Some physical constants

Quantity	Symbol	Value[a]
Atomic mass unit	u	$1.660\ 538\ 782\ (83) \times 10^{-27}$ kg 931.494 028 (23) MeV/c^2
Avogadro's number	N_A	$6.022\ 141\ 79\ (30) \times 10^{23}$ particles/mol
Bohr magneton	$\mu_B = \dfrac{e\hbar}{2m_e}$	$9.274\ 009\ 15\ (23) \times 10^{-24}$ J/T
Bohr radius	$a_0 = \dfrac{\hbar^2}{m_e e^2 k_e}$	$5.291\ 772\ 085\ 9\ (36) \times 10^{-11}$ m
Boltzmann's constant	$k_B = \dfrac{R}{N_A}$	$1.380\ 650\ 4\ (24) \times 10^{-23}$ J/K
Compton wavelength	$\lambda_C = \dfrac{h}{m_e c}$	$2.426\ 310\ 217\ 5\ (33) \times 10^{-12}$ m
Coulomb constant	$k_e = \dfrac{1}{4\pi\epsilon_0}$	$8.987\ 551\ 788 \ldots \times 10^{9}$ N•m²/C² (exact)
Deuteron mass	m_d	$3.343\ 583\ 20\ (17) \times 10^{-27}$ kg 2.013 553 212 724 (78) u
Electron mass	m_e	$9.109\ 382\ 15\ (45) \times 10^{-31}$ kg $5.485\ 799\ 094\ 3\ (23) \times 10^{-4}$ u 0.510 998 910 (13) MeV/c^2
Electron volt	eV	$1.602\ 176\ 487\ (40) \times 10^{-19}$ J
Elementary charge	e	$1.602\ 176\ 487\ (40) \times 10^{-19}$ C
Gas constant	R	8.314 472 (15) J/mol•K
Gravitational constant	G	$6.674\ 28\ (67) \times 10^{-11}$ N•m²/kg²
Neutron mass	m_n	$1.674\ 927\ 211\ (84) \times 10^{-27}$ kg 1.008 664 915 97 (43) u 939.565 346 (23) MeV/c^2
Nuclear magneton	$\mu_n = \dfrac{e\hbar}{2m_p}$	$5.050\ 783\ 24\ (13) \times 10^{-27}$ J/T
Permeability of free space	μ_0	$4\pi \times 10^{-7}$ T•m/A (exact)
Permittivity of free space	$\epsilon_0 = \dfrac{1}{\mu_0 c^2}$	$8.854\ 187\ 817 \ldots \times 10^{-12}$ C²/N•m² (exact)
Planck's constant	h	$6.626\ 068\ 96\ (33) \times 10^{-34}$ J•s
	$\hbar = \dfrac{h}{2\pi}$	$1.054\ 571\ 628\ (53) \times 10^{-34}$ J•s
Proton mass	m_p	$1.672\ 621\ 637\ (83) \times 10^{-27}$ kg 1.007 276 466 77 (10) u 938.272 013 (23) MeV/c^2
Rydberg constant	R_H	$1.097\ 373\ 156\ 852\ 7\ (73) \times 10^{7}$ m^{-1}
Speed of light in vacuum	c	$2.997\ 924\ 58 \times 10^{8}$ m/s (exact)

Note: These constants are the values recommended in 2006 by CODATA, based on a least-squares adjustment of data from different measurements. For a more complete list, see P. J. Mohr, B. N. Taylor, and D. B. Newell, CODATA Recommended Values of the Fundamental Physical Constants: 2006. *Rev. Mod. Phys.* **80**(2), 633–730, 2008.

[a]The numbers in parentheses for the values represent the uncertainties of the last two digits.

Solar system data

Body	Mass (kg)	Mean radius (m)	Period (s)	Mean distance from the Sun (m)
Mercury	3.30×10^{23}	2.44×10^{6}	7.60×10^{6}	5.79×10^{10}
Venus	4.87×10^{24}	6.05×10^{6}	1.94×10^{7}	1.08×10^{11}
Earth	5.97×10^{24}	6.37×10^{6}	3.156×10^{7}	1.496×10^{11}
Mars	6.42×10^{23}	3.39×10^{6}	5.94×10^{7}	2.28×10^{11}
Jupiter	1.90×10^{27}	6.99×10^{7}	3.74×10^{8}	7.78×10^{11}
Saturn	5.68×10^{26}	5.82×10^{7}	9.29×10^{8}	1.43×10^{12}
Uranus	8.68×10^{25}	2.54×10^{7}	2.65×10^{9}	2.87×10^{12}
Neptune	1.02×10^{26}	2.46×10^{7}	5.18×10^{9}	4.50×10^{12}
Pluto[a]	1.25×10^{22}	1.20×10^{6}	7.82×10^{9}	5.91×10^{12}
Moon	7.35×10^{22}	1.74×10^{6}	—	—
Sun	1.989×10^{30}	6.96×10^{8}	—	—

[a]In August 2006, the International Astronomical Union adopted a definition of a planet that separates Pluto from the other eight planets. Pluto is now defined as a 'dwarf planet' (like the asteroid Ceres).

Physical data often used

Average Earth–Moon distance	3.84×10^{8} m
Average Earth–Sun distance	1.496×10^{11} m
Average radius of the Earth	6.37×10^{6} m
Density of air (20°C and 1 atm)	1.20 kg/m^3
Density of air (0°C and 1 atm)	1.29 kg/m^3
Density of water (20°C and 1 atm)	1.00×10^{3} kg/m^3
Free-fall acceleration	9.80 m/s^2
Mass of the Earth	5.97×10^{24} kg
Mass of the Moon	7.35×10^{22} kg
Mass of the Sun	1.99×10^{30} kg
Standard atmospheric pressure	1.013×10^{5} Pa

Note: These values are the ones used in the text.

Some prefixes for powers of ten

Power	Prefix	Abbreviation	Power	Prefix	Abbreviation
10^{-24}	yocto	y	10^{1}	deka	da
10^{-21}	zepto	z	10^{2}	hecto	h
10^{-18}	atto	a	10^{3}	kilo	k
10^{-15}	femto	f	10^{6}	mega	M
10^{-12}	pico	p	10^{9}	giga	G
10^{-9}	nano	n	10^{12}	tera	T
10^{-6}	micro	μ	10^{15}	peta	P
10^{-3}	milli	M	10^{18}	exa	E
10^{-2}	centi	c	10^{21}	zetta	Z
10^{-1}	deci	d	10^{24}	yotta	Y

SERWAY JEWETT
WILSON WILSON ROWLANDS

PHYSICS

FOR GLOBAL SCIENTISTS AND ENGINEERS

VOLUME 1

2ND EDITION

CENGAGE

Physics for global scientists and engineers, Volume 1

2nd Edition

Raymond A Serway

John W Jewett

Adapted by:-

Kate Wilson

Anna Wilson

Wayne Rowlands

Publishing manager: Dorothy Chiu
Senior publishing editor: Fiona Hammond
Senior project editor: Nathan Katz
Developmental editor: Lydia Crisp
Cover design: Chris Starr
Text design: Norma Van Rees
Editor: Stephanie Ayres
Permissions/Photo researcher: Helen Mammides
Indexer: Russell Brooks
Art direction: Olga Lavecchia
Cover: Getty Images/Tara Moore
Typeset by MPS Limited

First local edition published in 2013
[Adapted from *Physics for scientists and engineers* (8th edition), by Raymond A Serway and John W Jewett, Jr, published by Brooks/Cole Cengage Learning, 2010]

Second edition published in 2017
[Adapted from *Physics for scientists and engineers* (9th edition), by Raymond A Serway and John W Jewett, Jr, published by Brooks/Cole Cengage Learning, 2014]

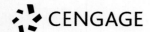

For product information and technology assistance,
in Australia call **1300 790 853**;
in New Zealand call **0800 449 725**

For permission to use material from this text or product, please email
aust.permissions@cengage.com

National Library of Australia Cataloguing-in-Publication Data
Creator: Serway, Raymond A., author.
Title: Physics for global scientists and engineers. Volume 1 / Raymond A Serway; John W Jewett; Adapted by: Kate Wilson; Anna Wilson; Wayne Rowlands.
Edition: 2nd edition.
ISBN: 9780170355513 (paperback)
Notes: Includes index.
Subjects: Physics--Textbooks. Physics--Study and teaching.
Other Creators/Contributors: Jewett, John W., author. Wilson, Kate, adapting author. Wilson, Anna, adapting author. Rowlands, Wayne, adapting author.
Dewey Number: 530

Cengage Learning Australia
Level 7, 80 Dorcas Street
South Melbourne, Victoria Australia 3205

Cengage Learning New Zealand
Unit 4B Rosedale Office Park
331 Rosedale Road, Albany, North Shore 0632, NZ

For learning solutions, visit **cengage.com.au**

Printed in China by 1010 Printing International Limited
2 3 4 5 6 7 21 20 19

Brief contents

Volume 1

Volume 2

Contents

Volume 1

Preface

This second edition of *Physics for global scientists and engineers* is an adaptation of the classic text *Physics for scientists and engineers* by Serway and Jewett to better suit students and instructors outside of the US. The language used has been modified, examples and case studies from local regions have been included, and quantities are given in SI units rather than imperial, except where a unit conversion is part of the learning objective for the problem. Uncertainty analysis is an integrated part of the text, in keeping with the empirical nature of the subject, and to help support students' learning in laboratories and reduce the 'disconnect' that sometimes occurs between the laboratory component of a course and the lecture/tutorial components. We have retained the excellent features of the original text, such as Pitfall Preventions and the selection of Quick Quizzes, conceptual and quantitative questions, as well as pedagogical features such as the *Try this* experiments.

The sequence of content reflects the ongoing development of physics. Rather than dividing the content into classical and modern, with the modern physics section largely consisting of discoveries and theories now about 100 years old, we instead divide the material by topic. Hence, we include the material on relativity in the first section on mechanics, where it is integrated with Newtonian mechanics that gives students an early introduction into what many find to be one of the more exciting aspects of physics. This arrangement also allows a stronger focus on quantum physics as the unifying theory that describes the physics of atoms, molecules and nuclei in the final chapters.

Objectives

This introductory physics textbook has two main objectives: to provide the student with a clear and logical presentation of the basic concepts and principles of physics and to strengthen an understanding of the concepts and principles through a broad range of interesting real-world applications. To meet these objectives, we emphasise sound physical arguments and problem-solving methodology. At the same time, we attempt to motivate the student through case studies and practical examples that demonstrate the role of physics in other disciplines, including engineering, chemistry, biology and medicine.

Changes from the ninth edition of Serway and Jewett's *Physics for scientists and engineers* and an overview of the second edition

A number of changes and improvements were made for the first edition of this text and these have been built upon for this second edition. The new features are based on our experiences and on current trends in science education. Other changes were incorporated in response to comments and suggestions offered by reviewers of the manuscript and our colleagues.

Line-by-line revision of the examples, questions and problems set. Each example, question and problem has been reviewed and many have been revised, to improve both readability and appeal to an international student cohort. Except in a few cases where a unit conversion is a deliberate element of a problem, all quantities are given in SI units. We have made careful revisions to worked examples so that the use of *Analysis models* and *Problem-solving strategies* are made more explicit and followed more consistently. The use of diagrams of various sorts to represent the situation, and as a first step in understanding the physical situation, is used in all but very simple mathematical problems. Solutions are presented symbolically as far as possible, and dimension checking is performed *before* numbers are substituted at the end. This approach helps students to think symbolically when they solve problems, and to check that their analysis is at least plausible instead of automatically looking to insert numbers into an equation to solve a problem.

Changes to and re-ordering of content. For the first edition, the material on relativity was placed in the mechanics section, giving students an early introduction to an area of physics that many find exciting and that is comparatively modern. A new chapter on the mechanical properties of solids was added, and the material on mechanical properties of fluids was expanded into two chapters including both static and dynamic properties. These chapters are grouped together in Part 2, where they follow logically from the mechanics introduced in Part 1. The section on X-ray diffraction was expanded, as this is a significant technique in analysis of materials, and is one often encountered in undergraduate teaching laboratories. Part 7, on Quantum Physics, groups together our treatment of all those physical systems that are described by quantum, rather than classical, mechanics.

For this second edition, Chapter 4, which describes Newton's laws, has been substantially revised based on recent physics education research. The concepts of 'weight' and 'apparent weight' are now dealt with far more explicitly and the language used to describe these concepts is discussed in detail. The section on Newton's third law has been expanded and is now applied more explicitly in worked examples.

The problem solving strategy has been updated for this edition to reflect best practice in pedagogy. The second step, previously 'Categorise', has been replaced with 'Model', which asks students to consider what assumptions and approximations they can make, and what existing models they can apply. This better reflects problem solving strategies used by experts.

Integration of uncertainties. Uncertainty, or the degree to which you can be confident in a measurement or other experimental result, is a critical part of any empirical science. An understanding of the role of uncertainty is perhaps most important in physics, which relies on quantitative measurements to develop and test mathematically expressed theories and laws. Physics courses generally include a laboratory component in which students meet uncertainties, but they are generally missing from other contexts such as lectures, textbooks and homework problems. Because of this, students may have little practice at uncertainty analysis, and may see it as something that only ever needs to be considered in the lab. The new edition integrates uncertainty analysis into the text, beginning with a section in Chapter 1 on 'Uncertainties in Measurement' contributed by Associate Professor Les Kirkup, followed by the inclusion of uncertainties in at least one worked example per chapter and several end-of-chapter problems. In these examples and problems the uncertainty analysis is an integral part of the problem, and a range of techniques for calculating the final uncertainty are demonstrated. The number of significant figures shown in the final answer depends upon the uncertainty, rather than being fixed. In some examples and problems the uncertainties are expressed as tolerances, for example for electronic components, as is typical in engineering.

Focus questions. Each chapter begins with a question designed to engage the interest of the student in the material within the chapter. These questions use a range of contexts including historical, such as the discovery of the shape of the DNA molecule and the use of bubble chambers; everyday, such as rainbows and colour-travel paint on cars; and technological examples, such as solar panels, lasers and reinforced concrete. These questions are answered at the end of the chapter, drawing together ideas from within the chapter into an answer and an explanation.

'Try this' examples. Each chapter includes *Try this* examples in which students are instructed to perform a simple experiment, using everyday items they are likely to have at hand in an office or kitchen, and to observe and explain the results. Research has shown that when students are actively engaged, particularly by 'doing' as well as thinking, deeper learning is likely to result.

Case studies highlighting interesting and significant local and international research. This new edition contains nine case studies, four of which are new, written by scientists, including physiologists, chemists, biologists and physicists as well as engineers, from around the world. The case studies highlight the application of physics to disciplines such as ergonomics (ergonomics of sheep shearing) and medicine (fibre optics and the human body) and important developments such as the discovery of the Higgs boson. A number of additional case studies can be found online.

Expansion of the analysis model approach. The analysis model approach used in the previous edition is used in this version of *Physics* and is expanded to include dimension checking as an explicit step in problem solving. It lays out a standard set of situations that appear in most physics problems. These situations are based on four simplification models: particle, system, rigid object and wave. The student thinks about what the entity is doing and how it interacts with its environment and what assumptions and approximations can reasonably be made. This leads the student to identify a particular analysis model for the problem. As the student gains more experience, he or she will lean less on the analysis model approach and begin to identify fundamental principles directly, more like a physicist does. This approach is further reinforced in the end-of-chapter summary under the heading *Analysis models for problem solving.*

Content

The material in this book provides an introduction to physics at a level appropriate for first year calculus-based university physics courses. It will also be a useful reference for students continuing their physics studies beyond first year. The book is divided into seven parts. Part 1 (Chapters 1 to 12) deals with the fundamentals of Newtonian mechanics and introduces students to relativity; Part 2 (Chapters 13 to 15) introduces the mechanical properties of fluids and solids; Part 3 (Chapters 16 to 18) covers oscillations, mechanical waves and sound; Part 4 (Chapters 19 to 22) addresses heat and thermodynamics; Part 5 (Chapters 23 to 34) treats electricity and magnetism; Part 6 (Chapters 35 to 38) covers light and optics; and Part 7 (Chapters 39 to 44) introduces the concepts of quantum mechanics needed to describe the physics of atoms, molecules and nuclei.

Helpful features

Pedagogical use of colour. Readers should consult the **pedagogical colour chart** (inside the front cover) for a listing of the colour-coded symbols used in the text diagrams. This system is followed consistently throughout the text.

Use of calculus. We have introduced calculus gradually, keeping in mind that students often take introductory courses in calculus and physics concurrently. Most steps are shown when basic equations are developed, and reference is often made to mathematical appendices.

Appendices and endpapers. Several appendices are provided. Most of the appendix material represents a review of mathematical concepts and techniques used in the text, including scientific notation, algebra, geometry, trigonometry, vector algebra and calculus. Reference to these appendices is made throughout the text, and where this is done an icon appears in the margin to highlight the link. In addition to the mathematical reviews, the appendices contain tables of physical data, conversion factors, and the SI units of physical quantities as well as a periodic table of the elements. Other useful information – fundamental constants and physical data, planetary data, a list of standard prefixes, mathematical symbols, the Greek alphabet, and standard abbreviations are given on the endpapers for quick access.

About the authors

Raymond A. Serway received his doctorate at Illinois Institute of Technology and is Professor Emeritus at James Madison University. In 2011, he was awarded with an honorary doctorate degree from his alma mater, Utica College. He received the 1990 Madison Scholar Award at James Madison University, where he taught for 17 years. Dr Serway began his teaching career at Clarkson University, where he conducted research and taught from 1967 to 1980. He was the recipient of the Distinguished Teaching Award at Clarkson University in 1977 and the Alumni Achievement Award from Utica College in 1985. As Guest Scientist at the IBM Research Laboratory in Zurich, Switzerland, he worked with K. Alex Müller, 1987 Nobel Prize recipient. Dr Serway also was a visiting scientist at Argonne National Laboratory, where he collaborated with his mentor and friend, the late Dr Sam Marshall. Dr Serway is the co-author of *College Physics,* Ninth Edition; *Principles of Physics,* Fifth Edition; *Essentials of College Physics; Modern Physics,* Third Edition; and the high school textbook *Physics,* published by Holt McDougal. In addition, Dr Serway has published more than 40 research papers in the field of condensed matter physics and has given more than 60 presentations at professional meetings. Dr Serway and his wife, Elizabeth, enjoy traveling, playing golf, fishing, gardening, singing in the church choir, and especially spending quality time with their four children, ten grandchildren, and a recent great grandson.

John W. Jewett, Jr. earned his undergraduate degree in physics at Drexel University and his doctorate at Ohio State University, specialising in optical and magnetic properties of condensed matter. Dr Jewett began his academic career at Richard Stockton College of New Jersey, where he taught from 1974 to 1984. He is currently Emeritus Professor of Physics at California State Polytechnic University, Pomona. Through his teaching career, Dr Jewett has been active in promoting effective physics education. In addition to receiving four National Science Foundation grants in physics education, he helped found and direct the Southern California Area Modern Physics Institute (SCAMPI) and Science IMPACT (Institute for Modern Pedagogy and Creative Teaching). Dr Jewett's honours include the Stockton Merit Award at Richard Stockton College in 1980, selection as Outstanding Professor at California State Polytechnic University for 1991–92, and the Excellence in Undergraduate Physics Teaching Award from the American Association of Physics Teachers (AAPT) in 1998. In 2010, he received an Alumni Lifetime Achievement Award from Drexel University in recognition of his contributions in physics education. He has given more than 100 presentations both domestically and abroad, including multiple presentations at national meetings of the AAPT. He has also published 25 research papers in condensed matter physics and physics education research. Dr Jewett is the author of *The World of Physics: Mysteries, Magic, and Myth,* which provides many connections between physics and everyday experiences. In addition to his work as the co-author for *Physics for Scientists and Engineers,* he is also the co-author on *Principles of Physics,* Fifth Edition, as well as *Global Issues,* a four-volume set of instruction manuals in integrated science for high school. Dr Jewett enjoys playing keyboard with his all-physicist band, travelling, underwater photography, learning foreign languages, and collecting antique quack medical devices that can be used as demonstration apparatus in physics lectures. Most importantly, he relishes spending time with his wife, Lisa, and their children and grandchildren.

Kate Wilson has a PhD in computational physics from Monash University and a Graduate Diploma in Secondary Teaching from the University of Canberra. She is a senior lecturer at UNSW Canberra (UNSW@ADFA) in the School of Engineering and Information Technology where she teaches in Civil Engineering, and is a member of the Learning and Teaching Group where she teaches the Graduate Teaching Program. Kate has been a member of the Sydney University Physics Education Research group, an Innovative Teaching and Educational Technology Fellow at the University of New South Wales, first year coordinator in physics at the Australian National University and director of the Australian Science Olympiads Physics Program. She has taught physics from first year algebra-based courses to third year condensed matter physics. She has published research on neural networks, magnetism, ballast water pumping and physics education. Her recent research looks at students' understanding of Newtonian mechanics, and this book is informed by that research. She is an author of the resource set 'Workshop Tutorials for Physics' and 'Nelson Physics for the Australian Curriculum Units 1 & 2' and 'Nelson Physics for the Australian Curriculum Units 3 & 4'.

Anna Wilson has a BSc(Hons) from the University of Bristol and obtained a PhD in nuclear physics from the University of Liverpool. She also has a Master of Higher Education from the Australian National University. She has worked at universities in the UK, the US, France and Australia. She has taught physics at all levels of the undergraduate degree, including algebra-based first year courses, quantum mechanics, and nuclear and particle physics, and is the recipient of teaching awards including an Australian Learning and Teaching Council Citation for Outstanding Contribution to Student Learning and an Award for Teaching Excellence. She has published research in the fields of optics, nuclear structure physics and higher education. While working on this book, Anna divided her time between the University of Canberra's Teaching and Learning Centre and the Research School of Physics and Engineering at the Australian National University. She is currently undertaking a second PhD in Education at the University of Stirling, UK.

Wayne Rowlands is a Senior Lecturer in the Department of Physics and Astronomy at Swinburne University of Technology. He has a PhD in laser atomic physics from the University of Melbourne, and a Graduate Certificate in Learning and Teaching from Swinburne University of Technology. His interests cover fundamental experimental research, science education and outreach. Wayne was a Chief Investigator in the ARC Centre of Excellence for Quantum-Atom Optics, with a particular interest in Bose-Einstein condensation. He is an active member of the Engineering and Science Education Research Group at Swinburne, has presented at education research conferences, and was invited to deliver the Australian Institute of Physics 'Youth Lecture' series of talks by the Victorian Branch (in 2002) and the Queensland Branch (in 2006). Wayne has been the editor of 'AOS News', the journal of the Australian Optical Society, and also served as a long-term presenter on the 3RRR radio science show 'Einstein A Go Go'.

Acknowledgements

We would like to thank all the reviewers, listed below, for their helpful feedback on the manuscript. In particular, we would like to thank Alix and Matthew Verdon who read the entire mechanics section as well as various other chapters and made many helpful comments, greatly improving the rigour of the text, and Darren Goossens who assisted with the revision of the material on x-ray diffraction.

We greatly appreciate the case study authors, listed below, for their contribution not only in explaining their work at an appropriate level and in an engaging way, but for collaborating with us in ensuring that each case study was integrated well into the text.

- Case study 1: Jack Harvey, John Culvenor, Warren Payne, Steve Cowley, Michael Lawrance and Robyn Williams
- Case study 2: Will Featherstone
- Case study 3: Elizabeth Angstmann
- Case study 4: Nicoleta Gacieu
- Case study 5: David Low
- Case study 6: Jeff Tallon
- Case study 7: Stephan Winkler
- Case study 8: John Love
- Case study 9: Geoffrey Taylor

We would like to thank the case study authors from the previous edition, which can now be found online:

- Online Case Study 1: Chiara Neto
- Online Case Study 2: Joe Wolfe
- Online Case Study 3: Mark Boland
- Online Case Study 4: Tony Irwin

We would also like to thank all our colleagues who contributed photos, including David Low, Matthew Verdon, Patrick Helean, Steve Keough, Joe Wolfe, George Hatsidimitris, Les Kirkup, Elizabeth Angstmann, Paul Davidson and Darren Goossens. Andrew Papworth helped with laboratory equipment and access to students. We are particularly grateful to the many ANU, UNSW, ADFA, Monash, UTS and Physics Olympiad students who allowed us to take photos of them while they were trying to perform laboratory work. Laurence and Marcus also stood still long enough for a few photos to be taken, and we appreciate the important role Questacon has in encouraging their, and many other children's, interest in science.

The staff at Cengage Learning were always a pleasure to work with, and we would particularly like to thank Fiona Hammond, Lydia Crisp and Nathan Katz for their work on the project. The help from our editor, Stephanie Ayres, was also crucial to the production of this book.

Finally, we would like to thank our families, friends, colleagues and especially our students for their support and encouragement.

Cengage Learning would also like to thank the following reviewers for their insightful and helpful feedback:

- Wayne Hutchison, University of New South Wales at the Australian Defence Force Academy
- Jim Webb, Griffith University
- David Parlevliet, Murdoch University
- Helen Georgiou, University of Sydney
- Dr Barry G Blundell, Auckland University of Technology
- Patrick Bowman, Massey University
- Joachim Brand, Massey University
- Matthew Collins, Charles Sturt University
- Allan Ernest, Charles Sturt University

- David Hoxley, La Trobe University
- Stephen Hughes, Queensland University of Technology
- Peter Killen, University Of Sunshine Coast
- Les Kirkup, University of Technology Sydney
- David Low, University of New South Wales at Australian Defence Force Academy
- Brendan McGann, Curtin University
- Nick Mermelengas, University of South Australia
- David Mills, Monash University
- Don Neely, University of Western Sydney
- John O'Byrne, University of Sydney
- Geoffrey Pang, Central Queensland University
- Darren Pearce, Queensland University of Technology
- Jamie Quinton, Flinders University
- Anton Rayner, University of Queensland
- Ben Ruck, Victoria University of Wellington
- Andrew Smith, Monash University
- Geoff Swan, Edith Cowan University
- M.N. Thomson, University of Melbourne
- Matthew Verdon, Australian Science Innovations
- Margaret Wegener, University of Queensland
- Jim Woolnough, University of Canberra
- Marjan Zadnik, Curtin University
- William Zealey, University of Wollongong

As well as several other anonymous reviewers.

Every attempt has been made to trace and acknowledge copyright holders. Where the attempt has been unsuccessful, the publishers welcome information that would redress the situation.

To the student

How to study

The most effective way to learn physics, or any other subject, is to be as active in your learning as possible.

Before going to lectures, read any notes provided in advance and any relevant sections of the text book. **During** lectures, pay attention and try to fit what is being discussed in lectures into your existing understanding and knowledge. Identify anything that doesn't seem to fit, and ask questions. If there are questions or activities in lectures, participate. **After** lectures, do any assigned homework problems, and review your understanding of the lecture content. Use resources including the textbook to help you. Work through example problems, don't just read them. Do the *Try this* examples – and try to predict what will happen before you do the experiment. When you try to explain what you observed, pay particular attention to any mismatch between your predictions and observations. Use other resources, such as websites, other books, articles and other students.

Find colleagues to study with. Sometimes an explanation from a friend will be easier to understand than one from a lecturer or tutor. Explaining things yourself to colleagues is also a great way to learn because it forces you to put into clear terms what you know, and can help you identify when you don't really understand a concept or principle as well as you thought you did. Anyone who has taught will tell you that teaching a subject is the best way to learn it yourself.

Participate in tutorials and laboratory classes. Ask lots of questions, think about what you are doing, and *ask yourself* lots of questions to make sure you understand. Use the opportunities to interact with teaching staff and other students. To really learn how to do something, you need to practice doing it – this is what laboratory and tutorial classes are for – for you to apply what you have learnt in lectures and from other resources and *do* some physics.

Use this book. Don't just read it, **do** the examples, problems, quiz questions and *Try this* experiments. We have tried to provide lots of opportunities for you to practice using the concepts and principles. When you do read, think about what you are reading, make notes of anything that you don't understand and ask questions of your lecturers and tutors, and other students. Often you will need to read a section more than once so that you understand it, especially when it is material that is new to you.

Work through the examples yourself, without looking at the solutions, then check your solution. Make sure you think carefully about any differences in your solution and the one given. Have you made a mistake? Have you made different simplifying assumptions? Follow the general problem-solving strategy, particularly ensuring that you understand the physical situation and which concepts and principles can be applied *before* you look to any equations. When you have found a solution, check that it is dimensionally correct and physically sensible. This is a good habit to get into, and will serve you well in exams. Few things annoy a marker more than dimensionally incorrect answers (or answers with missing units), or answers that are physically silly. Each chapter has many problems for you to practice on, as well as conceptual questions that will help you apply the concepts and principles covered in the chapter. It is important to be able to apply them without always resorting to an equation. Answers to *Quick quizzes* are given at the end of the textbook, and solutions to selected end-of-chapter questions and problems are provided in the accompanied *Student Solutions Manual.*

Do the *Try this* experiments. They have been designed to use only simple equipment that you can find in an office or kitchen, with very few exceptions. Use a *predict – observe – explain* strategy. Think about what you expect to happen before you do the experiment, based on your existing knowledge. Observe carefully what does happen, and repeat the experiment if necessary. Ask anyone else who is working with you to also observe, or to do the experiment while you watch. Then explain your observations, comparing them with your initial predictions. Was your prediction correct, and if not, why not? Think about similar situations you may have observed that can be explained by the same principles.

Use the online resources. Use the online tutorial material and Enhanced WebAssign content. Work through the tutorials, and use the Active Figures. Use a *predict – observe – explain* strategy with the Active Figures.

Set up a regular study schedule, and spend some time at least a few times each week on your study, making it as active as possible – doing, not just reading. Do not wait until just before the exam and then try to 'cram' a semester's worth of material.

Finally, and most importantly, think! Focus on understanding and applying the concepts and principles, rather than memorising equations.

We wish you great enjoyment and success in your studies.

Kate Wilson, Anna Wilson and Wayne Rowlands

Guide to the text

As you read this text you will find a number of features in every chapter to enhance your study of physics and help you understand how the theory is applied in the real world.

PART OPENING FEATURES

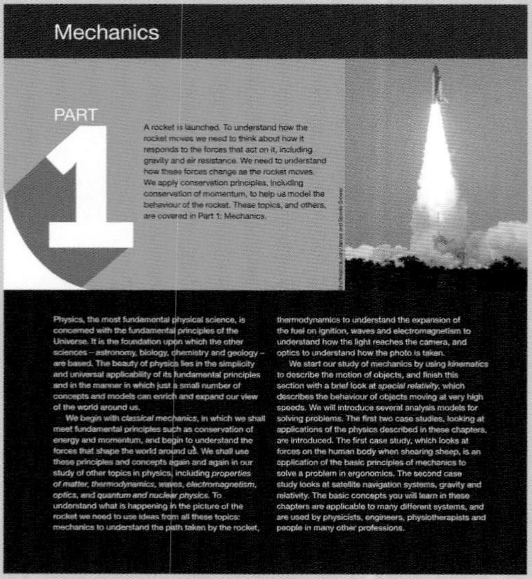
Mechanics

PART 1

A rocket is launched. To understand how the rocket moves we need to think about how it responds to the forces that act on it, including gravity and air resistance. We need to understand how these forces change as the rocket moves. We apply conservation principles, including conservation of momentum, to help us model the behaviour of the rocket. These topics, and others, are covered in Part 1: Mechanics.

Physics, the most fundamental physical science, is concerned with the fundamental principles of the Universe. It is the foundation upon which the other sciences – astronomy, biology, chemistry and geology – are based. The beauty of physics lies in the simplicity and universal applicability of its fundamental principles and in the manner in which just a small number of concepts and models can enrich and expand our view of the world around us.

We begin with *classical mechanics*, in which we shall meet fundamental principles such as conservation of energy and momentum, and begin to understand the forces that shape the world around us. We shall use these principles and concepts again and again in our study of other topics in physics, including *properties of matter, thermodynamics, waves, electromagnetism, optics*, and *quantum and nuclear physics*. To understand what is happening in the picture of the rocket we need to use ideas from all these topics: mechanics to understand the path taken by the rocket, thermodynamics to understand the expansion of the fuel on ignition, waves and electromagnetism to understand how the light reaches the camera, and optics to understand how the photo is taken.

We start our study of mechanics by using *kinematics* to describe the motion of objects, and finish this section with a brief look at *special relativity*, which describes the behaviour of objects moving at very high speeds. We will introduce several analysis models for solving problems. The first two case studies, looking at applications of the physics described in these chapters, are introduced. The first case study, which looks at forces on the human body when shearing sheep, is an application of the basic principles of mechanics to solve a problem in ergonomics. The second case study looks at satellite navigation systems, gravity and relativity. The basic concepts you will learn in these chapters are applicable to many different systems, and are used by physicists, engineers, physiotherapists and people in many other professions.

The part opener introduces the branch of physics to be covered in the following chapters, providing an overview of how the chapters relate to each other. Each part opener has a **vignette** that includes a real-world scenario and visual, providing context to the concepts to be covered.

CHAPTER OPENING FEATURES

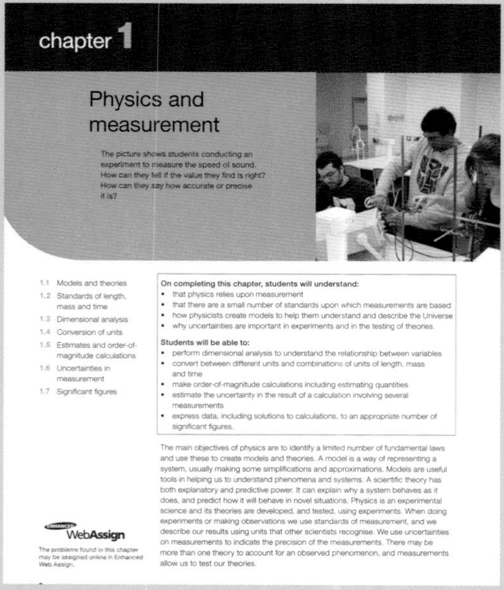
chapter 1

Physics and measurement

The picture shows students conducting an experiment to measure the speed of sound. How can they tell if the value they find is right? How can they say how accurate or precise it is?

1.1 Models and theories
1.2 Standards of length, mass and time
1.3 Dimensional analysis
1.4 Conversion of units
1.5 Estimates and order-of-magnitude calculations
1.6 Uncertainties in measurement
1.7 Significant figures

On completing this chapter, students will understand:
- that physics relies upon measurement
- that there are a small number of standards upon which measurements are based
- how physicists create models to help them understand and describe the Universe
- why uncertainties are important in experiments and in the testing of theories.

Students will be able to:
- perform dimensional analysis to understand the relationship between variables
- convert between different units and combinations of units of length, mass and time
- make order-of-magnitude calculations including estimating quantities
- estimate the uncertainty in the result of a calculation involving several measurements
- express data, including solutions to calculations, to an appropriate number of significant figures.

The main objectives of physics are to identify a limited number of fundamental laws and use these to create models and theories. A model is a way of representing a system, usually making some simplifications and approximations. Models are useful tools in helping us to understand phenomena and systems. A scientific theory has both explanatory and predictive power. It can explain why a system behaves as it does, and predict how it will behave in novel situations. Physics is an experimental science and its theories are developed, and tested, using experiments. When doing experiments or making observations we use standards of measurement, and we describe our results using units that other scientists recognise. We use uncertainties on measurements to indicate the precision of the measurements. There may be more than one theory to account for an observed phenomenon, and measurements allow us to test our theories.

WebAssign

The problems found in this chapter may be assigned online in Enhanced Web Assign.

Gain an insight into how physics theories relate to the real world through the **chapter opening vignette** with focus questions at the beginning of each chapter. The vignette is then revisited at the end of each chapter.

The **learning objectives** give you a clear sense of the topics covered in each chapter and what you should be able to do after reading the chapter.

FEATURES WITHIN CHAPTERS

Key Equations

$$\Delta\theta = \theta_f - \theta_i$$ ◄ Angular displacement

The rate at which this angular displacement occurs can be quantified by defining the **average angular speed** ω_{avg} (Greek letter omega) as the ratio of the angular displacement of a rigid object to the time interval Δt during which the displacement occurs:

$$\omega_{avg} = \frac{\theta_f - \theta_i}{t_f - t_i} = \frac{\Delta\theta}{\Delta t} \quad (9.2)$$ ◄ Average angular speed

In analogy to translational speed, the **instantaneous angular speed** ω is defined as the limit of the average angular speed as Δt approaches zero:

$$\omega = \lim_{\Delta t \to 0} \frac{\Delta\theta}{\Delta t} = \frac{d\theta}{dt} \quad (9.3)$$ ◄ Instantaneous angular speed

Key equations, **concepts** and **laws** are highlighted to help you identify important information.

Key equations are also numbered for easy reference.

Quick **Quiz**

Quick Quiz 5.5

Imagine the person shown in **Figure 5.24** is holding a ball in their hand. They gently let go of the ball (not throw it) when they are at the position shown in **Figure 5.24**. What is the path of the ball as seen by this person? (a) It stays where they released it, floating next to their hand. (b) It falls to land approximately at their feet. (c) It falls to land behind them, approximately where their feet were when they released it. Hint: look at **Figure 4.24** and draw a similar pair of diagrams for this situation.

Test your progress through each section by answering the **Quick Quiz** questions as you progress through the chapter.

Pitfall Prevention

Pitfall Prevention boxes give tips to help you avoid common physics mistakes and misconceptions.

Pitfall Prevention 9.1

The radian is an unusual unit. An angle expressed in radians is a pure number. It is a ratio of two lengths, so it is dimensionless. In rotational equations, you *must* use angles expressed in radians. Don't fall into the trap of using angles measured in degrees in rotational equations. Be careful how you use your calculator – make sure you know what units *it* is working in!

TRY THIS

TRY THIS

Stand next to one friend (A) and get two friends (B and C) to stand next to each other facing you, so they are the same distance from you and the person (A) standing next to you. You and your friend (A) throw a ball to the person (B or C) opposite you at exactly the same time, so they can catch them. The balls must travel the same horizontal distance, but you can throw them with different maximum heights. Does the ball with the higher or lower maximum height reach the person it is thrown to first? Does this happen every time?

Try this boxes provide examples of simple experiments using everyday items that you can easily try at home.

Active figures

Active Figure 2.3 (a) Graph representing the motion of the car in Active Figure 2.1. (b) An enlargement of the upper-left-hand corner of the graph.

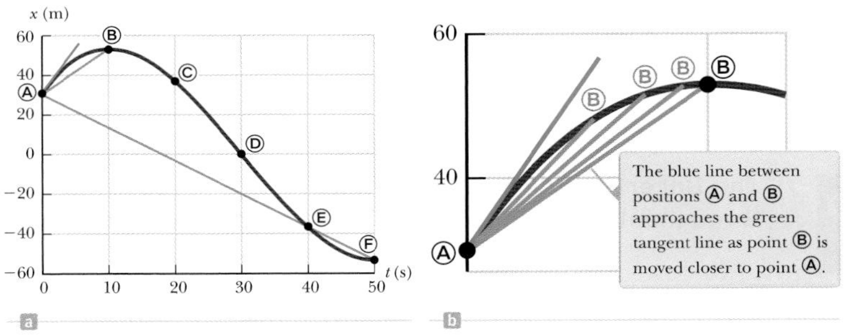

Active Figures indicate concepts that are supported by interactive animated presentations in the Physics companion website.

Worked Example

Example 5.12

Imagine a space station in the shape of a wheel like that shown in **Figure 5.24**, with a radius to the external wall of 250 m. If the inhabitants are to feel as if they are subject to normal Earth surface gravity, what linear speed must the 'floor' of the space station be moving at? What period of rotation does this correspond to?

Solution

Conceptualise Examine **Figure 5.24**. The rotational acceleration is caused by the force of the floor acting on the person. This acceleration needs to be the same as that due to the Earth's gravitational field at the surface of the Earth. This will give a normal force acting on the person's feet equal to the normal force they would experience standing on the ground on Earth.

Model We will treat the situation from the point of view of an observer outside the space station, not rotating but otherwise moving with it.

Analyse We want the centripetal force (the normal force) to be equal to mg where g is the magnitude of the acceleration due to gravity on Earth's surface.

$$n = \frac{mv^2}{r} = mg$$

so:

$$\frac{v^2}{r} = g$$

Rearranging for v:

$$v = \sqrt{rg}$$

Check dimensions:

$$[LT^{-1}] = ([L][LT^{-2}])^{\frac{1}{2}} = [LT^{-1}] \quad ☺$$

Substitute values:

$$v = \sqrt{250 \text{ m} \times 9.8 \text{ m.s}^{-2}} = 2450 \text{ m.s}^{-1}$$

Any point on the floor is travelling at 2450 $m.s^{-1}$, and travels a distance of $d = 2\pi r = 1570$ m in each rotational period. Therefore the period of rotation is $T = \frac{d}{v} = 0.64$ s.

What If? What if the space station had concentric floors and was spinning to give an 'artificial gravity' equal to g on the outermost floor? What would happen to an inhabitant's sensation of weight as they moved to a more inner floor?

Answer The period of rotation would be the same, but the radius would be smaller. As $T = \frac{2\pi r}{v}$, the velocity must be directly proportional to the radius. The centripetal acceleration, which is the 'artificial gravity' is $a_c = \frac{v^2}{r}$. So if $r \rightarrow \frac{1}{2} r$, then $v \rightarrow \frac{1}{2} v$ and $v^2 \rightarrow \frac{1}{4} v^2$, so $a_c \rightarrow \frac{1}{2} a_c$. Hence the artificial gravity decreases as you move closer to the centre of the space station.

Finalise The linear speed scales with the square root of the radius, as does the rotational period. So the larger the spacecraft, the greater the linear speed must be, but also the greater the rotational period. Hence the frequency of rotation is smaller for larger spacecraft.

Worked Examples provide conceptual explanations along with the calculations for every step. The examples closely follow the authors' proven **General Problem Solving Strategy**, which is introduced in Chapter 2 to reinforce good problem solving habits. About one-third of the worked examples include **What If?** extensions, which further strengthens conceptual understanding.

ICONS

Wherever you see the **Go online!** icon you will find additional relevant material, including Active Figures, on the book's website at http://www.cengagebrain.com.

±?

The **Uncertainty** icon highlights coverage of uncertainty, which is integrated throughout the book to help you understand this important concept in context.

NEW

Σ

The NEW **Maths icon** highlights mathematical concepts that are covered in Appendix B, directing you to the relevant content in the appendix for revision.

END-OF-CHAPTER FEATURES

At the end of each chapter you'll find several tools to help you to review, practise and extend your knowledge of the key learning objectives.

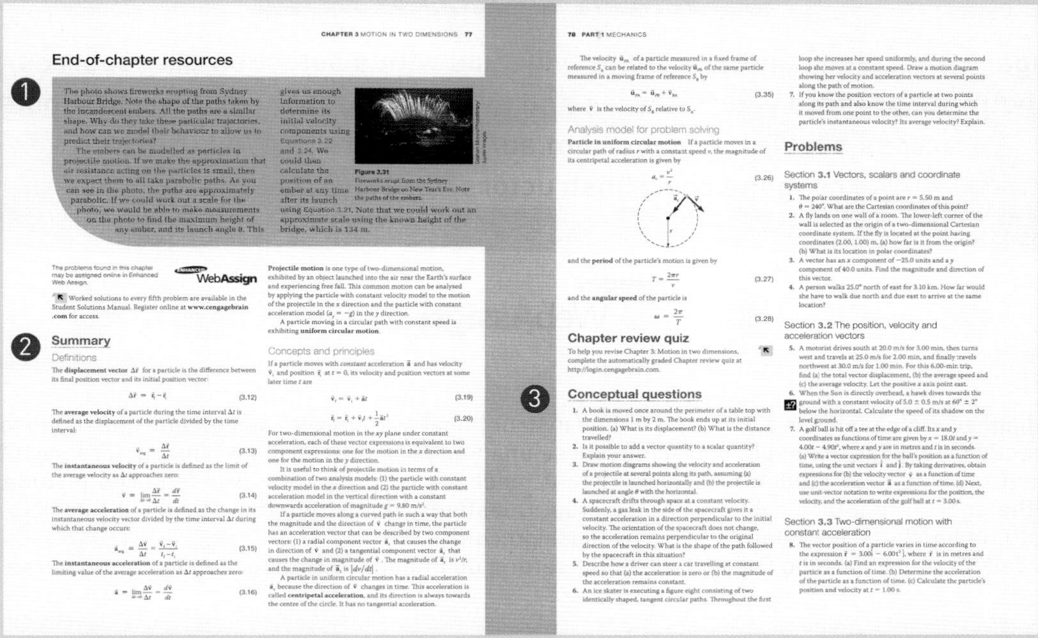

1. Revisit the chapter opening **vignette** to see how the chapter has helped you to understand the concepts involved.
2. **Definitions**, **Concepts and principles** and **Analysis for problem-solving sections** complete the **Summary** at the end of every chapter.
3. **Conceptual questions** and an extensive set of **Problems** are also included at the end of each chapter. About two thirds of the problems are keyed to specific sections of the chapter.

The **Additional problems** and **Challenge problems** will require you to synthesise key ideas from several sections.

CASE STUDIES

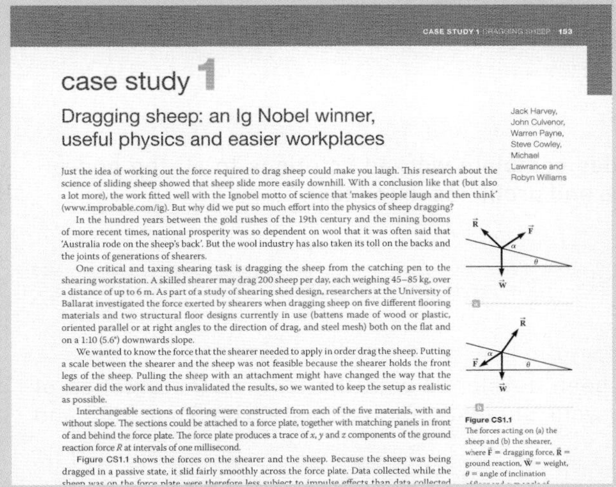

International and regional **Case studies** have been written by practitioners from a wide range of disciplines and cover relevant applications and research in physics.

Guide to the online resources

FOR THE INSTRUCTOR

Cengage Learning is pleased to provide you with a selection of resources that will help you prepare your lectures and assessments. These teaching tools are accessible via cengage.com.au/instructors for Australia or cengage.co.nz/instructors for New Zealand.

ENHANCED WebAssign

Enhanced Web Assign is a powerful online instructional system with assignable questions taken directly from your textbook, powerful course analytics and a Gradebook. Instructors save time spent grading, and students receive instant feedback on problems. Key features include: all of the end-of-chapter Problems, Conceptual Questions, Quick Quizzes, Master It tutorials and multimedia Watch Its. Learn more at webassign.com.

Talk to your Learning Consultant about setting up Enhanced Web Assign for your course.

PowerLecture DVD

The **PowerLecture Instructor's Resource DVD** provides everything you need for Physics. Key content includes art and images from the text, PowerPoint lectures, ExamView test generator software with questions, instructor's manual, solutions to all questions and problems in the text, animated **Active Figure simulations**, and a **physics movie** library.

INSTRUCTOR'S MANUAL

The Instructor's Manual includes:

- learning objectives
- suggestions for lecture demonstrations
- example tutorial and lab class activities.

SOLUTIONS MANUAL

The Solutions Manual includes complete worked solutions to all the end-of-chapter conceptual questions and problems in the text.

WORD-BASED TEST BANK

This bank of questions has been developed with the text for the creation of quizzes, tests and exams for your students. Deliver tests from your LMS and your classroom.

POWERPOINT™ PRESENTATIONS

Use the chapter-by-chapter **PowerPoint presentations** to enhance your lectures and handouts in order to reinforce the key principles of your subject.

ARTWORK FROM THE TEXT

Add the digital files of figures, graphs and pictures into your course management system, use them in student handouts, or copy them into your lecture presentations.

FOR THE STUDENT

STUDENT COMPANION WEBSITE

Visit the Physics companion website. You'll find:

- Revision quizzes
- Active Figures
- Solutions to selected questions from the text
- Extra case studies
- Chapter summaries, and
- Useful weblinks.

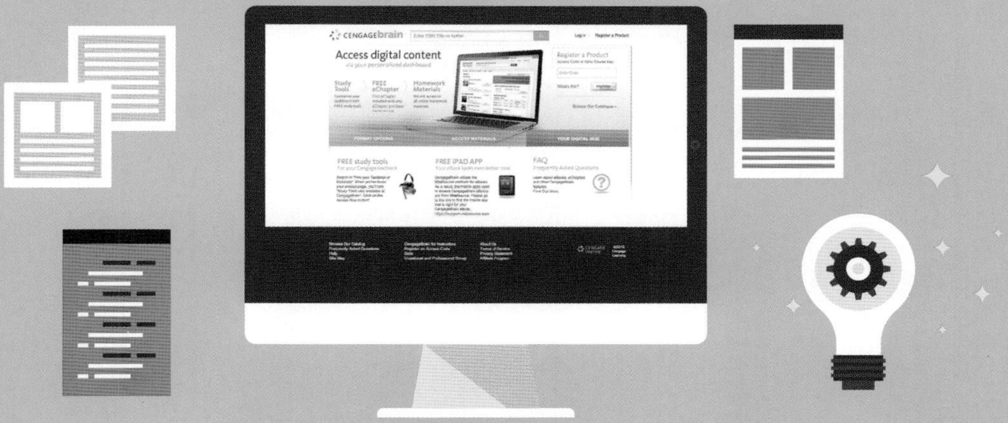

Enhanced Web Assign has assignable online questions taken directly from your textbook including all of the end-of-chapter Problems, Conceptual Questions, Quick Quizzes, Master It tutorials, and multimedia Watch Its. Ask your Instructor for details on how to access activities in Enhanced Web Assign.

Mechanics

PART 1

A rocket is launched. To understand how the rocket moves we need to think about how it responds to the forces that act on it, including gravity and air resistance. We need to understand how these forces change as the rocket moves. We apply conservation principles, including conservation of momentum, to help us model the behaviour of the rocket. These topics, and others, are covered in Part 1: Mechanics.

Shutterstock.com/James and Bonnie Grower

Physics, the most fundamental physical science, is concerned with the fundamental principles of the Universe. It is the foundation upon which the other sciences – astronomy, biology, chemistry and geology – are based. The beauty of physics lies in the simplicity and universal applicability of its fundamental principles and in the manner in which just a small number of concepts and models can enrich and expand our view of the world around us.

We begin with *classical mechanics*, in which we shall meet fundamental principles such as conservation of energy and momentum, and begin to understand the forces that shape the world around us. We shall use these principles and concepts again and again in our study of other topics in physics, including *properties of matter*, *thermodynamics*, *waves*, *electromagnetism*, *optics*, and *quantum and nuclear physics*. To understand what is happening in the picture of the rocket we need to use ideas from all these topics: mechanics to understand the path taken by the rocket, thermodynamics to understand the expansion of the fuel on ignition, waves and electromagnetism to understand how the light reaches the camera, and optics to understand how the photo is taken.

We start our study of mechanics by using *kinematics* to describe the motion of objects, and finish this section with a brief look at *special relativity*, which describes the behaviour of objects moving at very high speeds. We will introduce several analysis models for solving problems. The first two case studies, looking at applications of the physics described in these chapters, are introduced. The first case study, which looks at forces on the human body when shearing sheep, is an application of the basic principles of mechanics to solve a problem in ergonomics. The second case study looks at satellite navigation systems, gravity and relativity. The basic concepts you will learn in these chapters are applicable to many different systems, and are used by physicists, engineers, physiotherapists and people in many other professions.

chapter 1

Physics and measurement

The picture shows students conducting an experiment to measure the speed of sound. How can they tell if the value they find is right? How can they say how accurate or precise it is?

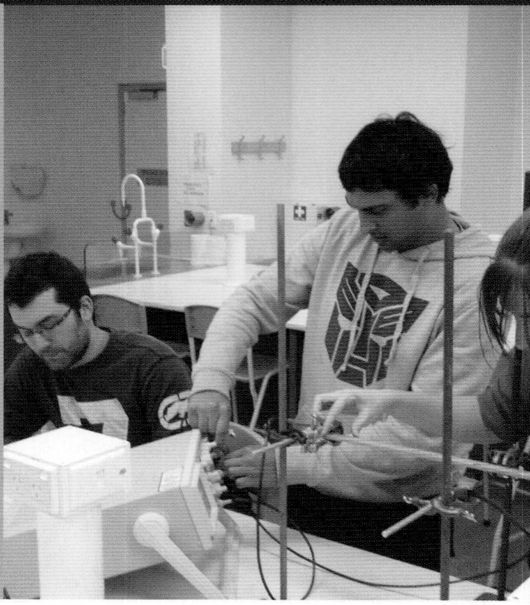

On completing this chapter, students will understand:

- that physics relies upon measurement
- that there are a small number of standards upon which measurements are based
- how physicists create models to help them understand and describe the Universe
- why uncertainties are important in experiments and in the testing of theories.

Students will be able to:

- perform dimensional analysis to understand the relationship between variables
- convert between different units and combinations of units of length, mass and time
- make order-of-magnitude calculations including estimating quantities
- estimate the uncertainty in the result of a calculation involving several measurements
- express data, including solutions to calculations, to an appropriate number of significant figures.

The main objectives of physics are to identify a limited number of fundamental laws and use these to create models and theories. A model is a way of representing a system, usually making some simplifications and approximations. Models are useful tools in helping us to understand phenomena and systems. A scientific theory has both explanatory and predictive power. It can explain why a system behaves as it does, and predict how it will behave in novel situations. Physics is an experimental science and its theories are developed, and tested, using experiments. When doing experiments or making observations we use standards of measurement, and we describe our results using units that other scientists recognise. We use uncertainties on measurements to indicate the precision of the measurements. There may be more than one theory to account for an observed phenomenon, and measurements allow us to test our theories.

ENHANCED WebAssign

The problems found in this chapter may be assigned online in Enhanced Web Assign.

1.1 Models and theories

We use models in physics to describe, explain and predict the behaviour of systems. These models may be conceptual models expressed in words and diagrams, mathematical models, computational models or even physical models. We use models because they allow us to generalise, and when we cannot interact directly with a system, such as individual atoms. Once we have a good model for how a particular type of system works, we can apply it to many systems of that type to explore how generally useful it is. We will develop and use several **analysis models** (described below) in this book, which can be applied to solve a large number of different problems. Models may be more or less accurate, they may change in time, and we may choose different models for the same system depending on the situation we are concerned with.

© The University of Sydney

Figure 1.1
The physics building at The University of Sydney was built in 1924. The names of some of the greatest physicists are carved on the outside of the building.

Consider the behaviour of matter. As an example, imagine a sample of solid gold. The simplest model of this piece of gold is as a uniform, rigid solid. If the sample is cut in half, the two pieces still retain their chemical identity as solid gold. What if the pieces are cut again and again, indefinitely? Will the smaller and smaller pieces always be gold? Early Greek philosophers developed a model for matter by speculating that the process ultimately must end when it produces a particle that can no longer be cut. In Greek, *atomos* means 'not sliceable'. From this Greek term comes our English word *atom*. The Greek model of the structure of matter stated that all ordinary matter consists of atoms. Beyond that, no additional structure was specified in the model. Then in 1897, J. J. Thomson identified the electron as a constituent of the atom. This led to the first atomic model that contained internal structure. Following the discovery of the nucleus in 1911, an atomic model was developed in which each atom is made up of electrons surrounding a central nucleus. This model leads, however, to a new question: Does the nucleus have structure? Is it a single particle or a collection of particles? By the early 1930s, a model evolved that described two basic entities in the nucleus: protons and neutrons. Is that, however, where the process of breaking down stops? Our current model of matter says that protons, neutrons and a host of other exotic particles are composed of six different varieties of particles called **quarks**, which have been given the names of *up, down, strange, charmed, bottom* and *top*.

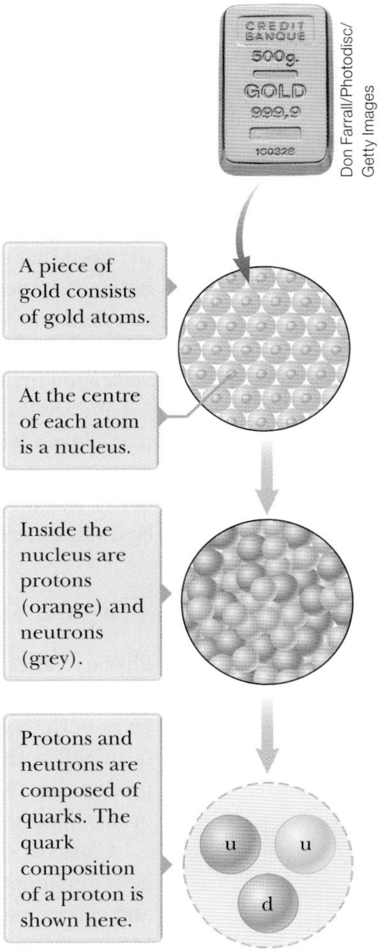
CREDIT BANQUE
500g.
GOLD
999,9
Don Farrall/Photodisc/Getty Images
A piece of gold consists of gold atoms.
At the centre of each atom is a nucleus.
Inside the nucleus are protons (orange) and neutrons (grey).
Protons and neutrons are composed of quarks. The quark composition of a proton is shown here.
u
u
d

Figure 1.2
Different ways of modelling the structure of a sample of gold, showing different levels of organisation

We model real systems all the time in physics – with conceptual models such as that described above, with mathematical models, with physical models sometimes and increasingly with simulations or computer models. Every time you solve a problem using physics you are using or creating a model, usually both. You draw on the existing mathematical models, such as the kinematic equations that describe motion or Newton's laws, and you apply them to a mental model of the situation which you create when you decide what the important features of the problem are, and what can be ignored.

When analysing a given situation, you choose the most appropriate model. Each of the models of gold described above is an approximation to the real system. If you are considering how the sample behaves when forces are applied, then modelling it as a uniform rigid body is likely to be an adequate model. If you are considering chemical reactions then you need to consider the atomic structure but not the details of the nucleus. If you are doing experiments in bombarding gold atoms with high energy particles, then you would probably use the quark model. As you learn more physics and practise problem solving, you will improve your skills at developing models and choosing the appropriate model for the situation.

When we choose a particular model to apply to a given situation, the model is generally based on a particular scientific theory. A scientific theory is a well substantiated explanation, which gives good predictive and explanatory power. Theories are developed by performing experiments and making observations, constructing and refining models to explain the observations, and then generalising them so that they can describe a wide range of systems and phenomena. Theories may include laws, such as conservation principles.

Like models, theories are generally satisfactory only under limited conditions. For example, classical mechanics, described in the first 11 chapters of this book, and largely developed by Isaac Newton (1642–1727) accurately describes the motion of objects moving at familiar everyday speeds but must be modified to more precisely model objects moving at speeds comparable to the speed of light. The special theory of relativity developed by Albert Einstein (1879–1955) predicts the same results as Newton's laws for objects moving at low speeds but also correctly describes the motion of objects at speeds approaching the speed of light. Hence, Einstein's special theory of relativity is a more general theory of motion based on, but modifying and extending, classical mechanics. This does not mean that we no longer use classical mechanics, or that classical mechanics is 'wrong' and relativity is 'right'. The special theory of relativity itself does not correctly describe the behaviour of very small objects, such as electrons. So it is also limited. The theory of quantum mechanics gives a better description of the behaviour of atomic and subatomic particles.

The theories and models that we construct are all approximations. They give us predictive and explanatory power, but they all have limitations.

One of the most important things to learn in physics is how to choose the appropriate model to describe a system. The model needs to include the important elements of the system. For an object moving very fast, relativistic effects are important and so we would use a model based on special relativity. For an object moving at more familiar speeds, like a car, we would use classical mechanics as relativistic effects are extremely small at the velocities a car can attain and classical mechanics is much simpler. In general, we use the simplest model that captures the behaviour of the system. Often, we start with a model that is too simple as a first approximation, and then use increasingly complex models until we have one that captures the behaviour of the system to the level of precision that we want. Using *analysis models* will give you practice in using models and recognising when simplifying approximations are appropriate.

Analysis models

In this book we use simplified mathematical models of systems which we call *analysis models.* Analysis models help us analyse common situations by making *appropriate* simplifications and approximations. An **analysis model** is a description of either (1) the behaviour of some physical entity or (2) the interaction between that entity and the environment. When you encounter a new problem, you should try to distinguish between important and unimportant details. For example, suppose a car is moving along a straight freeway at a constant speed. Is it important that it is a car? Is it important that it is a freeway? What about quantum and relativistic effects? If the answers to these questions are no, we model the car as a *particle with constant velocity,* which we will discuss in Chapter 2 as our first *analysis model.*

We shall generate analysis models based on four common approximations. The first of the four is the particle model discussed in Chapter 2. We will look at a particle under various behaviours and environmental interactions. Further analysis models are introduced in later chapters based on the ideas of a *system,* a *rigid object* and a *wave.* Once we have introduced these analysis models, we shall see that they appear again and again in different situations.

Theory and experiment

Physics relies on evidence obtained from experimental observations and quantitative measurements. Measurements and observations are used to formulate laws and theories. A scientific theory is one which gives both explanatory and predictive power. Because a scientific theory allows us to make predictions, it is falsifiable – that means that it can be tested and shown to be not true by performing experiments to test its predictions. When there is a discrepancy between the prediction of a theory and experimental results, new or modified theories are formulated.

Classical physics includes the principles of classical mechanics, thermodynamics, optics, and electromagnetism. Much of mechanics was developed in the 17th and 18th centuries, but thermodynamics and electromagnetism were not developed until the latter part of the 19th century, because before that time the apparatus for controlled experiments in these areas was either too crude or unavailable. The uncertainties in the measurements were too great to test competing theories. The improvement of experiments, which often means reducing uncertainties in measurements, has often led to the need for a new or refined theory.

A major revolution in physics began near the end of the 19th century because many physical phenomena were observed that could not be explained by classical physics. These observations were only possible because of the development of new technologies, which allowed new experiments to be performed. The two most important developments in this era were the theories of relativity and quantum mechanics, sometimes called 'modern physics'. Quantum mechanics was formulated to model physical phenomena at the atomic level, because the observed phenomena could not be explained by classical mechanics. The quantum theory of matter describes particles and waves very differently to the classical theories, as you will see in later chapters. Many practical devices have been developed using the principles of quantum mechanics; in fact, all modern electronics relies on it. Chances are you have such a device in your pocket – a smart phone, USB memory stick or MP3 player.

There is a constant interplay between theory, experiment and technology. New technologies lead to new and more precise experiments, which in turn lead to new observations and new theories. The new theories lead us to better understanding, and hence the development of new technologies.

Physicists continually strive to improve our understanding of fundamental laws. The new ways of understanding lead to changes in the way we look at and think about our world. The new technologies lead to changes in the way we interact with the world and each other, and how we live our lives.

1.2 Standards of length, mass and time

To describe natural phenomena, we make measurements of various aspects of nature. Each measurement is associated with a physical quantity, such as the length of an object. The laws of physics are expressed as mathematical relationships between physical quantities that we will introduce and discuss throughout this book. In mechanics, the three fundamental quantities are length, mass and time. All other quantities in mechanics can be expressed in terms of these three.

If we are to report the results of a measurement to someone who wishes to reproduce this measurement, a *standard* must be defined. It would be meaningless if a visitor from another planet were to talk to us about a length of 8 'glitches' if we do not know the meaning of the unit glitch. On the other hand, if someone familiar with our system of measurement reports that a wall is 2 metres high and our unit of length is defined to be 1 metre, we know that the height of the wall is twice our basic length unit. Whatever is chosen as a standard must be readily accessible and must possess some property that can be measured reliably. Measurement standards used by different people in different places must yield the same result, and standards used for measurements must not change with time.

In 1960, an international committee established a set of standards for the fundamental quantities of science. It is called the International System of Units (abbreviated to **SI** from the French *Système International d'unités*). Its fundamental units of length, mass and time are *metre*, *kilogram* and *second* respectively. Other standards for SI fundamental units established by the committee are those for temperature (*kelvin*), electric current (*ampere*), luminous intensity (*candela*) and the amount of substance (*mole*). These are the units we shall generally use throughout this book.

Each country has a laboratory or institute that is responsible for maintaining standards for that country, for example the National Measurement Institute in Australia and the Measurement Standards Laboratory in New Zealand.

Length

We can identify **length** as the distance between two points in space. In 1120, King Henry I of England decreed that the standard of length in his country would be named the *yard* and would be precisely equal to the distance from the tip of his nose to the end of his outstretched arm. Similarly, the original standard for the foot adopted by the French was the length of the royal foot of King Louis XIV. Neither of these standards were constant in time; when a new king took the throne, length measurements changed. The French standard prevailed until 1799, when the legal standard of length in France became the **metre** (m), defined as one ten-millionth of the distance from the equator to the North Pole along one particular longitudinal line that passes through Paris.

As recently as 1960, the length of the metre was defined as the distance between two lines on a specific platinum–iridium bar stored under controlled conditions in a laboratory in France. Current requirements of science and technology, however, necessitate more accuracy and precision than that with which the separation between the lines on the bar can be determined. In the 1960s and 1970s, the metre was defined as 1 650 763.73 wavelengths of orange-red light emitted from a krypton-86 lamp. In October 1983, however, the metre was redefined as **the distance travelled by light in a vacuum during a time of** $\frac{1}{299\,792\,458}$ **second**. In effect, this latest definition establishes that the speed of light in vacuum is precisely 299 792 458 metres per second. This definition of the metre is valid and measurable throughout the Universe based on our assumption that light is the same everywhere.

Pitfall Prevention 1.1

Generating intuition about typical values of quantities when solving problems is important because you must think about your end result and determine if it seems reasonable. For example, if you are calculating the mass of a housefly and arrive at a value of 100 kg, this answer is *unreasonable* and there is a mistake somewhere in your calculations.

Table 1.1 lists approximate values of some measured lengths.

Table 1.1
Approximate values of some measured lengths

Length (m)	
Distance from the Earth to the most remote known quasar	1.4×10^{26}
Distance from the Earth to the most remote normal galaxies	9×10^{25}
Distance from the Earth to the nearest large galaxy (Andromeda)	2×10^{22}
Distance from the Sun to the nearest star (Proxima Centauri)	4×10^{16}
One light-year	9.46×10^{15}
Mean orbit radius of the Earth about the Sun	1.496×10^{11}
Mean distance from the Earth to the Moon	3.84×10^{8}
Distance from the equator to the North Pole	1.00×10^{7}
Mean radius of the Earth	6.37×10^{6}
Typical altitude (above the surface) of a satellite orbiting the Earth	2×10^{5}
Length of a tennis court	2.4×10^{1}
Length of a fly	5×10^{-3}
Size of the smallest dust particles	$\sim 10^{-4}$
Size of cells of most living organisms	$\sim 10^{-5}$
Diameter of a hydrogen atom	$\sim 10^{-10}$
Diameter of an atomic nucleus	$\sim 10^{-14}$
Diameter of a proton	$\sim 10^{-15}$

Figure 1.3
(a) The National Standard Kilogram No. 20, an accurate copy of the International Standard Kilogram kept at Sèvres, France, is housed under a double bell jar in a vault at the National Institute of Standards and Technology. (b) A caesium fountain atomic clock. The clock will neither gain nor lose a second in 20 million years.

Mass

The SI fundamental unit of **mass**, the **kilogram** (kg), is defined as **the mass of a specific platinum-iridium alloy cylinder kept at the International Bureau of Weights and Measures at Sèvres, France**. This mass standard was established in 1887 and has not been changed since. A duplicate of the Sèvres cylinder is kept at the National Institute of Standards and Technology in Gaithersburg, Maryland (Figure 1.3a). Table 1.2 lists approximate values of the masses of various objects.

Time

Before 1967, the standard of **time** was defined in terms of the *mean solar day*. (A solar day is the time interval between successive appearances of the Sun at the highest point it reaches in the sky each day.) The fundamental unit of a **second** (s) was defined as $\left(\frac{1}{60}\right)\left(\frac{1}{60}\right)\left(\frac{1}{24}\right)$ of a mean solar day. Also, the time for one rotation of the Earth is changing, albeit slowly, making it unsatisfactory as a standard.

In 1967, the second was redefined to take advantage of the high precision attainable in a device known as an *atomic clock* (Figure 1.3b), which measures vibrations of caesium atoms. One second is now defined as **9 192 631 770 times the period of vibration of radiation from the caesium-133 atom**. Approximate values of some time intervals are presented in Table 1.3.

In addition to the fundamental SI units of metre, kilogram and second, we can also use other fractions or multiples of the fundamental units, such as millimetres and nanoseconds, where the prefixes *milli-* and *nano-* denote multipliers of the basic units based on various powers of ten. Prefixes for the various powers of ten and their abbreviations are listed in Table 1.4. For example, 10^{-3} m is equal to 1 millimetre (mm), and 10^{3} m corresponds to 1 kilometre (km). Likewise, 1 kilogram (kg) is 10^{3} grams (g), and 1 millisecond (ms) is 10^{-3} seconds (s).

Table 1.2
Approximate masses of various objects

Mass (kg)	
Observable Universe	$\sim 10^{52}$
Milky Way galaxy	$\sim 10^{42}$
Sun	1.99×10^{30}
Earth	5.98×10^{24}
Moon	7.36×10^{22}
Elephant	$\sim 10^{3}$
Human	$\sim 10^{2}$
Frog	$\sim 10^{-1}$
Mosquito	$\sim 10^{-5}$
Bacterium	$\sim 1 \times 10^{-15}$
Hydrogen atom	1.67×10^{-27}
Electron	9.11×10^{-31}

Table 1.3
Approximate values of some time intervals

Time interval (s)	
Age of the Universe	4×10^{17}
Age of the Earth	1.3×10^{17}
Average age of a university student	6.3×10^{8}
One year	3.2×10^{7}
One day	8.6×10^{4}
One class period	3.0×10^{3}
Time interval between normal heartbeats	8×10^{-1}
Period of audible sound waves	$\sim 10^{-3}$
Period of typical radio waves	$\sim 10^{-6}$
Period of vibration of an atom in a solid	$\sim 10^{-13}$
Period of visible light waves	$\sim 10^{-15}$
Duration of a nuclear collision	$\sim 10^{-22}$

Table 1.4
Prefixes for powers of ten

Power	Prefix	Abbreviation	Power	Prefix	Abbreviation
10^{-24}	yocto	y	10^{3}	kilo	k
10^{-21}	zepto	z	10^{6}	mega	M
10^{-18}	atto	a	10^{9}	giga	G
10^{-15}	femto	f	10^{12}	tera	T
10^{-12}	pico	p	10^{15}	peta	P
10^{-9}	nano	n	10^{18}	exa	E
10^{-6}	micro	μ	10^{21}	zetta	Z
10^{-3}	milli	m	10^{24}	yotta	Y
10^{-2}	centi	c			
10^{-1}	deci	d			

The variables length, time and mass are examples of *fundamental quantities.* Most other variables are *derived quantities,* those that can be expressed as a mathematical combination of fundamental quantities. Common examples are *area* (a product of two lengths) and *speed* (a ratio of a length to a time interval).

Another example of a derived quantity is **density**. The density ρ (Greek letter rho) of any substance is defined as its *mass, m, per unit volume, V*:

$$\rho = \frac{m}{V} \tag{1.1}$$

◀ Definition of density

In terms of fundamental quantities, density is a ratio of a mass to a product of three lengths.

Quick Quiz 1.1

In a machine shop, two cams are produced, one of aluminium (density 2.70×10^3 kg/m^3) and one of iron (density 7.86×10^3 kg/m^3). Both cams have the same mass. Which cam is larger? **(a)** The aluminium cam is larger. **(b)** The iron cam is larger. **(c)** Both cams have the same size.

TRY THIS

Walk around your house (or flat, or college) and see how many measuring devices you can find. What do they measure, and in what units? Do they measure fundamental quantities or some quantity made up of a combination of these?

1.3 Dimensional analysis

In physics, the word *dimension* denotes the physical nature of a quantity. The distance between two points, for example, can be measured in units of feet or metres, both of which have the dimension length. A given *dimension* can have many different *units*.

The symbols we use in this book to specify the dimensions of length, mass and time are L, M and T respectively. The *dimensions* of a quantity will be represented by capitalised, non-italic letters such as L or T. The *algebraic symbol* for the quantity itself will be an italicised letter such as L for the length of an object or t for time. We use square brackets [ ] to denote the dimensions of a physical quantity. For example, the symbol we use for speed in this book is v, and in our notation, the dimensions of speed are written $[v] = \mathrm{L/T}$. As another example, the dimensions of area A are $[A] = \mathrm{L}^2$. The dimensions and SI units of area, volume, speed and acceleration are listed in **Table 1.5**. The dimensions of other quantities, such as force and energy, will be described as they are introduced in the text.

Pitfall Prevention 1.2
The same symbol may be used to represent different quantities; for example, *t*, which we often use for time, can also be used for thickness. A particular quantity may be represented by more than one symbol, for example, distance may be represented by *x*, *s*, *y*, *l*, etc. Be careful to make sure you know what a symbol represents, and when you assign symbols to quantities yourself, make sure that you write down clearly exactly what each symbol represents.

Table 1.5
Dimensions and units of four derived quantities

Quantity	Area (*A*)	Volume (*V*)	Speed (*v*)	Acceleration (*a*)
Dimensions	L^2	L^3	L/T	$\mathrm{L/T}^2$
SI units	m^2	m^3	m/s	$\mathrm{m/s}^2$

It is always useful to check a derived equation to see if it matches your expectations. A useful procedure for doing that, called **dimensional analysis**, can be used because dimensions can be treated as algebraic quantities. Every time you solve a problem that requires you to derive an equation, you should check that equation using dimensional analysis. Quantities can be added or subtracted only if they have the same dimensions and the terms on both sides of an equation must have the same dimensions. By following these simple rules, you can use dimensional analysis to determine whether an expression has the correct form.

Pitfall Prevention 1.3
When solving problems, it is good practice to find the solution completely in algebraic form and do a dimension check before entering numerical values into the final symbolic expression. There are several advantages to this: first, you can do a dimension check; second, some quantities may cancel out and not need to be used at all; third, it saves keystrokes and hence opportunities for error with your calculator; and fourth, it decreases rounding errors. *Never* 'plug in' the numbers until you have your final algebraic expression and have done a dimension check.

To illustrate this procedure, suppose you are interested in an equation for the position x of a car at a time t if the car starts from rest at $x = 0$ and moves with constant acceleration a. The correct expression for this situation is $x = \frac{1}{2}at^2$, as we show in Chapter 2. The quantity x on the left side has the dimension of length. For the equation to be dimensionally correct, the quantity on the right side must also have the dimension of length. We can perform a dimensional check by substituting the dimensions for acceleration, $\mathrm{L/T}^2$ (**Table 1.5**), and time, T, into the equation. That is, the dimensional form of the equation $x = \frac{1}{2}at^2$ is

$$\mathrm{L} = \frac{\mathrm{L}}{\cancel{\mathrm{T}^2}} \cdot \cancel{\mathrm{T}^2} = \mathrm{L} ☺$$

The dimensions of time cancel as shown, leaving the dimension of length on the right-hand side to match that on the left. We use the symbol ☺ to show that we have performed a dimension check and that our expression is dimensionally correct.

The constant $\frac{1}{2}$ is a pure number and hence has no units or dimensions, so it does not appear in the dimensional analysis above.

Imagine now that you are solving a problem involving an accelerating car and have derived the equation, *incorrectly*, as $x = \frac{1}{2}at$. If you check the dimensions as above, you will find that:

$$\mathrm{L} = \frac{\mathrm{L}}{\mathrm{T}^2}\mathrm{T} = \frac{\mathrm{L}}{\mathrm{T}} ☹$$

At this point you would go back and check your working, rather than substituting in values.

It is also good practice to *always* include units when you substitute values, as incorrect final units will alert you to dimensionally incorrect equations.

Dimensional analysis is also useful when we don't have an equation and we want to know how one quantity might depend on another. We can apply a more general procedure using dimensional analysis. We start by writing an expression of the form

$$x \propto a^n t^m$$

where n and m are exponents that must be determined and the symbol $\propto$ indicates a proportionality. This relationship is correct only if the dimensions of both sides are the same. Because the dimension of the left side is length, the dimension of the right side must also be length. That is

$$\left[a^n t^m\right] = \mathrm{L} = \mathrm{L}^1\mathrm{T}^0$$

Because the dimensions of acceleration are $\mathrm{L/T^2}$ and the dimension of time is T, we have

$$\left(\frac{\mathrm{L}}{\mathrm{T}^2}\right)^n \mathrm{T}^m = \mathrm{L}^1\mathrm{T}^0 \rightarrow \left(\mathrm{L}^n\mathrm{T}^{m-2n}\right) = \mathrm{L}^1\mathrm{T}^0$$

The exponents of L and T must be the same on both sides of the equation. From the exponents of L, we see immediately that $n = 1$. From the exponents of T, we see that $m - 2n = 0$, which, once we substitute for n, gives us $m = 2$. Returning to our original expression $x \propto a^n t^m$, we conclude that $x \propto at^2$.

Quick Quiz 1.2

True or False: Dimensional analysis can give you the numerical value of constants of proportionality that may appear in an algebraic expression.

Example 1.1

Analysis of a power law

Suppose we are told that the acceleration a of a particle moving with uniform speed v in a circle of radius r is proportional to some power of r, say r^n, and some power of v, say v^m. Determine the values of n and m and write the simplest form of an equation for the acceleration.

Solution

Write an expression for a with a dimensionless constant of proportionality k: $a = kr^n v^m$

Substitute the dimensions of a, r, and v: $\frac{\mathrm{L}}{\mathrm{T}^2} = \mathrm{L}^n\left(\frac{\mathrm{L}}{\mathrm{T}}\right)^m = \frac{\mathrm{L}^{n+m}}{\mathrm{T}^m}$

Equate the exponents of L and T so that the dimensional equation is balanced: $n + m = 1$ and $m = 2$

Solve the two equations for n: $n = -1$

Write the acceleration expression: $a = kr^{-1}v^2 = k\frac{v^2}{r}$

In Section 3.5 on uniform circular motion, we show that $k = 1$ if a consistent set of units is used. The constant k would not equal 1 if, for example, v were in km/h and you wanted a in $\mathrm{m/s^2}$.

Pitfall Prevention 1.4

When performing calculations, include the units for every quantity and carry the units through the entire calculation. Avoid the temptation to drop the units early and then attach the expected units once you have an answer. By including the units in every step, you can detect errors if the units for the answer turn out to be incorrect.

1.4 Conversion of units

Sometimes it is necessary to convert units from one measurement system to another or convert within a system (for example, from kilometres to metres). Like dimensions, units can be treated as algebraic quantities that can cancel each other. For example, to convert from km/h to m/s:

$$\frac{1\text{ km}}{1\text{ h}} = \frac{1\text{ km}}{1\text{ h}} \times \frac{1000\text{ m}}{1\text{ km}} = \frac{1000\text{ m}}{1\text{ h}} = \frac{1000\text{ m}}{1\text{ h}} \times \frac{1\text{ h}}{3600\text{ s}} = \frac{1}{3.6}\text{ m/s}$$

So if you have a speed in km/h, you simply divide by 3.6 to convert to m/s.

In each step in the example above we are only multiplying by 1, for example, 1000 m = 1 km, so 1000 m/km = 1. So, in the first step we can cancel km to get a measurement in m/h, then in the next step we can cancel h to get m/s. Always write in the units carefully like this when you convert units, and check that your answer makes physical sense.

A list of conversion factors can be found in Appendix A.

Example 1.2

Is he speeding?

A car passes through a school zone at 8 a.m. on a school day in 2.0 minutes. The speed limit in the zone is 40 km/h. If the school zone is 1.5 km long, is the driver speeding?

Solution

We need to calculate the speed of the car in km/h.

The speed is: $v = x/t = 0.70\text{ km}/1.0\text{ min} = 0.70\text{ km/min}$

Convert this to km/h: $0.75\text{ km/min} = (0.70\text{ km/min}) \times (60\text{ min/h})$
$= 42\text{ km/h}$

The driver is speeding.

Note that we have calculated the *average* speed. The instantaneous speed at any moment may be higher or lower than this. If it is lower than this at any time, then it must be even higher than 2 km/h over the speed limit at some other time.

Paul Rands (http://www.expressway.online)

Figure 1.4
Safe-T-Cams are used on Australian highways to monitor heavy vehicle trip times. Photos are taken, and the time recorded. This allows the average speed between cameras to be calculated. The main purpose of the system is to reduce fatigue-related accidents by ensuring heavy vehicle drivers take adequate rest breaks.

TRY THIS

Measure your walking pace in m/s by timing how long it takes you to walk 100 m. Convert that to km/h. Check that your answer makes physical sense. What is your walking pace in ft/s and mi/h?

1.5 Estimates and order-of-magnitude calculations

Suppose someone asks you how big an image file on your computer or phone is. In response, you probably wouldn't give the exact number of bytes, but rather an estimate. You would probably say about 1.1 MB, rather than 1 132 437 B. The estimate may be made even more approximate by expressing it as an order of magnitude, which is a power of ten determined as follows.

1. Express the number in scientific notation, with the multiplier of the power of ten between 1 and 10. Make sure to include a unit!
2. If the multiplier is less than 3 (approximately the square root of ten), the order of magnitude of the number is the power of ten in the scientific notation. If the multiplier is greater than 3, the order of magnitude is one larger than the power of ten in the scientific notation.

   For example, $0.0086\ \text{m} = 8.6 \times 10^{-3}\ \text{m} \sim 10^{-2}\ \text{m}$.
   We use the symbol $\sim$ for 'is of the order of'.

Quick **Quiz 1.3**

What is the order of magnitude of each of the following? 0.0021 m, 72.0 m, 356 251 m

Usually, when an order-of-magnitude estimate is made, the results are reliable to within about a factor of ten. If a quantity increases in value by three orders of magnitude, its value increases by a factor of about $10^3 = 1000$. Inaccuracies caused by guessing too low for one number are often cancelled by other guesses that are too high. You will find that with practice your estimates become better and better. Estimation problems can be fun to work because you freely drop digits, venture reasonable approximations for unknown numbers, make simplifying assumptions, and turn the question around into something you can answer in your head or with minimal mathematical manipulation on paper. Because of the simplicity of these types of calculations, they can be performed on a small scrap of paper and are often called 'back-of-the-envelope calculations'.

TRY THIS

Pick up this book, and estimate how heavy it is. Then estimate its volume and do a 'back-of-the-envelope calculation' to find its density.

Example **1.3**

Breaths in a lifetime

Estimate the number of breaths taken during an average human lifetime without using your calculator.

Solution

We start by guessing that the typical human lifetime is about 80 years. Think about the average number of breaths that a person takes in 1 min. This number varies depending on whether the person is exercising, working on a difficult physics problem, excited or relaxed, and so forth. To the nearest order of magnitude, we shall choose 10 breaths per minute as our estimate. (This estimate is certainly closer to the true average value than an estimate of 1 breath per minute or 100 breaths per minute.)

Find the approximate number of minutes in 80 years:

$$80\ \text{yrs}\left(\frac{400\ \text{days}}{1\ \text{yr}}\right)\left(\frac{25\ \text{h}}{1\ \text{day}}\right)\left(\frac{60\ \text{min}}{1\ \text{h}}\right) = 5 \times 10^7\ \text{min}$$

Find the approximate number of breaths in a lifetime:

$$\text{number of breaths} = (10\ \text{breaths/min})(5 \times 10^7\ \text{min}) = 5 \times 10^8\ \text{breaths}$$

Therefore, a person takes of the order of 10^9 breaths in a lifetime. Notice how much simpler it is in the first calculation above to multiply 400×25 than it is to work with the more accurate 365×24 when you don't have a calculator.

What If? What if the average lifetime were estimated as 70 years instead of 80? Would that change our final estimate?

Answer We could claim that $(70\ \text{yr}) \approx 4 \times 10^7$ min, so our final estimate should be 4×10^8 breaths. This answer is still of the order of 10^9 breaths (recall that we round up from 4), so an order-of-magnitude estimate would be unchanged.

1.6 Uncertainties in measurement

This section was kindly contributed by Professor Les Kirkup

Measurement shapes the theories scientists use to describe physical phenomena and assists them to challenge those theories. As a consequence a theory may be replaced when it is found to be inadequate. Einstein's theory of gravitation predicts that the gravitational pull of the Sun will deviate light from distant stars passing close to the Sun by about 8.48 μrad (8.48×10^{-6} rad or $(4.86 \times 10^{-4})°$). Newton's theory of gravitation also predicts that light will be deviated by the Sun's gravity, but by only about 4.22 μrad. So, which theory is better? In 1919 a group of scientists led by Sir Arthur Eddington, who studied the bending of light during a total eclipse, confirmed Einstein's prediction. This profoundly affected our view of the Universe and brought considerable fame to Einstein.

What this success story disguises is that the measurement of the deviation of light by the Sun was extremely challenging and though the values obtained by Eddington and his co-workers broadly supported Einstein's theory, there was an amount of *uncertainty* in the best value they obtained for the deviation of the light. Eddington wrote the best value as 9.60 μrad with an uncertainty of 0.58 μrad. The value of the uncertainty is extremely important, for if it had been ten times larger (i.e. 5.8 μrad), no decision could have been made as to whether Einstein's or Newton's theory was the better. To justify this, we subtract 5.8 μrad from 9.60 μrad to give a low end of the possible deviation. This indicates that, as a consequence of the uncertainty, the deviation of light could have been as small as 3.8 μrad, in good agreement with Newton's theory.

This is a particularly striking example of where uncertainty matters. However, the simple fact is that irrespective of the sophistication of the measuring equipment, the expertise of the experimenters, and the constancy of the quantity being measured, there is *always* some uncertainty in values obtained through measurement. By carefully designing an experiment, the uncertainty can be reduced – sometimes to the extent where it can be neglected – but the uncertainty must be estimated first before we can be confident that its presence is not a cause for concern. For example, if your experiment requires that you use the value of the charge carried by an electron in your calculations, you can be assured that the value is well known and its uncertainty is small; many workers have measured the charge carried by an electron, and the current best value is $1.602\,176\,487 \times 10^{-19}$ C with an uncertainty of $0.000\,000\,040 \times 10^{-19}$ C (which is a percentage uncertainty of about 2.5×10^{-6}!). Few experiments manage to reduce uncertainty to this degree and due to its omnipresence and its potential effect on any conclusions drawn from data, it is a foolhardy scientist who does not consider the impact of uncertainty in measurement.

It is sometimes helpful to imagine that a quantity being measured during an experiment has a 'true value', which *would* be obtained if all influences, such as the limited resolution of instruments used during the experiment or the varying laboratory conditions (such as ambient temperature), could be eliminated. The variability in values obtained through experiment leads us naturally to being able to determine the uncertainty in the best estimate of the quantity that is the focus of the experiment. Suppose repeat measurements are made of a quantity. We first calculate the mean of those values. This gives us the best estimate of the true value under most conditions. There are several ways to calculate the uncertainty in the best estimate, and you can refer to books on experimental methods or uncertainties in measurements for other methods. A simple and convenient way to do this is to find the maximum value and minimum value in the set of repeat measurements, then calculate:

$$\text{uncertainty} = \frac{\text{max value} - \text{min value}}{n} \tag{1.2}$$

where n is the number of repeat measurements. (This equation works well for n up to 10, but is not good for larger n.) For example, suppose we measure the time for a ball to fall 10 m and obtain the following values (in seconds): 1.25, 1.55, 1.45, 1.56, 1.33, 1.45, 1.39. The mean is 1.426 s and the uncertainty, using **Equation 1.2**. is 0.044 s. Note that it is international convention to express the uncertainty to two significant figures, although it is also common to express it to only a single significant figure. See Section 1.7 for a discussion on significant figures.

It is very common for an experiment to require the measurement of several quantities. Those quantities are then brought together so that the particular quantity of interest (sometimes called the *measurand*) can be determined. If each quantity has some uncertainty, how do we calculate the uncertainty in the measurand? There are several ways to do this. We consider here the simplest, which is good enough for many purposes. Suppose a and b are quantities which have uncertainty Δa and Δb respectively. We might want to add, subtract, multiply or divide a and b. Let us use the symbol, z, to represent the measurand and Δz to represent the uncertainty in z.

If:

$$z = a + b, \text{ then } \Delta z = \Delta a + \Delta b \tag{1.3}$$

$$z = a - b, \text{ then } \Delta z = \Delta a + \Delta b \tag{1.4}$$

Hence whenever we add *or* subtract measured values, we must always *add* their uncertainties. Note that z and Δz have the same units, and Δz is known as the *absolute* uncertainty in z.

If:

$$z = ab, \text{ then } \frac{\Delta z}{z} = \frac{\Delta a}{a} + \frac{\Delta b}{b} \tag{1.5}$$

$$z = \frac{a}{b}, \text{ then } \frac{\Delta z}{z} = \frac{\Delta a}{a} + \frac{\Delta b}{b} \tag{1.6}$$

The quantity $\frac{\Delta z}{z}$ is dimensionless and has no units; it is known as the fractional or relative uncertainty in z, and is often expressed as a percentage.

As an example, suppose we wish to determine the density of a copper block. Through measurement we find that the best estimate of the mass, m, of the block is 250.5 g with an uncertainty of 2.5 g and the best estimate of volume, V, of the block is 28.33 cm^3 with an uncertainty of 0.18 cm^3. To find the density, ρ, we use the relationship (**Equation 1.1**):

$$\rho = \frac{m}{V} = \frac{250.5 \text{ g}}{28.33 \text{ cm}^3} = 8.842 \text{ g/cm}^3$$

Using **Equation 1.5** and substituting ρ, m and V for z, a and b respectively, we have:

$$\frac{\Delta \rho}{\rho} = \frac{\Delta m}{m} + \frac{\Delta V}{V} = \frac{2.5 \text{ g}}{250.5 \text{ g}} + \frac{0.18 \text{ cm}^3}{28.33 \text{ cm}^3} = 0.016\,33$$

It follows that the uncertainty, $\Delta \rho$, in the density is $0.016\,33 \times \rho = 0.016\,33 \times 8.84 \text{ g/cm}^3 = 0.14 \text{ g/cm}^3$. So we can say that the best estimate of the density of the copper is 8.84 g/cm^3 and the uncertainty in the best estimate is 0.14 g/cm^3. We write this as $\rho = (8.84 \pm 0.14)$ g/cm^3. We may also say that the uncertainty in the density is $0.016 \times 100\% = 1.6\%$.

Another simple method of finding the uncertainty when combining measured values is to estimate the maximum and minimum possible values. Such a method is useful if you have an equation which includes (as examples) the logarithmic or exponential function. Each quantity in the formula is modified by an amount equal to its uncertainty to produce the largest value and smallest value of the final calculated quantity. The uncertainty in this final calculated quantity is then given by half the difference between these values:

$$\Delta z = \frac{z_{max} - z_{min}}{2} \tag{1.7}$$

For example, suppose we wish to find the cross-sectional area of a large copper pipe with inner diameter $d = (0.10 \pm 0.01)$ m and outer diameter $D = (0.15 \pm 0.01)$ m.

The best estimate of the cross-sectional area is

$$A = \frac{\pi(D^2 - d^2)}{4} = \frac{\pi((0.15 \text{ m})^2 - (0.1 \text{ m})^2)}{4} = 0.009\,817 \text{ m}^2$$

To find the maximum cross-sectional area we substitute in the maximum value of D, 0.16 m, and the minimum value of d, 0.09 m, to get:

$$A_{max} = \frac{\pi(D^2 - d^2)}{4} = \frac{\pi((0.16 \text{ m})^2 - (0.09 \text{ m})^2)}{4} = 0.013\,74 \text{ m}^2$$

To find the minimum cross-sectional area we substitute in the minimum value of D, 0.14 m, and the maximum value of d, 0.11 m, to get:

$$A_{min} = \frac{\pi\left(D^2 - d^2\right)}{4} = \frac{\pi((0.14\text{ m})^2 - (0.11\text{ m})^2)}{4} = 0.005\,890\text{ m}^2$$

Now the uncertainty can be found from Equation 1.7:

$$\Delta A = \frac{(A_{max} - A_{min})}{2} = \frac{(0.01374\text{ m}^2 - 0.005890\text{ m}^2)}{2} = 0.0039\text{ m}^2$$

We can write the area with its uncertainty as $(0.0098 \pm 0.0039)\text{ m}^2$ or $(9.8 \pm 3.9) \times 10^{-3}\text{ m}^2$.

Problems and worked examples that include uncertainties are highlighted with this symbol in the margin. ▶ ±?

Because physics is an experimental science, and an understanding of measurement and uncertainty is important in all experimental sciences, we will use uncertainties in selected worked examples and end-of-chapter problems throughout this book. If your physics course includes a laboratory component, then you may need to know how to deal with more complicated combinations of uncertainties and how to apply statistical analysis to large data sets. You should refer to a text on experimental methods for more information.

TRY THIS

Get as many people as you can, each with a stopwatch, to time how long it takes for a ball dropped from 2 m to hit the ground after being released. Ask them to all start timing when you say 'go' (and drop the ball), and to stop when the ball hits the ground. What is the time taken? What is the uncertainty in the time?

1.7 Significant figures

When quantities are measured, the measured values are known only to within the limits of the experimental uncertainty as described above. The number of **significant figures** in a measurement can be used to infer the uncertainty in the measurement. The number of significant figures is related to the number of numerical digits used to express the measurement.

Zeroes may or may not be significant figures. Those used to position the decimal point in such numbers as 0.03 and 0.0075 are not significant. Therefore, there are one and two significant figures respectively in these two values. When the zeroes come after other digits, however, there is the possibility of misinterpretation. For example, suppose the mass of an object is given as 1500 g. This value is ambiguous because we do not know whether the last two zeros are being used to locate the decimal point or whether they represent significant figures in the measurement. To remove this ambiguity, it is common to use scientific notation to indicate the number of significant figures. In this case, we would express the mass as 1.5×10^3 g if there are two significant figures in the measured value, 1.50×10^3 g if there are three significant figures, and 1.500×10^3 g if there are four.

The same rule holds for numbers less than 1, so 2.3×10^{-4} has two significant figures (and therefore could be written 0.000 23) and 2.30×10^{-4} has three significant figures (also written as 0.000 230).

Quick Quiz 1.4

How many significant figures do each of these values have? 0.0021 m, 72.0 m, 356 251 m
Remember that the number of significant figures is not the same as the number of decimal places.

Pitfall Prevention 1.5

Read Carefully Notice that the rule for addition and subtraction is different from that for multiplication and division. For addition and subtraction, the important consideration is the number of *decimal places*, not the number of *significant figures*.

When no uncertainty in a value is given, it is usual to assume that the uncertainty is in the last digit of the value. Hence the number of significant figures to which an answer is given indicates the uncertainty in that answer. In problem solving, we often combine quantities mathematically through multiplication, division, addition, subtraction, and so forth. When doing so, you must make sure that the result has the appropriate number of significant figures. A good rule of thumb to use in determining the number of significant figures that can be claimed in a multiplication or a division is as follows:

When multiplying several quantities, the number of significant figures in the final answer is the same as the number of significant figures in the quantity having the smallest number of significant figures. The same rule applies to division.

For example, if you calculate a distance using the equation $x = \frac{1}{2}at^2$, where $a = 1.422\ \text{m/s}^2$ and $t = 2.5\ \text{s}$, then your answer will be 4.443 75 m, which should be rounded to 4.4 m because our least precise piece of data had only 2 significant figures. The factor of $\frac{1}{2}$ is not a value determined through experiment, it is an *exact* value, with no uncertainty, even though it is written to only one significant figure. This is also the case for the exponent, 2.

For addition and subtraction, you must consider the number of decimal places when you are determining how many significant figures to report:

When numbers are added or subtracted, the number of decimal places in the result should equal the smallest number of decimal places of any term in the sum or difference.

As an example of this rule, consider the sum

$$23.2 + 5.174 = 28.4$$

Notice that we do not report the answer as 28.374 because the lowest number of decimal places is one, for 23.2. Therefore, our answer must have only one decimal place.

The rule for addition and subtraction can often result in answers that have a different number of significant figures than the quantities with which you start. For example, consider these operations that satisfy the rule:

$$1.000\ 1 + 0.000\ 3 = 1.000\ 4$$

$$1.002 - 0.998 = 0.004$$

In the first example, the result has five significant figures even though one of the terms, 0.000 3, has only one significant figure. Similarly, in the second calculation, the result has only one significant figure even though the numbers being subtracted have four and three, respectively.

Note that while doing any calculation you should keep all significant figures of any value you use, and only round the answer at the end, especially where there are multiple steps. In all problems, you should obtain an algebraic solution and check its dimensions before performing any numerical calculations. This not only allows you to check the form of your solution, but also reduces errors due to accumulated rounding of values.

If the number of significant figures in the result of a calculation must be reduced, there is a general rule for rounding numbers: the last digit retained is increased by 1 if the last digit dropped is greater than 5. (For example, 1.346 becomes 1.35.) If the last digit dropped is less than 5, the last digit retained remains as it is. (For example, 1.343 becomes 1.34.) If the last digit dropped is equal to 5, the remaining digit should be rounded to the nearest even number. (This rule helps avoid accumulation of errors in long arithmetic processes.)

Example 1.4

Installing a carpet

A carpet is to be installed in a rectangular room whose length is measured to be 12.71 m and whose width is measured to be 3.46 m. Find the area of the room.

Solution

If you multiply 12.71 m by 3.46 m on your calculator, you will see an answer of 43.9766 m^2. How many of these numbers should you claim? Our rule of thumb for multiplication tells us that you can claim only the number of significant figures in your answer as are present in the measured quantity having the lowest number of significant figures. In this example, the lowest number of significant figures is three in 3.46 m, so we should express our final answer as 44.0 m^2.

End-of-chapter resources

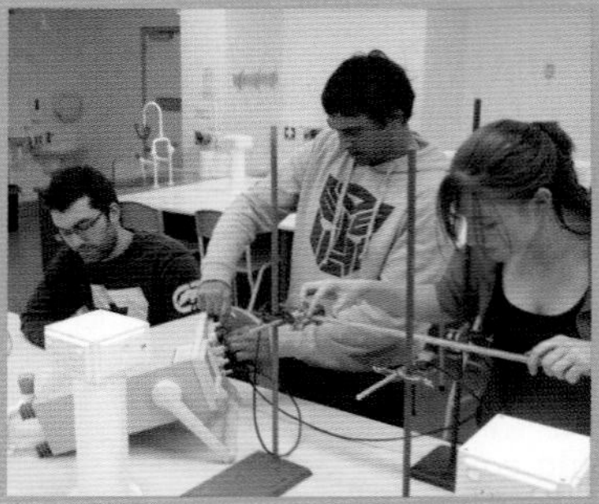

We started this chapter with a photo of students doing an experiment to measure the speed of sound in air, using a standing waves method (see Chapter 17). If they measured a value of 324.6 m/s, with an uncertainty of 32 m/s, are they 'right'?

If they write their value with the correct number of significant figures and include the uncertainty with the value, $v_{sound} = (320 \pm 30)$ m/s.

This is easier to compare with other measured values. For example, in 1738 members of the Paris Academy fired cannons situated at the two ends of a base line 29 km long and measured the interval between seeing the flash from the cannon and hearing the explosion. They calculated a value for the speed of sound as 332 m/s for dry air at 0°C. The students' value has a range from 290 m/s to 350 m/s, which includes the value of 332 m/s measured in 1738. The currently accepted value of 331 m/s (dry air, 0°C) also falls within this range, so the value measured by the students is reasonable in that it agrees with the currently accepted value. We do not usually ask whether an answer is 'right', as we do not know what the 'right' or 'true' answer is. Rather, we ask whether it is reasonable, and whether it agrees with experimental observations. In this case the students' answer is reasonable, and does agree with previously measured values.

The problems found in this chapter may be assigned online in Enhanced Web Assign.

Worked solutions to every fifth problem are available in the Student Solutions Manual. Register online at **www.cengagebrain.com** for access.

Summary

Definition

The three fundamental physical quantities of mechanics are **length**, **mass** and **time**, which in the SI system have the units **metre** (m), **kilogram** (kg) and **second** (s) respectively. These fundamental quantities cannot be defined in terms of more basic quantities.

The **density** of a substance is defined as its *mass per unit volume:*

$$\rho = \frac{m}{V} \quad (1.1)$$

Concepts and principles

Models are used in physics to represent objects and systems, and to help us understand, analyse and predict the behaviour of objects and systems. The model is a representation; it is not the object or system itself, so it is always an approximation. Different models may be chosen to represent a given system depending on the degree of accuracy required in the representation. Models are generally based on a particular theory.

Scientific theories are well substantiated explanations, which have both explanatory and predictive power. Scientific theories can be tested, and falsified, by experiment and observation.

There is an ongoing interplay between theory, experiment and technology, with advances in each contributing to advances in the other two.

The method of **dimensional analysis** is very powerful in solving physics problems. Dimensions can be treated as algebraic quantities. By making estimates and performing order-of-magnitude calculations, you should be able to approximate the answer to a problem when there is not enough information available to specify an exact solution. All measured quantities have an **uncertainty**, and any value derived from these also has an uncertainty that can be calculated by combining the uncertainties of the measured values in the appropriate way. The number of **significant figures** to which a value is written depends upon the uncertainty in that value.

When **multiplying** several quantities, the number of significant figures in the final answer is the same as the number of significant figures in the quantity having the smallest number of significant figures. The same rule applies to **division.** When numbers are **added** or **subtracted,** the number of decimal places in the result should equal the smallest number of decimal places of any term in the sum or difference.

When you compute a result from several measured numbers, each of which has a certain accuracy, you should give the result with the correct number of **significant figures**.

Chapter review quiz

To help you revise Chapter 1: Physics and measurement, complete the automatically graded Chapter review quiz at http://login.cengagebrain.com.

Conceptual questions

1. Suppose the three fundamental standards of the metric system were length, *density* and time rather than length, *mass* and time. The standard of density in this system is to be defined as that of water. What properties of water need to be considered to make sure that the standard of density is as accurate as possible?
2. Express the following quantities using the prefixes given in Table 1.4. (a) 3×10^{-4} m (b) 5×10^{-5} s (c) 72×10^{2} g
3. What natural phenomena could serve as alternative time standards? List at least 3.
4. ±? Two students perform experiments to measure the speed of light. One finds a value of $(3.0 \pm 0.1) \times 10^{8}$ m/s, the other finds a value of $(2.8 \pm 0.1) \times 10^{8}$ m/s. Do their values agree? Explain your answer.
5. ±? If the students in conceptual question 4 had found values of $(3.00 \pm 0.01) \times 10^{8}$ m/s, and $(2.97 \pm 0.01) \times 10^{8}$ m/s, would their experimental values agree?

Problems

Σ *Note*: You may wish to review your mathematics skills; the online resources contain mathematics review questions for this purpose. Consult the endpapers, appendices and tables in the text whenever necessary in solving problems. Answers to odd-numbered problems appear in the back of the book.

Section 1.1 Models and theories

1. Figure P.1.1 shows students studying the thermal conduction of energy into cylindrical blocks of ice. As we will see in Chapter 20, this process is described by the mathematical model:

$$\frac{Q}{\Delta t} = \frac{k\pi d^2 (T_h - T_c)}{4L}$$

For experimental control, in one set of trials all quantities except d and Δt are constant. (a) If d is made three times larger, does the equation predict that Δt will get larger or get smaller? By what factor? (b) What pattern of proportionality of Δt to d does the equation predict? (c) To display this proportionality as a straight line on a graph, what quantities should you plot on the horizontal and vertical axes? (d) What expression represents the theoretical gradient of this graph?

Alexandra Héder

Figure P1.1

2. In physics, it is often convenient to use mathematical approximations. For example, the small angle approximation says that:

$$\tan\alpha \approx \sin\alpha \approx \alpha = \frac{\pi\alpha'}{180°}$$

where α is in radians and α' is in degrees.

(a) On a single set of axes, plot $\tan\alpha$, $\sin\alpha$ and α as functions of α. (b) Find the largest value of α for which $\tan\alpha$ may be approximated as α with an error of less than 10%.

Section 1.2 Standards of length, mass and time

3. (a) Use information from the endpapers of this book to calculate the average density of the Earth. (b) Where does the value fit among those listed in Table 14.1 in Chapter 14? Look up the density of a typical surface rock such as granite in another source and compare it with the density of the Earth.
4. The standard kilogram (Figure 1.1a) is a platinum–iridium cylinder 39.0 mm in height and 39.0 mm in diameter. What is the density of the material?
5. Two spheres are cut from a certain uniform rock. One has radius 4.50 cm. The mass of the other is five times greater. Find its radius.

Section 1.3 Dimensional analysis

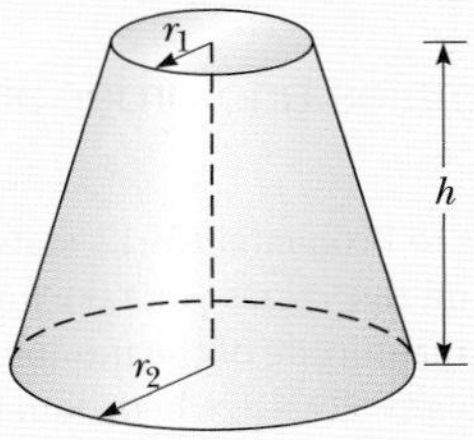

Figure P1.6

6. Figure P1.6 shows a *frustum of a cone*. Match each of the following expressions (a to c) with the quantity it describes (d to f):

(a) $\pi(r_1 + r_2)[h^2 + (r_2 - r_1)^2]^{\frac{1}{2}}$

(b) $2\pi(r_1 + r_2)$, and

(c) $\pi h(r_1^2 + r_1 r_2 + r_2^2)/3$

(d) the total circumference of the flat circular faces

(e) the volume

(f) the area of the curved surface.

7. Which of the following equations is dimensionally correct?

(a) $v_f = v_i + ax$ where v is measured in m/s, a in m/s^2 and x in m.

(b) $y = (2\text{ m})\cos(kx)$, where $k = 2\text{ m}^{-1}$

8. Newton's law of universal gravitation is represented by $F = \frac{GMm}{r^2}$ where F is the magnitude of the gravitational force exerted by one small object on another, M and m are the masses of the objects, and r is a distance. Force has the SI units kg m/s^2. What are the SI units of the proportionality constant G?

Section 1.4 Conversion of units

9. Hugh Jackman is looking at real estate on Sesame Street in New York and is thinking about buying an apartment that has a width of 75.0 ft and a length of 125 ft. Being an Australian he tends to think in SI units rather than imperial or 'US customary' units. Determine the area of this apartment in square metres.

10. Suppose your hair grows at the rate 0.079 cm per day. Find the rate at which it grows in nanometres per second. Because the distance between atoms in a molecule is of the order of 0.1 nm, your answer suggests how rapidly layers of atoms are assembled in this protein synthesis.

11. One gallon of paint covers an area of 25.0 m^2. What is the thickness of the fresh paint on the wall?

12. The dimensions of the 'Ark of the Covenant' as given in the bible are $1\frac{1}{2}$ cubits (height) $\times$ $1\frac{1}{2}$ cubits (width) $\times$ $2\frac{1}{2}$ cubits (length). Find the volume of the Ark in SI units. Do you think Indiana Jones would have been able to carry it?

13. One cubic metre (1.00 m^3) of aluminium has a mass of 2.70×10^3 kg, and the same volume of iron has a mass of 7.86×10^3 kg. Find the radius of a solid aluminium sphere that will balance a solid iron sphere of radius 2.00 cm on an equal-arm balance.

14. Let ρ_{Al} represent the density of aluminium and ρ_{Fe} that of iron. Find the radius of a solid aluminium sphere that balances a solid iron sphere of radius r_{Fe} on an equal-arm balance.

Section 1.5 Estimates and order-of-magnitude calculations

Note: In your solutions to Problems 15–17, state the quantities you measure or estimate and the values you take for them.

15. Find the order of magnitude of the number of table-tennis balls that would fit into a typical-size bedroom (without the balls being crushed).

16. A car tyre is rated to last for 80 000 kilometres. To an order of magnitude, through how many revolutions will it turn over its lifetime?

17. Bacteria and other prokaryotes are found deep underground, in water, and in the air. One micron (10^{-6} m) is a typical length scale associated with these microbes. (a) Estimate the total number of bacteria and other prokaryotes on the Earth. (b) Estimate the total mass of all such microbes.

Section 1.6 Uncertainties in measurement

18. ±? A rectangular plate has a length of (21.3 ± 0.2) cm and a width of (9.8 ± 0.1) cm. Calculate the area of the plate, including its uncertainty.

19. ±? What is the total perimeter of the plate described in Problem 18?

20. ±? The radius of a uniform solid sphere is measured to be (6.50 ± 0.20) cm, and its mass is measured to be (1.85 ± 0.02) kg. Determine the density of the sphere in kilograms per cubic metre and the uncertainty in the density.

21. ±? A boy measures his height as 1.45 m using a tape measure. The uncertainty in this measurement is 1 cm. A few weeks later he again measures his height and finds a value of 1.46 m this time, with the same uncertainty. By how much has he grown? Can he claim that he has definitely grown taller in this time period? Explain your answer.

22. ±? Students are doing an experiment to find the electrical resistivity of different materials. Resistivity, which we will learn about in Chapter 27, is a measure of how *resistant* a material is to the flow of electric current and is given by: $\rho = \frac{RA}{L}$, where R is the measured resistance of the sample of the material, A is its cross-sectional area and L is its length. For a certain piece of wire the students make the following measurements:

the length of the wire, L = (1.234 ± 0.002) m
the diameter of the wire, D = (0.0050 ± 0.0002) m
the resistance of the wire, R = (15.5 ± 0.1) Ω.

(a) What is the fractional uncertainty in each of the measurements the students have made? (b) What is the best value for the resistivity, ρ, from their measurements? (c) What is the uncertainty in this value of ρ?

23. ±? Students are doing an experiment to find the resistivity of different materials, as described in Problem 22 above. Instead of taking only a single measurement for the resistance, R, they make six measurements and record the following values: 15.5 Ω, 15.3 Ω, 15.6 Ω, 15.5 Ω, 15.4 Ω, 15.5 Ω. (a) What is the best value of R and its uncertainty from these data? (b) What is the best value for the resistivity, ρ, and its uncertainty?

Section 1.7 Significant figures

24. How many significant figures are in the following numbers? (a) 78.9 ± 0.2 (b) 3.788×10^9 (c) 2.46×10^{-6} (d) 0.0053

25. Carry out these arithmetic operations, giving your answers to the correct number of significant figures. (a) the sum of the measured values 756, 37.2, 0.83 and 2, (b) the product 0.0032 × 3356.3, and (c) the product 5.620 and π.

Additional problems

26. A typical bacterial length scale is 10^{-6} m. (a) Estimate the human intestinal volume and assume 1% of it is occupied by bacteria. What is the order of magnitude of the number of microorganisms in the human intestinal tract? (b) Does the number of bacteria suggest whether the bacteria are beneficial, dangerous, or neutral for the human body?

27. What distance differs from 100 m and from 1000 m by equal factors so that we could equally well choose to represent its order of magnitude as $\sim 10^2$ m or as $\sim 10^3$ m?

28. The diameter of our disc-shaped galaxy, the Milky Way, is about 1.0×10^5 light-years (ly). The distance to the Andromeda galaxy (**Figure P1.28**), which is the spiral galaxy nearest to the Milky Way, is about 2.0 million ly. If a scale model represents the Milky Way and Andromeda galaxies as dinner plates 25 cm in diameter, determine the distance between the centres of the two plates.

Robert Gendler/NASA

Figure P1.28
The Andromeda galaxy

29. The distance from the Sun to the nearest star is about 4×10^{16} m. The Milky Way (**Figure P1.29**) is roughly a disc of diameter $\sim 10^{21}$ m and thickness $\sim 10^{19}$ m. Find the order of magnitude of the number of stars in the Milky Way. Assume the distance between the Sun and our nearest neighbour is typical.

Richard Payne/NASA

Figure P1.29
The Milky Way galaxy

30. **Review.** A student is supplied with a stack of copy paper, a ruler, a compass, scissors, and a sensitive balance. He cuts out various shapes in various sizes, calculates their areas, measures their masses, and prepares the graph shown in **Figure P1.30**. (a) Consider the fourth experimental point from the top. How far is it from the best-fit straight line? Express your answer as a difference in vertical-axis coordinate. (b) Express your answer as a percentage. (c) Calculate the slope of the line. (d) State what the graph demonstrates, referring to the shape of the graph and the results of parts (b) and (c). (e) Describe whether this result should be expected theoretically. (f) Describe the physical meaning of the slope.

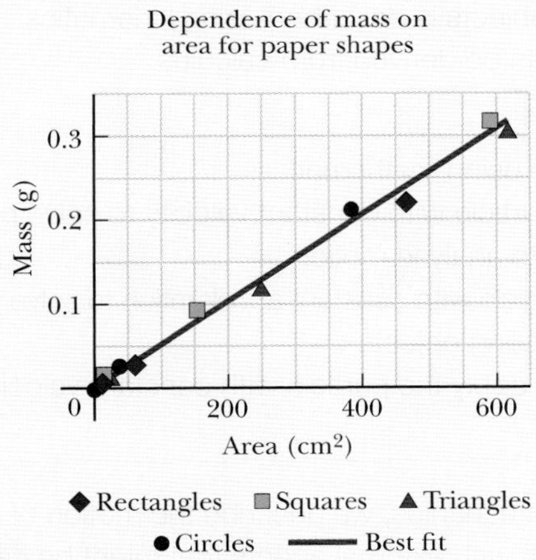

Figure P1.30

31. A high fountain of water is located at the centre of a circular pool as shown in **Figure P1.31**. A student walks around the pool and measures its circumference to be 15.0 m. Next, the student stands at the edge of the pool and uses a protractor to gauge the angle of elevation of the top of the fountain to be $\phi = 55.0°$. How high is the fountain?

Figure P1.31

Σ *See Appendix B.4 for help with trigonometry*

Challenge problems

32. You stand in a flat paddock and observe two cows (**Figure P1.32**). Cow A is due north of you and 15.0 m from your position. Cow B is 25.0 m from your position. From your point of view, the angle between cow A and cow B is 20.0°, with cow B appearing to the right of cow A. (a) How far apart are cow A and cow B? (b) Consider the view seen by cow A. According to this cow, what is the angle between you and cow B? (c) Consider the view seen by cow B. According to this cow, what is the angle between you and cow A? Hint: What does the situation look like to an eagle hovering above the meadow?

Figure P1.32
Your view of two cows in a paddock. Cow A is due north of you. You must rotate your eyes through an angle of 20.0° to look from cow A to cow B.

Before beginning the following chapters, you should prepare by revising your mathematics. There is a set of revision problems to help you check that you are ready to proceed at www.cengagebrain.com. If you can solve these problems, you are ready to proceed.

chapter 2

Motion in one dimension

Each year 'Race the Rattler' is held in Mary Valley, between the town of Dagun and the old Gympie Railway Station 18.5 km away. The race is a footrace against a C17 steam locomotive built in the 1920s. Who would you expect to win – a human or the train? Would it make any difference if the race was much shorter or much longer?

Newspix/News Ltd/Graeme Parkes

ENHANCED WebAssign

The problems found in this chapter may be assigned online in Enhanced Web Assign.

On completing this chapter, students will understand:

- how motion can be described in terms of position, velocity and acceleration
- the difference between speed and velocity
- how analysis models can be used to analyse motion
- how the motion of freely falling objects is described using kinematics
- that the kinematics equations can be derived from calculus.

Students will be able to:

- describe the motion of a particle using mathematics
- apply the analysis model for a particle with constant velocity to solve problems
- draw and use motion diagrams to analyse motion
- apply the analysis model for a particle with constant acceleration to solve problems
- solve problems involving freely falling objects
- identify when it is appropriate to apply the particle with constant velocity or constant acceleration model.

As a first step in studying classical mechanics, we describe the motion of an object while ignoring the interactions with external agents that might be causing or modifying that motion. This portion of classical mechanics is called *kinematics*. In this chapter, we consider only motion in one dimension, that is, motion of an object along a straight line.

In our study of translational motion, we use what is called the **particle model** and describe the moving object as a *particle* regardless of its size.

2.1 Position, velocity and speed

Remember our discussion of making models for physical situations in Section 1.1. In general, **a particle is a point-like object, that is, an object that has mass but is of infinitesimal size**. For example, if we wish to describe the motion of the Earth around the Sun, we can treat the Earth as a particle and obtain reasonably accurate data about its orbit. This approximation is justified because the radius of the Earth's orbit is large compared with the dimensions of the Earth and the Sun.

The use of a point to represent an extended object is one example of a common approximation used by physicists. This is called the particle approximation or particle model. In physics, it is often a good idea to look for ways to construct a simple model that captures the important features of a physical situation. However, when we do this, it is important to remember that we are dealing with a limited model and that what we have left out of that model will limit the accuracy of our results.

Position

A particle's **position** x is the location of the particle with respect to a chosen reference point that we can consider to be the origin of a coordinate system. The motion of a particle is completely known if the particle's position in space is known at all times.

Consider a car moving back and forth along the x axis as in **Active Figure 2.1a**. When we begin collecting position data, the car is 30 m to the right of the reference position $x = 0$. We will use the particle model by identifying some point on the car as a particle representing the entire car. Usually we choose either the geometric centre of the object or the average position of its mass, known as the centre of mass.

We start our clock and once every 10 s we note the car's position. As you can see from **Table 2.1**, the car moves to the right (which we have defined as the positive direction) during the first 10 s of motion, from position Ⓐ to position Ⓑ. After Ⓑ, the position values begin to decrease, suggesting the car is going in the opposite direction away from position Ⓑ to position Ⓕ. In fact, at Ⓓ, 30 s after we start measuring, the car is at the origin, $x = 0$ (see **Active Figure 2.1a**). It continues moving to the left and is more than 50 m to the left of $x = 0$ when we stop recording information after our sixth data point. A graphical representation of this information is presented in **Active Figure 2.1b**. Such a plot is called a *position–time graph*.

Table 2.1
Position of the car at various times

Position	t (s)	x (m)
Ⓐ	0	30
Ⓑ	10	52
Ⓒ	20	38
Ⓓ	30	0
Ⓔ	40	−37
Ⓕ	50	−53

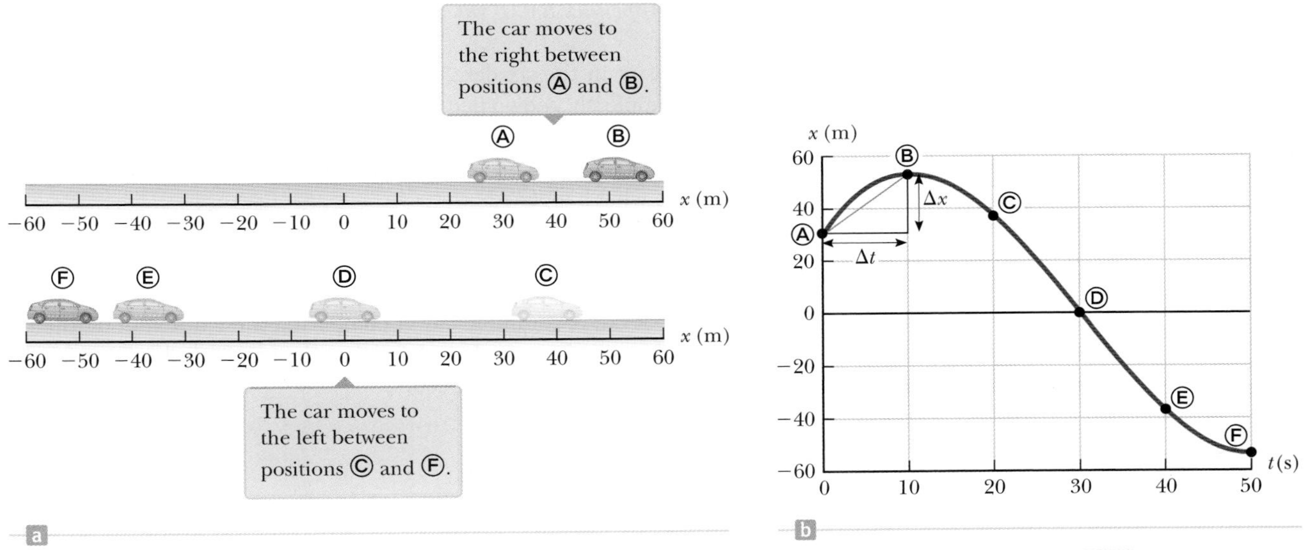

Active Figure 2.1

A car moves back and forth along a straight line. If we are interested only in the car's translational motion, and not, for example, in whether it is going forwards or in reverse, we can model it as a particle. Several representations of the information about the motion of the car can be used. **Table 2.1** is a tabular representation of the information. (a) A pictorial representation of the motion of the car (b) A graphical representation (position–time graph) of the motion of the car

Notice the *alternative representations* of information that we have used for the motion of the car. Active Figure 2.1a is a *pictorial representation*, whereas Active Figure 2.1b is a *graphical representation*. Table 2.1 is a *tabular representation* of the same information. We also have a *textual representation*, i.e., a description in words. Creating an alternative representation is often an excellent strategy for understanding the situation in a given problem. The ultimate goal in many problems is a *mathematical representation* that can be analysed to solve for some desired piece of information.

We do not give uncertainties here because we assume that any rounding errors in the simulation that produced this data are negligible compared to the precision to which the values are given. Note that this is rarely the case with real experiments, and not always the case with simulations.

Given the data in Table 2.1, we can determine the change in position of the car for various time intervals. The **displacement** Δx of a particle is defined as its change in position over some time interval. As the particle moves from an initial position x_i to a final position x_f, its displacement is given by

Displacement ▶

$$\Delta x \equiv x_f - x_i \tag{2.1}$$

Getty Images/Paul Kane

Figure 2.2
On this field, players run back and forth for the entire game. The distance that the players run over the duration of the game is non-zero. The displacement of the players over the duration of the game is approximately zero because they finish the game at approximately the same point at which they started it.

We use the capital Greek letter delta (Δ) to denote the *change* in a quantity. From this definition, we see that Δx is positive if x_f is greater than x_i and negative if x_f is less than x_i.

It is very important to recognise the difference between displacement and distance travelled. **Distance** is the length of a path followed by a particle. Consider, for example, the football players in Figure 2.2. If a player runs from the centre of the field to the goals and back again, the *displacement* of the player during this time interval is zero because he ended up at the same point as he started: $x_f = x_i$, so $\Delta x = 0$. During this time interval, however, he moved through a *distance* equal to the length of the football field. Distance is always represented as a positive number, whereas displacement can be either positive or negative.

Displacement is an example of a **vector quantity**. Many other physical quantities, including position, velocity and acceleration, also are vectors. In general, a vector quantity requires the specification of both direction and magnitude. By contrast, a **scalar quantity** has a numerical value and no direction. In this chapter, we use positive (+) and negative (−) signs to indicate vector direction. For example, for horizontal motion let us arbitrarily specify to the right as being the positive direction. It follows that any object always moving to the right undergoes a positive displacement $\Delta x > 0$, and any object moving to the left undergoes a negative displacement so that $\Delta x < 0$.

Σ *For more information about vectors, see Appendix B.6. If you have not studied vectors before you should review this Appendix before proceeding.*

Notice that the data in Table 2.1 only result in the six data points in the graph in Active Figure 2.1b. Therefore, the motion of the particle is not completely known because we don't know its position at *all* times. The smooth curve drawn through the six points in the graph is only a *possibility* of the actual motion of the car. We only have information about six instants of time; we have no idea what happened between the data points. The smooth curve is only a *guess* as to what happened.

It is often easier to see changes in position from a graph than from a verbal description or even a table of numbers. For example, it is clear that the car covers more ground during the middle of the 50 s interval than at the end. Between positions Ⓒ and Ⓓ, the car travels almost 40 m, but during the last 10 s, between positions Ⓔ and Ⓕ, it moves less than half that far. A common way of comparing these different motions is to consider the *velocity*. The **average velocity** $v_{x,\text{avg}}$ of a particle is defined as the particle's displacement Δx divided by the time interval Δt during which that displacement occurs:

Average velocity ▶

$$v_{x,\text{avg}} \equiv \frac{\Delta x}{\Delta t} \tag{2.2}$$

where the subscript x indicates motion along the x axis. From this definition we see that average velocity has dimensions of length divided by time [L/T], and SI units of metres per second. Average quantities are also sometimes written with a bar above them, for example, $\bar{v}_x = v_{x,\text{avg}}$.

The average velocity of a particle moving in one dimension can be positive or negative, depending on the sign of the displacement. (The time interval Δt is generally positive.) If the coordinate of the particle increases in time (that is, if $x_f > x_i$), then Δx is positive, $v_{x,\text{avg}} = \Delta x/\Delta t$ is also positive and the particle is moving in the positive x direction. If the coordinate of the particle decreases in time (that is, if $x_f < x_i$), Δx is negative and hence $v_{x,\text{avg}}$ is negative. Like displacement, velocity is a vector.

We can interpret average velocity geometrically by drawing a straight line between any two points on the position–time graph in **Active Figure 2.1b**. This line forms the hypotenuse of a right-angled triangle of height Δx and base Δt. The gradient of this line is the ratio $\Delta x/\Delta t$, which is what we have defined as average velocity in **Equation 2.2**. For example, the line between positions Ⓐ and Ⓑ in **Active Figure 2.1b** has a gradient equal to the average velocity of the car between those two times (52 m − 30 m)/(10 s − 0 s) = 2.2 m/s.

To review the concept of a gradient, see the section on linear equations in Appendix B.2. Σ

In everyday usage, the terms *speed* and *velocity* are interchangeable. In physics, however, there is a clear distinction between these two quantities. Consider a marathon runner who runs a distance s of more than 40 km and yet ends up at her starting point. Her total displacement is zero, so her average velocity is zero! Nonetheless, we need to be able to quantify how fast she was running. The **average speed** v_{avg} of a particle, a scalar quantity, is defined as the total distance s travelled divided by the total time interval required to travel that distance:

$$v_{avg} \equiv \frac{s}{\Delta t} \qquad (2.3)$$

◄ Average speed

The SI unit of average speed is the same as the unit of average velocity: metres per second. Unlike average velocity, however, average speed has no direction and is always expressed as a positive number. Notice the clear distinction between the definitions of average velocity and average speed: average velocity (**Equation 2.2**) is the *displacement* divided by the time interval, whereas average speed (**Equation 2.3**) is the *distance* divided by the time interval.

Knowledge of the average velocity or average speed of a particle does not provide information about the details of the trip. For example, suppose it takes you 45.0 s to travel 100 m down a long, straight hallway toward your departure gate at an airport. At the 100 m mark, you realise you missed the toilets, and you return back 25.0 m along the same hallway, taking 10.0 s to make the return trip. The magnitude of your average *velocity* is +75.0 m/55.0 s = +1.36 m/s. The average *speed* for your trip is 125 m/55.0 s = 2.27 m/s. You may have travelled at various speeds during the walk and, of course, you changed direction. Neither average velocity nor average speed on their own provides information about these details.

> **Pitfall Prevention 2.1**
> The magnitude of the average velocity is *not* the average speed. For example, consider the marathon runner discussed here. The magnitude of her average velocity is zero, but her average speed is clearly not zero.

Quick Quiz 2.1

Under which of the following conditions is the magnitude of the average velocity of a particle moving in one dimension smaller than the average speed over some time interval? **(a)** A particle moves in the $+x$ direction without reversing. **(b)** A particle moves in the $-x$ direction without reversing. **(c)** A particle moves in the $+x$ direction and then reverses the direction of its motion. **(d)** There are no conditions for which this is true.

Example 2.1

Calculating the average velocity and speed

Find the displacement, average velocity and average speed of the car in **Active Figure 2.1a** between positions Ⓐ and Ⓕ.

Solution

Consult **Active Figure 2.1** to form a mental image of the car and its motion. (In general, you will have to draw a diagram; in this case you are already given one.) We model the car as a particle. From the position–time graph given in **Active Figure 2.1b**, notice that $x_{Ⓐ} = 30$ m at $t_{Ⓐ} = 0$ s and that $x_{Ⓕ} = -53$ m at $t_{Ⓕ} = 50$ s.

Use **Equation 2.1** to find the displacement of the car: $\Delta x = x_{Ⓕ} - x_{Ⓐ} = -53\text{ m} - 30\text{ m} = -83\text{ m}$

This result means that the car ends up 83 m in the negative direction (to the left, in this case) from where it started. This number has the correct units and is of the same order of magnitude as the supplied data. A quick look at **Active Figure 2.1a** indicates that it is the correct answer.

Use **Equation 2.2** to find the car's average velocity:

$$v_{x,\text{avg}} = \frac{x_{Ⓕ} - x_{Ⓐ}}{t_{Ⓕ} - t_{Ⓐ}}$$

$$= \frac{-53\text{ m} - 30\text{ m}}{50\text{ s} - 0\text{ s}} = \frac{-83\text{ m}}{50\text{ s}} = -1.7\text{ m/s}$$

Example 2.1 cont.

We cannot unambiguously find the average speed of the car from the data in **Table 2.1** because we do not have information about the positions of the car between the data points. If we adopt the assumption that the details of the car's position are described by the curve in **Active Figure 2.1b**, the distance travelled is 22 m (from Ⓐ to Ⓑ) plus 105 m (from Ⓑ to Ⓕ), for a total of 127 m.

Use **Equation 2.3** to find the car's average speed:

$$v_{avg} = \frac{127\text{ m}}{50\text{ s}} = 2.5\text{ m/s}$$

Notice that the average speed is positive, as it must be.

What If? Suppose the red-brown curve in **Active Figure 2.1b** were different so that between 0 s and 10 s it went from Ⓐ up to 100 m and then came back down to Ⓑ. Would this change the values we calculate for average velocity and average speed?

Answer The average speed of the car would change because the distance is different, but the average velocity would not change.

2.2 Instantaneous velocity and speed

Pitfall Prevention 2.2
In any graph, the *gradient* represents the ratio of the change in the quantity represented on the vertical axis to the change in the quantity represented on the horizontal axis. Remember that *a gradient has units* (unless both axes have the same units). The units of the gradient in Active Figures 2.1b and 2.3 are metres per second, the units of velocity.

Often we need to know the velocity of a particle at a particular instant in time rather than the average velocity over a finite time interval. What does it mean to talk about how quickly something is moving if we 'freeze time' and talk only about a particular instant? In the late 1600s, with the invention of calculus, scientists began to understand how to describe an object's motion at any moment in time.

To see how that is done, consider **Active Figure 2.3a**, which is a reproduction of the graph in **Active Figure 2.1b**. We have already discussed the average velocity for the interval during which the car moved from position Ⓐ to position Ⓑ (given by the gradient of the short blue line) and for the interval during which it moved from Ⓐ to Ⓕ (represented by the gradient of the longer blue line and calculated in **Example 2.1**). The car starts out by moving to the right, which we defined to be the positive direction. The value of the average velocity during the interval from Ⓐ to Ⓑ is closer to the initial velocity than is the value of the average velocity during the interval from Ⓐ to Ⓕ. Now let us focus on the short blue line and slide point Ⓑ to the left along the curve, toward point Ⓐ, as in **Active Figure 2.3b**. As the two points become extremely close together, the line becomes a tangent line to the curve, indicated by the green line in **Active Figure 2.3b**. The gradient of this tangent line represents the velocity of the car at point Ⓐ. What we have done is determine the *instantaneous velocity* at that moment. In other words, the **instantaneous velocity** v_x equals the limiting value of the ratio $\frac{\Delta x}{\Delta t}$ as Δt approaches zero:

$$v_x \equiv \lim_{\Delta t \to 0} \frac{\Delta x}{\Delta t} \tag{2.4}$$

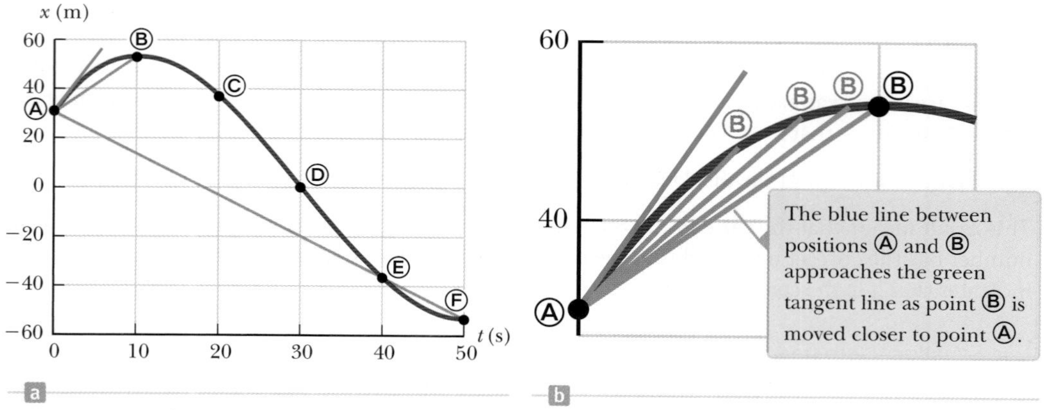

Active Figure 2.3
(a) Graph representing the motion of the car in **Active Figure 2.1**. (b) An enlargement of the upper-left-hand corner of the graph.

In calculus, this limit is called the *derivative* of x with respect to t, written $\frac{dx}{dt}$:

$$v_x \equiv \lim_{\Delta t \to 0} \frac{\Delta x}{\Delta t} = \frac{dx}{dt} \tag{2.5}$$

◀ Instantaneous velocity

Σ

Appendix B.7 summarises differential calculus and gives you the rules for taking derivatives of many different functions. If you have not studied calculus before you should review this Appendix before proceeding.

The instantaneous velocity can be positive, negative, or zero. When the gradient of the position–time graph is positive, such as at any time during the first 10 s in **Active Figure 2.3**, v_x is positive and the car is moving toward larger values of x. After point Ⓑ, v_x is negative because the gradient is negative and the car is moving toward smaller values of x. At point Ⓑ, the gradient and the instantaneous velocity are zero and the car is momentarily at rest.

From here on, we use the word *velocity* to designate instantaneous velocity. When we are interested in *average velocity,* we shall always use the adjective *average.*

The **instantaneous speed** of a particle is defined as the magnitude of its instantaneous velocity. As with average speed, instantaneous speed has no direction associated with it. Velocity is a vector; speed is a scalar. For example, if one particle has a velocity of +25 m/s along a given line and another particle has a velocity of −25 m/s along the same line, both have a speed of 25 m/s. As with velocity, we drop the adjective for instantaneous speed. 'Speed' means instantaneous speed.

In Pitfall Prevention 2.1, we said that the magnitude of the average velocity is not the average speed. The magnitude of the instantaneous velocity, however, *is* the instantaneous speed.

Quick Quiz 2.2

Is a police officer with a radar gun more interested in **(a)** your average speed or **(b)** your instantaneous speed as you drive past?

Conceptual Example 2.2

Average and instantaneous velocity I

Consider the following one-dimensional motions: **(A)** a ball thrown directly upwards rises to a highest point and falls back into the thrower's hand; **(B)** a race car starts from rest and speeds up to 100 m/s; and **(C)** a spacecraft drifts through space at constant velocity. Are there any points in the motion of these objects at which the instantaneous velocity has the same value as the average velocity over the entire motion? If so, identify the point(s).

Solution

(A) The average velocity for the thrown ball is zero because the ball returns to the starting point; therefore, its displacement is zero. There is one point at which the instantaneous velocity is zero: at the top of the motion.

(B) The car's average velocity cannot be evaluated unambiguously with the information given, but it must have some value between 0 and 100 m/s. Because the car will have every instantaneous velocity between 0 and 100 m/s at some time during the interval, there must be some instant at which the instantaneous velocity is equal to the average velocity over the entire motion.

(C) Because the spacecraft's instantaneous velocity is constant, its instantaneous velocity at any time and its average velocity over any time interval are the same.

Note that as long as the velocity is a continuous function, there must always exist at least one point at which the instantaneous velocity is equal to the average velocity.

Example 2.3

Average and instantaneous velocity II

A particle moves along the x axis. Its position varies with time according to the expression $x = -4t + 2t^2$, where x is in metres and t is in seconds.

We assume the coefficients, 4 and 2, have units which make the equation dimensionally consistent. We can determine what these are using dimensional analysis. In this case the coefficient in front of t must have units of m/s so that $4t$ has units of m, and the coefficient of t^2 must be m/s^2.

Because the position of the particle is given by a mathematical function, the motion of the particle is completely known, unlike that of the car in **Active Figure 2.1**. Notice that the particle moves in the negative x direction for the first second of motion, is momentarily at rest at the moment $t = 1$ s, and moves in the positive x direction at times $t > 1$ s.

(A) Determine the displacement of the particle in the time intervals $t = 0$ s to $t = 1$ s and $t = 1$ s to $t = 3$ s.

Solution

Always begin by drawing a diagram, as this is the first step in creating a model of the situation that you can then use to solve the problem. In this case there are two diagrams that are useful, a graph showing the position as a function of time, and a picture showing the car (as a particle) at different times. These are shown in **Figure 2.4**.

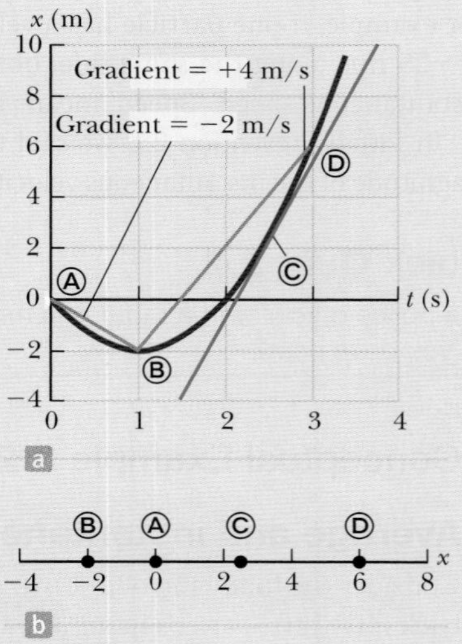

Figure 2.4
(Example 2.3) (a) Position–time graph for a particle having an x coordinate that varies in time according to the expression $x = -4t + 2t^2$. (b) The particle moves in one dimension along the x axis.

The graph in **Figure 2.4a** helps you to form a mental representation of the particle's motion. Keep in mind that the particle does not move in a curved path in space such as that shown by the red-brown curve in the graphical representation. The particle moves only along the x axis in one dimension as shown in **Figure 2.4b**.

During the first time interval, the gradient is negative and hence the average velocity is negative. Therefore we know that the displacement between Ⓐ and Ⓑ must be a negative number having units of metres. Similarly we expect the displacement between Ⓑ and Ⓓ to be positive.

In the first time interval, set $t_i = t_Ⓐ = 0$ and $t_f = t_Ⓑ = 1$ s and use **Equation 2.1** to find the displacement:

$$\Delta x_{Ⓐ \to Ⓑ} = x_f - x_i = x_Ⓑ - x_Ⓐ$$
$$= (-4t_Ⓑ + 2t_Ⓑ{}^2) - (-4t_Ⓐ + 2t_Ⓐ{}^2)$$
$$= [-4(1) + 2(1)^2] - [-4(0) + 2(0)^2] = -2 \text{ m}$$

We could also have read this directly off the graph, which we will do for the second time interval from $t = 1$ s to $t = 3$ s, giving a displacement of +8 m.

Both methods, using either the equation or reading directly from the graph, give the same answer. When there are two possible methods like this, as is often the case, it is a good idea to use one method to find your answer, then check your answer using the other method.

(B) Calculate the average velocity during these two time intervals.

Solution

In the first time interval, use **Equation 2.2** with $\Delta t = t_f - t_i = t_Ⓑ - t_Ⓐ = 1$ s:

$$v_{x,\,avg(Ⓐ\to Ⓑ)} = \frac{\Delta x_{Ⓐ\to Ⓑ}}{\Delta t} = \frac{-2 \text{ m}}{1 \text{ s}} = -2 \text{ m/s}$$

In the second time interval, $\Delta t = 2$ s:

$$v_{x,\,avg(Ⓑ\to Ⓓ)} = \frac{\Delta x_{Ⓑ\to Ⓓ}}{\Delta t} = \frac{8 \text{ m}}{2 \text{ s}} = +4 \text{ m/s}$$

Example 2.3 cont.

These values are the same as the gradients of the blue lines joining these points in Figure 2.4a.

(C) Find the instantaneous velocity of the particle at $t = 2.5$ s.

Solution

Measure the gradient of the green line at $t = 2.5$ s (point © in Figure 2.4a):

$$v_x = \frac{10\text{ m} - (-4\text{ m})}{3.8\text{ s} - 1.5\text{ s}} = +6\text{ m/s}$$

Notice that this instantaneous velocity is of the same order of magnitude as our previous results, that is, a few metres per second. Is that what you would have expected?

2.3 Analysis model: particle with constant velocity

In Section 1.1 we discussed the importance of making models. We now come to our first *analysis model*, where we apply the particle model to an object with constant velocity.

We use Equation 2.2 to build our first analysis model for solving problems. We imagine a particle moving with a constant velocity. The model of a **particle with constant velocity** can be applied in *any* situation in which an object that can be modelled as a particle is moving with constant velocity. This situation occurs frequently so this model is very useful.

If the velocity of a particle is constant over some interval, its instantaneous velocity at any instant is the same as the average velocity in that interval. That is, $v_x = v_{x,\text{ avg}}$. Therefore, Equation 2.2 gives us an equation to be used in the mathematical representation of this situation:

$$v_x = \frac{\Delta x}{\Delta t} \tag{2.6}$$

Remembering that $\Delta x = x_f - x_i$, we see that $v_x = \dfrac{x_f - x_i}{\Delta t}$, or

$$x_f = x_i + v_x \Delta t$$

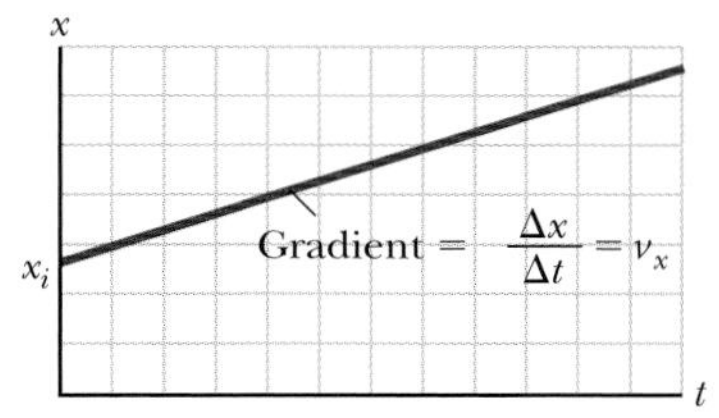

Figure 2.5
Position–time graph for a particle with constant velocity. The value of the constant velocity is the gradient of the line.

This equation tells us that the position of the particle is given by the sum of its original position x_i at time $t = 0$ plus the displacement $v_x \Delta t$ that occurs during the time interval Δt. In practice, we usually choose the time at the beginning of the interval to be $t_i = 0$ and the time at the end of the interval to be $t_f = t$, so our equation becomes

$$x_f = x_i + v_x t \quad (\text{for constant } v_x) \tag{2.7}$$

◀ Position as a function of time for the particle with constant velocity model

Equations 2.6 and 2.7 are the primary equations used in the model of a particle with constant velocity. Whenever you identify the analysis model in a problem to be the particle with constant velocity, you can immediately turn to these equations.

Figure 2.5 is a graphical representation of the particle with constant velocity. On this position–time graph, the gradient of the line representing the motion is constant and equal to the magnitude of the velocity. Equation 2.7, which is the equation of a straight line, is the mathematical representation of the particle with constant velocity model. The gradient of the straight line is v_x and the y intercept is x_i in both representations.

Example 2.4

Modelling a runner as a particle

A sports scientist at the Australian Institute of Sport in Canberra is studying the biomechanics of the human body. She determines the velocity of a runner while he runs along a straight line at a constant rate. The runner begins to run, quickly reaching a constant velocity. The scientist starts the stopwatch as he runs past her and then stops it after the runner has run a further distance of 20 m. The time interval indicated on the stopwatch is 4.0 s.

Getty Images/Howard Kingsnorth

(A) What is the runner's velocity?

Solution

We model the moving runner as a particle because the size of the runner and the movement of arms and legs are unnecessary details. Because the problem states that the subject runs at a constant rate, we can model him as a particle with constant velocity.

Having identified the model, we can use Equation 2.6 to find the constant velocity of the runner:

$$v_x = \frac{\Delta x}{\Delta t} = \frac{x_f - x_i}{\Delta t} = \frac{20\text{ m} - 0}{4.0\text{ s}} = 5.0\text{ m/s}$$

(B) If the runner continues his motion after the stopwatch is stopped, what is his position after 10 s has passed?

Solution

Use Equation 2.7 and the velocity found in part (A) to find the position of the particle at time $t = 10$ s:

$$v_f = v_i + v_x t = 0 + (5.0\text{ m/s})(10\text{ s}) = 50\text{ m}$$

What If? What if we doubled the value of t?

Answer As Equation 2.7 tells us that displacement is linearly proportional to time when acceleration is zero and $x_i = 0$, we expect that x will change by whatever factor t changes by. Hence doubling t will double x.

Particle with constant speed

A particle with constant velocity moves with a constant speed along a straight line. This situation can be represented with the model of a **particle with constant speed**. The primary equation for this model is Equation 2.3, with the average speed v_{avg} replaced by the constant speed v:

$$v = \frac{s}{\Delta t} \tag{2.8}$$

As an example, imagine a particle moving at a constant speed in a circular path. If the speed is 5.00 ± 0.05 m/s and the radius of the path is 10.0 ± 0.2 m, we can calculate the time interval required to complete one trip around the circle:

$$v = \frac{s}{\Delta t} \rightarrow \Delta t = \frac{s}{v} = \frac{2\pi r}{v} = \frac{2\pi(10.0\text{ m})}{5.00\text{ m/s}} = 12.6\text{ s}$$

with a relative uncertainty of $$\frac{\Delta r}{r} + \frac{\Delta v}{v} = \frac{0.2\text{ m}}{10\text{ m}} + \frac{0.05\text{ m/s}}{5.00\text{ m/s}} = 0.03 \text{ or } 3\%.$$

The absolute uncertainty in the time is $0.03 \times 12.6\text{ s} = 0.4$ s, so we can write $\Delta t = (12.6 \pm 0.4)$ s.

Analysis Model 2.1

Particle with constant velocity

Imagine a moving object that can be modelled as a particle. If it moves at a constant speed through a displacement Δx in a straight line in a time interval Δt, its constant velocity is

$$v_x = \frac{\Delta x}{\Delta t} \tag{2.6}$$

The position of the particle as a function of time is given by

$$x_f = x_i + v_x t \tag{2.7}$$

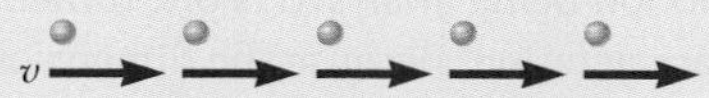

Examples

- a meteoroid travelling through gravity-free space
- a car travelling at a constant speed on a straight highway
- a runner travelling at constant speed on a perfectly straight path
- an object moving at terminal speed through a viscous medium (Chapter 5)

Analysis Model 2.2

Particle with constant speed

Imagine a moving object that can be modelled as a particle. If it moves at a constant speed through a distance d along a straight line or a curved path in a time interval Δt, its constant speed is

$$v = \frac{d}{\Delta t} \tag{2.8}$$

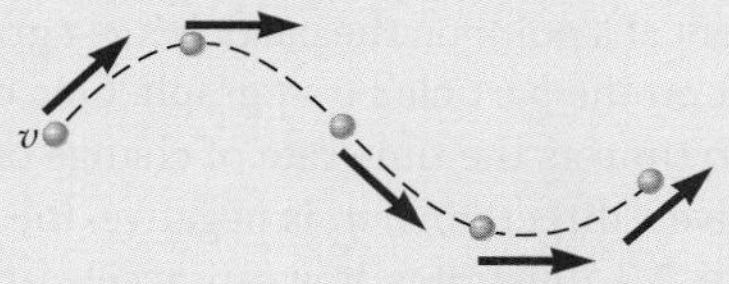

Examples

- a planet travelling around a perfectly circular orbit
- a car travelling at a constant speed on a curved racetrack
- a runner travelling at constant speed on a curved path
- a charged particle moving through a uniform magnetic field (Chapter 29)

These two analysis models, *particle with constant velocity* and *particle with constant speed*, are mathematical *models* which we use to describe and analyse the behaviour of *real* objects. They are approximations, with limitations. You should always think carefully when applying a model, and ensure that it is appropriate. These models are appropriate for the examples listed above, but they would not be appropriate, for example, to describe the motion of an object which was significantly changing its shape or size as it moved.

2.4 Acceleration

In **Example 2.3**, we worked with a common situation in which the velocity of a particle changes while the particle is moving. When the velocity of a particle changes with time, the particle is *accelerating*. For example, a car's velocity increases when you accelerate and decreases when you brake.

Suppose an object that can be modelled as a particle moving along the x axis has an initial velocity v_{xi} at time t_i at position Ⓐ and a final velocity v_{xf} at time t_f at position Ⓑ as in **Figure 2.6a**. The **average acceleration** $a_{x,avg}$ of the particle is defined as the *change* in velocity Δv_x divided by the time interval Δt during which that change occurs:

Average ▶ acceleration

$$a_{x,\text{avg}} \equiv \frac{\Delta v_x}{\Delta t} = \frac{v_{xf} - v_{xi}}{t_f - t_i} \tag{2.9}$$

As with velocity, when the motion being analysed is one dimensional, we can use positive and negative signs to indicate the direction of the acceleration. Because the dimensions of velocity are L/T and the dimension of time is T, acceleration has dimensions of length divided by time squared, or L/T^2. The SI unit of acceleration is metres per second squared (m/s^2). It might be easier to interpret these units if you think of them as metres per second. For example, suppose an object has an acceleration of $+2\ m/s^2$. You should form a mental image of the object having a velocity that is along a straight line and is increasing by 2 m/s every time an interval of 1 s passes. If the object starts from rest, you can picture it moving at a velocity of +2 m/s after 1 s, at +4 m/s after 2 s, and so on.

Often the value of the average acceleration is different over different time intervals. It is therefore useful to define the **instantaneous acceleration** as the limit of the average acceleration as Δt approaches zero. This concept is analogous to the definition of instantaneous velocity discussed in Section 2.2. If we imagine that point Ⓐ is brought closer and closer to point Ⓑ in **Figure 2.6a** and we take the limit of $\Delta v_x/\Delta t$ as Δt approaches zero, we obtain the instantaneous acceleration at point Ⓑ:

Instantaneous ▶ acceleration

$$a_x \equiv \lim_{\Delta t \to 0} \frac{\Delta v_x}{\Delta t} = \frac{dv_x}{dt} \tag{2.10}$$

Σ *For help with derivatives, see Appendix B.7.*

That is, the instantaneous acceleration equals the derivative of the velocity with respect to time, and by definition is the gradient of the velocity–time graph. The gradient of the green line in **Figure 2.6b** is equal to the instantaneous acceleration at point Ⓑ. Notice that **Figure 2.6b** is a *velocity–time* graph, not a *position–time* graph as in **Active Figures 2.1b** and **2.3**, and **Figures 2.4** and **2.5**. Therefore, we see that just as the velocity of a moving particle is the gradient at a point on the particle's x–t graph, the acceleration of a particle is the gradient at a point on the particle's v_x–t graph. One can interpret the derivative of the velocity with respect to time as the time rate of change of velocity. If a_x is positive, the acceleration is in the positive x direction; if a_x is negative, the acceleration is in the negative x direction.

Figure 2.7 illustrates how an acceleration–time graph is related to a velocity–time graph. The acceleration at any time is the gradient of the velocity–time graph at that time. Positive values of acceleration correspond to those points in **Figure 2.7a** where the velocity is increasing in the positive x direction. The acceleration reaches a maximum at time $t_Ⓐ$, when the gradient of the velocity–time graph is a maximum. The acceleration then goes to zero at time $t_Ⓑ$, when the velocity is a maximum (that is, when the gradient of the v_x–t

> **Pitfall Prevention 2.3**
> Keep in mind that *negative acceleration* does not necessarily mean that an object is *slowing down*. If the acceleration is negative and the velocity is negative, the object is speeding up!

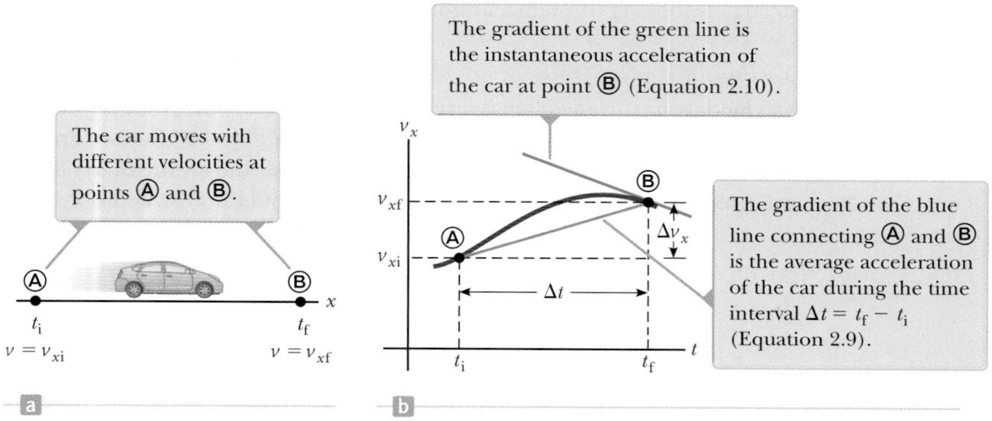

The car moves with different velocities at points Ⓐ and Ⓑ.
Ⓐ
Ⓑ
x
t_i
t_f
$v = v_{xi}$
$v = v_{xf}$
a
The gradient of the green line is the instantaneous acceleration of the car at point Ⓑ (Equation 2.10).
v_x
v_{xf}
v_{xi}
Δv_x
Δt
t
The gradient of the blue line connecting Ⓐ and Ⓑ is the average acceleration of the car during the time interval $\Delta t = t_f - t_i$ (Equation 2.9).
b

Figure 2.6
(a) A car, modelled as a particle, moving along the x axis from Ⓐ to Ⓑ, has velocity v_{xi} at $t = t_i$ and velocity v_{xf} at $t = t_f$. (b) Velocity–time graph (red-brown) for the particle moving in a straight line.

graph is zero). The acceleration is negative when the velocity is decreasing in the positive x direction, and it reaches its most negative value at time $t_{\text{Ⓒ}}$. The word *deceleration* has the common meaning of *slowing down.* We will not use this word in this book because it confuses the definition we have given for negative acceleration.

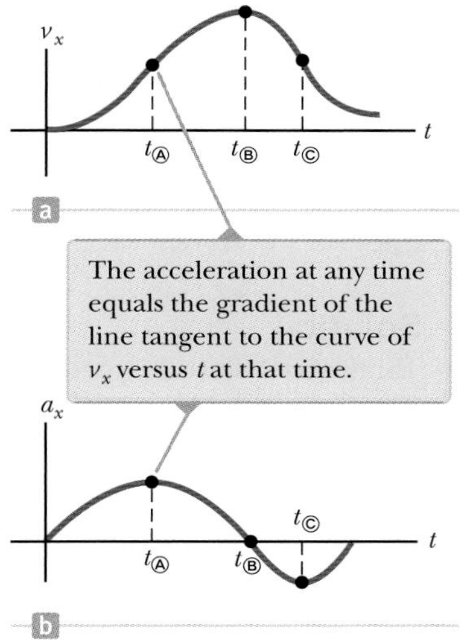

Figure 2.7
(a) The velocity–time graph for a particle moving along the x axis. **(b)** The instantaneous acceleration can be obtained from the velocity–time graph.

Quick Quiz 2.3

A car is moving in the negative x direction when the driver brakes. Is the car's velocity positive or negative? Is the car's acceleration positive or negative?

For the case of motion in a straight line, the direction of the velocity of an object and the direction of its acceleration are related as follows. When the object's velocity and acceleration are in the same direction, the object is speeding up. On the other hand, when the object's velocity and acceleration are in opposite directions, the object is slowing down.

From now on, we shall use the term *acceleration* to mean instantaneous acceleration. When we mean average acceleration, we shall always use the adjective *average*. Because $v_x = dx/dt$, the acceleration can also be written as

$$a_x = \frac{dv_x}{dt} = \frac{d}{dt}\left(\frac{dx}{\Delta dt}\right) = \frac{d^2x}{dt^2} \tag{2.11}$$

That is, in one-dimensional motion, the acceleration equals the *second derivative* of x with respect to time.

For help with derivatives, see Appendix B.7.

Conceptual Example 2.5

Graphical relationships between x, v_x, and a_x

The position of a jogger moving along the x axis varies with time as in **Figure 2.8**. Graph the velocity versus time and the acceleration versus time for the jogger.

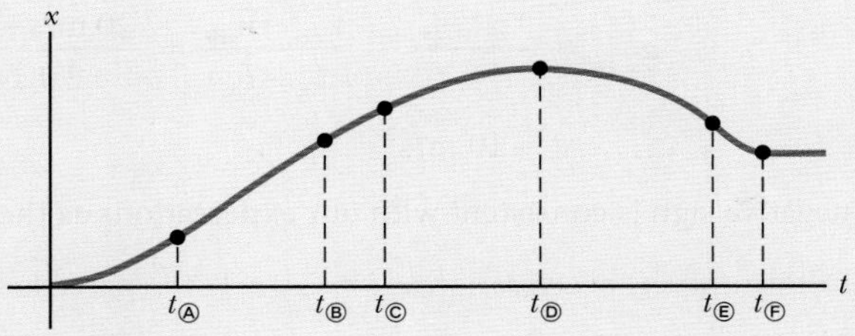

Figure 2.8
(Conceptual Example 2.5) Position–time graph for a jogger moving along the x axis.

Solution

The velocity at any instant is the gradient of the tangent to the x–t graph at that instant. Between $t = 0$ and $t = t_{\text{Ⓐ}}$, the gradient of the x–t graph increases uniformly, so the velocity increases linearly as shown in Figure 2.9a. Between $t_{\text{Ⓐ}}$ and $t_{\text{Ⓑ}}$, the gradient of the x–t graph is constant, so the velocity remains constant. Between $t_{\text{Ⓑ}}$ and $t_{\text{Ⓓ}}$, the gradient of the x–t graph decreases, so the value of the velocity in the v_x–t graph decreases. At $t_{\text{Ⓓ}}$, the gradient of the x–t graph is zero, so the velocity is zero at that instant. Between $t_{\text{Ⓓ}}$ and $t_{\text{Ⓔ}}$, the gradient of the x–t graph and therefore the velocity are negative and decrease uniformly in this interval. In the interval $t_{\text{Ⓔ}}$ to $t_{\text{Ⓕ}}$, the gradient of the x–t graph is still negative, and at $t_{\text{Ⓕ}}$ it goes to zero. Finally, after $t_{\text{Ⓕ}}$, the gradient of the x–t graph is zero, meaning that the object is at rest for $t > t_{\text{Ⓕ}}$. The acceleration at any instant is the gradient of the tangent to the v_x–t graph at that instant. The graph of acceleration versus time for this jogger is shown in Figure 2.9b. The acceleration is constant and positive between 0 and $t_{\text{Ⓐ}}$, where the gradient of the v_x–t graph is positive. It is zero between $t_{\text{Ⓐ}}$ and $t_{\text{Ⓑ}}$ and for $t > t_{\text{Ⓕ}}$ because the gradient of the v_x–t graph is zero at these times. It is negative between $t_{\text{Ⓑ}}$ and $t_{\text{Ⓔ}}$ because the gradient of the v_x–t graph is negative during this interval. Between $t_{\text{Ⓔ}}$ and $t_{\text{Ⓕ}}$, the acceleration is positive as it is between 0 and $t_{\text{Ⓐ}}$, but higher in value because the gradient of the v_x–t graph is steeper.

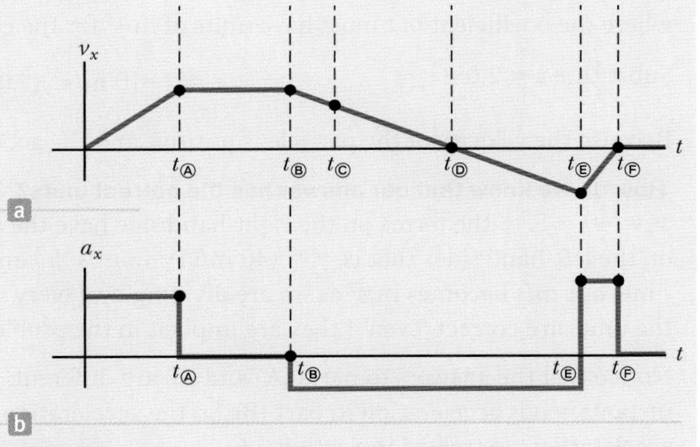

Figure 2.9
(Conceptual Example 2.5) (a) The velocity–time graph for the jogger is obtained by measuring the gradient of the position–time graph at each instant. (b) The acceleration–time graph for the jogger is obtained by measuring the gradient of the velocity–time graph at each instant.

Example 2.6

Average and instantaneous acceleration

The velocity of a particle moving along the x axis varies according to the expression $v_x = 40 - 5t^2$, where v_x is in metres per second and t is in seconds.

(A) Find the average acceleration in the time interval $t = 0$ to $t = 2.0$ s.

Solution

Think about what the particle is doing from the mathematical representation. As usual, you should start by drawing a diagram; in this case draw a graph (a rough sketch will usually do) showing how the velocity changes with time. Is it moving at $t = 0$? In which direction? Does it speed up or slow down? Thinking about what is happening at various times will help you. **Figure 2.10** shows the velocity–time graph for this situation. Because the gradient of the entire v_x–t curve is negative, we expect the acceleration to be negative, and larger at Ⓑ than at Ⓐ.

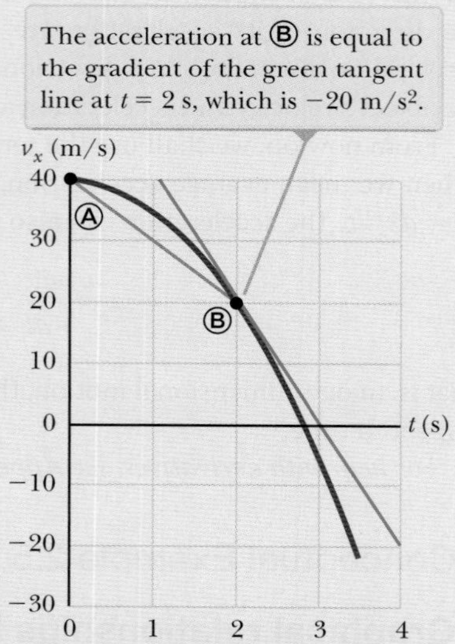

The acceleration at Ⓑ is equal to the gradient of the green tangent line at $t = 2$ s, which is -20 m/s².
v_x (m/s)
40
30
20
10
0
−10
−20
−30
Ⓐ
Ⓑ
t (s)
0
1
2
3
4

Figure 2.10
(Example 2.6) The velocity–time graph for a particle moving along the x axis according to the expression $v_x = 40 - 5t^2$.

The average acceleration is the gradient of the blue line joining points Ⓐ and Ⓑ on the graph.

Find the velocities at $t_i = t_Ⓐ = 0$ and $t_f = t_Ⓑ = 2.0$ s by substituting these values of t into the expression for the velocity: $v_{xⒶ} = 40 - 5(t_Ⓐ)^2 = 40 - 5(0)^2 = +40$ m/s

$$v_{xⒷ} = 40 - 5(t_Ⓑ)^2 = 40 - 5(2.0)^2 = +20 \text{ m/s}$$

Note that we could also read these directly from the graph.

Find the average acceleration in the specified time interval $\Delta t = t_Ⓑ - t_Ⓐ = 2.0$ s:

$$a_{x,\,avg} = \frac{v_{xf} - v_{xi}}{t_f - t_i} = \frac{v_{xⒷ} - v_{xⒶ}}{t_Ⓑ - t_Ⓐ} = \frac{20 \text{ m/s} - 40 \text{ m/s}}{2.0 \text{ s} - 0 \text{ s}}$$

$$= -10 \text{ m/s}^2$$

The negative sign is consistent with our expectations as the slope of the blue line is negative.

(B) Determine the acceleration at $t = 2.0$ s. Is the particle slowing down or speeding up?

Solution

We can apply **Equation 2.10** here, to obtain $a_x = \dfrac{dv_x}{dt} = \dfrac{d(40 - 5t^2)}{dt} = -10t$

where the coefficient of t must have units of m/s³ for the equation to be dimensionally correct.

Substitute $t = 2.0$ s: $a_x = (-10 \text{ m/s}^3)(2.0 \text{ s}) = -20 \text{ m/s}^2$

Because the velocity of the particle is positive and the acceleration is negative at this instant, the particle is slowing down.

How do we know that our answer has the correct units? We have assumed that in the original equation given, $v_x = 40 - 5t^2$, the terms on the right-hand side have the units of velocity, that is m/s, as they are equated to a velocity on the left-hand side. That is, $v_x = (40 \text{ m/s}) - (5 \text{ m/s}^3)t^2$ and t is in units of s. When we take the derivative with respect to time, our m/s becomes m/s² as we are dividing by a (very short) time, measured in s. It is always a good idea to check that the units are correct, even if they are implicit in the problem!

Notice that the answers to parts (A) and (B) are different. The average acceleration in part (A) is not equal to the instantaneous acceleration in part (B). So the acceleration is *not* constant in this example. Situations involving constant acceleration are treated in Section 2.6.

2.5 Motion diagrams

The concepts of velocity and acceleration are often confused with each other, but in fact they are quite different quantities. In forming a mental representation of a moving object, a pictorial representation called a *motion diagram* is sometimes useful to describe the velocity and acceleration while an object is in motion.

A motion diagram can be formed by imagining a *stroboscopic* photograph of a moving object, which shows several images of the object taken as the strobe light flashes at a constant rate. **Active Figure 2.1a** is a motion diagram for the car studied in Section 2.1. **Active Figure 2.11** represents three sets of strobe photographs of cars moving along a straight road in a single direction, from left to right. The time intervals between flashes of the stroboscope are equal in each part of the diagram. So as to not confuse the two vector quantities, we use red arrows for velocity and purple arrows for acceleration in **Active Figure 2.11**. The arrows are shown at several instants during the motion of the object. Let us describe the motion of the car in each diagram.

In **Active Figure 2.11a**, the images of the car are equally spaced, showing us that the car moves through the same displacement in each time interval. This equal spacing is consistent with the car moving with *constant positive velocity* and *zero acceleration*. We could model the car as a particle and describe it by using the particle with constant velocity analysis model.

In **Active Figure 2.11b**, the images become farther apart as time progresses. In this case, the velocity arrow increases in length with time because the car's displacement in consecutive time intervals is increasing. These features suggest the car is moving with a *positive velocity* and a *positive acceleration*. The velocity and acceleration are in the same direction.

In **Active Figure 2.11c**, we can tell that the car slows as it moves to the right because its displacement between adjacent images (consecutive time intervals) decreases with time. This suggests the car moves to the right with a negative acceleration. The length of the velocity arrow decreases in time and eventually reaches zero. From this diagram, we see that the acceleration and velocity arrows are *not* in the same direction. The car is moving with a *positive velocity*, but with a *negative acceleration*, and is slowing down. The velocity and acceleration are in opposite directions.

Each purple acceleration arrow in parts (b) and (c) of **Active Figure 2.11** is the same length. Therefore, these diagrams represent motion of a *particle with constant acceleration*. This important analysis model will be discussed in the next section.

Quick Quiz 2.4

Which one of the following statements is true? **(a)** If a car is travelling eastwards, its acceleration must be eastwards. **(b)** If a car is slowing down, its acceleration must be negative. **(c)** A particle with constant acceleration can never stop and stay stopped.

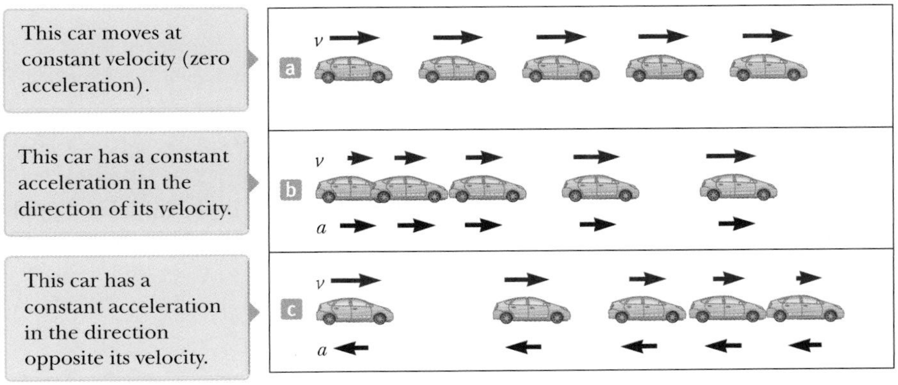

Active Figure 2.11

Motion diagrams of a car moving along a straight road in a single direction. The velocity at each instant is indicated by a red arrow, and the constant acceleration is indicated by a purple arrow.

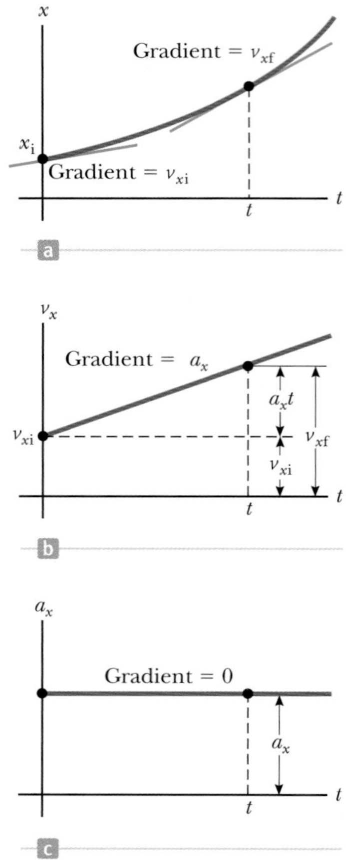

Active Figure 2.12

A particle with constant acceleration a_x moving along the x axis:
(a) the position–time graph,
(b) the velocity–time graph, and
(c) the acceleration–time graph.

2.6 Analysis model: particle with constant acceleration

If the acceleration of a particle varies in time, its motion can be complex and difficult to analyse. A very common and simple type of one-dimensional motion, however, is that in which the acceleration is constant; for example, the acceleration of falling objects close to the surface of the Earth can be approximated in many cases (but not all) as constant. (This case is discussed in detail in Section 2.8.) In such a case, the average acceleration $a_{x,\text{avg}}$ over any time interval is equal to the instantaneous acceleration a_x at any instant within the interval, and the velocity changes at the same rate throughout the motion. This situation occurs often enough that we identify it as an analysis model: the **particle with constant acceleration**. In the discussion that follows, we generate several equations that describe the motion of a particle for this model.

If we replace $a_{x,\text{avg}}$ by a_x in Equation 2.9 and take $t_i = 0$ and t_f to be any later time t, we find that

$$a_x = \frac{v_{xf} - v_{xi}}{t - 0}$$

or

$$v_{xf} = v_{xi} + a_x t \quad \text{(for constant } a_x\text{)} \tag{2.12}$$

This expression enables us to determine an object's velocity at *any* time t if we know the object's initial velocity v_{xi} and its (constant) acceleration a_x. A velocity–time graph for this constant-acceleration motion is shown in Active Figure 2.12b. The graph is a straight line, the gradient of which is the acceleration a_x; the (constant) gradient is consistent with $a_x = dv_x/dt$ being a constant. Notice that the gradient is positive, which indicates a positive acceleration. If the acceleration were negative, the gradient of the line in Active Figure 2.12b would be negative. When the acceleration is constant, the graph of acceleration versus time (Active Figure 2.12c) is a straight line with a gradient of zero.

Because velocity at constant acceleration varies linearly in time according to Equation 2.12, we can express the average velocity in any time interval as the arithmetic mean of the initial velocity v_{xi} and the final velocity v_{xf}:

$$v_{x,\text{avg}} = \frac{v_{xi} + v_{xf}}{2} \quad \text{for constant } a_x \tag{2.13}$$

Note that this expression for average velocity applies *only* in situations in which the acceleration is constant.

We can now use Equations 2.1, 2.2, and 2.13 to obtain the position of an object as a function of time. Recalling that Δx in Equation 2.2 represents $x_f - x_i$ and recognising that $\Delta t = t_f - t_i = t - 0 = t$, we find that

Position as a function of velocity and time for the particle with constant acceleration model ▶

$$x_f - x_i = v_{x,\text{avg}}t = \frac{1}{2}(v_{xi} + v_{xf})t$$

$$x_f = x_i + \frac{1}{2}(v_{xi} + v_{xf})t \quad \text{for constant } a_x \tag{2.14}$$

This equation provides the final position of the particle at time t in terms of the initial and final velocities.

We can obtain another useful expression for the position of a particle with constant acceleration by substituting Equation 2.12 into Equation 2.14:

Position as a function of time for the particle with constant acceleration model ▶

$$x_f = x_i + \frac{1}{2}[v_{xi} + (v_{xi} + a_x t)]t$$

$$x_f = x_i + v_{xi}t + \frac{1}{2}a_x t^2 \quad \text{for constant } a_x \tag{2.15}$$

This equation provides the final position of the particle at time t in terms of the initial position, the initial velocity, and the constant acceleration.

The position–time graph for motion at constant (positive) acceleration shown in Active Figure 2.12a is obtained from Equation 2.15. Notice that the curve is a parabola. The gradient of the tangent line to this curve at $t = 0$ equals the initial velocity v_{xi}, and the gradient of the tangent line at any later time t equals the velocity v_{xf} at that time.

Finally, we can obtain an expression for the final velocity that does not contain time as a variable by substituting the value of t from Equation 2.12 into Equation 2.14:

$$x_f = x_i + \frac{1}{2}(v_{xi} + v_{xf})\left(\frac{v_{xf} - v_{xi}}{a_x}\right) = x_i + \frac{(v_{xf})^2 - (v_{xi})^2}{2a_x}$$

$$(v_{xf})^2 = (v_{xi})^2 + 2a_x(x_f - x_i) \quad \text{(for constant } a_x\text{)} \qquad (2.16)$$

◀ Velocity as a function of position for the particle with constant acceleration model

This equation provides the final velocity in terms of the initial velocity, the constant acceleration, and the position of the particle.

For motion at *zero* acceleration, we see from Equations 2.12 and 2.15 that

$$\left.\begin{aligned} v_{xf} &= v_{xi} = v_x \\ x_f &= x_i + v_x t \end{aligned}\right\} \quad \text{when } a_x = 0$$

That is, when the acceleration of a particle is zero, its velocity is constant and its position changes linearly with time. In terms of models, when the acceleration of a particle is zero, the particle with constant acceleration model reduces to the particle with constant velocity model (Section 2.3).

Quick Quiz 2.5

In Active Figure 2.13, match each v_x–t graph on the top with the a_x–t graph on the bottom that best describes the motion.

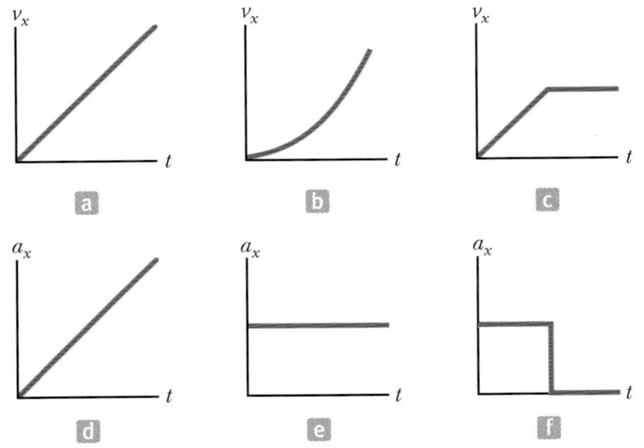

Active Figure 2.13

(Quick Quiz 2.5) Parts (a), (b) and (c) are v_x–t graphs of objects in one-dimensional motion. The possible accelerations of each object as a function of time are shown in scrambled order in (d), (e) and (f).

Equations 2.12 through 2.16 are **kinematic equations** that may be used to solve any problem involving a particle with constant acceleration in one dimension. The choice of which equation you use in a given situation depends on what you know beforehand. Sometimes it is necessary to use two of these equations to solve for two unknowns. You should recognise that the quantities that vary during the motion are position x_f, velocity v_{xf}, and time t. Remember that these equations can only be used when a_x is constant.

±? Example 2.7

Carrier landing

A pilot lands his jet on an aircraft carrier at a speed of 230 km/h (64 m/s), with an uncertainty in the speed of 10 km/h (3 m/s). You might have seen movies or television shows in which a jet lands on an aircraft carrier and is brought to rest surprisingly fast by an arresting cable, to prevent it falling off and into the sea. The idea of using arresting cables to slow down landing aircraft and enable them to land safely on ships originated at about the time of the First World War. The cables are still a vital part of the operation of modern aircraft carriers.

(A) What is the acceleration of the jet (assumed constant) if it stops in (2.2 ± 0.2) s after the arresting cable is engaged?

Solution

It is a good idea to draw a quick sketch for yourself showing the motion of the jet – either as a motion diagram as described in Section 2.5 or as a graph. This will help you visualise the problem. The diagram at right shows what is happening in the problem.

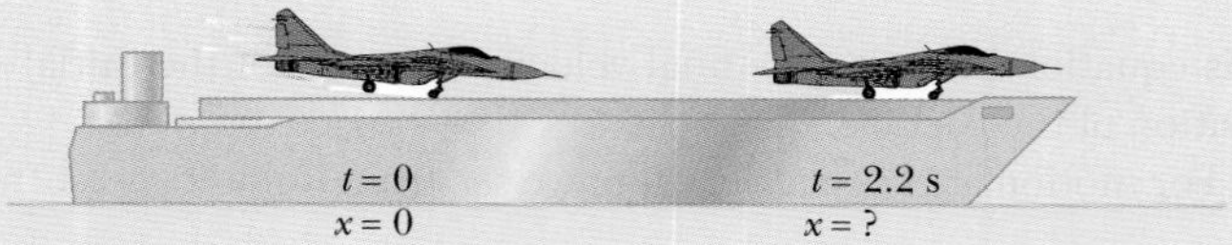

We are told both the initial speed, $v_{xi} = (64 \pm 3)$ m/s, and the final speed, $v_{xf} = 0$ (the jet must stop), and the time interval, $t = (2.2 \pm 0.2)$ s. We are making the approximation that the acceleration of the jet is constant, so that we can apply the *particle with constant acceleration* analysis model.

We define our x axis as the direction of motion of the jet. (Notice that we have no information about the change in position of the jet while it is slowing down.)

We can use **Equation 2.12** to find the acceleration of the jet, modelled as a particle:

$$a_x = \frac{v_{xf} - v_{xi}}{t}$$

Do a dimension check to ensure we have rearranged the equation correctly:

$$[LT^{-2}] = [LT^{-1}]/[T] = [LT^{-2}] \; ☺$$

Our dimensions are okay so we can proceed to substitute in the numbers:

$$a_x = \frac{0 - 64 \text{ m/s}}{2.2 \text{ s}} = -29 \text{ m/s}^2$$

The uncertainty in the acceleration is: $\frac{\Delta a_x}{a_x} = \frac{\Delta v_x}{v_x} + \frac{\Delta t}{t} = \frac{3 \text{ m/s}}{64 \text{ m/s}} + \frac{0.2 \text{ s}}{2.2 \text{ s}} = 0.14$ or 14%

so $\Delta a_x = 29 \text{ m/s} \times 0.14 = 4 \text{ m/s}^2$

Recall that uncertainties always have a positive value, so we multiply by the magnitude of a_x and ignore the negative sign.

So the acceleration is $a_x = (-29 \pm 4)$ m/s.

(B) If the jet touches down at position $x_i = 0$, what is its final position?

Solution

Use **Equation 2.14** to solve for the final position:

$$x_f = x_i + \frac{1}{2}(v_{xi} + v_{xf})t = 0 + \frac{1}{2}(64 \text{ m/s} + 0)2.2 \text{ s} = 70 \text{ m}$$

The uncertainty will be

$$\Delta x = x\left(\frac{\Delta v_{xi}}{v_{xi}} + \frac{\Delta t}{t}\right) = 70 \text{ m}\left(\frac{3 \text{ m/s}}{64 \text{ m/s}} + \frac{0.2 \text{ s}}{2.2 \text{ s}}\right) = 10 \text{ m}$$

so $x_f = (70 \pm 10)$ m.

Reality Check Given the size of aircraft carriers, a length of 70 m seems reasonable for stopping the jet. If we had made a calculation error and come up with an answer of 70 km, we would know we had made an error.

What If? Suppose the stopping distance before the edge of the carrier is 85 m. Can the pilot land safely? What if the speed is faster than 64 m/s?

Answer The calculated stopping distance was (70 ± 10) m, which has a maximum value of 80 m. So the jet can land safely if the available distance is 85 m. If the jet is travelling faster at the beginning, it will stop farther away from its starting point. Mathematically, we see in **Equation 2.14** that if v_{xi} is larger, then x_f will be larger. Given the uncertainty in stopping distance, only a small increase in speed could result in the jet falling into the ocean.

Example 2.8

Watch out for the speed limit!

A car travelling at a constant speed of 45.0 m/s passes a police officer on a motorcycle hidden behind a billboard. One second after the speeding car passes the billboard, the police officer sets out from the billboard to catch the car, accelerating at a constant rate of 3.00 m/s^2. How long does it take the officer on the motorcycle to overtake the car?

Solution

As usual, start by drawing a diagram to visualise what is happening. **Figure 2.14** helps clarify the sequence of events by showing the positions of the vehicles at three important times. We shall use two analysis models: particle with constant velocity and particle with constant acceleration. The car is modelled as a particle with constant velocity, and the police officer on the motorcycle is modelled as a particle with constant acceleration.

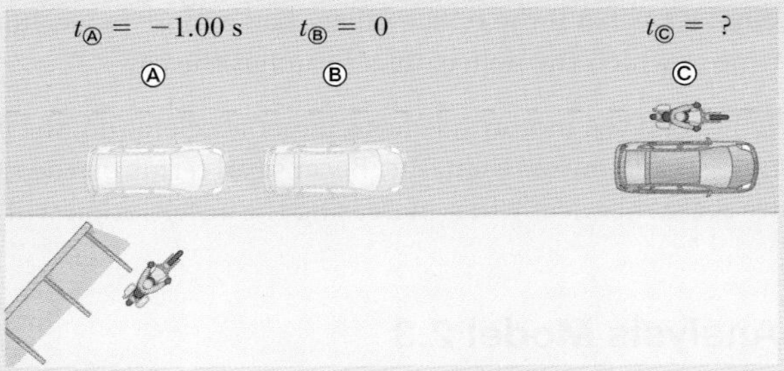

Figure 2.14
(Example 2.8) A speeding car passes a hidden motorcycle police officer.

First, we write expressions for the position of each vehicle as a function of time. It is convenient to choose the position of the billboard as the origin and to set $t_Ⓑ = 0$ as the time the officer begins moving. At that instant, the car has already travelled a distance of 45.0 m from the billboard because it has travelled at a constant speed of $v_x = 45.0$ m/s for 1 s. Therefore, the initial position of the speeding car is $x_Ⓑ = 45.0$ m.

Using the particle with constant velocity model, apply **Equation 2.7** to give the car's position at any time t:

$$x_{car} = x_Ⓑ + v_{x,car}t$$

A quick check shows that at $t = 0$, this expression gives the car's correct initial position when the police officer begins to move:

$$x_{car} = x_Ⓑ = 45.0 \text{ m}$$

The police officer starts from rest at $t_Ⓑ = 0$ and accelerates at $a_x = 3.00$ m/s^2 away from the origin. Use **Equation 2.15** to give the officer's position at any time t:

$$x_f = x_i + v_{xi}t + \frac{1}{2}a_x t^2$$

$$x_{police} = 0 + (0)t + \frac{1}{2}a_x t^2 = \frac{1}{2}a_x t^2$$

Set the positions of the car and police officer equal to represent the police officer overtaking the car at position Ⓒ:

$$x_{police} = x_{car}$$

$$\frac{1}{2}a_x t^2 = x_Ⓑ + v_{x,car}t$$

Rearrange to give a quadratic equation:

$$\frac{1}{2}a_x t^2 - v_{x,car}t - x_Ⓑ = 0$$

Solve the quadratic equation for the time at which the police officer catches the car (*For help with quadratic equations, see Appendix B.2.*):

$$t = \frac{v_{x,\,car} \pm \sqrt{(v_{x,\,car})^2 + 2a_x x_Ⓑ}}{a_x}$$

Σ

$$t = \frac{v_{x,\,car}}{a_x} \pm \sqrt{\frac{(v_{x,\,car})^2}{(a_x)^2} + \frac{2x_Ⓑ}{a_x}} \quad (1)$$

Evaluate the solution, choosing the positive root because that is the only choice consistent with a time $t > 0$:

$$t = \frac{45.0\text{ m/s}}{3.00\text{ m/s}^2} + \sqrt{\frac{(45.0\text{ m/s})^2}{(3.00\text{ m/s}^2)^2} + \frac{2(45.0\text{ m})}{3.00\text{ m/s}^2}} = 31.0\text{ s}$$

Why didn't we choose $t = 0$ as the time at which the car passes the police officer? If we did so, we would not be able to use the particle with constant acceleration model for the police officer. The officer's acceleration would be zero for the first second and then 3.00 m/s² for the remaining time. By defining the time $t = 0$ as when the police officer begins moving, we can use the particle with constant acceleration model for the officer's movement for all positive times.

What If? What if the police officer had a more powerful motorcycle with a larger acceleration? How would that change the time at which the police officer catches the car?

Answer If the motorcycle has a larger acceleration, the police officer should catch up to the car sooner, so the answer for the time should be less than 31 s. Because all terms on the right side of Equation (1) have the acceleration a_x in the denominator, we see symbolically that increasing the acceleration will decrease the time at which the police officer catches the car.

Analysis Model 2.3

Particle with constant acceleration

Imagine a moving object that can be modelled as a particle. If it begins from position x_i and initial velocity v_{xi} and moves in a straight line with a constant acceleration a_x, its subsequent position and velocity are described by the following kinematic equations:

$$v_{xf} = v_{xi} + a_x t \quad (2.12)$$

$$v_{x,\text{avg}} = \frac{v_{xi} + v_{xf}}{2} \quad (2.13)$$

$$x_f = x_i + \frac{1}{2}(v_{xi} + v_{xf})t \quad (2.14)$$

$$x_f = x_i + v_{xi}t + \frac{1}{2}a_x t^2 \quad (2.15)$$

$$v_{xf}^2 = v_{xi}^2 + 2a_x(x_f - x_i) \quad (2.16)$$

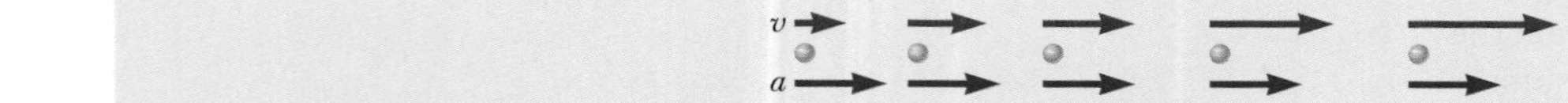

Examples

- a car accelerating at a constant rate along a straight freeway
- a dropped object in the absence of air resistance (Section 2.7)
- an object on which a constant net force acts (Chapter 4)
- a charged particle in a uniform electric field (Chapter 23)

You should think carefully about which model you choose to describe a given situation. Choose the simplest model that adequately describes the system. But remember that if the model chosen has too many approximations, it will not adequately describe or predict the behaviour of the system. Hence you should not generally use the particle with constant velocity analysis model to describe an accelerating particle, or the particle with constant acceleration analysis model to describe a particle with changing acceleration.

2.7 Freely falling objects

It is now well known that, in the absence of air resistance, all objects dropped near the Earth's surface fall towards the Earth with the same constant acceleration under the influence of the Earth's gravity. It was not until about 1600 that this conclusion was accepted. Before that time, the teachings of the Greek philosopher Aristotle (384–322 BC) had held that heavier objects fall faster than lighter ones.

The Italian Galileo Galilei (1564–1642) formulated our present-day ideas concerning falling objects. There is a legend that he demonstrated the behaviour of falling objects by observing that two different weights dropped simultaneously from the Leaning Tower of Pisa hit the ground at approximately the same time. Although there is some doubt that he carried out this particular experiment, it is well established that Galileo performed many experiments on objects moving on inclined planes. In his experiments, he rolled balls down a slight incline and measured the distances they covered in successive time intervals. The purpose of the incline was to reduce the acceleration, which made it possible for him to make accurate measurements of the time intervals. By gradually increasing the gradient of the incline, he was finally able to draw conclusions about freely falling objects because a freely falling ball is equivalent to a ball moving down a vertical incline.

Pitfall Prevention 2.4

Be sure not to confuse the italic symbol *g* for free-fall acceleration with the non-italic symbol g used as the abbreviation for the unit gram.

Pitfall Prevention 2.5

The convention we use in this book is that *g* is a *positive number*. It is tempting to substitute $-9.80\ \text{m/s}^2$ for *g*, but resist the temptation. Downwards gravitational acceleration is indicated explicitly by stating the acceleration as $a_y = -g$.

TRY THIS

Simultaneously drop a coin and a crumpled-up piece of paper from the same height. Try it with some other objects. What sort of objects fall faster or slower? Can you find any patterns?

If the effects of air resistance were negligible, a coin and crumpled-up piece of paper would have the same motion and would hit the floor at the same time. In the idealised case, in which air resistance is absent, such motion is referred to as *free-fall* motion. If this same experiment could be conducted in a vacuum, in which air resistance is truly negligible, the paper and the coin would fall with the same acceleration even when the paper is not crumpled. On 2 August 1971, astronaut David Scott conducted such a demonstration on the Moon. He simultaneously released a hammer and a feather, and the two objects fell together to the lunar surface. Of course on Earth, where air resistance is significant, the hammer hits the ground first. As always, we need to consider which simplifying approximations are reasonable, and which are not.

Galileo Galilei
Italian physicist and astronomer (1564–1642)
Galileo formulated the laws that govern the motion of objects in free fall and made many other significant discoveries in physics and astronomy. Galileo publicly defended Nicolaus Copernicus' assertion that the Sun is at the centre of the Universe (the heliocentric system). He published Dialogue Concerning Two New World Systems to support the Copernican model, a view that the Catholic Church declared to be heretical, and which he later recanted for his own safety.

When we use the expression *freely falling object*, we do not necessarily refer to an object dropped from rest. A freely falling object is any object moving freely under the influence of gravity alone, regardless of its initial motion. Objects thrown upwards or downwards and those released from rest are all falling freely once they are released (ignoring air resistance). Any freely falling object experiences an acceleration directed *downwards*, regardless of its initial motion.

We shall denote the magnitude of the *free-fall acceleration* by the symbol *g*. The value of *g* decreases with increasing altitude above the Earth's surface. Furthermore, slight variations in *g* occur with changes in latitude and composition of the Earth's crust. At the Earth's surface, the value of *g* is approximately $9.80\ \text{m/s}^2$. Unless stated otherwise, we shall use this value for *g* when performing calculations. For making quick estimates, use $g = 10\ \text{m/s}^2$.

If we neglect air resistance and assume the free-fall acceleration does not vary with altitude over short vertical distances, the motion of a freely falling object moving vertically is that of a particle with constant acceleration in one dimension. Therefore, the equations developed in Section 2.6 for the particle with constant acceleration model can be applied. The only modification for freely falling objects that we need to make in these equations is to note that the motion is in the vertical direction (the *y* direction) rather than in the horizontal direction (*x*) and that the acceleration is downwards and has a magnitude of $9.80\ \text{m/s}^2$. Therefore, we choose $a_y = -g = -9.80\ \text{m/s}^2$, where the negative sign means that the acceleration of a freely falling object is downwards. In Chapter 11, we shall study how to deal with variations in *g* with altitude. In Chapter 5 we look at the effect of air resistance.

Figure 2.15

TRY THIS

Hold a ruler vertically as shown in Figure 2.15, with some mark on the ruler between but not touching a friend's index finger and thumb. Without warning, release the ruler and see where your friend can catch it without moving their hand downwards. You can use the distance the ruler falls to estimate your friend's reaction time. Swap and see how yours compares.

Quick Quiz 2.6

Does the acceleration of a ball (i) (a) increase, (b) decrease, (c) increase and then decrease, (d) decrease and then increase or (e) remain the same after it is thrown up in the air? (ii) What about its speed?

Conceptual Example 2.9

The daring skydivers

A skydiver jumps out of a hovering helicopter. A few seconds later, another skydiver jumps out and they both fall along the same vertical line. Ignore air resistance so that both skydivers fall with the same constant acceleration. Does the difference in their speeds stay the same throughout the fall? Does the vertical distance between them stay the same throughout the fall?

Solution

At any given instant, the speeds of the skydivers are different because one had a head start. In any time interval Δt after this instant, however, the two skydivers increase their speeds by the same amount because they have the same acceleration. Therefore, the difference in their speeds remains the same throughout the fall.

The first jumper always has a greater speed than the second. Therefore, in a given time interval, the first skydiver covers a greater distance than the second. Consequently, the separation distance between them increases.

Drawing a diagram makes this easy to see, so always draw a diagram!

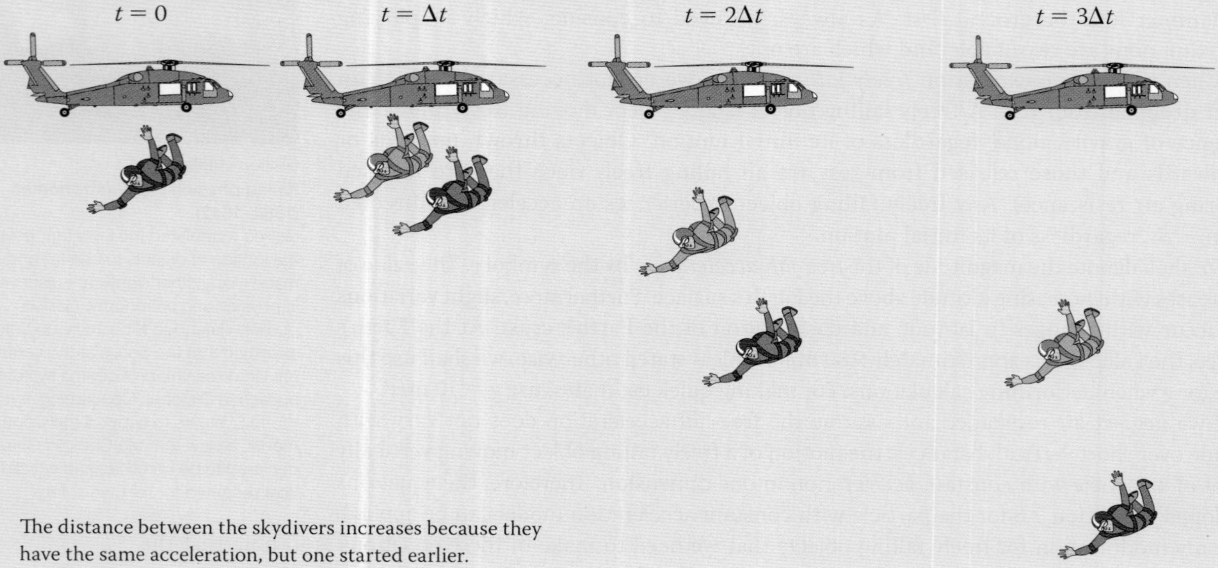

The distance between the skydivers increases because they have the same acceleration, but one started earlier.

TRY THIS

Get a friend to stand on a chair or table and drop one small object followed quickly by a second object. Observe what happens to the distance between them. Does it matter what sort of objects they are – try it with tennis balls and with crumpled up pieces of paper. You could also try doing the experiment with ball bearings in tall cylinders of oil – we will come back to this *Try this* experiment when we discuss friction in Chapter 5.

Pitfall Prevention 2.6

A common misconception is that the acceleration of a projectile at the top of its trajectory is zero. Although the velocity at the top of the motion of an object thrown upwards momentarily goes to zero, *the acceleration is still that due to gravity* at this point. If the velocity and acceleration were both zero, the projectile would stay at the top.

Example 2.10

Don't throw stones!

A stone thrown from the top of a building is given an initial velocity of 25.0 m/s straight upwards. The stone is launched 50.0 m above the ground, and the stone just misses the edge of the roof on its way down as shown in **Figure 2.16**.

(A) Using $t_{Ⓐ} = 0$ as the time the stone leaves the thrower's hand at position Ⓐ, determine the time at which the stone reaches its maximum height.

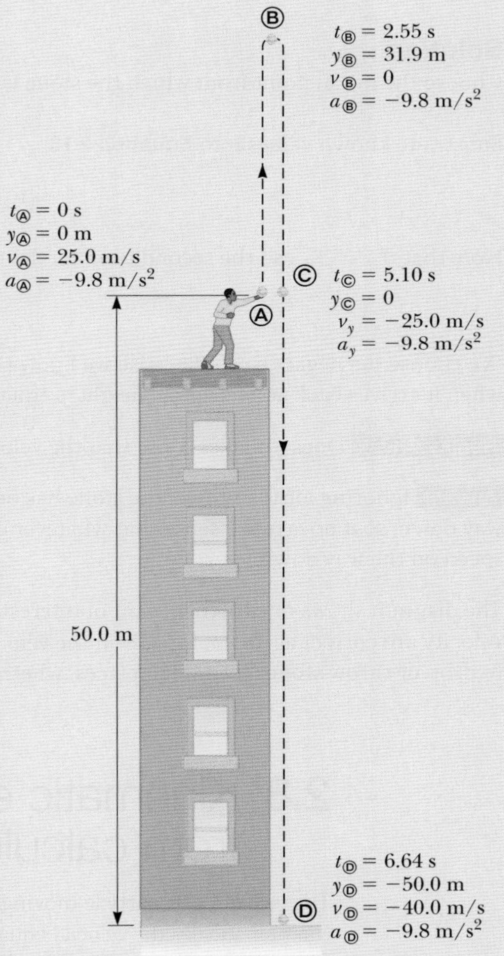

Figure 2.16
(Example 2.10) Position and velocity versus time for a freely falling stone initially thrown upwards with a velocity $v_{yi} = 25.0$ m/s. We calculate many of the quantities in the labels in the example. Can you verify the other values that are not?

Solution

You most likely have experience with dropping objects or throwing them upwards and watching them fall, so this problem should describe a familiar experience. Start by drawing a diagram, labelling all the points of interest. You can go back to your diagram and add extra data or labels as needed. The first point of interest, Ⓐ is just after the stone leaves the person's hand. Note that the initial velocity is positive because the stone is launched upwards. The velocity will change sign after the stone reaches its highest point (point Ⓑ), but the acceleration of the stone will *always* be downwards. Choose an initial point just after the stone leaves the person's hand and a final point at the top of its flight where the velocity is momentarily zero.

Use **Equation 2.12** to calculate the time at which the stone reaches its maximum height:

$$v_{yf} = v_{yi} + a_y t \rightarrow t = \frac{v_{yf} - v_{yi}}{a_y}$$

Do a quick dimension check to ensure you have rearranged the equation correctly: $[T] = [LT^{-1}]/[LT^{-2}] = [T]$ ☺

Then substitute numerical values:

$$t = t_{Ⓑ} = \frac{0 - 25.0 \text{ m/s}}{-9.80 \text{ m/s}^2} = 2.55 \text{ s}$$

Example 2.10 cont.

(B) Find the maximum height of the stone.

Solution

As in part (A), choose the initial and final points at the beginning and the end of the upwards flight. Set $y_{Ⓐ} = 0$ and substitute the time from part (A) into **Equation 2.15** to find the maximum height:

$$y_{max} = y_{Ⓑ} = y_{Ⓐ} + v_{xⒶ}t + \frac{1}{2}a_y t^2$$

$$y_{Ⓑ} = 0 + (25.0\ \text{m/s})(2.55\ \text{s}) + \frac{1}{2}(-9.80\ \text{m/s}^2)(2.55\ \text{s})^2 = 31.9\ \text{m}$$

(C) Determine the velocity of the stone when it returns to the height from which it was thrown.

Solution

Choose the initial point from which the stone is launched and the final point when it passes this position coming down.

Substitute known values into **Equation 2.16**:

$$(v_{yⒸ})^2 = (v_{yⒶ})^2 + 2a_y(y_{Ⓒ} - y_{Ⓐ})$$

Note that $y_{Ⓐ} = y_{Ⓒ}$, so the second term on the right hand side above must be zero, hence $(v_{yⒸ})^2 = (v_{yⒶ})^2$ and therefore

$$v_{yⒸ} = \pm v_{yⒶ} = -25.0\ \text{m/s}$$

We choose the negative root because we know that the stone is moving downwards at point Ⓒ. The velocity of the stone when it arrives back at its original height is equal in magnitude to its initial velocity but is opposite in direction.

Reality Check Does it makes sense that the velocity is the same?

Answer Ignoring air resistance, the stone has been accelerated for the same time and distance on the way up and on the way down, so it does make sense that whatever decrease in speed there was on the way up, we expect an equal increase in speed on the way down.

The diagram shows one further point of interest: the stone just as it hits the ground. Verify that the values for position and velocity are correct at this point. Given the velocity of the stone when it hits the ground, why do you think it is a bad idea to drop or throw stones from high places, whether you throw upwards or downwards first?

2.8 Kinematic equations derived from calculus

The velocity of a particle moving in a straight line can be obtained if its position as a function of time is known. Mathematically, the velocity equals the derivative of the position with respect to time. It is also possible to find the position of a particle if its velocity is known as a function of time using *integration.* Graphically, this is equivalent to finding the area under a curve.

Σ *If you are not comfortable with calculus you should review Appendices B.7 and B.8 before proceeding.*

Suppose the v_x–t graph for a particle moving along the x axis is as shown in **Figure 2.17**. Let us divide the time interval $t_f - t_i$ into many small intervals, each of duration Δt_n. From the definition of average velocity, we see that the displacement of the particle during any small interval, such as the one shaded in **Figure 2.17**, is given by $\Delta x_n = v_{xn,\text{avg}}\Delta t_n$, where $v_{xn,\text{avg}}$ is the average velocity in that interval. Therefore, the displacement during this small interval is simply the area of the shaded rectangle in **Figure 2.17**. The total displacement for the interval $t_f - t_i$ is the sum of the areas of all the rectangles from t_i to t_f:

$$\Delta x = \sum_n v_{xn,\text{avg}}\Delta t_n$$

where the symbol Σ (uppercase Greek sigma) signifies a sum over all terms, that is, over all values of n. Now, as the intervals are made smaller and smaller, the number of terms in the sum increases and the sum approaches a value equal to the area under the curve in the velocity–time graph. Therefore, in the limit $n \to \infty$, or $\Delta t_n \to 0$, the displacement is

$$\Delta x = \lim_{\Delta t_n \to 0} \sum_n v_{xn} \Delta t_n \tag{2.17}$$

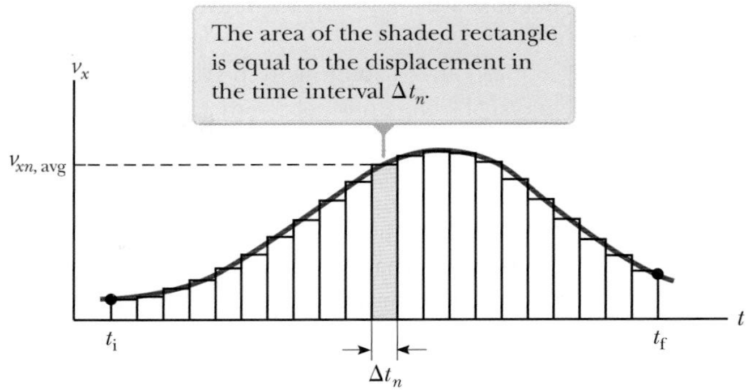

Figure 2.17
Velocity versus time for a particle moving along the x axis. The total area under the curve is the total displacement of the particle.

Notice that we have replaced the average velocity $v_{xn,\text{avg}}$ with the instantaneous velocity v_{xn} in the sum because the stepwise velocity $v_{xn,\text{avg}}$ approaches a continuous function v_{xn} as the time intervals shrink to zero. As you can see from **Figure 2.17**, this approximation is valid in the limit of very small intervals. Therefore, if we know the v_x–t graph for motion along a straight line, we can obtain the displacement during any time interval by measuring the area under the curve corresponding to that time interval.

The limit of the sum shown in **Equation 2.17** is called a **definite integral** and is written

$$\lim_{\Delta t_n \to 0} \sum_n v_{xn} \Delta t_n = \int_{t_i}^{t_f} v_x(t)\, dt \tag{2.18}$$

◀ Definite integral

where $v_x(t)$ denotes the velocity at any time t. If the explicit functional form of $v_x(t)$ is known and the limits are given, the integral can be evaluated. Sometimes the v_x–t graph for a moving particle has a shape much simpler than that shown in **Figure 2.17**. For example, suppose a particle moves at a constant velocity v_{xi}. In this case, the v_x–t graph is a horizontal line as in **Figure 2.18** and the displacement of the particle during the time interval Δt is simply the area of the shaded rectangle:

$$\Delta x = v_{xi} \Delta t \quad (\text{when } v_x = v_{xi} = \text{constant})$$

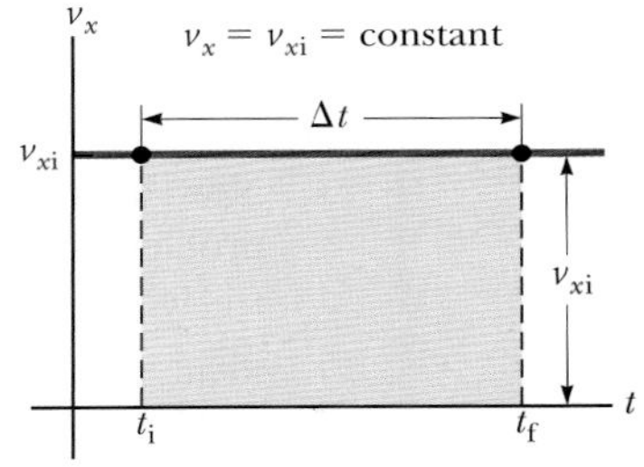

Figure 2.18
The velocity–time graph for a particle moving with constant velocity v_{xi}. The displacement of the particle during the time interval $t_f - t_i$ is equal to the area of the shaded rectangle.

Kinematic equations

We now use the defining equations for acceleration and velocity to derive two of our kinematic equations, **Equations 2.12** and **2.15**.

The defining equation for acceleration (**Equation 2.10**),

$$a_x = \frac{dv_x}{dt}$$

may be written as $dv_x = a_x\, dt$ or in terms of an integral as

$$v_{xf} - v_{xi} = \int_0^t a_x dt$$

For the special case in which the acceleration is constant, a_x can be removed from the integral to give

$$v_{xf} - v_{xi} = a_x \int_0^t dt = a_x(t - 0) = a_x t \tag{2.19}$$

which is **Equation 2.12**.

Now let us consider the defining equation for velocity (**Equation 2.5**):

$$v_x = \frac{dx}{dt}$$

We can write this equation as $dx = v_x\, dt$ or in integral form as

$$x_f - x_i = \int_0^t v_x dt$$

Because $v_x = v_{xf} = v_{xi} + a_x t$, this expression becomes

$$x_f - x_i = \int_0^t (v_{xi} + a_x t)dt = \int_0^t v_{xi}\,dt + a_x \int_0^t t\,dt = v_{xi}(t - 0) + a_x\left(\frac{t^2}{2} - 0\right)$$

$$x_f - x_i = v_{xi}t + \frac{1}{2}a_x t^2$$

which is **Equation 2.15**.

Besides what you might expect to learn about physics concepts, a very valuable skill you should hope to take away from your physics course is the ability to solve complicated problems. The way physicists approach complex situations and break them into manageable pieces is extremely useful. The following is a general problem-solving strategy that can be applied to any problem. You will find this strategy much more effective than 'formula hunting'.

General problem-solving strategy

Conceptualise

- The first thing to do when approaching a problem is to *think about* and *understand* the situation. Study carefully any representations of the information (for example, diagrams, graphs, tables or photographs) that accompany the problem. Imagine a movie running in your mind of what happens in the problem.
- If a pictorial representation is not provided, you should always make a quick drawing of the situation. Indicate any known values directly on your sketch. Even if a diagram is provided, it will help to draw another one of a different type, for example a graph or motion diagram, or some other type of diagram. We will meet other types, such as free-body diagrams and field diagrams, in later chapters.
- Now focus on what algebraic or numerical information is given in the problem. Carefully read the problem statement, looking for key phrases such as 'starts from rest' ($v_i = 0$), 'stops' ($v_f = 0$), or 'falls freely' ($a_y = -g = -9.80$ m/s^2).
- Read the question again. Make sure you know exactly what the question is asking you for. Will the final result be numerical or algebraic? Do you know what units to expect?
- Don't forget to use your own experiences and common sense. What should a reasonable answer look like? For example, you wouldn't expect to calculate the speed of a car to be 5×10^6 m/s.

Model

- Once you understand what the problem is about, and what you are trying to find, you should have a good mental model of the situation. You need to use that model to solve the problem. This will usually mean constructing a mathematical model of the situation.
- Think carefully about what approximations you can sensibly make. Can you use an analysis model such as the *particle with constant velocity* or the *particle with constant acceleration* analysis model? Are approximations such as no air resistance or no friction between surfaces reasonable? When you have decided what simplifications you can reasonably make then you can construct your mathematical model by finding or generating the equations you will use.
- Think about what general principles you can use. In later chapters, you will meet several conservation principles and force laws that are often used to solve problems. It is a good idea to write a few words saying what laws or principles you are using, and *then* write down the equations that represent these.
- Note that we are halfway through solving the problem *before* we write down any equations! It may seem quicker to simply write down the equations you think you will need right at the start, but you are much less likely to get the problem right if you do this, especially as the problems you deal with become more complex.

Analyse

- Now you analyse your model to find a mathematical solution. Because you have already modelled the situation in the problem, and identified relevant general laws or principles, and in some cases an analysis model that can be applied, you should be able to select relevant equations that apply. For example, if the problem involves a particle with constant acceleration, **Equations 2.12** to **2.16** are relevant.

- Use algebra and calculus as necessary to solve symbolically for the required variable in terms of what is given. Check the dimensions of your algebraic expression *before* substituting in any numbers. If the dimensions are wrong, the answer is wrong, and you will have to go back and look for the error. If the dimensions are correct, substitute in the appropriate numbers, calculate the result, and round it to the proper number of significant figures.

Finalise

- Examine your numerical answer. Does it have the correct units? Does it meet your expectations from your conceptualisation of the problem? What about the algebraic form of the result? Does it make sense? Do a 'reality check' – ask yourself if it makes sense given your understanding and experiences. Examine the variables in the problem to see whether the answer would change in a physically meaningful way if the variables were drastically increased or decreased or even became zero. Looking at limiting cases to see whether they yield expected values is a very useful way to make sure that you are obtaining reasonable results.
- When you have finished the problem, think about how this problem compared with others you have solved. How was it similar? In what critical ways did it differ? What have you learned by doing it? Can you use the same model again for solving similar problems in the future?

When solving complex problems, you may need to identify a series of sub-problems and apply the problem-solving strategy to each. When you are trying to solve a problem and you get stuck, go back to the steps in the strategy and use them as a guide. In the rest of this book, we will label these steps explicitly in the worked examples.

Several chapters in this book include a section labelled *Problem-solving strategy* that should help you through the rough spots. These sections are organised according to the *General problem-solving strategy* outlined above and are tailored to the specific types of problems addressed in that chapter.

To clarify how this strategy works, we repeat **Example 2.7** with the particular steps of the strategy identified.

Example 2.7

Carrier landing

A pilot lands his jet on an aircraft carrier at a speed of 230 km/h (64 m/s), with an uncertainty in the speed of 10 km/h (3 m/s). You might have seen movies or television shows in which a jet lands on an aircraft carrier and is brought to rest surprisingly fast by an arresting cable, to prevent it falling off and into the sea. The idea of using arresting cables to slow down landing aircraft and enable them to land safely on ships originated at about the time of World War I. The cables are still a vital part of the operation of modern aircraft carriers.

(A) What is the acceleration of the jet (assumed constant) if it stops in (2.2 ± 0.2) s after the arresting cable is engaged?

Solution

Conceptualise It is a good idea to draw a quick sketch for yourself showing the motion of the jet – either as a motion diagram as described in Section 2.5 or as a graph. This will help you visualise the problem. The diagram shows what is happening in the problem.

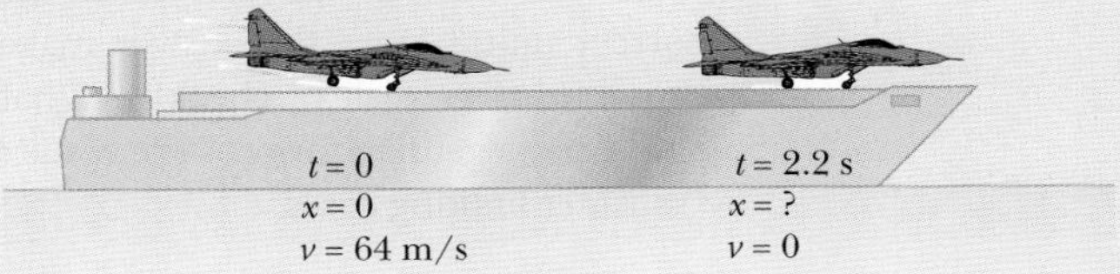

Conceptualise

When you conceptualise a problem, try to understand the situation that is presented in the problem statement. Study carefully any representations of the information (for example, diagrams, graphs, tables or photographs) that accompany the problem. Draw a diagram. Imagine a movie running in your mind of what happens in the problem.

Example 2.7 cont.

Model Because the acceleration of the jet is assumed constant, we model it as a particle with constant acceleration.

Model

Simplify the problem. Remove the details that are not important to the solution to model the situation. What general principles apply? In this case, identify the appropriate analysis model.

Analyse We define our x axis as the direction of motion of the jet. (Notice that we have no information about the change in position of the jet while it is slowing down.)
We are told both the initial speed, $v_{xi} = (64 \pm 3)$ m/s, and the final speed, $v_{xf} = 0$ (the jet must stop), and the time interval, $t = (2.2 \pm 0.2)$ s.

Analyse

Now analyse your model. Select relevant equations and solve symbolically for the required variable. Do a dimension check. Substitute in the appropriate numbers, calculate the result and round it to the proper number of significant figures.

We can use **Equation 2.12** to find the acceleration of the jet, modelled as a particle:

$$a_x = \frac{v_{xf} - v_{xi}}{t}$$

Do a dimension check to ensure we have rearranged the equation correctly: $[LT^{-2}] = [LT^{-1}]/[T] = [LT^{-2}]$ ☺

Our dimensions are okay so we can proceed to substitute in the numbers:

$$a_x = \frac{0 - 64\text{ m/s}}{2.2\text{ s}} = -29\text{ m/s}^2$$

The uncertainty in the acceleration is:

$$\frac{\Delta a_x}{a_x} = \frac{\Delta v_x}{v_x} + \frac{\Delta t}{t} = \frac{3\text{ m/s}}{64\text{ m/s}} + \frac{0.2\text{ s}}{2.2\text{ s}} = 0.14 \text{ or } 14\%$$

so $\Delta a_x = 29\text{ m/s} \times 0.14 = 4\text{ m/s}^2$.

Recall that uncertainties always have a positive value, so we multiply by the magnitude of a_x and ignore the negative sign.

So the acceleration is $a_x = (-29 \pm 4)$ m/s.

(B) If the jet touches down at position $x_i = 0$, what is its final position?

Solution

Use **Equation 2.14** to solve for the final position:

$$x_f = x_i + \frac{1}{2}(v_{xi} + v_{xf})t = 0 + \frac{1}{2}(64\text{ m/s} + 0)2.2\text{ s} = 70\text{ m}$$

The uncertainty will be $\Delta x = x\left(\frac{\Delta v_{xi}}{v_{xi}} + \frac{\Delta t}{t}\right) = 70\text{ m}\left(\frac{3\text{ m/s}}{64\text{ m/s}} + \frac{0.2\text{ s}}{2.2\text{ s}}\right) = 10\text{ m}$

so $x_f = (70 \pm 10)$ m.

Finalise Given the size of aircraft carriers, a length of 70 m seems reasonable for stopping the jet. If we had made a calculation error and come up with an answer of 70 km, we would know we had made an error.

Finalise

Finalise the problem. Examine the numerical answer. Does it have the correct units? Does the answer make sense (do a reality check)? What about the algebraic form of the result? Consider limiting cases.

Example 2.7 cont.

What If? Suppose the stopping distance before the edge of the carrier is 85 m. Can the pilot land safely? What if the speed is faster than 64 m/s?

What If?

What If? questions will appear in many examples in the text, and offer a variation on the situation just explored. This feature encourages you to think about the results of the example and assists in conceptual understanding of the principles.

Answer The calculated stopping distance was (70 ± 10) m, which has a maximum value of 80 m. So the jet can land safely if the available distance is 85 m. If the jet is travelling faster at the beginning, it will stop farther away from its starting point. Mathematically, we see in **Equation 2.14** that if v_{xi} is larger, then x_f will be larger. Given the uncertainty in stopping distance, only a small increase in speed could result in the jet falling into the ocean.

End-of-chapter resources

Each year the 'Race the Rattler' event is held in Mary Valley, between the town of Dagun and old Gympie Railway Station 18.5 km away. The race is a footrace against a C17 steam locomotive built in the 1920s. The winning times are typically an hour and a few minutes. The train has a top speed of around 19 km/h, but has quite small acceleration, taking minutes to obtain this maximum speed. Once at this speed, the train can continue to steam along at this pace as long as the coal and water last. In contrast, a human can accelerate to reach a top speed of around 35 km/h in only a few seconds, but cannot maintain this speed for more than short sprints. Hence if the race was much shorter the humans would win, and if the race was much longer the train would win. With a distance of 18.5 km, in some years the race is won by the train and in some by a human, by a margin of only a few minutes in either case.

Newspix/News Ltd/Graeme Parkes

The problems found in this chapter may be assigned online in Enhanced Web Assign.

Worked solutions to every fifth problem are available in the Student Solutions Manual. Register online at **www.cengagebrain.com** for access.

Summary

Definitions

When a particle moves along the x axis from some initial position x_i to some final position x_f, its **displacement** is

$$\Delta x \equiv x_f - x_i \quad (2.1)$$

The **average velocity** of a particle during some time interval is the displacement Δx divided by the time interval Δt during which that displacement occurs:

$$v_{x,\text{avg}} \equiv \frac{\Delta x}{\Delta t} \quad (2.2)$$

The **average speed** of a particle is equal to the ratio of the total distance it travels, s, to the total time interval during which it travels that distance:

$$v_{\text{avg}} \equiv \frac{s}{\Delta t} \quad (2.3)$$

The **instantaneous velocity** of a particle is defined as the limit of the ratio $\Delta x/\Delta t$ as Δt approaches zero. By definition, this limit equals the derivative of x with respect to t, or the time rate of change of the position:

$$v_x \equiv \lim_{\Delta t \to 0} \frac{\Delta x}{\Delta t} = \frac{dx}{dt} \quad (2.5)$$

The **instantaneous speed** of a particle is equal to the magnitude of its instantaneous velocity.

The **average acceleration** of a particle is defined as the ratio of the change in its velocity Δv_x divided by the time interval Δt during which that change occurs:

$$a_{x,\text{avg}} \equiv \frac{\Delta v_x}{\Delta t} = \frac{v_{xf} - v_{xi}}{t_f - t_i} \quad (2.9)$$

The **instantaneous acceleration** is equal to the limit of the ratio $\Delta v_x/\Delta t$ as Δt approaches 0. By definition, this limit equals the derivative of v_x with respect to t, or the time rate of change of the velocity:

$$a_x \equiv \lim_{\Delta t \to 0} \frac{\Delta v_x}{\Delta t} = \frac{dv_x}{dt} \quad (2.10)$$

Concepts and principles

When an object's velocity and acceleration are in the same direction, the object is speeding up. When the velocity and acceleration are in opposite directions, the object is slowing down.

An object falling freely in the presence of the Earth's gravity experiences free-fall acceleration directed towards the centre of the Earth. If air resistance is neglected, if the motion occurs near the surface of the Earth, and if the range of the motion is small compared with the Earth's radius, the free-fall acceleration $a_y = -g$ is constant over the range of motion, where g is equal to 9.80 m/s^2.

An important aid to problem solving is the use of **analysis models**. Analysis models can be applied to a range of different (but similar) situations. Each analysis model has one or more equations associated with it. When solving a new problem, identify the analysis model that corresponds to the problem. The model will tell you which equations to use. The first three analysis models introduced in this chapter are summarised below.

Complicated problems are best approached in an organised manner. Recall and apply the *Conceptualise, Model, Analyse* and *Finalise* steps of the **General problem-solving strategy** when you need them.

Analysis models for problem-solving

Particle with constant velocity If a particle moves in a straight line with a constant speed v_x, its constant velocity is given by

$$v_x = \frac{\Delta x}{\Delta t} \quad (2.6)$$

and its position is given by

$$x_f = x_i + v_x t \quad (2.7)$$

Particle with constant speed If a particle moves a distance s along a curved or straight path with a constant speed, its constant speed is given by

$$v = \frac{s}{\Delta t} \quad (2.8)$$

Particle with constant acceleration If a particle moves in a straight line with a constant acceleration a_x, its motion is described by the kinematic equations:

$$v_{xf} = v_{xi} + a_x t \quad (2.12)$$

$$v_{x,\text{avg}} = \frac{v_{xi} + v_{xf}}{2} \quad (2.13)$$

$$x_f = x_i + \frac{1}{2}(v_{xi} + v_{xf})t \quad (2.14)$$

$$x_f = x_i + v_{xi}t + \frac{1}{2}a_x t^2 \quad (2.15)$$

$$(v_{xf})^2 = (v_{xi})^2 + 2a_x(x_f - x_i) \quad (2.16)$$

Chapter review quiz

To help you revise Chapter 2: Motion in one dimension, complete the automatically graded Chapter review quiz at http://login.cengagebrain.com.

Conceptual questions

1. If the average velocity of an object is zero in some time interval, what can you say about the displacement of the object for that interval?
2. (a) Can the velocity of an object at an instant of time be greater in magnitude than the average velocity over a time interval containing the instant? (b) Can it be less?
3. If the velocity of a particle is non-zero, can the particle's acceleration be zero? Explain your answer. What about the other way around?
4. (a) Can the equations of kinematics (Equations 2.15–2.19) be used in a situation in which the acceleration varies in time? (b) Can they be used when the acceleration is zero?
5. If a car is traveling eastward, can its acceleration be westward? Explain.
6. A hard rubber ball, not affected by air resistance in its motion, is tossed upwards from shoulder height, falls to the footpath, rebounds to a smaller maximum height, and is caught on its way down again. This motion is represented in Figure CQ2.6, where the successive positions of the ball Ⓐ through Ⓔ are not equally spaced in time. At point Ⓓ the centre of the ball is at its lowest point in the motion. The motion of the ball is along a straight, vertical line, but the diagram shows successive positions offset to the right to avoid overlapping. Choose the positive y direction to be upwards. (a) Rank the situations Ⓐ through Ⓔ according to the speed of the ball $|v_y|$ at each point, with the largest speed first. (b) Rank the same situations according to the acceleration a_y of the ball at each point. (In both rankings, remember that zero is greater than a negative value. If two values are equal, show that they are equal in your ranking.)

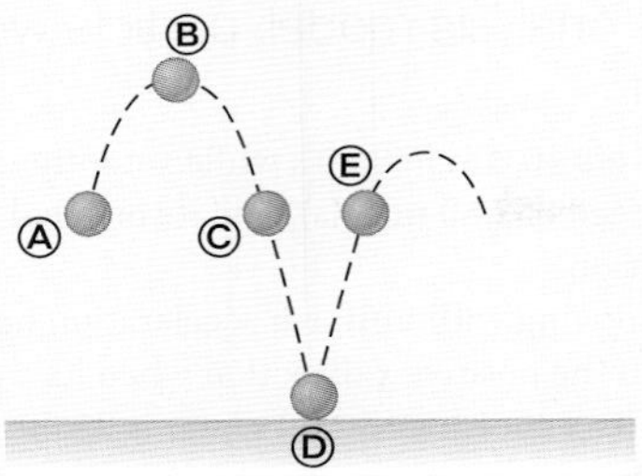

Figure CQ2.6

Problems

Section 2.1 Position, velocity and speed

1. The position versus time for a certain particle moving along the x axis is shown in Figure P2.1. Find the average velocity in the time intervals (a) 0 to 2 s, (b) 0 to 4 s, (c) 2 s to 4 s, (d) 4 s to 7 s, and (e) 0 to 8 s.

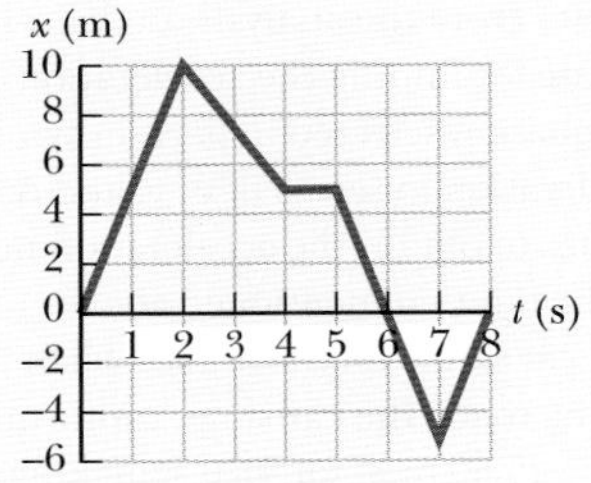

Figure P2.1

2. The position of a billy cart was observed at various times; the results are summarised in the following table. Find the average velocity of the billy cart for (a) the first second, (b) the last 3 s, and (c) the entire period of observation.

t (s)	0	1.0	2.0	3.0	4.0	5.0
x (m)	0	2.3	9.2	20.7	36.8	57.5

3. A person walks first at a constant speed of 5.00 m/s along a straight line from point Ⓐ to point Ⓑ and then back along the line from Ⓑ to Ⓐ at a constant speed of 3.00 m/s. (a) What is her average speed over the entire trip? (b) What is her average velocity over the entire trip? Remember to start by drawing a diagram!

Section 2.2 Instantaneous velocity and speed

4. A position–time graph for a particle moving along the x axis is shown in Figure P2.4. (a) Find the average velocity in the time interval $t = 1.50$ s to $t = 4.00$ s. (b) Determine the instantaneous velocity at $t = 2.00$ s by measuring the gradient of the tangent line shown in the graph. (c) At what value of t is the velocity zero?

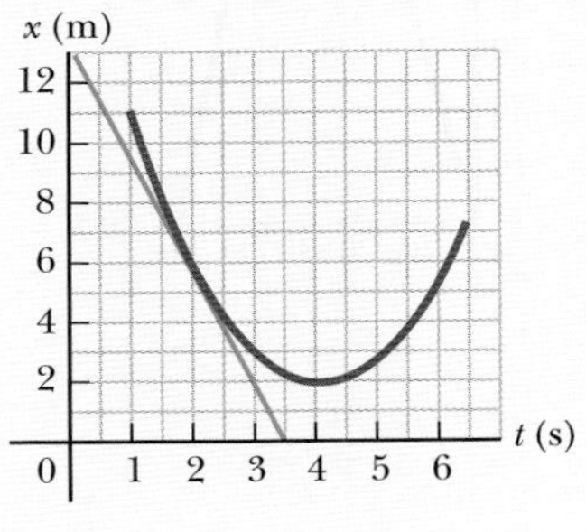

Figure P2.4

5. The position of a particle moving along the x axis varies in time according to the expression $x = 3t^2$, where x is in metres and t is in seconds. Evaluate its position (a) at $t = 3.00$ s and (b) at $t = 3.00\text{ s} + \Delta t$. (c) Evaluate the limit of $\Delta x/\Delta t$ as Δt approaches zero to find the velocity at $t = 3.00$ s.

Section 2.3 Analysis model: particle with constant velocity

6. ±? The North American and European plates of the Earth's crust are drifting apart with a relative speed of about 25 ± 5 mm/yr. Take the speed as constant and find when the rift between them started to open, to reach a current width of $(4.7 \pm 0.1) \times 10^3$ km.

7. A person takes a trip, driving with a constant speed of 89.5 km/h, except for a 22.0-min rest stop. If the person's average speed is 77.8 km/h, (a) how much time is spent on the trip and (b) how far does the person travel?

Section 2.4 Acceleration

8. ±? A (50 ± 2) g super ball travelling at (25 ± 1) m/s bounces off a brick wall and rebounds at (22 ± 2) m/s. A high-speed camera records this event. If the ball is in contact with the wall for (3.5 ± 0.1) ms, what is the magnitude of the average acceleration of the ball during this time interval?

9. (a) Use the data in Problem 2 to construct a smooth graph of position versus time. (b) By constructing tangents to the $x(t)$ curve, find the instantaneous velocity of the car at several instants. (c) Plot the instantaneous velocity versus time and, from this information, determine the average acceleration of the car. (d) What was the initial velocity of the car?

10. A child rolls a marble on a bent track that is 100 cm long as shown in Figure P2.10. We use x to represent the position of the marble along the track. On the horizontal sections from $x = 0$ to $x = 20$ cm and from $x = 40$ cm to $x = 60$ cm, the marble rolls with constant speed. On the sloping sections, the marble's speed changes steadily. At the places where the gradient changes, the marble stays on the track and does not undergo any sudden changes in speed. The child gives the marble some initial speed at $x = 0$ and $t = 0$ and then watches it roll to $x = 90$ cm, where it changes direction, eventually returning to $x = 0$ with the same speed with which the child released it. Draw graphs of (a) x versus t, (b) v_x versus t, and (c) a_x versus t, vertically aligned with their time axes identical, to show the motion of the marble. You will not be able to place numbers other than zero on the horizontal axis or on the velocity or acceleration axes, but show the correct graph shapes.

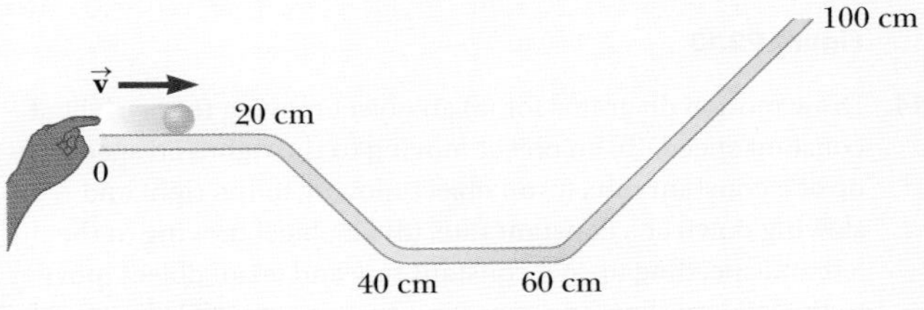

Figure P2.10

11. A particle starts from rest and accelerates as shown in **Figure P2.11**. Determine (a) the particle's speed at $t = 10.0$ s and at $t = 20.0$ s, and (b) the distance travelled in the first 20.0 s.

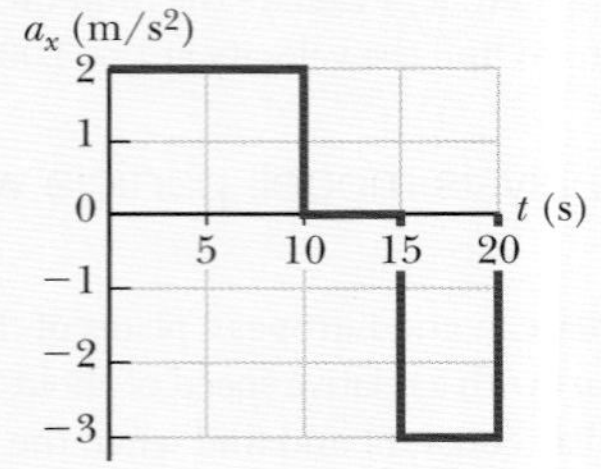

Figure P2.11

12. A particle moves along the x axis according to the equation $x = 2.00 + 3.00t - 1.00t^2$, where x is in metres and t is in seconds. At $t = 3.00$ s, find (a) the position of the particle, (b) its velocity and (c) its acceleration.

Section 2.5 Motion diagrams

13. Each of the strobe photographs (a), (b) and (c) in **Figure P2.13** was taken of a single disc moving toward the right, which we take as the positive direction. Within each photograph the time interval between images is constant. For each photograph, prepare graphs of x versus t, v_x versus t, and a_x versus t, vertically aligned with their time axes identical, to show the motion of the disc. You will not be able to place numbers other than zero on the axes, but show the correct shapes for the graph lines.

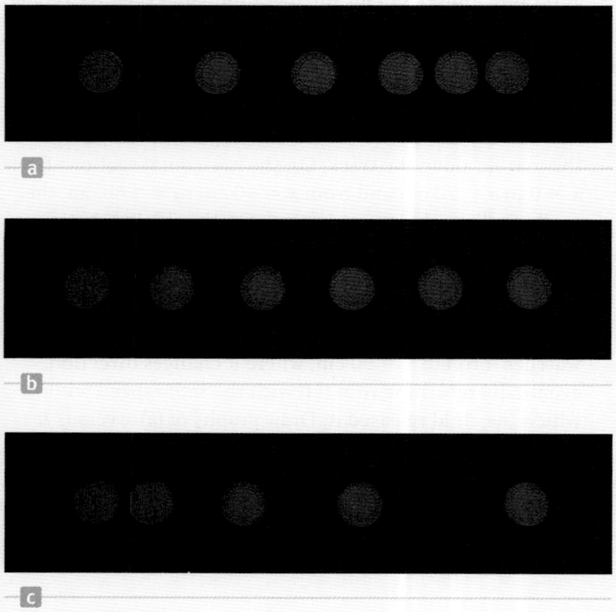

Figure P2.13

14. Draw motion diagrams for (a) an object moving to the right at constant speed, (b) an object moving to the right and speeding up at a constant rate, (c) an object moving to the right and slowing down at a constant rate, (d) an object moving to the left and speeding up at a constant rate and (e) an object moving to the left and slowing down at a constant rate. (f) How would your drawings change if the changes in speed were not uniform, that is, if the speed were not changing at a constant rate?

Section 2.6 Analysis model: particle with constant acceleration

15. A truck covers 40.0 m in 8.50 s while smoothly slowing down to a final speed of 2.80 m/s. (a) Find its original speed. (b) Find its acceleration.

16. An object moving with uniform acceleration has a velocity of 12.0 cm/s in the positive x direction when its x coordinate is 3.00 cm. If its x coordinate 2.00 s later is 25.00 cm, what is its acceleration?

17. Solve **Example 2.8** by a graphical method. On the same graph, plot position versus time for the car and the police officer. From the intersection of the two curves, read the time at which the police officer overtakes the car.

18. ±? A speedboat travels in a straight line and increases in speed uniformly from $v_i = (20.0 \pm 0.5)$ m/s to $v_f = (30.0 \pm 0.5)$ m/s in a displacement Δx of (200 ± 5) m. We wish to find the time interval required for the boat to move through this displacement. (a) Draw a diagram for this situation. (b) What analysis model is most appropriate for describing this situation? (c) From the analysis model, what equation is most appropriate for finding the acceleration of the speedboat? (d) Solve the equation selected in part (c) symbolically for the boat's acceleration in terms of v_i, v_f and Δx. (e) Substitute numerical values to obtain the acceleration numerically. (f) Find the time interval mentioned above.

19. The driver of a car slams on the brakes when he sees a tree blocking the road. The car slows uniformly with an acceleration of -25.60 m/s^2 for 4.20 s, making straight skid marks 62.4 m long, all the way to the tree. With what speed does the car then strike the tree?

20. In the particle with constant acceleration model, we identify the variables and parameters v_{xi}, v_{xf}, a_x, t, and $x_f - x_i$. Of the equations in the model, **Equations 2.13–2.17**, the first does not involve $x_f - x_i$, the second and third do not contain a_x, the fourth omits v_{xf}, and the last leaves out t. So, to complete the set, there should be an equation *not* involving v_{xi}. (a) Derive it from the others. (b) Use the equation in part (a) to solve Problem 19 in one step.

21. **Figure P2.21** represents part of the performance data of a car owned by a proud physics student. (a) Calculate the total distance travelled by computing the area under the red-brown graph line. (b) What distance does the car travel between the times $t = 10$ s and $t = 40$ s? (c) Draw a graph of its acceleration versus time between $t = 0$ and $t = 50$ s. (d) Write an equation for x as a function of time for each phase of the motion, represented by the segments $0a$, ab, and bc. (e) What is the average velocity of the car between $t = 0$ and $t = 50$ s?

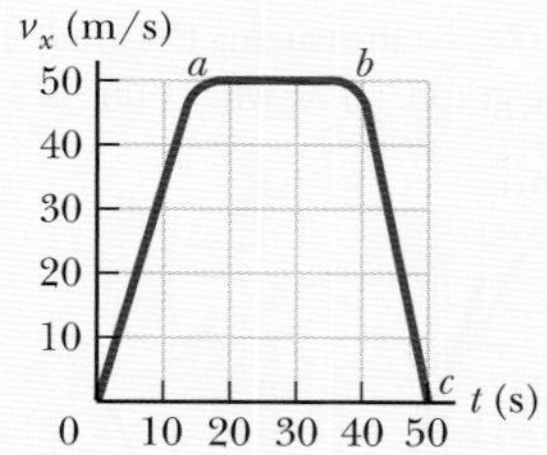

Figure P2.21

22. At $t = 0$, one toy car is set rolling on a straight track with initial position 15.0 cm, initial velocity –3.50 cm/s, and constant acceleration 2.40 cm/s^2. At the same moment, another toy car is set rolling on an adjacent track with initial position 10.0 cm, initial velocity 15.50 cm/s, and constant zero acceleration. (a) At what time, if any, do the two cars have equal speeds? (b) What are their speeds at that time? (c) At what time(s), if any, do the cars pass each other? (d) What are their locations at that time? (e) Explain the difference between question (a) and question (c) as clearly as possible.

Section 2.7 Freely falling objects

Note: In all problems in this section ignore the effects of air resistance.

23. An attacker at the base of a castle wall 3.65 m high throws a rock straight up with speed 7.40 m/s from a height of 1.55 m above the ground. (a) Will the rock reach the top of the wall? (b) If so, what is its speed at the top? If not, what initial speed must it have to reach the top? (c) Find the change in speed of a rock thrown straight down from the top of the wall at an initial speed of 7.40 m/s and moving between the same two points. (d) Does the change in speed of the downwards-moving rock agree with the magnitude of the speed change of the rock moving upwards between the same heights? (e) Explain physically why it does or does not agree.

24. ±? A ball is thrown directly downwards with an initial speed of (8.0 ± 0.1) m/s from a height of (30 ± 1) m. After what time interval does it strike the ground?

25. A bushranger hanging out of a hotel window wishes to drop vertically onto a horse galloping under the window. The constant speed of the horse is 10.0 m/s, and the distance from the window to the level of the saddle is 3.00 m. (a) What must be the horizontal distance between the saddle and window sill when the bushranger makes his move? (b) For what time interval is he in the air?

26. A package is dropped at time $t = 0$ from a helicopter that is descending steadily at a speed v_i. (a) What is the speed of the package in terms of v_i, g and t? (b) What vertical distance d is it from the helicopter in terms of g and t? (c) What are the answers to parts (a) and (b) if the helicopter is rising steadily at the same speed?

Section 2.8 Kinematic equations derived from calculus

27. Automotive engineers refer to the time rate of change of acceleration as the 'jerk'. Assume an object moves in one dimension such that its jerk J is constant. (a) Determine expressions for its acceleration $a_x(t)$, velocity $v_x(t)$ and position $x(t)$, given that its initial acceleration, velocity and position are a_{xi}, v_{xi} and x_i, respectively. (b) Show that $a_x^2 = a_{xi}^2 + 2J(v_x - v_{xi})$.

28. The speed of a bullet as it travels down the barrel of a rifle toward the opening is given by

$$v = (-5.00 \times 10^7)t^2 + (3.00 \times 10^5)t$$

where v is in metres per second and t is in seconds. The acceleration of the bullet just as it leaves the barrel is zero. (a) Determine the acceleration and position of the bullet as functions of time when the bullet is in the barrel. (b) Determine the time interval over which the bullet is accelerated. (c) Find the speed at which the bullet leaves the barrel. (d) What is the length of the barrel?

Additional problems

29. The froghopper *Philaenus spumarius* is supposedly the best jumper in the animal kingdom. To start a jump, this insect can accelerate at 4.00 km/s^2 over a distance of 2.00 mm as it straightens its specially adapted 'jumping legs'. Assume the acceleration is constant. (a) Find the upwards velocity with which the insect takes off. (b) In what time interval does it reach this velocity? (c) How high would the insect jump if air resistance were negligible? The actual height it reaches is about 70 cm, so air resistance must be a noticeable force on the leaping froghopper.

Getty Images/Stephen Dalton

Figure P2.29

30. In **Active Figure 2.12b**, the area under the velocity–time graph and between the vertical axis and time t (vertical dashed line) represents the displacement. As shown, this area consists of a rectangle and a triangle. (a) Calculate their areas. (b) Explain how the sum of the two areas compares with the expression on the right-hand side of **Equation 2.15**.

31. An inquisitive physics student and mountain climber climbs a 50.0 m cliff that overhangs a calm pool of water. He throws two stones vertically downwards, 1.00 s apart, and observes that they cause a single splash. The first stone has an initial speed of 2.00 m/s. (a) How long after release of the first stone do the two stones hit the water? (b) What initial velocity must the second stone have if the two stones are to hit the water simultaneously? (c) What is the speed of each stone at the instant the two stones hit the water?

32. *Why is the following situation impossible?* A freight train is lumbering along at a constant speed of 16.0 m/s. Behind the freight train on the same track is a passenger train travelling in the same direction at 40.0 m/s. When the front of the passenger train is 58.5 m from the back of the freight train, the engineer on the passenger train recognises the danger and hits the brakes of his train, causing the train to move with acceleration –3.00 m/s^2. Because of the engineer's action, the trains do not collide.

33. At $t = 0$, one athlete in a race running on a long, straight track with a constant speed v_1 is a distance d_1 behind a second athlete running with a constant speed v_2. (a) Under what circumstances is the first athlete able to overtake the second athlete? (b) Find the time t at which the first athlete overtakes the second athlete, in terms of d_1, v_1 and v_2. (c) At what minimum distance d_2 from the leading athlete must the finish line be located so that the trailing athlete can at least tie for first place? Express d_2 in terms of d_1, v_1 and v_2 by using the result of part (b).

34. Kathy tests her new sports car by racing with Stan, an experienced racer. Both start from rest, but Kathy leaves the starting line 1.00 s after Stan does. Stan moves with a constant acceleration of 3.50 m/s^2, while Kathy maintains an acceleration of 4.90 m/s^2. Find (a) the time at which Kathy overtakes Stan, (b) the distance she travels before she catches him, and (c) the speeds of both cars at the instant Kathy overtakes Stan.

35. Two objects, A and B, are connected by hinges to a rigid rod that has a length L. The objects slide along perpendicular guide rails as shown in Figure P2.35. Assume object A slides to the left with a constant speed v. (a) Find the velocity v_B of object B as a function of the angle θ. (b) Describe v_B relative to v. Is v_B always smaller than v, larger than v, or the same as v, or does it have some other relationship?

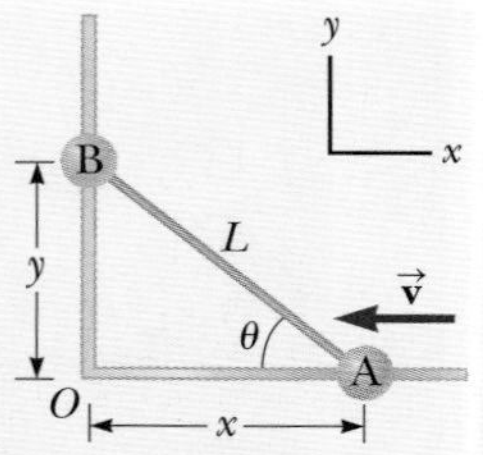

Figure P2.35

36. A commuter train travels between two city stations. Because the stations are only 1.00 km apart, the train never reaches its maximum possible cruising speed. During rush hour the engineer minimizes the time interval Δt between two stations by accelerating at a rate $a_1 = 0.100$ m/s^2 for a time interval Δt_1 and then immediately braking with acceleration $a_2 = -0.500$ m/s^2 for a time interval Δt_2. Find the minimum time interval of travel Δt and the time interval Δt_1.

Challenge problems

37. In a women's 100 m race, accelerating uniformly, Laura takes 2.00 s and Helen 3.00 s to attain their maximum speeds, which they each maintain for the rest of the race. They cross the finish line simultaneously, both setting a world record of 10.4 s. (a) What is the acceleration of each sprinter? (b) What are their respective maximum speeds? (c) Which sprinter is ahead at the 6.00 s mark, and by how much? (d) What is the maximum distance by which Helen is behind Laura, and at what time does that occur?

38. A man drops a rock into a well. (a) The man hears the sound of the splash 2.40 s after he releases the rock from rest. The speed of sound in air (at the ambient temperature) is 336 m/s. How far below the top of the well is the surface of the water? (b) **What If**? If the travel time for the sound is ignored, what percentage error is introduced when the depth of the well is calculated?

39. Two thin rods are fastened to the inside of a circular ring as shown in Figure P2.39. One rod of length D is vertical, and the other of length L makes an angle θ with the horizontal. The two rods and the ring lie in a vertical plane. Two small beads are free to slide without friction along the rods. (a) If the two beads are released from rest simultaneously from the positions shown, use your intuition and guess which bead reaches the bottom first. (b) Find an expression for the time interval required for the red bead to fall from point Ⓐ to point Ⓒ in terms of g and D. (c) Find an expression for the time interval required for the blue bead to slide from point Ⓑ to point Ⓒ in terms of g, L and θ. (d) Show that the two time intervals found in parts (b) and (c) are equal. Hint: What is the angle between the chords of the circle ⒶⒷ and ⒷⒸ? (e) Do these results surprise you? Was your intuitive guess in part (a) correct? This problem was inspired by an article by Thomas B. Greenslade, Jr., 'Galileo's Paradox', *Phys. Teach.* 46, 294 (May 2008).

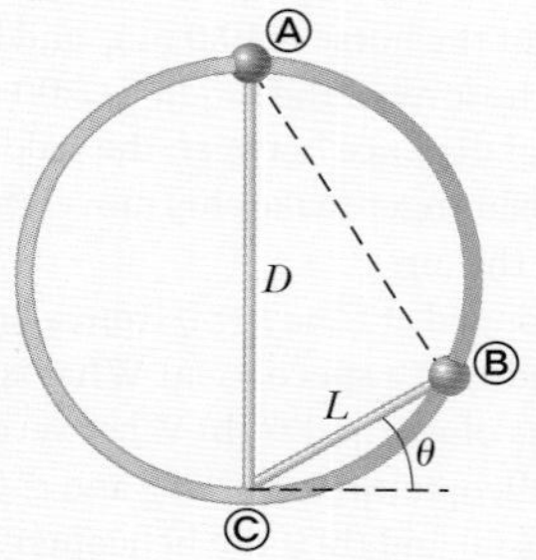

Figure P2.39

chapter 3

Motion in two dimensions

Graham Monro/Photolibrary/Jupiter Images

The photo shows fireworks erupting from Sydney Harbour Bridge. Note the shape of the paths taken by the incandescent embers. All the paths are a similar shape. Why do they take these particular trajectories, and how can we model their behaviour to allow us to predict their trajectories?

Fireworks erupt from the Sydney Harbour Bridge on New Year's Eve. Note the paths of the embers.

On completing this chapter, students will understand:
- how motion in two or more dimensions is described using vectors
- that uniform circular motion is due to a centripetal acceleration
- that motion is relative
- why observers in different reference frames observe different motion.

Students will be able to:
- distinguish between vectors and scalars
- represent a position in Cartesian or plane polar coordinates
- describe motion in two or more dimensions using the position, velocity and acceleration vectors
- solve problems to find times, positions, velocities and accelerations using the kinematic equations
- recognise and solve problems involving projectile motion
- solve problems using the analysis model 'particle in uniform circular motion'
- analyse non-uniform circular motion using tangential and radial acceleration
- apply Galilean transformations to calculate velocity and acceleration for different observers.

In this chapter, we explore the kinematics of a particle moving in two dimensions. Knowing the basics of two-dimensional motion will allow us to examine a variety of situations, including the motion of projectiles as shown in the picture and objects in circular motion. We also discuss the concept of relative motion, which shows why observers in different frames of reference may measure different positions and velocities for a given object.

ENHANCED WebAssign

The problems found in this chapter may be assigned online in Enhanced Web Assign.

3.1 Vectors, scalars and coordinate systems

In our study of physics, we often need to work with physical quantities that have both numerical and directional properties. Quantities of this nature are **vector** quantities.

In Chapter 2 we saw that the mathematical description of an object's motion requires a method for describing the object's position at various times. In two or more dimensions, this description is accomplished with the use of **vectors** and a **coordinate system**.

Σ *If you are not comfortable with vectors you should review Appendix B.6 before proceeding with this chapter.*

Vector and scalar quantities

When you want to know how much juice is in a bottle at the supermarket, you look at the volume shown on the packaging. All you need to know is the number and the units, for example 2 litres. Volume is therefore an example of a *scalar quantity*. If you are preparing to pilot a small plane and need to know the wind velocity, you must know both the speed of the wind and its direction. Because direction is important for its complete specification, velocity is a *vector quantity*.

A **scalar quantity** is completely specified by a single value with an appropriate unit and has no direction. Examples include volume, mass, speed and time. Some scalars are always positive, such as mass and speed. Others, such as temperature, can have either positive or negative values. The rules of ordinary arithmetic are used to manipulate scalar quantities.

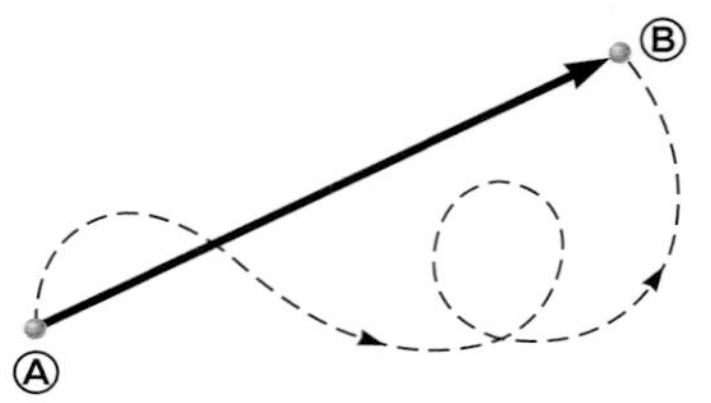

Figure 3.1
As a particle moves from Ⓐ to Ⓑ along an arbitrary path represented by the broken line, its displacement is a vector quantity shown by the arrow drawn from Ⓐ to Ⓑ.

A **vector quantity** is completely specified by a number with an appropriate unit plus a direction. Examples of vector quantities include displacement, acceleration and force. Suppose a particle moves from some point Ⓐ to some point Ⓑ along a straight path as shown in Figure 3.1. We represent this displacement by drawing an arrow from Ⓐ to Ⓑ. The direction of the arrowhead represents the direction of the displacement, and the length of the arrow represents the magnitude of the displacement. If the particle travels along some other path from Ⓐ to Ⓑ such as that shown by the dotted line in Figure 3.1, its displacement is still the arrow drawn from Ⓐ to Ⓑ. Displacement depends only on the initial and final positions, so the displacement vector is independent of the path taken by the particle between these two points.

In this text, we use a boldface letter with an arrow over the letter, such as $\vec{\mathbf{A}}$, to represent a vector. Another common notation for vectors with which you should be familiar is a simple boldface character: **A**. The magnitude of the vector $\vec{\mathbf{A}}$ is written either A or $|\vec{\mathbf{A}}|$. The magnitude of a vector has physical units, such as metres for displacement or metres per second for velocity. The magnitude of a vector is always a positive number.

Quick Quiz 3.1

Which of the following are vector quantities and which are scalar quantities? (a) your age (b) acceleration (c) velocity (d) speed (e) mass.

Coordinate systems

The most commonly used coordinate system is the Cartesian coordinate system, in which perpendicular axes intersect at a point defined as the origin (Figure 3.3). Cartesian coordinates are also called *rectangular coordinates*.

TRY THIS

Battleship is a two-player game in which each player has several ships on a Cartesian grid with one axis labelled with numbers and the other with letters. Players have to sink each other's ships by guessing the position of the ship and taking turns at firing shells at particular positions on the grid, by calling out the coordinates of the position. A ship may be one, two or three grid squares long, and aligned vertically or horizontally. The other player says hit or miss accordingly. The winner is the player who first sinks all the opponent's ships by hitting all coordinates at which there is a ship. Although there are many commercially available versions of this game, it can be played just as well with only pen and paper. Have a quiet game next time you are in a boring chemistry lecture.

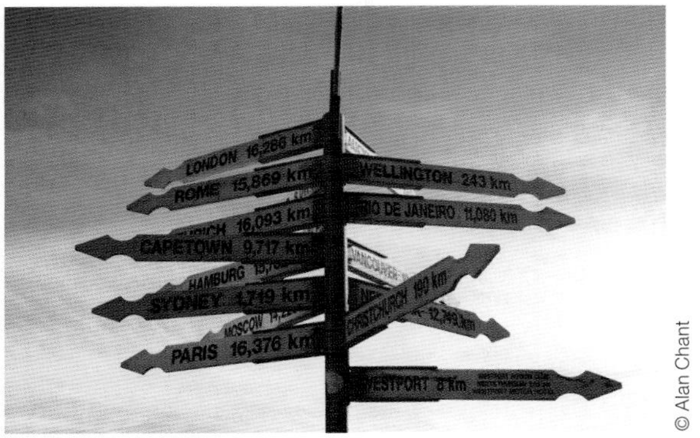

© Alan Chant

Figure 3.2
A signpost in Westport, New Zealand, shows the distance and direction to several cities. Quantities that are defined by both a magnitude and a direction are called *vector quantities.*

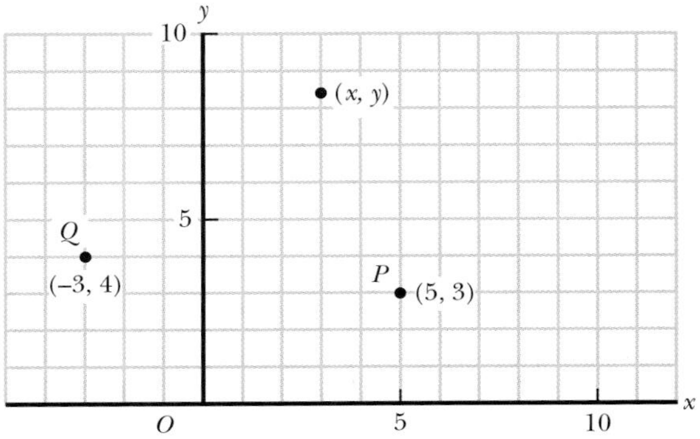

Figure 3.3
Designation of points in a Cartesian coordinate system. Every point is labelled with coordinates (x, y).

Sometimes it is more convenient to represent a point in a plane by its *plane polar coordinates* (r, θ) as shown in **Active Figure 3.4a**. In this *polar coordinate system*, r is the distance from the origin to the point having Cartesian coordinates (x, y) and θ is the angle between a fixed axis and a line drawn from the origin to the point. The fixed axis is often the positive x axis, and θ is usually measured anticlockwise from it. The angle θ may be measured in degrees or radians. Radians are defined as the ratio between the arc length, s, and the radius, r such that

$$\theta \equiv \frac{s}{r} \tag{3.1}$$

◀ Definition of radian

as shown in **Figure 3.5**, and there are 2π radians in a complete circle.

One radian is the angle subtended by an arc length equal to the radius of the arc. Remember that units have two purposes: they give dimensions and scale. The radian is an unusual unit in that it has no dimensions. Because the circumference of a circle is $2\pi r$, it follows from **Equation 3.1** that 360° corresponds to an angle of $(2\pi r/r)$ rad $= 2\pi$ rad. Hence, 1 rad $= 360°/2\pi \approx 57.3°$.

From the right-angled triangle in **Active Figure 3.4b**, we find that $\sin\theta = y/r$ and that $\cos\theta = x/r$. (A review of trigonometric functions is given in Appendix B.4.) Therefore, starting with the plane polar coordinates of any point, we can obtain the Cartesian coordinates by using the equations

$$x = r\cos\theta \tag{3.2}$$

$$y = r\sin\theta \tag{3.3}$$

Alternatively, if we know the Cartesian coordinates we can find the plane polar coordinates.

$$\tan\theta = \frac{y}{x} \tag{3.4}$$

$$r = \sqrt{x^2 + y^2} \tag{3.5}$$

Equation 3.5 is the familiar Pythagorean theorem.

Appendix B.4 provides a review of trigonometry and a summary of useful trigonometric identities.

These four expressions relating the coordinates (x, y) to the coordinates (r, θ) apply only when θ is defined as shown in **Active Figure 3.4a** – in other words, when positive θ is an angle measured anticlockwise from the positive x axis. (Some scientific calculators

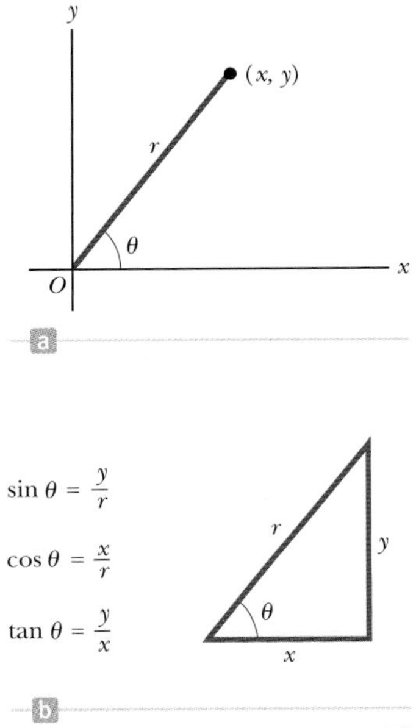

Active Figure 3.4
(a) The plane polar coordinates of a point are represented by the distance r and the angle θ where θ is measured anticlockwise from the positive x axis. (b) The right triangle used to relate (x, y) to (r, θ).

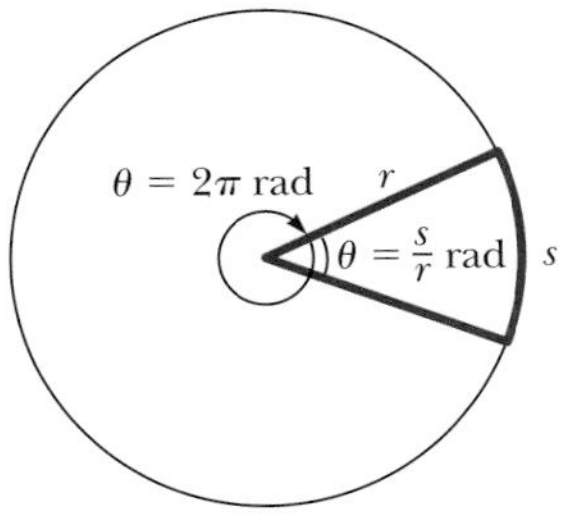

Figure 3.5
The angle, in radians, is the ratio of the arc length to the radius.

perform conversions between Cartesian and polar coordinates based on these standard conventions.) If the reference axis for the polar angle θ is chosen to be one other than the positive x axis or if the sense of increasing θ is chosen differently, the expressions relating the two sets of coordinates will change.

Example 3.1

Polar coordinates

The Cartesian coordinates of a point in the xy plane are $(x, y) = (-3.50, -2.50)$ m. Find the polar coordinates of this point.

Solution

Conceptualise Draw a diagram. The drawing in **Active Figure 3.6** helps us conceptualise the problem.

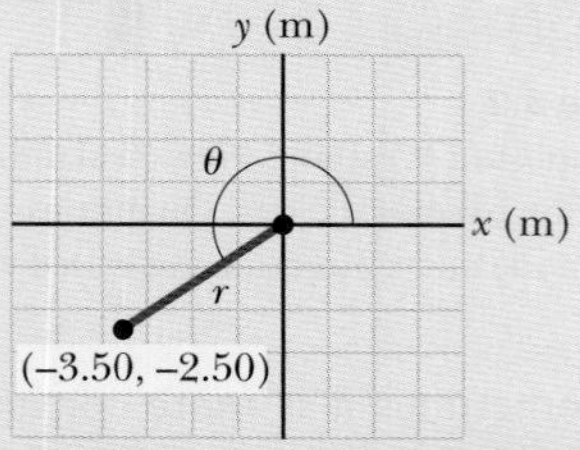

Active Figure 3.6
(Example 3.1) Finding polar coordinates when Cartesian coordinates are given

Model We are simply converting from Cartesian coordinates to polar coordinates. This is a substitution problem. Substitution problems generally do not have an extensive Analyse step other than the substitution of numbers into a given equation. Similarly, the Finalise step consists primarily of checking the units and making sure that the answer is reasonable. Therefore, for substitution problems, we will not label Analyse or Finalise steps.

Use **Equation 3.5** to find r:

$$r = \sqrt{x^2 + y^2} = \sqrt{(-3.50\text{ m})^2 + (-2.50\text{ m})^2} = 4.30\text{ m}$$

Use **Equation 3.4** to find θ:

$$\tan\theta = \frac{y}{x} = \frac{-2.50\text{ m}}{-3.50\text{ m}} = 0.714$$

$$\theta = 216° = 3.77\text{ rad}$$

Notice that you must use the signs of x and y to find that the point lies in the third quadrant of the coordinate system. That is, $\theta = 216°$ or 3.77 rad, not 35.5° or 0.610 rad, whose tangent is also 0.714.

Components of a vector

Any vector can be completely described by its components. The **components** of a vector are the projections of the vector onto Cartesian coordinates. Consider a vector $\vec{\mathbf{A}}$ lying in the xy plane and making an arbitrary angle θ with the positive x axis as shown in **Figure 3.7a**. This vector can be expressed as the sum of two other *component vectors*; $\vec{\mathbf{A}}_x$, which is parallel to the x axis, and $\vec{\mathbf{A}}_y$, which is parallel to the y axis. From **Figure 3.7b**, we see that the three vectors form a right triangle and that $\vec{\mathbf{A}} = \vec{\mathbf{A}}_x + \vec{\mathbf{A}}_y$. We shall often refer to the 'components of a vector $\vec{\mathbf{A}}$', written A_x and A_y (without the boldface notation). The component A_x represents the projection of $\vec{\mathbf{A}}$ along the x axis, and the component A_y represents the projection of $\vec{\mathbf{A}}$ along the y axis. These components can be positive or negative. The component A_x is positive if the component vector $\vec{\mathbf{A}}_x$ points in the positive x direction and is negative if $\vec{\mathbf{A}}_x$ points in the negative x direction. A similar statement can be made for the component A_y.

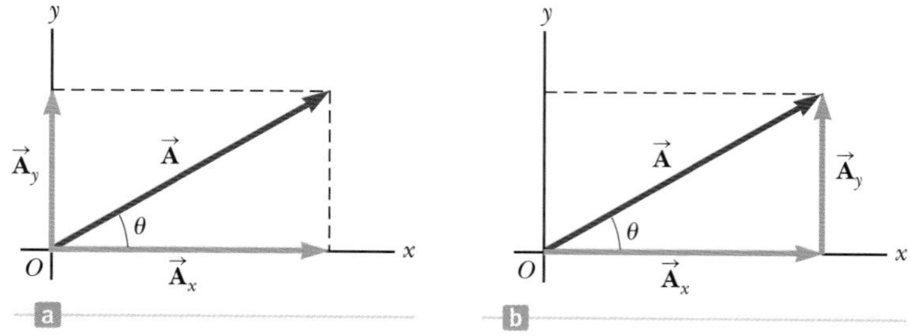

Figure 3.7
(a) A vector $\vec{\mathbf{A}}$ lying in the xy plane can be represented by its component vectors $\vec{\mathbf{A}}_x$ and $\vec{\mathbf{A}}_y$. (b) The y component vector $\vec{\mathbf{A}}_y$ can be moved to the right so that it adds to $\vec{\mathbf{A}}_x$. The vector sum of the component vectors is $\vec{\mathbf{A}}$. These three vectors form a right triangle.

From **Figure 3.7** and the definition of sine and cosine, we see that $\cos\theta = A_x/A$ and that $\sin\theta = A_y/A$. Hence, the components of $\vec{\mathbf{A}}$ are

$$A_x = A\cos\theta \tag{3.6}$$

$$A_y = A\sin\theta \tag{3.7}$$

The magnitudes of these components are the lengths of the two sides of a right triangle with a hypotenuse of length A. Therefore, the magnitude and direction of $\vec{\mathbf{A}}$ are related to its components through the expressions

$$A = \sqrt{A_x^2 + A_y^2} \tag{3.8}$$

$$\theta = \tan^{-1}\left(\frac{A_y}{A_x}\right) \tag{3.9}$$

Pitfall Prevention 3.1

Equations 3.6 and 3.7 associate the cosine of the angle with the *x* component and the sine of the angle with the *y* component. This association is true *only* because we measured the angle θ with respect to the *x* axis. Think about which side of the triangle is adjacent to the angle and which side is opposite and then identify the cosine and sine accordingly.

Remember, you can always look up trigonometric identities in Appendix B.4 if you need to.

When solving problems in two dimensions, you can specify a vector $\vec{\mathbf{A}}$ either with its components A_x and A_y or with its magnitude and direction A and θ.

Suppose you are working on a physics problem that requires resolving a vector into its components. In many applications, it is convenient to express the components in a coordinate system having axes that are not horizontal and vertical but that are still perpendicular to each other. For example, we will consider the motion of objects sliding down inclined planes. For these examples, it is often convenient to orient the x axis parallel to the plane and the *y* axis perpendicular to the plane.

Quick Quiz 3.2

Choose the correct response to make the sentence true: A component of a vector is **(a)** always, **(b)** never, or **(c)** sometimes larger than the magnitude of the vector.

Unit vectors

Vector quantities are often expressed in terms of unit vectors. A **unit vector** is a dimensionless vector having a magnitude of exactly 1. Unit vectors are used to specify a given direction and have no other physical significance. They are used solely as a bookkeeping convenience in describing a direction in space. We shall use the symbols $\hat{\mathbf{i}}$, $\hat{\mathbf{j}}$ and $\hat{\mathbf{k}}$ to represent unit vectors pointing in the positive *x*, *y* and *z* directions respectively. (The 'hats', or circumflexes, on the symbols are a standard notation for unit vectors.) The unit vectors $\hat{\mathbf{i}}$, $\hat{\mathbf{j}}$ and $\hat{\mathbf{k}}$ form a set of mutually perpendicular vectors in a right-handed coordinate system as shown in **Active Figure 3.8a**. The magnitude of each unit vector equals 1; that is, $|\hat{\mathbf{i}}| = |\hat{\mathbf{j}}| = |\hat{\mathbf{k}}| = 1$.

Consider a vector $\vec{\mathbf{A}}$ lying in the *xy* plane as shown in **Active Figure 3.8b**. The product of the component A_x and the unit vector $\hat{\mathbf{i}}$ is the component vector $\vec{\mathbf{A}}_x = A_x\hat{\mathbf{i}}$, which lies on the *x* axis and has magnitude $|A_x|$.

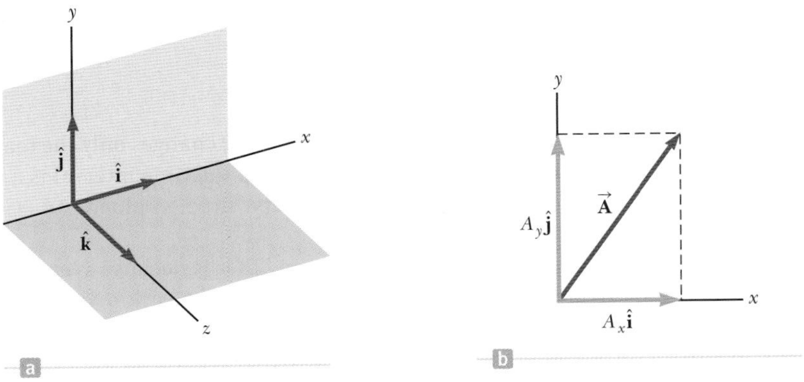

Active Figure 3.8

(a) The unit vectors $\hat{\mathbf{i}}$, $\hat{\mathbf{j}}$ and $\hat{\mathbf{k}}$ are directed along the *x*, *y* and *z* axes respectively. (b) Vector $\vec{\mathbf{A}} = A_x\hat{\mathbf{i}} + A_y\hat{\mathbf{j}}$ lying in the *xy* plane has components A_x and A_y.

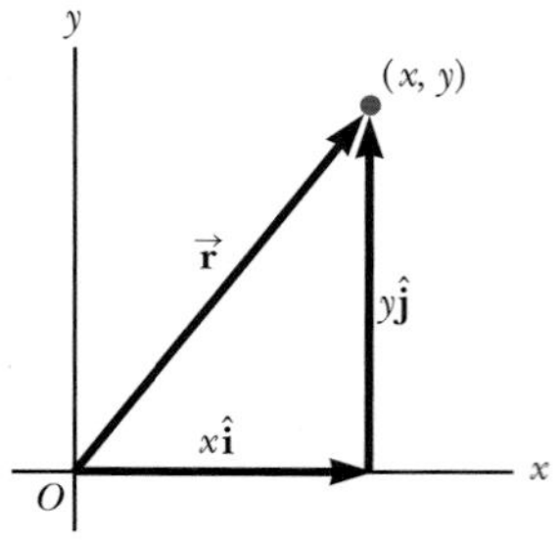

Figure 3.9
The point whose Cartesian coordinates are (x, y) can be represented by the position vector $\vec{r} = x\hat{i} + y\hat{j}$.

Likewise, $\vec{A}_y = A_y\hat{j}$ is the component vector of magnitude $|A_y|$ lying on the y axis. Therefore, the unit-vector notation for the vector $\vec{A}$ is

$$\vec{A} = A_x\hat{i} + A_y\hat{j} \tag{3.10}$$

For example, consider a point lying in the xy plane and having Cartesian coordinates (x, y) as in Figure 3.9. The point can be specified by the **position vector $\vec{r}$**, which in unit-vector form is given by

$$\vec{r} = x\hat{i} + y\hat{j} \tag{3.11}$$

This notation tells us that the components of $\vec{r}$ are the coordinates x and y.

The rules for calculating with vectors, including adding, subtracting, multiplying and dividing vectors, are given in Appendix B.6. If you are not already confident in using vectors, you should study this appendix carefully and do the practice problems, as we will be using vectors a lot in this book as many physical quantities have both magnitude and direction.

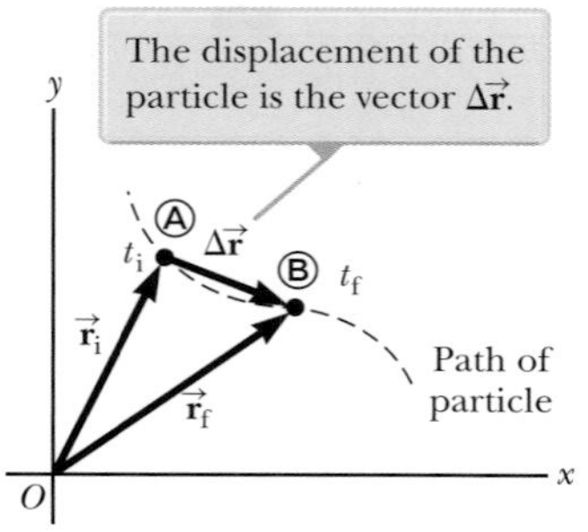

Figure 3.10
A particle moving in the xy plane is located with the position vector $\vec{r}$ drawn from the origin to the particle. The displacement of the particle as it moves from Ⓐ to Ⓑ in the time interval $\Delta t = t_f - t_i$ is equal to the vector $\Delta\vec{r} = \vec{r}_f - \vec{r}_i$, and hence the final position of the particle may be written as $\vec{r}_f = \vec{r}_i + \Delta\vec{r}$.

3.2 The position, velocity and acceleration vectors

In Chapter 2, we found that the motion of a particle along a straight line such as the x axis is completely known if its position is known as a function of time. Let us now extend this idea to two-dimensional motion of a particle in the xy plane. We begin by describing the position of the particle by its **position vector $\vec{r}$**, drawn from the origin of some coordinate system to the location of the particle in the xy plane as in Figure 3.10. At time t_i, the particle is at point Ⓐ, described by position vector $\vec{r}_i$. At some later time t_f, it is at point Ⓑ, described by position vector $\vec{r}_f$. The path from Ⓐ to Ⓑ is not necessarily a straight line. As the particle moves from Ⓐ to Ⓑ in the time interval $\Delta t = t_f - t_i$, its position vector changes from $\vec{r}_i$ to $\vec{r}_f$. As we learned in Chapter 2, displacement is a vector, and the displacement of the particle is the difference between its final position and its initial position. We now define the **displacement vector $\Delta\vec{r}$** for a particle such as the one in Figure 3.10 as being the difference between its final position vector and its initial position vector:

Displacement ► vector

$$\Delta\vec{r} \equiv \vec{r}_f - \vec{r}_i \tag{3.12}$$

The direction of $\Delta\vec{r}$ is indicated in Figure 3.10. As we see from the figure, the magnitude of $\Delta\vec{r}$ is *less* than the distance traveled along the curved path followed by the particle.

As we saw in Chapter 2, it is often useful to quantify motion by looking at the velocity. Two-dimensional (or three-dimensional) kinematics is similar to one-dimensional kinematics, but we now use full vector notation rather than positive and negative signs to indicate the direction of motion.

We define the **average velocity $\vec{v}_{avg}$** of a particle during the time interval Δt as the displacement of the particle divided by the time interval:

Average ► velocity

$$\vec{v}_{avg} \equiv \frac{\Delta\vec{r}}{\Delta t} \tag{3.13}$$

Multiplying or dividing a vector quantity by a positive scalar quantity such as Δt changes only the magnitude of the vector, not its direction. Because displacement is a vector quantity and the time interval is a positive scalar quantity, we conclude that the average velocity is a vector quantity directed along $\Delta\vec{r}$. Compare Equation 3.13 with its one-dimensional counterpart, Equation 2.2.

The average velocity between points is *independent of the path* taken. That is because average velocity is proportional to displacement, which depends only on the initial and final position vectors and not on the path taken. As with one-dimensional motion, we conclude that if a particle starts its motion at some point and returns to this point via any path, its average velocity is zero for this trip because its displacement is zero. Consider again our football players on the field in Figure 2.2. We previously

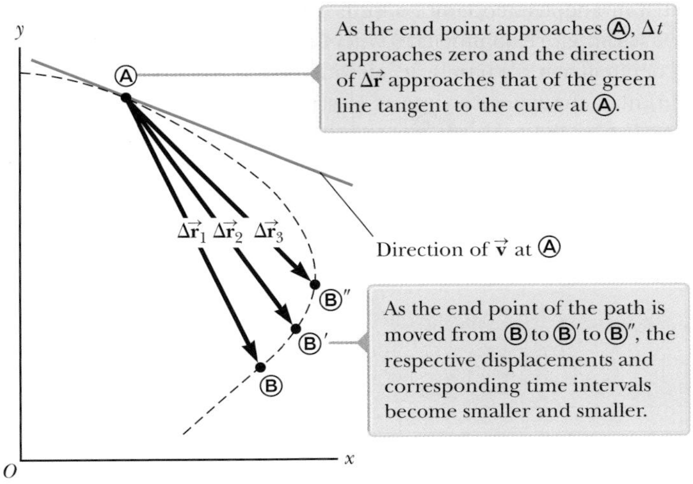

y
Ⓐ
As the end point approaches Ⓐ, Δt approaches zero and the direction of $\Delta\vec{\mathbf{r}}$ approaches that of the green line tangent to the curve at Ⓐ.
$\Delta\vec{\mathbf{r}}_1$ $\Delta\vec{\mathbf{r}}_2$ $\Delta\vec{\mathbf{r}}_3$
Direction of $\vec{\mathbf{v}}$ at Ⓐ
Ⓑ″
Ⓑ′
Ⓑ
As the end point of the path is moved from Ⓑ to Ⓑ′ to Ⓑ″, the respective displacements and corresponding time intervals become smaller and smaller.
O
x

Figure 3.11
As a particle moves between two points, its average velocity is in the direction of the displacement vector $\Delta\vec{\mathbf{r}}$. By definition, the instantaneous velocity at Ⓐ is directed along the line tangent to the curve at Ⓐ.

considered only their one-dimensional motion back and forth. In reality, however, they move over a two-dimensional surface, running back and forth between the goals as well as left and right across the width of the field. Starting from one end, a given player may follow a very complicated two-dimensional path. Upon returning to the original point, however, a player's average velocity is zero because the player's displacement for the whole trip is zero.

Consider again the motion of a particle between two points in the xy plane as shown in **Figure 3.11**. As the time interval over which we observe the motion becomes smaller and smaller – that is, as Ⓑ is moved to Ⓑ′ and then to Ⓑ″ and so on – the direction of the displacement approaches that of the line tangent to the path at Ⓐ. The **instantaneous velocity** $\vec{\mathbf{v}}$ is defined as the limit of the average velocity $\Delta\vec{\mathbf{r}}/\Delta t$ as Δt approaches zero:

$$\vec{\mathbf{v}} \equiv \lim_{\Delta t \to 0} \frac{\Delta\vec{\mathbf{r}}}{\Delta t} = \frac{d\vec{\mathbf{r}}}{dt} \qquad (3.14)$$

◀ Instantaneous velocity

That is, the instantaneous velocity equals the derivative of the position vector with respect to time. The direction of the instantaneous velocity vector at any point in a particle's path is along a line tangent to the path at that point and in the direction of motion. Compare **Equation 3.14** with the corresponding one-dimensional version, **Equation 2.5**.

Σ *A review of differential calculus, including taking limits, is given in Appendix B.7.*

The magnitude of the instantaneous velocity vector $v = |\vec{\mathbf{v}}|$ of a particle is called the *speed* of the particle, which is a scalar quantity.

As a particle moves from one point to another along some path, its instantaneous velocity vector changes from $\vec{\mathbf{v}}_i$ at time t_i to $\vec{\mathbf{v}}_f$ at time t_f. Knowing the velocity at these points allows us to determine the average acceleration of the particle. The **average acceleration** $\vec{\mathbf{a}}_{avg}$ of a particle is defined as the change in its instantaneous velocity vector $\Delta\vec{\mathbf{v}}$ divided by the time interval Δt during which that change occurs:

$$\vec{\mathbf{a}}_{avg} \equiv \frac{\Delta\vec{\mathbf{v}}}{\Delta t} = \frac{\vec{\mathbf{v}}_f - \vec{\mathbf{v}}_i}{t_f - t_i} \qquad (3.15)$$

◀ Average acceleration

Because $\vec{\mathbf{a}}_{avg}$ is the ratio of a vector quantity $\Delta\vec{\mathbf{v}}$ and a positive scalar quantity Δt, we conclude that average acceleration is a vector quantity directed along $\Delta\vec{\mathbf{v}}$. As shown in **Figure 3.12**, the direction of $\vec{\mathbf{a}}_{avg}$ is given by the direction of $\Delta\vec{\mathbf{v}} = \vec{\mathbf{v}}_f - \vec{\mathbf{v}}_i$. Compare **Equation 3.15** with **Equation 2.9**.

When the average acceleration of a particle changes during different time intervals, it is useful to define its instantaneous acceleration. The **instantaneous acceleration** $\vec{\mathbf{a}}$ is defined as the limiting value of the ratio $\Delta\vec{\mathbf{v}}/\Delta t$ as Δt approaches zero:

$$\vec{\mathbf{a}} \equiv \lim_{\Delta t \to 0} \frac{\Delta\vec{\mathbf{v}}}{\Delta t} = \frac{d\vec{\mathbf{v}}}{dt} \qquad (3.16)$$

◀ Instantaneous acceleration

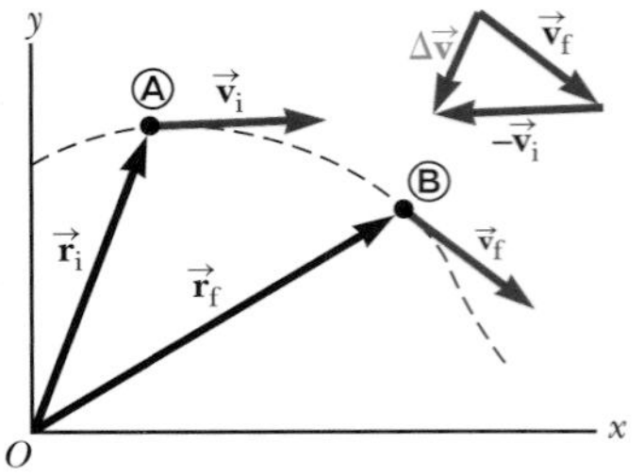

Figure 3.12
A particle moves from position Ⓐ to position Ⓑ. Its velocity vector changes from $\vec{v}_i$ to $\vec{v}_f$. The vector diagram at the upper right shows how to determine the vector $\Delta\vec{v}$ from the initial and final velocities.

In other words, the instantaneous acceleration equals the derivative of the velocity vector with respect to time. Compare Equation 3.16 with Equation 2.10.

Various changes can occur when a particle accelerates. First, the magnitude of the velocity vector (the speed) may change with time as in straight-line (one-dimensional) motion. Second, the direction of the velocity vector may change with time even if its magnitude (speed) remains constant as in two-dimensional motion along a curved path. Finally, both the magnitude and the direction of the velocity vector may change simultaneously.

Quick Quiz 3.3

Consider the following controls in a car: accelerator, brake, steering wheel. What are the controls in this list that cause an acceleration of the car? (a) all three controls (b) the accelerator and the brake (c) only the brake (d) only the accelerator (e) only the steering wheel.

3.3 Two-dimensional motion with constant acceleration

In Section 2.6, we investigated one-dimensional motion of a particle under constant acceleration. Let us now consider two-dimensional motion during which the acceleration of a particle remains constant in both magnitude and direction. As we shall see, this approach is useful for analysing some common types of motion.

Imagine a puck gliding along in a straight line on an air hockey table. We model the motion of the puck as a particle with constant velocity as the surface is frictionless. We will define our x axis as being in the direction of the motion. Now imagine puffing gently on the puck in the y direction as it goes past you, so it experiences a momentary acceleration perpendicular to its original motion. The puff of air does not affect the motion in the x direction, but after the puff the puck has a component of its velocity in the y direction. Figure 3.13 shows a motion diagram of the puck as seen from above before and after the puff. We can see that the x component of the velocity is unchanged, and the total velocity is the sum of the x and y components, which are independent.

In general, motion in two dimensions can be modelled as two independent motions in each of the two perpendicular directions associated with the x and y axes. That is, any influence in the y direction does not affect the motion in the x direction and vice versa.

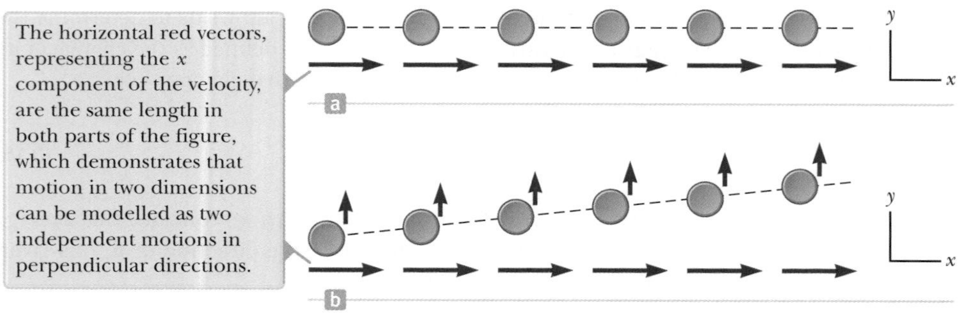

The horizontal red vectors, representing the x component of the velocity, are the same length in both parts of the figure, which demonstrates that motion in two dimensions can be modelled as two independent motions in perpendicular directions.
a
b
y
x

Figure 3.13
(a) A puck moves across a horizontal air hockey table at constant velocity in the x direction. (b) After a puff of air in the y direction is applied to the puck, the puck has gained a y component of velocity, but the x component is unaffected by the force in the perpendicular direction.

The position vector for a particle moving in the xy plane can be written

$$\vec{\mathbf{r}} = x\hat{\mathbf{i}} + y\hat{\mathbf{j}} \tag{3.17}$$

where x, y, and $\vec{\mathbf{r}}$ change with time as the particle moves while the unit vectors $\hat{\mathbf{i}}$ and $\hat{\mathbf{j}}$ remain constant. If the position vector is known, the velocity of the particle can be obtained from **Equations 3.14** and **3.17**, which give

$$\vec{\mathbf{v}} = \frac{d\vec{\mathbf{r}}}{dt} = \frac{dx}{dt}\hat{\mathbf{i}} + \frac{dy}{dt}\hat{\mathbf{j}} = v_x\hat{\mathbf{i}} + v_y\hat{\mathbf{j}} \tag{3.18}$$

Because the acceleration $\vec{\mathbf{a}}$ of the particle is assumed constant in this discussion, its components a_x and a_y are also constant. Therefore, we can model the particle as a particle under constant acceleration independently in each of the two directions and apply the equations of kinematics separately to the x and y components of the velocity vector. Substituting, from **Equation 2.12**, $v_{xf} = v_{xi} + a_x t$ and $v_{yf} = v_{yi} + a_y t$ into **Equation 3.18** to determine the final velocity at any time t, we obtain

$$\vec{\mathbf{v}}_f = (v_{xi} + a_x t)\,\hat{\mathbf{i}} + (v_{yi} + a_y t)\,\hat{\mathbf{j}} = (v_{xi}\hat{\mathbf{i}} + v_{yi}\hat{\mathbf{j}}) + (a_x\hat{\mathbf{i}} + a_y\hat{\mathbf{j}})t$$

$$\vec{\mathbf{v}}_f = \vec{\mathbf{v}}_i + \vec{\mathbf{a}}t \tag{3.19}$$

◀ Velocity vector as a function of time

This result states that the velocity of a particle at some time t equals the vector sum of its initial velocity $\vec{\mathbf{v}}_i$ at time $t = 0$ and the additional velocity $\vec{\mathbf{a}}t$ acquired at time t as a result of constant acceleration. **Equation 3.19** is the vector version of **Equation 2.12**.

Similarly, from **Equation 2.15** we know that the x and y coordinates of a particle moving with constant acceleration are

$$x_f = x_i + v_{xi}t + \frac{1}{2}a_x t^2 \qquad y_f = y_i + v_{yi}t + \frac{1}{2}a_y t^2$$

Substituting these expressions into **Equation 3.17** (and labelling the final position vector $\vec{\mathbf{r}}_f$) gives

$$\vec{\mathbf{r}}_f = \left(x_i + v_{xi}t + \frac{1}{2}a_x t^2\right)\hat{\mathbf{i}} + \left(y_i + v_{yi}t + \frac{1}{2}a_y t^2\right)\hat{\mathbf{j}}$$

$$= (x_i\hat{\mathbf{i}} + y_i\hat{\mathbf{j}}) + (v_{xi}\hat{\mathbf{i}} + v_{yi}\hat{\mathbf{j}})t + \frac{1}{2}(a_x\hat{\mathbf{i}} + a_y\hat{\mathbf{j}})t^2$$

$$\vec{\mathbf{r}}_f = \vec{\mathbf{r}}_i + \vec{\mathbf{v}}_i t + \frac{1}{2}\vec{\mathbf{a}}t^2 \tag{3.20}$$

◀ Position vector as a function of time

which is the vector version of **Equation 2.15. Equation 3.20** tells us that the position vector $\vec{\mathbf{r}}_f$ of a particle is the vector sum of the original position $\vec{\mathbf{r}}_i$, a displacement $\vec{\mathbf{v}}_i t$ arising from the initial velocity of the particle, and a displacement $\frac{1}{2}\vec{\mathbf{a}}t^2$ resulting from the constant acceleration of the particle.

Graphical representations of **Equations 3.19** and **3.20** are shown in **Active Figure 3.14**. The components of the position and velocity vectors are also illustrated in the figure. Notice from **Active Figure 3.14a** that $\vec{\mathbf{v}}_f$ is generally not along the direction of either $\vec{\mathbf{v}}_i$ or $\vec{\mathbf{a}}$ because the relationship between these quantities is a vector expression. For the same reason, from **Active Figure 3.14b** we see that $\vec{\mathbf{r}}_f$ is generally not along the direction of $\vec{\mathbf{r}}_i$, $\vec{\mathbf{v}}_i$, or $\vec{\mathbf{a}}$. Finally, notice that $\vec{\mathbf{v}}_f$ and $\vec{\mathbf{r}}_f$ are generally not in the same direction.

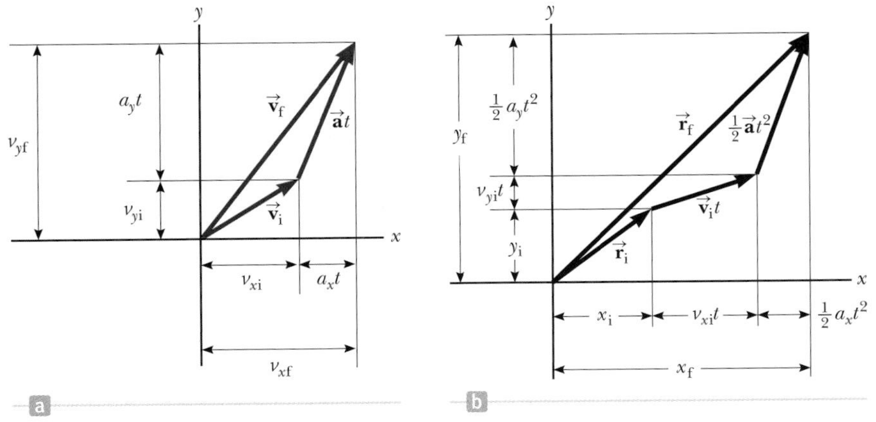

Active Figure 3.14

Vector representations and components of (a) the velocity and (b) the position of a particle moving with a constant acceleration $\vec{\mathbf{a}}$.

Example 3.2

Motion in a plane

A particle moves in the xy plane, starting from the origin at $t = 0$ with an initial velocity having an x component of 20 m/s and a y component of −15 m/s. The particle experiences an acceleration in the x direction, given by $a_x = 4.0\ \text{m/s}^2$.

(A) Determine the total velocity vector at any time.

Solution

Conceptualise The components of the initial velocity tell us that the particle starts by moving towards the right and downwards. We start by drawing a diagram showing this to help us visualise the problem: see **Figure 3.15**. The x component of velocity starts at 20 m/s and increases by 4.0 m/s every second. The y component of velocity does not change from its initial value of −15 m/s. Because the particle is accelerating in the $+x$ direction, its velocity component in this direction increases and the path curves as shown in the diagram. Notice that the spacing between successive images increases as time goes on because the speed is increasing. The placement of the acceleration and velocity vectors in **Active Figure 3.14** helps us further conceptualise the situation.

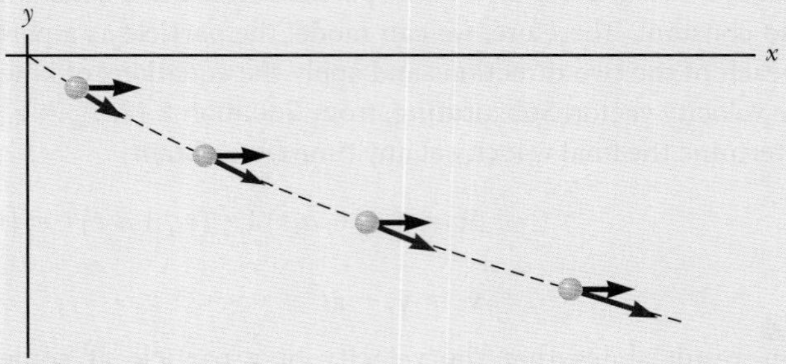

Figure 3.15
(Example 3.2) Motion diagram for the particle.

Model The initial velocity has components in both the x and y directions, so the particle is moving in two dimensions. Because the particle only has an x component of acceleration, we model it as a particle with constant acceleration in the x direction and a particle with constant velocity in the y direction.

Analyse To begin the mathematical analysis, we set $v_{xi} = 20\ \text{m/s}$, $v_{yi} = -15\ \text{m/s}$, $a_x = 4.0\ \text{m/s}^2$, and $a_y = 0$.

Use **Equation 3.19** for the velocity vector:

$$\vec{\mathbf{v}}_f = \vec{\mathbf{v}}_i + \vec{\mathbf{a}}t = (v_{xi} + a_x t)\hat{\mathbf{i}} + (v_{yi} + a_y t)\hat{\mathbf{j}}$$

Substitute numerical values with the velocity in metres per second and the time in seconds:

$$\vec{\mathbf{v}}_f = [20\ \text{m/s} + (4.0\ \text{m/s}^2)t]\,\hat{\mathbf{i}} + [-15\ \text{m/s} + (0)t]\,\hat{\mathbf{j}}$$

$$\vec{\mathbf{v}}_f = [20\ \text{m/s} + (4.0\ \text{m/s}^2 t)]\hat{\mathbf{i}} - 15\ \text{m/s}\,\hat{\mathbf{j}} \quad \textbf{(1)}$$

Finalise Notice that the x component of velocity increases in time while the y component remains constant; this result is consistent with our conceptualisation of the problem.

(B) Calculate the velocity and speed of the particle at $t = 5.0$ s and the angle the velocity vector makes with the x axis.

Solution

Analyse Evaluate the result from **Equation (1)** at t = 5.0 s:

$$\vec{\mathbf{v}}_f = [20\ \text{m/s} + 4.0\ \text{m/s}^2(5.0\text{s})]\hat{\mathbf{i}} - 15\hat{\mathbf{j}} = (40\hat{\mathbf{i}} - 15\hat{\mathbf{j}})\ \text{m/s}$$

Determine the angle θ that $\vec{\mathbf{v}}_f$ makes with the x axis at $t = 5.0$ s:

$$\theta = \tan^{-1}\left(\frac{v_{yf}}{v_{xf}}\right) = \tan^{-1}\left(\frac{-15\ \text{m/s}}{40\ \text{m/s}}\right) = -21°$$

Evaluate the speed of the particle as the magnitude of $\vec{\mathbf{v}}_f$:

$$v_f = |\vec{\mathbf{v}}_f| = \sqrt{v_{xf}^2 + v_{yf}^2} = \sqrt{(40\ \text{m/s})^2 + (-15\ \text{m/s})^2} = 43\ \text{m/s}$$

Finalise The negative sign for the angle θ indicates that the velocity vector is directed at an angle of 21° below the positive x axis.

Reality Check Notice that if we calculate v_i from the x and y components of $\vec{\mathbf{v}}_i$, we find that $v_f > v_i$. Is that consistent with our prediction?

Example 3.2 cont.

(C) Determine the x and y coordinates of the particle at any time t and its position vector at this time.

Solution

Analyse Use the components of Equation 3.20 with $x_i = y_i = 0$ at $t = 0$ with x and y in metres and t in seconds:

$$x_f = v_{xi}t + \frac{1}{2}a_x t^2 = 20\text{m/s } t + 2.0\text{m/s}^2 t^2$$

$$y_f = v_{yi}t = -15t$$

Express the position vector of the particle at any time t:

$$\vec{\mathbf{r}}_f = x_f\hat{\mathbf{i}} + y_f\hat{\mathbf{j}} = (20\text{m/s}t + 2.0\text{m/s}^2t^2)\hat{\mathbf{i}} - 15\text{m/s}t\hat{\mathbf{j}}$$

Finalise Let us now consider a limiting case for very large values of t.

What If? What if we wait a very long time and then observe the motion of the particle? How would we describe the motion of the particle for large values of the time?

Answer Looking at Figure 3.14, we see the path of the particle curving toward the x axis. If this continues, the path will become more and more parallel to the x axis as time increases. Mathematically, Equation (1) shows that the y component of the velocity remains constant while the x component increases linearly with t. Therefore, when t is very large, the x component of the velocity will be much larger than the y component, suggesting that the velocity vector becomes more and more parallel to the x axis. Both x_f and y_f continue to increase with time, although x_f increases much faster.

3.4 Projectile motion

Anyone who has observed a ball in motion has observed **projectile motion**. The ball moves in a curved path and returns to the ground. Projectile motion of an object is simple to model and analyse if we make two simplifying assumptions: (1) the free-fall acceleration is constant over the range of motion and is directed downwards, and (2) the effect of air resistance is negligible. With these assumptions, we find that the path of a projectile, which we call its *trajectory*, is *always* a parabola as shown in Active Figure 3.16.

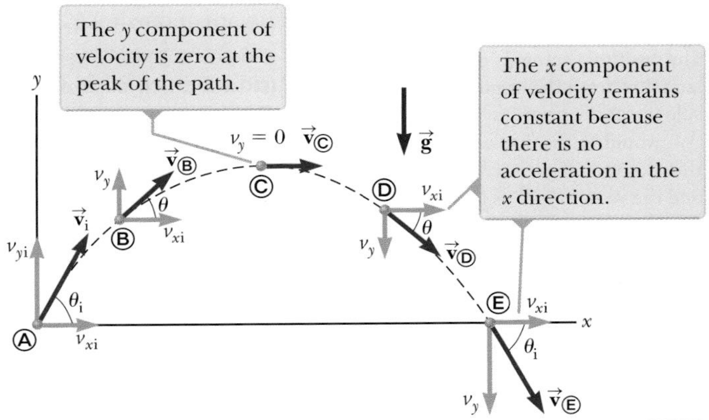

Active Figure 3.16

The parabolic path of a projectile that leaves the origin with a velocity $\vec{\mathbf{v}}_i$. The velocity vector $\vec{\mathbf{v}}$ changes with time in both magnitude and direction. This change is the result of acceleration $\vec{\mathbf{a}} = \vec{\mathbf{g}}$ in the negative y direction.

Our first assumption is reasonable as long as the range of motion is small compared with the radius of the Earth (6.4×10^6 m) and, in effect, this assumption is equivalent to assuming the Earth is flat over the range of motion considered. Our second assumption is often *not* justified, especially at high velocities or when the ball is spinning. Our model of projectile motion, with the approximations of constant acceleration and no air resistance, is appropriate for some projectiles but not others. Before you apply this model, you should think about whether those approximations are reasonable. If they are not, you can still use this model to get a first approximation of the behaviour of the projectile, but to analyse it properly you should use a model that includes air resistance (as in Chapter 5) and/or varying acceleration due to gravity (Chapter 11). However, we will use these simplifying assumptions for the moment.

The expression for the position vector of the projectile as a function of time follows directly from Equation 3.20, with its acceleration being that due to gravity, $\vec{\mathbf{a}} = \vec{\mathbf{g}}$:

$$\vec{\mathbf{r}}_f = \vec{\mathbf{r}}_i + \vec{\mathbf{v}}_i t + \frac{1}{2}\vec{\mathbf{g}}t^2 \quad (3.21)$$

where the initial x and y components of the velocity of the projectile are

$$v_{xi} = v_i \cos\theta_i \qquad v_{yi} = v_i \sin\theta_i \quad (3.22)$$

Pitfall Prevention 3.2

Whenever simplifying assumptions are made, it is important that the consequences of these assumptions are considered to make sure they are reasonable; sometimes it is reasonable to neglect air resistance, sometimes it is not.

Pitfall Prevention 3.3

As discussed in Pitfall Prevention 2.6, many people claim that the acceleration of a projectile at the topmost point of its trajectory is zero. This mistake arises from confusion between velocity and acceleration. The acceleration is not zero anywhere along the trajectory because the projectile is always subject to the gravitational force due to the Earth.

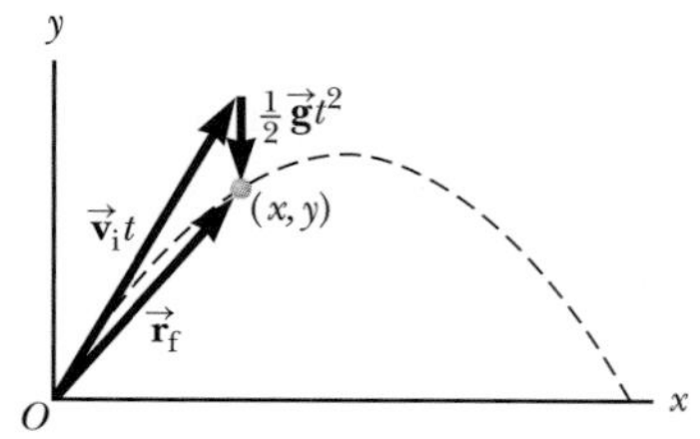

Figure 3.17
The position vector $\vec{r}_f$ of a projectile launched from the origin whose initial velocity at the origin is $\vec{v}_i$. The vector $\vec{v}_i t$ would be the displacement of the projectile if gravity were absent, and the vector $\frac{1}{2}\vec{g}t^2$ is its vertical displacement from a straight-line path due to its downwards gravitational acceleration.

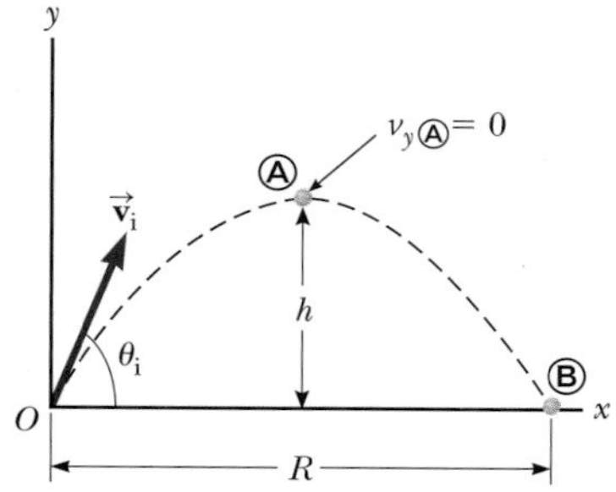

Figure 3.18
A projectile launched over a flat surface from the origin at $t_i = 0$ with an initial velocity $\vec{v}_i$. The maximum height of the projectile is h, and the horizontal range is R. At Ⓐ, the peak of the trajectory, the particle has coordinates $(R/2, h)$.

The expression in **Equation 3.21** is plotted in **Figure 3.17**, for a projectile launched from the origin, so that $\vec{r}_i = 0$. The final position of a particle can be considered to be the superposition of its initial position $\vec{r}_i$; the term $\vec{v}_i t$, which is its displacement if no acceleration were present; and the term $\frac{1}{2}\vec{g}t^2$ that arises from its acceleration due to gravity. In other words, if there were no gravitational acceleration, the particle would continue to move along a straight path in the direction of $\vec{v}_i$. Therefore, the vertical distance $\frac{1}{2}\vec{g}t^2$ through which the particle 'falls' off the straight-line path is the same distance that an object dropped from rest would fall during the same time interval.

In Section 3.3, we stated that two-dimensional motion with constant acceleration can be analysed as a combination of **two independent motions in the *x* and *y* directions**, with accelerations a_x and a_y. Projectile motion can also be handled in this way, with zero acceleration in the x direction and a constant acceleration in the y direction, $a_y = -g$. Therefore, when analysing projectile motion, we model it as the superposition of two motions: (1) motion of a particle with constant velocity in the horizontal direction and (2) motion of a particle with constant acceleration (free fall) in the vertical direction. The horizontal and vertical components of a projectile's motion are independent of each other and can be handled separately, with time t as the common variable for both components.

TRY THIS

Get a friend to run at a constant velocity while holding a ball, and *drop* the ball as they run past you. They should be careful to just let go of the ball, not to throw it forwards or backwards. Watch carefully – where does the ball land relative to your friend's feet? Where does it land relative to where it was released? You could do the same experiment by dropping a ball (preferably a light, soft ball) from a vehicle (skateboard, bike, etc.) moving in a straight line at constant speed, and compare its path as seen by an observer on the vehicle and one standing still and watching you go past. Try doing it with a target such as a bucket to drop the ball into. Where do you have to release the ball – before, exactly at, or after you pass the bucket?

Quick Quiz 3.4

(i) As a projectile thrown upwards moves in its parabolic path (as in **Figure 3.17**), at what point along its path are the velocity and acceleration vectors for the projectile perpendicular to each other? (a) nowhere (b) the highest point (c) the launch point **(ii)** From the same choices, at what point are the velocity and acceleration vectors for the projectile parallel to each other?

Horizontal range and maximum height of a projectile

Let us assume a projectile is launched from the origin at $t_i = 0$ with a positive v_{yi} component as shown in **Figure 3.18** and returns to the same horizontal level. This situation is common in sports, where cricket balls, footballs and golf balls often land at approximately the same level from which they were launched.

Two points in this motion are especially interesting to analyse: the peak point Ⓐ, which has Cartesian coordinates $(R/2, h)$, and the point Ⓑ, which has coordinates $(R, 0)$. The distance R is called the *horizontal range* of the projectile, and the distance h is its *maximum height*. Let us find h and R in terms of v_i, θ_i, and g.

We can determine h by noting that at the peak $v_{yⒶ} = 0$. Therefore, we can use the y component of **Equation 3.19** to determine the time $t_Ⓐ$ at which the projectile reaches the peak:

$$v_{yf} = v_{yi} + a_y t$$
$$0 = v_i \sin\theta_i - g t_Ⓐ$$
$$t_Ⓐ = \frac{v_i \sin\theta_i}{g} \qquad (3.23)$$

Substituting this expression for $t_Ⓐ$ into the y component of **Equation 3.20** and replacing $y = y_Ⓐ$ with h, we obtain an expression for h in terms of the magnitude and direction of the initial velocity vector:

$$h = (v_i \sin\theta_i)\frac{v_i \sin\theta_i}{g} - \frac{1}{2}g\left(\frac{v_i \sin\theta_i}{g}\right)^2$$

$$h = \frac{v_i^2 \sin^2\theta_i}{2g} \qquad (3.24)$$

The range R is the horizontal position of the projectile at a time that is twice the time at which it reaches its peak, that is, at time $t_Ⓑ = 2t_Ⓐ$. Using the x component of **Equation 3.20**, noting that $v_{xi} = v_{xⒷ} = v_i \cos\theta_i$, and setting $x_Ⓑ = R$ at $t = 2t_Ⓐ$, we find that

$$R = v_{xi}t_Ⓑ = (v_i \cos\theta_i)2t_Ⓐ$$

$$= (v_i \cos\theta_i)\frac{2v_i \sin\theta_i}{g} = \frac{2v_i^2 \sin\theta_i \cos\theta_i}{g}$$

Using the identity $\sin 2\theta = 2 \sin\theta \cos\theta$ (see Appendix B.4), we can write R in the more compact form

$$R = \frac{v_i^2 \sin 2\theta_i}{g} \qquad (3.25)$$

The maximum value of R from **Equation 3.24** is $R_{max} = v_i^2/g$. This result makes sense because the maximum value of $\sin 2\theta_i$ is 1, which occurs when $2\theta_i = 90°$. Therefore, R is a maximum when $\theta_i = 45°$.

Active Figure 3.19 illustrates various trajectories for a projectile having a given initial speed but launched at different angles. As you can see, the range is a maximum for $\theta_i = 45°$. In addition, for any θ_i other than 45°, a point having Cartesian coordinates $(R, 0)$ can be reached by using either one of two complementary values of θ_i, such as 75° and 15°. Of course, the maximum height and time of flight for one of these values of θ_i are different from the maximum height and time of flight for the complementary value.

Useful trigonometric identities are given in Appendix B.4.

Σ

Pitfall Prevention 3.4

Equation 3.24 is useful for calculating R only for a symmetrical path as shown in **Active Figure 3.19**. If the path is not symmetrical, *do not use this equation*. The general expressions given by **Equations 3.19** and **3.20** are the *more important* results because they give the position and velocity components of *any* particle moving in two dimensions at *any* time t.

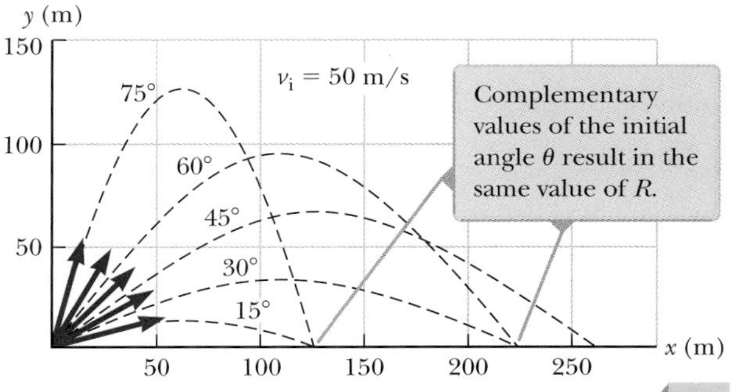

Active Figure 3.19

A projectile launched over a flat surface from the origin with an initial speed of 50 m/s at various angles of projection.

TRY THIS

Stand next to one friend (A) and get two friends (B and C) to stand next to each other facing you, so they are the same distance from you and the person (A) standing next to you. You and your friend (A) throw a ball to the person (B or C) opposite you at exactly the same time, so they can catch them. The balls must travel the same horizontal distance, but you can throw them with different maximum heights. Does the ball with the higher or lower maximum height reach the person it is thrown to first? Does this happen every time?

Quick **Quiz 3.5**

Rank the launch angles for the five paths in **Active Figure 3.19** with respect to time of flight from the shortest time of flight to the longest.

Problem-solving strategy

Projectile motion

We suggest you use the following approach when solving projectile motion problems.

1. Conceptualise

Think about what is going on physically in the problem. Imagine the projectile moving along its trajectory and draw a diagram. Label any important events along the trajectory. Consider what simplifying assumptions you can make, such as neglecting air resistance and assuming g is constant.

2. Model

If appropriate, model the projectile as a particle in free fall with negligible air resistance. Select a coordinate system with x in the horizontal direction and y in the vertical direction. Draw a set of *xy coordinate arrows* in the corner of your diagram.

3. Analyse

If the initial velocity vector is given, resolve it into x and y components. Treat the horizontal motion and the vertical motion independently. Analyse the horizontal motion of the projectile using the particle with constant velocity model. Analyse the vertical motion of the projectile using the particle with constant acceleration model.

4. Finalise

Once you have determined your result, check to see if your answers are consistent with your mental and pictorial representations and that your results are realistic.

Example 3.3

The long jump

A long jumper (**Figure 3.20**) leaves the ground at an angle of 20.0° above the horizontal and at a speed of 11.0 m/s.

(A) How far does he jump in the horizontal direction?

Solution

Conceptualise The arms and legs of a long jumper move in a complicated way, but we will ignore this motion. We conceptualise the motion of the long jumper as equivalent to that of a simple projectile. Draw a diagram showing the motion.

Model We will model the jumper as a particle in projectile motion, and ignore the motion of the arms and legs while he is airborne. We can reasonably make the approximation that g is constant over the trajectory, and we will further make the approximation that air resistance is negligible. This approximation is justified by the low speed of the jumper.

Sipa via AP Images

Figure 3.20
(Example 3.3) Romain Barras of France competes in the men's decathlon long jump at the 2008 Beijing Olympic Games.

Analyse His initial and final heights are the same, so we can use **Equation 3.26** to find the range of the jump. We could also use the more general analysis described in **Equations 3.17** to **3.20**.

$$R = \frac{v_i^2 \sin 2\theta_i}{g} = \frac{(11.0\,\text{m/s})^2 \sin 2(20.0°)}{9.80\,\text{m/s}^2} = \mathbf{7.94\ m}$$

(B) What is the maximum height reached?

Solution

Analyse Calculate the maximum height using **Equation 3.25**.

$$h = \frac{v_i^2 \sin^2 \theta_i}{2g} = \frac{(11.0\,\text{m/s})^2 (\sin 20.0°)^2}{2(9.80\,\text{m/s}^2)} = \mathbf{0.722\ m}$$

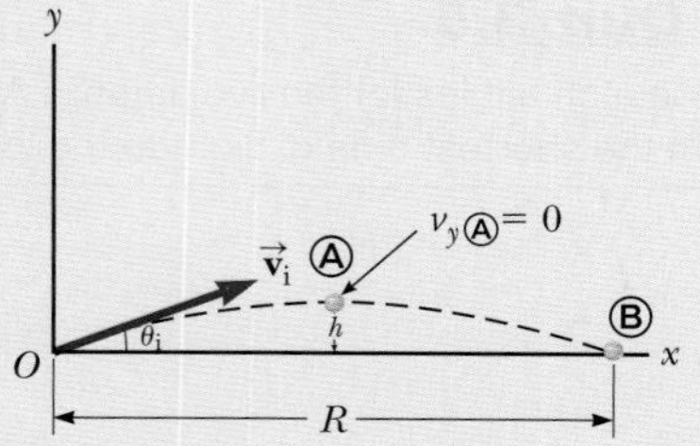

Finalise Find the answers to parts (A) and (B) using the general method. The results should agree. Treating the long jumper as a particle is an oversimplification. Nevertheless, the values obtained are consistent with experience in sports. We can model a complicated system such as a long jumper as a particle and still obtain reasonable results.

Example 3.4

A bullseye every time

In a popular lecture demonstration, a projectile is fired at a target in such a way that the projectile leaves the gun at the same time the target is dropped from rest. Show that if the gun is initially aimed at the stationary target, the projectile hits the falling target as shown in **Figure 3.21a**.

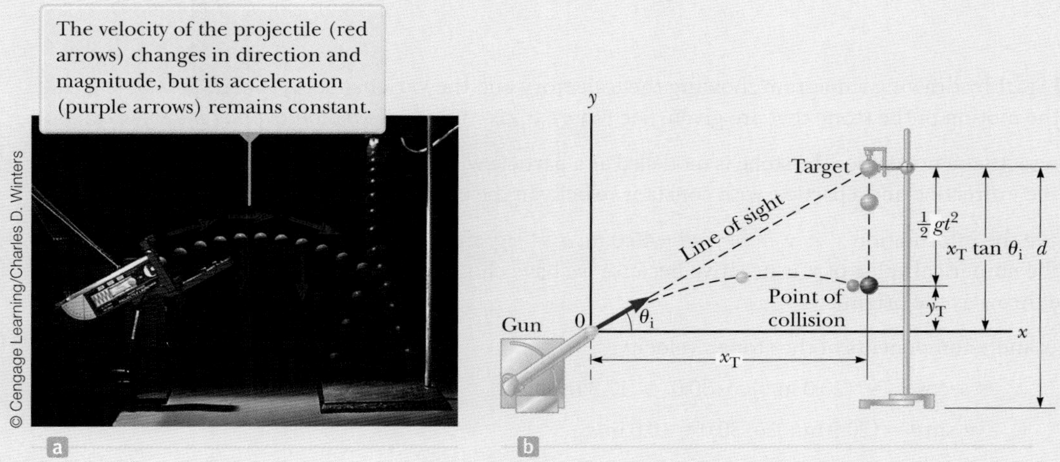

Figure 3.21
(Example 3.4) (a) Multiflash photograph of the projectile-target demonstration. If the gun is aimed directly at the target and is fired at the same instant the target begins to fall, the projectile will hit the target. (b) Schematic diagram of the projectile–target demonstration.

Solution

Conceptualise We can visualise the problem by studying **Figure 3.21**. You may also want to draw a diagram of your own; if we did not have **Figure 3.21b** we would begin by drawing such a diagram.

Model Because both objects are subject only to gravity, we model them as objects in free fall; the target moving in one dimension and the projectile moving in two.

Analyse The target T is modelled as a particle with constant acceleration in one dimension. **Figure 3.21b** shows that the initial y coordinate y_{iT} of the target is $x_T \tan\theta_i$ and its initial velocity is zero. It falls with acceleration $a = -g$. The projectile P is modelled as a particle with constant acceleration in the y direction and a particle with constant velocity in the x direction.

Write an expression for the y coordinate of the target at any moment after release, noting that its initial velocity is zero:

$$y_T = y_{iT} + (0)t - \frac{1}{2}gt^2 = x_T \tan\theta_i - \frac{1}{2}gt^2 \quad (1)$$

Write an expression for the y coordinate of the projectile at any moment:

$$y_P = y_{iP} + v_{yiP}t - \frac{1}{2}gt^2 = 0 + (v_{iP}\sin\theta_i)t - \frac{1}{2}gt^2 = (v_{iP}\sin\theta_i)t - \frac{1}{2}gt^2 \quad (2)$$

Write an expression for the x coordinate of the projectile at any moment:

$$x_P = x_{iP} + v_{xiP}t = 0 + (v_{iP}\cos\theta_i)t = (v_{iP}\cos\theta_i)t$$

Solve this expression for time as a function of the horizontal position of the projectile:

$$t = \frac{x_P}{v_{iP}\cos\theta_i}$$

Substitute this expression into **Equation (2)**:

$$y_P = (v_{iP}\sin\theta_i)\left(\frac{x_P}{v_{iP}\cos\theta_i}\right) - \frac{1}{2}gt^2 = x_P \tan\theta_i - \frac{1}{2}gt^2 \quad (3)$$

Compare **Equations (1)** and **(3)**. We see that when the x coordinates of the projectile and target are the same – that is, when $x_T = x_P$ – their y coordinates given by **Equations (1)** and **(3)** are the same and a collision results.

Finalise Note that a collision can result only when $v_{iP}\sin\theta_i \geq \sqrt{gd/2}$, where d is the initial elevation of the target above the floor. If $v_{iP}\sin\theta_i$ is less than this value, the projectile strikes the floor before reaching the target.

Example 3.5

Don't throw stones (again)!

A stone is thrown from the top of a building upwards at an angle of 30.0° to the horizontal with an initial speed of 20.0 m/s. The height from which the stone is thrown is 45.0 m above the ground.

(A) How long does it take the stone to reach the ground?

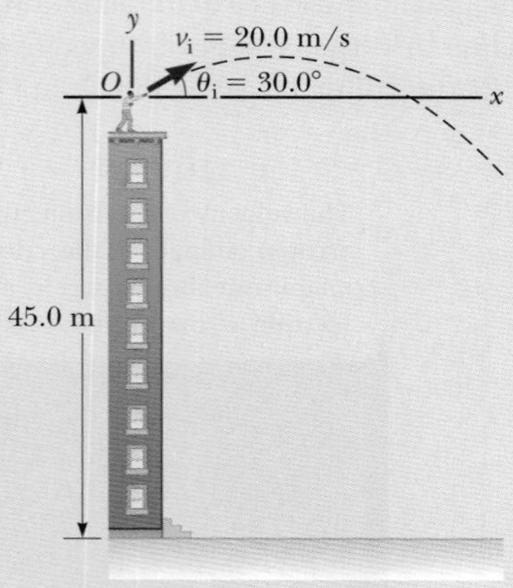

Figure 3.22 (Example 3.5) A stone is thrown from the top of a building.

Solution

Conceptualise Start by drawing a diagram showing the trajectory and the various parameters of the motion of the stone that are given. See Figure 3.22.

Model The stone acts as a projectile. The stone is modelled as a particle with constant acceleration in the y direction and a particle with constant velocity in the x direction.

Analyse We have the information $x_i = y_i = 0$, $y_f = -45.0$ m, $a = -g$, and $v_i = 20.0$ m/s (the numerical value of y_f is negative because we have chosen the point of the throw as the origin).

Find the initial x and y components of the stone's velocity:

$$v_{xi} = v_i \cos\theta_i = (20.0 \text{ m/s})\cos 30.0° = 17.3 \text{ m/s}$$

$$v_{yi} = v_i \sin\theta_i = (20.0 \text{ m/s})\sin 30.0° = 10.0 \text{ m/s}$$

Express the vertical position of the stone from the vertical component of Equation 3.20:

$$y_f = y_i + v_{yi}t + \frac{1}{2}a_y t^2$$

Substitute numerical values:

$$-45.0 \text{ m} = 0 + (10.0 \text{ m/s})t + \frac{1}{2}(-9.80 \text{ m/s}^2)t^2$$

Solve the quadratic equation for t: $t = 4.22$ s

(B) What is the speed of the stone just before it strikes the ground?

Solution

Analyse Use the y component of Equation 3.19 to obtain the y component of the velocity of the stone just before it strikes the ground:

$$v_{yf} = v_{yi} + a_y t$$

Substitute numerical values, using $t = 4.22$ s:

$$v_{yf} = 10.0 \text{ m/s} + (-9.80 \text{ m/s}^2)(4.22 \text{ s}) = -31.3 \text{ m/s}$$

Use this component with the horizontal component $v_{xf} = v_{xi} = 17.3$ m/s to find the speed of the stone at $t = 4.22$ s:

$$v_f = \sqrt{v_{xf}^2 + v_{yf}^2} = \sqrt{(17.3 \text{ m/s})^2 + (-31.3 \text{ m/s})^2} = 35.8 \text{ m/s}$$

Finalise Note that we could just as easily have set the initial position y_i to be +45 m, and our final position to be 0 m, and we would have arrived at the same final velocity. Repeat the calculation and check this for yourself.

3.5 Analysis model: particle in uniform circular motion

Figure 3.23a shows two physicists moving in a circular path on the Curve Ball exhibit at Questacon in Canberra; we describe this motion as **circular motion**. If a person is moving on this path with *constant speed* v, we call it **uniform circular motion**. Because it occurs so often, we use an analysis model called the **particle in uniform circular motion** to describe and analyse this type of motion.

An object that moves at a constant speed in a circular path *still has an acceleration*. To see why, consider the defining equation for acceleration, $\vec{a} = d\vec{v}/dt$ (Equation 3.16). Notice that the acceleration depends on the change in the *velocity*. Because velocity is a vector quantity, an acceleration can occur in two ways as mentioned in Section 3.2: by a change in the *magnitude* of the velocity and by a change in the *direction* of the velocity. The latter situation occurs for an object moving with constant speed in a circular path.

The constant-magnitude velocity vector is always tangent to the path of the object and perpendicular to the radius of the circular path.

> **Pitfall Prevention 3.5**
> Remember that in physics acceleration is defined as a change in the *velocity,* not a change in the *speed* (contrary to the everyday interpretation). In circular motion, the velocity vector is changing in direction, so there is an acceleration.

The acceleration vector in uniform circular motion is always perpendicular to the path and always points toward the centre of the circle. If that were not true, there would be a component of the acceleration parallel to the path and therefore parallel to the velocity vector. Such an acceleration component would lead to a change in the speed of the particle along the path. This situation, however, is inconsistent with our description of the situation: the particle moves with constant speed along the path. Therefore, for *uniform* circular motion, the acceleration vector can only have a component perpendicular to the path, which is towards the centre of the circle.

Let us now find the magnitude of the acceleration of the particle. Consider the diagram of the position and velocity vectors in **Figure 3.23c**. The figure also shows the vector representing the change in position $\Delta\vec{\mathbf{r}}$ for an arbitrary time interval. The particle follows a circular path of radius *r*, part of which is shown by the dashed curve. The particle is at Ⓐ at time t_i, and its velocity at that time is $\vec{\mathbf{v}}_i$; it is at Ⓑ at some later time t_f, and its velocity at that time is $\vec{\mathbf{v}}_f$. We assume $\vec{\mathbf{v}}_i$ and $\vec{\mathbf{v}}_f$ differ only in direction; their magnitudes are the same (that is, $v_i = v_f = v$ because it is *uniform* circular motion).

In **Figure 3.23d**, the velocity vectors in **Figure 3.23c** have been redrawn tail to tail. The vector $\Delta\vec{\mathbf{v}}$ connects the tips of the vectors, representing the vector addition $\vec{\mathbf{v}}_f = \vec{\mathbf{v}}_i + \Delta\vec{\mathbf{v}}$.

For help with vector addition, refer to Appendix B.6. For help with geometry, refer to Appendix B.3. Σ

In both **Figures 3.23c** and **3.23d**, we can identify triangles that help us analyse the motion. The angle $\Delta\theta$ between the two position vectors in **Figure 3.23c** is the same as the angle between the velocity vectors in **Figure 3.23d** because the velocity vector $\Delta\vec{\mathbf{v}}$ is always perpendicular to the position vector $\vec{\mathbf{r}}$. Therefore, the two triangles are *similar.* (Two triangles are similar if the angle between any two sides is the same for both triangles and if the ratio of the lengths of these sides is the same.) We can now write a relationship between the lengths of the sides for the two triangles in **Figures 3.23c** and **3.23d**:

$$\frac{|\Delta\vec{\mathbf{v}}|}{v} = \frac{|\Delta\vec{\mathbf{r}}|}{r}$$

where $v = v_i = v_f$ and $r = r_i = r_f$. This equation can be solved for $|\Delta\vec{\mathbf{v}}|$, and the expression obtained can be substituted into **Equation 3.15**, $\vec{\mathbf{a}}_{avg} = \Delta\vec{\mathbf{v}}/\Delta t$, to give the magnitude of the average acceleration over the time interval for the particle to move from Ⓐ to Ⓑ:

$$|\vec{\mathbf{a}}_{avg}| = \frac{|\Delta\vec{\mathbf{v}}|}{|\Delta t|} = \frac{v|\Delta\vec{\mathbf{r}}|}{r\Delta t}$$

Now imagine that points Ⓐ and Ⓑ in **Figure 3.23c** become extremely close together. As Ⓐ and Ⓑ approach each other, Δt approaches zero, $|\Delta\vec{\mathbf{r}}|$ approaches the distance travelled by the particle along the circular path, and the ratio $|\Delta\vec{\mathbf{r}}|/\Delta t$ approaches the speed *v*. In addition, the average acceleration becomes the instantaneous acceleration at point Ⓐ. Hence, in the limit $\Delta t \to 0$, the magnitude of the acceleration is

$$a_c = \frac{v^2}{r} \qquad (3.26)$$

◀ Centripetal acceleration

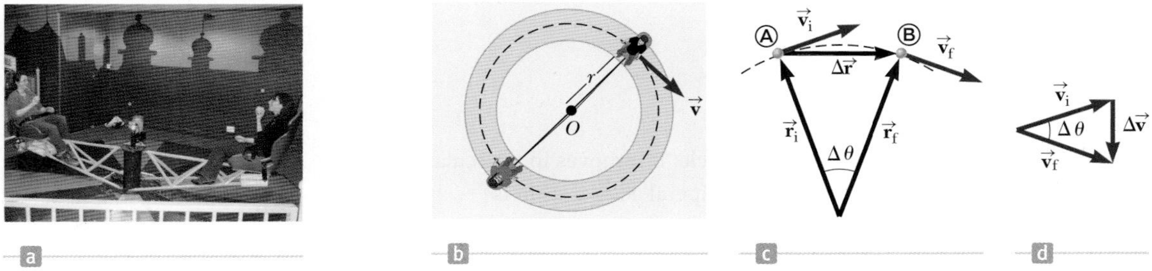

Figure 3.23
(a) Two physicists, Anna and Kate, riding the curve ball at Questacon. (b) A physicist moving along a circular path at constant speed experiences uniform circular motion. (c) As the physicist moves along a portion of a circular path from Ⓐ to Ⓑ, their velocity vector changes from $\vec{\mathbf{v}}_i$ to $\vec{\mathbf{v}}_f$. (d) The construction for determining the direction of the change in velocity $\Delta\vec{\mathbf{v}}$, which is toward the centre of the circle.

An acceleration of this nature is called a **centripetal acceleration** (*centripetal* means *centre-seeking*). The subscript on the acceleration symbol reminds us that the acceleration is centripetal.

In many situations, it is convenient to describe the motion of a particle moving with constant speed in a circle of radius r in terms of the **period** T, which is defined as the time interval required for one complete revolution of the particle. In the time interval T, the particle moves a distance of $2\pi r$, which is equal to the circumference of the particle's circular path. Therefore, because its speed is equal to the circumference of the circular path divided by the period, or $v = 2\pi r / T$, it follows that

Period of circular motion ▶

$$T = \frac{2\pi r}{v} \tag{3.27}$$

The period of a particle in uniform circular motion is a measure of the number of seconds for one revolution of the particle around the circle. The inverse of the period is the *rotation rate* and is measured in revolutions per second. Because one full revolution of the particle around the circle corresponds to an angle of 2π radians, the product of 2π and the rotation rate gives the **angular speed** ω of the particle, measured in radians/s or s^{-1}:

$$\omega = \frac{2\pi}{T} \tag{3.28}$$

Combining this equation with **Equation 3.27**, we find a relationship between angular speed and the translational speed with which the particle travels in the circular path:

$$\omega = 2\pi\left(\frac{v}{2\pi r}\right) = \frac{v}{r} \rightarrow v = r\omega \tag{3.29}$$

Pitfall Prevention 3.6
The magnitude of the centripetal acceleration vector is constant for uniform circular motion, but *the centripetal acceleration vector is not constant.* It always points towards the centre of the circle, but it changes direction as the object moves around the circular path.

Equation 3.29 demonstrates that, for a fixed angular speed, the translational speed becomes larger as the radial position becomes larger. Therefore, for example, if a merry-go-round rotates at a fixed angular speed ω, a rider at an outer position at large r will be travelling through space faster than a rider at an inner position at smaller r. We will investigate rotational motion more deeply in Chapter 10.

We can express the centripetal acceleration of a particle in uniform circular motion in terms of angular speed by combining **Equations 3.26** and **3.29**:

$$a_c = \frac{(r\omega)^2}{r}$$

$$a_c = r\omega^2 \tag{3.30}$$

Equations 3.27–3.30 can be used when the particle in uniform circular motion model is appropriate for a given situation.

Quick Quiz 3.6

A particle moves in a circular path of radius r with speed v. It then increases its speed to $2v$ while traveling along the same circular path. (i) The centripetal acceleration of the particle has changed by what factor? Choose one: (a) 0.25 (b) 0.5 (c) 2 (d) 4 (e) impossible to determine (ii) From the same choices, by what factor has the period of the particle changed?

Analysis Model 3.1

Particle in uniform circular motion

Imagine a moving object that can be modelled as a particle. If it moves in a circular path of radius r at a constant speed v, the magnitude of its centripetal acceleration is

$$a_c = \frac{v^2}{r} \tag{3.26}$$

and the period of the particle's motion is given by

$$T = \frac{2\pi r}{v} \tag{3.27}$$

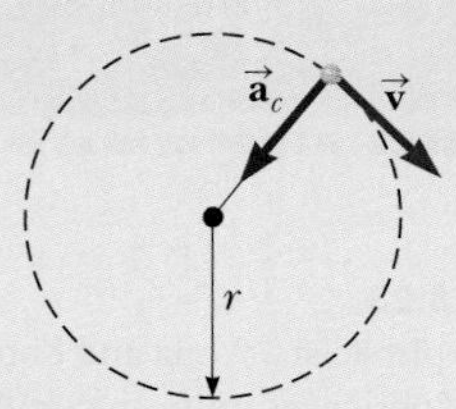

The angular speed of the particle is

$$\omega = \frac{2\pi}{T} \tag{3.28}$$

Examples

- a rock twirled in a circle on a string of constant length
- a planet travelling around a perfectly circular orbit (Chapter 11)
- a charged particle moving in a uniform magnetic field (Chapter 30)
- an electron in orbit around a nucleus in the Bohr model of the hydrogen atom (Chapter 41)

Example 3.6

The centripetal acceleration of the Earth

What is the centripetal acceleration of the Earth as it moves in its orbit around the Sun?

Solution

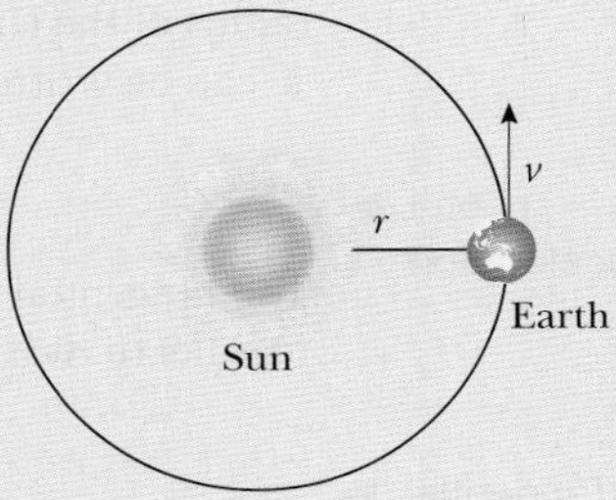

Conceptualise Draw a diagram of the Earth in a circular orbit around the Sun. The Earth is very small compared to its orbital radius.

Model We will model the Earth as a particle and approximate the Earth's orbit as circular (it's actually slightly elliptical). We apply the particle in uniform circular motion analysis model.

Analyse We do not know the orbital speed of the Earth to substitute into **Equation 3.26**. With the help of **Equation 3.27**, however, we can recast **Equation 3.26** in terms of the period of the Earth's orbit, which we know is one year, and the radius of the Earth's orbit around the Sun, which is 1.496×10^{11} m as given in the front endpaper.

Combine **Equations 3.26** and **3.27**:

$$a_c = \frac{v^2}{r} = \frac{\left(\dfrac{2\pi r}{T}\right)^2}{r} = \frac{4\pi^2 r}{T^2}$$

Do a dimension check, noting that 4 and π have no dimensions.

$$[LT^{-2}] = [L]/[T^2] = [LT^{-2}]$$ ☺

Substitute numerical values:

$$a_c = \frac{4\pi^2(1.496 \times 10^{11}\text{ m})}{(1\text{ yr})^2}\left(\frac{1\text{ yr}}{3.156 \times 10^7\text{ s}}\right)^2 = 5.93 \times 10^{-3}\text{ m/s}^2$$

Finalise This acceleration is much smaller than the free-fall acceleration on the surface of the Earth. We do not usually notice this acceleration because we are also accelerating at the same rate. Note that here we wrote the speed v in **Equation 3.25** in terms of the period T. In tables of planetary or astronomical data it is more common that T is given rather than v, so this is a useful technique.

3.6 Non-uniform circular motion: Tangential and radial acceleration

Let us consider a more general motion than that presented in Section 3.5. A particle moves to the right along a curved path, and its velocity changes both in direction and in magnitude as described in **Active Figure 3.24**. In this situation, the velocity vector is always tangent to the path; the acceleration vector $\vec{a}$, however, is at some angle to the path. At each of three points Ⓐ, Ⓑ and Ⓒ in **Active Figure 3.24**, the dashed blue circles represent the curvature of the actual path at each point. The radius of each circle is equal to the path's radius of curvature at each point.

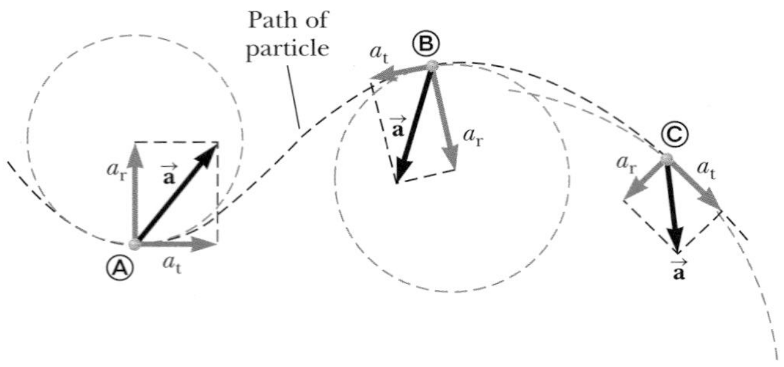

Active Figure 3.24

The motion of a particle along an arbitrary curved path lying in the *xy* plane. If the velocity vector $\vec{\mathbf{v}}$ (always tangent to the path) changes in direction and magnitude, the components of the acceleration $\vec{\mathbf{a}}$ are a tangential component a_t and a radial component a_r.

Σ *For help with vectors, see Appendix B.6.*

As the particle moves along the curved path in **Active Figure 3.24**, the direction of the total acceleration vector $\vec{\mathbf{a}}$ changes from point to point. At any instant, this vector can be resolved into two components based on an origin at the centre of the dashed circle corresponding to that instant: a radial component a_r along the radius of the circle and a tangential component a_t perpendicular to this radius. The *total* acceleration vector $\vec{\mathbf{a}}$ can be written as the vector sum of the component vectors:

Total acceleration ▶

$$\vec{\mathbf{a}} = \vec{\mathbf{a}}_r + \vec{\mathbf{a}}_t \tag{3.31}$$

The tangential acceleration component causes a change in the speed v of the particle. This component is parallel to the instantaneous velocity, and its magnitude is given by

Tangential acceleration ▶

$$a_t = \left|\frac{dv}{dt}\right| \tag{3.32}$$

The radial acceleration component arises from a change in direction of the velocity vector and is given by

Radial acceleration ▶

$$a_r = -a_c = -\frac{v^2}{r} \tag{3.33}$$

where r is the radius of curvature of the path at the point in question. We recognise the magnitude of the radial component of the acceleration as the centripetal acceleration discussed in Section 3.5. The negative sign in **Equation 3.29** indicates that the direction of the centripetal acceleration is toward the centre of the circle representing the radius of curvature. The direction is opposite that of the radial unit vector $\hat{\mathbf{r}}$, which always points away from the origin at the centre of the circle.

Because $\vec{\mathbf{a}}_r$ and $\vec{\mathbf{a}}_t$ are perpendicular component vectors of $\vec{\mathbf{a}}$, it follows that the magnitude of $\vec{\mathbf{a}}$ is $a = \sqrt{a_r^2 + a_t^2}$. At a given speed, a_r is large when the radius of curvature is small (as at points Ⓐ and Ⓑ in **Active Figure 3.24**) and small when r is large (as at point Ⓒ). The direction of $\vec{\mathbf{a}}_t$ is either in the same direction as $\vec{\mathbf{v}}$ (if v is increasing) or opposite $\vec{\mathbf{v}}$ (if v is decreasing, as at point Ⓑ).

In uniform circular motion, where v is constant, $a_t = 0$ and the acceleration is always completely radial as described in Section 3.5. In other words, uniform circular motion is a special case of motion along a general curved path. Furthermore, if the direction of $\vec{\mathbf{v}}$ does not change, there is no radial acceleration and the motion is one dimensional (in this case, $a_r = 0$, but a_t may not be zero).

Quick Quiz 3.7

A particle moves along a path and its speed increases with time. **(i)** In which of the following cases are its acceleration and velocity vectors parallel? (a) when the path is circular (b) when the path is straight (c) when the path is a parabola (d) never **(ii)** From the same choices, in which case are its acceleration and velocity vectors perpendicular everywhere along the path?

±? Example **3.7**

Over the hill

A car starts from rest and exhibits a constant acceleration of $(0.30 \pm 0.01)\ \text{m/s}^2$ parallel to the road. The car passes over a hill such that the top of the hill is shaped like a circle of radius (500 ± 5) m. At the moment the car is at the top of the hill, its velocity vector is horizontal and has a magnitude of (6.0 ± 0.2) m/s. What are the magnitude and direction of the total acceleration vector for the car at this instant?

Solution

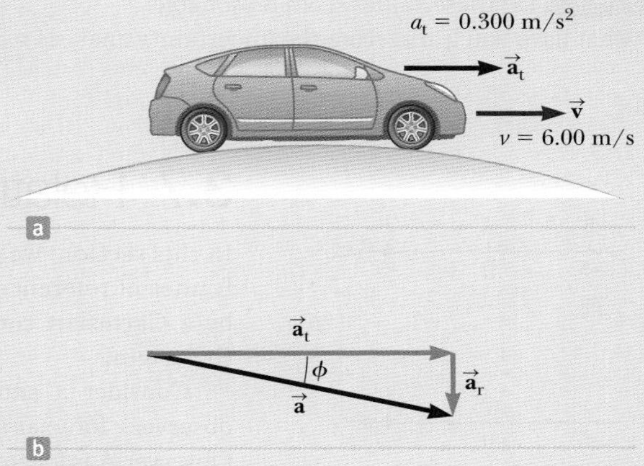

Figure 3.25
(Example 3.7) (a) A car passes over a rise that is shaped like a circle. (b) The total acceleration vector $\vec{\mathbf{a}}$ is the sum of the tangential and radial acceleration vectors $\vec{\mathbf{a}}_t$ and $\vec{\mathbf{a}}_r$.

Conceptualise Draw a diagram to help visualise the situation and recall any experiences you have had in driving (or riding) over hills to help you understand the problem.

Model Because the accelerating car is moving along a curved path, we model the car as a particle experiencing both tangential and radial acceleration.

Analyse The radial acceleration is given by **Equation 3.26**. The radial acceleration vector is directed straight downwards, and the tangential acceleration vector has magnitude $0.30\ \text{m/s}^2$ and is horizontal.

Evaluate the radial acceleration:

$$a_r = -\frac{v^2}{r} = -\frac{(6.0\ \text{m/s})^2}{500\ \text{m}} = -0.072\ \text{m/s}^2$$

The uncertainty in the radial acceleration is:

$$\Delta a_r = a_r\left(2\frac{\Delta v}{v} + \frac{\Delta r}{r}\right) = 0.072\ \text{m/s}^2\left(2\frac{0.2\ \text{m}}{6.0\ \text{m}} + \frac{5\ \text{m}}{500\ \text{m}}\right) = 0.006\ \text{m/s}^2$$

Find the magnitude of $\vec{\mathbf{a}}$:

$$\sqrt{a_r^2 + a_t^2} = \sqrt{(-0.072\ \text{m/s}^2)^2 + (0.30\ \text{m/s}^2)^2} = 0.31\ \text{m/s}^2$$

A quick way of finding the uncertainty here, as described in Chapter 1, is to find the maximum and minimum values, which gives us a range of values and hence the uncertainty.

The minimum value is:

$$a_{\text{min}} = \sqrt{(-0.066\ \text{m/s}^2)^2 + (0.29\ \text{m/s}^2)^2} = 0.30\ \text{m/s}^2$$

and the maximum is:

$$a_{\text{max}} = \sqrt{(-0.078\ \text{m/s}^2) + (0.31\ \text{m/s}^2)} = 0.32\ \text{m/s}^2$$

so

$$\Delta a = \frac{a_{\text{max}} - a_{\text{min}}}{2} = 0.01\ \text{m/s}^2$$

Note that as the tangential component of the acceleration is so much larger than the radial component, it would have been reasonable to make the approximation that $\Delta a \approx \Delta a_t = 0.01\ \text{m/s}^2$, which is in fact what we found.

So now we can write

$$\vec{\mathbf{a}} = (0.31 \pm 0.01)\ \text{m/s}^2$$

Find the angle ϕ (see **Figure 3.25b**) between $\vec{\mathbf{a}}$ and the horizontal:

$$\phi = \tan^{-1}\left(\frac{a_r}{a_t}\right) = \tan^{-1}\left(\frac{-0.072\ \text{m/s}^2}{0.30\ \text{m/s}^2}\right) = -13.5°$$

Again we will use the range method for finding the uncertainty.

Example 3.7 cont.

The minimum value is: $\phi_{min} = \tan^{-1}\left(\frac{0.066 \text{ m/s}^2}{0.31 \text{ m/s}^2}\right) = 12.0°$

The maximum value is: $\phi_{max} = \tan^{-1}\left(\frac{0.078 \text{ m/s}^2}{0.29 \text{ m/s}^2}\right) = 15.0°$

Hence the uncertainty is 1.5° and we can write $\phi = -13.5° \pm 1.5°$.

Finalise Do these values seem reasonable?
What have you learnt from the uncertainty analysis required in this problem?

3.7 Relative velocity and relative acceleration

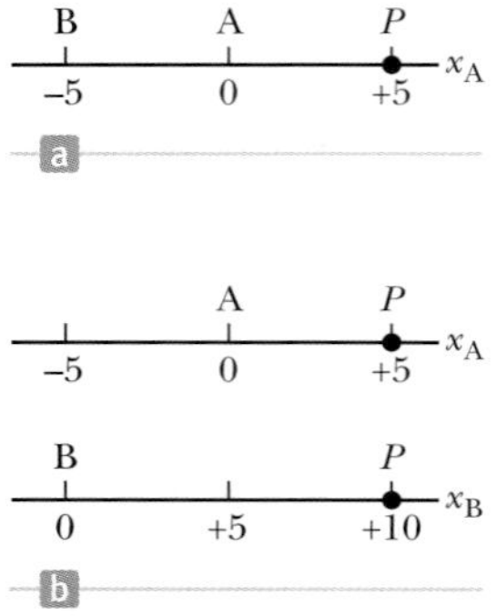

Figure 3.26
Different observers make different measurements. (a) Observer A is located at the origin and Observer B is at a position of −5. Both observers measure the position of a particle at *P*. (b) If both observers see themselves at the origin of their own coordinate system, they disagree on the value of the position of the particle at *P*.

In this section, we describe how observations made by different observers in different frames of reference are related to one another. A frame of reference can be described by a Cartesian coordinate system for which an observer is at rest with respect to the origin.

Consider a situation in which there will be different observations for different observers; for example, the two observers A and B along the number line in **Figure 3.26a.** Observer A is located at the origin of a one-dimensional x_A axis, while observer B is at the position $x_A = -5$. We denote the position variable as x_A because observer A is at the origin of this axis. Both observers measure the position of point *P*, which is located at $x_A = +5$. Suppose observer B decides that he is located at the origin of an x_B axis as in **Figure 3.26b.** Notice that the two observers disagree on the value of the position of point *P*. Observer A claims point *P* is located at a position with a value of +5, whereas observer B claims it is located at a position with a value of +10. Both observers are correct, even though they make different measurements. Their measurements differ because they are making the measurement from different frames of reference.

Now consider two observers watching a man walking on a travelator at an airport in **Figure 3.27.** The woman standing on the travelator sees the man moving at a normal walking speed. The woman observing from the stationary floor sees the man moving with a higher speed because the travelator speed combines with his walking speed. Both observers look at the same man and arrive at different values for his speed. Both are correct; the difference in their measurements results from the relative velocity of their frames of reference.

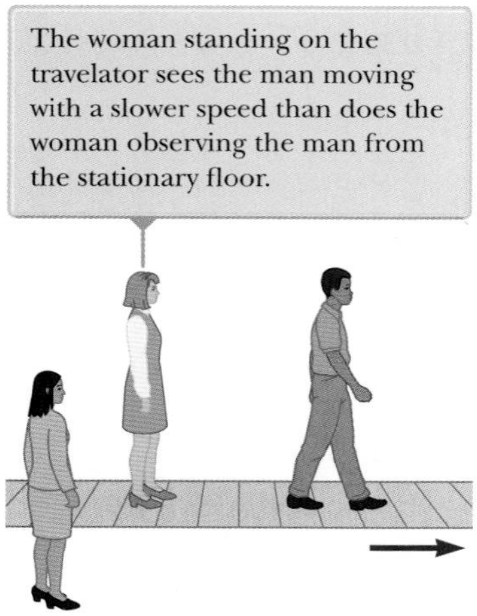

The woman standing on the travelator sees the man moving with a slower speed than does the woman observing the man from the stationary floor.

Figure 3.27
Two observers measure the speed of a man walking on a travelator (moving walkway).

In a more general situation, consider a particle located at point *P* in **Figure 3.28.** Imagine that the motion of this particle is being described by two observers, observer A in a reference frame S_A fixed relative to the Earth and a second observer B in a reference frame S_B moving to the right relative to S_A (and therefore relative to the Earth) with a constant velocity $\vec{\mathbf{v}}_{BA}$. In this discussion of relative velocity, we use a double-subscript notation; the first subscript represents what is being observed, and the second represents who is doing the observing. Therefore, the notation $\vec{\mathbf{v}}_{BA}$ means the velocity of observer B (and the attached frame S_B) as measured by observer A. With this notation, observer B measures A to be moving to the left with a velocity $\vec{\mathbf{v}}_{AB} = -\vec{\mathbf{v}}_{BA}$. For purposes of this discussion, let us place each observer at her or his respective origin.

We define the time $t = 0$ as the instant at which the origins of the two reference frames coincide in space. Therefore, at time *t*, the origins of the reference frames will be separated by a distance $v_{BA}t$. We label the position *P* of the particle relative to observer A with the position vector $\vec{\mathbf{r}}_{PA}$ and that relative to observer B with the position vector $\vec{\mathbf{r}}_{PB}$, both at time *t*. From **Figure 3.28,** we see that the vectors $\vec{\mathbf{r}}_{PA}$ and $\vec{\mathbf{r}}_{PB}$ are related to each other through the expression

$$\vec{\mathbf{r}}_{PA} = \vec{\mathbf{r}}_{PB} + \vec{\mathbf{v}}_{BA}t \qquad (3.34)$$

By differentiating **Equation 3.34** with respect to time, noting that $\vec{\mathbf{v}}_{BA}$ is constant, we obtain

$$\frac{d\vec{\mathbf{r}}_{PA}}{dt} = \frac{d\vec{\mathbf{r}}_{PB}}{dt} + \vec{\mathbf{v}}_{BA}$$

$$\vec{\mathbf{u}}_{PA} = \vec{\mathbf{u}}_{PB} + \vec{\mathbf{v}}_{BA} \quad (3.35)$$

◀ Galilean velocity transformation

where $\vec{\mathbf{u}}_{PA}$ is the velocity of the particle at P measured by observer A and $\vec{\mathbf{u}}_{PB}$ is its velocity measured by B. (We use the symbol $\vec{\mathbf{u}}$ for particle velocity rather than $\vec{\mathbf{v}}$, which we have already used for the relative velocity of two reference frames.) **Equations 3.34** and **3.35** are known as **Galilean transformation equations**. They relate the position and velocity of a particle as measured by observers in relative motion. Notice the pattern of the subscripts in **Equation 3.35**. When relative velocities are added, the inner subscripts (B) are the same, and the outer ones (P, A) match the subscripts on the velocity on the left of the equation.

Σ *For help with taking derivatives, see Appendix B.7.*

Although observers in two frames measure different velocities for the particle, they measure the *same acceleration* when $\vec{\mathbf{v}}_{BA}$ is constant. We can verify that by taking the time derivative of **Equation 3.35**:

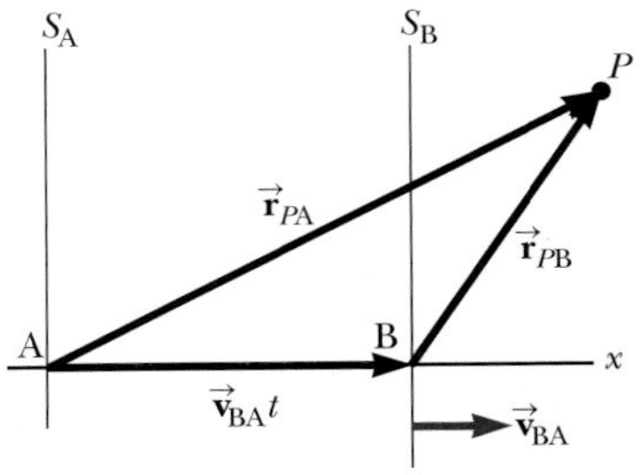

Figure 3.28
A particle located at P is described by two observers, one in the fixed frame of reference S_A and the other in the frame S_B, which moves to the right with a constant velocity $\vec{\mathbf{v}}_{BA}$. The vector $\vec{\mathbf{r}}_{PA}$ is the particle's position vector relative to S_A, and $\vec{\mathbf{r}}_{PB}$ is its position vector relative to S_B.

$$\frac{d\vec{\mathbf{u}}_{PA}}{dt} = \frac{d\vec{\mathbf{u}}_{PB}}{dt} + \frac{d\vec{\mathbf{v}}_{BA}}{dt} = \frac{d\vec{\mathbf{u}}_{PB}}{dt}$$

Because $\vec{\mathbf{v}}_{BA}$ is constant, $d\vec{\mathbf{v}}_{BA}/dt = 0$. Therefore, we conclude that $\vec{\mathbf{a}}_{PA} = \vec{\mathbf{a}}_{PB}$ because $\vec{\mathbf{a}}_{PA} = d\vec{\mathbf{u}}_{PA}/dt$ and $\vec{\mathbf{a}}_{PB} = d\vec{\mathbf{u}}_{PB}/dt$. That is, the acceleration of the particle measured by an observer in one frame of reference is the same as that measured by any other observer moving with constant velocity relative to the first frame.

Example 3.8

A boat crossing a river

A boat crossing a wide river moves with a speed of 10.0 km/h relative to the water. The water in the river has a uniform speed of 5.00 km/h due east relative to the Earth.

(A) If the boat heads due north, determine the velocity of the boat relative to an observer standing on the bank.

Solution

Conceptualise Imagine moving in a boat across a river while the current pushes you down the river. You will not move directly across the river, but will end up downstream. Draw a diagram showing the boat crossing the river, and include the data given in the question on your diagram. You should look again at **Figure 3.28** to help you draw your diagram.

Model We can model the river and the Earth as two reference frames moving at constant velocity relative to each other and apply the Galilean transformations to find the velocity of the boat in each frame.

Analyse We know $\vec{\mathbf{v}}_{br}$, the velocity of the *boat* relative to the *river*, and $\vec{\mathbf{v}}_{rE}$, the velocity of the *river* relative to the *Earth*. What we must find is $\vec{\mathbf{v}}_{bE}$, the velocity of the *boat* relative to the *Earth*. The relationship between these three quantities is $\vec{\mathbf{v}}_{bE} = \vec{\mathbf{v}}_{br} + \vec{\mathbf{v}}_{rE}$. The terms in the equation are vector quantities and should be manipulated as such. The vectors are shown in **Figure 3.29**. The quantity $\vec{\mathbf{v}}_{br}$ is due north; $\vec{\mathbf{v}}_{rE}$ is due east; and the vector sum of the two, $\vec{\mathbf{v}}_{bE}$, is at an angle θ as defined in **Figure 3.29**.

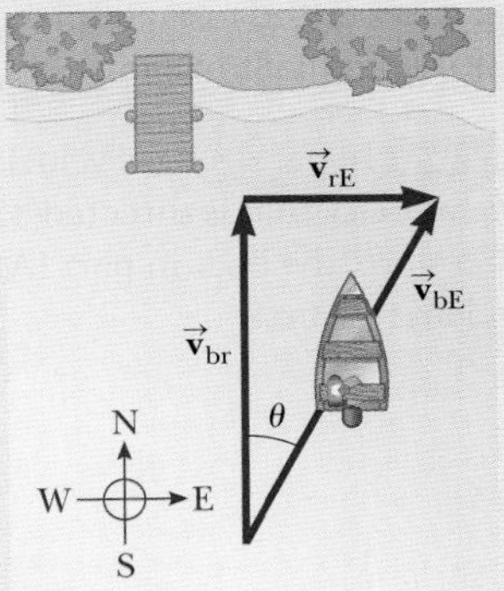

Figure 3.29
(Example 3.8) A boat aims directly across a river and ends up downstream.

Find the speed v_{bE} of the boat relative to the Earth using the Pythagorean theorem:

$$v_{bE} = \sqrt{v_{br}^2 + v_{rE}^2} = \sqrt{(10.0\ \text{km/h})^2 + (5.00\ \text{km/h})^2}$$
$$= 11.2\ \text{km/h}$$

Find the direction of $\vec{\mathbf{v}}_{bE}$: $$\theta = \tan^{-1}\left(\frac{v_{rE}}{v_{br}}\right) = \tan^{-1}\left(\frac{5.00}{10.0}\right) = 26.6°$$

Finalise The boat is moving at a speed of 11.2 km/h in the direction 26.6° east of north relative to the Earth. Notice that the speed of 11.2 km/h is faster than your boat speed of 10.0 km/h. The current velocity adds to yours to give you a higher speed. Notice in **Figure 3.29** that your resultant velocity is at an angle to the direction straight across the river, so you will end up downstream, as we predicted.

(B) If the boat travels with the same speed of 10.0 km/h relative to the river and is to travel due north as shown in **Figure 3.30**, what should its heading be?

Solution

Conceptualise In this case, we must aim the boat upstream so as to go straight across the river. Draw a new diagram showing the boat travelling at an angle to its previous path, aiming upstream.

Model We apply the same model as in (A).

Analyse The analysis now involves the new triangle shown in **Figure 3.30**. As in part (A), we know $\vec{\mathbf{v}}_{rE}$ and the magnitude of the vector $\vec{\mathbf{v}}_{br}$, and we want $\vec{v}_{bE}$ to be directed across the river. Notice the difference between the triangle in **Figure 3.29** and the one in **Figure 3.30**: the hypotenuse in **Figure 3.30** is no longer $\vec{\mathbf{v}}_{bE}$.

Use the Pythagorean theorem to find v_{bE}:

$$v_{bE} = \sqrt{v_{br}^2 - v_{rE}^2} = \sqrt{(10.0\ \text{km/h})^2 - (5.00\ \text{km/h})^2} = 8.66\ \text{km/h}$$

Find the direction in which the boat is heading: $$\theta = \tan^{-1}\left(\frac{v_{rE}}{v_{bE}}\right) = \tan^{-1}\left(\frac{5.00}{8.66}\right) = 30.0°$$

Finalise The boat must head upstream so as to travel directly northwards across the river. For the given situation, the boat must steer a course 30.0° west of north. For faster currents, the boat must be aimed upstream at larger angles.

What If? Imagine that the two boats in parts (A) and (B) are racing across the river. Which boat arrives at the opposite bank first?

Answer In part (A), the velocity of 10 km/h is aimed directly across the river. In part (B), the velocity that is directed across the river has a magnitude of only 8.66 km/h. Therefore, the boat in part (A) has a larger velocity component directly across the river and arrives first.

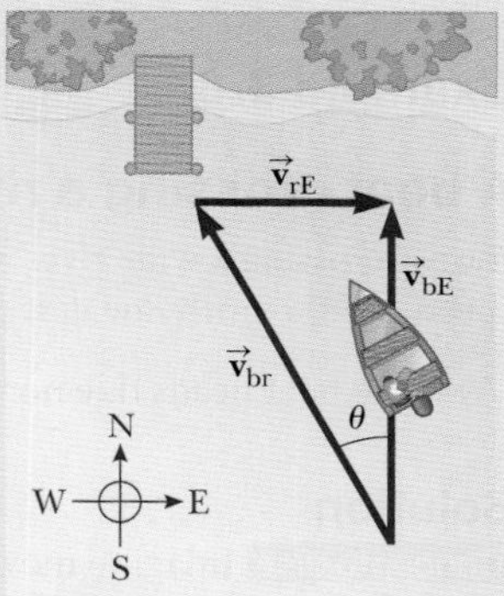

Figure 3.30 (Example 3.8) To move directly across the river, the boat must aim upstream.

End-of-chapter resources

The photo shows fireworks erupting from Sydney Harbour Bridge. Note the shape of the paths taken by the incandescent embers. All the paths are a similar shape. Why do they take these particular trajectories, and how can we model their behaviour to allow us to predict their trajectories?

The embers can be modelled as particles in projectile motion. If we make the approximation that air resistance acting on the particles is small, then we expect them to all take parabolic paths. As you can see in the photo, the paths are approximately parabolic. If we could work out a scale for the photo, we would be able to make measurements on the photo to find the maximum height of any ember, and its launch angle θ. This gives us enough information to determine its initial velocity components using Equations 3.22 and 3.24. We could then calculate the position of an ember at any time after its launch using Equation 3.21. Note that we could work out an approximate scale using the known height of the bridge, which is 134 m.

Graham Monro/Photolibrary/Jupiter Images

Figure 3.31
Fireworks erupt from the Sydney Harbour Bridge on New Year's Eve. Note the paths of the embers.

The problems found in this chapter may be assigned online in Enhanced Web Assign.

ENHANCED WebAssign

Worked solutions to every fifth problem are available in the Student Solutions Manual. Register online at **www.cengagebrain.com** for access.

Summary

Definitions

The **displacement vector** $\Delta\vec{r}$ for a particle is the difference between its final position vector and its initial position vector:

$$\Delta\vec{r} \equiv \vec{r}_f - \vec{r}_i \tag{3.12}$$

The **average velocity** of a particle during the time interval Δt is defined as the displacement of the particle divided by the time interval:

$$\vec{v}_{avg} \equiv \frac{\Delta\vec{r}}{\Delta t} \tag{3.13}$$

The **instantaneous velocity** of a particle is defined as the limit of the average velocity as Δt approaches zero:

$$\vec{v} \equiv \lim_{\Delta t \to 0} \frac{\Delta\vec{r}}{\Delta t} = \frac{d\vec{r}}{dt} \tag{3.14}$$

The **average acceleration** of a particle is defined as the change in its instantaneous velocity vector divided by the time interval Δt during which that change occurs:

$$\vec{a}_{avg} \equiv \frac{\Delta\vec{v}}{\Delta t} = \frac{\vec{v}_f - \vec{v}_i}{t_f - t_i} \tag{3.15}$$

The **instantaneous acceleration** of a particle is defined as the limiting value of the average acceleration as Δt approaches zero:

$$\vec{a} \equiv \lim_{\Delta t \to 0} \frac{\Delta\vec{v}}{\Delta t} = \frac{d\vec{v}}{dt} \tag{3.16}$$

Projectile motion is one type of two-dimensional motion, exhibited by an object launched into the air near the Earth's surface and experiencing free fall. This common motion can be analysed by applying the particle with constant velocity model to the motion of the projectile in the x direction and the particle with constant acceleration model ($a_y = -g$) in the y direction.

A particle moving in a circular path with constant speed is exhibiting **uniform circular motion**.

Concepts and principles

If a particle moves with *constant* acceleration $\vec{a}$ and has velocity $\vec{v}_i$ and position $\vec{r}_i$ at $t = 0$, its velocity and position vectors at some later time t are

$$\vec{v}_f = \vec{v}_i + \vec{a}t \tag{3.19}$$

$$\vec{r}_f = \vec{r}_i + \vec{v}_i t + \frac{1}{2}\vec{a}t^2 \tag{3.20}$$

For two-dimensional motion in the xy plane under constant acceleration, each of these vector expressions is equivalent to two component expressions: one for the motion in the x direction and one for the motion in the y direction.

It is useful to think of projectile motion in terms of a combination of two analysis models: (1) the particle with constant velocity model in the x direction and (2) the particle with constant acceleration model in the vertical direction with a constant downwards acceleration of magnitude $g = 9.80\ \text{m/s}^2$.

If a particle moves along a curved path in such a way that both the magnitude and the direction of $\vec{v}$ change in time, the particle has an acceleration vector that can be described by two component vectors: (1) a radial component vector $\vec{a}_r$ that causes the change in direction of $\vec{v}$ and (2) a tangential component vector $\vec{a}_r$ that causes the change in magnitude of $\vec{v}$. The magnitude of $\vec{a}_r$ is v^2/r, and the magnitude of $\vec{a}_t$ is $|dv/dt|$.

A particle in uniform circular motion has a radial acceleration $\vec{a}_r$ because the direction of $\vec{v}$ changes in time. This acceleration is called **centripetal acceleration**, and its direction is always towards the centre of the circle. It has no tangential acceleration.

The velocity $\vec{\mathbf{u}}_{PA}$ of a particle measured in a fixed frame of reference S_A can be related to the velocity $\vec{\mathbf{u}}_{PB}$ of the same particle measured in a moving frame of reference S_B by

$$\vec{\mathbf{u}}_{PA} = \vec{\mathbf{u}}_{PB} + \vec{\mathbf{v}}_{BA} \quad (3.35)$$

where $\vec{\mathbf{v}}$ is the velocity of S_B relative to S_A.

Analysis model for problem solving

Particle in uniform circular motion If a particle moves in a circular path of radius r with a constant speed v, the magnitude of its centripetal acceleration is given by

$$a_c = \frac{v^2}{r} \quad (3.26)$$

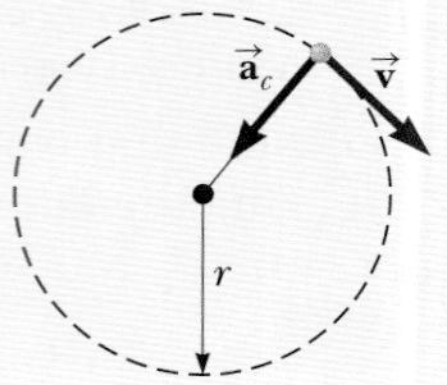

and the **period** of the particle's motion is given by

$$T = \frac{2\pi r}{v} \quad (3.27)$$

and the **angular speed** of the particle is

$$\omega = \frac{2\pi}{T} \quad (3.28)$$

Chapter review quiz

To help you revise Chapter 3: Motion in two dimensions, complete the automatically graded Chapter review quiz at http://login.cengagebrain.com.

Conceptual questions

1. A book is moved once around the perimeter of a table top with the dimensions 1 m by 2 m. The book ends up at its initial position. (a) What is its displacement? (b) What is the distance travelled?
2. Is it possible to add a vector quantity to a scalar quantity? Explain your answer.
3. Draw motion diagrams showing the velocity and acceleration of a projectile at several points along its path, assuming (a) the projectile is launched horizontally and (b) the projectile is launched at angle θ with the horizontal.
4. A spacecraft drifts through space at a constant velocity. Suddenly, a gas leak in the side of the spacecraft gives it a constant acceleration in a direction perpendicular to the initial velocity. The orientation of the spacecraft does not change, so the acceleration remains perpendicular to the original direction of the velocity. What is the shape of the path followed by the spacecraft in this situation?
5. Describe how a driver can steer a car travelling at constant speed so that (a) the acceleration is zero or (b) the magnitude of the acceleration remains constant.
6. An ice skater is executing a figure eight consisting of two identically shaped, tangent circular paths. Throughout the first loop she increases her speed uniformly, and during the second loop she moves at a constant speed. Draw a motion diagram showing her velocity and acceleration vectors at several points along the path of motion.
7. If you know the position vectors of a particle at two points along its path and also know the time interval during which it moved from one point to the other, can you determine the particle's instantaneous velocity? Its average velocity? Explain.

Problems

Section 3.1 Vectors, scalars and coordinate systems

1. The polar coordinates of a point are $r = 5.50$ m and $\theta = 240°$. What are the Cartesian coordinates of this point?
2. A fly lands on one wall of a room. The lower-left corner of the wall is selected as the origin of a two-dimensional Cartesian coordinate system. If the fly is located at the point having coordinates (2.00, 1.00) m, (a) how far is it from the origin? (b) What is its location in polar coordinates?
3. A vector has an x component of -25.0 units and a y component of 40.0 units. Find the magnitude and direction of this vector.
4. A person walks 25.0° north of east for 3.10 km. How far would she have to walk due north and due east to arrive at the same location?

Section 3.2 The position, velocity and acceleration vectors

5. A motorist drives south at 20.0 m/s for 3.00 min, then turns west and travels at 25.0 m/s for 2.00 min, and finally travels northwest at 30.0 m/s for 1.00 min. For this 6.00-min trip, find (a) the total vector displacement, (b) the average speed and (c) the average velocity. Let the positive x axis point east.
6. ±? When the Sun is directly overhead, a hawk dives towards the ground with a constant velocity of 5.0 ± 0.5 m/s at $60° \pm 2°$ below the horizontal. Calculate the speed of its shadow on the level ground.
7. A golf ball is hit off a tee at the edge of a cliff. Its x and y coordinates as functions of time are given by $x = 18.0t$ and $y = 4.00t - 4.90t^2$, where x and y are in metres and t is in seconds. (a) Write a vector expression for the ball's position as a function of time, using the unit vectors $\hat{\mathbf{i}}$ and $\hat{\mathbf{j}}$. By taking derivatives, obtain expressions for (b) the velocity vector $\vec{\mathbf{v}}$ as a function of time and (c) the acceleration vector $\vec{\mathbf{a}}$ as a function of time. (d) Next, use unit-vector notation to write expressions for the position, the velocity, and the acceleration of the golf ball at $t = 3.00$ s.

Section 3.3 Two-dimensional motion with constant acceleration

8. The vector position of a particle varies in time according to the expression $\vec{\mathbf{r}} = 3.00\hat{\mathbf{i}} - 6.00t^2\hat{\mathbf{j}}$, where $\vec{\mathbf{r}}$ is in metres and t is in seconds. (a) Find an expression for the velocity of the particle as a function of time. (b) Determine the acceleration of the particle as a function of time. (c) Calculate the particle's position and velocity at $t = 1.00$ s.

9. A fish swimming in a horizontal plane has velocity $\vec{\mathbf{v}}_i = (4.00\,\hat{\mathbf{i}} + 1.00\,\hat{\mathbf{j}})$ m/s at a point in the ocean where its position relative to a certain rock is $\vec{\mathbf{r}} = (10.0\,\hat{\mathbf{i}} - 4.00\,\hat{\mathbf{j}})$ m. After the fish swims with constant acceleration for 20.0 s, its velocity is $\vec{\mathbf{v}} = (20.0\,\hat{\mathbf{i}} - 5.00\,\hat{\mathbf{j}})$ m/s. (a) What are the components of the acceleration of the fish? (b) What is the direction of its acceleration with respect to unit vector $\hat{\mathbf{i}}$? (c) If the fish maintains constant acceleration, where is it at $t = 25.0$ s and in what direction is it moving?

10 **Review.** A snowmobile is originally at the point with position vector 29.0 m at 95.0° anticlockwise from the x axis, moving with velocity 4.50 m/s at 40.0°. It moves with constant acceleration 1.90 m/s^2 at 200°. After 5.00 s have elapsed, find (a) its velocity and (b) its position vector.

Section 3.4 Projectile motion

Note: Ignore air resistance in all problems and take $g = 9.80$ m/s^2 at the Earth's surface.

11. In a local bar, a customer slides an empty beer mug down the counter for a refill. The height of the counter is 1.22 m. The mug slides off the counter and strikes the floor 1.40 m from the base of the counter. (a) With what velocity did the mug leave the counter? (b) What was the direction of the mug's velocity just before it hit the floor?

12. A projectile is fired in such a way that its horizontal range is equal to three times its maximum height. What is the angle of projection?

13. Dolphins are able to move through water especially fast by jumping out of the water periodically. This behaviour is called *porpoising*. Suppose a dolphin swimming in still water jumps out of the water with velocity 6.26 m/s at 45.0° above the horizontal, sails through the air a distance L before returning to the water, and then swims the same distance L underwater in a straight, horizontal line with velocity 3.58 m/s before jumping out again. (a) Determine the average velocity of the dolphin for the entire process of jumping and swimming underwater. (b) Consider the time interval required to travel the entire distance of $2L$. By what percentage is this time interval reduced by the jumping–swimming process compared with simply swimming underwater at 3.58 m/s?

Figure P3.13

14. ±? A landscape architect is planning an artificial waterfall in a city park. Water flowing at (1.7 ± 0.2) m/s will leave the end of a horizontal channel at the top of a vertical wall h = (2.35 ± 0.01) m high, and from there it will fall into a pool (Figure P3.14). (a) Will the space behind the waterfall be wide enough for a pedestrian walkway? (b) To sell her plan to the city council, the architect wants to build a model to standard scale, which is one-twelfth actual size. How fast should the water flow in the channel in the model?

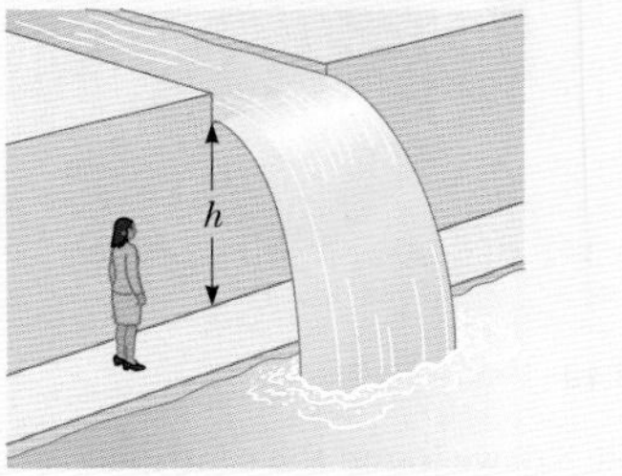

Figure P3.14

15. To start an avalanche on a mountain slope, an artillery shell is fired with an initial velocity of 300 m/s at 55.0° above the horizontal. It explodes on the mountainside 42.0 s after firing. What are the x and y coordinates of the shell where it explodes, relative to its firing point?

16. The record distance in the sport of throwing cowpats is 81.1 m. This record toss was set by Steve Uren of the United States in 1981. Assuming the initial launch angle was 45° and neglecting air resistance, determine (a) the initial speed of the projectile and (b) the total time interval the projectile was in flight. (c) How would the answers change if the range were the same but the launch angle were greater than 45°? Explain your answer.

17. A student tosses a coin into a large wishing well and wishes for good marks in physics. The coin is thrown from a height of 2.50 m above the water surface with a velocity of 4.00 m/s at an angle of 60.0° above the horizontal. As the coin strikes the water surface, it immediately slows down to exactly half the speed it had when it struck the water and maintains that speed while in the water. After the coin enters the water, it moves in a straight line in the direction of the velocity it had when it struck the water. If the well is 3.00 m deep, how much time elapses between when the coin is thrown and when it strikes the bottom of the well?

Section 3.5 Analysis model: particle in uniform circular motion

18. Paul and Darren are riding the Curve Ball shown in Figure 3.23 as Kate and Anna watch. Paul and Darren sit at the ends of the arms of the Curve Ball, 3.5 m apart. Kate times how long it takes for Darren to go past her five times, and measures a time of 17.5 s. (a) Find the velocity of Paul and Darren. (b) Find their centripetal acceleration. (c) What is the maximum velocity that the Curve Ball can spin Paul and Darren if they are not to lose consciousness? This typically happens when people are subjected to more than about $4g$ (4 times the acceleration due to gravity).

19. The astronaut orbiting the Earth in Figure P3.19 is preparing to dock with a Westar VI satellite. The satellite is in a circular orbit 600 km above the Earth's surface, where the free-fall acceleration is 8.21 m/s^2. Take the radius of the Earth as 6400 km. Determine the speed of the satellite and the time interval required to complete one orbit around the Earth, which is the period of the satellite.

NASA

Figure P3.19

20. In Example 3.6, we found the centripetal acceleration of the Earth as it revolves around the Sun. From information on the endpapers of this book, calculate the centripetal acceleration of a point on the surface of the Earth at the equator caused by the rotation of the Earth about its axis.

21. ±? The Australian Synchrotron, located in Melbourne, has a ring-shaped tunnel in which electrons travel at speeds close to the speed of light. The ring has a circumference of 130.2 m and the approximately circular beam of electrons that travels around it has a diameter of only a few millimetres. What *range* of centripetal accelerations can an electron travelling at 299 000 km/h have and still remain part of the beam of electrons circulating in the tunnel? Write your answer in the form acceleration ± uncertainty, and remember to give units and use the correct number of significant figures.

Photo courtesy of Australian Synchrotron

Figure P3.21

Section 3.6 Non-uniform circular motion: tangential and radial acceleration

22. A train slows down as it rounds a sharp horizontal turn, going from 90.0 km/h to 50.0 km/h in the 15.0 s it takes to round the bend. The radius of the curve is 150 m. Calculate the acceleration at the moment the train speed reaches 50.0 km/h. Assume the train continues to slow down at this time at the same rate.

23. Figure P3.23 represents the total acceleration of a particle moving clockwise in a circle of radius 2.50 m at a certain instant of time. For that instant, find (a) the radial acceleration of the particle, (b) the speed of the particle, and (c) its tangential acceleration.

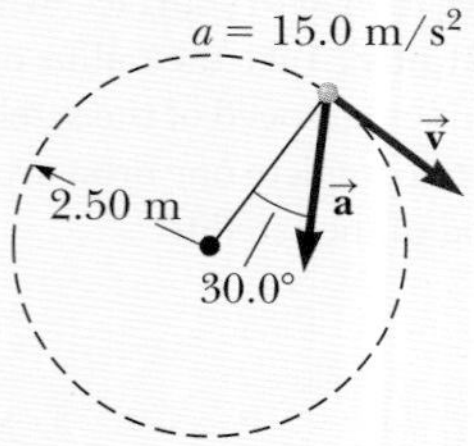

Figure P3.23

24. A poi (ball on a cord) is used in New Zealand in traditional Maori dances, as shown in Figure 3.24. It is also used for keeping hands flexible and strong, and improving coordination. In a particular Maori poi dance, a woman is swinging a poi such that it swings in a vertical circle at the end of a rope 50 cm long. When the ball is 39° past the lowest point on its way up, its total acceleration is $(-22.5\,\hat{\mathbf{i}} + 20.2\,\hat{\mathbf{j}})$ m/s². For that instant, (a) sketch a vector diagram showing the components of its acceleration, (b) determine the magnitude of its radial acceleration and (c) determine the speed and velocity of the poi.

Alamy/Photo Resource Hawaii

Figure P3.24

Section 3.7 Relative velocity and relative acceleration

25. A police car travelling at 95.0 km/h is travelling west, chasing a motorist travelling at 80.0 km/h. (a) What is the velocity of the motorist relative to the police car? (b) What is the velocity of the police car relative to the motorist? (c) If they are originally 250 m apart, in what time interval will the police car overtake the motorist?

26. ±? An aeroplane maintains a speed between 630 km/h and 660 km/h relative to the air it is flying through as it flies north from Wellington to Auckland, 460 km away. (a) What time interval is required for the trip if the plane flies through a headwind blowing at (35 ± 2) km/h toward the south? (b) What time interval is required if there is a tailwind with the same speed? (c) What time interval is required if there is a crosswind blowing at (35 ± 2) km/h to the east relative to the ground?

27. A car travels due east with a speed of 50.0 km/h. Raindrops are falling at a constant speed vertically with respect to the Earth. The traces of the rain on the side windows of the car make an angle of 60.0° with the vertical. Find the velocity of the rain with respect to (a) the car and (b) the Earth.

28. A science student is riding on a flatcar of a train traveling along a straight, horizontal track at a constant speed of 10.0 m/s. The student throws a ball into the air along a path that he judges to make an initial angle of 60.0° with the horizontal and to be in line with the track. The student's lecturer, who is standing on the ground nearby, observes the ball to rise vertically. How high does she see the ball rise?

Additional problems

29. The 'Vomit Comet'. In microgravity astronaut training and equipment testing, NASA flies a KC135A aircraft along a parabolic flight path. As shown in Figure P3.29, the aircraft climbs from 24 000 ft to 31 000 ft, where it enters a parabolic path with a velocity of 143 m/s nose high at 45.0° and exits with velocity 143 m/s at 45.0° nose low. During this portion of the flight, the aircraft and objects inside its padded cabin are in free fall; astronauts and equipment float freely as if there were no gravity. What are the aircraft's (a) speed and (b) altitude at the top of the manoeuvre? (c) What is the time interval spent in microgravity?

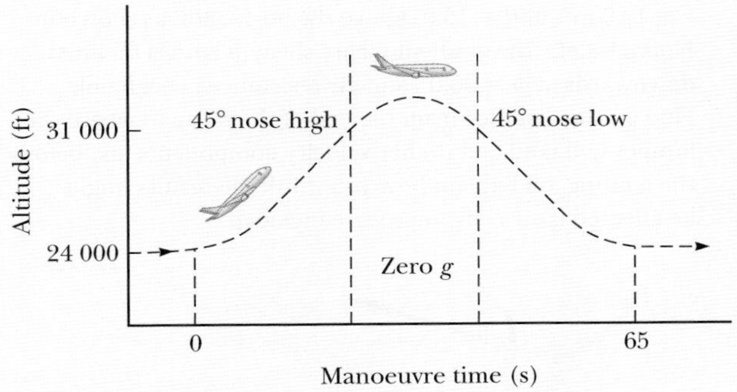

Figure P3.29

30. Lisa in her Lamborghini accelerates at the rate of $(3.00\,\hat{\mathbf{i}} - 2.00\,\hat{\mathbf{j}})$ m/s², while Jill in her Jaguar accelerates at $(1.00\,\hat{\mathbf{i}} + 3.00\,\hat{\mathbf{j}})$ m/s². They both start from rest at the origin of an xy coordinate system. After 5.00 s, (a) what is Lisa's speed with respect to Jill, (b) how far apart are they, and (c) what is Lisa's acceleration relative to Jill?

31. Anna and Kate are riding the Curve Ball at Questacon, as shown in Figure 3.23. They are seated 3.5 m apart at the ends of the two arms of the Curve Ball. Paul, who is standing at the side, measures the time taken for Anna go past him five times as 17.5 s. (a) If Kate is at point A when she drops a ball from 1.0 m above the floor, how far from the release point does the ball land? (b) How far from the centre of the Curve Ball does it land? Draw a diagram showing where it lands.

32. ±? Anna and Kate are riding the Curve Ball at Questacon again. Both are travelling at speed v, in a circle of radius r centred about the same point, and directly opposite each other. Paul, who is standing to the side and watching them, sees Anna throw the ball directly upwards. (a) With what horizontal speed has Anna thrown the ball? (b) Kate catches the ball as she swings around, coming directly beneath it. With what vertical speed has Anna thrown the ball? Write your answer in terms of v and r. (c) Write the velocity of the ball as it leaves Anna's hands as a vector.

33. A truck loaded with cannonball watermelons stops suddenly to avoid running over the edge of a washed-out bridge (Figure P3.33). The quick stop causes a number of melons to fly off the truck. One melon leaves the hood of the truck with an initial speed v_i = 10.0 m/s in the horizontal direction. A cross section of the bank has the shape of the bottom half of a parabola, with its vertex at the initial location of the projected watermelon and with the equation $y^2 = 16x$, where x and y are measured in meters. What are the x and y coordinates of the melon when it splatters on the bank?

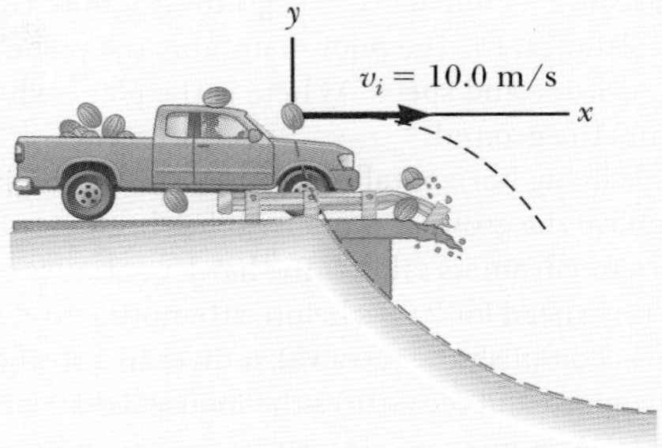

Figure P3.33

34. A pendulum with a cord of length r = 1.00 m swings in a vertical plane (Figure P3.34). When the pendulum is in the two horizontal positions θ = 90.0° and θ = 270°, its speed is 5.00 m/s. Find the magnitude of (a) the radial acceleration and (b) the tangential acceleration for these positions. (c) Draw vector diagrams to determine the direction of the total acceleration for these two positions. (d) Calculate the magnitude and direction of the total acceleration at these two positions.

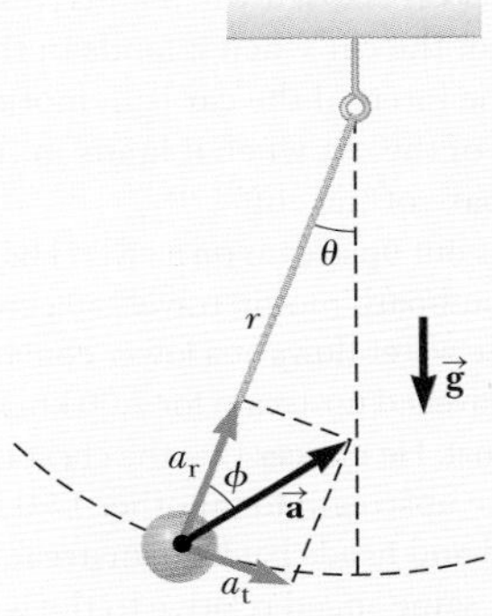

Figure P3.34

35. An astronaut on the surface of the Moon fires a cannon to launch an experiment package, which leaves the barrel moving horizontally. Assume the free-fall acceleration on the Moon is one-sixth of that on the Earth. (a) What must the muzzle speed of the package be so that it travels completely around the Moon and returns to its original location? (b) What time interval does this trip around the Moon require?

36. A projectile is launched from the point (x = 0, y = 0), with velocity $(12.0\,\hat{\mathbf{i}} + 49.0\,\hat{\mathbf{j}})$ m/s, at t = 0. (a) Make a table listing the projectile's distance $|\vec{\mathbf{r}}|$ from the origin at the end of each second thereafter for $0 \leq t \leq 10$ s. Tabulating the x and y coordinates and the components of velocity v_x and v will also be useful. (b) Notice that the projectile's distance from its starting point increases with time, goes through a maximum, and starts to decrease. Prove that the distance is a maximum when the position vector is perpendicular to the velocity. *Suggestion:* Argue that if $\vec{\mathbf{v}}$ is not perpendicular to $\vec{\mathbf{r}}$, then $|\vec{\mathbf{r}}|$ must be increasing or decreasing. (c) Determine the magnitude of the maximum displacement. (d) Explain your method for solving part (c).

37. ±? A Second World War bomber flies horizontally over level terrain with a speed of 275 m/s relative to the ground and at an altitude of 3.00 km. The bomber's altimeter has an uncertainty of 10 m and its speedometer has an uncertainty of 5%. (a) How far does the bomb travel horizontally between its release and its impact on the ground? Ignore the effects of air resistance. (b) The pilot maintains the plane's original course, altitude and speed. Where is the plane when the bomb hits the ground?

38. A hawk is flying horizontally at 10.0 m/s in a straight line, 200 m above the ground. A mouse it has been carrying struggles free from its talons. The hawk continues on its path at the same speed for 2.00 s before attempting to retrieve its prey. To accomplish the retrieval, it dives in a straight line at constant speed and recaptures the mouse 3.00 m above the ground. (a) Assuming no air resistance acts on the mouse, find the diving speed of the hawk. (b) What angle did the hawk make with the horizontal during its descent? (c) For what time interval did the mouse experience free fall?

39. A car is parked on a steep incline, making an angle of 37.0° below the horizontal and overlooking the ocean, when its brakes fail and it begins to roll. Starting from rest at $t = 0$, the car rolls down the incline with a constant acceleration of 4.00 m/s^2, travelling 50.0 m to the edge of a vertical cliff. The cliff is 30.0 m above the ocean. Find (a) the speed of the car when it reaches the edge of the cliff, (b) the time interval elapsed when it arrives there, (c) the velocity of the car when it lands in the ocean, (d) the total time interval the car is in motion, and (e) the position of the car when it lands in the ocean, relative to the base of the cliff.

40. A fisherman sets out upstream on a river. His small boat, powered by an outboard motor, travels at a constant speed v in still water. The water flows at a lower constant speed v_w. The fisherman has travelled upstream for 2.00 km when his ice chest falls out of the boat. He notices that the chest is missing only after he has gone upstream for another 15.0 min. At that point, he turns around and heads back downstream, all the time travelling at the same speed relative to the water. He catches up with the floating ice chest just as he returns to his starting point. How fast is the river flowing? Solve this problem in two ways. (a) First, use the Earth as a reference frame. With respect to the Earth, the boat travels upstream at speed $v - v_w$ and downstream at $v + v_w$. (b) A second much simpler and more elegant solution is obtained by using the water as the reference frame. This approach has important applications in many more complicated problems; examples are calculating the motion of rockets and satellites and analysing the scattering of subatomic particles from massive targets.

Challenge problems

41. Two swimmers, Chris and Sarah, start together at the same point on the bank of a wide river that flows with a speed v. Both move at the same speed c (where $c > v$) relative to the water. Chris swims downstream a distance L and then upstream the same distance. Sarah swims so that her motion relative to the Earth is perpendicular to the banks of the stream. She swims the distance L and then back the same distance, with both swimmers returning to the starting point. In terms of L, c and v, find the time intervals required (a) for Chris's round trip and (b) for Sarah's round trip. (c) Explain which swimmer returns first.

42. A skier leaves the ramp of a ski jump with a velocity of $v = 10.0$ m/s at $\theta = 15.0°$ above the horizontal as shown in Figure P3.42. The gradient where she will land is inclined downwards at $\phi = 50.0°$, and air resistance is negligible. Find (a) the distance from the end of the ramp to where the jumper will land and (b) her velocity components just before the landing. (c) Explain how you think the results might be affected if air resistance were included.

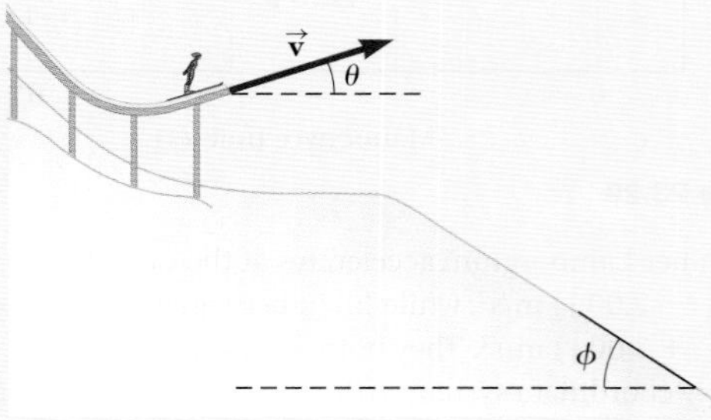

Figure P3.42

43. A fireworks rocket explodes at height h, the peak of its vertical trajectory. It throws out burning fragments in all directions, but all at the same speed v. Pellets of solidified metal fall to the ground without air resistance. Find the smallest angle that the final velocity of an impacting fragment makes with the horizontal.

44. An enemy ship is on the east side of a mountain island as shown in Figure P3.44. The enemy ship has manoeuvred to within 2500 m of the 1800-m-high mountain peak and can shoot projectiles with an initial speed of 250 m/s. If the western shoreline is horizontally 300 m from the peak, what are the distances from the western shore at which a ship can be safe from the bombardment of the enemy ship?

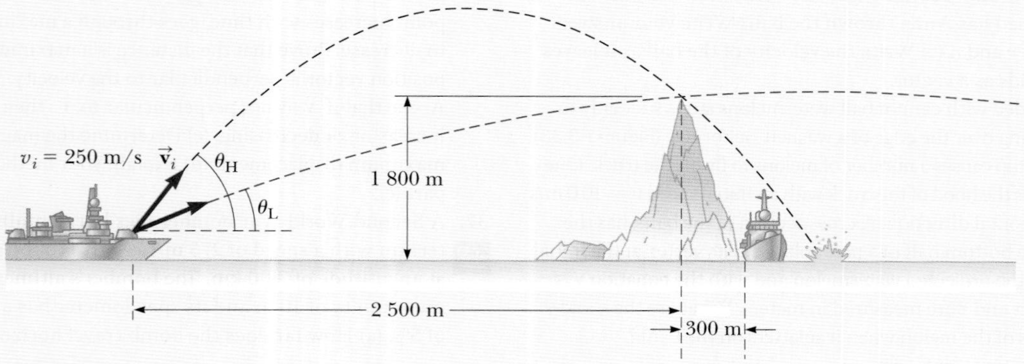

Figure P3.44

chapter 4

Forces and Newton's laws

The image shows astronauts floating, apparently weightless, in the International Space Station (ISS). The ISS orbits in a nearly circular orbit with an altitude of approximately 300 km from the surface of the Earth – a small fraction of the Earth's radius. What forces act on an astronaut in the ISS and are they actually weightless? If not, why do they float as if there were no gravity there?

On completing this chapter, students will understand:
- that forces cause accelerations and hence change the motion of objects
- the difference between inertial and non-inertial reference frames
- how the acceleration of an object due to an applied force is determined by its mass
- that mass is an intrinsic property of objects and weight is a force due to gravity
- that every interaction involves an equal force on each of the objects interacting
- why accelerations in non-inertial reference frames appear to be due to pseudoforces.

Students will be able to:
- describe the effects of forces, qualitatively and quantitatively
- identify and distinguish between inertial and non-inertial reference frames
- apply Newton's second law to solve problems involving forces
- draw and use free-body diagrams to analyse the effects of forces
- identify action–reaction force pairs by applying Newton's third law
- solve problems by applying analysis models using Newton's second law
- solve problems involving reference frames that are accelerating in one dimension
- identify and describe pseudoforces, including the 'centrifugal' force.

In Chapters 2 and 3, we *described* the motion of an object in terms of its position, velocity and acceleration without considering what might *influence* that motion. Now we consider that influence. The two main factors we need to consider are the *forces* acting on an object and the *mass* of the object.

In this chapter, we begin our study of *dynamics* by discussing the three basic laws of motion, which deal with forces and masses and were formulated more than three centuries ago by Isaac Newton.

ENHANCED WebAssign

The problems found in this chapter may be assigned online in Enhanced Web Assign.

Isaac Newton

English physicist and mathematician (1642–1727)

Isaac Newton was one of the most brilliant scientists in history. Before the age of 30, he formulated the basic concepts and laws of mechanics, discovered the law of universal gravitation, and contributed to the mathematical methods of calculus. As a consequence of his theories, Newton was able to explain the motions of the planets, the ebb and flow of the tides, and many special features of the motions of the Moon and the Earth.

4.1 The concept of force

Everyone has a basic understanding of the concept of force from everyday experience. When you push your empty dinner plate away, you exert a force on it. Similarly, you exert a force on a ball when you throw, kick or catch it. In these examples, the word *force* refers to an interaction with an object by means of muscular activity and some change in the object's velocity. Forces do not always cause motion, however. For example, when you are sitting, a gravitational force acts on your body and yet you remain stationary. As a second example, you can push (in other words, exert a force) on a large boulder and not be able to move it.

What force (if any) causes the Moon to orbit the Earth? Newton answered this, and related questions, by stating that any change in the velocity of an object is the result of a force. The Moon's velocity changes in direction as it moves in a nearly circular orbit around the Earth. This change in velocity is caused by the gravitational force exerted by the Earth on the Moon.

When a coiled spring is pulled, as in **Figure 4.1a**, the spring stretches. When a stationary cart is pulled, as in **Figure 4.1b**, the cart moves. When a football is kicked, as in **Figure 4.1c**, it is both deformed and set in motion. These situations are all examples of a class of forces called *contact forces*. That is, they involve physical contact between two objects. Other examples of contact forces are the force exerted by gas molecules on the walls of a container and the force exerted by your feet on the floor.

Another class of forces, known as *field forces*, does not involve physical contact between two objects. These forces can act through empty space. The gravitational force of attraction between two objects with mass, as in **Figure 4.1d**, is an example of this class of force. The gravitational force keeps objects bound to the Earth and keeps the planets in orbit around the Sun. Another common field force is the electric force that one electric charge exerts on another (**Figure 4.1e**), such as the charges of an electron and proton that are part of a hydrogen atom. A third example of a field force is the force a bar magnet exerts on a piece of iron (**Figure 4.1f**).

The distinction between contact forces and field forces is convenient on a macroscopic level, but when examined at the atomic level, all the forces we classify as contact forces turn out to be caused by electric (field) forces of the type illustrated in **Figure 4.1e**.

The fundamental forces ►

The only known *fundamental* forces in nature are all field forces: (1) *gravitational forces* between objects, (2) *electromagnetic forces* between charged objects, (3) *strong forces* between subatomic particles, and (4) *weak forces* that arise in certain radioactive decay processes.

You are subject to the gravitational force all the time, and all the contact forces you experience are ultimately due to the electromagnetic force. So we will focus on these two forces for most of this book. The normal and friction forces, which are components of the contact force between solid surfaces, are discussed in detail in Chapter 5. The gravitational force is discussed in detail in Chapter 11. The strong and weak forces and their role in interactions at the atomic and nuclear level are described in Part 7.

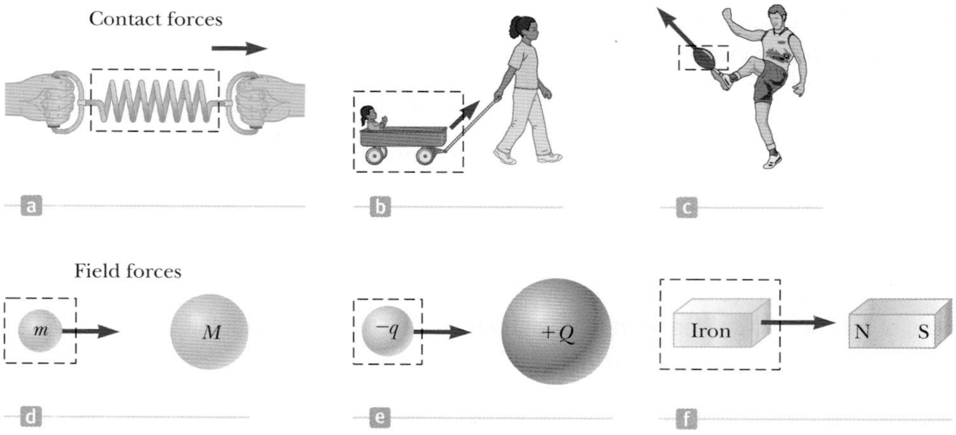

Figure 4.1
Some examples of applied forces. In each case, a force is exerted on the object within the boxed area. Some agent in the environment external to the boxed area exerts a force on the object.

The vector nature of force

Corbis/Colin Hawkins/cultura

We know from experience that if you apply a force to move an object it is not only the size of the force that is important, but the direction as well. Imagine two people pushing a trolley. They need to apply a force in the same direction to effectively move the trolley in a straight line. If you want to move a book to the left on a table, you need to apply a force to the left. Hence, to describe a force completely and understand its effects on motion we need to consider both the magnitude and direction of the force. Hence forces, like positions, velocities and accelerations, are vector quantities. When we do calculations with forces we need to remember that forces are vectors and apply the rules for manipulating vectors.

You may want to review Appendix B.6 before proceeding if you are not confident with vectors. Σ

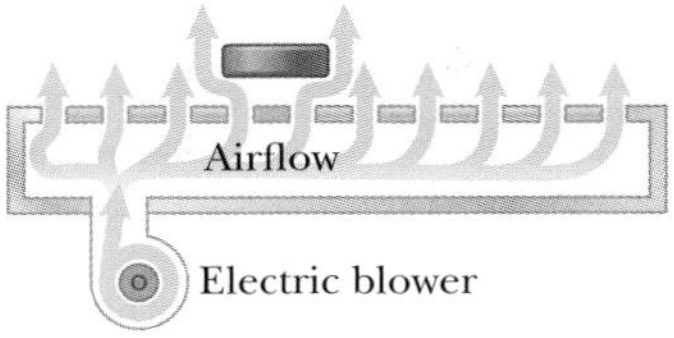

Figure 4.2
On an air hockey table, air blown through holes in the surface allows the puck to move almost without friction. If the table is not accelerating, a puck placed on the table will remain at rest.

4.2 Newton's first law and inertial frames

We begin our study of forces by imagining some physical situations involving a puck on a perfectly level air hockey table (**Figure 4.2**). You expect that the puck will remain stationary when it is placed gently at rest on the table. Now imagine your air hockey table is located on a train moving with constant velocity along a perfectly smooth track. If the puck is placed on the table, the puck again remains where it is placed. If the train were to accelerate, however, the puck would start moving along the table opposite the direction of the train's acceleration, just as an object on your dashboard falls onto the floor of your car when you accelerate rapidly.

As we saw in Section 3.7, a moving object can be observed from any number of reference frames. **Newton's first law of motion,** sometimes called the *law of inertia*, defines a special set of reference frames called *inertial frames*. This law can be stated as follows:

◀ Newton's first law

> If an object does not interact with other objects, it is possible to identify a reference frame in which the object has zero acceleration.

◀ Inertial frame of reference

Such a reference frame is called an **inertial frame of reference**. When the puck is on the air hockey table located on the ground, you are observing it from an inertial reference frame; there are no horizontal interactions of the puck with any other objects, and you observe it to have zero acceleration in that direction. Note that in the vertical direction the puck is subject to two external forces, the downwards gravitational force of the Earth and the upwards force of the air. This aspect of the situation is described by Newton's second law, which is discussed in Section 4.5. When you are on the train moving at constant velocity, you are also observing the puck from an inertial reference frame. Any reference frame that moves with constant velocity relative to an inertial frame is itself an inertial frame. When you and the train accelerate, however, you are observing the puck from a **non-inertial reference frame** because the train is accelerating relative to the inertial reference frame of the Earth's surface. While the puck appears to be accelerating according to your observations, a reference frame can be identified in which the puck has zero acceleration. For example, because there is almost no friction to hold the puck and the train together, an observer standing outside the train on the ground sees the puck sliding relative to the table but always moving with the same velocity with respect to the ground as the train had before it started to accelerate. Therefore, Newton's first law is still satisfied even though your observations as a rider on the train show an apparent acceleration relative to you.

A reference frame that moves with constant velocity relative to the distant stars is a good approximation of an inertial frame, and for our purposes we can consider the Earth as being such a frame. The Earth is not really an inertial frame because of its orbital motion around the Sun and its rotational motion about its own axis, both of which involve centripetal accelerations. These accelerations are small compared with g, however, and can often be neglected. For this reason, we can usually model the Earth as an inertial frame, along with any other frame attached to it. Inertial and non-inertial frames are defined and discussed in more detail when we look at relativity in Chapter 12.

Pitfall Prevention 4.1

Newton's first law does *not* say what happens for an object with *zero net force*, that is, multiple forces that cancel; it says what happens *in the absence of external forces*. This subtle but important difference allows us to define force as that which causes a change in the motion. The description of an object under the effect of forces that balance is covered by Newton's second law.

Pitfall Prevention 4.2

Force is the cause of changes in motion. An object can have motion in the absence of forces as described by Newton's first law. Therefore, don't interpret force as the cause of *motion*. Force is the cause of *changes in motion*.

Let us assume we are observing an object from an inertial reference frame. (We will return to observations made in non-inertial reference frames in Section 4.8.) Before about 1600, scientists believed that the natural state of matter was the state of rest. Observations showed that moving objects eventually stopped moving. Galileo was the first to take a different approach. He concluded that it is not the nature of an object to stop once set in motion: rather, it is its nature to *resist changes in its motion*. In his words, 'Any velocity once imparted to a moving body will be rigidly maintained as long as the external causes of retardation are removed'. For example, a spacecraft drifting through empty space with its engine turned off will keep moving forever. It would *not* seek a 'natural state' of rest.

Given our discussion of observations made from inertial reference frames, we can pose a more practical statement of Newton's first law of motion:

Another statement of Newton's first law ▶

In the absence of external forces and when viewed from an inertial reference frame, an object at rest remains at rest and an object in motion continues in motion with a constant velocity (that is, with a constant speed in a straight line).

In other words, **when no force acts on an object, the acceleration of the object is zero**. From the first law, we conclude that any *isolated object* (one that does not interact with its environment) is either at rest or moving with constant velocity. The tendency of an object to resist any attempt to change its velocity is called **inertia**. Given the statement of the first law above, we can conclude that an object that is accelerating must be experiencing a force. In turn, from the first law, **we can define force as that which causes a change in motion of an object**.

Definition of force ▶

Quick Quiz 4.1

Which of the following statements is correct? (a) It is possible for an object to have motion in the absence of forces on the object. (b) It is possible to have forces on an object in the absence of motion of the object. (c) Neither statement (a) nor statement (b) is correct. (d) Both statements (a) and (b) are correct.

4.3 Mass

Imagine playing catch with either a basketball or a bowling ball. Which ball is more likely to keep moving when you try to catch it? Which ball requires more effort to throw? In the language of physics, we say that the bowling ball is more resistant to changes in its velocity than the basketball. How can we quantify this concept?

Definition of mass ▶

Mass is that property of an object that specifies how much resistance an object exhibits to changes in its velocity. Experiments show that the greater the mass of an object, the less that object accelerates under the action of a given applied force.

To describe mass quantitatively, we conduct experiments in which we compare the accelerations a given force produces on different objects. Suppose a force acting on an object of mass m_1 produces a change in motion of the object that we can quantify with the object's acceleration $\vec{\mathbf{a}}_1$, and the *same force* acting on an object of mass m_2 produces an acceleration $\vec{\mathbf{a}}_2$. The ratio of the two masses is defined as the *inverse* ratio of the magnitudes of the accelerations produced by the force:

$$\frac{m_1}{m_2} \equiv \frac{a_2}{a_1} \tag{4.1}$$

For example, if a given force acting on a 3 kg object produces an acceleration of 4 m/s^2, the same force applied to a 6 kg object produces an acceleration of 2 m/s^2. According to a huge number of similar observations, we conclude that the magnitude of the acceleration of an object is inversely proportional to its mass when acted on by a given force. If one object has a known mass, the mass of the other object can be obtained from acceleration measurements.

Mass is an intrinsic property (sometimes called an inherent property) of an object and is independent of the object's surroundings and of the method used to measure it. Also, mass is a scalar quantity and thus obeys the rules of ordinary arithmetic. For example, if you combine a 3 kg mass with a 5 kg mass, the total mass is 8 kg. This result can be verified experimentally by comparing the acceleration that a known force gives to several objects separately with the acceleration that the same force gives to the same objects combined as one unit.

Mass should not be confused with **weight**. **Mass and weight are two different quantities**. The weight of an object is the gravitational force exerted on the object and varies with location (see Section 4.5). For example, your weight on the Moon is only around one sixth of your weight on the Earth, but the mass of an object is the same everywhere: a rock having a mass of 2 kg on the Earth also has a mass of 2 kg on the Moon. An astronaut kicking a rock on the Moon will hurt his foot as much as if he kicked it on the Earth. Weight, which is a force, is described in detail in the next section.

◀ Mass and weight are different quantities

4.4 The gravitational force and weight

All objects are attracted to the Earth. The attractive force exerted by the Earth on an object is called the **gravitational force** $\vec{\mathbf{F}}_g$. This force is directed towards the centre of the Earth.

We saw in Section 2.7 that a freely falling object experiences an acceleration $\vec{\mathbf{g}}$ acting towards the centre of the Earth. The gravitational force acting on an object close to the surface of the Earth is given by

$$\vec{\mathbf{F}}_g = m\vec{\mathbf{g}} \qquad (4.2)$$

where $\vec{\mathbf{g}}$ is the acceleration due to gravity.

As we shall see in Chapter 11, when we discuss gravity in detail, the gravitational force at any point depends on all masses in the region. Close to the Earth's surface, the gravitational force is dominated by the Earth. However, close to any other large body, such as the moon or another planet, the gravitational force experienced by an object will depend primarily on the mass of the moon or planet.

Remember also that *forces result from interactions between objects*, hence the gravitational force that acts on you due to the mass of the Earth, also depends on *your* mass. We can see this in **Equation 4.2**, where $\vec{\mathbf{F}}_g$ depends on both the local acceleration due to gravity which in turn depends on the mass of the Earth (see Chapter 11), and on the mass, m, of the body experiencing the force.

Pitfall Prevention 4.3

We are familiar with the everyday phrase, the 'weight of an object'. Weight, however, is not an intrinsic property of an object; it is a measure of the gravitational force between the object and the Earth (or other planet). Therefore, weight is a property of a *system* of items: the object and the Earth.

Pitfall Prevention 4.4

Kilogram is not a unit of weight. Despite popular statements of weights expressed in kilograms, the kilogram is not a unit of *weight*, it is a unit of *mass*. The unit of weight is the newton, N.

We define the weight of an object as the gravitational force acting on the object at a particular position. It is proportional to the mass of the object and the local acceleration due to gravity. It is a force so it has units of newtons (N).

◀ Definition of weight

Because it depends on g, weight varies with geographic location and decreases with increasing distance from the centre of the Earth, so objects weigh less at higher altitudes than at sea level. This difference is generally quite small however; for example, a 1000 kg load of steel used in the construction of the Burj Khalifa in Dubai, the world's tallest building at 828 m, weighed 9800 N at street level, but weighed only about 1 N less by the time it was lifted from ground level to the top of the building, a difference of about 0.01%.

Although this discussion has focused on the gravitational force on an object due to the Earth, the concept is generally valid on any planet. The value of g will vary from one planet to the next, but the magnitude of the gravitational force will always be given by the value of mg.

Note that we have modelled the Earth as having uniform, spherically distributed mass and hence g as being constant over the surface of the Earth. This model is an approximation that is adequate for solving many problems in physics, as variations in g across the surface of the Earth are very small. These small variations can be measured and are used by mining and resource companies to locate large mineral deposits, particularly oxides and sulfites, as g is slightly larger (by a few parts per million) above such deposits. The value of g also increases very slightly towards the poles because the Earth is not perfectly spherical.

Although we have so far focused on weight close to the surface of the Earth, any object with mass will have a weight given by **Equation 4.2** at any position in space, where $\vec{\mathbf{g}}$ is the local acceleration due to gravity. This local value depends on the size and distribution of masses in the region. For example, close to the surface of the moon the acceleration due to gravity is approximately one sixth that on Earth, so the weight of any object on the moon is approximately one sixth its weight on Earth. Jupiter has more than 300 times the mass of the Earth, but the local acceleration due to gravity on its surface is only around two and a half times that on Earth. As we shall see in Chapter 11, the gravitational force depends not only on the masses but also the

Figure 4.3
The life-support unit strapped to the back of astronaut Harrison Schmitt on the Moon had a mass of 136 kg. During his training on Earth, a mock-up with a mass of 23 kg was used. Although this strategy effectively simulated the reduced *weight* the unit would have on the Moon, it did not correctly mimic the unchanging *mass*.

distances involved. Hence an object with a mass of 1 kg on Earth has a weight of 9.8 N on Earth, and a mass of 1 kg and a weight of approximately 25 N on Jupiter. This is because while Jupiter is much more massive than the Earth, it is also much larger so the mass is more spread out.

Quick Quiz 4.2

Harrison Schmitt, pictured on the Moon in **Figure 4.3** carrying a 136 kg life-support unit, trained on Earth with a 23 kg life-support unit. Which of the following would be correctly simulated by the lighter unit on Earth? **(a)** Lifting the unit vertically upwards. **(b)** Pushing the unit horizontally on a low-friction surface. **(c)** Catching the unit when it was thrown to him.

Weight, apparent weight, scale forces and weightlessness

When you accelerate upwards or downwards in a lift, you feel an *apparent* change in your weight. This is because to make you accelerate upwards, the floor of the lift has to push upwards on your feet with more force than the downwards gravitational pull of the Earth. It is this large normal force on the bottoms of your feet that you feel as your *apparent weight*, making you feel heavier. However, your actual weight, the force exerted on you by the local gravitational field, changes negligibly in a lift, and neither has your mass changed.

apparent weight and scale force ▶

Your **apparent weight** is the force exerted upwards on you by the floor or any other supporting structure. A weighing scale measures this apparent weight. Hence this force is also sometimes referred to as a **scale force**.

Under some specific conditions this 'scale force', the force exerted on you by a set of scales that you stand on, is equal to your weight. However, in many situations, such as if you are accelerating upwards or downwards, if there is another force acting on you, or even if the surface on which the scale sits is not horizontal, then the scale force *is not* equal to your weight. Even the buoyant effects of the air around you (see Chapter 13), and the rotation of the Earth, have a small effect on the scale force, so that the gravitational force acting on an object (its weight) and the scale force are rarely precisely the same.

It is tempting to think of the force that a set of scales applies to you as being the same as your weight, and unfortunately the scale force is sometimes given as an 'operational' definition of weight, because it is how we often measure weight, as described in Section 4.7. However, **it is not the same**. Weight is the gravitational force exerted on you because of the gravitational interaction between you and other masses. The scale force is the normal force, which is electrostatic in nature, exerted by the scales on you. These are numerically equal when the scale is correctly calibrated, placed on a flat surface, there is no vertical acceleration and no other vertical forces act on the object on the scales. **Figure 4.4** shows how the scale force or apparent weight varies under different conditions.

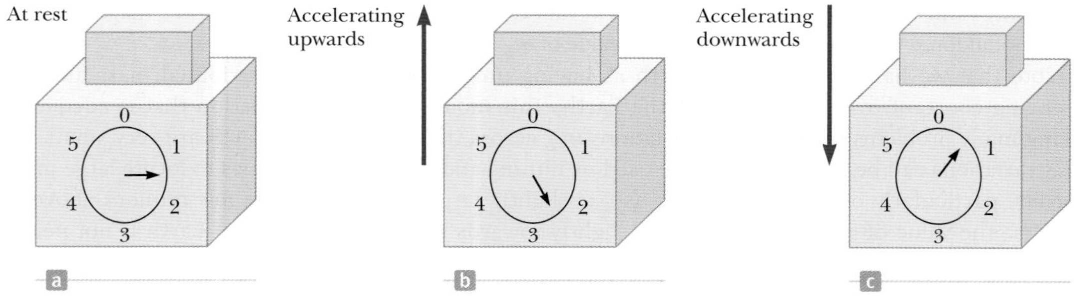

Figure 4.4
A scale is calibrated to read the mass of an object placed on top of it. (a) When at rest with no other forces acting, the scale exerts a force equal to the gravitational force acting on the object (the weight of the object) and the reading is equal to the mass of the object. (b) When accelerating upwards in a lift, the scale reading increases because it exerts a greater force than the gravitational force, and it reads a mass greater than the actual mass of the object. (c) When accelerating downwards the scale exerts a force less than the gravitational force on the object, and the mass reading is hence lower.

You can experience *apparent* weightlessness by reducing the contact force pushing up on you by the ground or floor to zero. This is called free fall, and happens whenever an object is allowed to fall subject only to the gravitational force. Skydiving is a good approximation to free fall (before the parachute is opened) particularly in the first part of the dive as air resistance is small. We feel 'weightless' when the only force acting on us *is* weight! You cannot actually be weightless unless you are either massless, or the local acceleration due to gravity is zero – that is, there is no gravitational force acting at your position.

The term weight is also used in engineering to refer to the load that an object places on a support. This is again the normal force that the object exerts on the support, like the scale force. In the absence of other vertical forces acting on the object and when the support and object are stationary, the load is equal to the gravitational force acting on the object (its weight). This usage can be particularly confusing when the load is held up by multiple supports. In this case the sum of the supporting forces is equal to the weight of the object, and this is often referred to as the 'weight' being 'spread' or distributed over the supports. However, the weight of the object is the gravitational force acting on the object, which is not necessarily the same as the load acting on any given support.

In general we will avoid the use of the term weight and use the term gravitational force to avoid confusion.

Pitfall Prevention 4.5

Weight, apparent weight, scale force and apparent weightlessness

Weight is the force that acts on an object due to the local gravitational field. It does not depend on the state of motion of the object. Apparent weight is the force acting on an object due to the surface supporting it, called a scale force if the supporting object is a set of scales. Apparent weight (and hence scale force) depends on the acceleration of the object, and on any other vertical forces acting on it. Apparent weightlessness occurs when there is no support exerting a normal force on an object. You **cannot actually be weightless** in a gravitational field unless you have zero mass.

TRY THIS

Place a bathroom scale next to a table and stand on the scale. Now push down on the top of the table. Has your mass changed? Has your weight changed? Has the reading on the scale changed? What does a bathroom scale actually measure? Experiment with pushing and pulling on the table. What do you do to increase the reading on the scale? What do you do to decrease it?

TRY THIS TOO

Get a bathroom scale and stand on it as you go up and down in a lift. An older style analogue scale with a moving needle works better than a digital one. What happens to the scale as you start to move up? As you start to move down? What do you feel?

Conceptual Example 4.1

How much do you weigh in a lift?

If you stand in a lift that is accelerating upwards you feel heavier. If you do the *Try this* experiment above you will find that the scale measures an increased weight. Are you heavier?

Solution

No; your weight is unchanged. Your experiences are due to your being in a non-inertial reference frame. To provide the acceleration upwards, the floor or scale must exert on your feet an upwards force that is greater in magnitude than your weight. It is this greater force you feel, which your body interprets as feeling heavier. The scale reads this upwards force, not your weight, and so its reading increases. This accelerating situation is an example of when the scale force is **not** equal to the weight of the object on the scale, and hence why we do not use the scale force as a definition of weight. If you did the previous *Try this* example of standing on a scale and pushing up or down on a table you would have noted that in this situation also your 'apparent weight' changes, so again the scale force is not a true measure of the weight force in this situation.

4.5 Newton's second law

Pitfall Prevention 4.6
An object can have motion in the absence of forces as described in Newton's first law. Therefore, don't interpret force as the *cause* of *motion*. Force is the cause of *changes* in motion as measured by acceleration.

Newton's first law explains what happens to an object when no forces act on it: it either remains at rest or moves in a straight line with constant speed. Newton's second law answers the question of what happens to an object when one or more forces act on it.

Imagine performing an experiment in which you push a block of mass m across a frictionless, horizontal surface. When you exert some horizontal force $\vec{\mathbf{F}}$ on the block, it moves with some acceleration $\vec{\mathbf{a}}$. If you apply a force twice as great on the same block, experimental results show that the acceleration of the block doubles; if you increase the applied force to $3\,\vec{\mathbf{F}}$, the acceleration triples; and so on. From such observations, we conclude that the acceleration of an object is directly proportional to the force acting on it: $\vec{\mathbf{F}} \propto \vec{\mathbf{a}}$. We also know that the magnitude of the acceleration of an object is inversely proportional to its mass: $|\vec{\mathbf{a}}| \propto 1/m$.

These experimental observations are summarised in **Newton's second law**:

When viewed from an inertial reference frame, the acceleration of an object is directly proportional to the net force acting on it and inversely proportional to its mass:

$$\vec{\mathbf{a}} \propto \frac{\sum \vec{\mathbf{F}}}{m}$$

If we choose a proportionality constant of 1, we can relate mass, acceleration and force through the following mathematical statement of Newton's second law:

Newton's second law ▶

$$\sum \vec{\mathbf{F}} = m\vec{\mathbf{a}} \tag{4.3}$$

Pitfall Prevention 4.7
Do *not* include an '$m\vec{\mathbf{a}}$ force' in your analysis of the forces on an object. The product $m\vec{\mathbf{a}}$ is the *net* force acting, and is the vector sum of all the individual forces. It is not another force.

In both the textual and mathematical statements of Newton's second law, we have indicated that the acceleration is due to the *net force* $\sum \vec{\mathbf{F}}$ acting on an object. The **net force** on an object is the vector sum of all forces acting on the object. (We sometimes refer to the net force as the *total force*, or the *resultant force*.) In solving a problem using Newton's second law, it is important to determine the correct net force on an object. **Many forces may be acting on an object, but there is only one acceleration.**

Note that Equation 4.3 is an approximation and is valid only when the speed of the object is much less than the speed of light. In this limit, that speeds are small compared to the speed of light, Newtonian mechanics is an excellent model that explains and allows us to predict the effect of forces. We treat the relativistic situation in Chapter 12.

Equation 4.3 is a vector expression and hence is equivalent to three component equations:

Newton's second law: component form ▶

$$\sum F_x = ma_x \qquad \sum F_y = ma_y \qquad \sum F_z = ma_z \tag{4.4}$$

Quick Quiz 4.3

An object experiences no acceleration. Which of the following *cannot* be true for the object? **(a)** A single force acts on the object. **(b)** No forces act on the object. **(c)** Forces act on the object, but the forces cancel.

Quick Quiz 4.4

You push an object, initially at rest, across a frictionless floor with a constant force for a time interval Δt, resulting in a final speed of v for the object. You then repeat the experiment, but with a force that is twice as large. What time interval is now required to reach the same final speed v? **(a)** $4\Delta t$ **(b)** $2\Delta t$ **(c)** Δt **(d)** $\Delta t/2$ **(e)** $\Delta t/4$

The SI unit of force is the **newton** (N). A force of 1 N is the force that, when acting on an object of mass 1 kg, produces an acceleration of 1 m/s². From this definition and Newton's second law, we see that the newton can be expressed in terms of the following fundamental units of mass, length and time:

$$1\ \text{N} \equiv 1\ \text{kg.m/s}^2 \qquad (4.5)$$

◀ Definition of the newton

and hence has the dimensions LMT^{-2}.

The mass, m, in Newton's second law (Equation 4.3) is sometimes referred to as the **inertial mass**, as distinct from the **gravitational mass**, m, in the equation for the gravitational force (Equation 4.2). We will not generally distinguish between the two as they have the same value. The conceptual distinction is discussed in Chapter 11 when we look at general relativity.

Conceptual Example 4.2

Forces between carriages in a train

Train carriages are connected by *couplers*, which are under tension as the locomotive pulls the train. Imagine you are on a train speeding up with a constant acceleration. As you move through the train from the locomotive to the last carriage, measuring the tension in each set of couplers, does the tension increase, decrease or stay the same? When the engineer applies the brakes, the couplers are under compression. How does this compression force vary from the locomotive to the last carriage? (Assume only the brakes on the wheels of the locomotive are applied.)

Solution

While the train is speeding up, tension decreases from the front of the train to the back. The coupler between the locomotive and the first carriage must apply enough force to accelerate all of the carriages. The coupler between the first and second carriages needs to apply enough force to accelerate all the carriages except the first carriage. As you move back along the train, each coupler is accelerating less mass behind it. The last coupler has to accelerate only the last carriage, and so it is under the least tension.

When the brakes are applied, the force again decreases from front to back. The coupler connecting the locomotive to the first carriage must apply a large force to slow down the rest of the carriages, but the final coupler must apply a force large enough to slow down only the last carriage.

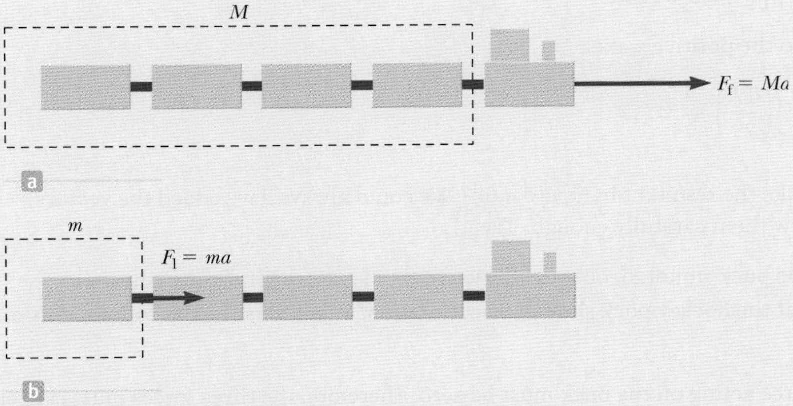

Figure 4.5
The train accelerating forwards. (a) The coupler from the locomotive to the first carriage must provide enough force, F_f, to accelerate the mass, M, of the entire train of carriages. (b) The coupler to the last carriage must only provide the force, F_l, required to accelerate the mass, m, of the last carriage.

Example 4.3

Appendix B.6 shows how to add and decompose vectors.

Σ

An accelerating ice hockey puck

A hockey puck having a mass of 0.30 kg slides on the frictionless, horizontal surface of an ice rink. Two ice hockey sticks strike the puck simultaneously. The first stick exerts a force $\vec{\mathbf{F}}_1$ with a magnitude of 5.0 N, directed at an angle of −20° to the x axis. The second stick exerts a force $\vec{\mathbf{F}}_2$ with a magnitude of 8.0 N at an angle of +60° to the x axis. Determine both the magnitude and the direction of the puck's acceleration.

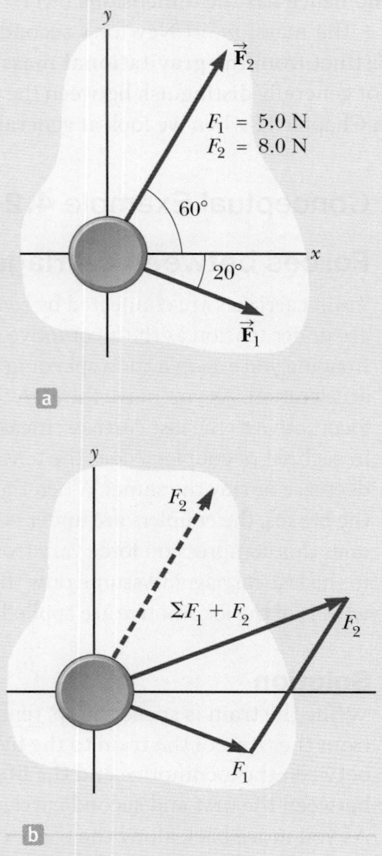

Figure 4.6
(Example 4.3) (a) A hockey puck moving on a frictionless surface is subject to two forces $\vec{\mathbf{F}}_1$ and $\vec{\mathbf{F}}_2$. (b) The vector addition of the forces $\vec{\mathbf{F}}_1$ and $\vec{\mathbf{F}}_2$ to give the resultant net force.

Solution

Conceptualise Draw diagrams showing the forces as vectors and the vector addition to find the net force. The acceleration of the puck will be in the same direction.

Model We make the approximation that the surface is frictionless, and we assume that there is no net force on the puck in the vertical direction. So we can model the puck as accelerating due to the effect of the two forces due to the sticks only, and apply Newton's second law.

Analyse Find the component of the net force acting on the puck in the x direction:

$$\sum F_x = F_{1x} + F_{2x} = F_1 \cos(-20°) + F_2 \cos 60°$$
$$= (5.0\text{ N})(0.940) + (8.0\text{ N})(0.500) = 8.7\text{ N}$$

Find the component of the net force acting on the puck in the y direction:

$$\sum F_y = F_{1y} + F_{2y} = F_1 \sin(-20°) + F_2 \sin 60°$$
$$= (5.0\text{ N})(-0.342) + (8.0\text{ N})(0.866) = 5.2\text{ N}$$

Use Newton's second law in component form (**Equation 4.3**) to find the x and y components of the puck's acceleration:

$$a_x = \frac{\sum F_x}{m} = \frac{8.7\text{ N}}{0.30\text{ kg}} = 29\text{ m/s}^2$$

$$a_y = \frac{\sum F_y}{m} = \frac{5.2\text{ N}}{0.30\text{ kg}} = 17\text{ m/s}^2$$

Find the magnitude of the acceleration:

$$a = \sqrt{(29\text{ m/s}^2)^2 + (17\text{ m/s}^2)^2} = 34\text{ m/s}^2$$

Find the direction of the acceleration relative to the positive x axis:

$$\theta = \tan^{-1}\left(\frac{a_y}{a_x}\right) = \tan^{-1}\left(\frac{17}{29}\right) = 31°$$

Finalise Does our resultant force vector look like the resultant in **Figure 4.6b**? We could always have added the vectors graphically instead of by components if we drew them carefully enough. Try it!

What If? Suppose three hockey sticks strike the puck simultaneously, with two of them exerting the forces shown in **Figure 4.6a**. The result of the three forces is that the hockey puck shows *no* acceleration. What must be the components of the third force?

Answer If there is zero acceleration, the net force acting on the puck must be zero. Therefore, the three forces must cancel. We have found the components of the combination of the first two forces. The components of the third force must be of equal magnitude and opposite sign so that all the components add to zero. Therefore, $F_{3x} = -8.7$ N and $F_{3y} = -5.2$ N.

Example 4.4

The runaway car

A car of mass m is on an icy driveway inclined at an angle θ as in **Figure 4.7**.

(A) Find the acceleration of the car, assuming the driveway is frictionless.

Figure 4.7
(Example 4.4) A car on a frictionless incline

Solution

Conceptualise Use **Figure 4.7** to conceptualise the situation. From everyday experience, we know that a car on an icy incline will accelerate down the incline. The same thing happens to a car on a hill if it is left in gear with the handbrake off.

Model We model the surface of the road as frictionless. Hence, the only forces acting on the car are the normal force $\vec{\mathbf{n}}$ exerted by the inclined plane, which acts perpendicular to the plane, and the gravitational force $\vec{\mathbf{F}}_g = m\vec{\mathbf{g}}$, which acts vertically downwards.

Analyse Begin by drawing a diagram showing the forces acting on the car (**Figure 4.8a**). Such a diagram is called a free-body diagram; these are described in detail in Section 4.7. For problems involving inclined planes, it is convenient to choose the coordinate axes with x along the incline and y perpendicular to it, as in **Figure 4.8b**. With these axes, we represent the gravitational force by a component of magnitude $mg \sin\theta$ along the positive x axis and one of magnitude $mg \cos\theta$ along the negative y axis.

The car can only move parallel to the road, so we know that the net force perpendicular to the road must be zero. The net force parallel to the road is not necessarily zero as the car may slide down the hill. Applying Newton's second law in these two directions gives:

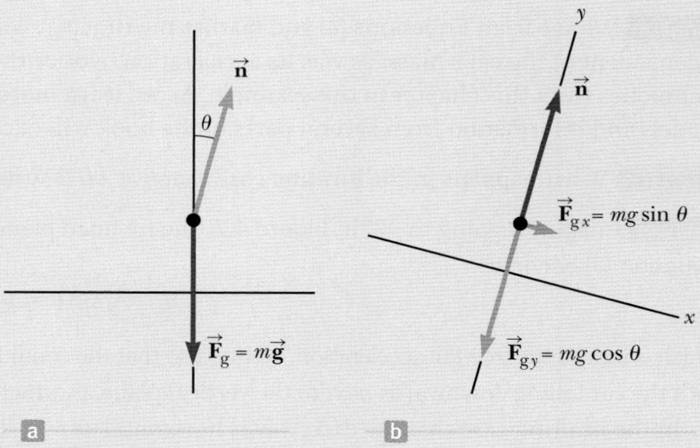

Figure 4.8
(Example 4.4) (a) The free-body diagram for the car. The black dot represents the position of the centre of mass of the car. (b) The free-body diagram with the forces resolved into components parallel and perpendicular to the plane.

$$\sum F_x = mg \sin\theta = ma_x \quad (1)$$

$$\sum F_y = n - mg \cos\theta = 0 \quad (2)$$

Solve **Equation (1)** for a_x:

$$a_x = g \sin\theta \quad (3)$$

Finalise Note that the acceleration component a_x is independent of the mass of the car! It depends only on the angle of inclination and on g.

From **Equation (2)**, we conclude that the component of $\vec{\mathbf{F}}_g$ perpendicular to the incline is balanced by the normal force; that is, $n = mg \cos\theta$. This situation is another case in which the normal force is *not* equal in magnitude to the weight of the object.

It is possible, although less convenient, to solve the problem with 'standard' horizontal and vertical axes. Try it.

(B) Suppose the car is released from rest at the top of the incline and the distance from the car's front bumper to the bottom of the incline is d. How long does it take the front bumper to reach the bottom of the hill, and what is the car's speed as it arrives there?

Solution

Conceptualise Review the diagram in part (A). Imagine that the car is sliding down the hill with acceleration $g \sin\theta$.

Model We can model the car as a particle with constant acceleration and apply the particle with constant acceleration analysis model (Chapter 2).

Example 4.4 cont.

Analyse Defining the initial position of the front bumper as $x_i = 0$ and its final position as $x_f = d$, and recognising that $v_{xi} = 0$, apply **Equation 2.15**, $x_f = x_i + v_{xi}t + \frac{1}{2}a_x t^2$:

$$d = \frac{1}{2}a_x t^2$$

Solve for t:

$$t = \sqrt{\frac{2d}{a_x}} = \sqrt{\frac{2d}{g\sin\theta}} \quad (4)$$

We do a quick dimension check: $[T] = ([L]/[LT^{-2}])^{1/2} = [T]$ ☺

Use **Equation 2.16**, with $v_{xi} = 0$, to find the final velocity of the car:

$$v_{xf}^2 = 2a_x d$$

$$v_{xf} = \sqrt{2a_x d} = \sqrt{2gd\sin\theta} \quad (5)$$

Again, we check the dimensions to make sure we haven't made an error: $[LT^{-1}] = ([LT^{-2}][L])^{1/2} = [LT^{-1}]$ ☺

Finalise We see from **Equations (4)** and **(5)** that the time t at which the car reaches the bottom and its final speed v_{xf} are independent of the car's mass, as was its acceleration. Notice that we have combined techniques from Chapter 2 with new techniques from this chapter in this example. As we learn more techniques in later chapters, this process of combining models and information from several parts of the book will occur more often.

What If? What happens in the limiting case that $\theta = 90°$? What if $\theta = 0°$?

Answer Imagine θ going to 90° in **Figure 4.7**. The inclined plane becomes vertical, and the car is an object in free fall! **Equation (3)** becomes

$$a_x = g\sin\theta = g\sin 90° = g$$

which is indeed the free-fall acceleration. Notice also that the condition $n = mg\cos\theta$ gives us $n = mg\cos 90° = 0$. That is consistent with the car falling downwards *next to* the vertical plane, in which case there is no contact force between the car and the plane.

In the limiting case that the driveway is horizontal, $\theta = 0°$, the acceleration and velocity become zero, and the time taken approaches infinite. This is what we expect – there is no reason for the car to begin sliding from rest on a horizontal surface if no other forces are exerted on it.

4.6 Newton's third law

If you press a finger against a corner of this textbook the book pushes back and makes a small dent in your skin. If you push harder, the book does the same and the dent in your skin is a little larger. This simple experiment illustrates that forces are *interactions* between two objects: when your finger pushes on the book, the book pushes back on your finger. This important principle is known as **Newton's third law:**

Newton's third law ▶

> If two objects interact, the force $\vec{\mathbf{F}}_{12}$ exerted by object 1 on object 2 is equal in magnitude and opposite in direction to the force $\vec{\mathbf{F}}_{21}$ exerted by object 2 on object 1:
>
> $$\vec{\mathbf{F}}_{12} = -\vec{\mathbf{F}}_{21} \quad (4.6)$$

When we designate forces as interactions between two objects we will use this subscript notation, where $\vec{\mathbf{F}}_{ab}$ means 'the force exerted *by* a *on* b'. The third law is illustrated in **Figure 4.9**. The force object 1 exerts on object 2 is often called the *action force*, and the force of object 2 on object 1 is called the *reaction force*. These italicised terms are not scientific terms, and either force can be labelled the action or reaction force. However, we will use these terms for convenience. In all cases, **the action and reaction forces act on *different* objects** and must be of the same type (gravitational, electrical, etc.). For example, the force acting on a freely falling projectile is the gravitational force exerted by the Earth on the projectile $\vec{\mathbf{F}}_g = \vec{\mathbf{F}}_{Ep}$ (E = Earth, p = projectile), and the magnitude of this force is mg. The reaction to this force is the gravitational force exerted by the projectile on the Earth $\vec{\mathbf{F}}_{pE} = -\vec{\mathbf{F}}_{Ep}$. The reaction force $\vec{\mathbf{F}}_{pE}$ must accelerate the Earth toward the projectile just as the action force $\vec{\mathbf{F}}_{Ep}$ accelerates the projectile toward the Earth. Because the Earth has such a large mass, however, its acceleration due to this reaction force is negligibly small.

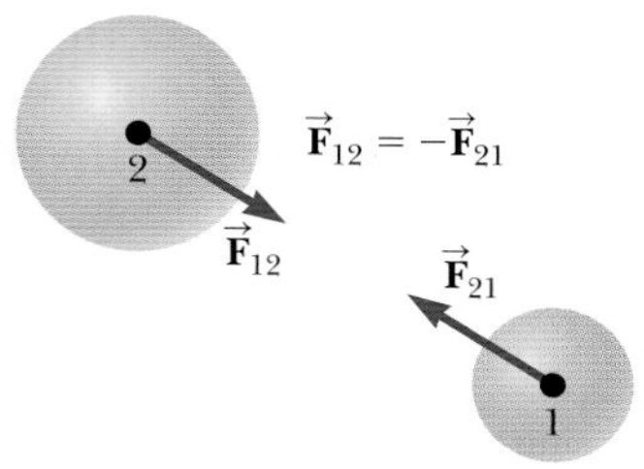

Figure 4.9
Newton's third law. The force $\vec{\mathbf{F}}_{12}$ exerted by object 1 on object 2 is equal in magnitude and opposite in direction to the force $\vec{\mathbf{F}}_{21}$ exerted by object 2 on object 1.

Quick Quiz 4.5

You sit down on a chair, compressing the cushion on the chair because of your weight. If the force that you exert on the cushion is the *action*, what is the *reaction* force according to Newton's third law? **(a)** The force of the Earth on the cushion **(b)** The force of the Earth on you **(c)** The force of the cushion on you **(d)** The force of you on the Earth

Consider a computer monitor at rest on a table as in **Figure 4.10**. The reaction force to the gravitational force $\vec{\mathbf{F}}_g = \vec{\mathbf{F}}_{Em}$ on the monitor is the force $\vec{\mathbf{F}}_{mE} = -\vec{\mathbf{F}}_{Em}$ exerted by the monitor on the Earth. The monitor does not accelerate because it is held up by the table. The table exerts on the monitor an upwards force $\vec{\mathbf{n}} = \vec{\mathbf{F}}_{tm}$, called the **normal force**. (*Normal* in this context means *perpendicular*.) This force, which prevents the monitor from falling through the table, can have any value needed, up to the point of breaking the table. Because the monitor has zero acceleration, Newton's second law applied to the monitor gives us $\Sigma\vec{\mathbf{F}} = \vec{\mathbf{n}} + m\vec{\mathbf{g}} = 0$, so $n\,\hat{\mathbf{j}} - mg\,\hat{\mathbf{j}} = 0$, or $n = mg$. The normal force balances the gravitational force on the monitor, so the net force on the monitor is zero. The reaction force to $\vec{\mathbf{n}}$ is the force exerted by the monitor downwards on the table, $\vec{\mathbf{F}}_{mt} = -\vec{\mathbf{F}}_{tm} = -\vec{\mathbf{n}}$. The only forces acting on the monitor are shown in **Figure 4.10a**; these forces and their reaction forces are shown in **Figure 4.10b**. We discuss normal force in more detail in Chapter 5.

Quick Quiz 4.6

(i) If a fly collides with the windshield of a fast-moving bus, which experiences an impact force with a larger magnitude? (a) The fly. (b) The bus. (c) The same force is experienced by both. **(ii)** Which experiences the greater acceleration? (a) The fly. (b) The bus. (c) The same acceleration is experienced by both.

TRY THIS

Stand on a skateboard or sit on a flat cart or trolley and push against a wall. What happens to the wall? What about you? Now try the same thing with a friend on a second skateboard or trolley. Now what happens when you push against them?

TRY THIS TOO

Inflate a balloon and hold it with its opening pinched shut. Let the balloon go and observe what it does. Explain this motion in terms of the forces acting on the balloon.

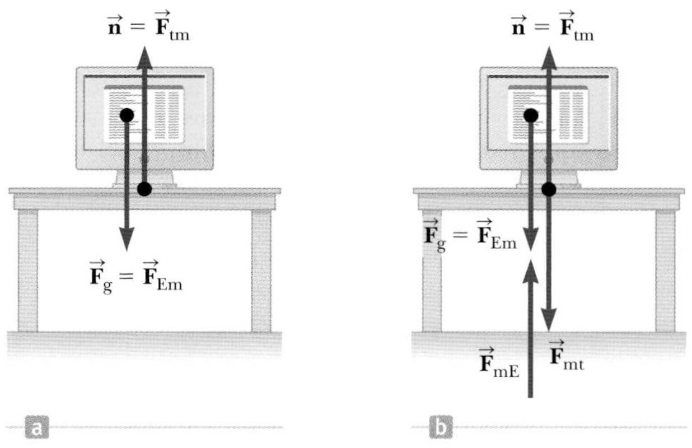

Figure 4.10
(a) When a computer monitor is at rest on a table, the forces acting on the monitor are the normal force $\vec{\mathbf{n}}$ and the gravitational force $\vec{\mathbf{F}}_g$. (b) The reaction to $\vec{\mathbf{n}}$ is the force $\vec{\mathbf{F}}_{mt}$ exerted by the monitor on the table. The reaction to $\vec{\mathbf{F}}_g$ is the force $\vec{\mathbf{F}}_{mE}$ exerted by the monitor on the Earth.

Pitfall Prevention 4.8

Remember that Newton's third-law action and reaction forces act on *different* objects. For example, in Figure 4.10, $\vec{\mathbf{F}}_{tm} = -m\vec{\mathbf{g}} = -\vec{\mathbf{F}}_{Em}$. The forces $\vec{\mathbf{n}}$ and $m\vec{\mathbf{g}}$ are equal in magnitude and opposite in direction, but they do not represent an action–reaction pair because both forces act on the *same* object, the monitor.

Pitfall Prevention 4.9

n does not always equal mg. In the situation shown in Figure 4.10 and in many others, we find that $n = mg$ (the normal force has the same magnitude as the gravitational force). This result, however, is *not* generally true. If an object is on an incline, if there are applied forces with vertical components, or if there is a vertical acceleration of the system, then $n \neq mg$. *Always* apply Newton's second law to find the relationship between n and mg.

Conceptual Example 4.5

Identifying action-reaction pairs

Consider the computer monitor shown in **Figure 4.10a**. A book is now placed on top of the monitor. Identify all the (action) forces acting on the book and on the monitor, and their reaction force pairs. Is the normal force of the table acting on the monitor still equal to the weight of the monitor?

Solution

Start by drawing a diagram. Our system now looks like that shown in **Figure 4.11**. **Figure 4.12** shows the forces acting on the book and the monitor. The forces acting on the book are the gravitational force of the Earth on the book, $\vec{F}_{Eb}$, and the normal force of the monitor on the book, $\vec{F}_{mb}$. Remembering that action-reaction pairs are of the form $\vec{F}_{a\text{ on }b} = -\vec{F}_{b\text{ on }a}$, the reaction forces must be the gravitational force of the book on the Earth, $\vec{F}_{bE}$, and the normal force of the book on the monitor, $\vec{F}_{bm}$ respectively.

The forces acting on the monitor are the gravitational force of the Earth on the monitor, $\vec{F}_{Em}$, the normal force of the table on the monitor, $\vec{F}_{tm}$, and the normal force of the book on the monitor, $\vec{F}_{bm}$. The reactions to these are, respectively, the gravitational force of the monitor on the Earth, $\vec{F}_{mE}$, the normal force of the monitor on the table, $\vec{F}_{mt}$, and the normal force of the monitor on the book, $\vec{F}_{mb}$.

If the monitor is still in equilibrium, that is, it is not accelerating, then by Newton's second law the net force acting on it must be zero. Look at the forces in **Figure 4.12b**. $\Sigma F = 0$ if $\vec{F}_{tm} = \vec{F}_{Em} + \vec{F}_{bm}$. Hence the normal force of the table on the monitor must now be larger than the gravitational force acting on (or the weight of) the monitor.

Figure 4.11

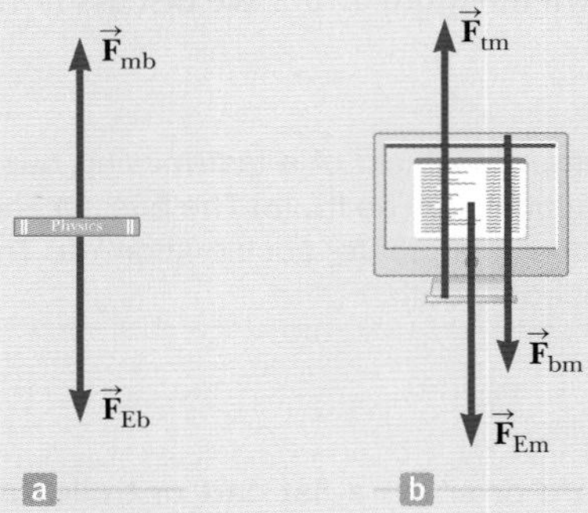

Figure 4.12

Example 4.6

One block pushes another

Two blocks of masses m_1 and m_2, with $m_1 > m_2$, are placed in contact with each other on a frictionless, horizontal surface as in **Active Figure 4.13a**. A constant horizontal force $\vec{F}$ is applied to m_1 as shown.

(A) Find the magnitude of the acceleration of the system.

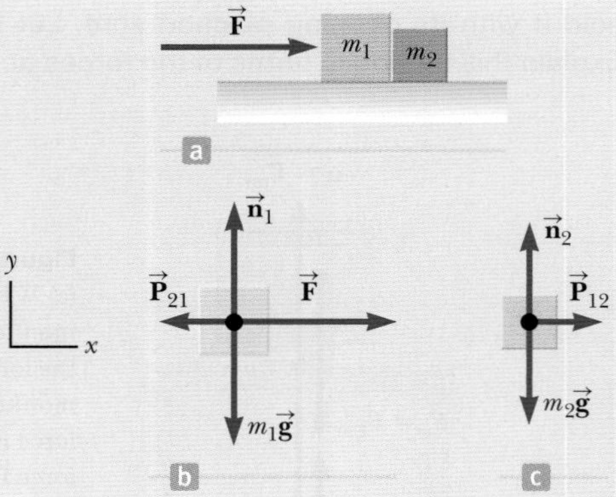

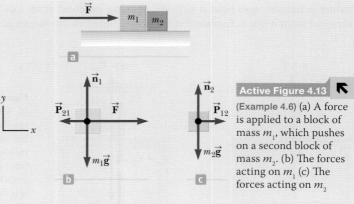

Active Figure 4.13 (Example 4.6) (a) A force is applied to a block of mass m_1, which pushes on a second block of mass m_2. (b) The forces acting on m_1 (c) The forces acting on m_2

Solution

Conceptualise Visualise the situation by using **Active Figure 4.13a** and note that both blocks must experience the *same* acceleration because they are in contact with each other and remain in contact throughout the motion. Note that if we did not have **Active Figure 4.13** we would start by drawing a diagram.

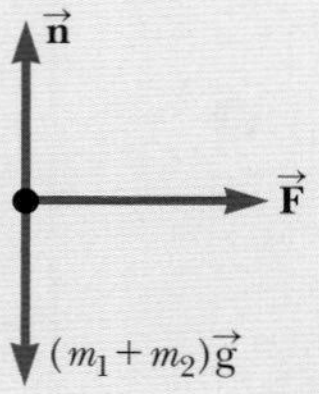

Model We make the approximation that the surface is frictionless. As both blocks will move together when the force is applied as shown to block 1, we can model the system of the two blocks as a single accelerating particle with mass $(m_1 + m_2)$. Draw a diagram showing the forces acting on the combination of blocks.

Example 4.6 cont.

Analyse Model the combination of two blocks as a single particle. Apply Newton's second law to the combination in the x direction to find the acceleration:

$$\sum F_x = F = (m_1 + m_2)a_x$$

$$a_x = \frac{F}{m_1 + m_2} \quad (1)$$

Finalise The acceleration given by **Equation (1)** is the same as that of a single object of mass $m_1 + m_2$ and subject to the same force.

(B) Determine the magnitude of the contact force that each blocks exerts upon the other.

Solution

Conceptualise The contact force is internal to the system of two blocks. Therefore, we cannot find this force by modelling the whole system (the two blocks) as a single particle. We also know from Newton's third law that the force that block 1 exerts on block 2 is equal in magnitude and opposite in direction to the force that block 2 exerts on block 1: $P_{12} = -P_{21}$.

Model We cannot treat the two blocks as a single object to determine the forces that they exert on each other. Block 1 has two forces acting on it, F and P_{21}, while block 2 only has P_{12} acting on it. Therefore we will consider block 2 and then apply Newton's third law.

Analyse **Active Figure 4.13c** shows that the only horizontal force acting on m_2 is the contact force $\vec{\mathbf{P}}_{12}$ (the force exerted by m_1 on m_2), which is directed to the right.

Apply Newton's second law to m_2:

$$\sum F_x = P_{12} = m_2 a_x \quad (2)$$

Substitute the value of the acceleration a_x given by **Equation (1)** into **Equation (2)**:

$$P_{12} = m_2 a_x = \left(\frac{m_2}{m_1 + m_2}\right)F \quad (3)$$

Now that we have an expression for F_{12} we can apply Newton's third law to write $P_{21} = -P_{12}$, so

$$P_{21} = -\left(\frac{m_2}{m_1 + m_2}\right)F$$

Finalise The contact forces, P_{12} and P_{21}, are internal forces to the system, and hence do not affect the acceleration of the overall system. However, internal forces, such as contact forces acting between two components of a system, *do* affect the acceleration of the components. In the *Try this* example, if you and a friend stand on skateboards and push against each other, you both accelerate.

Newton's third law and internal forces

> **Pitfall Prevention 4.10**
> Action–reaction pairs act on different objects. You cannot add or 'cancel' them when considering the forces on a single object.

When you consider a system of multiple objects, such as the two blocks in the worked example above, the Newton's third action–reaction pair of forces are internal forces to the system. This means that they do not affect the motion of the *system* overall. The average motion of the system, described by its centre of mass, is unchanged by internal forces. This is described in more detail in Chapter 8. Hence, while block 1 pushed on block 2, and block 2 pushed back on block 1, the acceleration of the block 1 plus block 2 system was determined solely by the external force, **F**. This does *not* mean that the motion of the individual blocks depends only on the external force **F**. Consider the situation if block 2 was not present. In this case block 1 would be subject only to force **F** in the horizontal direction and would have an acceleration equal to $\mathbf{F}/m_1$, which is greater than the acceleration when block 2 is present. Consider block 2. If block 1 was absent, then no horizontal force would act on block 2 and it would not accelerate at all.

It is important to remember that while internal forces within a system, which always occur in action-reaction pairs, can be 'cancelled out', you *cannot* cancel out action-reaction pairs when considering a single object. Action-reaction pairs act on separate objects. When you push a shopping trolley, you apply a force to it, and it exerts an equal and opposite force on you. The force that it exerts on you acts *on you* not *on the trolley*, so the reaction force to your push does not affect the trolley's motion, only yours.

Weight and scale force revisited

Imagine standing on a set of bathroom scales. The scales are calibrated to read your mass. But how do they do this? One simple mechanism is to use a spring attached to a pointer, which rotates and points to numbers on a scale as the spring is compressed. Digital scales use a similar mechanism, but with an electronic readout. When you stand on a set of scales, at rest on a flat horizontal floor, the only forces acting *on you* are the gravitational force F_g and the normal force, N, of the scales. Note that these cannot be an action-reaction pair because they act on the same object and they have different origins (gravitational and electrostatic). However, if we apply Newton's second law to you, then we can see that as $\Sigma\mathbf{F} = \mathbf{N} + \mathbf{F}_g = 0$ then $\mathbf{N} = -\mathbf{F}_g$.

The scale reads the compression of the spring, and converts this into a mass. Hence the scale reading depends upon the force that is exerted *on the scale*. According to Newton's third law, as the scale applies a force **N** to you, you must apply a force −**N** to the scale. In this case, the scale force, **N**, is equal to the gravitational force, $\mathbf{F}_g$, or your weight. However if there are other forces acting, as in the example below, this will not be the case.

Example 4.7

Tricking the scales

David has been eating lots of ice cream late at night while preparing lectures, and as a result his mass has increased. When he stands on his bathroom scales he is dismayed to see the increase. However, he notes that when he pushes down on a table the scale reading decreases to a value that he is happier with. If David has a mass of 84 kg, but only wants to see a reading of 80 kg, with what force must he push down on the table?

Conceptualise If you did the *Try this* earlier, you will already have some sense of how this works. If not, get a set of bathroom scales and stand on it. Then push up or down on a table and note how the reading changes. Remember that what the scale reads is the compressive force that is applied to it; this is the force that David exerts on the scales. Begin by drawing a diagram showing the situation (**Figure 4.14a**).

Model We will assume that the scales are correctly calibrated, the floor is horizontal and David is standing still (not accelerating). Hence the net force acting on David is zero.

Analyse Draw a diagram of the forces acting on the table (**Figure 4.14b**). The forces acting on the table are the downwards force due to gravity, $\vec{\mathbf{F}}_{Et}$, the upwards normal force of the floor on the table, $\vec{\mathbf{F}}_{ft}$, and the downwards force of David on the table, $\vec{\mathbf{F}}_{Dt}$.

Draw a diagram of the forces acting on David (**Figure 4.14c**). The forces acting on David are the downwards force due to gravity, $\vec{\mathbf{F}}_{ED}$, the upwards force of the table on David, $\vec{\mathbf{F}}_{tD}$, and the upwards force of the scales on David, $\vec{\mathbf{F}}_{sD}$.

Draw a diagram of the forces acting on the scales (**Figure 4.14d**). The forces acting on the scales are the downwards force due to gravity, $\vec{\mathbf{F}}_{Es}$, the downwards normal force of David on the scales, $\vec{\mathbf{F}}_{Ds}$, and the upwards normal force of the floor on the scales, $\vec{\mathbf{F}}_{fs}$.

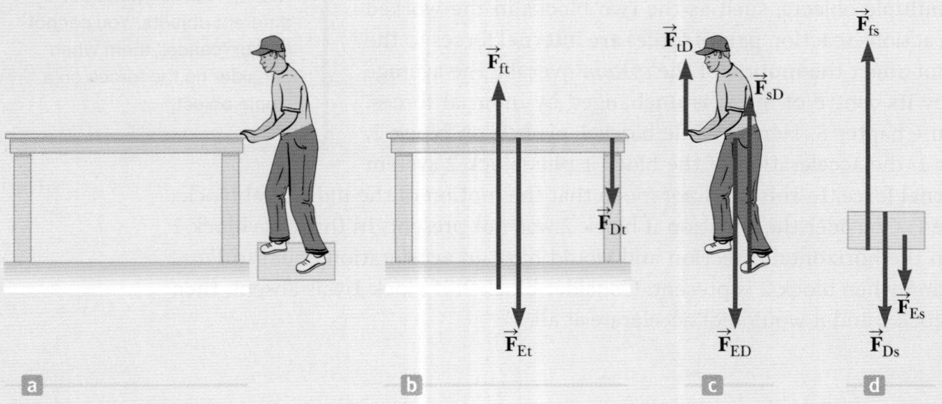

Figure 4.14
(a) David stands on the scales and pushes down on a table. (b) The forces acting on the table
(c) The forces acting on David
(d) The forces acting on the scales

Example 4.7 cont.

From Newton's third law we can say that $\vec{\mathbf{F}}_{tD} = -\vec{\mathbf{F}}_{Dt}$.

Applying Newton's second law to David, we can see that: $\sum\vec{\mathbf{F}} = \vec{\mathbf{F}}_{sD} + \vec{\mathbf{F}}_{tD} + \vec{\mathbf{F}}_{ED} = \vec{\mathbf{F}}_{sD} - \vec{\mathbf{F}}_{Dt} + \vec{\mathbf{F}}_{ED} = 0.$

so $\vec{\mathbf{F}}_{sD} = \vec{\mathbf{F}}_{Dt} - \vec{\mathbf{F}}_{ED}$.

The force exerted on the scales by David is, by Newton's third law, $\vec{\mathbf{F}}_{Ds} = -\vec{\mathbf{F}}_{sD} = -\vec{\mathbf{F}}_{Dt} + \vec{\mathbf{F}}_{ED}$.

Rearranging for the force David exerts on the table: $\vec{\mathbf{F}}_{Dt} = -\vec{\mathbf{F}}_{ED} + \vec{\mathbf{F}}_{sD}$

A scale reading of 80 kg corresponds to a force of 80 kg × 9.8 N/kg = 784 N, and a mass of 84 kg corresponds to a gravitational force (weight) of 823 N. $\vec{\mathbf{F}}_{Dt} = -823\text{N} + 784\text{N} = 39\text{N}.$

David must exert a force of 39 N downwards on the table to see a scale reading of 80 kg.

Finalise By pushing downwards on the table, the scale reading is reduced and the scale is 'tricked'. However, David's mass in unchanged, and hence the gravitational force he experiences (his weight) and his resistance to acceleration are also unchanged. To genuinely reduce his weight he needs to eat less ice cream or move somewhere with a lower gravitational field.

What If? What if David pushed upwards on the table instead?

Answer In this case the table would push downwards on him, and the force supplied by the scales would need to increase to keep him in equilibrium. If the scales apply a larger force to him, then he applies a larger force to the scales, and the reading increases.

4.7 Analysis models using Newton's laws

In this section we discuss two analysis models for solving problems in which objects are either in equilibrium ($\vec{\mathbf{a}} = 0$) or accelerating along a straight line under the action of constant external forces. Remember that when Newton's laws are applied to an object, we are interested only in external forces that act on that object. If the objects are modelled as particles, we need not worry about rotational motion. For now, we also neglect the effects of friction in those problems involving motion, and model surfaces as frictionless and any drag forces such as air resistance as negligible. Frictional forces are discussed in detail in Chapter 5.

We will make the approximation that any ropes, strings, or cables are light compared to objects suspended from them, and we treat their mass as negligible. We also assume that they do not stretch and hence that any force exerted on one end of a rope is transmitted undiminished to the other end. These assumptions describe an **ideal rope**. In this model, the magnitude of the force exerted by any element of the rope on the adjacent element is the same for all elements along the rope. When a rope attached to an object is pulling on the object, the rope exerts a force on the object in a direction away from the object, parallel to the rope. The magnitude T of that force is called the **tension** in the rope. Because it is the magnitude of a vector quantity, tension is a scalar quantity. Real ropes do stretch and often have significant mass, but this model is useful when these effects are small, which is often the case.

◀ Tension and ideal ropes

Free-body diagrams

When solving problems using Newton's laws it is a good idea to always start with a diagram showing the forces acting on the different objects involved. To work out the net force acting on, or the acceleration of, a particular object we draw a **free-body diagram**.

A free-body diagram is a type of force diagram that shows all the forces and only the forces acting on a single object. The object is represented as a point.

◀ Free-body diagrams

When analysing an object subject to forces, we are interested in the net force acting on one object, which we will model as a particle. Therefore, a free-body diagram helps us isolate only those forces on the object and eliminate the other forces from our analysis.

Pitfall Prevention 4.11

The *most important* step in solving a problem using Newton's laws is to draw a free-body diagram. Be sure to draw only those forces that act on the object you are considering. Be sure to draw *all* forces acting on the object, including any field forces, such as the gravitational force.

Conceptual Example 4.8

Forces acting on a computer monitor

Consider the computer monitor shown in Figure 4.10.

Draw a free-body diagram for this computer monitor.

Solution

We need to show only the forces acting on *one object,* the monitor. We use the particle model and represent the monitor as a dot and show the forces that act on the monitor as being applied to the dot. Figure 4.10a already gives us the forces, so we can easily draw the free-body diagram as shown in Figure 4.15.

Figure 4.15
(Conceptual Example 4.8) A free-body diagram showing the forces on the monitor.

Analysis model: the particle in equilibrium

If the acceleration of an object modelled as a particle is zero, the object is treated with the **particle in equilibrium** model. In this model, the net force on the object is zero. Newton's second law tells us that:

$$\sum \vec{F} = 0 \tag{4.7}$$

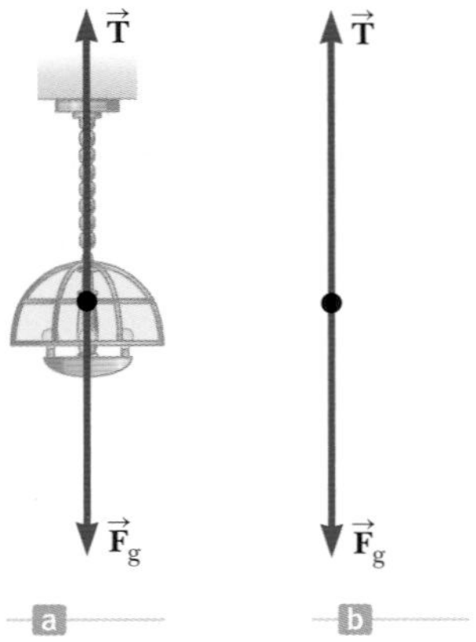

Figure 4.16
(a) A lamp suspended from a ceiling by a chain of negligible mass. (b) The forces acting on the lamp are the gravitational force $\vec{F}_g$ and the force $\vec{T}$ exerted by the chain.

Consider a lamp suspended from a light chain, which we can model as an ideal rope, fastened to the ceiling as in Figure 4.16a. The free-body diagram for the lamp (Figure 4.16b) shows that the forces acting on the lamp are the downwards gravitational force $\vec{F}_g$ and the upwards force $\vec{T}$ exerted by the chain. There are no forces in the x direction, so $\Sigma F_x = 0$ also. The condition $\Sigma F_y = 0$ gives

$$\sum F_y = T - F_g = 0 \quad \text{or} \quad T = F_g$$

Again, notice that $\vec{T}$ and $\vec{F}_g$ are *not* an action–reaction pair because they act on the same object, the lamp.

Newton's third law tells us about the forces that the lamp exerts on the chain and on the Earth. The reaction force to the gravitational force of the Earth on the lamp is a gravitational force of the lamp on the Earth. The reaction force to $\vec{T}$ is a downwards force exerted by the lamp on the chain.

Analysis model: the particle subject to a net force

If an object experiences an acceleration, its motion can be analysed with the **particle subject to a net force** model. The appropriate equation for this model is Newton's second law, Equation 4.3:

$$\sum \vec{F} = m\vec{a}$$

Consider a crate with a smooth base being pulled to the right on a smooth, horizontal floor as in Figure 4.17a. We model the surface of the floor as frictionless for the crate. (Of course, the floor directly under the boy must have friction; otherwise his feet would simply slip when he tries to pull on the crate!) Suppose you wish to find the acceleration of the crate and the force the floor exerts on it. The forces acting on the crate are illustrated in the free-body diagram in Figure 4.17b. Notice that the horizontal force $\vec{T}$ being applied to the crate acts through the rope. The magnitude of $\vec{\mathbf{T}}$ is equal to the tension in the rope. In addition to the force $\vec{\mathbf{T}}$, the free-body diagram for the crate includes the gravitational force $\vec{\mathbf{F}}_g$ and the normal force $\vec{\mathbf{n}}$ exerted by the floor on the crate.

We can now apply Newton's second law in component form to the crate. The only force acting in the x direction is $\vec{\mathbf{T}}$. Applying $\Sigma F_x = ma_x$ to the horizontal motion gives

$$\sum F_x = T = ma_x \quad \text{or} \quad a_x = \frac{T}{m}$$

See Appendix B.6 for help with decomposing vectors.

No acceleration occurs in the y direction because the crate moves only horizontally. Therefore, we use the particle in equilibrium model in the y direction. Applying the y component of **Equation 4.7** yields

$$\sum F_y = n + (-F_g) = 0 \quad \text{or} \quad n = F_g$$

That is, the normal force has the same magnitude as the gravitational force but acts in the opposite direction.

If $\vec{\mathbf{T}}$ is a constant force, the acceleration $a_x = T/m$ also is constant. Hence, the crate is also modelled as a particle with constant acceleration in the x direction, and the equations of kinematics from Chapter 2 can be used to obtain the crate's position x and velocity v_x as functions of time.

Newton's third law tells us about the force that the crate exerts on the rope. If the rope exerts a force $\vec{\mathbf{T}}$ on the crate, then the crate exerts a force $-\vec{\mathbf{T}}$ on the rope. Assuming the rope is ideal, this force is transmitted to the boy, pulling him back towards the crate. If there is no friction acting on the boy, then he will accelerate towards the crate with acceleration $\vec{\mathbf{a}} = -\vec{\mathbf{T}}/m_{boy}$.

In the situation just described, the magnitude of the normal force $\vec{\mathbf{n}}$ is equal to the magnitude of $\vec{F}_g$, but that is not always the case, as noted in Pitfall Prevention 4.9. (Remember that the normal force and the gravitational force are *not* an action–reaction pair.) For example, suppose a book is lying on a table and you push down on the book with a force $\vec{\mathbf{F}}$ as in **Figure 4.18**. Because the book is at rest and therefore not accelerating, $\Sigma F_y = 0$, which gives $n - F_g - F = 0$, or $n = F_g + F = mg + F$. In this situation, the normal force is *greater* than the gravitational force. Other examples in which $n \neq F_g$ are presented later.

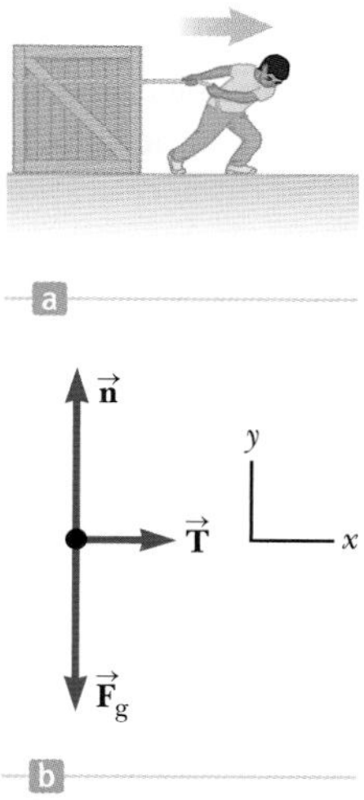

Figure 4.17
(a) A crate being pulled to the right on an approximately frictionless floor. (b) The free-body diagram representing the external forces acting on the crate.

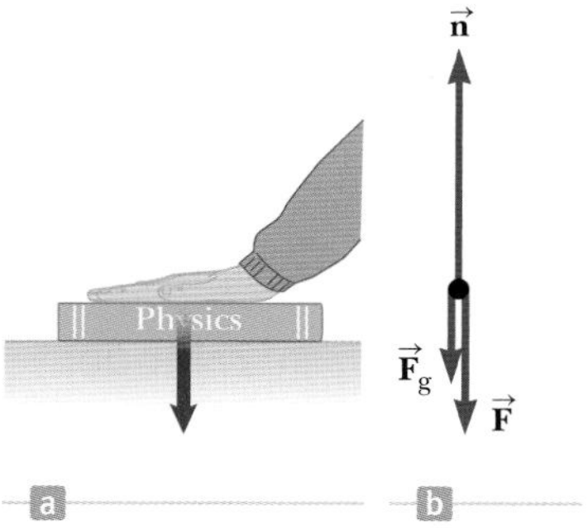

Figure 4.18
(a) When a force $\vec{\mathbf{F}}$ pushes vertically downwards on another object, the normal force $\vec{\mathbf{n}}$ on the object is greater than the gravitational force: $n = F_g + F$. (b) Free-body diagram for the book

Quick Quiz 4.7

If the boy in **Figure 4.17** applied the force to the crate at an angle above the horizontal, would the normal force acting on the crate be: (a) larger, (b) smaller or (c) the same as the gravitational force on the crate? What if the force was applied at an angle below the horizontal?

Analysis Model 4.1

Particle in equilibrium

Imagine an object that can be modelled as a particle. If it has several forces acting on it so that the forces all sum to zero, giving a net force of zero, the object will have an acceleration of zero. This condition is mathematically described as

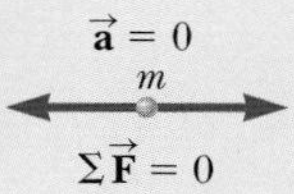

$$\sum \vec{\mathbf{F}} = 0 \qquad (4.7)$$

Examples

- a chandelier hanging over a dining room table
- an object moving at terminal speed through a viscous medium (Chapter 5)
- a steel beam in the frame of a building (Chapter 14)
- a boat floating on a body of water (Chapter 13)

Analysis Model 4.2

Particle subject to a net force

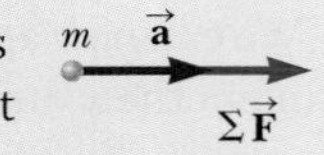

Imagine an object that can be modelled as a particle. If it has one or more forces acting on it so that there is a net force on the object, it will accelerate in the direction of the net force. The relationship between the net force and the acceleration is

$$\sum \vec{F} = m\vec{a} \tag{4.2}$$

Examples

- a crate pushed across a factory floor
- a falling object acted upon by a gravitational force (Chapter 11)
- a piston in an automobile engine pushed by hot gases (Chapter 22)
- a charged particle in an electric field (Chapter 23)

Problem-solving strategy

Applying Newton's laws

The following procedure is recommended when dealing with problems involving Newton's laws:

1. Conceptualise

Draw a simple, neat diagram of the system. The diagram helps establish the mental representation. Establish convenient coordinate axes for each object in the system. Draw separate free body diagrams for each object whose motion you want to analyse. Put on *all* the forces acting on the object. Do **not** draw on forces which are not acting on that object, such as forces it exerts on other objects or the net force. Do not add other vectors such as acceleration or velocity as this can cause confusion.

2. Model

Think about what approximations you can reasonably make. Can surfaces be treated as frictionless? Can ropes be modelled as ideal?

If an acceleration component for an object is zero, the object is modelled as a particle in equilibrium in this direction and $\Sigma F = 0$. If not, the object is modelled as a particle subject to a net force in this direction and $\Sigma F = ma$.

3. Analyse

Find the components of the forces along the coordinate axes. Use Newton's third law to identify action-reaction pairs on the free-body diagrams of interacting objects. These forces will be equal but act in opposite directions. (Note again that action–reaction force pairs act on different objects so will never be drawn on the same free-body diagram!). Apply the appropriate model from the *Model* step for each direction. Check your dimensions to make sure that all terms have units of force.

Solve the component equations for the unknowns. Remember that you generally must have as many independent equations as you have unknowns to obtain a complete solution.

4. Finalise

Make sure your results are consistent with the free-body diagram. Also consider limiting cases – check the predictions of your solutions for extreme values of the variables. By doing so, you can often detect errors in your results.

Example 4.9

A traffic light at rest

While roadworks are underway, a temporary traffic light weighing 122 N is hung from a cable tied to two other cables fastened to a support as in **Figure 4.19**. The upper cables make angles of 37.0° and 53.0° with the horizontal. These upper cables are not as strong as the vertical cable and the manufacturer has rated them to withstand a tension of 100 N with a tolerance (uncertainty) of 2%. The vertical cable can withstand a far greater tension. Does the traffic light remain hanging in this situation, or will one of the cables break?

Solution

Conceptualise Examine the drawing in **Figure 4.19**. Let us assume for the moment that the cables do not break and nothing is moving. We draw a free-body diagram for the traffic light (**Figure 4.20a**), the cable connecting the traffic light to the knot (**Figure 4.20b**) and the knot (**Figure 4.20c**).

This knot is a convenient object to choose because all the forces of interest act along lines passing through the knot.

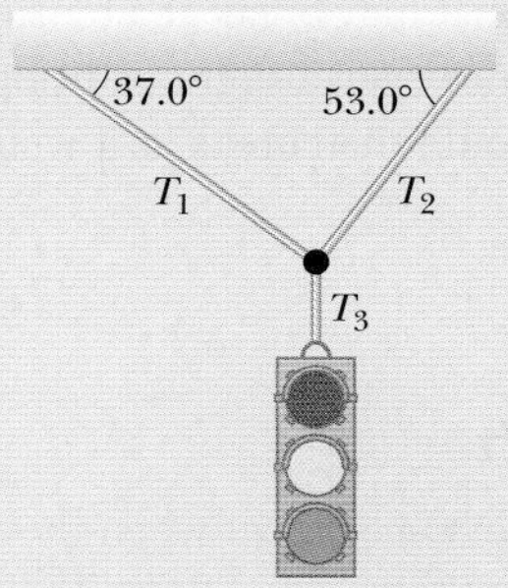

Figure 4.19
(Example 4.9) A traffic light suspended by cables

Model We will model the cables as ideal ropes. If we begin with the assumption that the light is not moving, then we can model it as a particle in equilibrium. The knot will also not be moving, so it too can be modelled as a particle in equilibrium.

Analyse Apply **Equation 4.7** for the traffic light in the y direction:

$$\sum F_y = 0 \rightarrow T_3 - F_g = 0$$
$$T_3 = F_g = 122 \text{ N}$$

We can now apply Newton's third law to the interaction between the cable and the traffic light. If the cable exerts force T_3 on the light, then the light exerts force $-T_3$ on the cable. If the cable is in equilibrium then the upwards force of the knot on the cable must be equal to T_3 (**Figure 4.20b**). Again applying Newton's third law, this time to the interaction between the knot and cable 3, we see that the force exerted by cable 3 on the knot must be $-T_3$.

Now consider **Figure 4.20c** and apply Newton's second law to the knot.

Choose the coordinate axes as shown in **Figure 4.20c** and resolve the forces acting on the knot into their components:

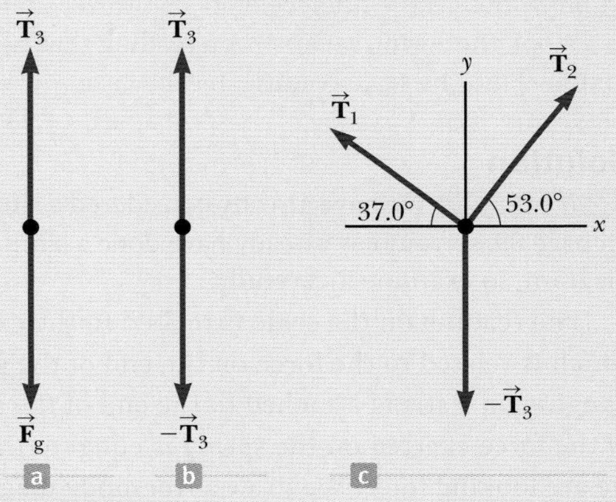

Figure 4.20
Free-body diagrams for the three objects of interest. (a) Free-body diagram for the traffic light. (b) Free-body diagram for cable 3. (c) Free-body diagram for the knot.

Force	x component	y component
$\vec{T}_1$	$-T_1 \cos 37.0°$	$T_1 \sin 37.0°$
$\vec{T}_2$	$T_2 \cos 53.0°$	$T_2 \sin 53.0°$
$\vec{T}_3$	0	−122 N

Apply the particle in equilibrium model to the knot:

$$\sum F_x = -T_1 \cos 37.0° + T_2 \cos 53.0° = 0 \quad (1)$$
$$\sum F_y = T_1 \sin 37.0° + T_2 \sin 53.0° + T_3 = 0 \quad (2)$$

Equation (1) shows that the horizontal components of $\vec{T}_1$ and $\vec{T}_2$ must be equal in magnitude, and **Equation (2)** shows that the sum of the vertical components of $\vec{T}_1$ and $\vec{T}_2$ must balance the downward force $\vec{T}_3$, which is equal in magnitude to the weight of the light.

Solve **Equation (1)** for T_2 in terms of T_1:

$$T_2 = T_1\left(\frac{\cos 37.0°}{\cos 53.0°}\right) = 1.33T_1 \quad (3)$$

Example 4.9 cont.

Substitute this value for T_2 and the value for T_3 into **Equation (2)**:

$$T_1 \sin 37.0° + (1.33T_1)(\sin 53.0°) - 122\text{ N} = 0$$
$$T_1 = 73.4\text{ N}$$
$$T_2 = 1.33T_1 = 97.4\text{ N}$$

Finalise Would you stand beneath the light?

Both values are less than 100 N (just barely for T_2), so the cables should not break. Allowing for a 2% variation in the strength of the cable, it may break at a tension of 98 N, which is only very slightly above the force exerted on cable 2 by the traffic light. Hence even a slight additional force, such as a bird landing on the lights, may cause the cable to break. The second cable would then break as it would have to take the full 122 N.

Example 4.10

Weight in a lift

A student is doing an experiment when they realise they need to use a piece of equipment in a lab on the second floor. Their experimental sample is attached to a spring balance as shown in **Figure 4.21**. They get in the lift and notice that their sample appears to change weight as they ascend.

(A) Show that as the lift accelerates upwards (starts to rise) and then downwards (slows to a stop), the spring scale gives a reading that is different from the weight of the sample. Find the scale reading in each case.

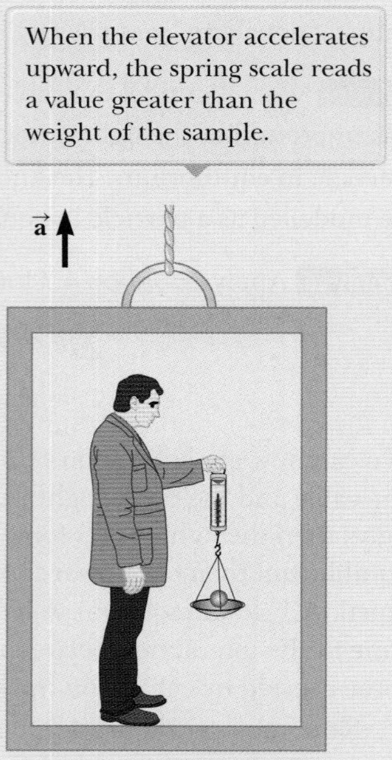

When the elevator accelerates upward, the spring scale reads a value greater than the weight of the sample.
$\vec{a}$

Figure 4.21
(Example 4.10) A sample attached to a spring balance in an accelerating lift.

Solution

Conceptualise We have already considered a similar situation in the *Try this* example on page 89, so you may already have done a similar experiment yourself! We have a diagram, so examine it carefully.

The reading on the scale is related to the extension of the spring in the scale, which is related to the force on the end of the spring. Imagine that the sample is hanging on a string attached to the end of the spring. In this case, the magnitude of the force exerted on the spring is equal to the tension T in the string. Therefore, we are looking for T. We draw a free-body diagram for the sample (**Figure 4.22**) for the two situations: accelerating upwards and accelerating downwards. The only forces acting in each case are the downwards gravitational force (the weight of the sample) and the upwards tension force (equal to the scale force or apparent weight). Note that if the lift was either at rest or moving at constant velocity then the sample would be in equilibrium, and according to Newton's second law these two forces would be equal.

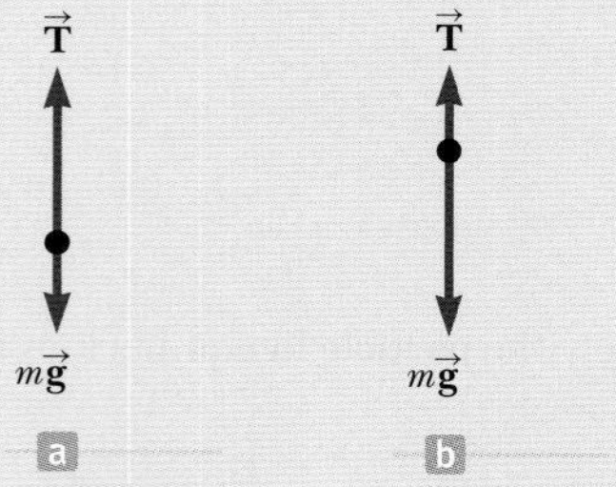

$\vec{T}$
$m\vec{g}$
a
$\vec{T}$
$m\vec{g}$
b

Figure 4.22
(Example 4.10) Free-body diagrams for the sample. (a) When the sample is being accelerated upwards, the tension, T, must be greater than F_g. (b) When the sample is being accelerated downwards, the tension, T, must be less than F_g.

Model We model the sample as a particle subject to a net force because it is accelerating. From Newton's second law, we know that the direction of the net force must be the same as the direction of acceleration. This is reflected in the relative lengths of the arrows in **Figure 4.22**.

Analyse Apply Newton's second law to the sample:

$$\sum F_y = T - mg = ma_y$$

Solve for T:

$$T = ma_y + mg = mg\left(\frac{a_y}{g} + 1\right) = F_g\left(\frac{a_y}{g} + 1\right) \quad \textbf{(1)}$$

Do a quick dimension check: $[MLT^{-2}] = [MLT^{-2}]([LT^{-2}]/[LT^{-2}]) = [MLT^{-2}]$ ☺

where we have chosen upwards as the positive y direction. We conclude from **Equation (1)** that the scale reading T is greater than the sample's weight mg if $\vec{a}$ is upwards, so a_y is positive (**Figure 4.22a**), and that the reading is less than mg if $\vec{a}$ is downwards, so a_y is negative (**Figure 4.22b**).

(B) Evaluate the scale readings for a 40.0 N sample if the lift moves with an acceleration $a_y = 2.00\ \text{m/s}^2$ up or down.

Solution

Evaluate the scale reading from **Equation (1)** if $\vec{a}$ is upwards:

$$T = 40.0\,\text{N}\left(\frac{2.00\ \text{m/s}^2}{9.80\ \text{m/s}^2} + 1\right) = 48.2\ \text{N}$$

Evaluate the scale reading from **Equation (1)** if $\vec{a}$ is downwards:

$$T = 40.0\,\text{N}\left(\frac{-2.00\ \text{m/s}^2}{9.80\ \text{m/s}^2} + 1\right) = 31.8\ \text{N}$$

Finalise The results agree with our experience from the *Try this* example – you feel lighter when accelerating downwards and heavier as you are pushed upwards by the floor of the lift. The same applies when the object is being pulled upwards on a string. In this case it is the tension in the string that supports the object, preventing it from falling due to gravity. Hence the apparent weight of the object is the tension in the string.

What If? Suppose the lift cable breaks, and the lift and its contents are in free fall. What happens to the reading on the scale?

Answer If the lift falls freely, its acceleration is $a_y = -g$. We see from **Equation (1)** that the scale reading T is zero in this case; that is, the sample *appears* to be weightless.

Example 4.11

Acceleration of two objects connected by a cord

A ball of mass m_1 and a block of mass m_2 are attached by a lightweight cord that passes over a frictionless pulley of negligible mass as in **Figure 4.23**. The block lies on a frictionless incline of angle θ. Find the magnitude of the acceleration of the two objects and the tension in the cord.

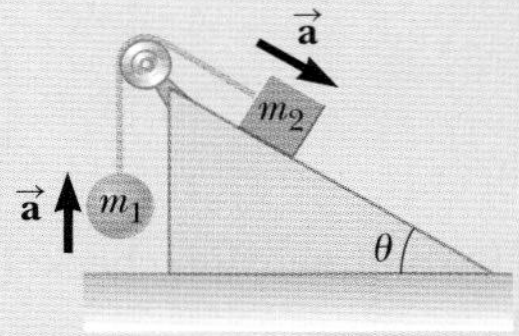

Figure 4.23
(Example 4.11) Two objects connected by a lightweight cord strung over a frictionless pulley

Solution

Conceptualise Imagine the objects in **Figure 4.23** in motion. If m_2 moves down the incline, then m_1 moves upwards. Because the objects are connected by a cord (which we assume does not stretch), their accelerations have the same magnitude.

Draw a free-body diagram for each object. It will be helpful to use a coordinate system for the block with directions parallel and perpendicular to the plane, so draw a second free-body diagram with forces resolved into components in these directions.

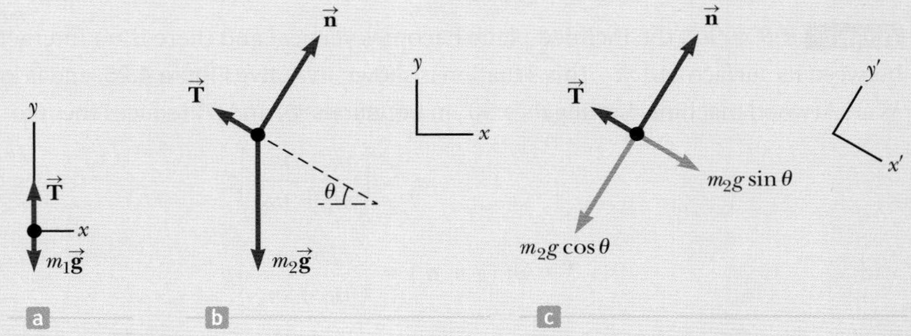

Figure 4.24
(Example 4.11) (a) The free-body diagram for the ball (b) The free-body diagram for the block (c) The free-body diagram for the block with the forces resolved into components parallel and perpendicular to the incline (The incline is frictionless.)

Model We model the two objects, the ball and the block, as particles subject to a net force. We model the slope as frictionless and the cord as an ideal rope which does not stretch and hence has the same tension at all points and transmits the force from one object to the other without diminishing it. Note, however, that in this case the *direction* of the force is changed because of the pulley. The ideal pulley exerts a force on the cable such that the direction of $\vec{T}$ but not its magnitude is changed.

Analyse Apply Newton's second law in component form to the ball, choosing the upwards direction as positive:

$$\sum F_x = 0 \qquad (1)$$

$$\sum F_y = T - m_1 g = m_1 a_y = m_1 a \qquad (2)$$

For the ball to accelerate upwards, it is necessary that $T > m_1 g$. In **Equation (2)**, we replaced a_y with a because the acceleration has only a y component.

For the block, it is convenient to choose the positive x' axis along the incline as in **Figure 4.24c**. For consistency with our choice for the ball, we choose the positive direction to be down the incline.

Apply Newton's second law in component form to the block:

$$\sum F_{x'} = m_2 g \sin\theta - T = m_2 a_{x'} = m_2 a \qquad (3)$$

$$\sum F_{y'} = n - m_2 g \cos\theta = 0 \qquad (4)$$

In **Equation (3)**, we replaced $a_{x'}$ with a because the two objects have accelerations of equal magnitude a.

Solve **Equation (2)** for T:

$$T = m_1(g + a) \qquad (5)$$

Substitute this expression for T into **Equation (3)**:

$$m_2 g \sin\theta - m_1(g + a) = m_2 a$$

Solve for a:

$$a = \left(\frac{m_2 \sin\theta - m_1}{m_1 + m_2}\right) g \qquad (6)$$

Do a dimension check: $[LT^{-2}] = ([M]/[M])[LT^{-2}] = [LT^{-2}]$ ☺

Substitute this expression for a into **Equation (5)** to find T:

$$T = \left(\frac{m_1 m_2(\sin\theta + 1)}{m_1 + m_2}\right) g \qquad (7)$$

Do a dimension check: $[MLT^{-2}] = ([M][M]/[M])[LT^{-2}] = [MLT^{-2}]$ ☺

Finalise The block accelerates down the incline only if $m_2 \sin\theta > m_1$. If $m_1 > m_2 \sin\theta$, the acceleration is up the incline for the block and downwards for the ball. Also notice that the result for the acceleration, **Equation (6)**, can be interpreted as the magnitude of the net external force acting on the ball-block system divided by the total mass of the system; this result is consistent with Newton's second law.

What If? What happens in this situation if $\theta = 90°$?

Answer If $\theta = 90°$, the inclined plane becomes vertical and there is no interaction between its surface and m_2. This situation is shown in **Active Figure 4.25**, and is known as an Atwood machine. Letting $\theta \to 90°$ in **Equations (6)** and **(7)** reduces them to

$$a_y = \left(\frac{m_2 - m_1}{m_1 + m_2}\right) g$$

and

$$T = m_1(g + a_y) = \left(\frac{2m_1 m_2}{m_1 + m_2}\right) g$$

Explore the effects of varying the masses by using **Active Figure 4.25**.

What If? What if $m_1 = 0$?

Answer If $m_1 = 0$, then m_2 is simply sliding down an inclined plane without interacting with m_1 through the string. Therefore, this problem becomes the sliding car problem in Example 4.4.

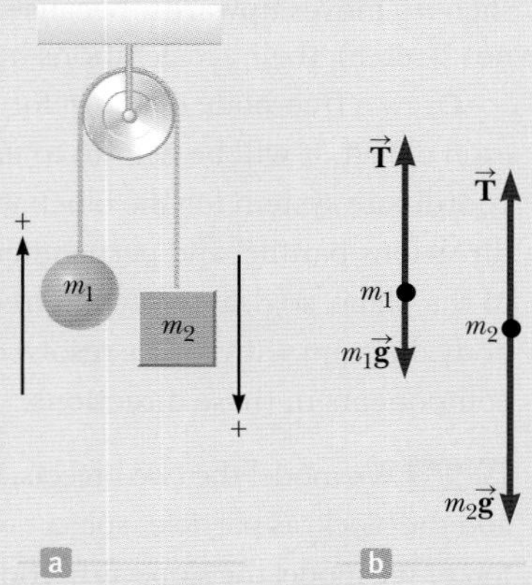

Active Figure 4.25

(Example 4.11) The Atwood machine. (a) Two objects connected by a massless inextensible cord over a frictionless pulley (b) The free-body diagrams for the two objects

Example 4.12

The Wellington cable car

The Wellington cable car carries passengers between Lambton Quay (at sea level) and the Wellington Botanic Gardens to a vertical height of 119 m, along 610 m of track. There are two cars joined by a heavy cable which goes over a pulley at the top of the tracks, and there is an electric motor to pull the cable, which allows one car to be drawn up the hill while the other goes down, irrespective of their weights. Without a motor, the heavier cars would tend to roll down the tracks, pulling the lighter car up at the same time. **Figure 4.26** shows how the system works.

Consider the following situation. Car 1 is at the bottom of the hill and empty and has a mass m_1 = 13 500 kg. Car 2 has several tourists in it and has a total mass m_2 = 14 500 kg. It is at the top of the tracks at the Botanic Gardens. The cable is light compared to the cars and as the cars will roll rather than slide we will ignore friction (in fact, friction is likely to be important, as discussed in the next section). If the brakes are all disengaged and the motor not turned on, how long does it take Car 2 to reach Lambton Quay?

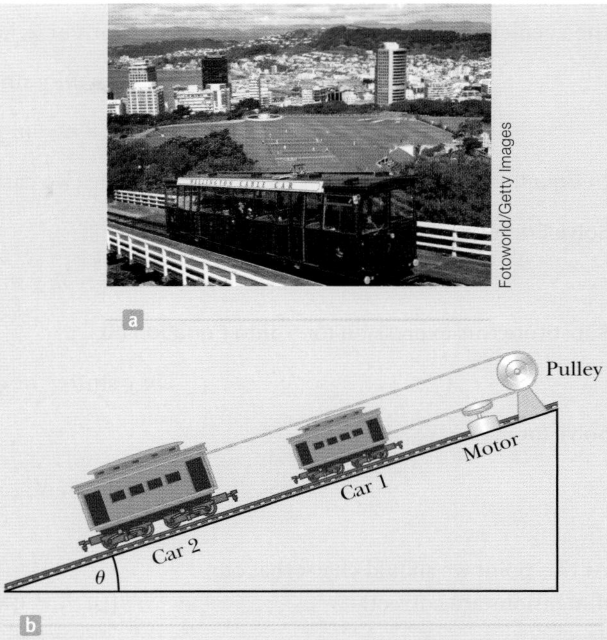

Figure 4.26
(Example 4.12) (a) The Wellington cable car (technically a funicular) carries passengers from Lambton Quay up a steep track to the Wellington Botanic Gardens. (b) A schematic diagram showing how the system works

Solution

Conceptualise The heavier car will tend to roll down the hill, pulling the lighter car up the hill. Because the cars are connected by a cable (which we assume does not stretch), their accelerations have the same magnitude. We start as usual by drawing a diagram showing the forces acting on the two cars. Note that in this problem we want to find a final time to travel some distance, starting from rest, so we will need to break this problem into parts to solve it. First we will need to find the acceleration, then we can use the acceleration to find the time taken to travel the distance to the bottom of the hill.

Model This is similar to the previous example. We can identify forces on each of the two cars and we are looking for an acceleration, so we model the cars as particles subject to a net force. We ignore friction, and model the cable as an ideal rope which does not stretch and hence has the same tension at all points. Once we have the acceleration we can use the analysis model from Chapter 2 for particles with constant acceleration to find the time taken.

Analyse Draw a free-body diagram for each car. Both cars must have the same acceleration, a, as they are connected by the cable. We choose our x direction for both cars to be parallel to the gradient of the hill, as this is the direction in which there will be an acceleration. In the y direction there is no acceleration and hence no net force.

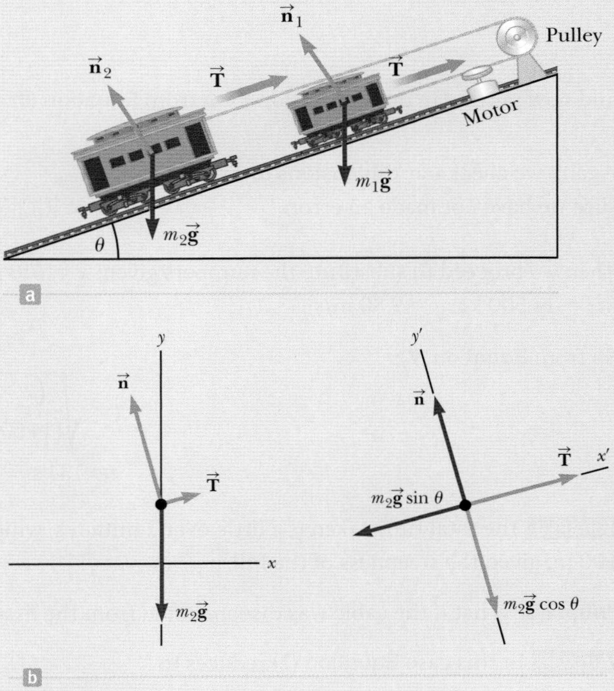

Figure 4.27
(Example 4.12) (a) Forces acting on the two cars (b) Free-body diagram for car 2 (c) Free-body diagram showing the components of all forces acting on the car parallel and perpendicular to the tracks

Apply Newton's second law in component form to car 1, choosing the upwards direction as positive:

$$\sum F_x = T - m_1 g \sin\theta = m_1 a_x = m_1 a \quad (1)$$

$$\sum F_y = n - m_1 g \cos\theta = 0 \quad (2)$$

For Car 1 to accelerate upwards, it is necessary that $T > m_1 g \sin\theta$. In **Equation (1)**, we replaced a_x with a because the acceleration has only an x component. For car 2, we can write:

$$\sum F_{x'} = m_2 g \sin\theta - T = m_2 a_{x'} = m_2 a \quad (3)$$

$$\sum F_{y'} = n - m_2 g \cos\theta = 0 \quad (4)$$

In **Equation (3)**, we replaced $a_{x'}$ with a because the two cars have accelerations of equal magnitude a.

Solve **Equation (1)** for T:

$$T = m_1 g \sin\theta + m_1 a \quad (5)$$

Substitute this expression for T into **Equation (3)**:

$$m_2 g \sin\theta - m_1 g \sin\theta - m_1 a = m_2 a$$

Solve for a:

$$a = \frac{(m_2 - m_1)(g \sin\theta)}{(m_1 + m_2)} \quad (6)$$

At this point we should check that our dimensions are correct:

$$[LT^{-2}] = ([M][LT^{-2}])/[M] = [LT^{-2}] \text{ ☺}$$

We can now either proceed to find a value for the acceleration or to find an expression for the time taken. It is generally preferable to find an algebraic expression for the final value before substituting any values.

Applying the analysis model for a particle with constant acceleration to car 2 we can now use **Equation 2.15**:

$x = v_i t + ½at^2$, with $v_i = 0$ and rearranging for t:

$$t = \sqrt{\frac{2x}{a}}$$

and now putting in our expression for a from **Equation (6)** above:

$$t = \sqrt{\frac{(2x)(m_1 + m_2)}{(m_2 - m_1)(g \sin\theta)}} \quad (7)$$

Again, we check our dimensions to make sure we have not made an error:

$$[T] = [([L][M])/([M][LT^{-2}])]^{1/2} = [T] \text{ ☺}$$

Then we proceed to substitute the numbers given: $x = 610$ m, $\sin\theta = \frac{119 \text{ m}}{610 \text{ m}} = \frac{119}{610}$, $m_1 = 13\,500$ kg, $m_2 = 14\,500$ kg, $g = 9.80$ m/s^2.

So from **Equation (7)**:

$$t = \sqrt{\frac{(2 \times 610 \text{ m})(13\,500 \text{ kg} + 14\,500 \text{ kg})}{(14\,500 \text{ kg} - 13\,500 \text{ kg})(9.80 \text{ m/s}^2 \times \frac{119}{610})}}$$

$$= 134 \text{ s}$$

Finalise The total time taken is a little over 2 minutes, which seems quite fast but not unreasonable for a distance of about 600 m, given the steepness of the hill.

What If? What if the cable was disconnected from the first car and the second car is simply allowed to roll down the hill?

Answer In this case **Equation (7)** reduces to

$$t = \sqrt{\frac{2x}{g \sin\theta}} = \sqrt{\frac{2 \times 610 \text{ m}}{9.8 \text{ m/s}^2 \times \frac{119}{610}}} = 25 \text{ s}$$

Note that in this case the mass of the object is no longer relevant, as acceleration due to gravity is independent of the mass of the object.

4.8 Motion in accelerating reference frames

Newton's laws of motion describe observations that are made in an inertial frame of reference. In this section, we analyse how Newton's laws are applied by an observer in a non-inertial frame of reference, that is, one that is accelerating. For example, recall the discussion of the air hockey table on a train in Section 4.2. The train moving at constant velocity represents an inertial frame. An observer on the train sees the puck at rest remain at rest, and Newton's first law appears to be obeyed.

The accelerating train is *not* an inertial frame. According to an observer on this train, there appears to be no force on the puck, yet it accelerates from rest towards the back of the train, appearing to violate Newton's first law. This property is a general property of observations made in non-inertial frames: there appear to be unexplained accelerations of objects that are not 'fastened' to the frame. Newton's first law is not, however, violated. It only appears to be violated because the observations are made from a non-inertial frame. In general, the direction of the unexplained acceleration is opposite the direction of the acceleration of the non-inertial frame.

On the accelerating train, as you watch the puck accelerating towards the back of the train, you might conclude, based on your belief in Newton's second law, that a force has acted on the puck to cause it to accelerate. We call an apparent force such as this one a **pseudoforce** (sometimes called a **fictitious force**) because it is due to observations made in an accelerating reference frame. A pseudoforce appears to act on an object in the same way as a real force; however, real forces are always interactions between two objects, and a second object cannot be identified for a pseudoforce. (What second object is interacting with the puck to cause it to accelerate?) In general, simple pseudoforces appear to act in the direction *opposite* that of the acceleration of the non-inertial frame, and are the evidence that you are in a non-inertial frame. For example, the train accelerates forwards and there appears to be a pseudoforce causing the puck to slide towards the back of the train. This acceleration of the puck, with no apparent cause, tells an observer that the train is accelerating even if they cannot feel it.

The train example describes a pseudoforce due to a change in the train's speed. Another pseudoforce is due to the change in the *direction* of the velocity vector. Consider a car travelling along a highway at a high speed and approaching a curved exit ramp on the left as shown in **Figure 4.28a**. As the car takes the sharp left turn on the ramp, the driver (on the right-hand side) leans or slides to the right and hits the door. At that point the force exerted by the door on the driver keeps her from being thrown from the car. What causes her to move towards the door? A popular but *incorrect* explanation is that a force, called the **centrifugal force**, acting towards the right in **Figure 4.28b** pushes the driver outwards from the centre of the circular path. The centrifugal force is a pseudoforce, and only appears to be acting because the car represents a non-inertial reference frame that has a centripetal acceleration towards the centre of its circular path. As a result, the driver feels an apparent force that is outwards from the centre of the circular path, or to the right in **Figure 4.28b**, in the direction opposite that of the acceleration.

Although we generally use inertial frames and avoid the use of pseudoforces in physics, you may come across them in other contexts. However, be careful that you understand when a pseudoforce such as the centrifugal force is being used as part of a model for forces in a non-inertial reference frame, and when it is used as an incorrect label for an apparent force in an inertial reference frame.

Let us consider this phenomenon in terms of Newton's laws. Before the car enters the ramp, the driver is moving in a straight-line path. As the car enters the ramp and travels a curved path, the driver tends to move along the original straight-line path, which is in accordance with Newton's first law: the natural tendency of an object is to continue moving in a straight line. If a sufficiently large force (towards the centre of curvature) acts on the driver as in **Figure 4.28c**, however, she moves in a curved path along with the car. This force is the force of friction between her and the car seat. If this friction force is not large enough, the seat follows a curved path while the driver tends to continue in the straight-line path of the car before the car began the turn. Therefore, from the point of view of an observer in the car, the driver leans or slides to the right relative to the seat. Eventually, she encounters the door, which provides a normal force large enough to enable her to follow the same curved path as the car.

a

From the driver's frame of reference, a force appears to push her towards the right door, but it is a fictitious force.

Fictitious force

b

Relative to the reference frame of the Earth, the car seat applies a real force (friction) towards the left on the driver, causing her to change direction along with the rest of the car.

Real force

c

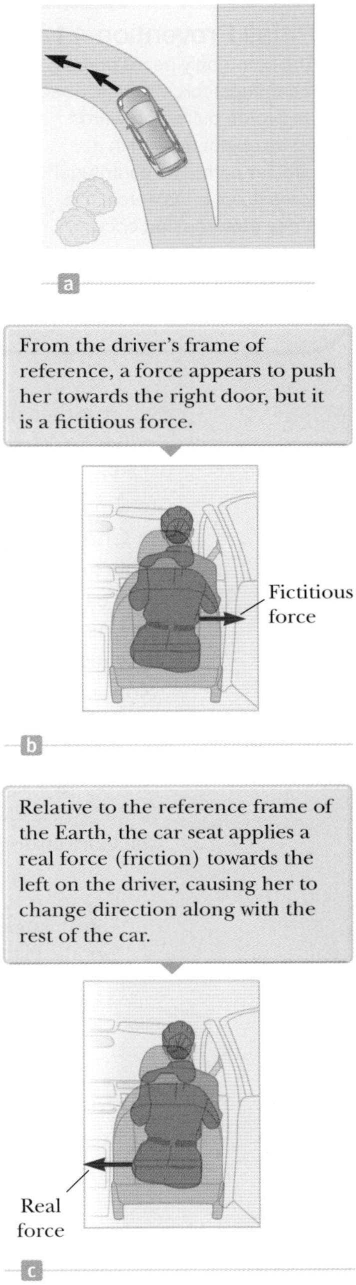

Figure 4.28
(a) A car approaching a curved exit ramp. What causes the driver to move towards the right-hand door? (b) Driver's frame of reference (c) Reference frame of the Earth

Pitfall Prevention 4.12

The commonly heard phrase 'centrifugal force' is described as a force pulling *outwards* on an object moving in a circular path. The centrifugal force is not a real force; it is a pseudoforce that occurs when you are in a non-inertial reference frame. This term is commonly used incorrectly.

From the point of view of an external observer, the person in the car is pushing against the door with a force N, because she is trying to continue in a straight line while the door (and the rest of the car) turns the corner. From Newton's third law, we know that the door must therefore exert a force of −N on her. In the car example, this force is horizontal and directed towards the centre of the curve.

Recall that your sensation of weight depends upon the supporting force that is applied to you by whatever surface you are standing on. In the absence of a gravitational force, or when an object is in orbit or freefall, the sensation of weight can be achieved by rotation. In a rotating reference frame, such as a rotating space station, a normal force provided by the internal surface of the space station, which provides the force necessary for its contents to stay inside the space station, just like the car door in the example above, gives the sensation of weight. Hence rotation has been proposed as a means of providing 'artificial gravity' in spacecraft. This is described in more detail in Chapter 5.

Another interesting pseudoforce is the '**Coriolis force**'. It is an apparent force caused by changing the radial position of an object in a rotating coordinate system.

For example, suppose you and a friend are on opposite sides of a rotating circular platform and you decide to throw a ball to your friend. This is just the same situation as the Curve Ball exhibit at Questacon, described in Chapter 3, in which Kate and Anna were trying to throw a ball to each other as the ride rotated. **Active Figure 4.29a** represents what an observer would see if the ball is viewed while the observer is hovering at rest above the rotating platform. According to this observer, who is in an inertial frame, the ball follows a straight line, as it must according to Newton's first law. At $t = 0$ you throw the ball towards your friend, but by the time t_f when the ball has crossed the platform, your friend has moved to a new position and can't catch the ball. Now, consider the situation from your friend's viewpoint. He is in a non-inertial reference frame because he is undergoing a centripetal acceleration and, therefore, is in a non-inertial (accelerating) reference frame. He starts off seeing the ball coming towards him, but as it crosses the platform, it veers to one side as shown in **Active Figure 4.29b**. Therefore, your friend on the rotating platform states that the ball does not obey Newton's first law and claims that a sideways force is causing the ball to follow a curved path. This pseudoforce is called the Coriolis force.

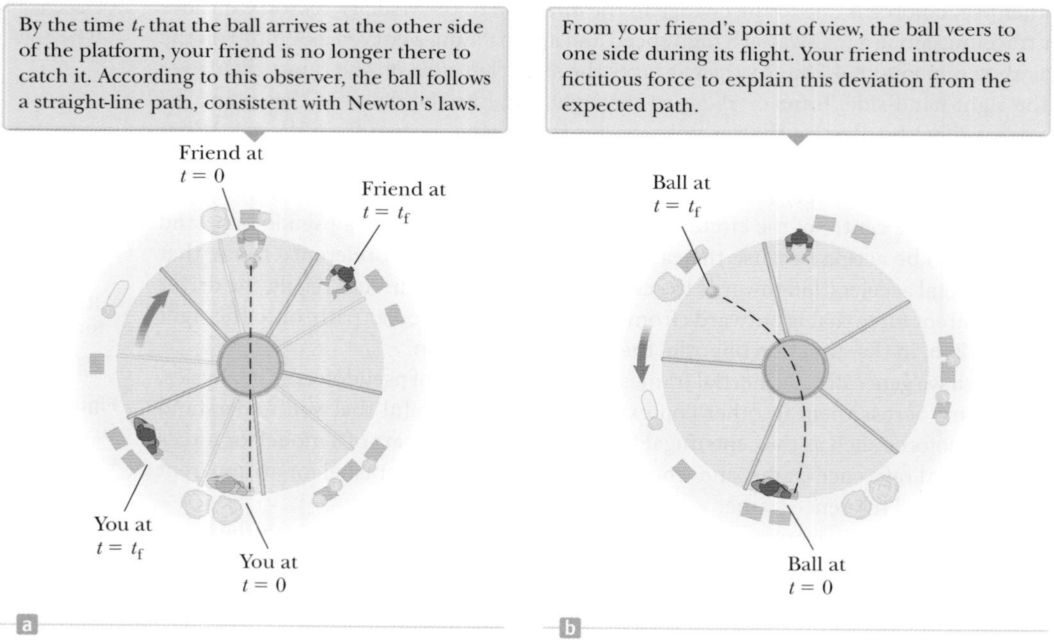
By the time t_f that the ball arrives at the other side of the platform, your friend is no longer there to catch it. According to this observer, the ball follows a straight-line path, consistent with Newton's laws.
Friend at $t = 0$
Friend at $t = t_f$
You at $t = t_f$
You at $t = 0$
a
From your friend's point of view, the ball veers to one side during its flight. Your friend introduces a fictitious force to explain this deviation from the expected path.
Ball at $t = t_f$
Ball at $t = 0$
b

Active Figure 4.29

You and your friend stand at the edge of a rotating circular platform. You throw the ball at $t = 0$ in the direction of your friend. (a) Overhead view observed by someone in an inertial reference frame attached to the Earth. The ground appears stationary, and the platform rotates clockwise. (b) Overhead view observed by someone in a non-inertial reference frame attached to the platform. The platform appears stationary, and the ground rotates anticlockwise.

We are frequently in non-inertial frames and observe the effects of these pseudoforces. For example, an object on your dashboard slides off if you press the accelerator of your car. As you ride on a merry-go-round, you feel pushed towards the outside as if due to the centrifugal force, and you are likely to fall over due to the Coriolis force if you walk along a radial line while a merry-go-round rotates. The Coriolis force due to the rotation of the Earth is responsible for rotations of cyclones and hurricanes and for large-scale ocean currents.

Figure 4.30
Cyclone Laurence crossing the north-west coast of Australia in December 2009. The rotation of the Earth causes cyclones in the southern hemisphere to rotate clockwise and hurricanes to rotate anticlockwise in the northern hemisphere, giving an impression of an additional or pseudoforce, in this case the Coriolis force.

Quick Quiz 4.8

Consider the driver in the car making a left turn in **Figure 4.27**. Which of the following is correct regarding the forces in the horizontal direction if she is making contact with the right-hand door? **(a)** She is in equilibrium due to real forces acting to the right and real forces acting to the left. **(b)** She is subject only to real forces acting to the right. **(c)** She is subject only to real forces acting to the left. **(d)** None of those statements is true.

TRY THIS

Hang an object from a piece of string and try running from rest while holding it. Watch what happens to the string – does it hang vertically while you accelerate? Ask a friend to hold the string and run past you, starting from rest and accelerating as fast as they can. You could also try this in a moving vehicle (with someone else driving).

Example 4.13

Pseudoforces in linear motion – the stroller mobile

A small toy of mass m hangs by a cord from the top of a stroller that is being pushed to the right and accelerating as shown in **Figure 4.31**. Both the child on the ground, an inertial observer, and the baby in the stroller (a non-inertial observer) notice that the cord makes an angle θ with respect to the vertical. The baby and his mother pushing the stroller, both non-inertial observers, believe that a force (which we know to be fictitious) causes the observed deviation of the cord from the vertical. How is the magnitude of this force related to the stroller's acceleration observed by the watching child in **Figure 4.31**?

Figure 4.31
(Example 4.13) A small toy suspended from the top of a stroller accelerating to the right is deflected as shown.

Solution

Conceptualise Place yourself in the role of each of the two observers in **Figure 4.31**. As the inertial observer on the ground, you see the stroller accelerating and know that the deviation of the cord is due to this acceleration. As the non-inertial observer in the stroller, imagine that you ignore any effects of the stroller's motion so that you are not aware of its acceleration. Because you are unaware of this acceleration, you claim that a force is pushing on the toy to cause the deviation of the cord from the vertical. If you did the *Try this* example above you may have observed this for yourself. Draw a free-body diagram showing the forces acting on the toy.

Model For the inertial observer, we model the toy as a particle subject to a net force in the horizontal direction and a particle in equilibrium in the vertical direction. For the non-inertial observer, the toy is modelled as a particle in equilibrium in both directions.

Analyse According to the watching child (an inertial observer at rest), the forces on the toy are the force $\vec{\mathbf{T}}$ exerted by the cord and the gravitational force. The inertial observer concludes that the toy's acceleration is the same as that of the stroller and that this acceleration is provided by the horizontal component of $\vec{\mathbf{T}}$.

Example 4.13 cont.

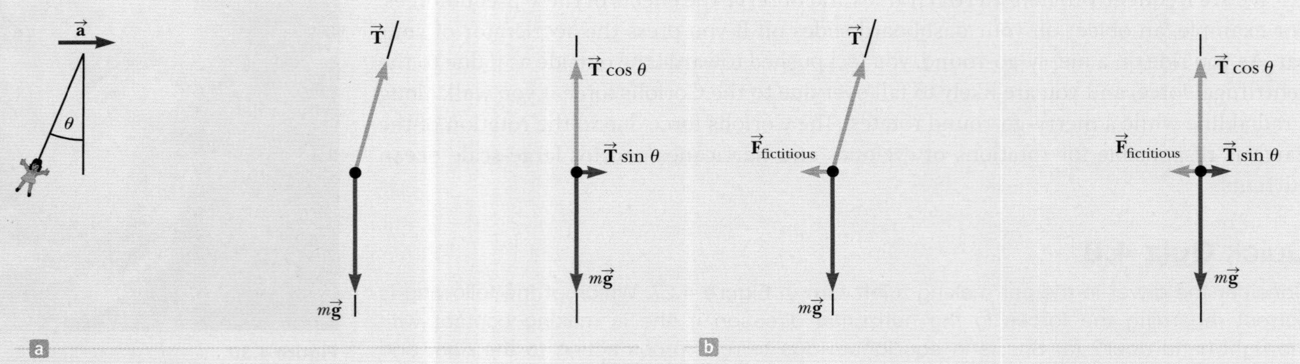

Figure 4.32
(Example 4.13) (a) The forces as seen by an inertial observer such as the watching child. (b) The forces as seen by an accelerating observer such as the baby in the stroller.

Apply Newton's second law in component form to the toy according to the inertial observer:

Inertial observer
$$\sum F_x = T\sin\theta = ma \quad (1)$$
$$\sum F_y = T\cos\theta - mg = 0 \quad (2)$$

According to the non-inertial observer riding in the stroller (the baby), the cord also makes an angle θ with the vertical; to that observer, however, the toy is at rest and so its acceleration is zero. Therefore, the non-inertial observer introduces a force (which we know to be fictitious) in the horizontal direction to balance the horizontal component of $\vec{\mathbf{T}}$ and claims that the net force on the toy is zero.

Apply Newton's second law in component form to the toy according to the non-inertial observer:

Non-inertial observer
$$\sum F'_x = T\sin\theta - F_{\text{fictitious}} = 0$$
$$\sum F'_y = T\cos\theta - mg = 0$$

These expressions are equivalent to **Equations (1)** and (2) if $F_{\text{fictitious}} = ma$, where a is the acceleration according to the inertial observer.

Finalise If we make this substitution in the equation for $\sum F'_x$ above, we obtain the same mathematical results as the inertial observer. The physical interpretation of the cord's deflection, however, differs in the two frames of reference.

What If? Suppose the inertial observer wants to measure the acceleration of the stroller by means of the pendulum (the toy hanging from the cord). How could he do so?

Answer Our intuition tells us that the angle θ the cord makes with the vertical should increase as the acceleration increases. By solving **Equations (1)** and (2) simultaneously for a, we find that $a = g\tan\theta$. Therefore, the magnitude of the stroller's acceleration can be found by measuring the angle θ. Because the deflection of the cord from the vertical serves as a measure of acceleration, a simple pendulum can be used as an accelerometer.

TRY THIS

You can make a simple accelerometer using a clear plastic bottle, some oil or water (or both) and some food colouring. Fill the bottle to about one-third full with water, add a centimetre of oil and a few drops of food colouring. The layer of oil helps make the surface more visible, and decreases 'sloshing' of the water. If you lie the bottle on its side on a cart or the dashboard of the car (while someone else drives) you can watch the shape of the water in the bottle change as the car accelerates. Try putting it on a spinning surface. What shape does it form, and why?

End-of-chapter resources

At the start of the chapter we asked what forces act on an astronaut in the International Space Station (ISS), and are they actually weightless? If not, why do they float as if there were no gravity there?

The ISS orbits in a nearly circular orbit at an altitude of approximately 300 km from the surface of the Earth – a small fraction of the Earth's radius of 6400 km. The gravity at that height is 88% that at the Earth's surface.

The ISS does not rotate to provide 'artificial gravity' for its occupants.

Recall that weight is the gravitational force acting on an object, $\vec{\mathbf{F}}_g = m\vec{\mathbf{g}}$. The mass of the astronaut does not change very much between leaving Earth and arriving on the ISS. However, the acceleration due to gravity, $\vec{\mathbf{g}}$, does change. Hence their weight (the gravitational force acting on them) is different.

We expect it will be less on the space station. We might even expect that it will be very small, as we have seen images of astronauts floating around, apparently weightless in the space station.

Let's do an example to check. If an astronaut has a mass of 80 kg on Earth, they will have a weight of 784 N. On the ISS their mass will still be 80 kg but their weight will be 88% of 784 N or 690 N. 690 N is certainly less than 784N, but only 12% less, so it doesn't account for the apparent weightlessness and floating. So why do astronauts appear weightless on the ISS?

The space station, and all its contents including astronauts, are in orbit around the Earth. An object in orbit, assuming no friction forces, can be modelled as subject *only* to the gravitational force exerted on it by the object it is orbiting, in the reference frame of the object it is orbiting. (We know of course that the ISS is subject to the gravitational force of not only the Earth, which it orbits, but also the Sun, which the Earth and the ISS orbit together.) This is equivalent to the free fall case of the broken lift cable in Example 4.10. The only difference is that the orbiting spacecraft also has a horizontal component to its motion, which is constant speed, while the lift and its occupants have zero horizontal velocity. In both cases, the only acceleration is towards the centre of the Earth.

Alamy Stock Photo/NASA Photo

Recall that your sensation of weight comes from the normal force exerted on you by a supporting surface. The scale force that a set of scales shows in an accelerating elevator is a measure of your perceived or 'apparent' weight. The scale force is *not*, however, equal to the gravitational force acting on you, which *is* your weight. If you accelerate downwards in a lift, your weight is the same but the scale force (and your sensation of weight) decreases. In the limit that your acceleration is equal to g the scale force reduces to zero, as does your sensation of weight, because (from Newton's third law) as you are no longer applying a force to the scale, it no longer applies a force to you.

The astronauts in the space station are accelerating all the time at 88% of g, along with all the rest of the space station, so they experience a feeling of 'weightlessness', although they feel just as much force from running into a wall at high speed as you would on Earth. The forces acting on the astronauts are gravity, and contact forces whenever they run into or push against something. These contact forces are the friction and normal forces, which are discussed in detail in the next chapter.

The problems found in this chapter may be assigned online in Enhanced Web Assign.

Worked solutions to every fifth problem are available in the Student Solutions Manual. Register online at **www.cengagebrain.com** for access.

Summary

Definitions

An **inertial frame of reference** is a frame in which an object that does not interact with other objects experiences zero acceleration. Any frame moving with constant velocity relative to an inertial frame is also an inertial frame.

We define **force** as **that which causes a change in the motion of an object**.

Weight is the gravitational force that acts on an object, $F_g = mg$, where m is the mass of the object and **g** is the acceleration due to gravity. **Mass** is an intrinsic property of an object and does not depend on its location. Weight is not an intrinsic property, it is a force, and depends on the location of the object.

Concepts and principles

Newton's first law states that it is possible to find an inertial frame in which an object that does not interact with other objects experiences zero acceleration or, equivalently, in the absence of an external force, when viewed from an inertial frame, an object at rest remains at rest and an object in uniform motion in a straight line maintains that motion.

Newton's second law states that the acceleration of an object is directly proportional to the net force acting on it and inversely proportional to its mass.

Newton's third law states that if two objects interact, the force exerted by object 1 on object 2 is equal in magnitude and opposite in direction to the force exerted by object 2 on object 1.

The **gravitational force** exerted on an object is equal to the product of its mass (a scalar quantity) and the free-fall acceleration: $\vec{\mathbf{F}}_g = m\vec{\mathbf{g}}$.

The **weight** of an object is the gravitational force acting on the object.

An **inertial reference frame** is one in which an object experiencing no force has no acceleration. An accelerating reference frame, such as an accelerating train or a rotating surface, is a **non-inertial reference frame**.

Pseudoforces, or fictitious forces, appear to act in **non-inertial reference frames**. They usually appear to act in the direction opposite to the acceleration of the reference frame. The **centrifugal force** is a pseudoforce which appears to act in non-inertial, rotating reference frames.

Analysis models for problem solving

Particle subject to a net force If a particle of mass m experiences a non-zero net force, its acceleration is related to the net force by Newton's second law:

$$\sum \vec{\mathbf{F}} = m\vec{\mathbf{a}} \tag{4.2}$$

Particle in equilibrium If a particle maintains a constant velocity (so that $\vec{\mathbf{a}} = 0$), which could include a velocity of zero, the forces on the particle balance and Newton's second law reduces to

$$\sum \vec{\mathbf{F}} = 0 \tag{4.7}$$

An observer in a non-inertial (accelerating) frame of reference introduces **pseudoforces** when applying Newton's second law in that frame.

Chapter review quiz

To help you revise Chapter 4: Forces and Newton's laws, complete the automatically graded Chapter review quiz at http://login.cengagebrain.com.

Conceptual questions

1. A person holds a ball in her hand. (a) Identify all the external forces acting on the ball and the Newton's third-law reaction force to each one. (b) If the ball is dropped, what force is exerted on it while it is falling? Identify the reaction force in this case. (Ignore air resistance.)
2. If a car is travelling due west with a constant speed of 20 m/s, what is the total force acting on it?
3. A passenger sitting in the rear of a bus claims that he was injured when the driver slammed on the brakes, causing a suitcase to come flying towards him from the front of the bus. If you were the judge in this case, what decision would you make? Why?
4. A spherical balloon inflated with air is held stationary, with its opening, on the west side, pinched shut. (a) Describe the forces exerted by the air inside and outside the balloon on sections of the rubber. (b) After the balloon is released, it takes off towards the east, gaining speed rapidly. Explain this motion in terms of the forces now acting on the balloon. (c) Explain how a skyrocket takes off from its launch pad.
5. If you hold a horizontal metal bar several centimetres above the ground and move it through grass, each leaf of grass bends out of the way. If you increase the speed of the bar, each leaf of grass will bend more quickly. How then does a rotary power lawn mower manage to cut grass? How can it exert enough force on a leaf of grass to shear it off?
6. A child tosses a ball straight up. She says that the ball is moving away from her hand because the ball feels an upwards 'force of the throw' as well as the gravitational force. Why does the ball move away from the child's hand?
7. The observer in the accelerating elevator of Example 4.10 would claim that the 'weight' of the sample is T, the scale reading, but this answer is obviously wrong. Why does this observation differ from that of a person outside the elevator, at rest with respect to the Earth?
8. Balancing carefully, three boys inch out onto a horizontal tree branch above a pond, each planning to dive in separately. The third boy in line notices that the branch is barely strong enough to support them. He decides to jump straight up and land back on the branch to break it, spilling all three into the pond. When he starts to carry out his plan, at what precise moment does the branch break? Explain your answer. *Suggestion*: Pretend to be the third boy and imitate what he does in slow motion. If you are still unsure, stand on a bathroom scale and repeat the suggestion.
9. A weightlifter stands on a bathroom scale. He pumps a barbell up and down. What happens to the reading on the scale as he does so? **What If?** What if he is strong enough to actually *throw* the barbell upwards? How does the reading on the scale vary now?
10. A bucket of water can be whirled in a vertical path such that no water is spilled. Why does the water stay in the bucket, even when the bucket is above your head?
11. Consider a tennis ball in contact with a stationary floor and with nothing else. (a) Can the normal force be different in magnitude from the gravitational force exerted on the ball? (b) Can the force exerted by the floor on the ball be different in magnitude from the force the ball exerts on the floor?
12. In **Figure CQ4.12**, the light, taut, unstretchable cord B joins block 1 and the larger-mass block 2. Cord A exerts a force on block 1 to make it accelerate forwards. (a) How does the magnitude of the force exerted by cord A on block 1 compare with the magnitude of the force exerted by cord B on block 2? Is it larger, smaller or equal? (b) How does the acceleration of

block 1 compare with the acceleration (if any) of block 2? (c) Does cord B exert a force on block 1? If so, is it forwards or backwards? Is it larger, smaller or equal in magnitude to the force exerted by cord B on block 2?

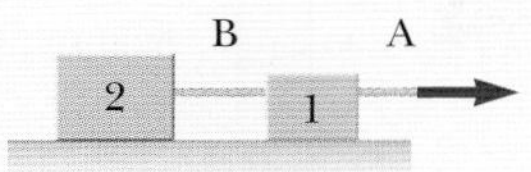

Figure CQ4.12

13. Identify action–reaction pairs in the following situations: (a) a man takes a step, (b) a snowball hits a girl in the back, (c) a ball thrown upwards slows down as it goes up, (d) a gust of wind strikes a window.
14. Twenty people participate in a tug-of-war. The two teams of ten people are so evenly matched that neither team wins. After the game they notice that a car is stuck in the mud. They attach the tug-of-war rope to the bumper of the car, and all the people pull on the rope. The heavy car has just moved a couple of decimetres when the rope breaks. Why did the rope break in this situation when it did not break when the same twenty people pulled on it in a tug-of-war?
15. The younger brother of a physics student asks her to tow him along in his wagon. The student explains that she can't because, whatever force she exerts on the wagon, the wagon, according to Newton's third law, will exert an equal and opposite force on her, preventing her from moving, and hence it is impossible for her to tow him along. What is wrong with the physics of this argument? Draw a diagram showing the forces acting on the wagon and the physics student to help explain your answer.
16. An athlete grips a light rope that passes over a low-friction pulley attached to the ceiling of a gym. A sack of sand precisely equal in weight to the athlete is tied to the other end of the rope. Both the sand and the athlete are initially at rest. The athlete climbs the rope, sometimes speeding up and slowing down as he does so. What happens to the sack of sand? Explain your answer.
17. A children's cartoon shows a person being thrown radially outwards from an out of control merry-go-round. This is explained by someone watching the show to be due to the centrifugal force. (a) Draw a diagram showing the path a person thrown from a rapidly spinning merry-go-round would actually follow. (b) Explain what role, if any, the centrifugal force plays in this situation.

Problems

Sections 4.1 through 4.6

1. The average speed of a nitrogen molecule in air is about 6.70×10^2 m/s, and its mass is 4.68×10^{-26} kg. (a) If it takes 3.00×10^{-13} s for a nitrogen molecule to hit a wall and rebound with the same speed but moving in the opposite direction, what is the average acceleration of the molecule during this time interval? (b) What average force does the molecule exert on the wall?
2. A toy rocket engine is securely fastened to a large puck that can glide with negligible friction over a horizontal air hockey table, taken as the xy plane. The 4.00 kg puck has a velocity of $3.00\hat{\mathbf{i}}$ m/s at one instant. Eight seconds later, its velocity is $(8\hat{\mathbf{i}} + 10\hat{\mathbf{j}})$ m/s. Assuming the rocket engine exerts a constant horizontal force, find (a) the components of the force and (b) its magnitude.
3. An orthodontist uses a wire brace to align a patient's crooked tooth as in **Figure P4.3**. The tension in the wire is adjusted to have a magnitude of 18.0 N. Find the magnitude of the net force exerted by the wire on the crooked tooth.

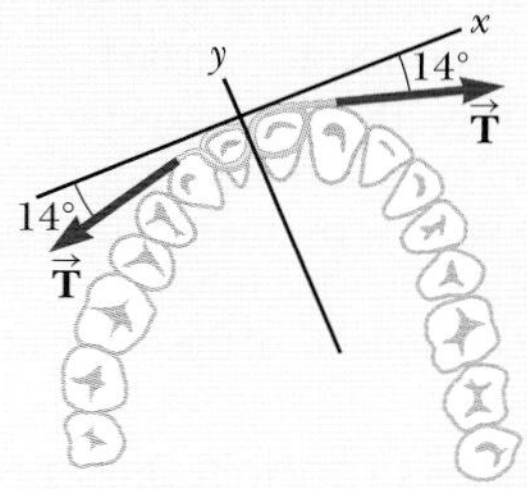

Figure P4.3

4. An electron of mass 9.11×10^{-31} kg has an initial speed of 3.00×10^5 m/s. It travels in a straight line, and its speed increases to 7.00×10^5 m/s in a distance of 5.00 cm. Assuming its acceleration is constant, (a) determine the magnitude of the force exerted on the electron and (b) compare this force with the weight of the electron, which we ignored.
5. One or more external forces, large enough to be easily measured, are exerted on each object enclosed in a dashed box shown in **Figure 4.1**. Identify the reaction to each of these forces.
6. A 3.00-kg object undergoes an acceleration given by $\vec{\mathbf{a}} = (2.00\hat{\mathbf{i}} + 5.00\hat{\mathbf{j}})$ m/s². Find (a) the resultant force acting on the object and (b) the magnitude of the resultant force.
7. A ball of mass m is dropped at $t = 0$ from the roof of a building of height h. While the ball is falling, a very strong wind blowing parallel to the face of the building exerts a constant horizontal force F on the ball. (a) At what time t does the object strike the ground? Express t in terms of g and h. State any assumptions you have made. (b) Find an expression in terms of m and F for the acceleration a_x of the object in the horizontal direction (taken as the positive x direction). (c) How far is the object displaced horizontally before hitting the ground? Answer in terms of m, g, F and h. (d) Find the magnitude of the object's acceleration while it is falling, using the variables F, m and g.
8. A force $\vec{\mathbf{F}}$ applied to an object of mass m_1 produces an acceleration of 3.00 m/s². The same force applied to a second object of mass m_2 produces an acceleration of 1.00 m/s². (a) What is the value of the ratio m_1/m_2? (b) If m_1 and m_2 are combined into one object, find its acceleration under the action of the force $\vec{\mathbf{F}}$.
9. ±? Two students are trying to move a heavy desk, mass = (50 ± 1) kg, across a smooth floor by applying forces $F_1 = (20.0 \pm 0.5)$ N and $F_2 = (15.0 \pm 0.5)$ N in a horizontal direction. They have put smooth cloth under the feet of the desk to stop it scratching the highly polished timber floor. Initially they apply the forces at right angles to each other. (a) Find the acceleration of the desk. Draw a diagram and write the acceleration as a vector. (b) If they push so that the angle between the direction of the forces is 60°, what is the acceleration of the desk? (c). What if? What if the forces are parallel? **What If** they are anti-parallel? Assume the uncertainties in any angles are negligible.

10. You stand on the seat of a chair and then hop off. (a) During the time interval you are in flight down to the floor, the Earth moves toward you with an acceleration of what order of magnitude? In your solution, explain your logic. Model the Earth as a perfectly solid object. (b) The Earth moves toward you through a distance of what order of magnitude?
11. Three forces acting on an object are given by $\vec{\mathbf{F}}_1 = (-22.00\hat{\mathbf{i}} + 2.00\hat{\mathbf{j}})$ N, $\vec{\mathbf{F}}_2 = (5.00\,\hat{\mathbf{i}} - 3.00\,\hat{\mathbf{j}})$ N and $\vec{\mathbf{F}}_3 = (-45.0\,\hat{\mathbf{i}})$ N. The object experiences an acceleration of magnitude 3.75 m/s². (a) What is the direction of the acceleration? (b) What is the mass of the object? (c) If the object is initially at rest, what is its speed after 10.0 s? (d) What are the velocity components of the object after 10.0 s?

Section 4.7 Analysis models using Newton's laws.

12. **Figure P4.12** shows a man poling a boat – a very efficient mode of transportation – across a shallow lake. He pushes parallel to the length of the light pole, exerting a force of magnitude 240 N on the bottom of the lake. Assume the pole lies in the vertical plane containing the keel of the boat. At one moment, the pole makes an angle of 35.0° with the vertical and the water exerts a horizontal drag force of 47.5 N on the boat, opposite to its forwards velocity of magnitude 0.857 m/s. The mass of the boat including its cargo and the worker is 370 kg. (a) The water exerts a buoyant force vertically upwards on the boat. Find the magnitude of this force. (b) Model the forces as constant over a short interval of time to find the velocity of the boat 0.450 s after the moment described.

AP Photo/Rebecca Blackwell

Figure P4.12

13. Assume the three blocks portrayed in **Figure P4.13** move on a frictionless surface and a 42-N force acts as shown on the 3.0-kg block. Determine (a) the acceleration given this system, (b) the tension in the cord connecting the 3.0-kg and the 1.0-kg blocks, and (c) the force exerted by the 1.0-kg block on the 2.0-kg block.

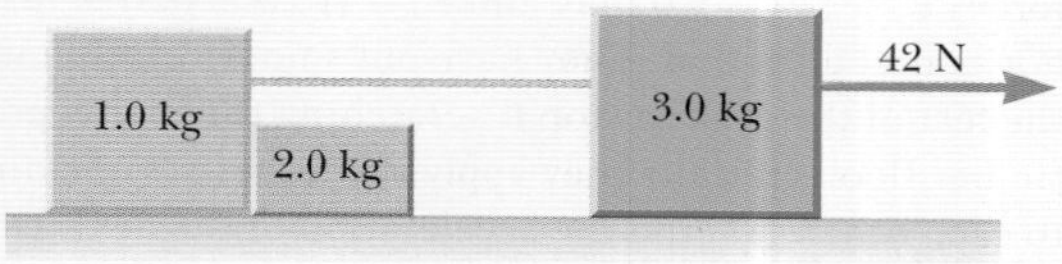

Figure P4.13

14. The systems shown in **Figure P4.14** are in equilibrium. If the spring scales are calibrated in newtons, what do they read? Ignore the masses of the pulleys and strings and assume the pulleys and the incline in **Figure P4.14d** are frictionless.

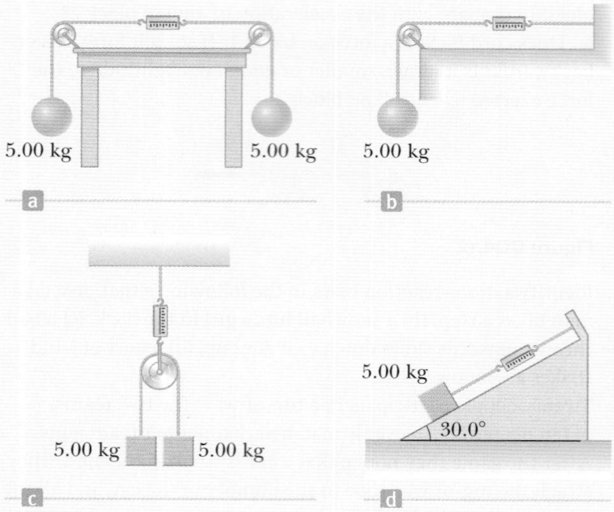

Figure P4.14

15. A child puts down their toboggan at the top of an icy hill, which has an inclination of $\theta = 15°$ and length of 20 m. The toboggan starts from rest at the top, and slides down the hill. (a) Draw a free-body diagram of the toboggan. Find (b) the acceleration of the toboggan and (c) its speed when it reaches the bottom of the incline.
16. A 3.00 kg object is moving in a plane, with its x and y coordinates given by $x = 5t^2 - 1$ and $y = 3t^3 + 2$, where x and y are in metres and t is in seconds. Find the magnitude of the net force acting on this object at $t = 2.00$ s.
17. The distance between two telephone poles is 50 m. When a 5 kg wedge-tailed eagle lands on the telephone wire midway between the poles, the wire sags 0.2 m. (a) Draw a free-body diagram for the bird. (b) Ignoring the weight of the wire, how much tension does the bird produce in the wire? (c) Do you think the approximation of the wire as an ideal rope is valid in this case? Explain your answer.
18. A beached whale is being lifted in a sling as shown in **Figure P4.18**. The whale, whose weight is F_g, hangs in equilibrium from three wires as shown in **Figure P4.18**. Two of the wires make angles θ_1 and θ_2 with the horizontal. Assuming the system is in equilibrium, show that the tension in the left-hand wire is

$$T_1 = \frac{F_g \cos \theta_2}{\sin \theta_1 + \theta_2}$$

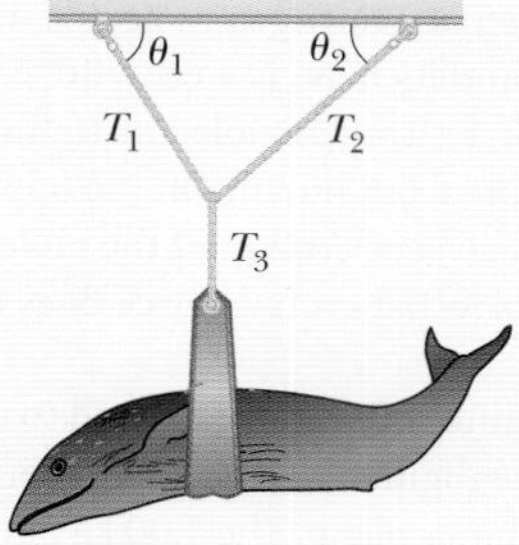

Figure P4.18
The forces involved in carrying a whale in a sling

19. A set-up similar to the one shown in **Figure P4.19** is often used in hospitals to support and apply a horizontal traction force to an injured leg. (a) Determine the force of tension in the rope supporting the leg. (b) What is the traction force exerted to the right on the leg?

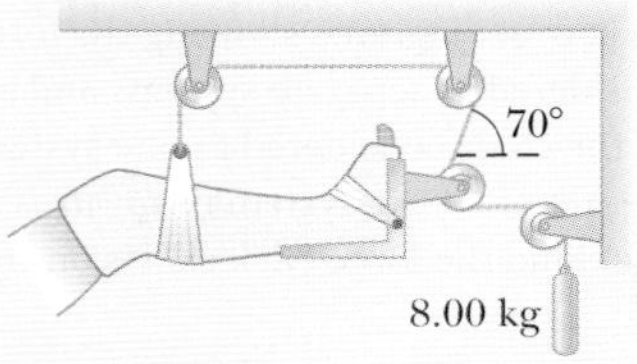

Figure P4.19

20. An object of mass $m = 1.00$ kg is observed to have an acceleration $\vec{\mathbf{a}}$ with a magnitude of 10.0 m/s² in a direction 60.0° east of north. **Figure P4.20** shows a view of the object from above. The force $\vec{\mathbf{F}}_2$ acting on the object has a magnitude of 5.00 N and is directed north. Determine the magnitude and direction of the horizontal force $\vec{\mathbf{F}}_1$ acting on the object.

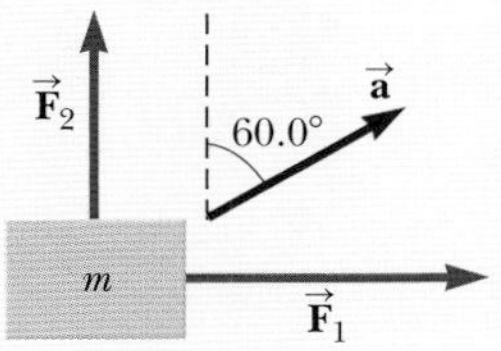

Figure P4.20

21. An object of mass $m_1 = 5.00$ kg placed on a frictionless, horizontal table is connected to a string that passes over a pulley and then is fastened to a hanging object of mass $m_2 = 9.00$ kg as shown in **Figure P4.21**. (a) Draw free-body diagrams of both objects. Find (b) the magnitude of the acceleration of the objects and (c) the tension in the string.

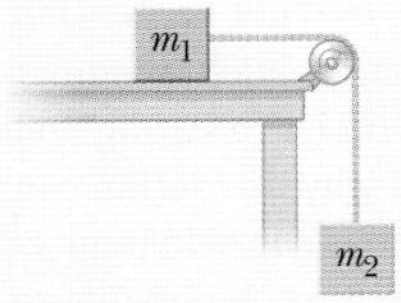

Figure P4.21

22. **Figure P4.22** shows the speed of a person's body as he does a chin-up. Assume the motion is vertical and the mass of the person's body is 64.0 kg. Determine the force exerted by the chin-up bar on his body at (a) $t = 0$, (b) $t = 0.5$ s, (c) $t = 1.1$ s and (d) $t = 1.6$ s.

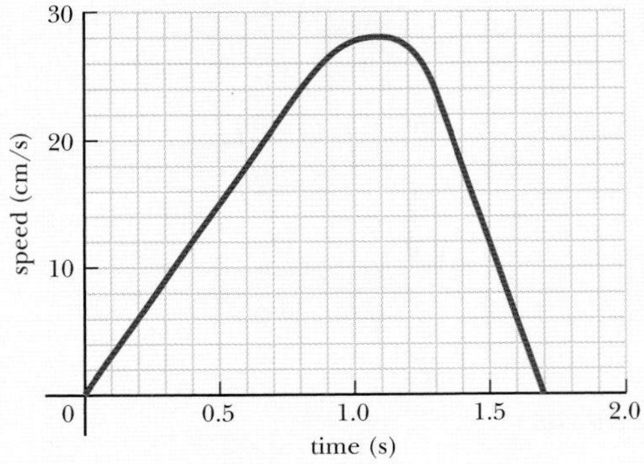

Figure P4.22

23. Two blocks, each of mass m, are hung from the ceiling of a lift as in **Figure P4.23**. The lift has an upwards acceleration a. The strings have negligible mass. (a) Find the tensions T_1 and T_2 in the upper and lower strings in terms of m, a and g. (b) Compare the two tensions and determine which string would break first if a is made sufficiently large. (c) What are the tensions if the cable supporting the lift breaks?

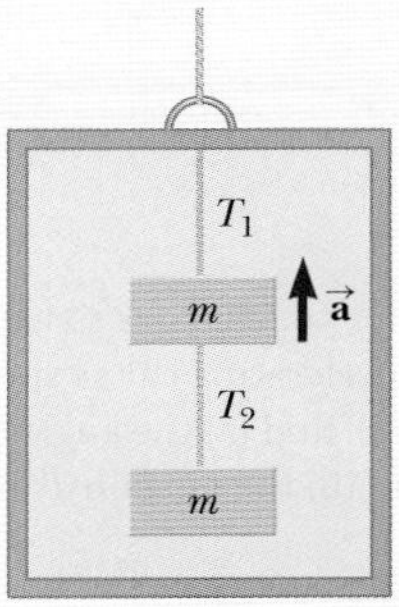

Figure P4.23

24. In the system shown in **Figure P4.24**, a horizontal force $\vec{\mathbf{F}}_x$ acts on an object of mass $m_2 = 8.00$ kg. The horizontal surface is frictionless. Consider the acceleration of the sliding object as a function of F_x. (a) For what values of F_x does the object of mass $m_1 = 2.00$ kg accelerate upwards? (b) For what values of F_x is the tension in the cord zero? (c) Plot the acceleration of the m_2 object versus F_x. Include values of F_x from −100 N to +100 N.

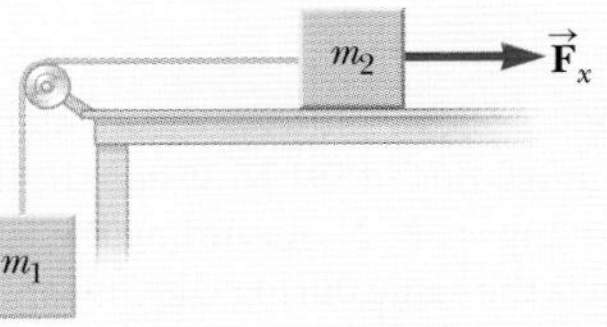

Figure P4.24

25. Consider a 72.0 kg man standing on a spring scale in a lift. Starting from rest, the lift ascends, attaining its maximum speed of 1.20 m/s in 0.800 s. It travels with this constant speed for the next 5.00 s. The lift then undergoes a uniform acceleration in the negative y direction for 1.50 s and comes to rest. What does the spring scale register (a) before the lift starts to move, (b) during the first 0.800 s, (c) while the lift is travelling at constant speed, and (d) during the time interval it is slowing down?

26. In the Atwood machine discussed in **Example 4.11** and shown in **Active Figure 4.20**, $m_1 = 2.00$ kg and $m_2 = 7.00$ kg. The masses of the pulley and string are negligible by comparison. The pulley turns without friction, and the string does not stretch. The lighter object is released with a sharp push that sets it into motion at $v_i = 2.40$ m/s downwards. (a) How far will m_1 descend below its initial level? (b) Find the velocity of m_1 after 1.80 s.

Section **4.8** Motion in accelerating reference frames

27. An object of mass $m = 5.00$ kg, attached to a spring scale, rests on a frictionless, horizontal surface as shown in **Figure P4.27**. The spring scale, attached to the front end of a boxcar, reads zero when the car is at rest. (a) Determine the acceleration of the car if the spring scale has a constant reading of 18.0 N when the car is in motion. (b) What constant reading will the spring scale show if the car moves with constant velocity? Describe the forces on the object as observed (c) by someone in the car and (d) by someone at rest outside the car.

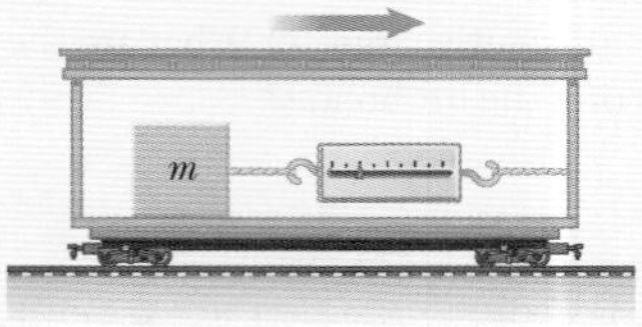

Figure P4.27

28. An object of mass $m = 0.500$ kg is suspended from the rear view mirror of an accelerating car as shown in **Figure P4.28**. Taking $a = 3.00$ m/s^2, find (a) the angle θ that the string makes with the vertical and (b) the tension T in the string.

Figure P4.28

29. A person stands on a scale in a lift. As the lift starts, the scale has a constant reading of 591 N. When the lift later stops, the scale reading is 391 N. Assuming the magnitude of the acceleration is the same during starting and stopping, determine (a) the weight of the person, (b) the person's mass, and (c) the acceleration of the lift.

30. A small container of water is placed on a turntable inside a microwave oven, at a radius of 12.0 cm from the centre. The turntable rotates steadily, turning one revolution in each 7.25 s. What angle does the water surface make with the horizontal?

Additional problems

31. A tyre of mass M is used as a swing and hangs from a length, L, of heavy rope with total mass m. (a) Find an expression for the tension, T, in the rope at a given point in the rope a length ℓ above the tyre. (b) Sketch a graph of the tension as a function of ℓ.

32. A rope with mass m_r is attached to a block with mass m_b as shown in **Figure P4.32**. The block rests on a frictionless, horizontal surface. The rope does not stretch. The free end of the rope is pulled to the right with a horizontal force $\vec{\mathbf{F}}$. (a) Draw force diagrams for the rope and the block, noting that the tension in the rope is not uniform. (b) Find the acceleration of the system in terms of m_b, m_r, and F. (c) Find the magnitude of the force the rope exerts on the block. (d) What happens to the force on the block as the rope's mass approaches zero? What can you state about the tension in a *light* cord joining a pair of moving objects?

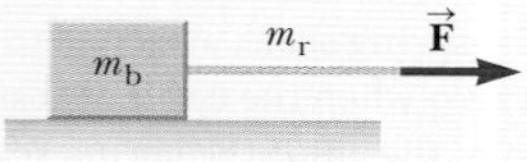

Figure P4.32

33. An inventive child named Nick wants to reach an apple in a tree without climbing the tree. Sitting in a chair connected to a rope that passes over a frictionless pulley (**Figure P4.33**), Nick pulls on the loose end of the rope with such a force that the spring scale reads 250 N. Nick's true weight is 320 N, and the chair weighs 160 N. Nick's feet are not touching the ground. (a) Draw one pair of diagrams showing the forces for Nick and the chair considered as separate systems and another diagram for Nick and the chair considered as one system. (b) Show that the acceleration of the system is *upwards* and find its magnitude. (c) Find the force Nick exerts on the chair.

Figure P4.33
Problems 33 and 34.

34. In the situation described in Problem 33 and shown in **Figure P4.33**, the masses of the rope, spring balance and pulley are negligible. Nick's feet are not touching the ground. (a) Assume Nick is momentarily at rest when he stops pulling down on the rope and passes the end of the rope to another child, of weight 440 N, who is standing on the ground next to him. The rope does not break. Describe the ensuing motion. (b) Instead, assume Nick is momentarily at rest when he ties the end of the rope to a strong hook projecting from the tree trunk. Explain why this action can make the rope break.

35. An object of mass M is held in place by an applied force $\vec{\mathbf{F}}$ and a pulley system as shown in **Figure P4.35**. The pulleys are massless and frictionless. (a) Draw diagrams showing the forces on each pulley. Find (b) the tension in each section of rope, T_1, T_2, T_3, T_4 and T_5 and (c) the magnitude of $\vec{\mathbf{F}}$.

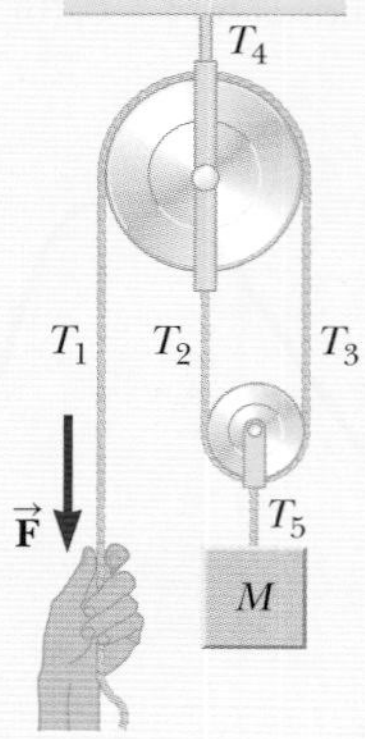

Figure P4.35

36. ±? A student is asked to measure the acceleration of a glider on a frictionless inclined plane, using an air track, a stopwatch and a metre ruler. The top of the track is measured to be (1.774 ± 0.005) cm higher than the bottom of the track, and the length of the track is $d = (127.1 \pm 0.5)$ cm. The cart is released from rest at the top of the incline, taken as $x = 0$, and its position x along the incline is measured as a function of time. For x values of (10.0 ± 0.5) cm, (20.0 ± 0.5) cm, (35.0 ± 0.5) cm, (50.0 ± 0.5) cm, (75.0 ± 0.5) cm and (100.0 ± 0.5) cm, the measured times at which these positions are reached (averaged over five runs) are (1.02 ± 0.02) s, (1.53 ± 0.05)s, (2.01 ± 0.06) s, (2.64 ± 0.10) s, (3.30 ± 0.14) s and (3.75 ± 0.16) s, respectively. (a) Construct a graph of x versus t^2, with a best-fit straight line to describe the data. (b) Determine the acceleration of the cart from the gradient of this graph. (c) Explain how your answer to part (b) compares with the theoretical value you calculate using $a = g \sin\theta$.

37. What horizontal force must be applied to a large block of mass M shown in **Figure P4.37** so that the tan blocks remain stationary relative to M? Assume all surfaces and the pulley are frictionless. Notice that the force exerted by the string accelerates m_2.

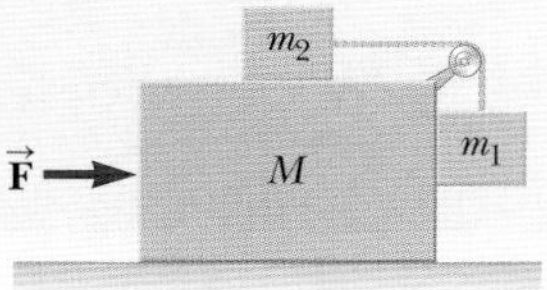

Figure P4.37

38. Because the Earth rotates about its axis, a point on the equator experiences a centripetal acceleration of 0.033 7 m/s², whereas a point at the poles experiences no centripetal acceleration. If a person at the equator has a mass of 75.0 kg, calculate (a) the gravitational force (true weight) on the person and (b) the normal force (apparent weight) on the person. (c) Which force is greater? Assume the Earth is a uniform sphere and take $g = 9.800$ m/s².

39. A car accelerates down a hill (**Figure P4.39**), going from rest to 30.0 m/s in 6.00 s. A toy with a mass of 100 g inside the car hangs by a string from the car's ceiling. The ball in the figure represents the toy. The acceleration is such that the string remains perpendicular to the ceiling. Determine (a) the angle θ and (b) the tension in the string.

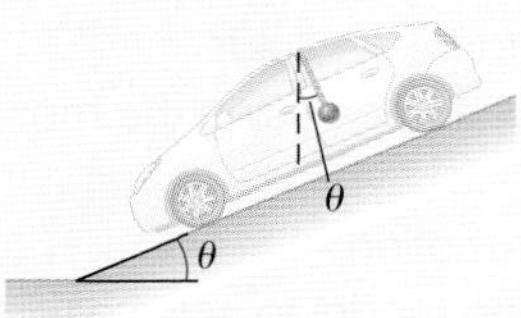

Figure P4.39

Challenge problems

40. A time-dependent force, $\vec{F} = (8.00\hat{\mathbf{i}} - 4.00t\hat{\mathbf{j}})$, where $\vec{F}$ is in newtons and t is in seconds, is exerted on a 2.00 kg object initially at rest. (a) At what time will the object be moving with a speed of 15.0 m/s? (b) How far is the object from its initial position when its speed is 15.0 m/s? (c) Through what total displacement has the object travelled at this moment?

41. A mobile is formed by supporting four metal butterflies of equal mass m from a string of length L. The points of support are evenly spaced a distance ℓ apart as shown in **Figure P4.41**. The string forms an angle θ_1 with the ceiling at each endpoint. The centre section of string is horizontal. (a) Find the tension in each section of string in terms of θ_1, m and g. (b) In terms of θ_1, find the angle θ_2 that the sections of string between the outside butterflies and the inside butterflies form with the horizontal. (c) Show that the distance D between the endpoints of the string is

$$D = \frac{L}{5}\left\{2\cos\theta_1 + 2\cos\left[\tan^{-1}\left(\tfrac{1}{2}\tan\theta_1\right)\right] + 1\right\}$$

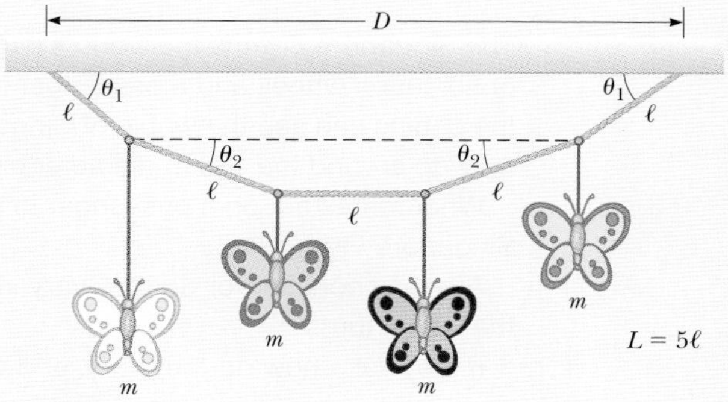

Figure P4.41

42. A block of mass $m = 2.00$ kg is released from rest at $h = 0.500$ m above the surface of a table, at the top of a $\theta = 30.0°$ incline as shown in **Figure P4.42**. The frictionless incline is fixed on a table of height $H = 2.00$ m. (a) Determine the acceleration of the block as it slides down the incline. (b) What is the velocity of the block as it leaves the incline? (c) How far from the table will the block hit the floor? (d) What time interval elapses between when the block is released and when it hits the floor? (e) Does the mass of the block affect any of the above calculations?

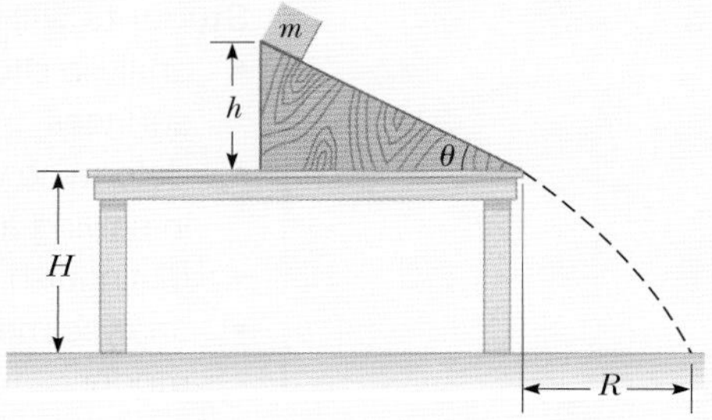

Figure P4.42
Problems 42 and 43.

43. In **Figure P4.42**, the incline has mass M and is fastened to the stationary horizontal tabletop. The block of mass m is placed near the bottom of the incline and is released with a quick push that sets it sliding upwards. The block stops near the top of the incline as shown in the figure and then slides down again, always without friction. Find the force that the tabletop exerts on the incline throughout this motion in terms of m, M, g and θ.

chapter 5

Further applications of Newton's laws

The Singapore Flyer is the world's largest observation wheel, and a passenger in one of the cabins can see the whole of Singapore as they go around the 165 m diameter circle.

What forces act on a passenger in the Singapore Flyer?

Are they constant or do they vary in time as the wheel turns?

If they vary, how do they vary, and why?

Corbis/Justin Guariglia

On completing this chapter, students will understand:

- how friction acts to oppose the sliding of solid surfaces relative to one another
- that resistive forces act to oppose motion relative to the medium in any fluid medium
- why and how we use simplifying assumptions to create models, such as those for analysing motion in the presence of resistive forces in fluids
- that non-uniform circular motion is due to a combination of radial and tangential forces.

Students will be able to:

- analyse situations and solve problems involving frictional forces between solid surfaces
- determine when resistive forces due to a fluid such as air or water are important in solving a problem
- distinguish between models for resistive forces in fluids
- analyse motion in the presence of resistive forces using the most appropriate model
- analyse uniform and non-uniform circular motion
- apply the particle in uniform circular motion model to solve a wide range of problems
- solve problems involving non-uniform circular motion.

ENHANCED WebAssign

The problems found in this chapter may be assigned online in Enhanced Web Assign.

In the preceding chapter we introduced Newton's laws of motion and used them in two analysis models for describing and analysing linear motion. Now we discuss motion that is slightly more complicated, such as motion affected by the resistive forces of friction and drag, and forces acting on objects following curved paths. This chapter illustrates the application of Newton's laws to a variety of new circumstances, showing how the idealised models of physical systems in Chapter 4 can be modified to better describe more complex, real-world situations.

5.1 Friction between solid surfaces

Newton's first law tells us that in the absence of external forces, an object in motion will continue in a straight line at a constant speed. Yet we know from everyday experience that anything we set in motion, such as a ball rolling on the floor or a book sliding across a table, will slow down and come to rest. This is because of the force of friction. Sometimes students get the impression that the physics they learn in a classroom doesn't actually apply to real life. This is because we often ignore frictional forces to simplify problems. It is important to be careful when we construct a model of a system, or choose an analysis model, to remember that we are making approximations, and to not make so many approximations that our model no longer reflects the real system!

Forces of friction are very important in our everyday lives. They allow us to walk or run and are necessary for the motion of cars and other wheeled vehicles. If there were no friction between your feet and the ground, your feet would slide out from beneath you when you tried to walk. On a slippery road, the wheels of a car may spin when the driver tries to accelerate, and the car stays on the spot unable to move.

Frictional forces act between solid surfaces and in any fluid medium such as air or water. In this section we shall look at friction between solid surfaces. We shall look at friction in viscous fluids in the next section.

Friction forces act to *resist* any *relative motion* between solid surfaces because of the interaction between those surfaces. Friction prevents your foot slipping against the floor, and allows you to walk. Tyres are designed to have as much friction between the tyre and the road as possible to prevent sliding, which is dangerous, and to allow maximum acceleration of the car. ◀ Friction forces

TRY THIS

Look at the soles of your shoes, and those of anyone else who will let you have a look. Try sliding (carefully) in different pairs of shoes. Why do sports shoes, such as running shoes, have ridged rubber soles, while dress shoes usually have quite smooth soles?

Imagine that you are cleaning out your room and have filled a rubbish bin with old food packaging, chemistry assignments and other rubbish. You then try to drag the bin across the surface of your concrete patio as in Active Figure 5.1a. This surface is *real*, not an idealised, frictionless surface. If we apply an external horizontal force $\vec{\mathbf{F}}$ to the rubbish bin, acting to the right, the rubbish bin remains stationary when $\vec{\mathbf{F}}$ is small. The force on the rubbish bin that counteracts $\vec{\mathbf{F}}$ and keeps it from moving acts towards the left and is called the **force of static friction $\vec{\mathbf{f}}_s$**. As long as the rubbish bin is not moving, $f_s = F$. Therefore, if $\vec{\mathbf{F}}$ is increased, $\vec{\mathbf{f}}_s$ also increases. Likewise, if $\vec{\mathbf{F}}$ decreases, $\vec{\mathbf{f}}_s$ also decreases. ◀ Force of static friction

Experiments show that the friction force arises from the nature of the two surfaces: because of their roughness, contact is made only at a few locations where peaks of the material touch. At these locations, the friction force arises in part because one peak physically blocks the motion of a peak from the opposing surface and in part from chemical bonding ('spot welds') of opposing peaks as they come into contact, as shown in Figure 5.2. Although the details of friction are quite complex at the atomic level, this force ultimately involves an electrical interaction between atoms or molecules – hence it is a manifestation of the *electromagnetic force*.

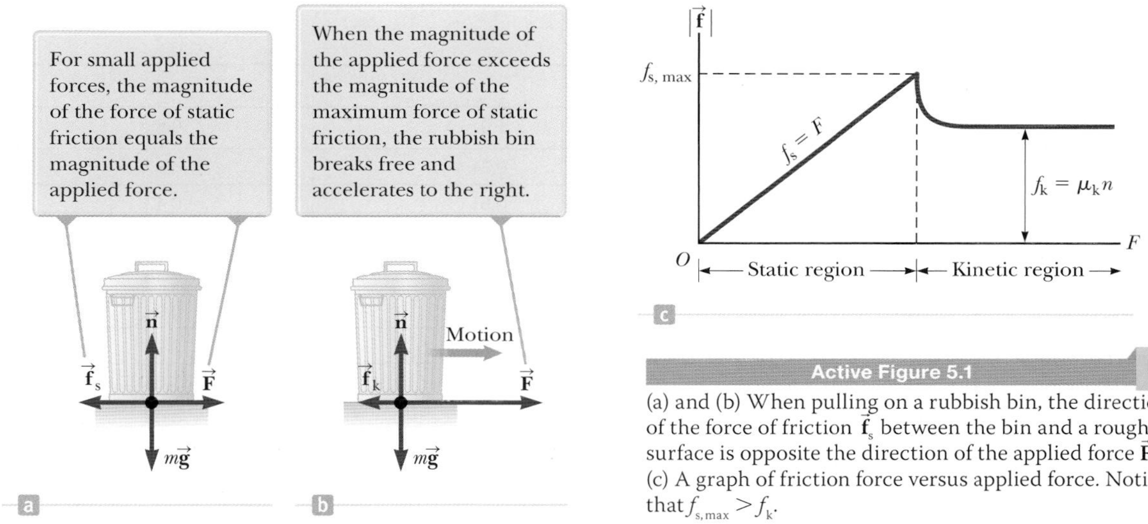

Active Figure 5.1

(a) and (b) When pulling on a rubbish bin, the direction of the force of friction $\vec{\mathbf{f}}_s$ between the bin and a rough surface is opposite the direction of the applied force $\vec{\mathbf{F}}$. (c) A graph of friction force versus applied force. Notice that $f_{s,\max} > f_k$.

If we increase the magnitude of $\vec{\mathbf{F}}$ as in **Active Figure 5.1b**, the rubbish bin eventually slips. When the rubbish bin is on the verge of slipping, f_s has its maximum value $f_{s,max}$ as shown in **Active Figure 5.1c**. When F exceeds $f_{s,max}$, the rubbish bin moves and accelerates to the right. We call the friction force acting on an object in motion the **force of kinetic friction** $\vec{\mathbf{f}}_k$. When the rubbish bin is in motion, the force of kinetic friction on the bin is less than $f_{s,max}$ (**Active Figure 5.1c**). The net force $F - f_k$ in the x direction produces an acceleration to the right, according to Newton's second law. If $F = f_k$, the acceleration is zero and the rubbish bin moves to the right with constant speed. If the applied force $\vec{\mathbf{F}}$ is removed from the moving bin, the friction force $\vec{\mathbf{f}}_k$ acting to the left provides an acceleration of the rubbish bin in the $-x$ direction and brings it to rest, again consistent with Newton's second law.

Force of kinetic friction ▶

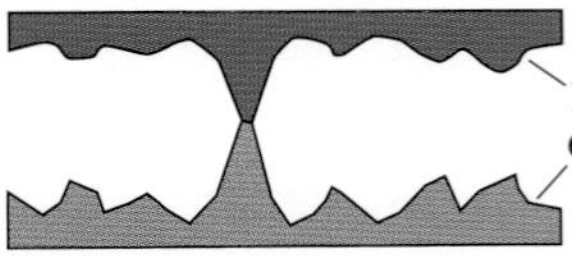

Figure 5.2
A simplified model of friction between two surfaces.

Experimentally we find that, to a good approximation, both $f_{s,max}$ and f_k are proportional to the magnitude of the normal force exerted on an object by the surface. The following descriptions of the force of friction are based on experimental observations and serve as the model we shall use for forces of friction between solid surfaces in problem solving.

The magnitude of the force of static friction between any two surfaces in contact can have the values

Force of static friction ▶

$$f_s \le \mu_s n \tag{5.1}$$

Coefficient of static friction ▶

where the dimensionless constant μ_s is called the **coefficient of static friction** and n is the magnitude of the normal force exerted by one surface on the other. The equality in **Equation 5.1** holds when the surfaces are on the verge of slipping, that is, when $f_s = f_{s,max} = \mu_s n$. This situation is called *impending* motion. The inequality holds when the surfaces are not on the verge of slipping.

The magnitude of the force of kinetic friction acting between two surfaces that are slipping relative to each other is

Force of kinetic friction ▶

$$f_k = \mu_k n \tag{5.2}$$

Coefficient of kinetic friction ▶

where μ_k is the **coefficient of kinetic friction**. Although the coefficient of kinetic friction can vary with speed, we shall usually neglect any such variations in this text.

The values of μ_k and μ_s depend on the nature of the surfaces, but μ_k is generally less than μ_s. Typical values range from about 0.03 to 1.0. **Table 5.1** lists some measured values.

The direction of the friction force on an object is parallel to the surface with which the object is in contact and opposite to the actual motion (kinetic friction) or the impending motion (static friction) of the object relative to the surface.

The coefficients of friction are nearly independent of the area of contact between the surfaces. We might expect that placing an object on the side having the most area might increase the friction force. Although this provides more points in contact, the force applied by the object (the load) is spread out over a larger area and the individual points are not pressed together as tightly. Because these effects approximately compensate for each other, the friction force can be modelled as independent of the surface area.

Pitfall Prevention 5.1

In Equation 5.1, the equals sign is used *only* in the case in which the surfaces are just about to break free and begin sliding. Do not fall into the common trap of using $f_s = \mu_s n$ in *any* static situation.

Pitfall Prevention 5.2

Equations 5.1 and 5.2 are *not* vector equations. They are relationships between the *magnitudes* of the vectors representing the friction and normal forces. Because the friction and normal forces are perpendicular to each other, the vectors cannot be related by a multiplicative constant.

Table 5.1
Coefficients of friction

	μ_s	μ_k
Rubber on concrete	1.0	0.8
Steel on steel	0.74	0.57
Aluminium on steel	0.61	0.47
Glass on glass	0.94	0.4
Copper on steel	0.53	0.36
Wood on wood	0.25–0.5	0.2
Metal on metal (lubricated)	0.15	0.06
Teflon on Teflon	0.04	0.04
Ice on ice	0.1	0.03
Synovial joints in humans	0.01	0.003

Note: All values are approximate. In some cases, the coefficient of friction can exceed 1.0.

Quick Quiz 5.1

You press your physics textbook flat against a vertical wall with your hand. What is the direction of the friction force exerted by the wall on the book? **(a)** downwards **(b)** upwards **(c)** out from the wall **(d)** into the wall

---

TRY THIS

Load some boxes onto a flat trolley or cart and accelerate it forwards rapidly. What happens to the boxes? What happens when you stop the trolley suddenly? (You can do this in the supermarket, but be careful not to run into anything.)

Pitfall Prevention 5.3

Friction does not 'oppose motion'. Friction opposes the *relative motion* of two surfaces; it opposes the sliding of one surface against another.

Conceptual Example 5.1

A crate on a ute

A ute is laden with a large crate as shown in **Figure 5.3**. The crate is resting on the tray of the ute and is not tied down.

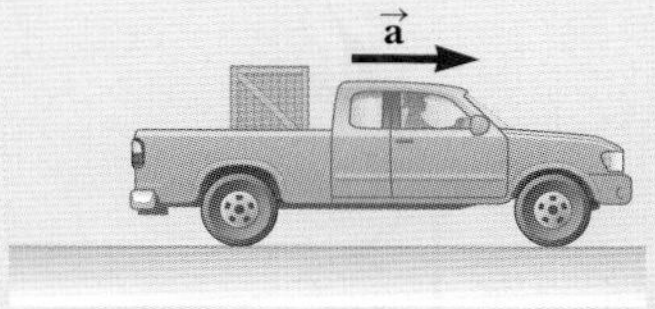

Figure 5.3
(Conceptual Example 5.1)

(A) Draw a free-body diagram showing the forces acting on the crate while the ute is accelerating forwards from rest, assuming the crate does not slip. In which direction is the frictional force between the ute tray and the crate acting?

Solution

The crate will be accelerating forwards with the ute, so the net force must be in the forwards direction. The only forces acting are the gravitational force down, the normal force up and the frictional force between the ute tray and the crate. There may also be a small force of air resistance or drag acting in the opposite direction to the motion of the ute and crate, but this will be very small at low speeds.

As we can see from the free-body diagram in **Figure 5.4b**, the frictional force between the ute's tray and the crate is in the direction of motion – i.e. forwards.

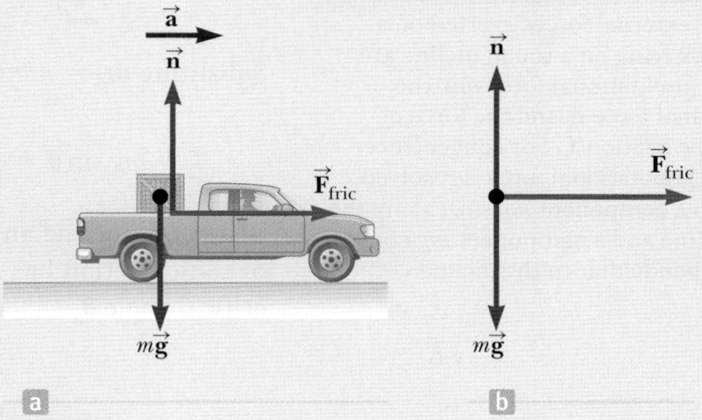

Figure 5.4
(Conceptual Example 5.1)

(B) Draw a free-body diagram showing the forces acting on the crate while the ute is braking, assuming the crate does not slip. In which direction is the frictional force between the ute tray and the crate acting?

Solution

The crate will be braking (accelerating backwards) now with the ute, so the net force must be in the backwards or negative direction. Again, it is the friction force that holds the crate to the tray, so this force causes the change in motion and is in the direction of the acceleration.

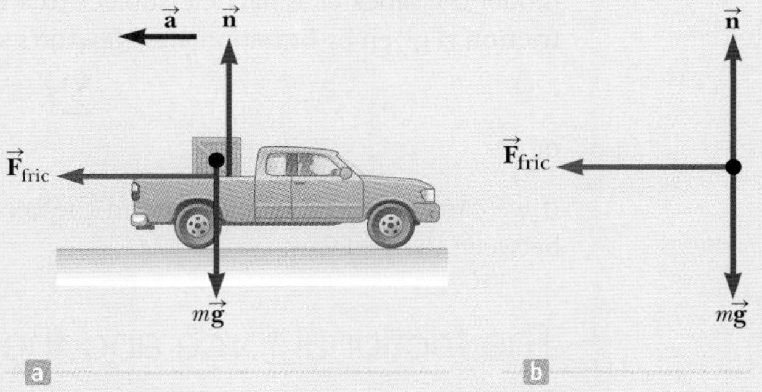

Figure 5.5
(Conceptual Example 5.1)

What If? What if the frictional force between the ute tray and the crate is not large enough to prevent the crate sliding? What will an observer on the ground see? What will an observer in the ute see?

Answer If the crate begins to slip while the ute is accelerating forwards then an observer on the ground sees the crate moving forwards, but not as fast as the ute. An observer on the ute sees the crate sliding backwards towards the back of the ute, and it may fall off the back. The opposite happens when the ute is braking and the box may slide forwards into the cabin of the ute. You may observe the same thing if you do the *Try this* example with boxes on a trolley.

TRY THIS

Place a block or coin on a flat object, such as a book, and gradually lift one edge of the book until the object begins to slide. By measuring the angle at which the object first begins to slide you can calculate the coefficient of static friction between the book and the object.

Experimental determination of μ_s and μ_k

For help with vector decomposition, see Appendix B.6.

As in the *Try this* example above, suppose a block is placed on a rough surface inclined relative to the horizontal as shown in **Active Figure 5.6**. The incline angle is increased until the block starts to move. The block tends to slide down the incline due to the gravitational force. The diagram in **Active Figure 5.6** shows the forces on the block: the gravitational force $m\vec{g}$, the normal force $\vec{n}$, and the force of static friction $\vec{f}_s$. We choose x to be parallel to the plane and y perpendicular to it.

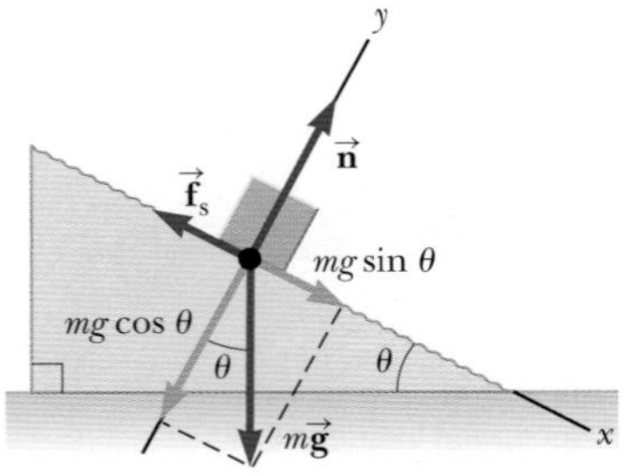

Active Figure 5.6

The external forces exerted on a block lying on a rough incline are the gravitational force $m\vec{g}$, the normal force $\vec{n}$, and the force of static friction $\vec{f}_s$. For convenience, the gravitational force is resolved into a component $mg\sin\theta$ along the incline and a component $mg\cos\theta$ perpendicular to the incline.

Notice how this differs from Example 4.4. When there is no friction on an incline, *any* angle of the incline will cause a stationary object to begin moving. When there is friction, however, there is no movement of the object for angles less than the critical angle, θ_c.

We apply Newton's second law to the block in both the x and y directions:

$$\sum F_x = mg\sin\theta - f_s = 0 \quad (1)$$

$$\sum F_y = n - mg\cos\theta = 0 \quad (2)$$

Substitute $mg = n/\cos\theta$ from **Equation (2)** into **Equation (1)**:

$$f_s = mg\sin\theta = \left(\frac{n}{\cos\theta}\right)\sin\theta = n\tan\theta \quad (3)$$

When the incline angle is increased until the block is on the verge of slipping, the force of static friction has reached its maximum value $\mu_s n$. The angle θ in this situation is the critical angle θ_c.

So $$f_s = \mu_s n = n\tan\theta_c$$

or $$\mu_s = \tan\theta_c$$

For example, if the block just slips at $\theta_c = 20.0°$, we find that $\mu_s = \tan 20.0° = 0.364$.

We can measure the coefficient of kinetic friction between two surfaces using a similar method. Consider the case when the block in **Active Figure 5.6** is sliding, so that it is now the kinetic friction force acting. We model the block as a particle subject to a net force in the direction parallel to the slope. The force due to friction is given by **Equation 5.2**. Newton's second law applied to the block in the direction of the slope gives:

$$\sum F = mg\sin\theta - \mu_k mg\cos\theta = ma$$

or $$a = g\sin\theta - \mu_k g\cos\theta$$

If we can measure the angle θ and the acceleration of the block we can find the coefficient of friction, μ_k, between the surfaces.

The frictional force and the normal force

The frictional force and normal force are both aspects of the contact force that acts between solid surfaces due to the electromagnetic force.

The frictional force acts to prevent relative movement of two surfaces in contact because of the microscopic bumps on the surfaces and the bonding of atoms on the two surfaces. The static friction force can supply a force equal to an applied force to keep the two surfaces together up to a certain maximum value, above which the effect of the bumps becomes less significant, the bonds break and the surfaces begin to slide. The friction force acts *parallel to the interface between the two surfaces.* The frictional force increases with the normal force.

The normal force acts to prevent one surface moving into another surface, also because of the interaction between atoms at the surfaces, and in particular the electrostatic repulsion of their electrons. The normal force can supply a force equal to an applied force up to some maximum value to prevent the surfaces moving into each other. Above this value bonds break between atoms and objects become damaged.

We can consider the friction and normal force to be the parallel and perpendicular components of the sum of all the microscopic forces acting between two surfaces because of the attractive and repulsive interactions of their atoms.

Note that we have been saying 'the friction force that acts between two surfaces'. This is because, as we know from Newton's third law, whatever force one object exerts on a second object, the same force (but in the opposite direction) is exerted on the first object. When you go to take a step forward, the friction force that your shoe exerts backwards on the floor is equal, but opposite, to the friction floor the force exerts on your shoe that moves you forwards. These forces are a Newton's third law action-reaction pair. When you push horizontally on a book sitting on a table, the reaction force is the book pushing back on you, not the friction force between the table and the book. Hence, the friction force is not necessarily equal to the applied force, and will be less than the applied force in many cases, as when the maximum static friction force is exceeded and a sliding object begins to accelerate. In the case of the normal force, the normal forces that two objects in contact exert on each other are also equal because they are an action-reaction pair. Even after the maximum normal force that one of the surfaces can exert is exceeded, the two normal forces are still equal. When the surface breaks, the force it exerts may even drop to zero, and at the same time the force exerted *on* it drops to zero as the other object passes through it.

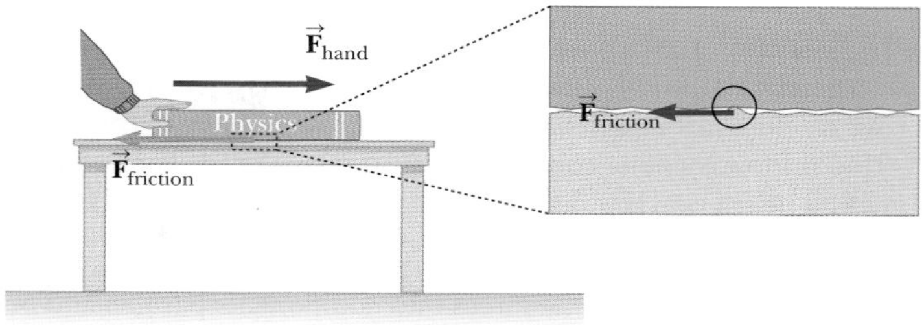

Figure 5.7
The forces acting on the book in the horizontal direction.

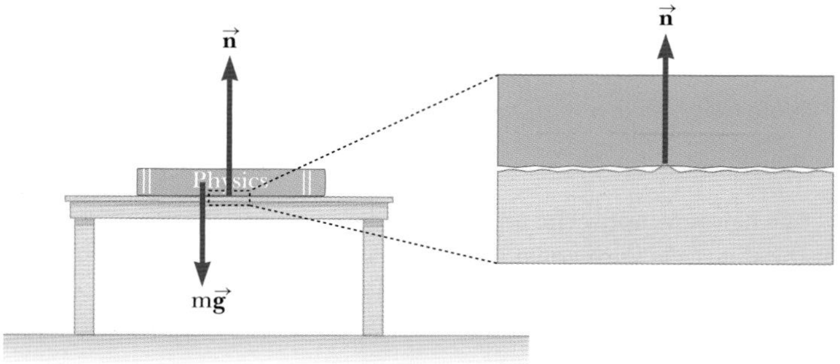

Figure 5.8
The forces acting on the book in the vertical direction.

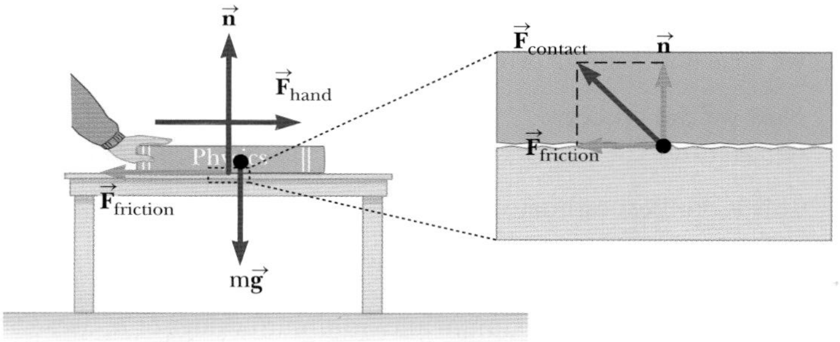

Figure 5.9
All forces acting on the book. The contact forces, $\vec{n}$ and $\vec{F}_{friction}$, act to oppose the applied forces.

Quick Quiz 5.2

A small child is trying to move a heavy bag of toys across a carpeted floor. Which is easier: **(a)** pushing the bag from behind at an angle below horizontal or **(b)** pulling the bag by a handle from in front at the same angle above horizontal?

Example 5.2

Marcus sits at the top of a slide as shown. The slide is (3.5 ± 0.1) m long at an angle of $30° \pm 1°$ to the horizontal. Starting from rest, Marcus slides down and reaches the bottom with a velocity of (4.4 ± 0.2) m/s. What is the coefficient of friction between Marcus's pants and the surface of the slide?

Solution

Conceptualise Draw a diagram showing the forces acting on Marcus. These will be gravity, acting to pull him down the slide; the normal force acting perpendicular to the surface of the slide and the friction force acting parallel to the surface in the opposite direction to his motion.

Example 5.2 cont.

Model The forces acting on Marcus are shown in **Figure 5.11a**. In the direction perpendicular to the slide we can model him as a particle subject to zero net force. We model Marcus as a particle subject to a net force in the direction of the slide. This net force will be constant because we make the approximation that the force of kinetic friction is independent of speed, as is the acceleration due to gravity, so we can also model Marcus as a particle with constant acceleration.

Analyse We break up the forces into components parallel and perpendicular to the slide, as shown in **Figure 5.11b**. Now we can write expressions for the force in each direction:

$$\sum F_y = n - mg\cos\theta = 0 \quad (1)$$

$$\sum F_x = mg\sin\theta - f_k = ma_x \quad (2)$$

Note that $f_k = \mu_k n = \mu_k mg\cos\theta$ so (2) becomes $\sum F_x = mg\sin\theta - \mu_k mg\cos\theta = ma_x$

Rearranging for the acceleration we find: $a = g\sin\theta - \mu_k g\cos\theta$ (3)

We can now apply the analysis model for a particle with constant acceleration and use **Equation 2.16** to relate the final velocity to the acceleration: $v_f^2 = v_i^2 + 2a(x_f - x_i)$

Taking $v_i = 0$ and $x_i = 0$ and putting in our expression for a from (3) we have $v_f^2 = 2ax = 2g(\sin\theta - \mu_k\cos\theta)x$

Rearranging for μ_k gives: $\mu_k = \tan\theta - [v_f^2/(2gx\cos\theta)]$

We do a dimension check to ensure that we have not made an error: $[\,] = [\,] - ([LT^{-1}]^2/[LT^{-2}][L]) = [\,]$ ☺

Now we can proceed to substitute the numbers: $\mu_k = \tan 30° - [(4.4\text{ m/s})^2/(2 \times 9.80\text{ m/s}^2 \times 3.5\text{ m} \times \cos 30°)] = 0.25$

This is our best estimate for μ_k.

To find the uncertainty, we will use the range method. The maximum and minimum values of μ_k are:

$$\mu_{k\,max} = \tan\theta_{max} - [v_{f,min}^2/(2gx_{max}\cos\theta_{max})]$$
$$= \tan 31° - [(4.2\text{ m/s})^2/(2 \times 9.80\text{ m/s}^2 \times 3.6\text{ m} \times \cos 31°)] = 0.31$$
$$\mu_{k\,min} = \tan\theta_{min} - [v_{f,max}^2/(2gx_{min}\cos\theta_{min})]$$
$$= \tan 29° - [(4.6\text{ m/s})^2/(2 \times 9.80\text{ m/s}^2 \times 3.4\text{ m} \times \cos 29°)] = 0.19$$
$$\Delta\mu_k = (0.31 - 0.19)/2 = 0.06.$$

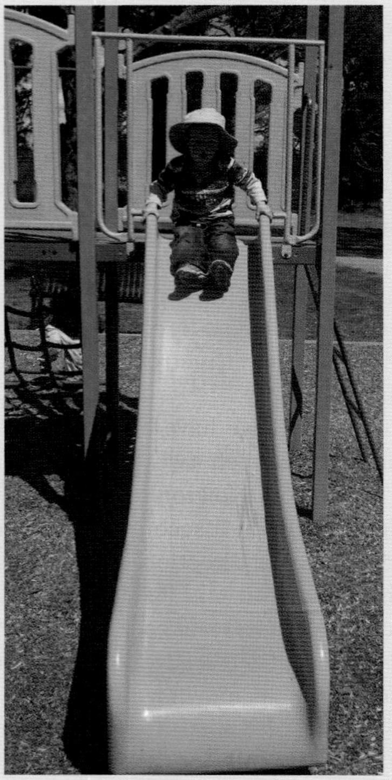

Figure 5.10
Marcus at the top of the slide

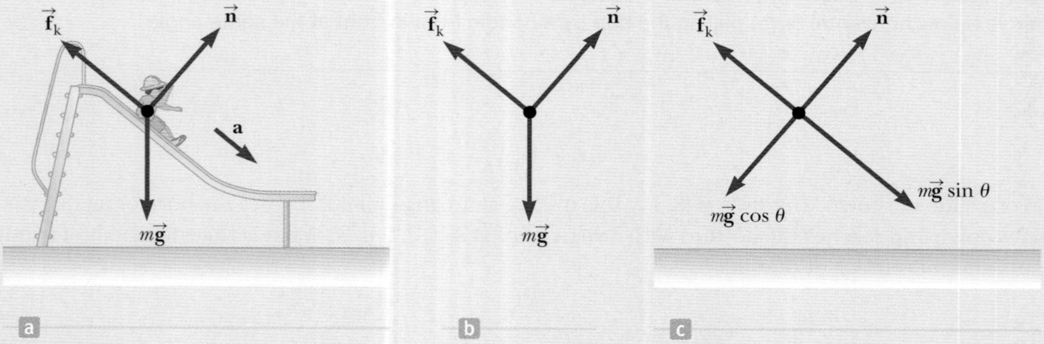

Figure 5.11
(a) Forces acting on Marcus as he accelerates down the slide. (b) Free-body diagram of the forces.
(c) Free-body diagram showing the forces parallel and perpendicular to the slide.

Worked Example 5.2 cont.

So we can write $\mu_k = 0.25 \pm 0.06$

Finalise What happens if the angle of the slide increases? The velocity will get larger due to the increase in the component of the gravitational force in the direction of motion, and also because the frictional force retarding the motion gets smaller. At an angle of 90°, or vertical, the frictional force becomes zero, and Marcus is in free fall. If the angle is reduced to 0°, so the slide is horizontal, the component of the gravitational force in the direction of the slide is zero, but the frictional force is a maximum, and Marcus will have to push himself along to move on the slide.

TRY THIS

Measure your 'weight' using a set of bathroom scales on a flat floor. Draw a free-body diagram showing the forces acting on you. Now draw a free-body diagram showing the forces acting on you if the scales were placed on an incline and you stood on them. Would the scales read higher or lower than when on the flat floor? Test your prediction.

Example 5.3

Acceleration of two connected objects when friction is present

A block of mass m_2 on a rough, horizontal surface is connected to a ball of mass m_1 by a lightweight cord over a lightweight, frictionless pulley as shown in **Figure 5.12a**. A force of magnitude F at an angle θ with the horizontal is applied to the block as shown, and the block slides to the right. The coefficient of kinetic friction between the block and surface is μ_k. Determine the magnitude of the acceleration of the two objects.

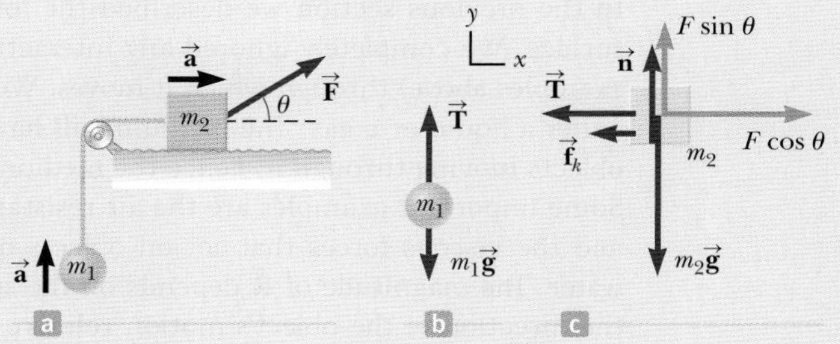

Figure 5.12
(Example 5.3) (a) The external force $\vec{F}$ applied as shown can cause the block to accelerate to the right. (b, c) Diagrams showing the forces on the two objects, assuming the block accelerates to the right and the ball accelerates upward.

Solution

Conceptualize Imagine what happens as $\vec{F}$ is applied to the block. Assuming $\vec{F}$ is large enough to break the block free from static friction but not large enough to lift the block, the block slides to the right and the ball rises.

Draw a free-body diagram showing the forces acting on each object. This is shown in **Figures 5.12b** and **5.12c**. Note that the force $\vec{F}$ has been decomposed into horizontal and vertical components in **Figure 5.12c**.

Model We can model the ball as a particle subject to a net force in the y direction. We can model the block as a particle subject to a net force in in the x direction, and a particle subject to a zero net force in the y direction, as we assume it has no acceleration in this direction. We make the approximation that the mass of the pulley and any friction acting at the pulley is negligible. We model the rope as ideal, so that the tension at all points in the rope is the same and it does not stretch.

Analyse As the ball and block are connected by an ideal rope, they each have a force T exerted on them, and they have the same magnitude acceleration, a.

Apply the particle subject to a net force model to the block in the horizontal direction:

$$\sum F_x = F\cos\theta - f_k - T = m_2 a_x = m_2 a \quad (1)$$

Because the block moves only horizontally, apply the particle in equilibrium model to the block in the vertical direction:

$$\sum F_y = n + F\sin\theta - m_2 g = 0 \quad (2)$$

Apply the particle subject to a net force model to the ball in the vertical direction:

$$\sum F_y = T - m_1 g = m_1 a_y = m_1 a \quad (3)$$

Solve **Equation (2)** for n:

$$n = m_2 g - F\sin\theta$$

Substitute n into $f_k = \mu_k n$ from **Equation 5.2**:

$$f_k = \mu_k(m_2 g - F\sin\theta) \quad (4)$$

Example 5.3 cont.

Substitute **Equation (4)** and the value of T from **Equation (3)** into **Equation (1)**:

$$F\cos\theta - \mu_k(m_2 g - F\sin\theta) - m_1(a+g) = m_2 a$$

Solve for a:

$$a = \frac{F(\cos\theta + \mu_k \sin\theta) - (m_1 + \mu_k m_2)g}{m_1 + m_2} \quad (5)$$

Check dimensions on our final Equation:

$$[\mathrm{LT^{-2}}] = ([\mathrm{MLT^{-2}}] - [\mathrm{M}][\mathrm{LT^{-2}}])/[\mathrm{M}] = [\mathrm{LT^{-2}}]$$ ☺

Finalise The acceleration of the block can be either to the right or to the left depending on the sign of the numerator in **Equation (5)**. If the velocity is to the left, we must reverse the sign of f_k in **Equation (1)** because the force of kinetic friction must oppose the motion of the block relative to the surface. In this case, the value of a is the same as in **Equation (5)**, with the two plus signs in the numerator changed to minus signs.

What does **Equation (5)** reduce to if the force $\vec{\mathbf{F}}$ is removed and the surface becomes frictionless? Call this expression **Equation (6)**. Does this algebraic expression match your intuition about the physical situation in this case? Now go back to Example 4.11 and let angle θ go to zero in **Equation (5)** of that example. How does the resulting equation compare with your **Equation (6)** here in Example 5.3? Is this what you expect?

5.2 Resistive forces in fluids

In the previous section we described the force of kinetic friction exerted on an object sliding across a surface. We completely ignored any interaction between the object and the fluid medium (the air in the examples above) through which it moves. We now consider the effect of that fluid medium, which can be either a liquid or a gas. The medium will have some viscosity, that is resistance to flow and resistance to objects moving through it, hence the medium exerts a **resistive force** $\vec{\mathbf{R}}$ on any object moving through it. Some important examples are the air resistance associated with moving vehicles (sometimes called *drag*) and the viscous forces that act on objects moving through a liquid, such as on a person swimming in water. The magnitude of $\vec{\mathbf{R}}$ depends on the speed of the object, and the direction of $\vec{\mathbf{R}}$ is always opposite the direction of the object's motion **relative to the medium**. For example, if a marble is dropped into a bottle of shampoo, the marble moves downwards and the resistive force is upwards, resisting the falling of the marble. This is also the case for skydivers, who reach a terminal velocity because of the drag force due to the air. The drag force due to air is also responsible for making streamers move in a breeze. In this case the air rather than the object (the streamer) is moving, but the force is still opposite to the direction of the motion of the streamer with respect to the air.

The magnitude of the resistive force can depend on speed in a complex way and here we consider only two simplified models. In the first model, we assume the resistive force is proportional to the speed of the moving object. This model is appropriate for objects falling slowly through a liquid and for very small objects, such as dust particles, moving through air. In the second model, we assume a resistive force that is proportional to the square of the speed of the moving object. Large objects, such as skydivers moving through air in free fall or cars driving at highway speeds, experience such a force.

Model 1: Resistive force proportional to object velocity

If we model the resistive force acting on an object moving through a liquid or gas as proportional to the object's velocity relative to the medium, the resistive force can be expressed as

$$\vec{\mathbf{R}} = -b\vec{\mathbf{v}} \quad (5.3)$$

where b is a constant whose value depends on the properties of the medium, such as its viscosity (which we shall discuss further in Section 14.3) and on the shape and size of the object; and $\vec{\mathbf{v}}$ is the velocity of the object relative to the medium. The negative sign indicates that $\vec{\mathbf{R}}$ is in the opposite direction to $\vec{\mathbf{v}}$.

Consider a small sphere, such as a ball bearing, of mass m released from rest in a liquid as in **Active Figure 5.13a**. The only forces acting on the sphere are the resistive force $\vec{\mathbf{R}} = -b\vec{\mathbf{v}}$, the gravitational force $\vec{\mathbf{F}}_g$ and a constant *buoyant force* which acts to reduce the weight of the ball. We shall ignore the buoyant force for the moment and deal with it in Chapter 13 when we study fluids. This is a reasonable approximation when the falling object has a density much greater than that of the fluid but it cannot be ignored when the densities are similar. Applying Newton's second law to the vertical motion, choosing the downwards direction to be positive and noting that $\sum F_y = mg - bv$, we obtain

$$mg - bv = ma \tag{5.4}$$

where the acceleration of the sphere is downwards. Noting that the acceleration a is equal to dv/dt gives

$$\frac{dv}{dt} = g - \frac{b}{m}v \tag{5.5}$$

This equation is called a *differential equation.*

For more information about differential equations, see Appendix B.7.

Σ

Notice that initially when $v = 0$, the magnitude of the resistive force is also zero and the acceleration of the sphere is simply g. As t increases, the magnitude of the resistive force increases and the acceleration decreases. The acceleration approaches zero when the magnitude of the resistive force approaches the gravitational force acting on the sphere. In this situation, the velocity of the sphere approaches its **terminal velocity** v_T.

◀ Terminal velocity

The terminal velocity is obtained from **Equation 5.5** by setting $dv/dt = 0$, which gives

$$mg - bv_T = 0 \quad \text{or} \quad v_T = \frac{mg}{b}$$

If $v = 0$ at $t = 0$, the solution to the differential **Equation 5.5** is:

$$v = \frac{mg}{b}\left(1 - e^{-bt/m}\right) = v_T\left(1 - e^{-t/\tau}\right) \tag{5.6}$$

This function is plotted in **Active Figure 5.13c**. The symbol e represents the base of the natural logarithm: $e = 2.718\,28$. The **time constant** $\tau = m/b$ (Greek letter tau) is the time at which the sphere released from rest at $t = 0$ reaches 63.2% of its terminal speed; when $t = \tau$, **Equation 5.6** yields $v = 0.632v_T$. (The number 0.632 is $1 - e^{-1}$.)

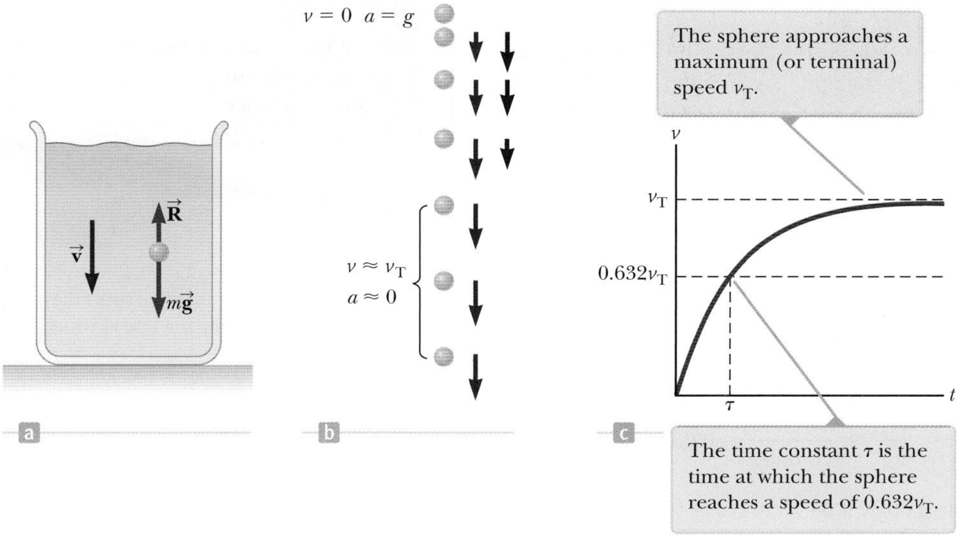

v = 0 a = g
The sphere approaches a maximum (or terminal) speed v_T.
v
v_T
0.632v_T
τ
t
R
v
mg
v ≈ v_T
a ≈ 0
a
b
c
The time constant τ is the time at which the sphere reaches a speed of 0.632v_T.

Active Figure 5.13

(a) A small sphere falling through a liquid (b) A motion diagram of the sphere as it falls. Velocity vectors (red) and acceleration vectors (violet) are shown for each image after the first one. (c) A speed–time graph for the sphere.

We can check that **Equation 5.6** is a solution to **Equation 5.5** by differentiation:

$$\frac{dv}{dt} = \frac{d}{dt}\left[\frac{mg}{b}(1 - e^{-bt/m})\right] = \frac{mg}{b}\left(0 + \frac{b}{m}e^{-bt/m}\right) = ge^{-bt/m}$$

Σ *Appendix B.7 gives a table of derivatives of common functions, including exponentials.*

Substituting into **Equation 5.5** both this expression for dv/dt and the expression for v given by **Equation 5.6** shows that our solution satisfies the differential equation.

TRY THIS

Fill a tall, clear container with a viscous (slow flowing) liquid such as oil or shampoo. Try dropping marbles or ball bearings or other small round objects gently into the liquid. What determines the rate at which they fall? What effect does size of the ball have? What about density?

±?

Example 5.4

Sphere falling in glycerine

The International Physics Olympiad competition includes a laboratory exam, in which the competitors have to design and perform an experiment to solve a particular problem. When the competition was held in Canberra in 1995, the laboratory problem was to find the viscosity of an unknown fluid (it was glycerine) using a tall cylinder of the fluid and a set of ball bearings of different sizes. The students needed to find the terminal velocities of the spheres so they could find the constant b and hence the viscosity of the medium.

A student doing the experiment releases a ball bearing of mass (2.00 ± 0.05) g from rest at the top of a tall cylinder filled with glycerine (**Figure 5.14**). The ball bearing falls through the fluid, where it experiences a resistive force proportional to its speed. The sphere reaches a terminal speed of (2.0 ± 0.1) cm/s. Determine the time constant τ and the time at which the sphere reaches 90% of its terminal speed.

Figure 5.14
A Physics Olympiad student measures the viscosity of glycerine by measuring the rate at which ball bearings fall.

Solution

Conceptualise With the help of **Active Figure 5.13**, imagine dropping the ball into the glycerine and watching it sink to the bottom of the vessel. You should also draw another diagram for yourself as part of your solution.

Model We model the sphere as a particle subject to a net force, with one of the forces being a resistive force that depends on the speed of the sphere.

Analyse The time constant τ is given by $\tau = m/b$, where, from **Equation 5.6**, $b = mg/v_T$. Hence (remembering to convert from cm/s to m/s) we can write

$$\tau = \frac{v_T}{g} = \frac{0.020 \text{ m/s}}{9.80 \text{ m/s}^2} = 2.04 \times 10^{-3} \text{ s}$$

The uncertainty is: $$\frac{\Delta\tau}{\tau} = \frac{\Delta v_T}{v_T} = \frac{0.1 \text{ cm/s}}{2.0 \text{ cm/s}} = 0.05$$

So $$\Delta\tau = 0.05 \times 2.04 \times 10^{-3} \text{ s} = 0.10 \times 10^{-3} \text{ s}$$

Example 5.4 cont.

So we can write $\tau = (2.0 \pm 0.1) \times 10^{-3}$ s.

Note that we did not need to use the mass of the ball bearing, because we found an algebraic expression for the time constant before substituting in any numbers. Had we found a numerical value for b and then used this numerical value to find the time constant we would have had to do more calculations and also substantially increased the uncertainty in our value for τ.

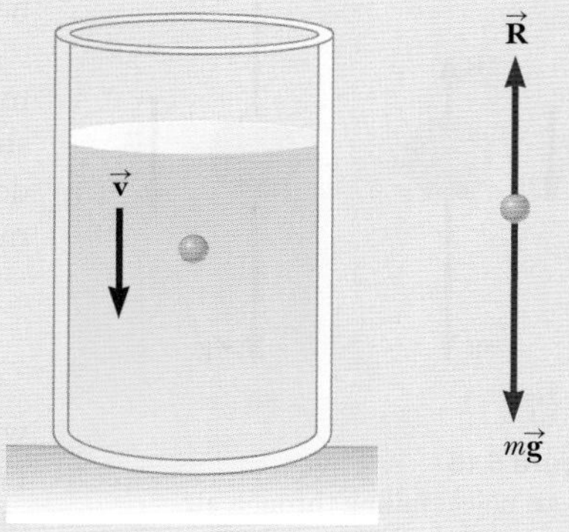

Find the time t at which the sphere reaches a speed of $0.90v_T$ by setting $v = 0.90v_T$ in **Equation 5.6** and solving for t:

$$0.90v_T = v_T(1 - e^{-t/\tau})$$
$$1 - e^{-t/\tau} = 0.90$$
$$e^{-t/\tau} = 0.10$$
$$-\frac{t}{\tau} = \ln 0.10 = -2.30$$
$$t = 2.30\tau = 2.30 \times 2.0 \times 10^{-3}\,\text{s} = 4.6 \times 10^{-3}\,\text{s}$$

The fractional uncertainty in t is the same as in τ, so $\Delta t/t = \Delta\tau/\tau = 0.05$ and hence $\Delta t = 0.05 \times 4.6 \times 10^{-3}\,\text{s} = 0.2 \times 10^{-3}$ s

The time taken for the ball to reach 90% of terminal velocity is $(4.6 \pm 0.2) \times 10^{-3}$ s.

Finalise The sphere reaches 90.0% of its terminal speed in a very short time interval. You may also have seen this behaviour if you tried the *Try this* activity. Note that we have ignored buoyant force in this example, although in fact in this case the buoyant force will be significant and students were expected to take it into account. You may wish to revisit this example after studying Chapter 13.

What If? What if the ball bearing didn't reach terminal velocity so quickly? How could you be sure when it was going at terminal velocity?

Answer In the Olympiad laboratory exam this was the case for many of the ball bearings and students had to make sure they timed the fall of the ball bearings through some distance for which they were moving at terminal velocity. They also had only five balls of each size, which could not be retrieved once dropped in the liquid, and to get maximum points in the exam they had to have repeated measurements and small uncertainties. Most students simply chose to assume that the balls would reach terminal velocity by about half of the way down the tube, and measured from there. One student dropped the first two balls of each size in quick succession and noted where the distance between them became constant. This allowed him to be sure that the subsequent three balls were indeed at terminal velocity when he began to time them, and that he could time over the maximum interval, thus minimising his uncertainties!

TRY THIS

Fill a tall, clear container with a viscous (slow flowing) liquid such as oil or shampoo. Try dropping marbles or ball bearings or other small round objects gently into the liquid. Can you tell when they reach terminal velocity? Try dropping two in quick succession and watching the gap between them.

Model 2: Resistive force proportional to object speed squared

For objects moving at high speeds through air, such as aeroplanes, skydivers, cars and cricket balls, the resistive force is reasonably well modelled as proportional to the square of the speed. In these situations, the magnitude of the resistive force can be expressed as

$$R = \frac{1}{2}D\rho A v^2 \qquad (5.7)$$

where D is a dimensionless empirical quantity called the *drag coefficient*, ρ is the density of air, and A is the cross-sectional area of the moving object measured in a plane perpendicular to its velocity (the area pushing

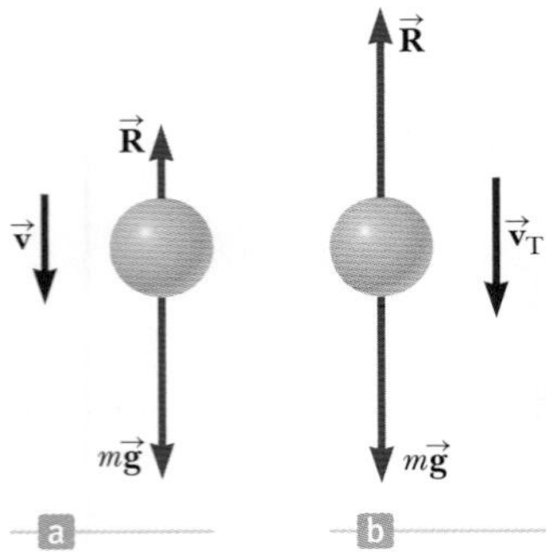

Figure 5.15
(a) An object falling through air experiences a resistive force $\vec{\mathbf{R}}$ and a gravitational force $\vec{\mathbf{F}}_g = m\vec{\mathbf{g}}$. (b) The object reaches terminal speed when the net force acting on it is zero, that is, when $\vec{\mathbf{R}} = -\vec{\mathbf{F}}_g$ or $R = mg$.

against the fluid). The drag coefficient has a value of about 0.5 for spherical objects but can have a value as great as 2 for irregularly shaped objects.

Let us analyse the motion of a falling object subject to an upwards resistive force of magnitude $R = \frac{1}{2}D\rho Av^2$. Suppose an object of mass m is released from rest. As **Figure 5.15** shows, the object experiences two external forces (again we ignore the buoyant force): the downwards gravitational force $\vec{\mathbf{F}}_g = m\vec{\mathbf{g}}$ and the upwards resistive force $\vec{\mathbf{R}}$. Hence, the magnitude of the net force is

$$\sum F = mg - \frac{1}{2}D\rho Av^2 \tag{5.8}$$

where we have taken downwards to be the positive vertical direction. Using the force in **Equation 5.8** in Newton's second law, we find that the object has a downwards acceleration of magnitude

$$a = g - \left(\frac{D\rho A}{2m}\right)v^2 \tag{5.9}$$

We can calculate the terminal speed v_T by noting that when the gravitational force is balanced by the resistive force, the net force on the object is zero and therefore its acceleration is zero. Setting $a = 0$ in **Equation 5.9** gives

$$g - \left(\frac{D\rho A}{2m}\right)v_T^2 = 0$$

so

$$v_T = \sqrt{\frac{2mg}{D\rho A}} \tag{5.10}$$

Table 5.2 lists the terminal speeds for some objects falling through air.

Note that a golf ball has a higher terminal velocity than we might expect for its weight and size. This is because the surface of a golfball is dimpled, which affects the airflow around the ball, reducing its drag. Early golfballs were not dimpled and could not be hit as far as modern dimpled ones can be.

Quick Quiz 5.3

A baseball and a basketball having the same mass are dropped through air from rest such that their bottoms are initially at the same height above the ground, on the order of 1 m or more. Which one strikes the ground first? **(a)** The baseball strikes the ground first. **(b)** The basketball strikes the ground first. **(c)** Both strike the ground at the same time.

Table 5.2
Terminal speed for various objects falling through air

Object	Mass (kg)	Cross-sectional area (m^2)	v_T (m/s)
Skydiver	75	0.70	60
Golf ball (radius 2.1 cm)	0.046	1.4×10^{-3}	44
Cricket ball (radius 3.5 cm)	0.160	4.0×10^{-3}	40
Hailstone (radius 0.50 cm)	4.8×10^{-4}	7.9×10^{-5}	14
Raindrop (radius 0.20 cm)	3.4×10^{-5}	1.3×10^{-5}	9.0

Conceptual Example 5.5

The skysurfer

Consider a skysurfer (**Figure 5.16**) who jumps from a plane with her feet attached firmly to her surfboard, does some tricks, and then opens her parachute. Describe the forces acting on her during these manoeuvres. Sketch a graph of the drag force acting on the skysurfer as a function of time.

Jump Run Productions/The Image Bank/Getty Images

Figure 5.16
(Conceptual Example 5.5) A skysurfer

Solution

When the surfer first steps out of the plane, she has no vertical velocity. The downwards gravitational force causes her to accelerate towards the ground. As her downwards speed increases, so does the upwards resistive force exerted by the air on her body and the board. This upwards force reduces her acceleration, and so her speed increases more slowly. Eventually, she is going so fast that the upwards resistive force matches the downwards gravitational force. Now the net force is zero and she no longer accelerates – she has reached her terminal speed. At some point after reaching terminal speed, she opens her parachute, resulting in a drastic increase in the upwards resistive force. The net force (and thus the acceleration) is now upwards, in the direction opposite the direction of the velocity. The downwards velocity therefore decreases rapidly, and the resistive force on the parachute also decreases. Eventually, the upwards resistive force and the downwards gravitational force balance each other and a much smaller terminal speed is reached, permitting a safe landing.

Note that the velocity vector of a skydiver never points upwards. You may have seen videos of skydivers where they appear to suddenly go upwards at high speed when the parachute opens. This is because the camera is continuing to fall at the same rate, but the skydiver has slowed down. Hence the diver is moving upwards *relative to the camera*, but they are still moving downwards relative to the Earth.

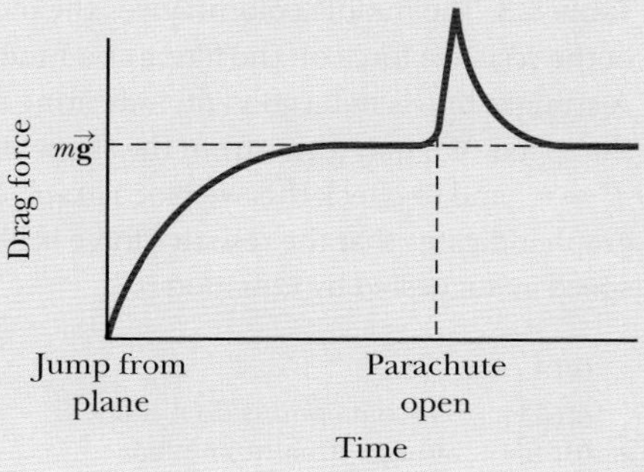

TRY THIS

Get some coffee filters or aluminium pie pans (paper patty pans also work well). Hold one flat and drop it carefully so that it falls directly down. Time how long it takes to fall. Hold two stacked together and drop them from the same height. How long do two together take? Now try it with three, four, five and ten. What happens to the time taken as you increase the number of filters or pie pans? Explain why.

±?

Example 5.6

Falling coffee filters

The dependence of resistive force on the square of the speed is a model. Let's test the model for a specific situation. A student performs an experiment in which she drops a series of bowl-shaped, pleated coffee filters and measures their terminal speeds. Each filter has a mass of (1.64 ± 0.02) g. The time constant t is small, so a dropped filter quickly reaches terminal speed. When the filters are nested together, they combine in such a way that the front-facing surface area does not increase. **Table 5.3** presents her measured data. Note that measuring speed is less precise than measuring mass in this experiment (as is often the case). Determine the relationship between the resistive force exerted by the air and the speed of the falling filters.

Example 5.6 cont.

Solution

Conceptualise Imagine dropping the coffee filters through the air. If you did the *Try this* example then you have a good idea of what happens. Because of the relatively small mass of the coffee filter, you probably won't notice the time interval during which there is an acceleration. The filters will appear to fall at constant velocity immediately upon leaving your hand. Draw a free-body diagram showing the forces acting on the filters at terminal velocity.

Model A filter moves at constant velocity, so we model it as a particle in equilibrium.

Analyse At terminal speed, the upwards resistive force on the filter balances the downwards gravitational force so that $R = mg$.

For a single coffee filter this will be

$$R = (1.64 \pm 0.02) \times 10^{-3}\ \text{kg} \times 9.80\ \text{m/s}^2 = (16.1 \pm 0.2) \times 10^{-3}\ \text{N}$$

Likewise, two filters nested together experience 0.0322 ± 0.0004 N of resistive force, and so forth. The relative uncertainty in each of the values will be the same, and will be approximately 1%. These values of resistive force are shown in the far right column of **Table 5.3**. The middle column gives the measured values of the terminal speed. A graph of the resistive force on the filters as a function of terminal speed is shown in **Figure 5.17a**. A straight line is not a good fit, indicating that the resistive force is *not* proportional to the speed. We may guess from the shape of the graph that the relationship is of the form $R \propto v_T^2$ and to check this we plot a graph of this form, as shown in **Figure 5.17b**. This graph indicates that the resistive force is likely to be proportional to the *square* of the speed as suggested by **Equation 5.7**.

Table 5.3
Terminal speed and resistive force for nested coffee filters

Number of filters	v_T (m/s) ± 10%	R (N) ± 1%
1	1.1	0.0161
2	1.4	0.0322
3	1.6	0.0483
4	2.0	0.0644
5	2.2	0.0805
6	2.3	0.0966
7	2.6	0.1127
8	2.8	0.1288
9	3.1	0.1449
10	3.2	0.1610

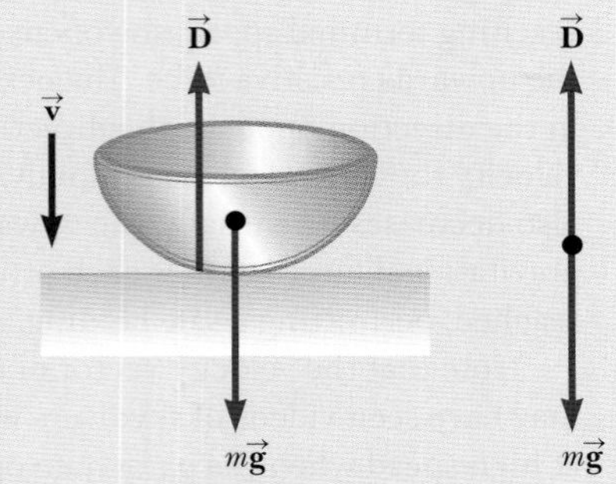

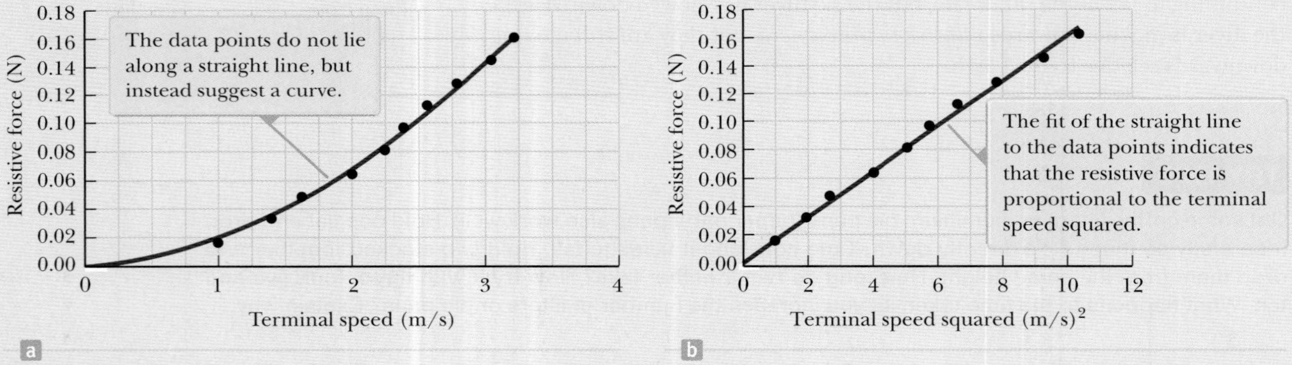

Figure 5.17
(Example 5.6) (a) Relationship between the resistive force acting on falling coffee filters and their terminal speed (b) Graph relating the resistive force to the square of the terminal speed

Finalise What if our graph of R vs v_T^2 had not given us a straight line? When we are unsure of the form of a relationship between two variables, but suspect that it is a power law of the form $y = ax^N$, then a plot of ln x vs ln y will tell us what the exponent N is. If $y = ax^N$, then (taking the natural logarithm of both sides) $\ln y = \ln(ax^N) = \ln a + \ln(x^N) = \ln a + N\ln x$.

Hence a plot of ln$y$ vs ln$x$ will have a gradient of N, and an intercept of ln$a$. **Figure 5.18** shows a plot of ln$R$ vs lnv_T. Note that it has a gradient of 2.0, indicating that our R $\propto v_T^2$ squared model is likely to be appropriate for the falling coffee filters. We can also read the intercept from the graph, and given the area of the coffee filters and the density of air, we could calculate the drag coefficient of the filters. Try it! Estimate the size of a typical coffee filter and calculate the drag coefficient from **Figure 5.18**.

For help with logarithms and linear equations see Appendix B.2.

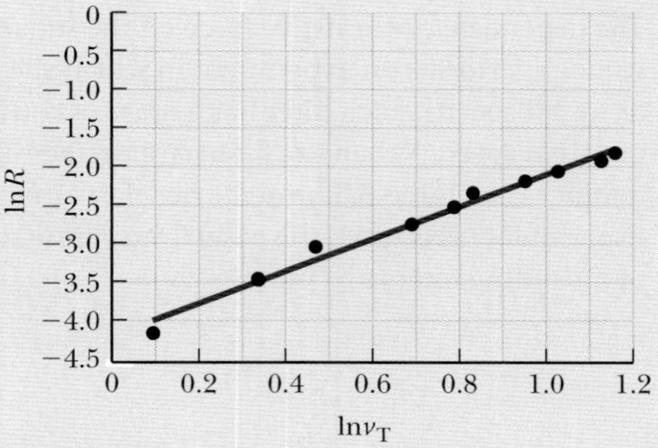

Figure 5.18
A graph of ln$R$ vs lnv_T. Note that it is a straight line with gradient 2, indicating that the resistive force is proportional to the square of the terminal velocity.

Example 5.7

Resistive force exerted on a cricketball – order of magnitude estimate

What is the drag force acting on a cricket ball bowled by a fast bowler? Make an order of magnitude estimate.

Solution

Conceptualise We are asked for an order of magnitude estimate, so let's make some simplifying assumptions. We will assume that the ball is travelling horizontally and only under the influence of gravity and the resistive force. The resistive force causes the ball to slow down and gravity causes its trajectory to curve downwards. We will need to estimate the ball's velocity; the other required data can be found in Table 5.2.

Model We model the ball as a projectile subject to two forces: gravity and the resistive force of the air. We model the resistive force using the model for drag given in Equation 5.7.

Analyse New Zealand cricketer Shane Bond's fastest bowled balls move at about 150 km/h, which is about 42 m/s. Note that this is almost the same as the terminal velocity of a cricket ball given in Table 5.2. We can dispense with a lot of substitution of numbers, as we are only after an order of magnitude estimate – if the velocity the ball is bowled at in the horizontal direction is the same as its vertical terminal velocity, then the drag force in each case will be the same. The drag force at terminal velocity in the vertical direction is equal to the gravitational force (weight). When Shane Bond (or another fast bowler) bowls a ball at about terminal velocity, it must experience a drag force roughly equal to its weight, or about 1 N.

Finalise How could we have solved this problem if the bowling speed was much lower or higher?

Answer We would have to rearrange Equation 5.7 to calculate the drag coefficient D of the ball using data from Table 5.2 and consider the ball in vertical motion. Then we could consider the situation of the ball in horizontal motion and again use Equation 5.7 but now solving for R. Try it for a spin bowler with a bowling speed of 25 m/s. Do you think this model for drag is appropriate for a spinning ball?

5.3 Uniform circular motion revisited

In Section 3.5, we discussed the analysis model of a particle in uniform circular motion, in which a particle moves with constant speed v in a circular path having a radius r. The particle experiences an acceleration that has a magnitude

$$a_c = \frac{v^2}{r}$$

The acceleration is called **centripetal acceleration** because $\vec{\mathbf{a}}_c$ is directed toward the centre of the circle. Furthermore, $\vec{\mathbf{a}}_c$ is *always* perpendicular to $\vec{\mathbf{v}}$. (If there were a component of acceleration parallel to $\vec{\mathbf{v}}$, the particle's *speed* would be changing.)

Let us now extend the particle in uniform circular motion model from Section 3.5 by incorporating the concept of force. Consider an air hockey puck of mass m that is tied to a string of length r and moves at constant speed in a horizontal, circular path as illustrated in Figure 5.19. Its weight is supported by a frictionless table, and the string is anchored to a peg at the centre of the circular path of the puck. Why does the puck move in a circle? According to Newton's first law, the puck would move in a straight line if there were no force on it; the string, however, prevents motion along a straight line by exerting on the puck a radial force $\vec{\mathbf{F}}_r$ that makes it follow the circular path. This force is directed along the string towards the centre of the circle as shown in Figure 5.19.

If Newton's second law is applied along the radial direction, the net force causing the centripetal acceleration can be related to the acceleration as follows:

$$\sum F = ma_c = m\frac{v^2}{r} \tag{5.11}$$

◀ Force causing centripetal acceleration

A force causing a centripetal acceleration acts towards the centre of the circular path and causes a change in the direction of the velocity vector. If that force should vanish, the object would no longer move in its circular path; instead, it would move along a straight-line path tangent to the circle. This idea is illustrated

Pitfall Prevention 5.4

Study Active Figure 5.20 very carefully. Many students (wrongly) think that the puck will move *radially* away from the centre of the circle when the string is cut. The velocity of the puck is *tangent* to the circle. By Newton's first law, the puck continues to move in the same direction in which it is moving when the force from the string disappears.

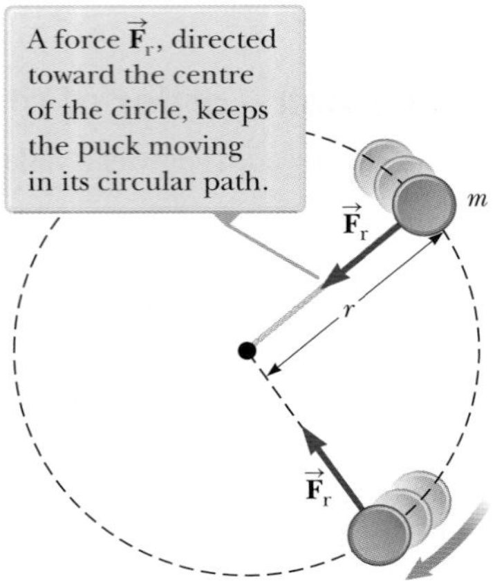

A force $\vec{F}_r$, directed toward the centre of the circle, keeps the puck moving in its circular path.
$\vec{F}_r$
m
r
$\vec{F}_r$

Figure 5.19
An overhead view of a puck moving in a circular path in a horizontal plane

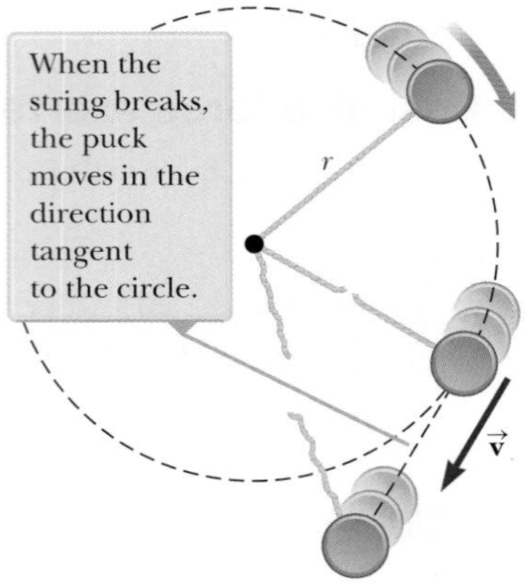

When the string breaks, the puck moves in the direction tangent to the circle.
r
$\vec{v}$

Active Figure 5.20
The string holding the puck in its circular path breaks.

in **Active Figure 5.20** for the puck moving in a circular path at the end of a string in a horizontal plane. If the string breaks at some instant, the puck moves along the straight-line path that is tangent to the circle at the position of the puck at this instant.

Quick Quiz 5.4

You are riding on a Ferris wheel that is rotating with constant speed. The car in which you are riding always maintains its correct upwards orientation; it does not invert. **(i)** What is the direction of the normal force on you from the seat when you are at the top of the wheel? (a) upwards (b) downwards (c) impossible to determine **(ii)** From the same choices, what is the direction of the net force on you when you are at the top of the wheel?

Analysis Model 5.1

Particle in uniform circular motion (extension)

Imagine a moving object that can be modelled as a particle. If it moves in a circular path of radius r at a constant speed v, it experiences a centripetal acceleration. Because the particle is accelerating, there must be a net force acting on the particle. That force is directed toward the centre of the circular path and is given by

$$\sum F = ma_c = m\frac{v^2}{r} \qquad (5.11)$$

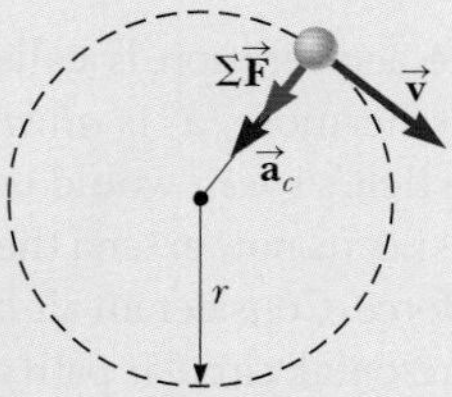

$\Sigma\vec{F}$
$\vec{v}$
$\vec{a}_c$
r

Examples

- the tension in a string of constant length acting on a rock twirled in a circle
- the gravitational force acting on a planet travelling around the Sun in a perfectly circular orbit (Chapter 11)
- the magnetic force acting on a charged particle moving in a uniform magnetic field (Chapter 30)
- the electric force acting on an electron in orbit around a nucleus in the Bohr model of the hydrogen atom (Chapter 41)

Example 5.8

The conical pendulum

A small ball of mass m is suspended from a string of length L. The string makes an angle θ with the vertical and sweeps out the surface of a cone as the ball revolves with constant speed v in a horizontal circle of radius r. This system is known as a *conical pendulum*. Find an expression for v.

Example 5.8 cont.

Solution

Conceptualise Imagine the motion of the ball and draw a diagram showing that the string sweeps out a cone and that the ball moves in a horizontal circle. Label your diagram using the variables given: r, L and θ. See **Figure 5.21a**.

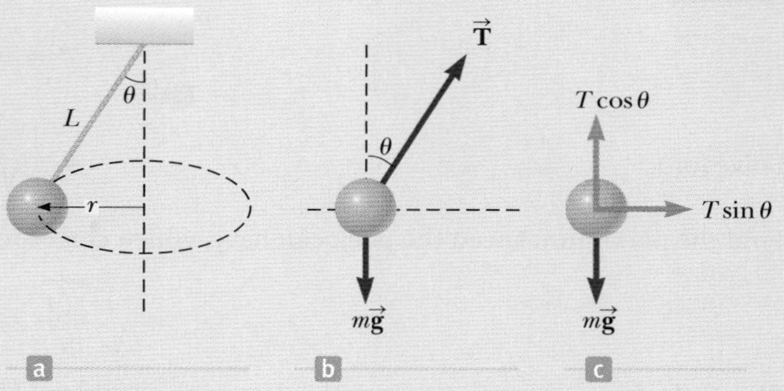

Figure 5.21
(Example 5.8) (a) A conical pendulum. The path of the ball is a horizontal circle. (b) A free-body diagram showing the forces acting on the ball (c) The horizontal and vertical components of the forces acting on the ball

Model The ball in **Figure 5.21a** does not accelerate vertically. Therefore, we model it as a particle in equilibrium in the vertical direction. It experiences a centripetal acceleration in the horizontal direction, so it is modelled as a particle in uniform circular motion in this direction.

Analyse We need to draw a free-body diagram showing the forces acting on the ball: see **Figure 5.21b**. We then need to resolve these forces into their horizontal and vertical components, as in **Figure 5.21c**. The force $\vec{T}$ exerted by the string on the ball is resolved into a vertical component $T\cos\theta$ and a horizontal component $T\sin\theta$ acting towards the centre of the circular path.

Apply the particle in equilibrium model in the vertical direction: $\sum F_y = T\cos\theta - mg = 0$

$$T\cos\theta = mg \quad (1)$$

Use **Equation 5.11** from the particle in uniform circular motion model in the horizontal direction:

$$\sum F_x = T\sin\theta = ma_c = \frac{mv^2}{r} \quad (2)$$

Divide **Equation (2)** by **Equation (1)** and use $\sin\theta/\cos\theta = \tan\theta$: $\tan\theta = \dfrac{v^2}{rg}$

Solve for v: $v = \sqrt{rg\tan\theta}$

Incorporate $r = L\sin\theta$ from the geometry in **Figure 5.21a**: $v = \sqrt{Lg\sin\theta\tan\theta}$

And finally, do a dimension check: $[\mathrm{LT}^{-1}] = ([\mathrm{L}][\mathrm{LT}^{-2}])^{1/2} = [\mathrm{LT}^{-1}]$ ☺

Finalise Notice that the speed is independent of the mass of the ball. Consider what happens when θ is 90° so that the string is horizontal. Because the tangent of 90° is infinite, the speed v is infinite, which tells us the string cannot possibly be horizontal. If it were, there would be no vertical component of the force $\vec{T}$ to balance the gravitational force on the ball. Compare this with **Figure 5.19** in which the puck's weight is supported by a frictionless table.

±? Example 5.9

How fast can it spin?

A puck of mass (0.50 ± 0.05) kg is attached to the end of a cord (1.5 ± 0.1) m long. The puck moves in a horizontal circle on a frictionless table as shown in **Figure 5.19**. The cord is rated to withstand a maximum tension of 50 N, with a tolerance (uncertainty) of 5%. What is the maximum speed at which the puck can move before the cord breaks? Assume the string remains horizontal during the motion.

Solution

Conceptualise The force required to keep the puck going around increases with the square of the velocity, so it makes sense that the stronger the cord, the faster the puck can move before the cord breaks. Also, we expect a more massive puck to break the cord at a lower speed. (Imagine whirling a bowling ball on the cord!) We already have a diagram of the system and the forces acting in **Figure 5.19** and **Active Figure 5.20**, otherwise we would start by drawing a diagram to help us conceptualise the problem. You should review these figures, particularly **Active Figure 5.20**.

Model As the puck moves in a circular path, we model it as a particle in uniform circular motion.

Analyse Incorporate the tension and the centripetal acceleration into Newton's second law as described by **Equation 5.11**:

$$T = m\frac{v^2}{r}$$

Solve for v:

$$v = \sqrt{\frac{Tr}{m}} \qquad (1)$$

Find the maximum speed the puck can have, which corresponds to the maximum tension the string can withstand:

$$v_{max} = \sqrt{\frac{T_{max}r}{m}} = \sqrt{\frac{(50\text{ N})(1.50\text{ m})}{0.500\text{ kg}}} = 12\text{ m/s}$$

This is the best value of v_{max} given the data in the problem. However, we know that the cord may actually break at a tension of only 95% of its rated strength of 50 N, or 47.5 N. The length of the cord could be as short as 1.4 m and the mass of the puck up to 0.55 kg.

Hence the maximum velocity for which we can be confident that the cord will not break is:

$$v = \sqrt{\frac{T_{max}r}{m}} = \sqrt{\frac{47.5\text{ N} \times 1.4\text{ m}}{0.55\text{ kg}}} = 11\text{ m/s}$$

Finalise **Equation (1)** shows that v increases with T and decreases with larger m, as we expected from our conceptualisation of the problem.

What If? Suppose the puck moves in a circle of larger radius at the same speed v. Is the cord more likely or less likely to break?

Answer The larger radius means that the change in the direction of the velocity vector will be smaller in a given time interval. Therefore, the acceleration is smaller and the required tension in the string is smaller. As a result, the string is less likely to break when the puck travels in a circle of larger radius.

Example 5.10

What is the maximum speed of the car?

A 1500 kg car moving on a flat, horizontal road goes around a curve with a radius of 35.0 m. The coefficient of static friction between the tyres and the dry road is 0.523. Find the maximum speed the car can have and still make the turn without sliding off the road.

Solution

Conceptualise Imagine that the curved roadway is part of a large circle so that the car is moving in a circular path. Draw a diagram showing this, as in **Figure 5.22a**. Think about your experience driving or being a passenger. When you turn the steering wheel, the car's front wheels turn so the plane of the wheels is at an angle to the direction the car is pointing. The wheels push against the road with a component in the direction perpendicular to the velocity of the car. From Newton's third law, the road must therefore push against the wheels (and via the wheels the rest of the car) in the opposite direction. This push by the road against the wheels provides the centripetal force to turn the car. The greater the velocity, the greater the force needed to turn in a circle. It is the force of *static* friction that makes the car turn the corner as, ideally, no slipping occurs at the point of contact between the road and the tyres.

Model Based on the *Conceptualise* step of the problem, we model the car as a particle in uniform circular motion in the horizontal direction. The car is not accelerating vertically, so it is modelled as a particle in equilibrium in the vertical direction.

Analyse As discussed in Section 5.1, it is the force of static friction that allows a car to move along the road and in this case stops it from sliding off the road on the corners. The maximum speed v_{max} the car can have around the curve is the speed at which it is on the verge of skidding outwards. At this point, the friction force has its maximum value $f_{s,max} = \mu_s n$.

Apply **Equation 5.11** in the radial direction for the maximum speed condition:

$$f_{s,max} = \mu_s n = m\frac{v_{max}^2}{r} \qquad (1)$$

Apply the particle in equilibrium model to the car in the vertical direction: $\sum F_y = 0 \rightarrow n - mg = 0 \rightarrow n = mg$

Solve **Equation (1)** for the maximum speed and substitute for n: $v_{max} = \sqrt{\frac{\mu_s n r}{m}} = \sqrt{\frac{\mu_s m g r}{m}} = \sqrt{\mu_s g r}$ (2)

Do a dimension check, remembering that μ_s is dimensionless:

$$[LT^{-1}] = ([LT^{-2}][L])^{1/2} = [LT^{-1}]$$ ☺

Substitute numerical values: $v_{max} = \sqrt{(0.523)(9.80 \text{ m/s}^2)(35.0 \text{ m})} = 13.4 \text{ m/s}$

Finalise This speed is equivalent to 50.0 km/h. Therefore, if the speed limit on this road is higher than 50 km/h this road could benefit greatly from some banking, as in the next example. Many roads have advisory speed signs before corners, to help drivers negotiate them safely. Notice that the maximum speed does not depend on the mass of the car, which is why highways do not need multiple speed limits to cover the various masses of vehicles using the road.

What If? Suppose a car travels this curve on a wet day and begins to skid on the curve when its speed reaches only 8.00 m/s. What can we say about the coefficient of static friction in this case?

Answer The coefficient of static friction between the tyres and a wet road should be smaller than that between the tyres and a dry road. This expectation is consistent with experience with driving – a skid is more likely on a wet road than a dry road and there are more accidents on wet days as a result.

To check our prediction, we can solve **Equation (2)** for the coefficient of static friction: $\mu_s = \frac{v_{max}^2}{gr}$

Substituting the numerical values gives $\mu_s = \frac{v_{max}^2}{gr} = \frac{(8.00 \text{ m/s})^2}{(9.80 \text{ m/s}^2)(35.0 \text{ m})} = 0.187$

which is indeed smaller than the coefficient of 0.523 for the dry road.

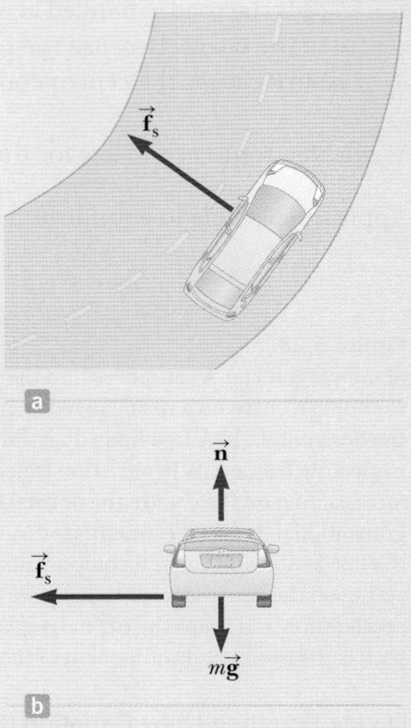

Figure 5.22
(Example 5.10) (a) The force of static friction directed towards the centre of the curve keeps the car moving in a circular path.
(b) A free-body diagram showing the forces acting on the car

Example 5.11

The banked road

A civil engineer wishes to redesign the curved road in Example 5.10 in such a way that a car will not have to rely on friction to round the curve without skidding. In other words, a car moving at the recommended speed can negotiate the curve even when the road is covered with ice. Such a curve is usually 'banked', which means that the road surface is tilted toward the inside of the curve. Suppose the designated speed for the curve is to be 13.4 m/s (50.0 km/h) and the radius of the curve is 35.0 m. At what angle should the curve be banked?

Solution

Conceptualise The difference between this example and Example 5.10 is that the car is no longer moving on a flat road. **Figure 5.23** shows the banked road, with the centre of the circular path of the car far to the left of the figure. Draw a free-body diagram for the car. Notice that the horizontal component of the normal force provides the car's centripetal acceleration.

Model As in Example 5.10, the car is modelled as a particle in equilibrium in the vertical direction and a particle in uniform circular motion in the horizontal direction. We take the limiting case of zero friction as we wish the road to be usable under icy conditions.

Analyse If the road is banked at an angle θ as in **Figure 5.23**, the normal force $\vec{\mathbf{n}}$ has a horizontal component towards the centre of the curve. Because the road is to be designed so that the force of static friction can be zero, only the component $n_x = n \sin\theta$ causes the centripetal acceleration.

Write Newton's second law for the car in the radial direction, which is the x direction: $\sum F_r = n\sin\theta = \dfrac{mv^2}{r}$ (1)

Apply the particle in equilibrium model to the car in the vertical direction: $\sum F_y = n\cos\theta - mg = 0$

$n\cos\theta = mg$ (2)

Figure 5.23
(Example 5.10) (a) A car moves into the page and is rounding a curve on a road banked at an angle θ to the horizontal. (b) A free-body diagram for the car, neglecting friction. When friction is neglected, the only forces acting on the car are the normal force and the gravitational force. (c) A free-body diagram showing the horizontal and vertical components of the forces acting on the car. The force that causes the centripetal acceleration and keeps the car moving in its circular path is the horizontal component of the normal force.

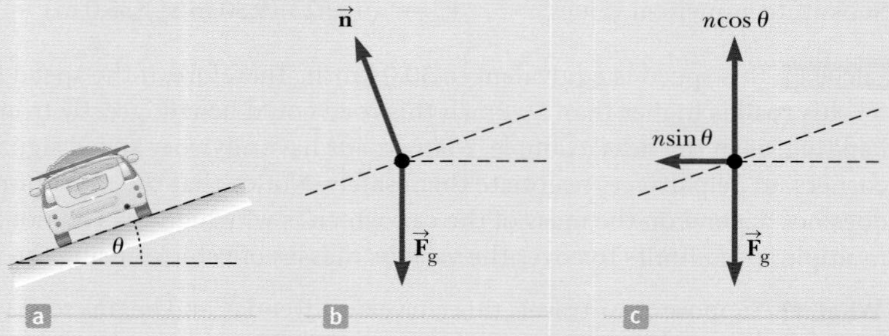

Divide **Equation (1)** by **Equation (2)**: $\tan\theta = \dfrac{v^2}{rg}$ (3)

Check dimensions, remembering that $\tan\theta$ is dimensionless: $[] = [LT^{-1}]^2/([L][LT^{-2}]) = []$ ☺

Solve for the angle θ: $\theta = \tan^{-1}\left[\dfrac{13.4\ \text{m/s}^2}{(35.0\ \text{m})(9.80\ \text{m/s}^2)}\right] = 27.6°$

Finalise **Equation (3)** shows that the banking angle is independent of the mass of the vehicle negotiating the curve.

What If? What if a car rounds the corner at a speed greater or less than 50 km/h?

Answer If a car rounds the curve at a speed less than 50 km/h, friction is needed to keep it from sliding down the bank (to the left in **Figure 5.23**). A driver attempting to negotiate the curve at a speed greater than 50 km/h has to depend on friction to keep from sliding up the bank (to the right in **Figure 5.23**). Fortunately a completely frictionless road is very unusual and the friction force is generally a major component of the centripetal force that makes cars go around corners.

If you have been on a carnival ride that rotates rapidly, or even been in a car going around a corner at high speed, you will have felt yourself being pushed against the inner surface of the ride or car. As described in Chapter 4 when we discussed the concept of 'apparent weight', when a surface presses against us in the vertical direction, we experience this as a sensation of weight.

Imagine being in a spacecraft that is in orbit. Every object inside the spacecraft, including you, is accelerating with $a_c = v^2/r$ where v is the orbital speed of the craft and r is the orbital radius. No surface presses against you unless you push against it, and any object held at rest (relative to the spacecraft) and released will stay in that position unless affected by other forces such as air currents. Hence there is no sensation of weight, and no sense of up or down.

This lack of a sense of weight can be problematic in several ways – it can be disorientating and cause nausea, and over extended time periods can cause muscle wastage as there is no need to support the body against the gravitational force. In addition, unsecured items tend to drift around and particles and liquids can drift about and interfere with equipment or be breathed in.

One way of addressing the apparent lack of weight is to rotate the spacecraft, as shown in **Figure 5.24**, so that an internal surface presses against the people and objects inside, providing the centripetal force to keep them rotating with the spacecraft in the same way the car door stops you sliding out of the car on a sharp corner. This is sometimes called 'artificial gravity' because the experience is much like being in a gravitational field. In fact the experiences of being in a gravitational field and being in an accelerating reference frame are indistinguishable. This is discussed in more detail in Chapter 12 when we discuss the general theory of relativity. The person shown in

Figure 5.24 experiences a normal force pushing up on their feet in the direction shown. Their body interprets this as a gravitational force (weight) in the opposite direction, just as when you accelerate upwards in a lift the increased normal force on your feet gives the sensation of increased weight. Note that Figure 5.24 is drawn from the point of view of someone in the same orbit as the space craft, but external to the craft. Hence it does not show the velocity, acceleration or (gravitational) force associated with the orbital motion. If the craft was in deep space, so far from any large body that the gravitational field was zero, the situation would be exactly the same as that shown in Figure 5.24.

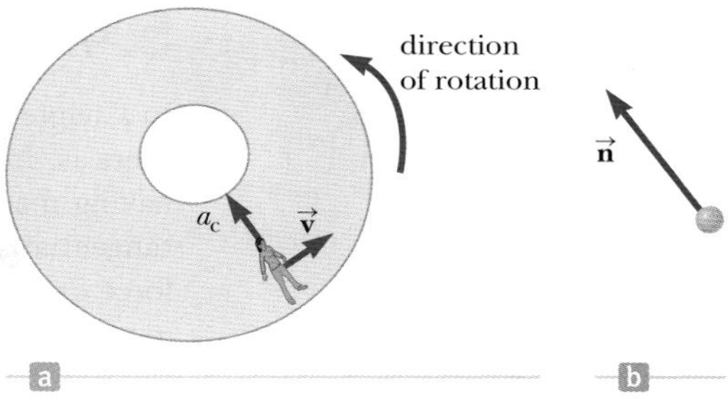

Figure 5.24
A rotating spacecraft as seen from the reference frame of a person in the same orbit but external to the craft (a) A person on the inside of the craft must have the velocity and acceleration as shown if they are to rotate with the craft. (b) Free-body diagram for the person. The contact (normal) force of the wall of the craft pushes against them to provide the necessary centripetal force.

Quick Quiz 5.5

Imagine the person shown in Figure 5.24 is holding a ball in their hand. They gently let go of the ball (not throw it) when they are at the position shown in Figure 5.24. What is the path of the ball as seen by this person? (a) It stays where they released it, floating next to their hand. (b) It falls to land approximately at their feet. (c) It falls to land behind them, approximately where their feet were when they released it. Hint: look at Figure 4.24 and draw a similar pair of diagrams for this situation.

Example 5.12

Imagine a space station in the shape of a wheel like that shown in Figure 5.24, with a radius to the external wall of 250 m. If the inhabitants are to feel as if they are subject to normal Earth surface gravity, what linear speed must the 'floor' of the space station be moving at? What period of rotation does this correspond to?

Solution

Conceptualise Examine Figure 5.24. The rotational acceleration is caused by the force of the floor acting on the person. This acceleration needs to be the same as that due to the Earth's gravitational field at the surface of the Earth. This will give a normal force acting on the person's feet equal to the normal force they would experience standing on the ground on Earth.

Model We will treat the situation from the point of view of an observer outside the space station, not rotating but otherwise moving with it.

Analyse We want the centripetal force (the normal force) to be equal to mg where g is the magnitude of the acceleration due to gravity on Earth's surface.

$$n = \frac{mv^2}{r} = mg$$

so:

$$\frac{v^2}{r} = g$$

Rearranging for v: $v = \sqrt{rg}$

Check dimensions: $[LT^{-1}] = ([L][LT^{-2}])^{\frac{1}{2}} = [LT^{-1}]$ ☺

Substitute values: $v = \sqrt{250\text{ m} \times 9.8\text{ m.s}^{-2}} = 2450\text{ m.s}^{-1}$

Any point on the floor is travelling at 2450 $m.s^{-1}$, and travels a distance of $d = 2\pi r = 1570$ m in each rotational period. Therefore the period of rotation is $T = \frac{d}{v} = 0.64$ s.

What If? What if the space station had concentric floors and was spinning to give an 'artificial gravity' equal to g on the outermost floor? What would happen to an inhabitant's sensation of weight as they moved to a more inner floor?

Answer The period of rotation would be the same, but the radius would be smaller. As $T = \frac{2\pi r}{v}$, the velocity must be directly proportional to the radius. The centripetal acceleration, which is the 'artificial gravity' is $a_c = \frac{v^2}{r}$. So if $r \rightarrow \frac{1}{2} r$, then $v \rightarrow \frac{1}{2} v$ and $v^2 \rightarrow \frac{1}{4} v^2$, so $a_c \rightarrow \frac{1}{2} a_c$. Hence the artificial gravity decreases as you move closer to the centre of the space station.

Finalise The linear speed scales with the square root of the radius, as does the rotational period. So the larger the spacecraft, the greater the linear speed must be, but also the greater the rotational period. Hence the frequency of rotation is smaller for larger spacecraft.

The net force exerted on the particle is the vector sum of the radial force and the tangential force.

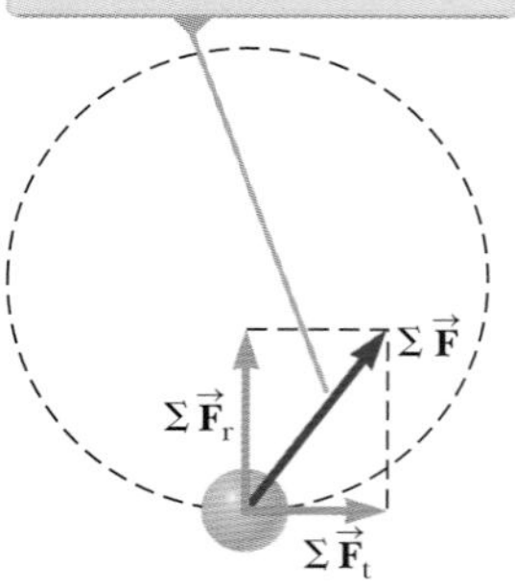

Active Figure 5.25

When the net force acting on a particle moving in a circular path has a tangential component $\Sigma\vec{F}_t$, the particle's speed changes.

5.4 Non-uniform circular motion

In Chapter 3, we found that if a particle moves with varying speed in a circular path, there is, in addition to the radial component of acceleration, a tangential component having magnitude $|dv/dt|$. Therefore, the force acting on the particle must also have a tangential and a radial component. Because the total acceleration is $\vec{\mathbf{a}} = \vec{\mathbf{a}}_r + \vec{\mathbf{a}}_t$, the total force exerted on the particle is $\sum\vec{\mathbf{F}} = \sum\vec{\mathbf{F}}_r + \sum\vec{\mathbf{F}}_t$ as shown in **Active Figure 5.25**. We express the radial and tangential forces as net forces with the summation notation because there could be multiple forces that combine. The vector $\sum\vec{\mathbf{F}}_r$ is directed towards the centre of the circle and is responsible for the centripetal acceleration. The vector $\sum\vec{\mathbf{F}}_t$ tangent to the circle is responsible for the tangential acceleration, which represents a change in the particle's speed with time.

Quick Quiz 5.6

A bead slides freely along a curved wire lying on a horizontal surface at constant speed as shown in **Figure 5.26**. **(a)** Draw the vectors representing the force exerted by the wire on the bead at points Ⓐ, Ⓑ and Ⓒ. **(b)** Suppose the bead in **Figure 5.26** speeds up with constant tangential acceleration as it moves towards the right. Draw the vectors representing the force on the bead at points Ⓐ, Ⓑ and Ⓒ.

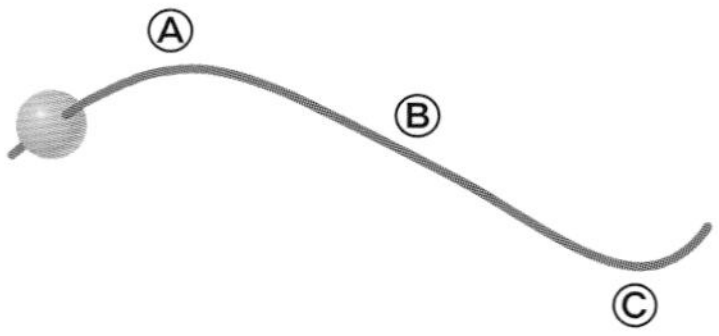

Figure 5.26
(Quick Quiz 5.6) A bead slides along a curved wire

Example 5.13

Poi swinging in a vertical circle

A girl is swinging a poi, a pair of balls on each end of a cord, so that one ball, with mass m, swings in a *vertical* circle of radius R about a fixed point O as shown in **Figure 5.28**. The girl starts the poi swinging and then allows it to swing without exerting any extra force other than to hold the end of the cord still at the centre. Determine the tangential acceleration of the poi and the tension in the cord at any instant when the speed of the poi is v and the cord makes an angle θ with the vertical.

Alamy/Photo Resource Hawaii

Figure 5.27

Solution

Conceptualise The poi will not move at a constant velocity because a tangential component of acceleration arises from the gravitational force exerted on the ball. We start by drawing a diagram showing the forces acting on the poi. We need to consider the forces for any arbitrary angle θ.

Categorise We model the ball as a particle subject to a net force and moving in a circular path, but it is not a particle in *uniform* circular motion. We need to use the techniques discussed in this section on non-uniform circular motion.

Analyse From the force diagram in **Figure 5.28**, we see that the only forces acting on the ball are the gravitational force $\vec{\mathbf{F}}_g = m\vec{\mathbf{g}}$ exerted by the Earth and the force $\vec{\mathbf{T}}$ exerted by the cord. We resolve $\vec{\mathbf{F}}_g$ into a tangential component $mg\sin\theta$ and a radial component $mg\cos\theta$.

Apply Newton's second law to the sphere in the tangential direction:

$$\sum F_t = mg\sin\theta = ma_t$$
$$a_t = g\sin\theta$$

Example 5.13 cont.

Apply Newton's second law to the forces acting on the sphere in the radial direction, noting that both $\vec{\mathbf{T}}$ and $\vec{\mathbf{a}}_r$ are directed towards O:

$$\sum F_T = T - mg\cos\theta = \frac{mv^2}{R}$$

$$T = mg\left(\frac{v^2}{Rg} + \cos\theta\right)$$

We check our dimensions, remembering that $\cos\theta$ is dimensionless and T is a force:

$[MLT^{-2}] = [M][LT^{-2}]([LT^{-1}]^2/[L][LT^{-2}] + []) = [M][LT^{-2}]([] + []) = [MLT^{-2}]$ ☺

Finalise Let us evaluate this result at the top and bottom of the circular path (**Figure 5.28**):

$$T_{top} = mg\left(\frac{v_{top}^2}{Rg} - 1\right) \qquad T_{bot} = mg\left(\frac{v_{bot}^2}{Rg} + 1\right)$$

We can see that the tension is at a minimum when the ball is at the top of the circle and a maximum when it is at the bottom.

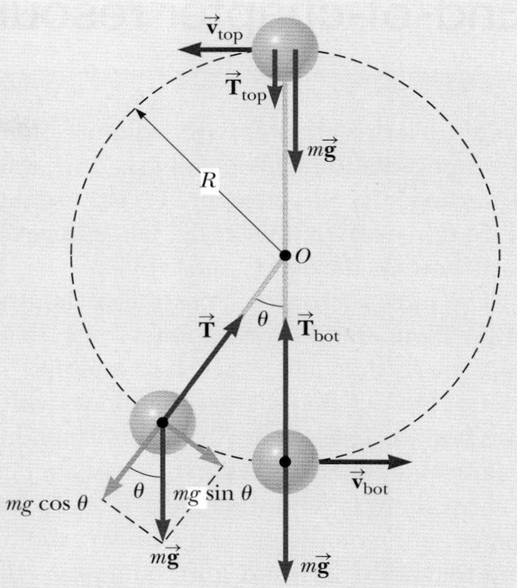

Figure 5.28
(Example 5.13) The forces acting on a ball of mass m connected to a cord of length R and rotating in a vertical circle centred at O. Forces acting on the sphere are shown when the ball is at the top and bottom of the circle and at an arbitrary location.

What If? What if the ball is set in motion with a slower speed?

(A) What speed would the ball have as it passes over the top of the circle if the tension in the cord goes to zero instantaneously at this point?

Answer Let us set the tension equal to zero in the expression for T_{top}:

$$0 = mg\left(\frac{v_{top}^2}{Rg} - 1\right) \rightarrow v_{top} = \sqrt{gR}$$

(B) What if the ball is set in motion such that the speed at the top is less than this value? What happens?

Answer In this case, the ball does not reach the top of the circle. At some point on the way up, the tension in the string becomes zero and the ball becomes a projectile. It follows a segment of a parabolic path over the top of its motion, rejoining the circular path on the other side when the tension becomes non-zero again.

Throughout this chapter we have introduced models that are less idealised, and have fewer simplifying assumptions and approximations, than those introduced in earlier chapters. For example, if we wish to model the behaviour of a car moving around a corner with a varying speed, we can now take into account friction due to skidding on the road and drag due to air resistance acting on the car, to analyse the radial and tangential acceleration. However, it is important to remember that these models are still just that – models. They still provide only an approximation to the forces acting on a car moving around a corner. In practice, a combination of theoretical modelling and empirical data from experiments is usually used to develop more precise models. For example, engineers use Computer Aided Engineering (CAE) tools which produce simulations based on kinematics, Newton's second law, conservation of energy and momentum (Chapters 7 and 8) and fluid dynamics (Chapter 14) as well as empirical data to model the complex behaviour and interactions of real surfaces and structures.

End-of-chapter resources

We started this chapter with a picture of the Singapore Flyer observation wheel, and asked what forces act on a passenger in the Singapore Flyer? Are they constant or do they vary in time as the wheel turns? If they vary, how do they vary, and why?

The forces acting on a passenger standing in a cabin of the Flyer are the gravitational force, which will be very nearly constant, and the normal force and the frictional force due to the cabin floor (which remains horizontal). These three forces must combine to give a centripetal force to make the passenger go around in a vertical circle with the Flyer. Figure 5.29 shows the forces acting on the passenger at an arbitrary position, and free-body diagrams for the forces acting at the top, bottom and halfway points. Note that the sum of the three forces acting will always equal mv^2/r as long as the Flyer rotates with constant speed.

Let's consider each of these positions. At the bottom, the acceleration is directed upwards only, and the frictional force will be zero. Applying Newton's second law in the vertical direction and taking upwards as positive, we see that:

$$n - mg = mv^2/r$$

So

$$n = m(g + v^2/r)$$

So the passenger feels heavier at the bottom because the floor is pushing up on them more than if they were still. At the top we find the opposite:

$$n - mg = -mv^2/r$$

So

$$n = m(g - v^2/r)$$

So the passenger feels lighter.

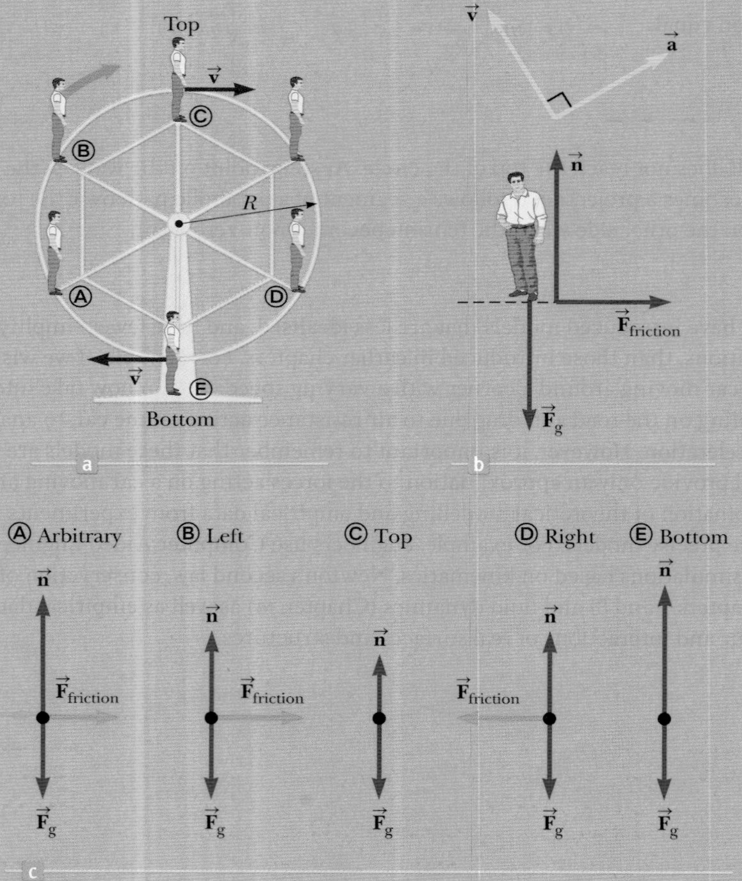

Figure 5.29
(a) A passenger in a cabin at various positions during a ride on the Singapore Flyer (b) The forces acting on a passenger at position Ⓐ (c) Free-body diagrams for the passenger at positions Ⓐ to Ⓔ as shown in (a)

The normal component of the force varies over time between these two extreme values. Considering the halfway positions, the centripetal force is horizontal and due entirely to the frictional force between the passenger's shoes and the floor. At each halfway position the frictional force must be $F = mv^2/r$. The normal force, which acts in the vertical direction, balances the gravitational force, hence $n = mg$ at these positions. Figure 5.30 shows a sketch of the gravitational, frictional and normal forces as a function of time.

Figure 5.29b shows the forces acting on the passenger at position A. At this position the net (centripetal) force is pointing at an angle to the horizontal, and is due to both the normal force and the frictional force.

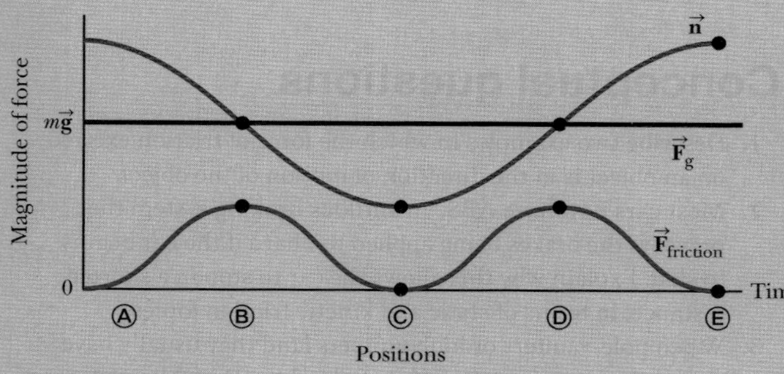

Figure 5.30
The forces (magnitude only) acting on a passenger in the Singapore Flyer as function of time

The problems found in this chapter may be assigned online in Enhanced Web Assign.

ENHANCED WebAssign

Worked solutions to every fifth problem are available in the Student Solutions Manual. Register online at **www.cengagebrain.com** for access.

Summary

Definitions

Resistive forces or **friction forces** act to resist the *relative motion between surfaces.* Note that friction forces do not oppose motion – they oppose the *relative motion of the surfaces.*

The resistive force between the surface of a solid object and a fluid in which it moves is called the **drag force**.

Concepts and principles

The maximum **force of static friction** $\vec{\mathbf{f}}_{s,\max}$ between an object and a surface is proportional to the normal force acting on the object. In general, $f_s \le \mu_s n$, where μ_s is the **coefficient of static friction** and n is the magnitude of the normal force.

When an object slides over a surface, the magnitude of the **force of kinetic friction** $\vec{\mathbf{f}}_k$ is given by $f_k = \mu_k n$, where μ_k is the **coefficient of kinetic friction**.

An object moving through a liquid or gas experiences a speed-dependent **resistive force**. This resistive force is in a direction opposite that of the velocity of the object relative to the medium and generally increases with speed. The magnitude of the resistive force depends on the object's size and shape, and on the properties of the medium through which the object is moving. In the limiting case for a falling object, when the magnitude of the resistive force equals the object's weight, the object reaches its **terminal velocity**.

A particle moving in uniform circular motion has a centripetal acceleration; this acceleration must be provided by a net force directed towards the centre of the circular path.

When an object moves in non-uniform circular motion such that its speed is varying, there must be a net force in both the radial and tangential directions. The net force in the radial direction acts to provide the centripetal acceleration; the net force in the tangential direction acts to change the speed of the object.

Analysis models for problem-solving

Friction between solid surfaces: resistive force not proportional to velocity The magnitude of the force of static friction between any two surfaces in contact can have the values

$$f_s \le \mu_s n \qquad (5.1)$$

The magnitude of the force of kinetic friction acting between two surfaces that are slipping relative to each other is

$$f_k = \mu_k n \qquad (5.2)$$

Resistive forces in fluids 1: resistive force proportional to velocity If we model the resistive force acting on an object moving through a liquid or gas as proportional to the object's velocity, the resistive force can be expressed as

$$\vec{\mathbf{R}} = -b\vec{\mathbf{v}} \qquad (5.3)$$

This is a good approximation for objects moving in fluids such as water.

Resistive forces in fluids 2: resistive force proportional to velocity squared For objects moving at high speeds through air, the resistive force is reasonably well modelled as proportional to the square of the speed:

$$R = \frac{1}{2}D\rho A v^2 \qquad (5.7)$$

Particle in uniform circular motion (extension) Newton's second law applied to a particle moving in uniform circular motion states that the net force causing the particle to undergo a centripetal acceleration (Equation 3.26) is related to the acceleration according to

$$\sum F = ma_c = m\frac{v^2}{r} \qquad (5.11)$$

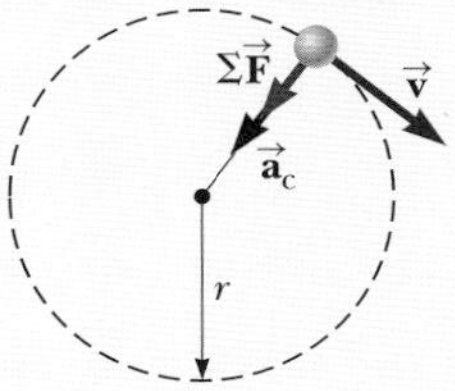

Chapter review quiz

To help you revise Chapter 5: Further applications of Newton's laws, complete the automatically graded Chapter review quiz at http://login.cengagebrain.com.

Conceptual questions

1. Describe two examples in which the force of friction exerted on an object is in the direction of motion of the object.
2. Most cars now have ABS, an antilock braking system that prevents the brakes being applied too hard if the car begins to skid. Explain why this allows the car to stop in a shorter distance, in terms of static and kinetic friction forces.
3. When pole-vaulters or high jumpers land they usually have a mattress or other soft surface to land on. Draw diagrams showing the direction of acceleration and the forces acting on a pole-vaulter (a) just before they hit the mattress, (b) as they sink into the mattress but before they come to rest and (c) just as they have come to rest. (d) Sketch a graph showing how the normal force of the mattress on the pole-vaulter changes with time, and mark in the times described in parts (a), (b) and (c).
4. A falling skydiver reaches terminal speed with her parachute closed. After the parachute is opened, what parameters change to decrease this terminal speed?
5. (a) Can a normal force be horizontal? (b) Can a normal force be directed vertically downwards? (c) Consider a tennis ball in contact with a stationary floor and with nothing else. Can the normal force be different in magnitude from the gravitational force exerted on the ball? (d) Can the force exerted by the floor on the ball be different in magnitude from the force the ball exerts on the floor? Explain your answer to each question and give an example.
6. Consider a small raindrop and a large raindrop falling through the atmosphere. (a) Compare their terminal speeds. (b) What are their accelerations when they reach terminal speed?
7. Describe the path of a moving body in the event that (a) its acceleration is constant in magnitude at all times and perpendicular to the velocity, and (b) its acceleration is constant in magnitude at all times and parallel to the velocity. Draw a diagram to illustrate your answers.
8. An object executes circular motion with constant speed whenever a net force of constant magnitude acts perpendicular to the velocity. What happens to the speed if the force is not perpendicular to the velocity?
9. A child is practising for a BMX race. His speed remains constant as he goes anticlockwise around a level track with two straight sections and two nearly semicircular sections as shown in the aerial view of Figure CQ5.9. (a) Rank the magnitudes of his acceleration at the points *A*, *B*, *C*, *D* and *E* from largest to smallest. If his acceleration is the same size at two points, display that fact in your ranking. If his acceleration is zero, display that fact. (b) What are the directions of his velocity at points *A*, *B* and *C*? For each point, choose one of north, south, east, west or non-existent. (c) What are the directions of his acceleration at points *A*, *B* and *C*?

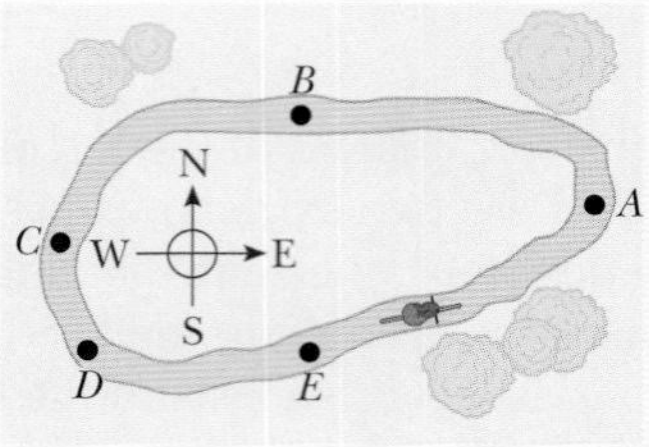

Figure CQ5.9

10. It has been suggested that rotating cylinders about 20 km in length and 8 km in diameter be placed in space and used as colonies. The purpose of the rotation is to simulate gravity for the inhabitants. Explain this concept for producing an effective imitation of gravity.
11. Why do fighter pilots tend to black out when pulling out of a steep dive?
12. A bucket of water can be whirled in a vertical path such that no water is spilled. Why does the water stay in the bucket, even when the bucket is above your head?

Problems

Section 5.1 Friction between solid surfaces

1. The Menzies building at Monash University has a long pavement of polished marble along the front, which becomes very slippery when it is wet. The slab shape of the building and its surroundings also tend to cause high wind speeds along the front of the building, making it an exciting place to walk in rainy, windy weather. A student of mass m stands at one end of the slippery pavement, holding their coat open like wings to catch as much wind as possible. The coefficient of kinetic friction between the student's shoes and the paving is μ_k. The wind applies a horizontal force of F_w to the student parallel to the front of the building. (a) Draw a free-body diagram showing all the forces acting on the student. (b) Identify the reaction force to each of these forces. (c) Find an expression for the speed, v, of the student sliding along the pavement as a function of the distance, x, they have travelled along it.
2. A small child with mass M rides down an icy slope of height h and inclination θ on a toboggan of mass m. At the bottom of the slope they continue to slide on flat ground, slowing and coming to a halt due to friction. (a) Draw a free-body diagram showing all the forces acting on the toboggan as it slides along the level ground at the bottom of the slope. (b) Identify the reaction forces to each of these forces on the toboggan. (c) The coefficient of friction between the toboggan and the ground is μ_k. Find an expression for the distance the child and toboggan travel before stopping.
3. ±? A (25 ± 1) kg box is initially at rest on a horizontal surface. A student measures the force required to set the block in motion, repeating the measurement five times, and finds the following values: 75.0 N, 75.2 N, 75.4 N, 74.9 N, and 74.8 N. She then measures the force required to keep the box moving with constant speed and finds values of 60.0 N, 60.1 N, 60.5 N, 58.9 N, and 59.6 N. Find (a) the coefficient of static friction and (b) the coefficient of kinetic friction between the block and the surface.

4. To meet a US Postal Service minimum requirement, employees' footwear must have a coefficient of static friction of 0.5 or more on a specified tile surface. A typical athletic shoe has a coefficient of static friction of 0.800. In an emergency, what is the minimum time interval in which a person starting from rest can move 3.00 m on the tile surface if she is wearing (a) footwear meeting the Postal Service minimum requirement and (b) a typical athletic shoe?

5. Before 1960, people believed that the maximum attainable coefficient of static friction for a car tyre on a road was $\mu_s = 1$. Around 1962, three companies independently developed racing tyres with coefficients of 1.6. Tyres have improved further since then. The shortest time interval in which a piston-engine car initially at rest has covered a distance of one-quarter of a mile (402 m) is about 4.43 s. (a) Assume the car's rear wheels lift the front wheels off the pavement as shown in **Figure P5.5**. What minimum value of μ_s is necessary to achieve the record time? (b) Suppose the driver were able to increase his or her engine power, keeping other things equal. How would this change affect the elapsed time?

Jamie Squire/Allsport/ Getty Images

Figure P5.5

6. At a biscuit factory a 3.00 kg box of biscuits starts from rest at the top of a 30.0° chute and slides a distance of 2.00 m down the chute in 1.50 s to land in a truck, ready to go to the supermarket. Find (a) the magnitude of the acceleration of the box, (b) the coefficient of kinetic friction between the box and the chute, (c) the friction force acting on the box and (d) the speed of the box after it has slid 2.00 m.

7. Three objects are connected on a table as shown in **Figure P5.7**. The coefficient of kinetic friction between the block of mass m_2 and the table is 0.350. The objects have masses of $m_1 = 4.00$ kg, $m_2 = 1.00$ kg and $m_3 = 2.00$ kg, and the pulleys are frictionless. (a) Draw a free-body diagram of each object. (b) Determine the acceleration of each object, including its direction. (c) Determine the tensions in the two cords. (d) **What If?** If the tabletop were smooth, would the tensions increase, decrease, or remain the same? Explain your answer.

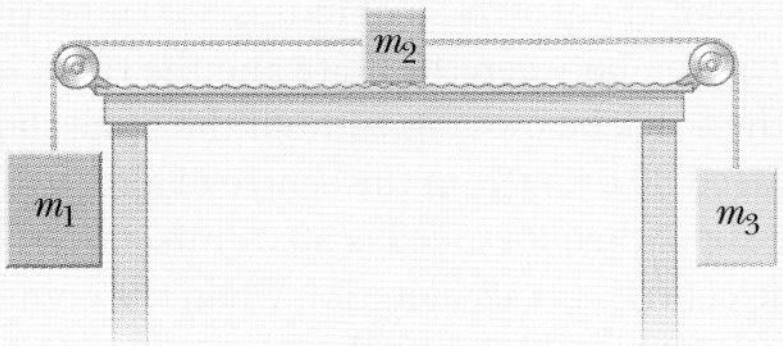

Figure P5.7

8. A small child has made himself a toy train by tying two cardboard boxes full of toys together by a rope of negligible mass and he drags it along with a horizontal force. Suppose he applies a force $F = 68.0$ N, the first box has a mass $m_1 = 12.0$ kg, the second box has a mass $m_2 = 18.0$ kg, and the coefficient of kinetic friction between each box and the surface is 0.100. (a) Draw a free-body diagram for each box. Determine (b) the acceleration of the system and (c) the tension T in the rope joining the boxes.

9. A block of mass 3.00 kg is pushed up against a wall by a force $\vec{\mathbf{P}}$ that makes an angle of $\theta = 50.0°$ with the horizontal as shown in **Figure P5.9**. The coefficient of static friction between the block and the wall is 0.250. (a) Determine the possible values for the magnitude of $\vec{\mathbf{P}}$ that allow the block to remain stationary. (b) Describe what happens if $|\vec{\mathbf{P}}|$ has a larger value and what happens if it is smaller. (c) Repeat parts (a) and (b) assuming the force makes an angle of $\theta = 13.0°$ with the horizontal.

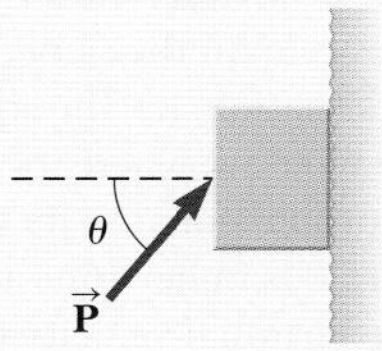

Figure P5.9

10. One side of the roof of a house slopes up at 37.0°. A roofer kicks a round, flat rock that has been thrown onto the roof by a neighbourhood child. The rock slides straight up the incline with an initial speed of 15.0 m/s. The coefficient of kinetic friction between the rock and the roof is 0.400. The rock slides 10.0 m up the roof to its peak. It crosses the ridge and goes into free fall, following a parabolic trajectory above the far side of the roof, with negligible air resistance. Determine the maximum height the rock reaches above the point where it was kicked.

11. A magician pulls a tablecloth from under a 200 g mug located 30.0 cm from the edge of the cloth. The cloth exerts a friction force of 0.100 N on the mug, and the cloth is pulled with a constant acceleration of 3.00 m/s². How far does the mug move relative to the horizontal tabletop before the cloth is completely out from under it? Note that the cloth must move more than 30.0 cm relative to the tabletop during the process.

Section **5.2** Resistive forces in fluids

12. A skydiver of mass 80.0 kg jumps from a slow-moving aircraft and reaches a terminal speed of 50.0 m/s. (a) What is her acceleration when her speed is 30.0 m/s? What is the drag force on the skydiver when her speed is (b) 50.0 m/s and (c) 30.0 m/s?

13. Assume the resistive force acting on a speed skater is proportional to the square of the skater's speed v and is given by $f = -kmv^2$, where k is a constant and m is the skater's mass. The skater crosses the finish line of a straight-line race with speed v_i and then slows down by coasting on his skates. Show that the skater's speed at any time t after crossing the finish line is $v(t) = v_i/(1 + ktv_i)$.

14. A small piece of styrofoam packing material is dropped from a height of 2.00 m above the ground. Until it reaches terminal speed, the magnitude of its acceleration is given by $a = g - Bv$. After falling 0.500 m, the styrofoam effectively reaches terminal speed and then takes 5.00 s more to reach the ground. (a) What is the value of the constant B? (b) What is the acceleration at $t = 0$? (c) What is the acceleration when the speed is 0.150 m/s?

15. ±? A small, spherical bead of mass (3.00 ± 0.02) g is released from rest at $t = 0$ from a point under the surface of a viscous liquid. The terminal speed is measured to be

$v_T = (2.0 \pm 0.2)$ cm/s. Find (a) the value of the constant b that appears in **Equation 5.3**, (b) the time t at which the bead reaches $0.632v_T$, and (c) the value of the resistive force when the bead reaches terminal speed.

16. A window washer pulls a rubber squeegee down a very tall vertical window. The squeegee has mass 160 g and is mounted on the end of a light rod. The coefficient of kinetic friction between the squeegee and the dry glass is 0.900. The window washer presses it against the window with a force having a horizontal component of 4.00 N. (a) If she pulls the squeegee down the window at constant velocity, what vertical force component must she exert? (b) The window washer increases the downwards force component by 25.0%, while all other forces remain the same. Find the squeegee's acceleration in this situation. (c) The squeegee is moved into a wet portion of the window, where its motion is resisted by a fluid drag force R proportional to its velocity according to $R = -20.0v$, where R is in newtons and v is in metres per second. Find the terminal velocity that the squeegee approaches, assuming the window washer exerts the same force described in part (b).

17. At international cricket games, it is commonplace to display on the scoreboard a speed for each bowl. Because the ball is subject to a drag force due to air proportional to the square of its speed given by $R = kmv^2$, it slows as it travels 20 m toward the wicket according to the formula $v = v_i e^{-kx}$. Suppose the ball leaves the bowler's hand at 150 km/h. Ignore its vertical acceleration, and determine the speed of the ball when it hits the wicket.

18. A motorboat cuts its engine when its speed is 10.0 m/s and then coasts to rest. The equation describing the motion of the motorboat during this period is $v = v_i e^{-ct}$, where v is the speed at time t, v_i is the initial speed at $t = 0$ and c is a constant. At $t = 20.0$ s, the speed is 5.00 m/s. (a) Find the constant c. (b) What is the speed at $t = 40.0$ s? (c) Differentiate the expression for $v(t)$ and thus show that the acceleration of the boat is proportional to the speed at any time.

19. You can feel a force of air drag on your hand if you stretch your arm out of the open window of a moving car. *Note*: Do not endanger yourself. What is the order of magnitude of this force? In your solution, state the quantities you measure or estimate and their values.

Section 5.3 Uniform circular motion revisited

20. A light string can support a stationary hanging load of 25.0 kg before breaking. An object of mass $m = 3.00$ kg attached to the string rotates on a frictionless, horizontal table in a circle of radius $r = 0.800$ m, while the other end of the string is held fixed as in **Figure P5.20**. What range of speeds can the object have before the string breaks?

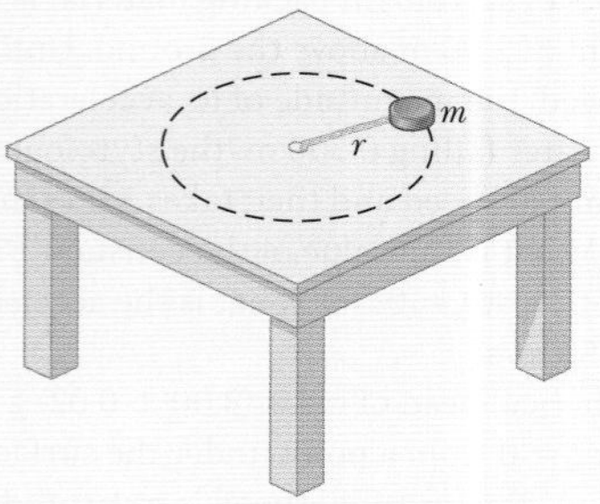

Figure P5.20

21. In the Bohr model of the hydrogen atom, an electron moves in a circular path around a proton. The speed of the electron is approximately 2.20×10^6 m/s. Find (a) the force acting on the electron as it revolves in a circular orbit of radius 0.529×10^{-10} m and (b) the centripetal acceleration of the electron.

22. Whenever two *Apollo* astronauts were on the surface of the Moon, a third astronaut orbited the Moon. Assume the orbit to be circular and 100 km above the surface of the Moon, where the acceleration due to gravity is 1.52 m/s². The radius of the Moon is 1.70×10^6 m. Determine (a) the astronaut's orbital speed and (b) the period of the orbit.

23. A car initially travelling eastwards turns north by travelling in a circular path at uniform speed as shown in **Figure P5.23**. The length of the arc ABC is 235 m, and the car completes the turn in 36.0 s. (a) What is the acceleration when the car is at B, located at an angle of 35.0°? Express your answer in terms of the unit vectors $\hat{\mathbf{i}}$ and $\hat{\mathbf{j}}$. Determine (b) the car's average speed and (c) its average acceleration during the 36.0 s interval.

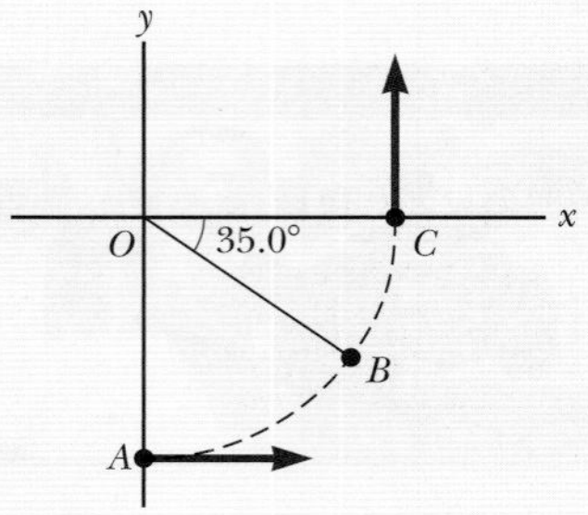

Figure P5.23

24. Consider a conical pendulum (**Figure P5.24**) with a bob of mass $m = 80.0$ kg on a string of length $L = 10.0$ m that makes an angle of $\theta = 5.00°$ with the vertical. Determine (a) the horizontal and vertical components of the force exerted by the string on the pendulum and (b) the radial acceleration of the bob.

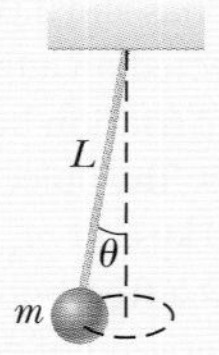

Figure P5.24

25. A crate of eggs is located in the middle of the flatbed of a ute as the ute negotiates a curve in the flat road. The curve may be regarded as an arc of a circle of radius 35.0 m. If the coefficient of static friction between the crate and the truck is 0.600, how fast can the truck be moving without the crate sliding?

26. A coin placed 30.0 cm from the centre of a rotating, horizontal turntable slips when its speed is 50.0 cm/s. (a) What force causes the centripetal acceleration when the coin is stationary relative to the turntable? (b) What is the coefficient of static friction between the coin and the turntable?

Section 5.4 Non-uniform circular motion

27. A bucket of water is rotated in a vertical circle of radius 1.00 m. (a) What two external forces act on the water in the bucket? (b) Which of the two forces is more important in causing the water to move in a circle? (c) What is the bucket's minimum speed at the top of the circle if no water is to spill out? (d) Assume the bucket with the speed found in part (c) were to suddenly disappear at the top of the circle. Describe the subsequent motion of the water. Would it differ from the motion of a projectile?

28. A 40.0 kg child swings in a swing supported by two chains, each 3.00 m long. The tension in each chain at the lowest point is 350 N. Find (a) the child's speed at the lowest point and (b) the force exerted by the seat on the child at the lowest point. (Ignore the mass of the seat.)

29. A roller-coaster car (**Figure P5.29**) has a mass of 500 kg when fully loaded with passengers. The path of the coaster from its initial point shown in the figure to point Ⓑ involves only up-and-down motion (as seen by the riders), with no motion to the left or right. (a) If the vehicle has a speed of 20.0 m/s at point Ⓐ, what is the force exerted by the track on the car at this point? (b) What is the maximum speed the vehicle can have at point Ⓑ and still remain on the track? Assume the roller-coaster tracks at points Ⓐ and Ⓑ are parts of vertical circles of radius r_1 = 10.0 m and r_2 = 15.0 m respectively.

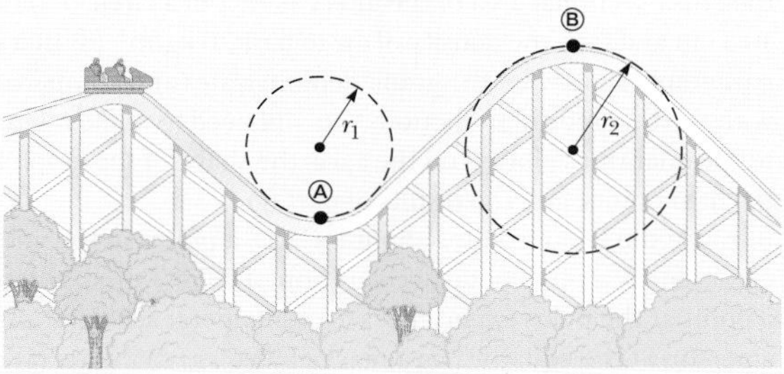

Figure P5.29

Problems 29 and 34

30. ±? An adventurous archeologist (m = (85 ± 1) kg) tries to cross a river by swinging from a vine. The vine is (10.0 ± 0.1) m long and his speed at the bottom of the swing is (8.0 ± 0.5) m/s. The archeologist doesn't know that the vine has a breaking strength of 1000 N. Does he make it across the river without falling in?

31. The Corkscrew roller-coaster at Rainbow's End theme park in Auckland has a teardrop-shaped vertical loop 30 m high. Suppose the speed at the top of the loop is 11.0 m/s and the corresponding centripetal acceleration of the riders is $2g$. (a) What is the radius of the arc of the teardrop at the top? (b) If the total mass of a car plus the riders is M, what force does the rail exert on the car at the top? (c) Suppose the roller-coaster had a circular loop of radius 20.0 m. If the cars have the same speed, 11.0 m/s at the top, what is the centripetal acceleration of the riders at the top? (d) Comment on the normal force at the top in the situation described in part (c) and on the advantages of having teardrop-shaped loops.

Additional problems

32. Two blocks of masses m_1 and m_2 are placed on a table in contact with each other as shown in **Figure P5.32**. The coefficient of kinetic friction between the block of mass m_1 and the table is μ_1, and that between the block of mass m_2 and the table is μ_2. A horizontal force of magnitude F is applied to the block of mass m_1. We wish to find P, the magnitude of the contact force between the blocks. (a) Draw diagrams showing the forces on each block. (b) What is the net force on the system of two blocks? (c) What is the net force acting on m_1? (d) What is the net force acting on m_2? (e) Write Newton's second law in the x direction for each block. (f) Solve the two equations in two unknowns for the acceleration a of the blocks in terms of the masses, the applied force F, the coefficients of friction and g. (g) Find the magnitude P of the contact force between the blocks in terms of the same quantities.

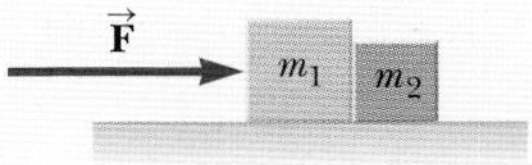

Figure P5.32

33. A crate of weight F_g is pushed by a force $\vec{\mathbf{P}}$ on a horizontal floor. The coefficient of static friction is μ_s, and $\vec{\mathbf{P}}$ is directed at angle θ below the horizontal. (a) Show that the minimum value of P that will move the crate is given by

$$P = \frac{\mu_s F_g \sec\theta}{1 - \mu_s \tan\theta}$$

(b) Find the condition on θ in terms of μ_s for which motion of the crate is impossible for any value of P.

34. ±? The mass of a roller-coaster car, including its passengers, is (500 ± 50) kg. Its speed at the bottom of the track in **Figure P5.29** is (19 ± 1) m/s. The radius of this section of the track is r_1 = (25.0 ± 0.1) m. Find the force that a seat in the roller-coaster car exerts on a (50.0 ± 0.5) kg passenger at the lowest point.

35. A string under a tension of 50.0 N is used to whirl a rock in a horizontal circle of radius 2.50 m at a speed of 20.4 m/s on a frictionless surface as shown in **Figure P5.35**. As the string is pulled in, the speed of the rock increases. When the string is 1.00 m long and the speed of the rock is 51.0 m/s, the string breaks. What is the breaking strength (the maximum tension the rope can withstand), in newtons, of the string?

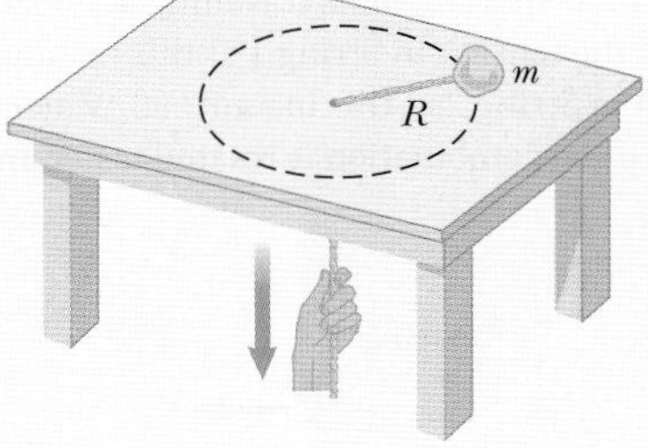

Figure P5.35

36. Disturbed by speeding cars outside his workplace, Nobel laureate Arthur Holly Compton designed a speed bump (called the 'Holly hump') and had it installed. A car of mass m passes over such a hump in a road that follows the arc of a circle of radius R. (a) If the car travels at a speed v, what force does the road exert on the car as the car passes the highest point of the hump? (b) **What If?** What is the maximum speed the car can have without losing contact with the road as it passes this highest point?

37. A child's toy consists of a small wedge that has an acute angle θ (Figure P5.37). The sloping side of the wedge is frictionless and an object of mass m on it remains at constant height if the wedge is spun at a certain constant speed. The wedge is spun by rotating, as an axis, a vertical rod that is firmly attached to the wedge at the bottom end. Show that, when the object sits at rest at a point at distance L up along the wedge, the speed of the object must be $v = (gL \sin\theta)^{1/2}$.

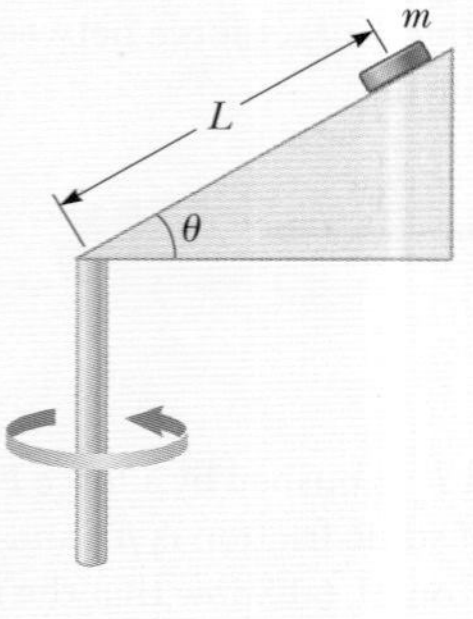

Figure P5.37

38. A seaplane of total mass m lands on a lake with initial speed $v_i\hat{\mathbf{i}}$. The only horizontal force on it is a resistive force on its pontoons from the water. The resistive force is proportional to the velocity of the seaplane: $\vec{R} = -b\vec{v}$. Newton's second law applied to the plane is $-bv\hat{\mathbf{i}} = m(dv/dt)\hat{\mathbf{i}}$. This differential equation implies that the speed changes according to

$$\int_{v_i}^{v} \frac{dv}{v} = -\frac{b}{m}\int_0^t dt$$

(a) Carry out the integration to determine the speed of the seaplane as a function of time. (b) Sketch a graph of the speed as a function of time. (c) Does the seaplane come to a complete stop after a finite interval of time? (d) Does the seaplane travel a finite distance in stopping?

39. An object of mass $m_1 = 4.00$ kg is tied to an object of mass $m_2 = 3.00$ kg with String 1 of length $\ell = 0.500$ m. The combination is swung in a vertical circular path on a second string, String 2, of length $\ell = 0.500$ m. During the motion, the two strings are collinear at all times as shown in Figure P5.39. At the top of its motion, m_2 is travelling at $v = 4.00$ m/s. (a) What is the tension in String 1 at this instant? (b) What is the tension in String 2 at this instant? (c) Which string will break first if the combination is rotated faster and faster?

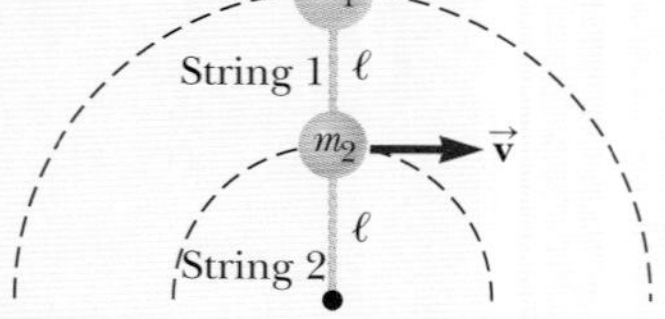

Figure P5.39

40. ±? In a tumble dryer a cylindrical tub containing wet clothes is rotated steadily about a horizontal axis as shown in Figure P5.40. So that the clothes will dry uniformly, they are made to tumble. The rate of rotation of the smooth-walled tub is chosen so that a small piece of cloth will lose contact with the tub when the cloth is at an angle of $\theta = 68 \pm 5°$ above the horizontal. If the radius of the tub is $r = 33$ cm, what rate of revolution is needed?

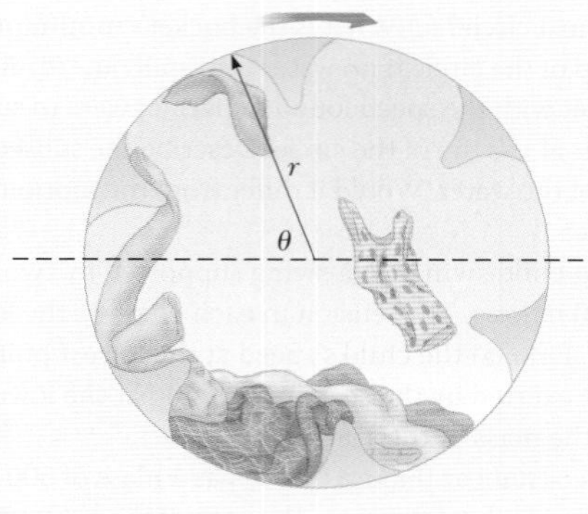

Figure P5.40

41. Interpret the graph in Figure 5.17b, which describes the results for falling coffee filters discussed in Example 5.6. Proceed as follows. (a) Find the gradient of the straight line, including its units. (b) From Equation 5.7, $R = \frac{1}{2}D\rho Av^2$, identify the theoretical gradient of a graph of resistive force versus squared speed. (c) Set the experimental and theoretical gradients equal to each other and calculate the drag coefficient of the filters. Model the cross-sectional area of the filters as that of a circle of radius 10.5 cm and take the density of air to be 1.20 kg/m³. (d) In a short paragraph, state what the graph demonstrates and compare it with the theoretical prediction. You will need to make reference to the quantities plotted on the axes, the shape of the graph line, the data points, and the results of part (c).

42. Because the Earth rotates about its axis, a point on the equator experiences a centripetal acceleration of 0.0337 m/s², whereas a point at the poles experiences no centripetal acceleration. If a person at the equator has a mass of 75.0 kg, calculate (a) the gravitational force (weight) on the person and (b) the normal force (apparent weight) on the person. (c) Which force is greater? Assume the Earth is a uniform sphere and take $g = 9.800$ m/s².

43. A student builds and calibrates an accelerometer and uses it to determine the speed of her car around a certain unbanked highway curve. The accelerometer is a plumb bob with a protractor that she attaches to the roof of her car. A friend riding in the car with the student observes that the plumb bob hangs at an angle of 15.0° from the vertical when the car has a speed of 23.0 m/s. (a) What is the centripetal acceleration of the car rounding the curve? (b) What is the radius of the curve? (c) What is the speed of the car if the plumb bob deflection is 9.00° while rounding the same curve?

44. While learning to drive, you are in a 1200 kg car moving at 20.0 m/s across a large, vacant, level carpark. Suddenly you realise you are heading straight towards the brick wall of a large supermarket and are in danger of running into it. The pavement can exert a maximum horizontal force of 7000 N on the car. (a) Explain why you should expect the force to have a well-defined maximum value. (b) Suppose you apply the brakes and do not turn the steering wheel. Find the minimum distance you must be from the wall to avoid a collision. (c) If you do not brake but instead maintain constant speed and turn the steering wheel, what is the minimum distance you must be from the wall to avoid a collision? (d) Of the two methods in parts (b) and (c), which is better for avoiding a collision? Or should you use both the brakes and the steering wheel, or neither? Explain your answer. (e) Does the conclusion in part (d) depend on the numerical values given in this problem, or is it true in general?

45. Figure P5.45 shows a swing ride at an amusement park. The structure consists of a horizontal, rotating circular platform of diameter D from which seats of mass m are suspended at the end of light chains of length d. When the system rotates at constant speed, the chains swing out and make an angle θ with the vertical. Consider such a ride with the following parameters: $D = 8.00$ m, $d = 2.50$ m, $m = 10.0$ kg, and $\theta = 28.0°$. (a) What is the speed of each seat? (b) Draw a diagram of forces acting on a 40.0 kg child riding in a seat and (c) find the tension in the chain.

Figure P5.45

46. An amusement park ride called the Gravitron consists of a large vertical cylinder that spins about its axis fast enough that any person inside is held up against the wall when the floor drops away (Figure P5.46). The coefficient of static friction between person and wall is μ_s, and the radius of the cylinder is R. (a) Show that the maximum period of revolution necessary to keep the person from falling is $T = (4\pi^2 R\mu_s/g)^{1/2}$. (b) If the rate of revolution of the cylinder is made to be somewhat larger, what happens to the magnitude of each one of the forces acting on the person? What happens to the motion of the person? (c) If the rate of revolution of the cylinder is instead made to be somewhat smaller, what happens to the magnitude of each one of the forces acting on the person? How does the motion of the person change?

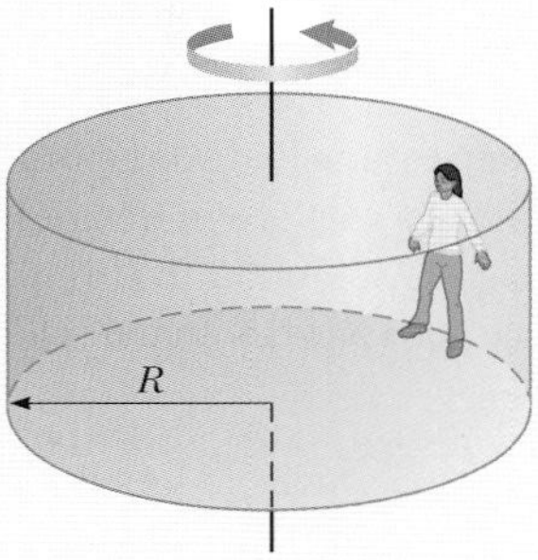

Figure P5.46

47. Members of a skydiving club were given the following data to use in planning their jumps. In the table below, d is the distance fallen from rest by a skydiver in a 'free-fall stable spread position' versus the time of fall t. (a) Convert the distances in feet into metres. (b) Graph d (in metres) versus t. (c) Determine the value of the terminal speed v_T by finding the gradient of the straight portion of the curve.

t (s)	*d* (ft)	*t* (s)	*d* (ft)	*t* (s)	*d* (ft)
0	0	7	652	14	1831
1	16	8	808	15	2005
2	62	9	971	16	2179
3	138	10	1138	17	2353
4	242	11	1309	18	2527
5	366	12	1483	19	2701
6	504	13	1657	20	2875

48. A car rounds a banked curve as discussed in Example 5.11 and shown in Figure 5.23. The radius of curvature of the road is R, the banking angle is θ, and the coefficient of static friction is μ_s. (a) Determine the range of speeds the car can have without slipping up or down the road. (b) Find the minimum value for μ_s such that the minimum speed is zero.

49. A model aeroplane of mass 0.750 kg flies with a speed of 35.0 m/s in a horizontal circle at the end of a 60.0 m control wire as shown in Figure P5.49a. The forces exerted on the aeroplane are shown in Figure P5.49b: the tension in the control wire, the gravitational force and the aerodynamic lift that acts at $\theta = 20.0°$ inwards from the vertical. Calculate the tension in the wire, assuming it makes a constant angle of $\theta = 20.0°$ with the horizontal.

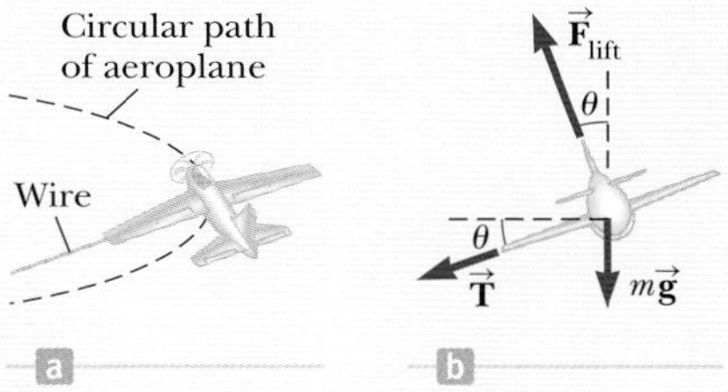

Figure P5.49

Challenge problems

50. The board sandwiched between two other boards in Figure P5.50 weighs 95.5 N. If the coefficient of static friction between the boards is 0.663, what must be the magnitude of the compression forces (assumed horizontal) acting on both sides of the middle board to keep it from slipping?

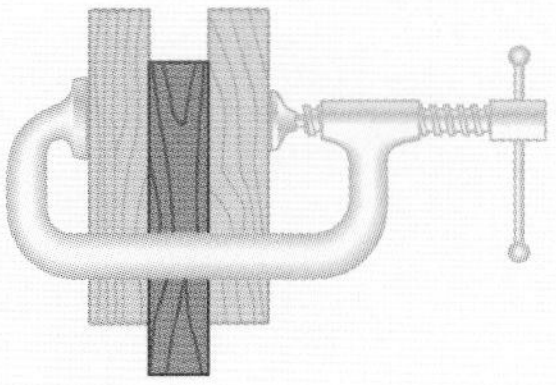

Figure P5.50

51. Because of the Earth's rotation, a plumb bob does not hang exactly along a line directed to the centre of the Earth. How much does the plumb bob deviate from a radial line at 35.0° north latitude? Assume the Earth is spherical.

52. A 9.00 kg object starting from rest falls through a viscous medium and experiences a resistive force given by Equation 5.3. The object reaches one-half its terminal speed in 5.54 s. (a) Determine the terminal speed. (b) At what time is the speed of the object three-fourths the terminal speed? (c) How far has the object travelled in the first 5.54 s of motion?

53. For $t < 0$, an object of mass m experiences no force and moves in the positive x direction with a constant speed v_i. Beginning at $t = 0$, when the object passes position $x = 0$, it experiences a net resistive force proportional to the square of its speed: $\vec{\mathbf{F}}_{net} = -mkv^2\hat{\mathbf{i}}$ where k is a constant. The speed of the object after $t = 0$ is given by $v = v_i/(1 + kv_i t)$. (a) Find the position x of the object as a function of time. (b) Find the object's velocity as a function of position.

54. A single bead can slide with negligible friction on a stiff wire that has been bent into a circular loop of radius 15.0 cm as shown in Figure P5.54. The circle is always in a vertical plane and rotates steadily about its vertical diameter with a period of 0.450 s. The position of the bead is described by the angle μ that the radial line from the centre of the loop to the bead makes with the vertical. (a) At what angle up from the bottom of the circle can the bead stay motionless relative to the turning circle? (b) **What If?** Repeat the problem, this time taking the period of the circle's rotation as 0.850 s. (c) Describe how the solution to part (b) is different from the solution to part (a). (d) For any period or loop size, is there always an angle at which the bead can stand still relative to the loop? (e) Are there ever more than two angles? Arnold Arons suggested the idea for this problem.

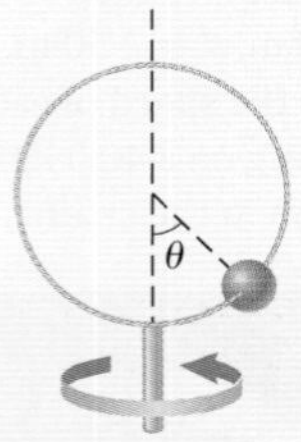

Figure P5.54

55. The expression $F = arv + br^2v^2$ gives the magnitude of the resistive force (in newtons) exerted on a sphere of radius r (in metres) by a stream of air moving at speed v (in metres per second), where a and b are constants with appropriate SI units. Their numerical values are $a = 3.10 \times 10^{-4}$ and $b = 0.870$. Using this expression, find the terminal speed for water droplets falling under their own weight in air, taking the following values for the drop radii: (a) 10.0 μm, (b) 100 μm, (c) 1.00 mm. For parts (a) and (c), you can obtain accurate answers without solving a quadratic equation by considering which of the two contributions to the air resistance is dominant and ignoring the lesser contribution.

case study 1

Dragging sheep: an Ig Nobel winner, useful physics and easier workplaces

Jack Harvey, John Culvenor, Warren Payne, Steve Cowley, Michael Lawrance and Robyn Williams

Just the idea of working out the force required to drag sheep could make you laugh. This research about the science of sliding sheep showed that sheep slide more easily downhill. With a conclusion like that (but also a lot more), the work fitted well with the Ignobel motto of science that 'makes people laugh and then think' (www.improbable.com/ig). But why did we put so much effort into the physics of sheep dragging?

In the hundred years between the gold rushes of the 19th century and the mining booms of more recent times, national prosperity was so dependent on wool that it was often said that 'Australia rode on the sheep's back'. But the wool industry has also taken its toll on the backs and the joints of generations of shearers.

One critical and taxing shearing task is dragging the sheep from the catching pen to the shearing workstation. A skilled shearer may drag 200 sheep per day, each weighing 45–85 kg, over a distance of up to 6 m. As part of a study of shearing shed design, researchers at the University of Ballarat investigated the force exerted by shearers when dragging sheep on five different flooring materials and two structural floor designs currently in use (battens made of wood or plastic, oriented parallel or at right angles to the direction of drag, and steel mesh) both on the flat and on a 1:10 (5.6°) downwards slope.

We wanted to know the force that the shearer needed to apply in order drag the sheep. Putting a scale between the shearer and the sheep was not feasible because the shearer holds the front legs of the sheep. Pulling the sheep with an attachment might have changed the way that the shearer did the work and thus invalidated the results, so we wanted to keep the setup as realistic as possible.

Interchangeable sections of flooring were constructed from each of the five materials, with and without slope. The sections could be attached to a force plate, together with matching panels in front of and behind the force plate. The force plate produces a trace of x, y and z components of the ground reaction force R at intervals of one millisecond.

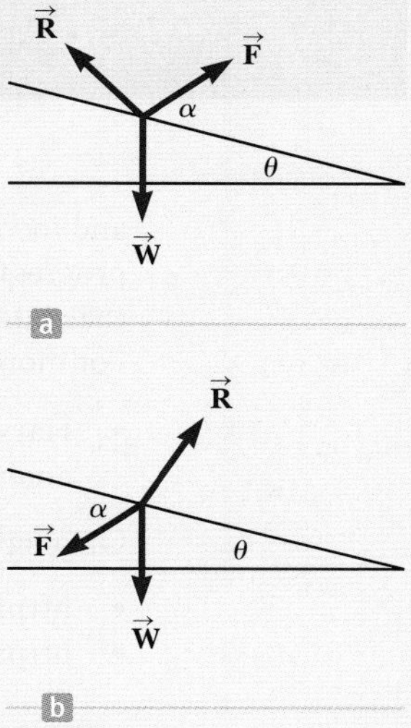

Figure CS1.1
The forces acting on (a) the sheep and (b) the shearer, where $\vec{F}$ = dragging force, $\vec{R}$ = ground reaction, $\vec{W}$ = weight, θ = angle of inclination of floor and α = angle of inclination of drag to floor

Figure CS1.1 shows the forces on the shearer and the sheep. Because the sheep was being dragged in a passive state, it slid fairly smoothly across the force plate. Data collected while the sheep was on the force plate were therefore less subject to impulse effects than data collected while the shearer was on the force plate. For this reason, sheep-based data were used to estimate a steady state average dragging force.

The dragging force F was calculated as:

$$F = \sqrt{R_x^2 + R_y^2 + (W - R_z)^2}$$

where W is the weight of the sheep.

Table CS1.1 shows results for the mean dragging force F by texture and slope. Statistical analysis showed that there were significant differences between most textures (the mean for wooden battens at right angles did not differ significantly from either of the two means for plastic battens). There was also a significant difference between the slopes. The combination that produced the lowest force was sloping wood battens oriented parallel to the direction of drag.

When compared with published maximum acceptable limits for sustained pulling forces, these mean forces were found to be close to or above the maximum figures for the most capable 10% of men. Thus dragging sheep repetitively is a difficult task that is within the

Table CS1.1
Mean dragging force on sheep (N): by texture and slope

	Slope		
Texture	**0°**	**5.6°**	**Both**
Wood: Parallel	388.2	359.2	373.7
Wood: At right angles	400.4	376.4	388.4
Plastic: Parallel	395.6	370.1	382.9
Plastic: At right angles	405.4	378.9	392.2
Steel mesh	422.8	394.1	408.4
All textures	402.5	375.7	

Newspix/Mark Calleja

capability of only the most able men, and likely to be too physically demanding for almost all women. The difference of about 15% in dragging force between the best floor tested and the worst floor tested is thus of great practical importance for the occupational health and safety and the productivity of shearers.

We also investigated the direction of the force: was it more of a lifting force, or more of a dragging force? The force F, as well as the angles θ and α and the typical shearer's dragging posture determined from videotapes of the experiment, were input to biomechanical modelling software to produce estimates of forces in the bodies of the shearers, including spine, knees and elbows. We were also able to determine the coefficient of friction for the various surfaces.

After testing the findings in a number of shearing sheds, the ideas were widely promoted and incorporated in government safety guidelines. Another benefit reported anecdotally as implementation progressed was that on a sloping floor the sheep tend to face up the slope and thus away from the shearer, so that catching them and tipping them over to drag becomes easier.

For more information see:

- Harvey, J. T., Culvenor, J., Payne, W. R., Cowley, S., Lawrance, M., Stuart, D. and Williams, R., 2002, 'An analysis of the forces required to drag sheep over various surfaces', *Applied Ergonomics*, 33(6): 523–31.

On a lighter note, this work was awarded the 2003 Ig Nobel Prize for Physics. See:

- http://en.wikipedia.org/wiki/List_of_Ig_Nobel_Prize_winners#2003
- http://improbable.com/ig/ig-pastwinners.html#ig2003

Dr Jack Harvey is a statistician and Senior Research Fellow in the School of Human Movement and Sport Sciences at the University of Ballarat.

Dr John F. Culvenor is a Melbourne-based independent consultant in engineering and a specialist in industrial ergonomics.

Professor Warren Payne is an exercise physiologist in the Institute of Sport, Physical Activity and Active Living at Victoria University.

Dr Steve Cowley is an occupational health and safety practitioner and industrial hygienist and Director of Steve Cowley Health & Safety Consulting.

Michael Lawrance is an occupational therapist.

Robyn Williams is a physiotherapist.

At the time of the research, they worked as a multidisciplinary research team based at the University of Ballarat, focusing on occupational health and safety issues in the wool industry.

chapter 6

Getty Images/Victor Fraile/Stringer

Work, force and energy

Competitors in the 'World's Strongest Man' truck-pull challenge need to drag a 24-tonne truck a distance of 25 metres in less than a minute. How much work does the strongman have to do on the truck?

On completing this chapter, students will understand:

- how force and work are related
- that systems contain kinetic and potential energy
- that there are several forms of potential energy
- how force and potential energy are related
- what equilibrium is in terms of potential energy
- that power is the rate of energy transfer.

Students will be able to:

- calculate the work done by a constant or varying force
- apply the work–kinetic energy theorem
- solve problems involving spring forces
- calculate changes in gravitational potential energy
- distinguish between stable, unstable and neutral equilibrium
- solve problems to find power usage by systems in different situations.

The concept of energy is one of the most important topics in science and engineering. In everyday life we think of energy in terms of fuel for transport and heating, electricity for lights and appliances and food for our bodies. These ideas tell us that fuels are needed to do a job and that those fuels provide us with something we call energy.

Every physical process that occurs in the universe involves energy transfers and transformations. Hence the models that we create to help us explain and predict physical processes need to include the concept of energy. We will find that using the concept of energy, and the principle of conservation of energy described in the next chapter, we are able to solve a wider range of problems and model the behaviour of many different systems.

ENHANCED WebAssign

The problems found in this chapter may be assigned online in Enhanced Web Assign.

6.1 Systems and environments

Energy is present in the Universe in various forms. *Every* physical process that occurs in the Universe involves energy and energy transfers or transformations. Unfortunately, despite its extreme importance, energy cannot be easily defined. The variables in previous chapters were relatively concrete; we have everyday experience with velocities and forces, for example. Although we have *experiences* with energy, such as running out of petrol or using electricity to run a computer, the *notion* of energy is more abstract. But it is a very powerful idea and allows us to solve complex problems without using Newton's laws. This is a big advantage when we are dealing with a complex system of many particles.

Our analysis models presented in earlier chapters were based on the motion of a *particle* or an object that could be modelled as a particle. We begin our new approach by focusing our attention on a *system*. In the system model, we focus our attention on a small portion of the Universe – the **system** – and ignore details of the rest of the Universe outside of the system. A critical skill in applying the system model to problems is *identifying the system*. A valid system:

- might be a single object or particle
- might be a collection of objects or particles
- might be a region of space (such as the interior of a car engine combustion cylinder)
- might vary with time in size and shape (such as a rubber ball, which deforms upon striking a wall).

Identifying the need for a system approach to solving a problem (as opposed to a particle approach) is part of the *Model* step in the general problem-solving strategy outlined in Chapter 2. Identifying the particular system is the second part of this step.

No matter what the particular system is in a given problem, we identify a **system boundary** (an imaginary surface) that may or may not coincide with a physical surface and which divides the Universe into the system and the **environment** surrounding the system.

As an example, imagine a force applied to an object in empty space. We can define the object as the system and its surface as the system boundary. The force applied to the object is an influence on the system from the environment that acts across the system boundary.

Another example was seen in Example 4.11, in which the system can be defined as the combination of the ball, the block and the cord. The influence from the environment includes the gravitational forces on the ball and the block, the normal and friction forces on the block, and the force exerted by the pulley on the cord. The forces exerted by the cord on the ball and the block are internal to the system and therefore are not included as an influence from the environment.

> **Pitfall Prevention 6.1**
> The most important *first* step to take in solving a problem using the energy approach is to identify the appropriate system of interest and decide where the boundary of the system is.

In general a system contains both objects and energy. The energy of a system is of two types – kinetic and potential. The kinetic energy of a system is due to the movement of the particles it contains. The potential energy of a system depends upon the forces acting within the system and acting on the system due to external objects. We shall define these in more detail in this chapter.

There are a number of mechanisms by which the energy of a system can be influenced by its environment. The first one we shall investigate is *work*.

6.2 Work done by a constant force

Almost all the terms we have used so far – velocity, acceleration, force and so on – convey a similar meaning in physics as they do in everyday life. Work, however, has a very specific meaning in physics. Work is **the amount of energy transferred to an object or system by the action of a force on that system through some distance**.

Definition of work ▶

For a single constant force acting on an object

Work done by a constant force ▶

$$W = \vec{\mathbf{F}} \cdot \Delta\vec{\mathbf{r}} \tag{6.1}$$

where Δr is the displacement of the object while the force F is acting.

Σ *For help with vector dot products see Appendix B.6.*

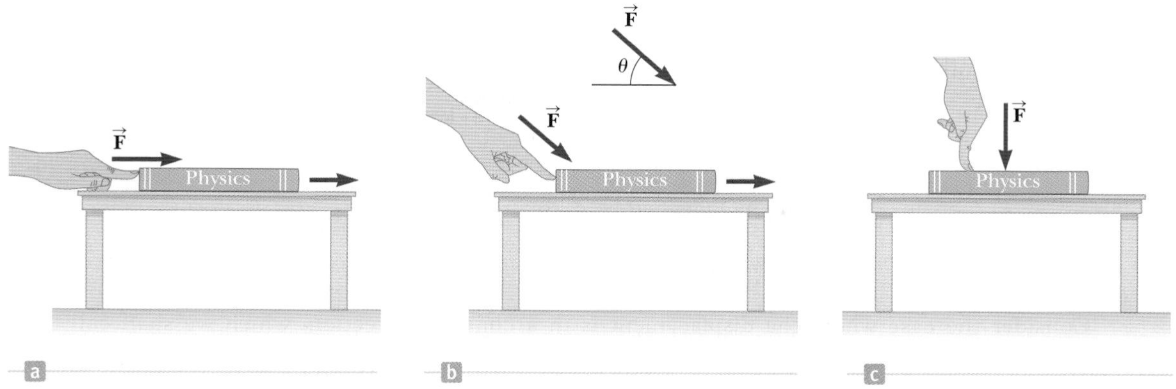

Figure 6.1
(a) A force is applied to a book on a table. The direction of the force is parallel to the surface and the direction of motion. (b) The force applied at an angle θ to the surface and direction of motion
(c) The force is applied perpendicular to the surface and any possible direction of motion

Equation 6.1 is a vector equation. It tells us that the work done is the **vector dot product** or **scalar product** of the two vectors, force and displacement. Work, like any other energy, is a scalar. Work has units of N.m or kg.m^2/s^2, called the joule, J, and, like energy, has dimensions of ML^2/T^2.

Consider the situation illustrated in Figure 6.1. A force $\vec{\mathbf{F}}$ is applied to a book, which we identify as the system, and the book slides along the table. If we want to know how effective the force is in moving the book, we must consider not only the magnitude of the force but also its direction. Notice that the finger in Figure 6.1 applies forces in three different directions to the book. Assuming the magnitude of the applied force is the same in all three diagrams, the push applied in Figure 6.1a does more to move the book than the push in Figure 6.1b. Figure 6.1c shows a situation in which the applied force does not move the book at all, regardless of how hard it is pushed (unless, of course, we apply a force so great that we break the table!). These results suggest that when analysing forces to determine the influence they have on the system, we must consider the vector nature of forces.

Notice also that the displacement in Equation 6.1 is that of *the point of application of the force.* If the force is applied to a particle or a rigid object that can be modelled as a particle, this displacement is the same as that of the particle. For a deformable system, however, these displacements are not the same. For example, imagine pressing in on the sides of a balloon with both hands. The centre of the balloon moves through zero displacement. The points of application of the forces from your hands on the sides of the balloon, however, do indeed move through a displacement as the balloon is compressed, and that is the displacement to be used in Equation 6.1. We will see other examples of deformable systems, such as springs and samples of gas contained in a vessel.

We said that the work done by the force is the **scalar product** or **vector dot product** of the force and the displacement. The scalar product of any two vectors $\vec{\mathbf{A}}$ and $\vec{\mathbf{B}}$ is defined as a scalar quantity equal to the product of the magnitudes of the two vectors and the cosine of the angle θ between them:

$$\vec{\mathbf{A}} \cdot \vec{\mathbf{B}} \equiv AB\cos\theta \qquad (6.2)$$

Figure 6.2 shows two vectors $\vec{\mathbf{A}}$ and $\vec{\mathbf{B}}$ and the angle θ between them used in the definition of the dot product. In Figure 6.2, $B\cos\theta$ is the projection of $\vec{\mathbf{B}}$ onto $\vec{\mathbf{A}}$. Therefore, Equation 6.2 means that $\vec{\mathbf{A}} \cdot \vec{\mathbf{B}}$ is the product of the magnitude of $\vec{\mathbf{A}}$ and the projection of $\vec{\mathbf{B}}$ onto $\vec{\mathbf{A}}$.

Sometimes it is simpler, particularly when working with forces and displacements in three dimensions, to perform the dot product in Cartesian form.

First we write our vectors $\vec{\mathbf{A}}$ and $\vec{\mathbf{B}}$ as $\vec{\mathbf{A}} = A_x\hat{\mathbf{i}} + A_y\hat{\mathbf{j}} + A_z\hat{\mathbf{k}}$ and $\vec{\mathbf{B}} = B_x\hat{\mathbf{i}} + B_y\hat{\mathbf{j}} + B_z\hat{\mathbf{k}}$.

Pitfall Prevention 6.2

Work is a scalar. Although Equation 6.1 defines the work in terms of two vectors, *work is a scalar;* there is no direction associated with it. *All* types of energy and energy transfer are scalars. This fact is a major advantage of the energy approach because we don't need vector calculations!

Pitfall Prevention 6.3

Work is done by … on … Not only must you identify the system, you must also identify what agent in the environment is doing work on the system. When discussing work, always use the phrase, 'the work done by … on …'. After 'by' insert the part of the environment that is interacting directly with the system. After 'on' insert the system. For example, 'the work done by the hammer on the nail' identifies the nail as the system, and the force from the hammer represents the influence from the environment.

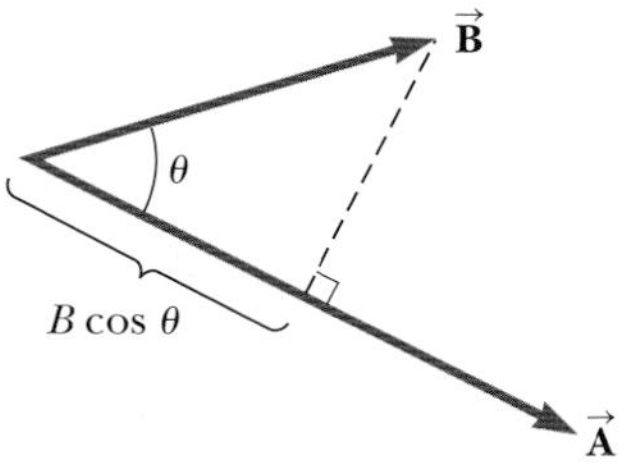

Figure 6.2
The scalar product $\vec{\mathbf{A}} \cdot \vec{\mathbf{B}}$ equals the magnitude of A multiplied by $B\cos\theta$, which is the projection of $\vec{\mathbf{B}}$ onto $\vec{\mathbf{A}}$.

Recall the dot product of perpendicular vectors is zero, and the dot product of parallel vectors is the algebraic product of the magnitudes. Hence the dot product of any unit vector with itself gives 1, and the dot product of any other unit vector with a different (and hence perpendicular) unit vector is 0:

$$\hat{\mathbf{i}}\cdot\hat{\mathbf{i}}=\hat{\mathbf{j}}\cdot\hat{\mathbf{j}}=\hat{\mathbf{k}}\cdot\hat{\mathbf{k}}=1 \text{ and } \hat{\mathbf{i}}\cdot\hat{\mathbf{j}}=\hat{\mathbf{i}}\cdot\hat{\mathbf{k}}=\hat{\mathbf{j}}\cdot\hat{\mathbf{k}}=0$$

Therefore

$$\vec{\mathbf{A}}\cdot\vec{\mathbf{B}}=A_xB_x+A_yB_y+A_zB_z \tag{6.3}$$

Appendix B.6 demonstrates how to find the vector dot and cross products in Cartesian coordinates and gives their properties.

Consider the case where the displacement is in a constant direction, which we shall call the x direction, so that $\Delta\vec{\mathbf{r}}$ in **Equation 6.1** becomes $\Delta\vec{\mathbf{x}}$. When the force and displacement are parallel, as in **Figure 6.1a**, the work done has its maximum value which is:

$$W = \vec{\mathbf{F}}\cdot\Delta\vec{\mathbf{x}} = F\Delta x$$

When the force and displacement are perpendicular, as in **Figure 6.1c**, the work done has its minimum value, which is:

$$W = \vec{\mathbf{F}}\cdot\Delta\vec{\mathbf{x}} = 0$$

For any angle, θ, in between, as in **Figure 6.1b**, the dot product can be written as:

$$W = \vec{\mathbf{F}}\cdot\Delta\vec{\mathbf{x}} = F_x\Delta x = F\Delta x\cos\theta \tag{6.4}$$

If you are not familiar with the vector dot product and how to calculate it you should refer to Appendix B.6. There are also practice problems on vector multiplication in the online resources.

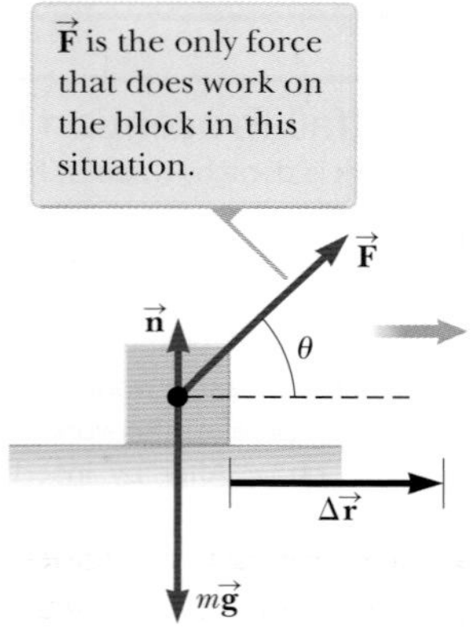

Figure 6.3
An object is displaced on a frictionless, horizontal surface. The normal force $\vec{\mathbf{n}}$ and the gravitational force $m\vec{\mathbf{g}}$ do no work on the object.

TRY THIS

Ask a friend to hold a heavy object, such as a large physics textbook, at arm's length for as long as they can. See if they can hold it steady for the time it takes you to explain to them why they are doing no actual work on the book in holding it there. You should include in your explanation the fact that **Equation 6.1** says that the work done is proportional to the distance through which the object or system to which the force is applied moves, and if they are holding the book steady then it has a zero displacement.

Pitfall Prevention 6.4

We can calculate the work done by a force on an object but that force is *not* necessarily the cause of the object's displacement. For example, if you lift an object, (negative) work is done on the object by the gravitational force, although gravity is not the cause of the object moving upwards!

Let us examine the situation in **Figure 6.3**, in which the object (the system) undergoes a displacement along a straight line while acted on by a constant force of magnitude F that makes an angle θ with the direction of the displacement.

In the *Try this* example above, no work is done on the book because it has no displacement. In fact, even if the book was being displaced horizontally by the person holding it, the work done would be given by the product of the displacement and the *horizontal* component of the force, as the vertical component of the force is still doing no work. The work done by a force on a moving object is zero when the force applied is perpendicular to the displacement of its point of application. That is, if $\theta = 90°$, then $W = 0$ because $\cos 90° = 0$. This is also the case in **Figure 6.3**; the work done by the normal force on the object and the work done by the gravitational force on the object are both zero because both forces are perpendicular to the displacement and have zero components along an axis in the direction of $\Delta\vec{\mathbf{r}}$.

The sign of the work also depends on the direction of $\vec{\mathbf{F}}$ relative to $\Delta\vec{\mathbf{r}}$. The work done by the applied force on a system is positive when the projection of $\vec{\mathbf{F}}$ onto $\Delta\vec{\mathbf{r}}$ is in the same direction as the displacement. For example, when an object is lifted, the work done by the applied force on the object is positive because the direction of that force is upwards, in the same direction as the displacement of its point of application. When the projection of $\vec{\mathbf{F}}$ onto $\Delta\vec{\mathbf{r}}$ is in the direction opposite the displacement, W is negative. For example, as an object is lifted, the work done by the gravitational force on the object is negative. The dot product in **Equation 6.1** or the factor $\cos\theta$ in **Equation 6.4** automatically takes care of the sign.

An important consideration for a system approach to problems is that **work is an energy transfer.** If W is the work done on a system and W is positive, energy is transferred *to* the system; if W is negative, energy is transferred *from* the system. Therefore, if a system interacts with its environment, this interaction can be described as a transfer of energy across the system boundary. The result is a change in the energy stored in the system. We will learn about the first type of energy storage in Section 6.5, after we investigate more aspects of work.

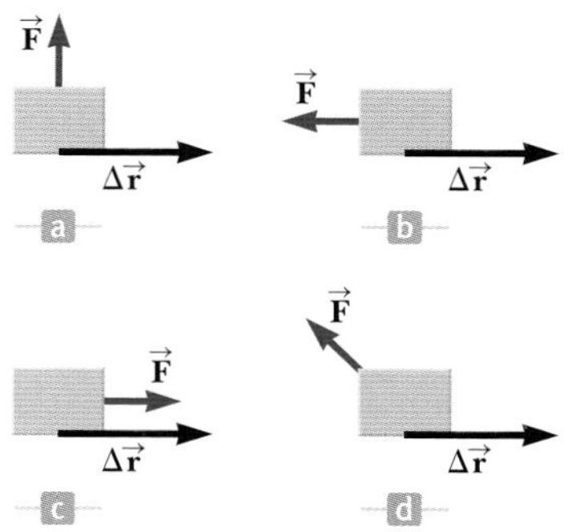

Figure 6.4
(Quick Quiz 6.2) A block is pulled by a force in four different directions. In each case, the displacement of the block is to the right and of the same magnitude.

Quick Quiz 6.1

The gravitational force exerted by the Sun on the Earth holds the Earth in an orbit around the Sun. Let us make the approximation that the orbit is perfectly circular. The work done by this gravitational force during a short time interval in which the Earth moves through a displacement in its orbital path is **(a)** zero **(b)** positive **(c)** negative **(d)** impossible to determine.

Quick Quiz 6.2

Figure 6.4 shows four situations in which a force is applied to an object. In all four cases, the force has the same magnitude, and the displacement of the object is to the right and of the same magnitude. Rank the situations in order of the work done by the force on the object, from most positive to most negative.

Example 6.1

Mr Clean

A man cleaning a floor pulls a vacuum cleaner with a force of magnitude $F = 50.0$ N at an angle of 30.0° to the horizontal (**Figure 6.5**). Calculate the work done by the force on the vacuum cleaner as the vacuum cleaner is displaced 3.00 m to the right.

Figure 6.5
(Example 6.1) A vacuum cleaner being pulled at an angle of 30° from the horizontal

Solution

Conceptualise Think about an experience in your life in which you pulled an object across the floor with a rope or cord. **Figure 6.5** helps conceptualise the situation, but we will need to draw another diagram – a free-body diagram – showing the forces acting on the vacuum cleaner.

Model We identify the vacuum cleaner as the system and draw a free-body diagram showing the forces acting on it. The only force acting in the direction of the displacement is the horizontal component of the force being applied by the man.

Analyse Use the definition of work (**Equation 6.4**):

$$W = F\Delta r\cos\theta = (50.0\text{ N})(3.00\text{ m})(\cos 30.0°) = 130\text{ J}$$

Finalise Notice in this situation that the normal force $\vec{n}$ and the gravitational force $\vec{F}_g = m\vec{g}$ do no work on the vacuum cleaner because these forces are perpendicular to its displacement. Note that the presence or absence of friction does not affect our calculation here. If friction is present, then it also does work. If the vacuum slides along the floor then the work done by friction is negative. Note that the work done does not depend on how long it took to move the vacuum through the displacement, or whether it moved at constant velocity or acceleration.

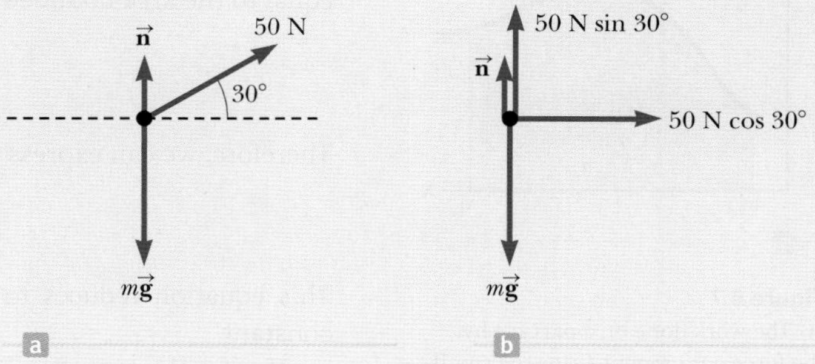

Figure 6.6
(a) A free-body diagram showing the forces acting on the vacuum cleaner
(b) A free-body diagram showing the forces acting broken into components parallel and perpendicular to the horizontal displacement

Example **6.2**

Work done by a constant force–the vector dot product

A particle moving in the *xy* plane undergoes a displacement given by $\Delta \vec{r} = (2.0\hat{i} + 3.0\hat{j})$ m as a constant force $\vec{F} = (5.0\hat{i} + 2.0\hat{j})$ N acts on the particle. Calculate the work done by $\vec{F}$ on the particle.

Solution

Conceptualise We are given force and displacement vectors and asked to find the work done by this force on the particle. This problem is really only asking us to find the dot product of two vectors, so this is a mathematics problem. The vectors are given in Cartesian form, so the best way to carry out the dot product is in Cartesian form, as in **Equation 6.3**.

Substitute the expressions for $\vec{F}$ and $\Delta \vec{r}$ into **Equation 6.1** and use **Equation 6.3** to solve the equation:

$$W = \vec{F} \cdot \Delta \vec{r} = [(5.0\hat{i} + 2.0\hat{j})\, N] \cdot [(2.0\hat{i} + 3.0\hat{j})\, m]$$

$$= (5.0\hat{i} \cdot 2.0\hat{i} + 5.0\hat{i} \cdot 3.0\hat{j} + 2.0\hat{j} \cdot 2.0\hat{i} + 2.0\hat{j} \cdot 3.0\hat{j})\, N.m$$

$$= [10 + 0 + 0 + 6]\, N.m = 16\, J$$

6.3 Work done by a varying force

Consider a particle being displaced along the *x* axis under the action of a force that varies with position. The particle is displaced in the direction of increasing *x* from $x = x_i$ to $x = x_f$. In such a situation, we cannot use $W = F\Delta r \cos\theta$ to calculate the work done by the force because this relationship applies only when $\vec{F}$ is constant in magnitude and direction. If, however, we imagine that the particle undergoes a very small displacement Δx, shown in **Figure 6.7a**, the *x* component F_x of the force is approximately constant over this small interval; for this small displacement, we can approximate the work done on the particle by the force as

$$W \approx F_x \Delta x$$

which is the area of the shaded rectangle in **Figure 6.7a**. If we imagine the F_x versus *x* curve divided into a large number of such intervals, the total work done for the displacement from x_i to x_f is approximately equal to the sum of a large number of such terms:

$$W \approx \sum_{x_i}^{x_f} F_x \Delta x$$

If the size of the small displacements is allowed to approach zero, the number of terms in the sum increases without limit but the value of the sum approaches a definite value equal to the area bounded by the F_x curve and the *x* axis:

$$\lim_{\Delta x \to 0} \sum_{x_i}^{x_f} F_x \Delta x = \int_{x_i}^{x_f} F_x \, dx$$

Therefore, we can express the work done by F_x on the particle as it moves from x_i to x_f as

$$W = \int_{x_i}^{x_f} F_x \, dx \tag{6.5}$$

This equation reduces to **Equation 6.1** when the component $F_x = F \cos\theta$ remains constant.

If more than one force acts on a system *and the system can be modelled as a particle,* the total work done on the system is just the work done by the net force. If we express the net force in the *x* direction as ΣF_x, the total work, or *net work,* done as the particle moves from x_i to x_f is

$$\sum W = W_{ext} = \int \sum F_x \, dx \quad \text{(particle)}$$

The total work done for the displacement from x_i to x_f is approximately equal to the sum of the areas of all the rectangles.

Area $= \Delta A \approx F_x \Delta x$

The work done by the component F_x of the varying force as the particle moves from x_i to x_f is *exactly* equal to the area under the curve.

Work

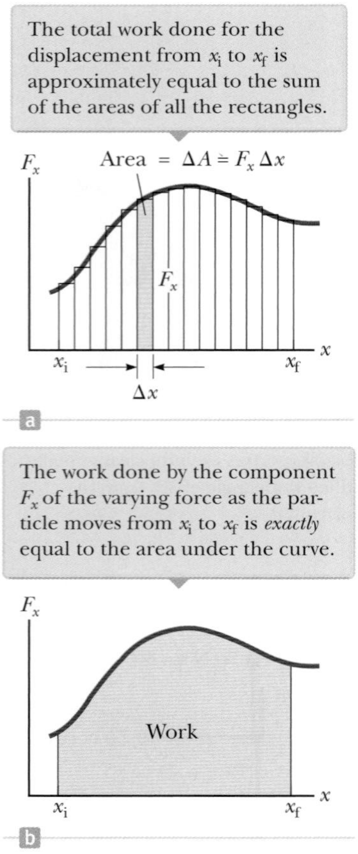

Figure 6.7

(a) The work done on a particle by the force component F_x for the small displacement Δx is $F_x \Delta x$, which equals the area of the shaded rectangle.

(b) The width Δx of each rectangle is shrunk to zero.

For the general case of a net force $\Sigma\vec{\mathbf{F}}$ whose magnitude and direction may vary, we use the scalar product,

$$\sum W = W_{\text{ext}} = \int \sum \vec{\mathbf{F}} \cdot d\vec{\mathbf{r}} \quad \text{(particle)} \tag{6.6}$$

where the integral is calculated over the path that the particle takes through space. The subscript 'ext' on work reminds us that the net work is done by an *external* agent on the system.

Example 6.3

Calculating total work done from a graph

A force acting on a particle varies with x as shown in **Figure 6.8**. Calculate the work done by the force on the particle as it moves from $x = 0$ to $x = 6.0$ m.

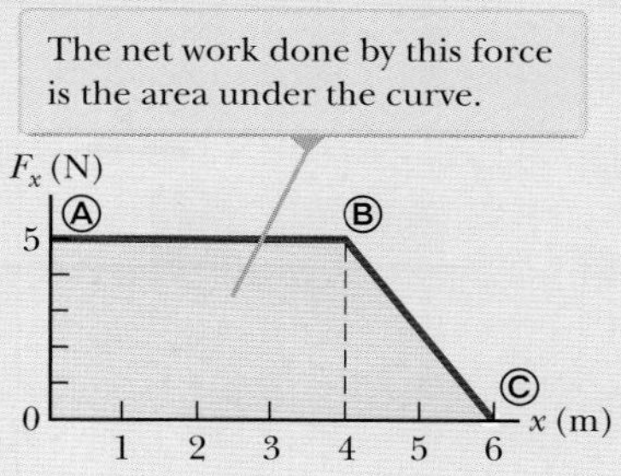

Figure 6.8
(Example 6.3) The force acting on a particle is constant for the first 4.0 m of motion and then decreases linearly with x from $x_{Ⓑ} = 4.0$ m to $x_{Ⓒ} = 6.0$ m.

Solution

Conceptualise Imagine a particle subject to the force in **Figure 6.8**. Notice that the force remains constant as the particle moves through the first 4.0 m and then decreases linearly to zero at 6.0 m.

Model We are considering a particle subject to a varying force.

Analyse We use the techniques for work done by varying forces. In this case, the graphical representation in **Figure 6.8** can be used to evaluate the work done.

The work done by the force is equal to the area under the curve from $x_{Ⓐ} = 0$ to $x_{Ⓒ} = 6.0$ m. This area is equal to the area of the rectangular section from Ⓐ to Ⓑ plus the area of the triangular section from Ⓑ to Ⓒ.

Evaluate the area of the rectangle: $W_{Ⓐ\text{ to }Ⓑ} = (5.0 \text{ N})(4.0 \text{ m}) = 20 \text{ J}$

Evaluate the area of the triangle: $W_{Ⓑ\text{ to }Ⓒ} = \frac{1}{2}(5.0 \text{ N})(2.0 \text{ m}) = 5.0 \text{ J}$

Find the total work done by the force on the particle: $W_{Ⓐ\text{ to }Ⓒ} = W_{Ⓐ\text{ to }Ⓑ} + W_{Ⓑ\text{ to }Ⓒ} = 20 \text{ J} + 5.0 \text{ J} = 25 \text{ J}$

Finalise As the graph of the force consists of straight lines, we can use rules for finding the areas of simple geometric shapes to evaluate the total work done in this example. In a case in which the force does not vary linearly, such rules cannot be used and the force function must be integrated as in **Equation 6.5** or **6.6**.

Work done by a spring

Many objects in everyday life behave like springs when small forces are applied to them. Try pushing a large ball against the floor – the force it applies back to you increases the further you push it and the more you deform it.

A model in which the force varies with position is shown in **Active Figure 6.9**. The system is a block, on a frictionless, horizontal surface, connected to a spring. For many springs, if the spring is either stretched or compressed a small distance from its unstretched (equilibrium) configuration, it exerts on the block a force that can be mathematically modelled as

◀ Spring force

$$F_s = -kx \tag{6.7}$$

where x is the position of the block relative to its equilibrium ($x = 0$) position and k is a positive constant called the **force constant** or the **spring constant** of the spring. In other words, the force required to stretch or compress a spring is proportional to the amount of stretch or compression x. This force law for springs is known as **Hooke's law**.

◀ Hooke's Law

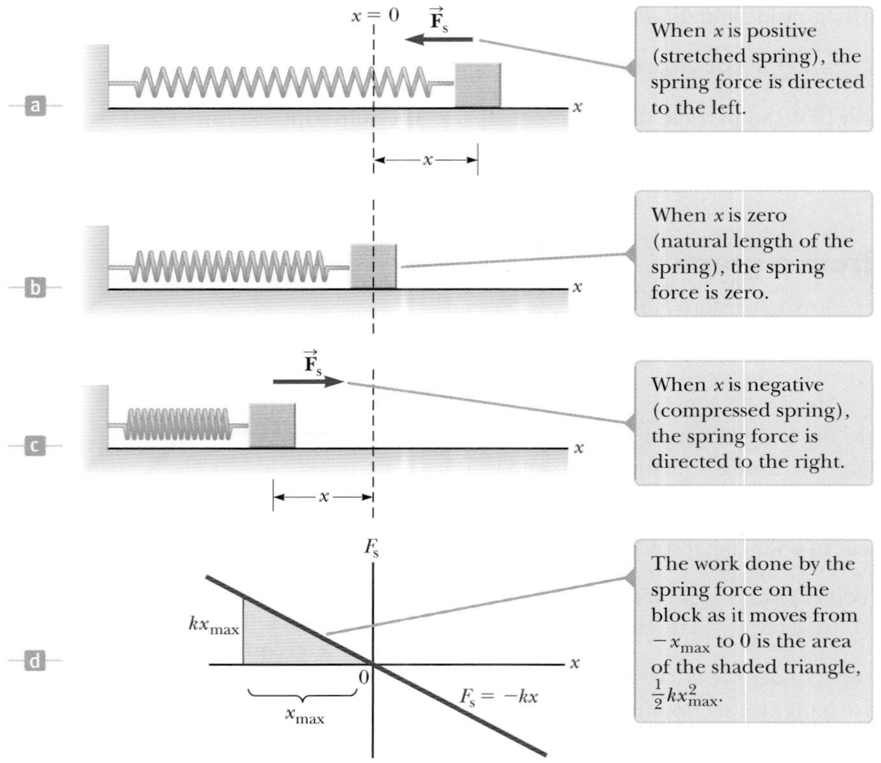

a
b
c
d
x = 0
$\vec{F}_s$
x
When x is positive (stretched spring), the spring force is directed to the left.
When x is zero (natural length of the spring), the spring force is zero.
When x is negative (compressed spring), the spring force is directed to the right.
F_s
kx_{max}
x_{max}
0
$F_s = -kx$
The work done by the spring force on the block as it moves from $-x_{max}$ to 0 is the area of the shaded triangle, $\frac{1}{2}kx^2_{max}$.

Active Figure 6.9 The force exerted by a spring on a block varies with the block's position x relative to the equilibrium position $x = 0$. (a) x is positive. (b) x is zero. (c) x is negative. (d) Graph of F_s versus x for the block–spring system.

The value of k is a measure of the *stiffness* of the spring. Stiff springs have large k values and soft springs have small k values. As can be seen from **Equation 6.7**, the units of k are N/m. The vector form of **Equation 6.7** is

$$\vec{\mathbf{F}}_s = F_s\hat{\mathbf{i}} = -kx\hat{\mathbf{i}} \tag{6.8}$$

where we have chosen the x axis to lie along the direction the spring extends or compresses.

Quick Quiz 6.3

What are the fundamental units and dimensions of the spring constant k?

The negative sign in **Equations 6.7** and **6.8** signifies that the force exerted by the spring is always directed *opposite* the displacement from equilibrium. When $x > 0$ as in **Active Figure 6.9a** so that the block is to the right of the equilibrium position, the spring force is directed to the left, in the negative x direction. When $x < 0$ as in **Active Figure 6.9c**, the block is to the left of the equilibrium position and the spring force is directed to the right, in the positive x direction. When $x = 0$ as in **Active Figure 6.9b**, the spring is unstretched and $F_s = 0$. The spring force is sometimes called a *restoring force* because it always acts towards the equilibrium position ($x = 0$).

If the spring is compressed until the block is at the point $-x_{max}$ and then released, the block can be observed to move from $-x_{max}$ through zero to $+x_{max}$. It then reverses direction, returns to $-x_{max}$, and continues oscillating back and forth. In **Active Figure 6.9** this oscillatory motion continues indefinitely. A real block on a spring would oscillate for a finite time, with decreasing amplitude, due to friction forces.

Suppose the block has been pushed to the left to a position $-x_{max}$ and is then released. We identify the block as our system and calculate the work W_s done by the spring force on the block as the block moves from $x_i = -x_{max}$ to $x_f = 0$. Applying **Equation 6.6** and assuming the block may be modelled as a particle, we obtain

Work done by ▶ a spring

$$W_s = \int \vec{\mathbf{F}}_s \cdot d\vec{\mathbf{r}} = \int_{x_i}^{x_f} (-kx\hat{\mathbf{i}}) \cdot (dx\hat{\mathbf{i}}) = \int_{-x_{max}}^{0} (-kx)dx = \frac{1}{2}kx^2_{max} \tag{6.9}$$

Σ *Appendix B.6 demonstrates how to find the vector dot product. Appendix B.8 shows how to integrate polynomials.*

The work done by the spring force is positive because the force is in the same direction as its displacement (both are to the right). Because the block arrives at $x = 0$ with some speed, it will continue moving until it reaches a position $+x_{max}$. The work done by the spring force on the block as it moves from $x_i = 0$ to $x_f = x_{max}$ is $W_s = -\frac{1}{2}kx^2_{max}$. The work is negative because for this part of the motion the spring force is to the left and its displacement is to the right. Therefore, the *net* work done by the spring force on the block as it moves from $x_i = -x_{max}$ to $x_f = x_{max}$ is *zero.*

Active Figure 6.9d is a plot of F_s versus x. The work calculated in **Equation 6.8** is the area of the shaded triangle, corresponding to the displacement from $-x_{max}$ to 0. Because the triangle has base x_{max} and height kx_{max}, its area is $\frac{1}{2}kx^2_{max}$, agreeing with the work done by the spring as given by **Equation 6.9**.

If the block undergoes an arbitrary displacement from $x = x_i$ to $x = x_f$, the work done by the spring force on the block is

$$W_s = \int_{x_i}^{x_f} (-kx)\, dx = \frac{1}{2}kx_i^2 - \frac{1}{2}kx_f^2 \tag{6.10}$$

From **Equation 6.10**, we see that the work done by the spring force is zero for any motion that ends where it began ($x_i = x_f$). We shall make use of this important result in Chapter 7 when we describe the motion of this system in greater detail.

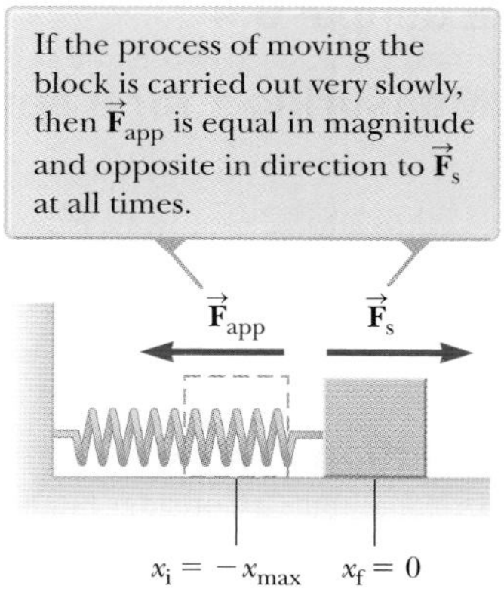
If the process of moving the block is carried out very slowly, then $\vec{F}_{app}$ is equal in magnitude and opposite in direction to $\vec{F}_s$ at all times.
$\vec{F}_{app}$
$\vec{F}_s$
$x_i = -x_{max}$
$x_f = 0$

Figure 6.10
A block moves from $x_i = -x_{max}$ to $x_f = 0$ on a frictionless surface as a force $\vec{F}_{app}$ is applied to the block.

Equations 6.9 and **6.10** describe the work done by the spring on the block. Now let us consider the work done on the block by an *external agent* as the agent applies a force on the block and the block moves *very slowly* from $x_i = -x_{max}$ to $x_f = 0$ as in **Figure 6.10**. We can calculate this work by noting that at any value of the position, the *applied force* $\vec{F}_{app}$ is equal in magnitude and opposite in direction to the spring force $\vec{F}_s$, so $\vec{F}_{app} = F_{app}\hat{\mathbf{i}} = -\vec{F}_s = -(-kx\,\hat{\mathbf{i}}) = kx\,\hat{\mathbf{i}}$. Therefore, the work done by this applied force (the external agent) on the system of the block is

$$W_{app} = \int \vec{F}_{app} \cdot d\vec{r} = \int_{x_i}^{x_f} (kx\hat{\mathbf{i}}) \cdot (dx\hat{\mathbf{i}}) = \int_{-x_{max}}^{0} kx\, dx = -\frac{1}{2}kx^2_{max}$$

This work is equal to the negative of the work done by the spring force for this displacement (**Equation 6.9**). The work is negative because the external agent must push inwards on the spring to prevent it from expanding and this direction is opposite the direction of the displacement of the point of application of the force as the block moves from $-x_{max}$ to 0.

For an arbitrary displacement of the block, the work done on the system by the external agent is

$$W_{app} = \int_{x_i}^{x_f} kx\, dx = \frac{1}{2}kx_f^2 - \frac{1}{2}kx_i^2 \tag{6.11}$$

Notice that this equation is the negative of **Equation 6.10**.

Quick Quiz 6.4

A dart is inserted into a spring-loaded dart gun by pushing the spring in by a distance *x*. For the next loading, the spring is compressed a distance 2*x*. How much work is required to load the second dart compared with that required to load the first? **(a)** four times as much **(b)** two times as much **(c)** the same **(d)** half as much **(e)** one-fourth as much

TRY THIS

Get a rubber band and hang it over something, such as a door handle or a hook. Use a paperclip to hang an object on the other end of the rubber band. What happens? Increase the mass hanging from the rubber band and observe what happens. If you can, try increasing the mass by uniform amounts, for example by hanging one, then two then three identical objects. Does the rubber band stretch by a uniform amount each time? Does it continue to do so as you add larger weights? If you remove the weights one by one does it go back to the same extension? Remember that Hooke's law is a *model* for elastic behaviour, with simplifying assumptions. Even the behaviour of simple things such as rubber bands is not completely explained by this model!

Example 6.4

Measuring *k* for a spring

A student measures the force constant of a spring using the setup in **Figure 6.11**. The spring is hung vertically and an object of mass m is attached to its lower end. Under the action of the load mg, the spring stretches a distance d from its equilibrium position.

(A) If the spring is stretched (2.0 ± 0.1) cm by a suspended object having a mass of (0.55 ± 0.05) kg, what is the force constant of the spring?

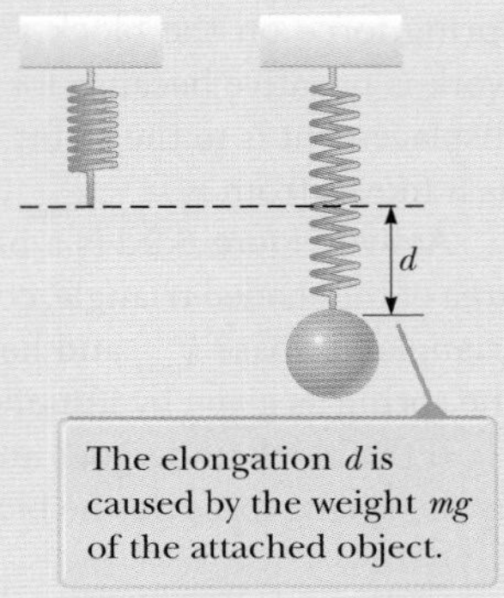
d
The elongation d is caused by the weight mg of the attached object.

Figure 6.11
(Example 6.4) Determining the force constant k of a spring

Solution

Conceptualise Consider **Figure 6.11**, which shows what happens to the spring when the object is attached to it. Draw a free-body diagram for the mass, showing the forces acting on it.

Model The object in **Figure 6.11** is not accelerating, so it is modelled as a particle in equilibrium.

Analyse As the object is in equilibrium, the net force on it is zero and the upwards spring force balances the downwards gravitational force $m\vec{g}$ (**Figure 6.12**).

$\vec{F}_s$
$m\vec{g}$

Figure 6.12
Free-body diagram showing the forces acting on the mass

Apply the particle in equilibrium model to the object:

$$\vec{F}_s + m\vec{g} = 0 \rightarrow F_s - mg = 0 \rightarrow F_s = mg$$

Apply Hooke's law to give $F_s = kd$ and solve for k:

$$k = \frac{mg}{d}$$

Do a quick dimension check:

$$[MT^{-2}] = [M][LT^{-2}]/[L] = [MT^{-2}]$$ ☺

Now substitute the numbers:

$$k = \frac{mg}{d} = \frac{(0.55\text{ kg})(9.80\text{ m/s}^2)}{2.0\times10^{-2}\text{ m}} = 2.7\times10^2\text{ N/m}$$

Note that the units for k can also be written as kg/s^2, which is the same as N/m.

With an uncertainty of

$$\Delta k = k\left(\frac{\Delta m}{m}+\frac{\Delta d}{d}\right) = 2.7\times10^2\text{ N}\left(\frac{0.05\text{ kg}}{0.55\text{ kg}}+\frac{0.01}{0.2}\right) = 38\text{ N/m}$$

So we can write the spring constant as (270 ± 40) N/m.

(B) How much work is done by the spring on the object as it stretches through this distance?

Solution

Use **Equation 6.10** to find the work done by the spring on the object:

$$W_s = 0-\frac{1}{2}kd^2 = -\frac{1}{2}(2.7\times10^2\text{ N/m})(2.0\times10^{-2}\text{ m}^2)$$
$$= -5.4\times10^{-2}\text{ J}$$

with an uncertainty of

$$\Delta W = W\left(\frac{\Delta k}{k}+2\frac{\Delta d}{d}\right) = 5.4\times10^{-2}\text{ J}\left(\frac{38\text{ N/m}}{270\text{ N/m}}+2\left(\frac{0.1\text{ cm}}{2.0\text{ cm}}\right)\right) = 1.3\times10^{-2}\text{ J}$$

So we write $W_s = (-5.4 \pm 1.3)\times10^{-2}$ J.

Note that we have used the relative uncertainty in displacement twice in our uncertainty calculation because we have multiplied by the displacement twice to get our value for work.

Finalise You may have done an experiment like this to measure the spring constant of a spring. A better method is to use multiple weights and construct a graph of displacement of the end of the spring as a function of applied force. The spring constant can then be found from the gradient. A different method again, described in Chapter 16, uses oscillating weights on springs. This is also a common physics experiment that you may have done.

Example 6.4 cont.

(C) How much work is done by gravity on the object?

Solution

In this case the work done is: $W_g = F_g d = mgd = (0.55\text{ kg})(9.8\text{ m/s}^2)(0.02\text{ m}) = 0.1078\text{ J}$

with an uncertainty of $\Delta W_g = W_g\left(\dfrac{\Delta m}{m} + \dfrac{\Delta d}{d}\right) = 0.1078\text{ J}\left(\dfrac{0.05\text{ kg}}{0.55\text{ kg}} + \dfrac{0.1\text{ cm}}{2.0\text{ cm}}\right) = 0.015\text{ J}$

where we have taken the uncertainty in g to be negligible.

So we can now write $W_g = (0.108 \pm 0.015)\text{ J}$

Note that this work is positive and is *not* the same as the work done by the spring, which we calculated in part B.

TRY THIS

Get some rubber bands, ideally identical ones from the same packet. Hold one so it loops around two fingers and stretch it. (Be careful not to accidentally flick it at someone.) Now repeat the experiment using two rubber bands together, then three, then four. What do you notice about the force you have to use to get the same stretch? Now try tying the rubber bands together to form a long chain. What do you notice about the force needed to achieve a given extension as a function of how many rubber bands you have tied together?

6.4 Kinetic energy and the work–kinetic energy theorem

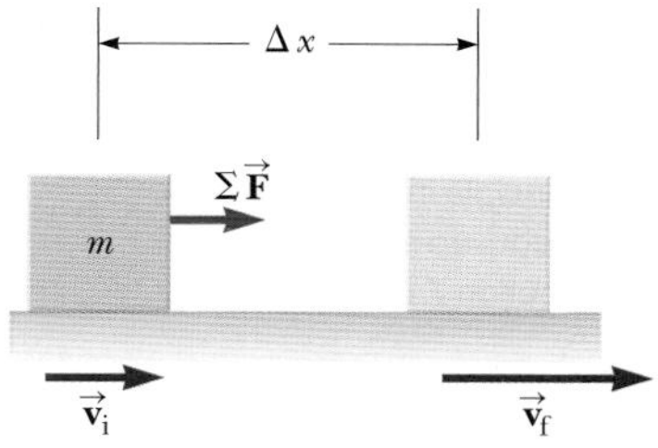

Figure 6.13
An object undergoing a displacement $\Delta\vec{r} = \Delta x\hat{i}$ and a change in velocity under the action of a constant net force $\Sigma\vec{F}$.

We have investigated work and identified it as a mechanism for transferring energy into a system. We have stated that work is an influence on a system from the environment, but we have not yet discussed the *result* of this influence on the system. One possible result of doing work on a system is that the system changes its speed. In this section, we investigate this situation and introduce our first type of energy that a system can possess, called *kinetic energy*.

Consider a system consisting of a single object. **Figure 6.13** shows a block of mass m moving through a displacement directed to the right under the action of a net force $\Sigma\vec{F}$, also directed to the right. We know from Newton's second law that the block moves with an acceleration $\vec{a}$. If the block moves through a displacement $\Delta\vec{r} = \Delta x\hat{i} = (x_f - x_i)\hat{i}$, the net work done on the block by the external net force $\Sigma\vec{F}$ is

$$W_{ext} = \int_{x_i}^{x_f} \sum F\, dx$$

Using Newton's second law, we substitute for the magnitude of the net force $\Sigma F = ma$ and then perform the following chain-rule manipulations on the integrand:

$$W_{ext} = \int_{x_i}^{x_f} ma\, dx = \int_{x_i}^{x_f} m\frac{dv}{dt}dx = \int_{x_i}^{x_f} m\frac{dv}{dx}\frac{dx}{dt}dx = \int_{v_i}^{v_f} mv\, dv$$

$$W_{ext} = \frac{1}{2}mv_f^2 - \frac{1}{2}mv_i^2 \qquad (6.12)$$

where v_i is the speed of the block at $x = x_i$ and v_f is its speed at $x = x_f$.

Appendix B.7 gives the chain rule and other useful properties of the derivative. Σ

Equation 6.12 was generated for the specific situation of one-dimensional motion, but it is a general result. It tells us that the work done by the net force on a particle of mass m is equal to the difference between the initial and final values of a quantity $\frac{1}{2}mv^2$. This quantity is so important that it has been given a special name, **kinetic energy**:

$$K \equiv \frac{1}{2}mv^2 \qquad (6.13)$$

◀ Kinetic energy

Kinetic energy represents the energy associated with the motion of the particle. Kinetic energy is a scalar quantity and has the same units as work.

Equation 6.12 states that the work done on a particle by a net force $\sum\vec{\mathbf{F}}$ acting on it equals the change in kinetic energy of the particle. It is often convenient to write Equation 6.12 in the form

$$W_{ext} = K_f - K_i = \Delta K \qquad (6.14)$$

Another way to write it is $K_f = K_i + W_{ext}$, which tells us that the final kinetic energy of an object is equal to its initial kinetic energy plus the change in energy due to the net work done on it.

We have generated Equation 6.14 by imagining doing work on a particle. We could also do work on a deformable system, in which parts of the system move with respect to one another. In this case, we also find that Equation 6.14 is valid as long as the net work is found by adding up the work done by each force.

Equation 6.14 is an important result known as the **work–kinetic energy theorem**:

Work–kinetic energy theorem ▶ *When work is done on a system and the only change in the system is in its speed, the net work done on the system equals the change in the kinetic energy of the system.*

> **Pitfall Prevention 6.5**
> The work–kinetic energy theorem is important but limited in its application; it is not a general principle. In many situations, other changes in the system occur besides its speed and there are other interactions with the environment besides work. A more general principle involving energy is *conservation of energy* in Section 7.1.

> **Pitfall Prevention 6.6**
> The work–kinetic energy theorem relates work to a change in the *speed* of a system, not a change in its velocity. For example, if an object is in uniform circular motion, its speed is constant. Even though its velocity is changing, no work is done on the object by the force causing the circular motion.

The work–kinetic energy theorem indicates that the speed of a system *increases* if the net work done on it is *positive* because the final kinetic energy is greater than the initial kinetic energy. The speed *decreases* if the net work is *negative* because the final kinetic energy is less than the initial kinetic energy.

Because we have so far only investigated translational motion through space, we arrived at the work–kinetic energy theorem by analysing situations involving translational motion. Another type of motion is *rotational motion,* in which an object spins about an axis. We will look at this in Chapter 9. The work–kinetic energy theorem is also valid for systems that undergo a change in rotational speed due to work done on the system.

The quantity ΔK in the work–kinetic energy theorem refers only to the initial and final points for the speeds; it does not depend on details of the path followed between these points. Recall that in Section 6.3 we arrived at a result of zero net work done when we let a spring push a block from $x_i = -x_{max}$ to $x_f = x_{max}$. As the speed is zero at both the initial and final points of the motion, the net work done on the block is zero; the details of how the speed varies in time do not matter. We will often see this concept of **path independence** used in solving problems.

Let us return to Example 6.4. Why was the work done by gravity not just the value of the work done by the spring with a positive sign? Notice that the work done by gravity is larger than the magnitude of the work done by the spring. Therefore, the total work done by all forces on the object is positive. Imagine now how to create the situation in which the *only* forces on the object are the spring force and the gravitational force. You must support the object at the highest point and then remove your hand and let the object fall. If you do so, you know that when the object reaches a position 2.0 cm below your hand, it will be *moving,* which is consistent with Equation 6.14. Positive net work is done on the object and the result is that it has kinetic energy as it passes through the 2.0 cm point.

Earlier, we indicated that work can be considered as a mechanism for transferring energy into a system. Equation 6.14 is a mathematical statement of this concept. When work W_{net} is done on a system, the result is a transfer of energy across the boundary of the system. The result on the system in the case of Equation 6.14 is a change ΔK in kinetic energy. In the next section, we investigate another type of energy that can be stored in a system.

Quick Quiz 6.5

A dart is inserted into a spring-loaded dart gun by pushing the spring in by a distance *x*. For the next loading, the spring is compressed a distance 2*x*. How much faster does the second dart leave the gun compared with the first? **(a)** four times as fast **(b)** two times as fast **(c)** the same **(d)** half as fast **(e)** one-fourth as fast

Example 6.5

Towing a trailer

A car tows a 60 kg trailer initially at rest with a constant horizontal force of 120 N. Find the trailer's speed after it has moved 3.0 m. Assume that the rotational kinetic energy of the trailer's wheels is negligible compared to its translational kinetic energy.

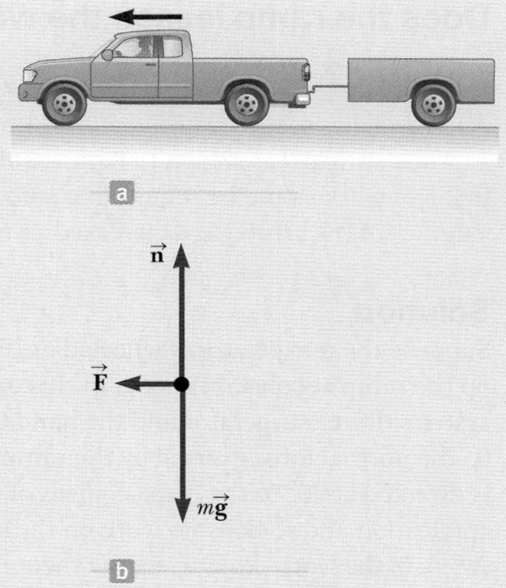

Figure 6.14
(Example 6.5) (a) A car towing a trailer (b) A free-body diagram showing the forces acting on the trailer

Solution

Conceptualise We start by drawing a diagram showing this situation, including a free-body diagram for the trailer (**Figure 6.14**).

Model The system of interest is the trailer. We are told that the rotational kinetic energy of the wheels is negligible compared to the translational kinetic energy of the trailer, so we ignore the rotational kinetic energy. It is unlikely that air resistance is significant at the low speed of the trailer (it starts from rest), and as it is rolling and not sliding we will make the approximation that any work done by frictional forces is negligible.

Three external forces act on the system. The normal force balances the gravitational force on the trailer and neither of these vertically acting forces does work on the trailer. Hence we only need to consider the work done by the car.

Analyse We will use an energy approach. The net external force acting on the trailer is the horizontal 120 N force.

Use the work–kinetic energy theorem for the trailer, noting that its initial kinetic energy is zero:

$$W_{ext} = K_f - K_i = \frac{1}{2}mv_f^2 - 0 = \frac{1}{2}mv_f^2$$

Solve for v_f and use **Equation 6.1** for the work done on the trailer by $\vec{\mathbf{F}}$:

$$v_f = \sqrt{\frac{2W_{ext}}{m}} = \sqrt{\frac{2F\Delta x}{m}}$$

Check dimensions:

$$[LT^{-1}] = ([MLT^{-2}][L]/[M])^{1/2} = [LT^{-1}]$$ ☺

Substitute numerical values:

$$v = \sqrt{\frac{2 \times 120\ \text{N} \times 3\ \text{m}}{60\ \text{kg}}} = 3.5\ \text{m/s}^2$$

Finalise It would be useful for you to solve this problem again by modelling the trailer as a particle subject to a net force to find its acceleration and then as a particle with constant acceleration to find its final velocity. Often there is more than one approach to solving a problem, which will give the same final answer, but usually one method is simpler than the other. Because energy is a scalar, it is often simpler to use an energy approach.

What If? Suppose the magnitude of the force in this example is doubled to $F' = 2F$. The 60 kg trailer accelerates to 3.5 m/s due to this applied force while moving through a displacement $\Delta x'$. How does the displacement $\Delta x'$ compare with the original displacement Δx?

Answer If we pull harder, the trailer should accelerate to a given speed in a shorter distance, so we expect that $\Delta x' < \Delta x$. In both cases, the block experiences the same change in kinetic energy ΔK. Mathematically, from the work–kinetic energy theorem, we find that

$$W_{ext} = F'\Delta x' = \Delta K = F\Delta x$$

$$\Delta x' = \frac{F}{F'}\Delta x = \frac{F}{2F}\Delta x = \frac{1}{2}\Delta x$$

and the distance is shorter, as suggested by our conceptual argument.

Conceptual Example 6.6

Does the ramp lessen the work required?

A man wishes to load a refrigerator onto a truck using a ramp at angle θ as shown in **Figure 6.15**. He claims that less work would be required to load the truck if the length L of the ramp were increased. Is his claim valid?

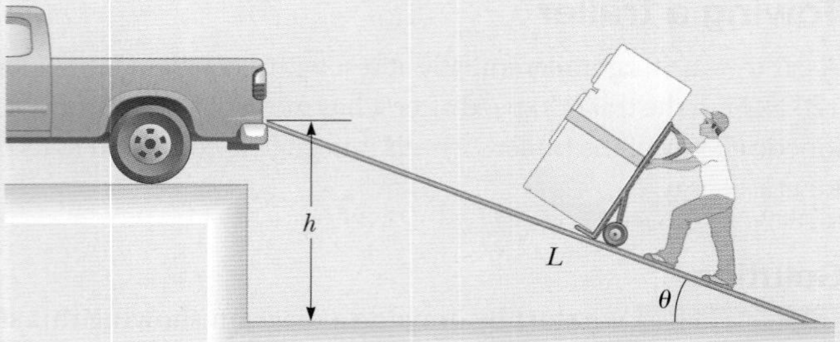

Figure 6.15
(Conceptual Example 6.6) A refrigerator attached to a frictionless, wheeled hand trolley is moved up a ramp at constant speed.

Solution

Suppose the refrigerator is wheeled on a hand trolley up the ramp at constant speed. In this case, for the system of the refrigerator and the hand trolley, $\Delta K = 0$. The normal force exerted by the ramp on the system is directed at 90° to the displacement of its point of application and so does no work on the system. Because $\Delta K = 0$, the work–kinetic energy theorem gives

$$W_{\text{ext}} = W_{\text{by man}} + W_{\text{by gravity}} = 0$$

The work done by the gravitational force equals the product of the weight mg of the system, the distance L through which the refrigerator is displaced and $\cos(\theta + 90°)$. Therefore,

$$W_{\text{by man}} = -W_{\text{by gravity}} = -(mg)\,(L)\,[\cos(\theta + 90°)]$$
$$= mgL \sin\theta = mgh$$

where $h = L \sin\theta$ is the height of the ramp. Therefore, the man must do the same amount of work mgh on the system *regardless* of the length of the ramp. The work depends only on the height of the ramp. Although less force is required with a longer ramp, the point of application of that force moves through a greater displacement.

6.5 Potential energy of a system

So far in this chapter, we have defined a system in general, but have focused our attention primarily on single particles or objects under the influence of external forces. Let us now consider systems of two or more particles or objects interacting via a force that is internal to the system. The kinetic energy of such a system is the algebraic sum of the kinetic energies of all members of the system. Sometimes one object in a system is so massive compared to all the others that we can simplify any calculations by taking that object to be stationary and everything else as moving relative to it. A common example is the Earth and small objects such as people, balls and cars moving around on its surface.

If we take our system to be the Earth and a book, for example, then if we want to know the total kinetic energy of the system we only need to consider the speed of the book relative to the Earth.

The book and the Earth interact via the gravitational force. We do some work on the system by lifting the book slowly from rest through a vertical displacement $\Delta\vec{\mathbf{r}} = (y_f - y_i)\hat{\mathbf{j}}$ as in **Active Figure 6.16**. According to our discussion of work as an energy transfer, this work done on the system must appear as an increase in energy of the system. The book is at rest before we perform the work and is at rest after we perform the work. Therefore, there is no change in the kinetic energy of the system.

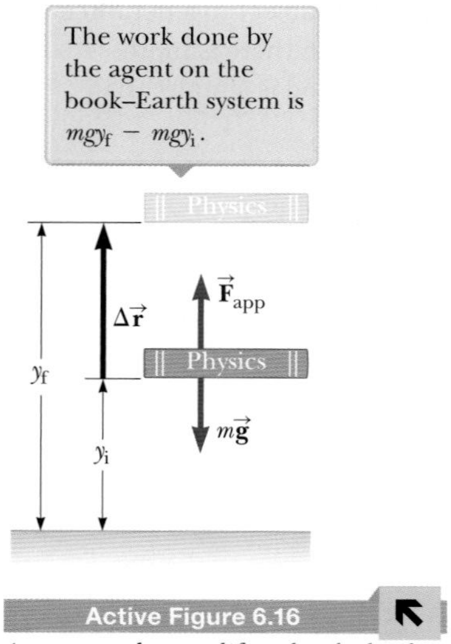

Active Figure 6.16
An external agent lifts a book slowly from a height y_i to a height y_f.

After lifting the book, we could release it and let it fall back to the position y_i. Notice that the book (and therefore, the system) now has kinetic energy and that its source is in the work that was done in lifting the book. While the book was at the highest point, the system had the *potential* to possess kinetic energy, but it did not do so until the book was allowed to fall. We call this type of stored energy

potential energy. The potential energy of a system can only be associated with forces acting between members of a system. The amount of potential energy in the system is determined by the *configuration* of the system. Moving members of the system to different positions or rotating them changes the configuration of the system and therefore may change its potential energy.

Let us now derive an expression for the potential energy associated with an object at a given location above the surface of the Earth. Consider an external agent lifting an object of mass m from an initial height y_i above the ground to a final height y_f as in **Active Figure 6.16**. We assume the lifting is done slowly, with no acceleration, so the applied force from the agent is equal in magnitude to the gravitational force on the object: the object is modelled as a particle in equilibrium moving at constant velocity. The work done by the external agent on the system is given by:

> ### Pitfall Prevention 6.7
> **Potential energy**
> The phrase *potential energy* does not refer to something that has the potential to become energy. Potential energy *is* energy.

$$W_{ext} = \vec{\mathbf{F}}_{app} \cdot \Delta\vec{\mathbf{r}} = (mg\hat{\mathbf{j}}) \cdot [(y_f - y_i)\hat{\mathbf{j}}] = mgy_f - mgy_i \quad (6.15)$$

where this result is the net work done on the system because the applied force is the only force on the system from the environment. (Remember that the gravitational force is *internal* to the system.) Notice the similarity between **Equation 6.15** and **Equation 6.12**. In each equation, the work done on a system equals a difference between the final and initial values of a quantity.

The vector dot product is reviewed in Appendix B.6.

Σ

We identify the quantity mgy as the **gravitational potential energy** U_g:

$$U_g \equiv mgy \quad (6.16)$$

◀ Gravitational potential energy

The units of gravitational potential energy are joules, the same as the units of work and kinetic energy. Potential energy, like work and kinetic energy, is a scalar quantity. Notice that **Equation 6.16** is valid only for objects near the surface of the Earth, where g is approximately constant.

Using our definition of gravitational potential energy, **Equation 6.15** can now be rewritten as

$$W_{ext} = \Delta U_g \quad (6.17)$$

which mathematically states that the net external work done on the system in this situation appears as a change in the gravitational potential energy of the system.

Gravitational potential energy depends only on the vertical height of the object above the surface of the Earth. The same amount of work must be done on an object–Earth system whether the object is lifted vertically from the Earth or is pushed up a frictionless incline, starting from the same point and ending up at the same height. We verified this statement for a specific situation of rolling a refrigerator up a ramp in Conceptual Example 6.6. This statement can be shown to be true in general by calculating the work done on an object by an agent moving the object through a displacement having both vertical and horizontal components:

> ### Pitfall Prevention 6.8
> Potential energy is always associated with a *system* of two or more interacting objects. When a small object moves near the surface of the Earth under the influence of gravity, we may sometimes refer to the potential energy 'associated with the object' rather than the more proper 'associated with the system' because the Earth does not move significantly. But you should remember that the energy belongs to the *system* – the object *and* the Earth.

$$W_{ext} = \vec{\mathbf{F}}_{app} \cdot \Delta\vec{\mathbf{r}} = (mg\hat{\mathbf{j}}) \cdot [(x_f - x_i)\hat{\mathbf{i}} + (y_f - y_i)\hat{\mathbf{j}}] = mgy_f - mgy_i$$

where there is no term involving x in the final result because $\hat{\mathbf{j}} \cdot \hat{\mathbf{i}} = 0$.

In solving problems, you must choose a reference configuration for which the gravitational potential energy of the system is set equal to some reference value, which is normally zero. The choice of reference configuration is completely arbitrary because the important quantity is the *difference* in potential energy and this difference is independent of the choice of reference configuration.

Sometimes it is convenient to choose as the reference configuration for zero gravitational potential energy the configuration in which an object is at the surface of the Earth, but sometimes a different configuration is more convenient. Often, the statement of the problem suggests a convenient configuration to use.

Quick Quiz 6.6

Choose the correct answer. The gravitational potential energy of a system **(a)** is always positive **(b)** is always negative **(c)** can be negative or positive.

Example 6.7

The proud athlete and the sore toe

A heavy trophy being shown off by a careless athlete slips from the athlete's hands and drops on his toe. The trophy has a mass m and the athlete (with height h_a) holds it at a height h_t above his head. Choosing floor level as the $y = 0$ point of your coordinate system, find the change in gravitational potential energy of the trophy–Earth system as the trophy falls. Repeat the problem, using the top of the athlete's head as the origin of coordinates.

Figure 6.17 (Example 6.7)

Solution

Conceptualise The trophy changes its vertical position with respect to the surface of the Earth. Associated with this change in position is a change in the gravitational potential energy of the trophy–Earth system. As always, it is helpful to draw a diagram and label it with the data given.

Model We wish to find the change in gravitational potential energy. The system is the trophy and the Earth. We model the trophy as a particle subject to a constant force, the gravitational force close to the Earth's surface. First, taking ground level as $y = 0$, use **Equation 6.16** for each case.

The gravitational potential energy of the system just before the trophy is released is:

$$U_i = mgy_i = mg(h_t + h_a)$$

The gravitational potential energy of the system when the trophy reaches the athlete's toe at ground level:

$$U_f = mgy_f = 0$$

The change in gravitational potential energy of the system is:

$$\Delta U_g = mg(h_t + h_a) - 0 = mg(h_t + h_a)$$

Second, taking the zero of potential energy as the athlete's head:

$$U_i = mgy_i = mgh_t$$

$$U_f = mgy_f = -mgh_a$$

so

$$\Delta U_g = mgh_t - (-mgh_a) = mg(h_t + h_a).$$

Hence, regardless of where we choose our zero of gravitational potential energy to be, the *change* in potential energy is the same. Note that in the second case when we took $y = h_a = 0$ we found that the final potential energy was negative. Unlike kinetic energy, which is always positive, potential energy can be positive or negative depending on how we choose our reference configuration. We will see that negative potential energies are very common.

Elastic potential energy

Members of a system can interact with one another by means of different types of forces, hence there are different types of potential energy in a system. When members of a system interact via the gravitational force, there is gravitational potential energy stored in the system. A second type of potential energy that a system can possess is elastic potential energy.

Consider a system consisting of a block and a spring as shown in **Active Figure 6.18**. In Section 6.4, we identified *only* the block as the system. Now we include both the block and the spring in the system and recognise that the spring force is the interaction between the two members of the system. The force that the spring exerts on the block is given by $F_s = -kx$ (**Equation 6.7**). The work done by an external applied force F_{app} on a system consisting of a block connected to the spring is given by **Equation 6.10**:

$$W_{app} = \frac{1}{2}kx_f^2 - \frac{1}{2}kx_i^2 \tag{6.18}$$

It is convenient to take as our reference coordinate for $x = 0$ the equilibrium position of the block. We then define the **elastic potential energy** of the spring–block system as

Elastic potential energy ▶

$$U_s \equiv \frac{1}{2}kx^2 \tag{6.19}$$

The elastic potential energy of the system can be thought of as the energy stored in the deformed spring (one that is either compressed or stretched from its equilibrium position). The elastic potential energy stored in a spring is zero whenever the spring is undeformed ($x = 0$). Energy is stored in the spring only when the spring is either stretched or compressed. Because the elastic potential energy is proportional to x^2, we see that U_s is always positive in a deformed spring. When the block is pushed against the spring by an external agent, the elastic potential energy and the total energy of the system increases. When the spring is compressed a distance x_{max} (**Active Figure 6.18**), the elastic potential energy stored in the spring is $\frac{1}{2}kx^2_{max}$. When the block is released from rest, the spring exerts a force on the block and pushes the block to the right.

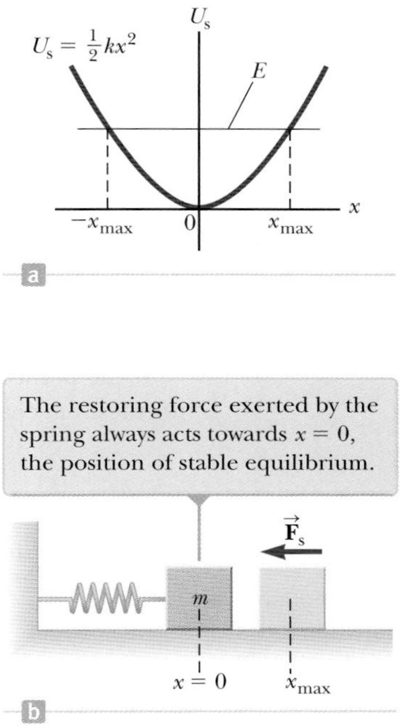

Active Figure 6.18
(a) Potential energy as a function of *x* for the frictionless block–spring system shown in (b). For a given energy *E* of the system, the block oscillates between the turning points, which have the coordinates $x = \pm x_{max}$.

TRY THIS

Hang an object from a spring or a rubber band and then pull the object down below its equilibrium point. Release the object and observe what happens. Why does the object oscillate up and down, and why does it come to a stop after a while? How much potential energy (elastic plus gravitational) was stored in the spring–mass system after you pulled down on the object compared to before? Was it more, less or the same?

Other forms of potential energy exist for other forces. We have now discussed gravitational and elastic potential energy; in later chapters we shall meet electrostatic potential energy, which is important in systems containing charged particles; nuclear potential energy, which is important in the nucleus; and magnetic potential energy, which is important when moving charged particles and magnetic materials interact.

Quick Quiz 6.7

A ball is connected to a light spring suspended vertically as shown in **Figure 6.19**. When pulled downwards from its equilibrium position and released, the ball oscillates up and down. **(i)** In the system of *the ball, the spring and the Earth*, what forms of energy are there during the motion? (a) kinetic and elastic potential (b) kinetic and gravitational potential (c) kinetic, elastic potential and gravitational potential (d) elastic potential and gravitational potential **(ii)** In the system of *the ball and the spring*, what forms of energy are there during the motion? Choose from the same possibilities (a) through (d).

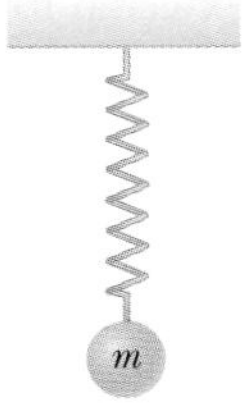

Figure 6.19
(Quick Quiz 6.7) A ball connected to a massless spring suspended vertically. What forms of potential energy are associated with the system when the ball is displaced downwards?

6.6 Energy diagrams and equilibrium of a system

The motion of a system can often be understood qualitatively through a graph of its potential energy versus the position of a member of the system. Consider the potential energy function for a block–spring system, given by $U_s = \frac{1}{2}kx^2$. This function is plotted versus x in **Active Figure 6.18a**, where x is the position of the block.

Recall that when a force acts on a system, the work done is given by **Equation 6.5**, and in the case of the elastic force, this work done appears as a change in elastic potential energy:

$$W = \int_{x_i}^{x_f} F_x\, dx = -\Delta U \tag{6.20}$$

The negative sign indicates that as the spring does work, pushing on the block, it acts to decrease the stored potential energy. This is because the force due to the spring acts to move the block back towards its equilibrium position (we call this a restoring force).

We can also write **Equation 6.20** as

$$\Delta U = U_f - U_i = -\int_{x_i}^{x_f} F_x dx \tag{6.21}$$

It is often convenient to establish some particular location x_i of one member of a system as representing a reference configuration and measure all potential energy differences with respect to it. We can then define the potential energy function as

$$U_f(x) = -\int_{x_i}^{x_f} F_x\,dx + U_i \tag{6.22}$$

Σ *Appendices B.7 and B.8 review differential and integral calculus.*

The value of U_i is often taken to be zero for the reference configuration. It does not matter what value we assign to U_i because any non-zero value merely shifts $U_f(x)$ by a constant amount and only the *change* in potential energy is physically meaningful.

If the point of application of the force undergoes an infinitesimal displacement dx, we can express the infinitesimal change in the potential energy of the system dU as

$$dU = -F_x\,dx$$

Therefore, the force is related to the potential energy through the relationship

Relationship between force and potential energy ▶

$$F_x = -\frac{dU}{dx} \tag{6.23}$$

In the case of the force F_s exerted by the spring on the block, we can say that:

$$F_s = -\frac{dU_s}{dx} = -kx$$

Therefore the x component of the force is equal to the negative of the gradient of the U versus x curve. When the block is placed at rest at the equilibrium position of the spring ($x = 0$), where $F_s = 0$, it will remain there unless some external force F_{ext} acts on it. If this external force stretches the spring from equilibrium, x is positive and the gradient dU/dx is positive; therefore, the force F_s exerted by the spring is negative and the block accelerates back towards $x = 0$ when released. If the external force compresses the spring, x is negative and the gradient is negative; therefore, F_s is positive and again the mass accelerates toward $x = 0$ upon release.

Pitfall Prevention 6.9

A common mistake is to think that potential energy on the graph in an energy diagram represents the height of some object. For example, that is not the case in Active Figure 6.18, where the block is only moving horizontally.

From this analysis, we conclude that the $x = 0$ position for a block–spring system is one of **stable equilibrium.** That is, any movement away from this position results in a force directed back towards $x = 0$. In general, configurations of a system in stable equilibrium correspond to those for which $U(x)$ for the system is a minimum.

If the block in **Active Figure 6.18** is moved to an initial position x_{max} and then released from rest, its total energy initially is the potential energy $\frac{1}{2}kx_{max}^2$ stored in the spring. As the block starts to move, the system acquires kinetic energy and loses potential energy. The block oscillates (moves back and forth) between the two points $x = -x_{max}$ and $x = +x_{max}$, called the *turning points.* In fact, if there is no friction, the block oscillates between $-x_{max}$ and $+x_{max}$ forever.

Another simple mechanical system with a configuration of stable equilibrium is a ball rolling about in the bottom of a bowl. Any time the ball is displaced from its lowest position, it tends to return to that position when released.

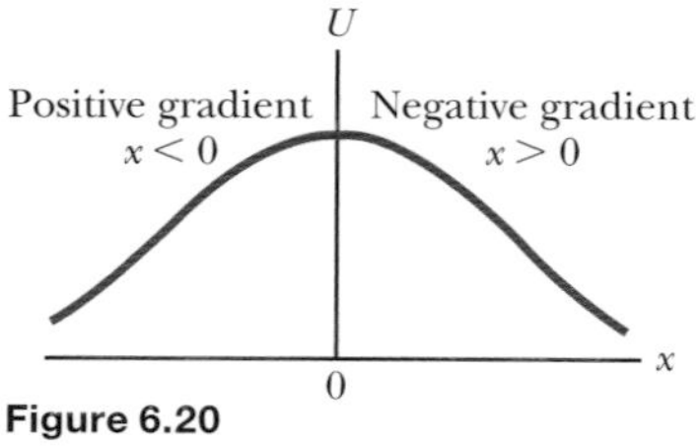

Figure 6.20
A plot of U versus x for a particle that has a position of unstable equilibrium located at $x = 0$. For any finite displacement of the particle, the force on the particle is directed away from $x = 0$.

Now consider a particle moving along the x axis under the influence of a force F_x, where the U versus x curve is as shown in **Figure 6.20**. Once again, $F_x = 0$ at $x = 0$ and so the particle is in equilibrium at this point. This position, however, is one of **unstable equilibrium** for the following reason. Suppose the particle is displaced to the right ($x > 0$). Because the gradient is negative for $x > 0$, $F_x = -dU/dx$ is positive and the particle accelerates away from $x = 0$. If instead the particle is at $x = 0$ and is displaced to the left ($x < 0$), the force is negative because the gradient is positive for $x < 0$ and the particle again accelerates away from the equilibrium position. The position $x = 0$ in this situation is one of unstable equilibrium because for any displacement from this point, the force pushes the particle farther away from equilibrium and towards a position of lower potential energy.

A pencil balanced on its point is in a position of unstable equilibrium. If the pencil is displaced slightly from its absolutely vertical position and then released, it will fall over. In general, configurations of a system in unstable equilibrium correspond to those for which $U(x)$ for the system is a maximum.

Finally, a configuration called **neutral equilibrium** arises when U is constant over some region. Small displacements of an object from a position in this region produce neither restoring nor disrupting forces. A ball lying on a flat, horizontal surface is an example of an object in neutral equilibrium.

Example 6.8

Force and energy on an atomic scale

The potential energy associated with the force between two neutral atoms in a molecule can be modelled by the Lennard-Jones potential energy function:

$$U(x) = 4\epsilon\left[\left(\frac{\sigma}{x}\right)^{12} - \left(\frac{\sigma}{x}\right)^{6}\right]$$

where x is the separation of the atoms. The function $U(x)$ contains two parameters σ and ϵ that are determined from experiments. Sample values for the interaction between two atoms in a molecule are $\sigma = 0.263$ nm and $\epsilon = 1.51 \times 10^{-22}$ J. Graph this function and find the most likely distance between the two atoms.

Solution

Conceptualise Based on our knowledge that stable molecules exist, we expect to find a stable equilibrium when the two atoms are separated by some equilibrium distance.

Model We identify the two atoms in the molecule as a system. Stable equilibrium exists for a separation distance at which the potential energy of the system of two atoms (the molecule) is a minimum, and the force is zero.

Analyse Equation 6.23 relates the force to the potential energy: $F = \frac{-dU(x)}{dx}$. Take the derivative of the function $U(x)$:

$$\frac{dU(x)}{dx} = 4\epsilon\frac{d}{dx}\left[\left(\frac{\sigma}{x}\right)^{12} - \left(\frac{\sigma}{x}\right)^{6}\right] = 4\epsilon\left[\frac{-12\sigma^{12}}{x^{13}} + \frac{6\sigma^{6}}{x^{7}}\right]$$

Minimise the function $U(x)$ by setting its derivative equal to zero:

$$4\epsilon\left[\frac{-12\sigma^{12}}{x_{eq}^{13}} + \frac{6\sigma^{6}}{x_{eq}^{7}}\right] = 0 \rightarrow x_{eq} = 2^{1/6}\sigma$$

Both sides have dimensions [L] as σ is in units of nm ☺.

Evaluate x_{eq}, the equilibrium separation of the two atoms in the molecule:

$$x_{eq} = (2)^{1/6}(0.263\text{ nm}) = 2.95 \times 10^{-10}\text{ m}$$

We graph the Lennard-Jones function on both sides of this critical value to create our energy diagram as shown in Figure 6.21.

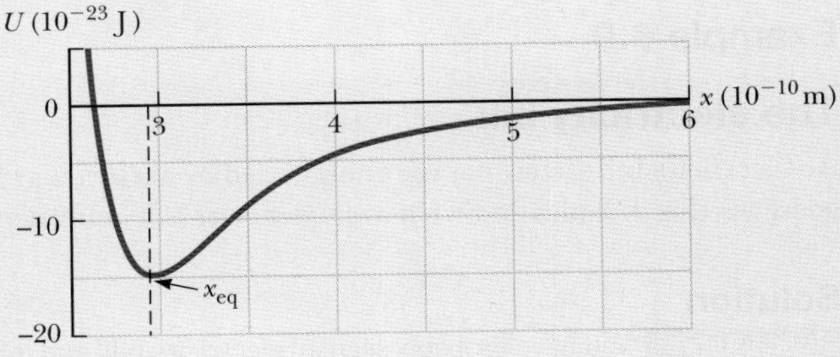

Figure 6.21
(Example 6.8) Potential energy curve associated with a molecule. The distance x is the separation between the two atoms making up the molecule.

Finalise Notice that $U(x)$ is extremely large when the atoms are very close together, is a minimum when the atoms are at their critical separation and then increases again as the atoms move apart. When $U(x)$ is a minimum, the atoms are in stable equilibrium, indicating that the most likely separation between them occurs at this point.

6.7 Power

Consider Conceptual Example 6.6 again, which involved rolling a refrigerator up a ramp into a truck. Suppose the man is not convinced the work is the same regardless of the ramp's length and sets up a long ramp with a gentle rise. Although he does the same amount of work as someone using a shorter ramp, he takes longer to do the work because he has to move the refrigerator over a greater distance. Although the work done on both ramps is the same, there is *something* different about the tasks: the *time interval* during which the work is done.

The time rate of energy transfer is called the **instantaneous power** P and is defined as

Definition of ▶ power

$$P \equiv \frac{dE}{dt} \tag{6.24}$$

where E is the energy being transferred.

The SI unit of power is joules per second (J/s), also called the **watt** (W) after James Watt:

The watt ▶

$$1 \text{ W} = 1 \text{ J/s} = 1 \text{ kg.m}^2/\text{s}^3$$

It has the dimensions $ML^2\,T^{-3}$.

> **Pitfall Prevention 6.10**
>
> **W, *W* and watts**
>
> Do not confuse the symbol W for the watt with the italic symbol *W* for work. Also, remember that the watt already represents a rate of energy transfer, so 'watts per second' does not make sense. The watt is *the same as* a joule per second.

We will focus on work as the energy transfer method in this discussion, but keep in mind that the notion of power is valid for *any* means of energy transfer. If an external force is applied to an object (which we model as a particle) and if the work done by this force on the object in the time interval Δt is W, the **average power** during this interval is

$$P_{\text{avg}} = \frac{W}{\Delta t}$$

Therefore, in Conceptual Example 6.6, although the same work is done in rolling the refrigerator up both ramps, less power is required for the longer ramp.

In a manner similar to the way in which we approached the definition of velocity and acceleration, the instantaneous power is the limiting value of the average power as Δt approaches zero:

$$P = \lim_{\Delta t \to 0} \frac{W}{\Delta t} = \frac{dW}{dt}$$

where we have represented the infinitesimal value of the work done by dW. We find from **Equation 6.4** that $dW = \vec{\mathbf{F}} \cdot d\vec{\mathbf{r}}$. Therefore, the instantaneous power can be written

$$P = \frac{dW}{dt} = \vec{\mathbf{F}} \cdot \frac{d\vec{\mathbf{r}}}{dt} = \vec{\mathbf{F}} \cdot \vec{\mathbf{v}} \tag{6.25}$$

where $\vec{\mathbf{v}} = d\vec{\mathbf{r}}/dt$ is the velocity of the object on which the force is acting.

Σ *See Appendix B.7 for the definition of the derivative.*

Example 6.9

The electricity bill

An electricity bill states that the energy used by a particular household is 1750 kWh over a 90 day period. How much energy is this in joules and what average power is it equivalent to?

Solution

Conceptualise You have probably seen an electricity bill, and if not, try to have a look at the one for the house in which you live. The energy usage above is a fairly typical electricity usage for a household of four people.

This is simply a numerical substitution problem, with a conversion of units required.

One kilowatt-hour (kWh) is the energy transferred in 1 h at the constant rate of 1 kW = 1000 J/s. Therefore the amount of energy represented by 1 kWh is

$$1 \text{ kWh} = (10^3 \text{ W})(3600 \text{ s}) = 3.60 \times 10^6 \text{ J}$$

$$1750 \text{ kWh} = 1750 \text{ kWh} \times 3.60 \times 10^6 \text{ J/kWh} = 6.3 \times 10^9 \text{ J}$$

The energy supplied is 6.3×10^9 J.

This is used over a period of 90 days: $90 \text{ days} \times 24 \text{ h/day} \times 60 \text{ min/h} \times 60 \text{ s/min} = 7.776 \times 10^6 \text{ s}$

Hence the average power is: $P = dE/dt = 6.3 \times 10^9 \text{ J} / 7.776 \times 10^6 \text{ s} = 810 \text{ W}$

Finalise Does this sound reasonable? A heater typically uses around 1000 W, and appliances such as clothes dryers and washing machines use even more. Some appliances such as refrigerators run all the time so even at night, when everyone is asleep, and lights, TVs and computers are mostly turned off, some energy is still being used. Try doing the same calculation for your own household!

Example 6.10

Power delivered by a lift motor

A lift has a mass of 1600 kg and is carrying passengers having a combined mass of 200 kg. A constant friction force of 4000 N retards its motion. How much power must a motor deliver to lift the lift and its passengers at a constant speed of 3.00 m/s?

Solution

Conceptualise Start by drawing a diagram showing the physical situation. Then draw a free-body diagram showing the forces acting on the system – in this case the lift and the passengers inside it. The motor must supply the force of magnitude T that pulls the elevator car upwards.

Model The friction force increases the power necessary to raise the lift. The problem states that the speed of the lift is constant, which tells us that $a = 0$, so we model the lift as a particle in equilibrium.

Analyse The free-body diagram in **Figure 6.22b** specifies the upwards direction as positive. The *total* mass M of the system (lift plus passengers) is equal to 1800 kg.

Using the particle in equilibrium model, apply Newton's second law to the system:

$$\sum F_y = T - f - Mg = 0$$

Solve for T: $T = f + Mg$

Use **Equation 6.25** and note that $\vec{\mathbf{T}}$ is in the same direction as $\vec{\mathbf{v}}$ to find the power:

$$P = \vec{\mathbf{T}} \cdot \vec{\mathbf{v}} = Tv = (f + Mg)v$$

Check dimensions: $[ML^2T^{-3}] = ([MLT^{-2}] + [M][LT^{-2}])[LT^{-1}] = [ML^2T^{-3}]$ ☺

Substitute numerical values:

$$P = [(4000 \text{ N}) + (1800 \text{ kg})(9.80 \text{ m/s}^2)](3.00 \text{ m/s}) = 6.49 \times 10^4 \text{ W}$$

Finalise This is 65 kW, which is large compared to the amount of power a person would require to simply walk up the stairs, which would also be better exercise. The velocity, however, at 3.00 m/s, is quite a lot faster than most people could go climbing up stairs.

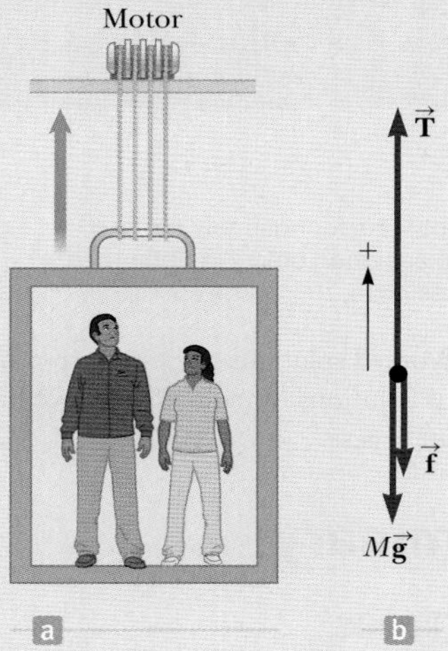

Figure 6.22
(Example 6.10) (a) The motor exerts an upwards force $\vec{\mathbf{T}}$ on the lift. The magnitude of this force is the tension T in the cable connecting the lift and motor. The downward forces acting on the car are a friction force $\vec{\mathbf{f}}$ and the gravitational force $\vec{\mathbf{F}}_g = M\vec{\mathbf{g}}$. (b) The free-body diagram for the lift

End-of-chapter resources

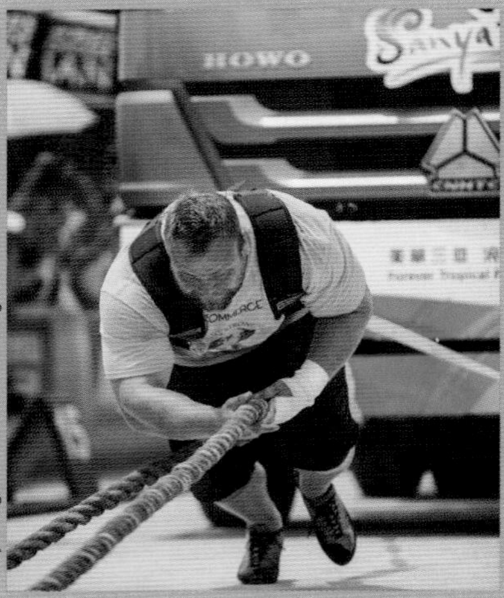

Getty Images/Victor Fraile/Stringer

At the beginning of this chapter we asked about how much work is done by a strongman in pulling a truck. At the very least, he needs to apply sufficient force to accelerate the truck up to a good speed; this is a race after all. This estimate ignores the work done in stretching the rope, heating the axle and the work done to overcome internal frictional forces, such as in bearings, and many other processes. Note that it is the friction between the road and tyres that makes the wheels rotate, hence this is not a force that needs to be 'overcome'. However, some energy is converted into the rotational kinetic energy of the wheels due to this force.

The competitor needs to keep his body as close to horizontal as possible so that the force he exerts on the truck is all in the horizontal component; he clearly understands that force is a vector. We can then make a rough estimate of the minimum amount of energy required by assuming the truck moves 25 metres starting from rest in 60 seconds. This gives an acceleration of 0.014 m.s^{-2}, and a final speed of 0.7 m.s^{-1}. Applying the work–kinetic energy theorem, this means the man has done around 6000 J of work on the truck.

The human body is thought to be about 25 per cent efficient in converting chemical energy to work, so he will have used at least four or five times this amount of energy in pulling the truck.

The problems found in this chapter may be assigned online in Enhanced Web Assign.

ENHANCED WebAssign

Worked solutions to every fifth problem are available in the Student Solutions Manual. Register online at **www.cengagebrain.com** for access.

Summary

Definitions

A **system** may be a single particle, a collection of particles or a region of space, and may vary in size and shape. A **system boundary** separates the system from the **environment**.

The **work** W done on a system by a constant applied force is the vector dot product of the force and the displacement of the point of application of the force

$$W = \vec{\mathbf{F}} \cdot \Delta\vec{\mathbf{r}} \quad (6.1)$$

The **scalar product** (dot product) of two vectors $\vec{\mathbf{A}}$ and $\vec{\mathbf{B}}$ is defined by the relationship

$$\vec{\mathbf{A}} \cdot \vec{\mathbf{B}} = AB\cos\theta \quad (6.2)$$

where the result is a scalar quantity and θ is the angle between the two vectors. The scalar product obeys the commutative and distributive laws.

If a varying force does work on a particle as the particle moves along the x axis from x_i to x_f, the work done by the force on the particle is given by

$$W = \int_{x_i}^{x_f} F_x\, dx \quad (6.5)$$

where F_x is the component of force in the x direction.

The **kinetic energy** of a particle of mass m moving with a speed v is

$$K \equiv \frac{1}{2}mv^2 \quad (6.13)$$

If a particle of mass m is at a distance y above the Earth's surface, the **gravitational potential energy** of the particle–Earth system is

$$U_g \equiv mgy \quad (6.16)$$

The **elastic potential energy** stored in a spring of force constant k is

$$U_s \equiv \frac{1}{2}kx^2 \quad (6.19)$$

The **instantaneous power** P is defined as the time rate of energy transfer:

$$P \equiv \frac{dE}{dt} \quad (6.24)$$

Concepts and principles

The **work–kinetic energy theorem** states that if work is done on a system by external forces and the only change in the system is in its speed,

$$W_{ext} = K_f - K_i = \Delta K = \frac{1}{2}mv_f^2 - \frac{1}{2}mv_i^2 \quad (6.12, 6.14)$$

A **potential energy function** U can be associated with a force. If a conservative force $\vec{\mathbf{F}}$ acts between members of a system while one member moves along the x axis from x_i to x_f, the change in the potential energy of the system equals the negative of the work done by that force:

$$U_f - U_i = -\int_{x_i}^{x_f} F_x\, dx \quad (6.21)$$

Force is related to potential energy through the relationship:

$$F_x = -\frac{dU}{dx}$$

Systems can be in three types of equilibrium configurations when the net force on a member of the system is zero. Configurations of **stable equilibrium** correspond to those for which $U(x)$ is a minimum.

Configurations of **unstable equilibrium** correspond to those for which $U(x)$ is a maximum.

Neutral equilibrium arises when U is constant as a member of the system moves over some region.

Chapter review quiz

To help you revise Chapter 6: Work, force and energy, complete the automatically graded Chapter review quiz at http://login.cengagebrain.com.

Conceptual questions

1. Give two examples in which a force is exerted on an object without doing any work on the object.
2. A certain uniform spring has spring constant k. Now the spring is cut in half. What is the relationship between k and the spring constant k' of each resulting smaller spring? Explain your reasoning.
3. (a) For what values of the angle θ between the two vectors force and displacement is their scalar product, the work done by the force, positive? (b) For what values of θ is the work negative?
4. Can kinetic energy be negative? Explain your answer.
5. Can a normal force do work? If not, why not? If so, give an example.
6. If only one external force acts on a particle, does it necessarily change the particle's (a) kinetic energy? (b) velocity?
7. Does the kinetic energy of an object depend on the frame of reference in which its motion is measured? Give an example to demonstrate your answer.
8. You are reshelving books in a library. You lift a book from the floor to the top shelf. The kinetic energy of the book on the floor was zero and the kinetic energy of the book on the top shelf is zero, so no change occurs in the kinetic energy, yet you did some work in lifting the book. Is the work–kinetic energy theorem violated? Explain your answer.
9. Preparing to clean them, you pop all the removable keys off a computer keyboard. Each key has the shape of a tiny box with one side open. By accident, you spill the keys onto the floor. Explain why many more keys land letter-side down than land open-side down.
10. What shape would the graph of U versus x have if a particle were in a region of neutral equilibrium? Sketch an example of this.
11. A bulldozer pushes on a stone. Assume the bulldozer does 15.0 J of work on the stone. Does the stone do work on the bulldozer? Explain your answer. If possible, determine how much work and explain your reasoning.
12. Express the units of the force constant of a spring in SI fundamental units.

Problems

Section 6.2 Work done by a constant force

1. A block of mass $m = 2.50$ kg is pushed a distance $d = 2.20$ m along a frictionless, horizontal table by a constant applied force of magnitude $F = 16.0$ N directed at an angle $\theta = 25.0°$ below the horizontal as shown in Figure P6.1. Determine the work done on the block by (a) the applied force, (b) the normal force exerted by the table, (c) the gravitational force and (d) the net force on the block.

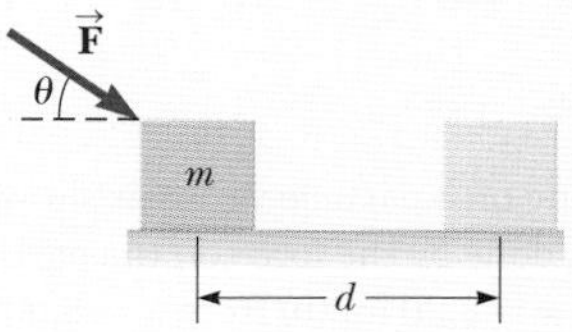

Figure P6.1

2. A group of students is doing an experiment to measure the drag force acting on raindrops of various sizes. A particular drop has a mass of $(3.4 \pm 0.1) \times 10^{-5}$ kg. It falls a distance of (100 ± 1) m at constant speed under the influence of gravity and air resistance. What is the work done on the raindrop (a) by the gravitational force and (b) by air resistance?
3. In 1990, Walter Arfeuille of Belgium lifted a 281.5 kg object through a distance of 17.1 cm using only his teeth. (a) How much work was done on the object by Arfeuille in this lift, assuming the object was lifted at constant speed? (b) What average force was exerted on Arfeuille's teeth during the lift?
4. A shopper in a supermarket pushes a trolley with a force of 35 N directed at an angle of 25° below the horizontal. The force is just sufficient to balance various friction forces, so the trolley moves at constant speed. (a) Find the work done by the shopper on the trolley as she moves down a 50.0 m aisle. (b) What is the net work done on the trolley by all forces? Why? (c) The shopper goes down the next aisle, pushing horizontally and maintaining the same speed as before. If the friction force doesn't change, would the shopper's applied force be larger, smaller or the same? (d) What about the work done on the trolley by the shopper?
5. Spiderman, whose mass is 80.0 kg, is dangling on the free end of a 12.0 m rope, the other end of which is fixed to a tree limb above. By repeatedly bending at the waist, he is able to get the rope in motion, eventually getting it to swing enough that he can reach a ledge when the rope makes a 60.0° angle with the vertical. How much work was done by the gravitational force on Spiderman in this manoeuvre?
6. A force $\vec{F} = 6\hat{i} - 2\hat{j}$ N acts on a particle that undergoes a displacement $\Delta\vec{r} = 3\hat{i} + \hat{j}$ m. Find (a) the work done by the force on the particle and (b) the angle between $\vec{F}$ and $\Delta\vec{r}$.

Section 6.3 Work done by a varying force

7. A particle is subject to a force F_x that varies with position as shown in Figure P6.7. Find the work done by the force on the particle as it moves (a) from $x = 0$ to $x = 5.00$ m, (b) from

$x = 5.00$ m to $x = 10.0$ m, and (c) from $x = 10.0$ m to $x = 15.0$ m. (d) What is the total work done by the force over the distance $x = 0$ to $x = 15.0$ m?

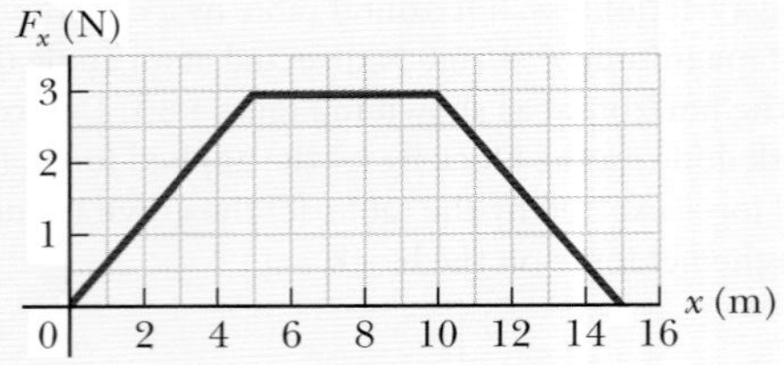

Figure P6.7

8. The force acting on a particle varies as shown in **Figure P6.8**. Find the work done by the force on the particle as it moves (a) from $x = 0$ to $x = 8.00$ m, (b) from $x = 8.00$ m to $x = 10.0$ m, and (c) from $x = 0$ to $x = 10.0$ m.

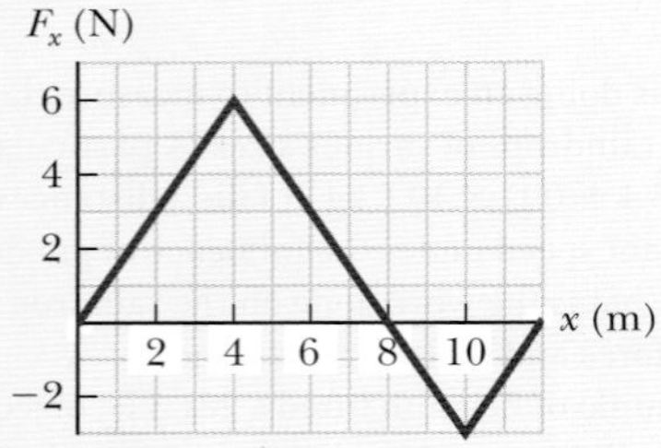

Figure P6.8

9. A group of students is doing an experiment to find a suitable spring to use in a model car's suspension system. When a (4.00 ± 0.01)-kg weight is hung vertically on a certain light spring, the spring stretches (2.5 ± 0.1) cm. If the weight is removed, (a) how far will the spring stretch if a 1.50 kg block is hung on it? (b) How much work must an external agent do to stretch the same spring 4.00 cm from its unstretched position?

10. A light spring with spring constant k_1 is hung from an elevated support. From its lower end a second light spring is hung, which has spring constant k_2. We describe these springs as 'in series'. An object of mass m is hung at rest from the lower end of the second spring. (a) Find the total extension distance of the pair of springs. (b) Find the effective spring constant of the pair of springs as a system.

11. When different loads hang on a spring, the spring stretches to different lengths as shown in the following table. (a) Make a graph of the applied force versus the extension of the spring. (b) Determine the straight line that best fits the data. (c) To complete part (b), do you want to use all the data points, or should you ignore some of them? Explain. (d) From the slope of the best-fit line, find the spring constant k. (e) If the spring is extended to 105 mm, what force does it exert on the suspended object?

F **(N)**	2.0	4.0	6.0	8.0	10	12	14	16	18	20	22
L **(mm)**	15	32	49	64	79	98	112	126	149	175	190

12. The force acting on a particle is $F_x = (8x - 16)$, where F is in newtons and x is in metres. (a) Make a plot of this force versus x from $x = 0$ to $x = 3.00$ m. (b) From your graph, find the net work done by this force on the particle as it moves from $x = 0$ to $x = 3.00$ m.

13. A force $\vec{\mathbf{F}} = 4x\hat{\mathbf{i}} + 3y\hat{\mathbf{j}}$, where $\vec{\mathbf{F}}$ is in newtons and x and y are in metres, acts on an object as the object moves in the x direction from the origin to $x = 5.00$ m. Find the work $W = \int \vec{\mathbf{F}} \cdot d\vec{\mathbf{r}}$ done by the force on the object.

14. A light spring with force constant 3.85 N/m is compressed by 8.00 cm as it is held between a 0.250 kg block on the left and a 0.500 kg block on the right, both resting on a horizontal surface. The spring exerts a force on each block, tending to push the blocks apart. The blocks are simultaneously released from rest. Find the acceleration with which each block starts to move, given that the coefficient of kinetic friction between each block and the surface is (a) 0, (b) 0.100 and (c) 0.462.

15. A 6000 kg freight car rolls along rails with negligible friction. The car is brought to rest by a combination of two coiled springs as illustrated in **Figure P6.15**. Both springs are described by Hooke's law and have spring constants $k_1 = 1600$ N/m and $k_2 = 3400$ N/m. After the first spring compresses a distance of 30.0 cm, the second spring acts with the first to increase the force as additional compression occurs as shown in the graph. The car comes to rest 50.0 cm after first contacting the two-spring system. Find the car's initial speed.

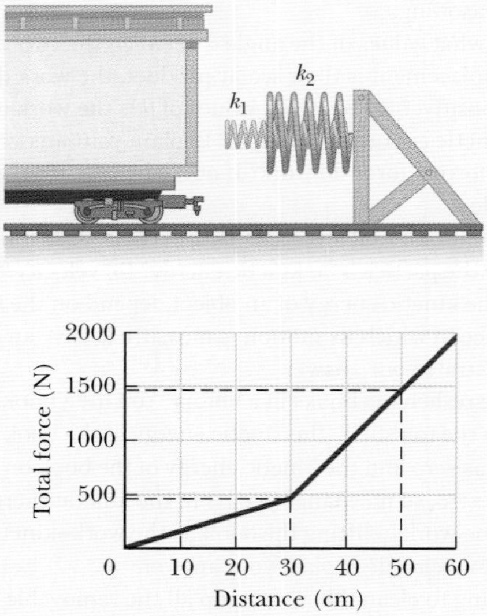

Figure P6.15

16. A small particle of mass m is pulled to the top of a frictionless half-cylinder of radius R by a light cord that passes over the top of the cylinder as illustrated in **Figure P6.16**. (a) Assuming the particle moves at a constant speed, show that $F = mg \cos \theta$. *Note:* If the particle moves at constant speed, the component of its acceleration tangent to the cylinder must be zero at all times. (b) By directly integrating $W = \int \vec{\mathbf{F}} \cdot d\vec{\mathbf{r}}$, find the work done in moving the particle at constant speed from the bottom to the top of the half-cylinder.

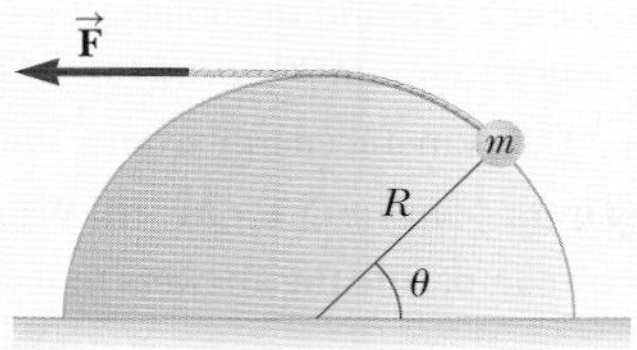

Figure P6.16

Section 6.4 Kinetic energy and the work–kinetic energy theorem

17. A 0.600-kg particle has a speed of 2.00 m/s at point Ⓐ and kinetic energy of 7.50 J at point Ⓑ. What is (a) its kinetic energy at Ⓐ, (b) its speed at Ⓑ, and (c) the net work done on the particle by external forces as it moves from Ⓐ to Ⓑ?
18. ±? A student is cleaning out their room and getting rid of their unwanted books. They push a (35 ± 1) kg box of chemistry books at a constant speed for (12.0 ± 0.5) m along a wood floor and do (350 ± 10) J of work by applying a constant horizontal force of magnitude F on the box. (a) Determine the value of F. (b) If the student now applies a force greater than F, describe the subsequent motion of the box. (c) Describe what would happen to the box if the applied force is less than F.
19. A 2100 kg pile driver is used to drive a steel I-beam into the ground. The pile driver falls 5.00 m before coming into contact with the top of the beam and it drives the beam 12.0 cm farther into the ground before coming to rest. Using energy considerations, calculate the average force the beam exerts on the pile driver while the pile driver is brought to rest.
20. In an electron microscope, there is an electron gun that contains two charged metallic plates 2.80 cm apart. An electric force accelerates each electron in the beam from rest to 9.60% of the speed of light over this distance. (a) Determine the kinetic energy of the electron as it leaves the electron gun. Electrons carry this energy to a phosphorescent viewing screen where the microscope's image is formed, making it glow. For an electron passing between the plates in the electron gun, determine (b) the magnitude of the constant electric force acting on the electron, (c) the acceleration of the electron and (d) the time interval the electron spends between the plates.
21. A 3.00-kg object has a velocity $(6.00\,\hat{\mathbf{i}} - 2.00\,\hat{\mathbf{j}})$ m/s. (a) What is its kinetic energy at this moment? (b) What is the net work done on the object if its velocity changes to $(8.00\,\hat{\mathbf{i}} + 4.00\,\hat{\mathbf{j}})$ m/s? (*Note:* From the definition of the dot product, $v^2 = \vec{\mathbf{v}} \cdot \vec{\mathbf{v}}$.)

Section 6.5 Potential energy of a system

22. A 0.20 kg stone is held 1.3 m above the top edge of a water well and then dropped into it. The well has a depth of 5.0 m. Relative to the configuration with the stone at the top edge of the well, what is the gravitational potential energy of the stone–Earth system (a) before the stone is released and (b) when it reaches the bottom of the well? (c) What is the change in gravitational potential energy of the system from release to reaching the bottom of the well?
23. ±? A (400 ± 20) N child is in a swing that is attached to a pair of ropes (2.0 ± 0.1) m long. Find the gravitational potential energy of the child–Earth system relative to the child's lowest position when (a) the ropes are horizontal, (b) the ropes make a 30° angle with the vertical and (c) the child is at the bottom of the circular arc.
24. A 'bungee fish' is a toy made by filling a round balloon with sand, adding eyes and a fish tail, and hanging the fish from a chain of tied together balloons. When the fish is pulled, it will oscillate up and down as if it were on a spring. Consider a bungee fish of mass 0.20 kg on a balloon bungee cord with spring constant 250 N.m^{-1}. The fish is pulled downwards from its equilibrium position, stretching the cord by a distance 2.0 cm. (a) What is the change in gravitational potential energy of the fish–cord–Earth system? (b) What is the change in elastic potential energy of the fish–cord–Earth system? (c) By how much would the gravitational and elastic potential energies of the same system change if the fish was pulled down by 4 cm?
25. Consider the bungee fish toy described in Problem 24. Sketch a graph of elastic potential energy and gravitational potential energy and the sum of these for the fish–cord–Earth system as a function of displacement of the fish from its equilibrium position.
26. The potential energy of a system of two particles separated by a distance r is given by $U(r) = A/r$, where A is a constant. Find the radial force $\vec{\mathbf{F}}_r$ that each particle exerts on the other.
27. A potential energy function for a system in which a two-dimensional force acts is of the form $U = 3x^3y - 7x$. Find the force that acts at the point (x, y).

Section 6.6 Energy diagrams and equilibrium of a system

28. For the potential energy curve shown in Figure P6.28, (a) determine whether the force F_x is positive, negative or zero at the five points indicated. (b) Indicate points of stable, unstable and neutral equilibrium. (c) Sketch the curve for F_x versus x from $x = 0$ to $x = 9.5$ m.

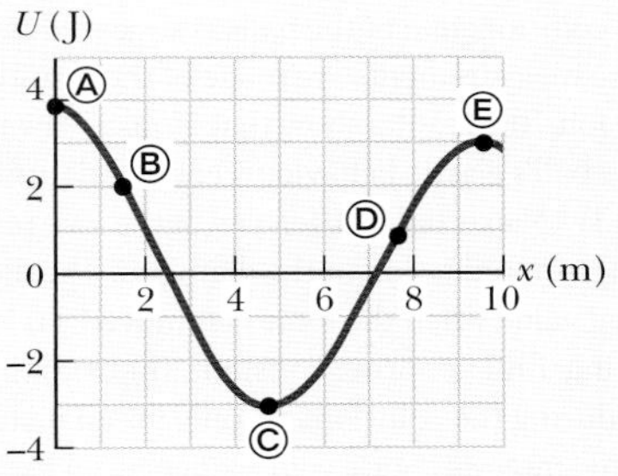

Figure P6.28

29. A right circular cone can theoretically be balanced on a horizontal surface in three different ways. Sketch these three equilibrium configurations and identify them as positions of stable, unstable or neutral equilibrium.

Section 6.7 Power

30. A certain rain cloud at an altitude of 1.75 km contains 3.20×10^7 kg of water vapour. How long would it take a 2.70 kW pump to raise the same amount of water from the Earth's surface to the cloud's position?
31. An 820-N Marine in basic training climbs a 12.0-m vertical rope at a constant speed in 8.00 s. What is his power output?
32. Make an order-of-magnitude estimate of the power a car engine contributes to speeding the car up to highway speed. In your solution, state the physical quantities you take as data and the values you measure or estimate for them.

33. A loaded ore car has a mass of 950 kg and rolls on rails with negligible friction. It starts from rest and is pulled up a mine shaft by a cable connected to a winch. The shaft is inclined at 30.0° above the horizontal. The car accelerates uniformly to a speed of 2.20 m/s in 12.0 s and then continues at constant speed. (a) What power must the winch motor provide when the car is moving at constant speed? (b) What maximum power must the winch motor provide? (c) What total energy has transferred out of the motor by work by the time the car moves off the end of the track, which is of length 1250 m?

Additional problems

34. A baseball outfielder throws a 0.150 kg baseball at a speed of 40.0 m/s and an initial angle of 30.0° to the horizontal. What is the kinetic energy of the baseball at the highest point of its trajectory?

35. The spring constant of a car's suspension system increases with increasing load due to a spring coil that is widest at the bottom, smoothly tapering to a smaller diameter near the top. The result is a softer ride on normal road surfaces from the wider coils, but the car does not bottom out on bumps because when the lower coils collapse, the stiffer coils near the top absorb the load. For such springs, the force exerted by the spring can be empirically found to be given by $F = ax^b$. For a tapered spiral spring that compresses 12.9 cm with a 1000 N load and 31.5 cm with a 5000 N load, (a) evaluate the constants a and b in the empirical equation for F and (b) find the work needed to compress the spring 25.0 cm.

36. ±? A group of students is investigating collisions between steel balls. Two identical steel balls, each of diameter (25.4 ± 0.1) mm and moving in opposite directions at (5.0 ± 0.2) m/s, run into each other head-on and bounce apart. Prior to the collision, one of the balls is squeezed in a vice while precise measurements are made of the resulting amount of compression. The results show that Hooke's law is a fair model of the ball's elastic behaviour. For one datum, a force of (16.0 ± 0.1) kN exerted by each jaw of the vice results in a (0.20 ± 0.01) mm reduction in the diameter. The diameter returns to its original value when the force is removed. (a) Modelling the ball as a spring, find its spring constant. (b) Does the interaction of the balls during the collision last only for an instant or for a non-zero time interval? State your evidence. (c) Estimate the kinetic energy of each of the balls before they collide. (d) Estimate the maximum amount of compression each ball undergoes when the balls collide. (e) Calculate an order-of-magnitude estimate for the time interval for which the balls are in contact.

37. The potential energy function for a system of particles is given by $U(x) = -x^3 + 2x^2 + 3x$, where x is the position of one particle in the system. (a) Determine the force F_x on the particle as a function of x. (b) For what values of x is the force equal to zero? (c) Plot $U(x)$ versus x and F_x versus x and indicate points of stable and unstable equilibrium.

38. Two constant forces act on an object of mass $m = 5.00$ kg moving in the xy plane as shown in **Figure P6.38**. Force $\vec{\mathbf{F}}_1$ is 25.0 N at 35.0°, and force $\vec{\mathbf{F}}_2$ is 42.0 N at 150°. At time $t = 0$, the object is at the origin and has velocity $(4.00\hat{\mathbf{i}} + 2.5\hat{\mathbf{j}})$ m/s. (a) Express the two forces in unit-vector notation. Use unit-vector notation for your other answers. (b) Find the total force exerted on the object. (c) Find the object's acceleration. Now, considering the instant $t = 3.00$ s, find (d) the object's velocity, (e) its position, (f) its kinetic energy from $\frac{1}{2}mv_f^{\,2}$, and (g) its kinetic energy from $\frac{1}{2}mv_i^2 + \Sigma\vec{\mathbf{F}}\cdot\Delta\vec{\mathbf{r}}$. (h) What conclusion can you draw by comparing the answers to parts (f) and (g)?

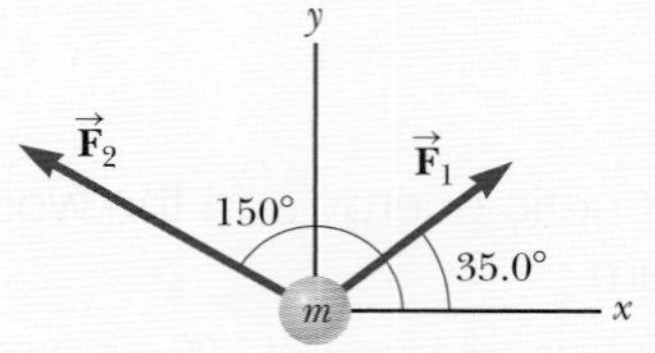

Figure P6.38

39. An inclined plane of angle θ has a spring of force constant k fastened securely at the bottom so that the spring is parallel to the surface. A block of mass m is placed on the plane at a distance d from the spring. From this position, the block is projected downwards towards the spring with speed v as shown in **Figure P6.39**. By what distance is the spring compressed when the block momentarily comes to rest?

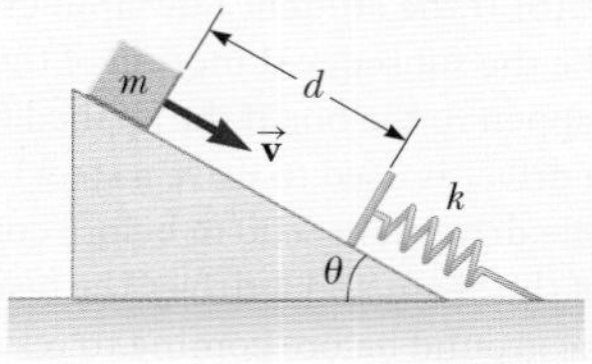

Figure P6.39

40. When an object is displaced by an amount x from stable equilibrium, a restoring force acts on it, tending to return the object to its equilibrium position. The magnitude of the restoring force can be a complicated function of x. In such cases, we can generally imagine the force function $F(x)$ to be expressed as a power series in x as $F(x) = -(k_1x + k_2x^2 + k_3x^3 + \ldots)$. The first term here is just Hooke's law, which describes the force exerted by a simple spring for small displacements. For small excursions from equilibrium, we generally ignore the higher-order terms, but in some cases it may be desirable to keep the second term as well. If we model the restoring force as $F = -(k_1x + k_2x^2)$, how much work is done on an object in displacing it from $x = 0$ to $x = x_{max}$ by an applied force $-F$?

41. When a car moves with constant speed down a highway, most of the power developed by the engine is used to compensate for the energy transformations due to friction forces exerted on the car by the air and the road. If the power developed by an engine is 175 hp, estimate the total friction force acting on the car when it is moving at a speed of 29 m/s. One horsepower (hp) equals 746 W.

42. An older-model family car accelerates from 0 to speed v in a time interval of Δt. A newer, more powerful sports car accelerates from 0 to $2v$ in the same time period. Assuming the energy coming from the engine appears only as kinetic energy of the cars, compare the power of the two cars.

Challenge problems

43. A particle of mass $m = 1.18$ kg is attached between two identical springs on a frictionless, horizontal tabletop. Both springs have spring constant k and are initially unstressed and the particle is at $x = 0$. (a) The particle is pulled a distance x along a direction perpendicular to the initial configuration of the springs as shown in **Figure P6.43**. Show that the force exerted by the springs on the particle is

$$\vec{\mathbf{F}} = -2kx\left(1 - \frac{L}{\sqrt{x^2 + L^2}}\right)\hat{\mathbf{i}}$$

(b) Show that the potential energy of the system is

$$U(x) = kx^2 + 2kL\,(L - \sqrt{x^2 + L^2}\,)$$

(c) Make a plot of $U(x)$ versus x and identify all equilibrium points. Assume $L = 1.20$ m and $k = 40.0$ N/m. (d) If the particle is pulled 0.500 m to the right and then released, what is its speed when it reaches $x = 0$?

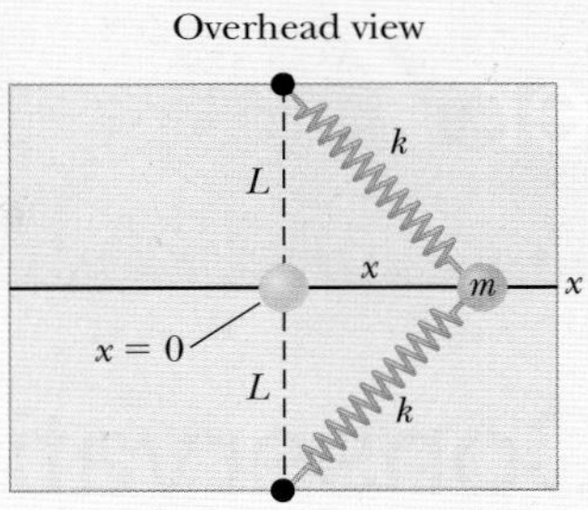

Figure P6.43

44. A light spring has unstressed length 15.5 cm. It is described by Hooke's law with spring constant 4.30 N/m. One end of the horizontal spring is held on a fixed vertical axle and the other end is attached to a puck of mass m that can move without friction over a horizontal surface. The puck is set into motion in a circle with a period of 1.30 s. (a) Find the extension of the spring x as it depends on m. Evaluate x for (b) $m = 0.0700$ kg, (c) $m = 0.140$ kg, (d) $m = 0.180$ kg, and (e) $m = 0.190$ kg. (f) Describe the pattern of variation of x as it depends on m.

chapter 7

Conservation of energy

AAP/Andrew Taylor

A wind farm near Lake George, close to Canberra in Australia.

Wind farms are increasingly being built to supply renewable energy to households and industry. What energy transformations are occuring here? If energy is conserved, then why is wind power considered renewable, while other power sources are not?

On completing this chapter, students will understand:

- how energy can be transferred into or out of a system
- that systems can be modelled as isolated or non-isolated
- that energy is conserved
- that conservative forces act to conserve mechanical energy
- that non-conservative forces act to increase internal energy in a system
- how energy bar charts can be used to represent changes in energy in a system.

Students will be able to:

- distinguish between isolated and non-isolated systems
- apply the analysis models for isolated and non-isolated systems
- solve problems using the principle of conservation of energy
- distinguish between conservative and non-conservative forces
- calculate changes in energy due to the actions of conservative and non-conservative forces
- use energy bar charts to represent changes in energy types in a system.

ENHANCED WebAssign

The problems found in this chapter may be assigned online in Enhanced Web Assign.

In Chapter 6 we introduced kinetic and potential energy, and described one way of transferring energy to a system: work. We now look at a fundamental property of energy: energy is conserved. This is the first of several conservation principles we shall describe and use.

7.1 Energy is conserved

We began the previous chapter with a discussion of systems. We looked at two ways in which energy can be stored in a system (kinetic and potential energy) and one way in which energy can be transferred to a system (work done by a force).

A third way in which a system can store energy is as *internal energy.* Internal energy is related to the temperature of the objects in a system and is a measure of the average kinetic energy of the individual particles (molecules, atoms, electrons etc.) that make up the system. We give the internal energy the symbol E_{int}. Note that internal energy is not the same as heat, which is a means of energy transfer as described below, although the word heat is often incorrectly used when internal energy is what is meant.

We now look again at systems and consider two types of systems: *isolated* and *non-isolated.*

The simplest example of a non-isolated system is a single object, modelled as a particle, which can be acted upon by various forces. This system, a single particle, interacts with its environment via forces and energy crosses the barrier of the system during these interactions.

The work–kinetic energy theorem from Chapter 6 is our first example of an energy equation appropriate for a non-isolated system. In the case of that theorem, the interaction of the system with its environment is the work done by the external force and the quantity in the system that changes is the kinetic energy.

More generally, **a non-isolated system is any system that interacts with its environment and is defined by a boundary through which energy can be transferred.**

◀ Non-isolated system

There are several mechanisms by which energy can be transferred into or out of a non-isolated system. These are shown in Figure 7.1.

Work, as we have learned in Chapter 6, is a method of transferring energy to a system by applying a force to the system such that the point of application of the force undergoes a displacement (Figure 7.1a).

Mechanical waves (Chapters 16–18) are a means of transferring energy by allowing a disturbance to propagate through air or another medium. It is the method by which energy (which you detect as sound) leaves a speaker and enters your ears to stimulate the hearing process (Figure 7.1b). Other examples of mechanical waves are seismic waves and ocean waves.

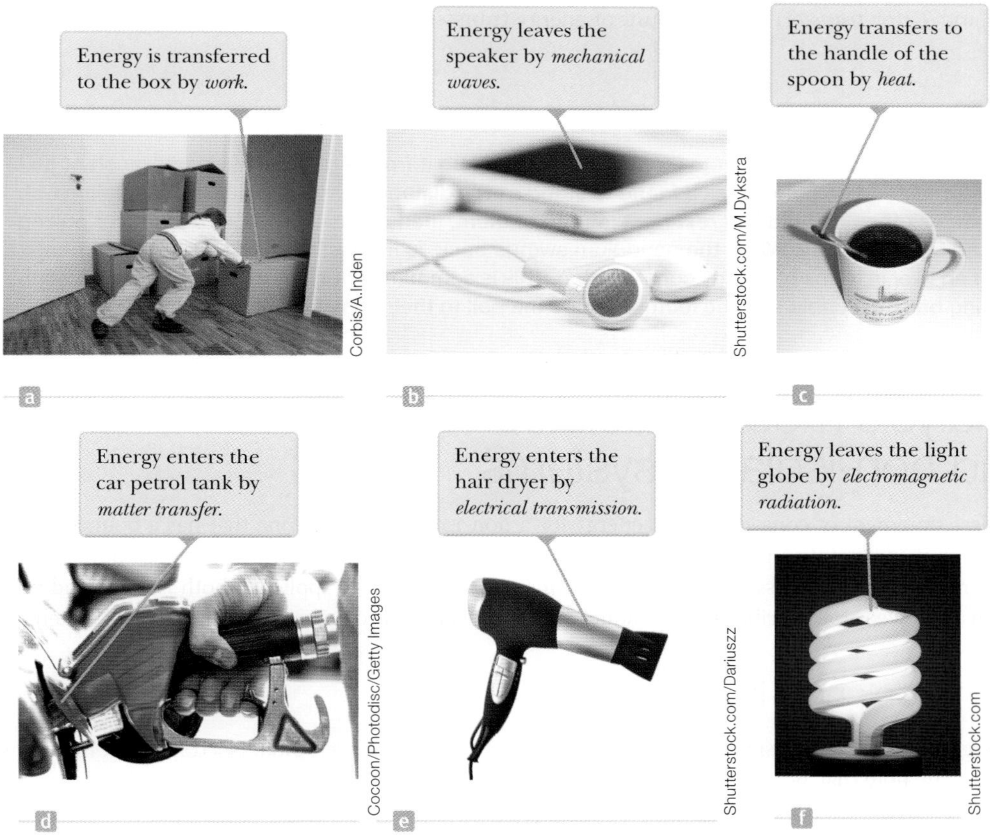
Energy is transferred to the box by *work.*
Corbis/A.Inden
a
Energy leaves the speaker by *mechanical waves.*
Shutterstock.com/M.Dykstra
b
Energy transfers to the handle of the spoon by *heat.*
c
Energy enters the car petrol tank by *matter transfer.*
Cocoon/Photodisc/Getty Images
d
Energy enters the hair dryer by *electrical transmission.*
Shutterstock.com/Dariuszz
e
Energy leaves the light globe by *electromagnetic radiation.*
Shutterstock.com
f

Figure 7.1
Energy transfer mechanisms. In each case, the system into which or from which energy is transferred is indicated.

Heat (Chapters 19 and 20) is a mechanism of energy transfer that is driven by a temperature difference between a system and its environment. For example, imagine dividing a metal spoon into two parts: the handle, which we identify as the system, and the portion submerged in a cup of coffee, which is part of the environment (Figure 7.1c). The handle of the spoon becomes hot because fast-moving electrons and atoms in the submerged portion bump into slower ones in the nearby part of the handle. These particles move faster because of the collisions and bump into the next group of slow particles. Therefore, the internal energy of the spoon handle rises as a result of energy transfer due to this collision process.

Matter transfer involves situations in which matter physically crosses the boundary of a system, carrying energy with it. Examples include filling a car's petrol tank (Figure 7.1d) or energy being carried to the rooms of a house with central heating by circulating warm air, a process called *convection* (Chapter 20).

Electrical transmission (Chapters 27 and 28) involves energy transfer into or out of a system by means of electric currents. It is how energy transfers into a hair dryer (Figure 7.1e), a stereo system or any other electrical device.

Electromagnetic radiation (Chapter 34) refers to electromagnetic waves such as light, microwaves and radio waves (Figure 7.1f) crossing the boundary of a system. Examples of this method of transfer include light energy travelling from the Sun to the Earth through space and the radio waves that carry the signal to a mobile phone.

A central feature of the energy approach is the notion that we can neither create nor destroy energy, that energy is always *conserved*. This feature has been tested in countless experiments and no experiment has ever shown this statement to be incorrect. As energy is conserved, **if the total amount of energy in a system changes, it can *only* be because energy has crossed the boundary of the system by a transfer mechanism such as one of the methods listed above**.

Energy is one of several quantities in physics that are conserved. We will see other conserved quantities in subsequent chapters. The general statement of the principle of **conservation of energy** can be described mathematically with the **conservation of energy equation** as follows:

Conservation of energy ▶

$$\Delta E_{\text{system}} = \sum T \tag{7.1}$$

where E_{system} is the total energy of the system, including all methods of energy storage (kinetic, potential, and internal) and T (for *transfer*) is the amount of energy transferred across the system boundary by some mechanism.

TRY THIS

Make a pendulum by tying an object to a cord and suspending it from a rafter, overhead branch or just standing on a chair and holding it. Ask a friend to stand a little distance away and pull the pendulum bob towards them. Ask them to hold it *at rest* right in front of them, then release it *without* pushing or pulling on it at all and not step away. When the pendulum swings back, is it possible that it will hit your friend? Explain your answer in terms of energy conservation. This is sometimes done as a lecture demonstration with a bowling ball on a cable and the lecturer holds the ball to their nose before releasing it.

7.2 Analysis models: isolated and non-isolated systems

Equation 7.1 is a statement of conservation of energy for a non-isolated system, one that has a boundary that allows the transfer of energy into and out of the system. Recall that an isolated system is one with a boundary that neither matter nor energy can cross. The principle of conservation of energy applies to both isolated and non-isolated systems, and we now describe analysis models useful for solving problems with either type of system.

Analysis model: non-isolated system

Two of the energy transfer mechanisms described in the previous section have well-established symbolic notations. For work, $T_{work} = W$; note that W is the external work, as described in Chapter 6. For internal work we shall use the symbol W_{int}. Heat, the energy transferred due to a temperature difference, is represented by Q, and described in detail in Chapter 20.

The other four members of our list do not have established symbols, so we will call them T_{MW} (mechanical waves), T_{MT} (matter transfer), T_{ET} (electrical transmission), and T_{ER} (electromagnetic radiation). Using K and U to represent kinetic and potential energy respectively (as in previous chapters), the full expansion of **Equation 7.1** is

$$\Delta K + \Delta U + \Delta E_{int} = W + Q + T_{MW} + T_{MT} + T_{ET} + T_{ER} \quad (7.2)$$

which is the primary mathematical representation of the energy version of the analysis model of the **non-isolated system**. In most cases, **Equation 7.2** reduces to a much simpler one because some of the terms are zero. If, for a given system, all terms on the right side of the conservation of energy equation are zero, the system is an *isolated system*.

Suppose a force is applied to a non-isolated system and the point of application of the force moves through a displacement. Then suppose the only effect on the system is to change its speed. In this case, the only transfer mechanism is work (so that the right side of **Equation 7.2** reduces to just W) and the only kind of energy in the system that changes is the kinetic energy (so that ΔE_{system} reduces to just ΔK). **Equation 7.2** then becomes

$$\Delta K = W$$

which is the work–kinetic energy theorem. This theorem is a special case of the more general principle of conservation of energy. We shall see several more special cases in future chapters.

Quick Quiz 7.1

By what transfer mechanisms does energy enter and leave **(a)** your television set? **(b)** a petrol-powered lawn mower? **(c)** a hand-cranked pencil sharpener?

Quick Quiz 7.2

Consider a block sliding over a horizontal surface with friction. Ignore any sound the sliding might make. **(i)** If the system is the *block*, this system is (a) isolated (b) non-isolated (c) impossible to determine. **(ii)** If the system is the *surface*, describe the system from the same set of choices. **(iii)** If the system is the *block and the surface*, describe the system from the same set of choices.

Analysis Model 7.1

Non-isolated system (energy)

Imagine you have identified a system to be analysed and have defined a system boundary. Energy can exist in the system in three forms: kinetic, potential and internal. The total of that energy can be changed when energy crosses the system boundary by any of six transfer methods shown in the diagram here. The total change in the energy in the system is equal to the total amount of energy that has crossed the system boundary. The mathematical statement of that concept is expressed in the **conservation of energy equation**:

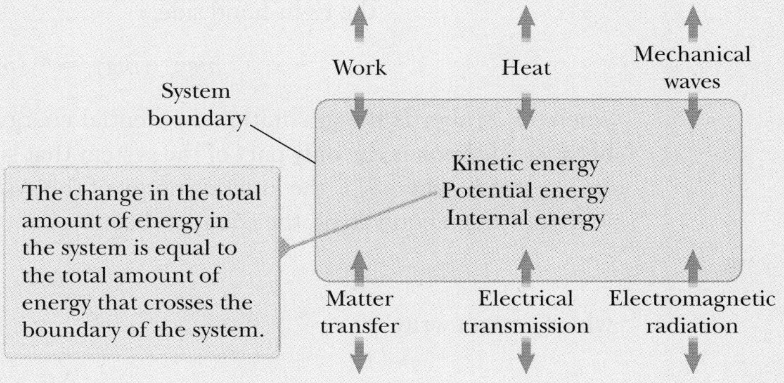
Work
Heat
Mechanical waves
System boundary
Kinetic energy
Potential energy
Internal energy
The change in the total amount of energy in the system is equal to the total amount of energy that crosses the boundary of the system.
Matter transfer
Electrical transmission
Electromagnetic radiation

$$\Delta E_{system} = \Sigma T \quad (7.1)$$

The full expansion of **Equation 7.1** shows the specific types of energy storage and transfer:

$$\Delta K + \Delta U + \Delta E_{int} = W + Q + T_{MW} + T_{MT} + T_{ET} + T_{ER} \quad (7.2)$$

For a specific problem, this equation is generally reduced to a smaller number of terms by eliminating the terms that are equal to zero because they are not appropriate to the situation.

Examples

- a force does work on a system of a single object, changing its speed: the work–kinetic energy theorem, $W = \Delta K$
- a gas contained in a vessel has work done on it and experiences a transfer of energy by heat, resulting in a change in its temperature: the first law of thermodynamics, $\Delta E_{int} = W + Q$ (Chapter 20)
- an incandescent light bulb is turned on, with energy entering the filament by electricity, causing its temperature to increase, and leaving by light: $\Delta E_{int} = T_{ET} + T_{ER}$ (Chapter 27)
- a photon enters a metal, causing an electron to be ejected from the metal: the photoelectric effect, $\Delta K + \Delta U = T_{ER}$ (Chapter 40)

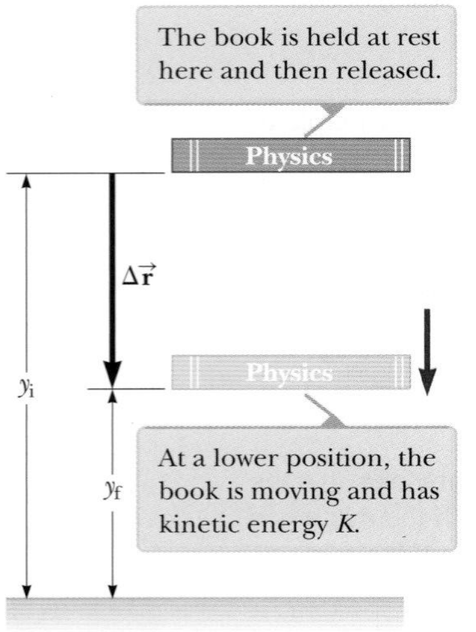
The book is held at rest here and then released.
Physics
$\Delta\vec{\mathbf{r}}$
Physics
y_i
y_f
At a lower position, the book is moving and has kinetic energy K.

Figure 7.2
A book is released from rest and falls due to work done by the gravitational force on the book

Analysis model: isolated system

Consider now an isolated system; such a system is chosen such that no energy crosses the system boundary by any method. Consider the book–Earth system in **Active Figure 6.16** in the preceding chapter. After we have lifted the book, there is gravitational potential energy stored in the system, which can be calculated from the work done by the external agent on the system, using $W = \Delta U_g$.

Let us now shift our focus to the work done *on the book alone* by the gravitational force (**Figure 7.2**) as the book falls back to its original height. As the book falls from y_i to y_f, the work done by the gravitational force on the book is

$$W_{\text{on book}} = (m\vec{\mathbf{g}}) \cdot \Delta\vec{\mathbf{r}} = (-mg\hat{\mathbf{j}}) \cdot [(y_f - y_i)\hat{\mathbf{j}}] = mgy_i - mgy_f \quad (7.3)$$

From the work–kinetic energy theorem, the work done on the book is equal to the change in the kinetic energy of the book:

$$W_{\text{on book}} = \Delta K_{\text{book}}$$

We can equate these two expressions for the work done on the book:

$$\Delta K_{\text{book}} = mgy_i - mgy_f \quad (7.4)$$

Let us now relate each side of this equation to the *system* of the book and the Earth. For the right-hand side,

$$mgy_i - mgy_f = -(mgy_f - mgy_i) = -\Delta U_g$$

where $U_g = mgy$ is the gravitational potential energy of the system. For the left-hand side of **Equation 7.4**, because the book is the only part of the system that is moving, if we treat the Earth as stationary, we see that $\Delta K_{\text{book}} = \Delta K$, where K is the kinetic energy of the system. Therefore, with each side of **Equation 7.4** replaced with its system equivalent, the equation becomes

$$\Delta K = -\Delta U_g \quad (7.5)$$

which we can write as:

$$\Delta K + \Delta U_g = 0$$

The left side represents a sum of changes of the energy stored in the system. The right-hand side is zero because there are no transfers of energy across the boundary of the system; the book–Earth system is *isolated* from the environment. We developed this equation for a gravitational system, but it can be shown to be valid for a system with any type of potential energy. Therefore, for an isolated system,

$$\Delta K + \Delta U = 0 \quad \text{or} \quad K_i + U_i = K_f + U_f \quad (7.6)$$

Quick Quiz 7.3

A rock of mass m is dropped to the ground from a height h. A second rock, with mass $2m$, is dropped from the same height. When the second rock strikes the ground, what is its kinetic energy? (a) twice that of the first rock (b) four times that of the first rock (c) the same as that of the first rock (d) half as much as that of the first rock (e) impossible to determine

Quick Quiz 7.4

Three identical balls are thrown from the top of a building, all with the same initial speed. As shown in **Active Figure 7.3**, the first is thrown horizontally, the second at some angle above the horizontal, and the third at some angle below the horizontal. Neglecting air resistance, rank the speeds of the balls at the instant each hits the ground.

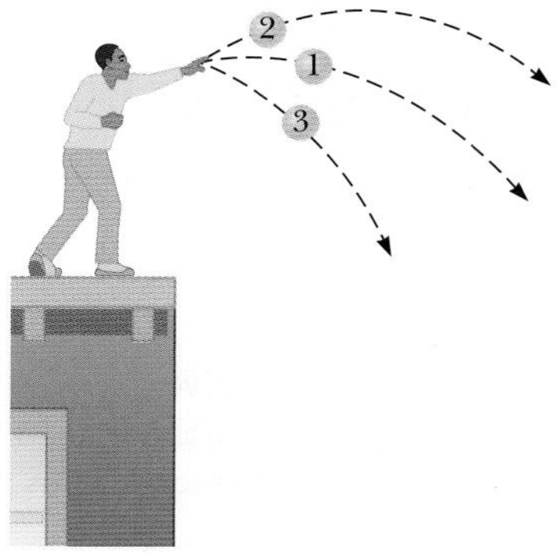

2
1
3

Active Figure 7.3
(Quick Quiz 7.4) Three identical balls are thrown with the same initial speed from the top of a building.

We define the sum of the kinetic and potential energies of a system as its mechanical energy:

$$E_{mech} = K + U \quad (7.7)$$

◀ Mechanical energy of a system

where U represents the total of *all* types of potential energy.

Note that there may be types of energy present other than mechanical energy; in particular a system will generally have some internal energy, related to the temperature of the components of the system. If frictional forces are acting then energy may be converted from mechanical energy to internal energy. For example, if you rub your hands together, friction between your hands acts to convert some kinetic energy into internal energy and your hands get warmer. If we define you as a system, then this friction force is internal to the system, and converts some mechanical energy into internal energy. So we must include the change of internal energy in our equation for conservation of energy for an isolated system:

$$\Delta K + \Delta U + \Delta E_{int} = 0 \quad (7.8)$$

These are all the different types of energy that a system can have, so we can summarise **Equation 7.8** as:

$$\Delta E_{system} = 0 \quad (7.9)$$

◀ The total energy of an isolated system is conserved.

which tells us that any increase in one type of energy must be accounted for by an equal decrease in another type.

Analysis Model 7.2

Isolated system (energy)

Imagine you have identified a system to be analysed and have defined a system boundary. Energy can exist in the system in three forms: kinetic, potential and internal. Imagine also a situation in which no energy crosses the boundary of the system by any method. Then, the system is isolated; energy transforms from one form to another and **Equation 7.8** becomes

$$\Delta E_{system} = 0 \quad (7.9)$$

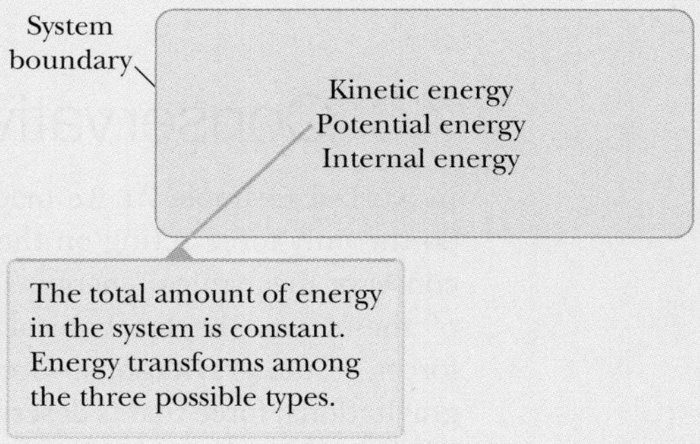

System boundary
Kinetic energy
Potential energy
Internal energy
The total amount of energy in the system is constant. Energy transforms among the three possible types.

Examples

- an object is in free fall; gravitational potential energy transforms to kinetic energy: $\Delta K + \Delta U = 0$
- a pendulum is raised and released with an initial speed; its motion eventually stops due to air resistance; gravitational potential energy and kinetic energy transform to internal energy, $\Delta K + \Delta U + \Delta E_{int} = 0$ (Chapter 16)
- a battery is connected to a resistor; chemical potential energy in the battery transforms to internal energy in the resistor: $\Delta U + \Delta E_{int} = 0$ (Chapter 27)

Example 7.1

Ball in free fall

A ball of mass m is dropped from a height h above the ground as shown in **Active Figure 7.4**.

(A) Neglecting air resistance, determine the speed of the ball when it is at a height y above the ground.

Solution

Conceptualise **Active Figure 7.4** and our everyday experience with falling objects allow us to conceptualise the situation. Although we can readily solve this problem with the techniques of Chapter 2, let us practise an energy approach.

Model We identify the system as the ball and the Earth. As we are ignoring air resistance and there are no other interactions between the system and the environment, the system is isolated and we use the isolated system model. We model the Earth as stationary, so only the ball moves, and we take the reference (zero) gravitational potential energy configuration as being the ball on the ground.

Analyse As the system is isolated, we apply the principle of conservation of energy to the ball–Earth system. At the instant the ball is released, its kinetic energy is $K_i = 0$ and the gravitational potential energy of the system is $U_{gi} = mgh$. When the ball is at a position y above the ground, its kinetic energy is $K_f = \frac{1}{2}mv_f^2$ and the potential energy of the system is $U_{gf} = mgy$.

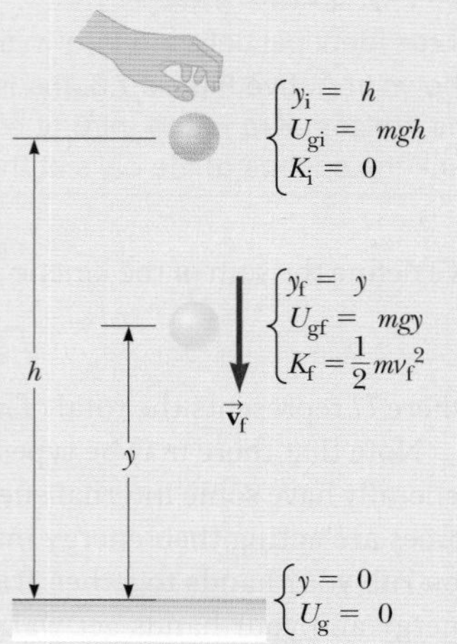

Active Figure 7.4

(Example 7.1) A ball is dropped from a height h above the ground. Initially, the total energy of the ball–Earth system is gravitational potential energy, equal to mgh relative to the ground. At the position y, the total energy is the sum of the kinetic and potential energies.

Apply **Equation 7.6**: $K_f + U_{gf} = K_i + U_{gi}$

$$\frac{1}{2}mv_f^2 + mgy = 0 + mgh$$

Solve for v_f: $v_f^2 = 2g(h-y) \Rightarrow v_f = \sqrt{2g(h-y)}$

Dimension check: $[LT^{-1}] = ([LT^{-2}][L])^{1/2} = [LT^{-1}]$ ☺

The speed is always positive. If you had been asked to find the ball's velocity, you would use the negative value of the square root to indicate the downwards motion.

Finalise This result for the final speed is consistent with the expression $v_{yf}^2 = v_{yi}^2 - 2g(y_f - y_i)$ from the particle with constant acceleration model for a falling object, where $y_i = h$.

7.3 Conservative and non-conservative forces

In worked Example 7.1 we made the approximation that air resistance was negligible on the falling ball. So the only force acting on the ball was gravity, which is a conservative force. Conservative forces act to conserve mechanical energy. They convert kinetic energy to potential energy or vice versa and the sum of these two, the mechanical energy, remains constant. We have already described two conservative forces – the gravitational force and the spring force. Field forces are conservative forces, including the gravitational force that is described in more detail in Chapter 10; the electrostatic force that we will study in Chapter 23; and nuclear forces that we will meet in Chapter 43. We often make the approximation that the only force acting when a spring or other elastic object is stretched or compressed is the spring force. In reality there is usually a non-conservative force, friction, also acting. The magnitude of the friction force may or may not be significant compared to the spring force.

Non-conservative forces, such as the friction force, act to convert some mechanical energy into thermal energy. Mechanical energy is not conserved by non-conservative forces, although *total energy* is.

Conservative forces have these two equivalent properties:

1. The work done by a conservative force on a particle moving between any two points is independent of the path taken by the particle.
2. The work done by a conservative force on a particle moving through any closed path is zero. (A closed path is one for which the beginning point and the endpoint are identical.)

◀ Properties of conservative forces

The gravitational force is one example of a conservative force; the force that an ideal spring exerts on any object attached to the spring is another. The work done by the gravitational force on an object moving between any two points near the Earth's surface is $W_g = (-mg\hat{\mathbf{j}}) \cdot [(y_f - y_i)\hat{\mathbf{j}}] = mgy_i - mgy_f$. From this equation, notice that W_g depends only on the initial and final y coordinates of the object and hence is independent of the path. Furthermore, W_g is zero when the object moves over any closed path (where $y_i = y_f$).

For the case of the object–ideal spring system, the work W_s done by the spring force is given by Equation 6.10: $W_s = \frac{1}{2}kx_i^2 - \frac{1}{2}kx_f^2$. We see that the spring force is conservative because W_s depends only on the initial and final x coordinates of the object and is zero for any closed path. Real springs generally have internal friction, but this ideal spring model is a good approximation to many systems, including the interaction between atoms on a molecular level.

Let us imagine a system of particles in which a conservative force $\vec{\mathbf{F}}$ acts between the particles. Imagine also that the configuration of the system changes due to the motion of one particle along the x axis. The work done by the force $\vec{\mathbf{F}}$ as the particle moves along the x axis is

$$W_{\text{int}} = \int_{x_i}^{x_f} F_x dx = -\Delta U \tag{7.10}$$

where F_x is the component of $\vec{\mathbf{F}}$ in the direction of the displacement. That is, the work done by a conservative force acting between members of a system equals the negative of the change in the potential energy of the system associated with that force when the system's configuration changes. We can also express Equation 7.10 as

$$\Delta U = U_f - U_i = -\int_{x_i}^{x_f} F_x dx \tag{7.11}$$

Therefore, ΔU is negative when F_x and dx are in the same direction, as when an object is lowered in a gravitational field or when a spring pushes an object towards equilibrium.

It is often convenient to establish some particular location x_i of one member of a system as representing a reference configuration and measure all potential energy differences with respect to it. We can then define the potential energy function as

$$U_f(x) = -\int_{x_i}^{x_f} F_x dx + U_i \tag{7.12}$$

The value of U_i is often taken to be zero for the reference configuration. It does not matter what value we assign to U_i because any non-zero value merely shifts $U_f(x)$ by a constant amount and only the *change* in potential energy is physically meaningful.

Appendices B.7 and B.8 review differential and integral calculus. Σ

If the point of application of the force undergoes an infinitesimal displacement dx, we can express the infinitesimal change in the potential energy of the system dU as

$$dU = -F_x dx$$

Therefore, the conservative force is related to the potential energy function through the relationship

$$F_x = -\frac{dU}{dx} \tag{7.13}$$

◀ Relation of force between members of a system to the potential energy of the system

That is, the x component of a conservative force acting on a member within a system equals the negative derivative of the potential energy of the system with respect to x.

We can easily check **Equation 7.13** for the two examples already discussed. In the case of the deformed spring, $U_s = \frac{1}{2}kx^2$; therefore,

$$F_s = -\frac{dU_s}{dx} = -\frac{d}{dx}\left(\frac{1}{2}kx^2\right) = -kx$$

which corresponds to the restoring force in the spring (Hooke's law). Because the gravitational potential energy function is $U_g = mgy$, it follows from **Equation 7.13** that $F_g = -mg$ when we differentiate U with respect to y instead of x.

Problem-solving strategy

Isolated systems with no non-conservative forces: conservation of mechanical energy

Many problems in physics can be solved using the principle of conservation of energy for an isolated system. The following procedure should be used when you apply this principle.

1. Conceptualise

Study the physical situation carefully and form a mental representation of what is happening. Draw a diagram. As you become more proficient working energy problems, you will begin to be comfortable imagining the types of energy that are changing in the system.

2. Model

Define your system, which might consist of more than one object and might or might not include springs or other means of storing potential energy. Determine if any energy transfers occur across the boundary of your system. If so, use the non-isolated system model, $\Delta E_{\text{system}} = \Sigma T$, from Section 7.1. If not, use the isolated system model, $\Delta E_{\text{system}} = 0$.

Determine whether any non-conservative forces are present and significant within the system. If so, use the techniques of the next section for non-conservative forces. If there are no non-conservative forces, or you can reasonably make the approximation that any non-conservative forces are negligible, use the principle of conservation of mechanical energy.

3. Analyse

Choose configurations to represent the initial and final conditions of the system. For each object that changes elevation, select a reference position for the object that defines the zero configuration of gravitational potential energy for the system. For an object on a spring, the zero configuration for elastic potential energy is when the object is at its equilibrium position. If there is more than one conservative force, write an expression for the potential energy associated with each force.

Write the total initial mechanical energy E_i of the system for some configuration as the sum of the kinetic and potential energies associated with the configuration. Then write a similar expression for the total mechanical energy E_f of the system for the final configuration that is of interest. Because mechanical energy is *conserved*, equate the two total energies and solve for the required quantity.

4. Finalise

Make sure your results are consistent with your mental representation. Also make sure the values of your results are reasonable and consistent with everyday experience.

Example 7.2

A grand entrance

You are designing an apparatus to support an actor of mass 65 kg who is to 'fly' down to the stage during the performance of a play. The play will run for at least a week, so the actor's mass may change by as much as 3 kg in this time, as he overeats at the opening night after-party and then goes on a strict diet afterwards. You attach the actor's harness to a 120 kg sandbag by means of a lightweight steel cable running smoothly over two frictionless pulleys as in **Figure 7.5**. You need 3.0 m of cable between the harness and the nearest pulley so that the pulley can be hidden behind a curtain. You can precisely measure and control the mass of the sandbag and length of the rope, and these do not change. For the apparatus to work successfully, the sandbag must never lift above the floor as the actor swings from above the stage to the floor. Let us call the initial angle that the actor's cable makes with the vertical θ. What is the maximum value θ can have before the sandbag lifts off the floor?

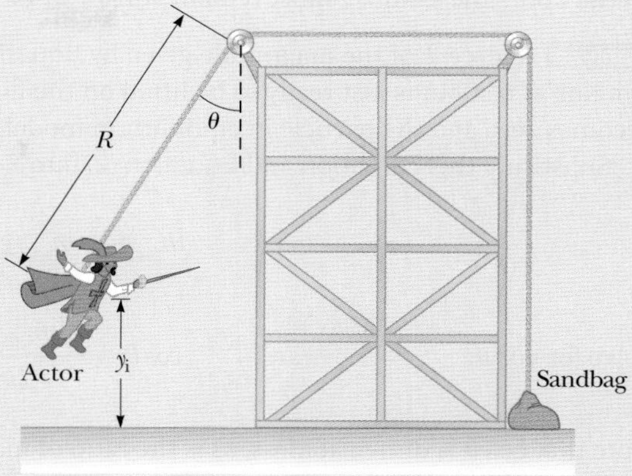

Figure 7.5
(Example 7.2) An actor uses some clever staging to make his entrance

Solution

Conceptualise We must use several concepts to solve this problem, including understanding the forces acting. Hence we begin by drawing free-body diagrams for the actor and the bag, as shown in **Figure 7.6**. Imagine what happens as the actor approaches the bottom of the swing. At the bottom, the cable is vertical and must support his weight as well as provide centripetal acceleration of his body in the upwards direction. At this point in his swing, the tension in the cable is the highest and the sandbag is most likely to lift off the floor.

Model Looking first at the swinging of the actor from the initial point to the lowest point, we model the actor and the Earth as an isolated system. We can reasonably make the approximation that air resistance is negligible, as the actor will not be swinging very fast, so there are no non-conservative forces acting. You might initially be tempted to model the system as non-isolated because of the interaction of the system with the cable, which is in the environment. The force applied to the actor by the cable, however, is always perpendicular to each element of the displacement of the actor and hence does no work. Therefore, in terms of energy transfers across the boundary, the system is isolated.

Analyse We first find the actor's speed as he arrives on the floor as a function of the initial angle θ and the radius R of the circular path through which he swings. From the isolated system model, apply conservation of mechanical energy to the actor–Earth system:

$$K_f + U_f = K_i + U_i$$

Let y_i be the initial height of the actor above the floor and v_f be his speed at the instant before he lands. $K_i = 0$ because the actor starts from rest and $U_f = 0$ because we define the configuration of the actor at the floor as having a gravitational potential energy of zero.

$$\frac{1}{2}m_{actor}v_f^2 + 0 = 0 + m_{actor}gy_i \qquad (1)$$

From the geometry in **Figure 7.5**, notice that $y_f = 0$, so $y_i = R - R\cos\theta = R(1 - \cos\theta)$. Use this relationship in **Equation (1)** and solve for v_f^2:

$$v_f^2 = 2gR(1 - \cos\theta) \qquad (2)$$

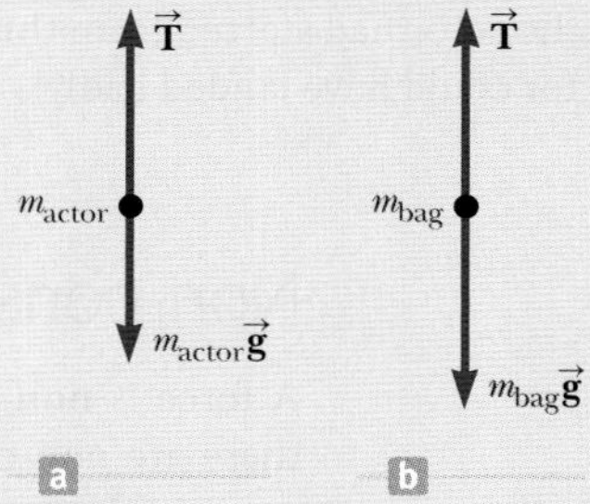

Figure 7.6
(Example 7.2) (a) The free-body diagram for the actor at the bottom of the circular path (b) The free-body diagram for the sandbag if the normal force from the floor goes to zero

Model Now focus on the instant the actor is at the lowest point. We model the actor at this instant as a particle in uniform circular motion because he moves along a circular arc. At the bottom of the swing he experiences a centripetal acceleration of v_f^2/r directed upwards.

Analyse Apply Newton's second law to the actor at the bottom of his path, using the free-body diagram in **Figure 7.6a** as a guide:

$$\sum F_y = T - m_{actor}g = m_{actor}\frac{v_f^2}{R} \qquad (3)$$

Model Finally, recognise that the sandbag lifts off the floor when the upwards force exerted on it by the cable exceeds the gravitational force acting on it; the normal force is zero when that happens. We do *not*, however, want the sandbag to lift off the floor. The sandbag must remain at rest, so we model it as a particle in equilibrium.

Analyse A force T of the magnitude given by **Equation (3)** is transmitted by the cable to the sandbag. If the sandbag remains at rest but is just ready to be lifted off the floor if any more force were applied by the cable, the normal force on it becomes zero and the particle in equilibrium model tells us that $T = m_{\text{bag}}g$ as in **Figure 7.6b**.

Substitute this condition and **Equation (2)** into **Equation (3)**:

$$m_{\text{bag}}g = m_{\text{actor}}g + m_{\text{actor}}\frac{2gR(1-\cos\theta)}{R}$$

Solve for $\cos\theta$:

$$\cos\theta = \frac{(3m_{\text{actor}} - m_{\text{bag}})}{2m_{\text{actor}}}$$

Note that $\cos\theta$ is dimensionless, as is the ratio of masses on the right-hand side, so our equation is dimensionally consistent. ☺

Substitute the given parameters, recalling that we need to find the *maximum* value for θ such that the bag will not lift off the floor, so we need to use the maximum value of the actor's mass:

$$\cos\theta_{\text{max allowed}} = \frac{3m_{\text{actor, max}} - m_{\text{bag}}}{2m_{\text{actor, max}}} = \frac{3(68\text{ kg}) - 120\text{ kg}}{2(68\text{ kg})} = 0.62$$

$$\theta = 52°$$

Finalise To solve this problem we had to decide what approximations were reasonable to make, decide how to define our system and make use of several different models. This is generally the case when solving real world problems in physics and engineering.

What If? What if we had neglected to consider the possible change in the actor's mass due to overindulgence? In this case we would have calculated the maximum allowed angle as:

$$\cos\theta_{\text{max allowed}} = \frac{3m_{\text{actor}} - m_{\text{bag}}}{2m_{\text{actor}}} = \frac{3(65\text{ kg}) - 120\text{ kg}}{2(65\text{ kg})} = 0.58$$

and $\theta = 55°$

If the actor had started from this angle, but actually weighed 68 kg, the sandbag would have lifted into the air and the actor could have landed badly!

Non-conservative forces

A force is **non-conservative** if it does not satisfy properties 1 and 2 for conservative forces (page 189). If there are non-conservative forces acting within the system, mechanical energy is transformed to internal energy. The total energy of the system is conserved although the mechanical energy is not. Friction and other resistive forces, as described in Chapter 5, act to convert kinetic or potential energy to internal energy, resulting in an increase in temperature of a system or components of a system.

For a book falling only under the action of the gravitational force, the mechanical energy of the book–Earth system remains fixed; gravitational potential energy transforms to kinetic energy, and the total energy of the system remains constant. Non-conservative forces acting within a system, however, cause a *change* in the mechanical energy of the system. For example, for a book sent sliding on a horizontal surface that is not frictionless, the mechanical energy of the book–surface system is transformed to internal energy. Only part of the book's kinetic energy is transformed to internal energy of the book. The rest appears as internal energy of the surface. Because the force of kinetic friction transforms the mechanical energy of a system into internal energy, it is a non-conservative force.

As an example of the path dependence of the work for a non-conservative force, consider **Figure 7.7**. Suppose you displace a book between two points on a table. If the book is displaced in a straight line along the blue path between points Ⓐ and Ⓑ in **Figure 7.7**, you do a certain amount of work against the kinetic friction force to keep the book moving at a constant speed. Now, imagine that you push the book along the brown semicircular path in **Figure 7.7**. You perform more work against friction along this curved path than along the straight path because the curved path is longer. The work done on the book depends on the path, so the friction force *cannot* be conservative.

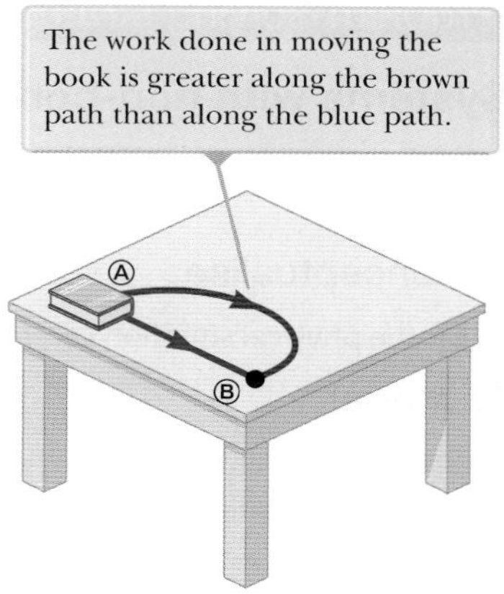

The work done in moving the book is greater along the brown path than along the blue path.
Ⓐ
Ⓑ

Figure 7.7
The work done against the force of kinetic friction depends on the path taken as the book is moved from Ⓐ to Ⓑ.

Imagine the book is given an initial push and then allowed to slide, coming to rest in some distance d. As the book moves through a distance d, the only force that does work on it is the force of kinetic friction. This force causes a change $-f_k d$ in the kinetic energy of the book.

Now, however, suppose the book is part of a system that also exhibits a change in potential energy. In this case, $-f_k d$ is the amount by which the mechanical energy of the system changes because of the force of kinetic friction. For example, if the book moves on an incline that is not frictionless, there is a change in both the kinetic energy and the gravitational potential energy of the book–Earth system. In general, if a friction force acts within an isolated system,

$$\Delta E_{\text{mech}} = \Delta K + \Delta U = -f_k d \qquad (7.14)$$

◀ Change in mechanical energy of a system due to friction within the system

where ΔU is the change in all forms of potential energy. Notice that **Equation 7.14** reduces to **Equation 7.8** when the friction force is zero.

If the system in which non-conservative forces act is non-isolated and the external influence on the system is by means of work,

$$\Delta E_{\text{mech}} = -f_k d + \sum W_{\text{other forces}} \qquad (7.15)$$

Equation 7.15, with the help of **Equation 7.7**, can be written as

$$\sum W_{\text{other forces}} = W = \Delta K + \Delta U + \Delta E_{\text{int}} \qquad (7.16)$$

This reduced form of **Equation 7.2** represents the non-isolated system model for a system that possesses potential energy and within which a non-conservative force acts. In practice, during problem solving you do not need to use equations such as **Equation 7.15**. You can simply use **Equation 7.2** and keep only those terms in the equation that correspond to the physical situation.

Pitfall Prevention 7.1

Conservative forces act such that mechanical energy, the sum of kinetic and potential energy, is conserved. When non-conservative forces act some energy is converted into internal energy, but *the total energy is still conserved.*

TRY THIS

Get a thick rubber band and hold it against your cheek so you can feel its temperature. Now rapidly stretch and release it 10 or 20 times and hold it against your cheek again. Can you feel any change in temperature? What forces were acting on and in the rubber band, and what energy transformations took place?

Quick **Quiz 7.5**

You are travelling along a freeway at 100 km/h. Your car has kinetic energy. You suddenly skid to a stop because of congestion in traffic. Where is the kinetic energy your car once had? **(a)** It is all in internal energy in the road. **(b)** It is all in internal energy in the tyres. **(c)** Some of it has transformed to internal energy and some of it transferred away by mechanical waves. **(d)** It is all transferred away from your car by various mechanisms.

Quick **Quiz 7.6**

A car travelling at an initial speed *v* slides a distance *d* to a halt after its brakes lock. If the car's initial speed is instead 2*v* at the moment the brakes lock, is the distance it slides equal to: (a) *d*, (b) 2*d*, (c) 4*d* or (d) 8*d*?

Problem-solving strategy

Systems with non-conservative forces

The following procedure should be used when you face a problem involving a system in which non-conservative forces act:

1. Conceptualise

Study the physical situation carefully and form a mental representation of what is happening. Draw a diagram.

2. Model

Define your system, which may consist of more than one object. Determine whether your system should be modelled as isolated or non-isolated. The system could include springs or other means of storing potential energy. Determine whether any non-conservative forces are significant. If not, use the principle of conservation of mechanical energy as outlined in Section 7.2. If so, use the procedure discussed below.

3. Analyse

Choose configurations to represent the initial and final conditions of the system. Draw diagrams representing these configurations. For each object that changes elevation, select a reference position for the object that defines the zero configuration of gravitational potential energy for the system. For an object on a spring, the zero configuration for elastic potential energy is when the object is at its equilibrium position. If there is more than one conservative force, write an expression for the potential energy associated with each force.

Use either **Equation 7.14** or **Equation 7.15** to establish a mathematical representation of the problem. Solve for the required quantity.

4. Finalise

Make sure your results are consistent with your mental representation. Also make sure the values of your results are reasonable and consistent with everyday experience.

±? Example **7.3**

Crate sliding down a ramp

A (3.0 ± 0.1) kg crate slides down a ramp. The ramp is (1.0 ± 0.1) m in length and inclined at an angle of $(30 \pm 3)°$. The crate starts from rest at the top, experiences a constant friction force of magnitude (5.0 ± 0.5) N, and continues to move a short distance on the horizontal floor after it leaves the ramp.

(A) Use energy methods to determine the speed of the crate at the bottom of the ramp.

Solution

Conceptualise Imagine the crate sliding down the ramp and draw a diagram showing the situation, as in **Figure 7.8**. The larger the friction force, the more slowly the crate will slide.

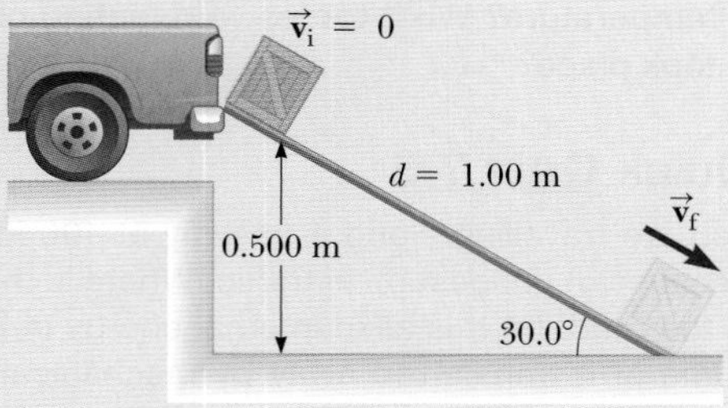

Figure 7.8
(Example 7.3) A crate slides down a ramp under the influence of gravity. The potential energy of the system decreases, whereas the kinetic energy increases.

Model We identify the crate, the surface and the Earth as the system. The system is modelled as isolated with a non-conservative force acting. The non-conservative force is the kinetic friction force. Air resistance will be negligible compared to the friction force of the ramp on the crate.

Analyse Because $v_i = 0$, the initial kinetic energy of the system when the crate is at the top of the ramp is zero. If the y coordinate is measured from the bottom of the ramp (the final position of the crate, for which we choose the gravitational potential energy of the system to be zero) with the upwards direction being positive, then $y_i = 0.50$ m.

Example 7.3 cont.

Write the expression for the total mechanical energy of the system when the crate is at the top:

$$E_i = K_i + U_i = 0 + U_i = mgy_i$$

Write an expression for the final mechanical energy: $E_f = K_f + U_f = \frac{1}{2}mv_f^2 + 0$

Apply **Equation 7.14**: $\Delta E_{mech} = E_f - E_i = \frac{1}{2}mv_f^2 - mgy_i = -f_k d$

Solve for v_f: $v_f = \sqrt{\frac{2}{m}(mgy_i - f_k d)}$ **(1)**

Check dimensions: $[LT^{-1}] = ([M^{-1}]([M][LT^{-2}][L] - [MLT^{-2}][L]))^{1/2} = ([M^{-1}][ML^2T^{-2}])^{1/2} = [LT^{-1}]$ ☺

Substitute numerical values: $v_f = \sqrt{\frac{2}{3.0\text{ kg}}[(3.0\text{ kg})(9.8\text{ m/s}^2)(0.50\text{ m}) - (5.0\text{ N})(1.0\text{ m})]} = 2.5\text{ m/s}$

We will use the range method to find the uncertainty:

$$\begin{aligned} v_{max} &= \sqrt{\frac{2}{m_{min}}(m_{min}gy_{max} - f_{k,min}d_{min})} \\ &= \sqrt{\frac{2}{2.9\text{ kg}}[(2.9\text{ kg})(9.8\text{ m/s}^2)(0.60\text{ m})] - (4.5\text{ N})(0.9\text{ m})} \\ &= 3.0\text{ m/s} \end{aligned}$$

$$\begin{aligned} v_{min} &= \sqrt{\frac{2}{m_{max}}(m_{max}gy_{min} - f_{k,max}d_{max})} \\ &= \sqrt{\frac{2}{3.1\text{ kg}}[(3.1\text{ kg})(9.8\text{ m/s}^2)(0.41\text{ m})] - (5.5\text{ N})(1.1\text{ m})} \\ &= 2.0\text{ m/s} \end{aligned}$$

So our uncertainty in v_f is 0.5 m/s and we can write $v_f = (2.5 \pm 0.5)$ m/s.

(B) How far does the crate slide on the horizontal floor if it continues to experience a friction force of (5.0 ± 0.5) N?

Solution

Analyse This part of the problem is handled in exactly the same way as part (A), but in this case we can consider the mechanical energy of the system to consist only of kinetic energy because the potential energy of the system remains fixed.

Write an expression for the mechanical energy of the system when the crate leaves the bottom of the ramp:

$$E_i = K_i = \frac{1}{2}mv_i^2$$

Apply **Equation 7.14** with $E_f = 0$: $E_f - E_i = 0 - \frac{1}{2}mv^2 = -f_k d \rightarrow \frac{1}{2}mv^2 = f_k d$

Solve for the distance d: $d = \frac{mv^2}{2f_k}$

Do a dimension check: $[L] = [M][LT^{-1}]^2/[MLT^{-2}] = [L]$ ☺

and substitute numerical values: $d = \frac{mv^2}{2f_k} = \frac{(3.00\text{ kg})(2.54\text{ m/s})^2}{2(5.0\text{ N})} = 1.94\text{ m}$

The uncertainty in d is:

$$\Delta d = d\left(\frac{\Delta m}{m} + \frac{\Delta f_k}{f_k} + 2\frac{\Delta v}{v}\right) = 1.94\text{ m}\left(\frac{0.1\text{ kg}}{3.0\text{ kg}} + \frac{0.5\text{ N}}{5.0\text{ N}} + 2\left(\frac{0.5}{2.5}\right)\right) = 1.0\text{ m}$$

Example 7.3 cont.

So the distance is $d = (1.9 \pm 1.0)$ m.

This is a very large fractional uncertainty, more than 50%, although the uncertainties in our initial data were never more than 10%. This illustrates the way uncertainties can compound when quantities are combined.

Finalise The increase in internal energy of the system as the crate slides down the ramp is $f_k d = (5.00\text{ N})(1.00\text{ m}) = 5.00$ J. This energy is shared between the crate and the surface, each of which is a bit warmer than before.

For comparison, calculate the speed at which the crate reaches the bottom on a frictionless surface. If the ground at the bottom really was frictionless, how far would the crate slide?

Example 7.4

A block pulled on a rough surface

A 6.0 kg block initially at rest is pulled to the right along a horizontal surface by a constant horizontal force of 12 N, as shown in **Active Figure 7.9**.

(A) Find the speed of the block after it has moved 3.0 m if the surfaces in contact have a coefficient of kinetic friction of 0.15.

Solution

Conceptualise The rough surface applies a friction force on the block opposite to the applied force.

Model The block is pulled by a force and the surface is rough, so we model the block–surface system as non-isolated with a non-conservative force acting.

Analyse **Active Figure 7.9a** illustrates this situation. Neither the normal force nor the gravitational force does work on the system because their points of application are displaced horizontally.

Find the work done on the system by the applied force:

$$\sum W_{\text{other forces}} = W_F = F\Delta x$$

Apply the particle in equilibrium model to the block in the vertical direction:

$$\sum F_y = 0 \rightarrow n - mg = 0 \rightarrow n = mg$$

Find the magnitude of the friction force:

$$f_k = \mu_k n = \mu_k mg = (0.15)(6.0\text{ kg})(9.80\text{ m/s}^2) = 8.82\text{ N}$$

Find the final speed of the block by rearranging **Equation 7.15**:

$$\frac{1}{2}mv_f^2 = \frac{1}{2}mv_i^2 - f_k d + W_F$$

$$v_f = \sqrt{v_i^2 + \frac{2}{m}(-f_k d + F\Delta x)}$$

where $v_i = 0$.

Check dimensions: $[LT^{-1}] = ([LT^{-1}]^2 + ([M^{-1}][MLT^{-2}][L]))^{1/2} = [LT^{-1}]$ ☺

Substitute numerical values: $v_f = \sqrt{0 + \frac{2}{6.0\text{ kg}}[-(8.82\text{ N})(3.0\text{ m}) + (12\text{N})(3.0\text{ m})]} = 1.8\text{ m/s}$

Finalise Compare this to the case with no friction acting. If $f_k = 0$, the final velocity will be 3.5 m/s. The difference in energy between the zero friction case and the block in this example is equal to the increase in internal energy of the block–surface system.

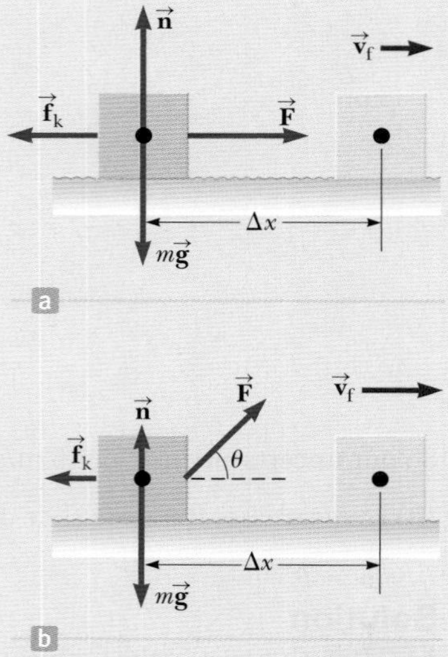

Active Figure 7.9

(Example 7.4) (a) A block pulled to the right on a rough surface by a constant horizontal force (b) The applied force is at an angle θ to the horizontal.

Example 7.4 cont.

(B) Suppose the force $\vec{\mathbf{F}}$ is applied at an angle θ as shown in **Active Figure 7.9b**. At what angle should the force be applied to achieve the largest possible speed after the block has moved 3.0 m to the right?

Solution

Conceptualise You might guess that $\theta = 0$ would give the largest speed because the force would have the largest component possible in the direction parallel to the surface. Think about $\vec{\mathbf{F}}$ applied at an arbitrary non-zero angle, however. Although the horizontal component of the force would be reduced, the vertical component of the force would reduce the normal force, in turn reducing the force of friction, which suggests that the speed could be maximised by pulling at an angle other than $\theta = 0$.

Model As in part (A), we model the block–surface system as non-isolated with a non-conservative force acting.

Analyse Find the work done by the applied force, noting that $\Delta x = d$ because the path followed by the block is a straight line:

$$\sum W_{\text{other forces}} = W_F = F\,\Delta x \cos\theta = Fd\cos\theta$$

Apply the particle in equilibrium model to the block in the vertical direction:

$$\sum F_y = n + F\sin\theta - mg = 0$$

Solve for n: $\quad n = mg - F\sin\theta$

Use **Equation 7.15** to find the final kinetic energy for this situation:

$$K_f = K_i - f_k d + W_F$$
$$= 0 - \mu_k n d + Fd\cos\theta = -\mu_k(mg - F\sin\theta)d + Fd\cos\theta$$

Maximising the speed is equivalent to maximising the final kinetic energy. Consequently, differentiate K_f with respect to θ and set the result equal to zero:

$$\frac{dK_f}{d\theta} = -\mu_k(0 - F\cos\theta)d - Fd\sin\theta = 0$$

$$\mu_k\cos\theta - \sin\theta = 0$$

$$\tan\theta = \mu_k$$

Note that $\tan\theta$ and μ_k are both dimensionless.

Evaluate θ for $\mu_k = 0.15$: $\quad \theta = \tan^{-1}(\mu_k) = \tan^{-1}(0.15) = 8.5°$

Finalise Notice that the angle at which the speed of the block is a maximum is indeed not $\theta = 0$. When the angle exceeds 8.5°, the horizontal component of the applied force is too small to be compensated by the reduced friction force and the speed of the block begins to decrease from its maximum value.

Example 7.5

Block–spring collision

A block having a mass of 0.80 kg is given an initial velocity $v_{Ⓐ} = 1.2$ m/s to the right and collides with a spring whose mass is negligible and whose force constant is $k = 50$ N/m.

(A) Assuming the surface to be frictionless, calculate the maximum compression of the spring after the collision.

Solution

Conceptualise Start as usual by drawing a diagram. In this case it will be helpful to show the system (block and spring) at various times, as shown in the various parts of **Figure 7.10**. All motion takes place in a horizontal plane, so we do not need to consider changes in gravitational potential energy.

Example 7.5 cont.

Model We identify the system to be the block and the spring. The block–spring system is modelled as an isolated system with no non-conservative forces acting.

Analyse Before the collision, when the block is at Ⓐ, it has kinetic energy and the spring is uncompressed, so the elastic potential energy stored in the system is zero. Therefore, the total mechanical energy of the system before the collision is just $\frac{1}{2}mv_Ⓐ^2$. After the collision, when the block is at Ⓒ, the spring is fully compressed; now the block is at rest and so has zero kinetic energy. The elastic potential energy stored in the system, however, has its maximum value $\frac{1}{2}kx^2 = \frac{1}{2}kx^2_{max}$, where the origin of coordinates $x = 0$ is chosen to be the equilibrium position of the spring and x_{max} is the maximum compression of the spring, which in this case happens to be $x_Ⓒ$. The total mechanical energy of the system is conserved because no non-conservative forces act on objects within the isolated system.

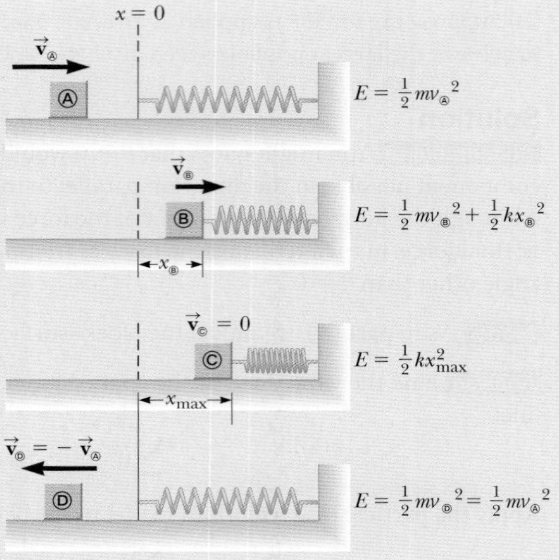

Figure 7.10
(Example 7.5) A block sliding on a frictionless, horizontal surface collides with a light spring. (a) Initially, the mechanical energy is all kinetic energy. (b) The mechanical energy is the sum of the kinetic energy of the block and the elastic potential energy in the spring. (c) The energy is entirely potential energy. (d) The energy is transformed back to the kinetic energy of the block. The total energy of the system remains constant throughout the motion.

Write a conservation of mechanical energy equation:

$$K_Ⓒ + U_{sⒸ} = K_Ⓐ + U_{sⒶ}$$

$$0 + \frac{1}{2}kx^2_{max} = \frac{1}{2}mv^2_Ⓐ + 0$$

Solve for x_{max}: $$x_{max} = \sqrt{\frac{m}{k}}\,v_Ⓐ$$

Check dimensions: $[L] = ([M]/[MT^{-2}])^{1/2}\,[LT^{-1}] = [L]$ ☺

Substitute in the numbers: $$x_{max} = \sqrt{\frac{0.80\text{ kg}}{50\text{ N/m}}}(1.2\text{ m/s}) = 0.15\text{ m}$$

(B) Suppose a constant force of kinetic friction acts between the block and the surface, with $\mu_k = 0.50$. If the speed of the block at the moment it collides with the spring is $v_Ⓐ = 1.2$ m/s, what is the maximum compression $x_Ⓒ$ in the spring?

Solution

Conceptualise Because of the friction force, we expect the compression of the spring to be smaller than in part (A) because some of the block's kinetic energy is transformed to internal energy in the block and the surface.

Model We identify the system as the block, the surface and the spring. This system is again modelled as isolated but now involves a non-conservative force.

Analyse In this case, the mechanical energy $E_{mech} = K + U_s$ of the system is *not* conserved because a friction force acts on the block. From the particle in equilibrium model in the vertical direction, we see that $n = mg$.

Evaluate the magnitude of the friction force: $f_k = \mu_k n = \mu_k mg$

Write the change in the mechanical energy of the system due to friction as the block is displaced from $x = 0$ to $x_Ⓒ$:

$$\Delta E_{mech} = -f_k x_Ⓒ$$

Substitute the initial and final energies: $$\Delta E_{mech} = E_f - E_i = \left(0 + \frac{1}{2}kx_Ⓒ^2\right) - \left(\frac{1}{2}mv_Ⓐ^2 + 0\right) = -f_k x_Ⓒ$$

$$\frac{1}{2}kx_Ⓒ^2 - \frac{1}{2}mv_Ⓐ^2 = -\mu_k mgx_Ⓒ$$

Example 7.5 cont.

This is a quadratic equation for x. Usually we rearrange for x before substituting in values, but in this case it is much simpler to substitute values first before rearranging to solve for x.

Appendix B.2 demonstrates how to solve a quadratic equation. Σ

Substitute numerical values:

$$\frac{1}{2}(50)x_{©}^{2} - \frac{1}{2}(0.80)(1.2)^2 = -(0.50)(0.80\text{ kg})(9.80\text{ m/s}^2)x_{©}$$

$$25x_{©}^{2} + 3.9x_{©} - 0.58 = 0$$

Solving the quadratic equation for $x_{©}$ gives $x_{©} = 0.093$ m and $x_{©} = -0.25$ m. The physically meaningful root is $x_{©} = 0.093$ m.

Finalise The negative root does not apply to this situation because the block must be to the right of the origin (positive value of x) when it comes to rest. Notice that the value of 0.093 m is less than the distance obtained in the frictionless case of part (A) as we expected.

TRY THIS

Get a tennis racquet or some other type of racquet and a tennis ball. Get a friend to hold the handle of the racquet and drop a ball onto the centre of the racquet. Observe how high the ball bounces. Now place the handle of the racquet on a table and get your friend to hold it as firmly as they can, by sitting on it or standing on it so the handle cannot move. Try dropping the ball again onto the racquet from the same height. Why does the ball bounce higher when the handle is held more firmly?

If you have a few different balls, try dropping them onto the floor one by one from the same height and note how high each one bounces. Why do some bounce higher than others?

Whenever a non-conservative force such as friction acts in a system, applying a force through some distance, then some mechanical energy is converted into internal energy and surfaces become warmer. For processes where we wish to convert potential energy, such as the chemical potential energy in fuel, into mechanical energy, such as the kinetic energy of a car or the gravitational potential energy of a lift, we want an *efficient* system, with as little energy 'lost' to internal energy as possible. Efficiency is a measure of how much energy is converted into the form we desire; it is the fraction of the energy converted that ends up in the desired form:

$$\varepsilon = \frac{\text{energy in desired form}}{\text{total energy converted}}$$

◀ Efficiency

where ε is the efficiency and is usually expressed as a percentage.

Example 7.6

Efficiency of the block–spring collision

In Example 7.5 we considered a block sliding on a rough surface and into a spring, compressing the spring. Energy was converted from kinetic energy into elastic potential energy in this process. How efficient was this process when the surface was rough (not frictionless)?

Conceptualise We have already conceptualised and modelled the problem in Example 7.5 and calculated the compression of the spring when the system (block, surface, spring) for the idealised situation with no non-conservative forces acting, and the real situation where there is a non-conservative force, friction, acting. We need to compare the energy stored in each case to find the efficiency.

Analyse We need to find the energy stored in each case from Example 7.5, then use the definition of efficiency.

The stored energy in the idealised system (part A of Example 7.5) is:

$$E_{\text{ideal}} = \frac{1}{2}kx_{\text{max}}^{2} = \frac{1}{2}mv_{\text{A}}^{2} = \frac{1}{2}(0.8\text{ kg})(1.2\text{ m/s})^2 = 0.576\text{ J}$$

The stored energy when friction acts (part B of Example 7.5) is:

$$E_{\text{real}} = \frac{1}{2}kx_{\text{max}}^{2} = \frac{1}{2}(50\text{ N/m})(0.093\text{ m})^2 = 0.216\text{ J}$$

Example 7.6 cont.

The efficiency of the process is:

$$\varepsilon = \frac{0.216\ \text{J}}{0.576\ \text{J}} = 0.375 \text{ or } 38\%$$

Finalise This process has an efficiency of 38%, which means that more than half the energy is 'lost' due to friction. This energy is lost in the sense that it is no longer stored as potential energy which is easily available for use; it has instead been converted to internal energy causing heating of the block and the surface.

Conceptual Example 7.7

Carrying a heavy bag

Imagine a student carrying a heavy bag of books across campus. If the student moves at approximately constant speed, the only work done on the bag of books is that required to initially accelerate it, and to lift it from the ground. Once the student is moving, no net work is being done on the bag. Yet the student knows they are doing work and using energy by carrying the bag, because it is heavy and tiring to carry.
So what *has* happened to the energy expended by the student?

Solution

Work is clearly done by the student to raise the bag from ground level to the student's back, as the gravitational potential energy of the bag-Earth system increases. However our muscles are not 100% efficient. You know from experience that if you lift a heavy weight, especially if you do it multiple times, you will start to feel hot. This is because of many internal forces of friction – including in the muscle fibres themselves – acting to convert potential and kinetic energy into internal energy. So much more energy is expended by the student in lifting the bag than is stored as gravitational potential energy.

Now consider walking. If we modelled walking as starting from rest and then moving at constant speed without friction, then the only work needed would be that required to accelerate us from rest to our final speed. This model is appropriate for sliding on a frictionless surface, and in some cases is a reasonable approximation for rolling. However, it is a very poor model for walking! As you walk, your speed constantly changes within each step, hence work is constantly being done to change your speed. (You apply a force to the ground, by Newton's third law it applies a force to you, and hence you accelerate.) In addition, different parts of your body are moving at different speeds at different times. All of the processes that are occurring to make you walk are less than 100% efficient, and most of the potential energy stored in your cells is actually converted into internal energy with only a small fraction converted into kinetic energy. When you carry a heavy bag of books on your back, you are constantly accelerating, and hence doing work on, the bag of books as well.

Hence, the student is doing plenty of work, but not very efficiently.

7.4 Energy bar charts

Consider **Active Figure 7.11**, which shows a spring on a frictionless, horizontal surface. When a block is pushed against the spring by an external agent, the elastic potential energy and the total energy of the system increase, as indicated in **Figure 7.11b**. When the spring is compressed a distance x_{max} (**Active Figure 7.11c**), the elastic potential energy stored in the spring is $\frac{1}{2}kx^2_{max}$. When the block is released from rest, the spring exerts a force on the block and pushes the block to the right. The elastic potential energy of the system decreases, whereas the kinetic energy increases and the total energy remains fixed (**Figure 7.11d**). When the spring returns to its original length, the stored elastic potential energy has been completely transformed into kinetic energy of the block (**Active Figure 7.11e**).

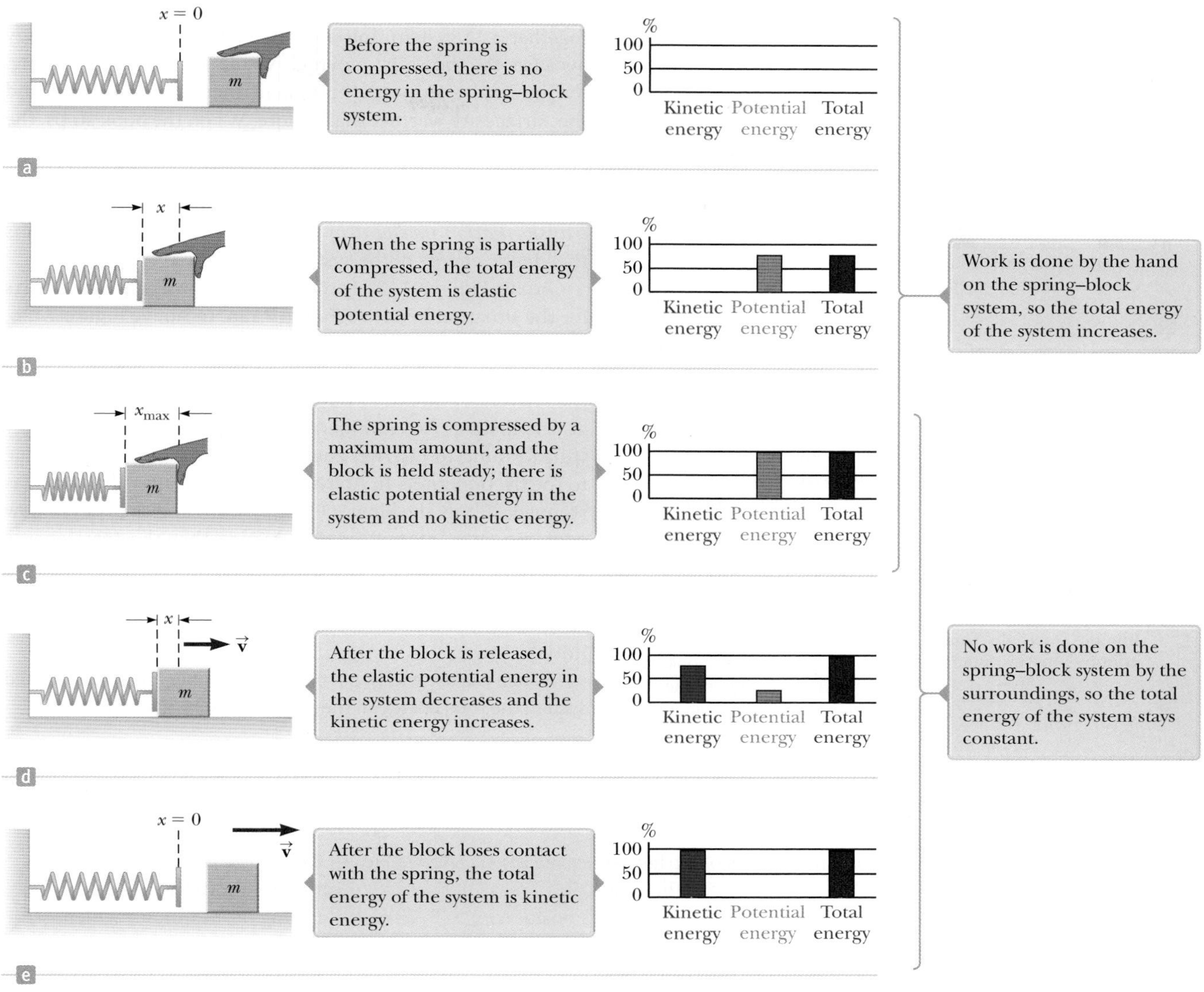

Active Figure 7.11

A spring on a frictionless, horizontal surface is compressed a distance x_{max} when a block of mass m is pushed against it. The block is then released and the spring pushes it to the right, where the block eventually loses contact with the spring. Parts (a) through (e) show various instants in the process. Energy bar charts on the right of each part of the figure help keep track of the energy in the system.

Active Figure 7.11 shows a useful graphical representation of information related to energy of systems called an **energy bar chart**. The vertical axis represents the amount of energy of a given type in the system. The horizontal axis shows the types of energy in the system. The bar chart in Active Figure 7.11a shows that the system contains zero energy because the spring is relaxed and the block is not moving. Between Active Figure 7.11a and Active Figure 7.11c, the hand does work on the system, compressing the spring and storing elastic potential energy in the system. In Active Figure 7.11d, the block has been released and is moving to the right while still in contact with the spring. The height of the bar for the elastic potential energy of the system decreases, the kinetic energy bar increases and the total energy bar remains fixed. In Active Figure 7.11e, the spring has returned to its relaxed length and the system now contains only kinetic energy associated with the moving block. Energy bar charts can be a very useful representation for keeping track of the various types of energy in a system. For practice, try making energy bar charts for the book–Earth system in Active Figure 6.16 when the book is dropped from the higher position.

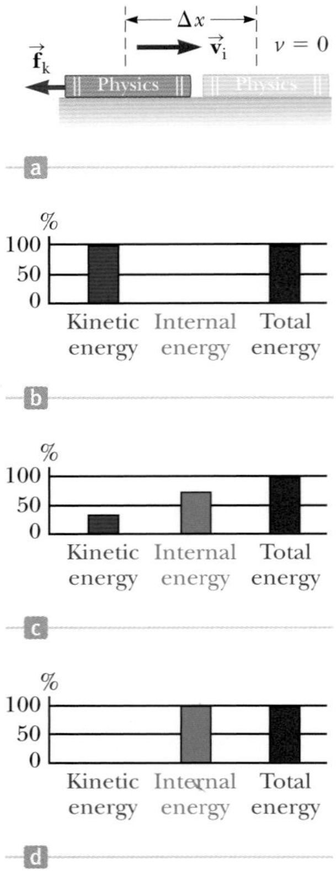

Active Figure 7.12

(a) A book sliding to the right on a horizontal surface slows down in the presence of a force of kinetic friction acting to the left. (b) An energy bar chart showing the energy in the system of the book and the surface at the initial instant of time. The energy of the system is all kinetic energy. (c) While the book is sliding, the kinetic energy of the system decreases as it is transformed to internal energy. (d) After the book has stopped, the energy of the system is all internal energy.

Consider the book and the surface in **Active Figure 7.12a** together as a system. Initially, the system has kinetic energy because the book is moving. While the book is sliding, the internal energy of the system increases: the book and the surface are warmer than before. When the book stops, the kinetic energy has been completely transformed to internal energy. We can consider the work done by friction within the system – that is, between the book and the surface – as a *transformation mechanism* for energy. This work transforms the kinetic energy of the system into internal energy.

Active Figures 7.12b through **7.12d** show energy bar charts for the situation in **Active Figure 7.12a**. In **Active Figure 7.12b**, the bar chart shows that the system contains kinetic energy at the instant the book is released by your hand. We define the reference amount of internal energy in the system as zero at this instant. **Active Figure 7.12c** shows the kinetic energy transforming to internal energy as the book slows down due to the friction force. In **Active Figure 7.12d**, after the book has stopped sliding, the kinetic energy is zero and the system now contains only internal energy. Notice that the total energy bar in red has not changed during the process. The amount of internal energy in the system after the book has stopped is equal to the amount of kinetic energy in the system at the initial instant.

When you use energy bar charts, remember that for an isolated system the total energy in the system must remain constant.

Example 7.8

The spring-loaded popgun

The launching mechanism of a popgun consists of a trigger-released spring (**Active Figure 7.13a**). The spring is compressed to a position $y_Ⓐ$ and the trigger is fired. The projectile of mass m rises to a position $y_Ⓒ$ above the position at which it leaves the spring, indicated in **Active Figure 7.13b** as position $y_Ⓑ = 0$. Consider a firing of the gun for which $m = 35.0$ g, $y_Ⓐ = -0.120$ m and $y_Ⓒ = 20.0$ m.

(A) Neglecting all resistive forces, determine the spring constant.

Solution

Conceptualise Imagine the process illustrated in parts (a) and (b) of **Active Figure 7.13**. The projectile starts from rest, speeds up as the spring pushes upwards on it, leaves the spring and then slows down as the gravitational force pulls downwards on it.

Model We identify the system as the projectile, the spring and the Earth. We ignore air resistance on the projectile and friction in the gun, so we model the system as isolated with no non-conservative forces acting.

Analyse Because the projectile starts from rest, its initial kinetic energy is zero. We choose the zero configuration for the gravitational potential energy of the system to be when the projectile leaves the spring. For this configuration, the elastic potential energy is also zero.

After the gun is fired, the projectile rises to a maximum height $y_Ⓒ$. The final kinetic energy of the projectile is zero.

From the isolated system model, write a conservation of mechanical energy equation for the system between points Ⓐ and Ⓒ:

$$K_Ⓒ + U_{gⒸ} + U_{sⒸ} = K_Ⓐ + U_{gⒶ} + U_{sⒶ}$$

Example 7.8 cont.

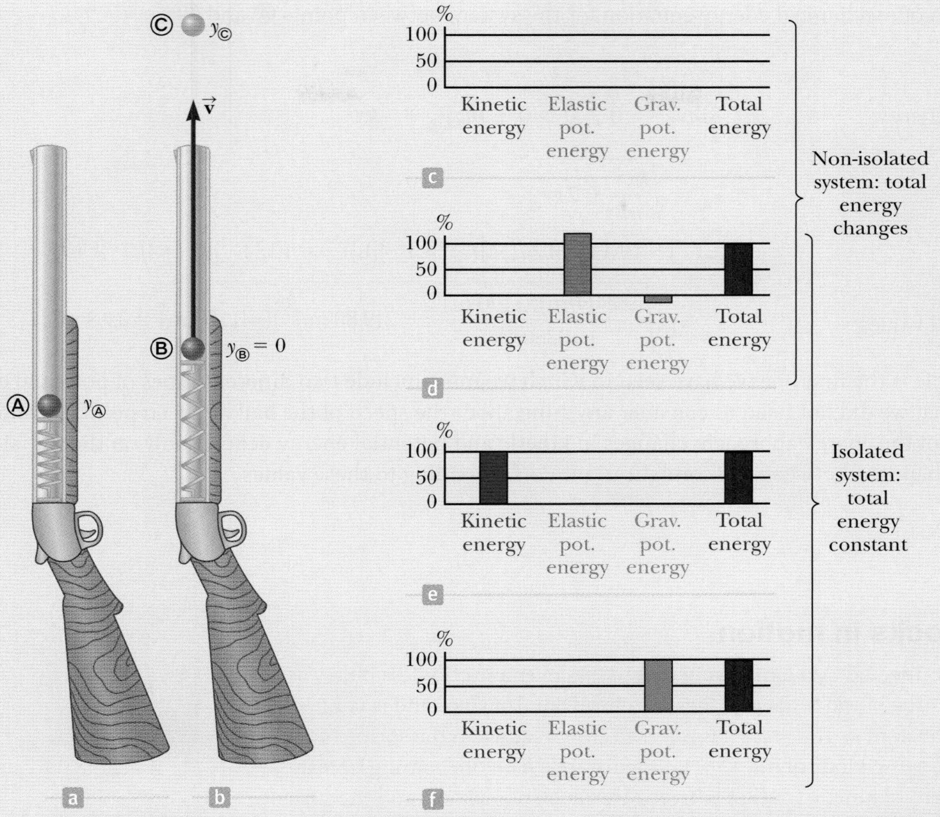

Active Figure 7.13

(Example 7.8) A spring-loaded popgun (a) before firing and (b) when the spring extends to its relaxed length (c) An energy bar chart for the popgun–projectile–Earth system before the popgun is loaded. The energy in the system is zero. (d) The popgun is loaded by means of an external agent doing work on the system to push the spring downwards. Therefore the system is non-isolated during this process. After the popgun is loaded, elastic potential energy is stored in the spring and the gravitational potential energy of the system is lower because the projectile is below point Ⓑ. (e) As the projectile passes through pointⒷ, all of the energy of the isolated system is kinetic. (f) When the projectile reaches point Ⓒ, all of the energy of the isolated system is gravitational potential.

Substitute for each energy: $$0 + mgy_{Ⓒ} + 0 = 0 + mgy_{Ⓐ} + \frac{1}{2}kx^2$$

Solve for k: $$k = \frac{2mg(y_{Ⓒ} - y_{Ⓐ})}{x^2}$$

Check dimensions: $$[MT^{-2}] = ([M][LT^{-2}][L])/[L]^2 = [MT^{-2}]$$ ☺

Substitute numerical values: $$k = \frac{2(0.0350\text{ kg})(9.80\text{ m/s})^2[20.0\text{ m} - (-0.120\text{ m})]}{(0.120\text{ m})^2} = 958\text{ N/m}$$

(B) Find the speed of the projectile as it moves through the equilibrium position Ⓑ of the spring as shown in **Active Figure 7.13b**.

Solution

Analyse The energy of the system as the projectile moves through the equilibrium position of the spring includes only the kinetic energy of the projectile $\frac{1}{2}mv_{Ⓑ}^2$. Both types of potential energy are equal to zero for this configuration of the system.

Example 7.8 cont.

Write a conservation of mechanical energy equation for the system between points Ⓐ and Ⓑ:

$$K_{Ⓑ} + U_{gⒷ} + U_{sⒷ} = K_{Ⓐ} + U_{gⒶ} + U_{sⒶ}$$

Substitute for each energy: $\frac{1}{2}mv_{Ⓑ}^2 + 0 + 0 = 0 + mgy_{Ⓐ} + \frac{1}{2}kx^2$

Solve for $v_{Ⓑ}$: $v_{Ⓑ} = \sqrt{\frac{kx^2}{m} + 2gy_{Ⓐ}}$

Check dimensions, $[LT^{-1}] = ([MT^{-2}][L^2]/[M]) + [LT^{-2}][L])^{1/2} = ([L^2T^{-2}])^{1/2} = [LT^{-1}]$ ☺

Substitute numerical values: $v_{Ⓑ} = \sqrt{\frac{(958\text{ N/m})(0.120\text{ m})^2}{(0.0350\text{ kg})} + 2(9.80\text{ m/s}^2)(-0.120\text{ m})} = 19.8\text{ m/s}$

Finalise This example is the first one we have seen in which we must include two different types of potential energy. Notice in part (A) that we did not need to consider anything about the speed of the ball between points Ⓐ and Ⓒ, which is part of the power of the energy approach: changes in kinetic and potential energy depend only on the initial and final values, not on what happens between the configurations corresponding to these values.

Example 7.9

Connected blocks in motion

(A) Two blocks are connected by a light string that passes over a frictionless pulley as shown in **Figure 7.14**. The block of mass m_1 lies on a horizontal surface and is connected to a spring of force constant k. The system is released from rest when the spring is unstretched. If the hanging block of mass m_2 falls a distance h before coming to rest, calculate the coefficient of kinetic friction between the block of mass m_1 and the surface.

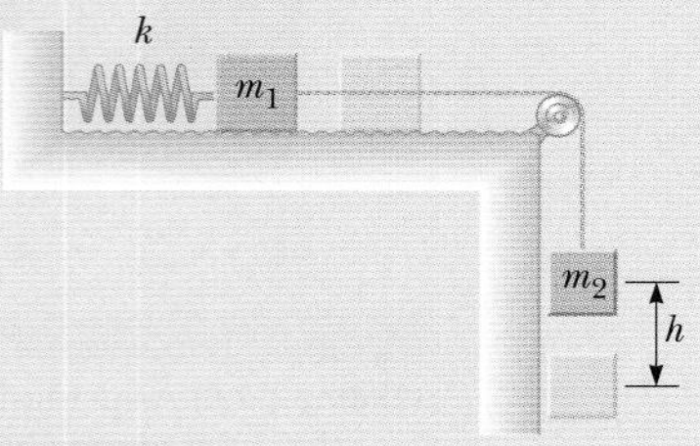

Figure 7.14
(Example 7.9) As the hanging block moves from its highest elevation to its lowest, the system loses gravitational potential energy but gains elastic potential energy in the spring. Some mechanical energy is transformed to internal energy because of friction between the sliding block and the surface.

Solution

Conceptualise There are two configurations in which the block is at rest. We choose these two configurations for the initial and final configurations because the kinetic energy of the system is zero for these configurations. Note that as we are given a diagram, we do not draw another one here, although it is good practice for you to draw a diagram even when one is given so that you can annotate it.

Model In this situation, the system consists of the two blocks, the spring, the surface and the Earth. The system is isolated with a non-conservative force acting. We model the sliding block as a particle in equilibrium in the vertical direction, leading to $n = m_1 g$ where n is the normal force acting on block 1.

Analyse We need to consider two forms of potential energy for the system, gravitational and elastic: $\Delta U_g = U_{gf} - U_{gi}$ is the change in the system's gravitational potential energy and $\Delta U_s = U_{sf} - U_{si}$ is the change in the system's elastic potential energy. The change in the gravitational potential energy of the system is associated with only the falling block because the vertical coordinate of the horizontally sliding block does not change. The initial and final kinetic energies of the system are zero, so $\Delta K = 0$.

For this example, let us start from **Equation 7.2** to show how this approach would work in practice. Because the system is isolated, the entire right side of **Equation 7.2** is zero. Based on the physical situation described in the problem, we see that there could be changes of kinetic energy, potential energy and internal energy in the system. Write the corresponding reduction of **Equation 7.2**:

$$\Delta K + \Delta U + \Delta E_{int} = 0$$

Incorporate into this equation that $\Delta K = 0$ and that there are two types of potential energy:

$$\Delta U_g + \Delta U_s + \Delta E_{int} = 0 \quad (1)$$

Example 7.9 cont.

Use **Equations 7.14** and **7.16** to find the change in internal energy in the system due to friction between the horizontally sliding block and the surface, noting that as the hanging block falls a distance h, the horizontally moving block moves the same distance h to the right:

$$\Delta E_{\text{int}} = f_k h = (\mu_k n)h = \mu_k m_1 gh \quad (2)$$

Evaluate the change in gravitational potential energy of the system, choosing the configuration with the hanging block at the lowest position to represent zero potential energy:

$$\Delta U_g = U_{gf} - U_{gi} = 0 - m_2 gh \quad (3)$$

Evaluate the change in the elastic potential energy of the system:

$$\Delta U_s = U_{sf} - U_{si} = \frac{1}{2}kh^2 - 0 \quad (4)$$

Substitute **Equations (2), (3),** and **(4)** into **Equation (1)**:

$$-m_2 gh + \frac{1}{2}kh^2 + \mu_k m_1 gh = 0$$

Solve for μ_k:

$$\mu_k = \frac{m_2 g - \frac{1}{2}kh}{m_1 g}$$

Check that both sides are dimensionless: $[\ ] = \dfrac{[\mathrm{M}][\mathrm{LT}^{-2}] - [\mathrm{MT}^{-2}][\mathrm{L}]}{[\mathrm{MLT}^{-2}]} = [\ ]$ ☺

Finalise This setup represents a method of measuring the coefficient of kinetic friction between an object and some surface. Notice that we do not need to remember which energy equation goes with which type of problem with this approach. You can always begin with **Equation 7.2** and then tailor it to the physical situation. This process may include deleting terms, such as the kinetic energy term and all terms on the right-hand side in this example. It can also include expanding terms, such as rewriting ΔU due to two types of potential energy in this example.

(B) The energy bar charts in **Figure 7.15** show three instants in the motion of the system in **Figure 7.14**. For each bar chart, identify the configuration of the system that corresponds to the chart.

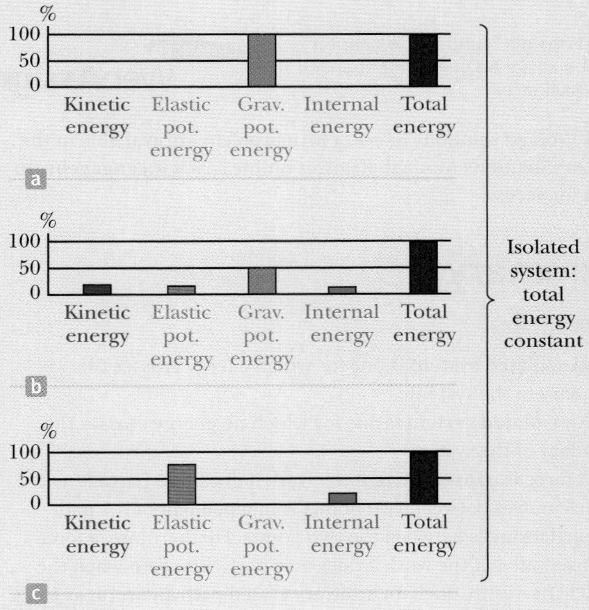

Figure 7.15
(Example 7.9) Three energy bar charts are shown for the system in **Figure 7.14**.

Solution

In **Figure 7.15a**, there is no kinetic energy in the system. Therefore, nothing in the system is moving. The bar chart shows that the system contains only gravitational potential energy and no internal energy yet, which corresponds to the configuration with the darker blocks in **Figure 7.14** and represents the instant just after the system is released.

In **Figure 7.15b**, the system contains four types of energy. The height of the gravitational potential energy bar is at 50%, which tells us that the hanging block has moved half-way between its position corresponding to **Figure 7.15a** and the position defined as $y = 0$. Therefore, in this configuration, the hanging block is between the dark and light images of the hanging block in **Figure 7.14**. The system has gained kinetic energy because the blocks are moving, elastic potential energy because the spring is stretching and internal energy because of friction between the block of mass m_1 and the surface.

In **Figure 7.15c**, the height of the gravitational potential energy bar is zero, telling us that the hanging block is at $y = 0$. In addition, the height of the kinetic energy bar is zero, indicating that the blocks have stopped moving momentarily. Therefore, the configuration of the system is that shown by the light images of the blocks in **Figure 7.14**. The height of the elastic potential energy bar is high because the spring is stretched its maximum amount. The height of the internal energy bar is higher than in **Figure 7.15b** because the block of mass m_1 has continued to slide over the surface.

End-of-chapter resources

AAP/Andrew Taylor

Wind farms are increasingly being built to supply renewable energy to households and industry.

A wind turbine, as shown, converts the kinetic energy of the air (the wind) into kinetic energy of itself as the drag force of the wind pushes on the blades. The wind itself is due to heating of the atmosphere by the Sun, which converts mass energy to internal energy during nuclear reactions as described in Chapter 43 and radiates some of this energy to the Earth as electromagnetic radiation, including visible light. This kinetic energy of the spinning blades, called rotational kinetic energy (Chapter 10), is in turn converted into electrical potential energy by a generator inside the turbine. This process will be discussed in detail in Chapter 31. Note that these processes are not 100% efficient, as friction forces act to convert some kinetic energy into internal energy, increasing the temperature of the wind turbine and the air.

Wind power is considered renewable because the energy ultimately comes from the Sun, and we do not 'use up' the Sun by collecting energy from it, although it will eventually use all its fuel and cease to radiate energy. In contrast, when we dig up fossil fuels, such as coal, and burn them, the coal is gone and cannot be used again. Regardless of the source, however, the *energy* is ultimately conserved.

Worked solutions to every fifth problem are available in the Student Solutions Manual. Register online at **www.cengagebrain.com** for access.

Summary

Definitions

A **non-isolated system** is one for which energy crosses the boundary of the system.

An **isolated system** is one for which no energy crosses the boundary of the system.

A force is **conservative** if the work it does on a particle as the particle moves between two points is independent of the path the particle takes between the two points. Furthermore, a force is conservative if the work it does on a particle is zero when the particle moves through an arbitrary closed path and returns to its initial position. A force that does not meet these criteria is said to be **non-conservative**.

The **total mechanical energy of a system** is defined as the sum of the kinetic energy and the potential energy:

$$E_{\text{mech}} \equiv K + U \quad (7.7)$$

Concepts and principles

For a non-isolated system, we can equate the change in the total energy stored in the system to the sum of all the transfers of energy across the system boundary, which is a statement of **conservation of energy**. For an isolated system, the total energy is constant.

If a system is isolated and if no non-conservative forces are acting on objects inside the system, the total mechanical energy of the system is constant:

$$K_f + U_f = K_i + U_i \quad (7.6)$$

If non-conservative forces (such as friction) act between objects inside a system, mechanical energy is not conserved. In these situations, the difference between the total final mechanical energy and the total initial mechanical energy of the system equals the energy transformed to internal energy by the non-conservative forces.

If a friction force acts within an isolated system, the mechanical energy of the system is reduced and the appropriate equation to be applied is

$$\Delta E_{\text{mech}} = \Delta K + \Delta U = -f_k d \quad (7.14)$$

If a friction force acts within a non-isolated system, the appropriate equation to be applied is

$$\Delta E_{\text{mech}} = -f_k d + \sum W_{\text{other forces}} \quad (7.15)$$

Analysis models for problem solving

Non-isolated system (energy) The most general statement describing the behaviour of a non-isolated system is the **conservation of energy equation:**

$$\Delta E_{\text{system}} = \sum T \quad (7.1)$$

Including the types of energy storage and energy transfer that we have discussed gives

$$\Delta K + \Delta U + \Delta E_{\text{int}} = W + Q + T_{\text{MW}} + T_{\text{MT}} + T_{\text{ET}} + T_{\text{ER}} \quad (7.2)$$

For a specific problem, this equation is generally reduced to a smaller number of terms by eliminating the terms that are not appropriate to the situation.

Isolated system (energy) The total energy of an isolated system is conserved, so

$$\Delta E_{\text{system}} = 0 \tag{7.9}$$

If no non-conservative forces act within the isolated system, the mechanical energy of the system is conserved.

Chapter review quiz

To help you revise Chapter 7: Conservation of energy, complete the automatically graded Chapter review quiz at http://login.cengagebrain.com.

Conceptual questions

1. Does everything have energy? Does it depend on the reference frame of the observer? Give the reasoning for your answer.
2. In Chapter 6, the work–kinetic energy theorem, $W_{\text{net}} = \Delta K$, was introduced. This equation states that work done on a system appears as a change in kinetic energy. It is a special-case equation, valid if there are no changes in any other type of energy such as potential or internal. Give three examples in which work is done on a system but the change in energy of the system is not a change in kinetic energy.
3. One person drops a ball from the top of a building while another person at the bottom observes its motion. Will these two people agree (a) on the value of the gravitational potential energy of the ball–Earth system? (b) on the change in potential energy? (c) on the kinetic energy of the ball at some point in its motion?
4. Can a force of static friction do work? If not, why not? If so, give an example.
5. A block is connected to a spring that is suspended from the ceiling. Assuming air resistance is negligible, describe the energy transformations that occur within the system consisting of the block, the Earth and the spring when the block is set into vertical motion. Draw energy bar charts to represent the different types of energy present (a) before the block is set into motion, (b) when it is displaced but not yet released, (c) when it has been released and returned to its equilibrium position, and (d) when it has reached its maximum height.
6. A bowling ball is suspended from the ceiling of a lecture hall by a strong cord. The ball is drawn away from its equilibrium position and released from rest at the tip of the demonstrator's nose as shown in Figure CQ7.6. The demonstrator remains stationary. (a) Explain why the ball does not strike her on its return swing. (b) Would this demonstrator be safe if the ball were given a push from its starting position at her nose?

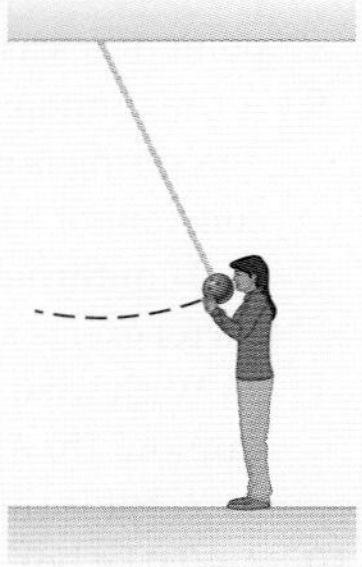

Figure CQ7.6

7. In the general conservation of energy equation, state which terms predominate in describing each of the following devices and processes. For a process going on continuously, you may consider what happens in a 10 s time interval. State which terms in the equation represent original and final forms of energy, which would be inputs and which outputs. (a) a slingshot firing a pebble (b) a fire burning (c) a portable radio operating (d) a car braking to a stop (e) the surface of the Sun shining visibly (f) a person jumping up onto a chair
8. A small boy climbs up a ladder at a playground to reach the top of a slide. He then slides down the slide and returns to the base of the ladder. Describe the forces that act in this process, and the energy transformations that occur as a result. Which of the forces are conservative, and which are non-conservative?
9. In a laboratory model of cars skidding to a stop, data are measured for four trials using two blocks. The blocks have identical masses but different coefficients of kinetic friction with a table: $\mu_k = 0.2$ and 0.8. Each block is launched with speed $v_i = 1$ m/s and slides across the level table as the block comes to rest. This process represents the first two trials. For the next two trials, the procedure is repeated but the blocks are launched with speed $v_i = 2$ m/s. Rank the four trials (a) through (d) according to the stopping distance from largest to smallest. If the stopping distance is the same in two cases, give them equal rank. (a) $v_i = 1$ m/s, $\mu_k = 0.2$ (b) $v_i = 1$ m/s, $\mu_k = 0.8$ (c) $v_i = 2$ m/s, $\mu_k = 0.2$ (d) $v_i = 2$ m/s, $\mu_k = 0.8$.
10. At time t_i, the kinetic energy of a particle is 30.0 J and the potential energy of the system to which it belongs is 10.0 J. At some later time t_f, the kinetic energy of the particle is 18.0 J. (a) If only conservative forces act on the particle, what are the potential energy and the total energy of the system at time t_f? (b) If the potential energy of the system at time t_f is 5.00 J, are any non-conservative forces acting on the particle? (c) Explain your answer to part (b).

Problems

Section 7.2 Analysis models: isolated and non-isolated systems

1. A block of mass 0.250 kg is placed on top of a light, vertical spring of force constant 5000 N/m and pushed downward so that the spring is compressed by 0.100 m. After the block is released from rest, it travels upward and then leaves the spring. To what maximum height above the point of release does it rise?
2. A block of mass m = 5.00 kg is released from point Ⓐ and slides on the frictionless track shown in Figure P7.2. Determine (a) the block's speed at points Ⓑ and Ⓒ and (b) the net work done by the gravitational force on the block as it moves from point Ⓐ to point Ⓒ.

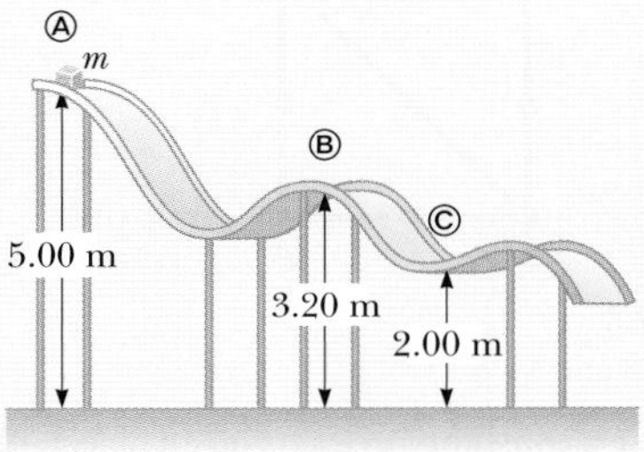

Figure P7.2

3. A bead slides without friction around a loop-the-loop (Figure P7.3). The bead is released from rest at a height $h = 3.50R$. (a) What is its speed at point Ⓐ? (b) How large is the normal force on the bead at point Ⓐ if its mass is 5.00 g?

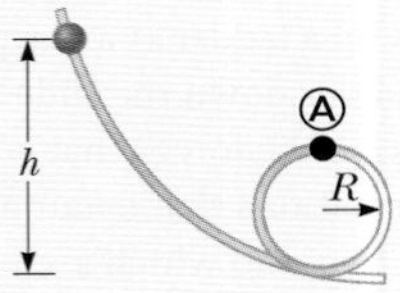

Figure P7.3

4. Two objects are connected by a light string passing over a light, frictionless pulley as shown in Figure P7.4. The object of mass $m_1 = 5.00$ kg is released from rest at a height $h = 4.00$ m above the table. Using the isolated system model, (a) determine the speed of the object of mass $m_2 = 3.00$ kg just as the 5.00 kg object hits the table and (b) find the maximum height above the table to which the 3.00 kg object rises.

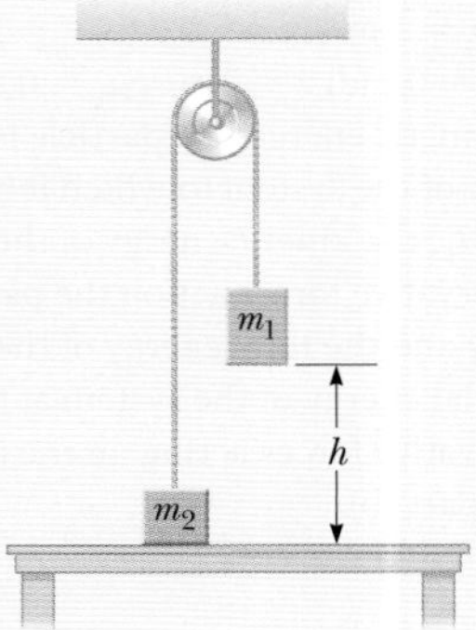

Figure P7.4

Section 7.3 Conservative and non-conservative forces

5. A 4.00 kg particle moves from the origin to position Ⓒ, having coordinates $x = 5.00$ m and $y = 5.00$ m (Figure P7.5). One force on the particle is the gravitational force acting in the negative y direction. Using Equation 6.1, calculate the work done by the gravitational force on the particle as it goes from O to Ⓒ along (a) the purple path, (b) the red path, and (c) the blue path. (d) Your results should all be identical. Why?

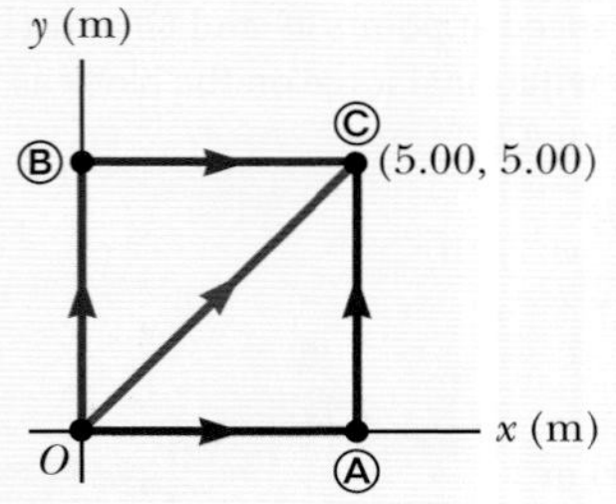

Figure P7.5
Problems 3 and 4

6. (a) Suppose a constant force acts on an object. The force does not vary with time or with the position or the velocity of the object. Start with the general definition for work done by a force

$$W = \int_i^f \vec{\mathbf{F}} \cdot d\vec{\mathbf{r}}$$

and show that the force is conservative. (b) As a special case, suppose the force $\vec{\mathbf{F}} = (3\hat{\mathbf{i}} + 4\hat{\mathbf{j}})$ N acts on a particle that moves from O to © in Figure P7.3. Calculate the work done by $\vec{\mathbf{F}}$ on the particle as it moves along each one of the three paths shown in the figure and show that the work done along the three paths is identical.

7. A child of mass m starts from rest and slides without friction from a height h along a slide next to a pool (Figure P7.7). She is launched from a height $h/5$ into the air over the pool. We wish to find the maximum height she reaches above the water in her projectile motion. (a) Is the child–Earth system isolated or non-isolated? Why? (b) Is there a non-conservative force acting within the system? (c) Define the configuration of the system when the child is at the water level as having zero gravitational potential energy. Express the total energy of the system when the child is at the top of the waterslide. (d) Express the total energy of the system when the child is at the launching point. (e) Express the total energy of the system when the child is at the highest point in her projectile motion. (f) From parts (c) and (d), determine her initial speed v_i at the launch point in terms of g and h. (g) From parts (d), (e), and (f), determine her maximum airborne height y_{max} in terms of h and the launch angle θ. (h) Would your answers be the same if the waterslide were not frictionless? Explain your answer.

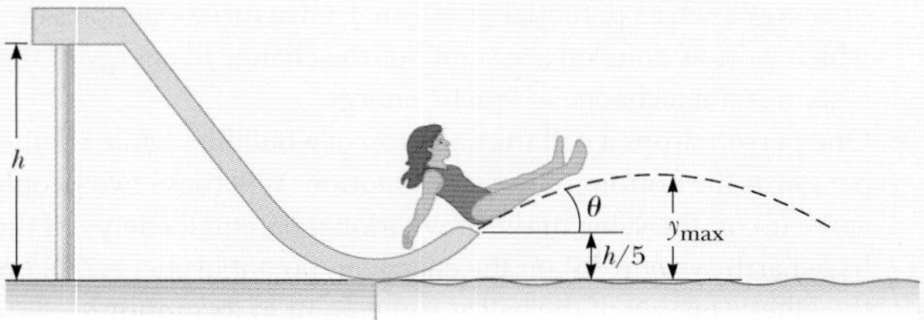

Figure P7.7

8. ±? A group of students is doing an experiment to investigate energy transformations, and very carefully launching blocks into the air using springs, ensuring they do not aim at any passing lecturers. A block of mass (0.250 ± 0.005) kg is placed on top of a light, vertical spring of force constant (5000 ± 500) N/m and pushed downwards so that the spring is compressed by (0.100 ± 0.002) m. Note that masses and distances can generally be measured more precisely than spring constants. After the block is released from rest, it travels upwards and then leaves the spring. To what maximum height above the point of release does it rise?

9. A toy cannon uses a spring to project a 5.30-g soft rubber ball. The spring is originally compressed by 5.00 cm and has a force constant of 8.00 N/m. When the cannon is fired, the ball moves 15.0 cm through the horizontal barrel of the cannon, and the barrel exerts a constant friction force of 0.0320 N on the ball. (a) With what speed does the projectile leave the barrel of the cannon? (b) At what point does the ball have maximum speed? (c) What is this maximum speed?

10. A 40.0 kg box initially at rest is pushed 5.00 m along a rough, horizontal floor with a constant applied horizontal force of 130 N. The coefficient of friction between box and floor is 0.300. Find (a) the work done by the applied force, (b) the increase in internal energy in the box–floor system as a result of friction, (c) the work done by the normal force, (d) the work done by the gravitational force, (e) the change in kinetic energy of the box and (f) the final speed of the box.

11. A block of mass $m = 2.00$ kg is attached to a spring of force constant $k = 500$ N/m as shown in **Figure P7.11**. The block is pulled to a position $x_i = 5.00$ cm to the right of equilibrium and released from rest. Find the speed the block has as it passes through equilibrium if (a) the horizontal surface is frictionless and (b) the coefficient of friction between block and surface is $\mu_k = 0.350$.

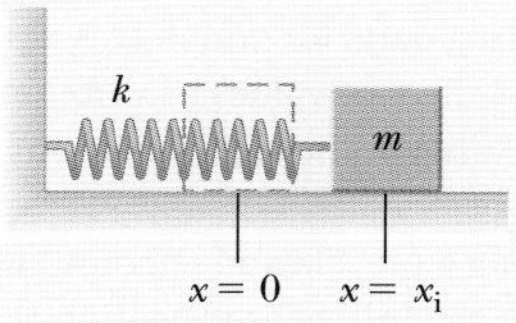

Figure P7.11

12. A crate of mass 10.0 kg is pulled up a rough incline with an initial speed of 1.50 m/s. The pulling force is 100 N parallel to the incline, which makes an angle of 20.0° with the horizontal. The coefficient of kinetic friction is 0.400, and the crate is pulled 5.00 m. (a) How much work is done by the gravitational force on the crate? (b) Determine the increase in internal energy of the crate–incline system owing to friction. (c) How much work is done by the 100 N force on the crate? (d) What is the change in kinetic energy of the crate? (e) What is the speed of the crate after being pulled 5.00 m?

13. As shown in **Figure P7.13**, a green bead of mass 25 g slides along a straight wire. The length of the wire from point Ⓐ to point Ⓑ is 0.600 m, and point Ⓐ is 0.200 m higher than point Ⓑ. A constant friction force of magnitude 0.0250 N acts on the bead. (a) If the bead is released from rest at point Ⓐ, what is its speed at point Ⓑ? (b) A red bead of mass 25 g slides along a curved wire, subject to a friction force with the same constant magnitude as that on the green bead. If the green and red beads are released simultaneously from rest at point Ⓐ, which bead reaches point Ⓑ with a higher speed? Explain.

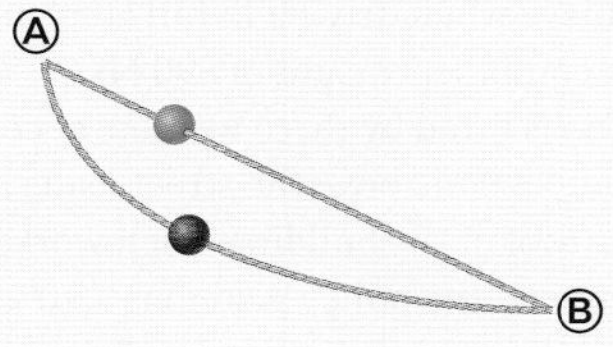

Figure P7.13

14. A 200 g block is pressed against a spring of force constant 1.40 kN/m until the block compresses the spring 10.0 cm. The spring rests at the bottom of a ramp inclined at 60.0° to the horizontal. Using energy considerations, determine how far up the incline the block moves from its initial position before it stops (a) if the ramp exerts no friction force on the block and (b) if the coefficient of kinetic friction is 0.400. (c) What is the efficiency of the process in part (b) for converting elastic potential energy to gravitational potential energy?

15. The coefficient of friction between the block of mass ±? $m_1 = (3.00 \pm 0.03)$ kg and the surface in **Figure P7.15** is $\mu_k = 0.40 \pm 0.04$. The system starts from rest. What is the speed of the ball of mass $m_2 = (5.00 \pm 0.05)$ kg when it has fallen a distance $h = 1.50$ m?

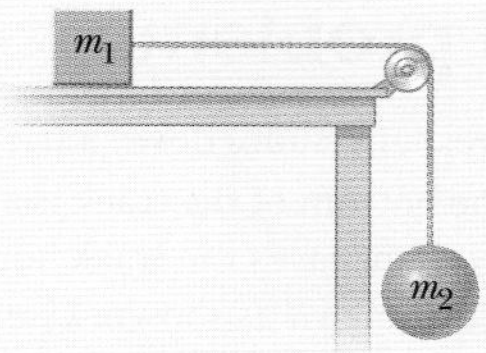

Figure P7.15

16. A 5.00 kg block is set into motion up an inclined plane with an initial speed of $v_i = 8.00$ m/s (**Figure P7.16**). The block comes to rest after travelling $d = 3.00$ m along the plane, which is inclined at an angle of $\theta = 30.0°$ to the horizontal. For this motion, determine (a) the change in the block's kinetic energy, (b) the change in the potential energy of the block–Earth system and (c) the friction force exerted on the block (assumed to be constant). (d) What is the coefficient of kinetic friction?

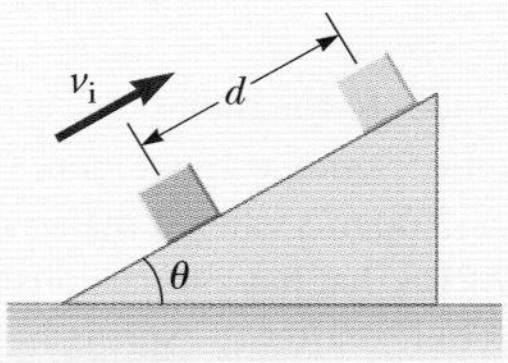

Figure P7.16

17. An 80.0 kg skydiver jumps out of a balloon at an altitude of 1000 m and opens his parachute at an altitude of 200 m. (a) Assuming the total retarding force on the skydiver is constant at 50.0 N with the parachute closed and constant at 3600 N with the parachute open, find the speed of the skydiver when he lands on the ground. (b) Do you think the skydiver will be injured? Explain your answer. (c) At what height should the parachute be opened so that the final speed of the skydiver when he hits the ground is 5.00 m/s? (d) How realistic is the assumption that the total retarding force is constant? Explain your answer.

Section 7.4 Energy bar charts

18. Consider the popgun in Example 7.8. Suppose the projectile mass, compression distance and spring constant remain the same as given or calculated in the example. Suppose, however, there is a friction force of magnitude 2.00 N acting on the projectile as it rubs against the interior of the barrel. The vertical length from point Ⓐ to the end of the barrel is 0.600 m. (a) After the spring is compressed and the popgun fired, to what height does the projectile rise above point Ⓑ? (b) Draw four energy bar charts for this situation, analogous to those in **Active Figures 7.13c–d**.

19. A boy in a wheelchair (total mass 47.0 kg) has speed 1.40 m/s at the crest of a slope 2.60 m high and 12.4 m long. At the bottom of the slope his speed is 6.20 m/s. Assume air resistance and rolling resistance can be modelled as a constant friction force of 41.0 N. (a) Find the work he did in pushing forwards on his wheels during the downhill ride. (b) Draw energy bar charts for the boy and his wheelchair at the crest and bottom of the slope.

Additional problems

20. A boy starts at rest and slides down a frictionless slide as in Figure P7.20. The bottom of the track is a height h above the ground. The boy then leaves the track horizontally, striking the ground at a distance d as shown. Using energy methods, determine the initial height H of the boy above the ground in terms of h and d.

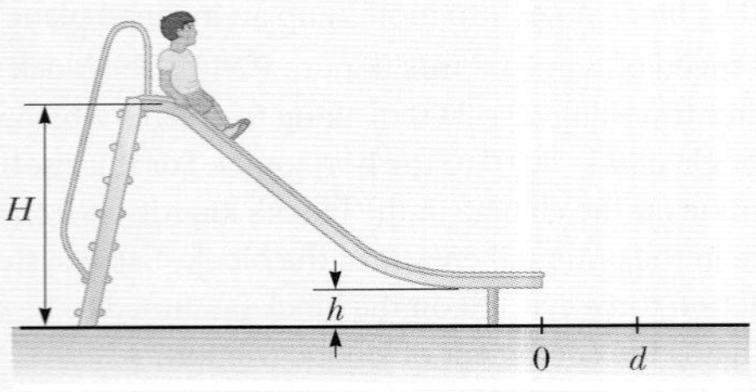

Figure P7.20

21. A 1.00-kg object slides to the right on a surface having a coefficient of kinetic friction 0.250 (Figure P7.21a). The object has a speed of $v_i = 3.00$ m/s when it makes contact with a light spring (Figure P7.21b) that has a force constant of 50.0 N/m. The object comes to rest after the spring has been compressed a distance d (Figure P7.21c). The object is then forced toward the left by the spring (Figure P7.62d) and continues to move in that direction beyond the spring's unstretched position. Finally, the object comes to rest a distance D to the left of the unstretched spring (Figure P7.21e). Find (a) the distance of compression d, (b) the speed v at the unstretched position when the object is moving to the left (Figure P7.21d), and (c) the distance D where the object comes to rest.

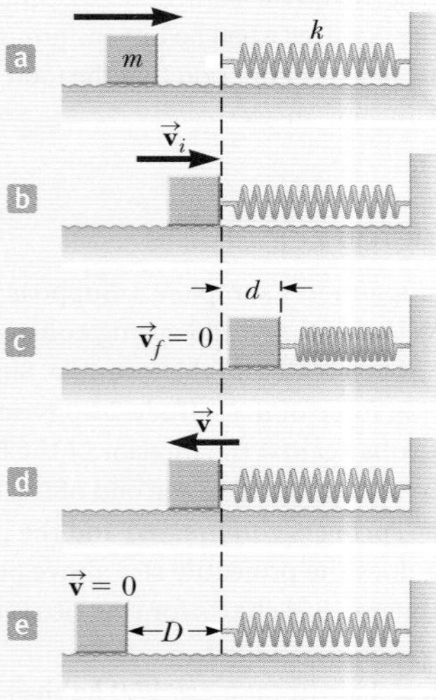

Figure P7.21

22. A 4.00 kg particle moves along the x axis. Its position varies with time according to $x = t + 2.0t^3$, where x is in metres and t is in seconds. Find (a) the kinetic energy of the particle at any time t, (b) the acceleration of the particle and the force acting on it at time t, (c) the power being delivered to the particle at time t, and (d) the work done on the particle in the interval $t = 0$ to $t = 2.00$ s.

23. Consider the energy conservation equation

$$\frac{1}{2}(46.0\text{ kg})(2.40\text{ m/s})^2 + (46.0\text{kg})(9.80\text{ m/s}^2)(2.80\text{m} + x)$$
$$= \frac{1}{2}(1.94 \times 10^4\text{ N/m})x^2$$

(a) Solve the equation for x. (b) Compose the statement of a problem, including data, for which this equation gives the solution. (c) Add the two values of x obtained in part (a) and divide by 2. (d) What is the significance of the resulting value in part (c)?

24. Jonathan is riding a bicycle and encounters a hill of height h. At the base of the hill, he is travelling at a speed v_i. When he reaches the top of the hill, he is travelling at a speed v_f. Jonathan and his bicycle together have a mass m. Ignore friction in the bicycle mechanism and between the bicycle tyres and the road. (a) What is the total external work done on the system of Jonathan and the bicycle between the time he starts up the hill and the time he reaches the top? (b) What is the change in potential energy stored in Jonathan's body during this process? (c) How much work does Jonathan do on the bicycle pedals within the Jonathan–bicycle–Earth system during this process?

25. A wind turbine on a wind farm turns in response to a force of high-speed air resistance, $R = \frac{1}{2}D\rho Av^2$. The power available is $P = Rv = \frac{1}{2}D\rho\pi r^2v^3$, where v is the wind speed and we have assumed a circular face for the wind turbine of radius r. Take the drag coefficient as $D = 1.00$. For a wind turbine having $r = 1.50$ m, calculate the power available with (a) $v = 8.00$ m/s and (b) $v = 24.0$ m/s. The power delivered to the generator is limited by the efficiency of the system, about 25%. For comparison, a house typically uses around 1 kW of power.

26. ±? An electric scooter has a battery that is capable of supplying (120 ± 10) Wh of energy. If friction forces and other losses account for (60 ± 5)% of the energy usage, what altitude change can a rider achieve when driving in hilly terrain if the rider and scooter have a combined weight of 890 N?

27. A 3.50 kN piano is lifted by three workers at constant speed to an apartment 25.0 m above the street using a pulley system fastened to the roof of the building. Each worker is able to deliver 165 W of power and the pulley system is 75.0% efficient (so that 25.0% of the mechanical energy is transformed to other forms due to friction in the pulley). Neglecting the mass of the pulley, find the time required to lift the piano from the street to the apartment.

28. A child's pogo stick (Figure P7.28) stores energy in a spring with a force constant of 2.50×10^4 N/m. At position Ⓐ ($x_{Ⓐ} = -0.100$ m), the spring compression is a maximum and the child is momentarily at rest. At position Ⓑ ($x_{Ⓑ} = 0$), the spring is relaxed and the child is moving upwards. At position Ⓒ, the child is again momentarily at rest at the top of the jump. The combined mass of the child and pogo stick is 25.0 kg. Although the boy must lean forward to remain balanced, the angle is small, so let's assume the pogo stick is vertical. Also assume the boy does not bend his legs during the motion. (a) Calculate the total energy of the child–stick–Earth system, taking both gravitational and elastic potential energies as zero for $x = 0$. (b) Determine $x_{Ⓒ}$. (c) Calculate the speed of the child at $x = 0$.

(d) Determine the value of x for which the kinetic energy of the system is a maximum. (e) Calculate the child's maximum upwards speed.

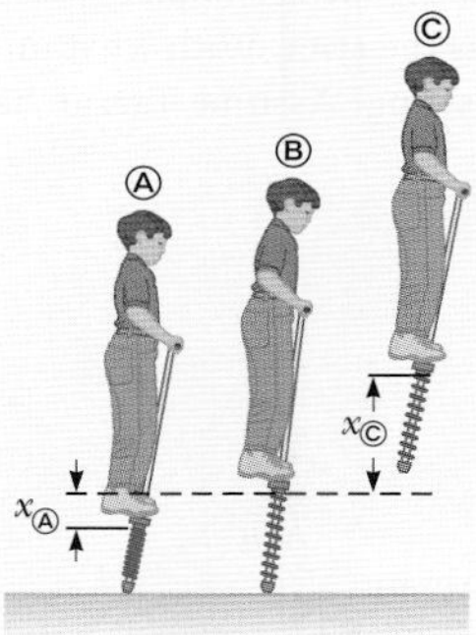

Figure P7.28

29. A block of mass $m_1 = 20.0$ kg is connected to a block of mass $m_2 = 30.0$ kg by a massless string that passes over a light, frictionless pulley. The 30.0 kg block is connected to a spring that has negligible mass and a force constant of $k = 250$ N/m, as shown in Figure P7.29. The spring is unstretched when the system is as shown in the figure, and the incline is frictionless. The 20.0 kg block is pulled a distance $h = 20.0$ cm down the incline of angle $\theta = 40.0°$ and released from rest. Find the speed of each block when the spring is again unstretched.

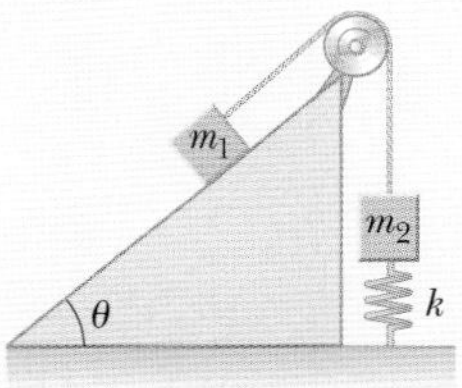

Figure P7.29

30. A 10.0-kg block is released from rest at point Ⓐ in Figure P7.30. The track is frictionless except for the portion between points Ⓑ and Ⓒ, which has a length of 6.00 m. The block travels down the track, hits a spring of force constant 2250 N/m, and compresses the spring 0.300 m from its equilibrium position before coming to rest momentarily. Determine the coefficient of kinetic friction between the block and the rough surface between points Ⓑ and Ⓒ.

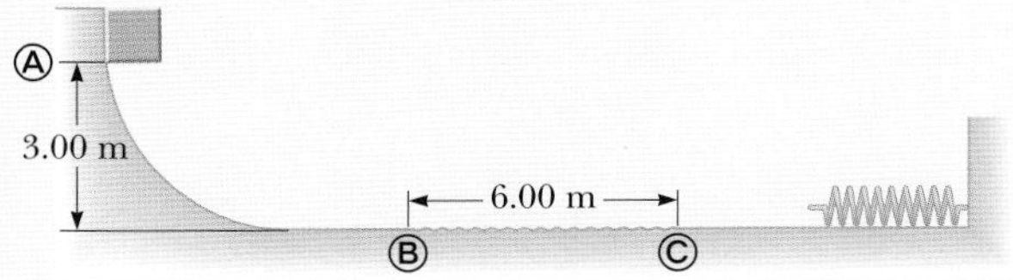

Figure P7.30

31. A pendulum, comprising a light string of length L and a small sphere, swings in the vertical plane. The string hits a peg located a distance d below the point of suspension (Figure P7.31). (a) Show that if the sphere is released from a height below that of the peg, it will return to this height after the string strikes the peg. (b) Show that if the pendulum is released from rest at the horizontal position ($\theta = 90°$) and is to swing in a complete circle centred on the peg, the minimum value of d must be $3L/5$.

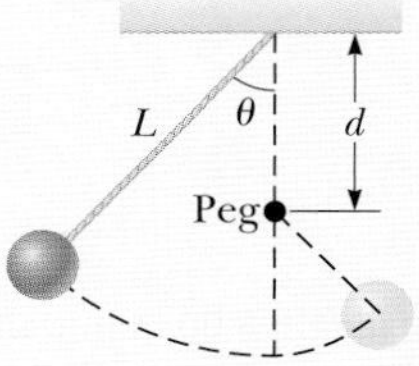

Figure P7.31

32. A ball whirls around in a *vertical* circle at the end of a string. The other end of the string is fixed at the centre of the circle. Assuming the total energy of the ball–Earth system remains constant, show that the tension in the string at the bottom is greater than the tension at the top by six times the ball's weight.

Challenge problems

33. Starting from rest, a 64.0 kg person bungee jumps from a tethered hot-air balloon 65.0 m above the ground. The bungee cord has negligible mass and unstretched length 25.8 m. One end is tied to the basket of the balloon and the other end to a harness around the person's body. The cord is modelled as a spring that obeys Hooke's law with a spring constant of 81.0 N/m, and the person's body is modelled as a particle. The hot-air balloon does not move. (a) Express the gravitational potential energy of the person–Earth system as a function of the person's variable height y above the ground. (b) Express the elastic potential energy of the cord as a function of y. (c) Express the total potential energy of the person–cord–Earth system as a function of y. (d) Plot a graph of the gravitational, elastic and total potential energies as functions of y. (e) Assume air resistance is negligible. Determine the minimum height of the person above the ground during his plunge. (f) Does the potential energy graph show any equilibrium position or positions? If so, at what elevations? Are they stable or unstable? (g) Determine the jumper's maximum speed.

34. A ball of mass $m = 300$ g is connected by a strong string of length $L = 80.0$ cm to a pivot and held in place with the string vertical. A wind exerts constant force F to the right on the ball as shown in Figure P7.34. The ball is released from rest. The wind makes it swing up to attain maximum height H above its starting point before it swings down again. (a) Find H as a function of F. Evaluate H for (b) $F = 1.00$ N and (c) $F = 10.0$ N. How does H behave (d) as F approaches zero and (e) as F approaches infinity? (f) Now consider the equilibrium height of the ball with the wind blowing. Determine it as a function of F. Evaluate the equilibrium height for (g) $F = 10$ N and (h) F going to infinity.

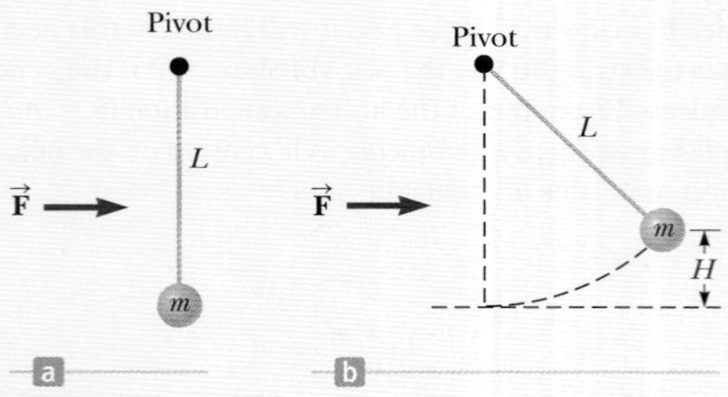

Figure P7.34

35. A uniform chain of length 8.00 m initially lies stretched out on a horizontal table. (a) Assuming the coefficient of static friction between chain and table is 0.600, show that the chain will begin to slide off the table if at least 3.00 m of it hangs over the edge of the table. (b) Determine the speed of the chain as its last link leaves the table, given that the coefficient of kinetic friction between the chain and the table is 0.400.

36. Jane, whose mass is 50.0 kg, needs to swing across a river (of width D) filled with person-eating crocodiles to save Tarzan from danger. She must swing into a wind exerting constant horizontal force $\vec{\mathbf{F}}$, on a vine of length L and initially making an angle θ with the vertical (**Figure P7.36**). Take $D = 50.0$ m, $F = 110$ N, $L = 40.0$ m and $\theta = 50.0°$. (a) With what minimum speed must Jane begin her swing to just make it to the other side? (b) Once the rescue is complete, Tarzan and Jane must swing back across the river. With what minimum speed must they begin their swing? Assume Tarzan has a mass of 80.0 kg.

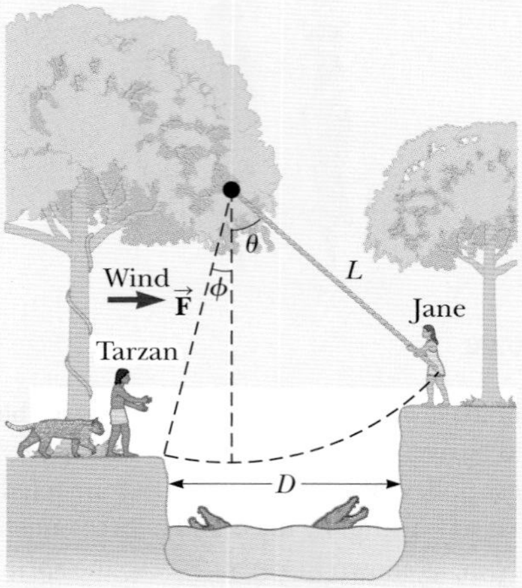

Figure P7.36

chapter 8

Linear momentum and collisions

Shutterstock.com/Celso Diniz

The picture shows a rocket blasting off. A rocket in space is an isolated system. So how does a rocket move in space, when it has nothing to push against? How can we analyse its motion?

On completing this chapter, students will understand:

- how linear momentum is related to mass and velocity
- that momentum is conserved for an isolated system
- that forces cause changes in momentum
- the significance of the centre of mass of a system of particles or extended object
- why objects undergoing collisions, such as cars or billiard balls, behave as they do
- how rockets are propelled.

Students will be able to:

- calculate the momentum of a moving particle
- apply the conservation of momentum principle to solve problems
- analyse problems involving collisions in one and two dimensions
- distinguish between isolated and non-isolated systems (momentum)
- choose and apply the appropriate analysis model for solving problems using a momentum approach
- find the centre of mass of a system of many particles or extended objects
- apply the principle of conservation of momentum to analyse the motion of a rocket.

In this chapter we introduce a second conservation law, that of conservation of momentum. The momentum of an object is related to both its mass and its velocity. We can identify new momentum versions of analysis models for isolated and non-isolated systems. These models are especially useful for treating problems that involve collisions between objects and for analysing rocket propulsion, and can be applied to objects on all scales – from galaxies to subatomic particles.

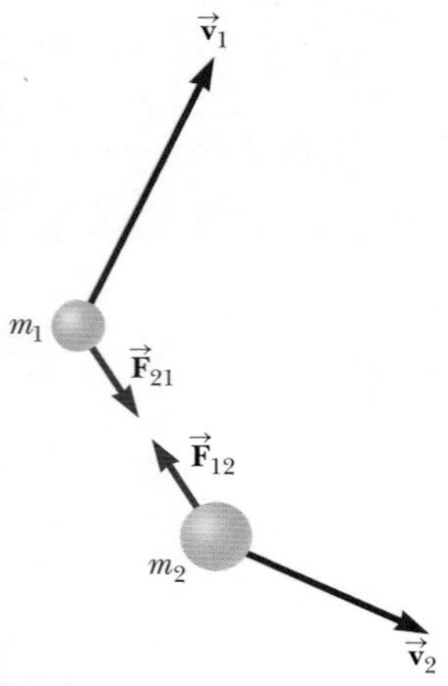

Figure 8.1
Two particles interact with each other. According to Newton's third law, we must have $\vec{F}_{12} = -\vec{F}_{21}$.

8.1 Linear momentum

In Chapters 6 and 7, we studied situations that are difficult to analyse with Newton's laws. We were able to solve problems involving these situations by identifying a system and applying a conservation principle: conservation of energy.

We now introduce a new quantity that describes motion, *linear momentum*. To generate this new quantity, consider an isolated system of two particles (**Figure 8.1**) with masses m_1 and m_2 moving with velocities $\vec{v}_1$ and $\vec{v}_2$ at an instant of time. Because the system is isolated, the only force on one particle is that from the other particle. If a force from particle 1 (for example, a gravitational force) acts on particle 2, there must be a second force – equal in magnitude but opposite in direction – that particle 2 exerts on particle 1. That is, the forces on the particles form a Newton's third law action–reaction pair, and $\vec{F}_{12} = -\vec{F}_{21}$. We can express this condition as

$$\vec{F}_{12} + \vec{F}_{21} = 0$$

Let us further analyse this situation by incorporating Newton's second law. At the instant shown in **Figure 8.1**, the interacting particles in the system have accelerations corresponding to the forces on them. Therefore, replacing the force on each particle with $m\vec{a}$ for the particle gives

$$m_1\vec{a}_1 + m_2\vec{a}_2 = 0$$

Now we replace each acceleration with its definition from **Equation 3.16**:

$$m_1\frac{d\vec{v}_1}{dt} + m_2\frac{d\vec{v}_2}{dt} = 0$$

If the masses m_1 and m_2 are constant, we can bring them inside the derivative operation, which gives

$$\frac{d(m_1\vec{v}_1)}{dt} + \frac{d(m_2\vec{v}_2)}{dt} = 0$$

$$\frac{d}{dt}(m_1\vec{v}_1 + m_2\vec{v}_2) = 0 \qquad (8.1)$$

We call the quantities $m_1\mathbf{v}_1$ and $m_2\mathbf{v}_2$ the **linear momentum** of particles 1 and 2.

Σ *The rules for manipulating derivatives are given in Appendix B.7.*

The *linear momentum* of a particle or an object that can be modelled as a particle of mass m moving with a velocity $\vec{v}$ is defined to be the product of the mass and velocity of the particle:

Definition of linear momentum of a particle ▶

$$\vec{p} \equiv m\vec{v} \qquad (8.2)$$

Linear momentum, or momentum as we shall refer to it in this chapter, is a vector quantity because it equals the product of a scalar quantity m and a vector quantity $\vec{v}$. Its direction is along $\vec{v}$, it has dimensions ML/T and its SI unit is kg.m/s.

If a particle is moving in an arbitrary direction, $\vec{p}$ has three components and **Equation 8.2** is equivalent to the component equations

$$p_x = mv_x \qquad p_y = mv_y \qquad p_z = mv_z$$

As you can see from its definition, the concept of momentum provides a quantitative distinction between heavy and light particles moving at the same velocity. For example, the momentum of a bowling ball is much greater than that of a tennis ball moving at the same speed. Newton called the product $m\vec{v}$ the *quantity of motion*; our present-day word *momentum* comes from the Latin word for movement.

Momentum is conserved ▶

Notice that in **Equation 8.1** the derivative of the sum $m_1\vec{v}_1 + m_2\vec{v}_2$ with respect to time is zero. Consequently, this sum must be constant. Hence **momentum is conserved in an isolated system where the masses of the individual components are constant.** This is true in general. Whenever two particles interact, the change in momentum of one is equal to the negative of the change in momentum of the other. Newton's third law is a statement of conservation of momentum.

Using Newton's second law of motion, we can relate the linear momentum of a particle to the resultant force acting on the particle. We start with Newton's second law and substitute the definition of acceleration:

$$\sum \vec{\mathbf{F}} = m\vec{\mathbf{a}} = m\frac{d\vec{\mathbf{v}}}{dt}$$

In Newton's second law, the mass m is assumed to be constant. Therefore, we can bring m inside the derivative operation to give us

$$\sum \vec{\mathbf{F}} = \frac{d(m\vec{\mathbf{v}})}{dt} = \frac{d\vec{\mathbf{p}}}{dt} \tag{8.3}$$

◀ Newton's second law for a particle

This equation shows that **the time rate of change of the linear momentum of a particle is equal to the net force acting on the particle.**

This alternative form of Newton's second law is the form in which Newton presented the law and it is actually more general than the form introduced in Chapter 3. In addition to situations in which the velocity vector varies with time, we can use **Equation 8.3** to study phenomena in which the mass changes. For example, the mass of a rocket changes as fuel is burned and ejected from the rocket. We cannot use $\sum \vec{\mathbf{F}} = m\vec{\mathbf{a}}$ to analyse rocket propulsion; we must use a momentum approach, as we will show in Section 8.8.

Quick Quiz 8.1

Two objects have equal kinetic energies. How do the magnitudes of their momenta compare? **(a)** $p_1 < p_2$ **(b)** $p_1 = p_2$ **(c)** $p_1 > p_2$ **(d)** not enough information to tell

Quick Quiz 8.2

Your physics lecturer throws a cricket ball to you at a certain speed and you catch it. The lecturer is next going to throw you a medicine ball whose mass is 10 times the mass of the cricket ball. She gives you the following choices: You can have the medicine ball thrown with **(a)** the same speed as the cricket ball, **(b)** the same momentum, or **(c)** the same kinetic energy. Rank these choices from easiest to hardest to catch.

8.2 Analysis model: isolated system (momentum)

Using the definition of momentum, **Equation 8.1** can be written

$$\frac{d}{dt}(\vec{\mathbf{p}}_1 + \vec{\mathbf{p}}_2) = 0$$

Because the time derivative of the total momentum $\vec{\mathbf{p}}_{\text{tot}} = \vec{\mathbf{p}}_1 + \vec{\mathbf{p}}_2$ is *zero*, we conclude that the *total* momentum of the isolated system of the two particles in **Figure 8.1** must remain constant:

$$\vec{\mathbf{p}}_{\text{tot}} = \text{constant} \tag{8.4}$$

or, equivalently,

$$\Delta\, \vec{\mathbf{p}}_{\text{tot}} = 0 \tag{8.5}$$

For the system of two particles, **Equation 8.5** can be written as

$$\vec{\mathbf{p}}_{1i} + \vec{\mathbf{p}}_{2i} = \vec{\mathbf{p}}_{1f} + \vec{\mathbf{p}}_{2f}$$

where $\vec{\mathbf{p}}_{1i}$ and $\vec{\mathbf{p}}_{2i}$ are the initial values and $\vec{\mathbf{p}}_{1f}$ and $\vec{\mathbf{p}}_{2f}$ are the final values of the momenta for the two particles for the time interval during which the particles interact. This equation in component form demonstrates that the total momenta in the x, y, and z directions are all independently conserved:

$$p_{1ix} + p_{2ix} = p_{1fx} + p_{2fx} \qquad p_{1iy} + p_{2iy} = p_{1fy} + p_{2fy} \qquad p_{1iz} + p_{2iz} = p_{1fz} + p_{2fz} \tag{8.6}$$

The ▶ momentum version of the isolated system model

Equation 8.5 is the mathematical statement of a new analysis model, the **isolated system (momentum).** It can be extended to any number of particles in an isolated system, as we show in Section 8.7. We studied the energy version of the isolated system model in Chapter 7 and now we have a momentum version. In general, Equation 8.5 can be stated in words as follows: ***Whenever two or more particles in an isolated system interact, the total momentum of the system remains constant.***

Pitfall Prevention 8.1

Momentum of an isolated *system* is conserved. Although the momentum of an isolated *system* is conserved, the momentum of one particle within an isolated system is not necessarily conserved because other particles in the system may be interacting with it. Do not apply conservation of momentum to a single particle.

This statement tells us that the total momentum of an isolated system at all times equals its initial momentum.

Notice that we have made no statement concerning the type of forces acting on the particles of the system. Furthermore, we have not specified whether the forces are conservative or non-conservative. We have also not indicated whether or not the forces are constant. The only requirement is that the forces must be *internal* to the system. This single requirement indicates the power of this new model.

TRY THIS

Stand on a skateboard (or similar) and throw a heavy ball away from you as hard as you can. In which direction do you move relative to the throw? Which moves faster, you or the ball?

Analysis Model 8.1

Isolated system (momentum)

Imagine you have identified a system to be analysed and have defined a system boundary. If there are no external forces on the system, the system is *isolated*. In that case, the total momentum of the system, which is the vector sum of the momenta of all members of the system, is conserved:

$$\Delta \vec{\mathbf{p}}_{\text{tot}} = 0 \qquad (8.5)$$

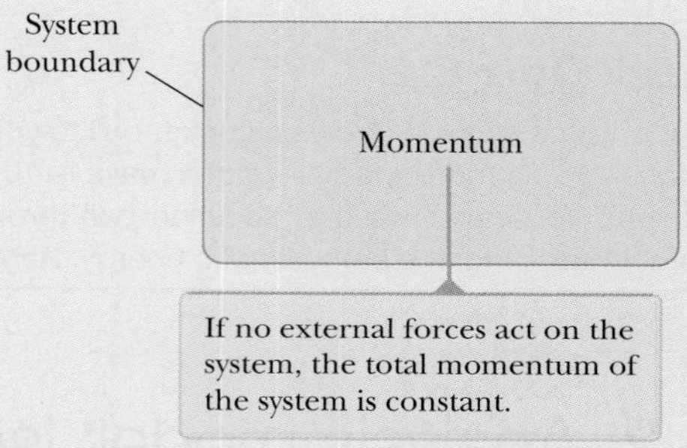
System boundary
Momentum
If no external forces act on the system, the total momentum of the system is constant.

Examples

- a cue ball strikes another ball on a pool table
- a spacecraft fires its rockets and moves faster through space
- molecules in a gas at a specific temperature move about and strike each other (Chapter 21)
- an incoming particle strikes a nucleus, creating a new nucleus and a different outgoing particle (Chapter 43)
- an electron and a positron annihilate to form two outgoing photons (Chapter 44)

Example 8.1

The archer

An archer stands at rest on frictionless ice and fires a 0.50 kg arrow horizontally at 50 m/s (Figure 8.2). The combined mass of the archer and the bow is 60 kg. With what velocity does the archer move across the ice after firing the arrow?

Figure 8.2
(Example 8.1) An archer fires an arrow horizontally to the right. Because he is standing on frictionless ice, he will begin to slide to the left across the ice.

Example 8.1 cont.

Solution

Conceptualise Imagine the arrow being fired one way and the archer recoiling in the opposite direction, as you would have experienced if you did the *Try this* example above. Take the system to be the archer, bow and arrow. Draw a diagram showing the two components of the system as the archer plus bow moving one way and the arrow moving the other.

Model We cannot easily solve this problem with models based on force or energy, but conservation of momentum offers a simple and elegant approach to problems such as this.

The system is not isolated because the gravitational force and the normal force from the ice act on the system. These forces, however, are vertical and perpendicular to the motion of the system. Therefore, there are no external forces in the horizontal direction and we can apply the isolated system (momentum) analysis model in terms of momentum components in this direction.

Analyse The total horizontal momentum of the system before the arrow is fired is zero because nothing in the system is moving. Therefore, the total horizontal momentum of the system after the arrow is fired must also be zero. We choose the direction of firing of the arrow as the positive x direction. Identifying the archer as particle 1 and the arrow as particle 2, we have $m_1 = 60$ kg, $m_2 = 0.50$ kg, and $\vec{\mathbf{v}}_{2f} = 50\hat{\mathbf{i}}$ m/s.

Using the isolated system (momentum) model, set the final momentum of the system equal to the initial value of zero:

$$m_1\vec{\mathbf{v}}_{1f} + m_2\vec{\mathbf{v}}_{2f} = 0$$

Solve this equation for $\vec{\mathbf{v}}_{1f}$ and substitute numerical values:

$$\vec{\mathbf{v}}_{1f} = -\frac{m_2}{m_1}\vec{\mathbf{v}}_{2f} = -\left(\frac{0.50\text{ kg}}{60\text{ kg}}\right)(50\hat{\mathbf{i}}\text{ m/s}) = -0.42\hat{\mathbf{i}}\text{ m/s}$$

Finalise The negative sign for $\vec{\mathbf{v}}_{1f}$ indicates that the archer is moving to the left in **Figure 8.2** after the arrow is fired, in the direction opposite the direction of motion of the arrow, in accordance with Newton's third law. Because the archer is much more massive than the arrow, his acceleration and consequent velocity are much smaller than the acceleration and velocity of the arrow.

What If? What if the arrow were fired in a direction that makes an angle θ with the horizontal? How will that change the recoil velocity of the archer?

Answer The recoil velocity should decrease in magnitude because only a component of the velocity of the arrow is in the x direction. Conservation of momentum in the x direction gives

$$m_1 v_{1f} + m_2 v_{2f}\cos\theta = 0$$

leading to

$$v_{1f} = -\frac{m_2}{m_1}v_{2f}\cos\theta$$

For $\theta = 0$, $\cos\theta = 1$ and the final velocity of the archer reduces to the value when the arrow is fired horizontally. For non-zero values of θ, the cosine function is less than 1 and the recoil velocity is less than the value calculated for $\theta = 0$. If $\theta = 90°$, then $\cos\theta = 0$ and $v_{1f} = 0$, so there is no recoil velocity. In this case, the archer is simply pushed downwards harder against the ice as the arrow is fired, and as long as the ice does not break he remains at equilibrium in the vertical direction.

Example 8.2

Can we really ignore the kinetic energy of the Earth?

In Section 6.5, we claimed that we can ignore the kinetic energy of the Earth when considering the energy of a system consisting of the Earth and a dropped ball. Verify this claim.

Example 8.2 cont.

Solution

Conceptualise Imagine dropping a ball at the surface of the Earth. From your point of view, the ball falls while the Earth remains stationary. By Newton's third law, however, the Earth experiences an upwards force and therefore an upwards acceleration while the ball falls. Can this motion be ignored?

Model We identify the system as the ball and the Earth. We assume there are no forces on the system from outer space, so the system is isolated. Let's use the momentum version of the isolated system analysis model.

Analyse We begin by setting up a ratio of the kinetic energy of the Earth to that of the ball. We identify v_E and v_b as the speeds of the Earth and the ball, respectively, after the ball has fallen through some distance.

Use the definition of kinetic energy to set up this ratio:

$$\frac{K_E}{K_b} = \frac{\frac{1}{2}m_E v_E^2}{\frac{1}{2}m_b v_b^2} = \left(\frac{m_E}{m_b}\right)\left(\frac{v_E}{v_b}\right)^2 \quad (1)$$

Apply the isolated system (momentum) model; the initial momentum of the system is zero, so set the final momentum equal to zero:

$$p_i = p_f \rightarrow 0 = m_b v_b + m_E v_E$$

Solve the equation for the ratio of speeds:

$$\frac{v_E}{v_b} = -\frac{m_b}{m_E}$$

Substitute this expression for v_E/v_b in **Equation (1)**:

$$\frac{K_E}{K_b} = \left(\frac{m_E}{m_b}\right)\left(-\frac{m_b}{m_E}\right)^2 = \frac{m_b}{m_E}$$

Substitute order-of-magnitude numbers for the masses:

$$\frac{K_E}{K_b} = \frac{m_b}{m_E} \sim \frac{1\text{ kg}}{10^{25}\text{ kg}} \sim 10^{-25}$$

Finalise The kinetic energy of the Earth is a very small fraction of the kinetic energy of the ball, so we are justified in ignoring it in the kinetic energy of the system.

8.3 Analysis model: non-isolated system (momentum)

In this section, we consider a *non-isolated system*. For energy considerations, a system is non-isolated if energy transfers across the boundary of the system by any of the means listed in Section 6.1. For momentum considerations, a system is non-isolated if a net force acts on the system for a time interval. In this case, we can imagine momentum being transferred to the system from the environment by means of the net force.

Assume a net force $\sum\vec{\mathbf{F}}$ acts on a particle and this force may vary with time. According to Newton's second law, $\sum\vec{\mathbf{F}} = d\vec{\mathbf{p}}/dt$, or

$$d\vec{\mathbf{p}} = \sum\vec{\mathbf{F}}\,dt \quad (8.7)$$

We can integrate this expression to find the change in the momentum of a particle when the force acts over some time interval. If the momentum of the particle changes from $\vec{\mathbf{p}}_i$ at time t_i to $\vec{\mathbf{p}}_f$ at time t_f, integrating **Equation 8.7** gives

$$\Delta\vec{\mathbf{p}} = \vec{\mathbf{p}}_f - \vec{\mathbf{p}}_i = \int_{t_i}^{t_f}\sum\vec{\mathbf{F}}\,dt \quad (8.8)$$

To evaluate the integral, we need to know how the net force varies with time. The quantity on the right side of this equation is a vector called the **impulse** of the net force $\sum\vec{\mathbf{F}}$ acting on a particle over the time interval $\Delta t = t_f - t_i$:

Impulse of a force ►

$$\vec{\mathbf{I}} \equiv \int_{t_i}^{t_f}\sum\vec{\mathbf{F}}\,dt \quad (8.9)$$

From its definition, we see that impulse $\vec{\mathbf{I}}$ is a vector quantity having a magnitude equal to the area under the force–time curve as described in **Figure 8.3a**. It is assumed the force varies in time in the general manner shown in the figure and is non-zero in the time interval $\Delta t = t_f - t_i$. The direction of the impulse vector is the same as the direction of the change in momentum. Impulse has the dimensions of momentum, that is, ML/T. Impulse is *not* a property of a particle; rather, it is a measure of the degree to which an external force changes the particle's momentum. Combining **Equations 8.8** and **8.9** gives us an important statement known as the **impulse–momentum theorem**:

The change in the momentum of a particle is equal to the impulse of the net force acting on the particle:

$$\Delta \vec{\mathbf{p}} = \vec{\mathbf{I}} \tag{8.10}$$

◀ Impulse–momentum theorem for a particle

This statement is equivalent to Newton's second law. When we say that an impulse is given to a particle, we mean that momentum is transferred from an external agent to that particle. **Equation 8.10** is the most general statement of the principle of **conservation of momentum** and is called the **conservation of momentum equation**. Note that **Equation 8.5** is a special case of **Equation 8.10** in which no external forces act. In practice, it is common to take **Equation 8.5** as a statement of conservation of momentum and choose the system such that there are no external forces.

The left side of **Equation 8.10** represents the change in the momentum of the system, which in this case is a single particle. The right side is a measure of how much momentum crosses the boundary of the system due to the net force being applied to the system. **Equation 8.10** is the mathematical statement of a new analysis model, the **non-isolated system (momentum)** model. Note that **Equation 8.10** is a vector equation, whereas **Equation 7.1** (conservation of energy) is a scalar equation. Therefore, directions are important for **Equation 8.10**. As there is only one type of momentum and one way of transferring it, there is no expansion of **Equation 8.10** analogous to **Equation 7.2**.

Because the net force imparting an impulse to a particle can generally vary in time, it is convenient to define a time-averaged net force:

$$\left(\sum \vec{\mathbf{F}}\right)_{\text{avg}} \equiv \frac{1}{\Delta t}\int_{t_i}^{t_f} \sum \vec{\mathbf{F}}\, dt \tag{8.11}$$

where $\Delta t = t_f - t_i$. (This equation is an application of the mean value theorem of calculus.) Therefore, we can express **Equation 8.9** as

$$\vec{\mathbf{I}} = \left(\sum \vec{\mathbf{F}}\right)_{\text{avg}} \Delta t \tag{8.12}$$

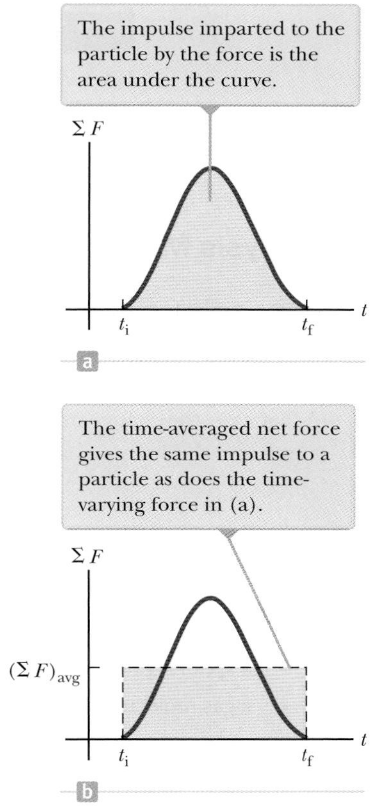
The impulse imparted to the particle by the force is the area under the curve.
ΣF
t_i
t_f
t
a
The time-averaged net force gives the same impulse to a particle as does the time-varying force in (a).
ΣF
$(\Sigma F)_{\text{avg}}$
t_i
t_f
t
b

Figure 8.3
(a) A net force acting on a particle may vary in time (b) The value of the constant force $(\Sigma F)_{\text{avg}}$ (horizontal dashed line) is chosen so that the area $(\Sigma F)_{\text{avg}} \Delta t$ of the rectangle is the same as the area under the curve in (a).

This time-averaged force, shown in **Figure 8.3b**, can be interpreted as the constant force that would give to the particle in the time interval Δt the same impulse that the time-varying force gives over this same interval.

In principle, if $\sum \vec{\mathbf{F}}$ is a known function of time, the impulse can be calculated from **Equation 8.9**. The calculation becomes especially simple if the force acting on the particle is constant. In this case, $\sum \vec{\mathbf{F}}_{\text{avg}} = \sum \vec{\mathbf{F}}$, where $\sum \vec{\mathbf{F}}$ is the constant net force, and **Equation 8.12** becomes

$$\vec{\mathbf{I}} = \sum \vec{\mathbf{F}}\, \Delta t \tag{8.13}$$

In many physical situations, we shall use what is called the **impulse approximation,** in which we assume one of the forces exerted on a particle acts for a short time but is much greater than any other force present. In this case, the net force $\sum \vec{\mathbf{F}}$ in **Equation 8.9** is replaced with a single force $\vec{\mathbf{F}}$ to find the impulse on the particle. This approximation is especially useful in treating collisions in which the duration of the collision is very short. For example, when a ball is struck with a bat, the time of the collision is about 0.01 s and the average force that the bat exerts on the ball in this time is typically several thousand newtons. Because this contact force is much greater than the magnitude of the gravitational force, the impulse approximation justifies our ignoring the gravitational forces exerted on the ball and bat.

Pitfall Prevention 8.2

When we use the impulse approximation, it is important to remember that $\vec{\mathbf{p}}_i$ and $\vec{\mathbf{p}}_f$ represent the momenta *immediately* before and after the interaction, respectively. In any situation in which it is reasonable to use the impulse approximation, the particle moves very little during the interaction.

Quick Quiz 8.3

A man pushes his twin sons in turn, each sitting on their own small cart, through a distance d, applying the same constant force to each son as he does so. Twin 2 has a greater mass than twin 1. Which statements are true? **(a)** $p_1 < p_2$ **(b)** $p_1 = p_2$ **(c)** $p_1 > p_2$ **(d)** $K_1 < K_2$ **(e)** $K_1 = K_2$ **(f)** $K_1 > K_2$. If he pushes the two boys with the same force for the same time, t, which of these statements are true?

Quick Quiz 8.4

Rank a car dashboard, seat belt, and air bag, each used alone in separate collisions from the same speed, in terms of (a) the impulse and (b) the average force each delivers to a front-seat passenger, from greatest to least.

TRY THIS

Get two friends to hold a sheet vertical and as taut as they can. Throw a raw egg into the sheet. See if you can throw the egg hard enough to break it against the sheet. (Do this over a soft surface so as not to waste too many eggs or over a clean surface so you can scramble them afterwards.) Is it possible break an egg against even a very taut sheet this way?

Analysis Model 8.2

Non-isolated system (momentum)

Imagine you have identified a system to be analysed and have defined a system boundary. If external forces are applied on the system, the system is *non-isolated.* In that case, the change in the total momentum of the system is equal to the impulse on the system, a statement known as the **impulse–momentum theorem**:

$$\Delta \vec{\mathbf{p}} = \vec{\mathbf{I}} \tag{8.10}$$

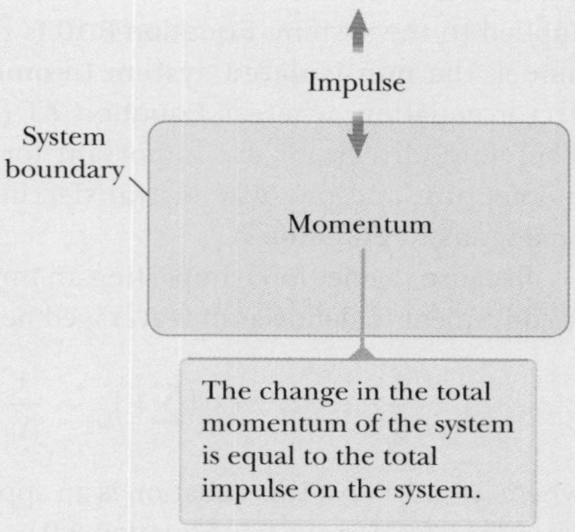
Impulse
System boundary
Momentum
The change in the total momentum of the system is equal to the total impulse on the system.

Examples

- a baseball is struck by a bat
- a spool is pulled by a string (Chapter 9)
- a gas molecule strikes the wall of the container holding the gas (Chapter 21)
- photons strike an absorbing surface and exert pressure on the surface (Chapter 34)

±? Example 8.3

How good are the bumpers?

In a particular crash test experiment, a car of mass (1500 ± 50) kg collides with a wall as shown. The initial and final velocities of the car are $\vec{\mathbf{v}}_i = (-15.0 \pm 0.1)$ m/s and $\vec{\mathbf{v}}_f = (2.6 \pm 0.2)$ m/s, respectively, both in the x direction. If the collision lasts (0.15 ± 0.01) s, find the impulse caused by the collision and the average net force exerted on the car.

Hyundai Motors/HO/Landov

Figure 8.4 (Example 8.3) In a crash test, much of the car's initial kinetic energy is transformed into energy associated with the damage to the car.

Solution

Conceptualise We start by drawing a diagram, showing the motion of the car before and after the collision. The collision time is short, so we can imagine the car being brought to rest very rapidly and then moving back in the opposite direction with a reduced speed.

Model Let us assume the net force exerted on the car by the wall and friction from the ground is large compared with other forces on the car (such as air resistance). The gravitational and normal forces on the car are perpendicular to

Example 8.3 cont.

the motion and therefore do not affect the horizontal momentum. The car's horizontal momentum changes due to an impulse from the environment so we apply the non-isolated system (momentum) analysis model.

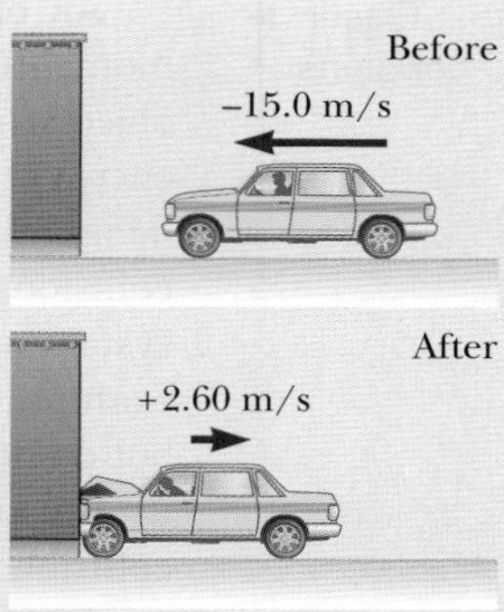

Figure 8.5
(Example 8.3) The car's momentum changes as a result of its collision with the wall.

Analyse Use **Equations 8.10** and **8.2** to find the impulse on the car:

$$\vec{\mathbf{I}} = \Delta\vec{\mathbf{p}} = \Delta\vec{\mathbf{p}}_f - \Delta\vec{\mathbf{p}}_i = m\vec{\mathbf{v}}_f - m\vec{\mathbf{v}}_i = m(\vec{\mathbf{v}}_f - \vec{\mathbf{v}}_i) = 1500\text{ kg }[2.6\text{ m/s} - (-15.0\text{ m/s})]$$

$$= 2.64 \times 10^4\text{ kg.m/s}$$

$$\Delta I = I\left(\frac{\Delta m}{m} + \frac{\Delta v_i}{v_i} + \frac{\Delta v_f}{v_f}\right)$$

$$= 2.64 \times 10^4\text{ kg.m/s}\left[\left(\frac{50\text{ kg}}{1500\text{ kg}}\right) + \left(\frac{0.2\text{ m/s}}{2.6\text{ m/s}}\right) + \left(\frac{0.1\text{ m/s}}{15.0\text{ m/s}}\right)\right]$$

$$= 0.31 \times 10^4\text{ kg.m/s}$$

so $\quad I = (2.64 \pm 0.31) \times 10^4\text{ kg.m/s}.$

Note that as the relative uncertainty in v_f is an order of magnitude greater than the uncertainty of any other data, we could have simply made the approximation $\Delta I/I \approx \Delta v_f/v_f$ and found the same result.

Use **Equation 8.12** to evaluate the average net force exerted on the car:

$$\sum\vec{\mathbf{F}}_{avg} = \frac{\vec{\mathbf{I}}}{\Delta t} = \frac{2.64 \times 10^4\hat{\mathbf{i}}\text{ kg.m/s}}{0.15\text{ s}} = 1.76 \times 10^5\hat{\mathbf{i}}\text{ N}$$

$$\Delta\sum F_{avg} = \sum F_{avg}\left(\frac{\Delta I}{I} + \frac{\Delta t}{t}\right) = 1.76 \times 10^5\text{ N}\left[\left(\frac{0.31\text{ kg.m/s}}{2.64\text{ kg.m/s}}\right) + \left(\frac{0.01\text{ s}}{0.15\text{ s}}\right)\right] = 0.32 \times 10^5\text{ N}$$

so $\quad \sum F_{avg} = (1.76 \pm 0.32) \times 10^5\text{ N}.$

Finalise The net force found above is a combination of the normal force on the car from the wall and any friction force between the tyres and the ground as the front of the car crumples. If the brakes are not operating while the crash occurs and the crumpling metal does not interfere with the free rotation of the tyres, this friction force will be relatively small due to the freely rotating wheels. Notice that the signs of the velocities in this example indicate the reversal of directions. What would the mathematics be describing if both the initial and final velocities had the same sign?

8.4 Collisions in one dimension

The term **collision** represents an event during which two particles come close to each other and interact by means of forces. The interaction forces are assumed to be much greater than any external forces present, so we can use the impulse approximation and treat the two particles as an isolated system during the collision. The forces may vary in time in complicated ways, such as that shown in **Figure 8.3**; however, this force is internal to the system of two particles and the momentum of the system must be conserved in any collision.

In contrast, the total kinetic energy of the system of particles may or may not be conserved, depending on the type of collision. Collisions are categorised as being either *elastic* or *inelastic* depending on whether or not kinetic energy is conserved.

A collision may involve physical contact between two macroscopic objects as described in **Active Figure 8.6a**, but on an atomic scale (**Active Figure 8.6b**), such as the collision of a proton with an alpha particle (the nucleus of a helium atom), there is no actual physical contact because the particles are both positively charged and repel each other due to the strong electrostatic force between them, as described in Chapter 23.

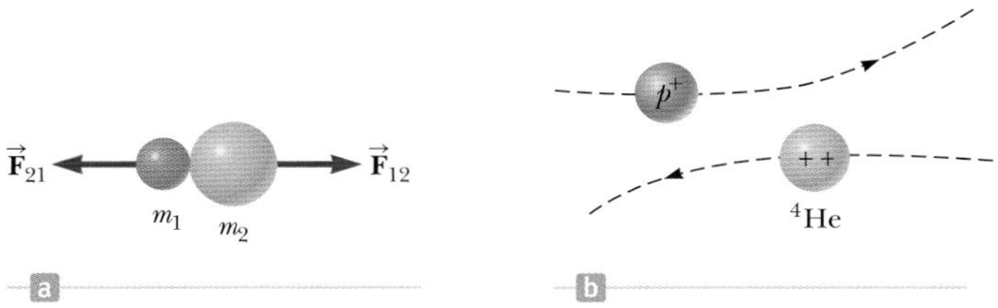

Active Figure 8.6
(a) The collision between two objects as the result of direct contact (b) The 'collision' between two charged particles

Elastic ▶ collisions

An **elastic collision** between two objects is one in which the total kinetic energy (as well as total momentum) of the system is the same before and after the collision. Collisions between macroscopic objects such as billiard balls are only *approximately* elastic because some deformation and loss of kinetic energy take place. For example, you can *hear* a billiard ball collision, so you know that some of the energy is being transferred away from the system by sound; an elastic collision must be perfectly silent. *Truly* elastic collisions do occur between atomic and subatomic particles. These collisions are described by the isolated system model for both energy and momentum. Very large-scale 'collisions' between galaxies are also an example of an elastic collsion, because the components of the two objects interact by the gravitational force but do not generally come into physical contact as they interact. Slingshot manoeuvres, such as have been used to accelerate the Voyager spacecraft on its way through the solar system, can also be modelled as elastic collisions.

Inelastic ▶ collisions

An **inelastic collision** is one in which the total kinetic energy of the system is not the same before and after the collision even though the momentum of the system is conserved. Some deformation of the objects takes place, and mechanical energy is transformed into sound and internal energy. If the objects stick together after they collide, as happens when a meteorite collides with the Earth, the collision is called **perfectly inelastic.** Inelastic collisions can be analysed using the momentum version of the isolated system model. Collisions between subatomic particles may be elastic or inelastic. For example, a neutron scattering off an atom in a solid may do so either elastically or inelastically. In the first case, the neutron scatters with the same speed as it had before hitting the atom. In the second case, the speed of the neutron can be either reduced or increased, and the change in speed tells us about the vibration of atoms in the material.

The principle of conservation of momentum applies at all scales, from subatomic particles to galaxies, and can be used to understand collisions of all sizes.

TRY THIS

Get a few different types of balls, including a ball of putty or Blu-tack, tennis balls, rubber balls, hard steel balls etc. Drop them on a hard surface and observe the collision. Which, if any, are perfectly inelastic? Which is closest to perfectly elastic?

Pitfall Prevention 8.3

Generally, inelastic collisions are hard to analyse without additional information. Lack of this information appears in the mathematical representation as having more unknowns than equations.

Perfectly inelastic collisions

Consider two particles of masses m_1 and m_2 moving with initial velocities $\vec{\mathbf{v}}_{1i}$ and $\vec{\mathbf{v}}_{2i}$ along the same straight line as shown in Active Figure 8.7. The two particles collide head-on, stick together, and then move with some common velocity $\vec{\mathbf{v}}_f$ after the collision. Because the momentum of an isolated system is conserved in *any* collision, we can say that the total momentum before the collision equals the total momentum of the composite system after the collision:

$$m_1\vec{\mathbf{v}}_{1i} + m_2\vec{\mathbf{v}}_{2i} = (m_1 + m_2)\vec{\mathbf{v}}_f \tag{8.14}$$

Solving for the final velocity gives

$$\vec{\mathbf{v}}_f = \frac{m_1\vec{\mathbf{v}}_{1i} + m_2\vec{\mathbf{v}}_{2i}}{m_1 + m_2} \tag{8.15}$$

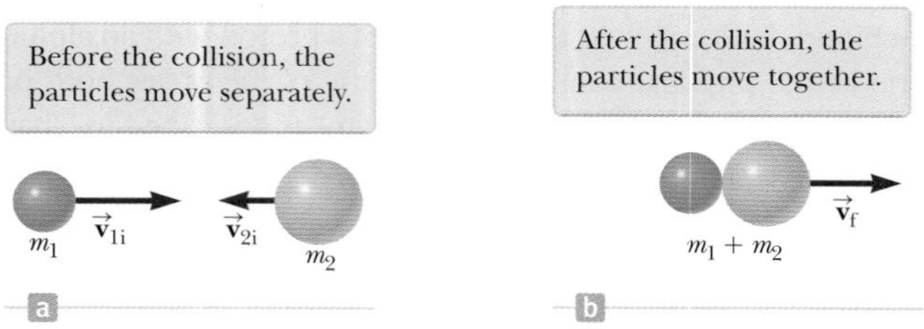

Before the collision, the particles move separately.
$\vec{v}_{1i}$
$\vec{v}_{2i}$
m_1
m_2
a
After the collision, the particles move together.
$\vec{v}_f$
$m_1 + m_2$
b

Active Figure 8.7
Schematic representation of a perfectly inelastic head-on collision between two particles

Elastic collisions

Consider two particles of masses m_1 and m_2 moving with initial velocities $\vec{\mathbf{v}}_{1i}$ and $\vec{\mathbf{v}}_{2i}$ along the same straight line as shown in **Active Figure 8.8**. The two particles collide head-on and then leave the collision site with different velocities, $\vec{\mathbf{v}}_{1f}$ and $\vec{\mathbf{v}}_{2f}$. In an elastic collision, both the momentum and the kinetic energy of the system are conserved. Therefore, considering velocities along the horizontal direction in **Active Figure 8.8**, we have

$$m_1 v_{1i} + m_2 v_{2i} = m_1 v_{1f} + m_2 v_{2f} \tag{8.16}$$

$$\frac{1}{2} m_1 v_{1i}^2 + \frac{1}{2} m_2 v_{2i}^2 = \frac{1}{2} m_1 v_{1f}^2 + \frac{1}{2} m_2 v_{2f}^2 \tag{8.17}$$

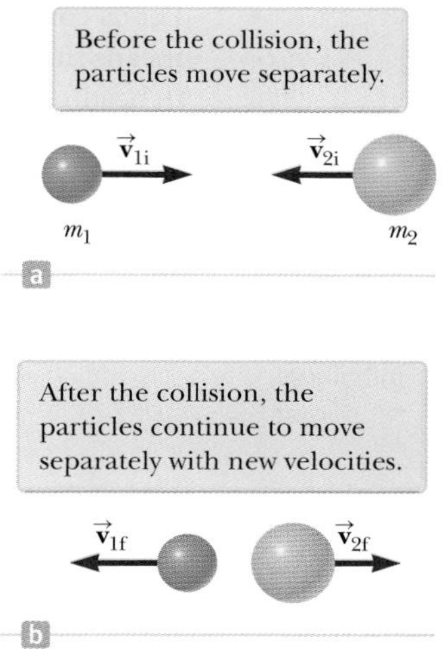

Before the collision, the particles move separately.
$\vec{\mathbf{v}}_{1i}$
$\vec{\mathbf{v}}_{2i}$
m_1
m_2
a
After the collision, the particles continue to move separately with new velocities.
$\vec{\mathbf{v}}_{1f}$
$\vec{\mathbf{v}}_{2f}$
b

Active Figure 8.8
Schematic representation of an elastic head-on collision between two particles

Because all velocities in **Active Figure 8.8** are either to the left or the right, they can be represented by the corresponding speeds along with algebraic signs indicating direction. We shall indicate v as positive if a particle moves to the right and negative if it moves to the left.

In a typical problem involving elastic collisions there are two unknown quantities and **Equations 8.16** and **8.17** can be solved simultaneously to find them.

We can also rearrange **Equations 8.16** and **8.17** to find expressions in terms of the velocities only. First we cancel the factor $\frac{1}{2}$ in **Equation 8.17** and rewrite the equation as

$$m_1(v_{1i}^2 - v_{1f}^2) = m_2(v_{2f}^2 - v_{2i}^2)$$

Factoring both sides of this equation gives

$$m_1(v_{1i} - v_{1f})(v_{1i} + v_{1f}) = m_2(v_{2f} - v_{2i})(v_{2f} + v_{2i}) \tag{8.18}$$

Next, let us separate the terms containing m_1 and m_2 in **Equation 8.16** to obtain

$$m_1(v_{1i} - v_{1f}) = m_2(v_{2f} - v_{2i}) \tag{8.19}$$

To obtain our final result, we divide **Equation 8.18** by **Equation 8.19** and obtain

$$v_{1i} + v_{1f} = v_{2f} + v_{2i}$$

$$v_{1i} - v_{2i} = -(v_{1f} - v_{2f}) \tag{8.20}$$

> **Pitfall Prevention 8.4**
> Equation 8.20 can only be used in a very *specific* situation, a one-dimensional, elastic collision between two objects. The *general* concept is conservation of momentum (and conservation of kinetic energy if the collision is elastic) for an isolated system.

Appendix B.2 reviews factoring and simultaneous equations.

This equation, in combination with **Equation 8.16**, can be used to solve problems dealing with elastic collisions. This pair of equations (**Equations 8.16** and **8.20**) is easier to handle than the pair of **Equations 8.16** and **8.17** because there are no quadratic terms as there are in **Equation 8.17**. According to **Equation 8.20**, the *relative* velocity of the two particles before the collision, $v_{1i} - v_{2i}$, equals the negative of their relative velocity after the collision, $-(v_{1f} - v_{2f})$.

Suppose the masses and initial velocities of both particles are known. **Equations 8.16** and **8.20** can be solved for the final velocities in terms of the initial velocities because there are two equations and two unknowns:

$$v_{1f} = \left(\frac{m_1 - m_2}{m_1 + m_2}\right) v_{1i} + \left(\frac{2m_2}{m_1 + m_2}\right) v_{2i} \tag{8.21}$$

$$v_{2f} = \left(\frac{2m_1}{m_1 + m_2}\right) v_{1i} + \left(\frac{m_2 - m_1}{m_1 + m_2}\right) v_{2i} \tag{8.22}$$

It is important to use the appropriate signs for v_{1i} and v_{2i} in **Equations 8.21** and **8.22**.

TRY THIS

Get some marbles and place one of them on a hard floor or a smooth table top. Roll a second identical marble directly towards it. What happens when they collide? Now try rolling one marble slowly forwards, then send a second identical marble directly after it from behind, at a higher speed. What happens this time? What happens if instead you use a large marble and a small marble? How good an approximation to elastic collisions are these?

Let us consider some special cases. If $m_1 = m_2$, **Equations 8.21** and **8.22** show that $v_{1f} = v_{2i}$ and $v_{2f} = v_{1i}$, which means that the particles exchange velocities if they have equal masses. That is approximately what one observes in head-on billiard ball collisions: the cue ball stops and the struck ball moves away from the collision with the same velocity the cue ball had.

If particle 2 is initially at rest, then $v_{2i} = 0$, and **Equations 8.21** and **8.22** become

Elastic collision: particle 2 initially at rest ▶

$$v_{1f} = \left(\frac{m_1 - m_2}{m_1 + m_2}\right)v_{1i} \quad (8.23)$$

$$v_{2f} = \left(\frac{2m_1}{m_1 + m_2}\right)v_{1i} \quad (8.24)$$

If m_1 is much greater than m_2 and $v_{2i} = 0$, we see from **Equations 8.23** and **8.24** that $v_{1f} \approx v_{1i}$ and $v_{2f} \approx 2v_{1i}$. That is, when a very heavy particle collides head-on with a very light one that is initially at rest, the heavy particle continues its motion unaltered after the collision and the light particle rebounds with a speed equal to about twice the initial speed of the heavy particle. An example of such a collision is that of a moving heavy atom, such as uranium, striking a light atom, such as hydrogen.

If m_2 is much greater than m_1 and particle 2 is initially at rest, then $v_{1f} \approx -v_{1i}$ and $v_{2f} \approx 0$. That is, when a very light particle collides head-on with a very heavy particle that is initially at rest, the light particle has its velocity reversed and the heavy one remains approximately at rest.

Note that we can always apply the case of either $v_{1i} = 0$ or $v_{2i} = 0$ by taking our reference frame as that in which one of the particles is initially stationary.

Quick Quiz 8.5

In a perfectly inelastic one-dimensional collision between two moving objects, what condition alone is necessary so that the final kinetic energy of the system is zero after the collision? **(a)** The objects must have initial momenta with the same magnitude but opposite directions. **(b)** The objects must have the same mass. **(c)** The objects must have the same initial velocity. **(d)** The objects must have the same initial speed, with velocity vectors in opposite directions.

Quick Quiz 8.6

A table-tennis ball is thrown at a stationary bowling ball. The table-tennis ball makes a one-dimensional elastic collision and bounces back along the same line. Compared with the bowling ball after the collision, does the table-tennis ball have **(a)** a larger magnitude of momentum and more kinetic energy, **(b)** a smaller magnitude of momentum and more kinetic energy, **(c)** a larger magnitude of momentum and less kinetic energy, **(d)** a smaller magnitude of momentum and less kinetic energy, or **(e)** the same magnitude of momentum and the same kinetic energy?

TRY THIS

Get a large ball such as a basketball and a small ball such as a tennis ball. Hold the two balls together with the large ball on top of the small ball and drop them. What happens? Now reverse the balls and put the small ball on top of the large ball and drop them together. What happens this time? Explain your observations in terms of conservation of momentum. Don't forget to include the Earth as part of the system. Note: You should do this outside.

Problem-solving strategy

One-dimensional collisions

You should use the following approach when solving collision problems in one dimension.

1. Conceptualise

Imagine the collision occurring. Draw simple diagrams of the particles before and after the collision and include appropriate velocity vectors. At first, you may have to guess at the directions of the final velocity vectors, and then draw them correctly once you have analysed the problem completely.

2. Model

Can you model the system of particles as isolated during the collision? If so, model the collision as elastic, inelastic or perfectly inelastic.

3. Analyse

Set up the appropriate mathematical representation for the problem. If the collision is perfectly inelastic, use **Equation 8.15**. If the collision is elastic, use **Equations 8.16** and **8.20**. If the collision is inelastic, use **Equation 8.16**. To find the final velocities in this case, you will need some additional information.

4. Finalise

Once you have determined your result, check to see if your answers are consistent with the mental and pictorial representations and that your results are realistic.

Example 8.4

Newton's cradle

A device called a 'Newton's cradle' that illustrates conservation of momentum and kinetic energy is shown in **Figure 8.9**. It consists of five identical hard balls supported by strings of equal lengths. When ball 1 is pulled out and released, there is an almost-elastic collision between it and ball 2, ball 1 stops and ball 5 moves out as shown in **Figure 8.9b**. If balls 1 and 2 are pulled out and released, they stop after the collision and balls 4 and 5 swing out. Is it ever possible that when ball 1 is released, it stops after the collision and balls 4 and 5 will swing out on the opposite side and travel with half the speed of ball 1?

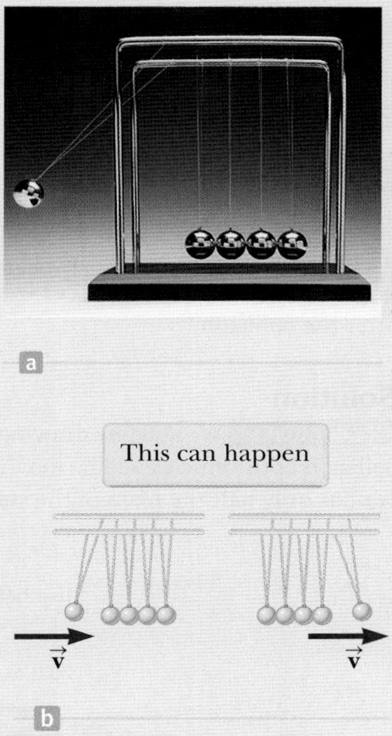

Figure 8.9
(Example 8.4) (a) A 'Newton's cradle' (b) If one ball swings down, we see one ball swing out at the other end.

Solution

Conceptualise Imagine one ball coming in from the left and two balls exiting the collision on the right. Draw a diagram showing this and include the velocity vectors. We want to know if this phenomenon is possible.

Model Because of the very short time interval between the arrival of the ball from the left and the departure of the ball(s) from the right, we can use the impulse approximation and ignore the gravitational forces on the balls, and model the system of five balls as isolated in terms of both momentum and energy. Because the balls are hard, we can model the collisions between them as elastic for purposes of calculation.

Analyse The momentum of the system before the collision is mv, where m is the mass of ball 1 and v is its speed immediately before the collision. After the collision, we imagine that ball 1 stops and balls 4 and 5 swing out, each moving with speed $v/2$. The total momentum of the system after the collision would be $m\left(\frac{v}{2}\right) + m\left(\frac{v}{2}\right) = mv$. Therefore, the momentum of the system is conserved.

The kinetic energy of the system immediately before the collision is $K_i = \frac{1}{2}mv^2$ and that after the collision is $K_f = \frac{1}{2}m\left(\frac{v}{2}\right)^2 + \frac{1}{2}m\left(\frac{v}{2}\right)^2 = \frac{1}{4}mv^2$. This shows that the kinetic energy of the system is *not* conserved, which is inconsistent with our assumption that the collisions are elastic.

Finalise Our analysis shows that it is not possible for balls 4 and 5 to swing out when only ball 1 is released. The only way to conserve both momentum and kinetic energy of the system is for one ball to move out when one ball is released, two balls to move out when two are released, and so on.

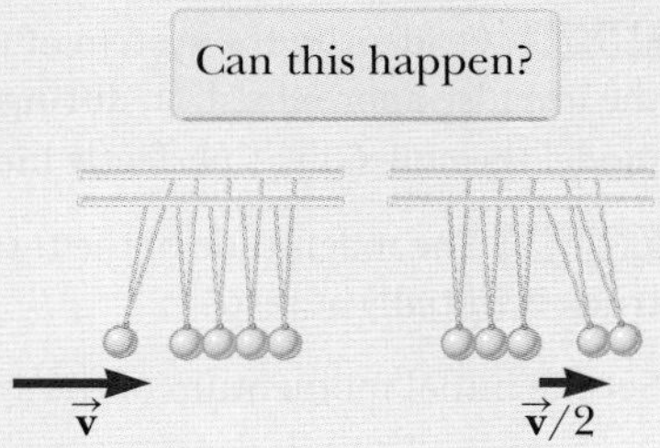

Figure 8.10
(Example 8.4) Is it possible for one ball to swing down and two balls to leave the other end with half the speed of the first ball?

What If? Consider what would happen if balls 4 and 5 are glued together. Now what happens when ball 1 is pulled out and released?

Example 8.4 cont.

Answer In this situation, balls 4 and 5 *must* move together as a single object after the collision. We have argued that both momentum and energy of the system cannot be conserved in this case. We assumed, however, ball 1 stopped after striking ball 2. What if we do not make this assumption? Consider the conservation equations with the assumption that ball 1 moves after the collision. For conservation of momentum,

$$p_i = p_f$$

$$mv_{1i} = mv_{1f} + 2mv_{4,5}$$

where $v_{4,5}$ refers to the final speed of the ball 4–ball 5 combination. Conservation of kinetic energy gives us

$$K_i = K_f$$

$$\frac{1}{2}mv_{1i}^{\ 2} = \frac{1}{2}mv_{1f}^{\ 2} + \frac{1}{2}(2m)v_{4,5}^{\ 2}$$

Combining these equations gives $$v_{4,5} = \frac{2}{3}v_{1i} \qquad v_{1f} = -\frac{1}{3}v_{1i}$$

Therefore, balls 4 and 5 move together as one object after the collision while ball 1 bounces back from the collision with one third of its original speed.

Example 8.5

Don't text and drive!

An 1800 kg truck stopped at a traffic light is struck from the rear by a 900 kg car whose driver is sending a text message instead of watching where they're going. The two vehicles become entangled, moving along the same path as that of the originally moving car. If the car was moving at 20.0 m/s before the collision, what is the velocity of the entangled vehicles after the collision?

Solution

Conceptualise We start by drawing a diagram showing the vehicles just before and just after the collision. After the collision both vehicles will be moving in the same direction as that of the initially moving car. Because the initially moving car has only half the mass of the stationary truck, we expect the final velocity of the combination to be relatively small.

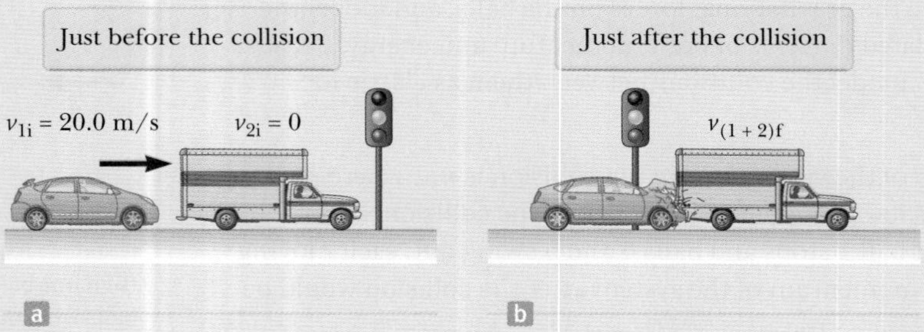

Model We identify the system of two vehicles as isolated in terms of momentum in the horizontal direction and apply the impulse approximation during the short time interval of the collision. The phrase 'become entangled' tells us to model the collision as perfectly inelastic.

Analyse The magnitude of the total momentum of the system before the collision is equal to that of the car because the truck is initially at rest.

Set the initial momentum of the system equal to the final momentum of the system: $$p_i = p_f \rightarrow m_1v_i = (m_1 + m_2)v_f$$

Example 8.5 cont.

Solve for v_f and substitute numerical values: $v_f = \dfrac{m_1 v_i}{m_1 + m_2} + \dfrac{(900\text{ kg})(20.0\text{ m/s})}{900\text{ kg} + 1800\text{ kg}} = 6.67\text{ m/s}$

Finalise Because the final velocity is positive, the direction of the final velocity of the combination is the same as the direction of the velocity of the initially moving car as predicted. The speed of the combination is also much lower than the initial speed of the moving car.

What If? Suppose we reverse the masses of the vehicles. What if a stationary 900 kg car is struck by a moving 1800 kg truck? Is the final speed the same as before?

Answer Intuitively, we can guess that the final speed of the combination is higher than 6.67 m/s if the initially moving vehicle is the more massive. Mathematically, that should be the case because the system has a larger momentum if the initially moving vehicle is the more massive one. Solving for the new final velocity, we find

$$v_f = \frac{m_1 v_i}{m_1 + m_2} = \frac{(1800\text{ kg})(20.0\text{ m/s})}{1800\text{ kg} + 900\text{ kg}} = 13.3\text{ m/s}$$

which is twice the previous final velocity.

Example 8.6

The ballistic pendulum

The ballistic pendulum (**Figure 8.11**) is an apparatus used to measure the speed of a fast-moving projectile such as a bullet and is sometimes used as a lecture demonstration. A projectile of mass m_1 is fired into a large block of wood of mass m_2 suspended from some light wires. The projectile embeds in the block, and the entire system swings through a height h. How can we determine the speed of the projectile from a measurement of h?

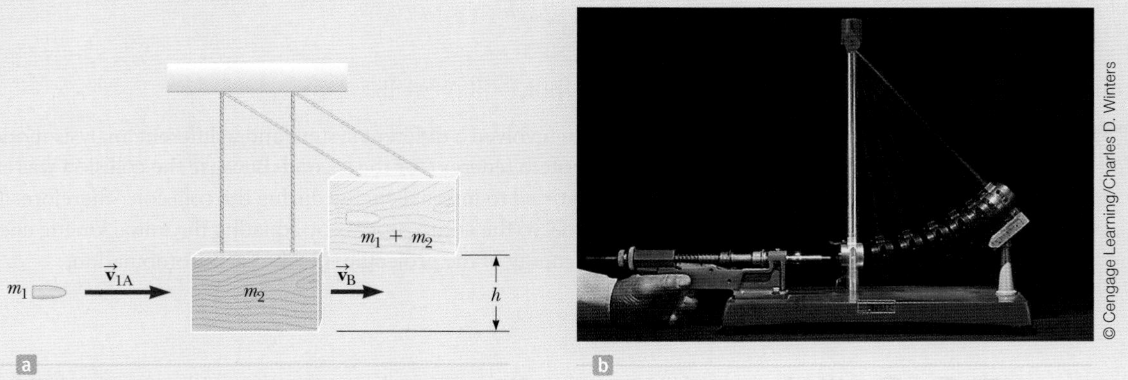

Figure 8.11
(Example 8.6) (a) Diagram of a ballistic pendulum. Notice that $\vec{v}_{1A}$ is the velocity of the projectile immediately before the collision and $\vec{v}_B$ is the velocity of the projectile–block system immediately after the perfectly inelastic collision.
(b) Multiflash photograph of a ballistic pendulum used in the laboratory

Solution

Conceptualise **Figure 8.11a** helps conceptualise the situation. Imagine the projectile entering the pendulum, which swings up to some height at which it momentarily comes to rest. We break up the problem into two parts – first we find the speed of the bullet–block system just after the collision, then we find the height to which the bullet–block system rises using an energy approach. We are given a diagram, although you may want to draw your own.

Model The projectile and the block form an isolated system in terms of momentum if we identify configuration A as immediately before the collision and configuration B as immediately after the collision. Because the projectile embeds in the block, we can model the collision between them as perfectly inelastic.

Example 8.6 cont.

Analyse To analyse the collision, we use **Equation 8.15**, which gives the speed of the system immediately after the collision when we assume the impulse approximation.

Noting that $v_{2A} = 0$, solve **Equation 8.15** for v_B:

$$v_B = \frac{m_1 v_{1A}}{m_1 + m_2} \quad (1)$$

Model For the process during which the projectile–block combination swings upwards to height h (ending at a configuration we'll call C), we focus on a *different* system, that of the projectile, the block and the Earth. We model this part of the problem as one involving an isolated system for energy with no non-conservative forces acting.

Analyse Write an expression for the total kinetic energy of the system immediately after the collision:

$$K_B = \frac{1}{2}(m_1 + m_2)v_B^{\,2} \quad (2)$$

Substitute the value of v_B from **Equation (1)** into **Equation (2)**: $K_B = \dfrac{m_1^2 v_{1A}^2}{2\,(m_1 + m_2)}$

Note that this kinetic energy of the system immediately after the collision is *less* than the initial kinetic energy of the projectile as is expected in an inelastic collision.

We define the gravitational potential energy of the system for configuration B to be zero. Therefore, $U_B = 0$, whereas $U_C = (m_1 + m_2)gh$.

Apply the conservation of mechanical energy principle to the system:

$$K_B + U_B = K_C + U_C$$

$$\frac{m_1^2 v_{1A}^2}{2\,(m_1 + m_2)} + 0 = 0 + (m_1 + m_2)\,gh$$

Solve for v_{1A}:

$$v_{1A} = \left(\frac{m_1 + m_2}{m_1}\right)\sqrt{2gh}$$

Do a dimension check: $[\mathrm{LT}^{-1}] = ([\mathrm{M}]/[\mathrm{M}])([\mathrm{LT}^{-2}][\mathrm{L}])^{1/2} = [\mathrm{LT}^{-1}]$ ☺

Finalise We had to solve this problem in two steps. Each step involved a different system and a different analysis model: isolated system (momentum) for the first step and isolated system (energy) for the second. Because the collision was assumed to be perfectly inelastic, some mechanical energy was transformed to internal energy during the collision. Therefore, it would have been *incorrect* to apply the isolated system (energy) model to the entire process by equating the initial kinetic energy of the incoming projectile with the final gravitational potential energy of the projectile–block–Earth combination.

8.5 Collisions in two dimensions

In Section 8.2, we showed that the momentum of a system of two particles is conserved when the system is isolated. For any collision of two particles, this result implies that the momentum in each of the directions x, y, and z is conserved. Many collisions take place in a two-dimensional plane – such as on the Earth's surface when distances are small compared to the size of the Earth. For such two-dimensional collisions, we obtain two component equations for conservation of momentum:

$$m_1 v_{1ix} + m_2 v_{2ix} = m_1 v_{1fx} + m_2 v_{2fx}$$

$$m_1 v_{1iy} + m_2 v_{2iy} = m_1 v_{1fy} + m_2 v_{2fy}$$

where the three subscripts on the velocity components in these equations represent, respectively, the identification of the object (1, 2), initial and final values (i, f), and the velocity component (x, y).

Consider a specific two-dimensional problem in which particle 1 of mass m_1 collides with particle 2 of mass m_2 initially at rest as in **Active Figure 8.12**. After the collision (**Active Figure 8.12b**), particle 1 moves at an angle θ with respect to the horizontal and particle 2 moves at an angle ϕ with respect to the horizontal. This event is called a *glancing* collision. Applying the law of conservation of momentum in component form and noting that the initial y component of the momentum of the two-particle system is zero gives

$$m_1v_{1i} = m_1v_{1f}\cos\theta + m_2v_{2f}\cos\phi \tag{8.25}$$

$$0 = m_1v_{1f}\sin\theta - m_2v_{2f}\sin\phi \tag{8.26}$$

where the minus sign in **Equation 8.26** is included because after the collision particle 2 has a y component of velocity that is downwards. (The symbols v in these particular equations are speeds, not velocity components. The direction of the component vector is indicated explicitly with plus or minus signs.) We now have two independent equations. As long as no more than two of the seven quantities in **Equations 8.25** and **8.26** are unknown, we can solve the problem.

If the collision is elastic, we can also use **Equation 8.17** (conservation of kinetic energy) with $v_{2i} = 0$:

$$\frac{1}{2}m_1v_{1i}^2 = \frac{1}{2}m_1v_{1f}^2 + \frac{1}{2}m_2v_{2f}^2 \tag{8.27}$$

Knowing the initial speed of particle 1 and both masses, we are left with four unknowns (v_{1f}, v_{2f}, θ, and ϕ). Because we have only three equations, one of the four remaining quantities must be measured to determine the motion after the elastic collision from conservation principles alone.

If the collision is inelastic, kinetic energy is *not* conserved and **Equation 8.27** does *not* apply.

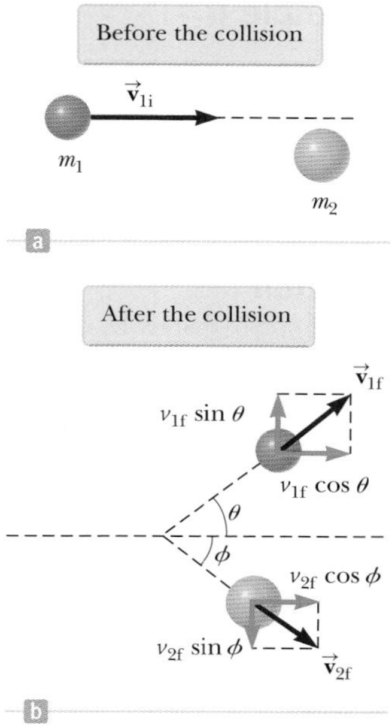

Active Figure 8.12
An elastic, glancing collision between two particles

Problem-solving strategy

Two-dimensional collisions

The following procedure is recommended when dealing with problems involving collisions between two particles in two dimensions.

1. Conceptualise

Imagine the collision occurring and predict the approximate directions in which the particles will move after the collision. Set up a coordinate system and define your velocities in terms of that system. It is convenient to have the x axis coincide with one of the initial velocities. Sketch the coordinate system, draw and label all velocity vectors, and include all the given information.

2. Model

Can you model the system of particles as isolated during the collision? If so, model the collision as elastic, inelastic or perfectly inelastic.

3. Analyse

Write expressions for the x and y components of the momentum of each object before and after the collision. Remember to include the appropriate signs for the components of the velocity vectors and pay careful attention to signs throughout the calculation.

Write expressions for the *total* momentum in the x direction *before* and *after* the collision and equate the two. Repeat this procedure for the total momentum in the y direction.

Solve the momentum equations for the unknown quantities. If the collision is inelastic, kinetic energy is *not* conserved and additional information is probably required. If the collision is perfectly inelastic, the final velocities of the two objects are equal. If the collision is elastic, kinetic energy is conserved and you can equate the total kinetic energy of the system before the collision to the total kinetic energy after the collision, providing an additional relationship between the velocity magnitudes.

4. Finalise

Once you have determined your result, check to see if your answers are consistent with your mental and pictorial representations and that your results are realistic.

Pitfall Prevention 8.5
Equation 8.20, relating the initial and final relative velocities of two colliding objects, is only valid for one-dimensional elastic collisions. Do not use this equation when analysing two-dimensional collisions.

Example 8.7

Collision at an intersection

A 1500 kg car travelling east with a speed of 25.0 m/s collides at an intersection with a 2500 kg four-wheel drive travelling north at a speed of 20.0 m/s. Find the direction and magnitude of the velocity of the wreckage, assuming the vehicles stick together after the collision.

Solution

Conceptualise Start by drawing a diagram (**Figure 8.13**) to help you conceptualise the situation before and after the collision. Let us choose east to be along the positive x direction and north to be along the positive y direction.

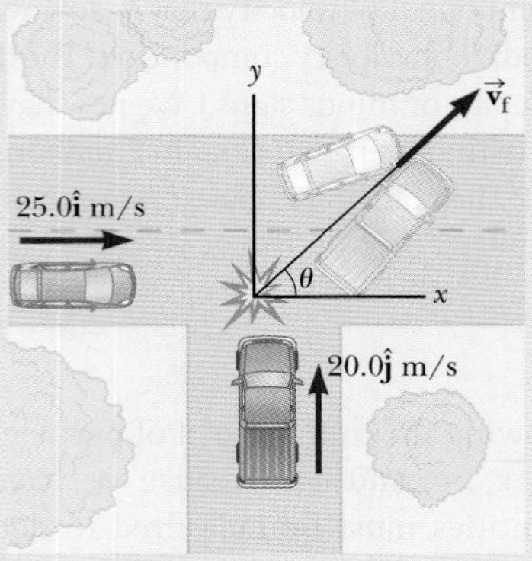

Figure 8.13
(Example 8.7) An eastbound car colliding with a northbound four-wheel drive

Model Because we consider moments immediately before and immediately after the collision as defining our time interval, we ignore the effect that friction would have on the vehicles and model the system of two vehicles as two particles isolated in terms of momentum. The collision is perfectly inelastic because the car and the four-wheel drive stick together after the collision.

Analyse Before the collision the only object having momentum in the x direction is the car. Therefore, the magnitude of the total initial momentum of the system (car plus four-wheel drive) in the x direction is that of only the car. Similarly, the total initial momentum of the system in the y direction is that of the four-wheel drive. After the collision, let us assume the wreckage moves at an angle θ with respect to the x axis with speed v_f.

Equate the initial and final momenta of the system in the x direction:

$$\sum p_{xi} = \sum p_{xf} \Rightarrow m_1 v_{1i} = (m_1 + m_2) v_f \cos\theta \quad (1)$$

Equate the initial and final momenta of the system in the y direction:

$$\sum p_{yi} = \sum p_{yf} \Rightarrow m_2 v_{2i} = (m_1 + m_2) v_f \sin\theta \quad (2)$$

Divide **Equation (2)** by **Equation (1)**

$$\frac{m_2 v_{2i}}{m_1 v_{1i}} = \frac{\sin\theta}{\cos\theta} = \tan\theta$$

where both sides are dimensionless ratios. ☺

Rearrange for θ and substitute numerical values:

$$\theta = \tan^{-1}\left(\frac{m_2 v_{2i}}{m_1 v_{1i}}\right) = \tan^{-1}\left[\frac{(2500\text{ kg})(20.0\text{ m/s})}{(1500\text{ kg})(25.0\text{ m/s})}\right] = 53.1°$$

Use **Equation (2)** to find the value of v_f and substitute numerical values:

$$v_f = \frac{m_2 v_{2i}}{(m_1 + m_2)\sin\theta} = \frac{(2500\text{ kg})(20.0\text{ m/s})}{(1500\text{ kg} + 2500\text{ kg})\sin 53.1°} = 15.6\text{ m/s}$$

Finalise Notice that the angle θ is qualitatively in agreement with **Figure 8.13**. Also notice that the final speed of the combination is less than the initial speeds of both cars. This result is consistent with the kinetic energy of the system being reduced in an inelastic collision.

Example 8.8

Proton–proton collision

Consider a glancing collision between two protons in a particle accelerator. A proton–proton collision can be modelled as elastic because when two protons come close together they exert a large electrostatic force on each other, and the electrostatic force is conservative. We model the situation in the rest frame of one of the protons (proton 2), so that $v_{2i} = 0$. Proton 1 has initial speed 3.50×10^5 m/s relative to proton 2. After the collision, proton 1 moves off at an angle of 37.0° to the original direction of motion and proton 2 deflects at an angle of ϕ to the same axis. Find the final speeds of the two protons and the angle ϕ.

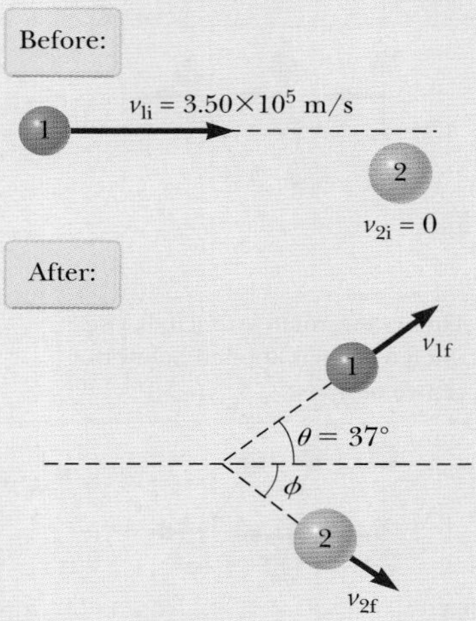

Solution

Conceptualise This collision is like that shown in **Active Figure 8.12**, which will help you conceptualise the behaviour of the system. We define the *x* axis to be along the direction of the velocity vector of proton 1. Draw a diagram showing the particles before and after the collision.

Model We model the two-proton system as isolated, and the collision as a two-dimensional elastic collision.

Analyse Both momentum and kinetic energy of the system are conserved in this glancing elastic collision.

Set up the mathematical representation with **Equations 8.25** through **8.27**:

$$v_{1f}\cos\theta + v_{2f}\cos\phi = v_{1i} \quad (1)$$

$$v_{1f}\sin\theta - v_{2f}\sin\phi = 0 \quad (2)$$

$$v_{1f}^2 + v_{2f}^2 = v_{1i}^2 \quad (3)$$

Rearrange **Equations (1)** and **(2)**:

$$v_{2f}\cos\phi = v_{1i} - v_{1f}\cos\theta$$

$$v_{2f}\sin\phi = v_{1f}\sin\theta$$

Square these two equations and add them:

$$v_{2f}^2\cos^2\phi + v_{2f}^2\sin^2\phi = v_{1i}^2 - 2v_{1i}v_{1f}\cos\theta + v_{1f}^2\cos^2\theta + v_{1f}^2\sin^2\theta$$

Incorporate that the sum of the squares of sine and cosine for *any* angle is equal to 1:

$$v_{2f}^2 = v_{1i}^2 - 2v_{1i}v_{1f}\cos\theta + v_{1f}^2 \quad (4)$$

Substitute **Equation (4)** into **Equation (3)**:

$$v_{1f}^2 + (v_{1i}^2 - 2v_{1i}v_{1f}\cos\theta + v_{1f}^2) = v_{1i}^2$$

$$v_{1f}^2 - v_{1i}v_{1f}\cos\theta = 0 \quad (5)$$

One possible solution of **Equation (5)** is $v_{1f} = 0$, which corresponds to a head-on, one-dimensional collision in which the first proton stops and the second continues with the same speed in the same direction. That is not the solution we want. Divide both sides of **Equation (5)** by v_{1f} and solve for the remaining factor of v_{1f}:

$$v_{1f} = v_{1i}\cos\theta = (3.50 \times 10^5 \text{ m/s})\cos 37.0° = 2.80 \times 10^5 \text{ m/s}$$

Use **Equation (3)** to find v_{2f}:

$$v_{2f} = \sqrt{v_{1i}^2 - v_{1f}^2} = \sqrt{(3.50 \times 10^5 \text{ m/s})^2 - (2.80 \times 10^5 \text{ m/s})^2} = 2.11 \times 10^5 \text{ m/s}$$

Use **Equation (2)** to find ϕ:

$$\phi = \sin^{-1}\left(\frac{v_{1f}\sin\theta}{v_{2f}}\right) = \sin^{-1}\left[\frac{(2.80 \times 10^5 \text{ m/s})\sin 37.0°}{2.11 \times 10^5 \text{ m/s}}\right] = 53.0°$$

Finalise Note that $\theta + \phi = 90°$. This result is *not* accidental. Whenever two objects of equal mass collide elastically in a glancing collision and one of them is initially at rest, their final velocities are perpendicular to each other. Collisions between subatomic particles are described in more detail in Chapter 43.

8.6 The centre of mass

In this section we describe the overall motion of a system in terms of a special point called the **centre of mass** of the system. The concept of centre of mass is particularly useful in analysing problems involving conservation of momentum. It is an example of a common approach in physics: we find a way to simplify a complex problem by modelling the motion of a group of connected particles as the motion of a single point. In this section we describe how to find that point within the system. The system can be either a group of particles, such as a collection of atoms in a container, or an extended object, such as a gymnast leaping through the air. The translational motion of the centre of mass of the system is the same as if all the mass of the system were concentrated at that point. That is, the system moves as if the net external force were applied to a single particle located at the centre of mass. This behaviour is independent of other motion, such as rotation or vibration of the system or deformation of the system (for instance, when a gymnast folds her body). This model, the *particle model*, was introduced in Chapter 2.

Consider a system consisting of a pair of particles that have different masses and are connected by a light, rigid rod (**Active Figure 8.14**). The position of the centre of mass of a system can be described as being the *average position* of the system's mass. The centre of mass of the system is located somewhere on the line joining the two particles and is closer to the particle having the larger mass. If a single force is applied at a point on the rod above the centre of mass, the system rotates clockwise (see **Active Figure 8.14a**). If the force is applied at a point on the rod below the centre of mass, the system rotates anticlockwise (see **Active Figure 8.14b**). If the force is applied at the centre of mass, the system moves in the direction of the force without rotating (see **Active Figure 8.14c**). The centre of mass of an object can be located experimentally with this procedure. We shall deal with what happens when the force is applied such that rotation is the result in the next chapter.

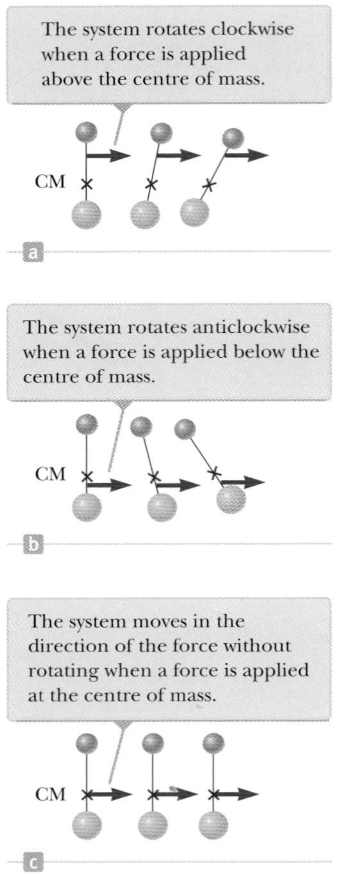

The system rotates clockwise when a force is applied above the centre of mass.
CM
a
The system rotates anticlockwise when a force is applied below the centre of mass.
CM
b
The system moves in the direction of the force without rotating when a force is applied at the centre of mass.
CM
c

Active Figure 8.14
A force is applied to a system of two particles of unequal mass connected by a light, rigid rod

The centre of mass of the pair of particles described in **Active Figure 8.15** is located on the x axis and lies somewhere between the particles. Its x coordinate is given by

$$x_{\text{CM}} \equiv \frac{m_1 x_1 + m_2 x_2}{m_1 + m_2} \tag{8.28}$$

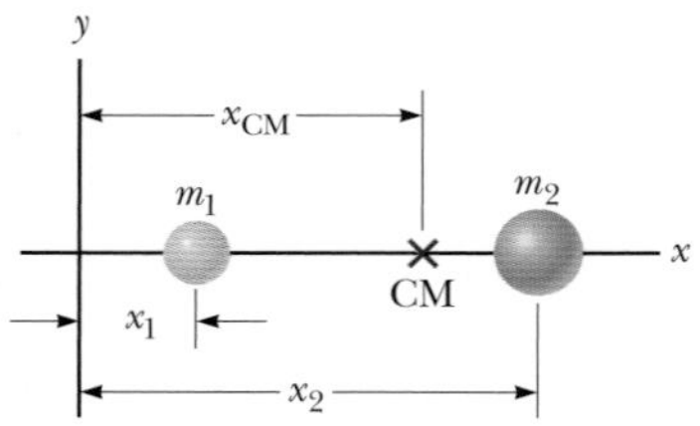

Active Figure 8.15
The centre of mass of two particles of unequal mass on the x axis is located at x_{CM}, a point between the particles, closer to the one having the larger mass.

For example, if $x_1 = 0$, $x_2 = d$, and $m_2 = 2m_1$, we find that $x_{\text{CM}} = \frac{2}{3}d$. That is, the centre of mass lies closer to the more massive particle. If the two masses are equal, the centre of mass lies midway between the particles.

We can extend this concept to a system of many particles with masses m_i in three dimensions. The x coordinate of the centre of mass of n particles is defined to be

$$x_{\text{CM}} \equiv \frac{m_1 x_1 + m_2 x_2 + m_3 x_3 + \cdots + m_n x_n}{m_1 + m_2 + m_3 + \cdots + m_n} = \frac{\sum_i m_i x_i}{\sum_i m_i} = \frac{\sum_i m_i x_i}{M} = \frac{1}{M}\sum_i m_i x_i \tag{8.29}$$

where x_i is the x coordinate of the ith particle and the total mass is $M \equiv \sum_i m_i$ where the sum runs over all n particles. The y and z coordinates of the centre of mass are similarly defined by the equations

$$y_{\text{CM}} \equiv \frac{1}{M}\sum_i m_i y_i \quad \text{and} \quad z_{\text{CM}} \equiv \frac{1}{M}\sum_i m_i z_i \tag{8.30}$$

The centre of mass can be located in three dimensions by its position vector $\vec{\mathbf{r}}_{\text{CM}}$. The components of this vector are x_{CM}, y_{CM}, and z_{CM}, defined in **Equations 8.29** and **8.30**. Therefore,

$$\vec{\mathbf{r}}_{\text{CM}} = x_{\text{CM}}\hat{\mathbf{i}} + y_{\text{CM}}\hat{\mathbf{j}} + z_{\text{CM}}\hat{\mathbf{k}} = \frac{1}{M}\sum_i m_i x_i \hat{\mathbf{i}} + \frac{1}{M}\sum_i m_i y_i \hat{\mathbf{j}} + \frac{1}{M}\sum_i m_i z_i \hat{\mathbf{k}}$$

$$\vec{\mathbf{r}}_{\text{CM}} \equiv \frac{1}{M}\sum_i m_i \vec{\mathbf{r}}_i \tag{8.31}$$

where $\vec{\mathbf{r}}_i$ is the position vector of the ith particle, defined by

$$\vec{\mathbf{r}}_i \equiv x_i\hat{\mathbf{i}} + y_i\hat{\mathbf{j}} + z_i\hat{\mathbf{k}}$$

Vectors are reviewed in Appendix B.6.

We use a similar procedure to find the centre of mass of an extended object with a continuous mass distribution. By dividing the object into elements of mass Δm_i as in Figure 8.16, with coordinates x_i, y_i, z_i, we see that the x coordinate of the centre of mass is:

$$x_{CM} = \lim_{\Delta m_i \to 0} \frac{1}{M}\sum_i x_i \Delta m_i = \frac{1}{M}\int x\, dm \quad (8.32)$$

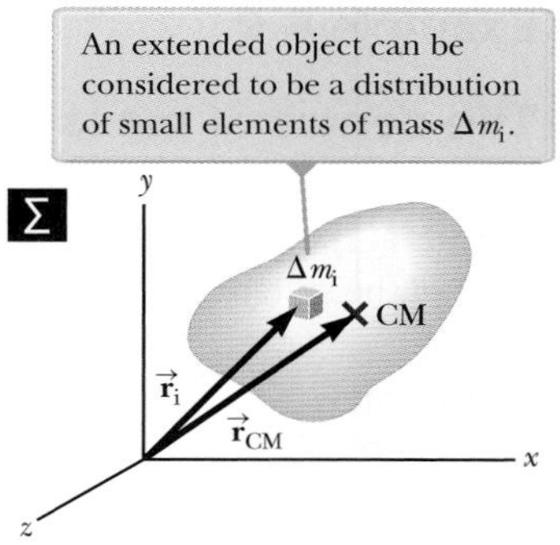
An extended object can be considered to be a distribution of small elements of mass Δm_i.
y
Δm_i
CM
$\vec{\mathbf{r}}_i$
$\vec{\mathbf{r}}_{CM}$
x
z

Figure 8.16
The centre of mass is located at the vector position $\vec{\mathbf{r}}_{CM}$, which has coordinates x_{CM}, y_{CM} and z_{CM}.

Likewise, for y_{CM} and z_{CM} we obtain

$$y_{CM} = \frac{1}{M}\int y\, dm \quad \text{and} \quad z_{CM} = \frac{1}{M}\int z\, dm \quad (8.33)$$

Integral calculus is reviewed in Appendix B.8.

We can express the vector position of the centre of mass of an extended object in the form

$$\vec{\mathbf{r}}_{CM} = \frac{1}{M}\int \vec{\mathbf{r}}\, dm \quad (8.34)$$

which is equivalent to the three expressions given by Equations 8.32 and 8.33.

The centre of mass of any symmetrical object of uniform density lies on an axis of symmetry and on any plane of symmetry. For example, the centre of mass of a uniform rod lies in the rod, midway between its ends. The centre of mass of a sphere or a cube lies at its geometric centre.

Because an extended object is a continuous distribution of mass, each small mass element is acted upon by the gravitational force. The net effect of all these forces is equivalent to the effect of a single force $M\vec{\mathbf{g}}$ acting through a special point, called the **centre of gravity**. If $\vec{\mathbf{g}}$ is constant over the mass distribution, the centre of gravity coincides with the centre of mass. If an extended object is pivoted at its centre of gravity, it balances in any orientation.

The centre of gravity of an irregularly shaped object such as a spanner can be determined by suspending the object first from one point and then from another. In Figure 8.17, a spanner is hung from point A and a vertical line AB (which can be established with a plumb bob) is drawn when the spanner has stopped swinging. The spanner is then hung from point C, and a second vertical line CD is drawn. The centre of gravity is halfway through the thickness of the spanner, under the intersection of these two lines. In general, if the spanner is hung freely from any point, the vertical line through this point must pass through the centre of gravity.

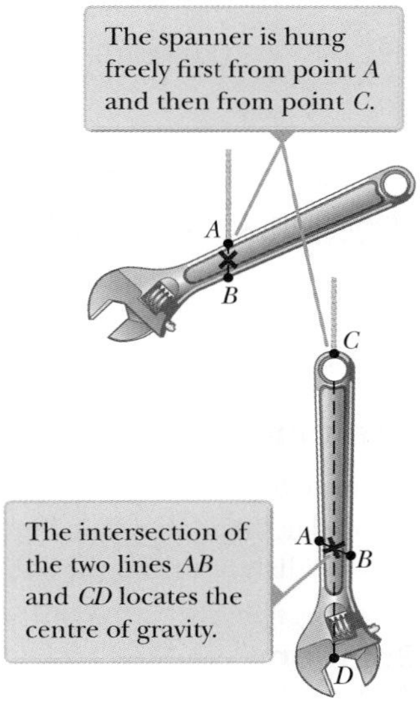
The spanner is hung freely first from point A and then from point C.
A
B
C
The intersection of the two lines AB and CD locates the centre of gravity.
A
B
D

Figure 8.17
An experimental technique for determining the centre of gravity of a spanner

TRY THIS

Get two bathroom scales and a wide plank of wood. Put the bathroom scales under the two ends of the plank, and put a mark halfway along the length of the plank. The scales should read the same if you have positioned the plank carefully. Now lie on the plank and get a friend to read the scales. If you lie such that the two scales have the same reading, your centre of mass in the (usually) vertical direction will be where the mark on the plank is. See if you can compare yours to that of your friend. Women usually have a lower centre of mass than men.

Example 8.9

The centre of mass of three particles

A system consists of three particles located at $\vec{r}_1 = 1\hat{i} + 0\hat{j}$, $\vec{r}_2 = 2\hat{i} + 0\hat{j}$ and $\vec{r}_3 = 0\hat{i} + 2\hat{j}$, with masses $m_1 = m_2 = 1.0$ kg and $m_3 = 2.0$ kg. Find the centre of mass of the system.

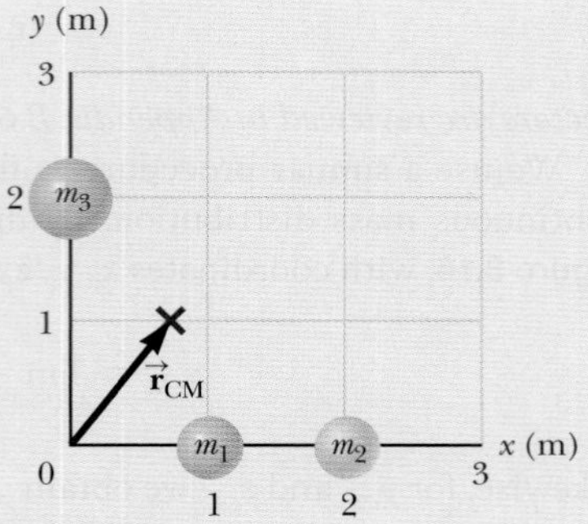

Figure 8.18
(Example 8.9) Two particles are located on the *x* axis, and a single particle is located on the *y* axis as shown. The vector indicates the location of the system's centre of mass.

Solution

Conceptualise Start by drawing a diagram showing the positions of the particles. Mark on the diagram approximately where the centre of mass should be, using your intuition.

Model This is a simple substitution problem using the equations for the centre of mass developed in this section in two dimensions.

$$x_{CM} = \frac{1}{M}\sum_i m_i x_i = \frac{m_1x_1 + m_2x_2 + m_3x_3}{m_1 + m_2 + m_3}$$

$$= \frac{(1.0\text{ kg})(1.0\text{ m}) + (1.0\text{ kg})(2.0\text{ m}) + (2.0\text{ kg})(0)}{1.0\text{ kg} + 1.0\text{ kg} + 2.0\text{ kg}} = \frac{3.0\text{ kg.m}}{4.0\text{ kg}} = 0.75\text{ m}$$

$$y_{CM} = \frac{1}{M}\sum_i m_i y_i = \frac{m_1y_1 + m_2y_2 + m_3y_3}{m_1 + m_2 + m_3}$$

$$= \frac{(1.0\text{ kg})(0) + (1.0\text{ kg})(0) + (2.0\text{ kg})(2.0\text{ m})}{4.0\text{ kg}} = \frac{4.0\text{ kg.m}}{4.0\text{ kg}} = 1.0\text{ m}$$

Write the position vector of the centre of mass: $\vec{r}_{CM} = x_{CM}\hat{i} + y_{CM}\hat{j} = (0.75\hat{i} + 1.0\hat{j})$ m

Finalise This position vector does correspond to our guess at approximately where the centre of mass should be.

Example 8.10

The centre of mass of a right triangle

You have been asked to hang a metal sign from a single vertical cable. The sign has the triangular shape shown in **Figure 8.19**. The bottom of the sign is to be parallel to the ground. At what distance from the left end of the sign should you attach the support cable?

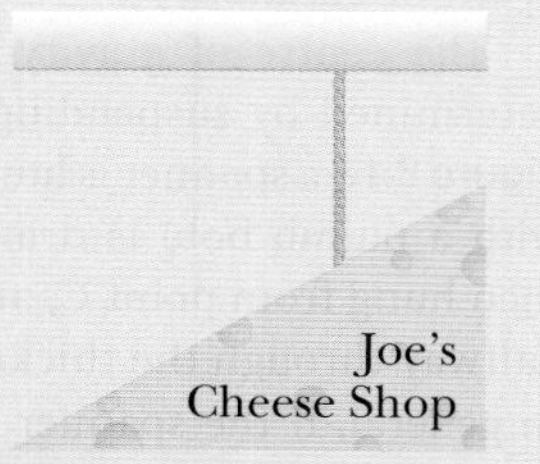

Figure 8.19
(Example 8.10) A triangular sign to be hung from a single string

Solution

Conceptualise **Figure 8.19** shows the sign hanging from the string. The string must be attached at a point directly above the centre of gravity of the sign, which is the same as its centre of mass because it is in a uniform gravitational field.

Model We model the sign as having a continuous mass distribution with uniform density. We also assume that the gravitational field is constant over the length and height of the sign, so that we can take the centre of gravity as being at the position of the centre of mass.

Analyse We use **Equation 8.32** to find the x coordinate of the centre of mass. We divide the triangle into narrow strips of width dx and height y as shown in **Figure 8.20**, where y is the height of the hypotenuse of the triangle above the x axis for a given value of x. The mass of each strip is the product of the volume of the strip and the density ρ of the material from which the sign is made: $dm = \rho yt\, dx$, where t is the thickness of the metal sign. The density of the material is the total mass of the sign divided by its total volume (area of the triangle times thickness).

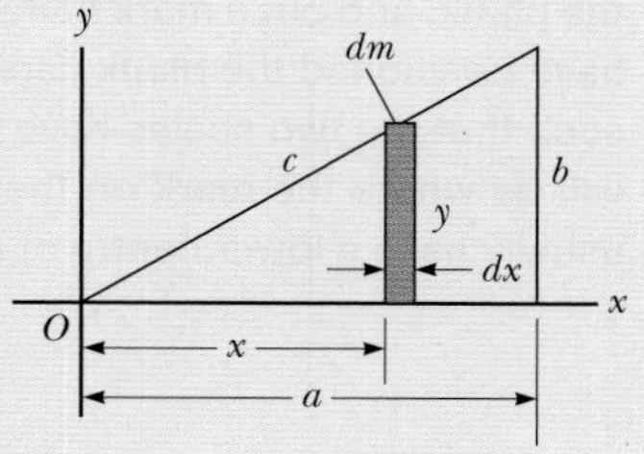

Figure 8.20
(Example 8.10) Geometric construction for locating the centre of mass

Evaluate dm: $$dm = \rho yt\,dx = \left(\frac{M}{\frac{1}{2}abt}\right)yt\,dx = \frac{2My}{ab}dx$$

Example 8.10 cont.

Use **Equation 8.32** to find the x coordinate of the centre of mass:

$$x_{CM} = \frac{1}{M}\int x\,dm = \frac{1}{M}\int_0^a x\frac{2My}{ab}dx = \frac{2}{ab}\int_0^a xy\,dx \quad (1)$$

To proceed further and evaluate the integral, we must express y in terms of x. The line representing the hypotenuse of the triangle in **Figure 8.20** has a gradient of b/a and passes through the origin, so the equation of this line is $y = (b/a)x$.

Substitute for y in **Equation (1)**:

$$x_{CM} = \frac{2}{ab}\int_0^a x\left(\frac{b}{a}x\right)dx = \frac{2}{a^2}\int_0^a x^2\,dx = \frac{2}{a^2}\left[\frac{x^3}{3}\right]_0^a$$

$$= \frac{2}{3}a$$

Therefore, the string must be attached to the sign at a distance two-thirds of the length of the bottom edge from the left.

Finalise For the triangular sign, the linear increase in height y with position x means that elements in the sign increase in mass linearly along the x axis. We could also find the y coordinate of the centre of mass of the sign, but that is not needed to determine where the string should be attached. Try cutting a right triangle out of cardboard and hanging it from a string so that the long base is horizontal. Try it with some other shapes.

8.7 Systems of many particles

Consider a system of two or more particles for which we have identified the centre of mass of the system. We can begin to understand the physical significance and utility of the centre of mass concept by taking the time derivative of the position vector for the centre of mass given by **Equation 8.31** to find the velocity vector. Assuming M remains constant for a system of particles – that is, no particles enter or leave the system – we obtain the following expression for the **velocity of the centre of mass** of the system:

$$\vec{\mathbf{v}}_{CM} = \frac{d\vec{\mathbf{r}}_{CM}}{dt} = \frac{1}{M}\sum_i m_i\frac{d\vec{\mathbf{r}}_i}{dt} = \frac{1}{M}\sum_i m_i\vec{\mathbf{v}}_i \quad (8.35)$$

where $\vec{\mathbf{v}}_i$ is the velocity of the ith particle. Rearranging **Equation 8.35** gives

$$M\vec{\mathbf{v}}_{CM} = \sum_i m_i\vec{\mathbf{v}}_i = \sum_i \vec{\mathbf{p}}_i = \vec{\mathbf{p}}_{tot} \quad (8.36)$$

◀ Total momentum of a system of particles

Therefore, the total linear momentum of the system equals the total mass multiplied by the velocity of the centre of mass. In other words, the total linear momentum of the system is equal to that of a single particle of mass M moving with a velocity $\vec{\mathbf{v}}_{CM}$.

Σ *You may find it helpful to review Appendices B.7: Differential Calculus and B.8: Integral calculus.*

Differentiating **Equation 8.35** with respect to time, we obtain the **acceleration of the centre of mass** of the system:

$$\vec{\mathbf{a}}_{CM} = \frac{d\vec{\mathbf{v}}_{CM}}{dt} = \frac{1}{M}\sum_i m_i\frac{d\vec{\mathbf{v}}_i}{dt} = \frac{1}{M}\sum_i m_i\vec{\mathbf{a}}_i \quad (8.37)$$

Rearranging this expression and using Newton's second law gives

$$M\vec{\mathbf{a}}_{CM} = \sum_i m_i\vec{\mathbf{a}}_i = \sum_i \vec{\mathbf{F}}_i \quad (8.38)$$

where $\vec{\mathbf{F}}_i$ is the net force on particle i.

The forces on any particle in the system may include both external forces (from outside the system) and internal forces (from within the system). By Newton's third law, however, the internal force exerted by particle 1 on particle 2, for example, is equal in magnitude and opposite in direction to the internal force exerted by particle 2 on particle 1. Therefore when we sum over all internal force vectors in **Equation 8.38**, they cancel in pairs and we find that the net force on the system is caused *only* by external forces. We can then write **Equation 8.38** in the form

◀ Newton's second law for a system of particles

$$\sum \vec{\mathbf{F}}_{ext} = M\vec{\mathbf{a}}_{CM} \quad (8.39)$$

That is, the net external force on a system of particles equals the total mass of the system multiplied by the acceleration of the centre of mass. Comparing **Equation 8.39** with Newton's second law for a single particle, we see that the particle model we have used in several chapters can be described in terms of the centre of mass:

> The centre of mass of a system of particles having combined mass M moves as an equivalent particle of mass M would move under the influence of the net external force on the system.

Integrating **Equation 8.39** over a finite time interval gives:

$$\int \sum \vec{\mathbf{F}}_{ext} dt = \int M\vec{\mathbf{a}}_{CM}\, dt = \int M \frac{d\vec{\mathbf{v}}_{CM}}{dt} dt = M\int d\vec{\mathbf{v}}_{CM} = M\Delta\vec{\mathbf{v}}_{CM}$$

Notice that this equation can be written as

Impulse–momentum theorem for a system of particles ▶

$$\Delta\vec{\mathbf{p}}_{tot} = \vec{\mathbf{I}} \tag{8.40}$$

where $\vec{\mathbf{I}}$ is the impulse imparted to the system by external forces and $\vec{\mathbf{p}}_{tot}$ is the momentum of the system. **Equation 8.40** is the generalisation of the impulse–momentum theorem for a single particle (**Equation 8.10**) to a system of many particles. It is also the mathematical representation of the non-isolated system (momentum) model for a system of many particles.

Finally, if the net external force on a system is zero, it follows from **Equation 8.39** that

$$M\vec{\mathbf{a}}_{CM} = M\frac{d\vec{\mathbf{v}}_{CM}}{dt} = 0$$

$$M\vec{\mathbf{v}}_{CM} = \vec{\mathbf{p}}_{tot} = \text{constant when } \sum \vec{\mathbf{F}}_{ext} = 0 \tag{8.41}$$

That is, the total linear momentum of a system of particles is conserved if no net external force is acting on the system. It follows that for an isolated system of particles, both the total momentum and the velocity of the centre of mass are constant in time. This statement is a generalisation of the isolated system (momentum) model for a many-particle system.

Suppose the centre of mass of an isolated system consisting of two or more members is at rest. The centre of mass of the system remains at rest if there is no net force on the system. In Example 8.1 the centre of mass of the arrow–archer system stays at rest, although both the arrow and the archer move. The arrow and archer move away from each other with equal and opposite momenta.

TRY THIS

Stand on a skateboard or low trolley and **very carefully** step off the front of the skateboard. What happens to the skateboard? Why is this a bad way to get on or off a skateboard?

Quick Quiz 8.7

A small boat glides through the water. A passenger stands up and runs aft towards the stern (back) of the boat. **(i)** While they are running toward the stern, is the speed of the boat (a) higher than it was before, (b) unchanged, (c) lower than it was before or (d) impossible to determine? **(ii)** The passenger stops running when they reach the stern of the boat. Is the speed of the boat now (a) higher than it was before they started running, (b) unchanged from what it was before they started running, (c) lower than it was before they started running or (d) impossible to determine?

Conceptual Example 8.11

Exploding projectile

A projectile fired into the air suddenly explodes into several fragments.

What can be said about the motion of the centre of mass of the system made up of all the fragments after the explosion?

Conceptual Example 8.11 cont.

Solution

Neglecting air resistance, the only external force on the projectile is the gravitational force. Therefore, if the projectile did not explode, it would continue to move along the parabolic path indicated by the dashed line in **Figure 8.21**. Because the forces caused by the explosion are internal, they do not affect the motion of the centre of mass of the system (the fragments). Therefore, after the explosion, the centre of mass of the fragments follows the same parabolic path the projectile would have followed if no explosion had occurred.

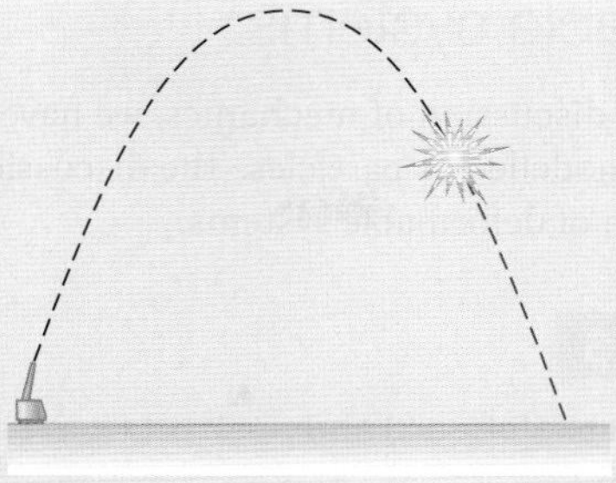

Figure 8.21
(Conceptual Example 8.11) When a projectile explodes into several fragments, the centre of mass of the system made up of all the fragments follows the same parabolic path the projectile would have taken had there been no explosion.

Example 8.12

The exploding rocket

A rocket is fired vertically upwards. At the instant it reaches an altitude of 1000 m and a speed of $v_i = 300$ m/s, it explodes into three fragments of equal mass. One fragment moves upwards with a speed of $v_1 = 450$ m/s following the explosion. The second fragment has a speed of $v_2 = 240$ m/s and is moving east right after the explosion. What is the velocity of the third fragment immediately after the explosion?

Solution

Conceptualise Picture the explosion in your mind, with one piece going upwards and a second piece moving horizontally towards the east. Draw a diagram showing the rocket before the explosion and the pieces and their velocities afterwards, using your intuition to guess the direction of the velocity of the third piece.

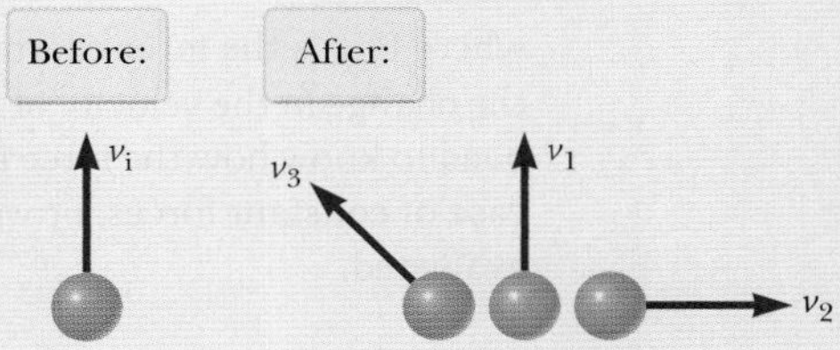

Model This example is a two-dimensional problem because we have two fragments moving in perpendicular directions after the explosion as well as a third fragment moving in an unknown direction in the plane defined by the velocity vectors of the other two fragments. We assume the time interval of the explosion is very short, so we use the impulse approximation in which we ignore the gravitational force and air resistance. Because the forces of the explosion are internal to the system (the rocket), the system is modelled as isolated in terms of momentum. Therefore, the total momentum $\vec{\mathbf{p}}_i$ of the rocket immediately before the explosion must equal the total momentum $\vec{\mathbf{p}}_f$ of the fragments immediately after the explosion.

Analyse As the three fragments have equal mass, the mass of each fragment is $M/3$, where M is the total mass of the rocket. We will let $\vec{\mathbf{v}}_3$ represent the unknown velocity of the third fragment.

Using the isolated system (momentum) model, equate the initial and final momenta of the system and express the momenta in terms of masses and velocities:

$$\vec{\mathbf{p}}_i = \vec{\mathbf{p}}_f \rightarrow M\vec{\mathbf{v}}_i = \frac{M}{3}\vec{\mathbf{v}}_1 + \frac{M}{3}\vec{\mathbf{v}}_2 + \frac{M}{3}\vec{\mathbf{v}}_3$$

Solve for $\vec{\mathbf{v}}_3$:

$$\vec{\mathbf{v}}_3 = 3\vec{\mathbf{v}}_i - \vec{\mathbf{v}}_1 - \vec{\mathbf{v}}_2$$

Substitute the numerical values:

$$\vec{\mathbf{v}}_3 = 3(300\hat{\mathbf{j}}\text{ m/s}) - (450\hat{\mathbf{j}}\text{ m/s}) - (240\hat{\mathbf{i}}\text{ m/s}) = (-240\hat{\mathbf{i}} + 450\hat{\mathbf{j}})\text{ m/s}$$

Finalise Notice that this event is the reverse of a perfectly inelastic collision. There is one object before the collision and three objects afterwards. Imagine running a movie of the event backwards: the three objects would come together and become a single object. In a perfectly inelastic collision, the kinetic energy of the system decreases. If you were to calculate the kinetic energy before and after the event in this example, you would find that the kinetic energy of the system increases (try it!). This increase in kinetic energy comes from the potential energy stored in whatever fuel exploded to cause the breakup of the rocket.

Deformable systems

So far in our discussion of mechanics, we have analysed the motion of particles or non-deformable systems that can be modelled as particles. The discussion of systems of many particles can be applied to an analysis of the motion of deformable systems.

TRY THIS

Stand on a skateboard and push off from a wall. What happens? Explain why in terms of impulse and momentum.

Consider the *Try this* example above. The force from the wall on your hands moves through no displacement; the force is always located at the interface between the wall and your hands. Therefore, this force does no work on the system, which is you and your skateboard. Pushing off the wall, however, does indeed result in a change in the kinetic energy of the system. Your body has deformed during this event: your arms were bent before the event, and they straightened out while you pushed off the wall. No work has been done on the system (you and your skateboard), so the kinetic energy has come from a transfer of energy internal to the system:

$$\Delta K + \Delta U = 0$$

where ΔK is the change in kinetic energy due to the increased speed of the system and ΔU is the decrease in potential energy stored in the body from previous meals. This equation tells us that the system transformed potential energy into kinetic energy by virtue of the muscular exertion necessary to push off the wall. Notice that the system is isolated in terms of energy but non-isolated in terms of momentum. Applying **Equation 8.40** to the system in this situation gives us

$$\Delta\vec{\mathbf{p}}_{\text{tot}} = \vec{\mathbf{I}} \rightarrow m\Delta\vec{\mathbf{v}} = \int \vec{\mathbf{F}}_{\text{wall}}\, dt$$

where $\vec{\mathbf{F}}_{\text{wall}}$ is the force exerted by the wall on your hands, m is the mass of you and the skateboard, and $\Delta\vec{\mathbf{v}}$ is the change in the velocity of the system during the event. To evaluate the right side of this equation, we would need to know how the force from the wall varies in time. In general, this process might be complicated. In the case of constant forces, or well-behaved forces, however, the integral on the right side of the equation can be evaluated.

Example 8.13

Pushing on a spring[1]

As shown in **Figure 8.22a**, two blocks are at rest on a frictionless, level table. Both blocks have the same mass m and they are connected by a spring of negligible mass. The separation distance of the blocks when the spring is relaxed is L. During a time interval Δt, a constant force of magnitude F is applied horizontally to the left block, moving it through a distance x_1 as shown in **Figure 8.22b**. During this time interval, the right block moves through a distance x_2. At the end of this time interval, the force F is removed.

(A) Find the resulting speed $\vec{\mathbf{v}}_{\text{CM}}$ of the centre of mass of the system.

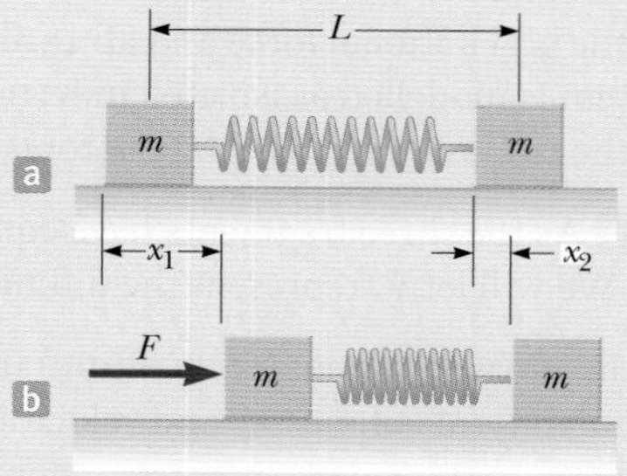

Figure 8.22
(Example 8.13) (a) Two blocks of equal mass are connected by a spring (b) The left block is pushed with a constant force of magnitude F and moves a distance x_1 during some time interval. During this same time interval, the right block moves through a distance x_2.

Solution

Conceptualise Imagine what happens as you push on the left block. It begins to move to the right in **Figure 8.22** and the spring begins to compress. As a result, the spring pushes to the right on the right block, which begins to move to the right. At any given time, the blocks are generally moving with different velocities. As the centre of mass of the system moves to the right after the force is removed, the two blocks oscillate back and forth with respect to the centre of mass.

[1] Example 8.13 was inspired in part by C. E. Mungan, 'A primer on work–energy relationships for introductory physics', *The Physics Teacher* 43:10, 2005.

Example 8.13 cont.

Model We apply three analysis models in this problem: the deformable system of two blocks and a spring is modelled as a non-isolated system in terms of energy because work is being done on it by the applied force. It is also modelled as a non-isolated system in terms of momentum because of the force acting on the system during a time interval. Since the applied force on the system is constant, the acceleration of its centre of mass is constant and the centre of mass is modelled as a particle with constant acceleration.

Analyse Using the non-isolated system (momentum) model, we apply the impulse–momentum theorem to the system of two blocks, recognising that the force F is constant during the time interval Δt while the force is applied.

Write **Equation 8.40** for the system: $F\Delta t = (2m)(v_{CM} - 0) = 2mv_{CM}$ **(1)**

During the time interval Δt, the centre of mass of the system moves a distance $\frac{1}{2}(x_1 + x_2)$. Use this fact to express the time interval in terms of $v_{CM,avg}$:

$$\Delta t = \frac{\frac{1}{2}(x_1 + x_2)}{v_{CM,avg}}$$

Because the centre of mass is modelled as a particle with constant acceleration, the average velocity of the centre of mass is the average of the initial velocity, which is zero, and the final velocity v_{CM}:

$$\Delta t = \frac{\frac{1}{2}(x_1 + x_2)}{\frac{1}{2}(0 + v_{CM})} = \frac{(x_1 + x_2)}{v_{CM}}$$

Substitute this expression into **Equation (1)**: $F\frac{(x_1 + x_2)}{v_{CM}} = 2mv_{CM}$

Solve for v_{CM}: $v_{CM} = \sqrt{F\frac{(x_1 + x_2)}{2m}}$

Check dimensions: $[LT^{-1}] = ([MLT^{-2}][L]/[M])^{1/2} = [LT^{-1}]$ ☺

(B) Find the total energy of the system associated with vibration relative to its centre of mass after the force F is removed.

Solution

Analyse The vibrational energy is all the energy of the system other than the kinetic energy associated with translational motion of the centre of mass. To find the vibrational energy, we apply the conservation of energy equation. The kinetic energy of the system can be expressed as $K = K_{CM} + K_{vib}$, where K_{vib} is the kinetic energy of the blocks relative to the centre of mass due to their vibration. The potential energy of the system is U_{vib}, which is the potential energy stored in the spring when the separation of the blocks is some value other than L.

From the non-isolated system (energy) model, express **Equation 7.2** for this system:

$$\Delta K_{CM} + \Delta K_{vib} + \Delta U_{vib} = W \qquad (2)$$

Express **Equation (2)** in an alternate form, noting that $K_{vib} + U_{vib} = E_{vib}$:

$$\Delta K_{CM} + \Delta E_{vib} = W$$

The initial values of the kinetic energy of the centre of mass and the vibrational energy of the system are zero. Use this fact and substitute for the work done on the system by the force F:

$$K_{CM} + E_{vib} = W = Fx_1$$

Where the displacement in the definition of work (**Equation 6.1**) is that of the point of application of the force.

Solve for the vibrational energy and use the result from part (A):

$$E_{vib} = Fx_1 - K_{CM} = Fx_1 - \frac{1}{2}(2m)v_{CM}^{\ 2} = F\frac{(x_1 - x_2)}{2}$$

Check dimensions to ensure there are no mistakes in rearranging the equations:

$$[ML^2T^{-2}] = [MLT^{-2}][L] = [ML^2T^{-2}]$$ ☺

Finalise Neither of the two answers in this example depends on the spring length, the spring constant or the time interval. Notice also that the magnitude x_1 of the displacement of the point of application of the applied force is different from the magnitude $\frac{1}{2}(x_1 + x_2)$ of the displacement of the centre of mass of the system.

End-of-chapter resources

8.8 Rocket propulsion

We started the chapter by asking how a rocket can move in space, when there is nothing for it to push against. When ordinary vehicles such as cars are propelled, the driving force for the motion is friction. In the case of the car, the driving force is the force exerted by the road on the car. A rocket moving in space, however, has no road to push against. The rocket is an isolated system in terms of momentum. Therefore, the source of the propulsion of a rocket must be something other than an external force. The operation of a rocket depends on the law of conservation of linear momentum as applied to an isolated system, where the system is the rocket plus its ejected fuel.

Shutterstock.com/Celso Diniz

As a rocket moves in free space, its linear momentum changes when some of its mass is ejected in the form of exhaust gases. Because the gases are given momentum when they are ejected out of the engine, the rocket receives a compensating momentum in the opposite direction, just like the archer on the ice when he fires an arrow. Therefore, the rocket is accelerated as a result of the push, or thrust, from the exhaust gases. In free space, the centre of mass of the system (rocket plus expelled gases) moves uniformly, independent of the propulsion process.

The rocket represents the reverse of a perfectly inelastic collision: momentum is conserved, but the kinetic energy of the rocket–exhaust gas system increases (at the expense of chemical potential energy in the fuel).

Suppose that at some time t the magnitude of the momentum of a rocket plus its fuel is $(M + \Delta m)v$, where v is the speed of the rocket relative to the Earth (Figure 8.23a). Over a short time interval Δt, the rocket ejects fuel of mass Δm. At the end of this interval, the rocket's mass is M and its speed is $v + \Delta v$, where Δv is the change in speed of the rocket (Figure 8.23b). If the fuel is ejected with a speed v_e relative to the rocket (the subscript e stands for *exhaust*, and v_e is usually called the *exhaust speed*), the velocity of the fuel relative to the Earth is $v - v_e$. Because the system of the rocket and the ejected fuel is isolated, we can equate the total initial momentum of the system to the total final momentum and obtain

$$(M + \Delta m)\, v = M(v + \Delta v) + \Delta m(v - v_e)$$

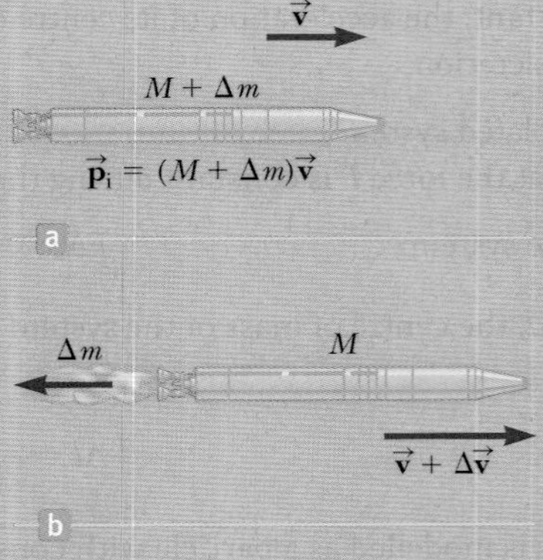

Figure 8.23 Rocket propulsion. (a) The initial mass of the rocket plus all its fuel is $M + \Delta m$ at a time t, and its speed is v. (b) At a time $t + \Delta t$, the rocket's mass has been reduced to M and an amount of fuel Δm has been ejected. The rocket's speed increases by an amount Δv.

Simplifying this expression gives

$$M\Delta v = v_e\, \Delta m$$

If we now take the limit as Δt goes to zero, we let $\Delta v \to dv$ and $\Delta m \to dm$. Furthermore, the increase in the exhaust mass dm corresponds to an equal decrease in the rocket mass, so $dm = -dM$. Notice that dM is negative because it represents a decrease in mass, so $-dM$ is a positive number. Using this fact gives

$$M\, dv = v_e\, dm = -v_e\, dM \qquad (8.42)$$

Now divide the equation by M and integrate, taking the initial mass of the rocket plus fuel to be M_i and the final mass of the rocket plus its remaining fuel to be M_f. The result is

$$\int_{v_i}^{v_f} dv = -v_e \int_{M_i}^{M_f} \frac{dM}{M} \qquad (8.43)$$

$$v_f - v_i = v_e \ln\left(\frac{M_i}{M_f}\right)$$

◀ Expression for rocket propulsion

which is the basic expression for rocket propulsion. First, Equation 8.43 tells us that the increase in rocket speed is proportional to the exhaust speed v_e of the ejected gases. Therefore, the exhaust speed should be very high. Second, the increase in rocket speed is proportional to the natural logarithm of the ratio M_i/M_f. Therefore, this ratio should be as large as possible; that is, the mass of the rocket without its fuel should be as small as possible and the rocket should carry as much fuel as possible.

The **thrust** on the rocket is the force exerted on it by the ejected exhaust gases. We obtain the following expression for the thrust from Newton's second law and Equation 8.42:

$$\text{Thrust} = M\frac{dv}{dt} = \left|v_e \frac{dM}{dt}\right| \qquad (8.44)$$

This expression shows that the thrust increases as the exhaust speed increases and as the rate of change of mass (called the *burn rate*) increases.

Worked solutions to every fifth problem are available in the Student Solutions Manual. Register online at **www.cengagebrain.com** for access.

Summary

Definitions

The **linear momentum** $\vec{\mathbf{p}}$ of a particle of mass m moving with a velocity $\vec{\mathbf{v}}$ is

$$\vec{\mathbf{p}} \equiv m\vec{\mathbf{v}} \quad (8.2)$$

The **impulse** imparted to a particle by a net force $\sum \vec{\mathbf{F}}$ is equal to the time integral of the force:

$$\vec{\mathbf{I}} \equiv \int_{t_i}^{t_f} \sum \vec{\mathbf{F}}\, dt \quad (8.9)$$

An **inelastic collision** is one for which the total kinetic energy of the system of colliding particles is not conserved. A **perfectly inelastic collision** is one in which the colliding particles stick together after the collision. An **elastic collision** is one in which the kinetic energy of the system is conserved.

The position vector of the **centre of mass** of a system of particles is defined as

$$\vec{\mathbf{r}}_{CM} \equiv \frac{1}{M}\sum_i m_i \vec{\mathbf{r}}_i \quad (8.31)$$

where $M = \sum_i m_i$ is the total mass of the system and $\vec{\mathbf{r}}_i$ is the position vector of the ith particle.

Concepts and principles

The position vector of the centre of mass of an extended object can be obtained from the integral expression

$$\vec{\mathbf{r}}_{CM} = \frac{1}{M}\int \vec{\mathbf{r}}\, dm \quad (8.34)$$

The velocity of the centre of mass for a system of particles is

$$\vec{\mathbf{v}}_{CM} = \frac{1}{M}\sum_i m_i \vec{\mathbf{v}}_i \quad (8.35)$$

The total momentum of a system of particles equals the total mass multiplied by the velocity of the centre of mass.

Newton's second law applied to a system of particles is

$$\sum \vec{\mathbf{F}}_{ext} = M\vec{\mathbf{a}}_{CM} \quad (8.39)$$

where $\vec{\mathbf{a}}_{CM}$ is the acceleration of the centre of mass and the sum is over all external forces. The centre of mass moves like an imaginary particle of mass M under the influence of the resultant external force on the system.

Analysis models for problem solving

Non-isolated system (momentum) If a system interacts with its environment in the sense that there is an external force on the system, the behaviour of the system is described by the **impulse–momentum theorem**:

$$\Delta\vec{\mathbf{p}}_{tot} = \vec{\mathbf{I}} \quad (8.40)$$

Isolated system (momentum) The principle of **conservation of linear momentum** indicates that the total momentum of an isolated system (no external forces) is conserved regardless of the nature of the forces between the members of the system:

$$M\vec{\mathbf{v}}_{CM} = \vec{\mathbf{p}}_{tot} = \text{constant when } \sum \vec{\mathbf{F}}_{ext} = 0 \quad (8.41)$$

In the case of a two-particle system, this principle can be expressed as

$$\vec{\mathbf{p}}_{1i} + \vec{\mathbf{p}}_{2i} = \vec{\mathbf{p}}_{1f} + \vec{\mathbf{p}}_{2f} \quad (8.5)$$

The system may be isolated in terms of momentum but non-isolated in terms of energy, as in the case of inelastic collisions.

Chapter review quiz

To help you revise Chapter 8: Linear momentum and collisions, complete the automatically graded Chapter review quiz at http://login.cengagebrain.com.

Conceptual questions

1. Does a larger net force exerted on an object always produce a larger change in the momentum of the object compared with a smaller net force? Explain your answer.
2. Does a larger net force always produce a larger change in kinetic energy than a smaller net force? Explain your answer.
3. While in motion, a cricket ball carries kinetic energy and momentum. (a) Can we say that it carries a force that it can exert on any object it strikes? (b) Can the ball deliver more kinetic energy to the bat and batter than the ball carries initially? (c) Can the ball deliver to the bat and batter more momentum than the ball carries initially? Explain each of your answers.
4. You are standing perfectly still and then take a step forwards. Before the step, your momentum was zero, but afterwards you have some momentum. Is the principle of conservation of momentum violated in this case? Explain your answer.
5. A sharpshooter fires a rifle while standing with the butt of the gun against her shoulder. If the forwards momentum of a bullet is the same as the backwards momentum of the gun, why isn't it as dangerous to be hit by the gun as by the bullet?
6. Each *Voyager* spacecraft was accelerated towards escape speed from the Sun by the gravitational force exerted by Jupiter on the spacecraft. (a) Is the gravitational force a conservative or a non-conservative force? (b) Does the interaction of the spacecraft with Jupiter meet the definition of an elastic collision? (c) How could the spacecraft be moving faster after the collision?
7. (a) Does the centre of mass of a rocket in free space accelerate? Explain your answer. (b) Can the speed of a rocket exceed the exhaust speed of the fuel? Why or why not?
8. An airbag in a car inflates when a collision occurs, which protects the passenger from serious injury. Why does the airbag soften the blow? Discuss the physics involved in this dramatic photograph (Figure CQ8.8).

Getty Images/Romilly Lockyer

9. In golf, novice players are often advised to be sure to 'follow through' with their swing. Why does this advice make the ball travel a longer distance? If a shot is taken near the green, very little follow-through is required. Why?
10. A glider with an open box on top glides along an air track. A student very carefully pours water vertically into the box. (a) What happens to the speed of the glider? Explain your answer. (b) The box has a small hole in the bottom, through which the water can fall. The student fills the box then starts the glider sliding on the air track and observes its motion. What happens to the speed of the glider as the water flows out? Explain your answer.

Problems

Section 8.1 Linear momentum

1. A particle of mass m moves with momentum of magnitude p. ±? (a) Show that the kinetic energy of the particle is $K = p^2/2m$. (b) Express the magnitude of the particle's momentum in terms of its kinetic energy and mass. (c) If the relative uncertainty in p is 5%, what is the relative uncertainty in K?
2. An object has a kinetic energy of 275 J and a momentum of magnitude 25.0 kg.m/s. Find the speed and mass of the object.
3. A student is doing an experiment to measure the frictional force ±? acting between surfaces. A (1.75 ± 0.05) kg block is accelerated until it is moving at 3.50 m/s. The accelerating force is then removed and the student measures the time taken for the block to come to a stop. The student repeats the measurement three times and obtains the results: 8.7 s, 8.9 s, 8.6 s. Use a momentum approach to find the coefficient of friction between the block and the surface.
4. A baseball approaches home plate at a speed of 45.0 m/s, moving horizontally just before being hit by a bat. The batter hits the ball such that after hitting the bat, the baseball is moving at 55.0 m/s straight up. The ball has a mass of 145 g and is in contact with the bat for 2.00 ms. What is the average vector force the ball exerts on the bat during their interaction?

Section 8.2 Analysis model: isolated system (momentum)

5. In research in cardiology and exercise physiology, it is often important to know the mass of blood pumped by a person's heart in one stroke. This information can be obtained by means of a *ballistocardiograph*. The instrument works as follows. The subject lies on a horizontal pallet floating on a film of air. Friction on the pallet is negligible. Initially, the momentum of the system is zero. When the heart beats, it expels a mass m of blood into the aorta with speed v, and the body and platform move in the opposite direction with speed V. The blood velocity can be determined independently (e.g., by observing the Doppler shift of ultrasound). Assume that it is 50.0 cm/s in one typical trial. The mass of the subject plus the pallet is 54.0 kg. The pallet moves 6.00×10^{-5} m in 0.160 s after one heartbeat. Calculate the mass of blood that leaves the heart. Assume that the mass of blood is negligible compared with the total mass of the person. (This simplified example illustrates the principle of ballistocardiography, but in practice a more sophisticated model of heart function is used.)
6. A 45.0 kg girl is standing on a 150 kg plank. Both are originally at rest on a frozen lake that constitutes a frictionless, flat surface. The girl begins to walk along the plank at a constant velocity of $1.50\hat{\mathbf{i}}$ m/s relative to the plank. (a) What is the velocity of the plank relative to the ice surface? (b) What is the girl's velocity relative to the ice surface?
7. When you jump straight up as high as you can, what is the order of magnitude of the maximum recoil speed that you give to the Earth? Model the Earth as a perfectly solid object. In your solution, state the physical quantities you take as data and the values you measure or estimate for them.
8. Two blocks of masses m and $3m$ are placed on a frictionless, horizontal surface. A light spring is attached to the more massive block, and the blocks are pushed together with the spring between them (**Figure P8.8**). A cord initially holding the blocks together is burned; after that happens, the block of mass $3m$ moves to the right with a speed of 2.00 m/s. (a) What is the velocity of the block of mass m? (b) Find the system's original elastic potential energy, taking m = 0.350 kg. (c) Is the original energy in the spring or in the cord? (d) Explain your answer to part (c). (e) Is the momentum of the system conserved in the bursting-apart process? Explain how that is possible considering (f) there are large forces acting and (g) there is no motion beforehand and plenty of motion afterwards?

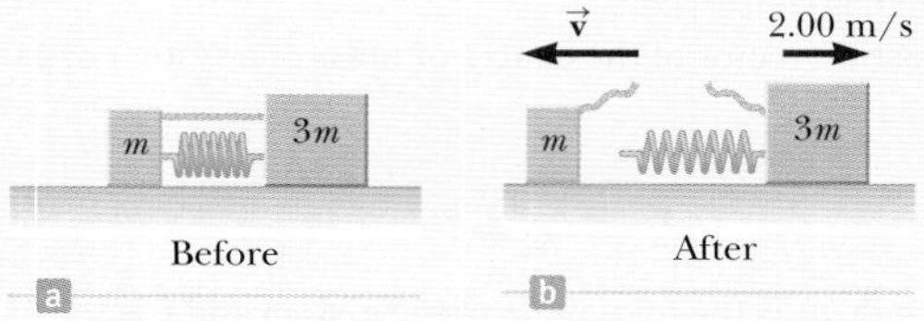

Figure P8.8

Section 8.3 Analysis model: non-isolated system (momentum)

9. The front 1.20 m of a 1 400-kg car is designed as a 'crumple zone' that collapses to absorb the shock of a collision. If a car travelling 25.0 m/s stops uniformly in 1.20 m, (a) how long does the collision last, (b) what is the magnitude of the average force on the car, and (c) what is the acceleration of the car? Express the acceleration as a multiple of the acceleration due to gravity.
10. An estimated force–time curve for a baseball struck by a bat is shown in **Figure P8.10**. From this curve, determine (a) the magnitude of the impulse delivered to the ball and (b) the average force exerted on the ball.

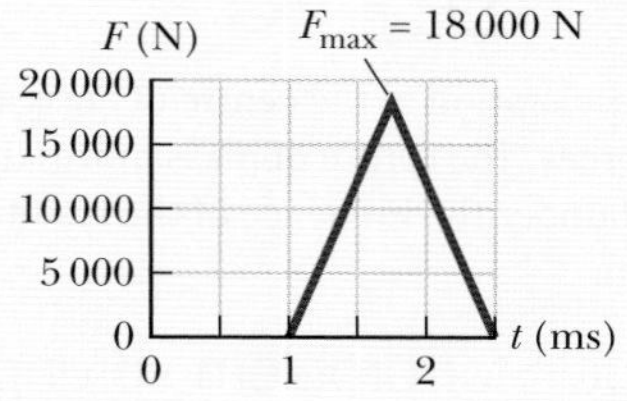

Figure P8.10

11. A glider of mass m is free to slide along a horizontal air track. It is pushed against a launcher at one end of the track. Model the launcher as a light spring of force constant k compressed by a distance x. The glider is released from rest. (a) Show that the glider attains a speed of $v = x(k/m)^{1/2}$. (b) Show that the magnitude of the impulse imparted to the glider is given by the

expression $I = x(km)^{1/2}$. (c) Is more work done on a cart with a large or a small mass?

12. A tennis player receives a shot with the ball (0.0600 kg) travelling horizontally at 50.0 m/s and returns the shot with the ball travelling horizontally at 40.0 m/s in the opposite direction. (a) What is the impulse delivered to the ball by the tennis racquet? (b) What work does the racquet do on the ball?

13. A sports scientist at the Australian Institute of Sport is using a *force platform* to analyse the performance of an athlete by measuring the vertical force the athlete exerts on the ground as a function of time. Starting from rest, a (65.0 ± 0.2) kg athlete jumps down onto the platform from a height of (0.60 ± 0.01) m. While he is in contact with the platform during the time interval $0 < t < 0.800$ s, the force he exerts on it is described by the function

Getty Images/Bongarts

$$F = 9200t - 11\,500t^2$$

where F is in newtons and t is in seconds. (a) What impulse did the athlete receive from the platform? (b) With what speed did he reach the platform? (c) With what speed did he leave it? (d) To what height did he jump upon leaving the platform?

14. Water falls without splashing at a rate of 0.250 L/s from a height of 2.60 m into a 0.750 kg bucket on a scale. If the bucket is originally empty, what does the scale read in newtons 3.00 s after water starts to accumulate in it?

Section 8.4 Collisions in one dimension

15. A 1200 kg car travelling initially at $v_{Ci} = 25.0$ m/s in an easterly direction crashes into the back of a 9000 kg truck moving in the same direction at $v_{Ti} = 20.0$ m/s (**Figure P8.15**). The velocity of the car immediately after the collision is $v_{Cf} = 18.0$ m/s to the east. (a) What is the velocity of the truck immediately after the collision? (b) What is the change in mechanical energy of the car–truck system in the collision? (c) Account for this change in mechanical energy.

Figure P8.15

16. A 10.0 g bullet is fired into a stationary block of wood having mass $m = 5.00$ kg. The bullet embeds into the block. The speed of the bullet-plus-wood combination immediately after the collision is 0.600 m/s. What was the original speed of the bullet?

17. A neutron in a nuclear reactor makes an elastic, head-on collision with the nucleus of a carbon atom initially at rest. (a) What fraction of the neutron's kinetic energy is transferred to the carbon nucleus? (b) The initial kinetic energy of the neutron is 1.60×10^{-13} J. Find its final kinetic energy and the kinetic energy of the carbon nucleus after the collision. (The mass of the carbon nucleus is nearly 12.0 times the mass of the neutron.)

18. A tennis ball of mass m_t is held just above a basketball of mass m_b, as described in a *Try this* example in this chapter and shown in **Figure P8.18**. With their centres vertically aligned, both are released from rest at the same moment so that the bottom of the basketball falls freely through a height h and strikes the floor. Assume an elastic collision with the ground instantaneously reverses the velocity of the basketball while the tennis ball is still moving down because the balls have separated very slightly while falling. Next, the two balls meet in an elastic collision. (a) To what height does the tennis ball rebound? (b) How do you account for the height in (a) being larger than h? Does that seem like a violation of conservation of energy?

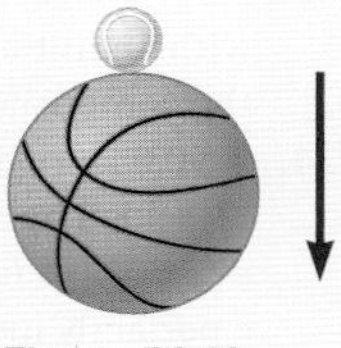

Figure P8.18

19. A 12.0 g wad of sticky clay is hurled horizontally at a 100 g wooden block initially at rest on a horizontal surface. The clay sticks to the block. After impact, the block slides 7.50 m before coming to rest. If the coefficient of friction between the block and the surface is 0.650, what was the speed of the clay immediately before impact?

20. Three carts of masses $m_1 = 4.00$ kg, $m_2 = 10.0$ kg and $m_3 = 3.00$ kg move on a frictionless, horizontal track with speeds of $v_1 = 5.00$ m/s to the right, $v_2 = 3.00$ m/s to the right and $v_3 = 4.00$ m/s to the left as shown in **Figure P8.20**. Velcro couplers make the carts stick together after colliding. (a) Find the final velocity of the train of three carts. (b) **What If?** Does your answer in part (a) require that all the carts collide and stick together at the same moment? What if they collide in a different order?

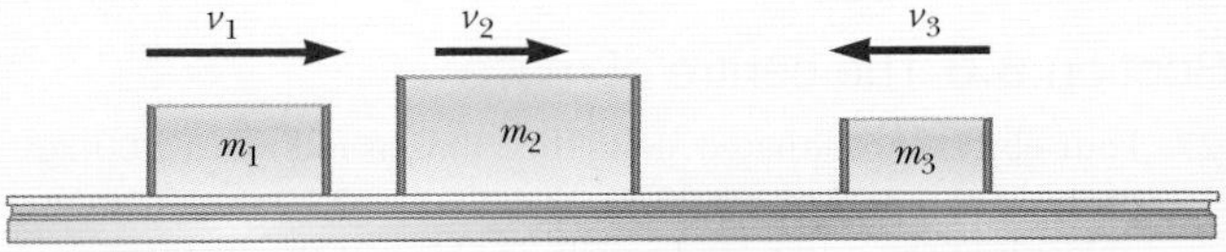

Figure P8.20

21. Two blocks are free to slide along the frictionless, wooden track shown in **Figure P8.21**. The block of mass $m_1 = 5.00$ kg is released from the position shown, at height $h = 5.00$ m above the flat part of the track. Protruding from its front end is the north pole of a strong magnet, which repels the north pole of an identical magnet embedded in the back end of the block of mass $m_2 = 10.0$ kg, initially at rest. The two blocks never touch. Calculate the maximum height to which m_1 rises after the elastic collision.

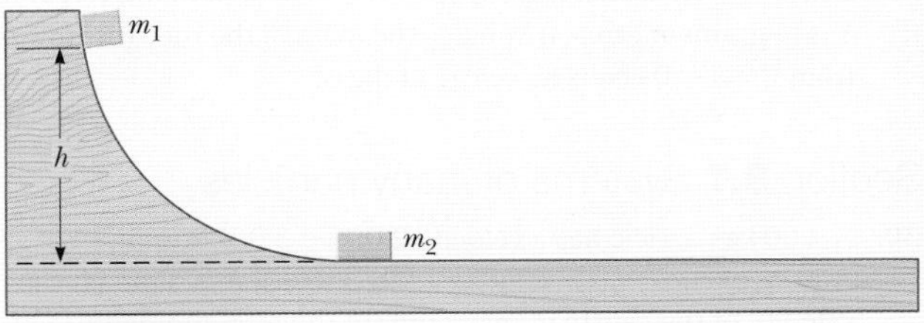

Figure P8.21

Section 8.5 Collisions in two dimensions

22. Two cars of equal mass approach an intersection. One vehicle is travelling with speed 13.0 m/s towards the east and the other is travelling north with speed v_{2i}. Neither driver sees the other. The vehicles collide in the intersection and stick together, leaving parallel skid marks at an angle of 55.0° north of east. The speed limit for both roads is 60 km/h and the driver of the northward-moving vehicle claims he was within the speed limit when the collision occurred. Is he telling the truth? Explain your reasoning.

23. An object of mass 3.00 kg, moving with an initial velocity of $5.00\hat{\mathbf{i}}$ m/s, collides with and sticks to an object of mass 2.00 kg with an initial velocity of $-3.00\hat{\mathbf{j}}$ m/s. Find the final velocity of the composite object.

24. A 90.0 kg rugby fullback running east with a speed of 5.00 m/s is tackled by a 95.0 kg opponent running north with a speed of 3.00 m/s. (a) Explain why the successful tackle constitutes a perfectly inelastic collision. (b) Calculate the velocity of the players immediately after the tackle. (c) Determine the mechanical energy that disappears as a result of the collision. Account for the missing energy.

25. A billiard ball moving at 5.00 m/s strikes a stationary ball of the same mass. After the collision, the first ball moves at 4.33 m/s at an angle of 30.0° with respect to the original line of motion. Assuming an elastic collision (and ignoring friction and rotational motion), find the struck ball's velocity after the collision.

26. An unstable atomic nucleus of mass 17.0×10^{-27} kg initially at rest disintegrates into three particles. One of the particles, of mass 5.00×10^{-27} kg, moves in the y direction with a speed of 6.00×10^{6} m/s. Another particle, of mass 8.40×10^{-27} kg, moves in the x direction with a speed of 4.00×10^{6} m/s. Find (a) the velocity of the third particle and (b) the total kinetic energy increase in the process.

Section 8.6 The centre of mass

27. Four objects are situated along the y axis as follows: a 2.00-kg object is at +3.00 m, a 3.00-kg object is at +2.50 m, a 2.50-kg object is at the origin, and a 4.00-kg object is at −0.500 m. Where is the centre of mass of these objects?

28. A uniform piece of sheet metal is shaped as shown in Figure P8.28. Calculate the x and y coordinates of the centre of mass of the piece.

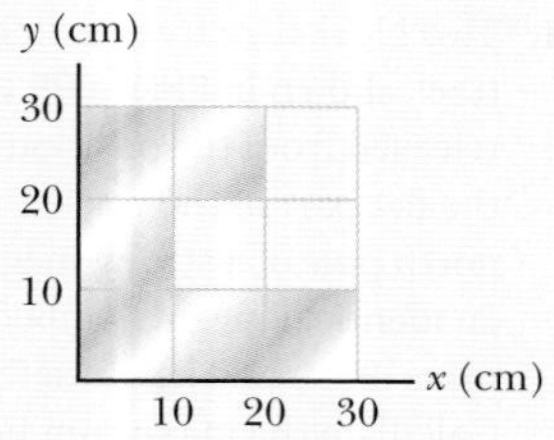

Figure P8.28

29. A rod of length 30.0 cm has linear density (mass per length) given by

$$\lambda = 50.0 + 20.0x$$

where x is the distance from one end, measured in metres, and λ is in grams/metre. (a) What is the mass of the rod? (b) How far from the $x = 0$ end is its centre of mass?

Section 8.7 Systems of many particles

30. A 2.00 kg particle has a velocity $2.00\hat{\mathbf{i}} - 3.00\hat{\mathbf{j}}$ m/s, and a 3.00 kg particle has a velocity $1.00\hat{\mathbf{i}} + 6.00\hat{\mathbf{j}}$ m/s. Find (a) the velocity of the centre of mass and (b) the total momentum of the system.

31. Romeo (77 ± 1) kg, entertains Juliet (55 ± 1) kg, by playing his guitar from the rear of their boat at rest in still water, 2.70 m away from Juliet, who is in the front of the boat. After the serenade, Juliet carefully moves to the rear of the boat to plant a kiss on Romeo's cheek. How far does the 80.0 kg boat move towards the shore it is facing?

32. The vector position of a 3.50-g particle moving in the xy plane varies in time according to $\vec{\mathbf{r}}_1 = (3\hat{\mathbf{i}} + 3\hat{\mathbf{j}})t + 2\hat{\mathbf{j}}t^2$, where t is in seconds and $\vec{\mathbf{r}}$ is in centimetres. At the same time, the vector position of a 5.50 g particle varies as $\vec{\mathbf{r}}_2 = 3\hat{\mathbf{i}} - 2\hat{\mathbf{i}}t^2 - 6\hat{\mathbf{j}}t$. At $t = 2.50$ s, determine (a) the vector position of the centre of mass, (b) the linear momentum of the system, (c) the velocity of the centre of mass, (d) the acceleration of the centre of mass, and (e) the net force exerted on the two-particle system.

33. For an engineering project a student has built a vehicle of total mass 6.00 kg, which moves itself. As shown in Figure P8.33, it runs on four light wheels. A reel is attached to one of the axles and a cord originally wound on the reel goes up over a pulley attached to the vehicle to support an elevated load. After the vehicle is released from rest, the load descends very slowly, unwinding the cord to turn the axle and make the vehicle move forward (to the left in Figure P8.33). Friction is negligible in the pulley and axle bearings. The wheels do not slip on the floor. The reel has been constructed with a conical shape so that the load descends at a constant low speed while the vehicle moves horizontally across the floor with constant acceleration, reaching a final velocity of $3.00\hat{\mathbf{i}}$ m/s. (a) Does the floor impart impulse to the vehicle? If so, how much? (b) Does the floor do work on the vehicle? If so, how much? (c) Does it make sense to say that the final momentum of the vehicle came from the floor? If not, where did it come from? (d) Does it make sense to say that the final kinetic energy of the vehicle came from the floor? If not, where did it come from? (e) Can we say that one particular force causes the forwards acceleration of the vehicle? What does cause it?

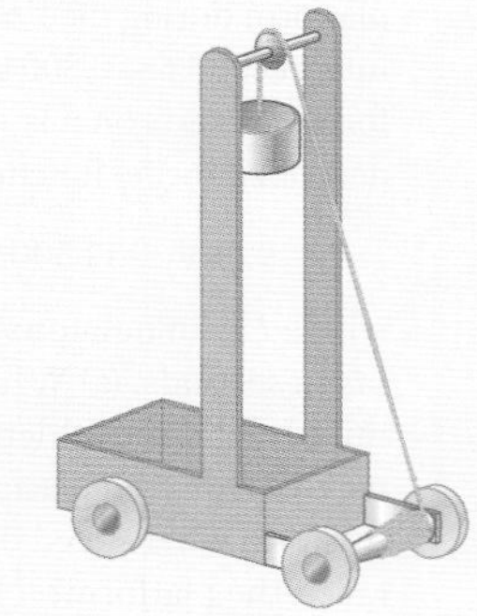

Figure P8.33

34. Figure P8.34a shows an overhead view of the initial configuration of two pucks of mass m on frictionless ice. The pucks are tied together with a string of length l and negligible mass. At time $t = 0$, a constant force of magnitude F begins to pull to the right on the centre point of the string. At time t, the moving pucks strike each other and stick together. At this time, the force has moved through a distance d, and the pucks have attained a speed v (Figure P8.34b). (a) What is v in terms of F, d, ℓ, and m? (b) How much of the energy transferred into the system by work done by the force has been transformed to internal energy?

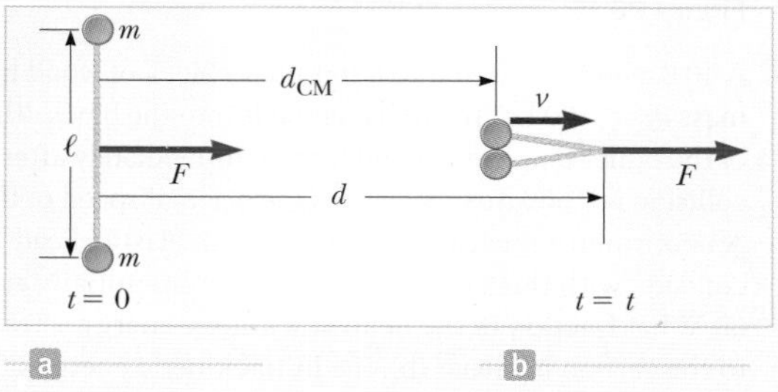

Figure P8.34

35. A 60.0 kg person bends his knees and then jumps straight up. After his feet leave the floor, his motion is unaffected by air resistance and his centre of mass rises by a maximum of 15.0 cm. Model the floor as completely solid and motionless. (a) Does the floor impart impulse to the person? (b) Does the floor do work on the person? (c) With what momentum does the person leave the floor? (d) Does it make sense to say that this momentum came from the floor? Explain your answer. (e) With what kinetic energy does the person leave the floor? (f) Does it make sense to say that this energy came from the floor? Explain your answer.

Section 8.8 Rocket propulsion

36. A garden hose is held as shown in Figure P8.36. The hose is originally full of motionless water. What additional force is necessary to hold the nozzle stationary after the water flow is turned on if the discharge rate is 0.600 kg/s with a speed of 25.0 m/s?

Figure P8.36

37. The first stage of a Saturn V space vehicle consumed fuel and oxidiser at the rate of 1.50×10^4 kg/s with an exhaust speed of 2.60×10^3 m/s. (a) Calculate the thrust produced by this engine. (b) Find the acceleration the vehicle had just as it lifted off the launch pad on the Earth, taking the vehicle's initial mass as 3.00×10^6 kg.

38. A rocket has total mass $M_i = 360$ kg, including $M_f = 330$ kg of fuel and oxidiser. In interstellar space, it starts from rest at the position $x = 0$, turns on its engine at time $t = 0$, and puts out exhaust with relative speed $v_e = 1500$ m/s at the constant rate $k = 2.50$ kg/s. The fuel will last for a burn time of $T_b = M_f/k = 330\text{ kg}/(2.5\text{ kg/s}) = 132$ s.

(a) Show that during the burn the velocity of the rocket as a function of time is given by

$$v(t) = -v_e \ln\left(1 - \frac{kt}{M_i}\right)$$

(b) Make a graph of the velocity of the rocket as a function of time for times running from 0 to 132 s.

(c) Show that the acceleration of the rocket is

$$a(t) = \frac{kv_e}{M_i - kt}$$

(d) Graph the acceleration as a function of time.

(e) Show that the position of the rocket is

$$x(t) = v_e\left(\frac{M_i}{k} - t\right)\ln\left(1 - \frac{kt}{M_i}\right) + v_e t$$

(f) Graph the position during the burn as a function of time.

Additional problems

39. A 2.00-g particle moving at 8.00 m/s makes a perfectly elastic head-on collision with a resting 1.00-g object. (a) Find the speed of each particle after the collision. (b) Find the speed of each particle after the collision if the stationary particle has a mass of 10.0 g. (c) Find the final kinetic energy of the incident 2.00-g particle in the situations described in parts (a) and (b). In which case does the incident particle lose more kinetic energy?

40. A 3.00 kg steel ball strikes a wall with a speed of 10.0 m/s at an angle of $\theta = 60.0°$ with the surface. It bounces off with the same speed and angle (Figure P8.40). If the ball is in contact with the wall for 0.200 s, what is the average force exerted by the wall on the ball?

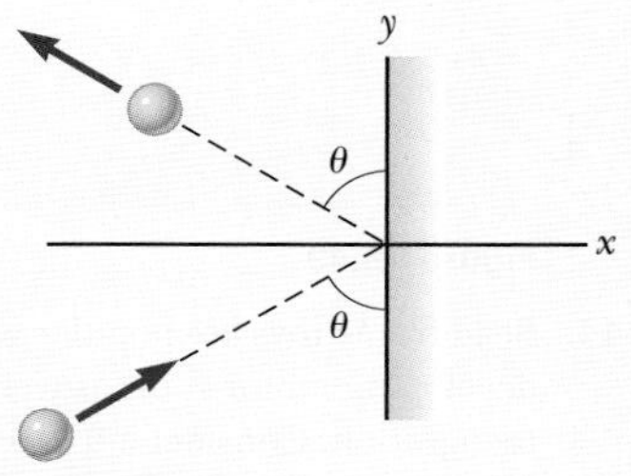

Figure P8.40

41. (a) Figure P8.41 shows three points in the operation of the ballistic pendulum discussed in Example 8.6 (and shown in Figure 8.11b). The projectile approaches the pendulum in Figure P8.41a. Figure P8.41b shows the situation just after the projectile is captured in the pendulum. In Figure P8.41c, the pendulum arm has swung upwards and come to rest at a height h above its initial position. Prove that the ratio of the kinetic energy of the projectile–pendulum system immediately after the collision to the kinetic energy immediately before it is $m_1/(m_1 + m_2)$. (b) What is the ratio of the momentum of the system immediately after the collision to the momentum immediately before it?

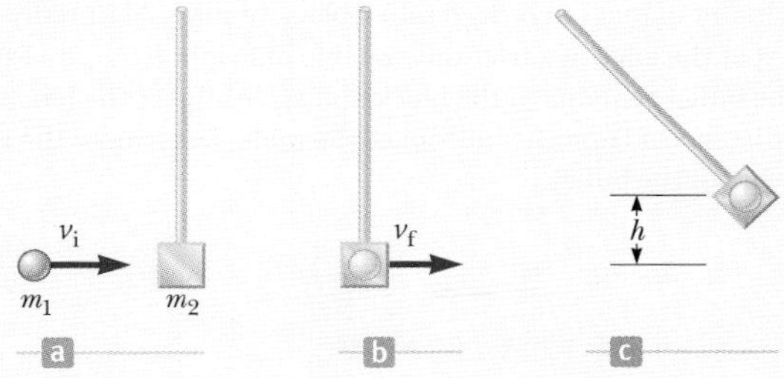

Figure P8.41

Problems 41 and 48. (a) A metal ball moves toward the pendulum. (b) The ball is captured by the pendulum. (c) The ball–pendulum combination swings up through a height h before coming to rest.

42. Two gliders are set in motion on a horizontal air track. A spring of force constant k is attached to the back end of the second glider. As shown in Figure P8.42, the first glider, of mass m_1, moves to the right with speed v_1, and the second glider, of mass m_2, moves more slowly to the right with speed v_2. When m_1 collides with the spring attached to m_2, the spring compresses by a distance x_{max}, and the gliders then move apart again. In terms of v_1, v_2, m_1, m_2, and k, find (a) the speed v at maximum compression, (b) the maximum compression x_{max}, and (c) the velocity of each glider after m_1 has lost contact with the spring.

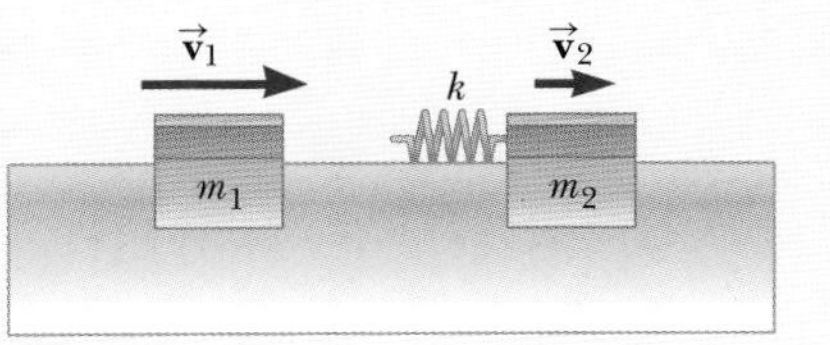

Figure P8.42

43. Two skateboarders of masses $m_1 = 48$ kg and $m_2 = 56$ kg start from rest at a height of $h = 5.00$ m on a ramp as shown in Figure P8.43, and roll down. When they meet on the level portion of the track, they undergo a head-on, elastic collision. Ignoring any friction, determine the maximum heights to which each skateboarder rises on the curved portion of the track after the collision.

Figure P8.43

44. Bead frame toys are popular with toddlers and good for developing motor skills as well as counting and colour recognition. Consider a 0.040 kg blue bead sliding on an approximately frictionless, curved wire, starting from rest at point Ⓐ in Figure P8.44, where $h = 1.50$ m. At point Ⓑ, the blue bead collides elastically with a 0.060 kg green bead at rest. Find the maximum height the green bead rises to as it moves up the wire.

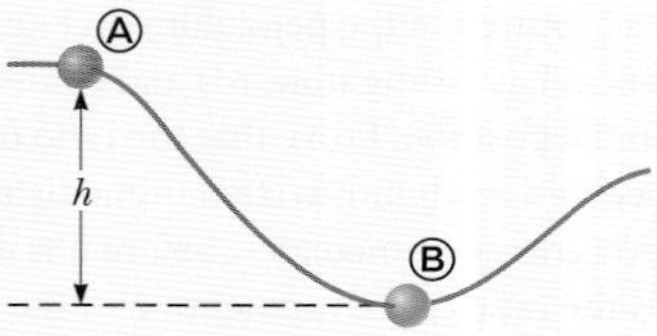

Figure P8.44

45. A bullet of mass m is fired into a block of mass M initially at rest at the edge of a frictionless table of height h (Figure P8.45). The bullet remains in the block and after impact the block lands a distance d from the bottom of the table. Determine the initial speed of the bullet.

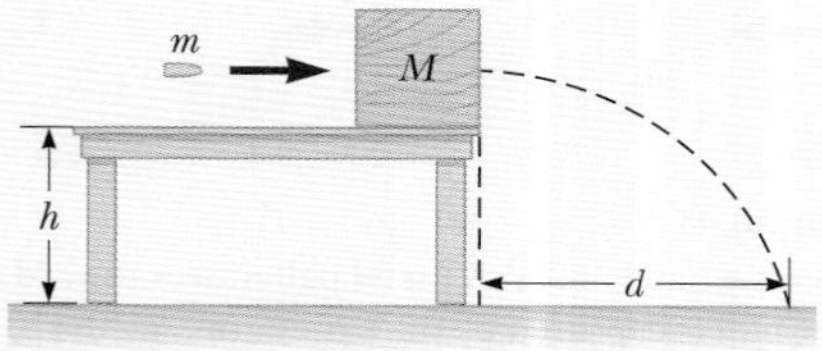

Figure P8.45

46. A small block of mass $m_1 = 0.500$ kg is released from rest at the top of a frictionless, curve-shaped wedge of mass $m_2 = 3.00$ kg, which sits on a frictionless, horizontal surface as shown in Figure P8.46a. When the block leaves the wedge, its velocity is measured to be 4.00 m/s to the right as shown in Figure P8.46b. (a) What is the velocity of the wedge after the block reaches the horizontal surface? (b) What is the height h of the wedge?

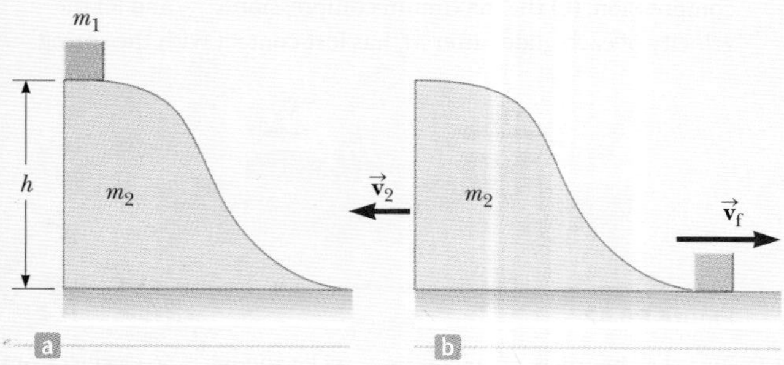

Figure P8.46

47. A 0.500 kg sphere moving with a velocity given by $(2.00\hat{\mathbf{i}} - 3.00\hat{\mathbf{j}} + 1.00\hat{\mathbf{k}})$ m/s strikes another sphere of mass 1.50 kg moving with an initial velocity of $(-1.00\hat{\mathbf{i}} + 2.00\hat{\mathbf{j}} - 3.00\hat{\mathbf{k}})$ m/s. (a) The velocity of the 0.500 kg sphere after the collision is $(-1.00\hat{\mathbf{i}} + 3.00\hat{\mathbf{j}} - 8.00\hat{\mathbf{k}})$ m/s. Find the final velocity of the 1.50 kg sphere and identify the kind of collision (elastic, inelastic or perfectly inelastic). (b) Now assume the velocity of the 0.500 kg sphere after the collision is $(-0.250\hat{\mathbf{i}} + 0.750\hat{\mathbf{j}} - 2.00\hat{\mathbf{k}})$ m/s. Find the final velocity of the 1.50 kg sphere and identify the kind of collision. (c) **What If?** Take the velocity of the 0.500 kg sphere after the collision as $(-1.00\hat{\mathbf{i}} + 3.00\hat{\mathbf{j}} + a\hat{\mathbf{k}})$ m/s. Find the value of a and the velocity of the 1.50 kg sphere after an elastic collision.

48. ±? A student performs a ballistic pendulum experiment using an apparatus similar to that discussed in Example 8.6 and shown in Figure P8.41. She obtains the following average data: $h = (8.68 \pm 0.02)$ cm, projectile mass $m_1 = (68.8 \pm 0.1)$ g, and pendulum mass $m_2 = (263 \pm 1)$ g. (a) Determine the initial speed v_{1A} of the projectile. (b) The second part of her experiment is to obtain v_{1A} by firing the same projectile horizontally (with the pendulum removed from the path) and measuring its final horizontal position x and distance of fall y (Figure P8.48). What numerical value does she obtain for v_{1A} based on her measured values of $x = (257.0 \pm 0.5)$ cm and $y = (85.3 \pm 0.1)$ cm? (c) What factors might account for the difference in this value compared with that obtained in part (a)?

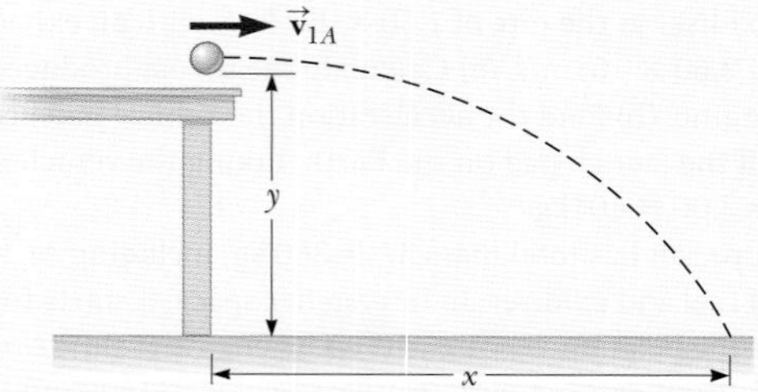

Figure P8.48

49. Consider as a system the Sun with the Earth in a circular orbit around it. Find the magnitude of the change in the velocity of the Sun relative to the centre of mass of the system over a six-month period. Ignore the influence of other celestial objects. You may obtain the necessary astronomical data from the back endpapers of this book.

50. A 5.00 g bullet moving with an initial speed of $v_i = 400$ m/s is fired into and passes through a 1.00 kg block as shown in Figure P8.50. The block, initially at rest on a frictionless, horizontal surface, is connected to a spring with force constant 900 N/m. The block moves $d = 5.00$ cm to the right after impact before being brought to rest by the spring. Find (a) the speed at which the bullet emerges from the block and (b) the amount of initial kinetic energy of the bullet that is converted into internal energy in the bullet–block system during the collision.

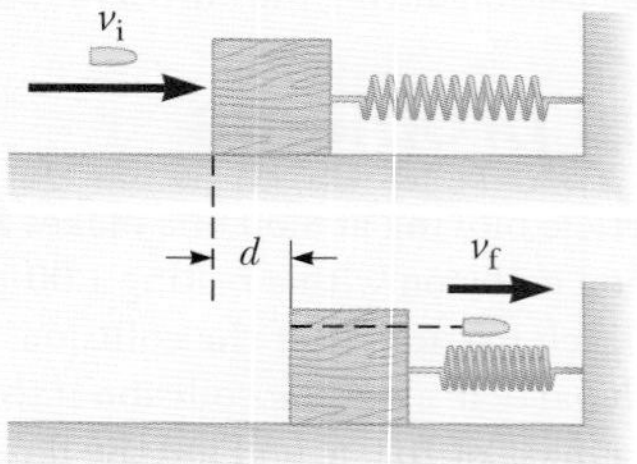

Figure P8.50

Challenge problems

51. In the 1968 Olympic games, University of Oregon jumper Dick Fosbury introduced a new technique of high jumping called the 'Fosbury flop'. It contributed to raising the world record by about 30 cm and is currently used by nearly every world-class jumper. In this technique, the jumper goes over the bar face up while arching her back as much as possible as shown in Figure P8.51a. This action places her centre of mass outside her body, below her back. As her body goes over the bar, her centre of mass passes below the bar. Because a given energy input implies a certain elevation for her centre of mass, the action of arching her back means that her body is higher than if her back were straight. As a model, consider the jumper as a thin uniform rod of length L. When the rod is straight, its centre of mass is at its centre. Now bend the rod in a circular arc so that it subtends an angle of 90.0° at the centre of the arc as shown in Figure P8.51b. In this configuration, how far outside the rod is the centre of mass?

Figure P8.51

52. On a horizontal air track, a glider of mass m carries a post supporting a small dense sphere, also of mass m, hanging just above the top of the glider on a cord of length L, as shown in Figure P8.52. The glider and sphere are initially at rest with the cord vertical. A constant horizontal force of magnitude F is applied to the glider, moving it through displacement x_1; then the force is removed. During the time interval when the force is applied, the sphere moves through a displacement with horizontal component x_2. (a) Find the horizontal component of the velocity of the centre of mass of the glider–sphere system when the force is removed. (b) After the force is removed, the glider continues to move on the track and the sphere swings back and forth, both without friction. Find an expression for the largest angle the cord makes with the vertical.

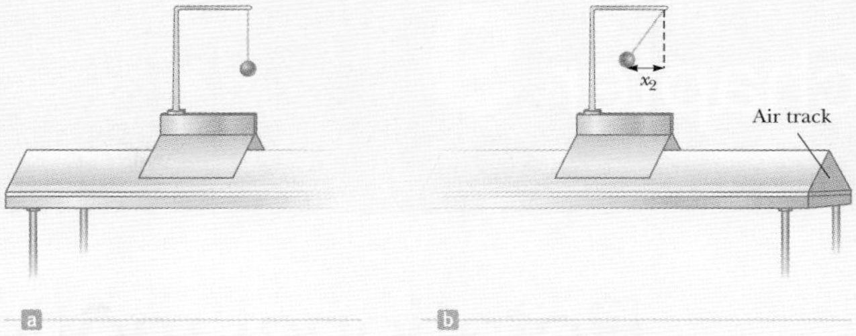

Figure P8.52

53. Sand from a stationary hopper falls onto a moving conveyor belt at the rate of 5.00 kg/s as shown in Figure P8.53. The conveyor belt is supported by frictionless rollers and moves at a constant speed of $v = 0.750$ m/s under the action of a constant horizontal external force $\vec{\mathbf{F}}_{ext}$ supplied by the motor that drives the belt. Find (a) the sand's rate of change of momentum in the horizontal direction, (b) the force of friction exerted by the belt on the sand, (c) the external force $\vec{\mathbf{F}}_{ext}$, (d) the work done by $\vec{\mathbf{F}}_{ext}$ in 1 s and (e) the kinetic energy acquired by the falling sand each second due to the change in its horizontal motion. (f) Why are the answers to parts (d) and (e) different?

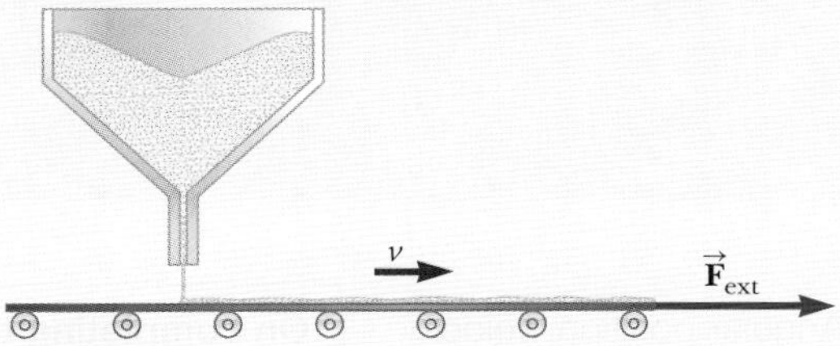

Figure P8.53

54. Two particles with masses m and $3m$ are moving towards each other along the x axis with the same initial speeds v_i. Particle m is travelling to the left, and particle $3m$ is travelling to the right. They undergo an elastic glancing collision such that particle m is moving in the negative y direction after the collision at a right angle from its initial direction. (a) Find the final speeds of the two particles in terms of v_i. (b) What is the angle θ at which the particle $3m$ is scattered?

chapter 9

Rotational motion

The wheelbarrow has been used all over the world for thousands of years. Why are wheelbarrows used so much today? Which part of a wheelbarrow is used as a lever system? What are the mechanical advantages of the wheel and axle? What changes could you make to a wheelbarrow to allow it to move even heavier loads without increasing the effort needed to lift the handles?

Shutterstock.com/Photographee.eu

On completing this chapter, students will understand:
- how rotational motion is described in terms of angular position, velocity and acceleration
- that rotational motion is described by rotational kinematics
- how linear and rotational kinematics are related
- that torque causes angular acceleration and hence changes the rotational motion of an object
- how an object's moment of inertia determines the rotational acceleration produced by an applied torque
- how tools such as hammers, spanners and wheelbarrows work.

Students will be able to:
- describe rotational motion in terms of angular position, velocity and acceleration
- calculate angular positions, velocities and accelerations using rotational kinematics
- describe the effects of torques, qualitatively and quantitatively
- apply Newton's second law to solve problems involving torques
- determine which analysis model is appropriate for a given situation involving torques and rotation
- apply analysis models for rigid objects to solve problems involving rigid objects in equilibrium and under net torques.

ENHANCED WebAssign

The problems found in this chapter may be assigned online in Enhanced Web Assign.

When an extended object such as a wheel rotates about its axis, the motion cannot be analysed by modelling the object as a particle because at any given time different parts of the object have different linear velocities and linear accelerations. In this chapter, we introduce another class of analysis models which describe the motion of rotating rigid objects such as wheels and explain how tools such as hammers, screwdrivers, spanners and wheelbarrows work.

9.1 Angular position, velocity and acceleration

We have already developed analysis models for systems of particles. However, the particle model is not suitable for rotating systems because all parts of a rotating system have different linear velocities and linear accelerations. In this chapter we develop new analysis models for dealing with such systems.

The analysis of a rotating object is greatly simplified by assuming it is rigid. A **rigid object** is one that is non-deformable; that is, the relative locations of all particles of which the object is composed remain constant. All real objects are deformable to some extent; our rigid-object model, however, is useful in many situations in which deformation is negligible and provides a good first approximation in situations where the deformation is small.

Figure 9.1 illustrates an overhead view of a rotating DVD. The disc rotates about a fixed axis perpendicular to the plane of the figure and passing through the centre of the disc at O. Every element of the disc undergoes circular motion about O. Consider a small element of the disc modelled as a particle at P, at a fixed distance r from the origin and rotating about it in a circle of radius r. We represent the position of P with its polar coordinates (r, θ), where r is the distance from the origin to P and θ is conventionally measured *anticlockwise* from some reference line fixed in space as shown in Figure 9.1a. The angle θ changes in time while r remains constant. As the particle moves along the circle from the reference line, which is at angle $\theta = 0$, it moves through an arc of length s as in Figure 9.1b. The arc length s is related to the angle θ through the relationship

$$s = r\theta \qquad \text{or} \qquad \theta = \frac{s}{r} \tag{9.1}$$

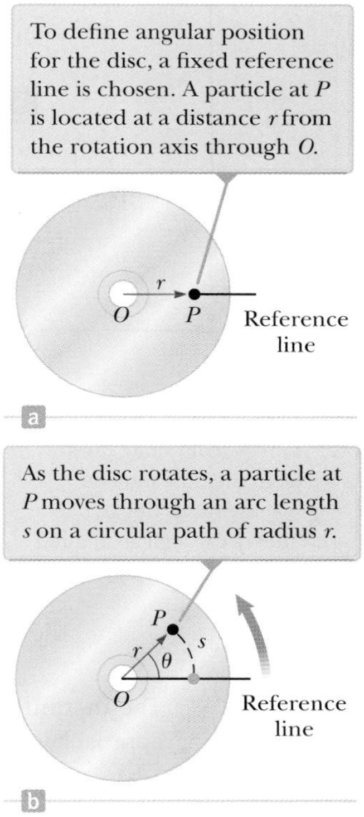

To define angular position for the disc, a fixed reference line is chosen. A particle at P is located at a distance r from the rotation axis through O.
r
O
P
Reference line
a
As the disc rotates, a particle at P moves through an arc length s on a circular path of radius r.
P
r
s
θ
O
Reference line
b

Figure 9.1
A DVD rotating about a fixed axis through O perpendicular to the plane of the figure

Recall from Chapter 3 that this is the definition of the radian, Equation 3.1. Because θ is the ratio of two lengths it is a pure number, so it has no dimensions. We give θ the artificial unit **radian** (rad), where one radian is the angle subtended by an arc length equal to the radius.

Because the disc in Figure 9.1 is a rigid object, as the particle moves through an angle θ from the reference line, every other particle on the object rotates through the same angle θ. Therefore, we can associate the angle θ with the entire rigid object as well as with an individual particle, which allows us to define the *angular position* of a rigid object in its rotational motion. We choose a reference line on the object, such as a line connecting O and a chosen particle on the object. The **angular position** of the rigid object is the angle θ between this reference line on the object and a fixed reference line in space, usually chosen as the x axis. This is similar to the way we define the position of an object in translational motion as the distance x between the object and a reference position such as the origin, $x = 0$. Therefore, **the angle θ plays the same role in rotational motion that the position x does in translational motion.**

> **Pitfall Prevention 9.1**
> The radian is an unusual unit. An angle expressed in radians is a pure number. It is a ratio of two lengths, so it is dimensionless. In rotational equations, you *must* use angles expressed in radians. Don't fall into the trap of using angles measured in degrees in rotational equations. Be careful how you use your calculator – make sure you know what units *it* is working in!

As the particle in question on our rigid object travels from position Ⓐ to position Ⓑ in a time interval Δt as in Figure 9.2, the reference line fixed to the object sweeps out an angle $\Delta\theta = \theta_f - \theta_i$. This quantity $\Delta\theta$ is defined as the **angular displacement** of the rigid object:

$$\Delta\theta \equiv \theta_f - \theta_i$$

◀ Angular displacement

The rate at which this angular displacement occurs can be quantified by defining the **average angular speed** ω_{avg} (Greek letter omega) as the ratio of the angular displacement of a rigid object to the time interval Δt during which the displacement occurs:

$$\omega_{avg} \equiv \frac{\theta_f - \theta_i}{t_f - t_i} = \frac{\Delta\theta}{\Delta t} \tag{9.2}$$

◀ Average angular speed

In analogy to translational speed, the **instantaneous angular speed** ω is defined as the limit of the average angular speed as Δt approaches zero:

$$\omega \equiv \lim_{\Delta t \to 0} \frac{\Delta\theta}{\Delta t} = \frac{d\theta}{dt} \tag{9.3}$$

◀ Instantaneous angular speed

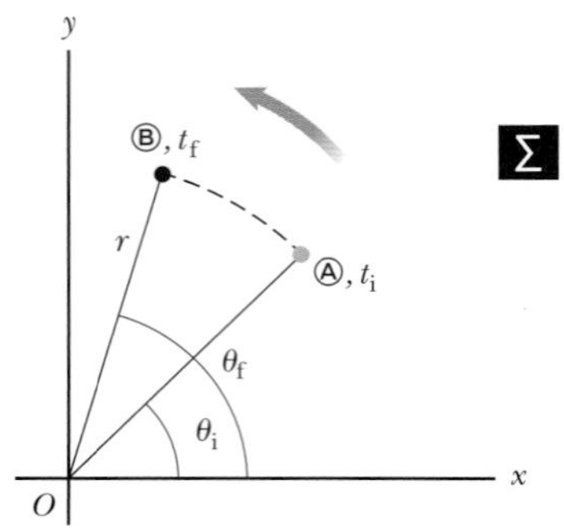

Figure 9.2
A particle on a rotating rigid object moves from Ⓐ to Ⓑ along the arc of a circle. In the time interval $\Delta t = t_f - t_i$, the radial line of length r moves through an angular displacement $\Delta\theta = \theta_f - \theta_i$.

Angular speed has *units* of radians per second (rad/s), and *dimensions* T^{-1}. We take ω to be positive when θ is increasing (anticlockwise motion in **Figure 9.2**) and negative when θ is decreasing (clockwise motion in **Figure 9.2**).

Σ *Differential calculus is summarised in Appendix B.7.*

Quick Quiz 9.1

A rigid object rotates in an anticlockwise direction around a fixed axis. Each of the following pairs of quantities represents an initial angular position and a final angular position of the rigid object. **(i)** Which of the sets can *only* occur if the rigid object rotates through more than 180°? (a) 3 rad, 6 rad (b) −1 rad, 1 rad (c) 1 rad, 5 rad **(ii)** Suppose the change in angular position for each of these pairs of values occurs in 1 s. Which choice represents the lowest average angular speed?

If the instantaneous angular speed of an object changes from ω_i to ω_f in the time interval Δt, the object has an angular acceleration. The **average angular acceleration** α_{avg} (Greek letter alpha) of a rotating rigid object is the ratio of the change in the angular speed to the time interval Δt during which the change in the angular speed occurs:

Average angular acceleration ▶

$$\alpha_{avg} \equiv \frac{\omega_f - \omega_i}{t_f - t_i} = \frac{\Delta\omega}{\Delta t} \qquad (9.4)$$

The **instantaneous angular acceleration** is defined as the limit of the average angular acceleration as Δt approaches zero:

Instantaneous angular acceleration ▶

$$\alpha \equiv \lim_{\Delta t \to 0} \frac{\Delta\omega}{\Delta t} = \frac{d\omega}{dt} \qquad (9.5)$$

Angular acceleration has units of radians per second squared (rad/s^2), and dimensions T^{-2}. Notice that α is positive when a rigid object rotating anticlockwise is speeding up or when a rigid object rotating clockwise is slowing down during some time interval.

When a rigid object is rotating about a *fixed* axis, every particle on the object rotates through the same angle in a given time interval and has the same angular speed and the same angular acceleration. Therefore, like the angular position θ, the quantities ω and α characterise the rotational motion of the entire rigid object as well as individual particles in the object.

TRY THIS

Look at a clock with an hour hand, a minute hand and a second hand. What is the angular velocity of each hand? What is the speed of each hand in revolutions per minute?

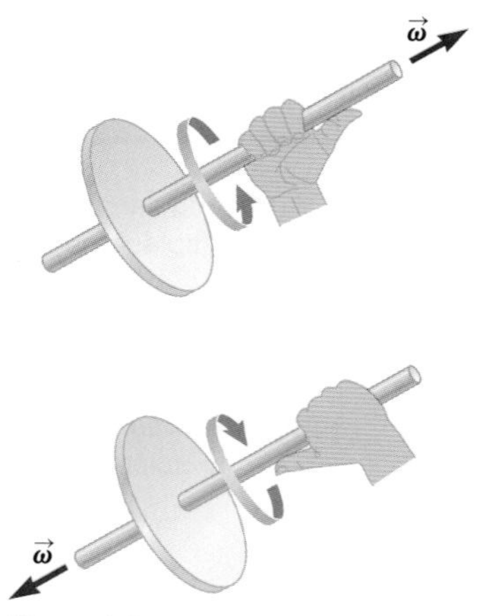

Figure 9.3
The right-hand rule for determining the direction of the angular velocity vector

Angular position (θ), angular speed (ω) and angular acceleration (α) are analogous to translational position (x), translational speed (v) and translational acceleration (a). The variables θ, ω, and α differ dimensionally from the variables x, v, and a by a factor having the dimension length.

The translational variables $\vec{\mathbf{x}}$, $\vec{\mathbf{v}}$ and $\vec{\mathbf{a}}$ are vectors, as we saw in Chapter 3. The rotational variables θ, ω and α also have direction. For rotation about a fixed axis, the only direction that uniquely specifies the rotational motion is the direction along the axis of rotation. Therefore, the directions of the vectors $\vec{\boldsymbol{\omega}}$ and $\vec{\boldsymbol{\alpha}}$ are along this axis. If a particle rotates in the xy plane as in **Figure 9.2**, the direction of $\vec{\boldsymbol{\omega}}$ for the particle is out of the plane of the diagram when the rotation is anticlockwise and into the plane of the diagram when the rotation is clockwise. To illustrate this convention, it is convenient to use the *right-hand rule* shown in **Figure 9.3**. When the four fingers of the right hand are wrapped in the direction of rotation, the extended right thumb points in the direction of $\vec{\boldsymbol{\omega}}$. The direction of $\vec{\boldsymbol{\alpha}}$ follows from its definition $\vec{\boldsymbol{\alpha}} \equiv d\vec{\boldsymbol{\omega}}/dt$. It is in the same direction as $\vec{\boldsymbol{\omega}}$ if the angular speed is increasing in time and it is antiparallel to $\vec{\boldsymbol{\omega}}$ if the angular speed is decreasing in time.

Strictly speaking, these rotational variables are pseudovectors because their signs depend on the choice of coordinate system (left handed or right handed), but in all other respects they behave as vectors and we treat them as such. You have already met pseudovectors in Chapter 3, probably without realising it – the unit vectors are also pseudovectors.

9.2 Analysis model: rigid object with constant angular acceleration

Imagine that a rigid object rotates about a fixed axis and that it has a constant angular acceleration. In this case, we generate a new analysis model for rotational motion called the **rigid object with constant angular acceleration**. This model is the rotational analogue to the particle with constant acceleration model. Writing **Equation 9.5** in the form $d\omega = \alpha\, dt$ and integrating from $t_i = 0$ to $t_f = t$ gives

$$\omega_f = \omega_i + \alpha t \quad \text{(for constant } \alpha\text{)} \tag{9.6}$$

where ω_i is the angular speed of the rigid object at time $t = 0$. **Equation 9.6** allows us to find the angular speed ω_f of the object at any later time t. Substituting **Equation 9.6** into **Equation 9.3** and integrating once more, we obtain

$$\theta_f = \theta_i + \omega_i t + \frac{1}{2}\alpha t^2 \quad \text{(for constant } \alpha\text{)} \tag{9.7}$$

where θ_i is the angular position of the rigid object at time $t = 0$. **Equation 9.7** allows us to find the angular position θ_f of the object at any later time t. Eliminating t from **Equations 9.6** and **9.7** gives

$$\omega_f^{\,2} = \omega_i^{\,2} + 2\alpha(\theta_f - \theta_i) \quad \text{(for constant } \alpha\text{)} \tag{9.8}$$

This equation allows us to find the angular speed ω_f of the rigid object for any value of its angular position θ_f. If we eliminate α between **Equations 9.6** and **9.7**, we obtain

$$\theta_f = \theta_i + \frac{1}{2}(\omega_i + \omega_f)t \quad \text{(for constant } \alpha\text{)} \tag{9.9}$$

Notice that these kinematic expressions for the rigid object with constant angular acceleration are of the same mathematical form as those for a particle with constant acceleration (Chapter 2). **Table 9.1** compares the kinematic equations for rotational and translational motion.

Integral calculus is summarised in Appendix B.8. Σ

Pitfall Prevention 9.2

Always specify your axis. In solving rotation problems, you must specify an axis of rotation. The choice is arbitrary, but once you make it, you must maintain that choice consistently throughout the problem. In some problems, the physical situation suggests a natural axis, such as through the centre of a car wheel. In other problems, there may not be an obvious choice and you must exercise judgement.

Pitfall Prevention 9.3

Just like translation
Equations 9.6 to 9.9 and Table 9.1 might suggest that rotational kinematics is just like translational kinematics. That is almost true, with two key differences. (1) In rotational kinematics, you must specify a rotation axis (per Pitfall Prevention 10.2). (2) In rotational motion, the object keeps returning to its original orientation; therefore, you may be asked for the number of revolutions made by a rigid object. This concept has no analogue in translational motion.

Quick Quiz 9.2

Consider again the pairs of angular positions for the rigid object in Quick Quiz 9.1. If the object starts from rest at the initial angular position, moves anticlockwise with constant angular acceleration and arrives at the final angular position with the same angular speed in all three cases, for which choice is the angular acceleration the highest?

Table 9.1
Kinematic equations for rotational and translational motion

Rigid body with constant angular acceleration	Particle with constant acceleration
$\omega_f = \omega_i + \alpha t$	$v_f = v_i + at$
$\theta_f = \theta_i + \omega_i t + \frac{1}{2}\alpha t^2$	$x_f = x_i + v_i t + \frac{1}{2}at^2$
$\omega_f^2 = \omega_i^2 + 2\alpha(\theta_f - \theta_i)$	$v_f^2 = v_i^2 + 2a(x_f - x_i)$
$\theta_f = \theta_i + \frac{1}{2}(\omega_i + \omega_f)t$	$x_f = x_i + \frac{1}{2}(v_i + v_f)t$

◀ Kinematics equations for rotation and translation

Analysis Model 9.1

Rigid object with constant angular acceleration

Imagine an object that undergoes a spinning motion such that its angular acceleration is constant. The equations describing its angular position and angular speed are analogous to those for the particle under constant acceleration model:

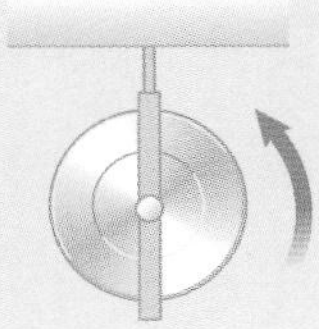

$$\omega_f = \omega_i + \alpha t \tag{9.6}$$

$$\theta_f = \theta_i + \omega_i t + \frac{1}{2}\alpha t^2 \tag{9.7}$$

$$\omega_f^2 = \omega_i^2 + 2\alpha(\theta_f - \theta_i) \tag{9.8}$$

$$\theta_f = \theta_i + \frac{1}{2}(\omega_i + \omega_f)t \tag{9.9}$$

Examples

- during its spin cycle, the tub of a clothes washer begins from rest and accelerates up to its final spin speed
- a workshop grinding wheel is turned off and comes to rest under the action of a constant friction force in the bearings of the wheel
- a gyroscope is powered up and approaches its operating speed (Chapter 10)
- the crankshaft of a diesel engine changes to a higher angular speed (Chapter 22)

Example 9.1

Rotating wheel

A wheel rotates with a constant angular acceleration of 3.50 rad/s^2.

(A) If the angular speed of the wheel is 2.00 rad/s at $t_i = 0$, through what angular displacement does the wheel rotate in 2.00 s? Give your answer in radians and degrees.

Solution

Conceptualise Draw a diagram to help you conceptualise the problem. Imagine that you start your stopwatch when the wheel is rotating at 2.00 rad/s, and count how many times it goes around in 2 seconds.

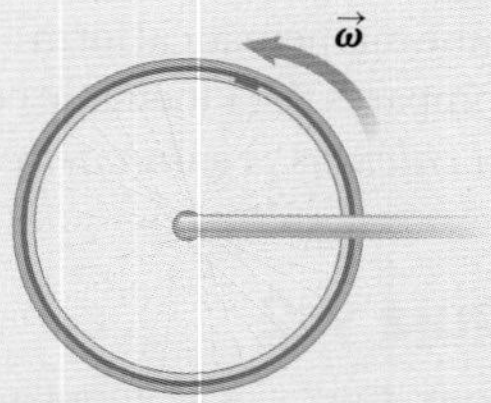

Figure 9.4
(Example 9.1) A wheel rotates with constant angular acceleration

Model The phrase 'with a constant angular acceleration' tells us we can use the rigid object with constant angular acceleration model.

Analyse Rearrange **Equation 9.7** so that it expresses the angular displacement of the object:

$$\Delta\theta = \theta_f - \theta_i = \omega_i t + \frac{1}{2}\alpha t^2$$

Substitute the known values to find the angular displacement at $t = 2.00$ s:

$$\Delta\theta = (2.00 \text{ rad/s})(2.00 \text{ s}) + \frac{1}{2}(3.50 \text{ rad/s}^2)(2.00 \text{ s})^2$$
$$= 11.0 \text{ rad}$$

Convert to degrees: $\Delta\theta = (11.0 \text{ rad})(180°/\pi \text{ rad}) = 630°$

(B) Through how many revolutions has the wheel turned during this time interval?

Solution

Multiply the angular displacement found in part (A) by the appropriate conversion factor to find the number of revolutions:

$$\Delta\theta = 630°\left(\frac{1 \text{ rev}}{360°}\right) = 1.75 \text{ rev}$$

Example 9.1 cont.

(C) What is the angular speed of the wheel at $t = 2.00$ s?

Solution

Use **Equation 9.6** to find the angular speed at $t = 2.00$ s:

$$\omega_f = \omega_i + \alpha t = 2.00 \text{ rad/s} + (3.50 \text{ rad/s}^2)(2.00 \text{ s})$$
$$= 9.00 \text{ rad/s}$$

What If? Suppose a particle moves along a straight line with a constant acceleration of 3.50 m/s². If the velocity of the particle is 2.00 m/s at $t_i = 0$, through what displacement does the particle move in 2.00 s? What is the velocity of the particle at $t = 2.00$ s?

Answer These questions are translational analogues to parts (A) and (C) of the original problem. The mathematical solution follows exactly the same form. For the displacement,

$$\Delta x = x_f - x_i = v_i t + \frac{1}{2} a t^2$$

$$= (2.00 \text{ m/s})(2.00 \text{ s}) + \frac{1}{2}(3.50 \text{ m/s}^2)(2.00 \text{ s})^2 = 11.0 \text{ m}$$

and for the velocity, $v_f = v_i + at = 2.00 \text{ m/s} + (3.50 \text{ m/s}^2)(2.00 \text{ s}) = 9.00 \text{ m/s}$

There is no translational analogue to part (B) because translational motion under constant acceleration is not cyclic.

9.3 Angular and translational quantities

In this section, we derive some useful relationships between the angular speed and acceleration of a rotating rigid object and the translational speed and acceleration of a point in the object. To do so, we must keep in mind that when a rigid object rotates about a fixed axis as in **Active Figure 9.5**, every particle of the object moves in a circle whose centre is on the axis of rotation.

Because point P in **Active Figure 9.5** moves in a circle, the translational velocity vector $\vec{\mathbf{v}}$ is always tangent to the circular path and hence is called *tangential velocity*. The magnitude of the tangential velocity of the point P is by definition the tangential speed $v = ds/dt$, where s is the distance travelled by this point measured along the circular path. Recalling that $s = r\theta$ (**Equation 9.1**) and noting that r is constant, we obtain

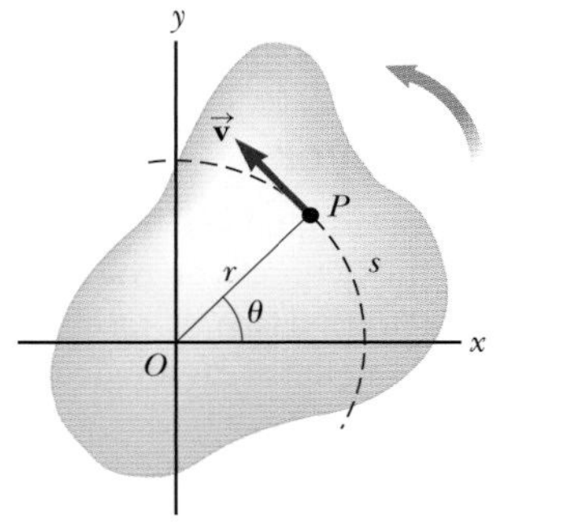

Active Figure 9.5
As a rigid object rotates about the fixed axis (the z axis) through O, the point P has a tangential velocity $\vec{\mathbf{v}}$ that is always tangent to the circular path of radius r.

$$v = \frac{ds}{dt} = r\frac{d\theta}{dt}$$

Because $d\theta/dt = \omega$ (see **Equation 9.3**), it follows that

$$v = r\omega \quad (9.10)$$

◀ Relation between tangential velocity and angular velocity

That is, the tangential speed of a point on a rotating rigid object is the product of the distance of that point from the axis of rotation and the angular speed. Therefore, although every point on the rigid object has the same *angular* speed, not every point has the same *tangential* speed because r is not the same for all points on the object. **Equation 9.10** shows that the tangential speed of a point on the rotating object increases as one moves outwards from the centre of rotation, as we would intuitively expect. For example, the outer end of a swinging golf club moves much faster than the handle.

We can relate the angular acceleration of the rotating rigid object to the tangential acceleration of the point P by taking the time derivative of v:

◀ Relation between tangential acceleration and angular acceleration

$$a_t = \frac{dv}{dt} = r\frac{d\omega}{dt}$$

$$a_t = r\alpha \quad (9.11)$$

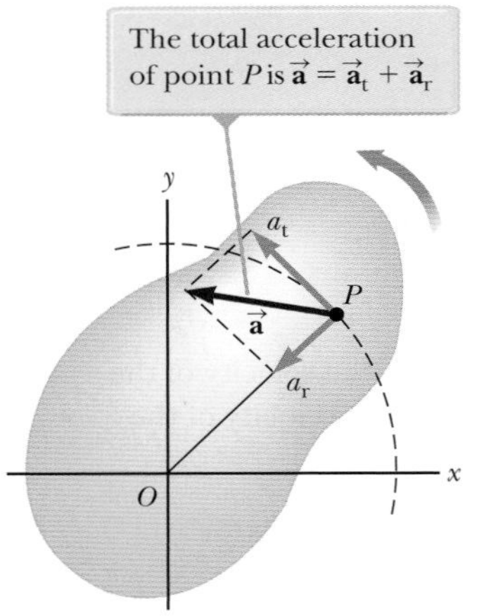

Figure 9.6
As a rigid object rotates about a fixed axis (the z axis) through O, the point P experiences a tangential component of translational acceleration a_t and a radial component of translational acceleration a_r.

That is, the tangential component of the translational acceleration of a point on a rotating rigid object equals the point's perpendicular distance from the axis of rotation multiplied by the angular acceleration.

In Section 3.5, we found that a point moving in a circular path undergoes a radial acceleration a_r directed towards the centre of rotation and whose magnitude is that of the centripetal acceleration v^2/r (**Figure 9.6**). Because $v = r\omega$ for a point P on a rotating object, we can express the centripetal acceleration at that point in terms of angular speed as

$$a_c = \frac{v^2}{r} = r\omega^2 \tag{9.12}$$

The total acceleration vector at the point is $\vec{a} = \vec{a}_t + \vec{a}_r$, where the magnitude of $\vec{a}_r$ is the centripetal acceleration a_c. Because $\vec{a}$ is a vector having a radial and a tangential component, the magnitude of $\vec{a}$ at the point P on the rotating rigid object is

$$a = \sqrt{a_t^2 + a_r^2} = \sqrt{r^2\alpha^2 + r^2\omega^4} = r\sqrt{\alpha^2 + \omega^4} \tag{9.13}$$

Quick Quiz 9.3

Marcus and Laurence are riding on a merry-go-round. Marcus rides on a horse at the outer rim of the circular platform, twice as far from the centre of the circular platform as Laurence, who rides on an inner horse. **(i)** When the merry-go-round is rotating at a constant angular speed, what is Marcus's angular speed? (a) twice Laurence's (b) the same as Laurence's (c) half of Laurence's (d) impossible to determine **(ii)** When the merry-go-round is rotating at a constant angular speed, describe Marcus's tangential speed from the same list of choices.

Example 9.2

DVD player

On a DVD (digital video disc) (**Figure 9.7**), information is stored digitally in a series of pits and flat areas on the surface of the disc. The alternations between pits and flat areas on the surface represent binary ones and zeros to be read by the DVD player. The pits and flat areas are detected by a system consisting of a laser and lenses. The length of a string of ones and zeroes representing one piece of information is the same everywhere on the disc, whether the information is near the centre of the disc or near its outer edge. So that this length of ones and zeroes always passes by the laser–lens system in the same time interval, the tangential speed of the disc surface at the location of the lens must be constant. According to **Equation 9.10**, the angular speed must therefore vary as the laser–lens system moves radially along the disc. In a typical DVD player, the constant speed of the surface at the point of the laser–lens system is 3.9 m/s.

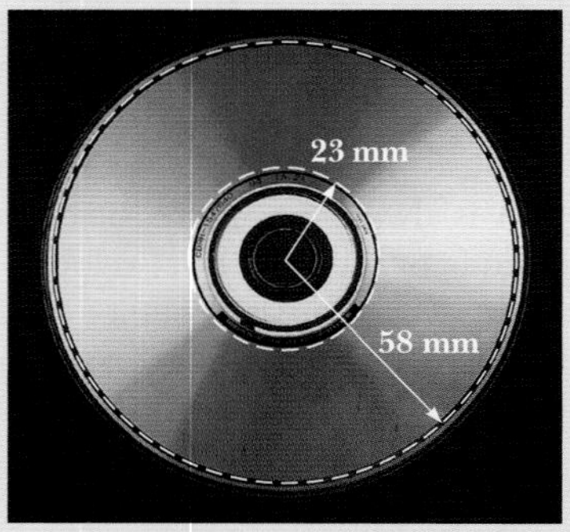

Figure 9.7
(Example 9.2) A DVD

(A) Find the angular speed of the disc in revolutions per minute when information is being read from the innermost part of the track ($r = 23$ mm) and the outermost part ($r = 58$ mm).

Solution

Conceptualise Figure 9.7 shows a photograph of a DVD. Trace your finger around the circle marked '23 mm' and mentally estimate the time interval to go around the circle once. Now trace your finger around the circle marked '58 mm', moving your finger across the surface of the page at the same speed as you did when tracing the smaller circle. Notice how much longer in time it takes your finger to go around the larger circle. If your finger represents the laser reading the disc, you can see that the disc rotates once in a longer time interval when the laser reads the information in the outer circle. Therefore, the disc must rotate more slowly when the laser is reading information from this part of the disc.

Model We model the disc as a rotating rigid object, with a changing angular speed.

Analyse Use **Equation 9.10** to find the angular speed that gives the required tangential speed at the position of the inner track:

$$\omega_i = \frac{v}{r_i} = \frac{3.9 \text{ m/s}}{2.3 \times 10^{-2} \text{ m}} = 170 \text{ rad/s}$$

$$= 170 \text{ rad/s}\left(\frac{1 \text{ rev}}{2\pi \text{ rad}}\right)\left(\frac{60 \text{ s}}{1 \text{ min}}\right) = 1.6 \times 10^3 \text{ rev/min}$$

Do the same for the outer track:

$$\omega_f = \frac{v}{r_f} = \frac{3.9 \text{ m/s}}{5.8 \times 10^{-2} \text{ m}} = 67 \text{ rad/s} = 6.4 \times 10^2 \text{ rev/min}$$

The DVD player adjusts the angular speed v of the disc within this range so that information moves past the laser–lens system at a constant rate.

(B) The playing time of the DVD is four hours. How many revolutions does the disc make during that time?

Solution

Model From part (A), the angular speed decreases as the disc plays. Let us assume it decreases steadily, with a constant α. We can then apply the rigid object with constant angular acceleration model to the disc.

Analyse If $t = 0$ is the instant the disc begins rotating, with angular speed of 170 rad/s, the final value of the time t is (4 h) (60 min/h) (60 s/min) = 14 400 s. We are looking for the angular displacement $\Delta\theta$ during this time interval.

Use **Equation 9.9** to find the angular displacement of the disc at $t = 14\,400$ s:

$$\Delta\theta = \theta_f - \theta_i = \frac{1}{2}(\omega_i + \omega_f)t$$

$$= \frac{1}{2}(170 \text{ rad/s} + 67 \text{ rad/s})(14\,400 \text{ s}) = 1.7 \times 10^6 \text{ rad}$$

Convert this angular displacement to revolutions:

$$\Delta\theta = (1.7 \times 10^6 \text{ rad})\left(\frac{1 \text{ rev}}{2\pi \text{ rad}}\right) = 2.7 \times 10^5 \text{ rev}$$

Finalise The disc is rotated more than 200 000 times during the play time. To maintain constant linear speed of the point on the disc being read, the disc must slow down gradually as it plays, but even at the end is moving at more than 600 rpm.

Conceptual Example 9.3

The rotating sprinkler and the Coriolis effect

The figure on the right shows a display at Questacon that demonstrates rotational motion. It is effectively a hand-driven rotating sprinkler. Water flows through the centre hub and into each of the radial pipes. Half of the pipes are bent backwards part way along their length. When you turn the handle and rotate the hub and attached pipes, the water flowing out of the straight pipes *lags* behind an imaginary line extending the pipe, and the water flowing from the bent – backwards pipes *leads* an imaginary line extended from the pipe towards the hub. Why?

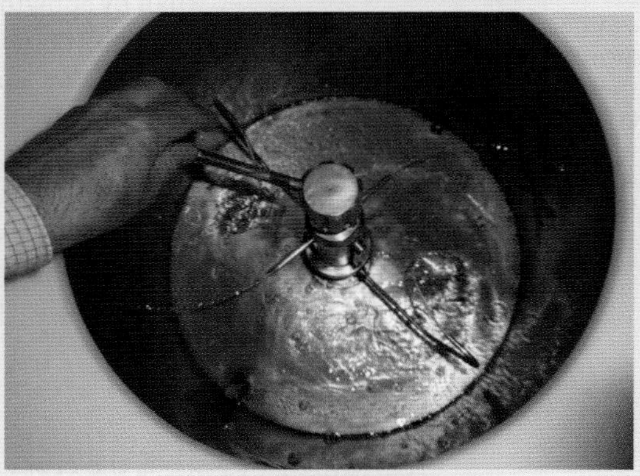

Solution

Consider the simple case of the sprinkler being turned at constant speed. The angular velocity of the system including the water *inside* the pipes is all moving at the same constant *angular* velocity. The *tangential* velocity of the water inside the pipes is greater the further out from the hub it is ($v = \omega r$). If you draw a line extending the length of the pipe, any point on the line must move with the same angular velocity, and hence if you extend the line outwards from the straight pipes, points on the line move faster than the end of the pipe, or the tangential velocity of the water exiting the pipe. When the water leaves the end of the pipe, there is no longer a centripetal force acting on it so it simply continues moving as a projectile subject only to gravity and air resistance. It moves with approximately constant horizontal velocity along a straight line

Conceptual Example 9.3 cont.

with the velocity at which it left the pipe. The velocity has both a radial and tangential component because the water was flowing outwards or inwards as well as moving in a circle. The tangential component, shown by the yellow arrows, is less than that of any point on the yellow dotted line extending from the pipe end, and so the water flow, shown in blue, lags behind this imaginary line.

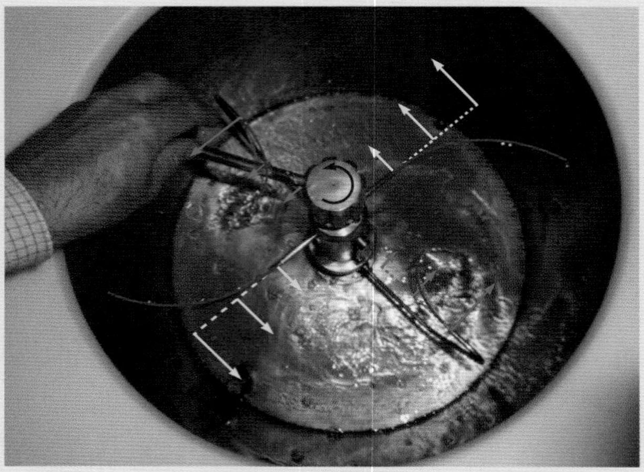

The opposite is seen for the bent pipes. Any point on the green dotted line extended from the mouth of the pipe towards the hub moves at lower tangential velocity, shown by the green arrows, than the water exiting the pipe end. Hence the water 'leads' the imaginary line.

Note that although the water appears to curve, any small element of water actually follows a straight path in the horizontal direction. Remember that what you are seeing as a curved stream is actually lots of molecules of water released at different moments and hence at different positions of the sprinkler. Any individual molecule follows a straight path from the viewpoint of an observer outside the system. Within the system, the individual water molecules appear to follow a curved path – this is the Coriolis effect as described in Chapter 5.

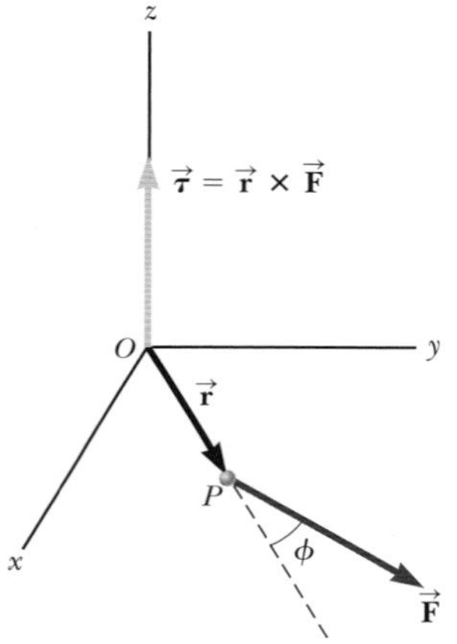

Active Figure 9.8

The torque vector $\vec{\tau}$ lies in a direction perpendicular to the plane formed by the position vector $\vec{r}$ and the applied force vector $\vec{F}$. In the situation shown, $\vec{r}$ and $\vec{F}$ lie in the xy plane, so the torque is along the z axis.

9.4 Torque

So far we have described rotational motion, and changes in rotational motion, without considering their cause. We now consider the analogue to force, which is called ***torque***.

TRY THIS

Push on a swinging door close to the edge with the hinges. Can you open the door this way? Now try pushing a little away from the hinges, towards the handle. Move your hand gradually across the door, pushing as you go. When does the door starts to open? Why does the force that you apply have more effect the further you get from the hinges?

As you will have found in the *Try this* example above, if you apply a force of magnitude F perpendicular to the surface of a door and near the hinges very little happens. You will achieve a more rapid rate of rotation for the door by applying the force near the doorknob than by applying it near the hinges. When a force is exerted on a rigid object pivoted about an axis, the object tends to rotate about that axis. The tendency of a force to rotate an object about some axis is measured by a quantity called **torque** $\vec{\tau}$ (Greek letter tau). Torque is a pseudovector, just as α and ω are. The direction of the applied torque is as important in determining the change in rotational motion as the direction of an applied force is in determining the change in translational motion.

Σ *You may want to review the vector cross product in Appendix B.6 before continuing.*

Consider a force $\vec{F}$ acting on a particle at the vector position $\vec{r}$ (Active Figure 9.8). The axis about which $\vec{F}$ tends to produce rotation is perpendicular to the plane formed by $\vec{r}$ and $\vec{F}$. The torque vector $\vec{\tau}$ is related to the two vectors $\vec{r}$ and $\vec{F}$:

$$\vec{\tau} \equiv \vec{r} \times \vec{F} \tag{9.14}$$

This says that torque is the **cross product**, or **vector product**, of position and force.

Torque has units of newton metres or kg.m^2/s^2 in SI units and dimensions of ML^2T^{-2}. Do not confuse torque and work, which have the same units but are very different concepts.

Given any two vectors $\vec{\mathbf{A}}$ and $\vec{\mathbf{B}}$, the vector product $\vec{\mathbf{A}} \times \vec{\mathbf{B}}$ is defined as a third vector $\vec{\mathbf{C}}$, which has a magnitude of $AB \sin\theta$, where θ is the angle between $\vec{\mathbf{A}}$ and $\vec{\mathbf{B}}$. That is, if $\vec{\mathbf{C}}$ is given by

$$\vec{\mathbf{C}} = \vec{\mathbf{A}} \times \vec{\mathbf{B}} \tag{9.15}$$

its magnitude is

$$C = AB\sin\theta \tag{9.16}$$

The quantity $AB \sin\theta$ is equal to the area of the parallelogram formed by $\vec{\mathbf{A}}$ and $\vec{\mathbf{B}}$ as shown in **Figure 9.9**. The *direction* of $\vec{\mathbf{C}}$ is perpendicular to the plane formed by $\vec{\mathbf{A}}$ and $\vec{\mathbf{B}}$, and the best way to determine this direction is to use the right-hand rule illustrated in **Figure 9.9**. The four fingers of the right hand are pointed along $\vec{\mathbf{A}}$ and then 'curled' in the direction that would rotate $\vec{\mathbf{A}}$ into $\vec{\mathbf{B}}$ through the angle θ. The direction of the upright thumb is the direction of $\vec{\mathbf{A}} \times \vec{\mathbf{B}} = \vec{\mathbf{C}}$. Because of the notation $\vec{\mathbf{A}} \times \vec{\mathbf{B}}$ is often read '$\vec{\mathbf{A}}$ cross $\vec{\mathbf{B}}$', the vector product is also called the **cross product**. Vector cross products point along the axis defined by a rotation from one vector to the other, as in **Active Figure 9.8**.

The cross product of any two vectors $\vec{\mathbf{A}}$ and $\vec{\mathbf{B}}$ can be expressed in determinant form:

$$\vec{\mathbf{A}} \times \vec{\mathbf{B}} = \begin{vmatrix} \hat{\mathbf{i}} & \hat{\mathbf{j}} & \hat{\mathbf{k}} \\ A_x & A_y & A_z \\ B_x & B_y & B_z \end{vmatrix} = \begin{vmatrix} A_y & A_z \\ B_y & B_z \end{vmatrix}\hat{\mathbf{i}} + \begin{vmatrix} A_z & A_x \\ B_z & B_x \end{vmatrix}\hat{\mathbf{j}} + \begin{vmatrix} A_x & A_y \\ B_x & B_y \end{vmatrix}\hat{\mathbf{k}}$$

Expanding these determinants gives the result

$$\vec{\mathbf{A}} \times \vec{\mathbf{B}} = (A_yB_z - A_zB_y)\hat{\mathbf{i}} + (A_zB_x - A_xB_z)\hat{\mathbf{j}} + (A_xB_y - A_yB_x)\hat{\mathbf{k}} \tag{9.17}$$

which implies that when crossing unit vectors:

$$\hat{\mathbf{i}} \times \hat{\mathbf{i}} = \hat{\mathbf{j}} \times \hat{\mathbf{j}} = \hat{\mathbf{k}} \times \hat{\mathbf{k}} = 0$$

and

$$\hat{\mathbf{i}} \times \hat{\mathbf{j}} = \hat{\mathbf{k}},\ \hat{\mathbf{j}} \times \hat{\mathbf{k}} = \hat{\mathbf{i}},\ \hat{\mathbf{k}} \times \hat{\mathbf{i}} = \hat{\mathbf{j}}$$

The properties of the vector cross product are given in Appendix B.6. Σ

Quick Quiz 9.4

Find the cross products: $\hat{\mathbf{j}} \times \hat{\mathbf{i}}$, $\hat{\mathbf{k}} \times \hat{\mathbf{j}}$ and $\hat{\mathbf{i}} \times \hat{\mathbf{k}}$.

We can now assign a direction to the torque vector. If the force lies in the xy plane as in **Active Figure 9.8**, the torque $\vec{\boldsymbol{\tau}}$ is represented by a vector parallel to the z axis. The force in **Active Figure 9.8** creates a torque that tends to rotate the object anticlockwise about the z axis; the direction of $\vec{\boldsymbol{\tau}}$ is towards increasing z, and $\vec{\boldsymbol{\tau}}$ is therefore in the positive z direction. If we reversed the direction of $\vec{\mathbf{F}}$ in **Active Figure 9.8**, $\vec{\boldsymbol{\tau}}$ would be in the negative z direction.

Torque depends on the applied force and the position of the application of the force. The vector that defines the position is measured from an *arbitrary* axis. For many situations there is an obvious rotation axis or pivot point from which we measure the position vector $\vec{\mathbf{r}}$. But this is not always the case. Hence there is no unique value of torque for an applied force on an object because the choice of axis is not unique. If we choose a different axis to measure the position vector $\vec{\mathbf{r}}$ from, then we will calculate a different value of torque. Hence we always need to specify the axis about which the torque is calculated. Sometimes subscript notation such as τ_O is used, where the subscript O indicates that the torque is about an axis through point O. Note that the term *moment* is often used in engineering rather than torque, and the moment about a point O is written M_O.

Pitfall Prevention 9.4

Torque and work both have units of N.m; however, they are different things and should not be confused. Torque is a vector, and it indicates the effect on rotational motion of an applied force. Work is a transfer of energy, and is a scalar. Units for work can be written as N.m or J. Units for torque are not written as J.

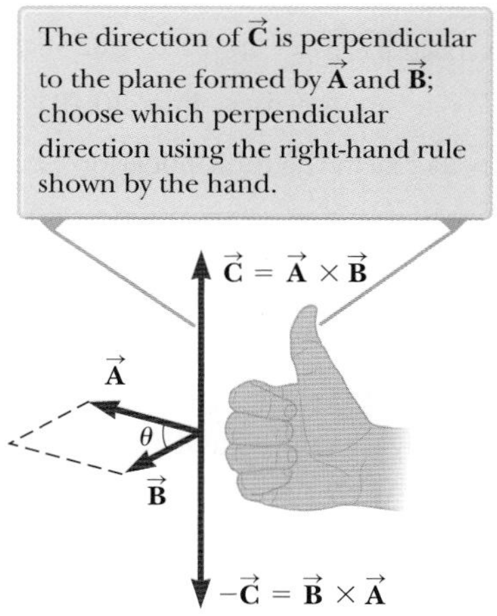

Figure 9.9
The vector product $\vec{\mathbf{A}} \times \vec{\mathbf{B}}$ is a third vector $\vec{\mathbf{C}}$ having a magnitude $AB \sin\theta$ equal to the area of the parallelogram shown.

Pitfall Prevention 9.5

Remember that the result of taking a vector product between two vectors is a *third vector*. Equation 9.16 gives only the magnitude of this vector; the direction is given by the right-hand rule.

Pitfall Prevention 9.6

Torque depends on your choice of axis. There is no unique value of the torque on an object. Its value depends on your choice of rotation axis. Often there is a natural choice, such as the axle about which a wheel rotates, but this is not always the case. Always remember to clearly indicate your choice of rotation axis, and to be consistent once you have chosen it.

Example 9.4

The torque vector

A force of $\vec{\mathbf{F}} = 2.00\,\hat{\mathbf{i}} + 3.00\,\hat{\mathbf{j}}$ N is applied to the handle of a spanner that is pivoted about a bolt aligned along the z coordinate axis. The force is applied at a point located at $\vec{\mathbf{r}} = 0.40\hat{\mathbf{i}} + 0.50\hat{\mathbf{j}}$ m. Find the torque $\vec{\tau}$ applied to the object.

Solution

Conceptualise Think about the directions of the force and position vectors. Draw a diagram to help visualise the problem. In what direction will the spanner and bolt turn?

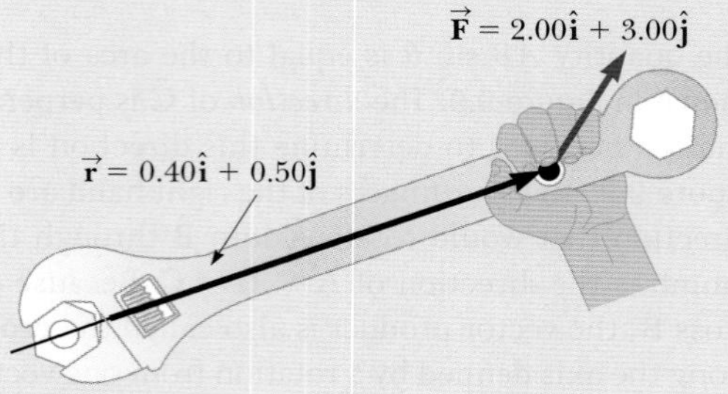

Figure 9.10
The force $\vec{\mathbf{F}}$ applied to a spanner at position $\vec{\mathbf{r}}$

Model We model the spanner as a rigid object and use the definition of torque.

Analyse Set up the torque vector using **Equation 9.14**:

$$\vec{\tau} = \vec{\mathbf{r}} \times \vec{\mathbf{F}} = [(0.40\,\hat{\mathbf{i}} + 0.50\,\hat{\mathbf{j}})\text{ m}] \times [(2.00\,\hat{\mathbf{i}} + 3.00\,\hat{\mathbf{j}})\text{ N}]$$

Perform the multiplication:

$$\vec{\tau} = [(0.40)(2.00)\,\hat{\mathbf{i}} \times \hat{\mathbf{i}} + (0.40)(3.00)\,\hat{\mathbf{i}} \times \hat{\mathbf{j}} + (0.50)(2.00)\,\hat{\mathbf{j}} \times \hat{\mathbf{i}} + (0.50)(3.00)\,\hat{\mathbf{j}} \times \hat{\mathbf{j}}]\text{ N.m}$$

$$\vec{\tau} = [0 + 1.20\,\hat{\mathbf{k}} - 1.0\,\hat{\mathbf{k}} + 0]\text{ N.m} = 0.20\,\hat{\mathbf{k}}\text{ N.m}$$

Finalise Notice that both $\vec{\mathbf{r}}$ and $\vec{\mathbf{F}}$ are in the xy plane. As expected, the torque vector is perpendicular to this plane, having only a z component. We have followed the rules for significant figures discussed in Section 1.7, which lead to an answer with two significant figures. We have lost some precision because we ended up subtracting two numbers that are close. What would the range of answers have been had there been a 10% uncertainty in the values given?

Consider the spanner in **Figure 9.11** that we wish to rotate around an axis that is perpendicular to the page and passes through the centre of the bolt. The applied force $\vec{\mathbf{F}}$ acts at an angle ϕ to the horizontal. The magnitude of the torque associated with the force $\vec{\mathbf{F}}$ around the axis passing through O is given by

$$\tau \equiv \vec{\mathbf{r}} \times \vec{\mathbf{F}} = rF\sin\phi = Fd \qquad (9.18)$$

where r is the distance between the rotation axis and the point of application of $\vec{\mathbf{F}}$, and d is the perpendicular distance from the rotation axis to the line of action of $\vec{\mathbf{F}}$. The *line of action* of a force is an imaginary line extending out both ends of the vector representing the force. The dashed line extending from the tail of $\vec{\mathbf{F}}$ in **Figure 9.11** is part of the line of action of $\vec{\mathbf{F}}$. From the right triangle in **Figure 9.11** that has the spanner as its hypotenuse, we see that $d = r\sin\phi$.

Lever arm ▶ The quantity d is called the **moment arm** or **lever arm** of $\vec{\mathbf{F}}$. The greater the lever arm, the greater the torque and hence the greater the effect of a given applied force. In general, the longer the handle on a tool such as a spanner, the more torque you can apply with it.

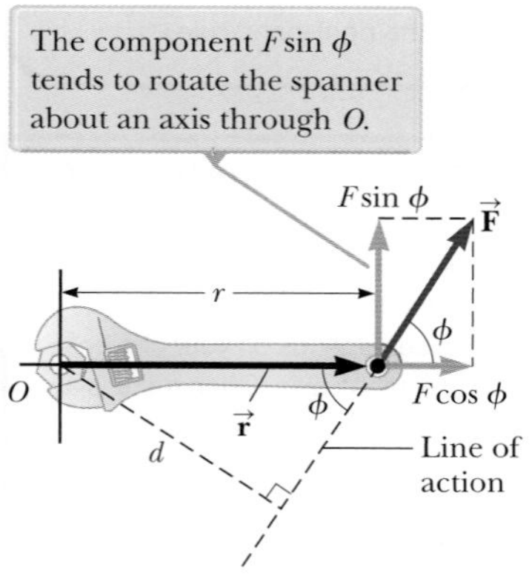

Figure 9.11
The force $\vec{\mathbf{F}}$ has a greater rotating tendency about an axis through O as F increases and as the moment arm d increases.

In **Figure 9.11**, the only component of $\vec{\mathbf{F}}$ that tends to cause rotation of the spanner around an axis through O is $F\sin\phi$, the component perpendicular to a line drawn from the rotation axis to the point of application of the force. The horizontal component, $F\cos\phi$, has no tendency to produce rotation about an axis passing through O because its line of action passes through O.

Hence when we open a door we want to apply our push as far from the hinges as possible and as closely perpendicular to the door as we can so that ϕ is close to 90°.

Levers

Consider again the spanner in **Figure 9.11**. To be able to turn a nut in the jaws of the spanner, the person using the spanner must apply a force such that the torque due to that force is greater than the torque due to the static friction force that holds the nut in place. The static friction force acts very close to the pivot point, the centre of the nut, while the force exerted by the person holding the spanner acts at a substantial distance. Recalling that torque is proportional to d, the greater the length of the lever arm (the longer the handle of the spanner) the greater the **mechanical advantage** that it gives. Mechanical advantage is a measure of how much load a lever can hold in equilibrium for a given applied force.

◀ Mechanical advantage

Consider the lever shown in Figure 9.12. The pivot point, also called the *fulcrum*, is shown by the triangle. The applied force or *effort* acts to balance the torque due to the *load*. The mechanical advantage is defined as

$$\text{Mechanical advantage} = \frac{\text{effort applied}}{\text{load balanced}}$$

which, using Equation 9.18, we can write as

$$\text{Mechanical advantage} = \frac{\text{effort arm}}{\text{load arm}}$$

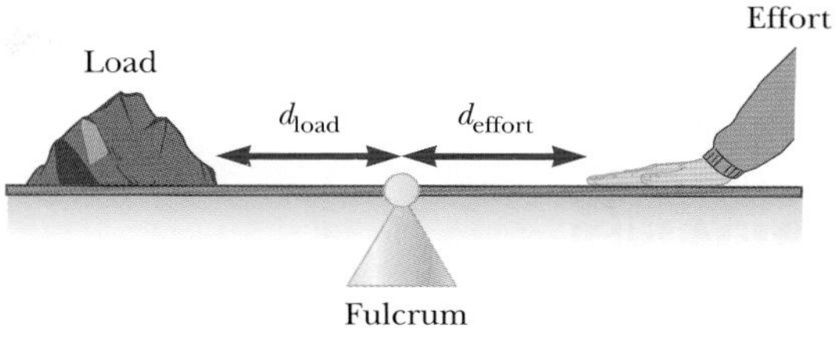

Figure 9.12
A class 1 lever, with the torque due to the load balanced by the effort

If the load is balanced then the torque due to the effort is equal to that due to the load, and we can write:

$$F_{load}\,d_{load} = F_{effort}\,d_{effort} \quad \text{or} \quad F_{load} = \frac{F_{effort}\,d_{effort}}{d_{load}} \tag{9.19}$$

This is the force exerted at the point of the load due to the effort force. We can see that the greater the ratio of d_{effort}/d_{load}, the greater the force that we can exert on the load for a given applied force. Many tools are just levers of one kind or another, using mechanical advantage to allow a small force to have a large effect.

Usually when we use a tool we wish to not only balance a load, but move it, such as the nut being turned by the spanner. So we need a lever arm long enough that our applied torque is greater than that applied by the load.

We can classify levers into three types, as shown in Figure 9.13.

A class 1 lever has the load and effort on opposite sides of the fulcrum, like a seesaw or a pair of scissors. A class two lever has the load between the fulcrum and the effort. A wheelbarrow is an example of a class two lever. A class three lever has the effort in the middle, between the fulcrum and the load, such as in a pair of tweezers. The expression for mechanical advantage above applies to all classes of levers.

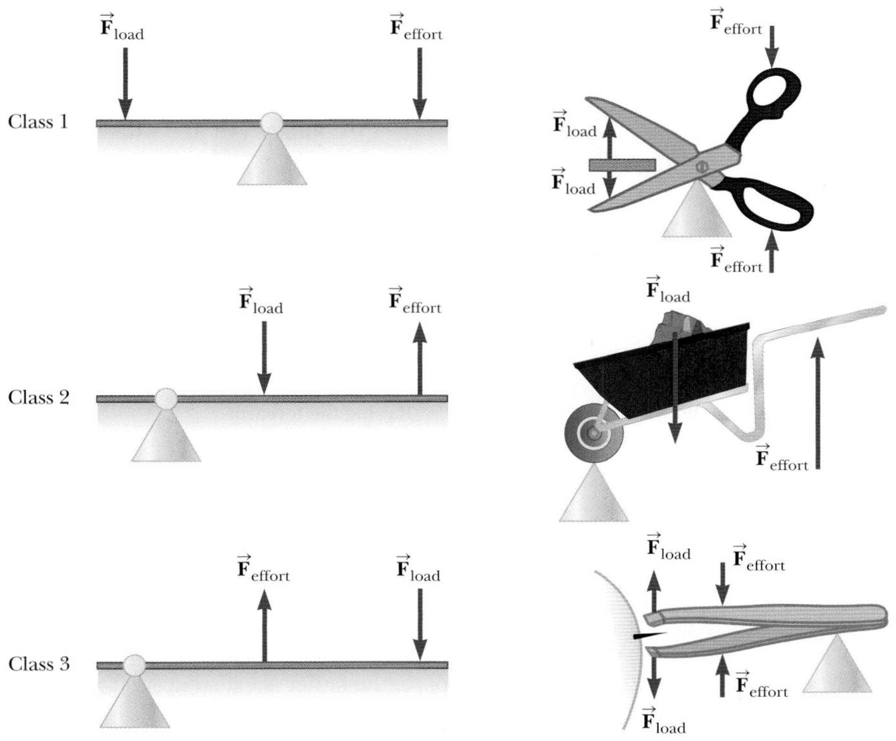

Figure 9.13
The three classes of levers and an example of each

TRY THIS

If you have your own toolbox, take out the tools and examine them. Otherwise look at someone else's toolbox, pay a visit to a hardware shop or see **Figure 9.14**. How many of the tools are really just levers? Can you identify what class of lever they are by locating the load, effort and fulcrum?

TRY THIS

Stand up straight and feel the muscles at the back of your thighs. Now lean forwards so your upper body pivots about your hips and feel your thigh muscles again. Stand on your toes and feel what happens to your calf muscles. Now hold a weight such as a heavy physics book in your hand and with your upper arm held vertical lift the book up and down. Feel what your bicep muscle does as you raise and lower the book. Can you identify the load, effort and fulcrum, and the class of lever in each of these exercises?

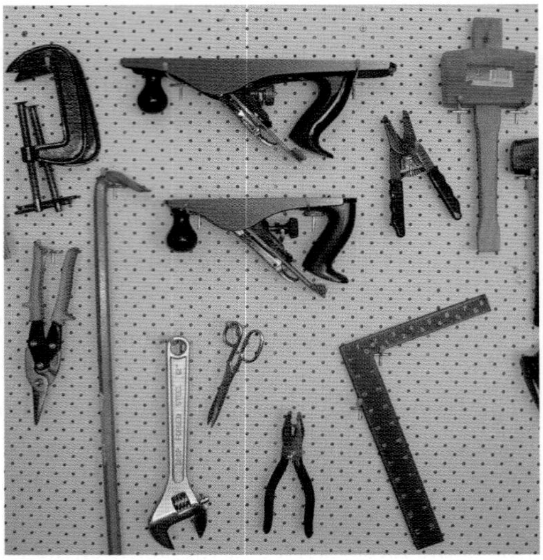

Figure 9.14
Can you identify the load, effort and fulcrum in these tools? What class of lever is each? Which ones are not levers?

Your body also contains levers of all three classes, as shown in **Figure 9.15**. Bending at the hips uses the muscles at the back of the thighs to provide the effort to balance the weight of your upper body which pivots about the hip, so that the load (your upper body's weight) is on the opposite side of the fulcrum to the effort (the force applied by the thigh muscles), as in a class 1 lever. When you stand up on your toes (the fulcrum), your weight (load) acts through your calf bone and the calf muscle, connecting to the back of your foot near the heel, supplies the effort. This is a class 2 lever. When you hold your upper arm vertical and your forearm horizontal with a weight in your hand, then lift the weight while keeping your upper arm vertical, your biceps provide the effort needed to raise the weight (load), as your arm pivots about your elbow joint (fulcrum). This is a class 3 lever. Levers are important simple machines, not just in physics but in engineering, medicine and everyday life.

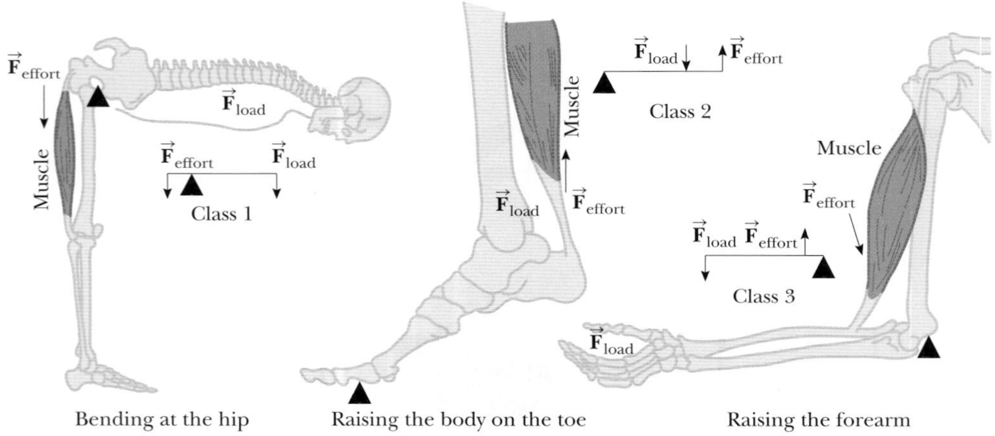

Figure 9.15
Three examples of levers in the body

Example 9.5

The seesaw

A seesaw consisting of a uniform board of mass M and length ℓ supports at rest a father and daughter with masses m_f and m_d, respectively. The support (the *fulcrum*) is under the centre of the board, the father is a distance d from the centre to the left, and the daughter is a distance $\ell/2$ from the centre to the right.

Determine where the father should sit to balance the system at rest.

Example 9.5 cont.

Solution

Conceptualise Draw a diagram showing the situation. Think about your experience on seesaws, and imagine where a heavy adult would need to sit to balance a child.

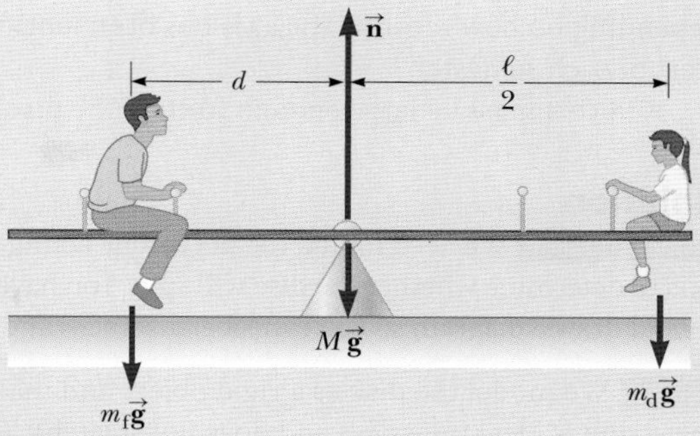

Figure 9.16
(Example 9.5) A balanced system

Model We can consider the seesaw as a class 1 lever with the daughter acting as the load to be balanced and the weight of the father providing the effort. We are told that the board is uniform and the fulcrum is at the centre, so we do not need to consider the weight of the board as it is evenly distributed about the fulcrum.

Analyse We want the torque due to the father to be equal to that of the daughter:

$$\tau_f = \tau_d$$

$$(m_f g)(d) = (m_d g)\frac{\ell}{2}$$

Solve for d:

$$d = \left(\frac{m_d}{m_f}\right)\frac{\ell}{2}$$

Do a dimension check: $[L] = ([M]/[M])[L]$ ☺

Finalise Check that this result makes physical sense by considering limiting cases. If the father is very heavy or the daughter is very light, then he needs to sit very close to the centre of the seesaw. If the daughter is heavier than he is, then he will not be able to balance her weight as long as she is sitting at the end of the seesaw; he will not be able to get enough mechanical advantage.

If two or more forces act on a rigid object as in **Active Figure 9.17**, each tends to produce rotation about the axis through O. In this example, $\vec{F}_2$ tends to rotate the object clockwise and $\vec{F}_1$ tends to rotate it anticlockwise. The direction of a torque can be found from the vector cross product, as shown in **Equation 9.14**, but often it is convenient to use the convention that the sign of the torque resulting from a force is positive if the turning tendency of the force is anticlockwise and negative if the turning tendency is clockwise. For example, in **Active Figure 9.17**, the torque resulting from $\vec{F}_1$, which has a moment arm d_1, is positive and equal to $+F_1d_1$; the torque from $\vec{F}_2$ is negative and equal to $-F_2d_2$. Hence, the *net* torque about an axis through O is

$$\sum \tau = \tau_1 + \tau_2 = F_1d_1 - F_2d_2$$

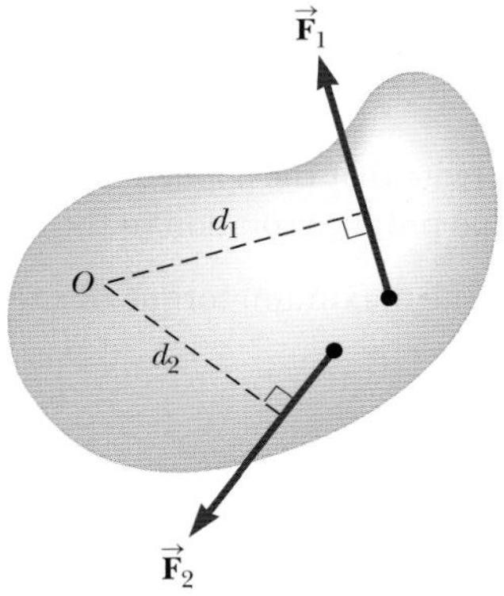

Active Figure 9.17
The force $\vec{F}_1$ tends to rotate the object anticlockwise about an axis through O, and $\vec{F}_2$ tends to rotate it clockwise.

Quick Quiz 9.5

(i) If you are trying to loosen a stubborn screw from a piece of wood with a screwdriver and fail, should you find a screwdriver for which the handle is (a) longer or (b) fatter? (ii) If you are trying to loosen a stubborn bolt from a piece of metal with a spanner and fail, should you find a spanner for which the handle is (a) longer or (b) fatter?

±? Example 9.6

Torque on a disc

In a rotational dynamics experiment, a group of first-year students attach a weight to a string that is wound around the small central hub of a large disc as shown. The disc is free to rotate about the central z axis as shown in the **Figure 9.19**. The hub has a diameter of (8.10 ± 0.05) cm and the students measure the mass of the attached weight as (50.0 ± 0.5) g. In addition, there is a frictional force acting at the axle, within the hub. The axle has a diameter of (1.00 ± 0.05) cm and the maximum static frictional force is approximately constant, varying between 0.2 N and 3.8 N

Example 9.6 cont.

depending on how recently the axle has been lubricated, which is done at the start of each semester.

Will this mass be large enough to start the disc rotating?

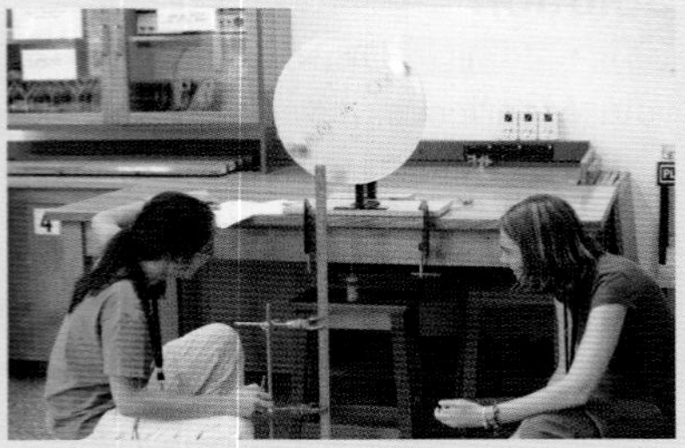

Figure 9.18
(Example 9.6) Students working on a rotational dynamics experiment at Monash University

Solution

Conceptualise We need to find the net torque acting on the disc about the rotation axis to determine whether the disc will spin. You have a photo of the system, but you should draw a diagram so you can label distances and directions of forces.

Model We model the disc as a rigid object and make the approximation that the string is ideal (massless and does not stretch). The two forces acting which produce torques are the tension in the string and the friction at the axle. We choose our rotation axis as the rotation axis of the disc.

Analyse We need to calculate the torque due to each force, noting that the friction force increases from a minimum value at the start of semester to a maximum at the end. We use **Equation 9.14**, with the forces perpendicular to the vector $\vec{\mathbf{r}}$ so that $\tau = rF$.

The tension in the string is equal to the applied weight, mg, when the weight is in equilibrium. This occurs if the friction force can balance the applied torque.

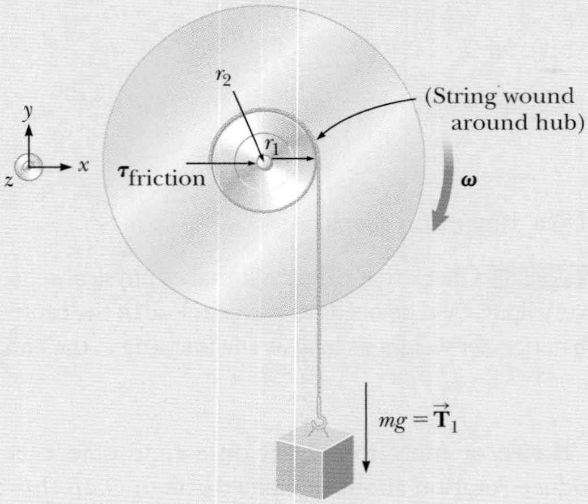

Figure 9.19
(Example 9.6) Forces acting in the rotational dynamics experiment

Allowing for the uncertainties in the measured values, the maximum torque due to the hanging weight is:

$$\tau_{1,\text{max}} = -r_{1,\text{max}} T_{1,\text{max}} = -r_{1,\text{max}} m_{\text{max}} g$$
$$= (0.04075 \text{ m})(0.00505 \text{ g})(9.80 \text{ m/s}^2) = 0.0202 \text{ N.m}$$

and the minimum applied torque is:

$$\tau_{1,\text{min}} = -r_{1,\text{min}} T_{1,\text{min}} = -r_{1,\text{min}} m_{\text{min}} g$$
$$= (0.04025 \text{ m})(0.00495 \text{ g})(9.80 \text{ m/s}^2) = 0.0195 \text{ N.m}$$

Note that the applied torque has a negative sign because it will tend to produce rotation in the clockwise direction.

The maximum torque due to the static friction force varies between

$$\tau_{2,\text{start}} = +r_2 F_{2,\text{start}} = (0.00525 \text{ m})(0.2 \text{ N}) = 0.00105 \text{ N} \quad \text{and}$$
$$\tau_{2,\text{end}} = +r_2 F_{2,\text{end}} = (0.00525 \text{ m})(3.8 \text{ N}) = 0.01995 \text{ N}$$

at the start and end of semester.

At the start of semester, the maximum static friction force provides a torque much less than that provided by the weight, and so the disc will begin to rotate.

At the end of semester, the maximum static friction force provides a torque that is within the range of that provided by the weight and so the disc may not rotate anymore unless additional weight is added or the axle is lubricated again.

Finalise The applied weight will provide easily enough torque to start the disc spinning at the start of semester, but by the final week the friction force will have increased enough that this weight may not cause the disc to spin anymore unless the axle is lubricated again.

9.5 Newton's Second Law for rotation and moments of inertia

Newton's second law states that the acceleration of an object due to an applied force is inversely proportional to the mass of the object – in fact, we can define mass as the constant of proportionality between force and acceleration (as in Chapter 4). The heavier the object, the more force is required to accelerate it by a given amount. Similarly, when we apply a torque to a rotating object, it experiences an angular acceleration. The angular acceleration, α, is proportional to the applied torque, τ. It also depends on the mass of the object and the way in which the mass is distributed.

TRY THIS

Sit right on the edge of a chair and, with your leg held straight, lift it in an arc from floor level as high as you can. (Hold on to the chair so you don't fall off.) Now do the same thing again with your leg bent at the knee. Which is easier and why?

The further from the axis of rotation that a mass is, the greater the torque required to achieve a given angular acceleration. The constant of proportionality linking torque and angular acceleration is called the ***moment of inertia***. Moment of inertia is a measure of the resistance of an object to changes in its rotational motion, just as mass is a measure of the tendency of an object to resist changes in its translational motion.

Before discussing the more complex case of rigid-object rotation, let us first consider the case of a particle moving in a circular path about some fixed point under the influence of an external force, such as a poi swung in a circle on the end of a string.

Consider a particle of mass m rotating in a circle of radius r under the influence of a tangential net force $\sum\vec{\mathbf{F}}_t$ and a radial net force $\sum\vec{\mathbf{F}}_r$ as shown in **Figure 9.20**. The radial net force causes the particle to move in the circular path with a centripetal acceleration. The tangential force provides a tangential acceleration $\vec{\mathbf{a}}_t$, and

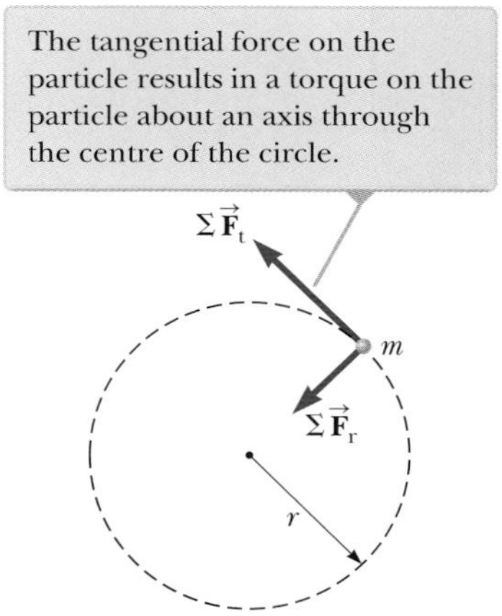
The tangential force on the particle results in a torque on the particle about an axis through the centre of the circle.
$\Sigma\vec{\mathbf{F}}_t$
m
$\Sigma\vec{\mathbf{F}}_r$
r

Figure 9.20
A particle rotating in a circle under the influence of a tangential net force $\sum\vec{\mathbf{F}}_t$. A radial net force $\sum\vec{\mathbf{F}}_r$ also must be present to maintain the circular motion.

$$\sum F_t = ma_t$$

The magnitude of the net torque due to $\sum\vec{\mathbf{F}}_t$ on the particle about an axis perpendicular to the page through the centre of the circle is

$$\sum\tau = \sum F_t\, r = (ma_t)r$$

Because the tangential acceleration is related to the angular acceleration through the relationship $a_t = r\alpha$ (**Equation 9.11**), the net torque can be expressed as

$$\sum\tau = (mr\alpha)r = (mr^2)\alpha$$

For a single particle, mr^2 is the moment of inertia of the particle about the z axis passing through the origin, and we can write

$$\sum\tau = I\alpha \qquad \textbf{(9.20)}$$

◀ Newton's second law for rotation

where I is the moment of inertia and is given by

$$I = mr^2 \qquad \textbf{(9.21)}$$

for a particle at a distance r from the axis of rotation. I has units of kg.m^2 and dimensions of ML^2.

Calculating the moment of inertia

Notice that $\sum\tau = I\alpha$ has the same mathematical form as Newton's second law of motion, $\sum F = ma$. We can extend this to a system of particles, with masses $m_1, m_2, m_3 \ldots$ at distances $r_1, r_2, r_3 \ldots$ from the centre of rotation, to give:

$$I \equiv \sum_i m_i r_i^2 \qquad \textbf{(9.22)}$$

◀ Moment of inertia

Σ

You may want to review integral calculus in Appendix B.8 before continuing.

We can evaluate the moment of inertia of a continuous rigid object by imagining the object to be divided into many small elements, each of which has mass Δm_i. We use the definition $I = \sum_i r_i^2 \Delta m_i$ and take the limit of this sum as $\Delta m_i \to 0$. In this limit, the sum becomes an integral over the volume of the object:

$$I = \lim_{\Delta m_i \to 0} \sum_i r_i^2 \Delta m_i = \int r^2\, dm \qquad \textbf{(9.23)}$$

◀ Moment of inertia of a rigid object

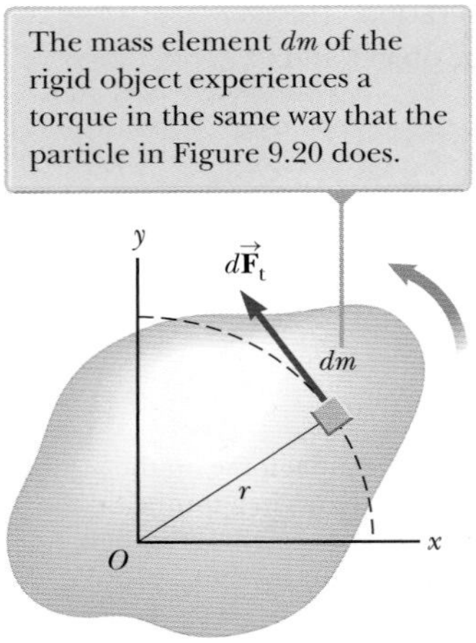

Figure 9.21
A rigid object rotating about an axis through O. Each mass element dm rotates about the axis with the same angular acceleration α.

It is usually easier to calculate moments of inertia in terms of the volume of the elements rather than their mass, and we can easily make that change by using **Equation 1.1**, $\rho \equiv m/V$, where ρ is the density of the object and V is its volume. From this equation, the mass of a small element is $dm = \rho dV$. Substituting this result into **Equation 9.23** gives

$$I = \int \rho r^2 dV$$

If the object is homogeneous, ρ is constant and the integral can be evaluated for a known geometry. If ρ is not constant, its variation with position must be known to perform the integration.

The density given by $\rho = m/V$ is sometimes referred to as *volumetric mass density* because it represents *mass per unit volume*. Often we use other ways of expressing density. For instance, when dealing with a sheet of uniform thickness t, we can define a *surface mass density* $\sigma = m/A = \rho t$, which represents *mass per unit area*. Or when mass is distributed along a rod or rope of uniform cross-sectional area A, we sometimes use *linear mass density* $\lambda = m/L = \rho A$, which is the *mass per unit length*.

Table 9.2 gives the moments of inertia for a number of objects about specific axes. The moments of inertia of rigid objects with simple geometry (high symmetry) are easy to calculate provided the rotation axis coincides with an axis of symmetry, as in the following examples.

Table 9.2
Moments of inertia of homogeneous rigid objects with different geometries

Hoop or thin cylindrical shell $I_{CM} = MR^2$	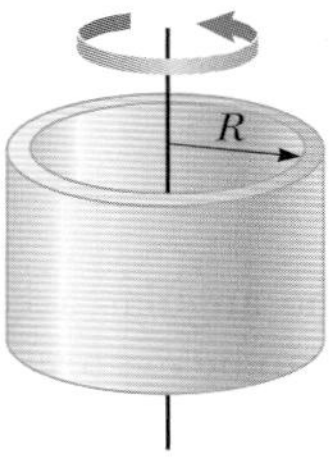	Hollow cylinder $I_{CM} = \frac{1}{2}M(R_1^2 + R_2^2)$	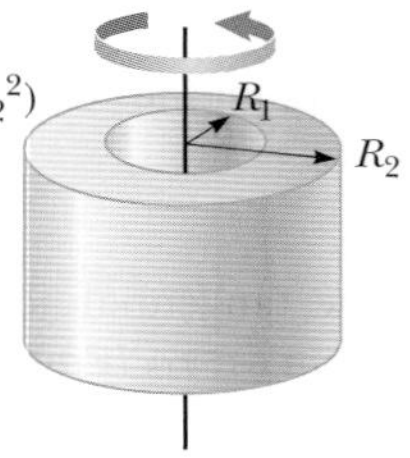
Solid cylinder or disc $I_{CM} = \frac{1}{2}MR^2$	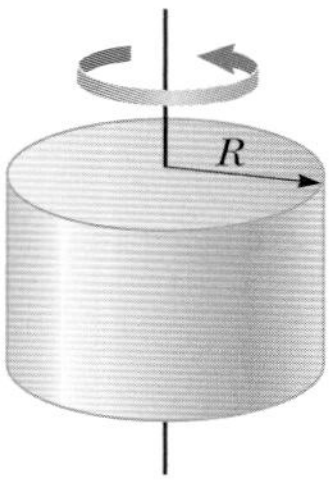	Rectangular plate $I_{CM} = \frac{1}{12}M(a^2 + b^2)$	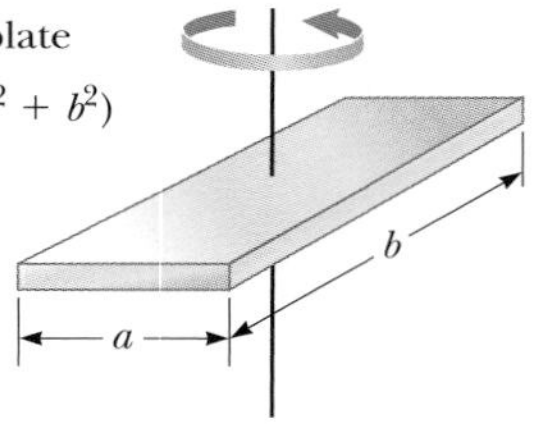
Long, thin rod with rotation axis through centre $I_{CM} = \frac{1}{12}ML^2$	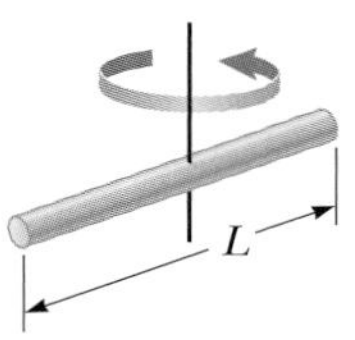	Long, thin rod with rotation axis through end $I = \frac{1}{3}ML^2$	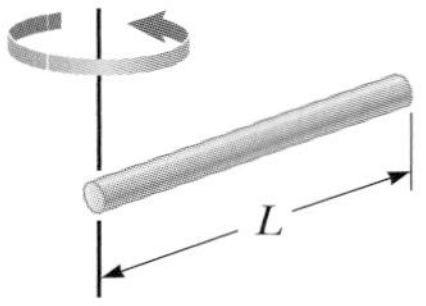
Solid sphere $I_{CM} = \frac{2}{5}MR^2$	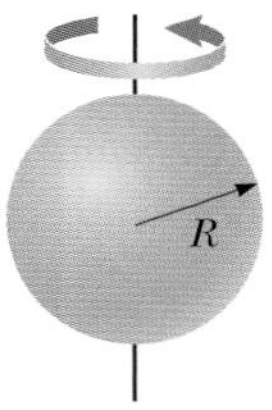	Thin spherical shell $I_{CM} = \frac{2}{3}MR^2$	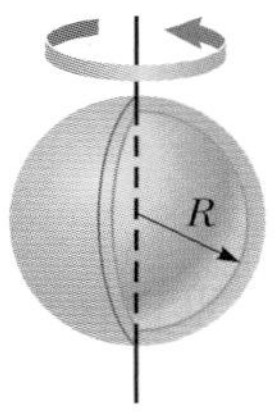

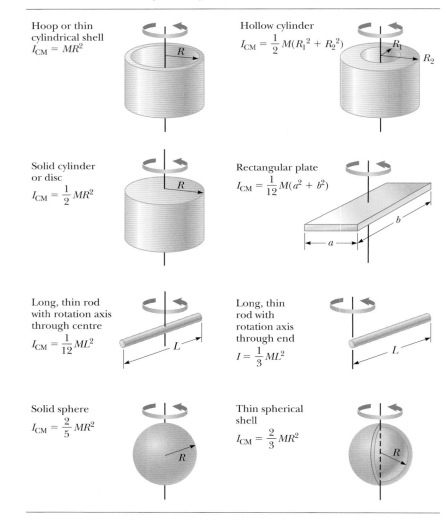

TRY THIS

Get a long rod, such as a ruler, and twirl it about its centre of mass as it lies on a table. Now try twirling it about one end. Which is easier, and why? Use the data in Table 9.2 to explain your answer.

Example 9.7

Uniform rigid rod

Calculate the moment of inertia of a uniform rigid rod of length L and mass M about an axis perpendicular to the rod (the y' axis) and passing through its centre of mass.

Solution

Conceptualise Draw a diagram and label it, showing the axis of rotation. Imagine twirling the rod in Figure 9.22 with your fingers around its midpoint. If you have a ruler handy, use it to simulate the spinning of a thin rod and feel the resistance it offers to being spun.

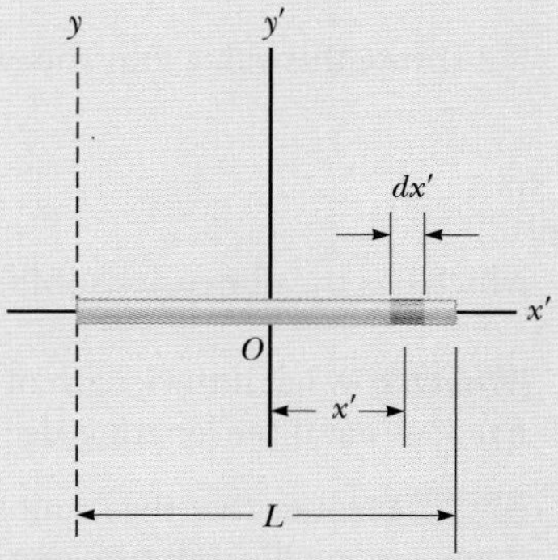

Figure 9.22
(Example 9.7) A uniform rigid rod of length L. The moment of inertia about the y' axis is less than that about the y axis. The latter axis is examined in Example 9.8.

Model We use the definition of moment of inertia in Equation 9.23. We simplify the problem by reducing the integrand to a single variable.

Analyse The shaded length element dx' in Figure 9.22 has a mass dm equal to the mass per unit length λ multiplied by dx'.

Express dm in terms of dx': $dm = \lambda dx' = \frac{M}{L}dx'$

Substitute this expression into Equation 9.23, with $r^2 = (x')^2$:

$$I_y = \int r^2\, dm = \int_{-L/2}^{L/2} (x')^2 \frac{M}{L} dx' = \frac{M}{L}\int_{-L/2}^{L/2} (x')^2\, dx'$$

$$= \frac{M}{L}\left[\frac{(x')^3}{3}\right]_{-L/2}^{L/2} = \frac{1}{12}ML^2$$

This has the correct dimensions, [M][L^2], and is the result given in Table 9.2.

Finalise We have found the moment of inertia of a rod about an axis perpendicular to the rod and passing through its centre of mass.

What If? How does this compare to an axis parallel to this one but passing through one end of the rod? What about an axis parallel to the rod and through the centre of the rod?

Answer The moment of inertia about one end of the rod is $\frac{1}{3}ML^2$, which is substantially larger. The moment of inertia about an axis coinciding with the rod is the same as that for a cylinder, $\frac{1}{2}MR^2$, where R is the radius of the rod. For a rod, R is typically much smaller than the length, L, so this will be a much smaller moment of inertia. Hence the moment of inertia can vary significantly with the axis of rotation.

Example 9.8

Uniform solid cylinder

A uniform solid cylinder has a radius R, mass M and length L. Calculate its moment of inertia about its central long axis.

Solution

Conceptualise Draw a diagram to visualise the problem, and label the axis as well as any dimensions given.

Model We again use the definition of moment of inertia. It is convenient to divide the cylinder into many cylindrical shells, each having radius r, thickness dr, and length L as shown in Figure 9.23.

Example 9.8 cont.

Analyse The density of the cylinder is ρ. The volume dV of each shell is its cross-sectional area multiplied by its length: $dV = L\,dA = L(2\pi r)\,dr$.

Express dm in terms of dr: $dm = \rho dV = \rho L(2\pi r)dr$

Substitute this expression into **Equation 9.23**:

$$I_z = \int r^2\,dm = \int r^2\,[\rho L(2\pi r)dr] = 2\pi\rho L\int_0^R r^3\,dr = \frac{1}{2}\pi\rho L R^4$$

Use the total volume $\pi R^2 L$ of the cylinder to express its density:

$$\rho = \frac{M}{V} = \frac{M}{\pi R^2 L}$$

Substitute this value into the expression for I_z:

$$I_z = \frac{1}{2}\pi\left(\frac{M}{\pi R^2 L}\right)LR^4 = \frac{1}{2}MR^2$$

which has the dimensions [M][L²] ☺

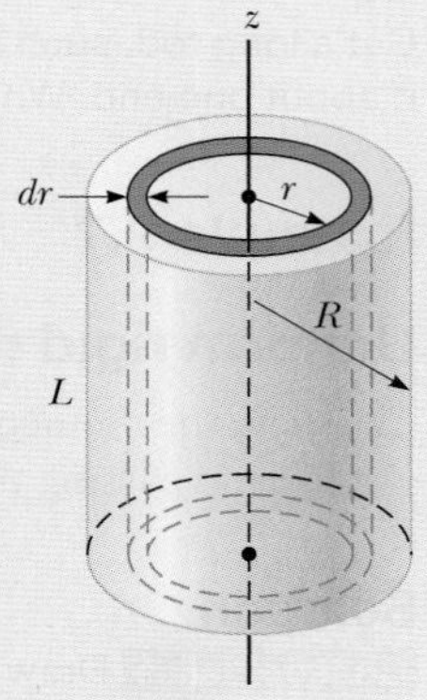

Figure 9.23 (Example 9.8) Calculating I about the z axis for a uniform solid cylinder

What If? What if the length of the cylinder in **Figure 9.23** is increased to $2L$, while the mass M and radius R are held fixed? What if the length is decreased to almost nothing so we have a thin disc?

Answer Notice that the result for the moment of inertia of a cylinder does not depend on L, the length of the cylinder. It applies equally well to a long cylinder and a flat disc having the same mass M and radius R. Therefore, the moment of inertia of the cylinder would not be affected by changing its length.

> **Pitfall Prevention 9.7**
> There is one major difference between mass and moment of inertia. Mass is an inherent property of an object. The moment of inertia of an object depends on your choice of rotation axis. Therefore, there is no single value of the moment of inertia for an object. There is a *minimum* value of the moment of inertia, which is that calculated about an axis passing through the centre of mass of the object.

The calculation of moments of inertia of an object about an arbitrary axis can be cumbersome, even for a highly symmetrical object. Fortunately, use of an important theorem, called the **parallel axis theorem**, often simplifies the calculation.

Suppose the object in **Figure 9.24a** rotates about the z axis. The moment of inertia does not depend on how the mass is distributed along the z axis; as we found in Example 9.8, the moment of inertia of a cylinder is independent of its length. Imagine collapsing the three-dimensional object into a planar object as in **Figure 9.24b**. In this imaginary process, all mass moves parallel to the z axis until it lies in the xy plane. The coordinates of the object's centre of mass are now x_{CM}, y_{CM} and $z_{CM} = 0$. Let the mass element dm have coordinates $(x, y, 0)$ as shown in the view down the z axis in **Figure 9.24c**. Because this element is a distance $r = \sqrt{x^2 + y^2}$ from the z axis, the moment of inertia of the entire object about the z axis is

$$I = \int r^2\,dm = \int (x^2 + y^2)\,dm$$

We can relate the coordinates x, y of the mass element dm to the coordinates of this same element located in a coordinate system having the object's centre of mass as its origin. If the coordinates of the centre of mass are x_{CM}, y_{CM} and $z_{CM} = 0$ in the original coordinate system centred on O, we see from **Figure 9.24c** that the relationships between the unprimed and primed coordinates are $x = x' + x_{CM}$, $y = y' + y_{CM}$ and $z = z' = 0$. Therefore,

$$\begin{aligned} I &= \int [(x' + x_{CM})^2 + (y' + y_{CM})^2]\,dm \\ &= \int [(x')^2 + (y')^2]\,dm + 2x_{CM}\int x'\,dm + 2y_{CM}\int y'\,dm + \left(x_{CM}^{\;2} + y_{CM}^{\;2}\right)\int dm \end{aligned}$$

The first integral is, by definition, the moment of inertia I_{CM} about an axis that is parallel to the z axis and passes through the centre of mass. The second two integrals are zero because, by definition

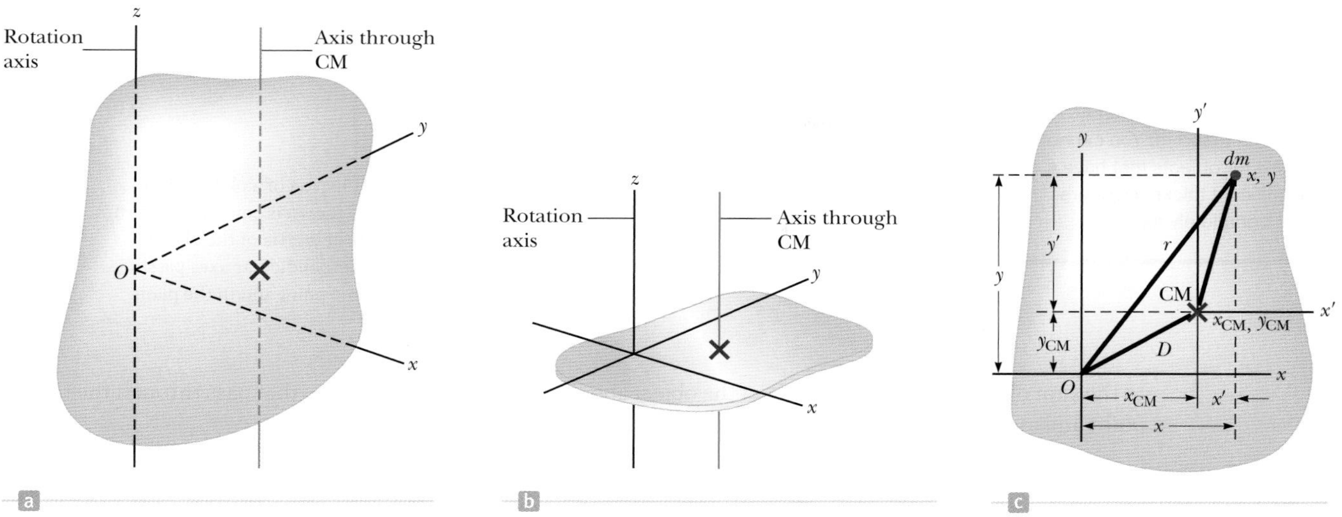

Figure 9.24
(a) An arbitrarily shaped rigid object. The origin of the coordinate system is not at the centre of mass of the object. Imagine the object rotating about the *z* axis. (b) All mass elements of the object are collapsed parallel to the *z* axis to form a planar object. (c) An arbitrary mass element *dm* is indicated in blue in this view down the *z* axis. The parallel axis theorem can be used with the geometry shown to determine the moment of inertia of the original object around the *z* axis.

of the centre of mass, $\int x'\,dm = \int y'\,dm = 0$. The last integral is simply MD^2 because $\int dm = M$ and $D^2 = x_{CM}^{\ 2} + y_{CM}^{\ 2}$ where D is the distance between the two axes. Therefore, we conclude that

◀ Parallel axis theorem

$$I = I_{CM} + MD^2 \quad (9.24)$$

Example 9.9

Applying the parallel axis theorem

Consider once again the uniform rigid rod of mass M and length L shown in **Figure 9.22**. Find the moment of inertia of the rod about an axis perpendicular to the rod through one end (the y axis in **Figure 9.22**).

Solution

Conceptualise Imagine twirling the rod around an endpoint rather than the midpoint. If you did the *Try this* example (page 269) you will already have noticed the degree of difficulty in rotating it around the end compared with rotating it around the centre. We have already have a diagram (**Figure 9.22**) but it is good practice to draw your own.

Categorise This example is a substitution problem, involving the parallel axis theorem.

We expect the moment of inertia to be greater than the result $I_{CM} = \frac{1}{12}ML^2$ from Example 9.7 because there is mass up to a distance of L away from the rotation axis, whereas the farthest distance in Example 9.7 was only $L/2$. The distance between the centre-of-mass axis and the y axis is $D = L/2$. If you did the *Try this* example you would also have noticed the difference between spinning the rod in the two different ways.

Use the parallel axis theorem: $$I = I_{CM} + MD^2 = \frac{1}{12}ML^2 + M\left(\frac{L}{2}\right)^2 = \frac{1}{3}ML^2$$

which is the result given in **Table 9.2**.

A second useful theorem for finding the moment of inertia of a planar object is the **perpendicular axis theorem**. For any planar object such as a round plate or square tile shown in **Figure 9.25**, the perpendicular axis theorem states that the moment of inertia about an axis perpendicular to the plane of the object is equal

Pitfall Prevention 9.8

The perpendicular axis theorem only applies to planar objects; it is not a general theorem. You cannot use it for objects that have a significant thickness or are not completely flat.

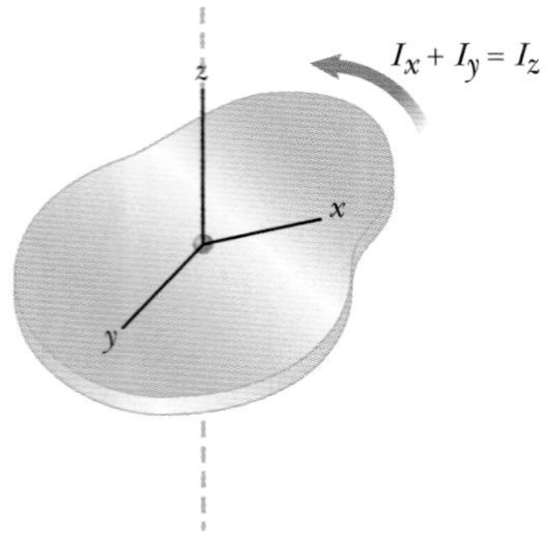

Figure 9.25
The moment of inertia of a planar object about the z axis shown is equal to the sum of the moments of inertia of the object about the two perpendicular axes passing through the same point and in the plane of the object.

to the sum of the moments of inertia of the object about any two axes in the plane that pass through the same point and are perpendicular to each other:

The perpendicular axis theorem ▶

$$I_z = I_x + I_y \tag{9.25}$$

The perpendicular axis theorem can also be used to find the moment of inertia of objects such as cylinders by breaking down the cylinder into a series of discs and summing over the moments of inertia of the individual discs.

Example 9.10

Spinning a frisbee

One student throws a frisbee to another student. It spins about the z axis, which is perpendicular to the plane of the frisbee and passes through its centre of mass as it flies through the air. The other student catches the frisbee and spins it, balanced vertically on their finger.

Find the moment of inertia of the frisbee in each case.

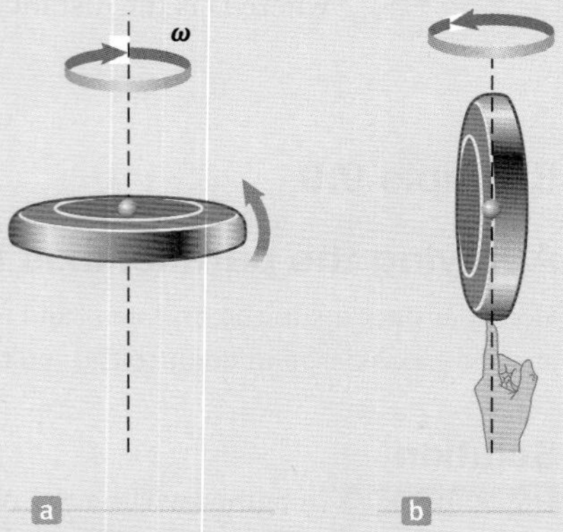

Figure 9.26
(Example 9.10) (a) A frisbee spinning about on an axis through its centre of mass and perpendicular to the plane of the frisbee (b) The frisbee spins about an axis in the plane passing through the centre of mass.

Solution

Conceptualise We start by drawing a diagram. You have probably played with a frisbee and sent it spinning through the air and you may be able to spin one balanced on your finger.

Model We model the frisbee as a flat disc with uniform mass distribution. This will allow us to apply the definition of moment of inertia and the perpendicular axis theorem.

Analyse We have already found the moment of inertia of a disc about an axis perpendicular to the plane of the disc and passing through the centre of mass in Example 9.8 (and it is also listed in **Table 9.2**).

$I_z = \frac{1}{2}MR^2$ where M is the mass of the frisbee and R is its radius.

To find the moment of inertia when the disc is spun about an axis in the plane of the disc and passing through the centre of mass, as when it is spun on a finger, we apply the perpendicular axis theorem. Because the disc is symmetrical about any axis in its plane, we can say that $I_x = I_y$, for any choice of x and y axes, so applying the perpendicular axis theorem gives:

$$I_z = I_x + I_y = 2I_x, \text{ therefore } I_x = \frac{1}{2}I_z = \frac{1}{2}\left(\frac{1}{2}MR^2\right) = \frac{1}{4}MR^2$$

Finalise We could also have done this problem by integrating over mass elements from one edge of the disc to the other, but that would take a lot longer.

9.6 Analysis model: rigid object subject to a net torque

We have seen that Newton's second law for a rotating system, **Equation 9.20**, is written in terms of torque and moment of inertia:

$$\sum \tau = I\alpha$$

That is, the net torque acting on the particle is proportional to its angular acceleration, and the proportionality constant is the moment of inertia.

Now let us extend this discussion to a rigid object of arbitrary shape rotating about a fixed axis as in **Figure 9.21**. The object can be regarded as an infinite number of mass elements dm of infinitesimal size. If we impose a Cartesian coordinate system on the object, each mass element rotates in a circle about the origin and each has a tangential acceleration $\vec{\mathbf{a}}_t$ produced by an external tangential force $d\vec{\mathbf{F}}_t$. For any given element, we know from Newton's second law that

$$dF_t = (dm)a_t$$

The external torque $d\tau_{ext}$ associated with the force $d\vec{\mathbf{F}}_t$ acts about the origin and its magnitude is given by

$$d\tau_{ext} = r\,dF_t = a_t r\,dm$$

Because $a_t = r\alpha$, the expression for $d\tau_{ext}$ becomes

$$d\tau_{ext} = \alpha r^2\,dm$$

Although each mass element of the rigid object may have a different translational acceleration $\vec{\mathbf{a}}_t$, they all have the *same* angular acceleration α. With that in mind, we can integrate the above expression to obtain the net external torque $\Sigma\tau_{ext}$ about an axis through O due to the external forces:

$$\sum \tau_{ext} = \int \alpha r^2\,dm = \alpha \int r^2\,dm$$

where α can be taken outside the integral because it is common to all mass elements. From **Equation 9.23**, we know that $\int r^2\,dm$ is the moment of inertia of the object about the rotation axis through O, and so the expression for $\Sigma\tau_{ext}$ becomes

$$\sum \tau_{ext} = I\alpha \qquad (9.26)$$

This equation for a rigid object is the same as that found for a particle moving in a circular path (**Equation 9.20**). The net torque about the rotation axis is proportional to the angular acceleration of the object, with the proportionality factor being the moment of inertia I, a quantity that depends on the axis of rotation and on the size and shape of the object. **Equation 9.26** is the mathematical representation of the analysis model of a **rigid object subject to a net torque**, the rotational analogue to the particle subject to a net force.

Finally, notice that the result $\Sigma\tau_{ext} = I\alpha$ also applies when the forces acting on the mass elements have radial components as well as tangential components. This is because the line of action of all radial components must pass through the axis of rotation; hence, all radial components produce zero torque about that axis.

Quick Quiz 9.6

You turn off your electric drill and find that the time interval for the rotating bit to come to rest due to frictional torque in the drill is Δt. You replace the bit with a larger one that results in a doubling of the moment of inertia of the drill's entire rotating mechanism. When this larger bit is rotated at the same angular speed as the first and the drill is turned off, the frictional torque remains the same as that for the previous situation. What is the time interval for this second bit to come to rest? **(a)** $4\Delta t$ **(b)** $2\Delta t$ **(c)** Δt **(d)** $0.5\Delta t$ **(e)** $0.25\Delta t$ **(f)** impossible to determine

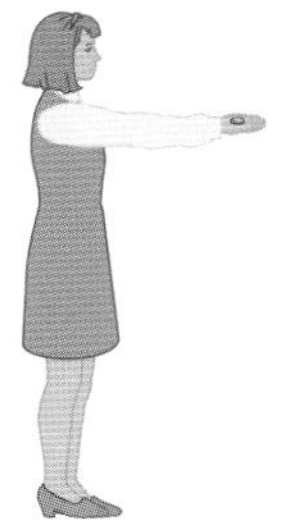

TRY THIS

Hold your arm out straight and horizontal and then let it fall without any resistance. Place a coin or other small object on the top your hand and repeat the experiment. Does the object stay against your hand as your arm swings? What if you put it on your arm? Be careful not to hit anyone with your arm or the object!

Analysis Model 9.2

Rigid object subject to a net torque

Imagine you are analysing the motion of an object that is free to rotate about a fixed axis. The cause of changes in rotational motion of this object is torque applied to the object and, in parallel to Newton's second law for translation motion, the torque is equal to the product of the moment of inertia of the object and the angular acceleration:

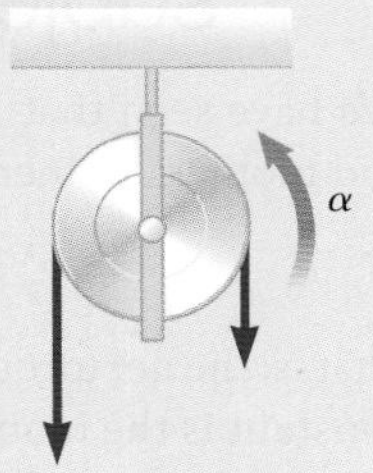

$$\sum \tau_{\text{ext}} = I\alpha \qquad (9.26)$$

The torque, the moment of inertia, and the angular acceleration must all be evaluated around the same rotation axis.

Examples

- a bicycle chain around the sprocket of a bicycle causes the rear wheel of the bicycle to rotate
- an electric dipole moment in an electric field rotates due to the electric force from the field (Chapter 23)
- a magnetic dipole moment in a magnetic field rotates due to the magnetic force from the field (Chapter 30)
- the armature of a motor rotates due to the torque exerted by a surrounding magnetic field (Chapter 31)

Example **9.11**

Rotating arm

We can model a rotating arm as in the *Try this* as a uniform rod of length L and mass M attached at one end to a frictionless pivot and free to rotate about the pivot in the vertical plane, as in **Figure 9.27**. The arm is released from rest in the horizontal position. What are the initial angular acceleration of the arm and the initial translational acceleration of its right end?

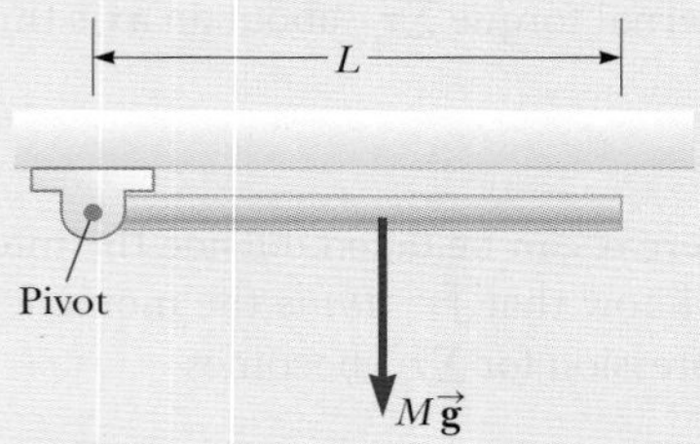

Figure 9.27
(Example 9.11) A rod is free to rotate around a pivot at the left end. The gravitational force on the rod acts at its centre of mass.

Solution

Conceptualise Imagine what happens to the rod in **Figure 9.27** when it is released. It rotates clockwise around the pivot at the left end. Recall what happened to your arm if you did the *Try this* example above.

Model The arm is modelled as a uniform rigid rod subject to a net torque. The torque is due only to the gravitational force on the rod if the rotation axis is chosen to pass through the pivot in **Figure 9.27**. We *cannot* model the arm as a rigid object with constant angular acceleration because the torque exerted on the rod and therefore the angular acceleration of the arm vary with its angular position.

Analyse The only force contributing to the torque about an axis through the pivot is the gravitational force $M\vec{\mathbf{g}}$ exerted on the arm. (The force exerted by the pivot on the arm has zero torque about the pivot because its moment arm is zero.) To determine the torque on the arm, we assume the gravitational force acts at the centre of mass of the arm as shown in **Figure 9.27**.

Write an expression for the magnitude of the net external torque due to the gravitational force about an axis through the pivot:

$$\sum \tau_{\text{ext}} = Mg\left(\frac{L}{2}\right)$$

Example 9.11 cont.

Use **Equation 9.26** to obtain the angular acceleration of the arm:

$$\alpha = \frac{\sum \tau_{ext}}{I} = \frac{Mg(L/2)}{\frac{1}{3}ML^2} = \frac{3g}{2L} \quad (1)$$

Use **Equation 9.11** with $r = L$ to find the initial translational acceleration of the right end of the arm:

$$a_t = L\alpha = \frac{3}{2}g$$

Finalise These values are the *initial* values of the angular and translational accelerations. Once the arm, which we modelled as a rod, begins to rotate, the gravitational force is no longer perpendicular to the arm and the values of the two accelerations decrease, going to zero at the moment the arm passes through the vertical orientation.

What If? What if we were to place a coin on the end of the arm and then release the arm? Would the coin stay in contact with the arm?

Answer The result for the initial acceleration of a point on the end of the arm shows that $a_t > g$. An unsupported coin falls at acceleration g. So, if we place a coin on the end of the arm and then release the arm, the hand at the end of the arm falls faster than the coin does!

The question now is to find the location on the arm at which we can place a coin that *will* stay in contact as both begin to fall. To find the translational acceleration of an arbitrary point on the arm at a distance $r < L$ from the pivot point, we combine **Equation (1)** with **Equation 9.11**:

$$a_t = r\alpha = \frac{3g}{2L}r$$

For the coin to stay in contact with the arm, the translational acceleration must be equal to or greater than that due to gravity. Taking the limiting case that they are equal:

$$a_t = g = \frac{3g}{2L}r$$

$$r = \frac{2}{3}L$$

Therefore, a coin placed closer to the pivot than two-thirds of the length of the arm stays in contact with the falling arm, but a coin farther out than this point loses contact. Is this what happens with your arm? What approximations have we made in this model that may significantly affect its accuracy?

Conceptual Example 9.12

Falling chimneys and tumbling blocks

When a tall chimney falls over, it often breaks somewhere along its length before it hits the ground, as shown in **Figure 9.28**. Why?

Solution

As the chimney rotates around its base, each higher portion of the chimney falls with a larger tangential acceleration than the portion below it according to **Equation 9.11**. The angular acceleration increases as the chimney tips further. Eventually, higher portions of the chimney experience an acceleration greater than the acceleration that could result from gravity alone; this situation is similar to that described in Example 9.11. It can happen only if these portions are being pulled downwards by a force in addition to the gravitational force. The force that causes that to occur is the **shear force** (which we shall discuss in Chapter 15) from lower portions of the chimney. Eventually, the shear force that provides this acceleration is greater than the chimney can withstand and the chimney breaks. The same thing happens with a tall tower of children's toy blocks.

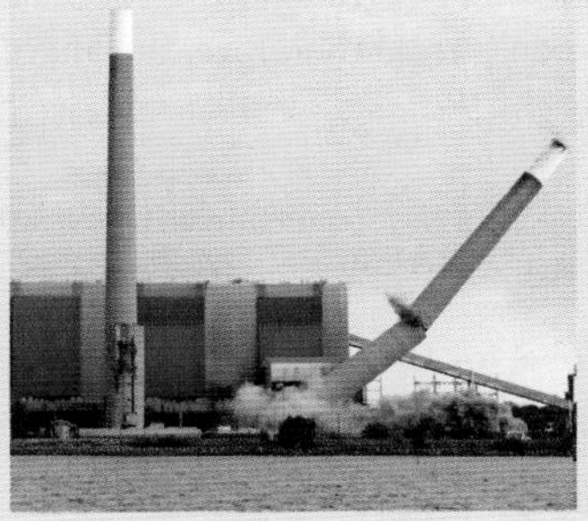

Kevin Spreekmeester/AGE fotostock

Figure 9.28
(Conceptual example 9.12) A falling chimney breaks at some point along its length.

TRY THIS

Borrow some blocks from a child and build a tower. Push it over and watch how it comes apart. How tall does it have to be to break apart at some point before it strikes the floor?

Example 9.13

Angular acceleration of a disc

Consider again the rotational dynamics experiment described in Example 9.6. A large metal disc of mass M and radius R is free to rotate about the central z axis. Imagine that some students have wrapped the string around the outside edge of the disc and attached a weight of mass m. Assume that the axle has just been lubricated so the frictional force is negligible.

When the disc is released, the weight accelerates downwards, the cord unwraps from the disc and the disc rotates with an angular acceleration. Find the angular acceleration of the disc, the translational acceleration of the object and the tension in the cord.

Solution

Conceptualise You may have done a similar experiment. As the cord unwinds from the disc, the tension in the string pulls on the disc, applying a torque and causing the disc to rotate. We need to consider the motion of both the falling weight and the spinning disc. We start by drawing a diagram showing the forces acting on the weight and the disc.

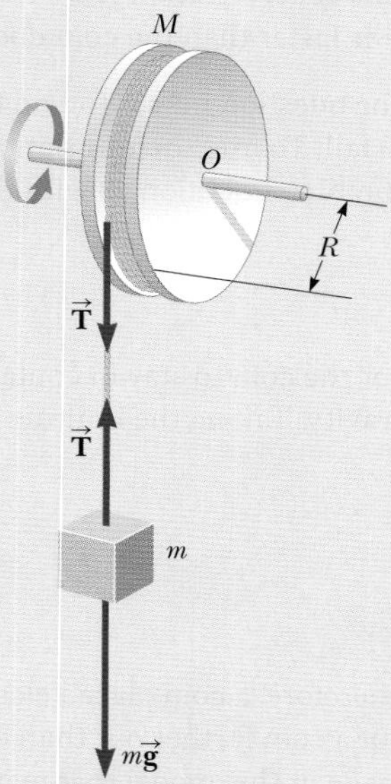

Figure 9.29
(Example 9.13) An object hangs from a cord wrapped around a wheel

Model We apply two analysis models here. The falling object is modelled as a particle subject to a net force. The wheel is modelled as a rigid object subject to a net torque.

Analyse The magnitude of the torque acting on the wheel about its axis of rotation is $\tau = TR$, where T is the force exerted by the cord on the rim of the wheel. (The gravitational force exerted by the Earth on the wheel and the normal force exerted by the axle on the wheel both pass through the axis of rotation and therefore produce no torque.)

Write **Equation 9.26**:
$$\sum \tau_{ext} = I\alpha$$

Solve for α and substitute the net torque:
$$\alpha = \frac{\sum \tau_{ext}}{I} = \frac{TR}{I} \quad (1)$$

Apply Newton's second law to the motion of the object, taking the downward direction to be positive:

$$\sum F_y = mg - T = ma$$

Solve for the acceleration a:
$$a = \frac{mg - T}{m} \quad (2)$$

Equations (1) and **(2)** have three unknowns: α, a and T. Because the object and wheel are connected by a cord that does not slip, the translational acceleration of the suspended object is equal to the tangential acceleration of a point on the wheel's rim. Therefore, the angular acceleration α of the wheel and the translational acceleration of the object are related by $a = R\alpha$.

Use this fact together with **Equations (1)** and **(2)**:
$$a = R\alpha = \frac{TR^2}{I} = \frac{mg - T}{m} \quad (3)$$

Solve for the tension T:
$$T = \frac{mg}{1 + (mR^2 / I)} \quad (4)$$

Substitute **Equation (4)** into **Equation (2)** and solve for a:
$$a = \frac{g}{1 + (I / mR^2)} \quad (5)$$

Use $a = R\alpha$ and **Equation (5)** to solve for α:
$$\alpha = \frac{a}{R} = \frac{g}{R + (I / mR)}$$

Example 9.13 cont.

The moment of inertia of a disc is given by $I = \frac{1}{2}MR^2$ (Table 9.2) so we can write our expressions for a, α and T in terms of the variables given in the question as:

$$T = \frac{(mg)}{1 + (2mR^2/MR^2)} = \frac{Mmg}{M + 2m}$$

$$a = \frac{g}{1 + (MR^2/2mR^2)} = \frac{2mg}{2m + M}$$

$$\alpha = \frac{a}{R} = \frac{2mg}{R(2m + M)}$$

Checking dimensions:

for T: $[\mathrm{MLT^{-2}}] = [\mathrm{M}][\mathrm{M}][\mathrm{LT^{-2}}]/[\mathrm{M}] = [\mathrm{MLT^{-2}}]$

for a: $[\mathrm{LT^{-2}}] = [\mathrm{M}][\mathrm{LT^{-2}}]/[\mathrm{M}] = [\mathrm{LT^{-2}}]$

and for α: $[\mathrm{T^{-2}}] = [\mathrm{M}][\mathrm{LT^{-2}}]/[\mathrm{L}][\mathrm{M}] = [\mathrm{T^{-2}}]$ ☺

What If? What if the wheel were to become very massive so that I becomes very large? What happens to the acceleration a of the object and the tension T?

Answer If the wheel becomes infinitely massive, we can imagine that the object of mass m will simply hang from the cord without causing the wheel to rotate.

We can show that mathematically by taking the limit $I \rightarrow \infty$. Equation (5) then becomes

$$a = \frac{g}{1 + (I/mR^2)} \rightarrow 0$$

which agrees with our conceptual conclusion that the object will hang at rest. Also, Equation (4) becomes

$$T = \frac{mg}{1 + (mR^2/I)} \rightarrow \frac{mg}{1 + 0} = mg$$

which is consistent, because the object simply hangs at rest in equilibrium between the gravitational force and the tension in the string.

9.7 Analysis model: rigid object in equilibrium

In Chapter 4 we discussed the particle in equilibrium model, in which a particle moves with constant velocity because the net force acting on it is zero. The situation with real (extended) objects is more complex because these objects often cannot be modelled as particles. For an extended object to be in equilibrium, a second condition must be satisfied. This second condition involves the rotational motion of the extended object.

Consider a single force $\vec{\mathbf{F}}$ acting on a rigid object as shown in Figure 9.30. Recall that the torque associated with the force $\vec{\mathbf{F}}$ about an axis through O is given by Equation 9.14:

$$\vec{\tau} = \vec{\mathbf{r}} \times \vec{\mathbf{F}}$$

The magnitude of $\vec{\tau}$ is Fd (see Equation 9.18), where d is the moment arm shown in Figure 9.30. According to Equation 9.26, the net torque on a rigid object causes it to undergo an angular acceleration.

We now consider those rotational situations in which the angular acceleration of a rigid object is zero. Such an object is in **rotational equilibrium**. Because $\sum \tau_{ext} = I\alpha$ for rotation about a fixed axis, the necessary condition for rotational equilibrium is that the net torque about any axis must be zero. We now have two necessary conditions for equilibrium of an object:

1. The net external force on the object must equal zero:

$$\sum \vec{\mathbf{F}}_{ext} = 0 \quad (9.27)$$

2. The net external torque on the object about *any* axis must be zero:

$$\sum \vec{\tau}_{ext} = 0 \quad (9.28)$$

These conditions describe the **rigid object in equilibrium** analysis model. The first condition is a statement of translational equilibrium; it states that the translational acceleration of the object's centre of mass must be zero when viewed from an inertial

Pitfall Prevention 9.9

Don't forget to perform dimension checks when you arrive at complicated algebraic expressions, especially when working with parameters (such as torque and moment of inertia) that you are not yet practised at using. Recall the earlier advice that there is no point in substituting numbers into an incorrect expression, so it is worth taking the time to check that it is at least dimensionally correct!

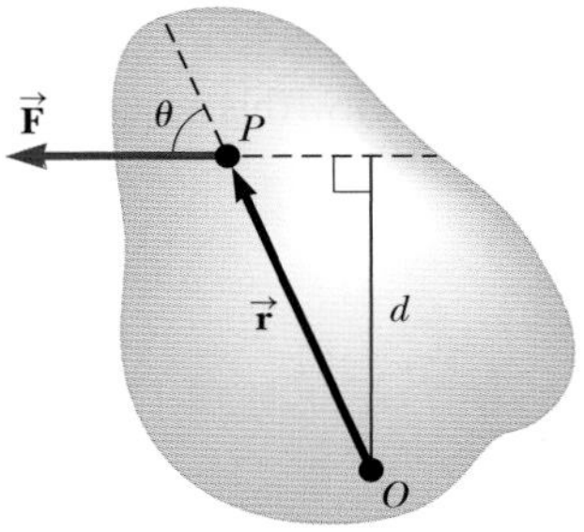

Figure 9.30
A single force $\vec{\mathbf{F}}$ acts on a rigid object at the point P.

> **Pitfall Prevention 9.10**
> Zero net torque does not mean an absence of rotational motion. An object that is rotating at a constant angular speed is subject to zero net torque. This is analogous to the translational situation: zero net force does not mean zero velocity.

reference frame. The second condition is a statement of rotational equilibrium; it states that the angular acceleration about any axis must be zero. In the special case of **static equilibrium**, the object in equilibrium is at rest relative to the observer and so has no translational or angular speed (that is, $v_{CM} = 0$ and $\omega = 0$).

Quick Quiz 9.7

Consider the object subject to the two forces of equal magnitude in **Figure 9.31**. Choose the correct statement with regard to this situation. **(a)** The object is in force equilibrium but not torque equilibrium. **(b)** The object is in torque equilibrium but not force equilibrium. **(c)** The object is in both force equilibrium and torque equilibrium. **(d)** The object is in neither force equilibrium nor torque equilibrium. The combination of two equal and opposite parallel forces acting on an object is known as a **couple** or **couple moment**.

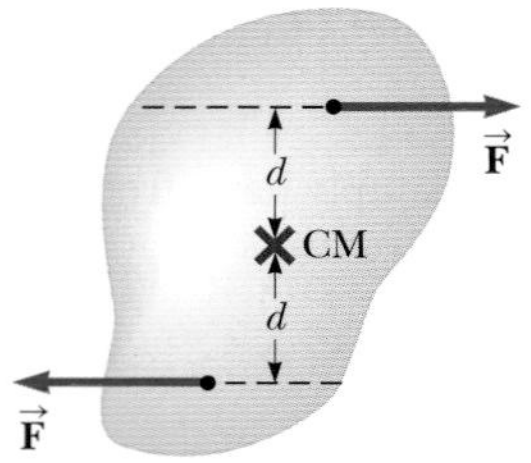

Figure 9.31
(Quick Quiz 9.7) Two forces of equal magnitude are applied at equal distances from the centre of mass of a rigid object.

The two vector expressions given by **Equations 9.27** and **9.28** are equivalent, in general, to six scalar equations: three from the first condition for equilibrium and three from the second (corresponding to x, y and z components). Hence, in a complex system involving several forces acting in various directions, you could be faced with solving a set of equations with many unknowns. Here, we restrict our discussion to situations in which all the forces lie in the xy plane. (Forces whose vector representations are in the same plane are said to be *coplanar*.) With this restriction, we must deal with only three scalar equations. Two come from balancing the forces in the x and y directions. The third comes from the torque equation, namely that the net torque about a perpendicular axis through *any* point in the xy plane must be zero. This perpendicular axis will necessarily be parallel to the z axis, so the two conditions of the rigid object in equilibrium model provide the equations

$$\sum F_x = 0 \qquad \sum F_y = 0 \qquad \sum \tau_z = 0 \tag{9.29}$$

where the location of the axis for the torque equation is arbitrary.

Σ *Appendix B.1 describes methods for solving simultaneous equations.*

The photograph of the one-bottle wine holder in **Figure 9.32** shows one example of a balanced mechanical system that seems to defy gravity. For the system (wine holder plus bottle) to be in equilibrium, the net external force must be zero (see **Equation 9.27**) and the net external torque must be zero (see **Equation 9.28**). The second condition can be satisfied only when the centre of gravity of the system is directly over the support point.

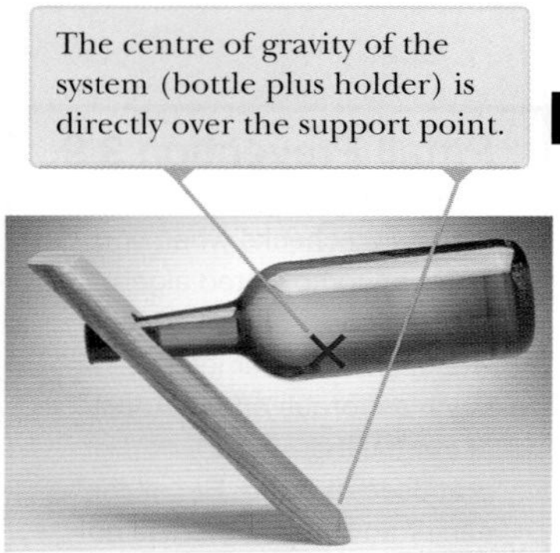

The centre of gravity of the system (bottle plus holder) is directly over the support point.

Figure 9.32
This one-bottle wine holder is a surprising display of static equilibrium.

Quick Quiz 9.8

Consider the object subject to the three forces in **Figure 9.33**. Choose the correct statement with regard to this situation. **(a)** The object is in force equilibrium but not torque equilibrium. **(b)** The object is in torque equilibrium but not force equilibrium. **(c)** The object is in both force equilibrium and torque equilibrium. **(d)** The object is in neither force equilibrium nor torque equilibrium.

Analysis Model 9.3

Rigid object in equilibrium

A rigid object that is in equilibrium has a constant translational velocity and a constant angular velocity. This occurs when no net torque and no net force act on the object.

$$\sum \vec{F}_{ext} = 0 \tag{9.27}$$

$$\sum \vec{\tau}_{ext} = 0 \tag{9.28}$$

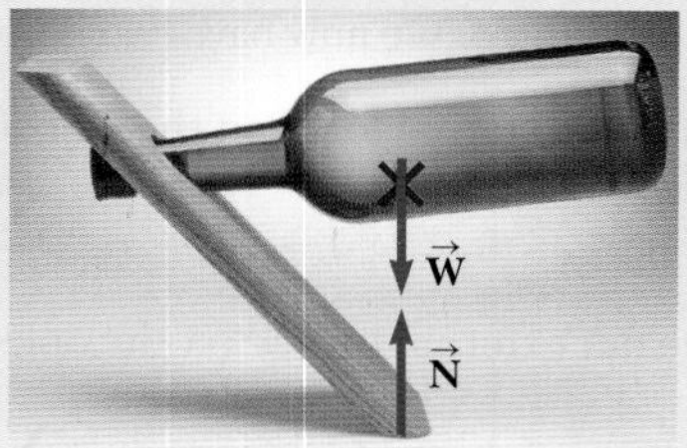

Examples

- a seesaw balanced by a person at either end (Example 9.5)
- an electric dipole aligned with the field in a uniform electric field (Chapter 23)
- a current loop aligned parallel to a uniform magnetic field (Chapter 30).

Problem-solving strategy

Rigid object in equilibrium

When analysing a rigid object in equilibrium under the action of several external forces, use the following procedure.

1. Conceptualise

Think about the object that is in equilibrium and identify all the forces acting on it. Imagine what effect each force would have on the rotation of the object if it were the only force acting. Draw diagrams showing the physical situation.

2. Model

Confirm that the object under consideration can be reasonably modelled as a rigid object in equilibrium. The object must have zero translational acceleration and zero angular acceleration.

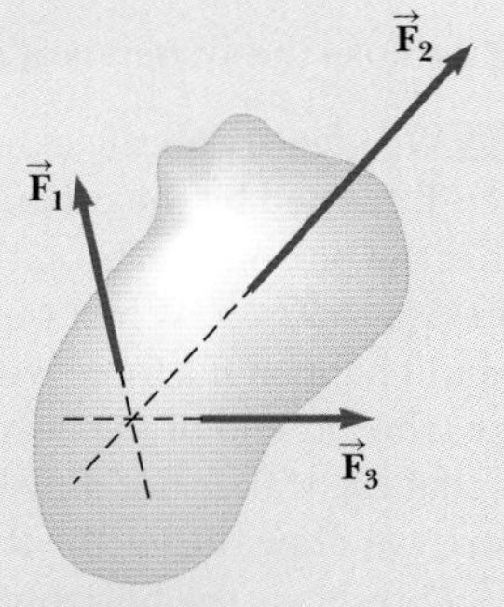

Figure 9.33
(Quick Quiz 9.8) Three forces act on an object. Notice that the lines of action of all three forces pass through a common point.

3. Analyse

Draw diagrams and label all external forces acting on the object. Try to guess the correct direction for any forces that are not specified. When using the particle subject to a net force model, the object on which forces act can be represented in a free-body diagram with a dot because it does not matter where on the object the forces are applied. When using the rigid object in equilibrium model, however, we cannot use a dot to represent the object because the location where forces act is important in the calculation. Therefore, in a diagram showing the forces on an object, we must show the actual object or a simplified version of it.

Resolve all forces into rectangular components, choosing a convenient coordinate system. Then apply the first condition for equilibrium, **Equation 9.27**. Remember to keep track of the signs of the various force components.

Choose a convenient axis for calculating the net torque on the rigid object. Remember that the choice of the axis for the torque equation is arbitrary; therefore, choose an axis that simplifies your calculation as much as possible. Usually, the most convenient axis for calculating torques is one through a point at which several forces act, so their torques around this axis are zero. If you do not know a force or do not need to know a force, it is often beneficial to choose an axis through the point at which this force acts. Apply the second condition for equilibrium, **Equation 9.28**.

Solve the simultaneous equations for the unknowns in terms of the known quantities.

4. Finalise

Make sure your results are consistent with your diagram. If you selected a direction that leads to a negative sign in your solution for a force, do not be alarmed; it merely means that the direction of the force is the opposite of what you guessed. Add up the vertical and horizontal forces on the object and confirm that each set of components adds to zero. Add up the torques on the object and confirm that the sum equals zero.

Example 9.14

Standing on a horizontal beam

A uniform horizontal beam with a length of $\ell = 8.00$ m and a weight of $W_b = 200$ N is attached to a wall by a pin connection. Its far end is supported by a cable that makes an angle of $\phi = 53.0°$ with the beam (**Figure 9.34**). A person of weight $W_p = 600$ N stands a distance $d = 2.00$ m from the wall. Find the tension in the cable as well as the magnitude and direction of the force exerted by the wall on the beam.

Solution

Conceptualise Imagine that the person in **Figure 9.34** moves outwards on the beam. He applies a load to the beam equal to his weight as he is not accelerating vertically. The further he moves outwards, the larger the torque he applies about the pivot and the larger the tension in the cable must be to balance this torque. We already have a diagram showing the situation, but we will need to draw a diagram showing the forces acting.

Model Because the system is at rest, we model the beam as a rigid object in static equilibrium.

Analyse We identify all the external forces acting on the beam: the 200 N gravitational force, the force $\vec{\mathbf{T}}$ exerted by the cable, the force $\vec{\mathbf{R}}$ exerted by the wall at the pivot and the 600 N force that the person exerts on the beam. These forces are shown in the force diagram for the beam in **Figure 9.35a**. When we assign directions for forces, it is sometimes helpful to imagine what would happen if a force were suddenly removed. For example, if the wall were to vanish suddenly, the left end of the beam would move to the left as it begins to fall. This scenario tells us that the wall is not only holding the beam up but is also pressing outwards against it. Therefore, we draw the vector $\vec{\mathbf{R}}$ in the direction shown in **Figure 9.35a**. **Figure 9.35b** shows the horizontal and vertical components of $\vec{\mathbf{T}}$ and $\vec{\mathbf{R}}$.

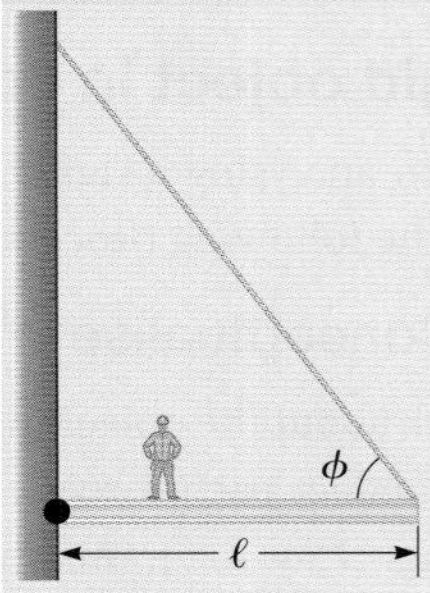

Figure 9.34
(Example 9.14) A uniform beam supported by a cable. A person walks outwards on the beam.

Substitute expressions for the forces on the beam into **Equation 9.29**:

$$\sum F_x = R\cos\theta - T\cos\phi = 0 \qquad (1)$$

$$\sum F_y = R\sin\theta + T\sin\phi - W_p - W_b = 0 \qquad (2)$$

where we have chosen rightwards and upwards as our positive directions. Because R, T and θ are all unknown, we cannot obtain a solution from these expressions alone. (To solve for the unknowns, the number of simultaneous equations must equal or exceed the number of unknowns.)

Now use the condition for rotational equilibrium. A convenient axis to choose for our torque equation is the one that passes through the pin connection. The feature that makes this axis so convenient is that the force $\vec{\mathbf{R}}$ and the horizontal component of $\vec{\mathbf{T}}$ both have a moment arm of zero; hence, these forces produce no torque about this axis.

Substitute expressions for the torques on the beam into **Equation 9.29**:

$$\sum \tau_z = (T\sin\phi)(\ell) - W_p d - W_b\left(\frac{\ell}{2}\right) = 0$$

This equation contains only T as an unknown because of our choice of rotation axis. Solve for T:

$$T = \frac{W_p d + W_b(l/2)}{\ell\sin\phi}$$

Check dimensions: $$[\mathrm{MLT^{-2}}] = \frac{[\mathrm{MLT^{-2}}]\,[\mathrm{L}]}{[\mathrm{L}]} = [\mathrm{MLT^{-2}}]$$ ☺

Substitute values: $$T = \frac{(600\ \mathrm{N})(2.00\ \mathrm{m}) + (200\ \mathrm{N})(4.00\ \mathrm{m})}{(8.00\ \mathrm{m})\sin 53.0^\circ} = 313\ \mathrm{N}$$

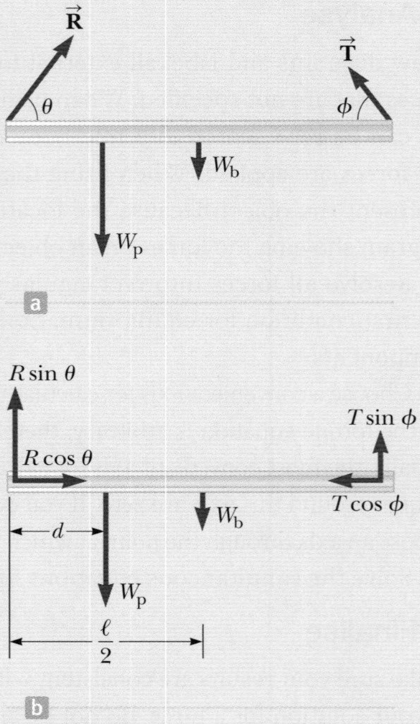

Figure 9.35
(Example 9.14) (a) The force diagram for the beam (b) The components of forces acting on the beam

Rearrange **Equations (1)** and **(2)** and then divide: $$\frac{R\sin\theta}{R\cos\theta} = \tan\theta = \frac{W_p + W_b - T\sin\phi}{T\cos\phi}$$

Solve for θ and substitute numerical values: $$\theta = \tan^{-1}\left(\frac{W_p + W_b - T\sin\phi}{T\cos\phi}\right)$$

$$= \tan^{-1}\left[\frac{600\ \mathrm{N} + 200\ \mathrm{N} - (313\ \mathrm{N})\sin 53.0^\circ}{(313\ \mathrm{N})\cos 53.0^\circ}\right] = 71.1^\circ$$

Note that the quantities in the argument all have units of force, and hence the argument is dimensionless, as we expect. ☺

Example 9.14 cont.

Solve **Equation (1)** for R and substitute numerical values:

$$R = \frac{T\cos\phi}{\cos\theta} = \frac{(313\ \text{N})\cos 53.0^\circ}{\cos 71.1^\circ} = 581\ \text{N}$$

Finalise The positive value for the angle θ indicates that $\vec{\mathbf{R}}$ does act outwards and upwards.

Had we selected some other axis for the torque equation, the solution might differ in the details but the answers would be the same. For example, had we chosen an axis through the centre of gravity of the beam, the torque equation would involve both T and R. This equation, coupled with **Equations (1)** and **(2)**, however, could still be solved for the unknowns. Try it!

What If? What if the person walks farther out on the beam? Does T change? Does R change? Does θ change?

Answer T must increase because the weight of the person exerts a larger torque about the pin connection, which must be countered by a larger torque in the opposite direction due to an increased value of T. If T increases, the vertical component of $\vec{\mathbf{R}}$ decreases to maintain force equilibrium in the vertical direction. Force equilibrium in the horizontal direction, however, requires an increased horizontal component of $\vec{\mathbf{R}}$ to balance the horizontal component of the increased $\vec{\mathbf{T}}$. This suggests that θ becomes smaller.

TRY THIS

Try leaning a ruler against a vertical surface. Does the ruler slip at some angles and stay up at others? Does it stay up when it is closer to vertical or closer to horizontal?

Example 9.15

The leaning ladder

A uniform ladder of length ℓ rests against a smooth, vertical wall (**Figure 9.36**). The mass of the ladder is m, and the coefficient of static friction between the ladder and the ground is $\mu_s = 0.40$. Find the minimum angle θ_{min} at which the ladder does not slip.

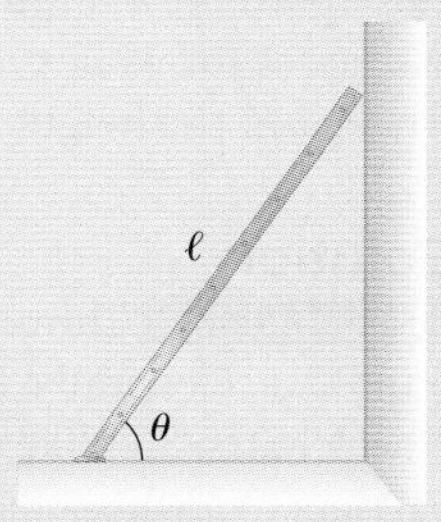

Figure 9.36 (Example 9.15) A uniform ladder at rest, leaning against a smooth wall. The ground is rough.

Solution

Conceptualise Think about any ladders you have climbed. Do you want a large friction force between the bottom of the ladder and the surface or a small one? If the friction force is zero, will the ladder stay up? Do the *Try this* example with the ruler above. Draw a diagram showing the forces acting on the ladder.

Model We do not want the ladder to slip, so we assume it is at rest and model it as a rigid object in equilibrium. We make the approximation that the wall is frictionless.

Analyse The external forces acting on the ladder are shown in **Figure 9.37**. The contact force exerted by the ground on the ladder is the vector sum of a normal force $\vec{\mathbf{n}}$ and the force of static friction $\vec{\mathbf{f}}_s$. The force $\vec{\mathbf{P}}$ exerted by the wall on the ladder is horizontal because the wall is frictionless.

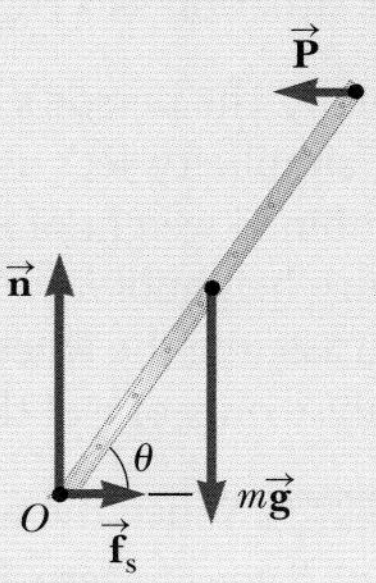

Figure 9.37 (Example 9.15) The forces on the ladder

Apply the first condition for equilibrium to the ladder:

$$\sum F_x = f_s - P = 0 \quad (1)$$

$$\sum F_y = n - mg = 0 \quad (2)$$

Solve **Equation (1)** for P: $\quad P = f_s \quad (3)$

Solve **Equation (2)** for n: $\quad n = mg \quad (4)$

When the ladder is on the verge of slipping, the force of static friction must have its maximum value, which is given by $f_{s,max} = \mu_s n$.

Example 9.15 cont.

Combine this equation with Equations (3) and (4):

$$P = f_{s,max} = \mu_s n = \mu_s mg \quad (5)$$

Apply the second condition for equilibrium to the ladder, taking torques about an axis through O:

$$\sum \tau_O = P\ell \sin \theta_{min} - mg\frac{\ell}{2}\cos \theta_{min} = 0$$

Solve for $\tan \theta_{min}$ and substitute for P from Equation (5):

$$\frac{\sin \theta_{min}}{\cos \theta_{min}} = \tan\theta_{min} = \frac{mg}{2P} = \frac{mg}{2\mu_s mg} = \frac{1}{2\mu_s}$$

Solve for the angle θ_{min}:

$$\theta_{min} = \tan^{-1}\left(\frac{1}{2\mu_s}\right) = \tan^{-1}\left[\frac{1}{2(0.40)}\right] = 51°$$

Recall that μ is dimensionless and note that both sides of our equation are dimensionless. ☺

Finalise Notice that the angle depends only on the coefficient of friction, not on the mass or length of the ladder. The shallower the angle is, the greater the coefficient of friction required to prevent the ladder sliding.

Example 9.16

Negotiating a curb

(A) Estimate the magnitude of the force $\vec{F}$ a person must apply to a wheelchair's main wheel to roll up over a footpath curb (Figure 9.38a). You will need to make estimates for values required such as the radius of the wheel and the height of the curb.

Solution

Conceptualise Think about wheelchair access to buildings. Generally, there are ramps built for people in wheelchairs. Steplike structures such as curbs are serious barriers to a wheelchair. We are given a diagram showing the situation and we will need to draw a diagram showing the forces acting.

Model Imagine that the person exerts enough force so that the bottom of the wheel just loses contact with the lower surface and hovers at rest. We model the wheel in this situation as a rigid object in equilibrium.

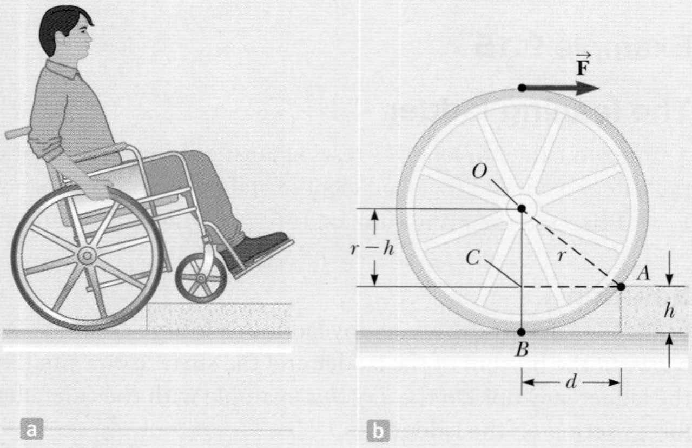

Figure 9.38
(Example 9.16) (a) A person in a wheelchair attempts to roll up over a curb. (b) Details of the wheel and curb. The person applies a force $\vec{F}$ to the top of the wheel.

Usually, the person's hands supply the required force to a slightly smaller wheel that is concentric with the main wheel. For simplicity, let's assume the radius of this second wheel is the same as the radius of the main wheel. Let's estimate a combined weight of $mg = 1400$ N for the person and the wheelchair and choose a wheel radius of $r = 30$ cm. We also pick a curb height of $h = 10$ cm. Note that our estimates for weights and heights are to only one or two significant figures. Let's also assume the wheelchair and occupant are symmetric and each wheel supports a load of 700 N. We then proceed to analyse only one of the wheels. Figure 9.38b shows the geometry for a single wheel.

Analyse When the wheel is just about to be raised from the street, the normal force exerted by the ground on the wheel at point B goes to zero. Hence, at this time only three forces act on the wheel as shown in the force diagram in Figure 9.39a. The force $\vec{R}$, which is the force exerted by the curb on the wheel, acts at point A, so if we choose to have our axis of rotation pass through point A, we do not need to include $\vec{R}$ in our torque equation. The moment arm of $\vec{F}$ relative to an axis through A is given by $2r - h$ (see Figure 9.39a).

Use the triangle OAC in **Figure 9.38b** to find the moment arm d of the gravitational force $m\vec{g}$ acting on the wheel relative to an axis through point A:

$$d = \sqrt{r^2 - (r-h)^2} = \sqrt{2rh - h^2} \quad \textbf{(1)}$$

Apply the second condition for equilibrium to the wheel, taking torques about an axis through A:

$$\sum \tau_A = mgd - F(2r - h) = 0 \quad \textbf{(2)}$$

Substitute for d from **Equation (1)**: $mg\sqrt{2rh - h^2} - F(2r-h) = 0$

Solve for F: $F = \dfrac{mg\sqrt{2rh - h^2}}{2r - h}$

Check dimensions: $[\mathrm{MLT}^{-2}] = \dfrac{[\mathrm{M}][\mathrm{LT}^{-2}]\sqrt{[\mathrm{L}^2]}}{[\mathrm{L}]} = [\mathrm{MLT}^{-2}]$ ☺

Substitute the known values:

$$F = \frac{(700\ \mathrm{N})\sqrt{2(0.3\ \mathrm{m})(0.1\ \mathrm{m}) - (0.1\ \mathrm{m})^2}}{2(0.3\ \mathrm{m}) - 0.1\ \mathrm{m}}$$

$$= 300\ \mathrm{N}$$

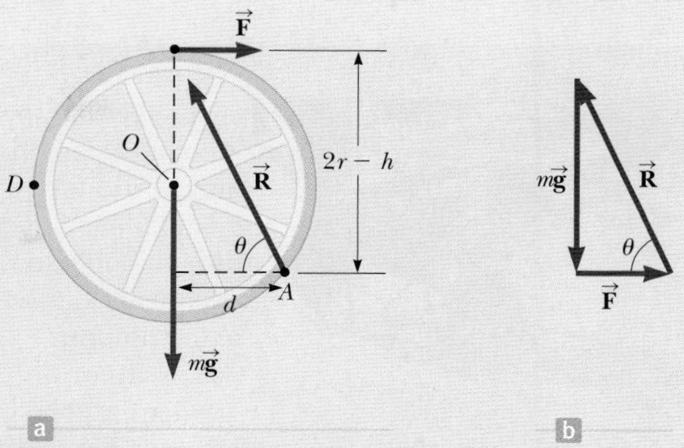

Figure 9.39
(Example 9.16) (a) A force diagram for the wheel when it is just about to be raised. Three forces act on the wheel at this instant: $\vec{\mathbf{F}}$, which is exerted by the hand; $\vec{\mathbf{R}}$, which is exerted by the curb; and the gravitational force $m\vec{g}$. We estimate that the centre of gravity of the system is above the centre of the wheel. (b) The vector sum of the three external forces acting on the wheel is zero.

(B) Determine the magnitude and direction of $\vec{\mathbf{R}}$.

Solution

Apply the first condition for equilibrium to the wheel:

$$\sum F_x = F - R\cos\theta = 0 \quad \textbf{(3)}$$

$$\sum F_y = R\sin\theta - mg = 0 \quad \textbf{(4)}$$

Divide **Equation (4)** by **Equation (3)**: $\dfrac{R\sin\theta}{R\cos\theta} = \tan\theta = \dfrac{mg}{F}$

Solve for the angle θ: $\theta = \tan^{-1}\left(\dfrac{mg}{F}\right) = \tan^{-1}\left(\dfrac{700\ \mathrm{N}}{300\ \mathrm{N}}\right) = 70°$

Note that the argument is dimensionless as required, because both numerator and denominator have dimensions $[\mathrm{MLT}^{-2}]$ ☺

Solve **Equation (4)** for R and substitute numerical values: $R = \dfrac{mg}{\sin\theta} = \dfrac{700\ \mathrm{N}}{\sin 70°} = 800\ \mathrm{N}$

Finalise Notice that we have kept only one digit as significant, because we had to estimate a lot of numerical values in the first place. The results indicate that the force that must be applied to each wheel is substantial. You may want to estimate the force required to roll a wheelchair up a typical footpath accessibility ramp for comparison.

End-of-chapter resources

Shutterstock.com/Photographee.eu

The wheelbarrow is one of the practical applications of a class 2 lever, and it is a valuable tool for use in agriculture, construction, landscaping and gardening. The first version of the wheelbarrow was developed in China in the BC era, and it then appeared in Europe in the late 12th century. We began the chapter by asking: What are the mechanical advantages of the wheel and axle?

As a class 2 lever, the wheelbarrow helps to reduce the effort needed to lift and transport heavy loads, because the load sits between the fulcrum and the wheelbarrow's effort arm. The axle of the wheel is the fulcrum. When the axle rotates for a short distance, the wheel moves a greater distance than the axle. For example, the mechanical advantage for the wheel and axle is computed as the ratio of the load to the effort (see Equation 9.19). The size ratio between the wheel and the axle corresponds to the amount of force applied to the axle and the distance covered by the wheel. The larger the ratio, the greater the torque created, or distance achieved. In addition, by lengthening the handles, a person can move even heavier loads without increasing the effort needed to lift the wheelbarrow (greater d_{effort}).

The problems found in this chapter may be assigned online in Enhanced Web Assign.

Worked solutions to every fifth problem are available in the Student Solutions Manual. Register online at **www.cengagebrain.com** for access.

Summary

Definitions

The **angular position** of a rigid object is defined as the angle θ between a reference line attached to the object and a reference line fixed in space. The **angular displacement** of a particle moving in a circular path or a rigid object rotating about a fixed axis is $\Delta\theta = \theta_f - \theta_i$.

The **instantaneous angular speed** of a particle moving in a circular path or of a rigid object rotating about a fixed axis is

$$\omega \equiv \frac{d\theta}{dt} \quad (9.3)$$

The **instantaneous angular acceleration** of a particle moving in a circular path or of a rigid object rotating about a fixed axis is

$$\alpha \equiv \frac{d\omega}{dt} \quad (9.5)$$

When a rigid object rotates about a fixed axis, every part of the object has the same angular speed and the same angular acceleration.

The **torque** $\vec{\tau}$ due to a force $\vec{F}$ about an axis through the origin in an inertial frame is defined to be

$$\vec{\tau} \equiv \vec{r} \times \vec{F} \quad (9.14)$$

The magnitude of the **torque** associated with a force $\vec{F}$ acting on an object at a distance r from the rotation axis is

$$\tau = rF\sin\phi = Fd \quad (9.18)$$

where ϕ is the angle between the position vector of the point of application of the force and the force vector, and d is the moment arm of the force, which is the perpendicular distance from the rotation axis to the line of action of the force.

The **moment of inertia of a system of particles** is defined as

$$I \equiv \sum_i m_i r_i^2 \quad (9.22)$$

where m_i is the mass of the ith particle and r_i is its distance from the rotation axis.

Concepts and principles

When a rigid object rotates about a fixed axis, the angular position, angular speed and angular acceleration are related to the translational position, translational speed and translational acceleration through the relationships

$$s = r\theta \quad (9.1)$$

$$v = r\omega \quad (9.10)$$

$$a_t = r\alpha \quad (9.11)$$

The **moment of inertia of a rigid object** is

$$I = \int r^2 dm \quad (9.23)$$

where r is the distance from the mass element dm to the axis of rotation.

Newton's second law for a rotation says that the angular acceleration of an object is proportional to the net torque acting on the object:

$$\sum \tau = I\alpha$$

Analysis models for problem solving

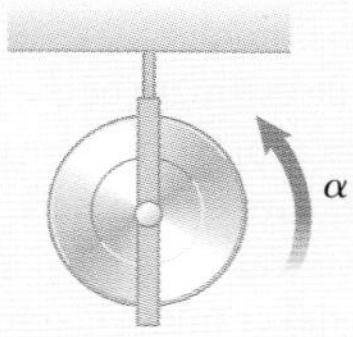

Rigid object with constant angular acceleration If a rigid object rotates about a fixed axis under constant angular acceleration, one can apply equations of kinematics that are analogous to those for translational motion of a particle with constant acceleration:

$$\omega_f = \omega_i + \alpha t \quad (9.6)$$

$$\theta_f = \theta_i + \omega_i t + \frac{1}{2}\alpha t^2 \quad (9.7)$$

$$\omega_f^{\,2} = \omega_i^2 + 2\alpha(\theta_f - \theta_i) \quad (9.8)$$

$$\theta_f = \theta_i + \frac{1}{2}(\omega_i + \omega_f)t \quad (9.9)$$

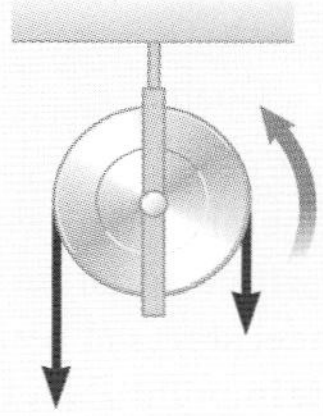

Rigid object subject to a net torque If a rigid object free to rotate about a fixed axis has a net external torque acting on it, the object undergoes an angular acceleration α, where

$$\sum \tau_{ext} = I\alpha \quad (9.26)$$

This equation is the rotational analogue to Newton's second law in the particle subject to a net force model.

Rigid object in equilibrium A rigid object in equilibrium exhibits no translational or angular acceleration. The net external force acting on it is zero, and the net external torque on it is zero about any axis:

$$\sum \vec{\mathbf{F}}_{ext} = 0 \quad (9.27)$$

$$\sum \vec{\tau}_{ext} = 0 \quad (9.28)$$

The first condition is the condition for translational equilibrium, and the second is the condition for rotational equilibrium.

Chapter review quiz

To help you revise Chapter 9: Rotational motion, complete the automatically graded Chapter review quiz at http://login.cengagebrain.com.

Conceptual questions

1. One blade of a pair of scissors rotates anticlockwise in the xy plane. (a) What is the direction of $\vec{\omega}$ for the blade? (b) What is the direction of $\vec{\alpha}$ if the magnitude of the angular velocity is decreasing in time?
2. Suppose just two external forces act on a stationary, rigid object and the two forces are equal in magnitude and opposite in direction. Under what condition does the object start to rotate?
3. Explain how you might use the apparatus shown in Figure CQ9.3 to determine the moment of inertia of the wheel. This is in fact the experiment the students were doing as described in Example 9.6. The wheel does not have a uniform mass density, so the moment of inertia is not necessarily equal to $\frac{1}{2}MR^2$.

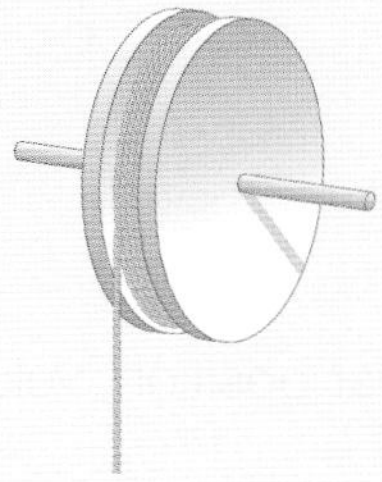

Figure CQ9.3

4. Which of the entries in Table 9.2 applies to finding the moment of inertia of (a) a long, hollow printing-press roller rotating about its axis of symmetry? (b) a hula hoop rotating about an axis through its centre and perpendicular to its plane? (c) a uniform door turning on its hinges? (d) a coin turning about an axis through its centre and perpendicular to its faces?
5. If you see an object rotating, is there necessarily a net torque acting on it? Can an object be in equilibrium if it is in motion?
6. A person balances a metre stick in a horizontal position on the extended index fingers of both hands. She slowly brings the two fingers together. If the stick remains balanced the two fingers always meet at the 50 cm mark regardless of their original positions. Why? (Try it!)
7. Stand with your back against a wall. Why can't you put your heels firmly against the wall and then bend forward without falling?
8. (a) Give an example in which the net force acting on an object is zero and yet the net torque is non-zero. (b) Give an example in which the net torque acting on an object is zero and yet the net force is non-zero.
9. A girl has a large, docile dog she wishes to weigh on a small bathroom scale. She reasons that she can determine her dog's weight with the following method. First she puts the dog's two front feet on the scale and records the scale reading. Then she places only the dog's two back feet on the scale and records the reading. She thinks that the sum of the readings will be the dog's weight. Is she correct? Explain your answer.
10. A ladder stands on the ground, leaning against a wall. Would you feel safer climbing up the ladder if you were told that the ground is frictionless but the wall is rough or if you were told that the wall is frictionless but the ground is rough? Explain your answer.
11. Figure CQ9.11 shows a side view of a child's tricycle with rubber tyres on a horizontal concrete footpath. (a) A string is attached to the lower pedal on the near side and pulled forwards horizontally

as shown by A. Will the tricycle start to roll? If so, which way? Answer the same questions if (b) the string is pulled forwards and upwards as shown by B, (c) the string is pulled straight down as shown by C and (d) the string is pulled forwards and downwards as shown by D. (e) Explain a pattern of reasoning, based on the figure, that makes it easy to answer questions such as these. What physical quantity must you evaluate?

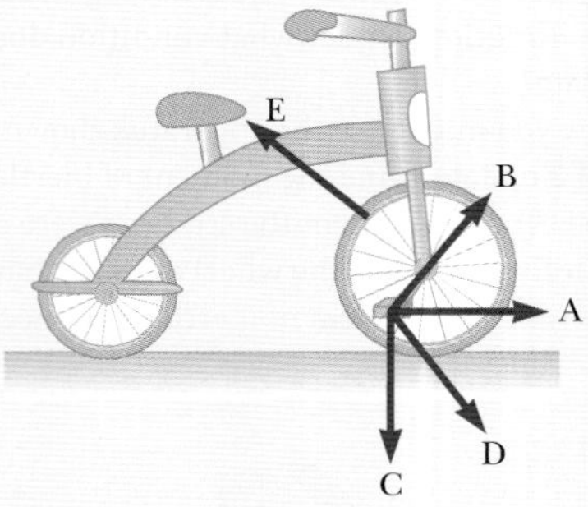

Figure CQ9.11

Problems

Section 9.1 Angular position, velocity and acceleration

1. Find the angular speed of the Earth's rotation about its axis.
2. A potter's wheel moves uniformly from rest to an angular speed of 1.00 rev/s in 30.0 s. (a) Find its average angular acceleration in radians per second per second. (b) Would doubling the angular acceleration during the given period have doubled the final angular speed?
3. During a certain time interval, the angular position of a swinging door is described by $\theta = 5.00 + 10.0t + 2.00t^2$, where θ is in radians and t is in seconds. Determine the angular position, angular speed and angular acceleration of the door (a) at $t = 0$ and (b) at $t = 3.00$ s.

Section 9.2 Analysis model: rigid object with constant angular acceleration

4. A wheel starts from rest and rotates with constant angular acceleration to reach an angular speed of 12.0 rad/s in 3.00 s. Find (a) the magnitude of the angular acceleration of the wheel and (b) the angle in radians through which it rotates in this time interval.
5. An electric motor rotating a workshop grinding wheel at 1.00×10^2 rev/min is switched off. Assume the wheel has a constant negative angular acceleration of magnitude 2.00 rad/s^2. (a) How long does it take the grinding wheel to stop? (b) Through how many radians has the wheel turned during the time interval found in part (a)?
6. A rotating wheel requires 3.00 s to rotate through 37.0 revolutions. Its angular speed at the end of the 3.00-s interval is 98.0 rad/s. What is the constant angular acceleration of the wheel?
7. ±? The tub of a washing machine is in its spin cycle, turning at (5.0 ± 0.5) rev/s. At this point, the person doing the laundry opens the lid and a safety switch turns off the washer. The machine is designed such that the tub smoothly slows to rest in (12 ± 1) s so that the operator is not injured. Through how many revolutions does the tub turn while it is in motion?

Section 9.3 Angular and translational quantities

8. A discus thrower accelerates a discus from rest to a speed of 25.0 m/s by whirling it through 1.25 rev. Assume the discus moves on the arc of a circle 1.00 m in radius. (a) Calculate the final angular speed of the discus. (b) Determine the magnitude of the angular acceleration of the discus, assuming it to be constant. (c) Calculate the time interval required for the discus to accelerate from rest to 25.0 m/s.
9. Make an order-of-magnitude estimate of the number of revolutions through which a typical car tyre turns in one year. State the quantities you measure or estimate and their values.
10. Figure P9.10 shows the drive train of a bicycle that has wheels 67.3 cm in diameter and pedal cranks 17.5 cm long. The cyclist pedals at a steady cadence of 76.0 rev/min. The chain engages with a front sprocket 15.2 cm in diameter and a rear sprocket 7.00 cm in diameter. Calculate (a) the speed of a link of the chain relative to the bicycle frame, (b) the angular speed of the bicycle wheels, and (c) the speed of the bicycle relative to the road. (d) What pieces of data, if any, are not necessary for the calculations?

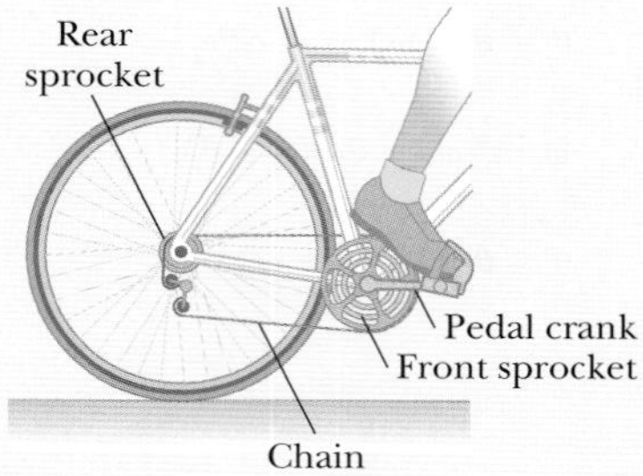

Figure P9.10

11. A wheel 2.00 m in diameter lies in a vertical plane and rotates about its central axis with a constant angular acceleration of 4.00 rad/s^2. The wheel starts at rest at $t = 0$, and the radius vector of a certain point P on the rim makes an angle of 57.3° with the horizontal at this time. At $t = 2.00$ s, find (a) the angular speed of the wheel and, for point P, (b) the tangential speed, (c) the total acceleration and (d) the angular position.
12. A disc 8.00 cm in radius rotates at a constant rate of 1200 rev/min about its central axis. Determine (a) its angular speed in radians per second, (b) the tangential speed at a point 3.00 cm from its centre, (c) the radial acceleration of a point on the rim and (d) the total distance a point on the rim moves in 2.00 s.
13. In a manufacturing process, a large, cylindrical roller is used to flatten material fed beneath it. The diameter of the roller is 1.00 m, and, while being driven into rotation around a fixed axis, its angular position is expressed as

$$\theta = 2.50t^2 - 0.600t^3$$

where θ is in radians and t is in seconds. (a) Find the maximum angular speed of the roller. (b) What is the maximum tangential speed of a point on the rim of the roller? (c) At what time t should the driving force be removed from the roller so that the roller does not reverse its direction of rotation? (d) Through how many rotations has the roller turned between $t = 0$ and the time found in part (c)?
14. A small object with mass 4.00 kg moves anticlockwise with constant angular speed 1.50 rad/s in a circle of radius 3.00 m centred at the origin. It starts at the point with position

vector $3.00\hat{\mathbf{i}}$ m. It then undergoes an angular displacement of 9.00 rad. (a) What is its new position vector? Use unit-vector notation for all vector answers. (b) In what quadrant is the particle located, and what angle does its position vector make with the positive x axis? (c) What is its velocity? (d) In what direction is it moving? (e) What is its acceleration? (f) Make a sketch of its position, velocity and acceleration vectors. (g) What total force is exerted on the object?

15. A car travelling on a flat (unbanked), circular track accelerates uniformly from rest with a tangential acceleration of a. The car makes it one quarter of the way around the circle before it skids off the track. From these data, determine the coefficient of static friction between the car and the track.

Section **9.4** Torque

16. Find the net torque on the wheel in **Figure P9.16** about the axle through O, taking a = 10.0 cm and b = 25.0 cm.

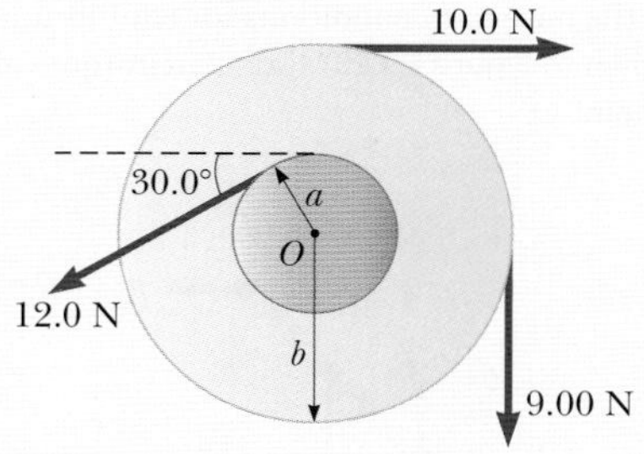

Figure P9.16

17. The fishing rod in **Figure P9.17** makes an angle of 20.0° with the horizontal. What is the torque exerted by the fish about an axis perpendicular to the page and passing through the angler's hand if the fish pulls on the fishing line with a force $\vec{\mathbf{F}}$ = 100 N at an angle 37.0° below the horizontal? The force is applied at a point 2.00 m from the angler's hands.

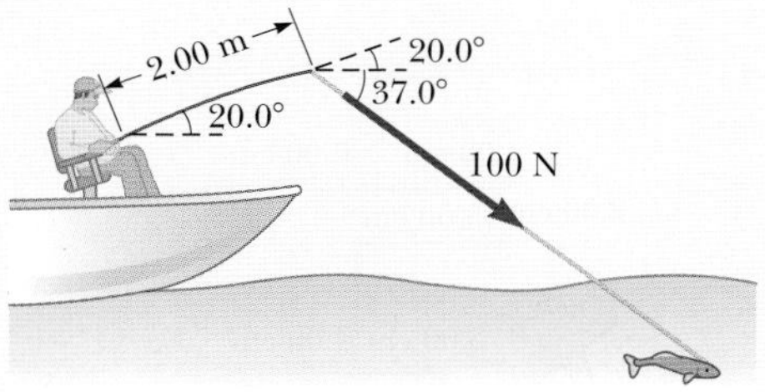

Figure P9.17

18. A man is using a hammer to remove a nail as shown in **Figure P9.18**. The man exerts a steadily increasing force in the horizontal direction, at the end of the handle which is 24 cm long. The nail, which is 2.0 cm from the pivot point on the head of the hammer, begins to move when the applied force reaches 35 N. If the coefficient of static friction between the wood and the nail is 0.4, what is the normal force acting on the nail due to the wood?

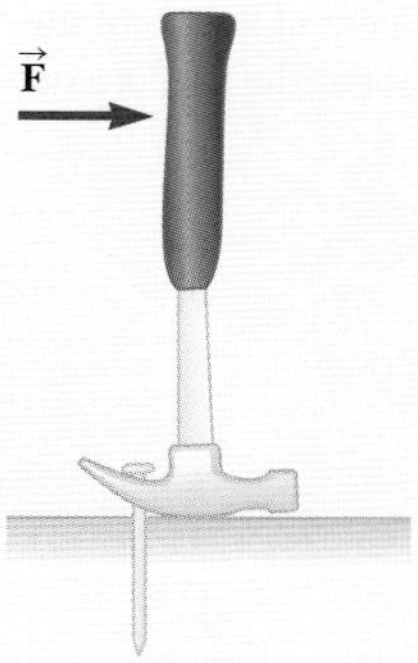

Figure P9.18

19. An engineering student is using a screwdriver with a thick handle to remove a screw from a 'black box' experiment in a physics lab, to see what is inside the black box. Consider the screwdriver as a lever. (a) Draw a diagram and identify the fulcrum, load and effort. (b) What class of lever is the screwdriver? (c) If the screwdriver has a handle 2 cm thick and the screw is a slotted type with just a single straight slit in the top with a length of 5 mm, what is the mechanical advantage of the screwdriver? Explain your reasoning. (d) If the engineer has to exert a force of 50 N to start the screw moving, what is the static frictional force acting on the screw?

Section **9.5** Newton's Second Law for rotation and moment of inertia

20. Imagine that you stand up straight and turn about a vertical axis through the top of your head and the point halfway between your ankles. Calculate an order-of-magnitude estimate for the moment of inertia of your body for this rotation. State the quantities you measure or estimate and their values.

21. **Figure P9.21** shows a side view of a car tyre before it is mounted on a wheel. Model it as having two sidewalls of uniform thickness 0.635 cm and a tread wall of uniform thickness 2.50 cm and width 20.0 cm. Assume the rubber has uniform density 1.10×10^3 kg/m^3. Find its moment of inertia about an axis perpendicular to the page through its centre.

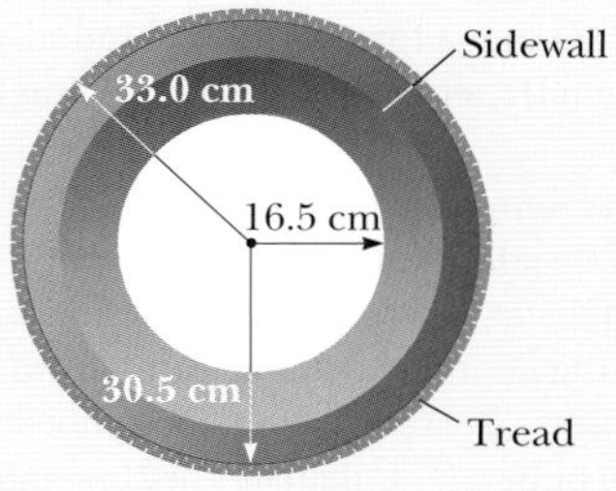

Figure P9.21

22. Car engines and other machines use cams for various purposes, such as opening and closing valves. **Figure P9.22** shows a segment of a camshaft. The cam is a circular disc of radius R with a hole of diameter R cut through it. As shown in the figure, the hole does not pass through the centre of the disc. The cam with the hole cut out has mass M. The cam is mounted on a uniform, solid, cylindrical shaft of diameter R and also of mass

M. What is the moment of inertia of the camshaft combination when it is rotating about the shaft's axis?

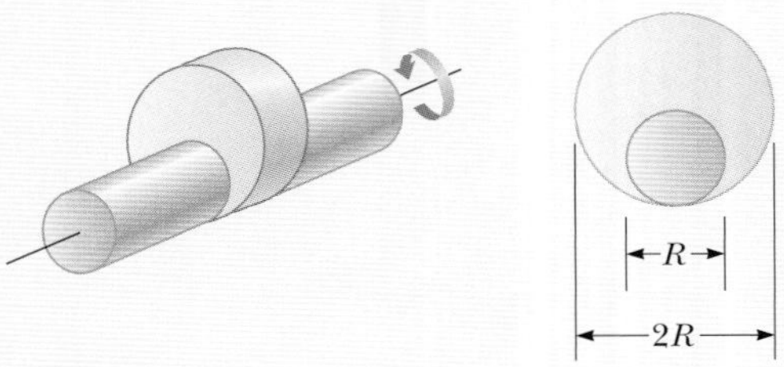

Figure P9.22

Section 9.6 Analysis model: rigid object subject to a net torque

23. A grinding wheel is in the form of a uniform solid disc of radius 7.00 cm and mass 2.00 kg. It starts from rest and accelerates uniformly under the action of the constant torque of 0.600 N.m that the motor exerts on the wheel. (a) How long does the wheel take to reach its final operating speed of 1200 rev/min? (b) Through how many revolutions does it turn while accelerating?
24. ±? A model aeroplane with mass 0.750 kg is tethered to the ground by a wire so that it flies in a horizontal circle 30.0 m in radius. The aeroplane engine provides a net thrust of (0.80 ± 0.02) N perpendicular to the tethering wire. (a) Find the torque the net thrust produces about the centre of the circle. (b) Find the angular acceleration of the aeroplane. (c) Find the translational acceleration of the aeroplane tangent to its flight path.
25. The combination of an applied force and a friction force produces a constant total torque of 36.0 N.m on a wheel rotating about a fixed axis. The applied force acts for 6.00 s. During this time, the angular speed of the wheel increases from 0 to 10.0 rad/s. The applied force is then removed, and the wheel comes to rest in 60.0 s. Find (a) the moment of inertia of the wheel, (b) the magnitude of the torque due to friction and (c) the total number of revolutions of the wheel during the entire interval of 66.0 s.
26. A potter's wheel – a thick stone disc of radius 0.500 m and mass 100 kg – is freely rotating at 50.0 rev/min. The potter can stop the wheel in 6.00 s by pressing a wet rag against the rim and exerting a force of 70.0 N radially inwards. Find the effective coefficient of kinetic friction between wheel and rag.
27. Consider the system shown in **Figure P9.27** with $m_1 = 20.0$ kg, $m_2 = 12.5$ kg, $R = 0.200$ m, and the mass of the pulley $M = 5.00$ kg. Object m_2 is resting on the floor and object m_1 is 4.00 m above the floor when it is released from rest. The pulley axis is frictionless. The cord is light, does not stretch and does not slip on the pulley. (a) Calculate the time interval required for m_1 to hit the floor. (b) How would your answer change if the pulley were massless?

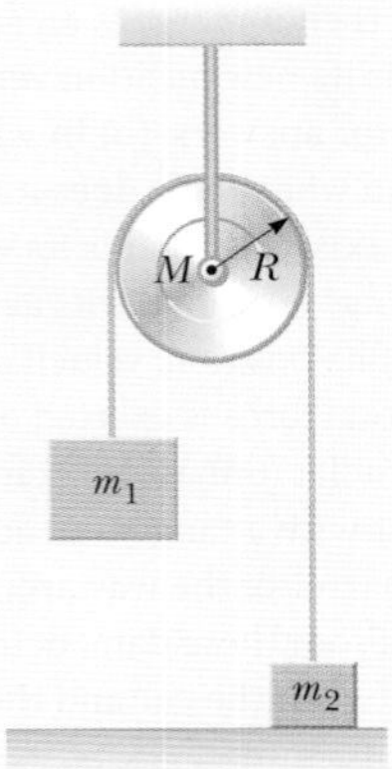

Figure P9.27

Section 9.7 Analysis model: rigid object in equilibrium

28. What are the necessary conditions for equilibrium of the object shown in **Figure P9.28**? Calculate torques about an axis through point O.

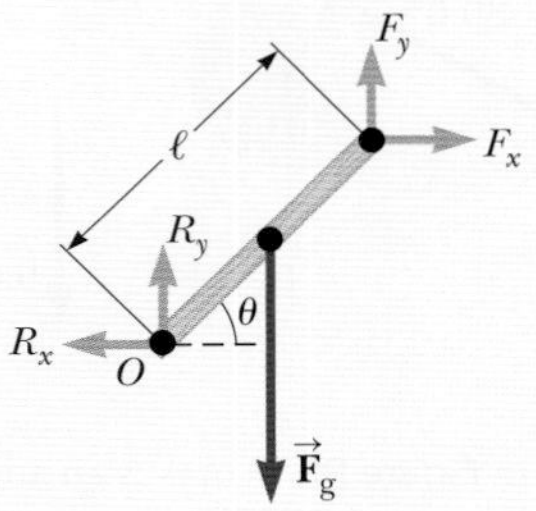

Figure P9.28

29. A mobile is constructed of light rods, light strings and beach souvenirs as shown in **Figure P9.29**. If $m_4 = 12.0$ g, find values for (a) m_1, (b) m_2 and (c) m_3.

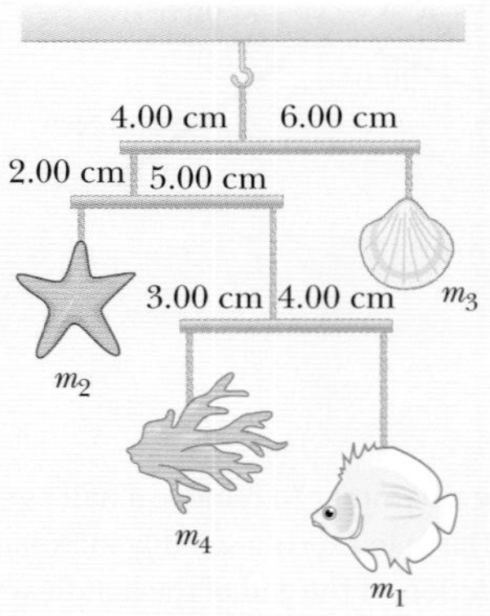

Figure P9.29

30. A 1500 kg car has a wheelbase (the distance between the axles) of 3.00 m. The car's centre of mass is on the centre line at a point 1.20 m behind the front axle. Find the force exerted by the ground on each wheel.

31. A 15.0 m uniform ladder weighing 500 N rests against a frictionless wall. The ladder makes a 60.0° angle with the horizontal. (a) Find the horizontal and vertical forces the ground exerts on the base of the ladder when an 800 N firefighter has climbed 4.00 m along the ladder from the bottom. (b) If the ladder is just on the verge of slipping when the firefighter is 9.00 m from the bottom, what is the coefficient of static friction between the ladder and the ground?

32. A 20.0 kg floodlight in a park is supported at the end of a horizontal beam of negligible mass that is hinged to a pole as shown in **Figure P9.32**. A cable at an angle of $\theta = 30.0°$ with the beam helps support the light. (a) Draw a force diagram for the beam. By computing torques about an axis at the hinge at the left-hand end of the beam, find (b) the tension in the cable, (c) the horizontal component of the force exerted by the pole on the beam and (d) the vertical component of this force. Now solve the same problem from the force diagram from part (a) by computing torques around the junction between the cable and the beam at the right-hand end of the beam. Find (e) the vertical component of the force exerted by the pole on the beam, (f) the tension in the cable, and (g) the horizontal component of the force exerted by the pole on the beam. (h) Compare the solutions to parts (b) through (d) with the solutions to parts (e) through (g). Is either solution more accurate?

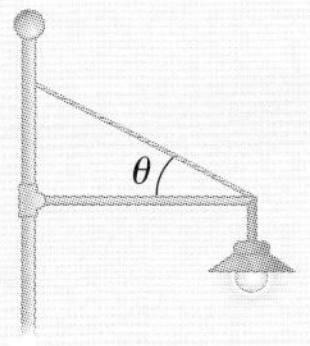

Figure P9.32

33. John is pushing his daughter Rachel in a wheelbarrow when it is stopped by a brick 8.00 cm high (**Figure P9.33**). The handles make an angle of $\theta = 15.0°$ with the ground. Due to the weight of Rachel and the wheelbarrow, a downwards force of 400 N is exerted at the centre of the wheel, which has a radius of 20.0 cm. (a) What force must John apply along the handles to just start the wheel over the brick? (b) What is the force (magnitude and direction) that the brick exerts on the wheel just as the wheel begins to lift over the brick? In both parts, assume the brick remains fixed and does not slide along the ground. Also assume the force applied by John is directed exactly toward the centre of the wheel.

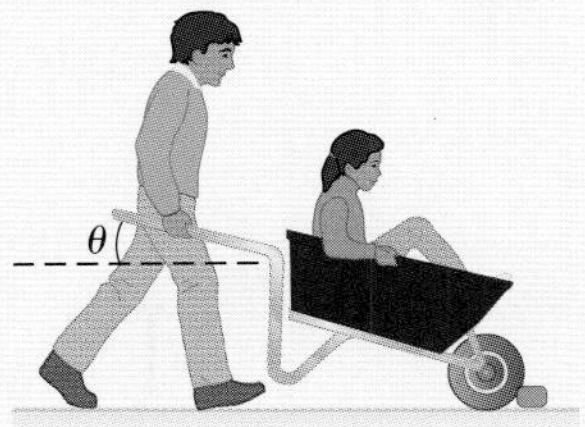

Figure P9.33

Additional problems

34. A flexible chain weighing 40.0 N hangs between two hooks located at the same height (**Figure P9.34**). At each hook, the tangent to the chain makes an angle $\theta = 42.0°$ with the horizontal. Find (a) the magnitude of the force each hook exerts on the chain and (b) the tension in the chain at its midpoint. *Suggestion*: For part (b), make a force diagram for half of the chain.

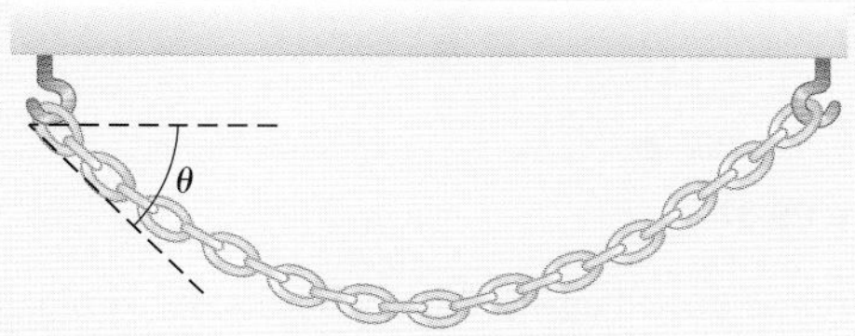

Figure P9.34

35. Three identical thin rods, each of length L and mass m, are welded perpendicular to one another as shown in **Figure P9.35**. The assembly is rotated about an axis that passes through the end of one rod and is parallel to another. Determine the moment of inertia of this structure about this axis.

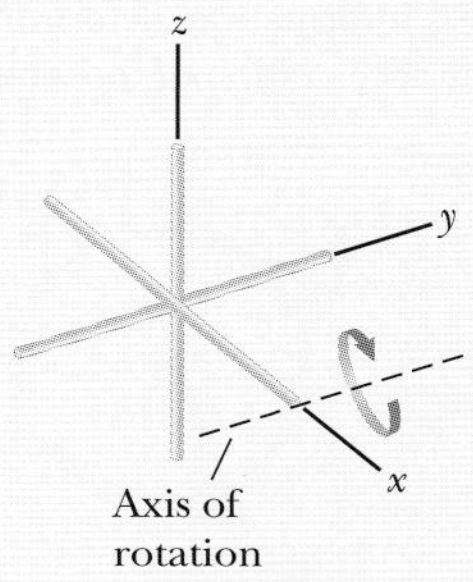

Figure P9.35

36. A block of mass $m_1 = 2.00$ kg and a block of mass $m_2 = 6.00$ kg are connected by a massless string over a pulley in the shape of a solid disc having radius $R = 0.250$ m and mass $M = 10.0$ kg. The fixed, wedge-shaped ramp makes an angle of $\theta = 30.0°$ as shown in **Figure P9.36**. The coefficient of kinetic friction is 0.360 for both blocks. (a) Draw force diagrams of both blocks and of the pulley. Determine (b) the acceleration of the two blocks and (c) the tensions in the string on both sides of the pulley.

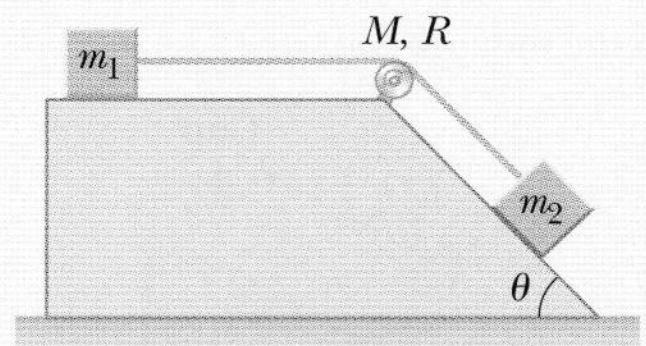

Figure P9.36

37. A long, uniform rod of length L and mass M is pivoted about a frictionless, horizontal pin through one end. The rod is nudged from rest in a vertical position as shown in **Figure P9.37**. At the instant the rod is horizontal, find (a) its angular speed, (b) the magnitude of its angular acceleration, (c) the x and y

components of the acceleration of its centre of mass and (d) the components of the reaction force at the pivot.

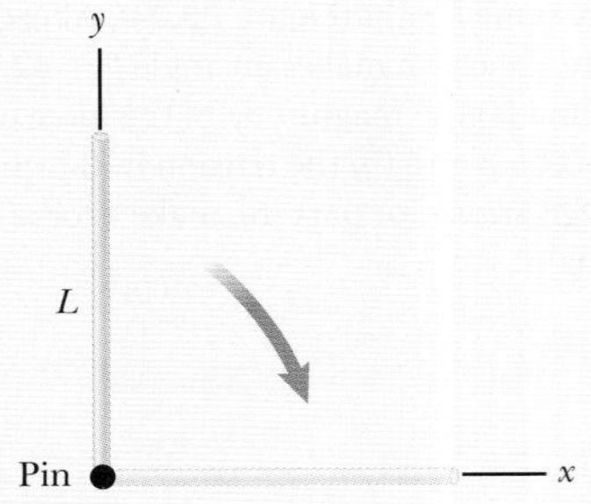

Figure P9.37

38. A bicycle is turned upside down while its owner repairs a flat tyre on the rear wheel. A friend spins the front wheel, of radius 0.381 m, and observes that drops of water fly off tangentially in an upwards direction when the drops are at the same level as the centre of the wheel. She measures the height reached by drops moving vertically (**Figure P9.38**). A drop that breaks loose from the tyre on one turn rises $h = 54.0$ cm above the tangent point. A drop that breaks loose on the next turn rises 51.0 cm above the tangent point. The height to which the drops rise decreases because the angular speed of the wheel decreases. From this information, determine the magnitude of the average angular acceleration of the wheel.

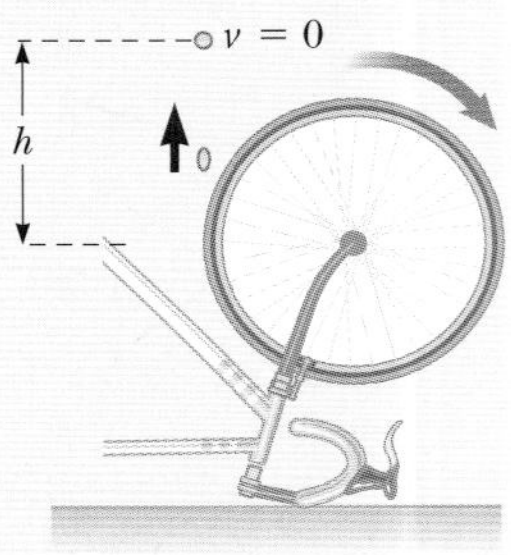

Figure P9.38

39. A common lecture demonstration, illustrated in **Figure P9.39**, consists of a ball resting at one end of a uniform board of length ℓ, which is hinged at the other end and elevated at an angle θ. A light cup is attached to the board at r_c so that it will catch the ball when the support stick is removed suddenly. (a) Show that the ball will lag behind the falling board when θ is less than 35.3°. (b) Assuming the board is 1.00 m long and is supported at this limiting angle, show that the cup must be 18.4 cm from the moving end.

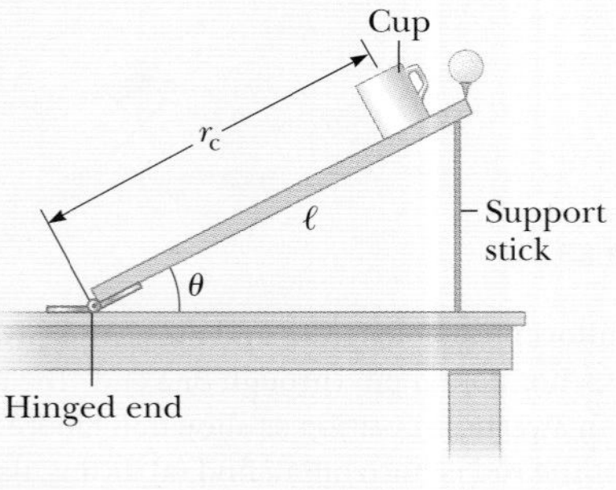

Figure P9.39

40. Consider a tall building located on the Earth's equator. As the Earth rotates, a person on the top floor of the building moves faster than someone on the ground with respect to an inertial reference frame because the person on the ground is closer to the Earth's axis. Consequently, if an object is dropped from the top floor to the ground a distance h below, it lands east of the point vertically below where it was dropped. (a) How far to the east will the object land? Express your answer in terms of h, g and the angular speed ω of the Earth. Ignore air resistance and assume the free-fall acceleration is constant over this range of heights. (b) Evaluate the eastwards displacement for $h = 50.0$ m. (c) In your judgment, were we justified in ignoring this aspect of the *Coriolis effect* in our previous study of free fall? (d) Suppose the angular speed of the Earth were to decrease due to tidal friction with constant angular acceleration. Would the eastwards displacement of the dropped object increase or decrease compared with that in part (b)?

41. ±? In a *Try this* example we described a method for finding your centre of mass. This can be done with the arrangement shown in **Figure P9.41**. A light plank rests on two scales, which read $M_1 = 38.0$ kg and $M_2 = 32.0$ kg. A distance of 1.65 m separates the scales. The scales have a precision of 0.5 kg and the uncertainty in the measured distance is 1 cm. How far from the woman's feet is her centre of mass?

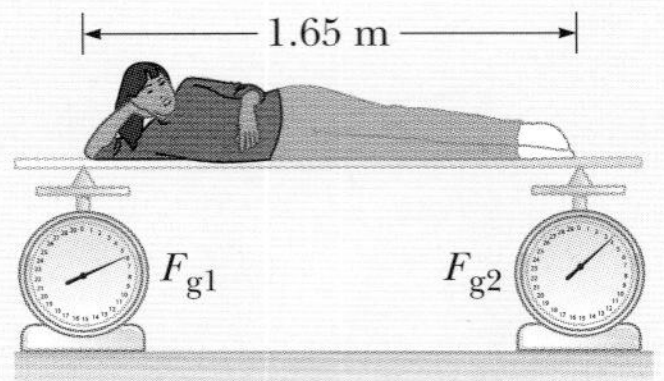

Figure P9.41

42. When a person stands on tiptoe on one foot, such as in various ballet positions, the position of the foot is as shown in **Figure P9.42a**. The total gravitational force $\vec{F}_g$ on the body is balanced by the normal force $\vec{n}$ exerted by the floor on the toes of one foot. A mechanical model of the situation is shown in **Figure P9.42b**, where $\vec{T}$ is the force exerted on the foot by the Achilles tendon and $\vec{R}$ is the force exerted on the foot by the tibia. Find the values of T, R and θ when $F_g = 700$ N.

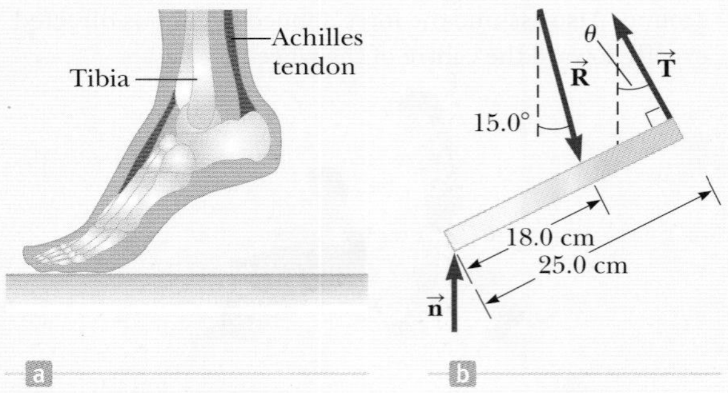

Figure P9.42

43. Find the mass m of the counterweight needed to balance a truck with mass $M = 1500$ kg on an incline of $\theta = 45°$ (**Figure P9.43**). Assume both pulleys are frictionless and massless.

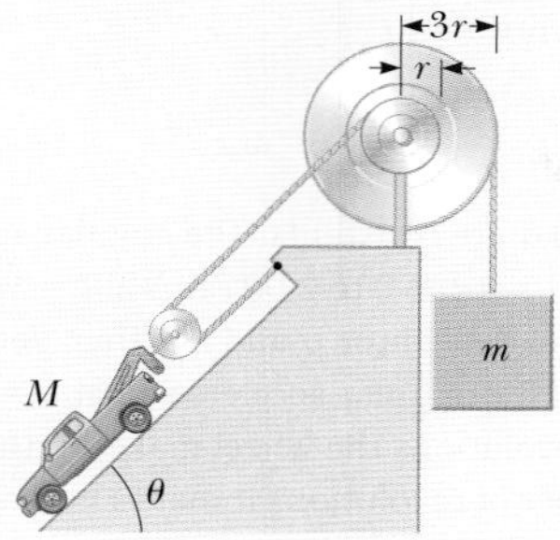

Figure P9.43

44. A crane of mass $m_1 = 3000$ kg supports a load of mass $m_2 = 10\,000$ kg as shown in Figure P9.44. The crane is pivoted with a frictionless pin at A and rests against a smooth support at B. Find the reaction forces at (a) point A and (b) point B.

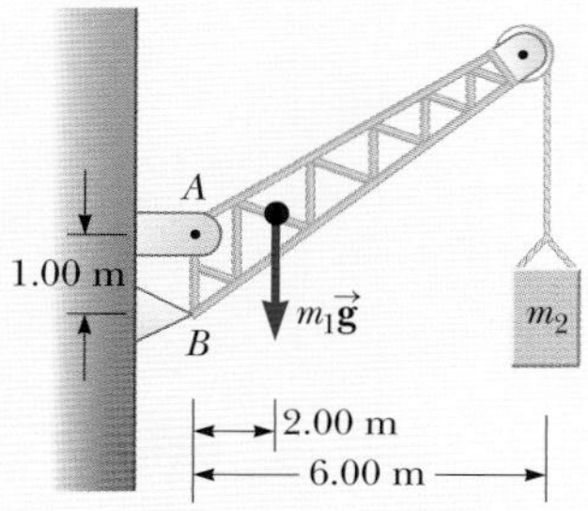

Figure P9.44

45. Assume a person bends forward to lift a load 'with his back' as shown in Figure P9.45a. The spine pivots mainly at the fifth lumbar vertebra, with the principal supporting force provided by the erector spinalis muscle in the back. To see the magnitude of the forces involved, consider the model shown in Figure P9.45b for a person bending forward to lift a 200 N object. The spine and upper body are represented as a uniform horizontal rod of weight 350 N, pivoted at the base of the spine. The erector spinalis muscle, attached at a point two-thirds of the way up the spine, maintains the position of the back. The angle between the spine and this muscle is $\theta = 12.0°$. Find (a) the tension T in the back muscle and (b) the compressional force in the spine. (c) Is this method a good way to lift a load? Explain your answer, using the results of parts (a) and (b). (d) Can you suggest a better method to lift a load?

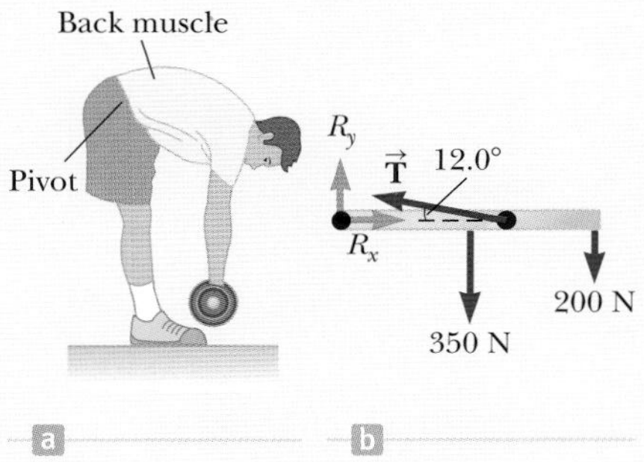

Figure P9.45

Challenge problems

46. A stepladder of negligible weight is constructed as shown in Figure P9.46, with $AC = BC = \ell = 4.00$ m. A painter of mass $m = 70.0$ kg stands on the ladder $d = 3.00$ m from the bottom. Assuming the floor is frictionless, find (a) the tension in the horizontal bar DE connecting the two halves of the ladder, (b) the normal forces at A and B and (c) the components of the reaction force at the single hinge C that the left half of the ladder exerts on the right half. *Suggestion*: Treat the ladder as a single object, but also treat each half of the ladder separately.

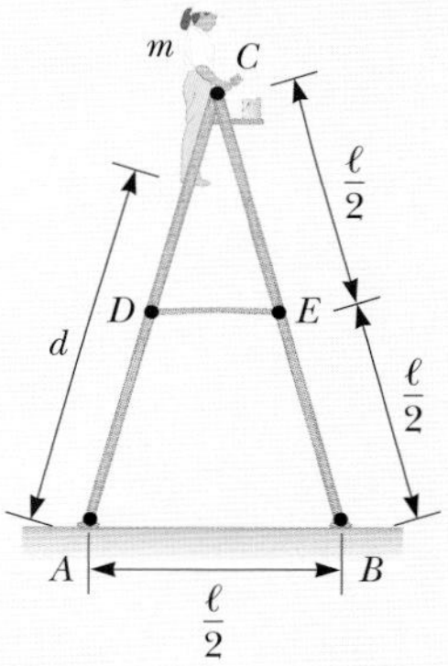

Figure P9.46

47. A merry-go-round is stationary. A dog is running around the merry-go-round on the ground just outside its circumference, moving with a constant angular speed of 0.750 rad/s. The dog does not change his pace when he sees what he has been looking for: a bone resting on the edge of the merry-go-round one-third of a revolution in front of him. At the instant the dog sees the bone ($t = 0$), the merry-go-round begins to move in the direction the dog is running, with a constant angular acceleration of 0.015 0 rad/s^2. (a) At what time will the dog first reach the bone? (b) The confused dog keeps running and passes the bone. How long after the merry-go-round starts to turn do the dog and the bone draw even with each other for the second time?

48. Consider the rectangular cabinet shown in Figure P9.48, with a force $\vec{\mathbf{F}}$ applied horizontally at the upper edge. (a) What is the minimum force required to start to tip the cabinet? (b) What is the minimum coefficient of static friction required for the cabinet not to slide with the application of a force of this magnitude? (c) Find the magnitude and direction of the minimum force required to tip the cabinet if the point of application can be chosen *anywhere* on the cabinet.

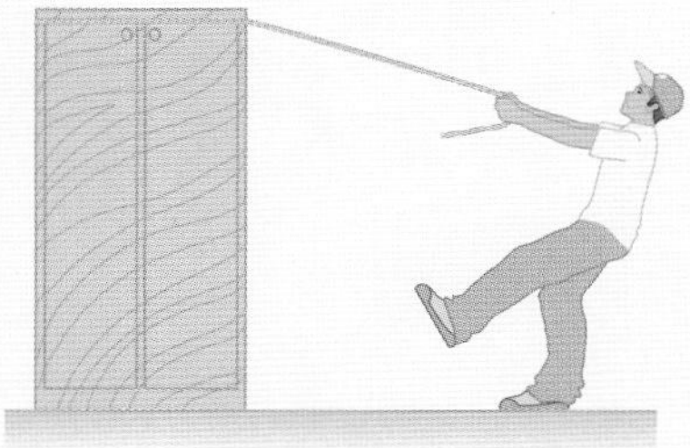

Figure P9.48

49. A uniform rod of weight F_g and length L is supported at its ends by a frictionless trough as shown in **Figure P9.49**. (a) Show that the centre of gravity of the rod must be vertically over point O when the rod is in equilibrium. (b) Determine the equilibrium value of the angle θ. (c) Is the equilibrium of the rod stable or unstable?

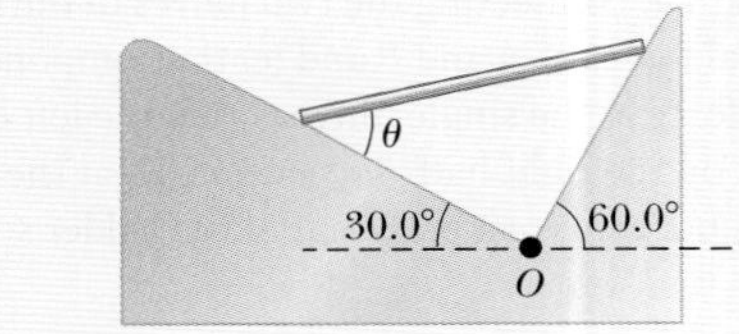

Figure P9.49

50. As a result of friction, the angular speed of a wheel changes with time according to

$$\frac{d\theta}{dt} = \omega_0 e^{-\sigma t}$$

where ω_0 and σ are constants. The angular speed changes from 3.50 rad/s at $t = 0$ to 2.00 rad/s at $t = 9.30$ s. (a) Use this information to determine σ and ω_0. Then determine (b) the magnitude of the angular acceleration at $t = 3.00$ s, (c) the number of revolutions the wheel makes in the first 2.50 s and (d) the number of revolutions it makes before coming to rest.

chapter **10**

Energy and momentum in rotating systems

Shutterstock.com/Lehrer

The Celtic stone is a fascinating spinning object and has been a physical curiosity since prehistoric times. When spun in a preferred direction, it will rotate on its axis. If it is spun in the opposite direction, it will become unstable, start to wobble and then reverse its spin to the preferred direction. Why does the stone have a preferred direction of spinning?

On completing this chapter, students will understand:

- how rotational and translational kinetic energies are related
- that work done on a system can result in change in translational and/or rotational kinetic energy
- that rolling is a combination of rotation and translation
- how linear and angular momenta are related
- that torque causes a change in angular momentum
- that angular momentum is conserved for an isolated system
- how spinning tops and gyroscopes work.

Students will be able to:

- describe rotational motion in terms of energy and angular momentum
- calculate rotational kinetic energy
- analyse rolling motion
- calculate angular momentum for systems of particles and extended objects
- determine which analysis model, isolated or non-isolated, is appropriate for a given situation
- define a system so that conservation of angular momentum can be applied
- apply analysis models for angular momentum to solve problems involving rotation.

The previous chapter looked at rotational motion and torque. In this chapter we consider the energy and momentum of rotating systems, known as rotational kinetic energy and angular momentum. The conservation laws for linear systems, conservation of energy and conservation of momentum that we have already discussed also apply to rotating systems. The law of conservation of angular momentum is a fundamental law of physics, equally valid for relativistic and quantum systems.

ENHANCED WebAssign

The problems found in this chapter may be assigned online in Enhanced Web Assign.

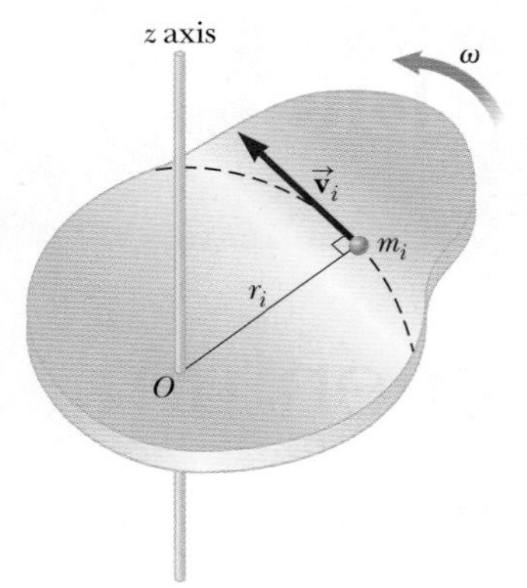

Figure 10.1
A rigid object rotating about the z axis with angular speed ω. The kinetic energy of the particle of mass m_i is $\frac{1}{2}m_i v_i^2$. The total kinetic energy of the object is called its rotational kinetic energy.

10.1 Rotational kinetic energy

In Chapter 6 we defined the kinetic energy of an object as the energy associated with its motion through space. An object rotating about a fixed axis remains stationary in space, so there is no kinetic energy associated with translational motion. However, the individual particles making up the rotating object are moving as they follow circular paths. Hence there is kinetic energy associated with rotational motion.

Consider an object as a system of particles and assume it rotates about a fixed z axis with an angular speed ω. **Figure 10.1** shows the rotating object and identifies one particle on the object located at a distance r_i from the rotation axis. If the mass of the ith particle is m_i and its tangential speed is v_i, its kinetic energy is

$$K_i = \frac{1}{2}m_i v_i^2$$

The individual tangential speeds of the particles depend on the distance r_i from the axis of rotation, however, every particle in the rigid object has the same angular speed ω, according to **Equation 9.10**. We can write the total kinetic energy of the rotating rigid object as the sum of the kinetic energies of the individual particles:

$$K_R = \sum_i K_i = \sum_i \frac{1}{2}m_i v_i^2 = \frac{1}{2}\sum_i m_i r_i^2 \omega^2$$

Recalling that the moment of inertia I of a rotating object is given by **Equation 9.22**

$$I \equiv \sum_i m_i r_i^2$$

we can write the kinetic energy as

Rotational ► kinetic energy

$$K_R = \frac{1}{2}\left(\sum_i m_i r_i^2\right)\omega^2 = \frac{1}{2}I\omega^2 \quad (10.1)$$

Although we commonly refer to the quantity $\frac{1}{2}I\omega^2$ as **rotational kinetic energy**, it is not a new form of energy. It is ordinary kinetic energy, derived from a sum over individual kinetic energies of the particles contained in the rigid object. Note the analogy between translational kinetic energy, $\frac{1}{2}mv^2$, and rotational kinetic energy, $\frac{1}{2}I\omega^2$. The quantities I and ω in rotational motion are analogous to m and v in translational motion, respectively. In fact, I takes the place of m and ω takes the place of v every time we compare a translational motion equation with its rotational counterpart. Recall that moment of inertia is a measure of the resistance of an object to changes in its rotational motion, just as mass is a measure of the tendency of an object to resist changes in its translational motion.

Example 10.1

Energy of rotation of a molecule

An oxygen molecule can be modelled as a stick (the bond) with a ball (an oxygen atom) at each end. An oxygen molecule has energy associated with its rotation, which can be about three different perpendicular axes, as shown in **Figure 10.2**.

If the bond length is L, the radius of the oxygen atom is r and the mass of each atom is m, what is the rotational kinetic energy for each of the rotational motions shown for a given ω?

Solution

Conceptualise We are already given a diagram in this case, otherwise we would draw one to help visualise the situation. We know that kinetic energy is proportional to the moment of inertia, and we can see that the mass is further from the axis of rotation in the second two cases, so we expect the kinetic energy to be larger in these cases.

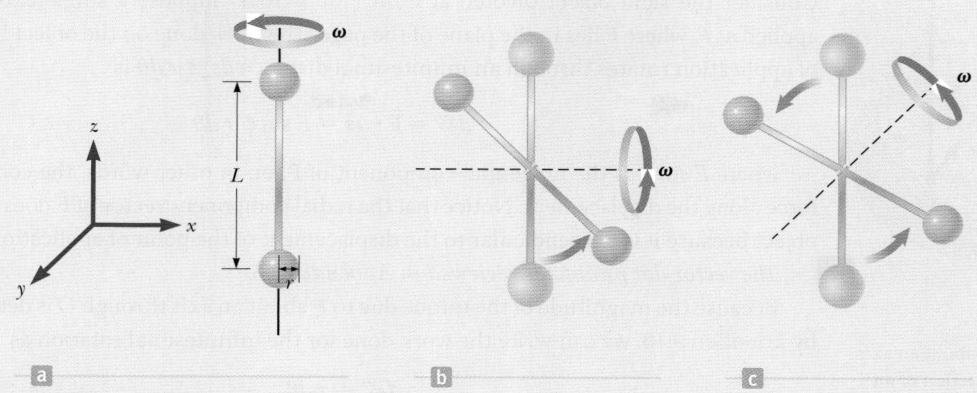

Figure 10.2
(Example 10.1) (a) An oxygen molecule rotating around the z axis (b) An oxygen molecule rotating around the x axis (c) An oxygen molecule rotating around the y axis

Model We will model the oxygen atoms as solid balls of uniform density and the molecule as rigid with the bond being massless. We will also make the approximation that the bond length is large compared to the size of the atoms.

We will need to use the definitions of kinetic energy and find the moment of inertia for each of these modes of rotation.

Analyse For the first case, the moment of inertia of each oxygen atom is that of a ball spinning about an axis passing through its centre of mass, which is given in **Table 9.2** as $\frac{2}{5}Mr^2$.

Evaluate the rotational kinetic energy using **Equation 10.1**:

$$K_x = \frac{1}{2}I\omega^2 = \frac{1}{2}\left(\frac{2}{5}mr^2\right)\omega^2 = \frac{1}{5}mr^2\omega^2$$

This is for a single ball, and there are two, so the total rotational kinetic energy is

$$K_x = \frac{2}{5}mr^2\omega^2$$

Check dimensions: $[ML^2T^{-2}] = [M][L]^2[T^{-1}]^2 = [ML^2T^{-2}]$ ☺

The second two cases are equivalent, each is a massless rod with a ball on each end. We will assume that the balls are small compared to the length L, that is, $r \ll L$, so that we can approximate them as particles of mass m at a distance $\frac{L}{2}$ from the axis of rotation. The moment of inertia of each ball (atom) is given by $m\left(\frac{L}{2}\right)^2$ and for the whole molecule (two atoms) the moment of inertia is $2 \times m\left(\frac{L}{2}\right)^2 = \frac{1}{2}mL^2$.

Again, evaluate the rotational kinetic energy using **Equation 10.1**:

$$K_y = K_z = \frac{1}{2}I\omega^2 = \frac{1}{2}\left(\frac{1}{2}mL^2\right)\omega^2 = \frac{1}{4}mL^2\omega^2$$

This again has the same dimensions as energy.

Finalise In the limiting case that the atoms are very small compared to the length of the bond between them, $r \ll L$, we can approximate r as 0. We see that $K_x \approx 0$ while K_y and K_z are unchanged. Most of the mass of an atom is concentrated in its nucleus (see Chapter 42), which is very tiny, so this is a good approximation. When we consider the rotational energy of molecules we usually ignore the rotational kinetic energy of modes such as the first case because they are many orders of magnitude smaller than the other possible modes.

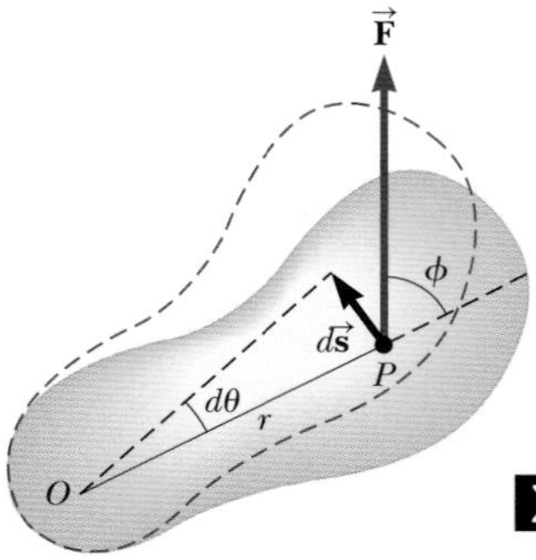

Figure 10.3
A rigid object rotates about an axis through O under the action of an external force $\vec{F}$ applied at P.

Consider the rigid object pivoted at O in **Figure 10.3**. Suppose a single external force $\vec{F}$ is applied at P, where $\vec{F}$ lies in the plane of the page. The work done on the object by $\vec{F}$ as its point of application rotates through an infinitesimal distance $ds = r\,d\theta$ is

$$dW = \vec{F} \cdot d\vec{s} = F \sin\phi \; r\,d\theta$$

where $F \sin\phi$ is the tangential component of $\vec{F}$, or, in other words, the component of the force along the displacement. Notice that the radial component vector of $\vec{F}$ does no work on the object because it is perpendicular to the displacement of the point of application of $\vec{F}$.

Σ *The vector dot product is reviewed in Appendix B.6.*

Because the magnitude of the torque due to $\vec{F}$ about an axis through O is defined as $rF \sin\phi$ by **Equation 9.18**, we can write the work done for the infinitesimal rotation as

$$dW = \tau\,d\theta \tag{10.2}$$

The rate at which work is being done by $\vec{F}$ as the object rotates about the fixed axis through the angle $d\theta$ in a time interval dt is

$$\frac{dW}{dt} = \tau \frac{d\theta}{dt}$$

Because dW/dt is the instantaneous power P delivered by the force (see Section 6.7) and $d\theta/dt = \omega$, this expression reduces to

Power ▶ delivered to a rotating rigid object

$$P = \frac{dW}{dt} = \tau\omega \tag{10.3}$$

This equation is analogous to $P = Fv$ in the case of translational motion, and **Equation 10.2** is analogous to $dW = F_x\,dx$.

When a symmetrical object rotates about a fixed axis, the work done by external forces equals the change in the rotational energy of the object. To show this, let us begin with $\Sigma\tau_{ext} = I\alpha$. Using the chain rule from calculus, we can express the net torque as

$$\sum \tau_{ext} = I\alpha = I\frac{d\omega}{dt} = I\frac{d\omega}{d\theta}\frac{d\theta}{dt} = I\frac{d\omega}{d\theta}\omega$$

Σ *Differential calculus and properties of the derivative, including the chain rule, are reviewed in Appendix B.7. Integral calculus, including the integration of polynomials, is reviewed in Appendix B.8.*

Rearranging this expression and noting that $\Sigma\tau_{ext}\,d\theta = dW$ gives

$$\sum \tau_{ext}\,d\theta = dW = I\omega\,d\omega$$

Integrating this expression, we obtain for the total work done by the net external force acting on a rotating system

Work–kinetic ▶ energy theorem for rotational motion

$$\sum W = \int_{\omega_i}^{\omega_f} I\omega\,d\omega = \frac{1}{2}I\omega_f^2 - \frac{1}{2}I\omega_i^2 \tag{10.4}$$

where the angular speed changes from ω_i to ω_f. **Equation 10.4** is the **work–kinetic energy theorem for rotational motion**. Similar to the work–kinetic energy theorem for translational motion, this theorem states that the net work done by external forces in rotating a symmetric rigid object about a fixed axis equals the change in the object's rotational energy. This theorem is a form of the non-isolated system (energy) model discussed in Chapter 7. Work is done on the system of the rigid object, which represents a transfer of energy across the boundary of the system that appears as an increase in the object's rotational kinetic energy.

In general, we can combine this theorem with the translational form of the work–kinetic energy theorem from Chapter 6. Therefore, the net work done by external forces on an object is the change in its *total* kinetic energy, which is the sum of the translational and rotational kinetic energies. For example, when a bowler throws a cricket ball the work done by the bowler's hands appears as translational kinetic energy associated with the ball moving through space as well as rotational kinetic energy associated with the spin of the ball.

In addition to the work–kinetic energy theorem, other energy principles can also be applied to rotational situations. For example, if a system involving rotating objects is isolated and no non-conservative forces act within the system, the isolated system model and the principle of conservation of mechanical energy can be used to analyse the system as in Example 10.2 below.

Example 10.2

Rotating rod

A uniform rod of length L and mass M is free to rotate on a frictionless pin passing through one end (**Figure 10.4**). The rod is released from rest in the horizontal position.

(A) What is its angular speed when the rod reaches its lowest position?

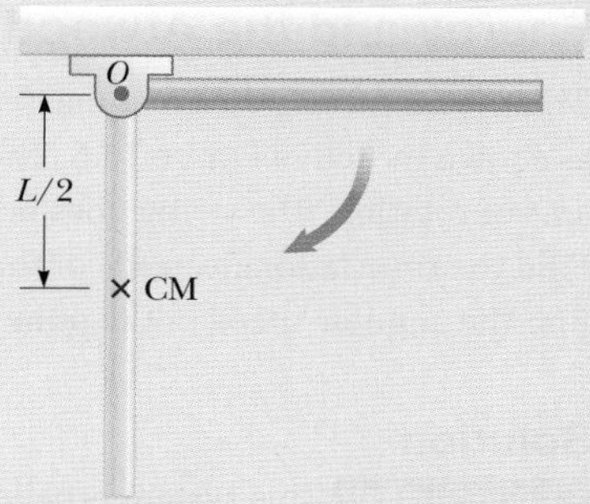

Figure 10.4
(Example 10.2) A uniform rigid rod pivoted at O rotates in a vertical plane under the action of the gravitational force.

Solution

Conceptualise Consider **Figure 10.4** and imagine the rod rotating downwards through a quarter turn about the pivot at the left end. Also look back at Example 9.11. This physical situation is the same.

Model As mentioned in Example 9.11, the angular acceleration of the rod is not constant. Therefore, the kinematic equations for rotation (Section 9.2) cannot be used to solve this example. As the pin is frictionless, we model the system of the rod and the Earth as an isolated system in terms of energy with no non-conservative forces acting and use the principle of conservation of mechanical energy.

Analyse We choose the configuration in which the rod is hanging straight down as the reference configuration for gravitational potential energy and assign a value of zero for this configuration. When the rod is in the horizontal position, it has no rotational kinetic energy. The potential energy of the system in this configuration relative to the reference configuration is $MgL/2$ because the centre of mass of the rod is at a height $L/2$ higher than its position in the reference configuration. When the rod reaches its lowest position, the energy of the system is entirely rotational energy $\frac{1}{2}I\omega^2$, where I is the moment of inertia of the rod about an axis passing through the pivot.

Using the isolated system (energy) model, write a conservation of mechanical energy equation for the system:

$$K_f + U_f = K_i + U_i$$

Substitute for each of the energies:

$$\frac{1}{2}I\omega^2 + 0 = 0 + \frac{1}{2}MgL$$

Solve for ω and use $I = \frac{1}{3}ML^2$ (see **Table 9.2**, page 264) for the rod:

$$\omega = \sqrt{\frac{MgL}{I}} = \sqrt{\frac{MgL}{\frac{1}{3}ML^2}} = \sqrt{\frac{3g}{L}}$$

Check dimensions:

$$[T^{-1}] = \left(\frac{[LT^{-2}]}{[L]}\right)^{1/2} = [T^{-1}]$$ ☺

(B) Determine the tangential speed of the centre of mass and the tangential speed of the lowest point on the rod when it is in the vertical position.

Solution

Use **Equation 9.10** and the result from part (A):

$$v_{CM} = r\omega = \frac{L}{2}\omega = \frac{1}{2}\sqrt{3gL}$$

Check dimensions:

$$[LT^{-1}] = ([LT^{-2}][L])^{1/2} = [LT^{-1}]$$ ☺

Because r for the lowest point on the rod is twice what it is for the centre of mass, the lowest point has a tangential speed twice that of the centre of mass:

$$v = 2v_{CM} = \sqrt{3gL}$$

Finalise The initial configuration in this example is the same as that in Example 9.11. In Example 9.11, however, we could only find the initial angular acceleration of the rod. Applying an energy approach in the current example allows us to find additional information, the angular speed of the rod at the lowest point. You could find the angular speed of the rod at any angular position by knowing the location of the centre of mass at this position.

Example 10.3

Energy and the Atwood machine

Two blocks having different masses m_1 and m_2 are connected by a string passing over a pulley as shown in **Active Figure 10.5**. The pulley has a radius R and moment of inertia I about its axis of rotation. The string does not slip on the pulley, and the system is released from rest. Find the translational speeds of the blocks after block 2 descends through a distance h and find the angular speed of the pulley at this time.

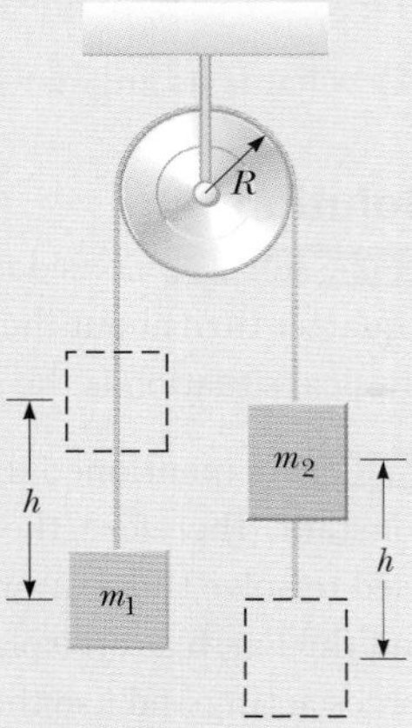

Figure 10.5 (Example 10.3) An Atwood machine with a massive pulley

Solution

Conceptualise This system is called an Atwood machine, and we have already seen it in Chapter 4 (Example 4.11). You can use **Active Figure 10.5** to help you visualise the system and understand how it works.

Model Because the string does not slip, the pulley rotates about the axle. We will neglect friction in the axle because the axle's radius is small relative to that of the pulley. Hence, the frictional torque is much smaller than the net torque applied by the two blocks provided that their masses are significantly different. Consequently, we model the system consisting of the two blocks, the pulley and the Earth as an isolated system in terms of energy with no non-conservative forces acting; therefore, the mechanical energy of the system is conserved.

Analyse We define the zero configuration for gravitational potential energy as that which exists when the system is released. From **Active Figure 10.5**, we see that the descent of block 2 is associated with a decrease in system potential energy and that the rise of block 1 represents an increase in potential energy. Using the isolated system (energy) model, write a conservation of mechanical energy equation for the system:

$$K_f + U_f = K_i + U_i$$

Substitute for each of the energies:

$$\left(\frac{1}{2}m_1v_f^{\,2} + \frac{1}{2}m_2v_f^{\,2} + \frac{1}{2}I\omega_f^{\,2}\right) + (m_1gh - m_2gh) = 0 + 0$$

Use $v_f = R\omega_f$ to substitute for ω_f:

$$\frac{1}{2}m_1v_f^{\,2} + \frac{1}{2}m_2v_f^{\,2} + \frac{1}{2}I\frac{v_f^2}{R^2} = m_2gh - m_1gh$$

$$\frac{1}{2}\left(m_1 + m_2 + \frac{I}{R^2}\right)v_f^2 = m_2gh - m_1gh$$

Solve for v_f:

$$v_f = \left[\frac{2(m_2 - m_1)gh}{m_1 + m_2 + I/R^2}\right]^{1/2} \qquad (1)$$

Use $v_f = R\omega_f$ to solve for ω_f:

$$\omega_f = \frac{v_f}{R} = \frac{1}{R}\left[\frac{2(m_2 - m_1)gh}{m_1 + m_2 + I/R^2}\right]^{1/2}$$

Example 10.3 cont.

Check dimensions, remembering that radians are dimensionless.

$$[T^{-1}] = [L^{-1}]([M][LT^{-2}][L]/([M]+[ML^2/L^2]))^{1/2} = [L]^{-1}([ML^2T^{-2}]/[M])^{1/2} = [T^{-1}] \; ☺$$

Hence it is unlikely that we have made an error in our algebraic manipulations and if we had numbers to substitute we would proceed to do so.

Finalise Compare this example to Example 4.11 where we ignored the rotation of the pulley. Does **Equation 1** reduce to the expression found in Example 4.11 in the limit that the pulley is massless? In this example we modelled the cable as massless. What effect would you expect the mass of the cable to have on **Equation 1**?

10.2 Rolling motion of a rigid object

In this section we treat the motion of a rigid object rolling along a flat surface. Suppose a cylinder is rolling on a straight path such that the axis of rotation remains parallel to its initial orientation in space. As **Figure 10.6** shows, a point on the rim of the cylinder moves in a complex path called a *cycloid*. We can simplify matters, however, by focusing on the centre of mass rather than on a point on the rim of the rolling object. As shown in **Figure 10.6**, the centre of mass moves in a straight line. If an object such as a cylinder rolls without slipping on the surface (called *pure rolling motion*), a simple relationship exists between its rotational and translational motions.

Consider a uniform cylinder of radius R rolling without slipping on a horizontal surface (**Figure 10.7**). As the cylinder rotates through an angle θ, its centre of mass moves a linear distance $s = R\theta$. Therefore, the translational speed of the centre of mass for pure rolling motion is given by

$$v_{CM} = \frac{ds}{dt} = R\frac{d\theta}{dt} = R\omega \tag{10.5}$$

where ω is the angular speed of the cylinder. **Equation 10.5** holds whenever a cylinder or sphere rolls without slipping and is the **condition for pure rolling motion**. The magnitude of the linear acceleration of the centre of mass for pure rolling motion is

$$a_{CM} = \frac{dv_{CM}}{dt} = R\frac{d\omega}{dt} = R\alpha \tag{10.6}$$

where α is the angular acceleration of the cylinder.

Imagine that you are moving along with a rolling object at speed v_{CM}, staying in a frame of reference at rest with respect to the centre of mass of the object, for example watching the wheel of a slowly moving car as you keep as you keep pace with it. As you observe the object, you will see the object in pure rotation around its centre of mass. **Figure 10.8a** shows the velocities of points at the top, centre and bottom of the object as observed by you. In addition to these velocities, every point on the object moves in the same direction with

> **Pitfall Prevention 10.1**
> Equation 10.5 looks familiar. It looks very similar to Equation 9.10, so be sure you're clear on the difference. Equation 9.10 gives the *tangential* speed of a point on a *rotating* object located a distance r from a fixed rotation axis if the object is rotating with angular speed v. Equation 10.5 gives the *translational* speed of the centre of mass of a *rolling* object of radius R rotating with angular speed ω.

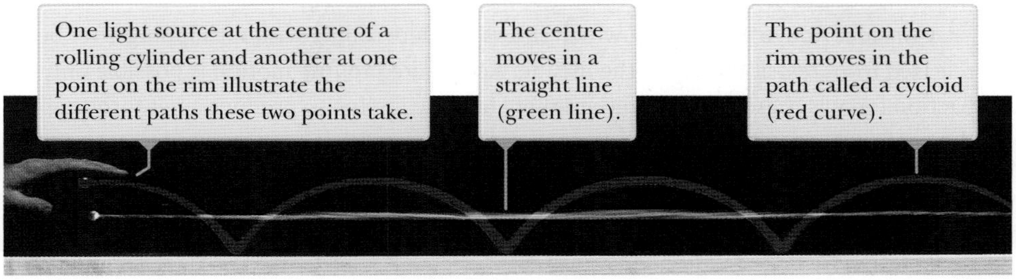

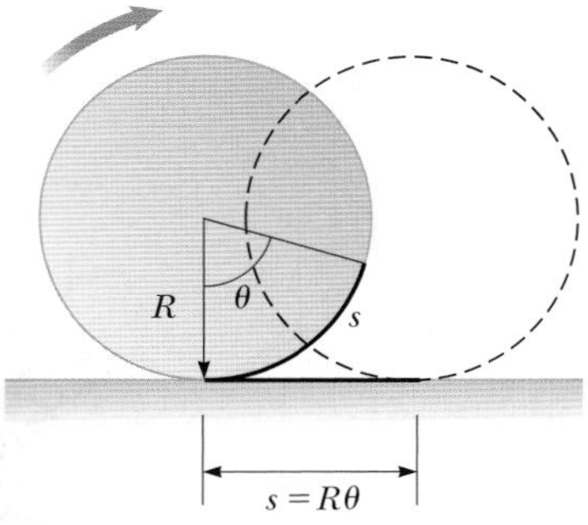

Figure 10.7
For pure rolling motion, as the cylinder rotates through an angle θ its centre moves a linear distance $s = R\theta$.

One light source at the centre of a rolling cylinder and another at one point on the rim illustrate the different paths these two points take.

The centre moves in a straight line (green line).

The point on the rim moves in the path called a cycloid (red curve).

Figure 10.6
Two points on a rolling object take different paths through space.

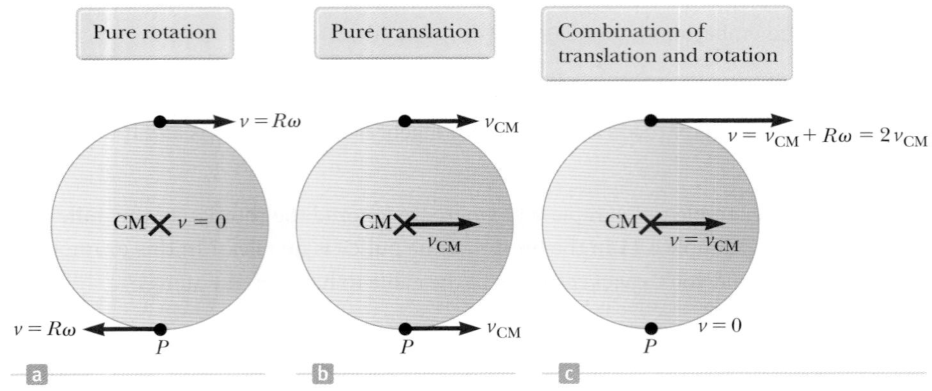

Figure 10.8
The motion of a rolling object can be modelled as a combination of pure translation and pure rotation.

speed v_{CM} relative to the surface on which it rolls. **Figure 10.8b** shows these velocities for a non-rotating object. In the reference frame at rest with respect to the surface, the velocity of a given point on the object is the sum of the velocities shown in **Figures 10.8a** and **10.8b**. **Figure 10.8c** shows the results of adding these velocities.

Notice that the contact point between the surface and object in **Figure 10.8c** has a translational speed of zero. At this instant, the rolling object is moving in exactly the same way as if the surface were removed and the object were pivoted at point P and spun about an axis passing through P. We can express the total kinetic energy of this imagined spinning object as

$$K = \frac{1}{2} I_P \omega^2 \qquad (10.7)$$

where I_P is the moment of inertia about a rotation axis through P.

Because the motion of the imagined spinning object is the same at this instant as our actual rolling object, **Equation 10.7** also gives the kinetic energy of the rolling object. Applying the parallel axis theorem, we can substitute $I_P = I_{CM} + MR^2$ into **Equation 10.7** to obtain

$$K = \frac{1}{2} I_{CM} \omega^2 + \frac{1}{2} MR^2 \omega^2$$

Using $v_{CM} = R\omega$, this equation can be expressed as

Total kinetic energy of a rolling object ▶

$$K = \frac{1}{2} I_{CM} \omega^2 + \frac{1}{2} M {v_{CM}}^2 \qquad (10.8)$$

The term $\frac{1}{2} I_{CM} \omega^2$ represents the rotational kinetic energy of the object about its centre of mass, and the term $\frac{1}{2} M {v_{CM}}^2$ represents the kinetic energy the object would have if it were just translating through space without rotating. Therefore, the total kinetic energy of a rolling object is the sum of the rotational kinetic energy about the centre of mass and the translational kinetic energy. This statement is consistent with the situation illustrated in **Figure 10.8**, which shows that the velocity of a point on the object is the sum of the velocity of the centre of mass and the tangential velocity around the centre of mass.

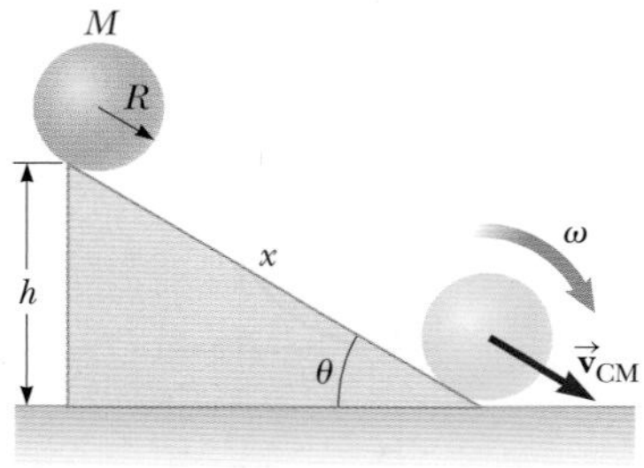

Active Figure 10.9
A sphere rolling down an incline. Mechanical energy of the sphere–Earth system is conserved if no slipping occurs.

Consider **Active Figure 10.9**, which shows a sphere rolling without slipping after being released from rest at the top of the incline. Accelerated rolling motion is possible only if a friction force is present between the sphere and the incline to produce a net torque about the centre of mass. Despite the presence of friction, no loss of mechanical energy occurs because the contact point is at rest relative to the surface at any instant. (On the other hand, if the sphere were to slip, mechanical energy of the sphere–incline–Earth system would decrease due to the non-conservative force of kinetic friction.)

In reality, *rolling friction* causes mechanical energy to transform to internal energy. Rolling friction is due to deformations of the surface and the rolling object. For example, car tyres flex as they roll on a road and mechanical energy is lost as internal energy.

The road surface also deforms slightly. Often we make the approximation that rolling friction is negligible and ignore it, but this is not always the case.

Using $v_{CM} = R\omega$ for pure rolling motion, we can express **Equation 10.8** as

$$K = \frac{1}{2} I_{CM} \left(\frac{v_{CM}}{R} \right)^2 + \frac{1}{2} M v_{CM}^2 \tag{10.9}$$

$$K = \frac{1}{2} \left(\frac{I_{CM}}{R^2} + M \right) v_{CM}^2$$

For the sphere–Earth system in **Active Figure 10.9**, we define the zero configuration of gravitational potential energy to be when the sphere is at the bottom of the incline. Therefore, conservation of mechanical energy gives

$$K_f + U_f = K_i + U_i$$

$$\frac{1}{2} \left(\frac{I_{CM}}{R^2} + M \right) v_{CM}^2 + 0 = 0 + Mgh$$

$$v_{CM} = \left[\frac{2gh}{1 + (I_{CM}/MR^2)} \right]^{1/2} \tag{10.10}$$

TRY THIS

Get a large can of dog food, a small of can of dog food (or some other solid material) and an empty can of the same size as one of the two full cans. Roll them down an incline together and see which one reaches the bottom first. Your incline will need to be rough enough that they do not slip. Explain your observations.

Quick Quiz 10.1

A ball rolls without slipping down incline A, starting from rest. At the same time, a box starts from rest and slides down incline B, which is identical to incline A except that it is frictionless. Which arrives at the bottom first? **(a)** The ball arrives first. **(b)** The box arrives first. **(c)** Both arrive at the same time. **(d)** It is impossible to determine.

Example 10.4

Ball rolling down an incline

For the solid sphere shown in **Active Figure 10.9**, calculate the translational speed of the centre of mass at the bottom of the incline and the magnitude of the translational acceleration of the centre of mass.

Solution

Conceptualise Imagine rolling the ball down the incline. You already have a diagram in **Active Figure 10.9**, otherwise you would start by drawing one.

Model We model the ball and the Earth as an isolated system in terms of energy with no non-conservative forces doing work. This model is the one that led to **Equation 10.10**, so we can use that result.

Analyse Evaluate the speed of the centre of mass of the sphere from **Equation 10.10**:

$$v_{CM} = \left[\frac{2gh}{1 + \left(\frac{2}{5}MR^2/MR^2\right)} \right]^{1/2} = \left(\frac{10}{7} gh \right)^{1/2} \tag{1}$$

To calculate the translational acceleration of the centre of mass, notice that the vertical displacement of the sphere is related to the distance x it moves along the incline through the relationship $h = x \sin\theta$.

Use this relationship to rewrite **Equation (1)**:

$$v_{CM}^2 = \frac{10}{7} gx \sin\theta$$

Example 10.4 cont.

Write **Equation 2.16** for an object starting from rest and moving through a distance x under constant acceleration:

$$v_{CM}{}^2 = 2a_{CM}x$$

Equate the preceding two expressions to find a_{CM}:

$$a_{CM} = \frac{5}{7}g\sin\theta$$

Finalise The speed is less than $\sqrt{2gh}$, which is the speed an object would have if it simply slid down the incline without rotating. The acceleration is also less than for a sliding object without friction.

Also note that both the speed and the acceleration of the centre of mass are *independent* of the mass and the radius of the sphere. That is, all homogeneous solid spheres experience the same speed and acceleration on a given incline. This is also the case for cylinders; if you did the *Try this* example with the different sized tins of dog food you would have noticed this.

If we were to repeat the acceleration calculation in Example 10.4 above for a hollow sphere, a solid cylinder or a hoop, we would obtain similar results in which only the factor in front of $g\sin\theta$ would differ. The constant factors that appear in the expressions for v_{CM} and a_{CM} depend only on the moment of inertia about the centre of mass for the specific object. In all cases, the acceleration of the centre of mass is *less than* $g\sin\theta$, the value the acceleration would have if the incline were frictionless and no rolling occurred.

TRY THIS

Get a 425 g can of condensed (solid) soup and a 425 g can of the same shape filled with a ready to eat (liquid) soup. (Any two cans of the same size, with one containing a solid and the other a liquid, will do just as well.) Roll them down an incline as in the previous example. Which one reaches the bottom first? Why? What happens to the gravitational potential energy of each can as it rolls down the hill? Write down the energy transformations that occur in each case.

Example 10.5

Pulling on a spool

A cylindrically symmetrical spool of cotton of mass m and radius R sits at rest on a horizontal table with friction. The cotton is wound around a central axle of radius r. You pull on the end of the cotton with a constant horizontal force of magnitude T to the right. As a result, the spool rolls without slipping a distance L along the table with no rolling friction.

(A) Find the final translational speed of the centre of mass of the spool.

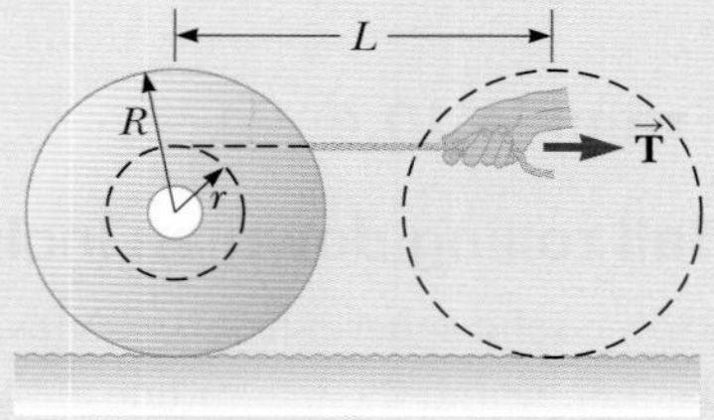

Figure 10.10
(Example 10.5) A spool rests on a horizontal table. A thread of cotton is wrapped around the axle and is pulled to the right by a hand.

Solution

Conceptualise Start by drawing a diagram to visualise the motion of the spool when you pull the cotton. For the spool to roll through a distance L, notice that your hand on the cotton must pull through a distance *different* from L.

Model The spool can be modelled as a rigid object subject to a net torque, but the net torque includes that due to the friction force at the bottom of the spool, about which we know nothing. Therefore we cannot use the rigid object subject to a net torque model. Work is done by your hand on the spool and cotton, which form a non-isolated system in terms of energy. Therefore we apply the non-isolated system (energy) model.

Analyse The only type of energy that changes in the system is the kinetic energy of the spool. We make the approximation that there is no rolling friction, so there is no change in internal energy. The only way that energy crosses the system's boundary is by the work done by your hand on the cotton. No work is done by the static force of friction on the bottom of the spool because the point of application of the force moves through no displacement.

Example 10.5 cont.

Write the appropriate reduction of the conservation of energy equation, **Equation 7.2**:

$$W = \Delta K = \Delta K_{trans} + \Delta K_{rot} \quad (1)$$

where W is the work done on the cotton by your hand. To find this work, we need to find the displacement of your hand during the process.

We first find the length of cotton that has unwound off the spool. If the spool rolls through a distance L, the total angle through which it rotates is $\theta = L/R$. The axle also rotates through this angle.

Use **Equation 9.1a** to find the total arc length through which the axle turns:

$$\ell = r\theta = r\frac{L}{R}$$

This result also gives the length of cotton pulled off the axle. Your hand will move through this distance *plus* the distance L through which the spool moves. Therefore, the magnitude of the displacement of the point of application of the force applied by your hand is $\ell + L = L(1 + r/R)$.

Evaluate the work done by your hand on the cotton:

$$W = TL\left(1 + \frac{r}{R}\right) \quad (2)$$

Substitute **Equation (2)** into **Equation (1)**:

$$TL\left(1 + \frac{r}{R}\right) = \frac{1}{2}mv_{CM}^2 + \frac{1}{2}I\omega^2$$

where I is the moment of inertia of the spool about its centre of mass, and v_{CM} and ω are the final values after the wheel rolls through the distance L.

Apply the non-slip rolling condition $\omega = \frac{v_{CM}}{R}$:

$$TL\left(1 + \frac{r}{R}\right) = \frac{1}{2}mv_{CM}^2 + \frac{1}{2}I\frac{v_{CM}^2}{R^2}$$

Solve for v_{CM}:

$$v_{CM} = \sqrt{\frac{2TL(1 + r/R)}{m(1 + I/mR^2)}} \quad (3)$$

This is a complicated-looking expression, so we do a dimension check to ensure there are no errors:

$$[LT^{-1}] = ([MLT^{-2}][L][]/[M]([]+[ML^2]/[M][L^2]))^{1/2}$$
$$= ([MLT^{-2}][L][]/[M][])^{1/2} = [LT^{-1}]$$ ☺

(B) Find the value of the friction force f.

Solution

Model As the friction force does no work, we cannot evaluate it from an energy approach. We model the spool as a non-isolated system, but this time in terms of momentum. The cotton applies a force across the boundary of the system, resulting in an impulse on the system. Because the forces on the spool are constant, we can model the spool's centre of mass as a particle with constant acceleration.

Analyse Write the impulse–momentum theorem (**Equation 8.40**) for the spool:

$$(T - f)\Delta t = m(v_{CM} - 0) = mv_{CM} \quad (4)$$

For a particle with constant acceleration starting from rest, **Equation 2.13** tells us that the average velocity of the centre of mass is half the final velocity.

Use **Equation 2.2** to find the time interval for the centre of mass of the spool to move a distance L from rest to a final speed v_{CM}:

$$\Delta t = \frac{L}{v_{CM,avg}} = \frac{2L}{v_{CM}} \quad (5)$$

Substitute **Equation (5)** into **Equation (4)**:

$$(T - f)\frac{2L}{v_{CM}} = mv_{CM}$$

Solve for the friction force f:

$$f = T - \frac{mv_{CM}^2}{2L}$$

Example 10.5 cont.

Substitute v_{CM} from **Equation (3)**:

$$f = T - \frac{m}{2L}\left[\frac{2TL(1 + r/R)}{m(1 + I/mR^2)}\right]$$

$$= T - T\frac{(1 + r/R)}{(1 + I/mR^2)} = T\left[\frac{I - mrR}{I + mR^2}\right]$$

We can see that this is dimensionally correct because the left-hand side and the quantity on the right outside the brackets are both forces. The quantity in the brackets is a dimensionless ratio. ☺

Finalise Notice that we could use the impulse–momentum theorem for the translational motion of the spool while ignoring that the spool is rotating! This fact demonstrates the power of our growing list of approaches to solving problems.

10.3 Angular momentum

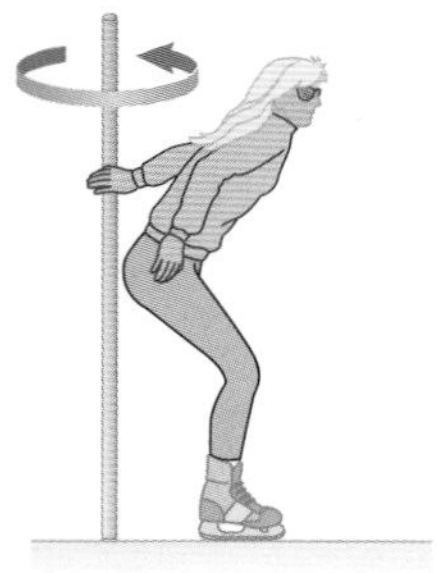

Active Figure 10.11
As the skater passes the pole she grabs hold of it, which causes her to swing rapidly around the pole in a circular path.

Imagine a rigid pole sticking up through the ice on a frozen pond (**Active Figure 10.11**). A skater glides rapidly toward the pole, aiming a little to the side so that she does not hit it. As she passes the pole, she reaches out to her side and grabs it, an action that causes her to move in a circular path around the pole. Just as the idea of linear momentum helps us analyse translational motion, a rotational analogue – *angular momentum* – helps us analyse the motion of this skater and other objects undergoing rotational motion.

Consider a particle of mass m located at the vector position $\vec{\mathbf{r}}$ and moving with linear momentum $\vec{\mathbf{p}}$ as in **Active Figure 10.12**. In describing translational motion, we found that the net force on the particle equals the time rate of change of its linear momentum, $\sum\vec{\mathbf{F}} = d\vec{\mathbf{p}}/dt$ (see **Equation 8.3**). Let us take the cross product of each side of **Equation 8.3** with $\vec{\mathbf{r}}$, which gives the net torque on the particle:

$$\sum\vec{\boldsymbol{\tau}} = \vec{\mathbf{r}} \times \sum\vec{\mathbf{F}} = \vec{\mathbf{r}} \times \frac{d\vec{\mathbf{p}}}{dt}$$

Σ *The vector cross product is reviewed in Appendix B.6.*

Now let's add to the right side the term $d\vec{\mathbf{r}}/dt \times \vec{\mathbf{p}}$, which is zero because $d\vec{\mathbf{r}}/dt = \vec{\mathbf{v}}$ and $\vec{\mathbf{v}}$ and $\vec{\mathbf{p}}$ are parallel. Therefore,

$$\sum\vec{\boldsymbol{\tau}} = \vec{\mathbf{r}} \times \frac{d\vec{\mathbf{p}}}{dt} + \frac{d\vec{\mathbf{r}}}{dt} \times \vec{\mathbf{p}}$$

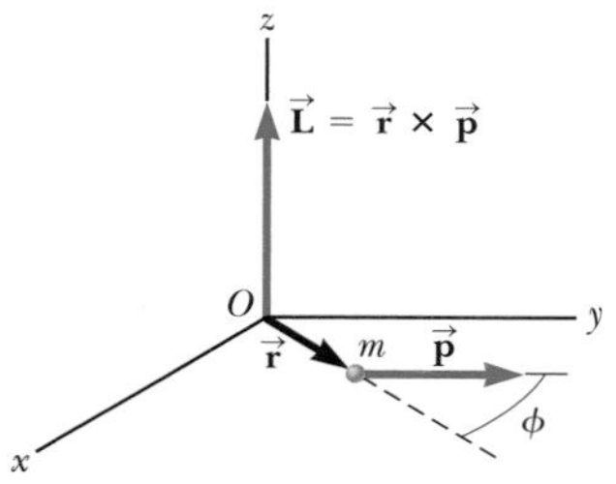

Active Figure 10.12
The angular momentum $\vec{\mathbf{L}}$ of a particle is a vector given by $\vec{\mathbf{L}} = \vec{\mathbf{r}} \times \vec{\mathbf{p}}$.

The right side of this equation is the derivative of $\vec{\mathbf{r}} \times \vec{\mathbf{p}}$ (see Appendix B.4). Therefore,

$$\sum\vec{\boldsymbol{\tau}} = \frac{d(\vec{\mathbf{r}} \times \vec{\mathbf{p}})}{dt} \quad (10.11)$$

which looks very similar in form to **Equation 8.3**, $\sum\vec{\mathbf{F}} = d\vec{\mathbf{p}}/dt$. Because torque plays the same role in rotational motion that force plays in translational motion, this result suggests that the combination $\vec{\mathbf{r}} \times \vec{\mathbf{p}}$ should play the same role in rotational motion that $\vec{\mathbf{p}}$ plays in translational motion. We call this combination the *angular momentum* of the particle.

The instantaneous angular momentum $\vec{\mathbf{L}}$ of a particle relative to an axis through the origin O is defined by the cross product of the particle's instantaneous position vector $\vec{\mathbf{r}}$ and its instantaneous linear momentum $\vec{\mathbf{p}}$:

Angular momentum of a particle ▶

$$\vec{\mathbf{L}} = \vec{\mathbf{r}} \times \vec{\mathbf{p}} \quad (10.12)$$

We can now write **Equation 10.11** as

Newton's Second Law for rotation ▶

$$\sum\vec{\boldsymbol{\tau}} = \frac{d\vec{\mathbf{L}}}{dt} \quad (10.13)$$

which is the rotational analogue of Newton's second law, $\sum\vec{\mathbf{F}} = d\vec{\mathbf{p}}/dt$. Torque causes the angular momentum $\vec{\mathbf{L}}$ to change just as force causes the linear momentum $\vec{\mathbf{p}}$ to change.

Notice that **Equation 10.13** is valid only if $\sum\vec{\boldsymbol{\tau}}$ and $\vec{\mathbf{L}}$ are measured about the same axis. Furthermore, the expression is valid for any axis fixed in an inertial frame.

The SI unit of angular momentum is kg.m^2/s and it has dimensions ML^2T^{-1}. Notice also that both the magnitude and the direction of $\vec{\mathbf{L}}$ depend on the choice of axis. Following the right-hand rule, we see that the direction of $\vec{\mathbf{L}}$ is perpendicular to the plane formed by $\vec{\mathbf{r}}$ and $\vec{\mathbf{p}}$. In **Active Figure 10.12**, $\vec{\mathbf{r}}$ and $\vec{\mathbf{p}}$ are in the xy plane, so $\vec{\mathbf{L}}$ points in the z direction. Because $\vec{\mathbf{p}} = m\vec{\mathbf{v}}$, the magnitude of $\vec{\mathbf{L}}$ is

$$L = mvr\sin\phi \qquad (10.14)$$

where ϕ is the angle between $\vec{\mathbf{r}}$ and $\vec{\mathbf{p}}$. It follows that L is zero when $\vec{\mathbf{r}}$ is parallel or antiparallel to $\vec{\mathbf{p}}$ ($\phi = 0$ or 180°). In other words, when the translational velocity of the particle is along a line that passes through the axis, the particle has zero angular momentum with respect to that axis. On the other hand, if $\vec{\mathbf{r}}$ is perpendicular to $\vec{\mathbf{p}}$ ($\phi = 90°$), then $L = mvr$.

Pitfall Prevention 10.2

We can define angular momentum even if the particle is not moving in a circular path. Even a particle moving in a straight line has angular momentum about any axis displaced from the path of the particle.

Quick Quiz 10.2

Recall the skater described at the beginning of this section. Let her mass be m. **(i)** What would be her angular momentum relative to the pole at the instant she is a distance d from the pole if she were skating directly toward it at speed v? **(a)** zero **(b)** mvd **(c)** impossible to determine **(ii)** What would be her angular momentum relative to the pole at the instant she is a distance d from the pole if she were skating at speed v along a straight path that is a perpendicular distance a from the pole? **(a)** zero **(b)** mvd **(c)** mva **(d)** impossible to determine

Example 10.6

Angular momentum in totem tennis

In a game of 'totem tennis' or 'swing ball' a tennis ball on the end of a light string is attached to a freely rotating bearing at the top of a vertical pole. A player hits the ball, causing it to move at a velocity $\vec{\mathbf{v}}$ in a circle of radius r about the pole. Find the magnitude and direction of its angular momentum about the pole.

Corbis/Ocean

Solution

Conceptualise We begin by drawing a diagram as usual. The axis of rotation (the pole) passes through point O at the centre of the circle. We model the ball as a particle.

Model We model the ball as a particle moving in uniform circular motion about the axis through O, and use the definition of angular momentum for a particle.

Analyse Use **Equation 10.14** to evaluate the magnitude of $\vec{\mathbf{L}}$:

$$L = mvr\sin 90° = mvr$$

This value of L is constant if all three factors on the right are constant. The direction of $\vec{\mathbf{L}}$ also is constant, even though the direction of $\vec{\mathbf{p}} = m\vec{\mathbf{v}}$ keeps changing. To verify this statement, apply the right-hand rule to find the direction of $\vec{\mathbf{L}} = \vec{\mathbf{r}} \times \vec{\mathbf{p}} = m\vec{\mathbf{r}} \times \vec{\mathbf{v}}$ in **Figure 10.13**. Your thumb points out of the page, so that is the direction of $\vec{\mathbf{L}}$. Hence, we can write the vector expression $\vec{\mathbf{L}} = mvr\hat{\mathbf{k}}$. If the particle were to move clockwise, $\vec{\mathbf{L}}$ would point downwards, and into the page and $\vec{\mathbf{L}} = -mvr\hat{\mathbf{k}}$. A particle in uniform circular motion has a constant angular momentum about an axis through the centre of its path.

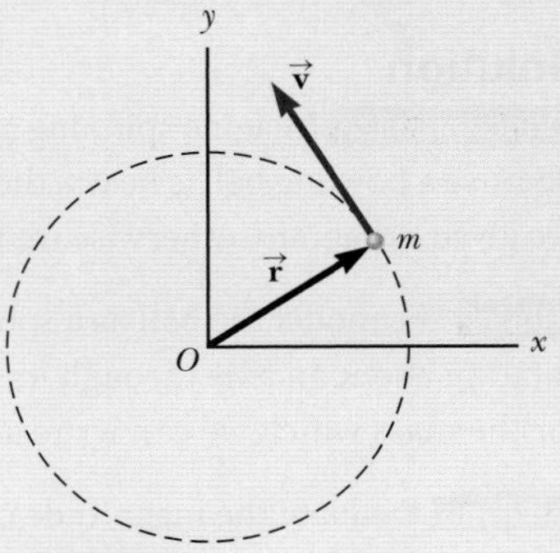

Figure 10.13 (Example 10.6) A particle moving in a circle of radius r has an angular momentum about an axis through O that has magnitude mvr. The vector $\vec{\mathbf{L}} = \vec{\mathbf{r}} \times \vec{\mathbf{p}}$ points *out* of the page.

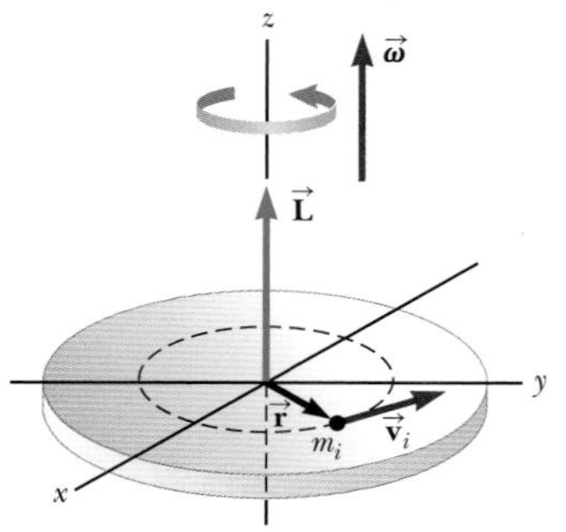

Figure 10.14
When a rigid object rotates about an axis, the angular momentum $\vec{L}$ is in the same direction as the angular velocity $\vec{\omega}$ according to the expression $\vec{L} = I\vec{\omega}$.

We now extend our discussion of angular momentum to extended objects. Consider a rigid object rotating about a fixed axis that coincides with the z axis of a coordinate system as shown in **Figure 10.14**. Let's determine the angular momentum of this object. Each *particle* of the object rotates in the xy plane about the z axis with an angular speed ω. The magnitude of the angular momentum of a particle of mass m_i about the z axis is $m_i v_i r_i$. Because $v_i = r_i\omega$ (**Equation 9.10**), we can express the magnitude of the angular momentum of this particle as

$$L_i = m_i r_i^2 \omega$$

The vector $\vec{L}_i$ for this particle is directed along the z axis, as is the vector $\vec{\omega}$.

We can now find the angular momentum (which in this situation has only a z component) of the whole object by taking the sum of L_i over all particles:

$$L_z = \sum_i L_i = \sum_i m_i r_i^2 \omega = \left(\sum_i m_i r_i^2\right)\omega$$

$$L_z = I\omega \qquad (10.15)$$

where $\sum_i m_i r_i^2$ is the moment of inertia I of the object about the z axis (**Equation 9.22**).

If a symmetrical object rotates about a fixed axis passing through its centre of mass, you can write **Equation 10.15** in vector form as $\vec{L} = I\vec{\omega}$, where $\vec{L}$ is the total angular momentum of the object measured with respect to the axis of rotation. Furthermore, this expression is valid for any object, regardless of its symmetry, if $\vec{L}$ represents the component of angular momentum along the axis of rotation.

> **Pitfall Prevention 10.3**
> Use an axis passing through the centre of mass. If a rigid object rotates about an *arbitrary* axis, then $\vec{L}$ and $\vec{\omega}$ may point in different directions. Strictly speaking, $\vec{L} = I\vec{\omega}$ applies only to rigid objects of any shape that rotate about an axis passing through the centre of mass.

Quick Quiz 10.3

A solid sphere and a hollow sphere have the same mass and radius. They are rotating with the same angular speed. Which one has the higher angular momentum? **(a)** the solid sphere **(b)** the hollow sphere **(c)** both have the same angular momentum **(d)** impossible to determine

±? Example 10.7

Bowling ball

Calculate the magnitude of the angular momentum of a bowling ball spinning at 10 rev/s as shown in **Figure 10.15**. An adult male typically uses a bowling ball with a mass between 6.4 kg and 7.2 kg, and a radius of 10.9 cm (regulation size).

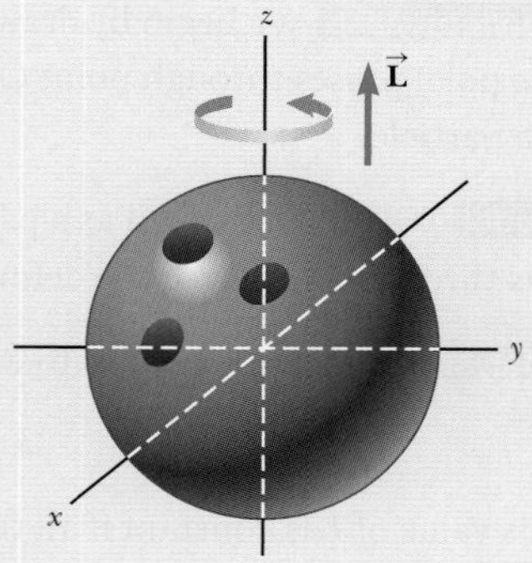

Figure 10.15
(Example 10.7) A ten-pin bowling ball that rotates about the z axis in the direction shown has an angular momentum $\vec{L}$ in the positive z direction. If the direction of rotation is reversed, then $\vec{L}$ points in the negative z direction.

Solution

Conceptualise Imagine spinning a bowling ball on the smooth floor of a bowling alley. Because a bowling ball is heavy, the angular momentum should be relatively large. We are given a diagram, otherwise we would begin by drawing one.

Model We model the ball as a sphere of uniform density (ignoring the finger holes), rotating about an axis through its centre of mass. We have a range of possible values for the mass which we can write as $M = (6.8 \pm 0.4\text{ kg})$.

Analyse Evaluate the magnitude of the angular momentum from **Equation 10.15**, using the moment of inertia of the ball about an axis through its centre from **Table 9.2**:

$$L_z = I\omega = \frac{2}{5}MR^2\omega = \frac{2}{5}(6.8\text{ kg})(0.109\text{ m})^2(10\text{ rev/s})(2\pi\text{ rad/rev}) = 2.03\text{ kg.m}^2\text{/s}$$

Example 10.7 cont.

with a range given by:

$$\Delta L_z = L_z\left(\frac{\Delta M}{M}\right) = 2.03\ \text{kg.m}^2\text{/s}\left(\frac{0.4\ \text{kg}}{6.8\ \text{kg}}\right) = 0.12\ \text{kg.m}^2\text{/s}$$

So we can write our value for angular momentum as

$$L_z = (2.03 \pm 0.12)\ \text{kg.m}^2\text{/s}$$

Table 10.1
Useful equations in rotational and translational motion

Rotational motion about a fixed axis	**Translational motion**
Angular speed $\omega = d\theta/dt$	Translational speed $v = dx/dt$
Angular acceleration $\alpha = d\omega/dt$	Translational acceleration $a = dv/dt$
Net torque $\sum\tau_{\text{ext}} = I\alpha$	Net force $\sum F = ma$
If $\alpha = \text{constant}\begin{cases}\omega_f = \omega_i + \alpha t\\ \theta_f = \theta_i + \omega_i t + \frac{1}{2}\alpha t^2\\ \omega_f^2 = \omega_i^2 + 2\alpha(\theta_f - \theta_i)\end{cases}$	If $a = \text{constant}\begin{cases}v_f = v_i + at\\ x_f = x_i + v_i t + \frac{1}{2}at^2\\ v_f^2 = v_i^2 + 2a(x_f - x_i)\end{cases}$
Work $W = \int_{\theta_i}^{\theta_f}\tau\, d\theta$	Work $W = \int_{x_i}^{x_f}F_x\, dx$
Rotational kinetic energy $K_R = \frac{1}{2}I\omega^2$	Kinetic energy $K = \frac{1}{2}mv^2$
Power $P = \tau\omega$	Power $P = Fv$
Angular momentum $L = I\omega$	Linear momentum $p = mv$
Net torque $\sum\tau = dL/dt$	Net force $\sum F = dp/dt$

We have now developed a rotational equivalent for all of the equations developed in Chapters 2 to 8 for translational motion. **Table 10.1** lists the various equations we have discussed pertaining to rotational motion together with the analogous expressions for translational motion. Notice the similar mathematical forms of the equations.

10.4 Analysis model: non-isolated system (angular momentum)

Newton's second law for a system of particles is

$$\sum\vec{\mathbf{F}}_{\text{ext}} = \frac{d\vec{\mathbf{p}}_{\text{tot}}}{dt}$$

This equation states that the net external force on a system of particles is equal to the time rate of change of the total linear momentum of the system.

The total angular momentum of a system of particles about some axis is defined as the vector sum of the angular momentum of the individual particles:

$$\vec{\mathbf{L}}_{\text{tot}} = \vec{\mathbf{L}}_1 + \vec{\mathbf{L}}_2 + \cdots \vec{\mathbf{L}}_n = \sum_i \vec{\mathbf{L}}_i$$

where the vector sum is over all n particles in the system.

Differentiating this equation with respect to time gives

$$\frac{d\vec{\mathbf{L}}_{\text{tot}}}{dt} = \sum_i \frac{d\vec{\mathbf{L}}_i}{dt} = \sum_i \vec{\boldsymbol{\tau}}_i$$

where we have used **Equation 10.13** to replace the time rate of change of the angular momentum of each particle with the net torque on the particle.

The torques acting on the particles of the system are those associated with internal forces between particles *and* those associated with external forces. The net torque associated with all internal forces, however, is zero. Recall that Newton's third law tells us that internal forces between particles of the system are equal in magnitude and opposite in direction. If we assume these forces lie along the line of separation of each pair of particles, the total torque around some axis passing through an origin O due to each action–reaction force pair is zero (that is, the lever arm d from O to the line of action of the forces is equal for both particles and the forces are in opposite directions). In the summation, therefore, the net internal torque is zero. We conclude that the total angular momentum of a system can vary with time only if a net external torque is acting on the system:

Net external ► torque on a system

$$\sum \vec{\boldsymbol{\tau}}_{\text{ext}} = \frac{d\vec{\mathbf{L}}_{\text{tot}}}{dt} \quad (10.16)$$

This equation is indeed the rotational analogue of $\sum \vec{\mathbf{F}}_{\text{ext}} = d\vec{\mathbf{p}}_{\text{tot}}/dt$ for a system of particles. **Equation 10.16** is the mathematical representation of the **angular momentum version of the non-isolated system model**. If a system is non-isolated in the sense that there is a net torque on it, the torque is equal to the time rate of change of angular momentum.

Although we do not prove it here, this statement is true regardless of the motion of the centre of mass. It applies even if the centre of mass is accelerating, provided the torque and angular momentum are evaluated relative to an axis through the centre of mass.

Equation 10.16 can be rearranged and integrated to give

$$\int_{t_i}^{t_f} \left(\sum \vec{\boldsymbol{\tau}}_{\text{ext}}\right) dt = \Delta \vec{\mathbf{L}}_{\text{tot}}$$

This equation is the rotational analogue to **Equation 8.40**. It represents the *angular impulse–angular momentum theorem.*

Consider again **Equation 10.15**, $L_z = I\omega$. If we differentiate **Equation 10.15** with respect to time, noting that I is constant for a rigid object, we get:

Rotational ► form of Newton's second law

$$\frac{dL_z}{dt} = I\frac{d\omega}{dt} = I\alpha \quad (10.17)$$

where α is the angular acceleration relative to the axis of rotation. Because dL_z/dt is equal to the net external torque (see **Equation 10.16**), we can express **Equation 10.17** as

$$\sum \tau_{\text{ext}} = I\alpha \quad (10.18)$$

That is, the net external torque acting on a rigid object rotating about a fixed axis equals the moment of inertia about the rotation axis multiplied by the object's angular acceleration relative to that axis. This result is the same as **Equation 9.27**, which was derived using a force approach, but we derived **Equation 10.18** using the concept of angular momentum. As we saw in Section 9.6, **Equation 10.18** is the mathematical representation of the rigid object subject to a net torque analysis model. This equation is also valid for a rigid object rotating about a moving axis, provided the moving axis (1) passes through the centre of mass and (2) is a symmetry axis.

Pitfall Prevention 10.4

We need to be careful to distinguish between internal and external torques. Equations 10.17 and 10.18 describe the effect of an *external* torque on a system. Internal torques may change the angular momentum of *parts* of a system, but the total angular momentum of a *system* remains constant unless there is an external torque acting.

Analysis Model 10.1

Nonisolated system (angular momentum)

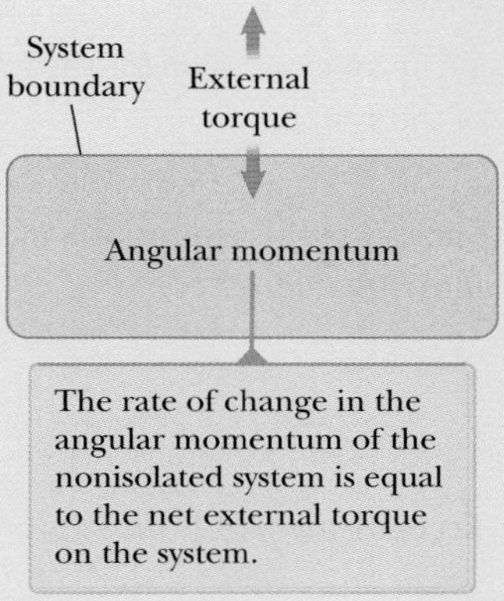
System boundary
External torque
Angular momentum
The rate of change in the angular momentum of the nonisolated system is equal to the net external torque on the system.

Imagine a system that rotates about an axis. If there is a net external torque acting on the system, the time rate of change of the angular momentum of the system is equal to the net external torque:

$$\sum \vec{\tau}_{ext} = \frac{d\vec{L}_{tot}}{dt} \quad (10.16)$$

A net external torque causes an angular acceleration given by:

$$\sum \tau_{ext} = I\alpha \quad (10.18)$$

Examples:

- a flywheel in an car engine increases its angular momentum when the engine applies torque to it
- the tub of a washing machine decreases in angular momentum due to frictional torque after the machine is turned off
- the axis of the Earth undergoes a precessional motion due to the torque exerted on the Earth by the gravitational force from the Sun
- the armature of a motor increases its angular momentum due to the torque exerted by a surrounding magnetic field (Chapter 31)

Example 10.8

A system of objects

A sphere of mass m_1 and a block of mass m_2 are connected by a light cord that passes over a pulley as shown in **Figure 10.16**. The radius of the pulley is R and the mass of the thin rim is M. The spokes of the pulley have negligible mass. The block slides on a smooth horizontal surface. Find an expression for the linear acceleration of the two objects, using the concepts of angular momentum and torque.

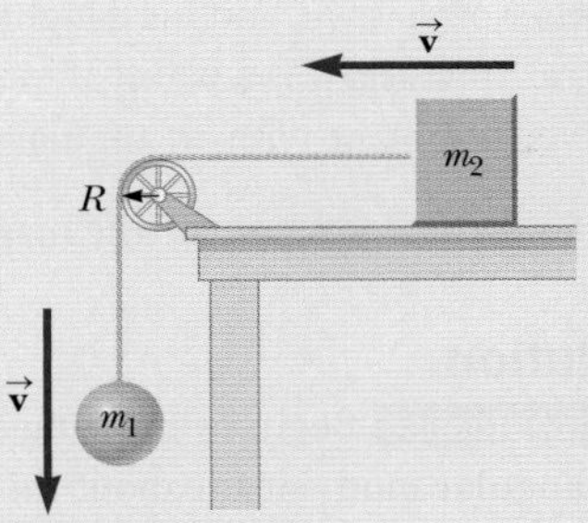
$\vec{v}$
m_2
R
$\vec{v}$
m_1

Figure 10.16
(Example 10.8) When the system is released, the sphere moves downwards and the block moves to the left.

Solution

Conceptualise When the system is released, the block slides to the left, the sphere drops downwards and the pulley rotates anticlockwise. This situation is similar to problems we have solved earlier except that now we want to use an angular momentum approach.

Model We model the block, pulley and sphere as a non-isolated system, subject to the external torque due to the gravitational force on the sphere. We model the pulley as a ring of radius R, and make the approximation that the smooth surface is frictionless. We calculate the angular momentum about an axis that coincides with the axle of the pulley. The angular momentum of the system includes that of two objects moving translationally (the sphere and the block) and one object undergoing pure rotation (the pulley).

Analyse At any instant of time, the sphere and the block have a common speed v, and the perpendicular distance between the vector $\vec{v}$ and the axis is R for both, so the angular momentum of the sphere is m_1vR and that of the block is m_2vR. At the same instant, all points on the rim of the pulley also move with speed v, so the angular momentum of the pulley is MvR.

Now consider the external torque acting on the system about the pulley axle. The force exerted by the axle on the pulley does not contribute to the torque, because it has a moment arm of zero. The normal force acting on the block is balanced by the gravitational force $m_2\vec{g}$, so these forces do not contribute to the torque. The gravitational force $m_1\vec{g}$ acting

Example 10.8 cont.

on the sphere produces a torque about the axle equal in magnitude to m_1gR, where R is the moment arm of the force about the axle. This result is the total external torque about the pulley axle, that is, $\sum\tau_{ext} = m_1gR$.

Write an expression for the total angular momentum of the system:

$$L = m_1vR + m_2vR + MvR = (m_1 + m_2 + M)vR \quad (1)$$

Substitute this expression and the total external torque into **Equation 10.16**:

$$\sum\tau_{ext} = \frac{dL}{dt}$$

$$m_1gR = \frac{d}{dt}[(m_1 + m_2 + M)\,vR]$$

$$m_1gR = (m_1 + m_2 + M)\,R\frac{dv}{dt} \quad (2)$$

Recognising that $dv/dt = a$, solve **Equation (2)** for a:

$$a = \frac{m_1g}{m_1 + m_2 + M} \quad (3)$$

Check dimensions:

$$[LT^{-2}] = [M][LT^{-2}]/[M] = [LT^{-2}]$$ ☺

Finalise When we evaluated the net torque about the axle, we did not include the forces that the cord exerts on the objects because these forces are internal to the system under consideration. Instead, we analysed the system as a whole. Only *external* torques contribute to the change in the system's angular momentum.

Consider an idealised, massless pulley, which is the limit that $M \to 0$. **Equation (3)** now gives $a = m_1g/(m_1 + m_2)$. Now go back to **Equation (6)** in Example 4.10 and let $\theta \to 0$. Does this limiting case give the expected result?

Example 10.9

The seesaw revisited

A father of mass m_f and his daughter of mass m_d sit on opposite ends of a seesaw at equal distances from the pivot at the centre. The seesaw is a plank of mass M and length ℓ and is smoothly pivoted. At a given moment, the combination rotates in a vertical plane with an angular speed ω.

(A) Find an expression for the magnitude of the system's angular momentum.

Solution

Conceptualise Begin by drawing a diagram and identify the z axis through O as the axis of rotation. The rotating system has angular momentum about this axis.

Model If we ignore any movement of arms or legs of the father and daughter we can model them both as particles. We model the plank as a rigid rod, and make the approximation that the pivot is frictionless. The system is therefore modelled as a rigid object, consisting of a rod with a particle at each end, rotating about the centre of the rod.

Analyse The moment of inertia of the system equals the sum of the moments of inertia of the three components: the seesaw and the two individuals. We can refer to **Table 9.2** to obtain the expression for the moment of inertia of the rod and use the particle expression $I = mr^2$ for each person.

Find the total moment of inertia of the system about the z axis through O:

$$I = \frac{1}{12}M\ell^2 + m_f\left(\frac{\ell}{2}\right)^2 + m_d\left(\frac{\ell}{2}\right)^2 = \frac{\ell^2}{4}\left(\frac{M}{3} + m_f + m_d\right)$$

Find the magnitude of the angular momentum of the system:

$$L = I\omega = \frac{\ell^2}{4}\left(\frac{M}{3} + m_f + m_d\right)\omega$$

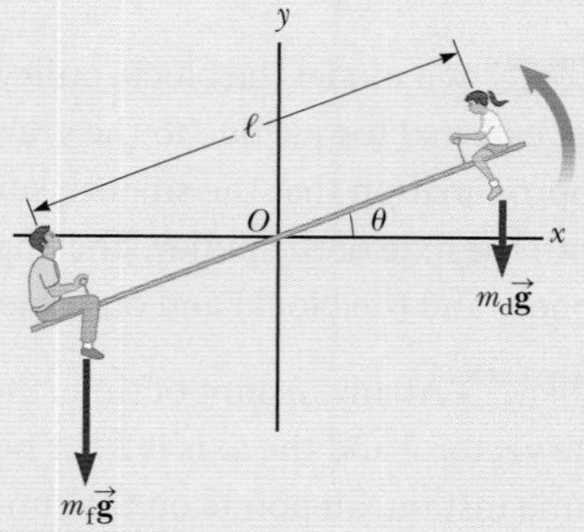

Figure 10.17
(Example 10.9) A father and daughter demonstrate angular momentum on a seesaw.

Example 10.9 cont.

(B) Find an expression for the magnitude of the angular acceleration of the system when the seesaw makes an angle θ with the horizontal.

Solution

Conceptualise Generally, fathers are more massive than daughters, so the system is not in equilibrium and has an angular acceleration. We expect the angular acceleration to be positive (anticlockwise) in Figure 10.17.

Model We model the system as non-isolated because of the external torque associated with the gravitational force. We again identify the axis of rotation as the z axis in Figure 10.17.

Analyse To find the angular acceleration of the system at any angle θ, we first calculate the net torque on the system and then use $\sum \tau_{ext} = I\alpha$ from the rigid object subject to a net torque model to obtain an expression for α.

Evaluate the torque due to the gravitational force on the father: $\tau_f = m_f g \frac{\ell}{2} \cos\theta$ ($\vec{\boldsymbol{\tau}}_f$ out of the page)

Evaluate the torque due to the gravitational force on the daughter: $\tau_d = -m_d g \frac{\ell}{2} \cos\theta$ ($\vec{\boldsymbol{\tau}}_d$ out of the page)

Evaluate the net external torque exerted on the system: $\sum \tau_{ext} = \tau_f + \tau_d = \frac{1}{2}(m_f - m_d) g \ell \cos\theta$

Use Equation 10.18 and I from part (A) to find a: $\alpha = \frac{\sum \tau_{ext}}{I} = \frac{2(m_f - m_d) g \cos\theta}{\ell[(M/3) + m_f + m_d]}$

Check dimensions: $[T^{-2}] = [M][LT^{-2}]/[L][M] = [T^{-2}]$ ☺

Finalise For a father more massive than his daughter, the angular acceleration is positive as expected. If the seesaw begins in a horizontal orientation ($\theta = 0$) and is released, the rotation is anticlockwise in Figure 10.17 and the father's end of the seesaw drops, which is consistent with everyday experience. If the daughter was heavier than her father then the rotation would be in the opposite direction.

10.5 Analysis model: isolated system (angular momentum)

In Chapter 8, we found that the total linear momentum of a system of particles remains constant if the system is isolated, that is, if the net external force acting on the system is zero. We have an analogous conservation law in rotational motion:

> The total angular momentum of a system is constant in both magnitude and direction if the net external torque acting on the system is zero, that is, if the system is isolated.

◀ Conservation of angular momentum

This statement is often called the principle of **conservation of angular momentum** and is the basis of the **angular momentum version of the isolated system model**. This principle follows directly from Equation 10.16, which indicates that if

$$\sum \vec{\boldsymbol{\tau}}_{ext} = \frac{d\vec{\mathbf{L}}_{tot}}{dt} = 0 \quad (10.19)$$

then

$$\vec{\mathbf{L}}_{tot} = \text{constant} \quad \text{or} \quad \vec{\mathbf{L}}_i = \vec{\mathbf{L}}_f \quad \text{or} \quad \Delta\vec{\mathbf{L}}_{tot} = 0 \quad (10.20)$$

In fact Equation 10.16 is the most general expression of conservation of angular momentum because it describes how the system interacts with its environment.

For an isolated system consisting of a number of particles, we write this conservation law as $\vec{\mathbf{L}}_{tot} = \sum \vec{\mathbf{L}}_n =$ constant, where the index n denotes the nth particle in the system.

If an isolated rotating system is deformable so that its mass undergoes redistribution in some way, the system's moment of inertia changes. Because the magnitude of the angular momentum of the system is $L = I\omega$ (Equation 10.15), conservation of angular momentum requires that the product of I and ω must remain constant. Therefore, a change in I for an isolated system requires a change in ω. In this case, we can express the principle of conservation of angular momentum as

$$I_i\omega_i = I_f\omega_f = \text{constant} \tag{10.21}$$

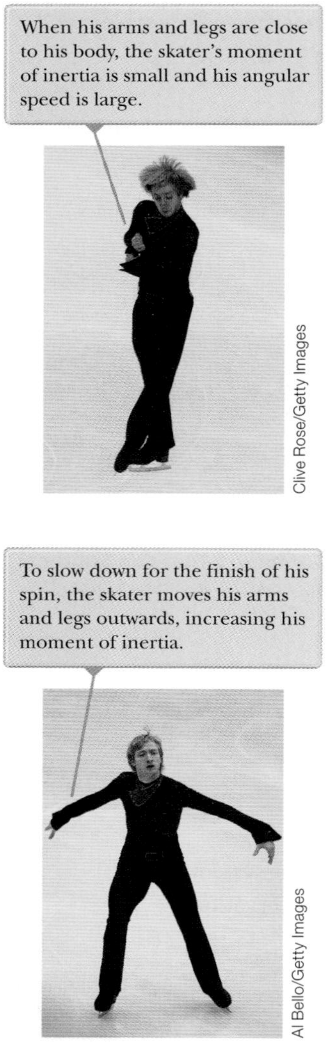

When his arms and legs are close to his body, the skater's moment of inertia is small and his angular speed is large.
Clive Rose/Getty Images
To slow down for the finish of his spin, the skater moves his arms and legs outwards, increasing his moment of inertia.
Al Bello/Getty Images

Figure 10.18
Angular momentum is conserved as Russian gold medallist Evgeni Plushenko performs during the Turin 2006 Winter Olympic Games.

This expression is valid both for rotation about a fixed axis and for rotation about an axis through the centre of mass of a moving system as long as that axis remains fixed in direction. We require only that the net external torque be zero.

Many examples demonstrate conservation of angular momentum for a deformable system. You may have observed a figure skater spinning (Figure 10.18). The angular speed of the skater is large when his hands and feet are close to his body. Ignoring friction between skater and ice, there are no external torques on the skater. The moment of inertia of his body increases as his hands and feet are moved away from his body at the finish of the spin. According to the principle of conservation of angular momentum, his angular speed must decrease. In a similar way, when divers or acrobats wish to make several somersaults, they pull their hands and feet close to their bodies to rotate at a higher rate. In these cases, the external force due to gravity acts through the centre of mass and hence exerts no torque about an axis through this point. Therefore, the angular momentum about the centre of mass must be conserved, that is, $I_i\omega_i = I_f\omega_f$.

In Equation 10.20, we have a third version of the isolated system model. We can now state that the energy, linear momentum and angular momentum of an isolated system are all constant:

$E_i = E_f$ (if there are no energy transfers across the system boundary)

$\vec{\mathbf{p}}_i = \vec{\mathbf{p}}_f$ (if the net external force on the system is zero)

$\vec{\mathbf{L}}_i = \vec{\mathbf{L}}_f$ (if the net external torque on the system is zero)

Energy, momentum and angular momentum are all conserved quantities and it is always possible to define the boundaries of a system such that it is isolated and these quantities are constant.

A system may be isolated in terms of one of these quantities but not in terms of another. If a system is non-isolated in terms of momentum or angular momentum, often it will be non-isolated also in terms of energy because the system has a net force or torque on it and the net force or torque will do work on the system. We can, however, identify systems that are non-isolated in terms of energy but isolated in terms of momentum. For example, imagine pushing inwards on a balloon (the system) between your hands. Work is done in compressing the balloon, so the system is non-isolated in terms of energy, but there is zero net force on the system, so the system is isolated in terms of momentum. A similar statement could be made about twisting the ends of a long, springy piece of metal with both hands. Work is done on the metal (the system), so energy is stored in the non-isolated system as elastic potential energy, but the net torque on the system is zero. Therefore, the system is isolated in terms of angular momentum.

TRY THIS

Sit on a rotating chair such as a computer chair, holding a heavy object in each hand – such as your copy of a physics textbook in one hand and a friend's copy in the other. Keep your arms tucked in. Get a friend to spin you slowly on the chair then let go. Slowly extend your arms out, then bring them back in again. What happens and why? Repeat the experiment, but this time drop the books when your arms are extended. Why does this make no difference to your angular velocity?
Caution: Only spin slowly and always start with your arms in; it is dangerous to start with your arms out and pull them in because you may fall off the chair and be injured!

Quick Quiz 10.4

A competitive diver leaves the diving board and falls towards the water with her body straight and rotating slowly. She pulls her arms and legs into a tight tuck position. What happens to her rotational kinetic energy? **(a)** it increases **(b)** it decreases **(c)** it stays the same **(d)** it is impossible to determine

Analysis Model 10.2

Isolated system (angular momentum)

Imagine a system rotates about an axis. If there is no net external torque on the system, there is no change in the angular momentum of the system:

$$\Delta\vec{L}_{tot} = 0 \quad (10.20)$$

Applying this law of conservation of angular momentum to a system whose moment of inertia changes gives

$$I_i\omega_i = I_f\omega_f = \text{constant} \quad (10.21)$$

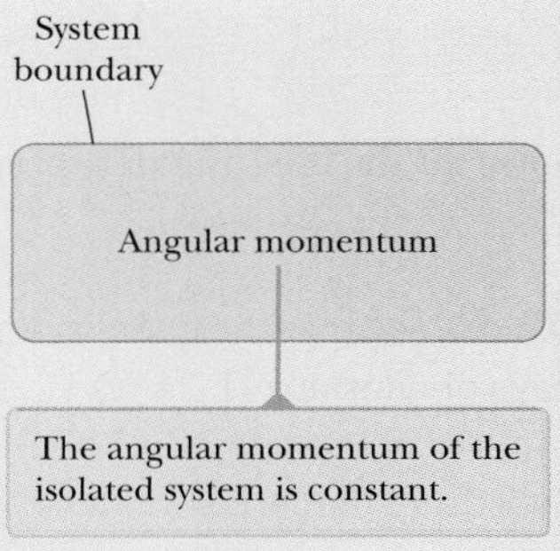
System boundary
Angular momentum
The angular momentum of the isolated system is constant.

Examples:

- after a supernova explosion, the core of a star collapses to a small radius and spins at a much higher rate
- the square of the orbital period of a planet is proportional to the cube of its semimajor axis; Kepler's third law (Chapter 11)
- in atomic transitions, selection rules on the quantum numbers must be obeyed in order to conserve angular momentum (Chapter 41)
- in beta decay of a radioactive nucleus, a neutrino must be emitted in order to conserve angular momentum (Chapter 43)

Example 10.10

The merry-go-round

A horizontal platform in the shape of a circular disc rotates freely in a horizontal plane about a frictionless, vertical axle (**Figure 10.19**). The platform has a mass $M = 100$ kg and a radius $R = 2.0$ m. A student whose mass is $m = 60$ kg walks slowly from the rim of the disc toward its centre. If the angular speed of the system is 2.0 rad/s when the student is at the rim, what is the angular speed when she reaches a point $r = 0.50$ m from the centre?

Solution

Conceptualise The speed change here is similar to those of the spinning skater and the *Try this* example above. This problem is different because part of the moment of inertia of the system changes (that of the student) while part remains fixed (that of the platform).

Example 10.10 cont.

Model We model the student as a particle. As the platform rotates on a frictionless axle, we model the system of the student and the platform as an isolated system in terms of angular momentum.

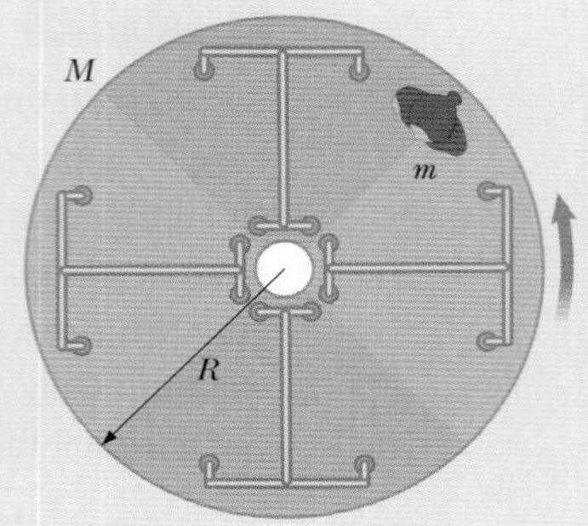

Figure 10.19
(Example 10.10) As the student walks toward the centre of the rotating platform, the angular speed of the system increases because the angular momentum of the system remains constant.

Analyse Let us denote the moment of inertia of the platform as I_p and that of the student as I_s.

Find the initial moment of inertia I_i of the system (student plus platform) about the axis of rotation:

$$I_i = I_{pi} + I_{si} = \frac{1}{2}MR^2 + mR^2$$

Find the moment of inertia of the system when the student walks to the position $r < R$:

$$I_f = I_{pf} + I_{sf} = \frac{1}{2}MR^2 + mr^2$$

Write **Equation 10.21** for the system:

$$I_i\omega_i = I_f\omega_f$$

Substitute the moments of inertia:

$$\left(\frac{1}{2}MR^2 + mR^2\right)\omega_i = \left(\frac{1}{2}MR^2 + mr^2\right)\omega_f$$

Solve for the final angular speed:

$$\omega_f = \left(\frac{\frac{1}{2}MR^2 + mR^2}{\frac{1}{2}MR^2 + mr^2}\right)\omega_i$$

We can see that the quantity in the brackets will be dimensionless, as it is a ratio of moments of inertia. Hence both sides have dimensions $[T^{-1}]$. ☺

Substitute numerical values:

$$\omega_f = \left[\frac{\frac{1}{2}(100\text{ kg})(2.0\text{ m})^2 + (60\text{ kg})(2.0\text{ m})^2}{\frac{1}{2}(100\text{ kg})(2.0\text{ m})^2 + (60\text{ kg})(0.5\text{ m})^2}\right](2.0\text{ rad/s}) = 4.1\text{ rad/s}$$

Finalise As expected, the angular speed increases. The fastest that this system could spin would be when the student moves to the centre of the platform.

What If? What if you measured the kinetic energy of the system before and after the student walks inwards? Are the initial kinetic energy and the final kinetic energy the same?

Answer You may be tempted to say yes because the system is isolated. Remember, however, that energy can be transformed among several forms, so we have to handle an energy question carefully.

Find the initial kinetic energy:

$$K_i = \frac{1}{2}I_i\omega_i^2 = \frac{1}{2}(440\text{ kg.m}^2)(2.0\text{ rad/s})^2 = 880\text{ J}$$

Find the final kinetic energy:

$$K_f = \frac{1}{2}I_f\omega_f^2 = \frac{1}{2}(215\text{ kg.m}^2)(4.1\text{ rad/s})^2 = 1.80\times10^3\text{ J}$$

Therefore, the kinetic energy of the system *increases*. The student must do work to move herself closer to the centre of rotation, so this extra kinetic energy comes from chemical potential energy in the student's body. The system is isolated in terms of energy, but a transformation process within the system changes potential energy to kinetic energy. This transformation is the result of internal forces in the system, so the total energy remains unchanged.

Example 10.11

Duck and stick collision

A 3.0 kg duck flies in to land on a wet lawn. The duck slides along the grass, striking a fallen totem-tennis pole. The pole is 2.0 m long, and has a mass of 2.0 kg. The pole is lying flat as shown in the overhead view of **Figure 10.20a**. The duck strikes at the endpoint of the pole, at a distance $r = 1.0$ m from the pole centre. Assume the collision is elastic and the duck does not deviate from its original line of motion. Find the translational speed of the duck, the translational speed of the pole and the angular speed of the pole after the collision. The moment of inertia of the pole about its centre of mass is 0.67 kg.m^2.

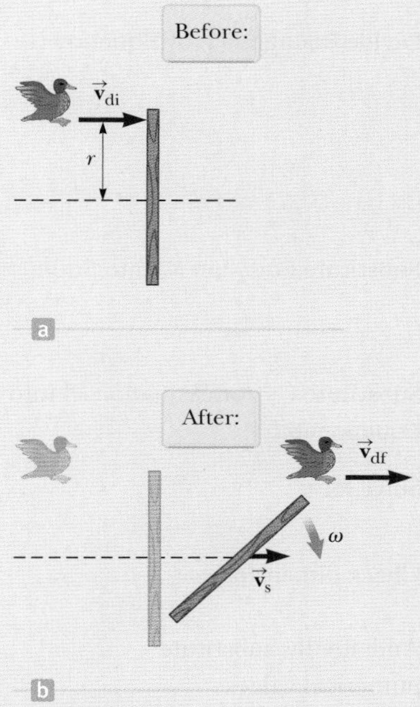

Figure 10.20
(Example 10.11) Overhead view of a duck striking a pole in an elastic collision (a) Before the collision, the duck moves toward the pole. (b) The collision causes the pole to rotate and move to the right.

Solution

Conceptualise Examine **Figure 10.20a** and imagine what happens after the duck hits the pole. **Figure 10.20b** shows what you might expect: the duck continues to move at a slower speed and the pole is in both translational and rotational motion. We assume the duck does not deviate from its original line of motion because the force exerted by the pole on the duck is parallel to the original path of the duck.

Model We model the pole as a uniform long, thin rod and the duck as a particle. We make the approximation that the wet grass is frictionless so that the duck and pole form an isolated system in terms of momentum and angular momentum. Ignoring the sound made in the collision, we also model the system as isolated in terms of energy. In addition, because the collision is assumed to be elastic, the kinetic energy of the system is constant.

Analyse First notice that we have three unknowns, so we need three equations to solve simultaneously. Apply the isolated system model for momentum to the system and then rearrange the result:

$$m_d v_{di} = m_d v_{df} + m_p v_p$$

$$m_d (v_{di} - v_{df}) = m_p v_p \qquad (1)$$

Apply the isolated system model for angular momentum to the system and rearrange the result. Use an axis passing through the centre of the pole as the rotation axis so that the path of the duck is a distance $r = 1.0$ m from the rotation axis:

$$-rm_d v_{di} = -rm_d v_{df} + I_p \omega$$

$$-rm_d (v_{di} - v_{df}) = I_p \omega \qquad (2)$$

Apply the isolated system model for energy to the system, rearrange the equation, and factor the left side:

$$\frac{1}{2} m_d v_{di}^2 = \frac{1}{2} m_d v_{df}^2 + \frac{1}{2} m_p v_p^2 + \frac{1}{2} I_p \omega^2$$

$$m_d (v_{di} - v_{df})(v_{di} + v_{df}) = m_p v_p^2 + I_p \omega^2 \qquad (3)$$

Multiply **Equation (1)** by r and add to **Equation (2)**:

$$rm_d (v_{di} - v_{df}) = rm_p v_p$$
$$-rm_d (v_{di} - v_{df}) = I_p \omega$$
$$0 = rm_p v_p + I_p \omega$$

Solve for ω:

$$\omega = -\frac{rm_p v_p}{I_p} \qquad (4)$$

Example 10.11 cont.

Check dimensions: $[T^{-1}] = [L][M][LT^{-1}]/[ML^2] = [T^{-1}]$ ☺

Divide **Equation (3)** by **Equation (1)**:

$$\frac{m_d(v_{di} - v_{df})(v_{di} + v_{df})}{m_d(v_{di} - v_{df})} = \frac{m_p v_p^2 + I_p\omega^2}{m_p v_p}$$

$$v_{di} + v_{df} = v_p + \frac{I_p\omega^2}{m_p v_p} \quad (5)$$

Substitute **Equation (4)** into **Equation (5)**:

$$v_{di} + v_{df} = v_p\left(1 + \frac{r^2 m_p}{I_p}\right) \quad (6)$$

Substitute v_{df} from **Equation (1)** into **Equation (6)**:

$$v_{di} + \left(v_{di} - \frac{m_p}{m_d}v_p\right) = v_p\left(1 + \frac{r^2 m_p}{I_p}\right)$$

Solve for v_p:

$$v_p = \frac{2v_{di}}{1 + (m_p/m_d) + (r^2 m_p/I)}$$

Check dimensions: $[LT^{-1}] = [LT^{-1}]/([M]/[M] + [L]^2[M]/[ML^2]) = [LT^{-1}]/([\,] + [\,]) = [LT^{-1}]$ ☺

And, finally, substitute numerical values:

$$v_p = \frac{2(3.0\text{ m/s})}{1 + (2.0\text{ kg}/3.0\text{ kg}) + [(1.0\text{ m})^2(1.0\text{ kg})/0.67\text{ kg.m}^2]} = 1.3\text{m/s}$$

Substitute numerical values into **Equation (4)**:

$$\omega = -\frac{(1.0\text{ m})(2.0\text{kg})(1.3\text{ m/s})}{0.67\text{ kg.m}^2} = -4.0\text{ rad/s}$$

Solve **Equation (1)** for v_{df} and substitute numerical values:

$$v_{df} = v_{di} - \frac{m_p}{m_d}v_p = 3.0\text{ m/s} - \frac{2.0\text{ kg}}{3.0\text{ kg}}(1.3\text{ m/s}) = 2.1\text{ m/s}$$

Finalise These values seem reasonable. The duck is moving more slowly after the collision than it was before the collision and the pole has a small translational speed.

10.6 The motion of gyroscopes and tops

Figure 10.21a shows a top spinning about its central axis. If the top spins rapidly, the symmetry axis rotates about the *z* axis, sweeping out a cone (see **Figure 10.21b**). The motion of the symmetry axis about the vertical – known as **precessional motion** – is usually slow relative to the spinning motion of the top. This is the motion shown in the photograph at the start of this chapter.

TRY THIS

If you do not have a spinning top, hold a coin vertically on a smooth table top and flick it horizontally at the edge with a finger so it spins about a vertical axis. (Let go of it as you flick it so it spins freely.) As it slows down you may be able to see it precessing before it falls over. How long does it take to fall down when you flick it hard so it is spinning fast? What about when you flick it gently so it spins only slowly? If you do have a spinning top, spin it and make the same observations.

Why doesn't the top fall over when it is spinning? Because the centre of mass is not directly above the pivot point *O*, a net torque is acting on the top about an axis passing through *O*, a torque resulting from the gravitational force $M\vec{\mathbf{g}}$. The top would certainly fall over if it were not spinning. Because it is spinning, however, it has an angular momentum $\vec{\mathbf{L}}$ directed along its symmetry axis. This symmetry axis moves about the z axis (precessional motion occurs) because the torque produces a change in the *direction* of the symmetry axis. This is an example of the importance of the vector nature of angular momentum.

The essential features of precessional motion can be illustrated by considering the simple gyroscope shown in **Figure 10.22a**. The two forces acting on the gyroscope are shown in **Figure 10.22b**: the downwards gravitational force $M\vec{\mathbf{g}}$ and the normal force $\vec{\mathbf{n}}$ acting upwards at the pivot point O. The normal force produces no torque about an axis passing through the pivot because its lever arm through that point is zero. The gravitational force, however, produces a torque $\vec{\boldsymbol{\tau}} = \vec{\mathbf{r}} \times M\vec{\mathbf{g}}$ about an axis passing through O, where the direction of $\vec{\boldsymbol{\tau}}$ is perpendicular to the plane formed by $\vec{\mathbf{r}}$ and $M\vec{\mathbf{g}}$. By necessity, the vector $\vec{\boldsymbol{\tau}}$ lies in a horizontal xy plane perpendicular to the angular momentum vector. The net torque and angular momentum of the gyroscope are related through **Equation 10.16**:

$$\sum \vec{\boldsymbol{\tau}}_{\text{ext}} = \frac{d\vec{\mathbf{L}}}{dt}$$

This expression shows that in the infinitesimal time interval dt, the non-zero torque produces a change in angular momentum $d\vec{\mathbf{L}}$, a change that is in the same direction as $\vec{\boldsymbol{\tau}}$. Therefore, like the torque vector, $d\vec{\mathbf{L}}$ must also be perpendicular to $\vec{\mathbf{L}}$. **Figure 10.22c** illustrates the resulting precessional motion of the symmetry axis of the gyroscope . In a time interval dt, the change in angular momentum is $d\vec{\mathbf{L}} = \vec{\mathbf{L}}_f - \vec{\mathbf{L}}_i = \vec{\boldsymbol{\tau}}dt$. Because $d\vec{\mathbf{L}}$ is perpendicular to $\vec{\mathbf{L}}$, the magnitude of $\vec{\mathbf{L}}$ does not change ($|\vec{\mathbf{L}}_i| = |\vec{\mathbf{L}}_f|$). Rather, what is changing is the *direction* of $\vec{\mathbf{L}}$. Because the change in angular momentum $d\vec{\mathbf{L}}$ is in the direction of $\vec{\boldsymbol{\tau}}$, which lies in the xy plane, the gyroscope undergoes precessional motion.

To simplify the description of the system, we assume the total angular momentum of the precessing wheel is the sum of the angular momentum $I\vec{\boldsymbol{\omega}}$ due to the spinning and the angular momentum due to the motion of the centre of mass about the pivot. In our treatment, we shall neglect the contribution from the centre of mass motion and take the total angular momentum to be simply $I\vec{\boldsymbol{\omega}}$. In practice, this approximation is good if $\vec{\boldsymbol{\omega}}$ is made very large. Hence we are using the case of a rapidly spinning object.

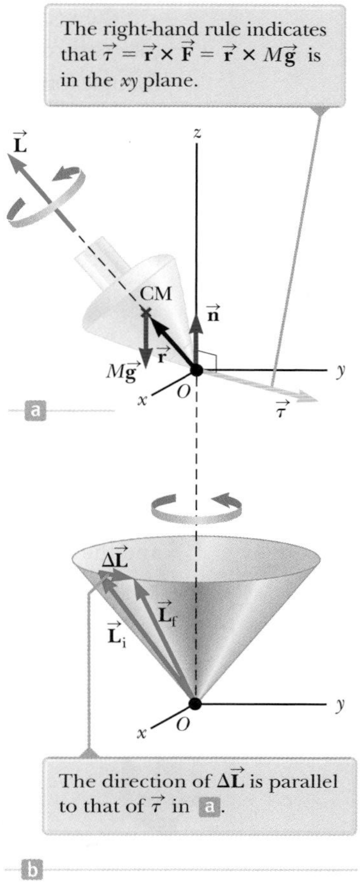
The right-hand rule indicates that $\vec{\tau} = \vec{\mathbf{r}} \times \vec{\mathbf{F}} = \vec{\mathbf{r}} \times M\vec{\mathbf{g}}$ is in the xy plane.
L
z
CM
n
r
Mg
y
x
O
τ
a
ΔL
Lf
Li
y
x
O
The direction of $\Delta\vec{\mathbf{L}}$ is parallel to that of $\vec{\tau}$ in a.
b

Figure 10.21
Precessional motion of a top spinning about its symmetry axis (a) The only external forces acting on the top are the normal force $\vec{\mathbf{n}}$ and the gravitational force $M\vec{\mathbf{g}}$. The direction of the angular momentum $\vec{\mathbf{L}}$ is along the axis of symmetry. (b) Because $\vec{\mathbf{L}}_f = \Delta\vec{\mathbf{L}} + \vec{\mathbf{L}}_i$, the top precesses about the z axis.

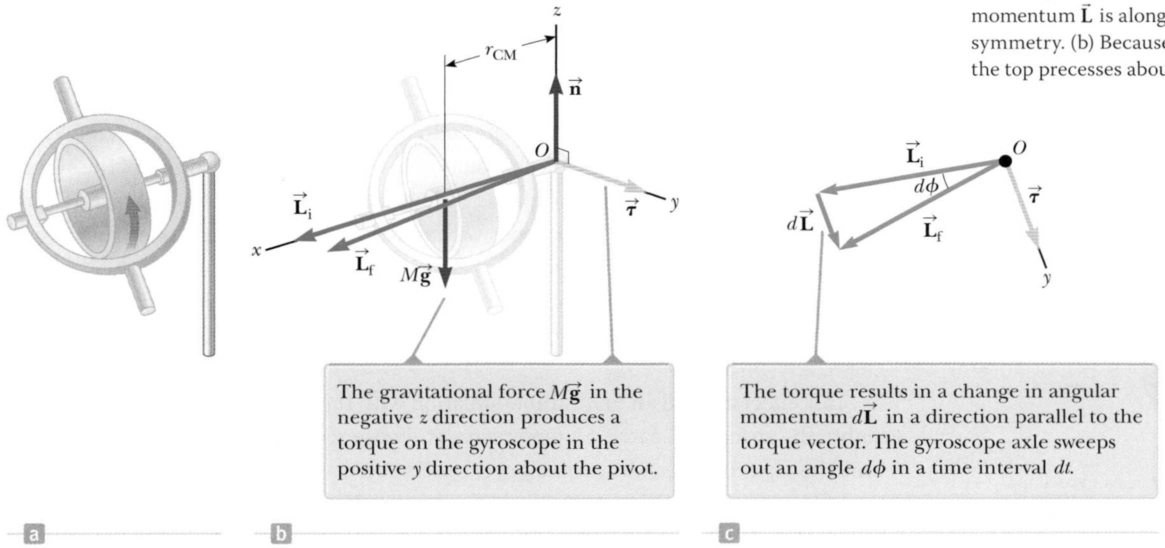
z
r_{CM}
n
O
Li
τ
y
x
Lf
Mg
The gravitational force $M\vec{\mathbf{g}}$ in the negative z direction produces a torque on the gyroscope in the positive y direction about the pivot.
Li
O
dφ
τ
dL
Lf
y
The torque results in a change in angular momentum $d\vec{\mathbf{L}}$ in a direction parallel to the torque vector. The gyroscope axle sweeps out an angle $d\phi$ in a time interval dt.
a
b
c

Figure 10.22
(a) A spinning gyroscope is placed on a pivot at the right end. (b) Diagram for the spinning gyroscope showing forces, torque, and angular momentum (c) Overhead view (looking down the z axis) of the gyroscope's initial and final angular momentum vectors for an infinitesimal time interval dt

The vector diagram in **Figure 10.22c** shows that in the time interval dt, the angular momentum vector rotates through an angle $d\phi$, which is also the angle through which the gyroscope axle rotates. From the vector triangle formed by the vectors $\vec{\mathbf{L}}_i$, $\vec{\mathbf{L}}_f$, and $d\vec{\mathbf{L}}$, we see that

$$d\phi = \frac{dL}{L} = \frac{\sum \tau_{\text{ext}} dt}{L} = \frac{(Mgr_{\text{CM}})dt}{L}$$

Dividing through by dt and using the relationship $L = I\omega$, we find that the rate at which the axle rotates about the vertical axis is

$$\omega_p = \frac{d\phi}{dt} = \frac{Mgr_{\text{CM}}}{I\omega} \qquad (10.22)$$

The angular speed ω_p is called the **precessional frequency**. This result is valid only when $\omega_p \ll \omega$. As you can see from **Equation 10.22**, the condition $\omega_p \ll \omega$ is met when ω is large, that is, when the wheel spins rapidly. This equation also tells us that the faster the spin, the slower the rate of precession of the wheel in a gyroscope. If you did the *Try this* example with a coin you would have noticed that as the coin slows down, the motion becomes unstable and the coin wobbles around and has a translational motion as well as the spin and the precession. The analysis of this part of the motion is more complex than we shall look at here.

Gyroscopes have many practical uses, particularly in navigation and aviation. The gyrocompass used in navigation on ships relies on a gyroscope to indicate direction. Gyrocompasses are used on most merchant and navy ships, and are preferable to magnetic compasses because they are not subject to errors due to local magnetic fields. They are also used in the heading indicator in aircraft, and to provide an artificial horizon indicator.

Gyroscopes are commonly used as stabilisers on autonomous vehicles, because they resist changes in orientation. They help a vehicle maintain a straight course, for example torpedoes and intercontinental ballistic missiles, and they help two-legged robots stay upright as they move.

As well as helping a vehicle stay on a straight course, a gyroscope can be used to change the direction of a vehicle's motion. Consider the spacecraft shown in **Figure 10.23**, with the gyroscope initially not rotating. In this case, the angular momentum of the spacecraft about its centre of mass is zero. Suppose the gyroscope is set into rotation, giving the gyroscope a non-zero angular momentum. There is no external torque on the isolated system (spacecraft and gyroscope), so the angular momentum of this system must remain zero according to the isolated system (angular momentum) model. The zero value can be satisfied if the spacecraft rotates in the direction opposite that of the gyroscope so that the angular momentum vectors of the gyroscope and the spacecraft cancel, resulting in no angular momentum of the system. The result of rotating the gyroscope, as in **Figure 10.23b**, is that the spacecraft turns around. By including three gyroscopes with mutually perpendicular axles, any desired rotation in space can be achieved.

Such a device is called a control moment gyroscope. Four of them are used on the international space station to hold the space station at a fixed attitude relative to the Earth.

This effect created an undesirable situation with the *Voyager 2* spacecraft during its flight. The spacecraft carried a tape recorder whose reels rotated at high speeds. Each time the tape recorder was turned on, the reels acted as gyroscopes and the spacecraft started an undesirable rotation in the opposite direction. This rotation had to be counteracted by Mission Control by using the sidewards-firing jets to *stop* the rotation!

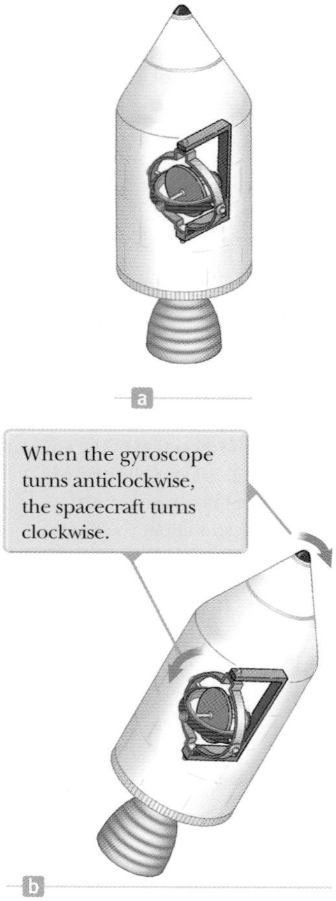

a
When the gyroscope turns anticlockwise, the spacecraft turns clockwise.
b

Figure 10.23
(a) A spacecraft carries a gyroscope that is not spinning. (b) The gyroscope is set into rotation.

End-of-chapter resources

The Celtic stone, also known as the wobble stone or rattleback toy, has an elongated semi-ellipsoidal rounded bottom and a flat top that will rotate on its axis, but only in a preferred direction. The distribution of mass is asymmetrical with respect to the axes of symmetry of the ellipsoid. When it is spun in a clockwise direction, it becomes unstable and starts to wobble, then stops and reverses the spin to its preferred anticlockwise direction. Although it seems, at first, that the Celtic stone defies conservation of angular momentum, it does not. This preferred direction of rotation is due to its offset centre of gravity and the frictional force from the table surface. The centre of mass is above the contact point on one side and below on the other, so the frictional force pushes to produce the same rotation about the centre on both sides. The friction from the table exerts a torque (Equation 10.13) on the Celtic stone that makes it rotate and acts in a preferred direction opposite to the direction of spinning. Energy is conserved and transformed into different forms. For example, the kinetic energy contained in the oscillations is transformed into rotational kinetic energy.

Shutterstock.com/Lehrer

The problems found in this chapter may be assigned online in Enhanced Web Assign.

Worked solutions to every fifth problem are available in the Student Solutions Manual. Register online at **www.cengagebrain.com** for access.

Summary

Definitions

The **angular momentum** $\vec{L}$ about an axis through the origin of a particle having linear momentum $\vec{p} = m\vec{v}$ is

$$\vec{L} = \vec{r} \times \vec{p} \qquad (10.12)$$

where $\vec{r}$ is the vector position of the particle relative to the origin.

Concepts and principles

If a rigid object rotates about a fixed axis with angular speed ω, its **rotational kinetic energy** can be written

$$K_R = \frac{1}{2}I\omega^2 \qquad (10.1)$$

where I is the moment of inertia of the object about the axis of rotation.

The rate at which work is done by an external force in rotating a rigid object about a fixed axis, or the **power** delivered, is

$$P = \tau\omega \qquad (10.3)$$

If work is done on a rigid object and the only result of the work is rotation about a fixed axis, the net work done by external forces in rotating the object equals the change in the rotational kinetic energy of the object:

$$\sum W = \frac{1}{2}I\omega_f^2 - \frac{1}{2}I\omega_i^2 \qquad (10.4)$$

The **total kinetic energy** of a rigid object rolling on a rough surface without slipping equals the rotational kinetic energy about its centre of mass plus the translational kinetic energy of the centre of mass:

$$K = \frac{1}{2}I_{CM}\omega^2 + \frac{1}{2}Mv_{CM}^2 \qquad (10.8)$$

The z component of angular momentum of a rigid object rotating about a fixed z axis is

$$L_z = I\omega \qquad (10.15)$$

where I is the moment of inertia of the object about the axis of rotation and ω is its angular speed.

Angular momentum is a conserved quantity, like energy and linear momentum.

Analysis models for problem solving

Non-isolated system (angular momentum) If a system interacts with its environment in the sense that there is an external torque on the system, the net external torque acting on a system is equal to the time rate of change of its angular momentum:

$$\sum \vec{\tau}_{ext} = \frac{d\vec{L}_{tot}}{dt} \qquad (10.16)$$

Isolated system (angular momentum) If a system experiences no external torque from the environment, the total angular momentum of the system is conserved:

$$\vec{L}_i = \vec{L}_f \text{ or } \Delta\vec{L} = 0 \qquad (10.20)$$

Applying this law of conservation of angular momentum to a system whose moment of inertia changes gives

$$I_i\omega_i = I_f\omega_f = \text{constant} \qquad (10.21)$$

Chapter review quiz

To help you revise Chapter 10: Energy and momentum in rotating systems, complete the automatically graded Chapter review quiz at http://login.cengagebrain.com.

Conceptual questions

1. Suppose you have two eggs, one hard boiled and the other raw. You wish to determine which is the hard boiled egg without breaking the eggs. This can be done by spinning the two eggs on the floor and comparing the rotational motions. (a) Which egg spins faster? Why? (b) Which egg rotates more uniformly? Why? (c) Which egg begins spinning again after being stopped and then immediately released? Why? (Hint: This is similar to the *Try this* experiment with the two cans of soup.)
2. Is it possible to change the translational kinetic energy of an object without changing its rotational energy?
3. Three objects of uniform density – a solid sphere, a solid cylinder and a hollow cylinder – are placed at the top of an incline (**Figure CQ10.3**). They are all released from rest at the same elevation and roll without slipping. (a) Which object reaches the bottom first? (b) Which reaches it last? If you did the *Try this* experiment with the full and empty cans you already know the answer to part of this question!

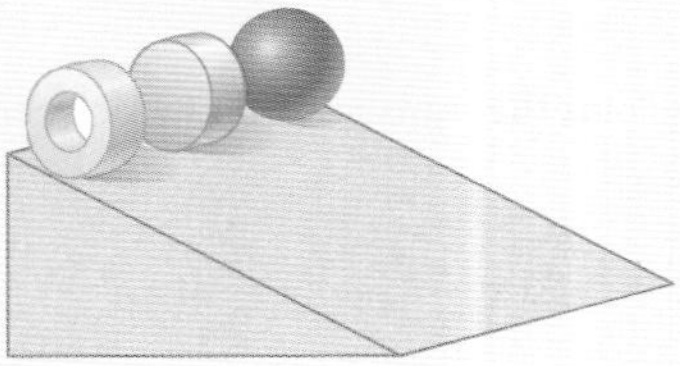

Figure CQ10.3

4. Suppose you set your textbook sliding across a smooth floor with a certain initial speed. It quickly stops moving because of a friction force exerted on it by the floor. Next, you start a basketball rolling with the same initial speed. It keeps rolling from one end of the room to the other. (a) Why does the basketball roll so far? (b) Does friction significantly affect the basketball's motion?
5. A ball is thrown in such a way that it does not spin about its own axis. Does this statement imply that the angular momentum is zero about an arbitrary axis? Explain.
6. Two naughty boys are playing with a roll of toilet paper. Laurence holds the roll between the index fingers of his hands so that it is free to rotate, and Marcus pulls at constant speed on the free end of the paper. As Marcus pulls the toilet paper, the radius of the roll of remaining paper decreases. (a) How does the torque on the roll change with time? (b) How does the angular speed of the roll change in time? (c) If Marcus suddenly jerks the end with a large force, is the sheet of toilet paper more likely to break from the others when it is being pulled from a nearly full roll or from a nearly empty roll?
7. In some motorcycle races, the riders drive over small hills and the motorcycle becomes airborne for a short time interval. If the motorcycle racer keeps the throttle open while leaving the hill and going into the air, the motorcycle tends to nose upwards. Why?
8. Stars originate as large bodies of slowly rotating gas. Because of gravity, these clumps of gas slowly decrease in size. What happens to the angular speed of a star as it shrinks? Explain why.
9. If global warming continues over the next one hundred years, it is likely that some polar ice will melt and the water will be distributed closer to the equator. (a) How would that change the moment of inertia of the Earth? (b) Would the duration of the day increase or decrease?
10. A cat usually lands on its feet regardless of the position from which it is dropped. A slow-motion film of a cat falling shows that the upper half of its body twists in one direction while the lower half twists in the opposite direction. (See **Figure CQ10.10**.) Why does this type of rotation occur?

Figure CQ10.10

Problems

Section 10.1 Rotational kinetic energy

1. The four particles in **Figure P10.1** are connected by rigid rods of negligible mass. The origin is at the centre of the rectangle. The system rotates in the xy plane about the z axis with an angular speed of 6.00 rad/s. Calculate (a) the moment of inertia of the system about the z axis and (b) the rotational kinetic energy of the system.

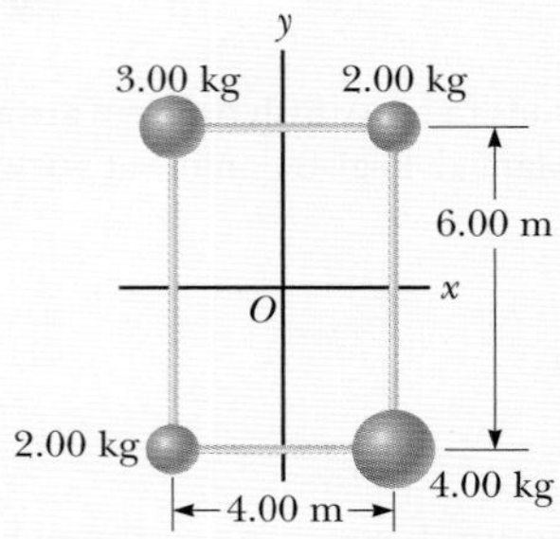

Figure P10.1

2. A *war-wolf* or *trebuchet* was a device used during the Middle Ages to throw rocks at castles and now sometimes built by engineering students just for fun. A simple trebuchet built by an engineering student is shown in **Figure P10.2**. Model it as a stiff rod of negligible mass, 3.00 m long, joining particles of mass m_1 = 0.120 kg and m_2 = 60.0 kg at its ends. It can turn on a frictionless, horizontal axle perpendicular to the rod and 14.0 cm from the large-mass particle. The operator releases the trebuchet from rest in a horizontal orientation. (a) Find the maximum speed that the small-mass object attains. (b) While the small-mass object is gaining speed, does it move with constant acceleration? (c) Does it move with constant tangential acceleration? (d) Does the trebuchet move with constant angular acceleration? (e) Does it have constant momentum? (f) Does the trebuchet–Earth system have constant mechanical energy?

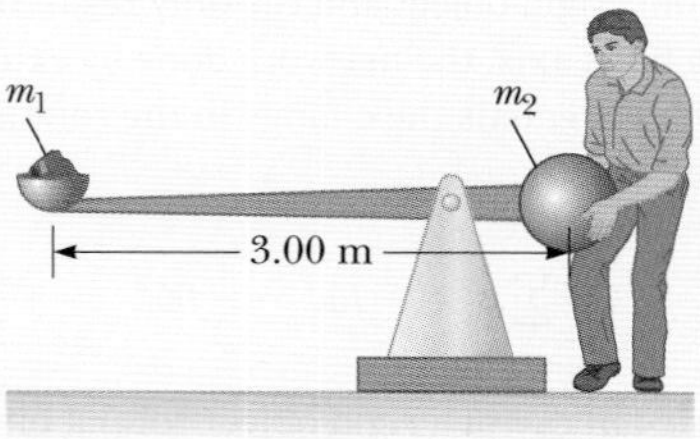

Figure P10.2

3. The Parliament tower clock in London, which houses the famous bell 'Big Ben', has an hour hand 2.70 m long with a mass of 60.0 kg and a minute hand 4.50 m long with a mass of 100 kg. Calculate the total rotational kinetic energy of the two hands about the axis of rotation. (Model the hands as long, thin rods rotating about one end.)

4. ±? In a physics experiment, an object with a mass of $m = (5.10 \pm 0.05)$ kg is attached to the free end of a light string wrapped around a reel of radius $R = (0.250 \pm 0.005)$ m and mass $M = (3.00 \pm 0.05)$ kg. The reel is a solid disc, free to rotate in a vertical plane about the horizontal axis passing through its centre as shown in **Figure P10.4**. The suspended object is released from rest (2.00 ± 0.02) m above the floor. Determine (a) the tension in the string, (b) the acceleration of the object, and (c) the speed with which the object hits the floor. (d) Verify your answer to part (c) by using the isolated system (energy) model.

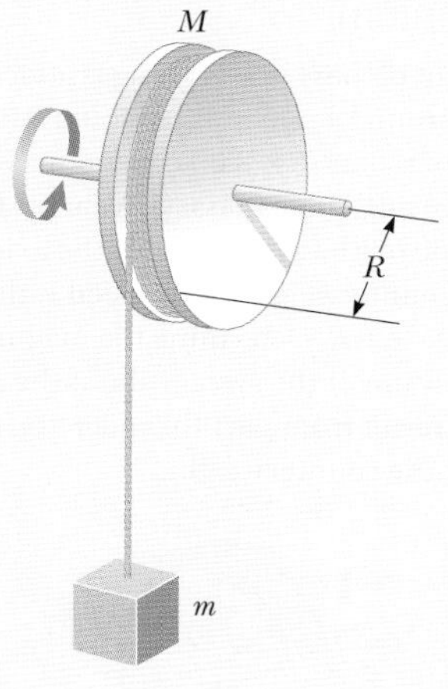

Figure P10.4

5. An experimental method for determining the moment of inertia of an irregularly shaped object or one with non-uniform density is shown in **Figure P10.5**. A counterweight of mass m is suspended by a cord wound around a spool of radius r, forming part of a turntable supporting the object. The turntable can rotate without friction. When the counterweight is released from rest, it descends through a distance h, acquiring a speed v. Show that the moment of inertia I of the rotating apparatus (including the turntable) is $mr^2(2gh/v^2 - 1)$. This experimental method was used as a part of an International Physics Olympiad experimental competition problem in South Korea in 2004, in which competitors had to work out the contents of a 'mechanical black box' containing connected springs and balls.

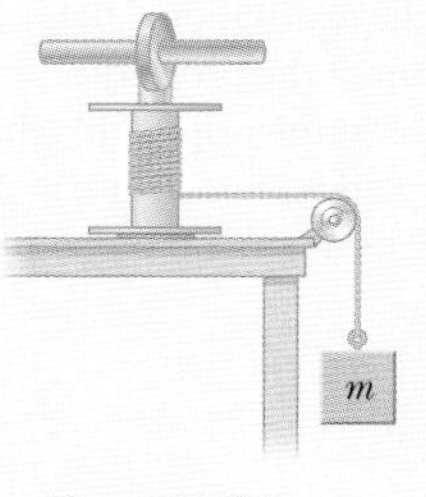

Figure P10.5

6. A uniform solid disc of radius R and mass M is free to rotate on a frictionless pivot through a point on its rim (**Figure P10.6**). If the disc is released from rest in the position shown by the copper-coloured circle, (a) what is the speed of its centre of mass when the disc reaches the position indicated by the dashed circle? (b) What is the speed of the lowest point on the disc in the dashed position? **What If?** (c) Repeat part (a) using a uniform hoop.

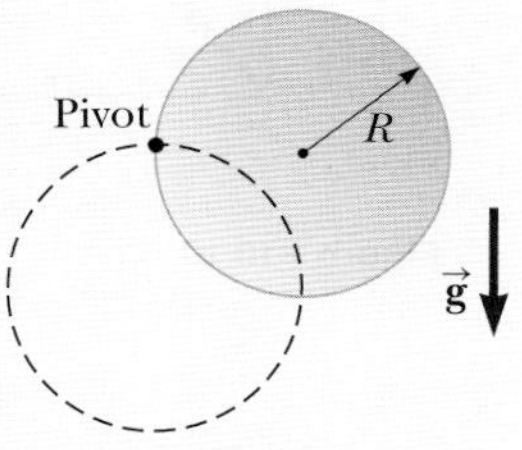

Figure P10.6

7. The head of a line trimmer or whipper-snipper has 100 g of cord wound in a light, cylindrical spool with inside diameter 3.00 cm and outside diameter 18.0 cm as shown in **Figure P10.7**. The cord has a linear density of 10.0 g/m. A single strand of the cord extends 16.0 cm from the outer edge of the spool. (a) When switched on, the trimmer speeds up from 0 to 2500 rev/min in 0.215 s. What average power is delivered to the head by the trimmer motor while it is accelerating? (b) When the trimmer is cutting grass, it spins at 2000 rev/min and the grass exerts an average tangential force of 7.65 N on the outer end of the cord, which is still at a radial distance of 16.0 cm from the outer edge of the spool. What is the power delivered to the head under load?

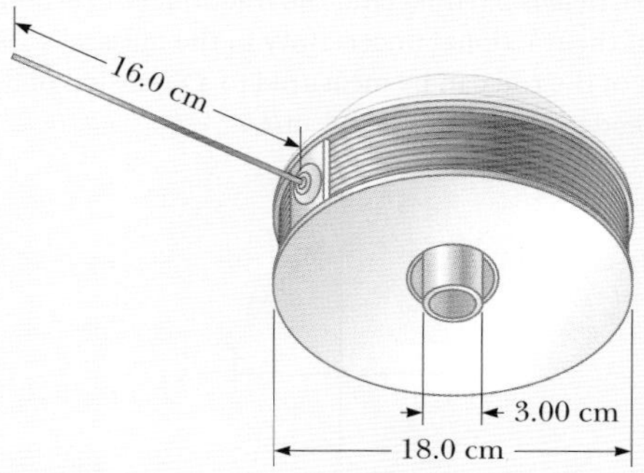

Figure P10.7

Section 10.2 Rolling motion of a rigid object

8. A cylinder of mass 10.0 kg rolls without slipping on a horizontal surface. At a certain instant, its centre of mass has a speed of 10.0 m/s. Determine (a) the translational kinetic energy of its centre of mass, (b) the rotational kinetic energy about its centre of mass and (c) its total energy.

9. (a) Determine the acceleration of the centre of mass of a uniform solid disc rolling down an incline making angle θ with the horizontal. (b) Compare the acceleration found in part (a) with that of a uniform hoop. (c) What is the minimum coefficient of friction required to maintain pure rolling motion for the disc?

10. A solid sphere is released from height h from the top of an incline making an angle θ with the horizontal. Calculate the speed of the sphere when it reaches the bottom of the incline (a) in the case that it rolls without slipping and (b) in the case that it slides frictionlessly without rolling. (c) Compare the time intervals required to reach the bottom in cases (a) and (b).

11. A tennis ball is a hollow sphere with a thin wall. It is set rolling without slipping at 4.03 m/s on a horizontal section of a track as shown in **Figure P10.11**. It rolls around the inside of a vertical circular loop of radius $r = 45.0$ cm. As the ball nears the bottom of the loop, the shape of the track deviates from a perfect circle so that the ball leaves the track at a point $h = 20.0$ cm below the horizontal section. (a) Find the ball's speed at the top of the loop. (b) Demonstrate that the ball will not fall from the track at the top of the loop. (c) Find the ball's speed as it leaves the track at the bottom. **What If?** (d) Suppose that static friction between the ball and the track were negligible so that the ball slid instead of rolling. Would its speed then be higher, lower or the same at the top of the loop? Explain your answer.

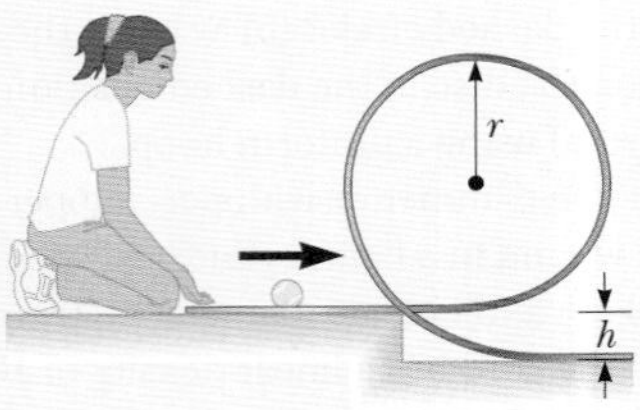

Figure P10.11

Section 10.3 Angular momentum

12. A light, rigid rod of length $\ell = 1.00$ m joins two particles, with masses $m_1 = 4.00$ kg and $m_2 = 3.00$ kg, at its ends. The combination rotates in the xy plane about a pivot through the centre of the rod (**Figure P10.12**). Determine the angular momentum of the system about the origin when the speed of each particle is 5.0 m/s. Find the fractional uncertainty in this value if the fractional uncertainty in the masses is 1%, and the length of the rod can be measured to a precision of 5 mm and the speed to a precision of 0.1 m/s.

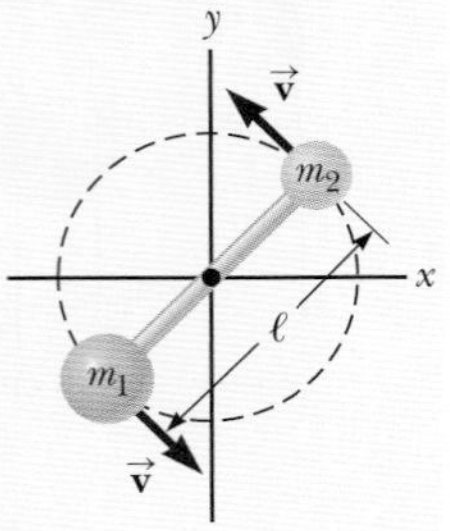

Figure P10.12

13. A 1.50-kg particle moves in the xy plane with a velocity of $\vec{v} = (4.20\,\hat{i} - 3.60\,\hat{j})$ m/s. Determine the angular momentum of the particle about the origin when its position vector is $\vec{r} = (1.50\,\hat{i} + 2.20\,\hat{j})$ m.

14. A conical pendulum consists of a bob of mass m in motion in a circular path in a horizontal plane as shown in **Figure P10.14**. During the motion, the supporting wire of length ℓ maintains a constant angle θ with the vertical. Show that the magnitude of the angular momentum of the bob about the vertical dashed line is

$$L = \left(\frac{m^2 g \ell^3 \sin^4 \theta}{\cos \theta}\right)^{1/2}$$

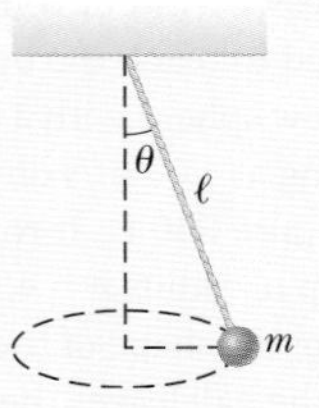

Figure P10.14

15. The position vector of a particle of mass 2.00 kg as a function of time is given by $\vec{r} = (6.00\hat{i} + 5.00t\hat{j})$, where $\vec{r}$ is in metres and t is in seconds. Determine the angular momentum of the particle about the origin as a function of time.

16. Model the Earth as a uniform sphere. (a) Calculate the angular momentum of the Earth due to its spinning motion about its axis. (b) Calculate the angular momentum of the Earth due to its orbital motion about the Sun. (c) Explain why the answer in part (b) is larger than that in part (a) even though it takes significantly longer for the Earth to go once around the Sun than to rotate once about its axis.

17. A particle of mass 0.400 kg is attached to the 100 cm mark of a metre stick of mass 0.100 kg. The metre stick rotates on the surface of a frictionless, horizontal table with an angular speed of 4.00 rad/s. Calculate the angular momentum of the system when the stick is pivoted about an axis (a) perpendicular to the table through the 50.0 cm mark and (b) perpendicular to the table through the 0 cm mark.

Section 10.4 Analysis model: non-isolated system (angular momentum)

18. A counterweight of mass $m = 4.00$ kg is attached to a light cord that is wound around a pulley as in **Figure P10.18**. The pulley is a thin hoop of radius $R = 8.00$ cm and mass $M = 2.00$ kg. The spokes have negligible mass. (a) What is the magnitude of the net torque on the system about the axle of the pulley? (b) When the counterweight has a speed v, the pulley has an angular speed $\omega = v/R$. Determine the magnitude of the total angular momentum of the system about the axle of the pulley. (c) Using your result from part (b) and $\vec{\tau} = d\vec{L}/dt$, calculate the acceleration of the counterweight.

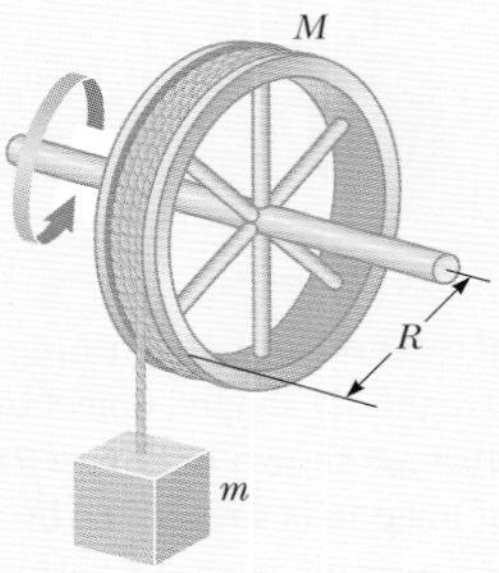

Figure P10.18

19. A projectile of mass m is launched with an initial velocity $\vec{v}_i$ making an angle θ with the horizontal as shown in **Figure P10.19**. The projectile moves in the gravitational field of the Earth. Find the angular momentum of the projectile about the origin (a) when the projectile is at the origin, (b) when it is at the highest point of its trajectory and (c) just before it hits the ground. (d) What torque causes its angular momentum to change?

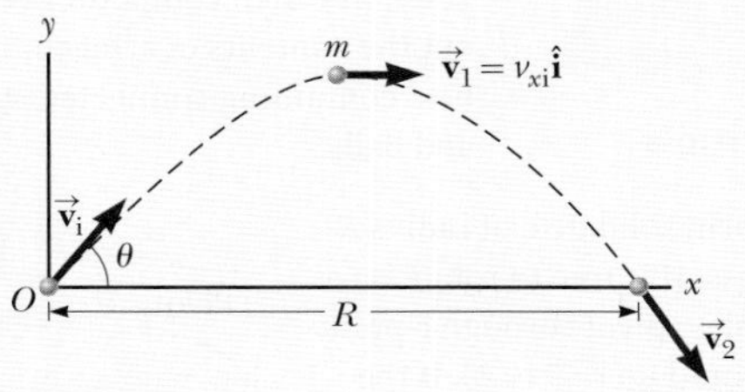

Figure P10.19

20. A ball of mass m is fastened at the end of a flagpole that is connected to the side of a tall building at point P as shown in **Figure P10.20**. The length of the flagpole is ℓ, and it makes an angle θ with the x axis. The ball becomes loose and starts to fall with acceleration $-g\hat{\mathbf{j}}$. (a) Determine the angular momentum of the ball about point P as a function of time. (b) For what physical reason does the angular momentum change? (c) What is the rate of change of the angular momentum of the ball about point P?

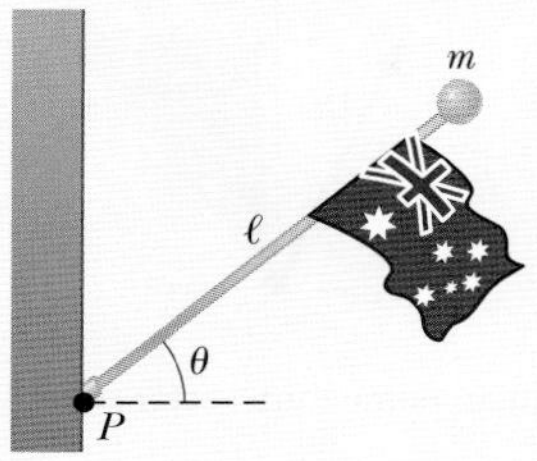

Figure P10.20

21. A space station is constructed in the shape of a hollow ring of mass 5.00×10^4 kg. (See **Figure P10.21**.) Members of the crew walk on a deck formed by the inner surface of the outer cylindrical wall of the ring, with radius $r = 100$ m. The ring is set rotating about its axis so that the people inside experience an effective free-fall acceleration equal to g. The rotation is achieved by firing two small rockets attached tangentially to opposite points on the rim of the ring. (a) What angular momentum does the space station acquire? (b) For what time interval must the rockets be fired if each exerts a thrust of 125 N?

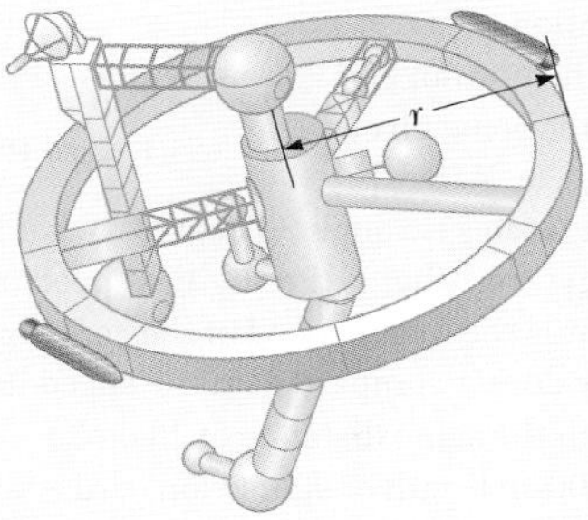

Figure P10.21

Section **10.5** Analysis model: isolated system (angular momentum)

22. A disc with moment of inertia I_1 rotates about a frictionless, vertical axle with angular speed ω_i. A second disc, this one having moment of inertia I_2 and initially not rotating, drops onto the first disc (**Figure P10.22**). Because of friction between the surfaces, the two eventually reach the same angular speed ω_f. (a) Calculate ω_f. (b) Calculate the ratio of the final to the initial rotational energy.

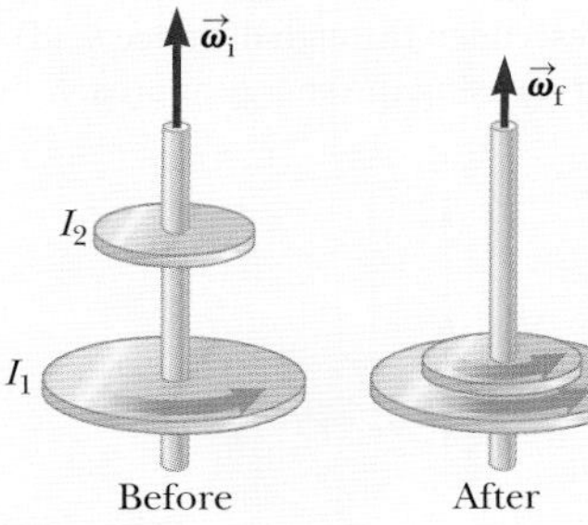

Figure P10.22

23. A student volunteer for a physics lecture demonstration sits on a freely rotating stool holding two dumbbells, each of mass 3.00 kg (**Figure P10.23**). When his arms are extended horizontally (**Figure P10.23a**), the dumbbells are 1.00 m from the axis of rotation and the student rotates with an angular speed of 0.750 rad/s. The moment of inertia of the student plus stool is 3.00 kg.m^2 and is assumed to be constant. The student pulls the dumbbells inwards horizontally to a position 0.300 m from the rotation axis (**Figure P10.23b**). (a) Find the new angular speed of the student. (b) Find the kinetic energy of the rotating system before and after he pulls the dumbbells inwards.

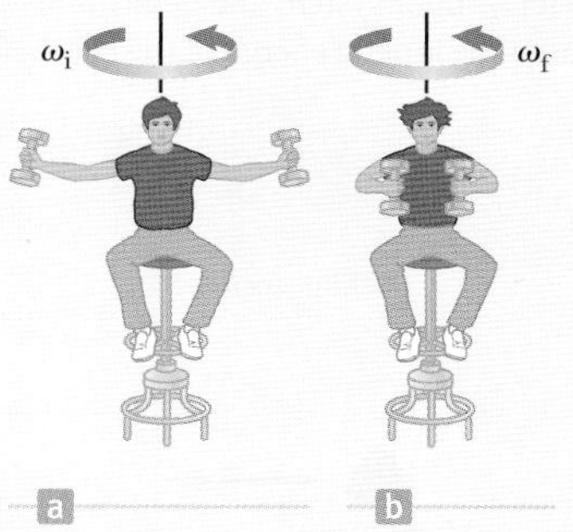

Figure P10.23

24. A 60.0 kg woman stands at the western rim of a merry-go-round in the shape of a horizontal turntable having a moment of inertia of 500 kg.m^2 and a radius of 2.00 m. The turntable is initially at rest and is free to rotate about a frictionless, vertical axle through its centre. The woman then starts walking around the rim clockwise (as viewed from above the system) at a constant speed of 1.50 m/s relative to the Earth. Consider the woman–turntable system as motion begins. (a) Is the mechanical energy of the system constant? (b) Is the momentum of the system constant? (c) Is the angular momentum of the system constant? (d) In what direction and with what angular speed does the turntable rotate? (e) How much chemical energy does the woman's body convert into mechanical energy of the woman–turntable system as the woman sets herself and the turntable into motion?

25. A wooden block of mass M resting on a frictionless, horizontal surface is attached to a rigid rod of length ℓ and of negligible mass (**Figure P10.25**). The rod is pivoted at the other end. A bullet of mass m travelling parallel to the horizontal surface and perpendicular to the rod with speed v hits the block and becomes embedded in it. (a) What is the angular momentum of the bullet–block system about a vertical axis through the pivot? (b) What fraction of the original kinetic energy of the bullet is converted into internal energy in the system during the collision?

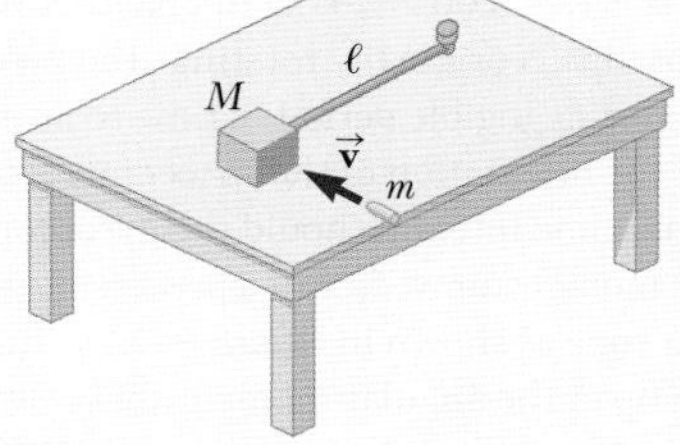

Figure P10.25

26. In a very rowdy art class a 0.50 kg lump of clay travelling horizontally with speed 15 m/s strikes an 18.0 kg door, sticking to it 10.0 cm from the side opposite the hinges as shown in Figure P10.26. The 1.00 m wide door is free to swing on its frictionless hinges. (a) Before it hits the door, does the clay have angular momentum relative to the door's axis of rotation? (b) If so, evaluate this angular momentum. If not, explain why there is no angular momentum. (c) Is the mechanical energy of the clay–door system constant during this collision? Answer without doing a calculation. (d) At what angular speed does the door swing open immediately after the collision? (e) Calculate the total energy of the clay–door system and determine whether it is less than or equal to the kinetic energy of the clay before the collision.

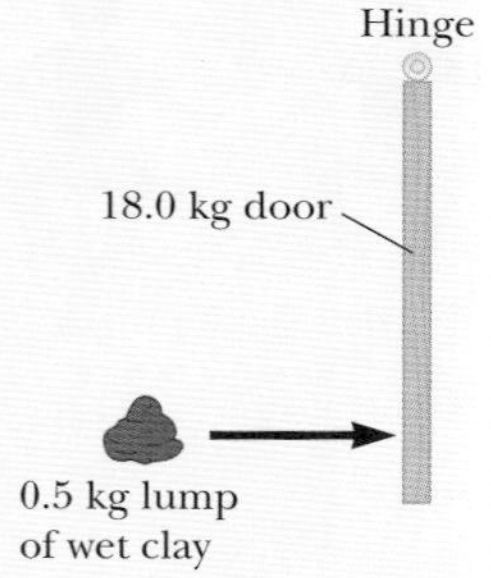

Figure P10.26
An overhead view of a lump of clay striking a door

27. In the same rowdy art class described in the previous question a wad of sticky clay with mass m and velocity $\vec{\mathbf{v}}_i$ is thrown at a fabric-printing machine, hitting a large roller which is a solid cylinder of mass M and radius R (Figure P10.27). The roller is initially at rest and is mounted on a fixed horizontal axle that runs through its centre of mass. The line of motion of the projectile is perpendicular to the axle and at a distance $d < R$ from the centre. (a) Find the angular speed of the system just after the clay strikes and sticks to the surface of the roller. (b) Is the mechanical energy of the clay–roller system constant in this process? Explain your answer. (c) Is the momentum of the clay–roller system constant in this process? Explain your answer.

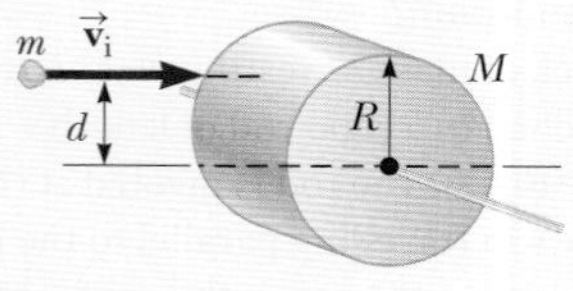

Figure P10.27

Section **10.6** The motion of gyroscopes and tops

28. A spacecraft is in empty space. It carries on board a gyroscope with a moment of inertia of $I_g = 20.0$ kg.m^2 about the axis of the gyroscope. The moment of inertia of the spacecraft around the same axis is $I_s = 5.00 \times 10^5$ kg.m^2. Neither the spacecraft nor the gyroscope is originally rotating. The gyroscope can be powered up in a negligible period of time to an angular speed of 100 rad/s. If the orientation of the spacecraft is to be changed by 30.0°, for what time interval should the gyroscope be operated?

29. The angular momentum vector of a precessing gyroscope sweeps out a cone as shown in Figure P10.29. The angular speed of the tip of the angular momentum vector, called its precessional frequency, is given by $\omega_p = \tau/L$, where τ is the magnitude of the torque on the gyroscope and L is the magnitude of its angular momentum. In the motion called *precession of the equinoxes,* the Earth's axis of rotation precesses about the perpendicular to its orbital plane with a period of 2.58×10^4 years. Model the Earth as a uniform sphere and calculate the torque on the Earth that is causing this precession.

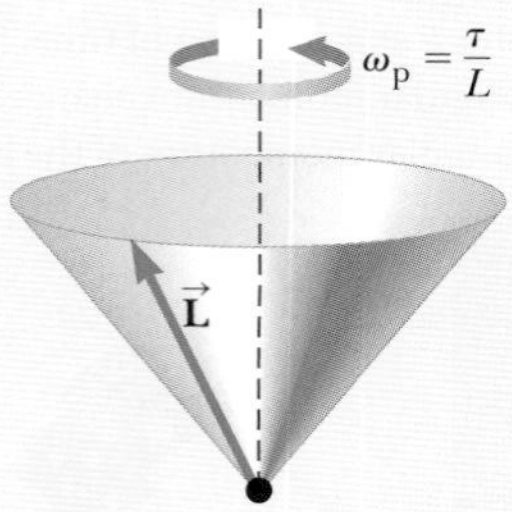

Figure P10.29
A precessing angular momentum vector sweeps out a cone in space.

Additional problems

30. **Review.** A thin, cylindrical rod $\ell = 24.0$ cm long with mass $m = 1.20$ kg has a ball of diameter $d = 8.00$ cm and mass $M = 2.00$ kg attached to one end. The arrangement is originally vertical and stationary, with the ball at the top as shown in Figure P10.30. The combination is free to pivot about the bottom end of the rod after being given a slight nudge. (a) After the combination rotates through 90 degrees, what is its rotational kinetic energy? (b) What is the angular speed of the rod and ball? (c) What is the linear speed of the centre of mass of the ball? (d) How does it compare with the speed had the ball fallen freely through the same distance of 28 cm?

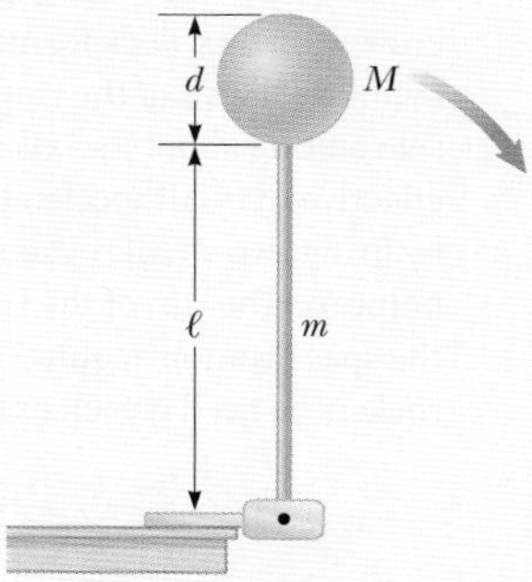

Figure P10.30

31. **Review.** A 4.00-m length of light nylon cord is wound around a uniform cylindrical spool of radius 0.500 m and mass 1.00 kg. The spool is mounted on a frictionless axle and is initially at rest. The cord is pulled from the spool with a constant acceleration of magnitude 2.50 m/s^2. (a) How much work has been done on the spool when it reaches an angular speed of 8.00 rad/s? (b) How long does it take the spool to reach this angular speed? (c) How much cord is left on the spool when it reaches this angular speed?

32. A string is wound around a uniform disc of radius R and mass M. The disc is released from rest with the string vertical and its top end tied to a fixed bar (Figure P10.32). Show that (a) the tension in the string is one-third of the weight of the disc, (b) the magnitude of the acceleration of the centre of mass is $2g/3$ and (c) the speed of the centre of mass is $(4gh/3)^{1/2}$ after the disc has descended through distance h. (d) Verify your answer to part (c) using the energy approach.

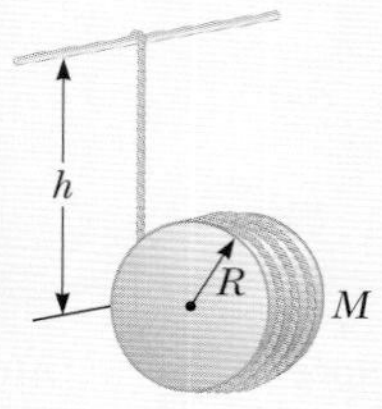

Figure P10.32

33. A uniform solid sphere of radius r is placed on the inside surface of a hemispherical bowl with radius R. The sphere is released from rest at an angle θ to the vertical and rolls without slipping (**Figure P10.33**). Determine the angular speed of the sphere when it reaches the bottom of the bowl.

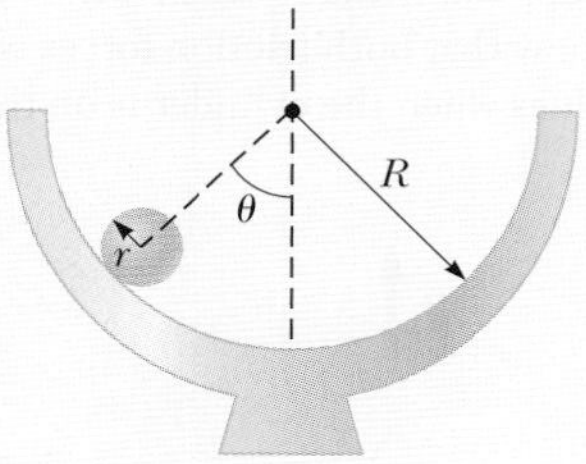

Figure P10.33

34. A spool of wire of mass M and radius R is unwound under a constant force $\vec{\mathbf{F}}$ (**Figure P10.34**). Assuming the spool is a uniform, solid cylinder that doesn't slip, show that (a) the acceleration of the centre of mass is $4\vec{\mathbf{F}}/3M$ and (b) the force of friction is to the *right* and equal in magnitude to $F/3$. (c) If the cylinder starts from rest and rolls without slipping, what is the speed of its centre of mass after it has rolled through a distance d?

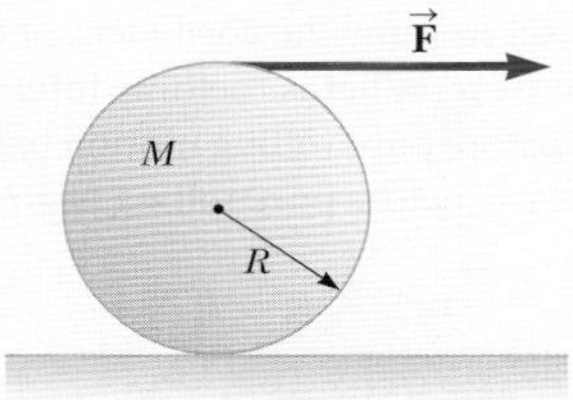

Figure P10.34

35. A light rope passes over a light, frictionless pulley. One end is fastened to a bunch of bananas of mass M, and a monkey of mass M clings to the other end (**Figure P10.35**). The monkey climbs the rope in an attempt to reach the bananas. (a) Treating the system as consisting of the monkey, bananas, rope and pulley, find the net torque on the system about the pulley axis. (b) Using the result of part (a), determine the total angular momentum about the pulley axis and describe the motion of the system. (c) Will the monkey reach the bananas?

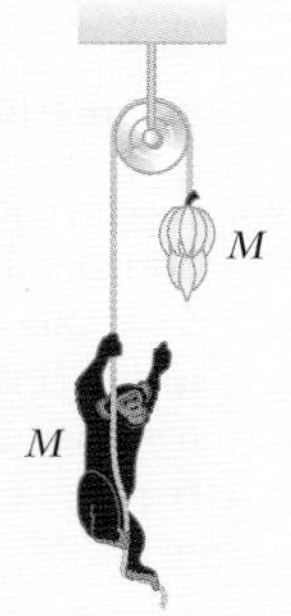

Figure P10.35

36. Halley's comet moves about the Sun in an elliptical orbit, with its closest approach to the Sun being about 0.590 AU and its greatest distance 35.0 AU (1 AU = the Earth–Sun distance). The angular momentum of the comet about the Sun is constant and the gravitational force exerted by the Sun has zero moment arm. The comet's speed at its closest approach is 54.0 km/s. What is its speed when it is farthest from the Sun?

37. A skateboarder with his board can be modelled as a particle of mass 76.0 kg, located at his centre of mass, 0.500 m above the ground. As shown in **Figure P10.37**, the skateboarder starts from rest in a crouching position at one lip of a half-pipe (point Ⓐ). The half-pipe forms one half of a cylinder of radius 6.80 m with its axis horizontal. On his descent, the skateboarder moves without friction and maintains his crouch so that his centre of mass moves through one-quarter of a circle. (a) Find his speed at the bottom of the half-pipe (point Ⓑ). (b) Find his angular momentum about the centre of curvature at this point. (c) Immediately after passing point Ⓑ, he stands up and raises his arms, lifting his centre of gravity to 0.950 m above the concrete (point Ⓒ). Explain why his angular momentum is constant in this manoeuvre, whereas the kinetic energy of his body is not constant. (d) Find his speed immediately after he stands up. (e) How much chemical energy in the skateboarder's legs was converted into mechanical energy in the skateboarder–Earth system when he stood up?

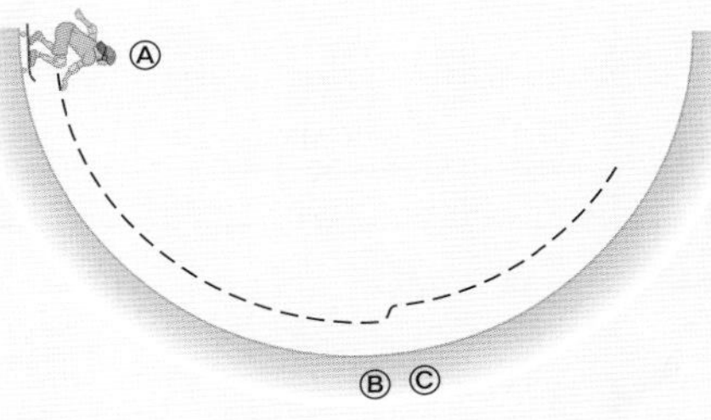

Figure P10.37

38. ±? A puck of mass m = (50.0 ± 0.1) g is attached to a taut cord passing through a small hole in a frictionless, horizontal surface (**Figure P10.38**). The puck is initially orbiting with speed v_i = (1.5 ± 0.1) m/s in a circle of radius r_i = (0.300 ± 0.005) m. The cord is then slowly pulled from below, decreasing the radius of the circle to r = (0.100 ± 0.05) m. (a) What is the puck's speed at the smaller radius? (b) Find the tension in the cord at the smaller radius. (c) How much work is done by the hand in pulling the cord so that the radius of the puck's motion changes from 0.300 m to 0.100 m?

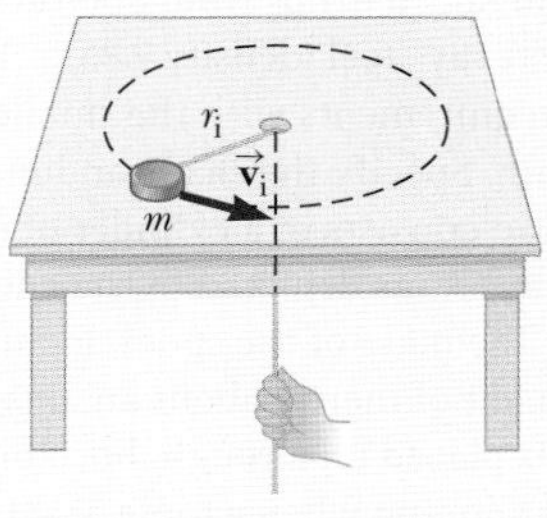

Figure P10.38

39. Two naughty children are playing on stools at a restaurant counter. Their feet do not reach the footrests and the tops of the stools are free to rotate without friction on pedestals fixed to the floor. Marcus catches a tossed apple, in a process described by the equation

$$(0.730\ \text{kg.m}^2)(2.40\hat{\jmath}\ \text{rad/s}) + (0.120\ \text{kg})(0.350\hat{\imath}\ \text{m}) \times (4.30\hat{\mathbf{k}}\ \text{m/s}) = [0.730\ \text{kg.m}^2 + (0.120\ \text{kg})(0.350\ \text{m})^2]\vec{\boldsymbol{\omega}}$$

(a) Solve the equation for the unknown $\vec{\boldsymbol{\omega}}$. (b) Complete the statement of the problem to which this equation applies. Your statement must include the given numerical information and specification of the unknown to be determined. (c) Could the equation equally well describe Marcus throwing the apple to Laurence? Explain your answer.

40. Two astronauts (**Figure P10.40**), each having a mass of 75.0 kg, are connected by a 10.0-m rope of negligible mass. They are isolated in space, orbiting their centre of mass at speeds of 5.00 m/s. Treating the astronauts as particles, calculate (a) the magnitude of the angular momentum of the two-astronaut system and (b) the rotational energy of the system. By pulling on the rope, one astronaut shortens the distance between them to 5.00 m. (c) What is the new angular momentum of the system? (d) What are the astronauts' new speeds? (e) What is the new rotational energy of the system? (f) How much chemical potential energy in the body of the astronaut was converted to mechanical energy in the system when he shortened the rope?

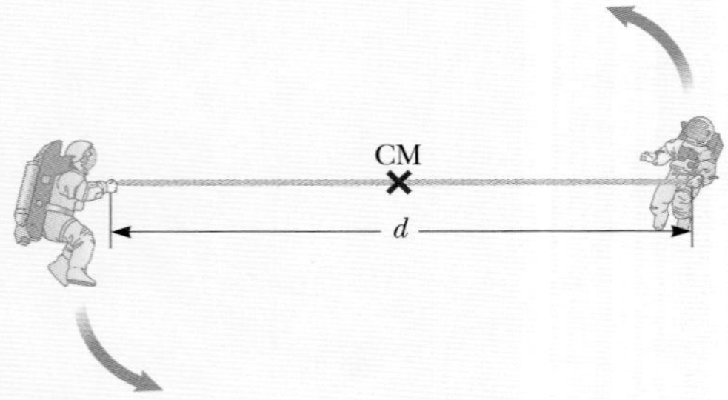

Figure P10.40

Challenge problems

41. As a petrol engine operates, a flywheel turning with the crankshaft stores energy after each fuel explosion, providing the energy required to compress the next charge of fuel and air. For the engine of a certain lawn mower, suppose a flywheel must be no more than 18.0 cm in diameter. Its thickness, measured along its axis of rotation, must be no larger than 8.00 cm. The flywheel must release energy 60.0 J when its angular speed drops from 800 rev/min to 600 rev/min. Design a sturdy steel (density 7.85×10^3 kg/m^3) flywheel to meet these requirements with the smallest mass you can reasonably attain. Specify the shape and mass of the flywheel.

42. A spool of thread consists of a cylinder of radius R_1 with end caps of radius R_2 as depicted in the end view shown in **Figure P10.42**. The mass of the spool, including the thread, is m, and its moment of inertia about an axis through its centre is I. The spool is placed on a rough, horizontal surface so that it rolls without slipping when a force $\vec{\mathbf{T}}$ acting to the right is applied to the free end of the thread. (a) Show that the magnitude of the friction force exerted by the surface on the spool is given by

$$f = \left(\frac{I + mR_1R_2}{I + mR_2^2}\right)T$$

(b) Determine the direction of the force of friction.

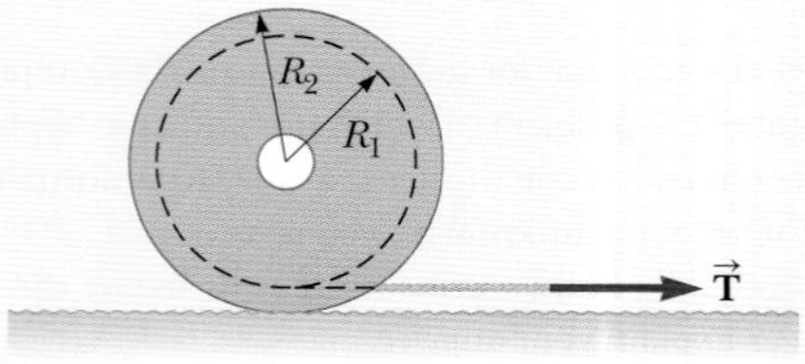

Figure P10.42

43. **Figure P10.43** shows a vertical force applied tangentially to a uniform cylinder of weight F_g. The coefficient of static friction between the cylinder and all surfaces is 0.500. The force $\vec{\mathbf{P}}$ is increased in magnitude until the cylinder begins to rotate. In terms of F_g, find the maximum force magnitude P that can be applied without causing the cylinder to rotate. *Suggestion*: Show that both friction forces will be at their maximum values when the cylinder is on the verge of slipping.

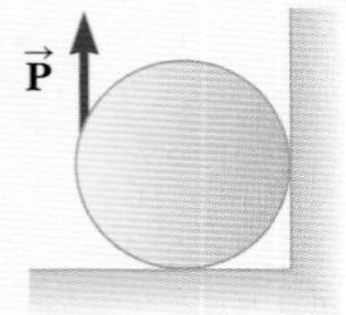

Figure P10.43

44. A uniform solid disc of radius R is set into rotation with an angular speed ω_i about an axis through its centre. While still rotating at this speed, the disc is placed into contact with a horizontal surface and immediately released as shown in **Figure P10.44**. (a) What is the angular speed of the disc once pure rolling takes place? (b) Find the fractional change in kinetic energy from the moment the disc is set down until pure rolling occurs. (c) Assume the coefficient of friction between disc and surface is μ. What is the time interval after setting the disc down before pure rolling motion begins? (d) How far does the disc travel before pure rolling begins?

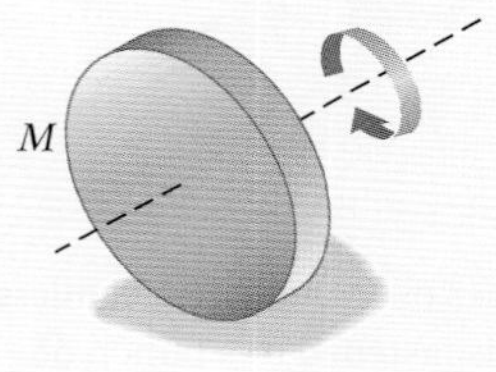

Figure P10.44

45. A solid cube of wood of side $2a$ and mass M is resting on a horizontal surface. The cube is constrained to rotate about a fixed axis AB (**Figure P10.45**). A bullet of mass m and speed v is shot at the face opposite $ABCD$ at a height of $4a/3$. The bullet becomes embedded in the cube. Find the minimum value of v required to tip the cube so that it falls on face $ABCD$. Assume $m << M$.

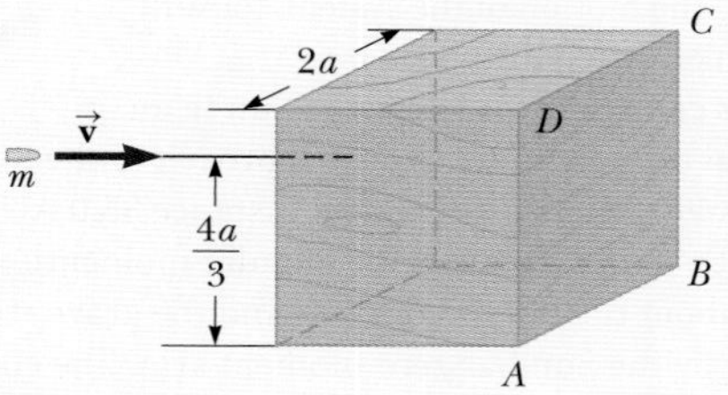

Figure P10.45

chapter 11

Gravity

Getty Images/Robin Smith

This image shows a nice moonrise above Parliament House in Canberra.

The Moon stays in orbit around the Earth due to the gravitational force. What effect does this force have on the Earth and why does this same force not simply pull the Moon in to collide with the Earth? In fact the Moon is getting further away by a few centimetres every year. Why is this happening?

On completing this chapter, students will understand:

- that all objects with mass produce a gravitational field
- that gravitational force decreases with the square of the distance between objects
- that the gravitational field is the gravitational force per unit mass on a test particle
- that there is a potential energy due to the gravitational force
- that gravitational potential is the potential energy per unit mass
- that an accelerating reference frame is indistinguishable from a reference frame in a gravitational field.

Students will be able to:

- calculate the gravitational force acting on an object in a gravitational field
- calculate the gravitational field due to a spherically symmetrical object
- draw simple field diagrams for gravitational fields
- analyse the motion of objects in a gravitational field
- apply analysis models for isolated systems (energy and angular momentum) to systems with gravitational fields
- determine when it is appropriate to apply the surface approximation for objects near the surface of the Earth
- apply Kepler's laws to analyse the motion of orbiting satellites and planets.

In this chapter, we introduce the first of the four fundamental forces: gravity. The idea of universal, fundamental forces or interactions is at the heart of how modern physics describes the world and Newton's development of his theory of gravity represents a critical change in the evolution of science. It was more than 300 years until any modification to Newton's law of universal gravitation was made, when Einstein introduced the general theory of relativity.

ENHANCED WebAssign

The problems found in this chapter may be assigned online in Enhanced Web Assign.

11.1 Newton's law of universal gravitation

In the 17th century a large amount of data had been collected on the motions of the Moon and the planets, but a clear understanding of the forces related to these motions was not available until Newton realised that the forces involved in the Earth–Moon attraction and in the Sun–planet attraction were not something special to those systems, but rather were particular cases of a general and universal attraction between objects – the force of gravity. The same underlying attraction that causes the Moon to follow its path around the Earth also causes an apple to fall from a tree. It was the first time that 'earthly' and 'heavenly' motions were unified and also the first time that the idea of a universal, fundamental force was suggested.

In previous chapters we have considered the changes in an object's motion and energy due to the presence of a non-zero net force. However, until now we have not considered the origins of and the laws that apply to the forces themselves. We mentioned in Chapter 1 that all observed forces are actually due to four fundamental interactions: gravity, electromagnetism, the weak nuclear force and the strong nuclear force. In this chapter we describe the first of these: gravity.

'Newton's apple tree' at Monash University. This tree was grown from a cutting of Newton's original apple tree, which according to Newton inspired his theory of gravity when he observed an apple falling from it.

In his memoirs, as recorded by William Stukely (*Memoirs of Sir Isaac Newton's Life*, 1752), Newton recounts watching an apple fall from a tree:

> *'why should that apple always descend perpendicularly to the ground,' thought he to him self: occasion'd by the fall of an apple, as he sat in a contemplative mood: 'why should it not go sideways, or upwards? but constantly to the earths centre? assuredly, the reason is, that the earth draws it. there must be a drawing power in matter. & the sum of the drawing power in the matter of the earth must be in the earths centre, not in any side of the earth. therefore dos this apple fall perpendicularly, or toward the centre. if matter thus draws matter; it must be in proportion of its quantity. therefore the apple draws the earth, as well as the earth draws the apple.'*

These few lines summarise most of the key features of **Newton's Law of Universal Gravitation**, as published in 1687 in his treatise *Mathematical Principles of Natural Philosophy*. Newton's law of universal gravitation states that every particle in the Universe attracts every other particle with a force that is directly proportional to the product of their masses (the quantity of matter as in the quote above) and inversely proportional to the square of the distance between them.

If the particles have masses m_1 and m_2 and are separated by a distance r, the magnitude of this gravitational force is

The law of universal gravitation ►

$$F_g = G\frac{m_1 m_2}{r^2} \quad \textbf{(11.1)}$$

where G is a constant, called the *universal gravitational constant* and $G = 6.674 \times 10^{-11}$ N.m²/kg².

TRY THIS

Get some coffee filters or aluminium pie pans (paper patty pans also work well) and hold one flat and drop it carefully so that it falls directly down. Time how long it takes to fall. Hold two stacked together and drop them from the same height. Recall that this was a *Try this* example in Chapter 5. How long do two together take? Now try it with 2, 3, 4, 5 and 10. What happens to the time taken as you increase the number of filters or pie pans? Draw a free-body diagram for the pie pans after they have reached terminal velocity. What forces are acting? How do they depend on the mass of the pie pans?

Quick Quiz 11.1

What are the units (in terms of m, kg and s only) of G? What are the dimensions of G?

The form of the force law given by **Equation 11.1** is often referred to as an **inverse-square law** because the magnitude of the force varies as the inverse square of the separation of the particles. Because the force varies as the inverse square of the distance between the particles, it decreases rapidly with increasing separation. We shall see other examples of this type of behaviour in subsequent chapters; for example, the electromagnetic force also follows an inverse square law. We can express this force in vector form by defining a unit vector $\hat{\mathbf{r}}_{12}$ (**Active Figure 11.1**). Because this unit vector is directed from particle 1 toward particle 2, the force exerted by particle 1 on particle 2 is

$$\vec{\mathbf{F}}_{12} = -G\frac{m_1 m_2}{r^2}\hat{\mathbf{r}}_{12} \quad (11.2)$$

where the negative sign indicates that particle 2 is attracted to particle 1; hence, the force on particle 2 must be directed towards particle 1. By Newton's third law, the force exerted by particle 2 on particle 1, designated $\vec{\mathbf{F}}_{21}$, is equal in magnitude to $\vec{\mathbf{F}}_{12}$ and in the opposite direction. That is, these forces form an action–reaction pair, and $\vec{\mathbf{F}}_{21} = -\vec{\mathbf{F}}_{12}$. As Newton said, the apple pulls the Earth as the Earth pulls the apple.

Equation 11.2 can also be used to show that the gravitational force exerted by a finite-size, spherically symmetrical mass distribution on a particle outside the distribution is the same as if the entire mass of the distribution were concentrated at the centre ('*the sum of the drawing power in the matter of the earth must be in the earths centre, not in any side of the earth*'). For example, the magnitude of the force exerted by the Earth on a particle of mass m near the Earth's surface is

$$F_g = G\frac{M_E m}{R_E^2} \quad (11.3)$$

where M_E is the Earth's mass and R_E its radius.

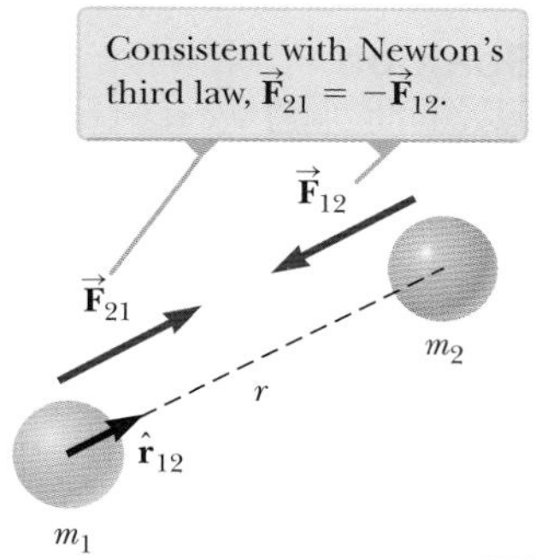

Consistent with Newton's third law, $\vec{\mathbf{F}}_{21} = -\vec{\mathbf{F}}_{12}$.
$\vec{\mathbf{F}}_{12}$
$\vec{\mathbf{F}}_{21}$
m_2
r
$\hat{\mathbf{r}}_{12}$
m_1

Active Figure 11.1
The gravitational force between two particles is attractive. The unit vector $\hat{\mathbf{r}}_{12}$ is directed from particle 1 towards particle 2.

Pitfall Prevention 11.1

Be clear on g and G. The symbol g represents the magnitude of the free-fall acceleration near a planet. At the surface of the Earth, g has an average value of 9.80 m/s^2. G is a universal constant that has the same value everywhere in the Universe, $G = 6.674 \times 10^{-11}$ N.m^2/kg^2.

Quick Quiz 11.2

Think about the expression for the force due to gravity we used in Chapter 4, $F = mg$. How does g relate to G, M_E and R_E?

Quick Quiz 11.3

A planet has two moons of equal mass. Moon 1 is in a circular orbit of radius r. Moon 2 is in a circular orbit of radius $2r$. What is the magnitude of the gravitational force exerted by the planet on moon 2? **(a)** four times as large as that on moon 1 **(b)** twice as large as that on moon 1 **(c)** equal to that on moon 1 **(d)** half as large as that on moon 1 **(e)** one-fourth as large as that on moon 1

±? Example 11.1

The Cavendish experiment

Henry Cavendish (1731–1810) measured the universal gravitational constant in an important experiment in 1798. Cavendish's apparatus consisted of two small spheres, each of mass m, fixed to the ends of a light, horizontal rod suspended by a fine fibre or thin metal wire as shown in **Figure 11.2**. When two large spheres, each of mass M, are placed near the smaller ones, the attractive force between smaller and larger spheres causes the rod to rotate and twist the wire suspension to a new equilibrium orientation. The angle of rotation is measured by the deflection of a light beam reflected from a mirror attached to the vertical suspension.

Consider a version of the Cavendish experiment in which students are attempting to measure the value of G. The mass of the small balls is $m = (20.0 \pm 0.1)$ kg, the mass of the large balls is (200.0 ± 0.5) kg, the horizontal rod joining the balls has a length of $l = (3.00 \pm 0.01)$ m, measured from the centres of the smaller balls, and the initial distance between the centre of each small ball and the large ball closest to it is $d = (0.300 \pm 0.001)$ m. The students measure the torque acting on the system consisting of the rod and small balls about the centre of the rod as $\vec{\tau} = (8.8 \pm 0.4) \times 10^{-6}$ N.m.

Find the constant G from this data.

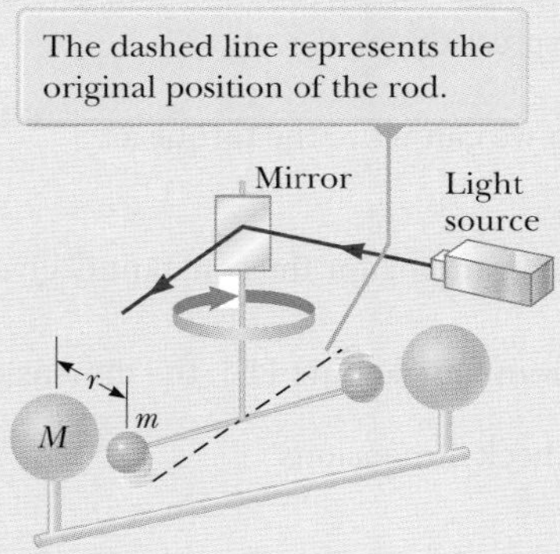

The dashed line represents the original position of the rod.
Mirror
Light source
r
m
M

Figure 11.2
Cavendish apparatus for measuring G

Example 11.1 cont.

Solution

Conceptualise We start by drawing a diagram showing the gravitational force acting on the small balls and relate this to the torque on the system.

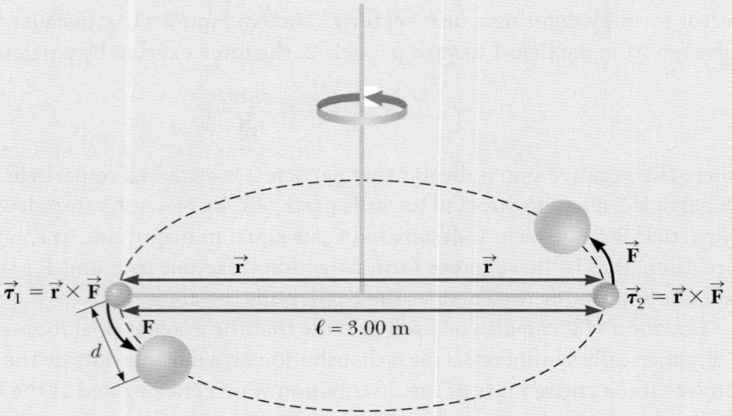

Model We can model the suspended rod and metal spheres assembly as a rigid object in equilibrium. The torque due to the gravitational force of the large spheres is balanced by the (measured) torque exerted by the wire. Note that the gravitational force due to the Earth is balanced by the tension in the wire, and neither of these forces exert a torque on the assembly in the horizontal plane. Hence we need to relate the measured torque to the gravitational force due to the large spheres.

Analyse As the system is symmetrical, the total torque due to the large spheres is twice that due to the force on one small sphere. As the object is in equilibrium,

$$\sum \tau = 0, \text{ so } \tau_{grav} = \tau_{wire}$$

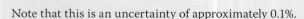

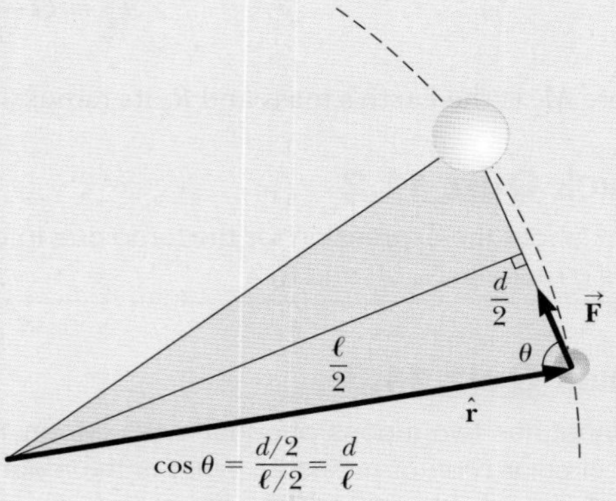

The gravitational torque on the ball–rod system (from **Equation 9.14**) will be

$$\vec{\tau} = 2(\vec{r} \times \vec{F}) = 2rF\sin\theta = \tau_{wire}$$

where $r = l/2$ and we find the angle θ from the construction shown: $\cos\theta = \dfrac{d}{l} = \dfrac{0.300\text{ m}}{3.00\text{ m}} = 0.1$, so $\theta = 84.3°$

We can find the uncertainty in this angle using the range method:

$$\theta_{min} = \cos^{-1}\left(\frac{d_{max}}{l_{min}}\right) = \cos^{-1}\left(\frac{0.301\text{ m}}{2.99}\right) = 84.2°$$

and $\theta_{max} = \cos^{-1}\left(\dfrac{d_{min}}{l_{max}}\right) = \cos^{-1}\left(\dfrac{0.299\text{ m}}{3.01\text{ m}}\right) = 84.3°$

so $\theta = 84.3° \pm 0.1°$

Note that this is an uncertainty of approximately 0.1%.

We now use **Equation 11.2** to give us an expression for F:

$$F = \frac{GmM}{d^2}$$

so we can write the torque as:

$$\tau_{wire} = 2r\left(\frac{GmM}{d^2}\right)\sin\theta$$

Rearranging for the constant G gives:

$$G = \frac{\tau_{wire} d^2}{2rmM\sin\theta}$$

From Quick Quiz 11.1, the dimensions of G are $[M^{-1}L^3T^{-2}]$

Check dimensions: $[M^{-1}L^3T^{-2}] = ([ML^2T^{-2}][L]^2)/([L][M][M]) = [M^{-1}L^3T^{-2}]$ ☺

Substituting the experimental values gives:

$$G = \frac{(8.8 \times 10^{-6}\text{ N.m})(0.300\text{ m}^2)}{2(1.50\text{ m})(20.0\text{ kg})(200.0\text{ kg})\sin 84.3°}$$

$$= 6.633 \times 10^{-11}\text{ N.m}^2/\text{kg}^2$$

Example 11.1 cont.

To find the uncertainty in G we use the sum of the relative uncertainties:

$$\frac{\Delta G}{G} = \frac{\Delta \tau}{\tau} + \frac{2\Delta d}{d} + \frac{\Delta l}{l} + \frac{\Delta r}{r} + \frac{\Delta M}{M} + \frac{\Delta m}{m} + \frac{\Delta(\sin\theta)}{\sin\theta}$$

This uncertainty is dominated by the uncertainty in τ, which is an order of magnitude or more larger than any other in the sum.

Hence we can say that $\frac{\Delta G}{G} \approx \frac{\Delta \tau}{\tau} = \frac{0.4}{8.8} = 5\%$

So the final experimental result is: $G = (6.6 \pm 0.3) \times 10^{-11}\ \text{N.m}^2/\text{kg}^2$

Finalise The result found in this experiment agrees with the accepted value for G. Also note that the size of the gravitational force acting between the balls is very small; this is generally the case for everyday objects, which is why we are not usually aware of it and only very precise experiments can measure it. Note that Cavendish did not set out to measure G, rather his aim was to 'weigh the world' by comparing the acceleration of a small ball due to a massive ball to the acceleration of the same small ball due to the Earth.

We estimated our uncertainty by making the approximation that $\frac{\Delta G}{G} \approx \frac{\Delta \tau}{\tau}$. How much does the uncertainty increase if we use the complete sum? Try it. You can use the range method to find the uncertainty in sin θ given the uncertainty in θ.

11.2 The gravitational field

When Newton published his theory of universal gravitation, it was considered a success because it satisfactorily explained the motion of the planets. It demonstrated that the same laws that explain phenomena on the Earth can be used to describe the behaviour of large objects such as planets. Nevertheless, both Newton's contemporaries and his successors found it difficult to accept the concept of a force that acts at a distance. They asked how it was possible for two objects such as the Sun and the Earth to interact when they were not in contact with each other. We use the concept of a **gravitational field** to explain this interaction.

Fields are a very useful concept in physics, and we shall meet them again in later chapters when we look at the electric field (Chapter 23) and the magnetic field (Chapter 29). A field is a way of representing a force that acts at a distance. When two objects interact without touching each other, we say that they interact via a force field or more usually we just say a field. Fields are defined in terms of the property of an object that creates the field. Every object with *mass* creates a gravitational field, via which it interacts with other objects with mass. All *charged* objects create an electric field, via which they interact with other charged objects. As we shall see below, the concept of a field is useful because it allows us to describe the effect one object will have on any arbitrary object, by defining a field around the object which depends upon the relevant property – for example mass or charge. The field concept is very useful and allows us to explain and predict the behaviour of many systems. Later we shall meet an alternative theory, that of particle exchange, when we study quantum physics (Chapter 44). The particle exchange theory provides an alternative way of understanding action at a distance.

We define the gravitational field in terms of the force exerted on a test particle. When a particle of mass m is placed at a point where the gravitational field is $\vec{\mathbf{g}}$, the particle experiences a force $\vec{\mathbf{F}}_g = m\vec{\mathbf{g}}$. In other words, we imagine that the field exerts a force on the particle rather than consider a direct interaction between two particles. The gravitational field $\vec{\mathbf{g}}$ is defined as

$$\vec{\mathbf{g}} \equiv \frac{\vec{\mathbf{F}}_g}{m} \quad (11.4)$$

◄ Gravitational field

That is, the gravitational field at a point in space equals the gravitational force experienced by a *test particle* placed at that point divided by the mass of the test particle; it is **the force per unit mass at that point in space**. Notice that the gravitational field is the same as the acceleration due to gravity and so has units of N/kg or m/s^2 and dimensions LT^{-2}.

Notice that the presence of the test particle is not necessary for the field to exist; the Earth creates a gravitational field whether or not we place a test particle near it.

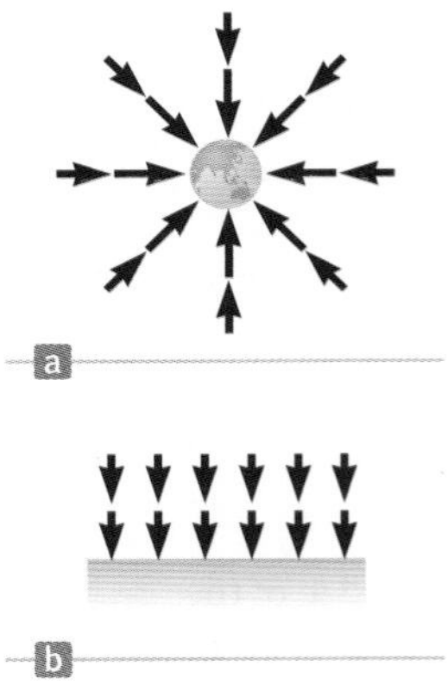

Figure 11.3
(a) The gravitational field vectors in the vicinity of a uniform spherical mass such as the Earth vary in both direction and magnitude. (b) The gravitational field vectors in a small region near the Earth's surface are uniform in both direction and magnitude.

Consider an object of mass m near the Earth's surface. Because the gravitational force acting on the object has a magnitude $GM_E m/r^2$ (see **Equation 11.3**), the field $\vec{\mathbf{g}}$ at a distance r from the centre of the Earth is

$$\vec{\mathbf{g}} = \frac{\vec{\mathbf{F}}_g}{m} = -\frac{GM_E}{r^2}\hat{\mathbf{r}} \tag{11.5}$$

where the negative sign indicates that the field points towards the centre of the Earth as illustrated in **Figure 11.3a** and $\hat{\mathbf{r}}$ is a unit vector pointing radially outwards from the Earth. The field vectors at different points surrounding the Earth vary in both direction and magnitude. In a small region near the Earth's surface, the downwards field $\vec{\mathbf{g}}$ is approximately constant and uniform as indicated in **Figure 11.3b**. **Equation 11.5** is valid at all points *outside* the Earth's surface, assuming the Earth is spherical. The unit N/kg is the same as m/s^2 and the value of the gravitational field, g, is the acceleration of the test particle due to the field, which at the surface of the Earth is just the free-fall acceleration of $9.80\ m/s^2$ which was introduced in Chapter 2.

We can represent gravitational fields (and in fact any other fields) using field diagrams. We construct these by drawing field vectors as shown in **Figure 11.3**, then joining up the vectors. We still need to indicate the direction of the field, which we do with an arrowhead as shown in **Figure 11.4**. We shall draw more field diagrams when we look at electric fields in Chapter 23.

TRY THIS

You can make your own gravitational field detector with a plumb-bob, which is just a heavy object attached to a piece of string. If you hold the end of the string, the plumb-bob always points in the direction of the gravitational field.

As you would have noted from Quick Quiz 11.2, at the Earth's surface, where $r = R_E$

Acceleration due to gravity at Earth's surface ▶

$$g = G\frac{M_E}{R_E^2} \tag{11.6}$$

and we find that $\vec{\mathbf{g}}$ has a magnitude of 9.80 N/kg.

Equation 11.6 relates the free-fall acceleration g to physical parameters of the Earth – its mass and radius – and explains the origin of the value of $9.80\ m/s^2$ that we have used in earlier chapters. The value of $g = 9.80\ m/s^2$ is the average value of the Earth's gravitational field at the surface of the Earth. The actual value varies from place to place, and even varies slightly with time. The value varies with position because of the shape of the Earth – the Earth is not perfectly spherical but is slightly oblate and the gravitational field is slightly lower at the poles. The field also varies due to the local composition of the Earth, and tiny variations in the field are used by the mining industry to locate large ore deposits. The Gravity Recovery and Climate Experiment (GRACE) launched by NASA in 1992 is measuring the variations in the Earth's gravitational field with position and time. The experiment is sensitive enough that it has been able to measure changes in the gravitational field of the Earth due to climate change. For example, changes due to the melting of ice sheets and even due to the loss of ground water during severe droughts have been measured.

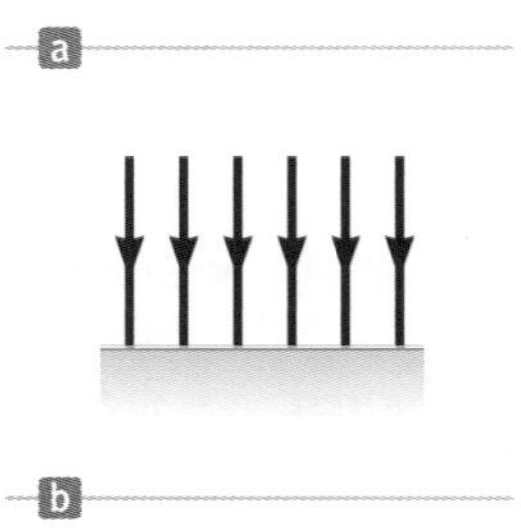

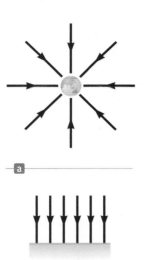

Figure 11.4
(a) The gravitational field in the vicinity of a uniform spherical mass such as the Earth (b) The gravitational field in a small region near the Earth's surface

Consider an object of mass m located a distance h above the Earth's surface or a distance r from the Earth's centre, where $r = R_E + h$. The magnitude of the gravitational force acting on this object is

$$F_g = G\frac{M_E m}{r^2} = G\frac{M_E m}{(R_E + h)^2}$$

The magnitude of the gravitational force acting on the object at this position is also $F = mg$, where g is the value of the free-fall acceleration at the altitude h. Substituting this expression for F_g into the last equation shows that g is given by

$$g = \frac{GM_E}{r^2} = \frac{GM_E}{(R_E + h)^2} \tag{11.7}$$

Therefore, it follows that *g decreases with increasing altitude.* Recall from Chapter 4 that the gravitational force that acts on any object is called the weight force or the weight of the object. Hence weight varies with position, and decreases with altitude. In the absence of a gravitational field, an object is weightless. Conversely, an object can never be weightless in the presence of a gravitational field. Remember that free fall, or 'apparent weightlessness' is not real weightlessness. In Example 11.2 we calculate the difference in weight of an astronaut on the surface of the Earth and in the International Space Station.

Quick Quiz 11.4

A group of students launch a water rocket with such force that it goes into a circular orbit around the Earth. While the rocket is in orbit, having expelled all the water to get there, what is the magnitude of the acceleration of the rocket? **(a)** It depends on how fast the rocket is thrown. **(b)** It is zero because the rocket does not fall to the ground. **(c)** It is slightly less than 9.80 m/s^2. **(d)** It is equal to 9.80 m/s^2.

Example 11.2

Variation of *g* with altitude *h* – weight in the International Space Station

The International Space Station (ISS) operates at an altitude of 350 km. What is the weight of a 75 kg astronaut in the ISS?

Solution

Conceptualise The mass of the astronaut is fixed; it is independent of his location. Based on the discussion in this section, we realise that the value of g will be reduced at the height of the space station's orbit. Therefore, his weight will be smaller than that at the surface of the Earth.

Model We model the astronaut as a particle subject to a gravitational field. His weight is the gravitational force exerted on him by the field, $w = mg$.

Analyse Use **Equation 11.7** with h = 350 km to find the gravitational field, g, at the orbital location:

$$g = \frac{GM_E}{(R_E + h)^2}$$

$$= \frac{(6.67 \times 10^{-11}\,\text{N.m}^2/\text{kg}^2)(5.97 \times 10^{24}\,\text{kg})}{(6.37 \times 10^6\text{ m} + 0.350 \times 10^6\text{ m})^2} = 8.82\text{ m/s}^2$$

Use this value of g to find the astronaut's weight in the ISS:

$$mg = (75\text{ kg})(8.82\text{ m/s}^2) = 660\text{ N}$$

Finalise The astronaut's weight on Earth is $w = mg = (75\text{ kg})(9.80\text{ m/s}^2) = 740$ N. So his weight on the space station is approximately 90% that on the surface of the Earth. He is in free fall, but he is *not* weightless.

Example 11.3

The density of the Earth

Using the known radius of the Earth and that the average value of g is 9.80 m/s^2 at the Earth's surface, find the average density of the Earth.

Solution

Conceptualise The gravitational field varies with mass and distance from the mass. For a uniform spherical distribution of mass the force is given by **Equation 11.1**, which can then be used to find the field. **Equation 11.6** for g at the Earth's surface was derived from **Equation 11.1**.

Model We model the Earth as a perfect sphere with uniform density. This density is the average density of the Earth. We can use **Equation 11.6** to find the mass of the Earth given the average value of g. To find the density we then need to calculate the volume of the Earth from the measured average radius, and use the definition of density as mass per volume.

Example 11.3 cont.

Analyse Solve **Equation 11.6** for the mass of the Earth:

$$M_E = \frac{gR_E^2}{G}$$

Substitute this mass into the definition of density (**Equation 1.1**):

$$\rho_E = \frac{M_E}{V_E} = \frac{gR_E^2/G}{\frac{4}{3}\pi R_E^3} = \frac{3}{4}\frac{g}{\pi G R_E}$$

Check dimensions:

$$[ML^{-3}] = [LT^{-2}]/[M^{-1}L^3T^{-2}][L] = [ML^{-3}] ☺$$

Then substitute the values:

$$\rho_E = \frac{3}{4}\frac{9.80 \text{ m/s}^2}{\pi(6.67\times10^{-11} \text{N.m}^2/\text{kg}^2)(6.37\times10^6 \text{ m})} = 5.51\times10^3 \text{ kg/m}^3$$

What If? What if you were told that a typical density of granite at the Earth's surface is 2.75×10^3 kg/m^3? What would you conclude about the density of the material in the Earth's interior?

Answer Because this value is about half the density we calculated as an average for the entire Earth, we would conclude that the inner core of the Earth has a density much higher than the average value. Hence, the Cavendish experiment combined with simple free-fall measurements of *g* provides information about the core of the Earth!

11.3 Gravitational potential energy

In Chapter 6 we introduced the concept of gravitational potential energy, which is the energy associated with the configuration of a system of objects interacting via the gravitational force. We emphasised that the gravitational potential energy function $U = mgy$ for a particle–Earth system is valid only when the particle is near the Earth's surface, where the gravitational force is constant. Because the gravitational force between two particles varies as $1/r^2$, we expect that a more general potential energy function – one that is valid without the restriction of the particle having to be near the Earth's surface – will be different from $U = mgy$.

Recall from **Equation 6.21** that the change in the potential energy of a system associated with a given displacement of part of the system is defined as

$$\Delta U = U_f - U_i = -\int_{r_i}^{r_f} F(r)\,dr \tag{11.8}$$

where the force acts on the part of the system being displaced.

Quick Quiz 11.5

Given that the force is proportional to $1/r^2$, how do you expect the potential energy to vary with distance from the Earth?

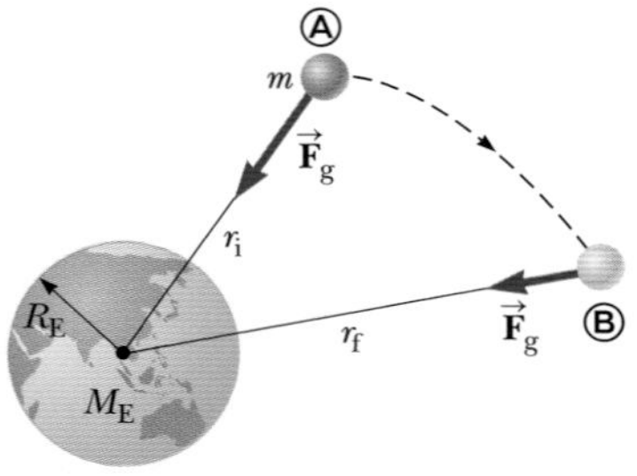

Figure 11.5
As a particle of mass *m* moves from Ⓐ to Ⓑ above the Earth's surface, the gravitational potential energy of the particle–Earth system changes according to **Equation 11.9**.

We can use this result to evaluate the general gravitational potential energy function. Consider a particle of mass *m* moving between two points Ⓐ and Ⓑ above the Earth's surface (**Figure 11.5**). The particle is subject to the gravitational force given by **Equation 11.1**:

$$F(r) = -\frac{GM_E m}{r^2}$$

Substituting this expression for *F*(*r*) into **Equation 11.8**, we find that:

$$U_f - U_i = GM_E m\int_{r_i}^{r_f}\frac{dr}{r^2} = GM_E m\left[-\frac{1}{r}\right]_{r_i}^{r_f}$$

$$U_f - U_i = -GM_E m\left(\frac{1}{r_f} - \frac{1}{r_i}\right) \tag{11.9}$$

The integration of polynomials is described in Appendix B.8.

As always, the choice of a reference configuration for the potential energy is completely arbitrary. It is customary to choose the reference configuration for zero potential energy to be such that the particles are infinitely separated, so that the force they exert on each other is zero. Taking $U_i = 0$ at $r_i = \infty$, we obtain the important result

$$U(r) = -\frac{GM_E m}{r} \quad (11.10)$$

◀ Gravitational potential energy of the Earth–particle system

This expression applies when the particle is separated from the centre of the Earth by a distance r, provided that $r \geq R_E$. The result is not valid for systems where the particle is inside the Earth, where $r < R_E$. Because of our choice of U_i, the function U is always negative (**Figure 11.6**).

Although **Equation 11.10** was derived for the particle–Earth system, it can be applied to any system of two particles. That is, the gravitational potential energy associated with any pair of particles of masses m_1 and m_2 separated by a distance r is

$$U = -\frac{Gm_1 m_2}{r} \quad (11.11)$$

This expression shows that the gravitational potential energy for any pair of particles varies as $1/r$, whereas the force between them varies as $1/r^2$. Furthermore, the potential energy is negative because the force is attractive and we have chosen the potential energy as zero when the particle separation is infinite. Because the force between the particles is attractive, an external agent must do positive work to increase the separation between the particles. The work done by the external agent produces an increase in the potential energy as the two particles are separated. That is, U becomes less negative as r increases. **Figure 11.6** shows the variation of gravitational potential energy with separation.

When two particles are at rest and separated by a distance r, an external agent has to supply an energy at least equal to $+Gm_1m_2/r$ to separate the particles to an infinite distance. It is therefore convenient to think of the absolute value of the potential energy as the *binding energy* of the system. If the external agent supplies an energy greater than the binding energy, the excess energy of the system is in the form of kinetic energy of the particles when the particles are at an infinite separation.

We can extend this concept to three or more particles. In this case, the total potential energy of the system is the sum over all pairs of particles. Each pair contributes a term of the form given by **Equation 11.11**. For example, if the system contains three particles as in **Figure 11.7**,

$$U_{total} = U_{12} + U_{13} + U_{23} = -G\left(\frac{m_1m_2}{r_{12}} + \frac{m_1m_3}{r_{13}} + \frac{m_2m_3}{r_{23}}\right) \quad (11.12)$$

The absolute value of U_{total} represents the work needed to separate the particles by an infinite distance.

In the same way that we defined the gravitational field, $\vec{\mathbf{g}}$, due to an object as the force per unit mass acting on a small test particle in the region of that object, we define the **gravitational potential** as the **potential energy per unit mass** of a test particle. Consider again the particle–Earth example. The gravitational potential energy of this system as given by **Equation 11.10** is $U(r) = -GM_E m/r$, from which we can find the gravitational potential due to the Earth at r by dividing the potential energy by the mass of the test particle, to give:

$$V_g \equiv \frac{U_g}{\mathrm{m}} \quad (11.13)$$

so

$$V_g(r) = \frac{-GM_E}{r} \quad (11.14)$$

Pitfall Prevention 11.2

A zero potential does not mean a zero force and vice versa. The gravitational field infinitely far from a particle is zero. We also choose this as our zero of gravitational potential. In general, however, remember that field depends on the rate of change of potential and that zero potential *at a point* does not imply a zero field. Nor does a zero field *at a point* imply a zero potential at that point.

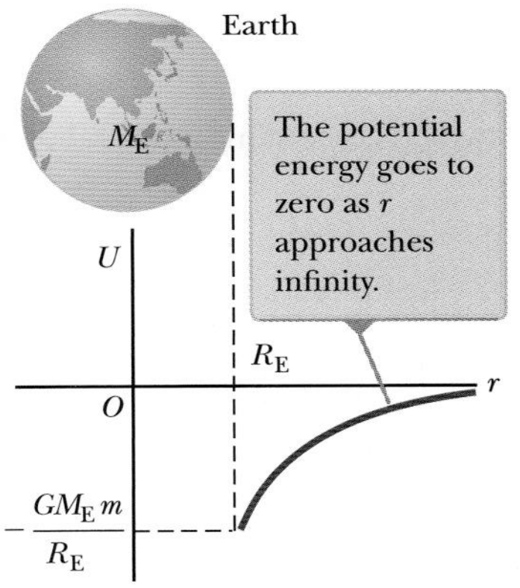

Figure 11.6
Graph of the gravitational potential energy U versus r for the system of an object above the Earth's surface

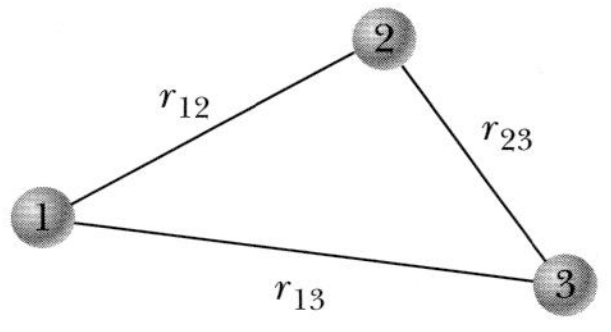

Figure 11.7
Three interacting particles

which can be generalised to give the gravitational potential at a distance r from any object of mass m:

$$V_g(r) = \frac{-Gm}{r} \tag{11.15}$$

Example 11.4

The change in potential energy

A student releases a weather balloon of mass m near the surface of the Earth and the balloon rises a small vertical distance Δy.

(A) Show that in this situation the general expression for the change in gravitational potential energy given by **Equation 11.9** reduces to the familiar relationship $\Delta U = mg\,\Delta y$.

Solution

Conceptualise We have developed two equations for gravitational potential energy: **Equation 11.9** for the case when two objects are far apart and **Equation 6.16** for an object close to the surface of a planet. We want to show that these equations are equivalent for a small object moving close to the surface of the Earth.

Model We are considering the case when the weather balloon is close to the Earth's surface. We model the balloon as a test particle and note that the change in height of the particle is small compared to the radius of the Earth.

Analyse We wish to show that these two expressions are equivalent. Combine the fractions in **Equation 11.9**:

$$\Delta U = -GM_E m\left(\frac{1}{r_f} - \frac{1}{r_i}\right) = GM_E m\left(\frac{r_f - r_i}{r_i r_f}\right) \tag{1}$$

Evaluate $r_f - r_i$ and $r_i r_f$ if both the initial and final positions of the particle are close to the Earth's surface:

$$r_f - r_i = \Delta y, \qquad r_i r_f \approx R_E^{\,2}$$

Substitute these expressions into **Equation (1)**:

$$\Delta U \approx \frac{GM_E m}{R_E^{\,2}}\Delta y = mg\Delta y$$

which is the same as **Equation 6.16**, where $g = \dfrac{GM_E}{R_E^2}$ (**Equation 11.6**) = 9.80 m/s².

(B) The balloon continues to rise into the upper atmosphere. Find the height at which the surface equation $\Delta U = mg\,\Delta y$ gives a 1.0% error in the change in the potential energy.

Analyse Set up a ratio reflecting a 1.0% error, noting that we expect the surface equation to give the larger value:

$$\frac{\Delta U_{\text{surface}}}{\Delta U_{\text{general}}} = 1.010$$

Substitute the expressions for each of these changes ΔU:

$$\frac{mg\Delta y}{GM_E m(\Delta y / r_i r_f)} = \frac{g r_i r_f}{GM_E} = 1.010$$

Substitute for r_i, r_f and g from **Equation 11.6**:

$$\frac{(GM_E/R_E^{\,2})R_E(R_E + \Delta y)}{GM_E} = \frac{R_E + \Delta y}{R_E} = 1 + \frac{\Delta y}{R_E} = 1.010$$

Solve for Δy:

$$\Delta y = 0.010R_E = 0.010(6.37 \times 10^6 \text{ m}) = 6.37 \times 10^4 \text{ m} = 63.7 \text{ km}$$

Finalise At this height the balloon would already be above 99.9% of the Earth's atmosphere, as air density decreases exponentially with altitude. Hence it is unlikely that the students will need to be concerned about correcting for the non-uniform nature of the Earth's gravitational field in this experiment; they can reasonably treat the Earth as flat and use the surface equation approximation.

11.4 Energy considerations in planetary and satellite motion

Consider an object of mass m moving with a speed v in the vicinity of a massive object of mass M, where $M \gg m$. The system might be a planet moving around the Sun, a satellite in orbit around the Earth as in **Figure 11.8** or a comet making a one-time flyby of the Sun. If we assume the object of mass M is at rest in an inertial reference frame, the total mechanical energy E of the two-object system when the objects are separated by a distance r is the sum of the kinetic energy of the object of mass m and the potential energy of the system given by **Equation 11.11**:

$$E = K + U$$

$$E = \frac{1}{2}mv^2 - \frac{GMm}{r} \quad (11.16)$$

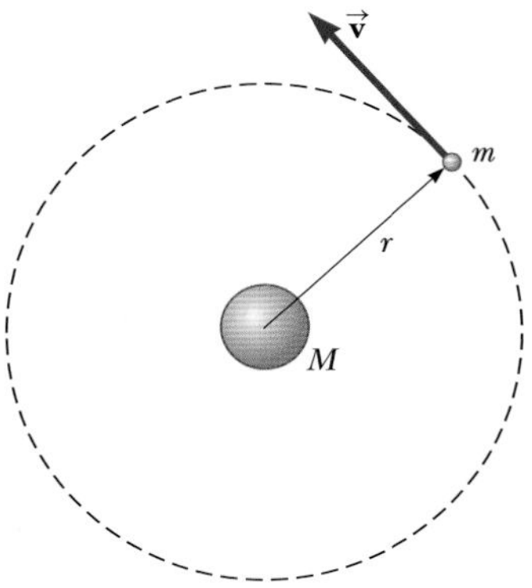

Figure 11.8
An object of mass m moving in a circular orbit about a much larger object of mass M

Equation 11.16 shows that E may be positive, negative or zero, depending on the value of v. For a bound system such as the Earth–Sun system, however, E is necessarily *less than zero* because we have chosen the convention that $U \to 0$ as $r \to \infty$.

In this case Newton's second law applied to the object of mass m gives

$$F_g = ma \Rightarrow \frac{GMm}{r^2} = \frac{mv^2}{r}$$

Multiplying both sides by r and dividing by 2 gives

$$\frac{1}{2}mv^2 = \frac{GMm}{2r} \quad (11.17)$$

Substituting this equation into **Equation 11.16**, we obtain

$$E = \frac{GMm}{2r} - \frac{GMm}{r}$$

$$E = -\frac{GMm}{2r} \quad \text{(circular orbits)} \quad (11.18)$$

◄ Total energy for circular orbits

This result shows that the total mechanical energy is negative in the case of circular orbits. Notice that the kinetic energy is positive and equal to half the absolute value of the potential energy. The absolute value of E is also equal to the binding energy of the system because this amount of energy must be provided to the system to move the two objects infinitely far apart.

The total mechanical energy is also negative in the case of elliptical orbits. The expression for E for elliptical orbits is the same as **Equation 11.18** with r replaced by the semimajor axis length a:

$$E = -\frac{GMm}{2a} \quad \text{(elliptical orbits)} \quad (11.19)$$

◄ Total energy for elliptical orbits

Furthermore, the total energy is constant if we assume the system is isolated. Therefore, as the object of mass m moves from Ⓐ to Ⓑ in **Figure 11.5**, the total energy remains constant and **Equation 11.16** gives

$$E = \frac{1}{2}mv_i^2 - \frac{GMm}{r_i} = \frac{1}{2}mv_f^2 - \frac{GMm}{r_f} \quad (11.20)$$

Because the total energy of an isolated gravitationally bound system is constant, regardless of the motion of the bound particles, the energy is called a *constant of the motion*. It is often useful when analysing a system to identify constants of the motion. As we shall see, angular momentum is also a constant of the motion for orbiting systems.

Quick Quiz 11.6

A comet moves in an elliptical orbit around the Sun. Which point in its orbit (closest or furthest from the Sun) represents the highest value of (a) the speed of the comet, (b) the potential energy of the comet–Sun system, (c) the kinetic energy of the comet and (d) the total energy of the comet–Sun system?

Example 11.5

Changing the orbit of a satellite

A space transportation vehicle releases a 470 kg communications satellite while in an orbit 280 km above the surface of the Earth. A rocket engine on the satellite is required to boost it into a geosynchronous orbit at 36 000 km. How much energy does the engine have to provide?

Solution

Conceptualise Energy must be expended to raise the satellite to this much higher position, to do work against the gravitational field of the Earth. We start by drawing a diagram.

Model We model the satellite as a particle in the Earth's gravitational field. The system of interest is the satellite and the Earth, which in this case is not an isolated system because work is being done to increase the potential energy of the system.

Analyse Use **Equation 11.8** to find the change in potential energy. This must be equal to the work done by the engine.

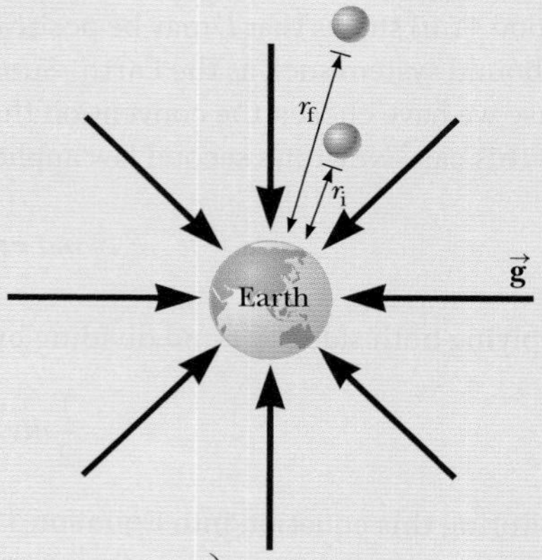

First, find the initial and final radius of the satellite's orbit:

$$r_i = R_E + 280\text{ km} = 6.37 \times 10^6\text{ m} + 280 \times 10^3\text{ m} = 6.65 \times 10^6\text{ m}$$

$$r_f = R_E + 36\,000\text{ km} = 6.37 \times 10^6\text{ m} + 3.60 \times 10^7\text{ m} = 4.22 \times 10^7\text{ m}$$

Use **Equation 11.18** to find the difference in energies for the satellite–Earth system with the satellite at the initial and final radii:

$$\Delta E = E_f - E_i = -\frac{GM_Em}{2r_f} - \left(-\frac{GM_Em}{2r_i}\right) = -\frac{GM_Em}{2}\left(\frac{1}{r_f} - \frac{1}{r_i}\right)$$

$$\Delta E = -\frac{(6.67 \times 10^{-11}\text{ N.m}^2/\text{kg}^2)(5.97 \times 10^{24}\text{ kg})(470\text{ kg})}{2} \times \left(\frac{1}{4.22 \times 10^7\text{ m}} - \frac{1}{6.65 \times 10^6\text{ m}}\right)$$

$$= 1.19 \times 10^{10}\text{ J}$$

Finalise This is the energy equivalent of 340 L of petrol. NASA engineers must account for the changing mass of the spacecraft as it ejects burned fuel, something we have not done here. Would you expect the calculation that includes the effect of this changing mass to yield a greater or a lesser amount of energy required from the engine?

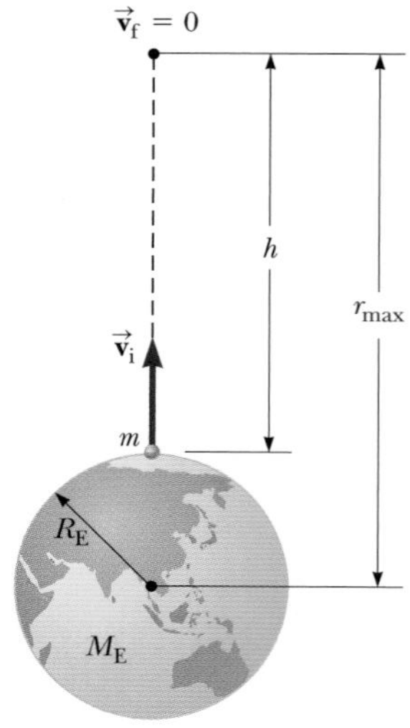

Figure 11.9
An object of mass m projected upwards from the Earth's surface with an initial speed v_i reaches a maximum altitude h.

Escape speed

Suppose an object of mass m is projected vertically upwards from the Earth's surface with an initial speed v_i as illustrated in **Figure 11.9**. We can use energy considerations to find the minimum value of the initial speed needed to allow the object to move infinitely far away from the Earth. **Equation 11.16** gives the total energy of the system for any configuration. As the object is projected upwards from the surface of the Earth, $v = v_i$ and $r = r_i = R_E$. When the object reaches its maximum altitude, $v = v_f = 0$ and $r = r_f = r_{max}$. Because the total energy of the isolated object–Earth system is constant, substituting these conditions into **Equation 11.20** gives

$$\frac{1}{2}mv_i^2 - \frac{GM_Em}{R_E} = -\frac{GM_Em}{r_{max}}$$

Solving for v_i^2 gives

$$v_i^2 = 2GM_E\left(\frac{1}{R_E} - \frac{1}{r_{max}}\right) \quad \textbf{(11.21)}$$

For a given maximum altitude $h = r_{max} - R_E$, we can use this equation to find the required initial speed.

We are now in a position to calculate **escape speed**, which is the minimum speed the object must have at the Earth's surface to approach an infinite separation distance from the Earth. Travelling at this minimum speed, the object continues to move further and further away from the Earth as its speed asymptotically approaches zero. Letting $r_{max} \to \infty$ in Equation 11.21 and taking $v_i = v_{esc}$ gives

$$v_{esc} = \sqrt{\frac{2GM_E}{R_E}} \quad (11.22)$$

◀ Escape speed from the Earth's surface

Notice that this expression for v_{esc} is independent of the mass of the object. In other words, a spacecraft has the same escape speed as a molecule. We have ignored air resistance and the rotation of the Earth here, both of which must be taken into account when launching a spacecraft.

If the object is given an initial speed equal to v_{esc}, the total energy of the system is equal to zero. Notice that when $r \to \infty$, the object's kinetic energy and the potential energy of the system are both zero. If v_i is greater than v_{esc}, the total energy of the system is greater than zero and the object has some residual kinetic energy as $r \to \infty$. This is called an **unbound** system because the two objects may become infinitely separated. If v_i is less than v_{esc} then the total energy is negative. This is called a **bound** system and the objects will always have a finite separation.

Example 11.6

Escape speed of a rocket

Calculate the escape speed from the Earth for a 5000 kg spacecraft and determine the kinetic energy it must have at the Earth's surface to move infinitely far away from the Earth.

Solution

Conceptualise Imagine projecting the spacecraft from the Earth's surface so that it moves further and further away, travelling more and more slowly, with its speed approaching zero. Its speed will never reach zero, however, so the object will never turn around and come back.

Model We model the spacecraft as a particle in the Earth's gravitational field. We will ignore air resistance, and the rotation of the Earth.

Analyse Use Equation 11.22 to find the escape speed: $v_{esc} = \sqrt{\frac{2GM_E}{R_E}} = \sqrt{\frac{2\,(6.67 \times 10^{-11}\,\text{N.m}^2/\text{kg}^2)(5.97 \times 10^{24}\,\text{kg})}{6.37 \times 10^6\,\text{m}}}$

$$= 1.12 \times 10^4\ \text{m/s}$$

Evaluate the kinetic energy of the spacecraft from Equation 6.13: $K = \frac{1}{2}mv_{esc}^2 = \frac{1}{2}(5.00 \times 10^3\ \text{kg})(1.12 \times 10^4\ \text{m/s})^2$

$$= 3.13 \times 10^{11}\ \text{J}$$

Finalise The calculated escape speed corresponds to about 40 000 km/h. Would the calculated speed be larger or smaller if we included air resistance in our model? What about the rotation of the Earth?

Conceptual Example 11.7

Launching a rocket

Rocket launch sites are generally close to the equator, for example the Vikram Sarabhai Space Centre in India is only 8.5° north of the equator, and the Kennedy Space Centre at Cape Canaveral in Florida, where the Apollo moon missions were launched, is 29° north of the equator. Why are launch sites close to the equator preferable to launch sites close to the poles? And which direction are rockets usually launched in and why?

Conceptual Example 11.7 cont.

Solution

The Earth rotates on its axis once every 24 hours. If you consider the Earth as a rotating rigid body then every point on the Earth rotates with the same angular velocity, ω. The linear velocity, $v = \omega r$, of a point on the Earth depends on the distance from the axis of rotation. This distance is greatest at the equator, where the linear speed is about 460m/s. (You should do the calculation to check!) So if a rocket is launched from the equator, it already has a speed of 460m/s, and not as much energy is needed to accelerate it up to the required escape speed. However, it needs to be launched in the right direction. The right direction to make use of the Earth's rotation is in the same direction as the rotation. As the Earth rotates eastwards, rockets are launched to the east.

In general, the escape speed from the surface of any planet of mass M and radius R is

$$v_{\text{esc}} = \sqrt{\frac{2GM}{R}} \tag{11.23}$$

Table 11.1
Escape speeds from the surfaces of the Moon, the Sun and the planets.

Planet	Mercury	Venus	Earth	Mars	Jupiter	Saturn	Uranus	Neptune	Moon	Sun
v_{esc} (km/s)	4.3	10.3	11.2	5.0	60	36	22	24	2.3	618

Escape speeds for the planets, the Moon and the Sun are provided in **Table 11.1**. The values vary from 2.3 km/s for the Moon to about 618 km/s for the Sun. These results, together with some ideas from the kinetic theory of gases (see Chapter 21), explain why some planets have atmospheres and others do not. As we shall see later, the average kinetic energy of a gas molecule depends only on the temperature. Hence lighter molecules, such as hydrogen and helium, have a higher average speed than heavier molecules at the same temperature. When the average speed of the lighter molecules is not much less than the escape speed of a planet, a significant fraction of them have a chance to escape.

This mechanism also explains why the Earth does not retain hydrogen molecules and helium atoms in its atmosphere but does retain heavier molecules, such as oxygen and nitrogen. Hydrogen bonds with other atoms such as oxygen, so is retained on Earth in molecular form. Helium is inert so is it not retained – it is a genuinely non-renewable resource! The helium present on the Earth was formed due to radioactive decays, and this process is far too slow to compensate for the rate at which helium is lost from the atmosphere after escaping from helium balloons and the like.

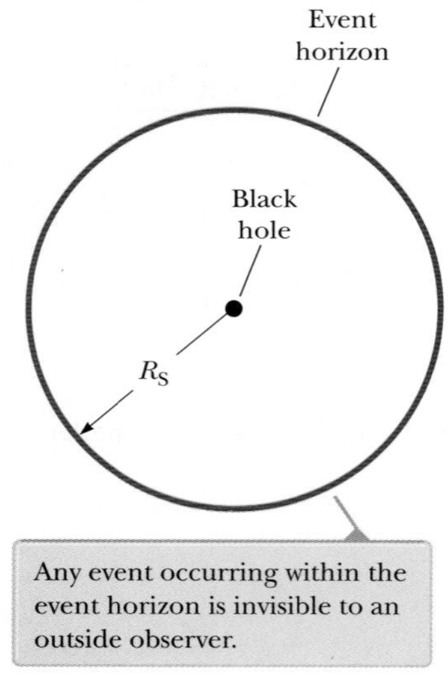
Event horizon
Black hole
R_S
Any event occurring within the event horizon is invisible to an outside observer.

Figure 11.10
A black hole. The distance R_S equals the Schwarzschild radius.

Black holes

A supernova is the catastrophic explosion of a very massive star. The material that remains in the central core of such an object continues to collapse, and the core's ultimate fate depends on its mass. If the core has a mass less than 1.4 times the mass of our Sun, it gradually cools down and ends its life as a white dwarf star. If the core's mass is greater than this value, it may collapse further due to gravitational forces and become a neutron star.

When the core has a mass greater than about three solar masses, the collapse may continue until the star becomes a very small object in space, commonly referred to as a **black hole**. In effect, black holes are remains of stars that have collapsed under their own gravitational force.

The escape speed for a black hole is very high because of the concentration of the star's mass into a sphere of very small radius (see **Equation 11.23**). If the escape speed exceeds the speed of light c, radiation from the object (such as visible light) cannot escape and the object appears to be black (hence 'black hole'). The critical radius R_S at which the escape speed is c is called the **Schwarzschild radius** (**Figure 11.10**).

The imaginary surface of a sphere of this radius surrounding the black hole is called the **event horizon**, which is the limit of how close you can approach the black hole and hope to escape.

There is evidence that supermassive black holes exist at the centres of galaxies, including a supermassive black hole of mass 2–3 million solar masses at the centre of our own galaxy.

11.5 Kepler's laws and the motion of planets

iStockphoto.com/GeorgiosArt

Johannes Kepler
German astronomer (1571–1630)
Kepler is best known for developing the laws of planetary motion based on the careful observations of Tycho Brahe.

Humans have observed the movements of the planets, stars and other celestial objects for thousands of years. In early history, these observations led scientists to regard the Earth as the centre of the Universe. This *geocentric model* was elaborated and formalised by the Greek astronomer Claudius Ptolemy (c. 100–c. 170) in the second century and was accepted for the next 1400 years. In 1543, Polish astronomer Nicolaus Copernicus (1473–1543) suggested that the Earth and the other planets revolved in circular orbits around the Sun (the *heliocentric model*).

Danish astronomer Tycho Brahe (1546–1601) wanted to determine how the heavens were constructed and made observations of the planets and 777 stars visible to the naked eye with only a large sextant and a compass. (The telescope had not yet been invented.) German astronomer Johannes Kepler was Brahe's assistant for a short while before Brahe's death, whereupon he acquired his mentor's astronomical data and spent 16 years trying to deduce a mathematical model for the motion of the planets.

Kepler's complete analysis of planetary motion is summarised in three statements known as **Kepler's laws:**

◀ Kepler's laws

1. All planets move in elliptical orbits with the Sun at one focus.
2. The radius vector drawn from the Sun to a planet sweeps out equal areas in equal time intervals.
3. The square of the orbital period of any planet is proportional to the cube of the semimajor axis of the elliptical orbit.

Kepler's first law

Kepler's first law indicates that the circular orbit is a special case and elliptical orbits are the general situation. This notion was difficult for scientists of the time to accept because they believed that perfect circular orbits of the planets reflected the perfection of heaven.

Active Figure 11.11 shows the geometry of an ellipse. An ellipse is mathematically defined by choosing two points F_1 and F_2, each of which is a called a **focus**, and then drawing a curve through points for which the sum of the distances r_1 and r_2 from F_1 and F_2 respectively is a constant. The longest distance through the centre between points on the ellipse (and passing through each focus) is called the **major axis**, and this distance is $2a$. In Active Figure 11.11, the major axis is drawn along the x direction. The distance a is called the **semimajor axis**. Similarly, the shortest distance through the centre between points on the ellipse is called the **minor axis** of length $2b$, where b is the **semiminor axis**. Each focus is located a distance c from the centre of the ellipse, where $a^2 = b^2 + c^2$. In the elliptical orbit of a planet around the Sun, the Sun is at one focus of the ellipse. There is nothing at the other focus.

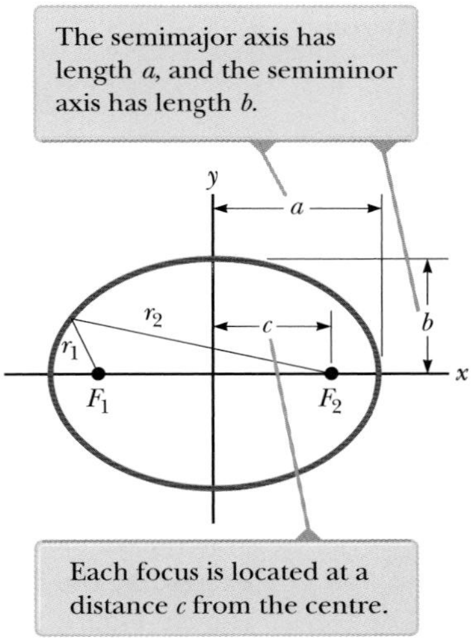
The semimajor axis has length a, and the semiminor axis has length b.
y
a
b
x
r_1
r_2
c
F_1
F_2
Each focus is located at a distance c from the centre.

Active Figure 11.11
Plot of an ellipse

The **eccentricity** of an ellipse is defined as $e = c/a$, and it describes the general shape of the ellipse. For a circle, $c = 0$, and the eccentricity is therefore zero. In Active Figure 11.11 as b decreases, c increases and the eccentricity e increases. Therefore, higher values of eccentricity correspond to longer and thinner ellipses. The range of values of the eccentricity for an ellipse is $0 < e < 1$.

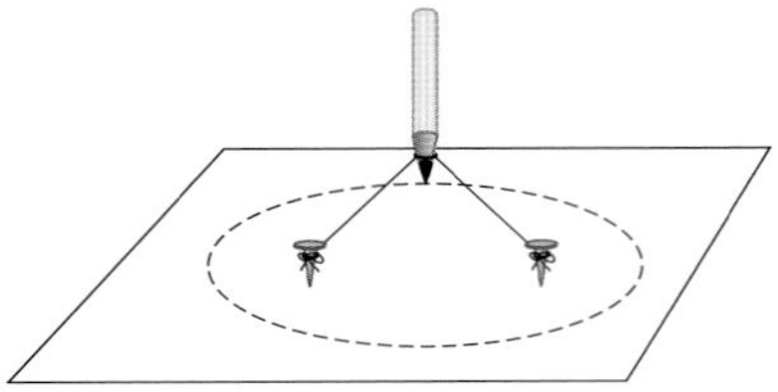

TRY THIS

You can draw an ellipse by using a piece of string, two pins and a pencil. Push the two pins into the surface you wish to draw an ellipse on. These will be the two foci. Tie one end of the string to each pin (or push the pins through the string). Place the pencil against the string and pull it taut. Holding the string taut with the pencil, use the string to guide the pencil to draw an arc. You will need to draw the upper and lower halves of the ellipse separately so the string doesn't catch on the pins.

Pitfall Prevention 11.3

The Sun is located at one focus of the elliptical orbit of a planet. It is *not* located at the centre of the ellipse.

Eccentricities for planetary orbits vary widely in the solar system. The eccentricity of the Earth's orbit is 0.017, which makes it nearly circular. In contrast, the eccentricity of Mercury's orbit is 0.21, the highest of the eight planets. **Figure 11.12a** shows an ellipse with an eccentricity equal to that of Mercury's orbit. Notice that even this highest-eccentricity orbit is difficult to distinguish from a circle. The eccentricity of the orbit of Comet Halley is 0.97, describing an orbit whose major axis is much longer than its minor axis, as shown in **Figure 11.12b**. As a result, Comet Halley spends much of its 76-year period far from the Sun and not visible from the Earth. It is only visible to the naked eye during a small part of its orbit when it is near the Sun.

Consider a planet in an elliptical orbit such as that shown in **Active Figure 11.11**, with the Sun at focus F_2. When the planet is farthest from the Sun, the distance between the planet and the Sun is $a + c$; this is called the *aphelion*. For an object in orbit around the Earth, this point is called the *apogee*. When the planet is at its closest to the Sun, the distance between the planet and the Sun is $a - c$, this point is called the *perihelion*, or for an Earth orbit, the *perigee*.

Kepler's first law is a direct result of the inverse-square nature of the gravitational force. Circular and elliptical orbits are the allowed shapes of orbits for objects that are *bound* to the gravitational force centre. These objects include planets, asteroids and comets that orbit the Sun, and moons orbiting a planet. There are also *unbound* objects, such as meteoroids from deep space that might pass by the Sun once and then never return. The allowed paths for these objects are parabolas ($e = 1$) and hyperbolas ($e > 1$).

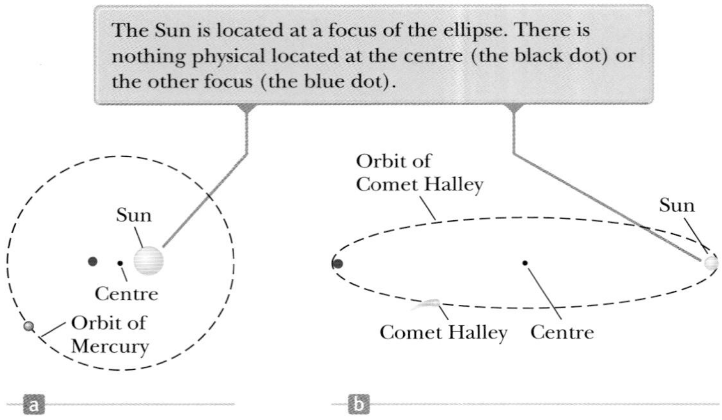

The Sun is located at a focus of the ellipse. There is nothing physical located at the centre (the black dot) or the other focus (the blue dot).
Sun
Centre
Orbit of Mercury
a
Orbit of Comet Halley
Sun
Comet Halley
Centre
b

Figure 11.12
(a) The shape of the orbit of Mercury, which has the highest eccentricity ($e = 0.21$) among the eight planets in the solar system (b) The shape of the orbit of Comet Halley. The shape of the orbit is correct but the comet and the Sun are shown larger than in reality for clarity.

Kepler's second law

Kepler's second law is a consequence of angular momentum conservation for an isolated system. **Active Figure 11.13a** shows a planet of mass M_p moving about the Sun in an elliptical orbit. Consider the planet as a system. We model the Sun to be much more massive than the planet and not moving. The gravitational force exerted by the Sun on the planet is a central force, always along the radius vector, directed toward the Sun (**Active Figure 11.13a**). The torque on the planet due to this central force is zero because $\vec{\mathbf{F}}_g$ is parallel to $\vec{\mathbf{r}}$. Because the external torque on the planet is zero, it is modelled as an isolated system for angular momentum, and the angular momentum $\vec{\mathbf{L}}$ of the planet is a constant of the motion:

$$\vec{\mathbf{L}} = \vec{\mathbf{r}} \times \vec{\mathbf{p}} = M_p \vec{\mathbf{r}} \times \vec{\mathbf{v}} = \text{constant}$$

Σ *The vector cross product is reviewed in Appendix B.6 and derivatives are reviewed in Appendix B.7.*

We can relate this result to the following geometric consideration. In a time interval dt, the radius vector $\vec{\mathbf{r}}$ in **Active Figure 11.13b** sweeps out the area dA, which equals half the area $|\vec{\mathbf{r}} \times d\vec{\mathbf{r}}|$ of the parallelogram formed by the vectors $\vec{\mathbf{r}}$ and $d\vec{\mathbf{r}}$. Because the displacement of the planet in the time interval dt is given by $d\vec{\mathbf{r}} = \vec{\mathbf{v}}\, dt$,

$$dA = \frac{1}{2}|\vec{\mathbf{r}} \times d\vec{\mathbf{r}}| = \frac{1}{2}|\vec{\mathbf{r}} \times \vec{\mathbf{v}}\, dt| = \frac{L}{2M_p}dt$$

$$\frac{dA}{dt} = \frac{L}{2M_p} \qquad (11.24)$$

where L and M_p are both constants. This result shows that that the radius vector from the Sun to any planet sweeps out equal areas in equal time intervals as stated in Kepler's second law.

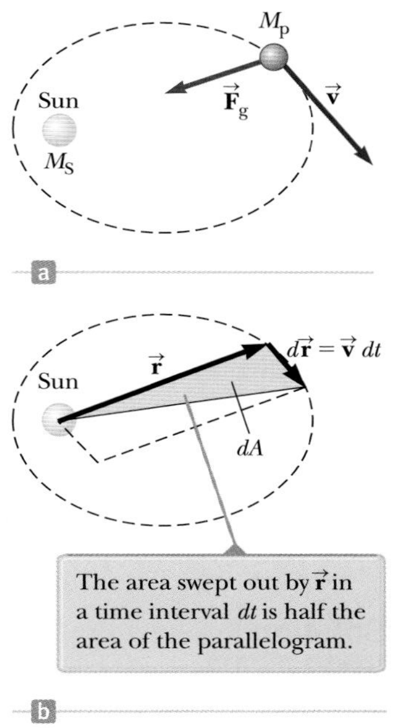

M_p
Sun
M_S
$\vec{F}_g$
$\vec{v}$
a
Sun
$\vec{r}$
$d\vec{r} = \vec{v}\,dt$
dA
The area swept out by $\vec{r}$ in a time interval dt is half the area of the parallelogram.
b

Active Figure 11.13
(a) The gravitational force acting on a planet is directed towards the Sun. (b) During a time interval dt, a parallelogram is formed by the vectors $\vec{\mathbf{r}}$ and $d\vec{\mathbf{r}} = \vec{\mathbf{v}}dt$.

Kepler's third law

Kepler's third law can be predicted from the inverse-square law for circular orbits. Consider a planet of mass M_p that is moving about the Sun (mass M_S) in a circular orbit as in **Figure 11.14**. Because the gravitational force provides the centripetal acceleration of the planet as it moves in a circle, we model the planet as a particle subject to a net force and as a particle in uniform circular motion and incorporate Newton's law of universal gravitation:

$$F_g = M_p a \rightarrow \frac{GM_S M_p}{r^2} = M_p\left(\frac{v^2}{r}\right)$$

The orbital speed of the planet is $2\pi r/T$, where T is the period; therefore:

$$\frac{GM_S}{r^2} = \frac{(2\pi r / T)^2}{r}$$

$$T^2 = \left(\frac{4\pi^2}{GM_S}\right)r^3 = K_S r^3$$

where K_S is a constant depending on the mass of the Sun.

This equation is also valid for elliptical orbits if we replace r with the length a of the semimajor axis (**Active Figure 11.11**):

$$T^2 = \left(\frac{4\pi^2}{GM_S}\right)a^3 = K_S a^3 \qquad (11.25)$$

◀ Kepler's third law

Equation 11.25 is Kepler's third law, which was stated in words at the beginning of this section. Because the semimajor axis of a circular orbit is its radius, this equation is valid for both circular and elliptical orbits. If we were to consider the orbit of a satellite such as the Moon about the Earth, the constant K_S would have a different value, with the Sun's mass replaced by the Earth's mass; that is, $K_E = 4\pi^2/GM_E$.

A table of useful data for objects in our solar system is given in the front endpapers.

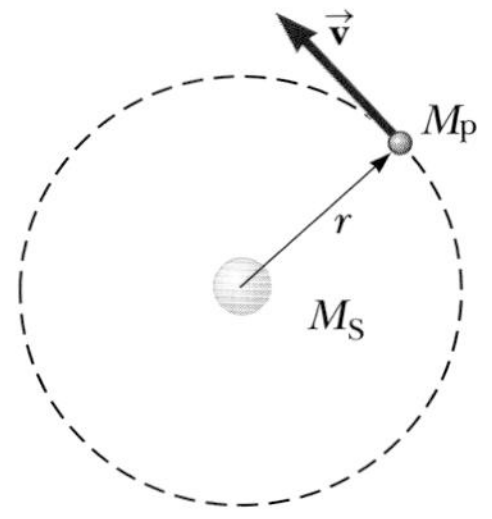

$\vec{v}$
M_p
r
M_S

Figure 11.14
A planet of mass M_p moving in a circular orbit around the Sun. The orbits of all planets except Mercury are nearly circular.

Quick Quiz 11.7

Using the table of planetary data in the endpapers find the value for T^2/r^3 for our solar system and verify that it is a constant for all the planets. Note that uncertainties in the measured period and radius will give slight variations in the value that you calculate.

Conceptual Example 11.8

GRACE and measurement of local gravity

The Gravity Recovery and Climate Experiment (GRACE) uses two closely spaced satellites in circular orbits, one following the other in the same orbit, to measure variations in the Earth's gravitational field by measuring the distance between the satellites. How does the distance between the satellites give information about the local gravitational field?

Solution

If the local gravitational field were constant throughout the orbit, which we expect for a spherically symmetric distribution of mass and a circular orbit, then both satellites would move at the same constant speed, and the distance between them would be constant. In fact the distance between them varies in time. Changes in the local gravitational field cause the satellites to accelerate. If the leading satellite starts to move faster because of an increase in the local field, then the distance between the satellites increases until the following satellite reaches that area and is also accelerated. Hence any change in their spacing indicates a changing value of the field, and hence of the distribution of the mass of the Earth.

Example 11.9

A geosynchronous satellite

Consider a satellite of mass m moving in a circular orbit around the Earth at a constant speed v and at an altitude h above the Earth's surface, such that the satellite is geosynchronous, that is, it remains above a fixed point on the surface of the Earth. Determine the speed of the satellite v and its height h above the Earth.

Solution

Conceptualise Imagine the satellite moving around the Earth in a circular orbit under the influence of the gravitational force. Start by drawing a diagram.

Model The satellite must have a centripetal acceleration. Therefore, we model the satellite as a particle subject to a net force and a particle in uniform circular motion. The satellite is well above the atmosphere, so the only external force acting on the satellite is the gravitational force, which acts towards the centre of the Earth and keeps the satellite in its circular orbit.

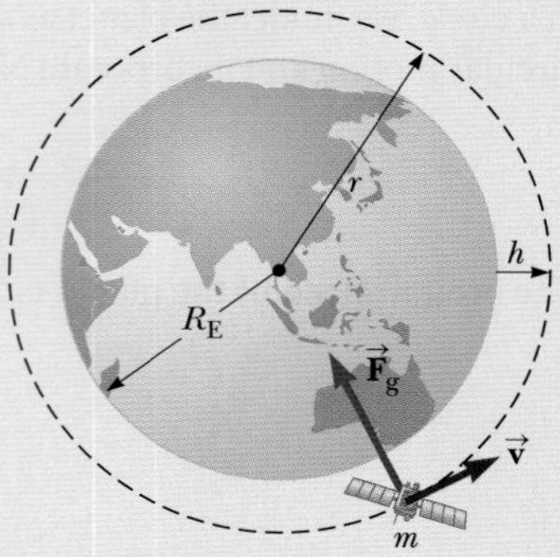

Figure 11.15 (Example 11.9) A satellite of mass m moving around the Earth in a circular orbit of radius r with constant speed v. The only force acting on the satellite is the gravitational force $\vec{\mathbf{F}}_g$. (Not drawn to scale)

Analyse Apply the particle subject to a net force and particle in uniform circular motion models to the satellite:

$$F_g = ma \rightarrow G\frac{M_E m}{r^2} = m\left(\frac{v^2}{r}\right)$$

Solve for v, noting that the distance r from the centre of the Earth to the satellite is $r = R_E + h$:

$$v = \sqrt{\frac{GM_E}{r}} = \sqrt{\frac{GM_E}{R_E + h}} \quad (1)$$

Check dimensions:

$$[LT^{-1}] = ([M^{-1}L^3T^{-2}][M]/[L])^{1/2} = [LT^{-1}]$$ ☺

To remain over a fixed position on the Earth, the period of the satellite must be 24 h = 86 400 s and the satellite must be in orbit directly over the equator.

Solve Kepler's third law (**Equation 11.25**, with $a = r$ and replacing M_S with M_E) for r:

$$r = \left(\frac{GM_E T^2}{4\pi^2}\right)^{1/3}$$

Substitute numerical values:

$$r = \left[\frac{(6.67 \times 10^{-11}\ \text{N.m}^2/\text{kg}^2)(5.67 \times 10^{24}\ \text{kg})(86\,400\ \text{s})^2}{4\pi^2}\right]^{1/3}$$

$$= 4.22 \times 10^7\ \text{m}$$

So $h = r - R_E = 4.22 \times 10^7\ \text{m} - 6.37 \times 10^6\ \text{m} = 3.58 \times 10^7\ \text{m}$

Use **Equation (1)** to find the speed of the satellite:

$$v = \sqrt{\frac{(6.67 \times 10^{-11}\ \text{N.m}^2/\text{kg}^2)(5.97 \times 10^{24}\ \text{kg})}{4.22 \times 10^7\ \text{m}}}$$

$$= 3.07 \times 10^3\ \text{m/s}$$

Finalise The satellite has a height above the surface of the Earth of almost 36 000 km. This is around ten times greater than the radius of the Earth, so our initial diagram is very far from a scale drawing.

What If? What if the satellite were orbiting at height h above the surface of another planet more massive than the Earth but of the same radius? Would the satellite be moving at a higher speed or a lower speed than it does around the Earth?

Answer If the planet exerts a larger gravitational force on the satellite due to its larger mass, the satellite must move with a higher speed to avoid moving towards the surface. This conclusion is consistent with the predictions of **Equation (1)**, which shows that because the speed v is proportional to the square root of the mass of the planet, the speed increases as the mass of the planet increases.

Dark matter

Kepler's third law, **Equation 11.25**, allows us to calculate the mass, M_s, of the large central body about which a planet or satellite orbits. It also applies to larger bodies, such as suns and solar systems about the galactic centre and entire galaxies within galaxy clusters. What has been observed is that stars and galaxies, particularly in larger orbits, have a shorter period than can be accounted for by the visible mass in the galaxy or galaxy cluster. The visible mass is matter which is either luminous, like stars, or reflects light, such as planets. This implies that there is mass which we *cannot* see, acting as a source of gravitational attraction. Because we cannot see this matter (it neither emits nor reflects light), it is called *dark matter*. Current theories suggest that the majority of the mass in the universe is dark matter. Dark matter and dark energy are discussed further in Chapter 44.

11.6 The general theory of relativity

Up to this point, we have sidestepped a curious puzzle. Mass has two seemingly different properties: a *gravitational attraction* for other masses and an *inertial* property that represents a resistance to acceleration. To designate these two attributes, we use the subscripts g and i and write

$$\text{Gravitational property:} \quad F_g = m_g g$$
$$\text{Inertial property:} \quad \sum F = m_i a$$

The value for the gravitational constant G was chosen to make the magnitudes of m_g and m_i numerically equal. Regardless of how G is chosen, however, the strict proportionality of m_g and m_i has been established experimentally to an extremely high degree: a few parts in 10^{12}. Therefore, it appears that gravitational mass and inertial mass may indeed be exactly equal.

Why, though? They seem to involve two entirely different concepts: a force of mutual gravitational attraction between two masses and the resistance of a single mass to being accelerated by any type of force. This question, which puzzled Newton and many other physicists over the years, was addressed by Einstein in 1916 when he published his theory of gravitation, known as the *general theory of relativity*. Because it is a mathematically complex theory, we shall discuss it in qualitative terms only.

In Einstein's view, the dual behaviour of mass was evidence for a very intimate and basic connection between the two behaviours. He pointed out that no mechanical experiment (such as dropping an object) could distinguish between the two situations illustrated in **Figures 11.16a** and **11.16b**. In **Figure 11.16a**, a person standing in a lift on the surface of a planet feels pressed into the floor due to the gravitational force. If he releases his briefcase, he observes it moving towards the floor with acceleration $\vec{\mathbf{g}} = -g\hat{\mathbf{j}}$. In **Figure 11.16b**, the person is in a lift in empty space accelerating upwards with $\vec{\mathbf{a}}_{el} = +g\hat{\mathbf{j}}$. The person feels pressed into the floor with the same force as in **Figure 11.16a**. If he releases his briefcase, he observes it moving towards the floor with acceleration g, exactly as in the previous situation. In each situation, an object released by the observer undergoes a downwards acceleration of magnitude g relative to the floor. In **Figure 11.16a**, the person is at rest in an

© Mary Evans Picture Library/Alamy

Albert Einstein
German physicist (1879–1955)
Einstein, one of the greatest physicists of all time, was born in Ulm, Germany. In 1905, at age 26, he published four scientific papers that revolutionised physics. Two of these papers were concerned with the special theory of relativity, the other two with quantum physics. In 1916 he published his work on the general theory of relativity.

The observer in the non-accelerating elevator drops his briefcase, which he observes to move downwards with acceleration g.

The observer in the accelerating elevator drops his briefcase, which he observes to move downwards with acceleration g.

In an accelerating elevator, the observer sees a light beam bend downwards.

Because of the equivalence in **a** and **b**, we expect a light ray to bend downwards in a gravitational field.

$\vec{v}_{el} = 0$
$\vec{a}_{el} = 0$

$\vec{a}_{el} = +g\hat{j}$

$\vec{a}_{el} = +g\hat{j}$

$\vec{v}_{el} = 0$
$\vec{a}_{el} = 0$

$\vec{g} = -g\hat{j}$

$\vec{g} = -g\hat{j}$

a   b   c   d

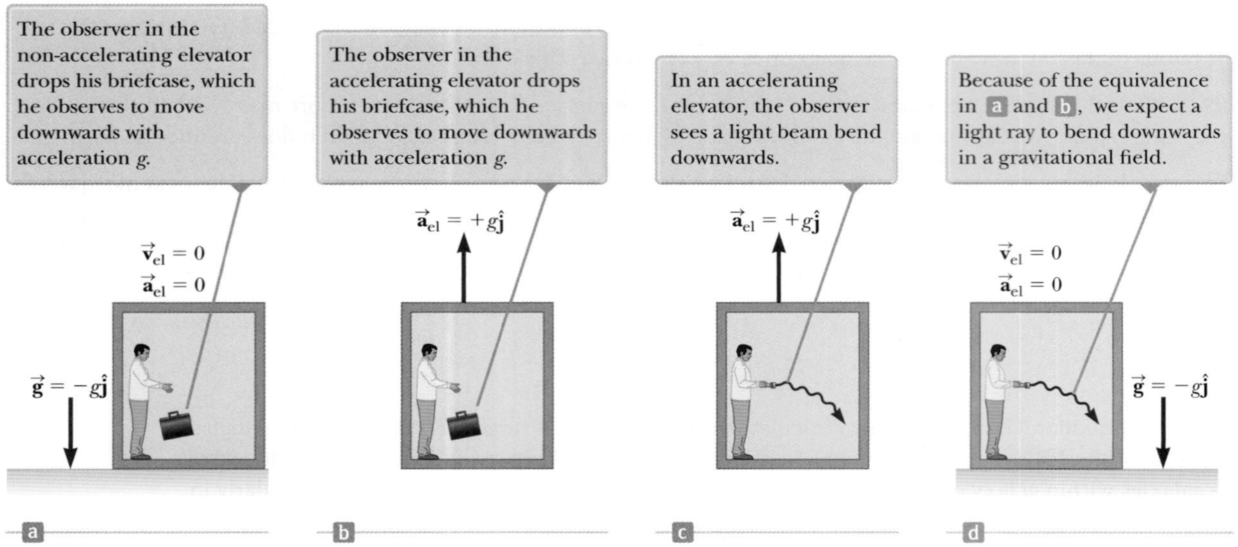

Figure 11.16
(a) The observer is at rest in a lift in a uniform gravitational field $\vec{g} = -g\hat{j}$, g directed downwards. (b) The observer is in a region where gravity is negligible, but the lift moves upwards with an acceleration $\vec{a}_{el} = +g\hat{j}$. According to Einstein, the frames of reference in (a) and (b) are equivalent in every way. No local experiment can distinguish any difference between the two frames. (c) An observer watches a beam of light in an accelerating lift. (d) Einstein's prediction of the behaviour of a beam of light in a gravitational field

inertial frame in a gravitational field due to the planet. In Figure 11.16b, the person is in a noninertial frame accelerating in gravity-free space. Einstein's claim is that these two situations are completely equivalent.

Einstein carried this idea further and proposed that *no* experiment, mechanical or otherwise, could distinguish between the two situations. This extension to include all phenomena (not just mechanical ones) has interesting consequences. For example, suppose a light pulse is sent horizontally across the lift as in Figure 11.16c, in which the lift is accelerating upwards in empty space. From the point of view of an observer in an inertial frame outside the lift, the light travels in a straight line while the floor of the lift accelerates upwards. According to the observer on the lift, however, the trajectory of the light pulse bends downwards as the floor of the lift (and the observer) accelerates upwards. Therefore, based on the equality of parts (a) and (b) of the figure, Einstein proposed that a beam of light should also be bent downwards by a gravitational field, as in Figure 11.16d. Experiments have verified the effect, although the bending is small. A laser aimed at the horizon falls less than 1 cm after travelling 6000 km. No such bending is predicted in Newton's theory of gravitation, because light has no mass so it should experience no gravitational force.

Einstein's **general theory of relativity** has two postulates:
- All the laws of nature have the same form for observers in any frame of reference, whether accelerated or not.
- In the vicinity of any point, a gravitational field is equivalent to an accelerated frame of reference in gravity-free space (the principle of equivalence).

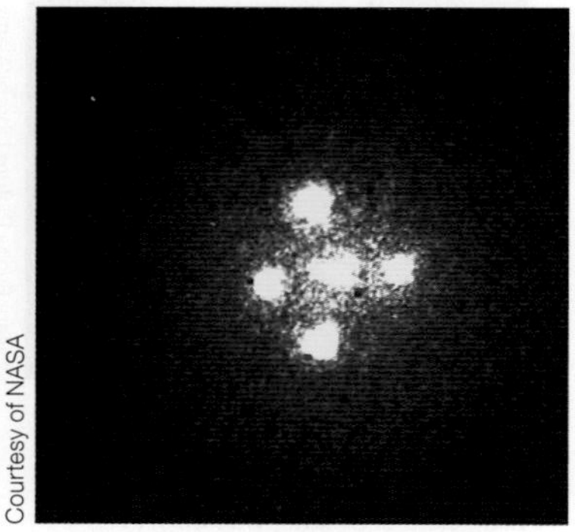

Einstein's cross. The four outer bright spots are images of the same galaxy that have been bent around a massive object located between the galaxy and the Earth. The massive object acts like a lens, causing the rays of light that were diverging from the distant galaxy to converge on the Earth. (If the intervening massive object had a uniform mass distribution, we would see a bright ring instead of four spots.)

TRY THIS

Fill a clear plastic bottle with water so only a small air bubble is present along with the water. The bubble sits at the top because of the buoyant force of the water acting on the air (see Chapter 13), which is due to the gravitational field acting on the water and creating a pressure gradient in the water. What happens if you accelerate the water bottle? Try accelerating it horizontally and vertically. The bubble acts an acceleration indicator. Can the bubble accelerometer distinguish between an acceleration and a gravitational field? Is there a preferred position for the bubble if you drop the bottle so the system is in free fall?

One interesting effect predicted by the general theory is that time is altered by gravity. A clock in the presence of gravity runs slower than one located where gravity is negligible. Consequently, the frequencies of radiation emitted by atoms in the presence of a strong gravitational field are *redshifted* to lower frequencies when compared with the same

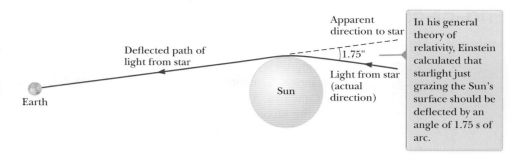

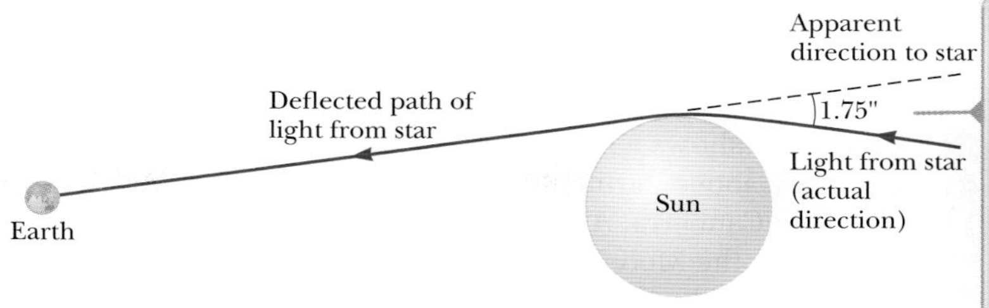

In his general theory of relativity, Einstein calculated that starlight just grazing the Sun's surface should be deflected by an angle of 1.75 s of arc.

Figure 11.17
Deflection of starlight passing near the Sun. Because of this effect, the Sun or some other remote object can act as a *gravitational lens.*

emissions in the presence of a weak field. This gravitational redshift has been detected in spectral lines emitted by atoms in massive stars. It has also been verified on the Earth by comparing the frequencies of gamma rays emitted from nuclei separated vertically by about 20 m.

The second postulate suggests that a gravitational field may be 'transformed away' at any point if we choose an appropriate accelerated frame of reference, a freely falling one. Einstein developed an ingenious method of describing the acceleration necessary to make the gravitational field 'disappear'. He introduced a concept, the *curvature of space–time,* that describes the gravitational effect at every point. According to Einstein, there is no such thing as a gravitational force. Rather, the presence of a mass causes a curvature of space–time in the vicinity of the mass, and this curvature dictates the path that all freely moving objects must follow.

As an example of the effects of curved space–time, imagine two travellers moving on parallel paths a few metres apart on the surface of the Earth and maintaining an exact northwards heading along two longitude lines. As they observe each other near the equator, they will claim that their paths are exactly parallel. As they approach the North Pole, however, they notice that they are moving closer together and will meet at the North Pole. Therefore, they claim that they moved along parallel paths, but moved toward each other, *as if there were an attractive force between them.* The travellers make this conclusion based on their everyday experience of moving on flat surfaces. From our mental model of the Earth as spherical, however, we realise they are walking on a curved surface and it is the geometry of the curved surface, rather than an attractive force, that causes them to converge. In a similar way, general relativity replaces the notion of forces with the movement of objects through curved space–time.

One prediction of the general theory of relativity is that a light ray passing near the Sun should be deflected in the curved space–time created by the Sun's mass. This prediction was initially believed to be confirmed when astronomers, led by Sir Arthur Eddington, attempted to measure the bending of starlight near the Sun during a total solar eclipse in 1919 (**Figure 11.17**). The results of these original experiments have since been shown to be highly ambiguous[1], and it is highly unlikely that such measurements could have been carried out with sufficient precision to distinguish between Newtonian and Einsteinian theories of gravity with the instruments available at that time. However, subsequent experiments and astronomical observations have confirmed that Einstein's theory correctly describes the behaviour of light in the presence of gravitational fields. If the concentration of mass becomes very great as is believed to occur when a large star exhausts its nuclear fuel and collapses to a very small volume, a black hole, as described in Section 11.4, may form. Here, the curvature of space–time is so extreme that within a certain distance from the centre of the black hole all matter and light become trapped.

While Einstein's postulates of general relativity have been borne out by many astronomical observations, they present a very different view of gravitational interactions from the field picture we used at the start of this chapter. Of the four fundamental forces, gravity is the only one that has this alternative description, and indeed the quest to combine general relativity with quantum field theory is one of the major challenges of modern physics.

In the next chapter we look in detail at a particular limiting case of the general theory of relativity, which is the case where there is no gravitational field and the reference frame is not accelerating. The special theory of relativity deals with the motion of objects in inertial reference frames, moving at very large speeds.

[1] Earman, J. and Glymour, C., 1980, 'Relativity and Eclipses: The British Eclipse Expeditions of 1919 and Their Predecessors', *Historical Studies in the Physical Sciences*, 11(1): 49.

End-of-chapter resources

Getty Images/Robin Smith

The image shows a nice moonrise above Parliament House in Canberra. As we have seen, the Moon stays in orbit around the Earth due to the gravitational force. The gravitational force of the Moon also acts upon the Earth, although because the Earth has a much greater mass, according to Newton's second law, the effect (acceleration) of the force is much smaller. The velocity of the Moon is perpendicular to the gravitational force, hence the Moon moves in an (approximately) circular orbit about the Moon–Earth system centre of mass. There is no net work done as the force and displacement are perpendicular and the Moon continues to move in circular motion. However, both the Earth and the Moon also experience tidal forces due to the small variation in gravitational force from one side of the Earth to the other, which cause the ocean tides on Earth and also cause a small distortion of the Moon. The energy loss due to friction, primarily because of the effects of viscosity of the water in the ocean, which we shall study in Chapter 14, results in a decrease in spin angular momentum of the Earth. However, we know from Chapter 10 that angular momentum is conserved, so something must be gaining angular momentum to compensate – the thing that is gaining angular momentum is the Moon! It does so by increasing the radius of its orbit (and hence its moment of inertia about the Earth), which is why it is slowly moving further away, by almost 4 cm per year.

The problems found in this chapter may be assigned online in Enhanced Web Assign.

Worked solutions to every fifth problem are available in the Student Solutions Manual. Register online at **www.cengagebrain.com** for access.

Summary

Definitions

The **gravitational field** at a point in space is defined as the gravitational force per unit mass experienced by a test particle at that point:

$$\vec{\mathbf{g}} \equiv \frac{\vec{\mathbf{F}}_g}{m} \quad (11.4)$$

The **gravitational potential** at a point in space is defined as the gravitational potential energy per unit mass at that point:

$$V_g \equiv \frac{U_g}{m} \quad (11.13)$$

Concepts and principles

Newton's law of universal gravitation states that the gravitational force of attraction between any two particles of masses m_1 and m_2 separated by a distance r has the magnitude

$$F_g = G\frac{m_1 m_2}{r^2} \quad (11.1)$$

where $G = 6.674 \times 10^{-11}$ N.m^2/kg^2 is the **universal gravitational constant**. This equation enables us to calculate the force of attraction between masses under many circumstances.

An object at a distance h above the Earth's surface experiences a gravitational force of magnitude mg, where g is the free-fall acceleration at that height:

$$g = \frac{GM_E}{r^2} = \frac{GM_E}{(R_E + h)^2} \quad (11.7)$$

In this expression, M_E is the mass of the Earth and R_E is its radius. Therefore, the weight of an object decreases as the object moves away from the Earth's surface.

The **gravitational potential energy** associated with a system of two particles separated by a distance r is

$$U = -\frac{Gm_1 m_2}{r} \quad (11.11)$$

where U is taken to be zero as $r \to \infty$.

The **gravitational potential** is the potential energy per unit mass:

$$V_g(r) = \frac{-Gm}{r} \quad (11.15)$$

If an isolated system consists of an object of mass m moving with a speed v in the vicinity of a massive object of mass M, the total energy E of the system is the sum of the kinetic and potential energies:

$$E = \frac{1}{2}mv^2 - \frac{GMm}{r} \quad (11.16)$$

The total energy of the system is a constant of the motion. If the object moves in an elliptical orbit of semimajor axis a around the massive object and $M \gg m$, the total energy of the system is

$$E = -\frac{GMm}{2a} \quad (11.19)$$

For a circular orbit, this same equation applies with $a = r$.

The **escape speed** for an object projected from the surface of a planet of mass M and radius R is

$$v_{esc} = \sqrt{\frac{2GM}{R}} \tag{11.23}$$

Kepler's laws of planetary motion state:

1. All planets move in elliptical orbits with the Sun at one focus.
2. The radius vector drawn from the Sun to a planet sweeps out equal areas in equal time intervals.
3. The square of the orbital period of any planet is proportional to the cube of the semimajor axis of the elliptical orbit.

Kepler's third law can be expressed as

$$T^2 = \left(\frac{4\pi^2}{GM_S}\right)a^3 \tag{11.25}$$

where M_S is the mass of the Sun and a is the semimajor axis. For a circular orbit, a can be replaced in Equation 11.25 by the radius r. Most planets have nearly circular orbits around the Sun.

Einstein's **general theory of relativity**:

- All the laws of nature have the same form for observers in any frame of reference, whether accelerated or not.
- In the vicinity of any point, a gravitational field is equivalent to an accelerated frame of reference in gravity-free space (the principle of equivalence).

Chapter review quiz

To help you revise Chapter 11: Gravity, complete the automatically graded Chapter review quiz at http://login.cengagebrain.com.

Conceptual questions

1. A satellite in low-Earth orbit is not truly travelling through a vacuum. Rather, it moves through very thin air. Does the resulting air friction cause the satellite to slow down?
2. Explain why it takes more fuel for a spacecraft to travel from the Earth to the Moon than for the return trip.
3. (a) Explain why the force exerted on a particle by a uniform sphere must be directed towards the centre of the sphere. (b) Would this statement be true if the mass distribution of the sphere were not spherically symmetrical? Explain your answer.
4. (a) At what position in its elliptical orbit is the speed of a planet a maximum? (b) At what position is the speed a minimum?
5. (a) If a hole could be dug to the centre of the Earth, would the force on an object of mass m still obey Equation 11.1 there? (b) What do you think the force on m would be at the centre of the Earth?
6. Sketch the gravitational potential due to the Earth as a function of distance from the Earth's surface. On the same axes, sketch the gravitational potential using the 'surface approximation', that is, the approximation derived from a potential energy of mgh.
7. Sketch the gravitational field for the Earth–Moon system. Hint: Start by considering the gravitational force on a small test particle at various positions and remember that fields add vectorially.
8. Two identical clocks are in the same house, one upstairs in a bedroom and the other downstairs in the kitchen. Which clock runs slower? Explain.
9. There are two equinoxes each year, one in spring and one in autumn. The equinoxes are associated with two points 180° apart in the Earth's orbit where the tilt of the Earth's orbit is neither towards nor away from the Sun and the centre of the Sun lies in the plane of the Earth's equator. Hence the Earth is on precisely opposite sides of the Sun when it passes through these two points. One equinox occurs on the 20th or 21st of March (the autumnal equinox in the southern hemisphere), the other on the 21st or 22nd of September (the spring equinox), 185 days later. Why is the interval from the March to the September equinox longer than the interval from the September to the March equinox rather than being equal to that interval?

Problems

Section 11.1 Newton's law of universal gravitation

1. A 200 kg object and a 500 kg object are separated by 4.00 m. (a) Find the net gravitational force exerted by these objects on a 50.0 kg object placed midway between them. (b) At what position (other than an infinitely remote one) can the 50.0 kg object be placed so as to experience a net force of zero from the other two objects?
2. During a solar eclipse, the Moon, the Earth and the Sun all lie on the same line, with the Moon between the Earth and the Sun. (a) What force is exerted by the Sun on the Moon? (b) What force is exerted by the Earth on the Moon? (c) What force is exerted by the Sun on the Earth? (d) Compare the answers to parts (a) and (b). Why doesn't the Sun capture the Moon away from the Earth?
3. In introductory physics laboratories, a typical Cavendish balance for measuring the gravitational constant G uses lead spheres with masses (1.50 ± 0.01) kg and (15.0 ± 0.1) g whose centres are separated by (4.50 ± 0.02) cm. Calculate the gravitational force between these spheres, treating each as a particle located at the sphere's centre.
4. Three uniform spheres of masses $m_1 = 2.00$ kg, $m_2 = 4.00$ kg and $m_3 = 6.00$ kg are placed at the corners of a right triangle as shown in Figure P11.4. Calculate the resultant gravitational force on the object of mass m_2, assuming the spheres are isolated from the rest of the Universe.

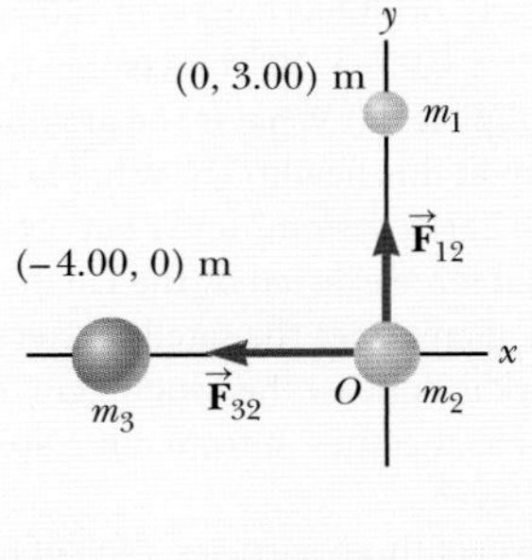

Figure P11.4

5. Two objects attract each other with a gravitational force of magnitude 1.00×10^{-8} N when separated by 20.0 cm. If the total mass of the two objects is 5.00 kg, what is the mass of each?

Section 11.2 The gravitational field

6. (a) Calculate the vector gravitational field at a point P on the perpendicular bisector of the line joining two objects of equal mass separated by a distance $2a$ as shown in Figure P11.6. (b) Explain physically why the field should approach zero as $r \to 0$. (c) Prove mathematically that the answer to part (a) behaves in this way. (d) Explain physically why the magnitude of the field should approach $2GM/r^2$ as $r \to \infty$. (e) Prove

mathematically that the answer to part (a) behaves correctly in this limit.

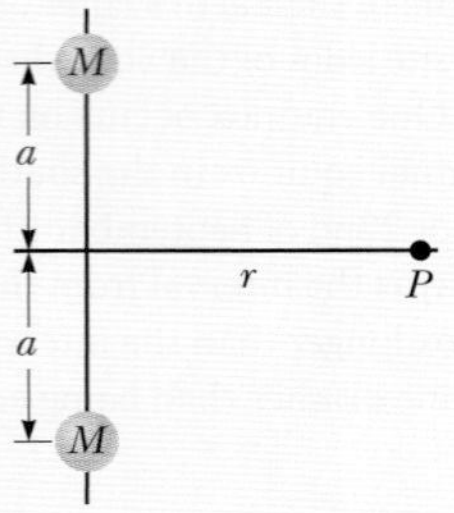

Figure P11.6

7. A spacecraft in the shape of a long cylinder has a length of 100 m, and its mass with occupants is 10 000 kg. It has strayed too close to a black hole having a mass 100 times that of the Sun (**Figure P11.7**). The nose of the spacecraft points toward the black hole, and the distance between the nose and the centre of the black hole is 10.0 km. (a) Determine the total force on the spacecraft. (b) What is the difference in the gravitational fields acting on the occupants in the nose of the ship and on those in the rear of the ship, furthest from the black hole?

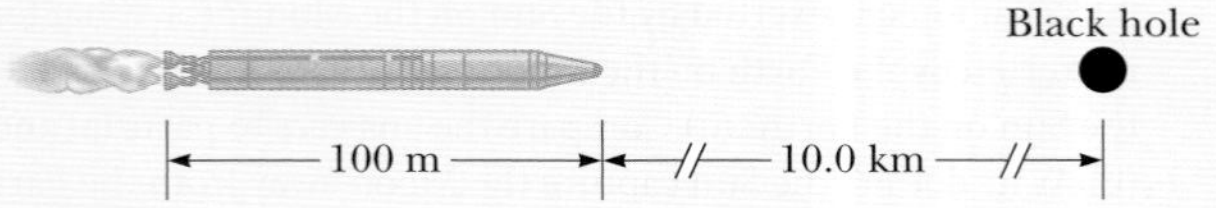

Figure P11.7

8. When a falling meteoroid is at a distance above the Earth's surface of 3.00 times the Earth's radius, what is its acceleration due to the Earth's gravitation?
9. The free-fall acceleration on the surface of the Moon is about one-sixth that on the surface of the Earth. The radius of the Moon is about $0.250R_E$ (R_E = Earth's radius = 6.37×10^6 m). Find the ratio of their average densities, ρ_{Moon}/ρ_{Earth}.

Section 11.3 Gravitational potential energy

10. ±? A satellite in Earth orbit has a mass of 100 kg and is at an altitude of $(2.00 \pm 0.05) \times 10^6$ m. Assume the uncertainty in the mass is negligible. (a) What is the gravitational potential due to the Earth at this height? (b) What is the potential energy of the satellite–Earth system? (c) What is the magnitude of the gravitational force exerted by the Earth on the satellite? (d) What force, if any, does the satellite exert on the Earth?
11. How much work is done by the Moon's gravitational field on a 1000 kg meteor as it comes in from outer space and impacts on the Moon's surface?
12. A system consists of three particles, each of mass 5.00 g, located at the corners of an equilateral triangle with sides of 30.0 cm. (a) Calculate the potential energy of the system. (b) Assume the particles are released simultaneously. Describe the subsequent motion of each. Will any collisions take place? Explain your answer.
13. After the Sun exhausts its nuclear fuel, its ultimate fate will be to collapse to a *white dwarf state*. In this state, it would have approximately the same mass as it has now, but its radius would be equal to the radius of the Earth. Calculate (a) the average density of the white dwarf, (b) the surface free-fall acceleration and (c) the gravitational potential at the surface of the white dwarf.
14. At the Earth's surface, a projectile is launched straight up at a speed of 10.0 km/s. To what height will it rise? Ignore air resistance and the rotation of the Earth. Are these approximations reasonable?

Section 11.4 Energy considerations in planetary and satellite motion

15. A space probe is fired as a projectile from the Earth's surface with an initial speed of 2.00×10^4 m/s. What will its speed be when it is very far from the Earth? Ignore atmospheric friction and the rotation of the Earth.
16. A 1000 kg satellite orbits the Earth at a constant altitude of 100 km. (a) How much energy must be added to the system to move the satellite into a circular orbit with altitude 200 km? What are the changes in the system's (b) kinetic energy and (c) potential energy?
17. A 'treetop satellite' moves in a circular orbit just above the surface of a planet, assumed to offer no air resistance. Show that its orbital speed v and the escape speed from the planet are related by the expression $v_{esc} = \sqrt{2}v$.
18. (a) What is the minimum speed, relative to the Sun, necessary for a spacecraft to escape the solar system if it starts at the Earth's orbit? (b) *Voyager 1* achieved a maximum speed of 125 000 km/h on its way to photograph Jupiter. Beyond what distance from the Sun is this speed sufficient to escape the solar system?
19. Ganymede is the largest of Jupiter's moons. Consider a rocket on the surface of Ganymede, at the point furthest from the planet (**Figure P11.19**). Model the rocket as a particle. (a) Does the presence of Ganymede make Jupiter exert a larger, smaller, or same size force on the rocket compared with the force it would exert if Ganymede were not interposed? (b) Determine the escape speed for the rocket from the planet–satellite system. The radius of Ganymede is 2.64×10^6 m and its mass is 1.495×10^{23} kg. The distance between Jupiter and Ganymede is 1.071×10^9 m, and the mass of Jupiter is 1.90×10^{27} kg. Ignore the motion of Jupiter and Ganymede as they revolve about their centre of mass.

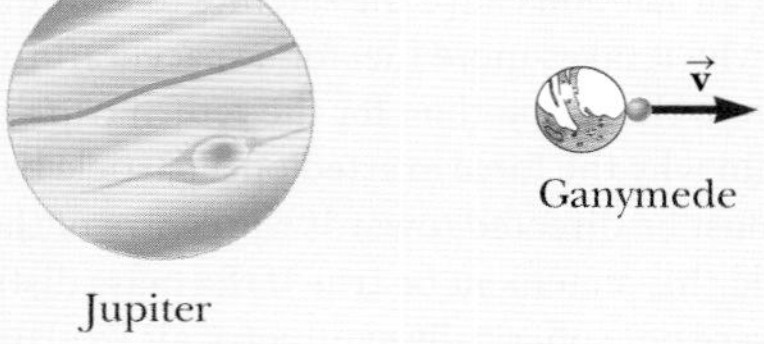

Figure P11.19

Section 11.5 Kepler's laws and the motion of planets

20. A satellite circles the Earth in a circular orbit at a location where the acceleration due to gravity is 9.00 m/s². Determine the orbital period of the satellite.
21. A minimum-energy transfer orbit to an outer planet consists of putting a spacecraft on an elliptical trajectory with the departure planet corresponding to the perihelion of the ellipse, or the closest point to the Sun, and the arrival planet at the aphelion, or the furthest point from the Sun. (a) Use Kepler's third law to calculate how long it would take to go from Earth to Mars on such an orbit as shown in **Figure P11.21**. (b) Can such an orbit be undertaken at any time? Explain your answer.

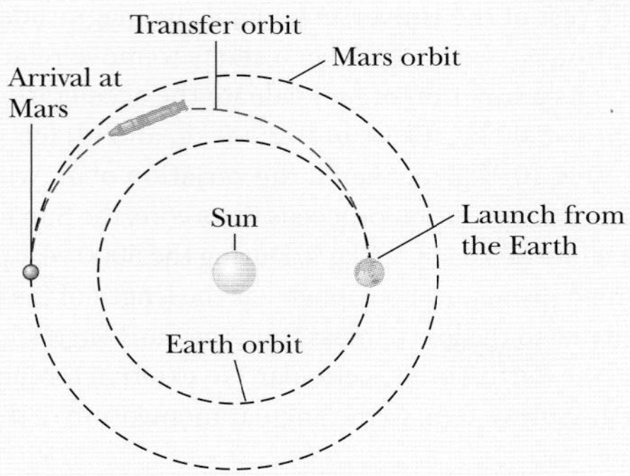

Figure P11.21

22. Plaskett's binary system consists of two stars that revolve in a circular orbit about a centre of mass midway between them. This statement implies that the masses of the two stars are equal (**Figure P11.22**). Assume the orbital speed of each star is $v = 220$ km/s and the orbital period of each is 14.4 days. Find the mass M of each star. Assume the uncertainty in each value is one in the last significant figure given. (For comparison, the mass of our Sun is 1.99×10^{30} kg.)

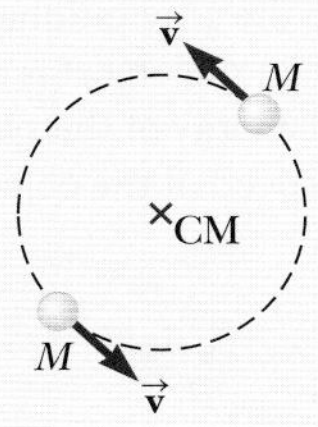

Figure P11.22

23. Comet Halley (**Figure P11.23**) approaches the Sun to within 0.570 AU, and its orbital period is 75.6 yr. (AU is the symbol for astronomical unit, where 1 AU = 1.50×10^{11} m is the mean Earth–Sun distance.) How far from the Sun will Halley's comet travel before it starts its return journey?

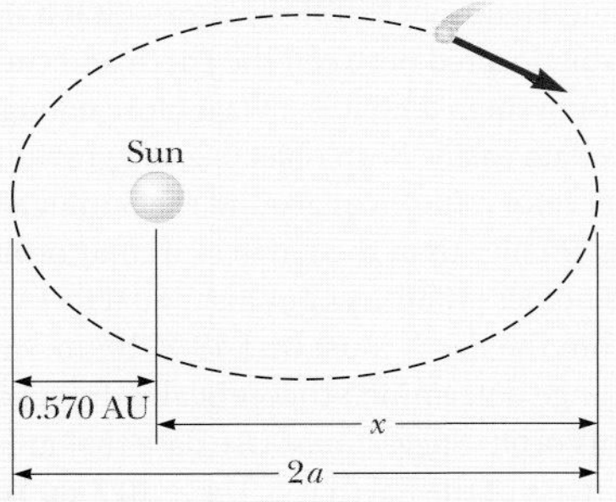

Figure P11.23
(Orbit is not drawn to scale.)

24. A synchronous satellite, which always remains above the same point on a planet's equator, is put in orbit around Jupiter to study that planet's famous red spot. Jupiter rotates once every 9.84 h. Use the data in the endpapers to find the altitude of the satellite above the surface of the planet.

25. Neutron stars are extremely dense objects formed from the remnants of supernova explosions. Many rotate very rapidly. Suppose the mass of a certain spherical neutron star is twice the mass of the Sun and its radius is 10.0 km. Determine the greatest possible angular speed it can have so that the matter at the surface of the star on its equator is just held in orbit by the gravitational force.

26. A satellite of mass 200 kg is placed into Earth orbit at a height of 200 km above the surface. (a) Assuming circular orbit, how long does the satellite take to complete one orbit? (b) What is the satellite's speed? (c) Starting from the satellite on the Earth's surface, what is the minimum energy input necessary to place this satellite in orbit? Ignore air resistance but include the effect of the planet's daily rotation.

Additional problems

27. A student proposes to study the gravitational force by suspending two 100.0 kg spherical objects at the lower ends of cables from the ceiling of a tall cathedral and measuring the deflection of the cables from the vertical. The 45.00 m cables are attached to the ceiling 1.000 m apart. The first object is suspended and its position is carefully measured. The second object is suspended and the two objects attract each other gravitationally. By what distance has the first object moved horizontally from its initial position due to the gravitational attraction to the other object? Hint: Keep in mind that this distance will be very small and make appropriate approximations. With what precision will the student be able to calculate the gravitational force if the smallest deflection he can measure is 0.1 μm? Assume the uncertainty in other quantities is negligible.

28. A satellite is in a circular orbit around the Earth at an altitude of 2.80×10^6 m. Find (a) the period of the orbit, (b) the speed of the satellite and (c) the acceleration of the satellite.

29. A satellite moves around the Earth in a circular orbit of radius r. (a) What is the speed v_i of the satellite? (b) Suddenly, an explosion breaks the satellite into two pieces with masses m and $4m$. Immediately after the explosion, the smaller piece of mass m is stationary with respect to the Earth and falls directly towards the Earth. What is the speed v of the larger piece immediately after the explosion? (c) Because of the increase in its speed, this larger piece now moves in a new elliptical orbit. Find its distance away from the centre of the Earth when it reaches the other end of the ellipse.

30. A rocket is fired straight up through the atmosphere from the South Pole, burning out at an altitude of 250 km when travelling at 6.00 km/s. (a) What maximum distance from the Earth's surface does it travel before falling back to the Earth? (b) Would its maximum distance from the surface be larger if the same rocket were fired with the same fuel load from a launch site on the equator? Why or why not?

31. Let Δg_M represent the difference in the gravitational fields produced by the Moon at the points on the Earth's surface nearest to and farthest from the Moon. Find the fraction $\Delta g_M/g$, where g is the Earth's gravitational field. (This difference is responsible for the occurrence of the *lunar tides* on the Earth.)

32. Two hypothetical planets of masses m_1 and m_2 and radii r_1 and r_2 respectively are nearly at rest when they are an infinite distance apart. Because of their gravitational attraction, they head toward each other on a collision course. (a) When their centre-to-centre separation is d, find expressions for the speed of each planet and for their relative speed. (b) Find the kinetic energy of each planet just before they collide, taking $m_1 = 2.00 \times 10^{24}$ kg, $m_2 = 8.00 \times 10^{24}$ kg, $r_1 = 3.00 \times 10^6$ m and $r_2 = 5.00 \times 10^6$ m. *Note:* Both the energy and momentum of the isolated two-planet system are constant.

33. (a) Show that the rate of change of the free-fall acceleration with vertical position near the Earth's surface is

$$\frac{dg}{dr} = -\frac{2GM_E}{R_E^3}$$

This rate of change with position is called a *gradient*. (b) Assuming h is small in comparison to the radius of the Earth, show that the difference in free-fall acceleration between two points separated by vertical distance h is

$$|\Delta g| = \frac{2GM_E h}{R_E^{\ 3}}$$

(c) Evaluate this difference for $h = 6.00$ m, a typical height for a two-storey building.

34. A certain quaternary star system consists of three stars, each of mass m, moving in the same circular orbit of radius r about a central star of mass M. The stars orbit in the same sense and are positioned one-third of a revolution apart from one another. Show that the period of each of the three stars is given by

$$T = 2\pi\sqrt{\frac{r^3}{G(M + m/\sqrt{3})}}$$

35. Studies of the relationship of the Sun to our galaxy – the Milky Way – have revealed that the Sun is located near the outer edge of the galactic disc, about 30 000 ly (1 ly = 9.46×10^{15} m) from the centre. The Sun has an orbital speed of approximately 250 km/s around the galactic centre. (a) What is the period of the Sun's galactic motion? (b) What is the order of magnitude of the mass of the Milky Way galaxy? (c) Suppose the galaxy is made mostly of stars of which the Sun is typical. What is the order of magnitude of the number of stars in the Milky Way?

36. The maximum distance from the Earth to the Sun (at aphelion) is 1.521×10^{11} m, and the distance of closest approach (at perihelion) is 1.471×10^{11} m. The Earth's orbital speed at perihelion is 3.027×10^4 m/s. Determine (a) the Earth's orbital speed at aphelion and the kinetic and potential energies of the Earth–Sun system (b) at perihelion and (c) at aphelion. (d) Is the total energy of the system constant? Explain your answer. Ignore the effect of the Moon and other planets.

37. Many people assume air resistance acting on a moving object will always make the object slow down. It can, however, actually be responsible for making the object speed up. Consider a 100 kg Earth satellite in a circular orbit at an altitude of 200 km. A small force of air resistance makes the satellite drop into a circular orbit with an altitude of 100 km. (a) Calculate the satellite's initial speed. (b) Calculate its final speed in this process. (c) Calculate the initial energy of the satellite–Earth system. (d) Calculate the final energy of the system. (e) Show that the system has lost mechanical energy and find the amount of the loss due to friction. (f) What force makes the satellite's speed increase? Hint: You will find a free-body diagram useful in explaining your answer.

38. Show that the minimum period for a satellite in orbit around a spherical planet of uniform density ρ is

$$T_{\min} = \sqrt{\frac{3\pi}{G\rho}}$$

and independent of the planet's radius.

39. Astronomers detect a distant meteoroid moving along a straight line that, if extended, would pass at a distance $3R_E$ from the centre of the Earth, where R_E is the Earth's radius. What minimum speed must the meteoroid have if it is *not* to collide with the Earth?

40. Consider an object of mass m, not necessarily small compared with the mass of the Earth, released at a distance of 1.20×10^7 m from the centre of the Earth. Assume the Earth and the object behave as a pair of particles, isolated from the rest of the Universe. (a) Find the magnitude of the acceleration a_{rel} with which each starts to move relative to the other as a function of m. Evaluate the acceleration (b) for $m = 5.00$ kg, (c) for $m = 2000$ kg, and (d) for $m = 2.00 \times 10^{24}$ kg. (e) Sketch the variation of a_{rel} with m.

41. As thermonuclear fusion proceeds in its core, the Sun loses mass at a rate of 3.64×10^9 kg/s. During the 5000 year period of recorded history, by how much has the length of the year changed due to the loss of mass from the Sun? *Suggestions*: Assume the Earth's orbit is circular. No external torque acts on the Earth–Sun system, so the angular momentum of the Earth is constant.

Challenge problems

42. A ring of matter is a familiar structure in planetary and stellar astronomy. Examples include Saturn's rings and a ring nebula. Consider a uniform ring of mass 2.36×10^{20} kg and radius 1.00×10^8 m. An object of mass 1000 kg is placed at a point A on the axis of the ring, 2.00×10^8 m from the centre of the ring (**Figure P11.42**). When the object is released, the attraction of the ring makes the object move along the axis toward the centre of the ring (point B). (a) Calculate the gravitational potential energy of the object–ring system when the object is at A. (b) Calculate the gravitational potential energy of the system when the object is at B. (c) Calculate the speed of the object as it passes through B.

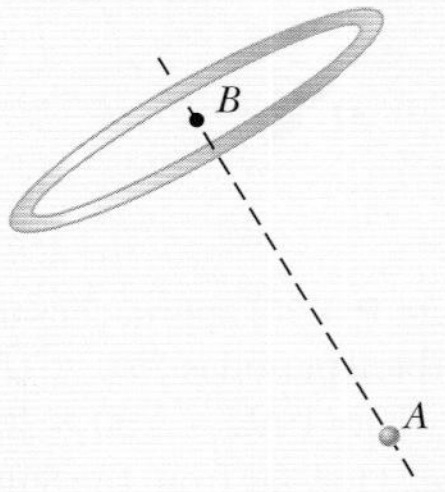

Figure P11.42

43. The Solar and Heliospheric Observatory (SOHO) spacecraft has a special orbit, located between the Earth and the Sun along the line joining them, and it is always close enough to the Earth to transmit data easily. Both objects exert gravitational forces on the observatory. It moves around the Sun in a near-circular orbit that is smaller than the Earth's circular orbit. Its period, however, is not less than 1 yr but just equal to 1 yr. Show that its distance from the Earth must be 1.48×10^9 m. In 1772, Joseph Louis Lagrange determined theoretically the special location allowing this orbit. *Suggestions*: Use data that are precise to four digits. The mass of the Earth is 5.974×10^{24} kg. You will not be able to easily solve the equation you generate; instead, use a computer to verify that 1.48×10^9 m is the correct value.

44. The oldest artificial satellite still in orbit is *Vanguard I*, launched on 3 March 1958. Its mass is 1.60 kg. Neglecting atmospheric drag, the satellite would still be in its initial orbit, with a minimum distance from the centre of the Earth of 7.02 Mm and a speed at this perigee point of 8.23 km/s. For this orbit, find (a) the total energy of the satellite–Earth system and (b) the magnitude of the angular momentum of the satellite. (c) At apogee, find the satellite's speed and its distance from the centre of the Earth. (d) Find the semimajor axis of its orbit. (e) Determine its period.

chapter 12

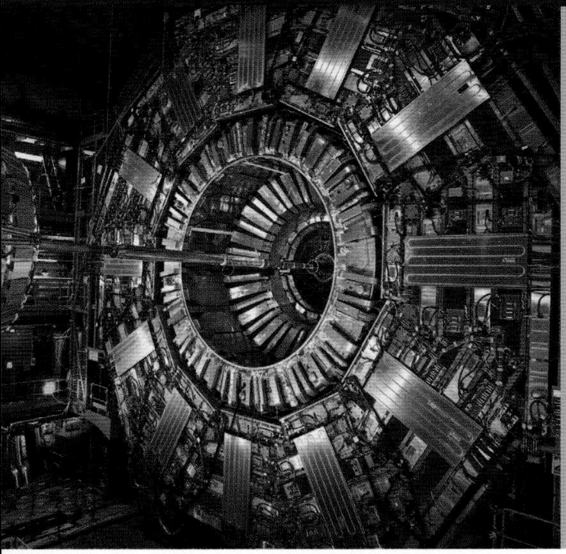
CERN

Special relativity

The Compact Muon Solenoid (CMS) Detector is part of the Large Hadron Collider at the European Laboratory for Particle Physics (CERN).

Collider experiments involve the acceleration of subatomic particles to speeds close to the speed of light. The kinetic energy carried by these particles is used to create new, exotic particles, such as the top quark and the Higgs boson. Where does the mass for these new particles come from?

On completing this chapter, students will understand:

- that the laws of physics are the same in all inertial reference frames
- that the speed of light is the maximum speed at which anything can travel
- that simultaneity, length and time are all relative and depend on the observer
- how velocity depends on an observer's reference frame
- how momentum and energy depend upon an observer's reference frame
- that mass is a form of potential energy.

Students will be able to:

- calculate the velocity of an object as measured by different observers
- apply the Lorentz transformations to calculate lengths and times in different reference frames
- analyse the motion of objects in different reference frames
- identify situations where special relativity needs to be used
- calculate the momentum and energy of objects moving at relativistic speeds
- distinguish between kinetic energy and rest energy of objects in motion in different reference frames.

Newtonian mechanics was formulated on the basis of observations of objects moving at speeds much less than that of light. However, in the early 1900s it became clear that it failed to describe properly the motion of objects whose speeds approach that of light.

In 1906, Einstein published the special theory of relativity. With this theory, experimental observations can be correctly predicted over the range of speeds from $v = 0$ to speeds approaching the speed of light. At low speeds, Einstein's theory reduces to Newtonian mechanics as a limiting situation.

ENHANCED WebAssign

The problems found in this chapter may be assigned online in Enhanced Web Assign.

12.1 The breakdown of Newtonian mechanics at high speeds

In the preceding chapters, we introduced the key concepts needed to describe the motion and mechanical interactions of objects – these concepts form the foundations of what we know as Newtonian mechanics. In Newtonian mechanics, positions and velocities are relative quantities, but acceleration is absolute. Much of the formalism is derived from the law relating force to acceleration, as expressed through Newton's second law (**Equation 4.3**):

$$\sum \vec{\mathbf{F}} = m\vec{\mathbf{a}}$$

As we have seen in Chapters 7 to 9, the other key ideas underpinning Newtonian mechanics are conservation of energy and conservation of momentum. We relate the change in an object's energy to the applied force through the work–energy relation (**Equation 6.5**),

$$W = \int_{x_i}^{x_f} F_x \, dx$$

Figure 12.1
The tower contains the Australian National University's Heavy Ion Accelerator, in which ions are accelerated up to a few percent of the speed of light.

and the change in an object's speed v to the change in its kinetic energy K through **Equation 6.13**, which can be re-written as

$$v = \sqrt{\frac{2K}{m}}$$

This equation indicates that as long as you keep providing an object with more kinetic energy, its speed will keep increasing, with no limit. These equations, and the associated ideas developed in Chapters 2 to 11, have provided highly reliable and accurate predictions of the motions and interactions of objects that we encounter in our everyday lives. But a little over 100 years ago, scientists started to make observations that suggested that Newtonian mechanics might break down in some limits. One of these limits was in the realm of the very small, which is now described using quantum mechanics (see Part 7). The other limit was the realm of the very fast, which we will consider in this chapter.

Equation 6.13 tells us how an object's speed should vary with kinetic energy if it behaves according to the laws of Newtonian mechanics. For example, as we will see in Chapter 23, when an electron experiences an electric field, it is accelerated by the field to higher speeds. It is this phenomenon that allows accelerators such as the one at the Australian National University (**Figure 12.1**) to accelerate beams of ions to be used in nuclear physics experiments (see Chapters 43 and 44). But when an object is accelerated to very high speeds, we observe something strange. **Figure 12.2** shows the speed of an electron as a function of its kinetic energy, up to the very high energies provided by accelerators.

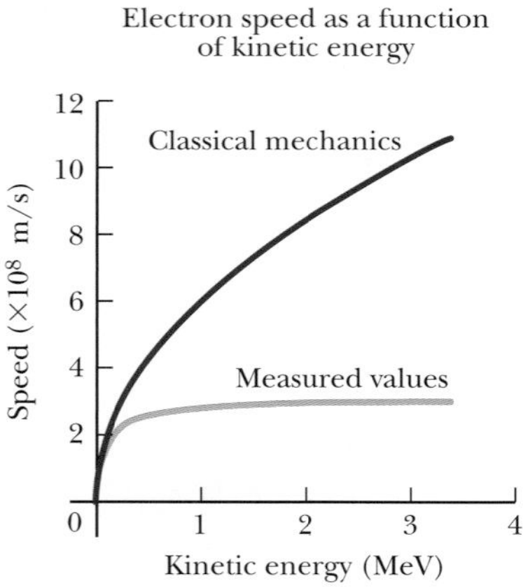

Figure 12.2
Newtonian mechanics predicts the speed of an accelerated electron will depend on the kinetic energy as shown by the red line. Observations show quite different behaviour, as shown by the green line.

We can see straight away from this graph that the electron stops following the predictions of Newtonian mechanics as it speed increases. One of the most striking features of the observed behaviour is that the speed of the electron approaches a maximum value, so that no matter how much energy it is given, it can never exceed this limit. It turns out that this limit is independent of the object being accelerated and that it is equal to the speed of light. This immediately poses a problem to Newtonian mechanics.

In fact, the observation (now extremely well confirmed experimentally) that the speed of light is a constant also poses a problem for Newtonian mechanics. In Chapter 34, we will see that James Clerk Maxwell (1831–1879) showed that the speed of light in free space is $c = 3.00 \times 10^8$ m/s. This speed is always measured to be the same, regardless of the relative velocities of the object emitting the light and the observer. To understand why this is a problem for Newtonian mechanics, we need to consider the concepts of frames of reference and *relativity*.

12.2 Relativity and Einstein's postulates

In our previous analysis of Newtonian mechanics, we made some unstated assumptions about any observers of these physical events. In fact, in order to describe a physical event, we must establish a frame of reference. Recall from Chapter 4 that an inertial frame of reference is one in which an object is observed to have no acceleration when no forces act on it. Any frame moving with constant velocity with respect to an inertial frame is also an inertial frame. Since all inertial frames are equally valid, there is no *absolute* inertial reference frame. This implies that the results of an experiment performed in a vehicle moving with uniform velocity must be identical to the results of the same experiment performed in a stationary vehicle. This idea is captured by the **principle of Galilean relativity**:

The laws of mechanics must be the same in all inertial frames of reference.

◀ Principle of Galilean relativity

In the following, we refer to any occurrence that can be observed and assigned a time and a location by an observer as an *event*. If the observer is at rest with respect to an inertial frame ('in a frame'), the event's location and time of occurrence can be specified by the four coordinates (x, y, z, t). In Section 3.7, we introduced the **Galilean transformation equations**, which allow us to transform between observations made by observers in different frames when their relative motion is at low speeds. Consider two inertial frames, labelled S and S′ (**Figure 12.3**). The S′ frame moves with a constant velocity $\vec{\mathbf{v}}$ along the common x and x' axes, where $\vec{\mathbf{v}}$ is measured relative to S. We assume the origins of S and S′ coincide at $t = 0$ and an event occurs at point P in space at some instant of time. For simplicity, we show the observer O in the S frame and the observer O' in the S′ frame as blue dots at the origins of their coordinate frames in **Figure 12.3**, but that is not necessary: either observer could be at any fixed location in his or her frame. Observer O describes the event with space–time coordinates (x, y, z, t), whereas observer O' in S′ uses the coordinates (x', y', z', t') to describe the same event. As we see from the geometry in **Figure 12.3**, the relationships among these various coordinates can be written

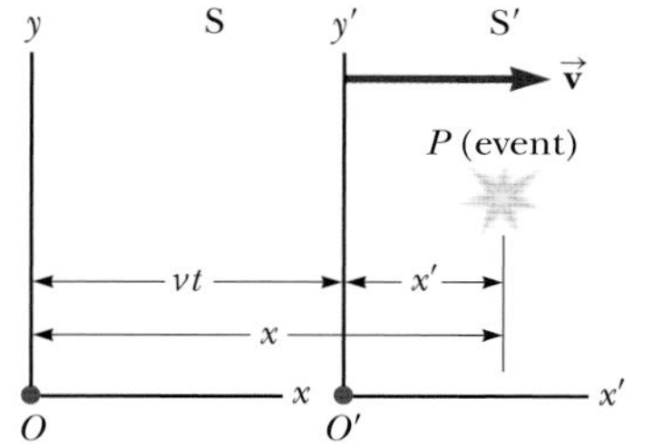

Figure 12.3
An event occurs at a point P. The event is seen by two observers in inertial frames S and S′, where S′ moves with a velocity v relative to S.

$$x' = x - vt \qquad y' = y \qquad z' = z \qquad t' = t \tag{12.1}$$

◀ Galilean transformation equations

These equations are the **Galilean space–time transformation equations**. Note that position is a relative quantity, depending on the frame of the observer, but time is assumed to be the same in both inertial frames. That is, in classical mechanics, all clocks run at the same rate regardless of their velocity, so the time at which an event occurs for an observer in S is the same as the time for the same event for an observer in S′. Consequently, the time interval between two successive events should be the same for both observers. (Note: This apparently natural assumption turns out to be incorrect, which becomes obvious in situations where v is comparable to the speed of light.)

The Galilean transformation equations naturally lead to the velocity transformation equation introduced in Section 3.7. Suppose a particle moves through a displacement of magnitude dx along the x axis in a time interval dt as measured by an observer in S. It follows from **Equation 12.1** that the corresponding displacement dx' measured by an observer in S′ is $dx' = dx - v\,dt$, where frame S′ is moving with speed v in the x direction relative to frame S. Because $dt = dt'$, we find that

$$\frac{dx'}{dt'} = \frac{dx}{dt} - v$$

or

$$u'_x = u_x - v \tag{12.2}$$

◀ Galilean velocity transformation equation

where u_x and u'_x are the x components of the velocity of the particle measured by observers in S and S′, respectively. (We use the symbol $\vec{\mathbf{u}}$ rather than $\vec{\mathbf{v}}$ for particle velocity because v is already used for the relative velocity of two reference frames.) **Equation 12.2** is the **Galilean velocity transformation equation**. It is consistent with our intuitive notion of time and space as well as with our discussions in Section 3.7.

Pitfall Prevention 12.1

It is important to be aware of the relationship between the S and S′ frames. Many of the mathematical representations in this chapter are true *only* for the specified relationship between the S and S′ frames. The x and x' axes coincide, except their origins are different. The y and y' axes (and the z and z' axes) are parallel, but they only coincide at one instant due to the time-varying displacement of the origin of S′ with respect to that of S. We choose the time $t = 0$ to be the instant at which the origins of the two coordinate systems coincide.

Quick Quiz 12.1

A boy throws a ball at 50 km/h while standing in a train moving at 110 km/h. The ball is thrown in the same direction as that of the velocity of the train. If you apply the Galilean velocity transformation equation to this situation, is the speed of the ball relative to the Earth **(a)** 50 km/h, **(b)** 110 km/h, **(c)** 40 km/h, **(d)** 160 km/h or **(e)** impossible to determine?

Although the Galilean transformations were formulated with reference to mechanical interactions, it is reasonable to expect the same behaviour to apply to light, which as we will see in Chapter 35 can be treated as particles (photons). However, as we have already noted, it is well established experimentally that the measured speed of light is independent of the relative motion of the observer and the light source. This was first established by A. A. Michelson and E. W. Morley in the latter part of the 19th century through a series of experiments that relied on the wave properties of light (see Chapter 37). To resolve this contradiction, we must conclude that either (1) the laws of electricity and magnetism, which indicate that light moves with constant speed, are not the same in all inertial frames or (2) the Galilean velocity transformation equation is incorrect. If we assume the first alternative, a preferred reference frame in which the speed of light has the value c must exist and the measured speed must be greater or less than this value in any other reference frame, in accordance with the Galilean velocity transformation equation. If we assume the second alternative, we must abandon the notions of absolute time and absolute length that form the basis of the Galilean space–time transformation equations. This second alternative is the basis of special relativity.

Einstein's postulates of special relativity

Einstein proposed a theory that boldly removed the difficulties arising from the failure of the Galilean velocity transformations in the case of light. In doing so, he completely altered our notion of space and time, ultimately leading to his theory of general relativity described in Chapter 11. He based his special theory of relativity on two postulates.

Einstein's postulates of the special theory of relativity ►

1. **The principle of relativity:** The laws of physics must be the same in all inertial reference frames.
2. **The constancy of the speed of light:** The speed of light in vacuum has the same value, $c = 3.00 \times 10^8$ m/s, in all inertial frames, regardless of the velocity of the observer or the velocity of the source emitting the light.

The first postulate asserts that *all* the laws of physics – those dealing with mechanics, electricity and magnetism, optics, thermodynamics, and so on – are the same in all reference frames moving with constant velocity relative to one another. This postulate is a generalisation of the principle of Galilean relativity, which refers only to the laws of mechanics. From an experimental point of view, Einstein's principle of relativity means that any kind of experiment (measuring the speed of light, for example) performed in a laboratory at rest must give the same result when performed in a laboratory moving at a constant velocity with respect to the first one. Hence, no preferred inertial reference frame exists and it is impossible to detect absolute motion.

Note that postulate 2 is required by postulate 1: if the speed of light were not the same in all inertial frames, measurements of different speeds would make it possible to distinguish between inertial frames. As a result, a preferred, absolute frame could be identified, in contradiction to postulate 1.

Many different experimental observations, from fields as different as radioactive β-decay studies and astrophysics, support Einstein's postulates. If we accept the special theory of relativity, we must conclude that relative motion is unimportant when measuring the speed of light. At the same time, we must alter our commonsense notion of space and time and be prepared for some counterintuitive consequences.

David Serway, son of one of the authors, watches over two of his children, Nathan and Kaitlyn, as they play on the statue of Albert Einstein at the Einstein memorial in Washington, D.C. (*Emily Serway*)

12.3 Consequences of the loss of simultaneity: measurements of time and length

A basic premise of Newtonian mechanics is that a universal time scale exists that is the same for all observers. Classical mechanics takes this *simultaneity* for granted. The special theory of relativity abandons this assumption.

Einstein devised the following thought experiment to illustrate this point. A railway carriage moves with uniform velocity, and two bolts of lightning strike its ends as illustrated in **Figure 12.4a**, leaving marks on the railway carriage and on the ground. The marks on the railway carriage are labelled A' and B', and those on the ground are labelled A and B. An observer O' moving with the railway carriage is midway between A' and B', and a ground observer O is midway between A and B. The events recorded by the observers are the striking of the railway carriage by the two lightning bolts.

The light signals emitted from A and B at the instant at which the two bolts strike later reach observer O at the same time as indicated in **Figure 12.4b**. This observer realises that the signals travelled at the same speed over equal distances and so concludes that the events at A and B occurred simultaneously. Now consider the same events as viewed by observer O'. By the time the signals have reached observer O, observer O' has moved as indicated in **Figure 12.4b**. Therefore, the signal from B' has already swept past O', but the signal from A' has not yet reached O'. In other words, O' sees the signal from B' before seeing the signal from A'. Remember that, according to the postulates of special relativity, *the two observers must find that light travels at the same speed, even though they are in different inertial frames.* Therefore, observer O' concludes that one lightning bolt strikes the front of the railway carriage *before* the other one strikes the back.

This thought experiment clearly demonstrates that the two events that appear to be simultaneous to observer O do *not* appear to be simultaneous to observer O'. Simultaneity is not an absolute concept but rather one that depends on the motion of the observer. The loss of an absolute timescale and hence of absolute simultaneity has some very interesting consequences: the results of measurements of lengths and time intervals depend on the reference frame of the observer making the measurement.

Pitfall Prevention 12.2

You might wonder which observer in Figure 12.4 is correct concerning the two lightning strikes. *Both are correct* because the principle of relativity states that *there is no preferred inertial frame of reference.* Although the two observers reach different conclusions, both are correct in their own reference frame because the concept of simultaneity is not absolute. That, in fact, is the central point of relativity: any uniformly moving frame of reference can be used to describe events.

Time dilation

Consider a vehicle moving to the right with a speed v, such as the railway carriage shown in **Active Figure 12.5a**. A mirror is fixed to the ceiling of the vehicle. Observer O' is at rest in the rest frame of the carriage. She holds a torch a distance d below the mirror. At some instant, the torch emits a pulse of light directed toward the mirror (event 1) and at some later time after reflecting from the mirror, the pulse arrives back at the torch (event 2). Observer O' carries a clock (which is therefore also in the rest frame of the carriage) and uses it to measure

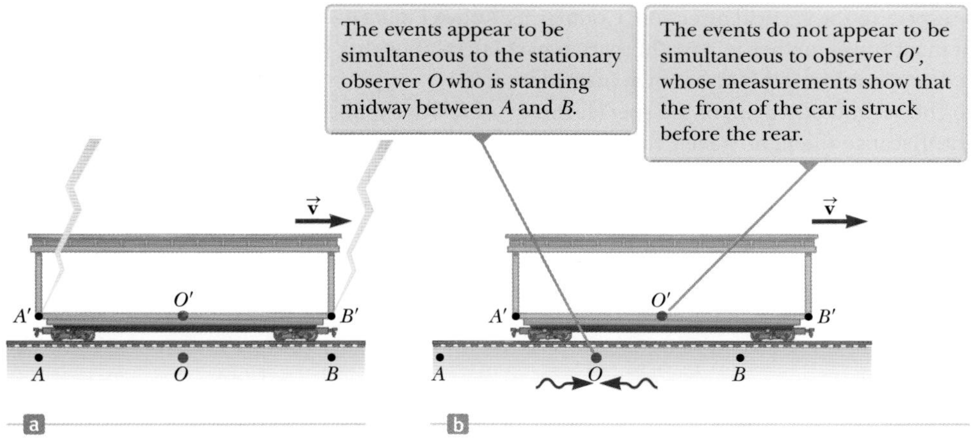
The events appear to be simultaneous to the stationary observer O who is standing midway between A and B.
The events do not appear to be simultaneous to observer O', whose measurements show that the front of the car is struck before the rear.
$\vec{v}$
A'
O'
B'
A
O
B
a
b

Figure 12.4
(a) Two lightning bolts strike the ends of a moving railway carriage. (b) The leftwards-travelling light signal has already passed O', but the rightwards-travelling signal has not yet reached O'.

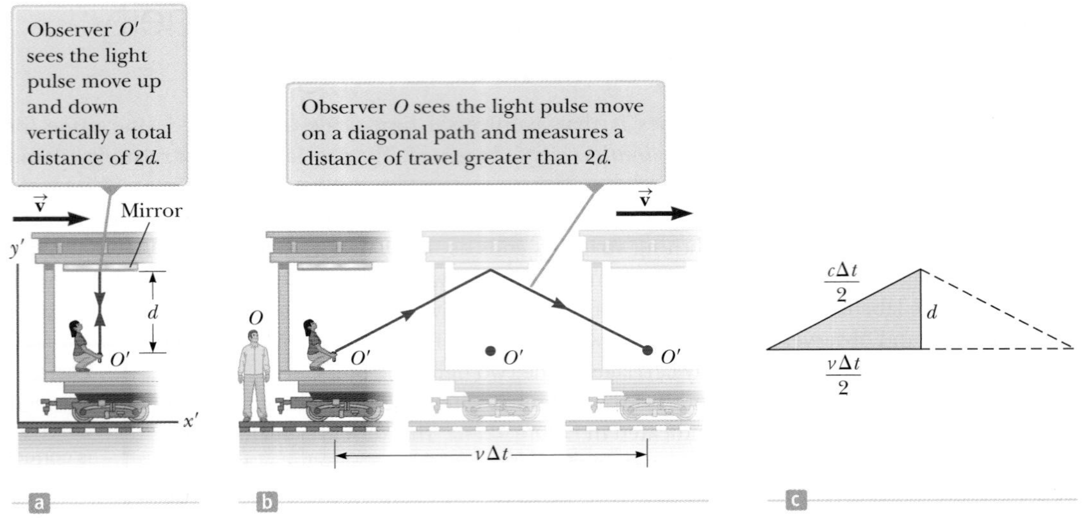

Active Figure 12.5

(a) A mirror is fixed to a moving vehicle and a light pulse is sent out by observer O' at rest in the vehicle.
(b) Relative to a stationary observer O standing alongside the vehicle, the mirror and O' move with a speed v.
(c) The right triangle for calculating the relationship between Δt and Δt_p.

the time interval Δt_p between these two events. The subscript p stands for *proper*, referring to the fact that this measurement of the time interval is being made in the frame in which the two events occur at the same location (the light pulse leaving and arriving at the torch).

We model the pulse of light from the torch as a particle with constant speed. (We shall see in Part 7 of this book why we can treat light as a particle.) Because the light pulse has a speed c, the time interval required for the pulse to travel from O' to the mirror and back is

$$\Delta t_p = \frac{\text{distance travelled}}{\text{speed}} = \frac{2d}{c} \tag{12.3}$$

This is the time interval between the torch emitting the pulse of light and the reflected pulse arriving back at the torch, as measured by observer O'.

Now consider the same pair of events as viewed by observer O in a second frame at rest with respect to the ground as shown in **Active Figure 12.5b**. According to this observer, the carriage, the mirror and the torch are all moving to the right with speed v. As a result, the sequence of events appears entirely different. By the time the light from the torch reaches the mirror, the mirror has moved to the right a distance $v\Delta t/2$, where Δt is the time interval required for the light to travel from O' to the mirror and back to O' as measured by O. Observer O concludes that because of the motion of the vehicle, if the light is to hit the mirror, it must travel at an angle with respect to the vertical direction. Comparing **Active Figure 12.5a** with **Active Figure 12.5b**, we see that the light must travel further in part (b) than in part (a). (Notice that neither observer 'knows' that he or she is moving. Each is at rest in his or her own inertial frame.)

According to the second postulate of the special theory of relativity, both observers must measure c for the speed of light. Because the light travels further according to O, the time interval Δt measured by O is longer than the time interval Δt_p measured by O'. To obtain a relationship between these two time intervals, let's use the right triangle shown in **Active Figure 12.5c**. The Pythagorean theorem gives

$$\left(\frac{c\Delta t}{2}\right)^2 = \left(\frac{v\Delta t}{2}\right)^2 + d^2$$

Solving for Δt gives

$$\Delta t = \frac{2d}{\sqrt{c^2 - v^2}} = \frac{2d}{c\sqrt{1 - \dfrac{v^2}{c^2}}} \tag{12.4}$$

Because $\Delta t_p = 2d/c$, we can express this result as

$$\Delta t = \frac{\Delta t_p}{\sqrt{1 - \frac{v^2}{c^2}}} \equiv \gamma\, \Delta t_p \qquad (12.5)$$

◄ Time dilation

where

$$\gamma = \frac{1}{\sqrt{1 - \frac{v^2}{c^2}}} \qquad (12.6)$$

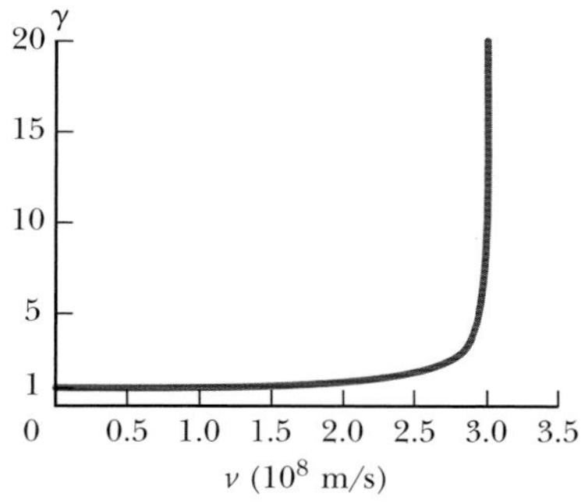

Figure 12.6
Graph of γ versus v. As the speed approaches that of light, γ increases rapidly.

The quantity γ is called the Lorentz factor. Because γ is always greater than unity, **Equation 12.5** shows that the time interval Δt measured by an observer moving with respect to a clock is longer than the time interval Δt_p measured by an observer at rest with respect to the clock. This effect is known as time dilation.

Time dilation is not observed in our everyday lives, which can be understood by considering the factor γ. This factor deviates significantly from a value of 1 only for very high speeds, as shown in **Figure 12.6** and **Table 12.1**. For example, for a speed of $0.1c$, the value of γ is 1.005. Therefore, there is a time dilation of only 0.5% even at 10% of the speed of light. Speeds encountered on an everyday basis are far slower than $0.1c$.

◄ Proper time

The time interval Δt_p in **Equations 12.3** and **12.5** is called the **proper time** interval. (Einstein used the German term *Eigenzeit*, which means 'own-time'.) The proper time interval is the time interval between two events measured by an observer who is at rest in the reference frame in which the two events occur at the same point in space. For example, the proper time interval between the creation of an unstable particle such as a muon (see below) and its subsequent decay is the time measured in the rest frame of the muon.

If a clock is moving with respect to you, the time interval between ticks of the moving clock is observed to be longer than the time interval between ticks of an identical clock in your reference frame. Therefore, a moving clock is measured to run more slowly than a clock in your reference frame by a factor γ. We can generalise this result by stating that all physical processes, including mechanical, chemical and biological ones, are measured to slow down when those processes occur in a frame moving with respect to the observer. For example, a clock inside a spacecraft is measured to slow down relative to a clock back on the Earth, as does the astronaut's heartbeat and other life processes.

Table 12.1
Approximate values for γ at various speeds

v/c	γ
0	1
0.0010	1.0000005
0.010	1.00005
0.10	1.005
0.20	1.021
0.50	1.155
0.80	1.667
0.90	2.294
0.94	2.931
0.99	7.089
0.999	22.37

Quick Quiz 12.2

Suppose the observer O' on the train in **Active Figure 12.5** aims her torch at the far wall of the railway carriage and turns it on and off, sending a pulse of light toward the far wall. Both O' and O measure the time interval between when the pulse leaves the torch and when it hits the far wall. Which observer measures the proper time interval between these two events? **(a)** O' **(b)** O **(c)** both observers **(d)** neither observer

Time dilation is a very real phenomenon that has been verified by various experiments involving natural clocks. One direct illustration of time dilation involves the observation of *muons*, unstable elementary particles that have a charge equal to that of the electron and a mass 207 times that of the electron. Muons can be produced by the collision of cosmic radiation with atoms high in the atmosphere. Slow-moving muons in the laboratory have a lifetime that is measured to be the proper time interval $\Delta t_p = 2.2\ \mu s$. If we take $2.2\ \mu s$ as the average lifetime of a muon, we find that muons travelling close to the speed of light can travel a distance of approximately $(3.0 \times 10^8\ m/s)(2.2 \times 10^{-6}\ s) \approx 6.6 \times 10^2$ m before they decay (**Figure 12.7a**). Hence, they are unlikely to reach the surface of the Earth from high in the atmosphere where they are produced. Experiments show, however, that a large number of muons *do* reach the surface. The phenomenon of time dilation explains this effect.

Pitfall Prevention 12.3

It is *very* important in relativistic calculations to correctly identify the proper time interval. The proper time interval between two events is always the time interval measured by an observer for whom the two events take place at the same position.

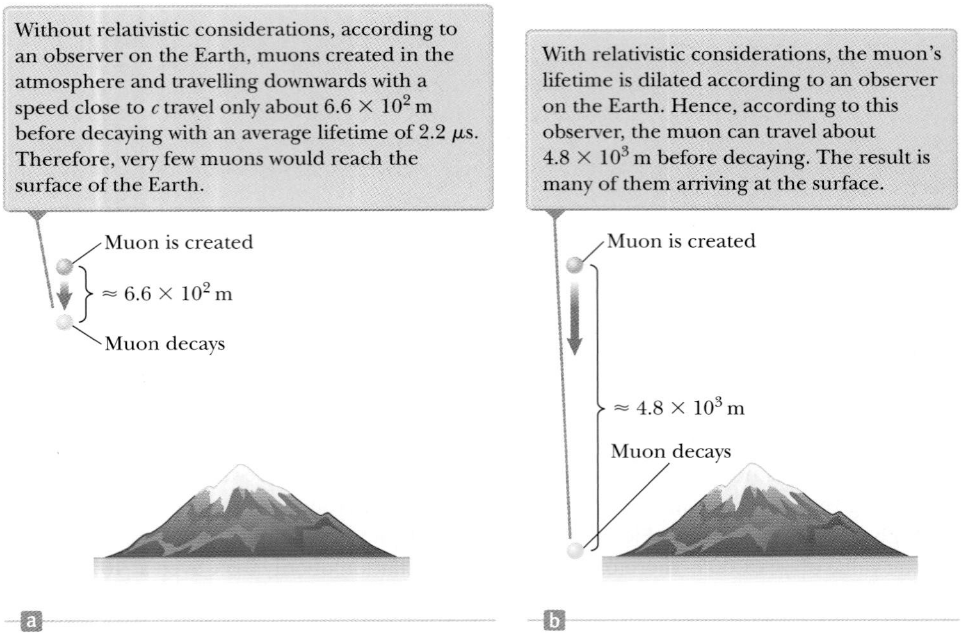
Without relativistic considerations, according to an observer on the Earth, muons created in the atmosphere and travelling downwards with a speed close to c travel only about 6.6×10^2 m before decaying with an average lifetime of 2.2 μs. Therefore, very few muons would reach the surface of the Earth.
Muon is created
$\approx 6.6 \times 10^2$ m
Muon decays
a
With relativistic considerations, the muon's lifetime is dilated according to an observer on the Earth. Hence, according to this observer, the muon can travel about 4.8×10^3 m before decaying. The result is many of them arriving at the surface.
Muon is created
$\approx 4.8 \times 10^3$ m
Muon decays
b

Figure 12.7
Travel of muons according to an Earth-based observer

As measured by an observer on the Earth, the muons have a dilated lifetime equal to $\gamma\Delta t_p$. For example, for $v = 0.99c$, $\gamma \approx 7.1$, and $\gamma\,\Delta t_p \approx 16\ \mu$s. Hence, the average distance travelled by the muons in this time interval as measured by an observer on the Earth is approximately $(0.99)(3.0 \times 10^8\ \text{m/s})(16 \times 10^{-6}\ \text{s}) \approx 4.8 \times 10^3$ m, as indicated in **Figure 12.7b**.

In 1976, at the laboratory of the European Council for Nuclear Research (CERN) in Geneva, muons injected into a large storage ring reached speeds of approximately 0.9994c. Electrons produced by the decaying muons were detected by counters around the ring, enabling scientists to measure the decay rate and hence the muon lifetime. The lifetime of the moving muons was measured to be approximately 30 times as long as that of the stationary muon, in agreement with the prediction of special relativity to within two parts in a thousand.

±? Example 12.1

What is the period of the pendulum?

The period of a pendulum is measured to be (3.00 ± 0.01) s in the reference frame of the pendulum. What is the period when measured by an observer moving at a speed of 0.50c relative to the pendulum?

Solution

Conceptualise Let's change frames of reference. Instead of the observer moving at 0.50c, we can take the equivalent point of view that the observer is at rest and the pendulum is moving at 0.50c past the stationary observer. Hence, the pendulum is an example of a clock moving at high speed with respect to an observer.

Model Based on the *Conceptualise* step, we can use the model of time dilation of a moving clock.

Analyse The proper time interval, measured in the rest frame of the pendulum where each swing occurs at the same location, is $\Delta t_p = 3.00$ s.

Use **Equation 12.5** to find the dilated time interval:

$$\Delta t = \gamma\Delta t_p = \frac{1}{\sqrt{1 - \dfrac{(0.50c)^2}{c^2}}}\Delta t_p = \frac{1}{\sqrt{1 - 0.25}}\Delta t_p$$

$$= 1.155[(3.00 \pm 0.01)\ \text{s}] = (3.47 \pm 0.01)\ \text{s}$$

Example 12.1 cont.

Finalise The moving pendulum is indeed measured to take longer to complete a period than does a pendulum at rest. The period increases by a factor of $\gamma = 1.155$.

What If What if the speed of the observer increases by 10%? Does the dilated time interval increase by 10%?

Answer Based on the highly nonlinear behaviour of γ as a function of v in **Figure 12.6**, we would guess that the increase in Δt would be different from 10%.

Find the new speed if it increases by 10%:

$$v_{\text{new}} = (1.10)(0.50c) = 0.55c$$

Perform the time dilation calculation again:

$$\Delta t = \gamma \Delta t_{\text{p}} = \frac{1}{\sqrt{1 - \dfrac{(0.55c)^2}{c^2}}}\Delta t_{\text{p}} = \frac{1}{\sqrt{1 - 0.3025}}\Delta t_{\text{p}}$$

$$= 1.43(3.00 \pm 0.01)\text{ s} = (4.29 \pm 0.01)\text{ s}$$

Therefore, the 10% increase in speed results in more than a 40% increase in the dilated time! As γ increases very rapidly with v, at speeds above $0.9c$ a 10% increase in speed gives many hundreds of times increase in time. Note that the uncertainty does not appear to change here because we are rounding to two decimal places to match the precision of the given data. In fact the uncertainty is increasing as the time increases.

Example 12.2

How long was your trip?

Suppose you are driving your car on a business trip and are travelling at 30 m/s. Your boss, who is waiting at your destination, expects the trip to take 5.0 h. When you arrive late, your excuse is that the clock in your car registered the passage of 5.0 h but that you were driving fast and so your clock ran more slowly than the clock in your boss' office. If your car clock actually did indicate a 5.0 h trip, how much time passed on your boss' clock, which was at rest on the Earth?

Solution

Conceptualise The observer is your boss standing stationary on the Earth. The clock is in your car, moving at 30 m/s with respect to your boss.

Model Given the everyday speed of 30 m/s, we would use a model with classical concepts and equations. Based on the problem statement that the moving clock runs more slowly than a stationary clock, however, we employ a relativistic model involving time dilation.

Analyse The proper time interval, measured in the rest frame of the car, is $\Delta t_{\text{p}} = 5.0$ h.

Use **Equation 12.6** to evaluate γ:

$$\gamma = \frac{1}{\sqrt{1 - \dfrac{v^2}{c^2}}} = \frac{1}{\sqrt{1 - \dfrac{(3.0 \times 10^1\text{ m/s})^2}{(3.0 \times 10^8\text{ m/s})^2}}} = \frac{1}{\sqrt{1 - 10^{-14}}}$$

If you try to determine this value on your calculator, you will probably obtain $\gamma = 1$. Instead, use a binomial expansion to first order:

$$\gamma = (1 - 10^{-14})^{-1/2} \approx 1 + \frac{1}{2}(10^{-14}) = 1 + 5.0 \times 10^{-15}$$

Use **Equation 12.5** to find the dilated time interval measured by your boss:

$$\Delta t = \gamma \Delta t_{\text{p}} = (1 + 5.0 \times 10^{-15})(5.0\text{ h})$$
$$= 5.0\text{ h} + 2.5 \times 10^{-14}\text{ h} = 5.0\text{ h} + 0.090\text{ ns}$$

Finalise Your boss' clock would be only 0.090 ns ahead of your car clock. You might want to think of another excuse. This example shows how tiny relativistic effects such as time dilation are in most circumstances and why Newtonian mechanics is an appropriate approximation for virtually all everyday situations.

The twin paradox

The famous *twin paradox* is often given as an example of the counterintuitive nature of special relativity. However, as we shall see in the following, to understand and resolve the paradox we must go beyond special relativity, which only applies to inertial frames. Consider an experiment involving twin brothers named Laurence and Marcus. When they are 20 years old, Laurence, the more adventurous of the two, sets out on an epic journey from the Earth to Planet X, located 20 light-years away. One light-year (ly) is the distance light travels through free space in 1 year. Laurence's spacecraft is capable of reaching a speed of $0.95c$ relative to the inertial frame of his twin brother back home on the Earth. After reaching Planet X, Laurence becomes homesick and immediately returns to the Earth at the same speed, $0.95c$. Upon his return, Laurence is shocked to discover that Marcus has aged 42 years and is now 62 years old. Laurence, on the other hand, has aged only 13 years.

The paradox is *not* that the twins have aged at different rates. Here is the apparent paradox. According to Marcus, he was at rest while his brother travelled at a high speed away from him and then came back. According to Laurence, however, he himself remained stationary while Marcus and the Earth raced away from him and then headed back. Therefore, we might expect Laurence to claim that Marcus ages more slowly than himself. The situation appears to be symmetrical from either twin's point of view. Which twin *actually* ages more slowly?

The situation is actually not symmetrical. Consider a third observer moving at a constant speed relative to Marcus. According to the third observer, Marcus does not change inertial frames. Marcus' speed relative to the third observer is always the same. The third observer notes, however, that Laurence accelerates during his journey when he slows down and starts moving back toward the Earth, *changing reference frames in the process.* From the third observer's perspective, there is something very different about the motion of Marcus when compared to that of Laurence. Einstein showed that acceleration can be treated as equivalent to gravitational free fall, and so while Laurence is accelerating back towards Earth, his motion needs to be considered in the framework of general, rather than special, relativity.

Quick Quiz 12.3

Suppose astronauts are paid according to the amount of time they spend travelling in space. After a long voyage travelling at a speed approaching c, would a crew rather be paid according to **(a)** an Earth-based clock, **(b)** their spacecraft's clock or **(c)** it doesn't make any difference?

Length contraction

The measured distance between two points in space also depends on the frame of reference of the observer.

Proper length ▶ The **proper length** L_p of an object is the length measured by an observer *at rest relative to the object.* The length of an object measured by someone in a reference frame that is moving with respect to the object is always less than the proper length. This effect is known as **length contraction**.

> **Pitfall Prevention 12.4**
> As with the proper time interval, it is *very* important in relativistic calculations to correctly identify the observer who measures the proper length. The proper length between two points in space is always the length measured by an observer at rest with respect to the points.

To understand length contraction, consider a spacecraft travelling with a speed v from one star to another. There are two observers: one on the Earth and the other in the spacecraft. The observer at rest on the Earth (and also assumed to be at rest with respect to the two stars) measures the distance between the stars to be the proper length L_p. According to this observer, the time interval required for the spacecraft to complete the voyage is $\Delta t = L_p/v$. The pass of the two stars by the spacecraft occurs at the same position for the space traveller; namely, at the location of the spacecraft. Therefore, the space traveller measures the proper time interval Δt_p. Because of time dilation, the proper time interval is related to the Earth-measured time interval by $\Delta t_p = \Delta t/\gamma$. Because the space traveller reaches the second star in the time Δt_p, he or she concludes that the distance L between the stars is

$$L = v\Delta t_p = v\frac{\Delta t}{\gamma}$$

Because the proper length is $L_p = v\,\Delta t$, we see that

$$L = \frac{L_p}{\gamma} = L_p\sqrt{1 - \frac{v^2}{c^2}} \qquad (12.7)$$

where $\sqrt{1 - v^2/c^2}$ is a factor less than one. If an object has a proper length L_p when it is measured by an observer at rest with respect to the object, its length L when it moves with speed v in a direction parallel to its length is measured to be shorter according to **Equation 12.7**.

For example, suppose a metre stick moves past a stationary Earth-based observer with speed v as in **Active Figure 12.8**. The length of the metre stick as measured by an observer in a frame attached to the stick is the proper length L_p shown in **Active Figure 12.8a**. The length of the stick L measured by the Earth observer is shorter than L_p by the factor $(1 - v^2/c^2)^{1/2}$ as suggested in **Active Figure 12.8b**. Notice that length contraction takes place only along the direction of motion.

The proper length and the proper time interval are defined differently. The proper length is measured by an observer for whom the endpoints of the length remain fixed in space. The proper time interval is measured by someone for whom the two events take place at the same position in space. As an example of this point, let's return to the decaying muons moving at speeds close to the speed of light. An observer in the muon's reference frame measures the proper lifetime, whereas an Earth-based observer measures the proper length (the distance between the creation point and the decay point in **Figure 12.7b**). In the muon's reference frame, there is no time dilation, but the distance of travel to the surface is shorter when measured in this frame. Likewise, in the Earth observer's reference frame, there is time dilation but the distance of travel is measured to be the proper length. Therefore, when calculations on the muon are performed in both frames, the outcome of the experiment in one frame is the same as the outcome in the other frame: more muons reach the surface than would be predicted without relativistic effects.

Quick **Quiz 12.4**

You are observing a spacecraft moving away from you. Before it set off, you measured its length and synchronised your watch with an on-board clock that you can still see through its window. Now that it is moving away from you, you measure the spacecraft to be shorter than when it was at rest on the ground next to you. By comparing the time displayed on the on-board clock with the time on your watch, you observe that time runs slower on the spacecraft. Compared with when the spacecraft was on the ground, what do you measure if the spacecraft turns around and comes *towards* you at the same speed? **(a)** The spacecraft is measured to be longer, and the clock runs faster. **(b)** The spacecraft is measured to be longer, and the clock runs slower. **(c)** The spacecraft is measured to be shorter, and the clock runs faster. **(d)** The spacecraft is measured to be shorter, and the clock runs slower.

Length contraction and time dilation raise the important question of what we mean when we talk about measurements of time and position. Since both time and space intervals depend on the frame of reference of the observer, it is usually sensible to define the proper length or time and then calculate the effect of a non-zero relative speed of an observer on the result of a measurement of that length or time interval.

World-lines and light cones

It is sometimes helpful to represent a physical situation with a **space–time graph**, in which ct is the ordinate and position x is the abscissa. The twin paradox is displayed on such a graph in **Figure 12.9** from Marcus' point of view. A path through space–time is called a **world-line**. At the origin, the world-lines of Laurence (blue) and Marcus (green) coincide because the twins are in the same location at the same time. After Laurence leaves on his trip, his world-line diverges from that of his brother. Marcus' world-line is vertical because he remains fixed in location in his own reference frame. At Marcus and Laurence's reunion, the two world-lines again come together. It would be impossible for Laurence to have a world-line that crossed the path of a light beam that left the Earth when he did. To do so would require him to have a speed greater than c (which, as shown in Sections 12.4 and 12.5, is not possible).

World-lines for light beams are diagonal lines on space–time graphs, typically drawn at 45° to the right or left of vertical (assuming the x and ct axes have the same scales), depending on whether the light beam is travelling in the direction of increasing or decreasing x. All possible future events for Marcus and Laurence lie above the x axis and between the red-brown lines in **Figure 12.9** because neither twin can travel faster than light. The only past events that Marcus and Laurence could have experienced occur between two similar 45° world-lines that approach the origin from below the x axis.

A metre stick measured by an observer in a frame attached to the stick has its proper length L_p.

A metre stick measured by an observer in a frame in which the stick has a velocity relative to the frame is measured to be shorter than its proper length.

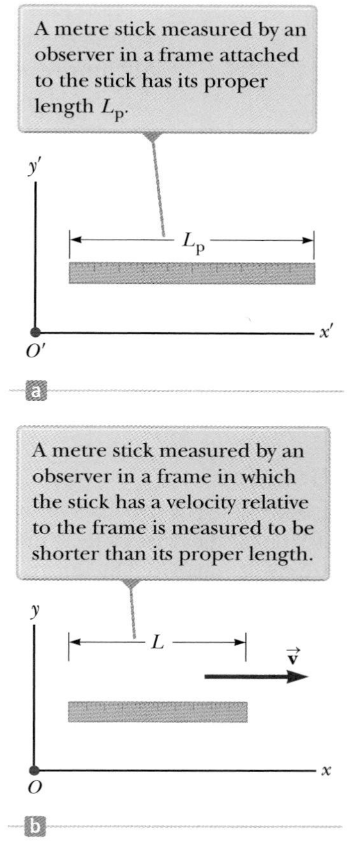

Active Figure 12.8

The length of a metre stick is measured by two observers.

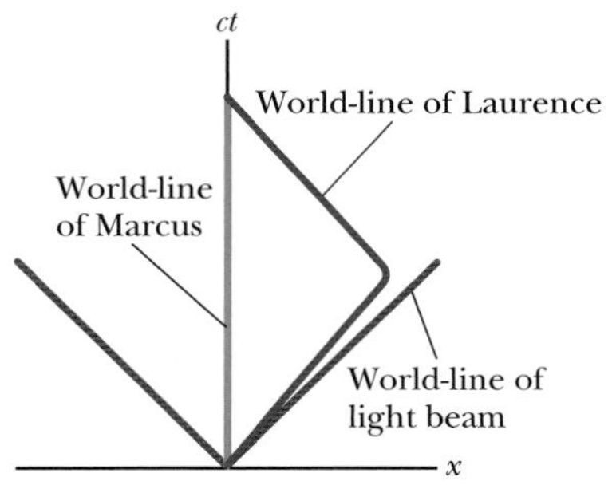

Figure 12.9

The twin paradox on a space–time graph. The twin who stays on the Earth has a world-line along the ct axis (green). The path of the travelling twin through space–time is represented by a world-line that changes direction (blue). The red-brown lines are world-lines for light beams travelling in the positive x direction (on the right) or the negative x direction (on the left).

If Figure 12.9 is rotated about the ct axis, the red-brown lines sweep out a cone, called the *light cone*, which generalises Figure 12.9 to two space dimensions. The y axis can be imagined coming out of the page. All future events for an observer at the origin must lie within the light cone. We can imagine another rotation that would generalise the light cone to three space dimensions to include z, but because of the requirement for four dimensions (three space dimensions and time), we cannot represent this situation in a two-dimensional drawing on paper.

Example 12.3

A voyage to Sirius

An astronaut takes a trip to Sirius, which is located a distance of 8 light-years from the Earth. The astronaut measures the time of the one-way journey to be 6 years in a spaceship moving at a constant speed of $0.8c$. Is the astronaut travelling faster than light?

Solution

Conceptualise An observer on the Earth measures light to require 8 years to travel between Sirius and the Earth. The astronaut measures a time interval for his travel of only 6 years. How can the 8-ly distance be reconciled with the 6-year trip time measured by the astronaut? We know already that time is dilated and length is contracted in a moving reference frame, so will apply these ideas.

Model We model the astronaut as a relativistic particle moving with constant velocity. We will need to use the ideas of length contraction and time dilation.

Analyse The distance of 8 ly represents the proper length from the Earth to Sirius measured by an observer on the Earth seeing both objects nearly at rest.

Calculate the contracted length measured by the astronaut using Equation 12.7:

$$L = \frac{8\text{ ly}}{\gamma} = (8\text{ ly})\sqrt{1 - \frac{v^2}{c^2}} = (8\text{ ly})\sqrt{1 - \frac{(0.8c)^2}{c^2}} = 5\text{ ly}$$

Use the particle with constant velocity model to find the travel time measured on the astronaut's clock:

$$\Delta t = \frac{L}{v} = \frac{5\text{ ly}}{0.8c} = \frac{5\text{ ly}}{0.8(1\text{ ly/yr})} = 6\text{ yr}$$

Finalise Notice that we have used the value for the speed of light as $c = 1$ ly/yr. The trip takes a time interval shorter than 8 years for the astronaut because, to her, the distance between the Earth and Sirius is measured to be shorter.

What If? What if this trip is observed with a very powerful telescope by a technician in Mission Control on the Earth? At what time will this technician see that the astronaut has arrived at Sirius, as measured by a clock on Earth?

Answer We have to consider both time dilation and the time taken for the light signal to travel back from Sirius to Earth. The duration of the astronaut's journey measured in the rest frame of the technician is

$$\Delta t = \frac{L_p}{v} = \frac{8\text{ ly}}{0.8c} = 10\text{ yr}$$

For the technician to *see* the arrival, the light from the scene of the arrival must travel back to the Earth and enter the telescope. This requires an additional time interval of

$$\Delta t = \frac{L_p}{v} = \frac{8\text{ ly}}{c} = 8\text{ yr}$$

Therefore, the technician sees the arrival after 10 yr + 8 yr = 18 yr. If the astronaut immediately turns around and comes back home, she arrives, according to the technician, 20 years after leaving, only 2 years after the technician saw her arrive!

Example 12.4

The pole-in-the-barn paradox

The twin paradox, discussed earlier, is a classic 'paradox' in relativity. Another classic paradox is as follows. Suppose a runner moving at $0.75c$ carries a horizontal pole 15 m long toward a barn that is 10 m long. The barn has front and rear doors that are initially open. An observer on the ground can instantly and simultaneously close and open the two doors by remote control. When the runner and the pole are inside the barn, the ground observer closes and then opens both doors so that the runner and pole are momentarily captured inside the barn and then proceed to exit the barn from the back doorway. Do both the runner and the ground observer agree that the runner makes it safely through the barn?

Solution

Conceptualise From your everyday experience, you would be surprised to see a 15 m pole fit inside a 10 m barn, but we are becoming used to surprising results in relativistic situations.

Model The pole is in motion with respect to the ground observer so that the observer measures its length to be contracted, whereas the stationary barn has a proper length of 10 m. We use the model of length contraction of moving objects.

Analyse Use Equation 12.7 to find the contracted length of the pole according to the ground observer:

$$L_{\text{pole}} = L_{\text{p}}\sqrt{1-\frac{v^2}{c^2}} = (15\text{ m})\sqrt{1-(0.75)^2} = 9.9\text{ m}$$

Therefore, the ground observer measures the pole to be slightly shorter than the barn and there is no problem with momentarily capturing the pole inside it. The 'paradox' arises when we consider the runner's point of view.

Use Equation 12.7 to find the contracted length of the barn according to the running observer:

$$L_{\text{barn}} = L_{\text{p}}\sqrt{1-\frac{v^2}{c^2}} = (10\text{ m})\sqrt{1-(0.75)^2} = 6.6\text{ m}$$

Because the pole is in the rest frame of the runner, the runner measures it to have its proper length of 15 m. Now the situation looks even worse: how can a 15 m pole fit inside a 6.6 m barn? Although this question is the classic one that is often asked, it is not the question we have asked because it is not the important one. We asked, *'Does the runner make it safely through the barn?'*

The resolution of the 'paradox' lies in the relativity of simultaneity. The closing of the two doors is measured to be simultaneous by the ground observer. Because the doors are at different positions, however, they do not close simultaneously as measured by the runner. The rear door closes and then opens first, allowing the leading end of the pole to exit. The front door of the barn does not close until the trailing end of the pole passes by.

We can analyse this 'paradox' using a space–time graph. Figure 12.10a is a space–time graph from the ground observer's point of view. We choose $x = 0$ as the position of the front doorway of the barn and $t = 0$ as the instant at which the leading end of the pole is located at the front doorway of the barn. The world-lines for the two doorways of the barn are separated by 10 m and are vertical because the barn is not moving relative to this observer. For the pole, we follow two tilted world-lines, one for each end of the moving pole. These world-lines are 9.9 m apart horizontally, which is the contracted length seen by the ground observer. As seen in Figure 12.10a, the pole is entirely within the barn at some time.

Figure 12.10b shows the space–time graph according to the runner. Here, the world-lines for the pole are separated by 15 m and are vertical because the pole is at rest in the runner's

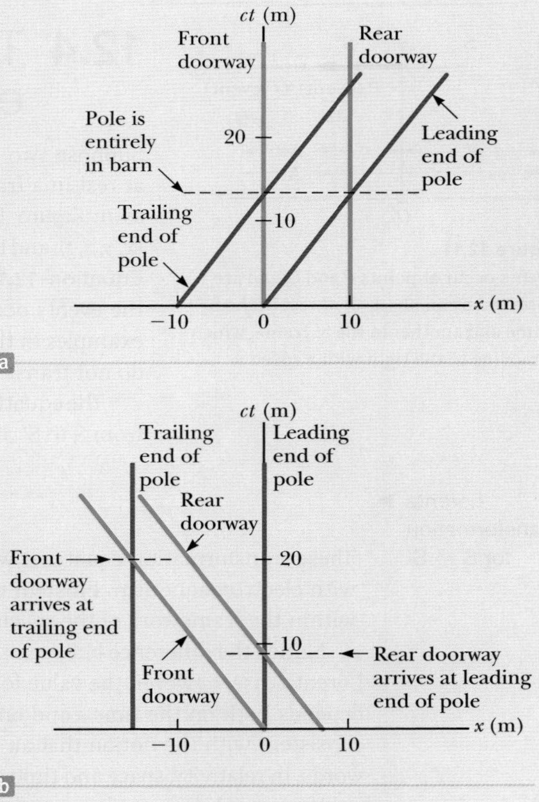

Figure 12.10
(Example 12.4) Space–time graphs for the pole-in-the-barn paradox (a) from the ground observer's point of view and (b) from the runner's point of view

Example 12.4 cont.

frame of reference. The barn is hurtling *towards* the runner, so the world-lines for the front and rear doorways of the barn are tilted to the left. The world-lines for the barn are separated by 6.6 m, the contracted length as seen by the runner. The leading end of the pole leaves the rear doorway of the barn long before the trailing end of the pole enters the barn. Therefore, the opening of the rear door occurs before the closing of the front door.

From the ground observer's point of view, use the particle with constant velocity model to find the time after $t = 0$ at which the trailing end of the pole enters the barn:

$$t = \frac{\Delta x}{v} = \frac{9.9\text{ m}}{0.75c} = \frac{13.2\text{ m}}{c} \quad (1)$$

From the runner's point of view, use the particle with constant velocity model to find the time at which the leading end of the pole leaves the barn:

$$t = \frac{\Delta x}{v} = \frac{6.6\text{ m}}{0.75c} = \frac{8.8\text{ m}}{c} \quad (2)$$

Find the time at which the trailing end of the pole enters the front door of the barn:

$$t = \frac{\Delta x}{v} = \frac{15\text{ m}}{0.75c} = \frac{20\text{ m}}{c} \quad (3)$$

Finalise From **Equation (1)**, the pole should be completely inside the barn at a time corresponding to $ct = 13.2$ m. This situation is consistent with the point on the ct axis in **Figure 12.10a** where the pole is inside the barn. From **Equation (2)**, the leading end of the pole leaves the barn at $ct = 8.8$ m. This situation is consistent with the point on the ct axis in **Figure 12.10b** where the rear doorway of the barn arrives at the leading end of the pole. **Equation (3)** gives $ct = 20$ m, which agrees with the instant shown in **Figure 12.10b** at which the front doorway of the barn arrives at the trailing end of the pole.

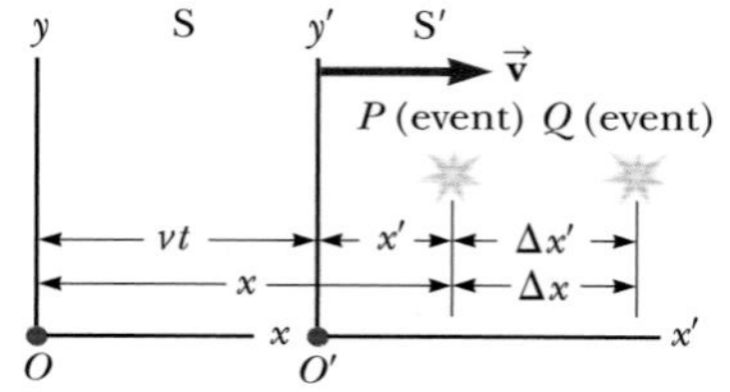

Figure 12.11
Events occur at points P and Q and are observed by an observer at rest in the S frame and another in the S′ frame, which is moving to the right with a speed v.

12.4 The Lorentz transformation equations

Suppose two events occur at points P and Q and are reported by two observers, one at rest in a frame S and another in a frame S′ that is moving to the right with speed v as in **Figure 12.11**. The observer in S reports the events with space–time coordinates (x, y, z, t), and the observer in S′ reports the same events using the coordinates (x', y', z', t'). **Equation 12.1** predicts that the distance between the two points in space at which the events occur does not depend on motion of the observer: $\Delta x = \Delta x'$. However, the examples in the previous section show that at very high relative speeds, time and length do not transform according to the Galilean transformation equations.

The equations that are valid for all speeds and that enable us to transform coordinates from S to S′ are the **Lorentz transformation equations**:

Lorentz ► transformation for S → S′

$$x' = \gamma(x - vt) \quad y' = y \quad z' = z \quad t' = \gamma\left(t - \frac{v}{c^2}x\right) \quad (12.8)$$

These transformation equations were developed by Hendrik A. Lorentz (1853–1928) in 1890 in connection with electromagnetism. Einstein recognised their more general physical significance and interpreted them within the framework of the special theory of relativity.

Notice the difference between the Galilean and Lorentz time equations. In the Galilean case, $t = t'$. In the Lorentz case, however, the value for t' assigned to an event by an observer O' in the S′ frame in **Figure 12.11** depends both on the time t and on the coordinate x as measured by an observer O in the S frame, which is consistent with the notion that an event is characterised by four space–time coordinates (x, y, z, t). In other words, in relativity, space and time are *not* separate concepts but rather are closely interwoven with each other.

If you wish to transform coordinates in the S′ frame to coordinates in the S frame, simply replace v by $-v$ and interchange the primed and unprimed coordinates in **Equations 12.8**:

Inverse Lorentz ► transformation for S′ → S

$$x = \gamma(x' + vt') \quad y = y' \quad z = z' \quad t = \gamma\left(t' + \frac{v}{c^2}x'\right) \quad (12.9)$$

When $v \ll c$, the Lorentz transformation equations should reduce to the Galilean equations. As v approaches zero, $v/c \ll 1$; therefore, $\gamma \to 1$ and **Equations 12.8** indeed reduce to the Galilean space–time transformation equations in **Equation 12.1**.

In many situations, we would like to know the difference in coordinates between two events or the time interval between two events as seen by observers O and O'. From **Equations 12.8** and **12.9**, we can express the differences between the four variables x, x', t and t' in the form

$$\left.\begin{aligned} \Delta x' &= \gamma(\Delta x - v\Delta t) \\ \Delta t' &= \gamma\left(\Delta t - \frac{v}{c^2}\Delta x\right) \end{aligned}\right\} S \to S' \qquad (12.10)$$

$$\left.\begin{aligned} \Delta x &= \gamma(\Delta x' - v\Delta t') \\ \Delta t &= \gamma\left(\Delta t' - \frac{v}{c^2}\Delta x'\right) \end{aligned}\right\} S' \to S \qquad (12.11)$$

where $\Delta x' = x'_2 - x'_1$ and $\Delta t' = t'_2 - t'_1$ are the differences measured by observer O' and $\Delta x = x_2 - x_1$ and $\Delta t = t_2 - t_1$ are the differences measured by observer O.

Example 12.5

Simultaneity and time dilation revisited

(A) Use the Lorentz transformation equations in difference form to show that simultaneity is not an absolute concept.

Solution

Conceptualise Imagine two events that are simultaneous and separated in space as measured in the S′ frame such that $\Delta t' = 0$ and $\Delta x' \neq 0$. These measurements are made by an observer O' who is moving with speed v relative to O.

Model The statement of the problem tells us to use the relativistic model based on the Lorentz transformation.

Analyse From the expression for Δt given in **Equation 12.11**, find the time interval Δt measured by observer O:

$$\Delta t = \gamma\left(\Delta t' + \frac{v}{c^2}\Delta x'\right) = \gamma\left(0 + \frac{v}{c^2}\Delta x'\right) = \gamma\frac{v}{c^2}\Delta x'$$

Finalise The time interval for the same two events as measured by O is non-zero, so the events do not appear to be simultaneous to O.

(B) Use the Lorentz transformation equations in difference form to show that a moving clock is measured to run more slowly than a clock that is at rest with respect to an observer.

SOLUTION

Conceptualise Imagine that observer O' carries a clock that he uses to measure a time interval $\Delta t'$. He finds that two events occur at the same place in his reference frame ($\Delta x' = 0$) but at different times ($\Delta t' \neq 0$). Observer O' is moving with speed v relative to O.

Model The statement of the problem tells us to use the relativistic model based on the Lorentz transformation.

Analyse From the expression for Δt given in **Equation 12.11**, find the time interval Δt measured by observer O:

$$\Delta t = \gamma\left(\Delta t' + \frac{v}{c^2}\Delta x'\right) = \gamma\left(\Delta t' + \frac{v}{c^2}(0)\right) = \gamma\Delta t'$$

Finalise This result is the equation for time dilation found earlier (**Equation 12.5**), where $\Delta t' = \Delta t_p$ is the proper time interval measured by the clock carried by observer O'. Therefore, O measures the moving clock to run slow.

12.5 Relativistic velocity, linear momentum and energy

The Lorentz velocity transformation equations

Suppose two observers in relative motion with respect to each other are both observing an object's motion. We know that the Galilean velocity transformation (**Equation 12.2**) is valid for low speeds. How do the observers' measurements of the velocity of the object relate to each other if the speed of the object or the relative speed of the observers is close to that of light? Once again, S′ is our frame moving in the x direction at a speed v relative to S. Suppose an object has a velocity component u'_x measured in the S′ frame, where

$$u'_x = \frac{dx'}{dt'} \tag{12.12}$$

Using **Equation 12.8**, we have

$$dx' = \gamma(dx - v\,dt)$$

$$dt' = \gamma\left(dt - \frac{v}{c^2}dx\right)$$

Substituting these values into **Equation 12.12** gives

$$u'_x = \frac{dx - v\,dt}{dt - \dfrac{v}{c^2}dx} = \frac{\dfrac{dx}{dt} - v}{1 - \dfrac{v}{c^2}\dfrac{dx}{dt}}$$

The term dx/dt, however, is simply the velocity component u_x of the object measured by an observer in S, so this expression becomes

Lorentz velocity transformation for S → S′ ►

$$u'_x = \frac{u_x - v}{1 - \dfrac{u_x v}{c^2}} \tag{12.13}$$

> **Pitfall Prevention 12.5**
> We have seen several measurements that the two observers O and O′ do *not* agree on: (1) the time interval between events that take place in the same position in one of their frames, (2) the distance between two points that remain fixed in one of their frames, (3) the velocity components of a moving particle, and (4) whether two events occurring at different locations in both frames are simultaneous or not. The two observers *can* agree on (1) their relative speed of motion v with respect to each other, (2) the speed c of any ray of light and (3) the simultaneity of two events that take place at the same position *and* time in some frame.

Although the Lorentz transformation equations show that coordinates perpendicular to the direction of relative motion are unchanged, the velocity components along the y and z axes are transformed due to the different time measurement. The relations given in **Equation 12.8** can be used to show that the components as measured by an observer in S′ are

$$u'_y = \frac{u_y}{\gamma\left(1 - \dfrac{u_x v}{c^2}\right)} \quad \text{and} \quad u'_z = \frac{u_z}{\gamma\left(1 - \dfrac{u_x v}{c^2}\right)}$$

Notice that the expressions for u'_y and u'_z do not contain the parameter v in the numerator because the relative velocity of S and S′ is along the x axis.

When v is much smaller than c (the nonrelativistic limit), the denominator of **Equation 12.13** approaches unity and so $u'_x \approx u_x - v$, which is the Galilean velocity transformation equation. In another extreme, when $u_x = c$, **Equation 12.13** becomes

$$u'_x = \frac{c - v}{1 - \dfrac{cv}{c^2}} = \frac{c\left(1 - \dfrac{v}{c}\right)}{1 - \dfrac{v}{c}} = c$$

This result shows that a speed measured as c by an observer in S is also measured as c by an observer in S′, independent of the relative motion of S and S′. This conclusion is consistent with Einstein's second postulate: the speed of light must be c relative to all inertial reference frames. Furthermore, we find that the speed of an object can never be measured as larger than c. That is, the speed of light is the ultimate speed. We shall return to this point later.

To obtain u_x in terms of u'_x, we replace v by $-v$ in **Equation 12.13** and interchange the roles of u_x and u'_x:

$$u_x = \frac{u'_x + v}{1 + \dfrac{u'v}{c^2}} \tag{12.14}$$

Quick Quiz 12.5

You are driving on a freeway at a relativistic speed. **(i)** Straight ahead of you, a police officer standing on the ground turns on a searchlight and a beam of light moves exactly vertically upwards as seen by the police officer. As you observe the beam of light, do you measure the magnitude of the vertical component of its velocity as (a) equal to c, (b) greater than c or (c) less than c? **(ii)** If the police officer aims the searchlight directly at you instead of upwards, do you measure the magnitude of the horizontal component of its velocity as (a) equal to c, (b) greater than c or (c) less than c?

Example 12.6

Relative velocity of two spacecraft

Two spacecraft A and B are moving in opposite directions. An observer on the Earth measures the speed of spacecraft A to be $0.750c$ and the speed of spacecraft B to be $0.850c$. Find the velocity of spacecraft B as measured by the crew on spacecraft A.

Solution

Conceptualise Start by drawing a diagram, showing the two reference frames of the two observers, one (O) on the Earth and one (O') on spacecraft A. The event is the motion of spacecraft B.

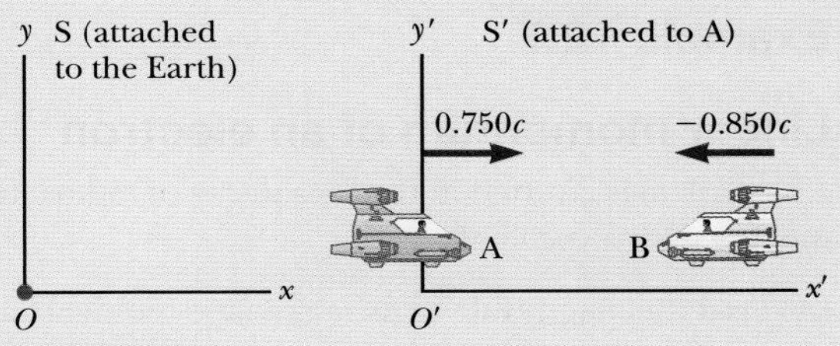

Figure 12.12

(Example 12.6) Two spacecraft A and B move in opposite directions. The speed of spacecraft B relative to spacecraft A is *less* than c and is obtained from the relativistic velocity transformation equation.

Model As the problem asks to find a velocity measured by a particular observer moving close to the speed of light, we use a relativistic model based on the Lorentz velocity transformation.

Analyse The Earth-based observer at rest in the S frame makes two measurements, one of each spacecraft. The velocity of spacecraft A is also the velocity of the observer at rest in spacecraft A (the S′ frame) relative to the observer at rest on the Earth. Therefore, $v = 0.750c$. We want to find u'_x, the velocity of spacecraft B as measured by the crew on spacecraft A, using the velocity measured by the Earth observer, $u_x = -0.850c$.

Obtain the velocity u_x of spacecraft B relative to spacecraft A using **Equation 12.13**:

$$u'_x = \frac{u_x - v}{1 - \dfrac{u_x v}{c^2}} = \frac{-0.850c - 0.750c}{1 - \dfrac{(-0.850c)(0.750c)}{c^2}} = -0.977c$$

Finalise The negative sign indicates that spacecraft B is moving in the negative x direction as observed by the crew on spacecraft A. Is that consistent with your expectation from **Figure 12.12**? Notice that the speed is less than c. That is, an object whose speed is less than c in one frame of reference must have a speed less than c in any other frame. (Had you used the Galilean velocity transformation equation in this example, you would have found that $u'_x = u_x - v = -0.850c - 0.750c = -1.60c$, which is impossible. The Galilean transformation equation does not work in relativistic situations.)

Relativistic linear momentum

In Chapter 7, we saw how important the principle of conservation of momentum is in Newtonian mechanics. Because the laws of physics must remain unchanged under the Lorentz transformation, we must generalise Newton's laws and the definitions of linear momentum and energy to conform to the Lorentz transformation equations and the principle of relativity. These generalised definitions should reduce to the classical (nonrelativistic) definitions for $v \ll c$*, and conservation of momentum and energy must still apply.*

First, recall from the isolated system model that when two particles (or objects that can be modelled as particles) collide, the total momentum of the isolated system of the two particles remains constant. Suppose we observe this collision in a reference frame S and confirm that the momentum of the system is conserved. Now imagine that the momenta of the particles are measured by an observer in a second reference frame S′ moving with velocity $\vec{\mathbf{v}}$ relative to the first frame. Using the Lorentz velocity transformation equation and the classical definition of linear momentum, $\vec{\mathbf{p}} = m\vec{\mathbf{u}}$ (where $\vec{\mathbf{u}}$ is the velocity of a particle), we find that linear momentum is *not* measured to be conserved by the observer in S′. However, because the laws of physics are the same in all inertial frames, linear momentum of the system *must* be conserved in all frames. We have a contradiction. In view of this contradiction and assuming the Lorentz velocity transformation equation is correct, we must modify the definition of linear momentum so that the momentum of an isolated system is conserved for all observers. For any particle, the correct relativistic equation for linear momentum that satisfies this condition is

Definition of relativistic linear momentum ▶

$$\vec{\mathbf{p}} \equiv \frac{m\vec{\mathbf{u}}}{\sqrt{1-\dfrac{u^2}{c^2}}} = \gamma m\vec{\mathbf{u}} \tag{12.15}$$

where m is the mass of the particle and $\vec{\mathbf{u}}$ is the velocity of the particle. When u is much less than c, $\gamma = (1 - u^2/c^2)^{-1/2}$ approaches unity and $\vec{\mathbf{p}}$ approaches $m\vec{\mathbf{u}}$. Therefore, the relativistic equation for $\vec{\mathbf{p}}$ reduces to the classical expression when u is much smaller than c, as it should.

Example 12.7

Linear momentum of an electron

An electron is emitted in the beta decay of radioactive caesium (see Chapter 43). It has a mass of 9.11×10^{-31} kg and is moving with a speed of $0.750c$.

(A) Find the magnitude of its relativistic momentum and compare this value with the momentum calculated from the classical expression.

Solution

Conceptualise Imagine an electron moving with high speed. The electron carries momentum, but the magnitude of its momentum is not given by $p = mu$ because the speed is relativistic.

Model We use the model of relativistic momentum.

Use **Equation 12.15** with $u = 0.750c$ to find the momentum:

$$p = \frac{m_e u}{\sqrt{1-\dfrac{u^2}{c^2}}}$$

$$p = \frac{(9.11\times10^{-31}\text{ kg})(0.750)(3.00\times10^{8}\text{ m/s})}{\sqrt{1-\dfrac{(0.750c)^2}{c^2}}}$$

$$= 3.10 \times 10^{-22}\text{ kg.m/s}$$

The classical expression (used incorrectly here) gives $p_{\text{classical}} = m_e u = 2.05 \times 10^{-22}$ kg.m/s. Hence, the correct relativistic result is 50% greater than the classical result!

Example 12.7 cont.

(B) At what speed does the relativistic momentum differ from that calculated classically by more than 1%?

Solution

Conceptualise We have seen from part (A) that the relativistic momentum is greater than the classical momentum.

Model In this example we are comparing the results of the classical and the relativistic models of momentum. We wish to find the value of u when $p_{\text{classical}} = 0.99p$.

Using **Equation 12.15** again, we find:

$$m_e u = 0.99\left(\frac{m_e u}{\sqrt{1 - \frac{u^2}{c^2}}}\right)$$

which we rearrange to give

$$u^2 = \sqrt{c^2(1 - 0.99^2)}$$

so $u = 4.2 \times 10^7$ m/s, or 152 million km/h.

Finalise As we have noted before, you have to be going very fast before relativistic effects become significant.

Force and acceleration

Newton's second law tells us that force is equal to the rate of change of momentum. For a particle moving at relativistic speeds, the relativistic force $\vec{\mathbf{F}}$ is thus defined as

$$\vec{\mathbf{F}} \equiv \frac{d\vec{\mathbf{p}}}{dt} \tag{12.16}$$

where $\vec{\mathbf{p}}$ is the relativistic momentum given by **Equation 12.15**. This expression, which is the relativistic form of Newton's second law, is reasonable because it preserves classical mechanics in the limit of low velocities and is consistent with conservation of linear momentum for an isolated system ($\vec{\mathbf{F}}_{\text{ext}} = 0$) both relativistically and classically.

Under relativistic conditions, the acceleration $\vec{\mathbf{a}}$ of a particle decreases with the action of a constant force, in which case $a \propto (1 - u^2/c^2)^{3/2}$. This proportionality shows that as the particle's speed approaches c, the acceleration caused by any finite force approaches zero. Hence, it is impossible to accelerate a particle from rest to a speed $u \geq c$. This argument reinforces that the speed of light is the maximum possible speed for energy transfer and for information transfer. Any object with mass must move at a slower speed.

Relativistic energy

To derive the relativistic form of the work–kinetic energy theorem, imagine a particle moving in one dimension along the x axis. A force in the x direction causes the momentum of the particle to change according to **Equation 12.16**. In what follows, we assume the particle is accelerated from rest to some final speed u. The work done by the force F on the particle is

$$W = \int_{x_1}^{x_2} F\,dx = \int_{x_1}^{x_2} \frac{dp}{dt}\,dx \tag{12.17}$$

To perform this integration and find the work done on the particle and the relativistic kinetic energy as a function of u, we first evaluate dp/dt:

$$\frac{dp}{dt} = \frac{d}{dt}\frac{mu}{\sqrt{1-\frac{u^2}{c^2}}} = \frac{m}{\left(1-\frac{u^2}{c^2}\right)^{3/2}}\frac{du}{dt}$$

Σ *For help with this derivation, see Appendix B.7 on differential calculus.*

Substituting this expression for dp/dt and $dx = u\,dt$ into **Equation 12.17** gives

$$W = \int_0^t \frac{m}{\left(1 - \frac{u^2}{c^2}\right)^{3/2}}\frac{du}{dt}(u\,dt) = m\int_0^u \frac{u}{\left(1 - \frac{u^2}{c^2}\right)^{3/2}}\,du$$

where we use the limits 0 and u in the integral because the integration variable has been changed from t to u. Evaluating the integral gives

$$W = \frac{mc^2}{\sqrt{1 - \frac{u^2}{c^2}}} - mc^2 \tag{12.18}$$

Recall from Chapter 6 that the work done by a force acting on a system consisting of a single particle equals the change in kinetic energy of the particle. Because we assumed the initial speed of the particle is zero, its initial kinetic energy is zero. Therefore, the work W in **Equation 12.18** is equivalent to the relativistic kinetic energy K:

Relativistic kinetic energy ▶

$$K = \frac{mc^2}{\sqrt{1 - \frac{u^2}{c^2}}} - mc^2 = \gamma mc^2 - mc^2 = (\gamma - 1)mc^2 \tag{12.19}$$

This equation is routinely confirmed by experiments using high-energy particle accelerators.

At low speeds, where $u/c \ll 1$, **Equation 12.19** should reduce to the classical expression $K = \frac{1}{2}mu^2$. We can check that by using the binomial expansion $(1 - \beta^2)^{-1/2} \approx 1 + \frac{1}{2}\beta^2 + \cdots$ for $\beta \ll 1$, where the higher-order powers of β are neglected in the expansion. (In treatments of relativity, β is a common symbol used to represent u/c or v/c.) In our case, $\beta = u/c$, so

$$\gamma = \frac{1}{\sqrt{1 - \frac{u^2}{c^2}}} = \left(1 - \frac{u^2}{c^2}\right)^{-1/2} \approx 1 + \frac{1}{2}\frac{u^2}{c^2}$$

Σ *See Appendix B.5 for a review of series expansions.*

Substituting this result into **Equation 12.19** gives

$$K \approx \left[\left(1 + \frac{1}{2}\frac{u^2}{c^2}\right) - 1\right]mc^2 = \frac{1}{2}mu^2 \quad (\text{for } u/c \ll 1)$$

which is the classical expression for kinetic energy. **Figure 12.2** showed the classical prediction and the observed relationship between speed and kinetic energy. Consistent with relativity, the observed particle speed never exceeds c, regardless of the kinetic energy. The two curves are in good agreement when $u \ll c$. Note also that the work required to accelerate a particle to close to the speed of light is extremely large – although we sometimes talk about relativistic spaceships and motorcycles, the energy required to accelerate a motorcycle to relativistic speeds is far more than could be provided by all the fuel reserves on Earth. We can, however, accelerate fundamental particles to relativistic speeds in particle accelerators, and it is in high-energy particle accelerators that most experiments that show relativistic effects have been performed. Indeed, the velocity of the accelerated electrons shown in **Figure 12.2** shows exactly the behaviour predicted by the equations above (check this for yourself!).

The constant term mc^2 in **Equation 12.19**, which is independent of the speed of the particle, is called the **rest energy** E_R of the particle:

Rest energy ▶

$$E_R = mc^2 \tag{12.20}$$

Equation 12.20 shows that mass is a form of energy, where c^2 is simply a constant conversion factor. This expression also shows that a small mass corresponds to an enormous amount of energy, a concept fundamental to nuclear and elementary-particle physics.

Energy–momentum relationship for a relativistic particle

The term γmc^2 in **Equation 12.19**, which depends on the particle speed, is the sum of the kinetic and rest energies. It is called the **total energy** E:

$$\text{Total energy} = \text{kinetic energy} + \text{rest energy}$$

$$E = K + mc^2 \tag{12.21}$$

Or

$$E = \frac{mc^2}{\sqrt{1 - \frac{u^2}{c^2}}} = \gamma mc^2 \tag{12.22}$$

◀ Total energy of relativistic particle

Note that the total relativistic energy is obtained by multiplying the rest energy by γ, just as the relativistic momentum was obtained by multiplying the classical momentum by γ.

In many situations, the linear momentum or energy of a particle rather than its speed is measured. It is therefore useful to have an expression relating the total energy E to the relativistic linear momentum p, which is accomplished by using the expressions $E = \gamma mc^2$ and $p = \gamma mu$. By squaring these equations and subtracting, we can eliminate u. The result, after some algebra, is

$$E^2 = p^2c^2 + (mc^2)^2 \tag{12.23}$$

When the particle is at rest, $p = 0$, so $E = E_R = mc^2$.

For particles that have zero mass, such as photons, we set $m = 0$ in **Equation 12.23** and find that

$$E = pc \tag{12.24}$$

This equation is an exact expression relating total energy and linear momentum for photons, which always travel at the speed of light (in a vacuum).

Finally, because the mass m of a particle is independent of its motion, m must have the same value in all reference frames. For this reason, m is often called the **invariant mass**. On the other hand, because the total energy and linear momentum of a particle both depend on velocity, these quantities depend on the reference frame in which they are measured.

When dealing with subatomic particles, it is convenient to express their energy in electron volts (eV). One eV is the energy obtained by an electron after being accelerated through a potential difference of 1 V (see Chapter 25). The conversion factor is

$$1 \text{ eV} = 1.602 \times 10^{-19} \text{ J}$$

For example, the mass of an electron is 9.109×10^{-31} kg. Hence, the rest energy of the electron is

$$m_e c^2 = (9.109 \times 10^{-31} \text{ kg})(2.998 \times 10^8 \text{ m/s})^2 = 8.187 \times 10^{-14} \text{ J}$$

$$= (8.187 \times 10^{-14} \text{ J})(1 \text{ eV}/1.602 \times 10^{-19} \text{ J}) = 0.511 \text{ MeV}$$

Quick Quiz 12.6

The following pairs of energies represent the rest energy and total energy of three different particles: Particle 1: E, $2E$; particle 2: E, $3E$; particle 3: $2E$, $4E$. Rank the particles from greatest to least according to their **(a)** mass, **(b)** kinetic energy and **(c)** speed.

Example 12.8

The energy of a speedy proton

(A) Find the rest energy of a proton in units of electron volts and joules.

Solution

Conceptualise Even if the proton is not moving, it has energy associated with its mass. If it moves, the proton possesses more energy, with the total energy being the sum of its rest energy and its kinetic energy.

Model The phrase 'rest energy' suggests we must use a relativistic rather than a classical model for this problem. We require a conversion of mass to energy.

Example 12.8 cont.

Analyse Use Equation 12.20 to find the rest energy:

$$E_R = m_p c^2 = (1.673 \times 10^{-27}\ \text{kg})(2.998 \times 10^8\ \text{m/s})^2 = 1.504 \times 10^{-10}\ \text{J}$$

$$= (1.504 \times 10^{-10}\ \text{J})\left(\frac{1.00\ \text{eV}}{1.602 \times 10^{-19}\ \text{J}}\right) = 938\ \text{MeV}$$

(B) If the total energy of a proton is three times its rest energy, what is the speed of the proton?

Solution

Use Equation 12.22 to relate the total energy of the proton to the rest energy:

$$E = 3m_p c^2 = \frac{m_p c^2}{\sqrt{1 - \frac{u^2}{c^2}}} \rightarrow 3 = \frac{1}{\sqrt{1 - \frac{u^2}{c^2}}}$$

Solve for u:

$$1 - \frac{u^2}{c^2} = \frac{1}{9} \rightarrow \frac{u^2}{c^2} = \frac{8}{9}$$

$$u = \frac{\sqrt{8}}{3} c = 0.943c = 2.83 \times 10^8\ \text{m/s}$$

(C) Determine the kinetic energy of the proton in units of electron volts.

Solution

Use Equation 12.21 to find the kinetic energy of the proton:

$$K = E - m_p c^2 = 3m_p c^2 - m_p c^2 = 2m_p c^2$$

$$= 2(938\ \text{MeV}) = 1.88 \times 10^3\ \text{MeV}$$

(D) What is the proton's momentum?

Solution

Use Equation 12.23 to calculate the momentum:

$$E^2 = p^2c^2 + (m_p c^2)^2 = (3m_p c^2)^2$$

$$p^2c^2 = 9(m_p c^2)^2 - (m_p c^2)^2 = 8(m_p c^2)^2$$

$$p = \sqrt{8}\frac{m_p c^2}{c} = \sqrt{8}\frac{938\ \text{MeV}}{c} = 2.65 \times 10^3\ \text{MeV}/c$$

Finalise The unit of momentum in part (D) is written MeV/c, which is a common unit in particle physics. Masses of subatomic particles are often given in units of MeV/c^2; this makes it very easy to find their rest energy.

12.6 Mass and energy

Equation 12.22, $E = \gamma mc^2$, represents the total energy of a particle. This important equation suggests that even when a particle is at rest ($\gamma = 1$), it still possesses enormous energy through its mass. The clearest experimental proof of the equivalence of mass and energy occurs in nuclear and elementary-particle interactions in which the conversion of mass into kinetic energy takes place. Consequently, we must modify the statement of the principle of conservation of energy we used in Chapter 7 to include rest energy.

For example, as we will see in Chapter 43, in a conventional nuclear reactor the uranium nucleus undergoes *fission*, a reaction that results in several lighter fragments that have considerable kinetic energy. In the case of ^{235}U, which is used as fuel in nuclear power plants, the fragments are two lighter nuclei and a few neutrons. The total mass of the fragments is less than that of the ^{235}U by an amount Δm. The corresponding energy Δmc^2 associated with this mass difference is exactly equal to the sum of the kinetic energies of the fragments. The kinetic energy is absorbed as the fragments move through water, raising the internal energy of the water. This internal energy is used to produce steam for the generation of electricity. This process provides much of the power generated in Europe and the United States. Australia has only a single nuclear reactor, which is not a power reactor but is used for research and production of radiopharmaceuticals, with an operating temperature well below the boiling point of water. New Zealand has no nuclear reactors.

Next, consider a basic *fusion* reaction in which two deuterium atoms combine to form one helium atom. The decrease in mass that results from the creation of one helium atom from two deuterium atoms is $\Delta m = 4.25 \times 10^{-29}$ kg. Hence, the corresponding energy that results from one fusion reaction is $\Delta mc^2 = 3.83 \times 10^{-12}$ J = 23.9 MeV. To appreciate the magnitude of this result, consider that if only 1 g of deuterium were converted to helium, the energy released would be on the order of 10^{12} J. This is the process that fuels stars, including our Sun.

Example 12.9

Mass change in a radioactive decay

The ^{216}Po nucleus is unstable and exhibits radioactivity (Chapter 43). It decays to ^{212}Pb by emitting an alpha particle, which is a helium nucleus, ^{4}He. The relevant masses are $m_i = m(^{216}Po) = 216.001\,915$ u and $m_f = m(^{212}Pb) + m(^{4}He) =$ 211 991 898 u + 4.002 603 u.

(A) Find the mass change of the system in this decay.

Solution

Conceptualise The initial system is the ^{216}Po nucleus. Imagine the mass of the system decreasing during the decay and transforming to kinetic energy of the alpha particle and the ^{212}Pb nucleus after the decay.

Model We use a relativistic model to convert between mass and energy. Calculate the mass change:

$$\Delta m = 216.001\,915 \text{ u} - (211.991\,898 \text{ u} + 4.002\,603 \text{ u})$$

$$= 0.007\,414 \text{ u} = 1.23 \times 10^{-29} \text{ kg}$$

where we have used the conversion factor for u to kg from the table of physical constants in the front of this book.

(B) Find the energy this mass change represents.

Solution

Use **Equation 12.20** to find the energy associated with this mass change:

$$E = \Delta mc^2 = (1.23 \times 10^{-29} \text{ kg})(3.00 \times 10^8 \text{ m/s})^2$$

$$= 1.11 \times 10^{-12} \text{ J} = 6.92 \text{ MeV}$$

Note that it is usual to express energies of particles in units of eV rather than joules. The conversion factor is given in the table of physical constants at the start of this book.

(C) What happens to this energy?

The principle of conservation of energy tells us that this energy cannot have been lost or destroyed in the decay process, therefore it must have been turned into another form. In fact, the mass energy difference is transformed into kinetic energy of the two final particles, the ^{212}Pb nucleus and the α particle. Since the total momentum in the rest frame of the initial ^{216}Po nucleus was zero, the kinetic energy must be shared between the ^{212}Pb nucleus and the α particle in a way that conserves momentum.

End-of-chapter resources

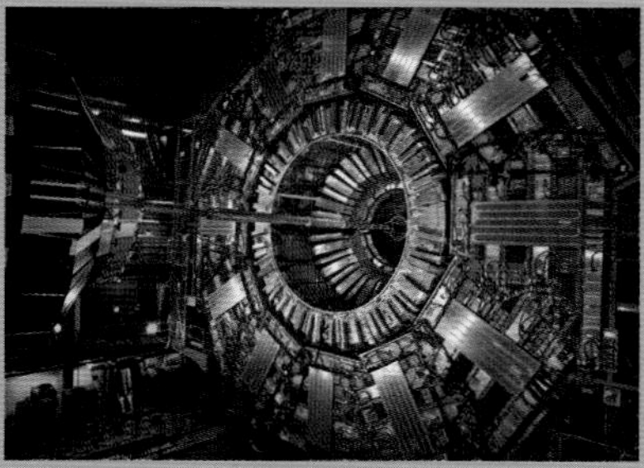

CERN

At the start of this chapter, we asked where the mass came from to create new particles in collider experiments such as those at CERN's Large Hadron Collider. We can use the principles of special relativity and the associated relativistic kinematics described above to answer this question. When particles are accelerated to very high speeds, the total kinetic energy they carry is given by

$$K = (p^2c^2 + m^2c^4)^{1/2} - mc^2$$

Because of the principle of mass–energy equivalence, this energy can be converted into mass during a collision. It is this that allows new particles, such as those described in Chapters 43 and 44, to be created and studied.

Worked solutions to every fifth problem are available in the Student Solutions Manual. Register online at **www.cengagebrain.com** for access.

Summary

Definitions

The Lorentz factor γ is defined as

$$\gamma \equiv \frac{1}{\sqrt{1-\frac{v^2}{c^2}}} \tag{12.6}$$

The relativistic expression for the **linear momentum** of a particle moving with a velocity $\vec{\mathbf{u}}$ is

$$\vec{\mathbf{p}} \equiv \frac{m\vec{\mathbf{u}}}{\sqrt{1-\frac{u^2}{c^2}}} = \gamma m\vec{\mathbf{u}} \tag{12.15}$$

The relativistic force $\vec{\mathbf{F}}$ acting on a particle whose linear momentum is $\vec{\mathbf{p}}$ is defined as

$$\vec{\mathbf{F}} \equiv \frac{d\vec{\mathbf{p}}}{dt} \tag{12.16}$$

Concepts and principles

The two basic postulates of the special theory of relativity are as follows.

1. The laws of physics must be the same in all inertial reference frames.
2. The speed of light in vacuum has the same value, $c = 3.00 \times 10^8$ m/s, in all inertial frames, regardless of the velocity of the observer or the velocity of the source emitting the light.

Three consequences of the special theory of relativity are as follows.

- Events that are measured to be simultaneous for one observer are not necessarily measured to be simultaneous for another observer who is in motion relative to the first.
- Clocks in motion relative to an observer are measured to run slower than stationary clocks by a factor $\gamma = (1 - v^2/c^2)^{-1/2}$. This phenomenon is known as **time dilation**.
- The lengths of objects in motion are measured to be contracted in the direction of motion by a factor $1/\gamma = (1 - v^2/c^2)^{1/2}$. This phenomenon is known as **length contraction**.

To satisfy the postulates of special relativity, the Galilean transformation equations must be replaced by the **Lorentz transformation equations**:

$$x' = \gamma(x - vt) \quad y' = y \quad z' = z \quad t' = \gamma\left(t - \frac{v}{c^2}x\right) \tag{12.8}$$

where $\gamma = (1 - v^2/c^2)^{-1/2}$ and the S′ frame moves in the x direction at speed v relative to the S frame.

The relativistic form of the **Lorentz velocity transformation equation** is

$$u'_x = \frac{u_x - v}{1 - \frac{u_x v}{c^2}} \tag{12.13}$$

where u'_x is the x component of the velocity of an object as measured in the S′ frame and u_x is its component as measured in the S frame.

The relativistic expression for the **kinetic energy** of a particle is

$$K = \frac{mc^2}{\sqrt{1-\frac{u^2}{c^2}}} - mc^2 = (\gamma - 1)mc^2 \tag{12.19}$$

The constant term mc^2 in Equation 12.19 is called the **rest energy** E_R of the particle:

$$E_R = mc^2 \tag{12.20}$$

The total energy E of a particle is given by

$$E = \frac{mc^2}{\sqrt{1-\frac{u^2}{c^2}}} = \gamma mc^2 \tag{12.22}$$

The relativistic linear momentum of a particle is related to its total energy through the equation

$$E^2 = p^2c^2 + (mc^2)^2 \tag{12.23}$$

Chapter review quiz

To help you revise Chapter 12: Special relativity, complete the automatically graded Chapter review quiz at http://login.cengagebrain.com.

Conceptual questions

1. The speed of light in water is 230 Mm/s. Suppose an electron is moving through water at 250 Mm/s. Does that violate the principle of relativity? Explain your answer.
2. Explain why, when defining the length of a rod, it is necessary to specify that the positions of the ends of the rod are to be measured simultaneously.
3. A train is approaching you at very high speed as you stand on an embankment. Just as an observer on the train passes you, you both begin to play the same recorded track on your MP3 players. (a) According to you, whose track finishes first? (b) According to the observer on the train, whose track finishes first? (c) Whose track actually finishes first?
4. Describe three ways our day-to-day lives would change if the speed of light were only 50 m/s.
5. How is acceleration indicated on a space–time graph?
6. A particle is moving at a speed less than $c/2$. If the speed of the particle is doubled, what happens to its momentum?
7. Give a physical argument that shows it is impossible to accelerate an object of mass m to the speed of light, even with a continuous force acting on it.
8. Describe what you would see in a mirror if you carried it in your hands and ran at a speed near that of light.
9. With regard to reference frames, how does general relativity differ from special relativity?

Problems

Section 12.3 Consequences of the loss of simultaneity: measurements of time and length

1. A metre ruler moving at $0.900c$ relative to the Earth's surface approaches an observer at rest with respect to the Earth's surface. The direction of the ruler's motion is parallel to its length. (a) What is the metre ruler's length as measured by the observer? (b) Qualitatively, how would the answer to part (a) change if the observer started running towards the metre ruler?
2. The average lifetime of a pi meson in its own frame of reference (i.e., the proper lifetime) is 2.6×10^{-8} s. If the meson moves with a speed of $0.98c$, what is (a) its mean lifetime as measured by an observer on Earth, and (b) the average distance it travels before decaying, as measured by an observer on Earth? (c) What distance would it travel if time dilation did not occur?
3. A star is 5.00 ly from the Earth. At what speed must a spacecraft travel on its journey to the star such that the Earth–star distance measured in the frame of the spacecraft is 2.00 ly?
4. An astronaut is travelling in a space vehicle moving at $0.500c$ relative to the Earth. The astronaut measures her pulse rate at 75.0 beats per minute. Signals generated by the astronaut's pulse are radioed to the Earth when the vehicle is moving in a direction perpendicular to the line that connects the vehicle with an observer on the Earth. (a) What pulse rate does the Earth-based observer measure? (b) **What If?** What would be the pulse rate if the speed of the space vehicle were increased to $0.990c$?
5. A fellow astronaut passes by you in a spacecraft travelling at a high speed. The astronaut tells you that his craft is 20.0 m long and that the identical craft you are sitting in is 19.0 m long. According to your observations, (a) how long is your craft, (b) how long is the astronaut's craft and (c) what is the speed of the astronaut's craft relative to your craft?
6. For what value of v does $\gamma = 1.0100$? Note that for speeds lower than this value, time dilation and length contraction are effects amounting to less than 1%.
7. A spacecraft with a proper length of 300 m passes by an observer on the Earth. According to this observer, it takes 0.750 ms for the spacecraft to pass a fixed point. Determine the speed of the spacecraft as measured by the Earth-based observer.
8. ±? The twins Laurence and Marcus join a migration from the Earth to Planet X, 20.0 ly away in a reference frame in which both planets are at rest. The twins, of the same age, depart at the same moment on different spacecraft. Laurence's spacecraft travels steadily at $(0.950 \pm 0.002)c$ and Marcus' at $(0.750 \pm 0.002)c$. (a) Calculate the age difference between the twins after Marcus' spacecraft reaches Planet X. (b) Which twin is older?
9. An atomic clock moves at 1000 km/h for 1.00 h as measured by an identical clock on the Earth. At the end of the 1.00 h interval, how many nanoseconds slow will the moving clock be compared with the Earth-based clock?
10. In 1963, astronaut Gordon Cooper orbited the Earth 22 times. The press stated that for each orbit, he aged two-millionths of a second less than he would have had he remained on the Earth. (a) Assuming Cooper was 160 km above the Earth in a circular orbit, determine the difference in elapsed time between someone on the Earth and the orbiting astronaut for the 22 orbits. You may use the approximation

$$\frac{1}{\sqrt{1-x}} \approx 1 + \frac{x}{2}$$

for small x. (b) Did the press report accurate information? Explain your answer.

Section 12.4 The Lorentz transformation equations

11. Shannon observes two light pulses to be emitted from the same location, but separated in time by 3.00 μs. Kim observes the emission of the same two pulses to be separated in time by 9.00 μs. (a) How fast is Kim moving relative to Shannon? (b) According to Kim, what is the separation in space of the two pulses?
12. A red light flashes at position $x_R = 3.00$ m and time $t_R = 1.00 \times 10^{-9}$ s, and a blue light flashes at $x_B = 5.00$ m and $t_B = 9.00 \times 10^{-9}$ s, all measured in the S reference frame. Reference frame S′ moves uniformly to the right and has its origin at the same point as S at $t = t' = 0$. Both flashes are observed to occur at the same place in S′. (a) Find the relative speed between S and S′. (b) Find the location of the two flashes in frame S′. (c) At what time does the red flash occur in the S′ frame?
13. Susan, in reference frame S, measures two events to be simultaneous. Event A occurs at the point (50.0 m, 0, 0) at the instant 9:00:00 Universal Time on 15 January 2010. Event B

occurs at the point (150 m, 0, 0) at the same moment. Robert, moving past with a velocity of $0.800c\hat{\mathbf{i}}$, also observes the two events. In his reference frame S′, which event occurred first and what time interval elapsed between the events?

14. A moving rod is observed to have a length of $\ell = 2.00$ m and to be oriented at an angle of $\theta = 30.0°$ with respect to the direction of motion as shown in Figure P12.14. The rod has a speed of $0.995c$. (a) What is the proper length of the rod? (b) What is the orientation angle in the reference frame in which the rod is at rest?

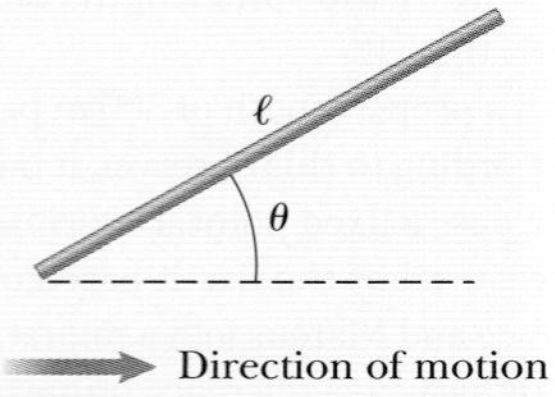

Figure P12.14

Section **12.5** Relativistic velocity, linear momentum and energy

15. Figure P12.15 shows a jet of material (at the upper right) being ejected by galaxy M87 (at the lower left). Such jets are believed to be evidence of supermassive black holes at the centre of a galaxy. Suppose two jets of material from the centre of a galaxy are ejected in opposite directions. Both jets move at $0.750c$ relative to the galaxy centre. Determine the speed of one jet relative to the other.

Figure P12.15

16. ±? A spacecraft moves away from the Earth at a speed of $v = 0.800c$ (Figure P12.16). A galactic patrol space vehicle pursues at a speed of $u = 0.900c$ relative to the Earth. Observers on the Earth measure the patrol vehicle to be overtaking the craft at a relative speed of $(0.100 \pm 0.005)c$. With what speed is the patrol vehicle overtaking the craft as measured by the patrol craft's crew?

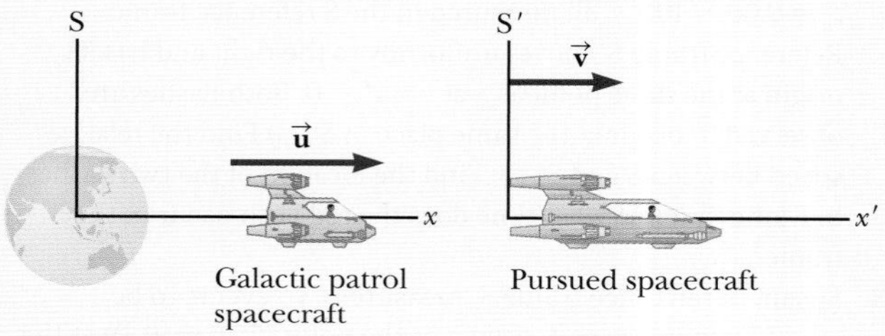

Figure P12.16

17. An electron has a relativistic momentum that is three times larger than its classical momentum. (a) Find the speed of the electron. (b) **What If?** How would your result change if the particle were a proton?

18. Calculate the momentum of an electron moving with a speed of (a) $0.0100c$, (b) $0.500c$ and (c) $0.900c$.

19. The non-relativistic expression for the momentum of a particle, $p = mu$, agrees with experiment if $u << c$. For what speed does the use of this equation give an error in the measured momentum of (a) 1.00% and (b) 10.0%?

20. An unstable particle at rest spontaneously breaks into two fragments of unequal mass. The mass of the first fragment is 2.50×10^{-28} kg and that of the other is 1.67×10^{-27} kg. If the lighter fragment has a speed of $0.893c$ after the breakup, what is the speed of the heavier fragment?

21. Protons in the Large Hadron Collider at CERN are accelerated to a total energy that is 7×10^6 times their rest energy. What is the speed of these protons in terms of c?

22. An electron in the Australian Synchrotron has a kinetic energy of 1 GeV. Calculate its (a) rest energy, (b) total energy and (c) speed.

23. An electron in the Australian Synchrotron moves with a speed of $c/2$. Use the work–kinetic energy theorem to find the work required to increase its speed to (a) $0.750c$ and (b) $0.995c$.

24. Show that for any object moving at less than 10% of the speed of light, the relativistic kinetic energy agrees with the result of the classical equation $K = \frac{1}{2}mu^2$ to within less than 1%.

25. The total energy of a proton is twice its rest energy. Find the momentum of the proton in units of MeV/c.

26. Show that the energy–momentum relationship in Equation 12.23, $E^2 = p^2c^2 + (mc^2)^2$, follows from the expressions $E = \gamma mc^2$ and $p = \gamma mu$.

27. A pion at rest ($m_\pi = 273m_e$) decays to a muon ($m_\mu = 207m_e$) and an antineutrino ($m_{\bar{\nu}} \approx 0$). The reaction is written $\pi^- \rightarrow \mu^- + \bar{\nu}$. Find (a) the kinetic energy of the muon and (b) the energy of the antineutrino in electron volts.

28. In Example 12.9, we saw that the mass–energy difference of 6.92 MeV between the parent nucleus ^{216}Po and its decay products ^{212}Pb and an α particle is shared among them as kinetic energy. Do the decay products need to be treated relativistically? What is the momentum of each outgoing particle, in units of MeV/c?

29. An unstable particle with mass $m = 3.34 \times 10^{-27}$ kg is initially at rest. The particle decays into two fragments that fly off along the x axis with velocity components $u_1 = 0.987c$ and $u_2 = -0.868c$. From this information, we wish to determine the masses of fragments 1 and 2. (a) Is the initial system of the unstable particle, which becomes the system of the two fragments, isolated or non-isolated? (b) Based on your answer to part (a), what two analysis models are appropriate for this situation? (c) Find the values of γ for the two fragments after the decay. (d) Using one of the analysis models in part (b), find a relationship between the masses m_1 and m_2 of the fragments. (e) Using the second analysis model in part (b), find a second relationship between the masses m_1 and m_2. (f) Solve the relationships in parts (d) and (e) simultaneously for the masses m_1 and m_2.

30. Massive stars ending their lives in supernova explosions produce the nuclei of all the atoms in the bottom half of the periodic table by fusion of smaller nuclei. This problem roughly models that process. A particle of mass m moving along the x axis with a velocity component $+u$ collides head-on and sticks to a particle of mass $m/3$ moving along the x axis with the velocity component $-u$. (a) What is the mass M of the resulting particle? (b) Evaluate the expression from part (a) in the limit $u \to 0$. (c) Explain whether the result agrees with what you should expect from nonrelativistic physics.

Section **12.6** Mass and energy

31. When 1.00 g of hydrogen combines with 8.00 g of oxygen, 9.00 g of water is formed. During this chemical reaction, 2.86×10^5 J of energy is released. (a) Is the mass of the water larger or smaller than the mass of the reactants? (b) What is the difference in mass? (c) Explain whether the change in mass is likely to be detectable.

32. ±? In a nuclear power plant, the fuel rods last 3 yr before they are replaced. The plant can transform energy at a maximum possible rate of 1.00 GW. Supposing it operates at $(80.0 \pm 0.5)\%$ capacity for 3.00 yr, what is the loss of mass of the fuel?

33. The power output of the Sun is 3.85×10^{26} W. By how much does the mass of the Sun decrease each second?

Additional problems

34. The net nuclear fusion reaction inside the Sun can be written as $4\,{}^1\text{H} \to {}^4\text{He} + E$. The rest energy of each hydrogen atom is 938.78 MeV, and the rest energy of the helium-4 atom is 3728.4 MeV. Calculate the percentage of the starting mass that is transformed to other forms of energy.

35. An astronaut wishes to visit the Andromeda galaxy, making a one-way trip that will take 30.0 yr in the spacecraft's frame of reference. Assume the galaxy is 2.00×10^6 ly away and the astronaut's speed is constant. (a) How fast must he travel relative to the Earth? (b) What will be the kinetic energy of his 1000 metric tonne spacecraft? (c) What is the cost of this energy if it is purchased at a cost of \$0.11/kWh?

36. The equation

$$K = \left(\frac{1}{\sqrt{1 - u^2/c^2}} - 1\right)mc^2$$

gives the kinetic energy of a particle moving at speed u. (a) Solve the equation for u. (b) From the equation for u, identify the minimum possible value of speed and the corresponding kinetic energy. (c) Identify the maximum possible speed and the corresponding kinetic energy. (d) Differentiate the equation for u with respect to time to obtain an equation describing the acceleration of a particle as a function of its kinetic energy and the power input to the particle. (e) Observe that for a non-relativistic particle we have $u = (2K/m)^{1/2}$ and that differentiating this equation with respect to time gives $a = P/(2mK)^{1/2}$. State the limiting form of the expression in part (d) at low energy. State how it compares with the nonrelativistic expression. (f) State the limiting form of the expression in part (d) at high energy. (g) Consider a particle with constant input power. Explain how the answer to part (f) helps account for the answer to part (c).

37. The cosmic rays of highest energy are protons that have kinetic energy on the order of 10^{13} MeV. (a) As measured in the proton's frame, what time interval would a proton of this energy require to travel across the Milky Way galaxy, which has a proper diameter $\sim 10^5$ ly? (b) From the point of view of the proton, how many kilometres across is the galaxy?

38. An object disintegrates into two fragments. One fragment has mass 1.00 MeV/c^2 and momentum 1.75 MeV/c in the positive x direction, and the other has mass 1.50 MeV/c^2 and momentum 2.00 MeV/c in the positive y direction. Find (a) the mass and (b) the speed of the original object.

39. Around the core of a nuclear reactor shielded by a large pool of water, Cerenkov radiation appears as a blue glow (**Figure P12.39**). Cerenkov radiation occurs when a particle travels faster through a medium than the speed of light in that medium. It is the electromagnetic equivalent of a bow wave or a sonic boom. An electron is travelling through water at a speed 10.0% faster than the speed of light in water. Determine the electron's (a) total energy, (b) kinetic energy and (c) momentum. (d) Find the angle between the shock wave and the electron's direction of motion.

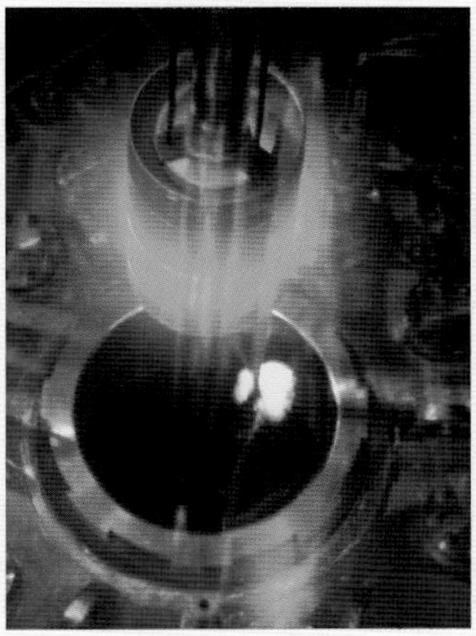

US Department of Energy/Photo Researchers Inc.

Figure P12.39

40. Imagine that the entire Sun, of mass M_S, collapses to a sphere of radius R_g such that the work required to remove a small mass m from the surface would be equal to its rest energy mc^2. This radius is called the *gravitational radius* for the Sun. (a) Use this approach to show that $R_g = GM_S/c^2$. (b) Find a numerical value for R_g.

41. An observer in a coasting spacecraft moves towards a mirror at speed v relative to the reference frame labelled S in **Figure P12.41**. The mirror is stationary with respect to S. A light pulse emitted by the spacecraft travels toward the mirror and is reflected back to the spacecraft. The spacecraft is a distance d from the mirror (as measured by observers in S) at the moment the light pulse leaves the spacecraft. What is the total travel time of the pulse as measured by observers in (a) the S frame and (b) the spacecraft?

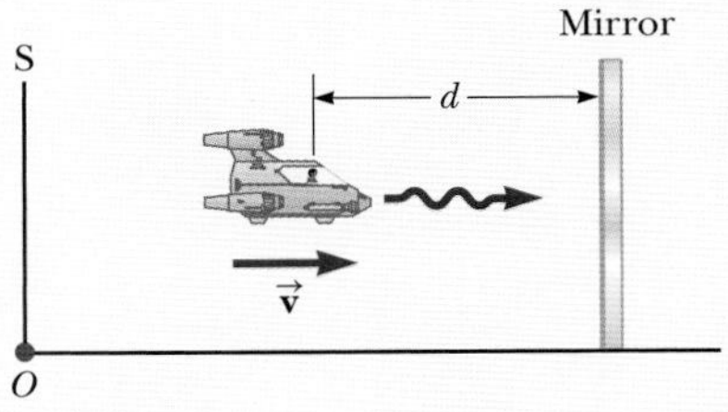

Figure P12.41

42. A ^{57}Fe nucleus at rest emits a 14.0 keV photon. Use conservation of energy and momentum to find the kinetic energy of the recoiling nucleus in electron volts. Use $Mc^2 = 8.60 \times 10^{-9}$ J for the final state of the ^{57}Fe nucleus.

43. Spacecraft I, containing students taking a physics exam, approaches the Earth with a speed of 0.600c (relative to the Earth), while spacecraft II, containing lecturers invigilating the exam, moves at 0.280c (relative to the Earth) directly toward the students. If the lecturers stop the exam after 50.0 min have passed on their clock, for what time interval does the exam last as measured by (a) the students and (b) an observer on the Earth?

Challenge problems

44. A global positioning system (GPS) satellite moves in a circular orbit with period 11 h 58 min. (a) Determine the radius of its orbit. (b) Determine its speed. (c) The non-military GPS signal is broadcast at a frequency of 1575.42 MHz in the reference frame of the satellite. When it is received on the Earth's surface by a GPS receiver (**Figure P12.44**), what is the fractional change in this frequency due to time dilation as described by special relativity? (d) The gravitational 'blueshift' of the frequency according to general relativity is a separate effect. It is called a blueshift to indicate a change to a higher frequency. The magnitude of that fractional change is given by

$$\frac{\Delta f}{f} = \frac{\Delta U_g}{mc^2}$$

where U_g is the change in gravitational potential energy of an object–Earth system when the object of mass m is moved between the two points where the signal is observed. Calculate this fractional change in frequency due to the change in position of the satellite from the Earth's surface to its orbital position. (e) What is the overall fractional change in frequency due to both time dilation and gravitational blueshift?

Figure P12.44

45. The creation and study of new and very massive elementary particles is an important part of contemporary physics. To create a particle of mass M requires an energy Mc^2. With enough energy, an exotic particle can be created by allowing a fast-moving proton to collide with a similar target particle. Consider a perfectly inelastic collision between two protons: an incident proton with mass m_p, kinetic energy K and momentum magnitude p joins with an originally stationary target proton to form a single product particle of mass M. Not all the kinetic energy of the incoming proton is available to create the product particle because conservation of momentum requires that the system as a whole still must have some kinetic energy after the collision. Therefore, only a fraction of the energy of the incident particle is available to create a new particle. (a) Show that the energy available to create a product particle is given by

$$Mc^2 = 2m_pc^2\sqrt{1+\frac{K}{2m_pc^2}}$$

This result shows that when the kinetic energy K of the incident proton is large compared with its rest energy m_pc^2, then M approaches $(2m_pK)^{1/2}/c$. Therefore, if the energy of the incoming proton is increased by a factor of 9, the mass you can create increases only by a factor of 3, not by a factor of 9 as would be expected. (b) This problem can be alleviated by using *colliding beams* as is the case in most modern accelerators. Here the total momentum of a pair of interacting particles can be zero. The centre of mass can be at rest after the collision, so, in principle, all the initial kinetic energy can be used for particle creation. Show that

$$Mc^2 = 2mc^2\left(1+\frac{K}{mc^2}\right)$$

where K is the kinetic energy of each of the two identical colliding particles. Here, if $K >> mc^2$, we have M directly proportional to K as we would desire.

46. A particle with electric charge q moves along a straight line in a uniform electric field $\vec{\mathbf{E}}$ with speed u. The electric force exerted on the charge is $q\vec{\mathbf{E}}$. The velocity of the particle and the electric field are both in the x direction. (a) Show that the acceleration of the particle in the x direction is given by

$$a = \frac{du}{dt} = \frac{qE}{m}\left(1-\frac{u^2}{c^2}\right)^{3/2}$$

(b) Discuss the significance of the dependence of the acceleration on the speed. (c) **What If?** If the particle starts from rest at $x = 0$ at $t = 0$, how would you proceed to find the speed of the particle and its position at time t?

case study 2

Geodesy and gravity can tell us about our changing planet

Will Featherstone

Geodesy is the study of the Earth's size, shape and gravity field, including temporal variations in these quantities. The sub-discipline of physical geodesy uses observations of gravity to infer the equipotential surface of the Earth's gravity field, known as the geoid (**Figure CS2.1**), which corresponds with undisturbed mean sea level. Disturbances from mean sea level are caused by variations in temperature, salinity, atmospheric pressure and currents.

Sensitive gravity measurements can be made by tracking the motion of satellites. A satellite orbiting in the Earth's gravitational field attempts to follow an equipotential surface. However, departures from the equipotential surface occur because of perturbing forces such as atmospheric drag, third-body gravitation (Sun, Moon and nearby planets), solar radiation pressure, and even the Earth's magnetic field, and these perturbing forces must be modelled and removed before we can determine the gravitational field.

In 2002, NASA and the German Space Agency launched the Gravity Recovery And Climate Experiment (GRACE) satellite mission. It comprises two near-identical satellites that are both tracked by the higher GPS (Global Positioning System) satellites, as well as K-band radar ranging between them. These measurements allow for time variations in the Earth's gravitational field to be determined on a monthly basis, or even more frequently. This has been one of the most exciting environmental applications of geodesy. The measurements allow us to monitor important changes to the Earth's system – from determining the amount of ice-melt to determining hydrological cycles in major river basins – all from space.

Our group has used GRACE to determine the amount of ice-melt over Greenland, showing it to have previously been overestimated by a factor of two! **Figure CS2.2** shows the inferred ice-mass loss over Greenland. Over the period from 2002 to 2008, the ice volume loss over Greenland was 177 ± 12 cubic kilometres per year.

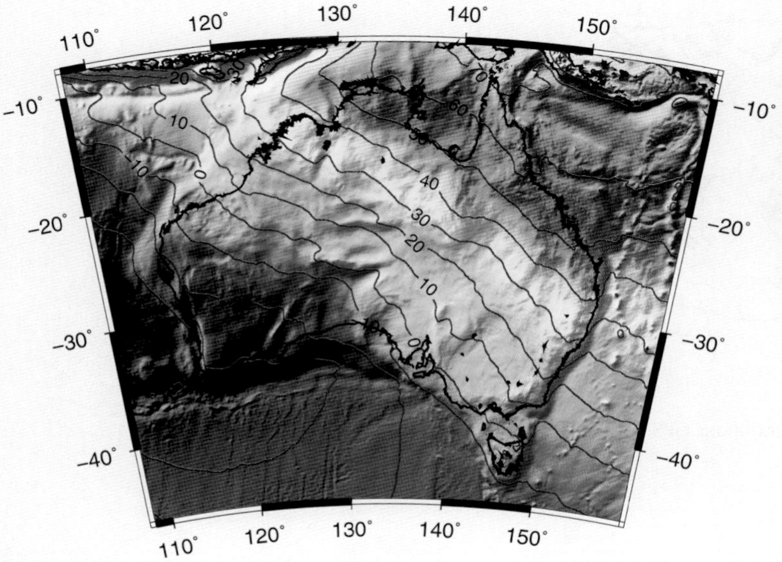

Figure CS2.1
The geoid over Australia slopes from −30 m in south-western Australia to + 70 m in northern Queensland.

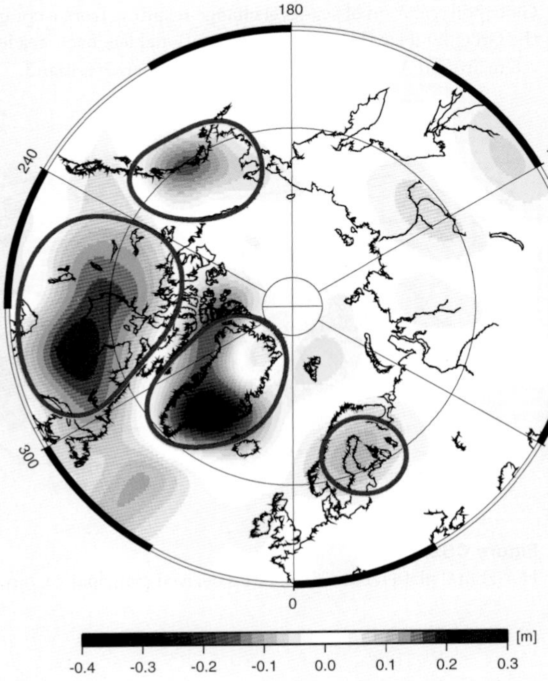

Figure CS2.2
GRACE-derived equivalent water thickness variations from August 2002 to July 2008 (inclusive) over the Arctic region. Red isolines enclose the areas of main mass variation.

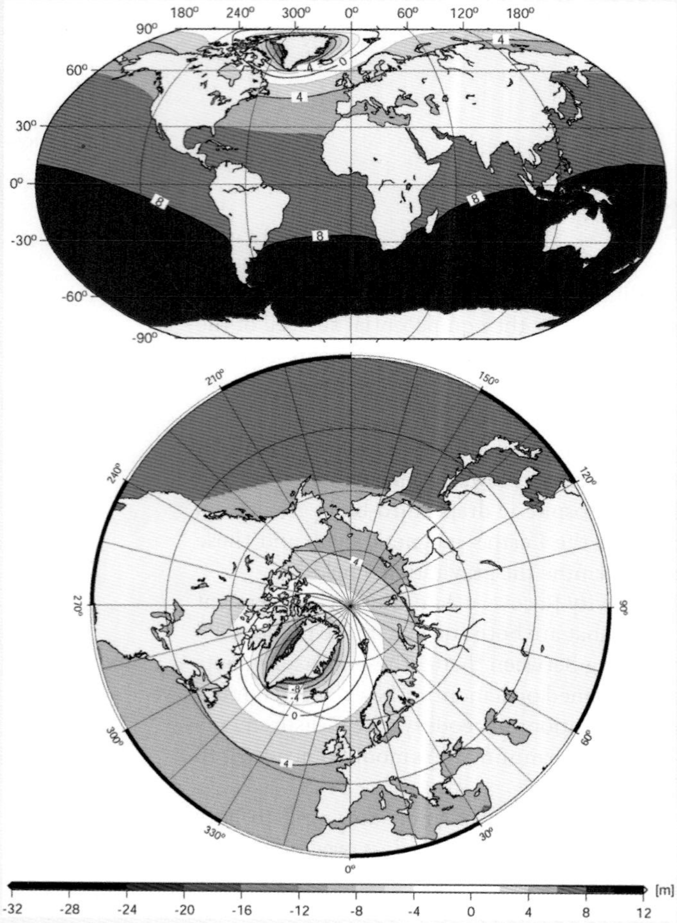

Figure CS2.3
Global distribution of sea-level change resulting from a complete melt of the Greenland ice sheet. Due to gravitational feedback, sea level will rise by >8 m around Australia, yet fall by >20 m near Greenland.

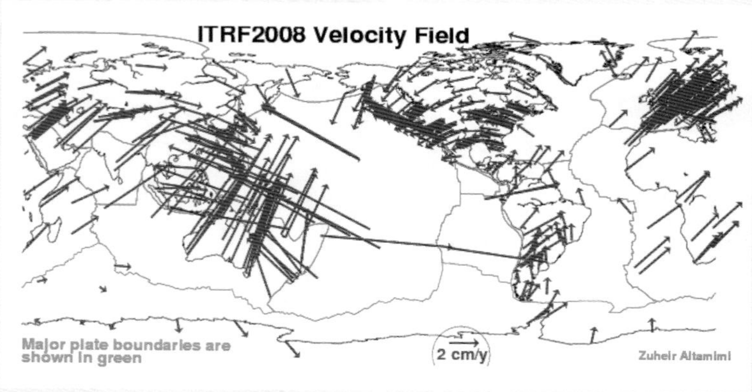

(http://itrf.ensg.ign.fr/ITRF_solutions/2008/)

Figure CS2.4
Horizontal plate tectonic velocities derived principally from continuous GPS

Why is the melting of Greenland of significance to Australia? Well, if the entire Greenland ice shield were to melt, the discharge of melt-water into the oceans would cause sea level to rise around Australia by over 8 metres, whereas it would fall by over 20 metres close to Greenland (**Figure CS2.3**). This is again an effect of gravitation. At present, the ice masses gravitationally attract the sea water towards Greenland. If they were to be removed, then this attraction vanishes and the sea level would fall in the vicinity. Consequently, the amount of sea-level rise is enhanced at the antipode. This is why Greenland ice-melt is of concern to Australia, and physical geodesy can play a useful role.

GPS geodesy can also be used to tell us about our changing Earth. By placing a continuously operating GPS receiver on the Earth's crust and processing the carrier-phase data in a scientific software package, plate tectonics can be measured and monitored. **Figure CS2.4** shows that Australia is drifting northwards at about 9 centimetres per year. There is a large amount of physics behind GPS, but one often forgotten element is the role of relativity. Since the satellite's clocks are moving at approximately 14 000 kilometres per hour and the satellite is orbiting at an altitude of approximately 20 000 km, special and general relativity have to be accounted for. This is done by 'tricking' the Earth-based user by letting the satellite clocks run at a slightly slower rate than their counterparts on Earth.

Will Featherstone is Professor of Geodesy at Curtin University, Western Australia. His degrees were in geophysics and planetary physics from Newcastle University, and geodesy from Oxford University. He computed the AUSGeoid98 and AUSGeoid09 models that are used by all high-precision GPS users in Australia.

Mechanical properties of solids and fluids

PART

2

The water flowing in this mountain stream appears opaque and white. But we know that water doesn't usually look this way; the film of water on the wet rocks and branches is clear, just as the water flowing from a tap in your house is transparent. The change in the water's appearance is a sign of turbulent flow.

In Part 1 of this book, we developed the framework we use to describe the motion of objects. In doing so, we tended to ignore the nature of the objects themselves. For example, our analysis models allowed us to focus on the motion of the centre of mass of an object, and when we considered the rotation of an extended object we did so assuming it was rigid. However, we have already seen that different objects respond to forces in different ways – for example, in elastic or inelastic collisions – and resistive forces depend on the medium an object is moving through. In Part 2, we begin to ask what creates these differences. We look at some of the properties of bulk matter and how they influence motion and response to forces.

Matter is normally classified as being in one of three states: solid, liquid or gas. From everyday experience we know that a solid has an approximately constant volume and definite shape, a liquid has an approximately constant volume but no definite shape, and an unconfined gas has neither a definite volume nor a definite shape. These descriptions help us picture the states of matter but they are somewhat artificial. For example, asphalt and plastics are normally considered solids but over long time intervals they tend to flow like liquids. Likewise, most substances can be a solid, a liquid or a gas (or a combination of any or all of these three), depending on the temperature and pressure.

chapter **13**

Fluid statics

Fish swim by pushing against the water so that the reaction force (Newton's third law) exerted by the water pushes back on them. But how do they control their movements up and down so that they do not either float to the top or sink? When fish die, they rise to the surface where they float for a few days before sinking to the bottom. What governs whether something floats or sinks?

Shutterstock.com/Pete Niesen

On completing this chapter, students will understand:

- that gases and liquids are types of fluids
- how pressure is generated by the motion of individual particles in a fluid
- why pressure varies with depth rather than volume
- how pressure is transmitted through fluids and how hydraulic tools work
- why things float (and why they don't!)
- that the buoyant force is independent of the object submerged in the fluid
- that there is an energy associated with a fluid interface
- why the pressure inside a bubble is different from the pressure outside.

Students will be able to:

- distinguish between force and pressure
- calculate the pressure in a fluid as a function of its depth
- distinguish between gauge and absolute pressures
- apply Pascal's law to relate input and output forces
- apply Archimedes' principle to determine whether an object will float in a fluid
- calculate the force on an object due to surface tension
- use the Young–Laplace equation to calculate pressure differences inside and outside bubbles and droplets.

A **fluid** is a collection of molecules that are randomly arranged and held together by weak cohesive forces and by forces exerted by the walls of a container. Both liquids and gases are fluids.

In our treatment of the mechanics of fluids, we will be applying concepts we have already discussed in the previous chapters, such as force, momentum and energy. We will also introduce some new concepts such as *pressure* and *buoyancy*. We'll start by considering the mechanics of a fluid at rest, *fluid statics*.

ENHANCED WebAssign

The problems found in this chapter may be assigned online in Enhanced Web Assign.

13.1 Pressure

While the concept of pressure is not confined to problems in fluid mechanics, it turns out to be particularly important in such situations. Mechanical pressure P relates the magnitude of an applied contact force F to the surface area A over which it is applied:

$$P \equiv \frac{F}{A} \tag{13.1}$$

Pressure is a scalar quantity: it is proportional to the magnitude of the applied force, but is not related to its direction. Although there is no direction associated with pressure, the direction of the force associated with the pressure is perpendicular to the surface on which the pressure acts.

If the pressure varies over an area, the infinitesimal force dF on an infinitesimal surface element of area dA is

$$dF = P\,dA \tag{13.2}$$

where P is the pressure at the location of the area dA. To calculate the total force exerted on a surface of a container holding a fluid, we must integrate **Equation 13.2** over the surface.

The units of pressure are newtons per square metre (N/m^2) in the SI system, which has dimensions MLT^{-2}. Another name for the SI unit of pressure is the **pascal** (Pa):

$$1\ \text{Pa} \equiv 1\ \text{N/m}^2 \tag{13.3}$$

Pitfall Prevention 13.1

Equations 13.1 and 13.2 make a clear distinction between force and pressure. Another important distinction is that *force is a vector and pressure is a scalar.* There is no direction associated with pressure, but the direction of the force associated with the pressure is perpendicular to the surface on which the pressure acts.

TRY THIS

Hold a pencil between your two thumbs so that the point of the pencil is against one thumb and the other end of the pencil is against the other thumb. Gently press your thumbs together. Why does one thumb begin to hurt but not the other?

Quick Quiz 13.1

Suppose you are standing directly behind someone who steps back and accidentally stamps on your foot with the heel of one shoe. Would you be better off if that person were (a) a large rugby player wearing trainers or (b) a small woman wearing stilettos?

±? Example 13.1

The water bed

The mattress of a waterbed is 2.00 ± 0.01 m long by 2.00 ± 0.01 m wide and 30 ± 1 cm deep.

(A) Find the weight of (the gravitational force acting on) the water in the mattress.

Solution

Conceptualise Think about carrying a jug of water and how heavy it is. Now imagine a sample of water the size of a waterbed. We expect the weight to be large.

Model The density of the water is approximately constant, and we will assume that it is fresh water (or at least has the same density as fresh water).

Analyse First, we need to find the volume of the water filling the mattress. We label the length ℓ, the width w and the depth d. Then:

$$V = \ell w d$$

Now use **Equation 1.1** to find the mass:

$$M = \rho V = \rho \ell w d$$

Then the weight of the water can be found using the usual relationship between mass and weight:

$$F = Mg = \rho V g = \rho \ell w d g$$

Do a dimension check to make sure the expression makes sense.

Example 13.1 cont.

Noting that density has dimensions [ML^{-3}], we have

$$[MLT^{-2}] = [ML^{-3}][L][L][L][LT^{-2}] = [MLT^{-2}]$$ ☺

Finally substitute in the numerical values for each term, using **Table 13.1** to find the density of fresh water:

$$F = (1.00 \times 10^3\,\text{kg/m}^3)(2.00 \times 2.00 \times 0.30\,\text{m}^3)(9.80\,\text{m/s}^2) = 1.18 \times 10^4\,\text{N}$$

As the fractional uncertainties in the length and width of the bed (1/200) are much smaller than the fractional uncertainty in the depth (1/30), we can assume that the uncertainty in the volume will be dominated by the uncertainty in the depth. Since $V = \ell wd$, the fractional uncertainty in the volume will also be 1 in 30. Looking at the rest of our solution, we can see that the weight F depends on the volume and two constants, the density of fresh water and the acceleration due to gravity. Thus the fractional uncertainty in the weight will *also* be 1 in 30, leading to a final answer of

$$F = (1.18 \pm 0.04) \times 10^4\,\text{N}$$

This is similar to the weight of a car! A normal double bed weighs less than 15% of this. Because this load is so great, it is best to place a water bed on a sturdy, well-supported floor.

(B) Find the pressure exerted by the waterbed on the floor when the bed rests in its normal position. Assume the entire lower surface of the bed makes contact with the floor.

Solution

The area in contact with the floor is $\ell w = 4.00\,\text{m}^2$.
Use **Equation 13.1** to find the pressure:

$$P = \frac{F}{A} = \frac{1.18 \times 10^4\,\text{N}}{4.00\,\text{m}^2} = 2.94 \times 10^3\,\text{Pa}$$

The uncertainty can be found by summing the fractional uncertainties:

$\Delta P/P = \Delta F/F + \Delta A/A = \Delta F/F + \Delta l/l + \Delta w/w = 0.04/1.18 + 0.01/2.00 + 0.01/2.00 = 0.04$ or 4%

So our final answer is $P = (2.94 \pm 0.13) \times 10^3$ Pa

What If? What if the waterbed is replaced by a 150 kg double bed that is supported by four legs? Each leg has a circular cross section of radius 2.00 cm. What pressure does this bed exert on the floor?

Answer The weight of the bed is distributed over four circular cross sections at the bottom of the legs. Therefore, the pressure is

$$P = \frac{F}{A} = \frac{mg}{(4\pi r^2)} = \frac{(150\,\text{kg})(9.80\,\text{m/s}^2)}{4\pi\,(0.0200\,\text{m})^2}$$

$$= 2.92 \times 10^5\,\text{Pa}$$

This result is almost 100 times larger than the pressure due to the waterbed. The weight of the double bed, even though it is much less than the weight of the waterbed, is applied over the very small cross-sectional area of the four legs. The high pressure on the floor at the feet of a regular bed can cause damage to the floor surface.

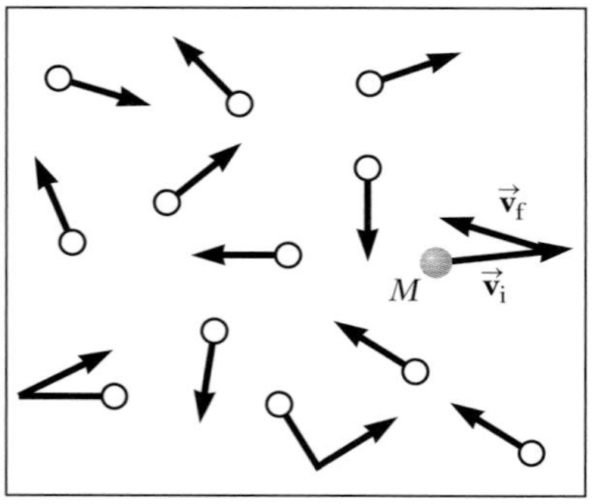

Figure 13.1
Molecules collide with the walls of a container.

13.2 Pressure in fluids

Imagine a container full of fluid as shown in **Figure 13.1**. The atoms or molecules in the fluid are in constant motion and move about in random directions. As we shall see in Chapter 21 their average velocities depend on their temperature.

As the molecules collide with the walls of the container, we know from Newton's laws (Chapter 4) that they exert a force:

$$F_{\text{molecule on container wall}} = -F_{\text{container wall on molecule}} = \frac{\Delta p}{\Delta t}$$

where Δp is the change in the momentum and Δt is the duration of the collision. The molecules collide with the walls from all possible angles, as shown in **Figure 13.1**. However, since the motion of the molecules is random, there is an equal probability of collisions occurring from angles less than or greater than 90° and the average direction of the collisions is perpendicular to the wall, regardless of whether the wall is the bottom, side or top of the container. Thus the molecules exert a total force F that is perpendicular to the surface area A, resulting in a pressure $P = F/A$. The same is true for an object submerged in a fluid; that is, the force exerted by a static fluid on an object is always perpendicular to the surfaces of the object, as shown in **Figure 13.2**.

The pressure in a fluid can be measured with the device pictured in **Figure 13.3**. The device consists of an evacuated cylinder that encloses a light piston connected to a spring. As the device is submerged in a fluid, the fluid presses on the top of the piston and compresses the spring until the inwards force exerted by the fluid is balanced by the outwards force exerted by the spring. The fluid pressure can be measured directly if the spring is calibrated in advance. If F is the magnitude of the force exerted on the piston and A is the surface area of the piston, the pressure P of the fluid at the level to which the device has been submerged is the ratio of the force to the area.

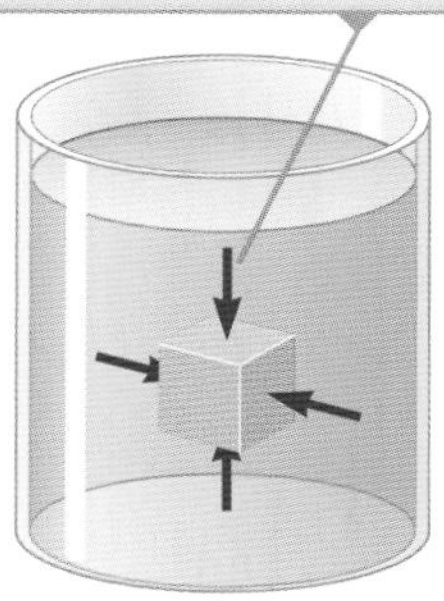

At any point on the surface of the object, the force exerted by the fluid is perpendicular to the surface of the object.

Figure 13.2
The forces exerted by a fluid on the surfaces of a submerged object (The forces on the front and back sides of the object are not shown.)

TRY THIS

Inflate a balloon. Is the pressure inside the balloon higher or lower than the pressure outside? How can you tell?

The downwards pressure exerted by a solid object on the surface it sits on depends on the object's density. Similarly, the pressure exerted by a fluid on an object inside the fluid, or on the walls of the container holding it, depends on the density of the fluid. The density of a substance is defined as its mass per unit volume (see **Equation 1.1**); **Table 13.1** lists the densities of various substances. These values vary slightly with temperature because the volume of a substance is dependent on temperature (as shown in Chapter 19). Under standard conditions (at 0°C and at atmospheric pressure), the densities of gases are two or three orders of magnitude lower than the densities of solids and liquids. This implies that gases occupy 100–1000 times the volume occupied by solids of the same material. If we model every atom or molecule as occupying a cube of volume $V = L^3$, this implies that the average molecular spacing in a gas under these conditions is about 10 times greater than that in a solid or liquid.

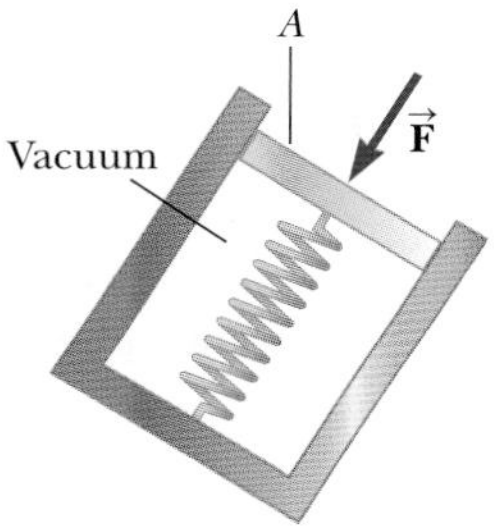

Figure 13.3
A simple device for measuring the pressure exerted by a fluid

Table 13.1
Densities of some substances at standard temperature (0°C) and pressure (atmospheric)

Substance	ρ (kg/m³)	Substance	ρ (kg/m³)
Air	1.29	Iron	7.86×10^3
Air (at 20°C and atmospheric pressure)	1.20	Lead	11.3×10^3
		Mercury	13.6×10^3
Aluminium	2.70×10^3	Nitrogen gas	1.25
Benzene	0.879×10^3	Oak	0.710×10^3
Brass	8.4×10^3	Osmium	22.6×10^3
Copper	8.92×10^3	Oxygen gas	1.43
Ethyl alcohol	0.806×10^3	Pine	0.373×10^3
Fresh water	1.00×10^3	Platinum	21.4×10^3
Glycerine	1.26×10^3	Seawater	1.03×10^3
Gold	19.3×10^3	Silver	10.5×10^3
Helium gas	1.79×10^{-1}	Tin	7.30×10^3
Hydrogen gas	8.99×10^{-2}	Uranium	19.1×10^3
Ice	0.917×10^3		

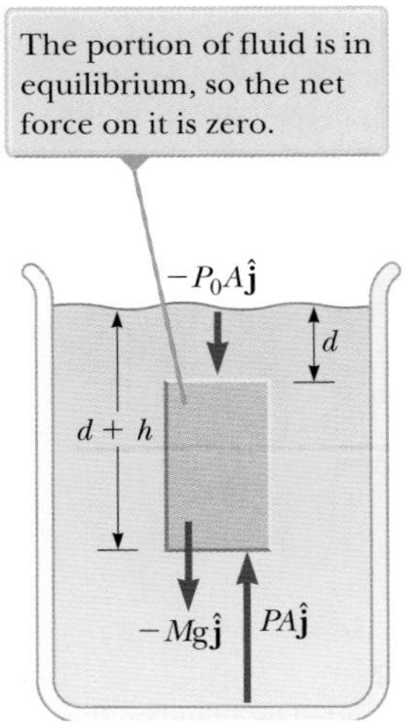

Figure 13.4
Variation of pressure with depth in an incompressible fluid

Variation of pressure with depth

As divers well know, water pressure increases with depth. Likewise, atmospheric pressure decreases with increasing altitude; for this reason, aircraft flying at high altitudes must have pressurised cabins.

Consider a liquid of density ρ at rest. We assume ρ is uniform throughout the liquid, which means the liquid is incompressible. Let us examine the portion of the liquid contained within a volume of cross-sectional area A extending from depth d to depth $d + h$ as shown in **Figure 13.4**. The liquid external to this exerts forces at all points on the surface of the volume, perpendicular to the surface. The pressure exerted by the liquid on the bottom face of the volume is P, and the pressure on the top face is P_0. Therefore, the upwards force exerted by the outside fluid on the bottom of the volume has a magnitude PA, and the downwards force exerted on the top has a magnitude P_0A. The mass of liquid in the portion is $M = \rho V = \rho Ah$; therefore, the weight of the liquid in the volume is $Mg = \rho Ahg$. Because the fluid in the volume is in equilibrium, the net force acting on it must be zero. Choosing upwards to be the positive y direction, we see that

$$\sum \vec{\mathbf{F}} = PA\hat{\mathbf{j}} - P_0A\hat{\mathbf{j}} - Mg\hat{\mathbf{j}} = 0$$

or

$$PA - P_0A - \rho Ahg = 0$$

Variation of pressure with depth ►

$$P = P_0 + \rho gh \qquad (13.4)$$

That is, the pressure P at a depth h below a point in the liquid at which the pressure is P_0 is greater by an amount ρgh. If the liquid is open to the atmosphere and P_0 is the pressure at the surface of the liquid, then P_0 is **atmospheric pressure**. We usually take atmospheric pressure to be

$$P_0 = 1.00 \text{ atm} = 1.013 \times 10^5 \text{ Pa}$$

Equation 13.4 implies that the pressure is the same at all points having the same depth, independent of the shape of the container.

TRY THIS

Take some empty plastic bottles and fill them with water. Make a hole in each one, at varying depths beneath the fill level. What do you observe? What does this tell you about the pressure in the bottles? Try to find bottles with different diameters – does the diameter affect what you see, or just the depth of the hole below the surface? Note: Do this outside, or put the bottles in your bath.

Quick Quiz 13.2

The pressure at the bottom of a filled glass of water is P. The water is poured out, and the glass is filled with vodka (40% ethyl alcohol and 60% water). What is the pressure at the bottom of the glass? **(a)** smaller than P **(b)** equal to P **(c)** larger than P **(d)** indeterminate

Example 13.2

A pain in your ear

Estimate the force exerted on your eardrum due to the water when you are swimming at a depth of 5.0 m.

Solution

Conceptualise As you descend in the water, the pressure increases; you can feel this most easily in your ears. This is because the air inside the middle ear is normally at atmospheric pressure P_0. The pressure difference on the two sides of the eardrum results in a net force.

Model To find the net force on the eardrum, we must find the difference between the total pressure at a depth of 5.0 m underwater and atmospheric pressure. We also need to estimate the surface area of the eardrum. A reasonable estimate is around $1 \text{ cm}^2 = 1 \times 10^{-4} \text{ m}^2$.

Example 13.2 cont.

Analyse Use **Equation 13.4** to find the pressure difference:

$$P - P_0 = (\rho gh + P_0) - P_0 = \rho gh = (1.00 \times 10^3 \text{ kg/m})(9.80 \text{ m/s}^2)(5.0 \text{ m}) = 4.9 \times 10^4 \text{ Pa}$$

Use **Equation 13.1** to find the net force on the ear: $F = (P - P_0)A = (4.9 \times 10^4 \text{ Pa})(1 \times 10^{-4} \text{ m}^2) \approx 5 \text{ N}$

Finalise Because a force of this magnitude on the eardrum is extremely uncomfortable, swimmers often 'pop their ears' while under water, an action that pushes air from the lungs into the middle ear. Using this technique equalises the pressure on the two sides of the eardrum and relieves the discomfort.

Example 13.3

The force on a dam

Water is filled to a height H behind a dam of width w (**Figure 13.5**). Determine the resultant force exerted by the water on the dam.

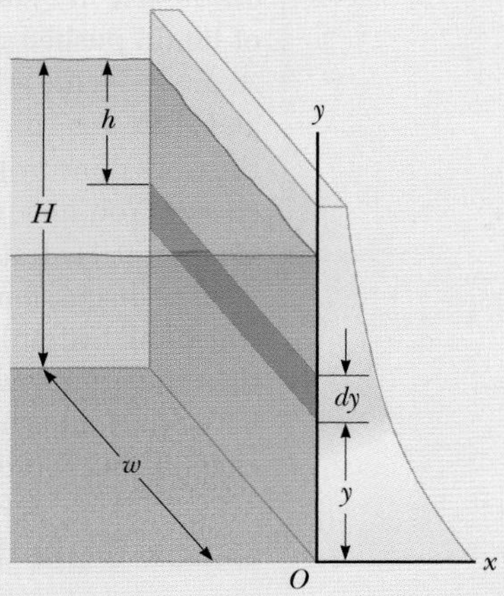

Figure 13.5
(Example 13.3) Water exerts a force on a dam.

Solution

Conceptualise Because pressure varies with depth, we cannot calculate the force simply by multiplying the area by the pressure. As the pressure in the water increases with depth, the force on the adjacent portion of the dam also increases.

Model We model the water as having uniform density, ρ, which does not vary with depth. The pressure at any depth is given by **Equation 13.4**, and force is related to pressure by **Equation 13.2**. Atmospheric pressure acts on both sides of the dam, so the resultant, or net, force is that due to the pressure by the water only.

Analyse We must integrate **Equation 13.2** to solve this example.

Let's imagine a vertical y axis, with $y = 0$ at the bottom of the dam. We divide the face of the dam into narrow horizontal strips at a distance y above the bottom, such as the red strip in **Figure 13.5**.

Use **Equation 13.4** to calculate the pressure due to the water at the depth h: $P = \rho gh = \rho g(H - y)$

Use **Equation 13.2** to find the force exerted on the shaded strip of area $dA = w\,dy$: $dF = P\,dA = \rho g(H - y)w\,dy$

Σ *For help with integration see Appendix B.8.*

Integrate to find the total force on the dam: $F = \int P\,dA = \int_0^H \rho g(H - y)w\,dy = \frac{1}{2}\rho g w H^2$

Finalise Notice that the thickness of the dam shown in **Figure 13.5** increases with depth. This design accounts for the greater force the water exerts on the dam at greater depths.

What If? What if you were asked to find this force without using calculus? How could you determine its value?

Answer We know from **Equation 13.4** that pressure varies linearly with depth. Therefore, the average pressure due to the water over the face of the dam is the average of the pressure at the top and the pressure at the bottom:

$$P_{avg} = \frac{P_{top} + P_{bottom}}{2} = \frac{0 + \rho gH}{2} = \frac{1}{2}\rho gH$$

The total force on the dam is equal to the product of the average pressure and the area of the face of the dam:

$$F = P_{avg}A = \left(\frac{1}{2}\rho gH\right)(Hw) = \frac{1}{2}\rho g w H^2$$

which (of course) is the same result we obtained using calculus.

13.3 Pascal's law and hydraulic tools

Because the pressure in a fluid depends on depth and on the value of P_0, any increase in pressure at the surface must be transmitted to every other point in the fluid. This concept was first recognised by French scientist Blaise Pascal (1623–1662) and is called **Pascal's law: a change in the pressure applied to a fluid is transmitted undiminished to every point of the fluid and to the walls of the container.**

Pascal's law ▶

An important application of Pascal's law is the hydraulic press illustrated in Figure 13.6a. A force of magnitude F_1 is applied to a small piston of surface area A_1. The pressure is transmitted through an incompressible liquid to a larger piston of surface area A_2. Because the pressure must be the same on both sides, $P = F_1/A_1 = F_2/A_2$. Therefore, the force F_2 is greater than the force F_1 by a factor of A_2/A_1. By designing a hydraulic press with appropriate areas A_1 and A_2, a large output force can be applied by means of a small input force. Hydraulic brakes on cars, car hoists and hydraulic jacks all make use of this principle (Figure 13.6b).

Because liquid is neither added to nor removed from the system, the volume of liquid pushed down on the left in Figure 13.6a as the piston moves downwards through a displacement Δx_1 equals the volume of liquid pushed up on the right as the right piston moves upwards through a displacement Δx_2. That is, $A_1\Delta x_1 = A_2\Delta x_2$; therefore, $A_2/A_1 = \Delta x_1/\Delta x_2$. We have already shown that $A_2/A_1 = F_2/F_1$, so $F_2/F_1 = \Delta x_1/\Delta x_2$, and $F_1\Delta x_1 = F_2\Delta x_2$. Each side of this equation is the work done by the force on its respective piston. Therefore, the work done by $\vec{\mathbf{F}}_1$ on the input piston equals the work done by $\vec{\mathbf{F}}_2$ on the output piston, as it must to conserve energy. You may recognise these equations as the same as those introduced in our discussion of levers in Chapter 9. Hydraulic devices such as the one shown in the figure operate on the principle of the lever.

The discussion above makes the assumption that the liquid is incompressible. For many liquids, including water and hydraulic oils such as brake fluids, this is a very good approximation. If the liquid was compressible, then the volume would not be constant as the force is applied. Compressibility is one of the main distinctions between liquids and gases. We will use this model of liquids as perfectly incompressible throughout this chapter, and discuss the compressibility of materials further in Chapter 15.

Quick Quiz 13.3

(i) If the fluid used in brakes was gas rather than liquid, such as in air brakes on a truck, would the force at the brakes piston be (a) greater than, (b) smaller than or (c) the same as, that predicted by the ratio of the piston areas as described above? (ii) What about the distance through which the brake piston moves? Will it be: (a) shorter, (b) longer or (c) the same?

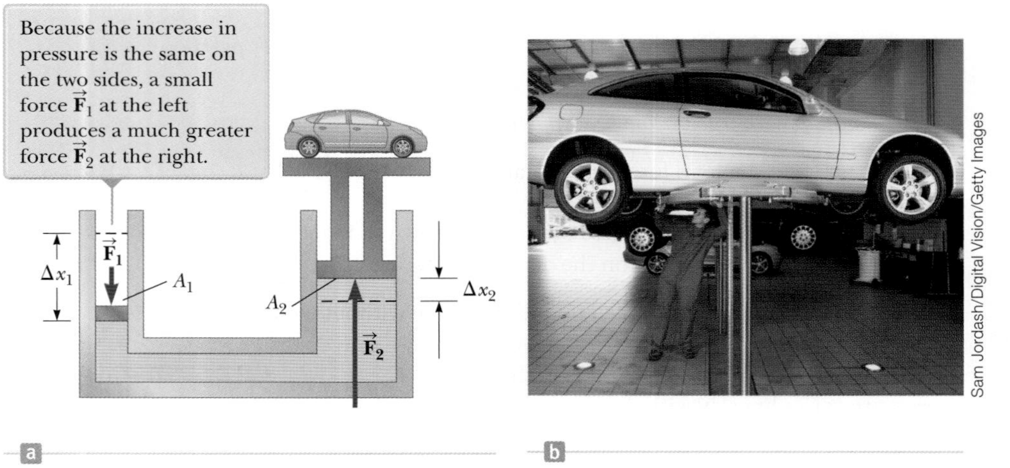

Figure 13.6
(a) Diagram of a hydraulic press
(b) A vehicle undergoing repair is supported by a hydraulic lift in a garage.

Example 13.4

The car hoist

In a car hoist used in a service station (see Figure 13.6), compressed air exerts a force on a small piston that has a circular cross section and a radius of 5.00 cm. This pressure is transmitted by a liquid to a piston that has a radius of 15.0 cm. What force must the compressed air exert to lift a car weighing 13 300 N? What air pressure produces this force?

Example 13.4 cont.

Solution

Conceptualise Our previous discussion of Pascal's law helps us understand how the car hoist operates. We can use the diagram in **Figure 13.6**, otherwise we would draw our own.

Model We model the liquid as incompressible, and use Pascal's law. The air is not incompressible, but we only need to find the pressure it exerts, not a distance or volume.

Analyse We label the forces and areas as in **Figure 13.6b**. Then we need to solve $F_1/A_1 = F_2/A_2$ for F_1:

$$F_1 = \left(\frac{A_1}{A_2}\right)F_2 = \left(\frac{\pi r_1^2}{\pi r_2^2}\right)F_2 = \frac{\pi\,(5.00 \times 10^{-2}\ \text{m})^2}{\pi\,(15.0 \times 10^{-2}\ \text{m}^2)}(1.33 \times 10^4\ \text{N})$$
$$= 1.48 \times 10^3\ \text{N}$$

Use **Equation 13.1** to find the air pressure that produces this force:

$$P = \frac{F_1}{A_1} = \frac{F_2}{A_2} = \frac{F_2}{\pi r_2^2} = \frac{13\,300\ \text{N}}{\pi(15 \times 10^{-2}\ \text{m})^2}$$
$$= 1.88 \times 10^5\ \text{Pa}$$

Finalise This pressure is approximately twice atmospheric pressure.

13.4 Pressure measurements

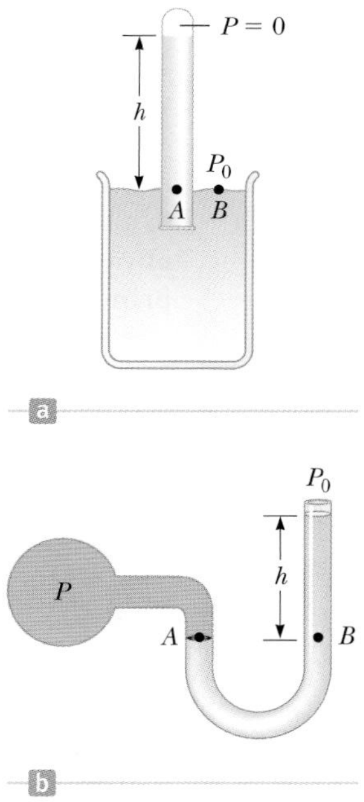

Figure 13.7
Two devices for measuring pressure: (a) a mercury barometer and (b) an open-tube manometer

Weather reports often give the *barometric pressure*. This reading is the current pressure of the atmosphere, which varies over a small range from the standard value provided earlier. How is this pressure measured?

One instrument used to measure atmospheric pressure is the common barometer, invented by Evangelista Torricelli (1608–1647). A long tube closed at one end is filled with mercury and then inverted into a dish of mercury (**Figure 13.7a**). The closed end of the tube is nearly a vacuum, so the pressure at the top of the mercury column can be taken as zero. In **Figure 13.7a**, the pressure at point A, due to the column of mercury, must equal the pressure at point B, due to the atmosphere. If that were not the case, there would be a net force that would move mercury from one point to the other until equilibrium is established. Therefore, $P_0 = \rho_{Hg}gh$, where ρ_{Hg} is the density of the mercury and h is the height of the mercury column. As atmospheric pressure varies, the height of the mercury column varies, so the height can be calibrated to measure atmospheric pressure. Let us determine the height of a mercury column for one atmosphere of pressure, $P_0 = 1\ \text{atm} = 1.013 \times 10^5$ Pa:

$$P_0 = \rho_{Hg}gh \;\rightarrow\; h = \frac{P_0}{\rho_{Hg}g} = \frac{1.013 \times 10^5\ \text{Pa}}{(13.6 \times 10^3\ \text{kg/m}^3)(9.80\ \text{m/s}^2)} = 0.760\ \text{m}$$

One atmosphere of pressure is defined to be the pressure equivalent of a column of mercury that is exactly 0.7600 m in height at 0°C. Pressures are often given in units of mm Hg, which refers to this definition; atmospheric pressure is thus 760 mm Hg in this system of units.

A device for measuring the pressure of a gas contained in a vessel is the open-tube manometer illustrated in **Figure 13.7b**. One end of a U-shaped tube containing a liquid is open to the atmosphere, and the other end is connected to a container of gas at pressure P. In an equilibrium situation, the pressures at points A and B must be the same (otherwise, the curved portion of the liquid would experience a net force and would accelerate), and the pressure at A is the unknown pressure of the gas. Therefore, equating the unknown pressure P to the pressure at point B, we see that $P = P_0 + \rho gh$. Again, we can calibrate the height h to the pressure P.

The difference in the pressures in each part of **Figure 13.7** (that is, $P - P_0$) is equal to ρgh. The pressure P is called the **absolute pressure** and the difference $P - P_0$ is called the **gauge pressure**. For example, the pressure you measure in your bicycle tyre is gauge pressure.

Although the SI units for pressure are N.m^2, there are many different units in use. Which system a pressure is given in depends on the context and the size of the pressure being measured. Common units include the pascal (1 Pa = 1 N/m^2), the atmosphere (1 atm = 101 325 N/m^2), millimetres of mercury (1 mm Hg = 133.32 N/m^2) and pounds per square inch (1 psi = 6894.8 N/m^2).

Pitfall Prevention 13.2
There are many different units in use in pressure measurements; it is important that you make sure you know which one is being used.

TRY THIS

Next time you see free blood pressure testing (for example at your local clinic or library), get yours tested. Ask the person who tests you what units they are using and whether the value they tell you is the gauge or absolute pressure.

Quick Quiz 13.4

Several common barometers are built, with a variety of fluids. For which of the following fluids will the column of fluid in the barometer be the highest? (a) mercury (b) water (c) ethyl alcohol (d) benzene

13.5 Buoyant forces and Archimedes' principle

Have you ever tried to push a beach ball down under water? It is extremely difficult to do because of the large upwards force exerted by the water on the ball. The upwards force exerted by a fluid on any immersed object is called a **buoyant force**. We can determine the magnitude of a buoyant force by applying some logic. Imagine a beach-ball-sized volume of water beneath the water surface (**Figure 13.8**). Because this volume is in equilibrium, there must be an upwards force that balances the downwards gravitational force due to its mass. This upwards force is the buoyant force and its magnitude is equal to the weight of the water in the volume. The buoyant force is the resultant force on the volume due to all forces applied by the fluid surrounding the volume.

Now imagine replacing the beach-ball-sized volume of water with a beach ball of the same size. The net force applied by the fluid surrounding the beach ball is the same, regardless of whether it is applied to a beach ball or to a volume of water. Consequently, **the magnitude of the buoyant force on an object always equals the weight of the fluid displaced by the object**. This statement is known as **Archimedes' principle**.

With the beach ball under water, the buoyant force is larger than the weight of the beach ball. Therefore, there is a net upwards force, which explains why it is so hard to hold the beach ball under the water.

To better understand the origin of the buoyant force, consider a cube of solid material immersed in a liquid as in **Figure 13.9**. According to **Equation 13.4**, the pressure P_{bot} at the bottom of the cube is greater than the pressure P_{top} at the top by an amount $\rho_{fluid}gh$, where h is the height of the cube and ρ_{fluid} is the density of the fluid. The pressure at the bottom of the cube causes an *upwards* force equal to $P_{bot}A$, where A is the area of the bottom face. The pressure at the top of the cube causes a *downwards* force equal to $P_{top}A$. The resultant of these two forces is the buoyant force $\vec{\mathbf{B}}$ with magnitude

$$B = (P_{bot} - P_{top})A = (\rho_{fluid}gh)A$$

$$B = \rho_{fluid}gV_{disp} \qquad (13.5)$$

Hulton Deutsch Collection/Corbis

Archimedes
Greek mathematician, physicist and engineer (c. 287–212 BC)
Archimedes was the first person to compute accurately the ratio of a circle's circumference to its diameter. He also showed how to calculate the volume and surface area of spheres, cylinders and other geometric shapes. He is well known for discovering the nature of the buoyant force and was also a gifted inventor. One of his practical inventions, still in use today, is the Archimedes' screw, an inclined, rotating, coiled tube used originally to lift water. He also invented the catapult and devised systems of levers, pulleys and weights for raising heavy loads.

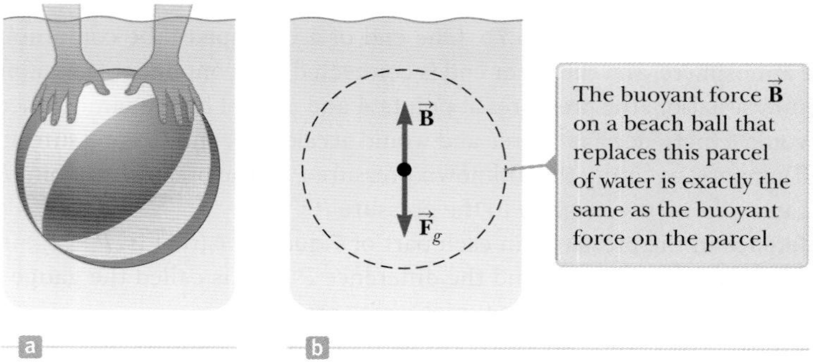
B
F g
The buoyant force B on a beach ball that replaces this parcel of water is exactly the same as the buoyant force on the parcel.
a
b

Figure 13.8
(a) A swimmer pushes a beach ball under water. (b) The forces on a beach ball–sized parcel of water.

where $V_{disp} = Ah$ is the volume of the fluid displaced by the cube. Because the product $\rho_{fluid}V_{disp}$ is equal to the mass of fluid displaced by the object,

$$B = Mg$$

where Mg is the weight of the fluid displaced by the cube.

TRY THIS

Take the lid of a pen or a similar object that can hold some trapped air when submerged in water. Submerge the pen lid in a bottle of water (you may need to use some putty to balance it so that it doesn't turn over). Close the lid of the bottle tightly: you have now made *a Cartesian diver*. Squeeze the sides of the bottle and observe what happens to your diver – can you explain your observations using the ideas of buoyancy and pressure we have just discussed? You may need to adjust the weight and balance of the pen lid 'diver' by adding putty to it.

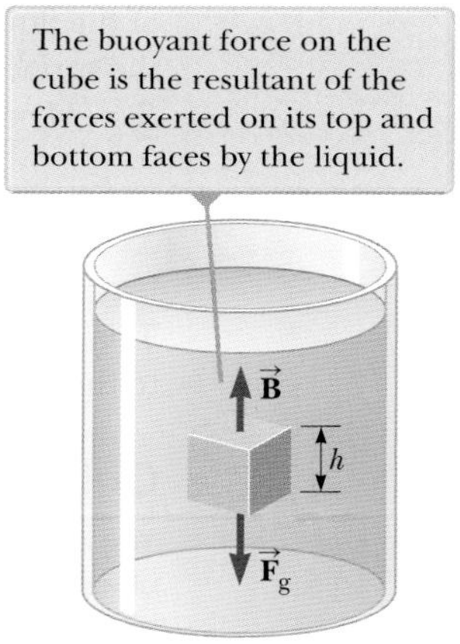
The buoyant force on the cube is the resultant of the forces exerted on its top and bottom faces by the liquid.
$\vec{B}$
h
$\vec{F}_g$

Figure 13.9
The external forces acting on an immersed cube are the gravitational force $\vec{F}_g$ and the buoyant force $\vec{B}$.

When applying Archimedes' principle, there are two common situations: an object may be totally submerged or it may be floating (partly submerged) on the surface of the fluid.

Case 1: Totally submerged object When an object is totally submerged in a fluid of density ρ_{fluid}, the volume V_{disp} of the displaced fluid is equal to the volume V_{obj} of the object; so, from Equation 13.5, the magnitude of the upwards buoyant force is $B = \rho_{fluid}\, gV_{obj}$. If the object has a mass M and density ρ_{obj}, its weight is equal to $F_g = Mg = \rho_{obj}\, gV_{obj}$, and the net force on the object is $B - F_g = (\rho_{fluid} - \rho_{obj})\, gV_{obj}$. Hence, if the density of the object is less than the density of the fluid, the downwards gravitational force is less than the buoyant force and the unsupported object accelerates upwards (Active Figure 13.10a). In this situation, the object is described as having **positive buoyancy**. A bubble of gas submerged in a glass of carbonated water, for example, has positive buoyancy, causing it to rise to the surface.

If the density of the object is greater than the density of the fluid, the upwards buoyant force is less than the downwards gravitational force and the unsupported object sinks (Active Figure 13.10b). In this situation, the object is described as having **negative buoyancy**. A coin dropped gently into water will sink, because the metal of the coin has a larger density than the water.

If the density of the submerged object equals the density of the fluid, the net force on the object is zero and the object remains in equilibrium. In this situation, the object is described as having **neutral buoyancy**. Fish have approximately neutral buoyancy, so are able to stay at the same depth with very little effort, as the buoyant force and gravitational forces are approximately equal.

Case 2: Floating object Consider an object of volume V_{obj} and density $\rho_{obj} < \rho_{fluid}$ in static equilibrium floating on the surface of a fluid, that is, an object that is only *partially* submerged (Active Figure 13.11). In this case, the upwards buoyant force is balanced by the downwards gravitational force acting on the object. If V_{disp} is the volume

Pitfall Prevention 13.3

Remember that **the buoyant force is exerted by the fluid.** It is not determined by the properties of the object except for the amount of fluid displaced by the object. Therefore, if several objects of different densities but the same volume are immersed in a fluid, they will all experience the same buoyant force. Whether they sink or float is determined by the relationship between the buoyant force and the gravitational force (the object's weight).

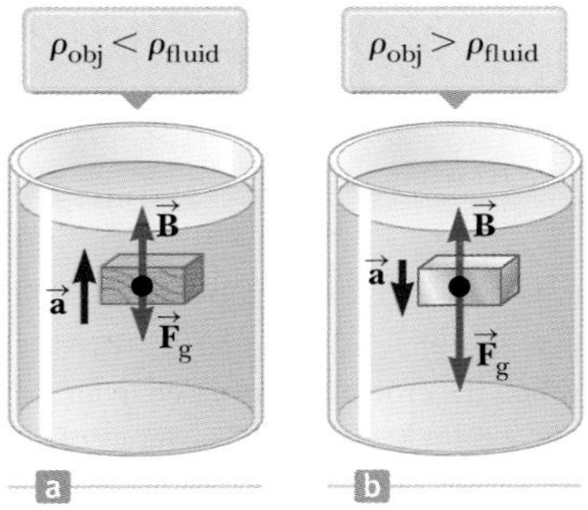
$\rho_{obj} < \rho_{fluid}$
$\rho_{obj} > \rho_{fluid}$
$\vec{B}$
$\vec{a}$
$\vec{F}_g$
a
b

Active Figure 13.10
(a) A totally submerged object that is less dense than the fluid in which it is submerged experiences a net upwards force and rises to the surface after it is released. (b) A totally submerged object that is denser than the fluid experiences a net downwards force and sinks.

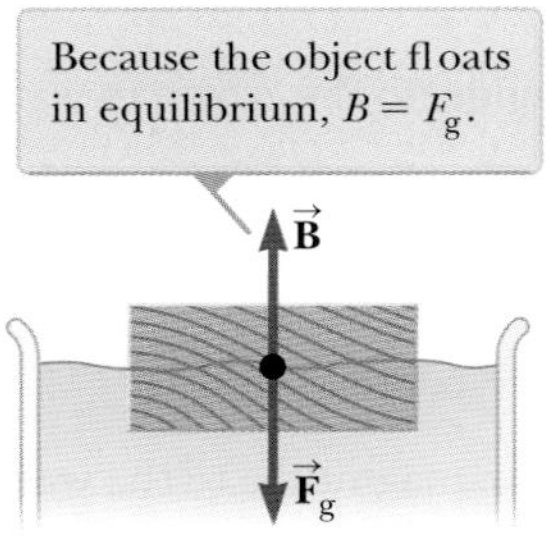
Because the object floats in equilibrium, $B = F_g$.
$\vec{B}$
$\vec{F}_g$

Active Figure 13.11
An object floating on the surface of a fluid experiences two forces, the gravitational force $\vec{F}_g$ and the buoyant force $\vec{B}$.

of the fluid displaced by the object (this volume is the same as the volume of that part of the object beneath the surface of the fluid), the buoyant force has a magnitude $B = \rho_{\text{fluid}} g V_{\text{disp}}$. Because the weight of the object is $F_g = M_g = \rho_{\text{obj}} g V_{\text{obj}}$ and because $F_g = B$, we see that $\rho_{\text{fluid}} g V_{\text{disp}} = \rho_{\text{obj}} g V_{\text{obj}}$, or

$$\frac{V_{\text{disp}}}{V_{\text{obj}}} = \frac{\rho_{\text{obj}}}{\rho_{\text{fluid}}} \tag{13.6}$$

This equation shows that the fraction of the volume of a floating object that is below the fluid surface is equal to the ratio of the density of the object to that of the fluid.

Quick Quiz 13.5

You are shipwrecked and floating in the middle of the ocean on a raft. Your cargo on the raft includes a treasure chest full of gold that you found before your ship sank and the raft is just barely afloat. To keep you floating as high as possible in the water, should you **(a)** leave the treasure chest on top of the raft, or **(b)** secure the treasure chest to the underside of the raft?

TRY THIS

Place two cans of soft drinks, one regular and one diet, in a container of water. What happens? Why?

Example 13.5

Eureka!

Archimedes supposedly was asked to determine whether a crown made for the king consisted of pure gold. According to legend, he solved this problem by weighing the crown first in air and then in water. Suppose he measured a weight of 7.84 N when the crown was in air and 6.84 N when it was in water. What should Archimedes have told the king?

Solution

Conceptualise Draw a diagram showing the two situations. Draw a free body diagram for the crown in each case.

Model This problem is an example of Case 1: the crown is completely submerged. The scale reading is a measure of one of the forces on the crown and the crown is stationary. We can model the crown as a particle in equilibrium.

Analyse When the crown is suspended in air, the scale reads the true weight $T_1 = F_g$ (neglecting the small buoyant force due to the surrounding air). When the crown is immersed in water, the buoyant force $\vec{\mathbf{B}}$ reduces the scale reading to an *apparent* weight of $T_2 = F_g - B$.

Apply the particle in equilibrium model to the crown in water:

$$\sum F = B + T_2 - F_g = 0$$

Solve for B and substitute the known values:

$$B = F_g - T_2 = 7.84\text{ N} - 6.84\text{ N} = 1.00\text{ N}$$

Because this buoyant force is equal in magnitude to the weight of the displaced water, $B = \rho_w g V_{\text{disp}}$, where V_{disp} is the volume of the displaced water and ρ_w is its density. Also, the volume of the crown V_c is equal to the volume of the displaced water because the crown is completely submerged, so $B = \rho_w g V_c$.

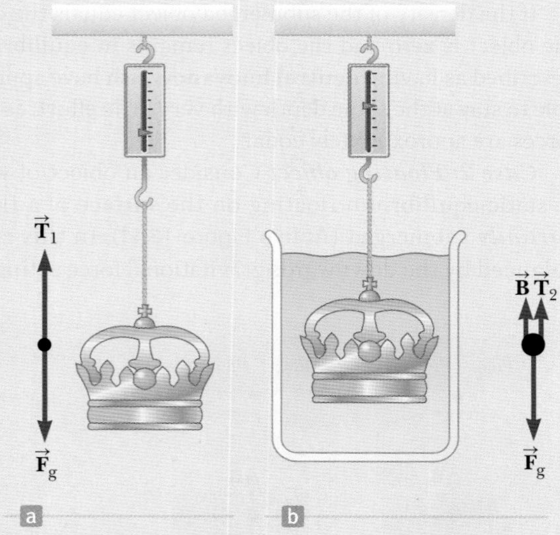

Figure 13.12
(Example 13.5) (a) When the crown is suspended in air, the scale reads its true weight because the buoyancy of air is negligible compared to the buoyancy of water: $T_1 = F_g$. (b) When the crown is immersed in water, the buoyant force $\vec{\mathbf{B}}$ changes the scale reading to a lower value $T_2 = F_g - B$.

Example 13.5 cont.

Find the density of the crown from **Equation 1.1**: $\rho_c = \frac{m_c}{V_c} = \frac{m_c g}{V_c g} = \frac{m_c g}{(B/\rho_w)} = \frac{m_c g \rho_w}{B}$

Check dimensions: $[ML^{-3}] = [M][LT^{-2}][ML^{-3}] / [MLT^{-2}] = [ML^{-3}]$ ☺

Substitute numerical values: $\rho_c = \frac{(7.84\ N)(1000\ kg/m^3)}{1.00\ N} = 7.84 \times 10^3\ kg/m^3$

Finalise From **Table 13.1**, we see that the density of gold is $19.3 \times 10^3\ kg/m^3$. Therefore, Archimedes should have told the king that either the crown was hollow, or it was not made of pure gold.

What If? Suppose the crown has the same weight but is indeed pure gold and not hollow. What would the scale reading be when the crown is immersed in water?

Answer Find the buoyant force on the crown: $B = \rho_w g V_w = \rho_w g V_c = \rho_w g \left(\frac{m_c}{\rho_c}\right) = \rho_w \left(\frac{m_c g}{\rho_c}\right)$

Substitute numerical values: $B = (1.00 \times 10^3 kg/m^3)\frac{7.84\ N}{19.3 \times 10^3 kg/m^3} = 0.406\ N$

Find the tension in the string hanging from the scale: $T_2 = F_g - B = 7.84\ N - 0.406\ N = 7.43\ N$

Example 13.6

A titanic surprise

An iceberg floating in seawater is extremely dangerous because most of the ice is below the surface. This hidden ice can damage a ship that is still a considerable distance from the visible ice. What fraction of the iceberg lies below the water level?

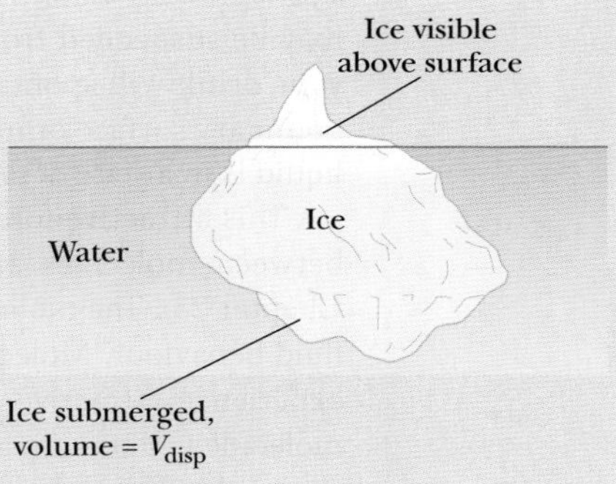

Figure 13.13
(Example 13.6) An iceberg floats in seawater.

Solution

Conceptualise Start by drawing a diagram to help visualise the problem.

Model This example corresponds to Case 2: the iceberg is not submerged.

Analyse Evaluate **Equation 13.6** using the densities of ice and seawater (**Table 13.1**):

$$f\% = \frac{V_{disp}}{V_{ice}} = \frac{\rho_{ice}}{\rho_{seawater}} = \frac{917\ kg/m^3}{1030\ kg/m^3} = 0.890 \text{ or } 89.0\%$$

Therefore, the visible fraction of ice above the water's surface is only about 11%.

Hydrometers and density measurements

Archimedes' principle is exploited in the design of **hydrometers**, devices that measure the relative densities or specific gravities of fluids. Hydrometers normally consist of a sealed cylinder of glass with a ballast material at one end (often mercury or lead) that ensures the cylinder floats upright, constructed so as to have the same density as a reference fluid (usually pure water). The hydrometer is gently lowered into the liquid being tested until it floats. Archimedes' principle tells us that a floating object displaces a volume of water with mass equal to its own mass; thus the hydrometer will sink deeper in less dense fluids and float

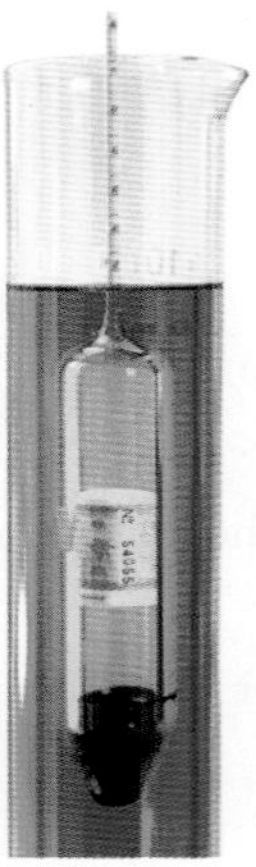

Figure 13.14
A hydrometer used to measure sugar content in beer

higher in denser fluids. The upper part of the hydrometer is marked with a scale that is calibrated so that the density of the liquid can be directly read from the level to which the hydrometer is submerged.

Hydrometers have many uses, including measuring the density of milk (which gives an indication of fat content) and identifying the perfect consistency for maple syrup. Hydrometers calibrated to the density of water at 15°C are used in the car industry to measure the density of the electrolyte (and hence the acid content and charge state) in lead–acid batteries. Special *proof and traille* hydrometers are used to measure alcohol content in spirits, which are less dense than water. You may have a hydrometer at home, if you or any of your family make home brew beer or wine, to let you assess the sugar content of your brew.

Quick **Quiz 13.6**

You may have noticed that when you swim in the sea or a saltwater pool, you float higher in the water than when you swim in a freshwater pool. Why is this?

13.6 Surface tension

Archimedes' principle explains why objects float if they are less dense than the fluid they displace. You may have observed that a sewing needle or paperclip placed gently on the surface of a glass of water will also float. If not, try it! But the density of the steel from which these objects are made is about eight times that of water, so what stops them sinking?

The answer to this question is **surface tension**. The same phenomenon is responsible for the fact that slowly flowing water leaves an almost closed tap as drops, rather than a steady stream; that a drop of water may be suspended from a surface for some time before falling, as if the water were inside a bag; and that your drink will spontaneously rise part way up a drinking straw. These phenomena occur when there is a boundary surface or interface between a liquid and another substance and they indicate that the surface of a liquid is in a state of tension: it exerts a force on the material at the boundary surface.

This attractive force can be understood by considering the microscopic structure of liquids. The forces between molecules are mainly due to electromagnetic interactions with neighbouring molecules (see Chapter 23). The molecules in a liquid experience strong cohesive forces, which are responsible for much of fluid behaviour. Molecules in the bulk of the liquid, beginning a few molecular diameters from the surface, experience attractive forces from neighbours on all sides – the *average* intermolecular force on such a molecule is thus zero and they are in a state of minimum (or low) potential energy. Those at the surface have no neighbours to bond with above them; if they undergo a small upwards displacement, they experience a downwards force due to the molecules below them in the bulk of the liquid. Thus moving molecules at the surface away from the fluid requires work to be done – this implies that these molecules have potential energy relative to those in the bulk of the fluid. There is thus an excess energy per unit area of surface – this excess energy is called the surface tension. It is measured in J/m^2 (or equivalently, N/m) and has dimensions MT^{-2}. The shape with the minimum surface-to-volume ratio (and hence the lowest energy) is a sphere, and so a droplet of liquid forms a spherical drop in the absence of any forces or contact surfaces. We can relate surface potential energy to the force it results in. The surface tension γ is defined as the ratio of the potential energy U of a surface to its area A:

$$\gamma = \frac{U}{A} \tag{13.7}$$

Applying the work-energy relation (**Equation 6.23**) we can find the force exerted on the surface:

$$F = -\frac{dU}{dx} = \gamma\frac{dA}{dx}$$

Consider a needle floating on water as shown in **Figure 13.15**. The force on the surface of the water due to the weight of the needle pushes the surface down, increasing its surface area. In order to minimise energy, the water exerts an upwards force on the needle. This additional force is sufficient to make the steel needle float.

When a droplet of liquid is stable, it is in equilibrium – this implies that its potential energy is a minimum. As the potential energy of the liquid increases with increasing surface area, this gives us an explanation of why small droplets tend to be almost perfectly spherical. Mathematically, it can be shown that the shape with the smallest surface-to-volume ratio is a sphere.

If you have ever blown bubbles using a soap solution, you have probably observed that they rapidly settle into a spherical shape – the weight of a bubble is so small that their shape is almost entirely caused by their surface tension. In contrast, drops of water about to fall from a tap are not perfectly spherical. This is because of the effects of gravity. If you suspend a droplet in a fluid that provides enough buoyant force to balance gravity (so that the droplet is neutrally buoyant), the droplets are spherical. The Belgian physicist Joseph Plateau demonstrated the spherical shape of oil drops suspended in a mixture of alcohol and water in 1873. The tendency of liquids to form spherical droplets was exploited in the manufacture of lead shot, which was made by spraying molten lead from a *shot tower*; as the droplets fell from the nozzle at the top of the tower, surface tension causes them to form spheres before they cool and harden into solid lead balls.

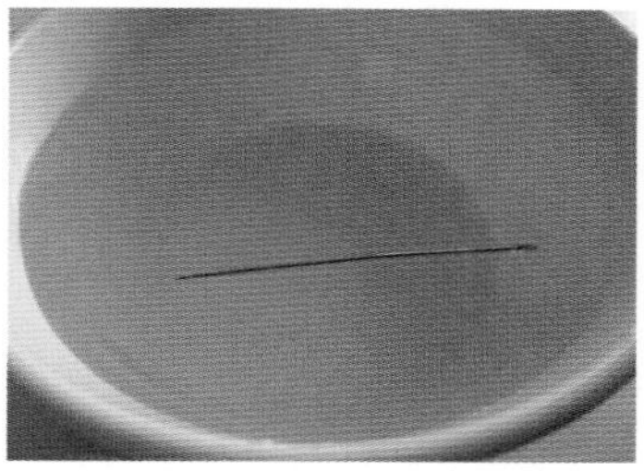

Figure 13.15
A steel needle floats on water, in apparent defiance of Archimedes' principle.

Figure 13.16
Clifton Hill shot tower, in Melbourne, is the tallest shot tower ever built in Australia.

Another way to think of surface tension is as a line tension. Imagine a soap film of surface tension γ on a rectangular frame, as illustrated in **Figure 13.17**. One side of the frame is a wire that is free to slide. If the wire is displaced by an amount dx, the surface area of the film changes by $dA = 2\ell dx$ and does work:

$$W = 2\gamma\ell\, dx$$

and so the force on the wire is

$$F = 2\gamma\ell \tag{13.8}$$

This shows that the surface tension can be thought of as a force per unit length along the wire's perimeter. Note that the factor of 2 comes from there being two faces to the surface, a front and a back.

TRY THIS

Make yourself a wire ring, or get a large plastic ring from a child's bubble-blowing kit. Tie a piece of thread to opposite sides of the ring, and then tie them together in the middle so that you have a loose loop of thread attached by two anchor lines to the wire. Dip the ring and thread in a bubble or soap solution to form a film. What happens to the loop of thread? Now pierce the film trapped inside the loop of thread. What happens now? Explain what you observe using the concept of surface tension.

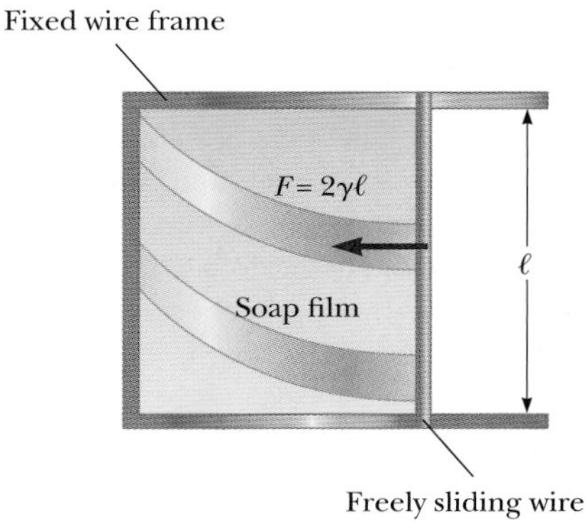

Figure 13.17
A soap film moves freely on a wire frame.

One way the molecules at the surface of a fluid can reduce their potential energy is to decrease the distance between molecules. This can be achieved by contracting the surface – the net effect of this energy minimisation is to produce a continual pressure due to an inwards force applied by the surface. This means that the interior of a droplet or bubble is at higher pressure than the exterior. We can relate the surface tension to this excess pressure as follows.

Consider a soap bubble in equilibrium. We will model the bubble as having negligible mass and ignore the gravitational force acting on it. The bubble consists of two spherical surface films (inside and outside surfaces) very close together. First, take a small element of the surface, dA. We label the air pressure outside the bubble as P_0 and the pressure inside as P_b. The normal force acting on the element is then

$$dF = (P_0 - P_b)\, dA \cos\theta$$

where θ is the angle between the surface element and the vertical.

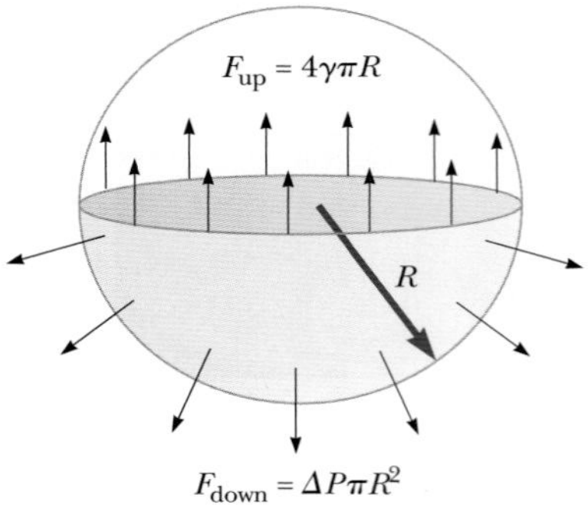

Figure 13.18
The forces acting on the lower half of a bubble are the downwards force due to the pressure inside the bubble, $F_{down} = \Delta P\pi R^2$, and the upwards force due to the surface tension of the film, $F_{up} = 4\gamma\pi R$.

Now consider the lower half of the bubble as shown in **Figure 13.18**. From **Equation 13.8**, the upwards force exerted by the top half of the bubble on the bottom half is:

$$F_{up} = 2\gamma\, 2\pi R$$

The downwards force on the lower half is the result of the pressure difference across the bubble surface and acts downwards on the entire lower half of the bubble. The net force due to the pressure difference will be downwards, as any horizontal components will cancel out. If we integrate our expression for dF, noting that when we do this what we are really finding is the product of the pressure difference $\Delta P = P_0 - P_b$ with the cross-sectional area $A = \pi R^2$, we get:

$$F_{down} = (P_0 - P_b)\,\pi R^2$$

As the bubble is in equilibrium, $F_{up} = F_{down}$ and so

$$(P_0 - P_b)\,\pi R^2 = 2\gamma\, 2\pi R$$

or

$$(P_0 - P_b) = \frac{4\gamma}{R} \tag{13.9}$$

For a liquid drop, which has only one surface rather than the two the bubble has, the pressure difference is halved, so that the pressure difference between the interior and exterior of a drop is given by the Young–Laplace equation:

The Young–Laplace ► equation

$$\Delta P = \frac{2\gamma}{R} \tag{13.10}$$

Quick Quiz 13.7

Two bubbles are connected via a pipe with a valve, which is initially closed. One bubble is bigger than the other. What will happen when the valve is opened?

Example 13.7

Droplet stability and gravity

Bubbles and droplets are only stable and spherical if they are smaller than a minimum size that depends on the surface tension: above a certain size, a droplet will break up into smaller droplets, or a bubble will deform and flatten and finally break. Although the stability of droplets and bubbles is governed by some quite complex physics, we can make reasonable estimates of the maximum stable size by comparing the internal pressure due to surface tension with the (gravitational) hydrostatic pressure at the bottom of the drop. If they are similar or if the hydrostatic pressure is bigger, the droplet is likely to be unstable.

Using this approach, calculate the maximum stable size for a water droplet in air, assuming the surface tension of water in air is 0.072 N/m.

Solution

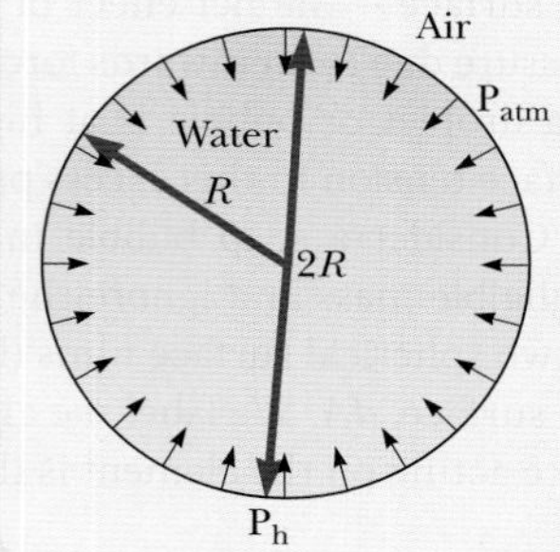

Conceptualise Start by drawing a diagram to help visualise the problem.

Model We are no longer making the approximation that the droplet has negligible mass. However we will still make the approximation that the droplet is spherical, and then test whether this approximation is reasonable. This problem requires us to calculate two different pressures and compare them.

Analyse Equation 13.4 tells us that the hydrostatic pressure at the bottom of a drop of water of radius R is

$$P_h = P_{atm} + \rho_{water}\, g\, 2R$$

Example 13.7 cont.

The difference between the external and internal pressures due to surface tension is

$$P_s = \frac{2\gamma}{R}$$

To find the maximum stable drop size, we need to solve for R_{max} such that $P_s = P_h - P_{atm}$

i.e.
$$\frac{2\gamma}{R_{max}} = 2\rho_{water} g R_{max}$$

Rearranging for R_{max} gives
$$R^2_{max} = \frac{\gamma}{\rho_{water} g}$$

Check the dimensions for consistency:
$$[L^2] = \frac{[MT^{-2}]}{[ML^{-3}][LT^{-2}]} = [L^2]$$ ☺

Substituting in appropriate values for the surface tension, density and acceleration due to gravity we obtain

$$R^2_{max} = \frac{0.072\ \text{N/m}}{(1000\ \text{kg/m}^3)(9.8\ \text{m/s}^2)}$$

$$= \frac{0.072\ \text{kg/s}^2}{(1000\ \text{kg/m}^3)(9.8\ \text{m/s}^2)} = 7.35 \times 10^{-6}\ \text{m}^2$$

Hence
$$R_{max} = 2.7 \times 10^{-3}\ \text{m}$$

Finalise Our simple approach to finding the maximum radius of a stable, spherical water drop gives a value of just less than 3 mm. For values of R larger than this the droplets will not be spherical.

±?

Example 13.8

A simple way to measure surface tension

The surface tension of a liquid can be measured using a sensitive balance and a microscope slide. A container of the liquid is placed on the balance and weighed. The microscope slide is then lowered long edge down onto the liquid until it is just touching the surface of the liquid. This causes a reduction in the scale reading, because the surface of the liquid in contact with the slide exerts an upwards force on the surface of the liquid below.

If the slide is 5.00 ± 0.05 cm long and 2.00 ± 0.05 mm thick, and the apparent mass of the container plus liquid decreases by 0.75 ± 0.05g, what is the surface tension of the liquid? Could the liquid be water?

Solution

Conceptualise When the slide is in contact with the surface of the liquid, the surface tension acts at the boundary of the slide to pull the slide downwards. By Newton's third law, the slide exerts an equal and opposite upwards force on the liquid. Draw a diagram showing the forces acting on the container of liquid.

Model We model the container as an object in equilibrium both before and after the slide is lowered to touch the surface, as shown. Initially, only the gravitational force and the normal force due to the scales are acting. The reading on the scales is therefore equal to the weight of the container plus liquid. When the slide is just in contact with the surface and the

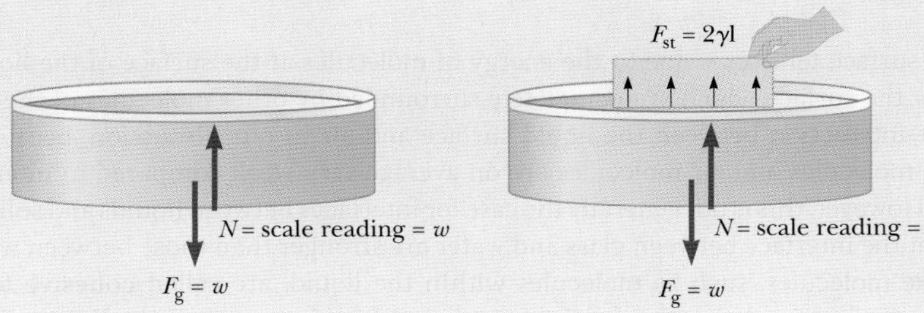

Example 13.8 cont.

system has come to equilibrium, there are three forces acting. These must again balance as the object is in equilibrium. The apparent weight change, Δw, must be equal to the upwards force due to surface tension.

Analyse This problem requires us to use the definition of surface tension in terms of force and length given in Equation 13.8.

We will first solve the problem symbolically and then substitute in numerical values. If the length of the slide is ℓ and the thickness t, the length of the boundary is $(2\ell + 2t)$. Equation 13.8 tells us that the resulting force is

$$F = \gamma(2\ell + 2t)$$

If the apparent weight change is Δw, then

$$\gamma(2\ell + 2t) = \Delta w$$

and so

$$\gamma = \frac{\Delta w}{(2\ell + 2t)}$$

We now substitute in the given values for the slide size and the mass, being careful to check that we are using consistent units.

$$\gamma = \frac{(0.75 \times 10^{-3} \times 9.80\ \text{m/s}^2)}{(2 \times 0.05\ \text{m} + 2 \times 0.002\ \text{m})} = 0.07067\ \text{N/m}$$

Finalise We cannot compare our answer with the expected value for water until we have calculated the uncertainty. The uncertainty in the length of the boundary is $2\Delta l + 2\Delta t = 2\ (0.05\ \text{cm}) + 2\ (0.05\ \text{mm}) = 1.1 \times 10^{-3}$ m. The boundary length is $2l + 2t = 2\ (5.00\ \text{cm}) + 2\ (2.00\ \text{mm}) = 10.4$ cm or 0.104 m.

This is a fractional uncertainty of 1.1×10^{-3} m / 0.104 m = 0.01 or 1%.

The fractional uncertainty in Δw is 0.05 g / 0.75 g = 0.067 or 6.7%.

The fractional uncertainty in γ is the sum of these: $\Delta\gamma/\gamma = 0.01 + 0.067 = 0.077$ or 7.7%.

So $\Delta\gamma = 0.077\ (0.07067\ \text{N/m}) = 0.005$ N/m.

So our final answer for the measured surface tension is $\gamma = (0.071 \pm 0.005)$ N/m, which is consistent with the value expected for water.

Contact angle and capillarity

You may have noticed that you can fill a glass with water so that the water is slightly higher than the rim – the water appears to curve upwards slightly from the edges of the glass (Figure 13.19a). Similarly, if you look closely at the surface of water in a half-filled glass, you can see a small rise in the surface at the edges. This effect is easiest to see in a narrow container, such as a measuring cylinder or test tube (Figure 13.19b).

TRY THIS

Fill a glass to the brim with water and look carefully at the surface. How much higher is the top of the water than the rim of the glass? How steeply does the water curve away from the rim? Now find a narrow glass tube (such as a measuring cylinder) and fill it part way with water. Look closely at the surface of the water. Try this with a few different liquids – for example, add table salt to make a solution of sodium chloride, add washing-up liquid to make a soap solution or try using cooking oil. Do they all behave in the same way?

Recall that surface tension is due to the energy of molecules at the surface of the liquid being higher than those below the surface which are completely surrounded by other molecules of the liquid. So far we have ignored the interaction between the liquid surface and air in our discussion, because the forces between the surface molecules and air molecules are on average very weak compared to interactions between fluid molecules. However, this is not generally the case for interfaces between liquids and solids. The forces between molecules at the interface between glass and water are stronger than those between water molecules. Forces between like molecules, such as molecules within the liquid, are called **cohesive** forces. Forces between different types of molecules, such as at the water–glass interface, are called **adhesive** forces. Hence, when we consider a liquid in a solid container we need to define two distinct surface tensions: γ_{LS}, which is the surface

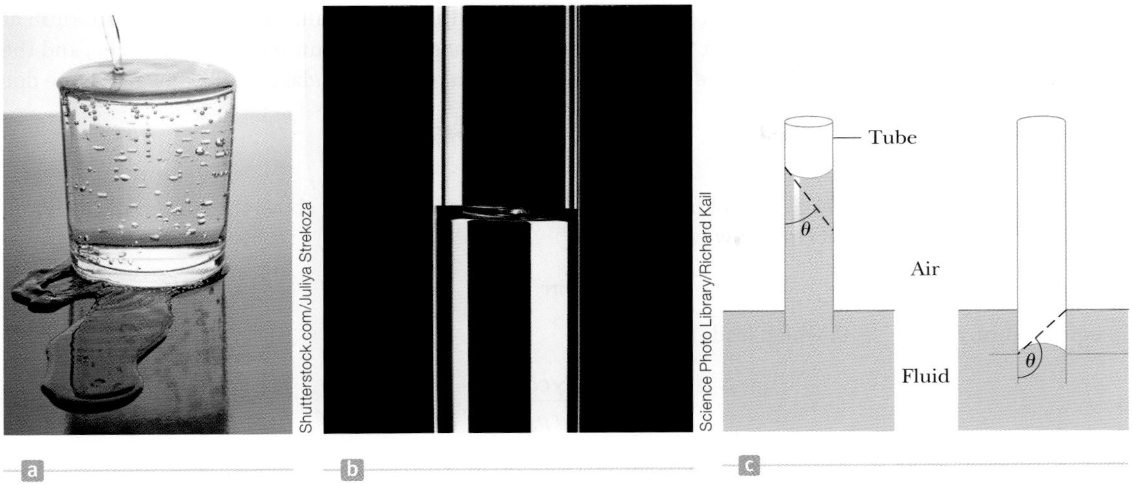

Figure 13.19
(a) An overfilled glass, (b) water in a test tube and (c) definition of the contact angle, θ.

tension for the liquid-solid interface, and γ_{LG}, which is the surface tension for the liquid–gas interface. γ_{LG} is always positive for liquids in air, so that liquids form droplets to minimise their surface energy by minimising their surface area. γ_{LS} can be positive or negative. The relative sizes of the surface tensions and the sign of γ_{LS} can be determined from the contact angle that the liquid makes with the solid:

$$\gamma_{LS} = -\gamma_{LG} \cos\theta \tag{13.11}$$

where the contact angle is measured through the liquid as shown in **Figure 13.19c**.

If the contact angle is less than 90°, as in the left hand tube in **Figure 13.19c**, then $\cos\theta$ is positive and γ_{LS} must be negative. This is the case for a water–glass interface, where the angle is small.

The tube on the right hand side of **Figure 13.19c** shows a contact angle which is greater than 90°, and hence this liquid–solid interface has a positive surface tension. This is the case for mercury in glass, which forms a convex meniscus. Remembering that the surface tension describes the excess energy per unit area of an interface, this tells us that the energy of the water is lowered by its contact with the glass. This is because the *adhesive* forces between the water molecules and the glass are stronger than the *cohesive* forces between water molecules. Water in a test tube or glass maximises its contact area with the glass by climbing up the sides of the container, thus minimising its energy, giving rise to a concave meniscus as shown in **Figure 13.19b**. This is also why water spilled on a clean glass slide will spread out in a single pool.

In contrast, a liquid such as mercury forms a convex meniscus when placed inside a glass tube – you may have noticed this if you have a mercury thermometer at home. A convex meniscus corresponds to an obtuse contact angle, meaning that the liquid–solid surface tension is positive and the force at the interface is repulsive. It is this positive liquid–solid surface tension that stops the mercury drop in a mercury switch from spreading out into a film.

If the adhesion between the liquid and solid is greater than the cohesion between molecules in the liquid, as is the case between water and glass, we observe another important phenomenon – that of **capillary action** or **capillarity**.

If you stand a narrow-bore tube such as a capillary tube upright in a beaker of water so that it extends above the water's surface, you will see that the water rises inside the tube above the surface of the water outside the tube. You can see this effect using a drinking straw as your narrow tube, although with a smaller-bore tube the rise is more dramatic. If you have ever walked through a puddle while wearing jeans you will already be familiar with this phenomenon, as the water soaks upwards through the material towards your knees, in apparent defiance of gravity. The same phenomenon lets candles burn: melted wax is drawn up into the wick by capillary action, feeding the flame. In each case, the liquid is drawn up into narrow tubes (either the capillary tube, or the space between fibres in your jeans or in a candle wick) because the liquid–solid surface tension is negative.

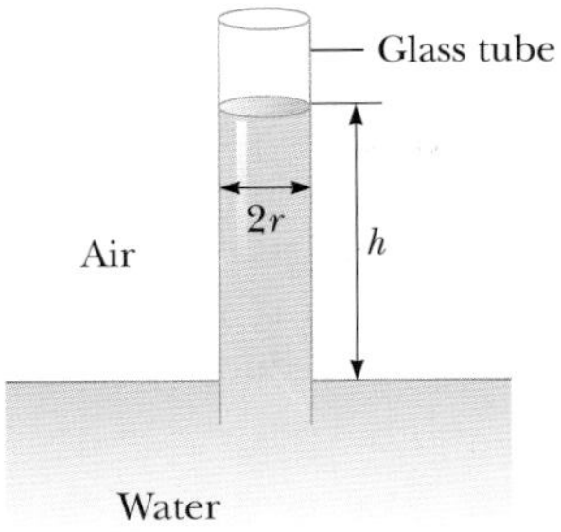

Figure 13.20
A liquid rises in a capillary tube.

We can use our understanding of surface tension to calculate how high a liquid will rise up a capillary tube. Suppose a tube of radius r has one end immersed in a liquid of surface tension $\gamma_{LG} = \gamma$ and density ρ as shown in **Figure 13.20**. The contact angle of the liquid at the tube walls is θ. The liquid rises to height h and then stays there, indicating that all relevant forces are in equilibrium at that point. The length of the boundary at the liquid–solid interface is the circumference of the tube, and the force due to surface tension acts along the direction perpendicular to the boundary, so the upwards force due to the surface tension is

$$F_{ST} = 2\pi r\gamma \cos\theta$$

This force supports the weight of the column of water of height h and radius r above the surface of the liquid outside the tube, which results in a downwards force:

$$F_g = -\pi r^2 h\rho g$$

In equilibrium, $F_{ST} + F_g = 0$, which leads to

$$h = \frac{2\gamma\cos\theta}{r\rho g} \tag{13.12}$$

That is, a liquid with a negative liquid–solid surface tension and acute contact angle such as water will be pulled up the tube. In contrast, if the tube were immersed in mercury, the mercury would be pushed down the tube to some depth below the surface.

TRY THIS

Run some water into a sink and hold some cloth, for example a pair of jeans, hanging into the water so that most of the cloth is still above the water. Watch what happens, and measure how high the water rises up the cloth. Try it with different sorts of cloth, and with paper towels. Use the measured heights to estimate the spacing between fibres in the cloth. What happens if you add detergent to the water first? Does it make any difference?

Example 13.9

How do trees drink?

Water and some nutrients are transported from the roots of trees to the rest of the plant through xylem, which act as capillary tubes. We can model the xylem as rigid tubes with inner diameter 20 μm. Calculate the maximum height to which water can rise through capillary action. Is this model sufficient to describe the movement of water through trees? What other factors might be important?

Solution

Conceptualise The xylem resembles a capillary tube immersed in water, as in **Figure 13.19c**.

Model We will model the xylem as capillary tubes, carrying water. The tubes have radius $d/2 = 10\ \mu$m, and fresh water has a density of 1000 kg/m^3. We do not know what the contact angle is, but we can define the limiting case as a contact angle of zero, which will give us the maximum possible value of h, as required.

Analyse We can use **Equation 13.12** to calculate the height.

$$h = \frac{2\gamma\cos\theta}{r\rho g} = \frac{2(0.072\,\text{N/m})\cos 0}{(10\times 10^{-6}\,\text{m})(1000\,\text{kg/m}^3)(9.80\,\text{m/s}^2)} = 1.5\ \text{m}$$

Finalise Modelling the water's behaviour through capillary action gives a maximum height of 1.5 m. This is far smaller than the height of most trees, so capillary action alone cannot explain the transport of water through trees. We know that liquid will rise up a tube if there is a larger pressure applied at the bottom than at the top, so we might look for possible reasons for reduced pressure at the top of a tree or increased pressure at the bottom. In fact, evaporation from leaves creates a reduced pressure at the top, which helps to draw water further up.

End-of-chapter resources

At the beginning of this chapter, we asked how fish control their movements, and why it is that they swim underwater while they are alive but float to the surface when they die, and then subsequently sink to the bottom. Under normal conditions, the weight of a fish is slightly greater than the buoyant force on the fish. Hence, the fish would sink if it did not have some mechanism for adjusting the buoyant force. The fish accomplishes that by internally regulating the size of its air-filled swim bladder and hence the magnitude of the buoyant force acting on it, according to Equation 13.5. In this manner, fish are able to swim to various depths and adjust their average density so that they are neutrally buoyant, allowing them to stay at the same depth. To swim higher, they must decrease their average density, and to descend lower they must increase it, before adjusting to become neutrally buoyant again. When fish die, as they decompose gases are released into the inner cavity, swelling the body and decreasing the average density so that the fish floats to the surface. After a few days, the gases escape from the fish, leaving it denser than the water and causing it to sink.

Shutterstock.com/Pete Niesen

The problems found in this chapter may be assigned online in Enhanced Web Assign.

ENHANCED WebAssign

Worked solutions to every fifth problem are available in the Student Solutions Manual. Register online at **www.cengagebrain.com** for access.

Summary

Definitions

The **pressure** P in a fluid is the force per unit area exerted by the fluid on a surface:

$$P \equiv \frac{F}{A} \tag{13.1}$$

In the SI system, pressure has units of newtons per square metre (N/m^2), and

$$1\ \text{N/m}^2 = 1\ \textbf{pascal}\ (\text{Pa}) \tag{13.3}$$

Surface tension γ is the potential energy per unit area or force per unit length at an interface:

$$\gamma = \frac{U}{A} \tag{13.7}$$

$$\gamma = \frac{F}{2\ell} \tag{13.8}$$

For a liquid in contact with both gas and solid, the surface tension at the liquid–solid interface is related to the surface tension at the liquid–gas interface through the contact angle θ.

$$\gamma_{LS} = -\gamma_{LG} \cos\theta \tag{13.11}$$

Concepts and principles

The pressure in a fluid at rest varies with depth h in the fluid according to the expression

$$P = P_0 + \rho g h \tag{13.4}$$

where P_0 is the pressure at $h = 0$ and ρ is the density of the fluid, assumed uniform.

Pascal's law states that when pressure is applied to an enclosed, incompressible fluid, the pressure is transmitted undiminished to every point in the fluid and to every point on the walls of the container.

When an object is partially or fully submerged in a fluid, the fluid exerts on the object an upwards force called the **buoyant force.** According to **Archimedes' principle,** the magnitude of the buoyant force is equal to the weight of the fluid displaced by the object:

$$B = \rho_{\text{fluid}} g V_{\text{disp}} \tag{13.5}$$

Whether an object is **negatively, positively or neutrally buoyant** depends on its relative density compared to the fluid it is submerged in.

Objects may also float if the **surface tension** of the fluid is sufficiently large. The surface tension is energy per unit area or force per unit length at an interface, and it results in an inwards pressure in bubbles and droplets. The Young–Laplace equation relates the difference in pressure on the interior and exterior of a droplet of liquid:

$$\Delta P = \frac{2\gamma}{R} \tag{13.10}$$

where R is the radius of the drop.

Negative surface tension at a liquid-solid interface is responsible for capillary action, where liquid rises to a higher level inside a tube than outside. The narrower the tube, the further the liquid rises. The height reached is

$$h = \frac{2\gamma \cos\theta}{r\rho g} \tag{13.12}$$

Chapter review quiz

To help you revise Chapter 13: Fluid statics, complete the automatically graded Chapter review quiz at http://login.cengagebrain.com.

Conceptual questions

1. Because atmospheric pressure is about 10^5 N/m^2 and the area of a person's chest is about 0.13 m^2, the force of the atmosphere on one's chest is about 13 000 N. In view of this enormous force, why don't our bodies collapse?
2. When an object is immersed in a liquid at rest, why is the net force on the object in the horizontal direction equal to zero?
3. A fish rests on the bottom of a bucket of water while the bucket is being weighed on a scale. When the fish begins to swim around, does the scale reading change? Explain your answer.
4. A typical silo on a farm has many metal bands wrapped around its perimeter for support as shown in Figure CQ13.4. Why is the spacing between successive bands smaller for the lower portions of the silo on the left, and why are double bands used at lower portions of the silo on the right?

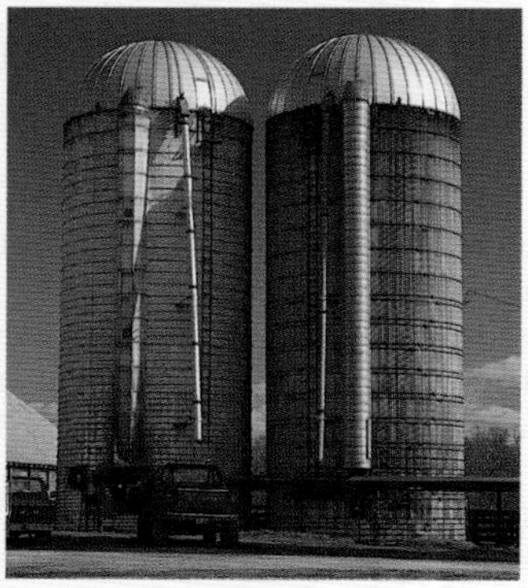
Henry Leap and Jim Lehman

Figure CQ13.4

5. Does a ship float higher in the water of an inland lake or in the ocean? Why?
6. In the movie 'Titanic', Rose and Jack are left floating in the ocean after the ship sinks. Jack helps Rose to climb onto a piece of floating debris. Why might it have made more sense for Jack to climb onto the debris rather than Rose?
7. (a) Is the buoyant force a conservative force? (b) Is a potential energy associated with the buoyant force? (c) Explain your answers to parts (a) and (b).
8. If you release a ball while inside a freely falling elevator, the ball remains in front of you rather than falling to the floor because the ball, the elevator and you all experience the same downward gravitational acceleration. What happens if you repeat this experiment with a helium-filled balloon?
9. How would you determine the density of an irregularly shaped rock?
10. Explain how the Cartesian diver in the *Try this* example in Section 13.5 works.
11. Is it possible for a needle that is initially wet to float on water?
12. If two soap bubbles of different sizes are connected via a tube, the smaller bubble will collapse and the larger one will grow. Why?
13. Many areas experiencing low rainfall have to draw water from the underground watertable through bores. The level of the watertable varies quite substantially even over short distances, depending on soil type among other factors. Why is the watertable higher in fine-grained soils than in coarse-grained soils?

Problems

In all problems, assume the density of air is the 20°C value from **Table 13.1**, 1.20 kg/m^3, unless noted otherwise.

Section 13.1 Pressure

1. The nucleus of an atom can be modelled as several protons and neutrons closely packed together. Each particle has a mass of 1.67×10^{-27} kg and radius of the order of 10^{-15} m. (a) Use this model and the data provided to estimate the density of the nucleus of an atom. (b) Compare your result with the density of a material such as iron. What do your result and comparison suggest concerning the structure of matter?
2. ±? A 50.0 kg woman enters a room that has a vinyl floor covering. The heel on each of her shoes is circular and has a radius of (0.500 ± 0.005 cm). (a) If the woman balances on one heel, what pressure does she exert on the floor? (b) Should the woman take her shoes off before entering the room? Explain your answer.

Section 13.2 Pressure in fluids

3. Suction cups are used in a variety of industrial and construction applications, including handling glass, stone and other smooth materials with no handles or grip points. Explain how they work.
4. If a giraffe is 5.0 m tall, with its heart at approximately half that height, what pressure does the heart need to produce to keep blood flowing to the brain? How does this required pressure change when the giraffe lowers its head to drink from a pool? Giraffes spread their front legs out when they drink – why?
5. ±? A swimming pool has dimensions (30.0 ± 0.1) m × (10.0 ± 0.1) m and a flat bottom. When the pool is filled to a depth of 2.00 m with fresh water, what is the force exerted by the water on (a) the bottom? (b) each end? (c) each side?
6. The spring of the pressure gauge shown in **Figure P13.6** has a spring constant of 1250 N/m, and the piston has a diameter of 1.20 cm. As the gauge is lowered into water in a lake, what change in depth causes the piston to move in by 0.750 cm?

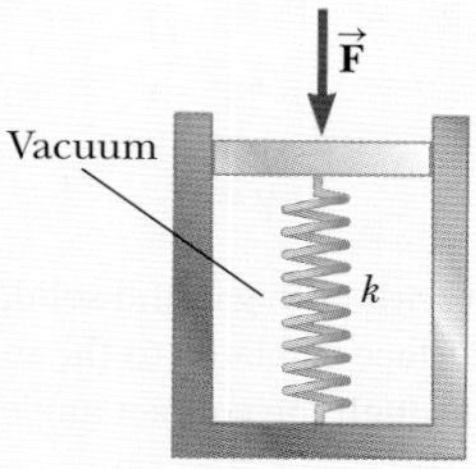

Figure P13.6

Section 13.3 Pascal's law and hydraulic tools

7. The small piston of a hydraulic hoist (**Figure P13.7**) has a cross-sectional area of 3.00 cm^2, and its large piston has a

cross-sectional area of 200 cm^2. What downwards force of magnitude F_1 must be applied to the small piston for the lift to raise a load whose weight is $F_g = 15.0$ kN?

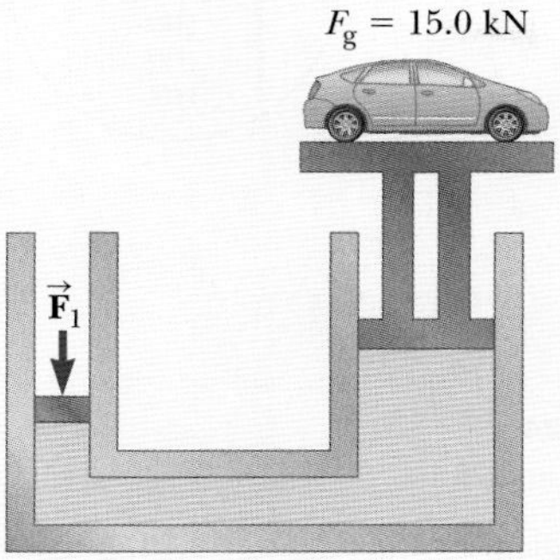

Figure P13.7

8. What must be the contact area between a suction cup (completely evacuated) and a ceiling if the cup is to support the weight of an 80.0 kg student?

9. (a) A very powerful vacuum cleaner has a hose 2.86 cm in diameter. With the end of the hose placed perpendicularly on the flat face of a brick, what is the weight of the heaviest brick that the cleaner can lift? (b) An octopus uses one sucker of diameter 2.86 cm on each of the two shells of a clam in an attempt to pull the shells apart. Find the greatest force the octopus can exert on a clamshell in salt water 32.3 m deep.

10. (a) Calculate the absolute pressure at the bottom of a freshwater lake at a point whose depth is 27.5 m. Assume the density of the water is 1.00×10^3 kg/m^3 and that the air above is at a pressure of 101.3 kPa. (b) What force is exerted by the water on the window of an underwater vehicle at this depth if the window is circular and has a diameter of 35.0 cm?

11. A tank is filled with water of depth d. At the bottom of one sidewall is a rectangular hatch of height h and width w that is hinged at the top of the hatch. (a) Determine the magnitude of the force the water exerts on the hatch. (b) Find the magnitude of the torque exerted by the water about the hinges.

Section 13.4 Pressure measurements

12. Blaise Pascal duplicated Torricelli's barometer using a red Bordeaux wine, of density 984 kg/m^3, as the working liquid. (a) What was the height h of the wine column for normal atmospheric pressure? (b) Would you expect the vacuum above the column to be as good as for mercury?

13. Normal atmospheric pressure is 1.013×10^5 Pa. The approach of a storm causes the height of a mercury barometer to drop by 20.0 mm from the normal height. What is the atmospheric pressure?

14. The human brain and spinal cord are immersed in the cerebrospinal fluid. The fluid is normally continuous between the cranial and spinal cavities and exerts a pressure of 100 to 200 mm of H_2O above the prevailing atmospheric pressure. In medicine pressures are often measured in units of millimetres of H_2O because body fluids, including the cerebrospinal fluid, typically have the same density as water. The pressure of the cerebrospinal fluid can be measured by means of a *spinal tap* as illustrated in **Figure P13.14**. A hollow tube is inserted into the spinal column, and the height to which the fluid rises is observed. If the fluid rises to a height of 160 mm, we write its gauge pressure as 160 mm H_2O. (a) Express this pressure in pascals, in atmospheres and in millimetres of mercury. (b) Some conditions that block or inhibit the flow of cerebrospinal fluid can be investigated by means of *Queckenstedt's test.* In this procedure, the veins in the patient's neck are compressed to make the blood pressure rise in the brain, which in turn should be transmitted to the cerebrospinal fluid. Explain how the level of fluid in the spinal tap can be used as a diagnostic tool for the condition of the patient's spine.

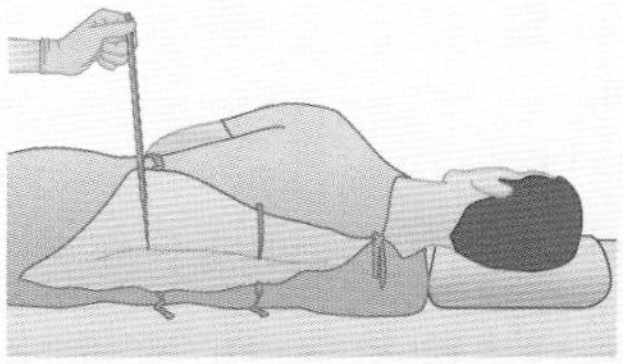

Figure P13.14

15. A tank with a flat bottom of area A and vertical sides is filled to a depth h with water. The pressure is P_0 at the top surface. (a) What is the absolute pressure at the bottom of the tank? (b) Suppose an object of mass M and density less than the density of water is placed into the tank and floats. No water overflows. What is the resulting increase in pressure at the bottom of the tank?

16. Mercury is poured into a U-tube as shown in **Figure P13.16a**. The left arm of the tube has cross-sectional area A_1 of 10.0 cm^2, and the right arm has a cross-sectional area A_2 of 5.00 cm^2. One hundred grams of water is then poured into the right arm as shown in **Figure P13.16b**. (a) Determine the length of the water column in the right arm of the U-tube. (b) Given that the density of mercury is 13.6 g/cm^3, what distance h does the mercury rise in the left arm?

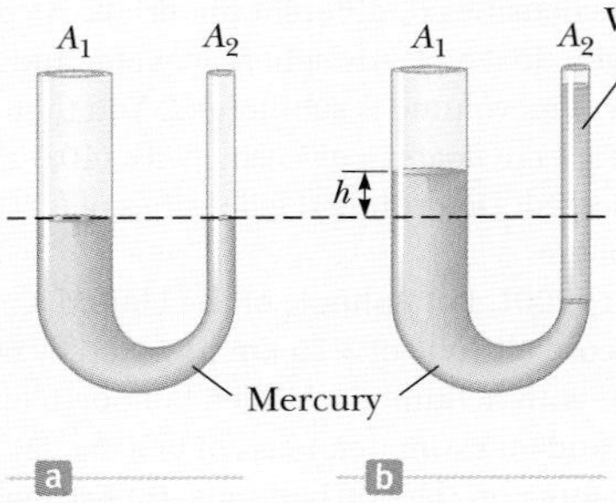

Figure P13.16

Section 13.5 Buoyant forces and Archimedes' principle

17. A light balloon is filled with 400 m^3 of helium at atmospheric pressure. (a) At 0°C, what is the maximum mass that the balloon can lift? (b) **What If?** The density of hydrogen is nearly half the density of helium. What load can the balloon lift if filled with hydrogen?

18. The weight of a solid object is 5.00 N. When the object is suspended from a spring scale and submerged in water, the scale reads 3.50 N. Find the density of the object.

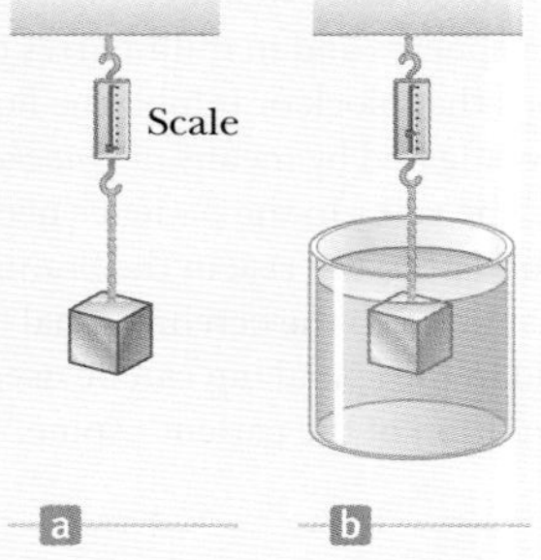

Figure P13.18

Problems 18 and 19

19. A 10.0 kg block of metal measuring 12.0 cm by 10.0 cm by 10.0 cm is suspended from a scale and immersed in water as shown in **Figure P13.18b**. The 12.0 cm dimension is vertical, and the top of the block is 5.00 cm below the surface of the water. (a) What are the magnitudes of the forces acting on the top and on the bottom of the block due to the surrounding water? (b) What is the reading of the spring scale? (c) Show that the buoyant force equals the difference between the forces at the top and bottom of the block.

20. A cube of wood having an edge dimension of 20.0 cm and a density of 650 kg/m^3 floats on water. (a) What is the distance from the horizontal top surface of the cube to the water level? (b) What mass of lead should be placed on the cube so that the top of the cube will be just level with the water surface?

21. ±? A spherical vessel used for deep-sea exploration has a radius of 1.50 m and a mass of (1.20×10^4) kg. To dive, the vessel takes on mass in the form of seawater. Determine the mass the vessel must take on if it is to descend at a constant speed of 1.20 m/s, when the resistive force on it is (1100 ± 50) N in the upwards direction. The density of seawater is equal to (1.03×10^3) kg/m^3.

22. ±? An experiment in the first-year physics labs of a university asks you to find the densities of different materials. As part of the experiment, you float a plastic sphere in water and find that (50.0 ± 0.5)% of its volume is submerged. You then observe that the same sphere floats in glycerine with (40.0 ± 0.5)% of its volume submerged. Determine the densities of (a) the glycerine and (b) the sphere.

23. On 21 October 2001, Ian Ashpole of the United Kingdom achieved a record altitude of 3.35 km powered by 600 toy balloons filled with helium. Each filled balloon had a radius of about 0.50 m and an estimated mass of 0.30 kg. (a) Estimate the total buoyant force on the 600 balloons. (b) Estimate the net upwards force on all 600 balloons. (c) Ashpole parachuted to the Earth after the balloons began to burst at the high altitude and the buoyant force decreased. Why did the balloons burst?

24. A hydrometer is an instrument used to determine liquid density. A simple one is sketched in **Figure P13.24**. The bulb of a syringe is squeezed and released to let the atmosphere lift a sample of the liquid of interest into a tube containing a calibrated rod of known density. The rod, of length L and average density ρ_0, floats partially immersed in the liquid of density ρ. A length h of the rod protrudes above the surface of the liquid. Show that the density of the liquid is given by

$$\rho = \frac{\rho_0 L}{L - h}$$

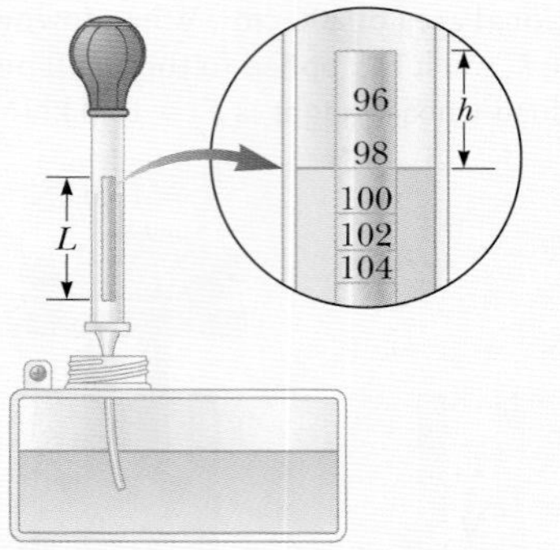

Figure P13.24

Problems 24 and 25

25. Refer to Problem 24 and **Figure P13.24**. A hydrometer is to be constructed with a cylindrical floating rod. Nine fiduciary marks are to be placed along the rod to indicate densities of 0.98 g/cm^3, 1.00 g/cm^3, 1.02 g/cm^3, 1.04 g/cm^3, . . . , 1.14 g/cm^3. The row of marks is to start 0.200 cm from the top end of the rod and end 1.80 cm from the top end. (a) What is the required length of the rod? (b) What must be its average density? (c) Should the marks be equally spaced? Explain your answer.

26. ±? How many cubic metres of helium are required to lift a balloon with a 400 kg payload to a height of (8.0 ± 0.1) km? Take $\rho_{He} = 0.179$ kg/m^3. Assume the balloon maintains a constant volume and the density of air decreases with the altitude z according to the expression $\rho_{air} = \rho_0 e^{-z/8\,000}$, where z is in metres and $\rho_0 = 1.20$ kg/m^3 is the density of air at sea level.

Section 13.6 Surface tension

27. A steel needle of length 3.0 cm and average diameter 0.5 mm is placed on the surface of a container of water. Assuming the surface tension of water is (7.28×10^{-2}) N/m, and taking the density of steel as 7.85 g/cm^3, calculate whether the needle will float. If the needle were twice as long, would that make it easier or harder for it to float?

28. The water in a fish tank is oxygenated by the introduction of a stream of bubbles through a nozzle in the side of the tank. The bubbles have a diameter of 3 mm. What must the gauge pressure be at the tip of the nozzle? (Assume the surface tension between air and water is 7.28×10^{-2} N/m.)

29. A child blows a soap bubble at the end of a straw. (a) Explain why it gets easier to blow as the bubble gets bigger. (b) The surface tension between the soap solution the child is using and air is 2.5×10^{-2} N/m. How much pressure must the child exert when the bubble diameter reaches 5 mm? What about when the bubble diameter reaches 5 cm?

30. When crude oil is initially pumped from oil fields, it often contains many very small droplets of water. It is important that these droplets are removed prior to the oil being transported and refined, and it is much easier to remove the water if the small droplets can be made to coalesce into larger ones. Droplet coalescence can be induced mechanically or electrostatically, but whatever process is used it results in the liberation of energy (which results in the oil heating up). Calculate the energy released when 1000 water droplets of radius 10^{-6} m coalesce to produce a single droplet. Take the surface tension at the water–oil interface to be 7.5×10^{-2} N/m.

Additional problems

31. (a) Calculate the absolute pressure at an ocean depth of 1000 m. Assume the density of seawater is 1030 kg/m^3 and the air above exerts a pressure of 101.3 kPa. (b) At this depth, what is the buoyant force on a spherical submarine having a diameter of 5.00 m?

32. In about 1657, Otto von Guericke, inventor of the air pump, evacuated a sphere made of two brass hemispheres (**Figure P13.32**). Two teams of eight horses each could pull the hemispheres apart only on some trials and then 'with greatest difficulty', with the resulting sound likened to a cannon firing. Find the force F required to pull the thin-walled evacuated hemispheres apart in terms of: R, the radius of the hemispheres; P, the pressure inside the hemispheres; and atmospheric pressure P_0.

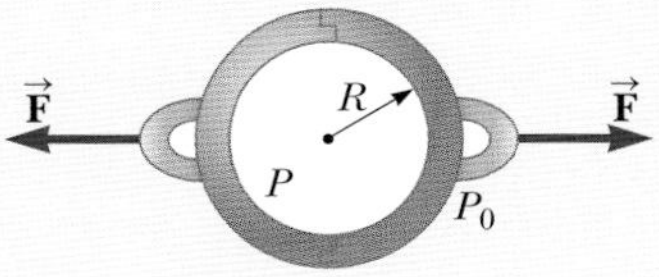

Figure P13.32

33. A spherical aluminium ball of mass 1.26 kg contains an empty spherical cavity that is concentric with the ball. The ball barely floats in water. Calculate (a) the outer radius of the ball and (b) the radius of the cavity.

34. A helium-filled balloon (whose envelope has a mass of $m_b = 0.250$ kg) is tied to a uniform string of length $\ell = 2.00$ m and mass $m = 0.050\,0$ kg. The balloon is spherical with a radius of $r = 0.400$ m. When released in air of temperature 20°C and density $\rho_{air} = 1.20$ kg/m^3, it lifts a length h of string and then remains stationary as shown in **Figure P13.34**. We wish to find the length of string lifted by the balloon. (a) When the balloon remains stationary, what is the appropriate analysis model to describe it? (b) Write a force equation for the balloon from this model in terms of the buoyant force B, the weight F_b of the balloon, the weight F_{He} of the helium, and the weight F_s of the segment of string of length h. (c) Make an appropriate substitution for each of these forces and solve symbolically for the mass m_s of the segment of string of length h in terms of m_b, r, ρ_{air}, and the density of helium ρ_{He}. (d) Find the numerical value of the mass m_s. (e) Find the numerical value of the length h.

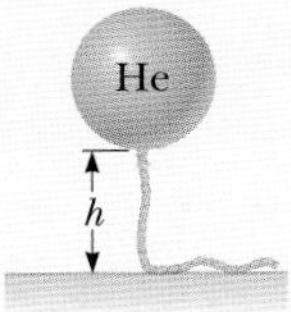

Figure P13.34

35. The true weight of an object can be measured in a vacuum, where buoyant forces are absent. A measurement in air, however, is disturbed by buoyant forces. An object of volume V is weighed in air on an equal arm balance with the use of counterweights of density ρ. Representing the density of air as ρ_{air} and the balance reading as F'_g show that the true weight F_g is

$$F_g = F'_g + \left(V - \frac{F'_g}{\rho g}\right)\rho_{air} g$$

36. Assume a certain liquid, with density 1230 kg/m^3, exerts no friction force on spherical objects. A ball of mass 2.10 kg and radius 9.00 cm is dropped from rest into a deep tank of this liquid from a height of 3.30 m above the surface. (a) Find the speed at which the ball enters the liquid. (b) Calculate the magnitudes of the two forces that are exerted on the ball as it moves through the liquid. (c) Explain why the ball moves down only a limited distance into the liquid and calculate this distance. (d) With what speed will the ball pop up out of the liquid? (e) How does the time interval Δt_{down}, during which the ball moves from the surface down to its lowest point, compare with the time interval Δt_{up} for the return trip between the same two points? (f) **What If?** Now modify the model to suppose the liquid exerts a small friction force on the ball, opposite in direction to its motion. In this case, how do the time intervals Δt_{down} and Δt_{up} compare? Explain your answer with a conceptual argument rather than a numerical calculation.

37. The United States possesses the ten largest warships in the world, aircraft carriers of the *Nimitz* class. Suppose one of the ships bobs up to float 11.0 cm higher in the ocean water when 50 fighters take off from it in a time interval of 25 min, at a location where the free-fall acceleration is 9.78 m/s^2. The planes have an average laden mass of 29 000 kg. Find the horizontal area enclosed by the waterline of the ship.

38. Decades ago, it was thought that huge herbivorous dinosaurs such as *Apatosaurus* and *Brachiosaurus* habitually walked on the bottom of lakes, extending their long necks up to the surface to breathe. *Brachiosaurus* had its nostrils on the top of its head. In 1977, Knut Schmidt-Nielsen pointed out that breathing would be too much work for such a creature. For a simple model, consider a sample consisting of 10.0 L of air at absolute pressure 2.00 atm, with density 2.40 kg/m^3, located at the surface of a freshwater lake. Find the work required to transport it to a depth of 10.3 m, with its temperature, volume, and pressure remaining constant. This energy investment is greater than the energy that can be obtained by metabolism of food with the oxygen in that quantity of air.

39. With reference to the dam studied in Example 13.3 and shown in **Figure 13.5**, (a) show that the total torque exerted by the water behind the dam about a horizontal axis through O is $\frac{1}{6}\rho g w H^3$. (b) Show that the effective line of action of the total force exerted by the water is at a distance $\frac{1}{3}H$ above O.

40. ±? Consider a rectangular wire frame of width $\ell = (10.0 \pm 0.2)$ cm, with a sliding bar. A film of soap with surface tension $\gamma = (2.5 \times 10^{-2})$ N/m fills the space enclosed by the frame and the bar. Show that the change in surface energy per unit surface area when the bar is moved is equal to the surface tension.

Challenge problems

41. Show that the variation of atmospheric pressure with altitude is given by $P = P_0 e^{-\alpha y}$, where $\alpha = \rho_0 g/\rho P_0$, P_0 is atmospheric pressure at some reference level $y = 0$, and ρ_0 is the atmospheric density at this level. Assume the decrease in atmospheric pressure over an infinitesimal change in altitude (so that the density is approximately uniform over the infinitesimal change) can be expressed from **Equation 13.4** as $dP = -\rho g\, dy$. Also assume the density of air is proportional to the pressure, which, as we will see in Chapter 20, is equivalent to assuming the temperature of the air is the same at all altitudes.

42. An ice cube whose edges measure 20.0 mm is floating in a glass of ice-cold water, and one of the ice cube's faces is parallel to the water's surface. (a) How far below the water's surface is the bottom face of the block? (b) Ice-cold ethyl alcohol is gently poured onto the water surface to form a layer 5.00 mm thick above the water. The alcohol does not mix with the water. When the ice cube again attains hydrostatic equilibrium, what is the distance from the top of the water to the bottom face of the block? (c) Additional cold ethyl alcohol is poured onto the water's surface until the top surface of the alcohol coincides with the top surface of the ice cube (in hydrostatic equilibrium). How thick is the required layer of ethyl alcohol?

case study 3

Measuring time by understanding fluids

Elizabeth Angstmann

The precision measurement of time is so ubiquitous that it is hard to imagine that this has not always been the case. The construction of a highly accurate clock requires detailed knowledge and control of the physics of a system. From a physical perspective a clock is simply a system whose state can be mapped onto a one-dimensional coordinate that we call 'time'. Clocks come in all shapes and sizes, from a state-of-art caesium atomic clock through to the motion of the stars through the sky. As our knowledge of physics has increased, so too has the complexity of the systems that we can use as clocks.

One of the earliest versions of a clock was a *water clock*, also known by the Greek name *clepsydra*. These were commonly used through much of the world. There are many different types of water clocks, but they all operate on the same basic physical mechanism. The passage of time is marked by water flowing from one container to another. The earliest known examples of water clocks were from ancient Egypt and consisted of a container with an outlet at the bottom. To make this simple device into an actual clock, another marked container was added that caught the water, measuring the passage of time.

There were two major issues with the use of water clocks in the ancient world. Firstly, the flow rate out of a water clock changed in proportion to the amount of water in the clock. As a result, the amount of time having passed is not simply proportional to the amount of water in the bottom container. Secondly, the clocks needed to match the time given by the other major clock of the time – the sundial. As the amount of daylight varies throughout the year, so too does the time measured by a sundial. Having an understanding of the physics of the water clock solved both of these issues, even though the ancient Greeks did not fully understand what was going on.

Thinking of an idealised water clock as having an opening in a container at some depth h, we know that the pressure at the opening will vary based on this depth (see **Equation 13.4**):

$$P = P_0 + \rho g h$$

The pressure difference between the outside of the container and the water is what drives the flow of the water out of the container. The easiest way to make this constant is by designing a container where the water height will remain constant. By adding another container to the water clock this can be achieved (see **Figure CS3.1**). The middle container has two outlets, an overflow outlet at the top and the water outlet at the bottom. Water flows at a high rate from the upper container to the middle container, keeping the height in this container fixed. With the height fixed the pressure at the outlet is now also fixed, and hence the flow rate out should be constant.

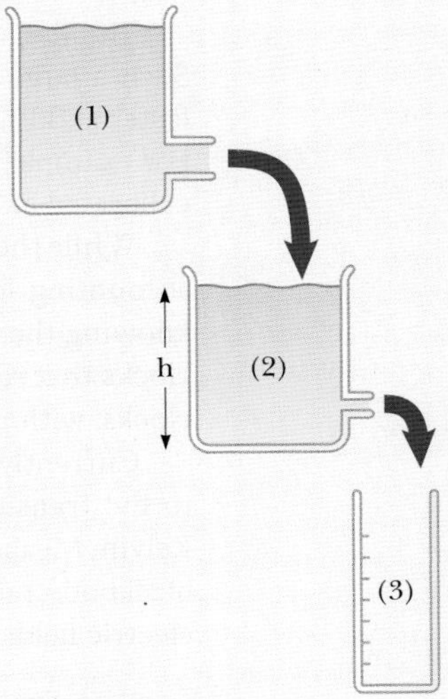

Figure CS3.1
Container (1) is kept filled. This releases water into container (2) at a rate greater than the rate water exits container (2), keeping container (2) constantly filled to height h. Container (3) measures the volume of water exiting container (2). This volume is proportional to the time.

Let us calculate the velocity of water from the container. The continuity equation (**Equation 14.1**) tells us that in a given time interval the same volume of water flows through cross-sectional areas A_1 and A_2 (see **Figure CS3.2**). Using Bernoulli's equation (**Equation 14.2**), we can see that

$$\rho g h + P_1 + \frac{1}{2}\rho v_1^2 = \rho g h + P_2 + \frac{1}{2}\rho v_2^2$$

We know that the pressures are related by

$$P_1 = P_2 = \rho g h$$

and from the continuity equation, there is a relationship between the velocities,

$$v_2 = v_1 \frac{A_1}{A_2}$$

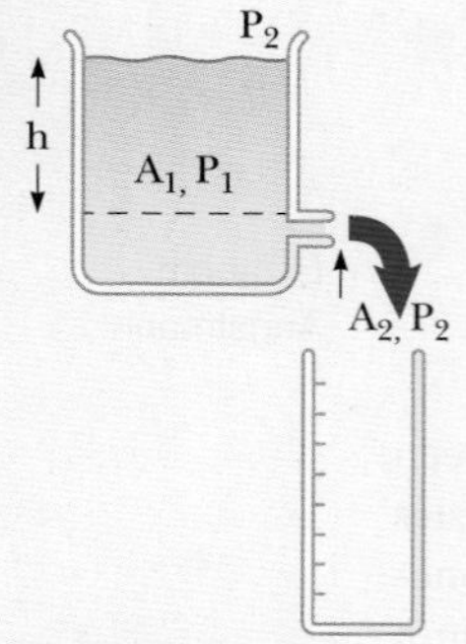

Figure CS3.2 The volume flow rate through surface A_1 is identical to that through A_2.

We can put this all back into Bernoulli's equation, which then gives

$$v_2 = \sqrt{\frac{2gh}{\left(1 - \frac{A_1^2}{A_2^2}\right)}}$$

for the velocity of water leaving the container. For the case when $A_1 > A_2$ – the cross-sectional area of the container is much larger than the cross-sectional area of the hole – we end up with Torricelli's law,

$$v_2 = \sqrt{2gh}$$

In reality there is a loss of energy due to the viscosity of the fluid, as the fluid close to the wall travels more slowly than the fluid at the centre of the hole. An interesting result is that the flow rate actually depends on the *shape* of the hole, as well as its size. We can account for this effect by writing the velocity as

$$v_2 = C\sqrt{2gh}$$

where C is a coefficient depending on the size of the container, and the size and shape of the hole.

In order to solve the problem of matching a sundial the flow rate out of the water clock needed to be changed daily. As the flow rate is given by the velocity multiplied by the cross-sectional area (see **Equation 14.1**), we can see that the easiest way to modify the flow rate is by changing the area of the outlet.

Even at their most accurate, water clocks could only have an accuracy of the order of 15 minutes per day. One of the largest factors in this was the effect of temperature variations on the viscosity. Poiseuille's equation (**Equation 14.8**) illustrates how flow rate depends upon the viscosity:

$$Q = Av = \frac{\pi \Delta P R^4}{8\eta l}$$

The dynamic viscosity of water changes from 1.787×10^{-3} Pa s at 0°C to 0.547×10^{-3} Pa s at 50°C. Such a large variation means that a water clock that was calibrated at 25°C would be running 1.12 times faster at 30°C. This limitation could be partially addressed by changing the fluid that was used in the clock. For example, Galileo used a mercury clepsydra when performing his motion experiments. A mercury clock calibrated at 25°C would only be running 1.02 times faster at 30°C.

While the role of water clocks has been relegated to artistic installations rather than precision timepieces, accounting for the effects of properties such as temperature on the accuracy of clocks is still reliant on knowing the physics of a system. Thermal variations still accounted for a lack of precision in the pendulum clocks that replaced water clocks. Even today temperature effects are a limitation on the accuracy of atomic clocks, with proper control and theoretical adjustments required.

Currently the exact definition of a metric second is proportional to the period of the radiation corresponding to the transition between the two hyperfine levels in the ground state of caesium–133 measured at zero Kelvin. Measurements of this transition are generally conducted at room temperature. At room temperature blackbody radiation subtly changes the frequency of this transition as it creates a temperature-dependent electric field. To accurately measure a second, this shift needs to be taken into account.

Dr Elizabeth Angstmann is a lecturer in the School of Physics, UNSW, Australia. Elizabeth has researched the effect of temperature on the frequency of caesium atomic clocks.

chapter 14

Shutterstock.com/Roman Sigaev

Fluid dynamics

The smoke particles moving with the hot gases above the candle in the picture initially rise in an orderly, neat line. A few centimetres above the hot end of the candle this behaviour suddenly changes and the smoke particles show that the gases are moving in complex, disordered paths. Why?

On completing this chapter, students will understand:

- that fluid flow may be laminar or turbulent
- why the velocity of an ideal fluid in a pipe depends on the pipe's diameter
- how Bernoulli's equation relates to conservation of energy
- that viscosity characterises the internal friction of a fluid
- why viscous fluids need a pressure drop to maintain flow
- why turbulence arises
- some of the factors that contribute to the lift force on an aeroplane wing.

Students will be able to:

- apply continuity principles to calculate volume rates of flow
- relate the velocity of a moving fluid to its pressure
- identify when it is appropriate to use an ideal fluid model
- calculate the viscous forces arising in some simple flow geometries
- identify when it is appropriate to apply Stokes's law and do so
- calculate the pressure drop needed to maintain steady flow through a pipe
- use the Reynolds number to determine whether flow through a pipe will be laminar or turbulent.

In the previous chapter we looked at the behaviour of fluids at rest. However, most of the fluids surrounding us, including the atmosphere and oceans, are moving. In the earlier chapters of this book, we developed analysis models to described the motion of particles under various conditions, but we know from experience that fluids do not behave like single particles. Fluids *flow*; they stick to surfaces as they trickle down or flow across them, they change shape depending on the surface they flow over and they wet absorbent materials. The physics that describes the motion of fluids is *fluid dynamics*.

14.1 Ideal fluid flow

Thus far, our study of fluids has been restricted to fluids at rest. We now turn our attention to fluids in motion. When fluid is in motion, its flow can be characterised as being one of two main types. The flow is said to be **steady**, or **laminar**, if each particle of the fluid follows a smooth path such that the paths of different particles never cross each other as shown in Figure 14.1a. In laminar flow, every fluid particle arriving at a given point has the same velocity.

Above a certain critical speed, characteristic of the fluid and the aperture it flows through or surfaces it flows around, fluid flow becomes **turbulent** (Figure 14.1b). We will examine turbulent flow in Section 14.5.

Because the motion of real fluids is very complex and not fully understood, we make some simplifying assumptions in our approach. To start with, we will develop the simplest model possible – that of **ideal fluid flow**. In this model, we make the following four assumptions.

1. **The flow is steady.** In steady (laminar) flow, all particles passing through a point have the same velocity.
2. **The flow is irrotational.** In irrotational flow, the fluid has no angular momentum about any point. If a small paddle wheel placed anywhere in the fluid does not rotate about the wheel's centre of mass, the flow is irrotational.
3. **The fluid is non-viscous.** In a non-viscous fluid, internal friction is neglected. An object moving through the fluid experiences no viscous force. We will develop a model including viscosity in Section 14.3.
4. **The fluid is incompressible.** The density of an incompressible fluid is constant.

The ideal fluid flow model is an approximation, which gives a useful description of many fluid flows. However, as with all models, we need to recognise that it is an approximation, and use it when appropriate.

Water has a very low compressibility, and a low viscosity, so the ideal fluid model provides a good description of the motion of water flowing steadily. The effects of non-negligible viscosity will be discussed in detail in Section 14.3. Compressibility is described in Chapter 15.

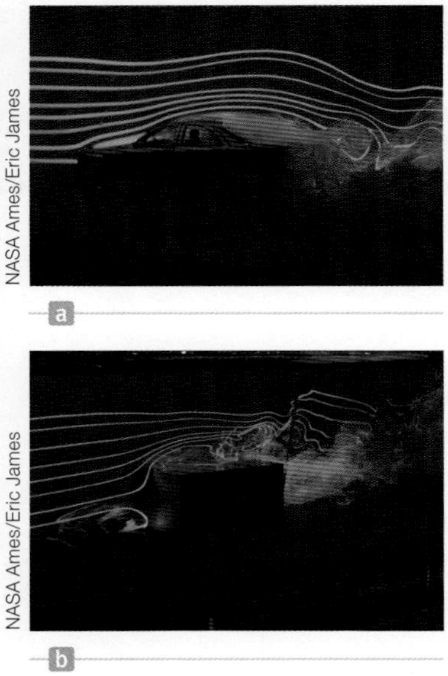

Figure 14.1
Air flow around a car in a test wind tunnel is made visible by streams of airborne particles. (a) Laminar flow around a car at low speeds, with a small amount of turbulence at the back of the car. (b) The flow around the boxy truck-like shape is turbulent.

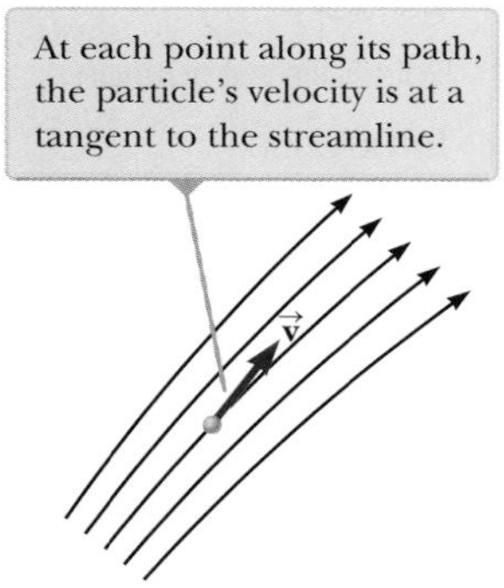

Figure 14.2
A particle in laminar flow follows a streamline.

Streamlines

The path taken by a fluid particle with steady flow is called a **streamline**. The velocity of the particle is always at a tangent to the streamline as shown in Figure 14.2. A set of streamlines like the ones shown in Figure 14.2 form a *tube of flow*. Fluid particles cannot flow into or out of the sides of this tube; if they could, the streamlines would cross one another. The ideal fluid model can help us to understand a variety of observations. For example, as illustrated in Figure 14.3, when a river crosses open countryside, where its bed is fairly wide, it flows fairly slowly. But when it flows through narrow constrictions such as gorges or canyons formed in harder rock, its velocity increases, sometimes to the point of creating the kind of extremely

Figure 14.3
A river runs through (a) a flood plain and (b) a narrow gorge.

turbulent flow referred to as rapids. To understand why the flow gets faster, we consider ideal fluid flow through a pipe of non-uniform size as illustrated in Figure 14.4.

Let's focus our attention on a segment of fluid in the pipe. Figure 14.4a shows the segment at time $t = 0$ consisting of the grey portion between point 1 and point 2 and the short blue portion to the left of point 1. At this time, the fluid in the short blue portion is flowing through a cross section of area A_1 at speed v_1. During the time interval Δt, the small length Δx_1 of fluid in the blue portion moves past point 1. During the same time, fluid moves past point 2 at the other end of the pipe. Figure 14.4b shows the situation at the end of the time interval Δt. The blue portion at the right end represents the fluid that has moved past point 2 through an area A_2 at a speed v_2.

The mass of fluid contained in the blue portion in Figure 14.4a is given by $m_1 = \rho A_1 \Delta x_1 = \rho A_1 v_1 \Delta t$, where ρ is the (unchanging) density of the ideal fluid. Similarly, the fluid in the blue portion in Figure 14.4b has a mass $m_2 = \rho A_2 \Delta x_2 = \rho A_2 v_2 \Delta t$. Because the fluid is incompressible and the flow is steady, however, the mass of fluid that passes point 1 in a time interval Δt must equal the mass that passes point 2 in the same time interval. That is, $m_1 = m_2$ or $\rho A_1 v_1 \Delta t = \rho A_2 v_2 \Delta t$, which means that

$$A_1 v_1 = A_2 v_2 = Q = \text{constant} \tag{14.1}$$

This expression is called the **equation of continuity for fluids**, and we call the constant Q the **volume rate of flow** or **flux**. Q has SI units of m^3/s, and hence dimensions $[L^3T^{-1}]$. The equation of continuity states that the product of the area and the fluid speed at all points along a pipe is constant for an incompressible fluid. Equation 14.1 shows that the speed is high where the tube is constricted (small A) and low where the tube is wide (large A). The volume rate of flow has the dimensions of volume per unit time and is often given in litres per second. The condition Q = constant is equivalent to the statement that the volume of fluid that enters one end of a tube in a given time interval equals the volume leaving the other end of the tube in the same time interval if no leaks are present.

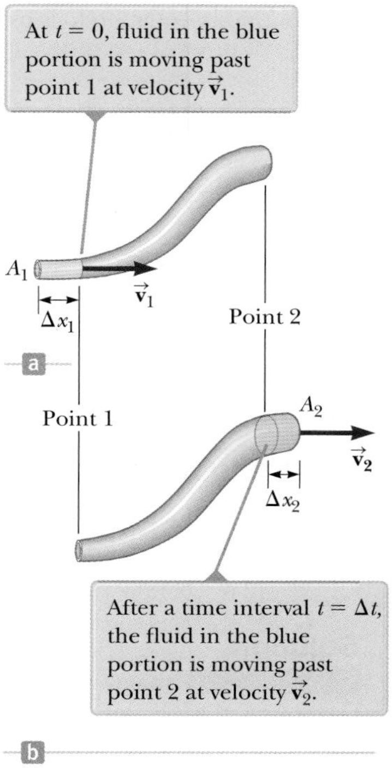

At $t = 0$, fluid in the blue portion is moving past point 1 at velocity $\vec{v}_1$.
A_1
$\vec{v}_1$
Δx_1
Point 2
a
Point 1
A_2
$\vec{v}_2$
Δx_2
After a time interval $t = \Delta t$, the fluid in the blue portion is moving past point 2 at velocity $\vec{v}_2$.
b

Figure 14.4
A fluid moving with steady flow through a pipe of varying cross-sectional area

TRY THIS

Next time you are watering the garden or cleaning your car (when there are no water restrictions, of course), partially block the opening of the hose with your thumb. What happens to the speed of the water spraying from the hose as you make the opening larger and smaller? Try to explain what you observe in terms of continuity.

Example 14.1

Narrowed arteries

Smoking and a high cholesterol diet can both cause arteriosclerosis, a condition in which fatty deposits build up on the inside of arteries and effectively reduce their diameter. If an artery's diameter is reduced from 3 mm to 1 mm by such a build up, what effect does this have on the velocity of the blood?

Solution

Conceptualise Figure 14.4 helps us visualise the situation, but with the flow going from the wider to the narrower region. From the equation for continuity, we expect the blood flow velocity to increase.

Model We model the artery as a pipe, and ignore any stretching of the walls. We will make the approximation that blood behaves as an ideal fluid.

Analyse We can apply the continuity equation, Equation 14.1, to find the ratio of the velocities. First we find expressions for the cross-sectional areas of the two regions in terms of the diameters:

$$A_1 = \frac{\pi d_1^{\,2}}{4} \text{ and } A_2 = \frac{\pi d_2^{\,2}}{4}$$

The continuity equation tells us that the volume rate of flow must remain constant, i.e.:

$$A_1 v_1 = A_2 v_2$$

Rearranging and substituting in the expressions for the areas, we find the ratio of the final velocity to the initial velocity is:

$$\frac{v_2}{v_1} = \frac{A_1}{A_2} = \frac{d_1^{\,2}}{d_2^{\,2}}$$

This equation shows that the velocity changes more rapidly than the diameter. If the diameter of the blood vessel is decreased by a factor of 3, the velocity of the blood will increase by a factor of $3^2 = 9$.

Example 14.2

Watering a garden

A gardener uses a hose to fill a 10 litre watering can. He notes that it takes 20 s to fill. The hose has a nozzle with an opening of cross-sectional area 0.50 cm². The nozzle is held so that water is projected horizontally from a point 1.0 m above the ground and aimed at a garden bed. Over what horizontal distance can the water be projected?

Solution

Conceptualise Imagine any past experience you have with projecting water from a horizontal hose or a pipe. The faster the water is travelling as it leaves the hose, the further from the end of the hose it will land on the ground. We draw a diagram (**Figure 14.5**) to help visualise the problem.

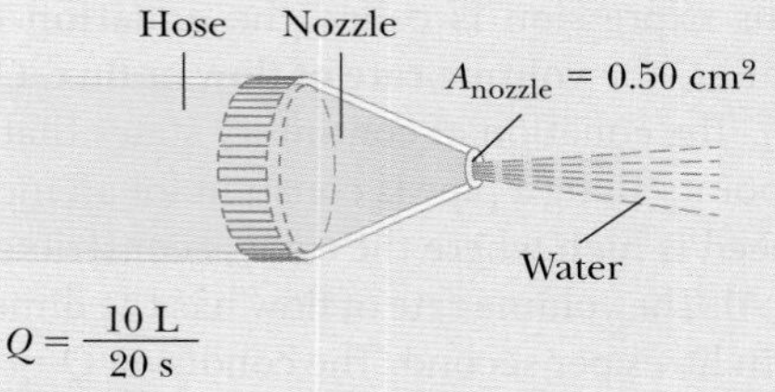

Figure 14.5
Water flows from a hose into a watering can.

Model We model the fluid flow as ideal while it is in the hose. Once the water leaves the hose, it is in free fall. Therefore, we model a given element of the water as a projectile with constant acceleration (due to gravity) in the vertical direction and a particle with constant velocity in the horizontal direction. The horizontal distance over which the element is projected depends on the speed with which it is projected.

Analyse First we find the horizontal velocity of the water leaving the hose. We identify a point at the end of the nozzle. We must find the speed v_{xi} with which the water exits the nozzle. The subscript i indicates that this is the initial value, and the subscript x indicates that the initial velocity vector of the projected water is horizontal.

The continuity equation tells us that the flow rate is a constant, so we can use it to find v_{xi}:

$$Q \equiv v_{xi} A_{\text{nozzle}} = A_{\text{hose}}\, v_{\text{hose}}$$

and therefore

$$v_{xi} = \frac{Q}{A_{\text{nozzle}}}$$

We now model the motion of the fluid using what we know about projectile motion. In the vertical direction, an element of the water starts from rest and falls through a vertical distance of 1.0 m.

Write **Equation 2.15** for the vertical position of an element of water, modelled as a particle with constant acceleration:

$$y_f = y_i + v_{yi} t - \frac{1}{2} g t^2$$

Setting the initial position at the exit point of the hose as the origin and remembering that, initially, the vertical component of the velocity is zero, this simplifies to

$$y_f = -\frac{1}{2} g t^2$$

Rearranging for t gives

$$t = \sqrt{-\frac{2 y_f}{g}}$$

Example 14.2 cont.

Use **Equation 2.7** to find the horizontal position of the element at this time, modelled as a particle with constant velocity:

$$x_f = x_i + v_{xi}t = v_{xi}\sqrt{-\frac{2y_f}{g}} = \frac{Q}{A_{\text{nozzle}}}\sqrt{-\frac{2y_f}{g}}$$

Check the dimensions:

$$[\text{L}] = \frac{[\text{L}^3\ \text{T}^{-1}]}{[\text{L}^2]}\sqrt{\frac{[\text{L}]}{[\text{LT}^{-2}]}} = [\text{L}]$$ ☺

Before we substitute numbers we must convert units. Noting that $1\ \text{l} = 1 \times 10^{-3}\text{m}^3$, and $1\ \text{cm}^2 = 1 \times 10^{-4}\text{m}^2$ we have:

$$Q = 10\ \text{l}/20\ \text{s} = 10 \times 10^{-3}\ \text{m}^3/20\ \text{s} = 5.0 \times 10^{-4}\ \text{m}^3/\text{s}$$

$$A_{\text{nozzle}} = 0.50\ \text{cm}^2 \times 1 \times 10^{-4}\ \text{m}^2/\text{cm}^2 = 5.0 \times 10^{-5}\ \text{m}^2.$$

Finally, substitute in numerical values to obtain

$$x_f = \frac{5.0 \times 10^{-4}\ \text{m}^3/\text{s}}{5.0 \times 10^{-5}\ \text{m}^2}\sqrt{\frac{-2 \times 1.0\ \text{m}}{9.8\ \text{m/s}^2}} = 4.5\ \text{m}$$

Finalise This distance will be enough to water a small garden. We have ignored air resistance – what effect will it have on the horizontal distance travelled by the water?

TRY THIS

Next time you're doing the washing up or are in the bath, grab a funnel and push it down into the water, wide end first. What do you observe? How does what you observe change depending on whether the fluid is plain water or has enough washing-up liquid or bubble bath dissolved in it to create bubbles? Why would washing-up liquid cause this difference?

14.2 The Bernoulli effect

Corbis/Bettmann

Daniel Bernoulli

Swiss physicist (1700–1782)

Bernoulli made important discoveries in fluid dynamics. His most famous work, *Hydrodynamica*, was published in 1738; it is both a theoretical and a practical study of equilibrium, pressure, and speed in fluids. He showed that as the speed of a fluid increases, its pressure decreases. Referred to as 'Bernoulli's principle', Bernoulli's work is used to produce a partial vacuum in chemical laboratories by connecting a vessel to a tube through which water is running rapidly.

You have probably experienced driving on a highway and having a large truck pass you at high speed. In this situation, you may have had the frightening feeling that your car was being pulled in towards the truck as it passed. We will investigate the origin of this effect in this section.

As a fluid moves through a region where its speed or elevation above the Earth's surface changes, the pressure in the fluid varies with these changes. The relationship between fluid speed, pressure and elevation was first derived in 1738 by Swiss physicist Daniel Bernoulli. Consider the flow of a segment of an ideal fluid through a non-uniform pipe in a time interval Δt as illustrated in **Figure 14.6**. This figure is very similar to **Figure 14.4**, which we used to develop the continuity equation. We have added two features: the forces on the outer ends of the blue portions of fluid and the heights of these portions above the reference position $y = 0$.

The force exerted by the fluid to the left of the blue portion in **Figure 14.6a** has magnitude P_1A_1. The work done by this force on the segment in a time interval Δt is $W_1 = F_1\,\Delta x_1 = P_1A_1\,\Delta x_1 = P_1V$, where V is the volume of the blue portion of fluid passing point 1 in **Figure 14.6a**. In a similar manner, the work done by the fluid to the right of the segment in the same time interval Δt is $W_2 = -P_2A_2\,\Delta x_2 = -P_2V$, where V is the volume of the blue portion of fluid passing point 2 in **Figure 14.6b**. (The volumes of the blue portions of fluid in **Figures 14.6a** and **14.6b** are equal because the fluid is incompressible.) This work is negative because the force on the segment of fluid is to the left and the displacement of the point of application of the force is to the right. Therefore, the net work done on the segment by these forces in the time interval Δt is

$$W = (P_1 - P_2)V$$

Part of this work goes into changing the kinetic energy of the segment of fluid, and part goes into changing the gravitational potential energy of the segment–Earth system. Because we are assuming streamline flow,

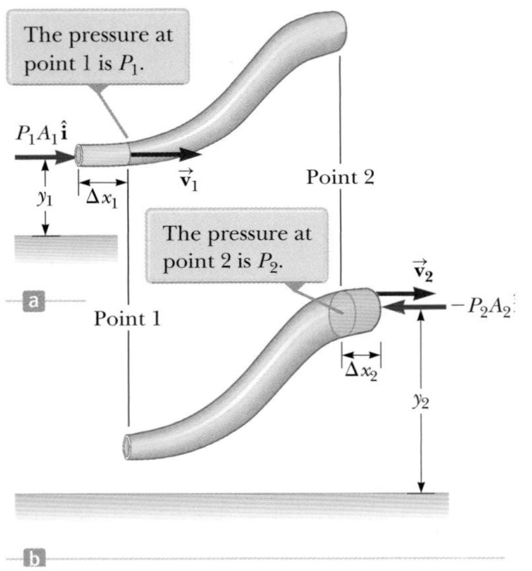

Figure 14.6
A fluid in laminar flow through a pipe. (a) A segment of the fluid at time $t = 0$. A small portion of the blue-coloured fluid is at height y_1 above a reference position. (b) After a time interval Δt, the entire segment has moved to the right. The blue-coloured portion of the fluid is at height y_2.

the kinetic energy K_{grey} of the grey portion of the segment is the same in both parts of **Figure 14.6**. Therefore, the change in the kinetic energy of the blue segment of fluid is

$$\Delta K = \left(\frac{1}{2}mv_2^2 + K_{grey}\right) - \left(\frac{1}{2}mv_1^2 + K_{grey}\right) = \frac{1}{2}mv_2^2 - \frac{1}{2}mv_1^2$$

where m is the mass of the blue portions of fluid in both parts of **Figure 14.6**. (Because the volumes of both portions are the same, they also have the same mass.)

Considering the gravitational potential energy of the segment–Earth system, once again there is no change during the time interval for the gravitational potential energy U_{grey} associated with the grey portion of the fluid. Consequently, the change in gravitational potential energy of the system is

$$\Delta U = (mgy_2 + U_{grey}) - (mgy_1 + U_{grey}) = mgy_2 - mgy_1$$

From **Equation 7.2**, the total work done on the system by the fluid outside the segment is equal to the change in mechanical energy of the system: $W = \Delta K + \Delta U$. Substituting for each of these terms gives

$$(P_1 - P_2)V = \frac{1}{2}mv_2^2 - \frac{1}{2}mv_1^2 + mgy_2 - mgy_1$$

If we divide each term by the portion volume V and recall that $\rho = m/V$, this expression reduces to

$$P_1 - P_2 = \frac{1}{2}\rho v_2^2 - \frac{1}{2}\rho v_1^2 + \rho g y_2 - \rho g y_1$$

Rearranging terms gives

$$P_1 + \frac{1}{2}\rho v_1^2 + \rho g y_1 = P_2 + \frac{1}{2}\rho v_2^2 + \rho g y_2 \qquad (14.2)$$

which is **Bernoulli's equation** as applied to an ideal fluid. This equation is often expressed as

Bernoulli's ▶ equation

$$P + \frac{1}{2}\rho v^2 + \rho g y = \text{constant} \qquad (14.3)$$

Bernoulli's equation is essentially an expression of conservation of energy for fluid flow. It shows that the pressure of a fluid decreases as the speed of the fluid increases. In addition, the pressure decreases as the elevation increases. This latter point explains why water pressure from taps on the upper floors of a tall building is weak unless measures are taken to provide higher pressure for these upper floors.

When the fluid is at rest, $v_1 = v_2 = 0$ and **Equation 14.2** becomes

$$P_1 - P_2 = \rho g(y_2 - y_1) = \rho g h$$

This result is in agreement with **Equation 13.4**.

Note that we have derived **Equation 14.3** assuming that mechanical energy is constant, in other words we have made the approximation that no non-conservative forces such as friction are acting.

Example 14.3

Flow to a bathroom

Water at an absolute pressure of 4.0×10^5 Pa enters a house at ground level through a 2.0 cm diameter pipe. It is carried to the bathroom on the first floor, 5.0 m above ground, through a pipe of diameter 1.0 cm to a toilet. What are the flow velocity and pressure in the bathroom when the flow velocity at the inlet pipe is 4.0 m/s?

Solution

Conceptualise First draw a diagram to help visualise the problem.

Model We model the flow as ideal. We know that the volume rate of flow of a fluid has to remain constant (more water cannot leave the pipe system than enters it and we will assume there are no leaks). We also know that we can apply conservation of energy via Bernoulli's equation to find the difference in pressure.

Example 14.3 cont.

Analyse We need to use both the continuity equation and the Bernoulli equation to solve this problem.

As the volume rate of flow must be constant throughout the pipe system, we know that

$$v_2 = \frac{A_1}{A_2} v_1 = \left(\frac{d_1}{d_2}\right)^2 v_1$$

where the subscripts 1 and 2 refer to the first and second sections of pipe respectively. As the diameter of the first pipe is twice that of the second, the velocity in the second pipe is 4 times larger, i.e. $v_2 = 16.0$ m/s.

We can now obtain the pressure using Bernoulli's equation:

$$P_2 = P_1 - \frac{\rho(v_2^2 - v_1^2)}{2} - \rho g(y_2 - y_1)$$

where ρ is the density of water.

Substitute the numerical values to obtain

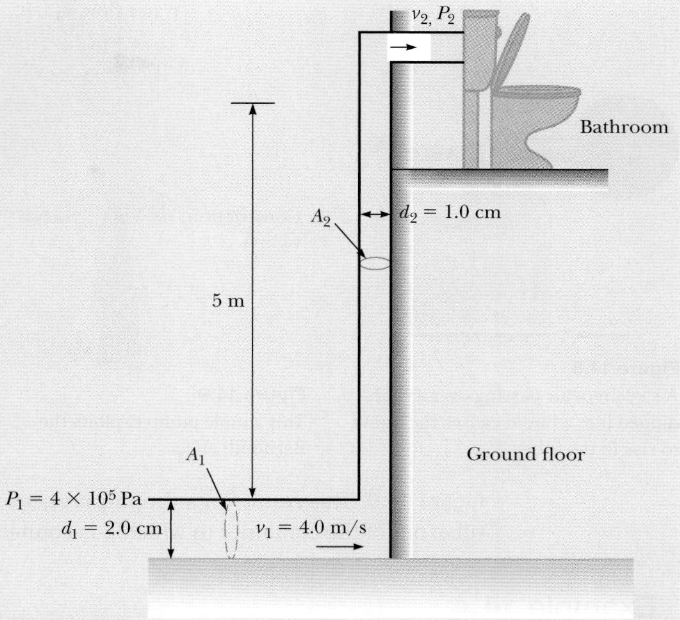

Figure 14.7
Water flows through pipes to a bathroom tap.

$$P_2 = 4.0 \times 10^5 \text{ Pa} - \frac{1}{2}(1.0 \times 10^3 \text{ kg/m}^3)[(16.0 \text{ m/s})^2 - (4.0 \text{ m/s})^2] - (1.0 \times 10^3 \text{ kg/m}^3)(9.8 \text{ m/s}^2)(5.0 \text{ m}) = 2.3 \times 10^5 \text{ Pa}$$

Finalise The velocity of the water at the top is much higher than at the bottom, because the pipe is narrower. The pressure is reduced, as energy is converted into kinetic and gravitational potential energy.

What If? What happens to the pressure when the valve at the cistern is closed?

Answer If the valve is closed, the velocities all go to zero, but there is still a decrease in pressure due to the difference in height. Then

$$P_2 = P_1 - \rho g(y_2 - y_1) = 3.5 \times 10^5 \text{ Pa}$$

That is, the pressure at the valve is higher when it is closed.

Although **Equation 14.3** was derived for an ideal, incompressible fluid, the general behaviour of pressure with speed is true for real fluids including gases: as the speed increases, the pressure decreases. This *Bernoulli effect* explains the experience with the truck on the highway mentioned at the opening of this section. As air passes between you and the truck, it must pass through a relatively narrow channel. According to the continuity equation, the speed of the air is higher. According to the Bernoulli effect, this higher speed air exerts less pressure on your car than the slower moving air on the other side of your car. Therefore, there is a net force pushing you towards the truck.

TRY THIS

Take two sheets of paper and tape their edges to the barrel of a hairdryer so that they are parallel to each other. Turn the hairdryer on. What happens and why? If you don't have a hairdryer handy you can hold the two sheets of paper close to each other and blow between them.

Quick Quiz 14.1

You observe two helium balloons floating next to each other at the ends of strings secured to a table. The facing surfaces of the balloons are separated by 1–2 cm. You blow through the small space between the balloons. What happens to the balloons? **(a)** They move toward each other. **(b)** They move away from each other. **(c)** They are unaffected.

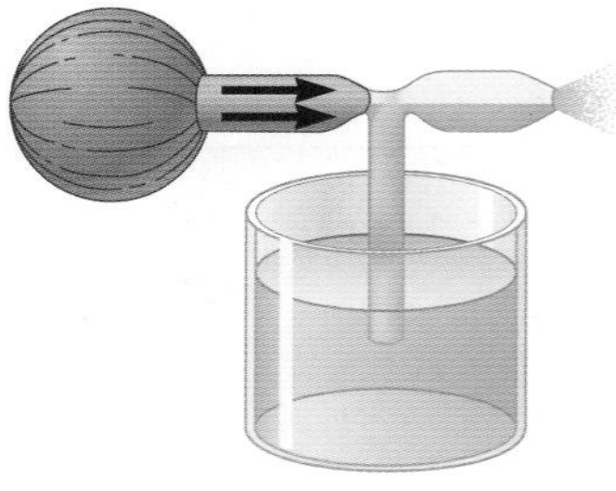

Figure 14.8
A stream of air passing over a tube dipped into a liquid causes the liquid to rise in the tube.

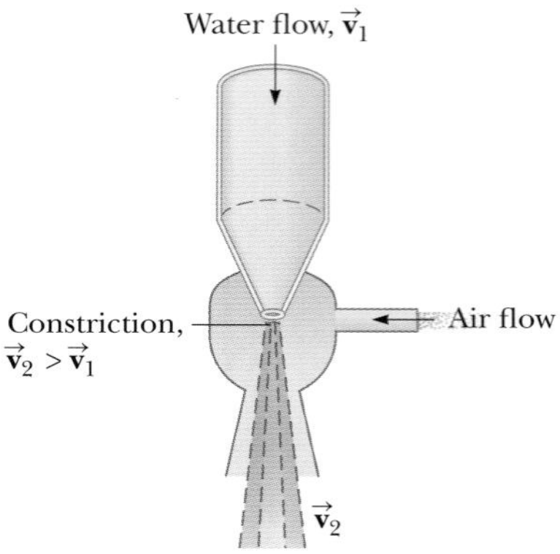

Figure 14.9
This simple pump exploits the Bernoulli effect.

A number of devices operate by means of the pressure differentials that result from differences in a fluid's speed. For example, a stream of air passing over one end of an open tube, the other end of which is immersed in a liquid, reduces the pressure above the tube as illustrated in **Figure 14.8**. This reduction in pressure causes the liquid to rise into the airstream. The liquid is then dispersed into a fine spray of droplets. You might recognise that this *atomiser* is used in perfume bottles and paint sprayers.

The Bernoulli effect is also exploited in the design of a type of very simple pump sometimes called a *filter pump*, which is used to reduce the gas pressure in a container. In these devices, water flows through a tapered tap so that its velocity is increased, producing a jet of water as shown in **Figure 14.9**. This results in a drop in pressure near the flowing water and gas is sucked in through the side tube, out of the container to which it is connected. Both gas and water leave the pump at the bottom.

Example 14.4

The Venturi tube

The horizontal constricted pipe illustrated in **Figure 14.10**, known as a *Venturi tube*, can be used to measure the flow speed of an incompressible fluid. Determine the flow speed at point 2 of **Figure 14.6a** if the pressure difference $P_1 - P_2$ is known.

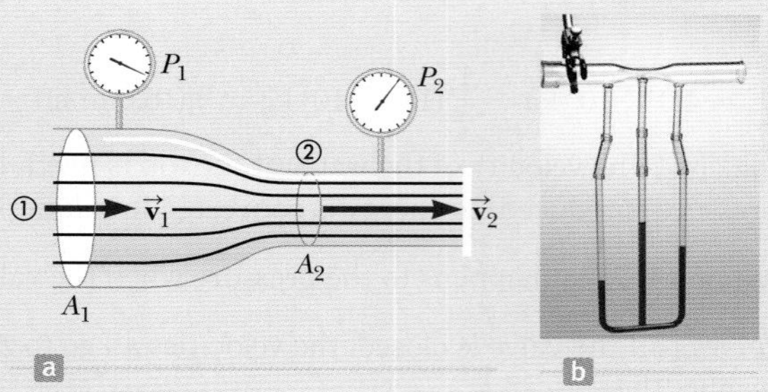

Figure 14.10
(Example 14.4) (a) Pressure P_1 is greater than pressure P_2 because $v_1 < v_2$. This device can be used to measure the speed of fluid flow. (b) A Venturi tube, located at the top of the photograph. The higher level of fluid in the middle column shows that the pressure at the top of the column, which is in the constricted region of the Venturi tube, is lower.

Solution

Conceptualise Bernoulli's equation shows how the pressure of a fluid decreases as its speed increases. Therefore, we should be able to calibrate a device to give us the fluid speed if we can measure pressure.

Model The problem states that the fluid is incompressible, so we can model the flow as ideal and use the equation of continuity for fluids and Bernoulli's equation.

Analyse Apply **Equation 14.2** to points 1 and 2, noting that $y_1 = y_2$ because the pipe is horizontal:

$$P_1 + \frac{1}{2}\rho v_1^2 = P_2 + \frac{1}{2}\rho v_2^2 \qquad \textbf{(1)}$$

Solve the equation of continuity for v_1:

$$v_1 = \frac{A_2}{A_1}v_2$$

Substitute this expression into **Equation (1)**:

$$P_1 + \frac{1}{2}\rho\left(\frac{A_2}{A_1}\right)^2 v_2^2 = P_2 + \frac{1}{2}\rho v_2^2$$

Solve for v_2:

$$v_2 = A_1\sqrt{\frac{2\,(P_1 - P_2)}{\rho\,(A_1^2 - A_2^2)}}$$

Check the dimensions of this expression are consistent.

$$[\mathrm{LT^{-1}}] = [\mathrm{L^2}]\sqrt{\frac{[\mathrm{ML^{-1}\,T^{-2}}]}{[\mathrm{ML^{-3}}]\,[\mathrm{L^2}]^2}} = [\mathrm{LT^{-1}}]$$ ☺

Finalise From the design of the tube (areas A_1 and A_2) and measurements of the pressure difference $P_1 - P_2$, we can calculate the speed of the fluid with this equation.

Example 14.5

Torricelli's law

An enclosed tank containing a liquid of density ρ has a hole in its side at a distance y_1 from the bottom of the tank (**Figure 14.11**). The hole is open to the atmosphere, and its diameter is much smaller than the diameter of the tank. The air above the liquid is maintained at a pressure P. Determine the speed of the liquid as it leaves the hole when the liquid's level is a distance h above the hole.

Solution

Conceptualise Use **Figure 14.11** to help you visualise the problem. When the hole is opened, liquid leaves the hole with a certain speed. If the pressure P at the top of the liquid is increased, the liquid leaves with a higher speed.

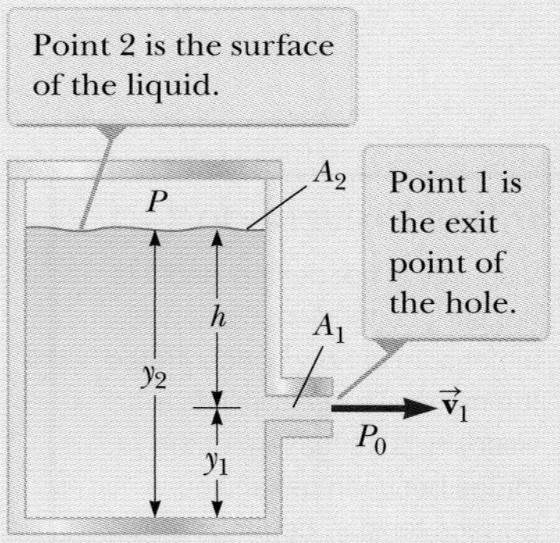

Point 2 is the surface of the liquid.
Point 1 is the exit point of the hole.
P
A_2
A_1
h
y_2
y_1
P_0
$\vec{v}_1$

Figure 14.11
(Example 14.5) A liquid leaves a hole in a tank at speed v_1.

Model We model the flow as ideal. We know the pressure at two points and the velocity at one of those points. We wish to find the velocity at the second point. Therefore, we can apply Bernoulli's equation. Note that A_2 is much greater than A_1, so continuity tells us that the velocity is very small at the top. We will make the approximation that the water is not moving at the top.

Analyse At the top of the tank, the pressure is P. At the hole, P_1 is equal to atmospheric pressure P_0.

Apply Bernoulli's equation between points 1 and 2: $$P_0 + \frac{1}{2}\rho v_1^2 + \rho g y_1 = P + \rho g y_2$$

Solve for v_1, noting that $y_2 - y_1 = h$: $$v_1 = \sqrt{\frac{2(P - P_0)}{\rho} + 2gh}$$

Check dimensions: $$[\mathrm{LT}^{-1}] = \sqrt{\frac{[\mathrm{ML}^{-1}\mathrm{T}^{-2}]}{[\mathrm{ML}^{-3}]} + [\mathrm{LT}^{-2}][\mathrm{L}]} = [\mathrm{LT}^{-1}]$$ ☺

Finalise Consider limiting cases. When P is much greater than P_0 (so that the term $2gh$ can be neglected), the exit speed of the water is mainly a function of P. If the tank is open to the atmosphere, then $P = P_0$ and $v_1 = \sqrt{2gh}$. In other words, for an open tank, the speed of the liquid leaving a hole a distance h below the surface is equal to that acquired by an object falling freely through a vertical distance h. This phenomenon is known as **Torricelli's law**.

What If? What if the position of the hole in **Figure 14.11** could be adjusted vertically? If the tank is open to the atmosphere and sitting on a table, what position of the hole would cause the water to land on the table at the furthest distance from the tank?

Answer Model a volume of water exiting the hole as a projectile. Find the time at which the volume strikes the table from a hole at an arbitrary position y_1:

$$y_f = y_i + v_{yi}t - \frac{1}{2}gt^2$$

$$0 = y_1 + 0 - \frac{1}{2}gt$$

$$t = \sqrt{\frac{2y_1}{g}}$$

Find the horizontal position of the volume at the time it strikes the table: $$x_f = x_i + v_{xi}t = 0 + \sqrt{2g(y_2 - y_1)}\sqrt{\frac{2y_1}{g}}$$
$$= 2\sqrt{(y_2 y_1 - y_1^2)}$$

Use the chain rule. See Appendix B.7.

Σ

Maximise the horizontal position by taking the derivative of x_f with respect to y_1 (because y_1, the height of the hole, is the variable that can be adjusted) and setting it equal to zero: $$\frac{dx_f}{dy_1} = \frac{1}{2}(2)(y_2 y_1 - y_1^2)^{-1/2}(y_2 - 2y_1) = 0$$

Example 14.5 cont.

Solve for y_1: $$y_1 = \frac{1}{2}y_2$$

Therefore, to maximise the horizontal distance, the hole should be halfway between the bottom of the tank and the upper surface of the water. Below this location, the water is projected at a higher speed but falls for a short time interval, reducing the horizontal range. Above this point, the water is in the air for a longer time interval but is projected with a smaller horizontal speed. Note that there is a second solution when $y_1 = y_2$, but this gives the minimum distance rather than the desired maximum.

14.3 Viscosity

Pitfall Prevention 14.1

Viscosity is not density and it is not surface tension. Density is a measure of how concentrated the material's mass is, while viscosity is a measure of the forces between molecules. Surface tension depends on the difference between the strength of the bonds within the liquid and those between the liquid and the solid or gas at the interface.

We have already seen in Chapter 5 that when an object falls through air or moves through water, it encounters a frictional force. The magnitude of that force depends on the viscosity of the fluid through which the object moves. If the frictional force is low, as in water or alcohol, then we say the viscosity of the fluid is low; if it is large, as in glue or glycerine, we say the viscosity is large. The viscosity of different liquids can be compared by observing the time it takes for a ball bearing or similar object to fall a fixed distance through the fluid. The ideal fluid model is only appropriate when the viscosity of the fluid is low.

TRY THIS

Fill some tumblers with different liquids such as water, runny honey and olive oil, making sure they are filled to the same level. Drop a marble into each one and see whether you can come up with a scale relating the viscosity of each fluid.

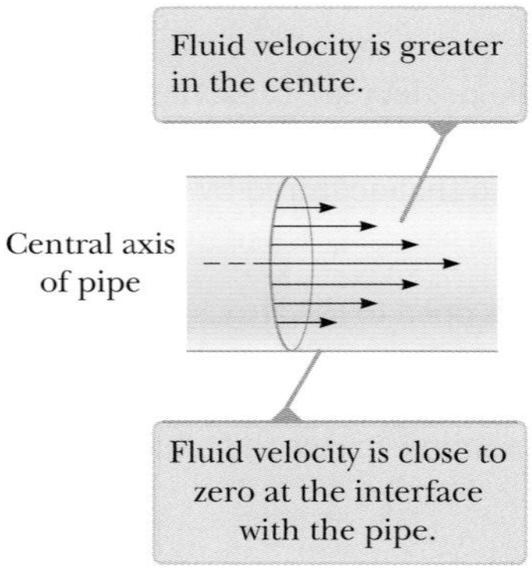

Fluid velocity is greater in the centre.
Central axis of pipe
Fluid velocity is close to zero at the interface with the pipe.

Figure 14.12
Schematic illustration of laminar flow of a viscous fluid through a pipe

The viscosity of a fluid governs how the fluid moves around an object, and is used to characterise the degree of internal friction in the fluid. When a liquid flows slowly and steadily through a pipe, the part of the liquid that is in contact with the pipe is almost stationary, while the central part moves relatively fast. The velocity profile is illustrated in **Figure 14.12**, where the lengths of the arrows indicate the changing magnitude of the fluid velocity at different positions in the pipe. Just as with solids moving against each other, the *viscous force* is an internal friction associated with the resistance that two adjacent layers of fluid have to moving relative to each other. (It is important to remember that these layers are not separated physically and are simply a tool to visualise the internal motion of the fluid.) Viscosity causes part of the fluid's kinetic energy to be converted to internal energy, raising the temperature of the fluid. It is important to realise that viscosity and density are not the same: in a liquid, viscosity is primarily due to the attraction between molecules, not the concentration of mass. In a fluid, the viscosity is largely due to transfer of momentum due to collisions between fluid particles. Surface tension is due to forces at the interface of a liquid and another material. Viscosity is due to forces throughout the bulk of the liquid or gas. So viscosity is not the same as, or directly proportional to, either density or surface tension.

The viscosity of a fluid depends on its temperature, as increased temperature corresponds to increased kinetic energy of the individual molecules. You have probably noticed the viscosity of cooking oil change with temperature; when the oil is cold, it flows slowly and clings to the surface of the pan. When it is hot, the oil flows easily.

TRY THIS

Put some runny honey in an empty bottle and place it in the fridge for a couple of hours. Take it out and observe how slowly the honey moves when you tilt the bottle. Let the honey warm up and observe its motion again – has its viscosity changed? What about its density? Now heat the honey in a microwave oven for a few seconds – have the viscosity or density changed?

A mathematical model for the frictional force in a fluid was first suggested by Newton in 1687. Newton noticed that the frictional force F was proportional to the surface area A of the liquid. He also suggested that it was proportional to the *velocity gradient*. In fact, this proportionality is only correct for a class of liquids now known as *Newtonian fluids*, while other liquids (such as toothpaste, paint and vegemite) behave somewhat differently. In the following, we restrict our discussion to Newtonian fluids.

The velocity gradient describes how fast the velocity changes from the edges to the middle of the pipe or other vessel through which the liquid flows. For the liquid flowing in the pipe illustrated in **Figure 14.12**, the velocity gradient w is given by the difference between the velocities at points A and B divided by the distance between them, i.e.

$$w = \frac{v_A - v_B}{d}$$

Note that the velocity gradient has units of s^{-1}. For a Newtonian fluid:

$$F = \eta A w$$

We call the constant of proportionality η the *coefficient of viscosity* of the fluid, which gives us a definition of viscosity for Newtonian fluids:

$$\eta = \frac{F}{Aw} \qquad (14.4)$$

◀ Coefficient of viscosity

The viscosity coefficient describes the frictional force per unit area and per unit velocity gradient. It has dimensions $ML^{-1}T^{-1}$ and is measured in units of Pa.s, N.s/m² or kg/(m.s) (check this for yourself).

Quick Quiz 14.2

Does the frictional force due to a liquid's viscosity depend on the flow rate?

Example 14.6

How big is the viscous force?

To give a sense of the magnitude of the viscous forces in different liquids, we compare identical flow conditions in water and olive oil, which have viscosities of 1.00×10^{-3} Pa.s and 0.081 Pa.s respectively at room temperature. The two liquids are flowing through separate but identical pipes. Calculate the frictional force on a 10 cm² area of liquid between two layers 1 mm apart, with the outer layer moving at 3 cm/s and the inner layer moving at 5 cm/s. What does this suggest to you about the ease with which these liquids flow and the amount of energy dissipated due to the viscosity of these fluids?

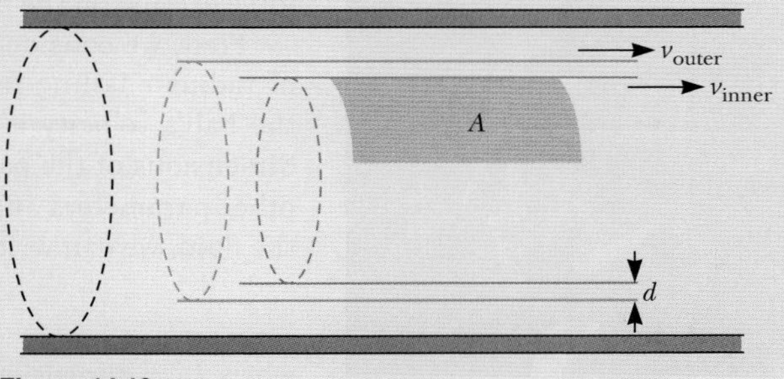

Figure 14.13
(Example 14.6) A viscous fluid flows through a pipe.

Solution

Conceptualise Start by drawing a diagram to help visualise the problem.

Model We assume the flow is laminar, and model the fluids as Newtonian so we can apply Newton's viscosity equation.

Analyse We use **Equation 14.4** to calculate the force on each area of liquid:

$$F = \eta A w = \eta A \frac{(v_{inner} - v_{outer})}{d}$$

Substituting in the numerical values for the velocities and distance between layers (remembering to be careful about the units) gives

$$F = \frac{\eta \times (10 \times 10^{-4}\ m^2) \times (0.02\ m/s)}{0.001\ m} = \eta \times 2 \times 10^{-2}\ Pa.m^2$$

giving

$$F_{water} = 2.00 \times 10^{-5}\ N \quad \text{and} \quad F_{oil} = 1.62 \times 10^{-4}\ N$$

Example 14.6 cont.

Finalise These forces are very small, indicating that these liquids flow fairly easily, with olive oil resisting flow a little more than water. This is consistent with our experiences when cooking. The results also show that only small amounts of mechanical energy are dissipated in the flow of low-viscosity liquids.

What If? What if the fluid was glycerine (a low-toxicity, sweet-tasting substance used as a filler in some low-fat foods and also used as antifreeze), which has a viscosity of 69 Pa.s at room temperature? The magnitude of the viscous force scales linearly with the coefficient of viscosity. As the viscosity of glycerine is 10^5 times larger than the viscosity of water, it experiences frictional forces 10^5 times larger, causing it to flow very slowly.

All real fluids (with the exception of superfluids, such as liquid helium at 2.17 K) have non-zero viscosity and so, although the ideal fluid model can provide a good approximation to some fluid flow, its application is limited. We can improve on this model by taking viscosity and the associated resistance to flow into account in our description of moving fluids. In Section 5.2, we described how an object falling through a fluid with non-zero viscosity reaches a terminal velocity and gave two models for the motion of such an object. In Model 1, we examined what would happen if the resistive force were proportional to the object's velocity; and in Model 2, we examined what would happen if the resistive force were proportional to the square of the object's velocity. Now that we have started to quantify a fluid's viscosity, we can understand Model 1 in more detail.

As we know that viscous forces in laminar flow depend on the fluid's velocity gradient, it seems reasonable to expect that when an object moves through a fluid, it will experience a resistive force that depends in some way on its velocity, as the fluid can be thought of as flowing past the object. Because we don't yet know how the force depends on the velocity, we can only write down the general expression,

$$F = k_0 + k_1 v + k_2 v^2 + \cdots \tag{14.5}$$

where the ks are constants that we need to find. We can identify the types of forces involved in the case of a falling object – gravity acting downwards on the object, the buoyant force acting upwards and the frictional force due to the fluid's viscosity opposing the object's motion relative to the fluid – and so we expect that the terminal velocity is reached when the viscous force combined with the upwards buoyant force exerted on the ball bearing by the liquid is equal to the weight acting downwards.

First, we consider the resistive force due to viscosity alone. Imagine a ball bearing of radius r falling from rest into a viscous liquid with viscosity η. At some instant the ball's velocity is v. The frictional force F can be partly found by examining the dimensions of the parameters involved in the problem. If we assume that there are no other parameters affecting the motion and that the viscosity is a constant property of the fluid, we can write

$$F = kr^a \eta^b v^c$$

where k is a dimensionless constant and the indices a, b and c remain to be found. Looking at the units of each of the parameters, we have

$$\begin{aligned} \text{kg m s}^{-2} &= \text{m}^a(\text{kg m}^{-1}\text{ s}^{-1})^b(\text{m s}^{-1})^c \\ &= \text{kg}^b\ \text{m}^{a(c-b)}\ \text{s}^{-(b+c)} \end{aligned}$$

We can make the mass on the right-hand side match up with the mass to the power one on the left-hand side if $b = 1$. Then to make the time on the right-hand side match up with the time to the power negative two on the left-hand side, we need $(b + c) = 2$ and therefore $c = 1$. Finally, to make the length on the right-hand side match up to the length to the power one on the right-hand side, we need $a(c - b) = 1$ and so $a = 1$. This leaves us with an expression

$$F = kr\eta v$$

Corbis/Bettmann

Baronet George Stokes
Irish physicist and mathematician (1819–1903)
Stokes worked at Cambridge University along with James Clerk Maxwell and William Thomson (Lord Kelvin). Together, these three scientists made extensive contributions to physics, mathematics and engineering. Stokes held the position of Lucasian professor of mathematics at Cambridge from 1847 to his death in 1903, was president of the UK's Royal Society and served as a member of parliament representing his University for five years. He is most famous for his work on light, and the motion of fluids.

which tells us that the resistive force experienced by a sphere moving through a fluid is directly proportional to its radius and velocity and also to the viscosity of the fluid. The value of the constant k can only be calculated using vector calculus, a branch of mathematics you may not yet be familiar with. It was shown to be 6π by George Stokes (1819–1903), leading to

$$F = 6\pi r\eta v \qquad (14.6)$$

◀ Stokes's law

This more complete version of **Equation 5.3**, which formed the basis for Model 1 in Section 5.2, is known as Stokes's law. It can be applied in the regime of *creeping flow*, when the velocity of the object is low and the flow of the fluid past the object is laminar, as illustrated in **Figure 14.14**.

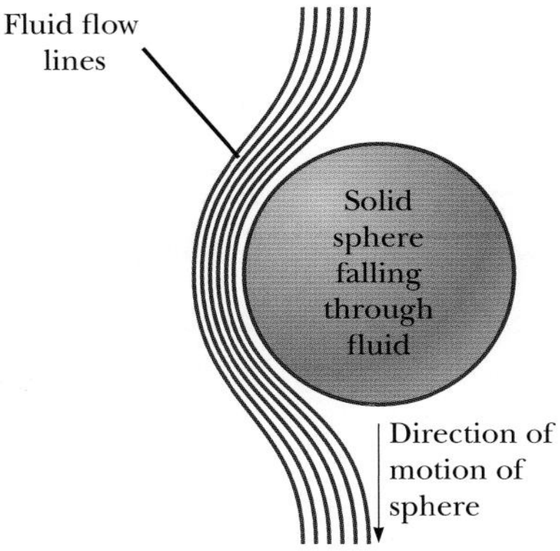

Figure 14.14
Creeping flow around a falling sphere

±? Example 14.7

Measuring viscosity

Stokes's law can be used to measure the coefficient of viscosity of very viscous liquids. Imagine you are doing a laboratory exam and have been given a tall glass cylinder of glycerine (density 1260 kg/m^3) and some small steel balls, and been asked to find the viscosity of the glycerine. You drop a small steel ball (density 7810 kg/m^3, radius (0.50 ± 0.01) cm) gently into the glycerine; it reaches its terminal velocity and falls through a distance of (10.0 ± 0.1) cm in (19.3 ± 0.1) s, which you time with a stopwatch. Find the coefficient of viscosity of the glycerine.

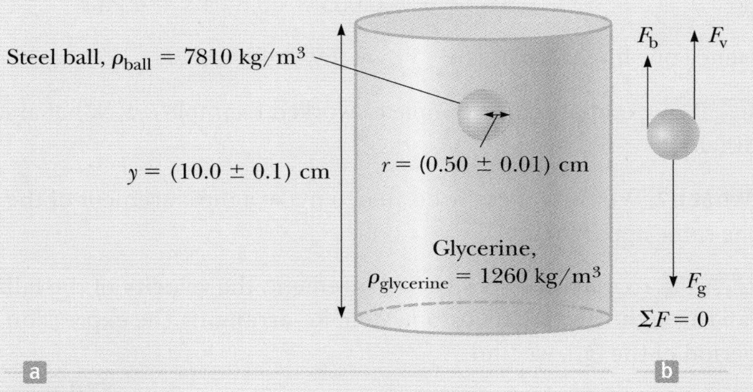

Figure 14.15
(Example 14.7) (a) A ball falls through glycerine. (b) Forces acting on the ball

Solution

Conceptualise Start by drawing a diagram to help visualise the problem. Identify the forces involved: we have to consider the viscous force, the upwards buoyant force and the weight of the ball bearing. At the terminal velocity, these forces must balance so that the net force is zero. Draw a free-body diagram showing the forces acting.

Model The ball has constant velocity so model it as a particle in equilibrium. This problem requires us to bring together our understanding of the buoyant force developed in the previous chapter with the discussion of viscosity above. We will assume that the flow around the ball bearing is laminar, so that Stokes's law can be applied.

Analyse First we find expressions for each of the forces.

The buoyant force is given by the mass of the glycerine displaced by the ball multiplied by g, i.e.

$$F_{buoyant} = m_{displaced}\, g = V_{ball}\rho_{glycerine}\, g = \frac{4}{3}\pi r^3_{ball}\rho_{glycerine}\, g$$

The weight of the ball is

$$F_{gravity} = m_{ball}\, g = V_{ball}\rho_{ball}\, g = \frac{4}{3}\pi r^3_{ball}\,\rho_{ball}\, g$$

and the viscous force is

$$F_{viscous} = 6\pi r_{ball}\eta_{glycerine}v_{ball} = 6\pi r_{ball}\eta_{glycerine}\left(\frac{y}{t}\right)$$

where y is the distance fallen in time t.

At the terminal velocity,

$$F_{viscous} = -\left(F_{gravity} - F_{buoyant}\right)$$

i.e.

$$6\pi r_{ball}\eta_{glycerine}\left(\frac{y}{t}\right) = \frac{4}{3}\pi r^3_{ball}\eta_{glycerine}\left(\rho_{ball} - \rho_{glycerine}\right)g$$

which rearranges to give $$\eta_{\text{glycerine}} = \frac{2r_{\text{ball}}^2(\rho_{\text{ball}} - \rho_{\text{glycerine}})gt}{9y}$$

Check the dimensions: $$[\text{ML}^{-1}\text{T}^{-1}] = \frac{[\text{L}^2][\text{ML}^{-3}][\text{LT}^{-2}][\text{T}]}{[\text{L}]} = [\text{ML}^{-1}\text{T}^{-1}]$$ ☺

Substitute values to obtain

$$\eta_{\text{glycerine}} = \frac{2 \times (0.005^2\ \text{m}^2) \times (7810 - 1260\ \text{kg/m}^3) \times (9.8\ \text{m/s}^2) \times (19.3\ \text{s})}{9 \times 0.10\ \text{m}} = 68.8\ \text{Pa.s}$$

Now we need to calculate the uncertainty in the viscosity. The uncertainty depends on the uncertainty in the three experimentally measured parameters: r_{ball}, y and t. Note that r_{ball} is squared in the expression for η, so its fractional uncertainty appears twice in the expression for the uncertainty:

$$\frac{\Delta\eta_{glycerine}}{\eta_{glycerine}} = 2\frac{\Delta r_{ball}}{r_{ball}} + \frac{\Delta y}{y} + \frac{\Delta t}{t} = 2\frac{0.01\ \text{cm}}{0.5\ \text{cm}} + \frac{0.1\ \text{cm}}{10.0\ \text{cm}} + \frac{0.1\ \text{s}}{19.3\ \text{s}} = 0.055 \text{ or } 5.5\%$$

So $\Delta\eta_{\text{glycerine}} = 0.055 \times 68.8\ \text{Pa.s} = 4\ \text{Pa.s}$

Hence our final answer is $\eta_{\text{glycerine}} = (69 \pm 4)$ Pa.s.

Finalise Compare this example to worked Example 5.4. What assumptions are made in the two models, and how are they different?

What If? Would you easily be able to make a measurement of the viscosity of olive oil ($\eta \approx 0.08$ Pa.s, $\rho \approx 900$ kg/m^3) with the same apparatus?

Answer To make a viscosity measurement, the velocity of the falling ball must be low enough for accurate timing of a fall through 10 cm with a stopwatch. Re-arranging the expression we derived for viscosity to give us an expression for the period of the fall, we find

$$t = \frac{9y\eta_{\text{oil}}}{2r_{\text{ball}}^2(\rho_{\text{ball}} - \rho_{\text{oil}})g} \approx 0.02\ \text{s}$$

which is probably too fast to time with a stopwatch.

14.4 Poiseuille's equation

Our understanding of viscosity needs to be included in our description of the flow of real fluids, as well as the movement of solid objects through fluids. We know from experience that fluids with a high viscosity move more slowly than those with a low viscosity. The steady flow of liquid through a pipe was first investigated thoroughly by Jean Poiseuille (1797–1869), a French physiologist and physician, who worked on methods of measuring blood pressure. Poiseuille derived an expression relating the rate at which a fluid flows through a pipe to its viscosity. Consider a fluid flowing steadily through a pipe of radius R. We will first examine the behaviour of a cylinder of the fluid within the pipe, having the same axis, but with radius $r < R$ and length ℓ (see **Figure 14.16**).

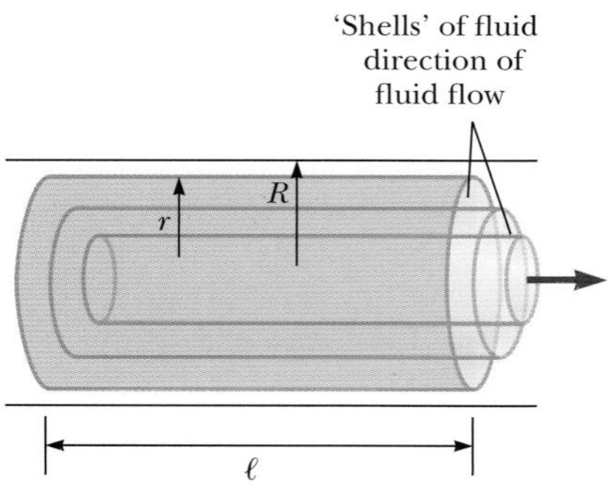

Figure 14.16
Elements of a fluid flowing through a pipe

Since the flow is steady, we know that there is no net force acting on the cylinder of fluid. We want to identify and balance the various forces that are present. First, there is a force due to the excess pressure, ΔP, which is responsible for the direction of flow. The magnitude of this force acting on the cylinder is

$$F_{\text{P}} = \Delta P\pi r^2$$

To calculate the viscous force, we can imagine the cylinder to be made up of concentric shells. Then the total viscous force on the cylinder is the sum of the viscous forces on each of the shells. The viscous force on one shell is

$$F_{\text{v}} = \eta A\frac{dv}{dr_i}$$

where dv/dr_i is the velocity gradient across the shell and A is the surface area of the shell. If a shell produces a force F_{vi} on the next shell, then it must experience a force $-F_{vi}$ from that shell, according to Newton's third law (Chapter 4). Hence as we add up all the forces acting on all the surfaces, the sum is zero apart from the forces acting on the very outermost and innermost surfaces of the flow. Hence the net force is:

$$F_v = \eta A \frac{dv}{dr}$$

where dv/dr is the velocity gradient at the surface of the cylinder.

Thus when we have steady, laminar flow, we know that

$$F_v = -F_P$$

so

$$\frac{dv}{dr} = -\frac{\Delta P \pi r^2}{\eta A}$$

The surface area of the cylinder is $2\pi r\ell$, giving

$$\frac{dv}{dr} = -\frac{\Delta P r}{2\eta\ell}$$

Integrating both sides gives

$$v = -\frac{\Delta P r^2}{4\eta\ell} + c$$

Refer to Appendix B.8 for help with integration.

The Bridgeman Art Library/Archives Charmet

Jean Louis Marie Poiseuille
French physiologist (1797–1869)
Poiseuille was the first scientist to thoroughly investigate laminar flow through pipes. The unit of viscosity is sometimes referred to as the poise, named in his honour. He was primarily interested in the flow of blood, leading him to investigate its flow through artificial capillaries and pipes. Poiseuille developed an improved method of measuring blood pressure using an instrument called a haemodynamometer, a mercury manometer with the connection to the artery filled with potassium carbonate to prevent clotting.

Σ

To find the constant of integration c, we think about what happens at the edge of the pipe. Earlier, we said that in laminar flow, the fluid adjacent to the pipe walls has negligible velocity. We therefore set $v = 0$ at $r = R$:

$$0 = -\frac{\Delta P R^2}{4\eta\ell} + c$$

or

$$c = \frac{\Delta P R^2}{4\eta\ell}$$

Substituting this back into the expression we have derived for velocity, we obtain

$$v = -\frac{\Delta P(R^2 - r^2)}{4\eta\ell} \quad (14.7)$$

Equation 14.7 gives us the velocity of the fluid in a pipe as a function of its distance from the central axis of the pipe.

We are often interested in the volume rate of flow of a fluid through a pipe or other container, rather than the velocity profile of the fluid. To find this, we again consider the fluid in the pipe as being a series of concentric shells. Consider a shell between radii r and $r + \delta r$. The fluid in this shell has a forwards velocity given by **Equation 14.7**. The volume per second flowing through this shell Q_{shell} is the product of the velocity and the cross-sectional area of the shell, that is

$$Q_{shell} = v \times 2\pi r \delta r$$

The total volume per second is obtained by integrating across the pipe,

$$Q = \int_0^R v 2\pi r\, dr$$
$$= \int_0^R -\frac{\Delta P(R^2 - r^2)\pi r}{2\eta\ell} dr$$

Evaluating this integral gives

$$Q = \frac{\pi \Delta P R^4}{8\eta\ell} \quad (14.8)$$

◄ Poiseuille's equation

Example 14.8

Domestic water supply

The water supply to a house consists of a horizontal mains pipe 20 cm in diameter and 5 km long that feeds another horizontal pipe with a diameter of 15 mm, which carries the water 10 m into the house. If someone in this house (and nobody else) draws water, what fraction of the total pressure drop occurs between the ends of the shorter, narrower pipe?

Solution

Conceptualise Start by drawing a diagram to help visualise the problem.

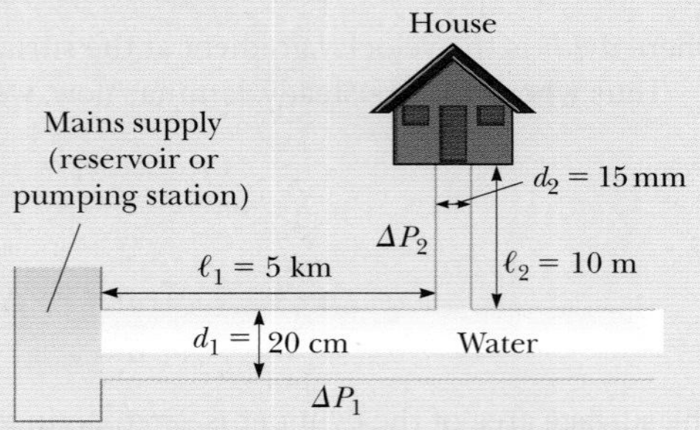

Figure 14.17
(Example 14.8) Mains and domestic pipes

Model If we assume that the volume rate of flow is small, we can treat the flow as laminar and use Poiseuille's equation.

Analyse Equation 14.8 tells us that the volume rate of flow in a pipe is

$$Q = \frac{\pi \Delta P R^4}{8\eta\ell}$$

If nobody else is drawing water from the mains supply, the rates of flow in the two pipes must be equal, so we can write

$$\frac{\pi \Delta P_1 R_1^{\,4}}{8\eta\ell_1} = \frac{\pi \Delta P_2 R_2^{\,4}}{8\eta\ell_2}$$

where the subscripts 1 and 2 refer to the water in the mains and domestic pipes respectively. Cancelling common factors and rearranging, we find

$$\Delta P_2 = \left(\frac{R_1}{R_2}\right)^4 \left(\frac{\ell_2}{\ell_1}\right) \Delta P_1$$

$$= \left(\frac{0.10}{0.0075}\right)^4 \left(\frac{10}{500}\right) \Delta P_1 = 63.2 \Delta P_1$$

Thus the total pressure drop $(\Delta P_1 + \Delta P_2) = 64.2\ \Delta P_1$ and the fraction along the domestic pipe is 0.984; that is, 98.4% of the pressure drop occurs in the final 10 m.

Finalise Is this realistic? If you have ever had a water mains burst near your home, you will know that the pressure in the mains pipe is very high, so it makes sense that most of the pressure drop will occur over the short distance to a house. In addition, if the pressure dropped significantly along the mains pipe, houses drawing water near the start of the pipe would find that water emerged from their taps dangerously fast, while houses at the far end of the pipe would only get a slow trickle when they turned on their taps.

Example 14.9

Angioplasty

As described in Example 14.1, blood vessels can become blocked due to smoking or a high-cholesterol diet. Angioplasty is a medical procedure that increases the blood flow in blocked arteries by dilating the blood vessel. By what factor must an artery's diameter change to increase blood flow by a factor of five? Take the viscosity of blood to be 3.5×10^{-3} Pa.s.

Solution

Conceptualise Figure 14.18 shows the situation. We want to compare the flow rate before and after the angioplasty, and find the required ratio of final artery radius to initial artery radius to increase flow rate by a factor of 5.

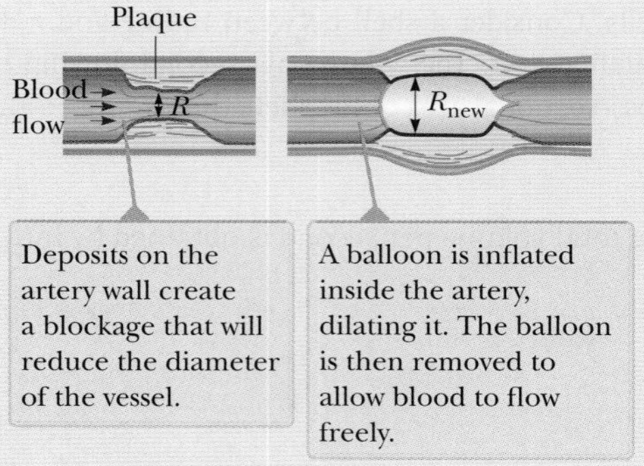

Figure 14.18
Angioplasty increases the volume of a partially blocked blood vessel.

Example 14.9 cont.

Model The pressure difference is supplied by the heart. We will assume that this pressure is the same before and after the angioplasty. We model the flow as laminar flow through a pipe, so we can use Poiseuille's equation.

Analyse Equation 14.8 tells us how the volume rate of flow depends on the radius:

$$Q = \frac{\pi \Delta P R^4}{8\pi \ell}$$

Because the only thing changing on the right-hand side of this equation is R, we can simplify by writing

$$Q = kR^4, \text{ which leads to } R = \left(\frac{Q}{k}\right)^{1/4}$$

where k is constant. Thus to obtain a flow rate $Q_{new} = 5Q$, we need

$$R_{new} = \left(\frac{Q_{new}}{k}\right)^{1/4} = \left(\frac{5Q}{k}\right)^{1/4} = 5^{1/4} R = 1.495R$$

Finalise To increase the flow rate by a factor of five, the blood vessel diameter needs to be increased by just under 50%.

14.5 Turbulence

Poiseuille's equation holds as long as the velocity vector of each layer of fluid is parallel to the axis of the vessel through which it flows and the flow remains steady. If the pressure difference along the pipe is increased, at some point a critical velocity is reached and the motion within the fluid changes from orderly to turbulent. Turbulent flow is irregular flow characterised by apparently random or chaotic motion disturbing the forwards motion of the fluid and resulting in eddies of all sizes. Common examples of turbulent flow include movement in the Earth's atmosphere and oceans, fluid flow around a turbine, fast-moving water through rapids and the movement of smoke, such as that from the volcano shown in Figure 14.19.

In turbulent flow, the average motion is in one direction, but within the flow there are irregularities and random movements. If you were to track the movement of just one smoke particle emerging from the erupting volcano, you would see it jitter about, moving backwards and forwards and only moving upwards *on the average.* Similarly, if you were to watch the movement of a small volume of water as it flows through rapids – perhaps by watching the motion of a leaf floating on the surface – you would see it swirl around in whirlpool-like patterns, only sometimes moving in the overall direction of the river.

Corbis

Figure 14.19
Smoke and ash billow during Ngauruhoe's 1974 eruption, giving a beautiful illustration of turbulent flow.

TRY THIS

Turn on a tap and watch the water come out. First, open the tap a small amount. The water looks glassy and clear; it is hard to see that it is moving at all. If you take a photo of the water and then take another a few seconds (or even hours) later, it will look exactly the same. Of course you shouldn't leave a tap running for hours to check this though! Now fully open the tap. The water is no longer clear; the flow pattern appears to be changing all the time. Turn off the tap slowly – is the transition from orderly (laminar) flow to turbulent flow gradual or sudden?

We often associate turbulence with rapid flow, but slow-moving fluids can also be turbulent – for example, the smoke above the volcano in Figure 14.19 is quite slow moving. Turbulence was first investigated in detail by Osborne Reynolds (1842–1912), a British physicist who is most famous for his work on fluids and thermodynamics. Reynolds found that the onset of turbulence does not depend on velocity alone, but instead depends on the velocity, viscosity and density of the fluid and the diameter of the pipe (or other vessel) it flows through. Using careful experimental observations of the behaviours of a range of liquids flowing at different velocities through different diameter pipes, he discovered that he could characterise the onset of turbulence using a dimensionless constant now known as the Reynolds number,

◀ Definition of the Reynolds number

$$Re = \frac{\rho v d}{\eta} \quad (14.9)$$

where ρ is the liquid's density, v its velocity, η its viscosity and d the diameter of the pipe it flows through. (Check that this number is dimensionless for yourself.) Reynolds found that as long as the quantity Re is less than about 2300, the flow remains laminar, no matter which parameters are individually varied. Thus a high-velocity fluid can maintain orderly flow if it is passing through a pipe of narrow diameter and even a high-viscosity fluid may experience turbulence if it is flowing through a pipe of large diameter. Above $Re = 2300$, the flow may be partly laminar and partly turbulent; above $Re = 4000$, the flow is completely turbulent.

Pitfall Prevention 14.2

Kinematic viscosity is not velocity. It is unfortunate that the most common letter used to represent kinematic viscosity is the Greek letter nu, which looks very much like the italic *v* for velocity which appears in the definition of the Reynolds number. Make sure you know which is which.

TRY THIS

Fill a cup with water (or tea or coffee) and stir rapidly. Is the movement of the liquid laminar or turbulent? Now replace the water with honey or golden syrup and stir at the same speed (you may find this takes a bit more effort – why?). Is the flow laminar or turbulent?

You may have noticed that the Reynolds number depends on two inherent properties of the material, the density and viscosity. Because of this, we sometimes make use of what is known as the kinematic viscosity, ν (Greek letter nu):

Definition of ► kinematic viscosity

$$\nu = \frac{\eta}{\rho} \tag{14.10}$$

and so we can also write the Reynolds number as

$$Re = \frac{vd}{\nu} \tag{14.11}$$

Kinematic viscosity has units m^2/s.

Example 14.10

Stickier than you think?

It is important to remember that the kinematic viscosity of a liquid is not the same as the normal (sometimes called dynamic) viscosity. The dynamic viscosity and density of air at 20°C are $\eta = 18.21 \times 10^{-6}$ Pa.s and $\rho = 1.205$ kg/m^3. The dynamic viscosity and density of water at 20°C are $\eta = 1.002 \times 10^{-3}$ Pa.s and $\rho = 0.998 \times 10^3$ kg/m^3. Calculate and compare their kinematic viscosities. Comment on your results.

Solution

Conceptualise Think about what you expect to find: which is more viscous, air or water?

Model Calculating the kinematic viscosities is a simple substitution problem.

Analyse We use **Equation 14.10**, the definition of kinematic viscosity: $\nu = \frac{\eta}{\rho}$

$$\nu_{air} = \eta_{air}/\rho_{air} = (18.21 \times 10^{-6}\ \text{Pa.s})/(1.205\ \text{kg/m}^3) = 15.11 \times 10^{-6}\ \text{m}^2/\text{s}$$

$$\nu_{water} = \eta_{water}/\rho_{water} = (1.002 \times 10^{-3}\ \text{Pa.s})/(0.998 \times 10^3\ \text{kg/m}^3) = 1.004 \times 10^{-6}\ \text{m}^2/\text{s}$$

Finalise It may seem counterintuitive that the kinematic viscosity of air is higher than that of water – after all, water feels more viscous than air when you move your fingers through it, for example when you push against the water when swimming (the feeling is particularly obvious if you are swimming breaststroke). Part of the resistance you feel is due to the inertial mass of the water, rather than its viscosity. If you were moving through air compressed to the same density as water, it would feel more viscous than water does (although the pressure required to compress the air this much would mean you would be crushed).

When the flow of a liquid is turbulent, the drag force is higher than when the flow is laminar. In laminar flow, the dissipation of energy due to viscosity is limited to surface areas parallel to the axis of flow. In contrast, in turbulent flow there are many more surfaces of relative motion due to the creation of eddies that move counter to or across the average direction of flow, giving rise to a much greater dissipation of energy due to resistive forces within the fluid. The resulting larger drag forces mean that the volume rate of flow along a pipe with a given pressure difference between its ends is less for a turbulent fluid than for an orderly fluid. The higher energy dissipation can cause unwanted heating, which can, in turn, lower the viscosity of the fluid and make turbulence even more likely. This is an important factor in the design of pipelines in the oil and gas industry and for water supply companies – it costs a lot less energy (and so a lot less money) to pump large volumes of fluid over long distances when there is little or no turbulence. The lower temperatures and decreased mechanical wear associated with laminar flow also lead to lower operation costs.

Example 14.11

Pumping oil

At a particular oil well, heavy (unrefined) crude (dynamic viscosity 0.083 Pa.s, density 935 kg/m^3) is pumped 2 km through a 10 cm diameter pipe to a *battery*, where the oil enters a larger diameter pipe fed by oil from other wells. The oil is then pumped through a further 8 km of pipeline of diameter 30 cm to the island, where it is collected in tankers and taken to a refinery.

(A) What is the maximum velocity of the oil in the narrower pipeline in order to ensure the flow remains laminar?

Figure 14.20
An oil rig at the Barrow Island fields, Western Australia

Solution

Conceptualise Start by drawing a diagram to help clarify the problem.

Model We want the flow to be laminar. We will need to use the Reynolds number for the two pipelines to find a safe non-turbulent velocity, remembering that flow in a pipe will be laminar if $Re < 2300$. We will then need to use the velocity to find the volume rate of flow in each case.

Analyse First, we rearrange the definition of the Reynolds number to give the velocity:

$$v = \frac{Re\eta}{\rho d}$$

Now substituting the values given in the question for the viscosity and density and setting $Re = 2300$ to ensure the flow is laminar,

$$v_1 = \frac{2300 \times 0.083 \text{ Pa.s}}{935 \text{ kg/m}^3 \times 0.1 \text{ m}} = 2.03 \text{ m/s}$$

where the subscript 1 refers to the first stage of the pipeline.

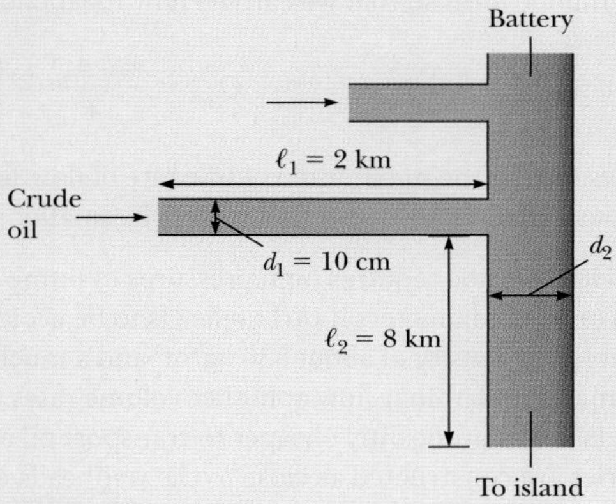

Figure 14.21
(Example 14.11) Oil pipelines

(B) What is the rate of flow in litres per second in this part of the pipeline at the maximum velocity?

Solution

We need to relate the velocity we found in part (A) to the volume rate of flow.

To obtain the volume rate of flow, we use the definition

$$Q = Av$$

Example 14.11 cont.

Assuming the pipeline is cylindrical and has constant diameter, we can write

$$Q_1 = \frac{\pi d_1^{\,2} v_1}{4} = \frac{\pi(0.01\text{ m})^2(2.03\text{ m/s})}{4} = 0.0159\text{ m}^3/\text{s} = 15.9\text{ L/s}$$

(C) What is the pressure drop required along this part of the pipeline?

Solution

The pressure drop is related to the volume rate of flow through Poiseuille's equation.

We rearrange Poiseuille's equation to find the pressure drop along the initial 2 km length:

$$\Delta P = \frac{8\eta\ell Q}{\pi R^4} = \frac{8 \times 0.083\text{ Pa.s} \times 2000\text{ m} \times 0.0159\text{ m}^3/\text{s}}{\pi \times (0.05\text{ m})^4}$$
$$= 1.08 \times 10^6\text{ Pa}$$

This is about 10 times atmospheric pressure.

(D) What is the maximum velocity in the second part of the pipeline needed to maintain laminar flow?

Solution

As the only parameter that has changed in the question is the diameter of the pipe (which has tripled) and as velocity is inversely proportional to the diameter, the maximum velocity for reliably laminar flow in the larger pipe is one-third that in the narrower pipe, i.e. $v_2 = v_1/3 = 0.68$ m/s.

(E) How many 10 cm pipelines, all operating at maximum velocity, can feed into the larger pipeline without causing turbulence?

Solution

Conceptualise As volume rate of flow is a constant, we need to find the volume rate of flow corresponding to the maximum velocity in the larger pipe and compare it to that in the smaller pipes.

Analyse Although we could use the numbers calculated in the previous answers, it is worth re-writing the volume rate of flow in symbolic form so that we can see how it depends on changes in diameter:

$$Q_{\text{max}} = \frac{\pi d^2 v_{\text{max}}}{4} = \left(\frac{\pi d^2}{4}\right)\left(\frac{Re\eta}{\rho d}\right) = \left(\frac{\pi d}{4}\right)\left(\frac{Re\eta}{\rho}\right)$$

This shows us that the maximum volume rate of flow for laminar flow depends *linearly* on the diameter. The larger pipeline has a diameter three times that of the smaller pipeline, so three smaller pipes can safely feed it.

Finalise Heavy crude requires high pressures to pump it along pipelines and it can only be pumped at relatively low velocities or small diameters if turbulence is to be avoided. Once it has been refined into light crude, which has a somewhat lower density of about 830 kg/m^3 and a much lower dynamic viscosity of about 0.007 Pa.s, it becomes much easier to maintain laminar flow at higher volume rates of flow and requires a lower pressure difference for any given rate of flow. It is thus significantly cheaper to transport oil after it has been refined than before. This is one of the reasons why oil refineries are constructed as close to the wellheads as possible.

14.6 Other applications of fluid dynamics

Consideration of the Bernoulli effect, viscosity and turbulence play an important role in the design of many different objects, from golf balls to aeroplanes and from engine design to pipeline construction. We have already seen in Example 14.11 how the properties of the fluid intended to flow through the pipeline affect the diameters of the pipes and the pressures required to drive the flow. Similarly, the lower frictional forces associated with fluid flow compared to the movement of one solid object against another are exploited in the use of oils as lubricants to reduce wear and tear on machine components. The combined effects of Bernoulli's principle, viscosity and turbulence are critical factors in situations where an object's aerodynamics need to be optimised.

In general, an object moving through a fluid experiences an upwards force referred to as **lift** as the result of any effect that causes the fluid to change its direction as it flows past the object. Some factors that influence lift are the shape of the object, its orientation with respect to the fluid flow, any spinning motion it might have, the texture of its surface and the viscosity of the fluid it moves through. For example, a golf ball struck with a club is given a rapid backspin due to the slant of the club. The dimples on the ball increase the friction force between the ball and the air so that air adheres to the ball's surface. Figure 14.22 shows air adhering to the ball and being deflected downwards as a result. Because the ball pushes the air down, the air must push up on the ball. Without the dimples, the friction force is lower and the golf ball does not travel as far. It may seem counterintuitive to increase the range by increasing the friction force, but the lift gained by spinning the ball more than compensates for the loss of range due to the effect of friction on the horizontal translational motion of the ball.

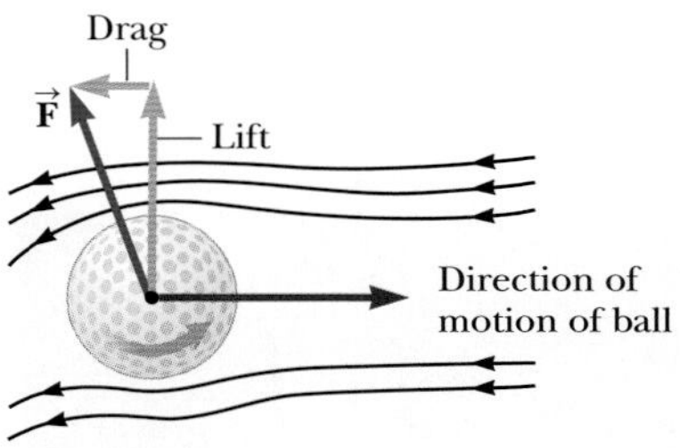
Drag
$\vec{F}$
Lift
Direction of motion of ball

Figure 14.22
Due to the deflection of air, a spinning golf ball experiences a lifting force that allows it to travel much farther than it would if it were not spinning.

Consider the streamlines that flow around an aeroplane wing as shown in Figure 14.23. If the wing is moving to the left with a velocity v_1, we can treat the problem as if the wing were stationary and the airstream approached the wing horizontally from the right with a velocity $\vec{\mathbf{v}}_1$. The tilt of the wing causes the airstream to be deflected downwards with a velocity $\vec{\mathbf{v}}_2$. As the airstream is deflected by the wing, the wing must exert a force on the airstream. According to Newton's third law, the airstream exerts a force $\vec{\mathbf{F}}$ on the wing that is equal in magnitude and opposite in direction. This force has a vertical component (the lift force) and a horizontal component (the drag force). The lift depends on several factors, such as the speed of the plane, the area of the wing, the curvature of the wing and the angle between the wing and the direction of motion (the *angle of attack*). The speed of the plane and the viscosity of the air contribute to whether or not the flow around the wing is laminar. In the laminar regime, the air flows smoothly along the surface of the wing, as illustrated schematically in Figure 14.23. The viscous forces between the wing and the air oppose their relative motion, giving rise to both horizontal and vertical (upwards) components of force; the horizontal force is the main component of the drag force, while the vertical component contributes to the lift. The curvature of the wing surfaces causes the pressure above the wing to be lower than that below the wing due to the Bernoulli effect. This pressure difference also contributes to the lift force. As long as the flow around the wing remains laminar, the Bernoulli effect takes place; however, if the plane's speed decreases too much or the angle of attack increases above about 15°, turbulent flow can set in above the wing to reduce the lift and the plane can stall.

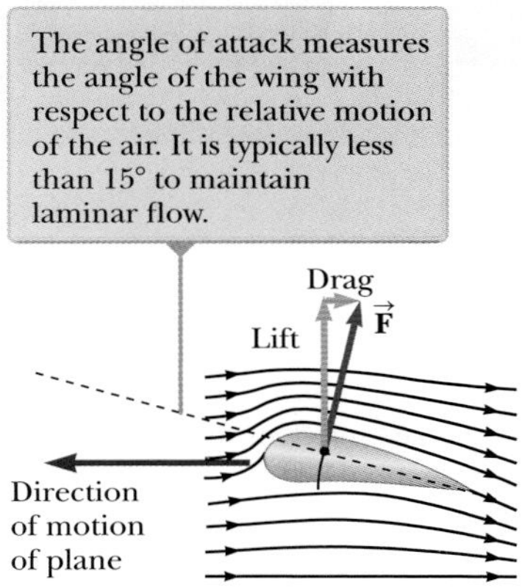
The angle of attack measures the angle of the wing with respect to the relative motion of the air. It is typically less than 15° to maintain laminar flow.
Drag
Lift
$\vec{F}$
Direction of motion of plane

Figure 14.23
Streamline flow around a moving aeroplane wing

While the factors contributing to successful flight include some of the ideas discussed in this chapter, accurate modelling of the behaviour of air around an aerofoil is a complex activity, requiring understanding of the behaviour of fluids at boundary layers. Aerofoil design is important in a wide range of applications, including wind turbines for electricity production and ships' propellers.

End-of-chapter resources

At the start of the chapter, we asked why the smoke particles moving with the hot gases above the candle in the picture initially rise in an orderly, neat line, but at some point start moving in complex, disordered paths. We can now understand this behaviour in terms of the different types of flow discussed in this chapter. Close to the burning candle, the flow is laminar. The smoke particles are carried upwards along streamlines, with velocity vectors parallel to the direction of motion. In this region, the flow can be described using Poiseuille's equation. As the smoke and hot gases move further upwards, they begin to spread out, occupying a wider and wider region and so increasing their Reynolds number. At a critical point, the Reynolds number becomes large enough for turbulent motion to start. This transition is clearly visible in the picture.

The problems found in this chapter may be assigned online in Enhanced Web Assign.

Worked solutions to every fifth problem are available in the Student Solutions Manual. Register online at **www.cengagebrain.com** for access.

Summary

Definitions

Laminar flow in a pipe is flow in which the velocity vectors of all elements of the fluid are parallel to the direction of flow.

Viscosity is a measure of the internal friction of a fluid.

Dynamic viscosity η has units of Pascal seconds.

Kinematic viscosity ν is defined as the ratio of the dynamic viscosity to the density:

$$\nu = \frac{\eta}{\rho} \qquad (14.10)$$

and has units metres squared per second.

The Reynolds number is a dimensionless constant that can be used to determine whether flow in a pipe is laminar or turbulent:

$$Re = \frac{\rho v d}{\eta} \qquad (14.9)$$

Concepts and principles

The volume rate of flow rate through a pipe that varies in cross-sectional area is constant; that is equivalent to stating that the product of the cross-sectional area A and the speed v at any point is a constant. This result is expressed in the **equation of continuity for fluids:**

$$A_1v_1 = A_2v_2 = Q = \text{constant} \qquad (14.1)$$

The sum of the pressure, kinetic energy per unit volume, and gravitational potential energy per unit volume has the same value at all points along a streamline for an ideal fluid. This result is summarised in **Bernoulli's equation**:

$$P + \frac{1}{2}\rho v^2 + \rho g y = \text{constant} \qquad (14.3)$$

As a fluid flows, its viscosity results in viscous forces that oppose the fluid's motion. The viscous force depends on the surface area of relative motion, the viscosity and the velocity gradient:

$$\eta = \frac{F}{Aw} \qquad (14.4)$$

If an object falls through a viscous fluid, at some point the upwards forces due to buoyancy and viscosity balance with the downwards force due to gravity, resulting in a terminal velocity. The viscous force experienced by such an object is given by Stokes's law:

$$F = 6\pi r \eta v \qquad (14.6)$$

Because of the resistance due to viscosity, a pressure drop is required to maintain a constant flow rate. For a fluid flowing through a pipe, the volume rate of flow is related to the pressure drop by Poiseuille's equation:

$$Q = \frac{\pi \Delta P R^4}{8\eta\ell} \qquad (14.8)$$

If the Reynolds number of a particular fluid flowing through a pipe is less than 2300, the flow is laminar. If it is greater than 4000, it is wholly turbulent. In between, it will be in a transition state, partially laminar and partially turbulent.

Chapter review quiz

To help you revise Chapter 14: Fluid dynamics, complete the automatically graded Chapter review quiz at http://login.cengagebrain.com.

Conceptual questions

1. The water supply for a city is often provided from reservoirs built on high ground. Water flows from the reservoir, through pipes and into your home when you turn a tap. Why does water flow more rapidly out of a tap on the ground floor of a building than from a tap on a higher floor?
2. If the airstream from a hair dryer is directed over a table-tennis ball, the ball can be levitated. Explain why this happens, using a diagram to illustrate the forces involved.
3. When ski jumpers are airborne (**Figure CQ14.3**), they bend their bodies forward and keep their hands at their sides. Why?

Figure CQ14.3

4. Why do pilots prefer to take off with the plane facing into the wind?
5. Rabbits ventilate their burrows by building a mound around one entrance, which is open to a stream of air when wind blows from any direction. A second entrance at ground level is open to almost stagnant air. How does this construction create an airflow through the burrow?
6. In **Figure CQ14.6**, an airstream moves from right to left through a tube that is constricted at the middle. Three table-tennis balls are levitated in equilibrium above the vertical columns through which the air escapes. (a) Why is the ball at the right higher than the one in the middle? (b) Why is the ball at the left lower than the ball at the right even though the horizontal tube has the same dimensions at these two points?

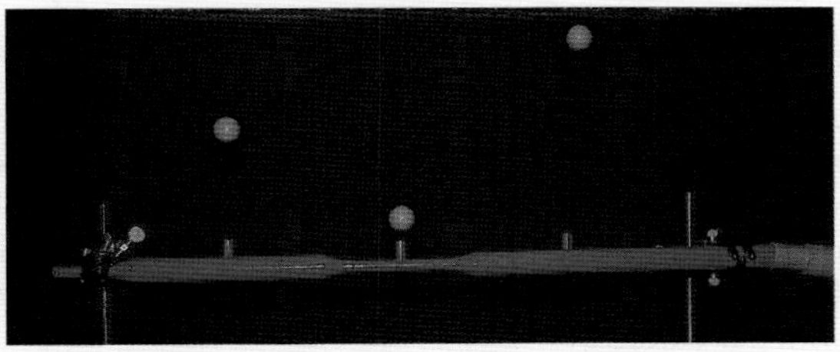

Figure CQ14.6

7. Why is it a good idea to open the windows of your house on the side not facing into the wind during gale force winds?
8. Water has a lower viscosity than oil. Why is it not used as a lubricant in machinery, car engines etc.?
9. Hot water running into a sink splashes less than cold water running at the same rate. Why?
10. Why do shower curtains blow into the shower when the water is flowing?
11. Airport runways are usually built along the prevailing wind direction, and take-off and landing are both usually in the same direction. Why?
12. Does the lift force on an aeroplane wing depend on altitude? Explain your answer.
13. Why do jet planes fly at altitudes of about 10 km, despite the fact it uses a lot of fuel to climb that high?

Problems

Section **14.1** Ideal fluid flow

1. For a healthy person at rest, the heart pumps blood at a rate of about 5 litres per minute. All of the blood eventually passes through capillaries, which have typical diameters of about 5×10^{-6} m. Blood flows through the capillaries at about 1 mm/s. Using this information, estimate the number of capillaries.
2. Water flowing through a hosepipe (diameter 2.0 ± 0.2 cm) fills a 300 litre paddling pool in (6.0 ± 0.5) minutes. What is the speed of the water in the hosepipe? ±?
3. Water flowing through a garden hose of diameter 2.74 cm fills a 25 litre container in 1.50 min. (a) What is the speed of the water leaving the end of the hose? (b) A nozzle is now attached to the end of the hose. If the nozzle diameter is one-third the diameter of the hose, what is the speed of the water leaving the nozzle?

Section **14.2** The Bernoulli effect

4. A large storage tank, open at the top and filled with water, develops a small hole in its side at a point 16.0 m below the water level. The rate of flow from the leak is found to be 2.50×10^{-3} m³/min. Determine (a) the speed at which the water leaves the hole and (b) the diameter of the hole.
5. Water moves through a constricted pipe in steady, ideal flow. At the lower point shown in **Figure P14.5**, the pressure is $P_1 = 1.75 \times 10^4$ Pa and the pipe diameter is 6.00 cm. At another point $y = 0.250$ m higher, the pressure is $P_2 = 1.20 \times 10^4$ Pa and the pipe diameter is 3.00 cm. Find the speed of flow (a) in the lower section and (b) in the upper section. (c) Find the volume flow rate through the pipe.

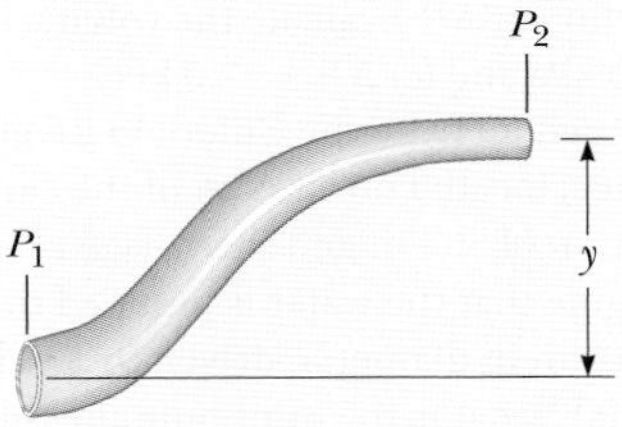

Figure P14.5

6. A pitot tube is a device used in aeroplanes to measure airspeed. It usually consists of a single tube with two holes, one at the front of the tube pointing in the direction of motion of the aircraft and one on the side of the tube. Airflow through the hole at the front is used to measure the total pressure on the plane P_t, while airflow through the side is used to measure the *static* (ambient) pressure P_s. Derive an expression for the airspeed in terms of the total and static pressures and the density of air. Explain why such a device might not work in icy conditions.
7. **Figure P14.7** shows a stream of water in steady flow from a kitchen tap. At the tap, the diameter of the stream is 0.960 cm. The stream fills a 125 cm³

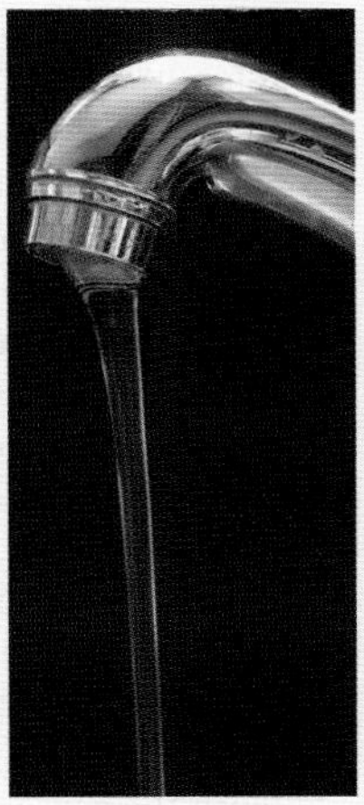

Figure P14.7

container in 16.3 s. Find the diameter of the stream 13.0 cm below the opening of the tap.

8. Water falls over a dam of height h with a mass flow rate of R, in units of kilograms per second. (a) Show that the power available from the water is $P = Rgh$ where g is the free-fall acceleration. (b) Each hydroelectric unit at Lake Jindabyne takes in water at a rate of 8.50×10^5 kg/s from a height of 87.0 m. The power developed by the falling water is converted to electric power with an efficiency of 85.0%. How much electric power does each hydroelectric unit produce?

9. **Review.** Figure P14.9 shows a valve separating a reservoir from a water tank. If this valve is opened, what is the maximum height above point B attained by the water stream coming out of the right side of the tank? Assume $h = 10.0$ m, $L = 2.00$ m, and $\theta = 30.0°$, and assume the cross-sectional area at A is very large compared with that at B.

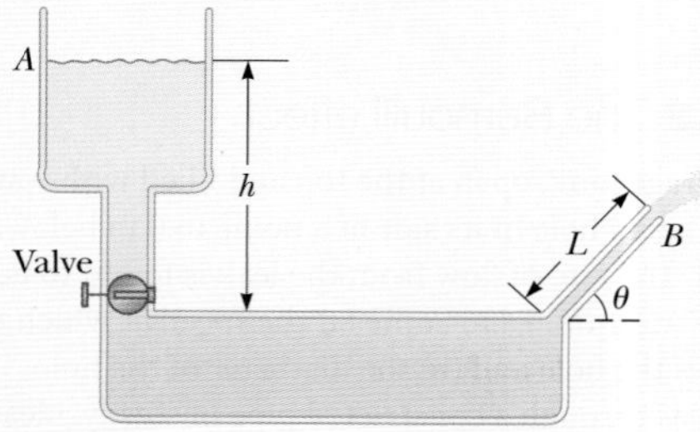

Figure P14.9

10. In ideal flow, a liquid of density 850 kg/m³ moves from a horizontal tube of radius 1.00 cm into a second horizontal tube of radius 0.500 cm at the same elevation as the first tube. The pressure between the liquid in one tube and the liquid in the second tube differs by ΔP. (a) Find the volume flow rate as a function of ΔP. Evaluate the volume flow rate for (b) $\Delta P = 6.00$ kPa and (c) $\Delta P = 12.0$ kPa.

11. Water is pumped up from the Colorado River to supply Grand Canyon Village, located on the rim of the canyon. The river is at an elevation of 564 m and the village is at an elevation of 2096 m. Imagine that the water is pumped through a single long pipe 15.0 cm in diameter, driven by a single pump at the bottom end. (a) What is the minimum pressure at which the water must be pumped if it is to arrive at the village? (b) If 4500 m³ of water are pumped per day, what is the speed of the water in the pipe? *Note*: Assume the free-fall acceleration and the density of air are constant over this range of elevations. The pressures you calculate are too high for an ordinary pipe. The water is actually lifted in stages by several pumps through shorter pipes.

12. When some of the geysers in New Zealand's Whakarewarewa geyser field erupt, the height of the water column reaches 40.0 m. (Figure P14.12). (a) Model the rising stream as a series of separate droplets. Analyse the free-fall motion of one of the droplets to determine the speed at which the water leaves the ground. (b) **What If?** Model the rising stream as an ideal fluid in streamline flow. Use Bernoulli's equation to determine the speed of the water as it leaves ground level. (c) How does the answer from part (a) compare with the answer from part (b)? (d) What is the pressure (above atmospheric) in the heated underground chamber if its depth is 175 m? Assume the chamber is large compared with the geyser's vent.

Bildagentur Rm/Photolibrary

Figure P14.12

13. ±? The Venturi tube discussed in Example 14.4 and shown in Figure P14.13 may be used as a fluid flowmeter. Suppose the device is used at a service station to measure the flow rate of petrol ($\rho = 7.00 \times 10^2$ kg/m³) through a hose having an outlet radius of 1.20 cm. If the difference in pressure is measured to be $P_1 - P_2 = (1.20 \pm 0.05)$ kPa and the radius of the inlet tube to the metre is (2.40 ± 0.01) cm, find (a) the speed of the petrol as it leaves the hose and (b) the fluid flow rate in cubic metres per second.

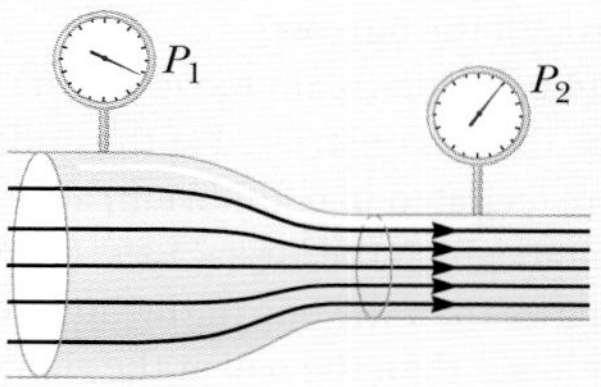

Figure P14.13

Section 14.3 Viscosity

14. A flat plate of area 0.1 m² is placed on a flat surface and is separated from it by a film of oil 10^{-5} m thick with coefficient of viscosity 1.5 Pa.s. Calculate the force needed to slide the plate on the surface at a constant speed of 1 mm/s. (Assume that the flow is laminar and that the oil adjacent to each surface moves with that surface.)

15. A rotating axle (diameter 5 cm) is supported in a metal sleeve (length 50 cm). The axle rotates at a constant speed of 4 revolutions per second. A layer of oil 10^{-4} m thick protects the axle and the sleeve. What is the magnitude of the force driving the axle? Explain any assumptions you make.

16. Two small boys, Marcus and Laurence, are playing with a set of marbles in the kitchen. Marcus decides that he wants to keep the marbles to himself, so he looks for a good place to hide them. His eyes light on the jar of jam his mother has left on the table. Assuming that the jam has a coefficient of viscosity $\eta = 20$ Pa.s and that the flow of the jam past the marble can be treated as creeping, what will be the terminal velocity reached by the marble?

17. In an undergraduate laboratory, Stokes's law is applied to measure the viscosity of a few different liquids as described in Example 14.7. Estimate the minimum viscosity that can be measured with this apparatus, explaining any assumptions that you make.

18. A student performs a viscosity measurement on golden ±? syrup using the equipment described in Example 14.7. The student makes one measurement and says that the ball takes (2.45 ± 0.05) s to fall through the 10 cm drop. What is the viscosity of the syrup? Is 0.05 s a reasonable estimate of the uncertainty for a single measurement of the time? What could be done to decrease this uncertainty?

19. In Section 14.3, we found that Stokes's law could be used to explain the resistive force experienced by an object falling at low speed through a fluid. However, the situation changes at high velocities, where the density of the fluid (and thus the buoyant force) becomes more important than the viscosity. We can try to find how the force depends on the velocity in this regime by checking the dimensions of the parameters involved. Starting from the expression $F = k\,r^a \rho^b v^c$, where k is a dimensionless constant, r and v are the radius and velocity of the falling object and ρ is the density of the fluid, find the values of a, b and c.

Section 14.4 Poiseuille's equation

20. An elephant with a trunk 2.0 m long squirts water ($\eta = 1.005 \times 10^{-3}$ Pa.s) at its calf. Inside the trunk are two nostrils that can be modelled as tubes the length of the trunk and with internal diameter 2.0 cm. If water leaves the trunk at a rate of 5.0 ± 0.1 litres per second, (a) what is the average velocity of the water inside the trunk? (b) What pressure is the elephant applying to squirt the water? (c) How would the pressure change if the elephant were squirting runny honey ($\eta \approx 5$ Pa.s)?

21. A cylindrical fuel injector injects fuel into an engine at a rate of 300 mL per minute. If the injector has a diameter of 1.0 mm and is 5.0 mm long, and the coefficient of viscosity of the fuel is $\eta = 0.65 \times 10^{-3}$ Pa.s, what is the pressure drop across the injector?

22. What pressure must be applied by a vet in order to inject 1 cm^3 of anaesthetic into a cat in 3 s, through a needle of diameter 0.1 cm and length 2.5 cm? Assume the viscosity and density of the anaesthetic are the same as water, and that the pressure inside the blood vessel is approximately 10% greater than atmospheric pressure. (b) The volume of anaesthetic a vet needs to apply to render an animal unconscious depends on the size of the animal – it requires a lot more anaesthetic for a horse than for a guinea pig. This means that vets have a range of syringes, which have different sized needles as well as different volume reservoirs. What would the diameter of the needle need to be for the vet to inject 5 cm^3 of anaesthetic into a horse in the same time, applying the same pressure?

23. For a healthy person at rest, the heart pumps blood at a rate of about 5 litres per minute. All of the blood eventually passes through capillaries, which have typical diameters of about 5×10^{-6} m and lengths of about 1 mm. Blood flows through the capillaries at about 1 mm/s. What is the pressure drop from one end of a capillary to the other?

24. Consider the laminar flow of a viscous fluid through a pipe. Sketch the variation in the volume rate of flow as a function of (a) the pressure drop across the pipe, (b) the radius of the pipe and (c) the length of the pipe.

25. What is the gauge pressure needed in the mains water in order that a stream of water emerging from a hose on a fire engine can reach a roof of height 15 m?

26. A large rainwater tank is filled with water to a depth of 1.5 m. An outlet of area 5 cm^2 in the bottom allows the water to drain out in a continuous stream. What is the volume rate at which water flows out of the tank?

27. A siphon is used to drain water from a tank as illustrated in Figure P14.27. Assume steady flow without friction. (a) If $h = 1.00$ m, find the speed of outflow at the end of the siphon. (b) **What If?** What is the limitation on the height of the top of the siphon above the end of the siphon? *Note*: For the flow of the liquid to be continuous, its pressure must not drop below its vapour pressure. Assume the water is at 20.0°C, at which the vapour pressure is 2.3 kPa.

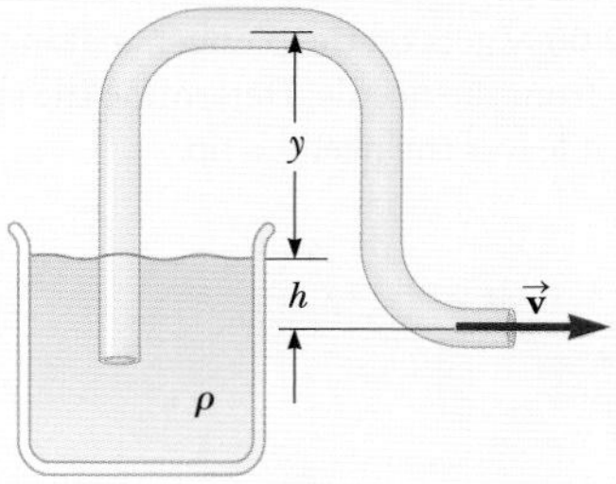

Figure P14.27

Section 14.5 Turbulence

28. Calculate the maximum velocity for laminar flow for water flowing through (a) mains pipes, diameter 20.0 cm and (b) domestic pipes, diameter 2.5 cm. If 10 houses draw water at the same time, is the maximum laminar volume rate of flow limited by the domestic pipes or the mains pipe? What is it?

29. Explain why a boat's propeller is more efficient at low rotational speed than at high speed.

30. Show that the pressure drop required to maintain a flow rate through a pipe of length ℓ and diameter d at the maximum velocity before turbulence occurs is $\Delta P = 32\eta^2 Re\,\ell/\rho d^3$.

Section 14.6 Other applications of fluid dynamics

31. An aeroplane has a mass of 1.60×10^4 kg, and each wing has an area of 40.0 m^2. During level flight, the pressure on the lower wing surface is 7.00×10^4 Pa. (a) Suppose the lift on the aeroplane were due to a pressure difference alone. Determine the pressure on the upper wing surface. (b) More realistically, a significant part of the lift is due to deflection of air downwards by the wing. Does the inclusion of this force mean that the pressure in part (a) is higher or lower? Explain your answer.

32. The Bernoulli effect can have important consequences for the design of buildings. For example, wind can blow around a skyscraper at remarkably high speed, creating low pressure. The higher atmospheric pressure in the still air inside the buildings can cause windows to pop out. (a) Suppose a horizontal wind blows with a speed of 11.2 m/s outside a large pane of plate glass with dimensions 4.00 m × 1.50 m. Assume the density of the air to be constant at 1.20 kg/m^3. The air inside the building is at atmospheric pressure. What is the total force exerted by air on the windowpane? (b) **What If?** If a second skyscraper is built nearby, the airspeed can be especially high where wind passes through the narrow separation between the buildings. Solve part (a) again with a wind speed of 22.4 m/s, twice as high.

Additional problems

33. An aeroplane is cruising at altitude 10 km. The pressure outside the craft is 0.287 atm; within the passenger compartment, the pressure is 1.00 atm and the temperature is 20°C. A small leak occurs in one of the window seals in the passenger compartment. Model the air as an ideal fluid to estimate the speed of the airstream flowing through the leak.

34. A hypodermic syringe contains a medicine with the density of water (Figure P14.34). The barrel of the syringe has a cross-sectional area $A = 2.50 \times 10^{-5}$ m², and the needle has a cross-sectional area $a = 1.00 \times 10^{-8}$ m². In the absence of a force on the plunger, the pressure everywhere is 1.00 atm. A force $\vec{F}$ of magnitude 2.00 N acts on the plunger, making medicine squirt horizontally from the needle. Determine the speed of the medicine as it leaves the needle's tip.

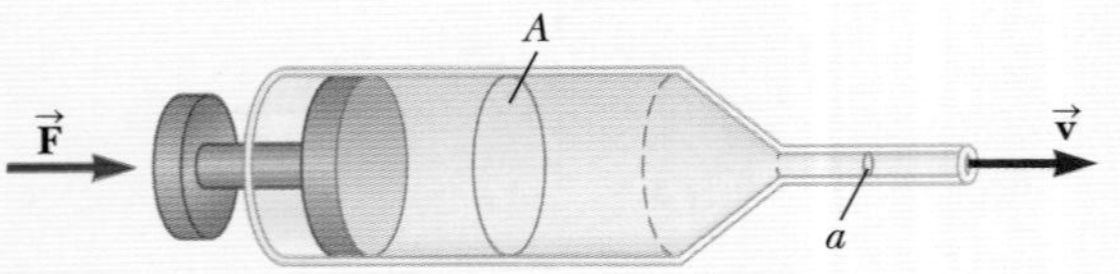

Figure P14.34

35. Assume a certain liquid, with density 1230 kg/m³, exerts no friction force on spherical objects. A ball of mass 2.10 kg and radius 9.00 cm is dropped from rest into a deep tank of this liquid from a height of 3.30 m above the surface. (a) Find the speed at which the ball enters the liquid. (b) Evaluate the magnitudes of the two forces that are exerted on the ball as it moves through the liquid. (c) Explain why the ball moves down only a limited distance into the liquid and calculate this distance. (d) With what speed will the ball pop up out of the liquid? (e) How does the time interval Δt_{down}, during which the ball moves from the surface down to its lowest point, compare with the time interval Δt_{up} for the return trip between the same two points? (f) **What If?** Now modify the model to suppose the liquid exerts a small friction force on the ball, opposite in direction to its motion. In this case, how do the time intervals Δt_{down} and Δt_{up} compare? Explain your answer with a conceptual argument rather than a numerical calculation.

36. Evangelista Torricelli was the first person to realise that we live at the bottom of an ocean of air. He correctly surmised that the pressure of our atmosphere is attributable to the weight of the air. The density of air at 0°C at the Earth's surface is 1.29 kg/m³. The density decreases with increasing altitude (as the atmosphere thins). On the other hand, if we assume the density is constant at 1.29 kg/m³ up to some altitude h and is zero above that altitude, then h would represent the depth of the ocean of air. (a) Use this model to determine the value of h that gives a pressure of 1.00 atm at the surface of the Earth. (b) Would the peak of Mount Everest rise above the surface of such an atmosphere?

37. In a water pistol, a piston drives water through a large tube of area A_1 into a smaller tube of area A_2 as shown in Figure P14.37. The radius of the large tube is 1.00 cm and that of the small tube is 1.00 mm. The smaller tube is 3.00 cm above the larger tube. (a) If the pistol is fired horizontally at a height of 1.50 m, determine the time interval required for the water to travel from the nozzle to the ground. Neglect air resistance and assume atmospheric pressure is 1.00 atm. (b) If the desired range of the stream is 8.00 m, with what speed v_2 must the stream leave the nozzle? (c) At what speed v_1 must the plunger be moved to achieve the desired range? (d) What is the pressure at the nozzle? (e) Find the pressure needed in the larger tube. (f) Calculate the force that must be exerted on the trigger to achieve the desired range. (The force that must be exerted is due to pressure over and above atmospheric pressure.)

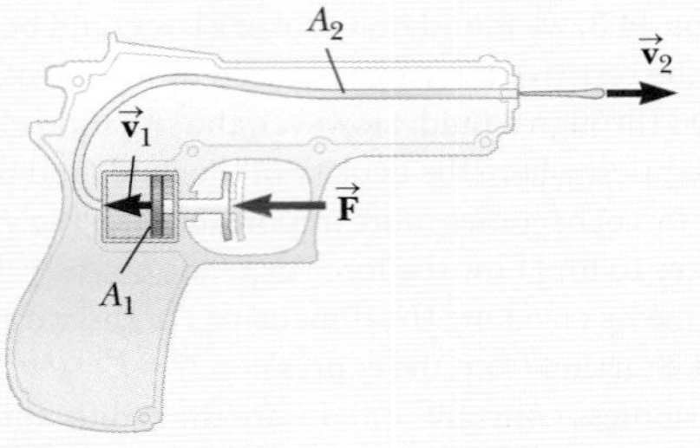

Figure P14.37

38. An incompressible, non-viscous fluid is initially at rest in the vertical portion of the pipe shown in Figure P14.38a, where $L = 2.00$ m. When the valve is opened, the fluid flows into the horizontal section of the pipe. What is the fluid's speed when all the fluid is in the horizontal section as shown in Figure P14.38b? Assume the cross-sectional area of the entire pipe is constant.

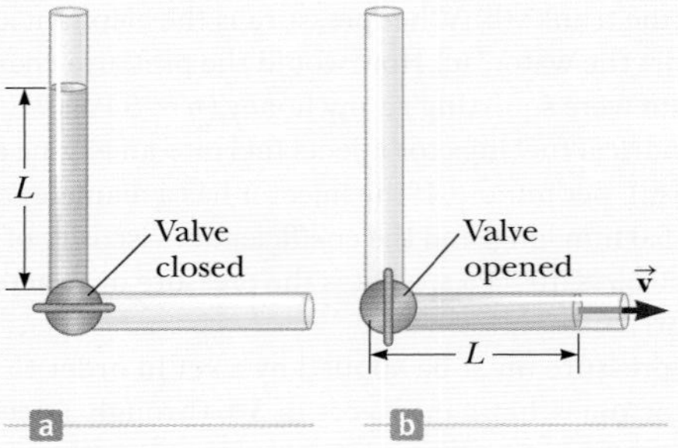

Figure P14.38

39. Imagine you are in the physics lab at university, trying to measure viscosities of different fluids by measuring the terminal velocity of spheres falling through them. Sketch the graph of velocity versus time you expect to observe for such an experiment and explain the features you have included.

40. A jet of water squirts out horizontally from a hole near the bottom of the tank shown in Figure P14.40. If the hole has a diameter of 3.50 mm, what is the height h of the water level in the tank?

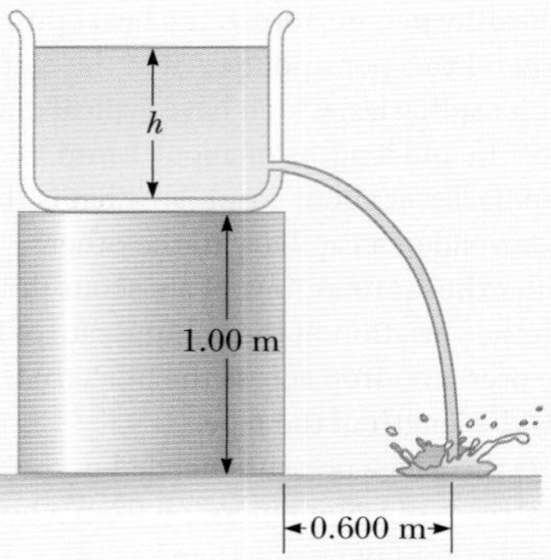

Figure P14.40

Challenge problems

41. A certain physics lecturer likes to have baths rather than showers but she is keen to conserve water. To make herself feel better about the amount of water she uses, she siphons the water from her bath into the garden. Model the bath as a rectangular volume with footprint area A and depth h. The garden is located a distance d below the surface of the water in the bath, where $d >> h$. The cross-sectional area of the siphon tube is A'. Model the water as flowing without friction. Show that the time interval required to empty the tank is given by

$$\Delta t = \frac{Ah}{A'\sqrt{2gd}}$$

How does this change if friction is taken into account?

42. The hull of an experimental boat is to be lifted above the water by a hydrofoil mounted below its keel as shown in Figure P14.42. The hydrofoil has a shape like that of an aeroplane wing. Its area projected onto a horizontal surface is A. When the boat is towed at sufficiently high speed, water of density ρ moves in streamline flow so that its average speed at the top of the hydrofoil is n times larger than its speed v_b below the hydrofoil. (a) Ignoring the buoyant force, show that the upwards lift force exerted by the water on the hydrofoil has a magnitude

$$F \approx \frac{1}{2}(n^2 - 1)\rho v_b^2 A$$

(b) The boat has mass M. Show that the lift-off speed is given by

$$v \approx \sqrt{\frac{2Mg}{(n^2 - 1)A\rho}}$$

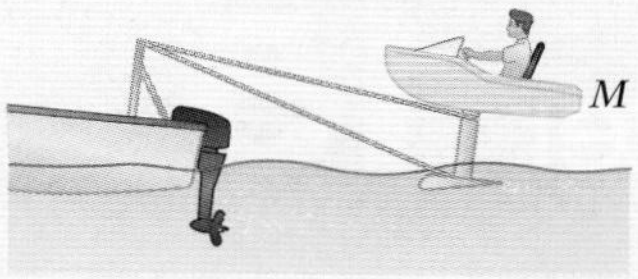

Figure P14.42

43. A U-tube open at both ends is partially filled with water (Figure P14.43a). Oil having a density 750.0 kg/m³ is then poured into the right arm and forms a column $L = (5.00 \pm 0.05)$ cm high (Figure P14.43b). (a) Determine the difference h in the heights of the two liquid surfaces. (b) The right arm is then shielded from any air motion while air is blown across the top of the left arm until the surfaces of the two liquids are at the same height (Figure P14.43c). Determine the speed of the air being blown across the left arm. Take the density of air as constant at 1.20 kg/m³.

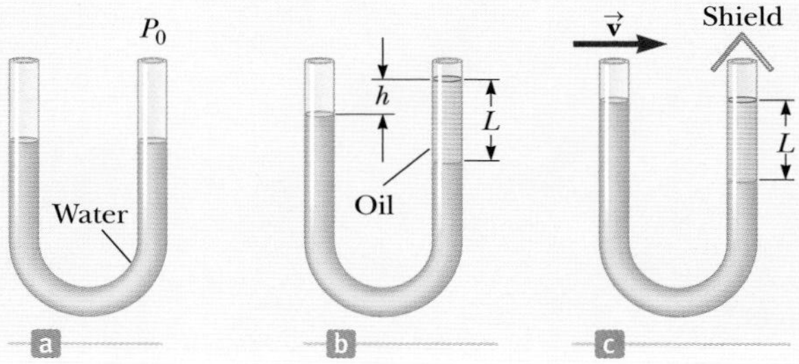

Figure P14.43

chapter 15

Solids

The pre-stressed concrete in the photograph is reinforced with steel rods, which apply a compressive force on the concrete itself. Why does this make concrete stronger?

Crobis/David Sailors

On completing this chapter, students will understand:

- that the properties of a material depend on the microscopic structure and the interactions between molecules or ions inside the material
- how the Lennard-Jones potential models interactions between neutral molecules
- why the interactions between ions and neutral molecules are different
- how Hooke's law relates to the intermolecular potential
- why solid materials break when a stress greater than some critical value is applied
- that ions in crystals vibrate like harmonic oscillators.

Students will be able to:

- calculate the average spacing between molecules in different materials
- use the Lennard-Jones potential to estimate the potential energy of molecules in a solid
- calculate the force exerted on one molecule or ion by another
- distinguish between different types of stress
- identify and use the elastic modulus that is relevant to a particular problem
- calculate the energy stored in a compressed or stretched wire or bar
- analyse bending and twisting in terms of the three simple types of deformation.

In this chapter, we look at some of the observable properties of solids and see how they relate to underlying structure. The properties of solid materials are critical to a wide range of applications, from design and safety in the construction industry to the creation of new high-strength materials such as carbon nanotubes. We look at how solid objects deform under load conditions and how this relates to our microscopic model.

15.1 Characteristics of solids and intermolecular potentials

In the previous two chapters, we examined the properties of fluids, such as density, viscosity and surface tension, and saw how they determine the behaviour of liquids and gases under static and laminar flow conditions. In this chapter, we focus on the properties of solids. (We will not consider the fourth state of matter, plasma, which is common throughout the galaxy but only forms a small proportion of matter on Earth and tends to be ephemeral, occurring in, for example, lightning, flames and the aurorae.)

First let's think about what we mean by *solid*. Some substances exist in three clearly distinct *phases* or states; water is a familiar example, with obvious differences between its gas, liquid and solid phases. A solid phase can be clearly defined for any substance with a sharply defined melting point.

However, there are many materials (such as glass or glue) which do not have sharp transitions from solid to liquid, but which we classify as solids under some conditions. In fact, solid and liquid phases of the same substance generally have the same compressibility and density. Because of this, it is difficult to provide a clear definition of what a solid is. Instead, we must be satisfied with a rough division based on properties such as rigidity and elasticity – that is, based on a substance's response to an applied force. We have already seen that fluids flow in response to an applied force or pressure drop. Normally, we call a substance a solid if it does not easily change its shape under the action of small forces. Even this is not a perfect distinction; as you may know, glaciers flow slowly downhill, metals such as indium and lead are of high density but are soft and malleable and will also change their shape under the effects of gravity, and if indium is subjected to high pressures it flows more like a low-viscosity fluid. In fact, solids may be better defined as things that do not easily change their shape under the action of small forces on short timescales and at normal pressures.

As you might expect, the nature of the transition from liquid to solid – sharp or gradual – depends on the microscopic structure of the material. Those materials that do have a well-defined melting point have **crystalline** structures; those that solidify gradually are **amorphous**. In general, solids are made up of neutral atoms, molecules or electrically charged ions (fragments of molecules, or atoms from which an electron has been removed), which experience attractive forces due to their neighbouring molecules that cause them to cohere. The molecules or ions in a crystalline solid are arranged in regularly-spaced, geometric patterns; they are described as occupying *sites* on a *lattice*. In bulk, this leads to geometric shapes, and such solids are bounded by plane surfaces and have characteristic angles. If a substance with a sharply-defined melting point solidifies slowly, its crystal structures can be quite large, indicating a large-scale ordering of the molecules or ions within the material. More rapid solidification can result in the creation of many small crystals, arranged at different angles with respect to each other, so that the crystalline nature is not obvious. Examples of crystalline materials include obviously crystalline substances such as diamond, quartz and ice, as well as materials such as metals and insulin. Amorphous materials, in contrast, do not adopt geometric, regular shapes, no matter how slowly they solidify. Their constituent molecules show neither long-range nor short-range order – they resemble liquids that have been frozen in time. Examples of amorphous materials include rubber, glass, wax and most common types of glue.

TRY THIS

Have a look around your house (particularly in your kitchen). Which materials are you certain you would classify as solids? Which are you sure are crystalline? Are there any solids you are sure are amorphous? What ways can you think of to test whether a substance is crystalline or amorphous?

Shutterstock.com/N.Ratanavijittakorn

Shutterstock.com/Graeme Knox

Shutterstock.com/Rossillicon Photos

Figure 15.1
Water in its three different phases

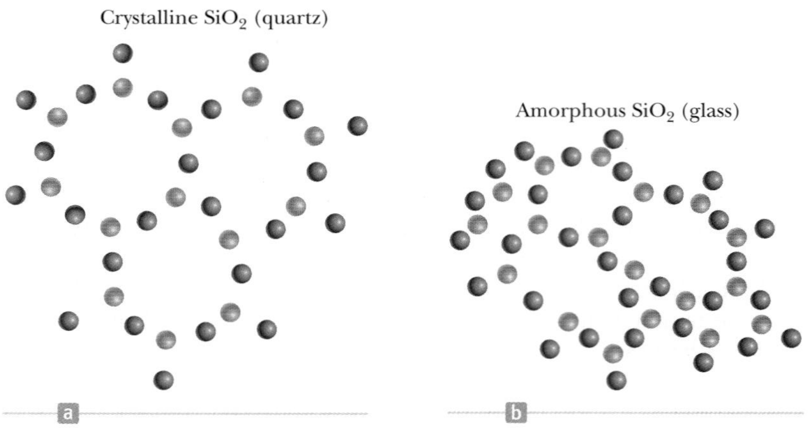

Figure 15.2
Schematic illustration of the microscopic structure of (a) crystalline and (b) amorphous materials. Individual molecules or ions are represented by circles.

Starting from the loose definition of a solid described above, we can use some of the simple observable properties of solids to learn something about the atoms and molecules that make up matter and of the forces between them. First, we can use what we know about the masses and densities of solids, which we can measure at the macroscopic or bulk scale, to find average intermolecular spacings in different materials.

Example 15.1

Estimating the distance between molecules

In which solid material are the molecules or ions closer together, ice or tin?

Solution

Conceptualise Matter is made up of molecules, atoms or ions. We start by drawing a diagram to help visualise the problem. Each of these particles has a fixed mass, so the total mass of a sample of material depends on the number of particles present, while the density of a material indicates how spread out the mass is. If both of these properties are known, we should be able to find the volume occupied by a single molecule or ion.

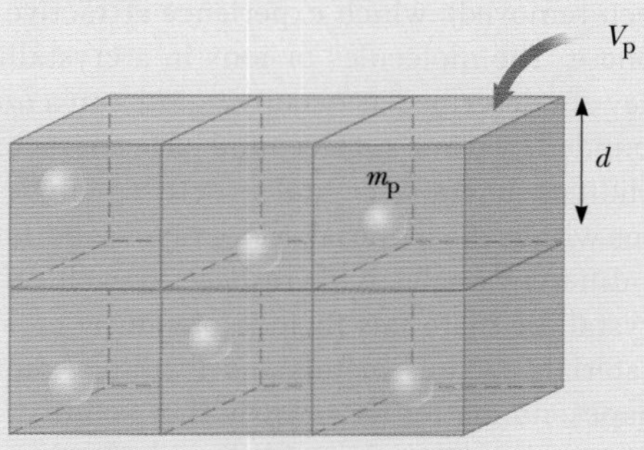

Figure 15.3
A simple representation of particles in matter

Model We model the material in each case as made of particles, either tin atoms or water molecules, which each occupy a cube as shown in Figure 15.3. We can then use the definition of density (Equation 1.1) and tables of data to find the atomic and molecular masses and densities of the two materials.

Analyse The molar mass M of a substance tells us the mass of a sample of the substance containing Avogadro's number N_A of particles and so is related to the mass of a single particle m_p by:

$$M = N_A m_p$$

The density of a substance is the mass of a number of particles n_p divided by the volume of space they occupy,

$$\rho = \frac{m}{V} = \frac{n_p m_p}{n_p V_p}$$

and so

$$\frac{M}{\rho} = \frac{N_A m_p V_p}{m_p} = N_A V_p$$

where V_p is the volume occupied by a single particle. That is:

$$V_p = \frac{M}{\rho N_A}$$

Example 15.1 cont.

If each particle occupies a cube-shaped volume of space, the average distance between the particles is the side length of the cube:

$$d_{\text{cube}} = V_{\text{p}}^{1/3} = \left(\frac{M}{\rho N_{\text{A}}}\right)^{1/3}$$

Check dimensions: $$[\text{L}] = \left(\frac{[\text{M}]}{[\text{ML}^{-3}][\,]}\right)^{1/3} = [\text{L}] ☺$$

We can look up the various constants needed to solve the problem in a handbook of physical constants or on the Internet. If you use the Internet make sure you take your data from a reputable source. The molar mass of water is 18.015 g and the density of ice at 0°C is 0.9167 g/cm³; the molar mass of tin is 118.71 g and the density of tin is 5.77 g/cm³; Avogadro's number is 6.022×10^{23}. Hence:

$$d_{\text{ice}} = \left[\frac{18.015\text{g}}{0.9167\text{g/cm}^3 \times 6.022 \times 10^{23}}\right]^{1/3} = 3.196 \times 10^{-8}\,\text{cm} = 0.3196\text{ nm}$$

$$d_{\text{tin}} = \left[\frac{118.71\text{g}}{5.77\text{g/cm}^3 \times 6.022 \times 10^{23}}\right]^{1/3} = 3.24 \times 10^{-8}\,\text{cm} = 0.324\text{ nm}$$

Our calculations suggest that the average distance between ions in tin is only slightly larger than the average distance between molecules in water. They also show that materials with quite different densities can have very similar interatomic distances.

Finalise You may worry that our assumption that the particles occupy cubic volumes of space is unjustified. To see whether this assumption has a significant effect, we can repeat the calculation assuming a different shape, for example spheres instead of cubes. Then the average separation is equal to the diameter of the sphere. The volume of a sphere is related to its diameter by

$$V = \frac{4}{3}\pi\left(\frac{d}{2}\right)^2 = \frac{\pi d^3}{6}$$

Therefore

$$d_{\text{sphere}} = \left(\frac{6}{\pi}\right)^{1/3} d_{\text{cube}} \approx 1.2 d_{\text{cube}}$$

This doesn't make much difference to the magnitude of the intermolecular or interionic distances; however, it does show that a comparison between different materials, in which the molecules or ions may be packed in different ways, is sensitive to the shape we assume, so we cannot really say whether the ions in tin are closer together than the molecules in water without knowing something about the way they are organised or their packing distribution.

What If? What if we had compared the separation for tin with that of water in its liquid state?

Answer You may know that water at temperatures close to 0°C is actually denser than ice, so the ions in tin are likely to be more separated than the molecules in liquid water. This is a good illustration of the fact that we cannot come up with a simple definition of a solid in terms of how dense it is from the average molecular separation.

The estimate of average molecular separation r_0, derived in Example 15.1,

$$r_0 \approx \left(\frac{M}{\rho N_{\text{A}}}\right)^{1/3} \quad (15.1)$$

can be applied to any material. Such estimates suggest that most solids have roughly similar spacings between molecules or ions, of the order of 0.1–1 nm. Indeed, the densest naturally occurring material, osmium, has spacing $r_0 \sim 0.2$ nm, compared to the least dense synthetic solid, aerogel, which has typical intermolecular spacings of about 2 nm. These intermolecular distances also provide an upper limit to the size of the molecules and ions that constitute the material.

The observable properties of solids can tell us much more than typical molecular spacings. We have already said that solids (like liquids) are difficult to compress. But they also resist expansion. You may have noticed that the air trapped in a plastic bottle of water expands when it is subjected to lower pressure (this is most obvious if the bottle is taken on a plane, but can also be observed between sea level and higher-altitude land), but neither the water nor the plastic of the bottle changes volume. Such resistance to expansion is observed over much larger pressure drops. For example, pressures in the beamlines at the Australian Synchrotron are typically as low as 10^{-13} atmospheres, but the metal used in constructing the beamlines undergoes negligible expansion.

These two observations – that solids (and liquids) resist both compression and expansion – provide a basis from which to develop a model of the forces that molecules and ions exert on each other. Let's consider two *electrically neutral* (uncharged) molecules in a solid. If we try to compress the solid, we are in effect trying to push the molecules closer together. When we say that the solid resists compression, we are saying that we have to do work to bring the molecules closer together, indicating that there is a repulsive force between them. However, resistance to expansion tells us there is also an attractive force between molecules. The net force between the molecules is thus repulsive at relatively short distances and attractive at relatively long distances. The 'natural' distance indicates an equilibrium separation where there is no net force. We therefore arrive at the conclusion that the force between two neutral molecules combines a *short-range* repulsive element with a *long-range* attractive element, which balance at the equilibrium distance. The size of this attractive force at large distances cannot be too great, however, or we would see an attractive force between solid objects placed next to each other that is greater than their gravitational attraction. This allows us to make an additional requirement: that the attractive force *decreases* in magnitude for large separations.

One mathematically simple form of a force with these properties is

$$F = A\left(\frac{r_0}{r}\right)^m - B\left(\frac{r_0}{r}\right)^n$$

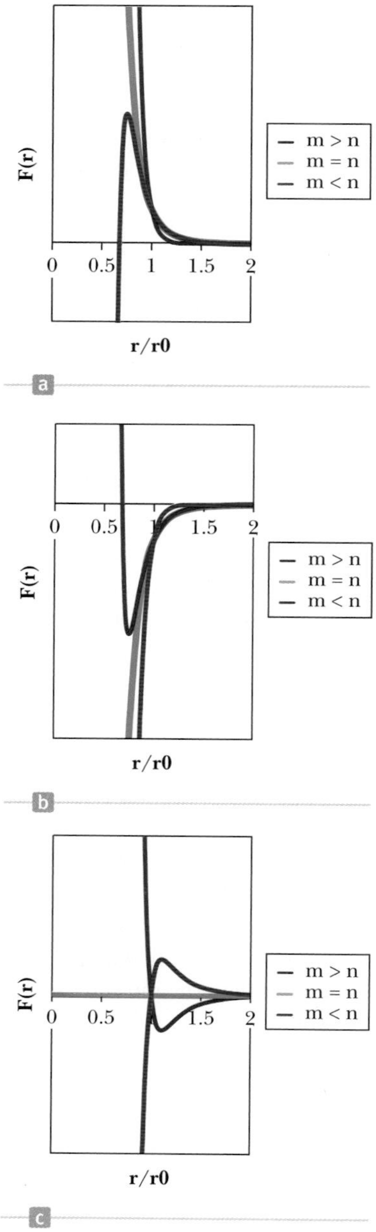

Figure 15.4
Forces of the type described by **Equation 15.1** with the following relationships between the parameters: (a) $A > B$, (b) $A < B$, (c) $A = B$. In all panels, red lines represent $m > n$, green $m = n$ and blue $m < n$.

where r represents the distance between the two molecules, r_0 is the equilibrium separation and m, $n > 0$. The first term provides a positive, repulsive force that decreases with increasing separation, while the second term provides a negative, attractive force that also decreases with increasing separation. **Figure 15.4** shows how such forces might vary with increasing molecular separation. If we think about what we require of the intermolecular force to reproduce experimental observations, we can narrow down the form of the force even further. None of the three curves in **Figure 15.4a** for which A > B provides a negative force at values of r greater than r_0, meaning that these equations do not produce the attraction we need at larger distances. The red curve in panel **Figure 15.4b** for B < A gives a positive (repulsive) force at short distances and a negative (attractive) force at longer distances and looks at first like it might be a good candidate. However, we need the force to be zero at $\frac{r}{r_0} = 1$, since there should be no net force at the equilibrium position. The only version of **Equation 15.1** that satisfies all the required conditions is the red curve in **Figure 15.4c** where A = B. We can therefore write our mathematical model for the intermolecular force as

$$F = A\left[\left(\frac{r_0}{r}\right)^m - \left(\frac{r_0}{r}\right)^n\right] \qquad (15.2)$$

where $m > n$ and A is an overall scaling factor.

Experimental data on responses to applied forces, melting and boiling points (and more recently direct measurements of intermolecular forces) suggest that setting $m = 13$ and $n = 7$ provides a reasonable approximation for many materials. That is:

$$F = A\left[\left(\frac{r_0}{r}\right)^{13} - \left(\frac{r_0}{r}\right)^{7}\right] \tag{15.3}$$

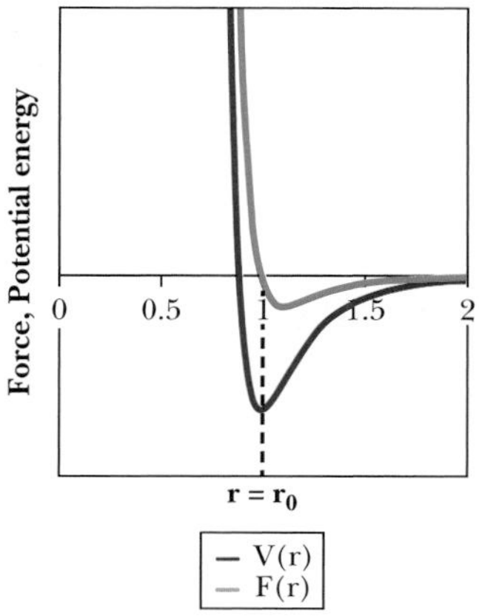

Figure 15.5
The Lennard-Jones potential and associated intermolecular force. The red line shows the potential energy of one molecule with respect to the other. The green line shows the force exerted on one molecule by the other.

Quick **Quiz 15.1**

What units must the constant A appearing in **Equation 15.3** have? **(a)** joules **(b)** metres **(c)** newtons **(d)** none, it's dimensionless

As we have seen with gravity and will see again with the electromagnetic force, it is frequently useful to describe an interaction between two objects in terms of potential energy rather than the force – this is as true for interactions between two molecules in a solid as it is for the Earth–Moon system or electrons orbiting a positively charged nucleus. Thinking about the relationship between force and work described in Chapter 6, we can write

$$F(r) = -\frac{d}{dr}U(r)$$

As we have a suggested form for $F(r)$, we rearrange the equation to find $U(r)$:

$$U(r) = -\int_{\infty}^{r} F(r)dr$$

where we have set the potential energy to be zero at infinity so that $U(\infty) = 0$. *Integral calculus, including the integration of polynomials, is reviewed in Appendix B.8.* Evaluating this for a force of the form given in **Equation 15.3**, we obtain

Σ

$$U_{LJ}(r) = \epsilon\left[\left(\frac{r_0}{r}\right)^{12} - 2\left(\frac{r_0}{r}\right)^{6}\right] \tag{15.4}$$

◀ Lennard-Jones potential

where ϵ is a scaling constant with units of energy. This form of intermolecular potential is called the Lennard-Jones potential. It is shown in **Figure 15.5**, along with the associated force. The potential energy is a minimum at the equilibrium separation $r = r_0$, increasing faster as the separation is decreased and more slowly as it is increased. A minimum in a potential such as that shown in the figure is often called a *potential well*, since it takes work to remove something from the minimum just as it takes work to lift a bucket out of a well. At $r = r_0$, $U_{LJ} = -\epsilon$, thus ϵ corresponds to the depth of the potential well.

John Lennard-Jones
British mathematician and theoretical physicist (1894–1954)
Lennard-Jones is most famous for his work on the interactions between atoms and molecules. He was born John Jones and adopted his wife's maiden name of Lennard as part of his own surname on their marriage. He undertook his PhD at Cambridge after serving as a pilot in the First World War, and later became a professor first at the University of Bristol and then at Cambridge. He made significant contributions to the development of molecular orbital theory and formulated intermolecular potentials of the type given in **Equation 15.4**.

Quick **Quiz 15.2**

Two molecules, initially at their equilibrium separation, interact via a Lennard-Jones potential as shown in **Figure 15.5**. Does it require more energy to **(a)** halve their separation, **(b)** double it or **(c)** neither – both require the same?

It should be stressed that the same form of intermolecular potential can be deduced from the properties of liquids; the Lennard-Jones and similar potentials can be used to describe interactions between neutral molecules in any state, not just solids.

Example 15.2

Squeezing ice molecules 1

Consider two adjacent water molecules in an ice crystal. How much work must be done to reduce the distance to 5% less than the equilibrium distance? Take the depth of the potential well as $\epsilon = 9.6 \times 10^{-21}$ J.

Solution

Conceptualise Start by drawing a diagram to help visualise the problem. We have two molecules, A and B, separated by a distance $r = 0.95r_0$. If work is done on a molecule to move it away from its equilibrium position, the molecule acquires potential energy equal to that work.

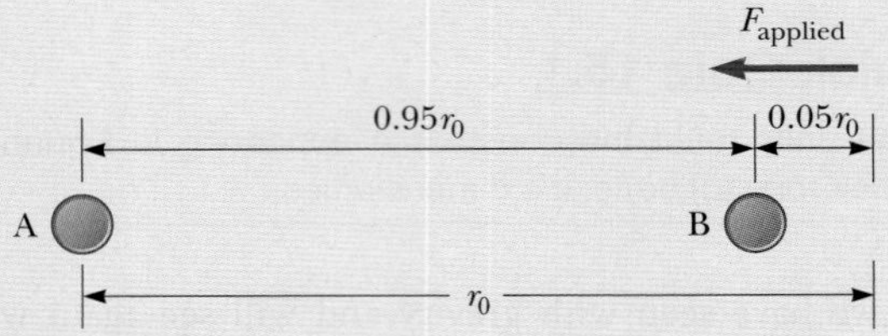

Figure 15.6
(Example 15.2) Molecules A and B, separated by distance r

Model We can use the Lennard-Jones potential to estimate the potential energy of one molecule at a distance r from another; the work done in moving a molecule from r_0 to $0.95r_0$ is then the difference in potential energy between these two points.

Analyse **Equation 15.4** gives the potential energy at separation r as:

$$U_{LJ}(r) = \epsilon\left[\left(\frac{r_0}{r}\right)^{12} - 2\left(\frac{r_0}{r}\right)^{6}\right]$$

The change in potential energy as the separation changes from r_0 to r is then

$$\Delta U_{LJ} = \epsilon\left[\left(\frac{r_0}{r}\right)^{12} - 2\left(\frac{r_0}{r}\right)^{6}\right] - \epsilon\left[\left(\frac{r_0}{r_0}\right)^{12} - 2\left(\frac{r_0}{r_0}\right)^{6}\right]$$

$$= \epsilon\left[\left(\frac{r_0}{0.95r_0}\right)^{12} - 2\left(\frac{r_0}{0.95r_0}\right)^{6} + 1\right]$$

$$= \epsilon[(0.95)^{-12} - 2(0.95)^{-6} + 1]$$

$$= 0.130\epsilon = 0.130 \times 9.6 \times 10^{-21}\ \text{J} = 1.2 \times 10^{-21}\ \text{J}$$

Finalise The increased potential energy was introduced into the system by the work done in moving the molecules closer together, so our final answer is that 1.2×10^{-21} J of work is done to move two water molecules closer together. This may seem like a tiny amount, but consider that in order to compress the ice cube containing the two molecules by 5%, the distances between all pairs of molecules must be reduced by this amount. As an ice cube may contain of the order of 10^{23} molecules, the *total* amount of energy needed is *not* tiny.

Example 15.3

Squeezing ice molecules 2

Consider the two water molecules in the previous example. If molecule A is held fixed and molecule B is released from an initial position a distance $0.95r_0$ from molecule A, what will its inital acceleration be? Take the mass of a water molecule as 2.99×10^{-26} kg.

Solution

Conceptualise To find the initial acceleration of molecule B, we need to find the repulsive force exerted on it by molecule A.

Model We model the force between the atoms as due to the Lennard-Jones potential, so we can use **Equation 15.3**. When we have the force, we can apply Newton's second law to find the acceleration. Note, however, that we are given a value for ϵ, but we need a value for the constant A. Hence we first need to find a relationship between ϵ and A.

Example 15.3 cont.

Analyse Equations 15.3 and 15.4 are related since $F = -\dfrac{dU}{dr}$

that is:

$$A\left[\left(\frac{r_0}{r}\right)^{13} - \left(\frac{r_0}{r}\right)^{7}\right] = -\epsilon\left[\frac{d}{dr}\left(\frac{r_0}{r}\right)^{12} - 2\frac{d}{dr}\left(\frac{r_0}{r}\right)^{6}\right]$$

$$= -\epsilon\left[\left(r_0^{12}\frac{dr^{-12}}{dr}\right) - \left(2r_0^{6}\frac{dr^{-6}}{dr}\right)\right]$$

$$= \epsilon\left[12r_0^{12}r^{-13} - 12r_0^{6}r^{-7}\right]$$

$$= \frac{12\epsilon}{r_0}\left[\left(\frac{r_0}{r}\right)^{13} - \left(\frac{r_0}{r}\right)^{7}\right]$$

Hence:

$$A = \frac{12\epsilon}{r_0}$$

and

$$F = \frac{12\epsilon}{r_0}\left[\left(\frac{r_0}{r}\right)^{13} - \left(\frac{r_0}{r}\right)^{7}\right]$$

Thus the force on molecule B at a distance $0.95r_0$ from molecule A is

$$F = \frac{12\epsilon}{r_0}[(0.95)^{-13} - (0.95)^{-7}] = \frac{6.19\epsilon}{r_0}$$

The resulting acceleration is

$$a = \frac{F}{m} = \frac{6.19\epsilon}{r_0 m}$$

Check dimensions, noting that ϵ must have dimensions of energy:

$$[LT^{-2}] = \frac{[ML^2\,T^{-2}]}{([L][M])} = [LT^{-2}]\ ☺$$

Using the values given in this example and the average intermolecular distance calculated in Example 15.1 we find

$$a = \frac{6.19 \times 9.6 \times 10^{-21}\,J}{0.3196 \times 10^{-9}\,m \times 2.99 \times 10^{-26}\,kg}$$

$$= 6.2 \times 10^{15}\ m/s^2$$

Finalise This is an enormous acceleration!

The force associated with the Lennard-Jones potential is frequently given in the form derived in Example 15.3, i.e.

$$F = \frac{12\epsilon}{r_0}\left[\left(\frac{r_0}{r}\right)^{13} - \left(\frac{r_0}{r}\right)^{7}\right] \tag{15.5}$$

The Lennard-Jones potential provides an estimate of the forces between two *neutral* molecules. However, we know that while some solids are made of bound neutral molecules, others are made up of electrically charged ions, e.g. metals (which are composed of a crystal lattice of positive ions through which 'free' electrons wander) and ionic solids such as sodium chloride, in which positive and negative ions occupy adjacent crystal lattice sites.

If a substance is composed of electrically charged ions, rather than neutral molecules, the potential has to be modified to take into account the *electrostatic* interaction between them. The electrostatic interaction

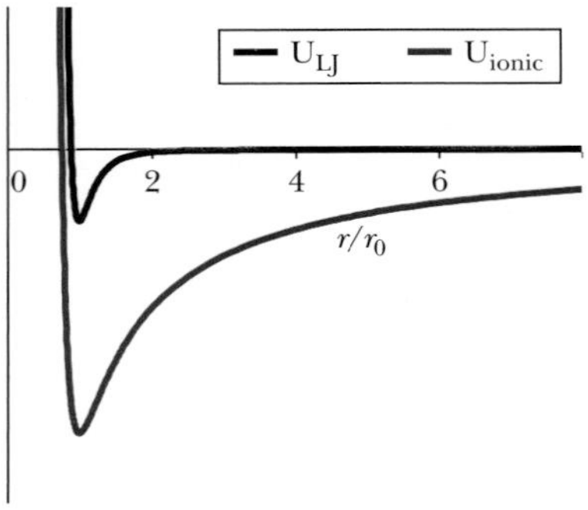

Figure 15.7
The potential between ions with unlike charges (red line), compared to the Lennard-Jones potential between neutral molecules (blue line)

between two charged particles will be discussed in detail in Chapter 23. The effect of this extra interaction depends on whether the ions have like or unlike charges.

When the ions have like charges, as in a metal, the electrostatic interaction is repulsive, increasing the short-range repulsion and dominating the long-range attraction of **Equation 15.5** so that the net force is *always* repulsive. At first glance you might think that this means that metals shouldn't be able to cohere and yet of course we know they do. This is because the electrons tend to concentrate between the ions so that the charges cancel out (this minimises their potential energy), meaning that in any small region metals tend to be electrically neutral. Metallic bonding is discussed in more detail in Chapter 42.

If the ions have unlike charges the short-range repulsive part of the interaction is slightly reduced and the long-range attractive part becomes much stronger. In an ionic solid such as table salt, the potential can be written in the form

$$U_{\text{ionic}} = \frac{a}{r^9} - \frac{b}{r} \tag{15.6}$$

where the constant b is typically about two orders of magnitude larger than the constant a.

This kind of potential is illustrated in **Figure 15.7**. The Lennard-Jones potential is included on the same axes. The depth of the potential described by **Equation 15.6** is much greater than that described by **Equation 15.5**, indicating that ionic bonds are stronger than those between neutral molecules. The bonds between ions in metals (see Chapter 43) are weaker still.

Quick Quiz 15.3

Which of the graphs shown in **Figure 15.8a–d** best represents the force on an ion due to the potential given in **Equation 15.6**?

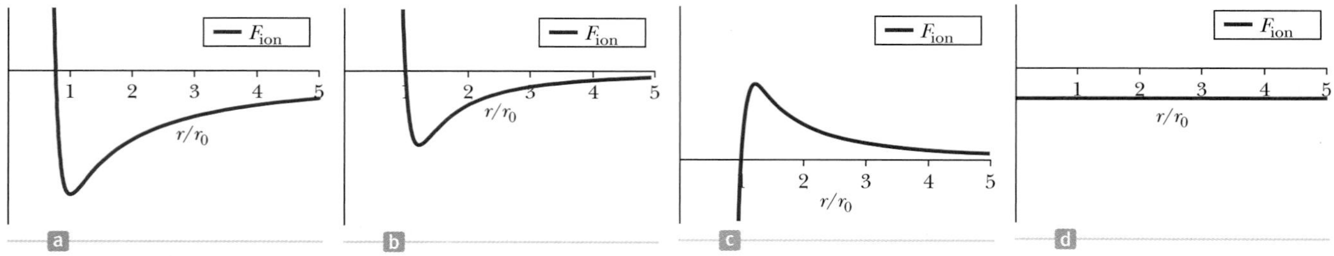

Figure 15.8
Quick quiz 15.3

15.2 Elasticity, stress and strain

Several physical properties of bulk materials can be explained using the intermolecular potentials described in the previous section. In Part 4: Thermodynamics, we shall see how thermal properties such as latent heat and thermal expansion can be related to the microscopic structure of materials. In Chapters 27 and 43 we shall see how electrical properties such as conductivity can also be related to microscopic structure. In this section, we look at the elastic properties of solids.

Look back at **Figure 15.5**. In the region around the equilibrium position where $r = r_0$ and $F(r) = 0$, the force decreases approximately linearly with increasing separation; that is, the force is approximately proportional to the separation,

$$F = -k(r - r_0) \tag{15.7}$$

You may recognise this as Hooke's law (**Equation 6.7**). It means that for small applied forces, we expect the intermolecular separation to scale linearly with the force. This is the origin of the behaviour we observe in bulk solids that obey Hooke's law (see for example the discussion of springs in Section 6.3).

Consider a chain of N molecules as shown in Figure 15.9. When there is no external force applied, the molecules are separated by their equilibrium spacing ℓ_0 and the length of the chain is $L_0 = N\ell_0$. When a force is applied that increases the molecular separation by an amount $\Delta\ell$, the total extension is $\Delta L = N\Delta\ell$. We define the **tensile strain** resulting from a stretching applied force as the ratio of the extension to the original length:

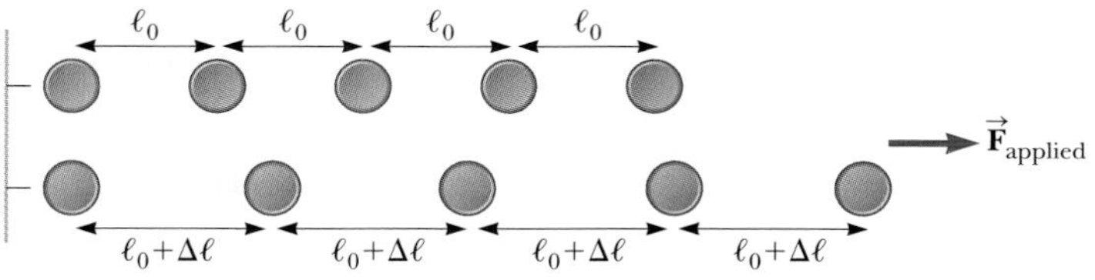

Figure 15.9
A force is applied to one end of a chain of molecules in a solid (the other end is held fixed).

$$\text{Tensile strain} = \frac{\Delta L}{L_0} \tag{15.8}$$

◀ Definition of tensile strain

In this case, we can see that

$$\frac{\Delta L}{L_0} = \frac{N\Delta\ell}{N\ell_0} = \frac{\Delta\ell}{\ell_0}$$

Thus if the extension of the molecular bond is directly proportional to the applied force, the extension of the whole chain of molecules will be proportional to the applied force too.

Quick Quiz 15.4

Which will stretch more easily: **(a)** a molecular solid for which the potential is closely approximated by the Lennard-Jones potential or **(b)** an ionic solid for which the potential is similar to that shown in Figure 15.7?

We can deduce something else about the response of a solid made of neutral molecules to an increasing force from Figure 15.5. The figure shows that the attractive force between molecules that are separated by a distance greater than the equilibrium separation increases up to some separation R, where the force is $F = -F_R$. The point at which $F = -F_R$ in Figure 15.5 is the turning point in the force, where it has its most negative value, and hence this is the point at which the attraction is strongest. This is the separation at which the magnitude of the restoring force is a maximum. Beyond this, the restoring force *decreases* with increasing distance.

Let's consider the implications of this. If a force $F \leq +F_R$ is applied to separate the molecules, the restoring force will balance it and the molecules will remain bound (if further apart than normal). Once the external force is removed, the molecules will relax back to the equilibrium separation. In bulk, this results in **elasticity**, the property of returning to the original size and shape after a deforming force is removed. You can see this kind of behaviour when you compress a spring or rubber ball, or stretch an elastic band (as long as you don't stretch it too far).

If the applied force is greater than $+F_R$, there will be a net positive force and the molecules will accelerate away from each other. Thus the distance R indicates the maximum separation at which the molecules are bound; beyond this point they are not. In bulk, this corresponds to the point at which a solid permanently deforms or breaks when forces are applied to stretch it. When this occurs, the material is said to have exceeded its **elastic limit**. **Brittle** materials are those that break just after the elastic limit is exceeded; amorphous materials (such as glass) tend to be brittle, as do ionic crystalline materials (such as sodium chloride and flint). **Ductile** materials (such as mild steel or iron) elongate more rapidly after the elastic limit is exceeded. Ductility is usually a property of a subset of crystalline materials (particularly metals) whose planes of molecules can 'glide' past one another. The deformations induced beyond the elastic limit are permanent or **plastic**.

So far, we have discussed the response of a single molecular bond or a linear chain of bonds to an applied force. For a bulk solid, we find it more useful to talk about the applied force per unit area. Consider the chain of molecules shown in Figure 15.9. In reality, the bulk material is composed of many such chains, linked to each other, each of which must be subject to a particular force F to result in an elongation ΔL. If we assume the chains of molecules are evenly distributed throughout the material, the number of chains will be proportional to the cross-sectional area of the solid. The *total* force that must be applied to the end of the solid to stretch all of the chains it contains is therefore also proportional to the cross-sectional area; that is, twice the force is needed to stretch a solid bar of cross-sectional area $2A$ than to stretch a solid bar of the same material of cross-sectional area A. We call the force per unit area **stress**. Stress results in **strain**. In general, strain is the response of the material to the applied stress.

Experimentally it is found that, for sufficiently small stresses, stress is proportional to strain, as we would expect from our understanding of molecular bonds. The constant of proportionality or **elastic modulus** depends on the material being deformed and the way in which it is being deformed. In general it is defined as

$$\text{Elastic modulus} \equiv \frac{\text{stress}}{\text{strain}} \qquad (15.9)$$

The elastic modulus relates what is done to a solid object (a force is applied) to how that object responds (it deforms to some extent). As well as a characteristic elastic modulus each material has its own characteristic **breaking stress** – that is, the stress that if applied causes the material to fracture. It is important to note that fracture depends on applied stress rather than applied force, so that two wires of the same material but different cross-sections will support different maximum loads before snapping.

An understanding of how materials respond to stress is of great importance in engineering and construction. For example, a road bridge is subjected to loads of varying magnitude as the volume of traffic varies and carries a significant base load due to the weight of the concrete, road surface etc. If the design is not adequate or the material flawed, a bridge can collapse under its own weight, as shown in Figure 15.10. Before such a bridge is erected, samples of the steel to be used are tested to find out whether they will withstand likely loads without buckling or breaking.

Richard Wood/Flickr

Figure 15.10
A bridge in Canberra collapses when concrete is poured onto the structure.

In the following sections, we consider three types of deformation: change in length only, change in pressure, and shearing, that is, opposing motion of planes in the solid.

TRY THIS

Take an eraser and try to stretch it and compress it longitudinally. Now try to compress it by pressing on all sides at once. Now twist it. Which is easier? Now take a piece of chalk and try the same. Can you break the chalk? If so, which type of deforming stress leads to it breaking most easily?

Getty Images/Graham Monro/gm photographics

Figure 15.11
Cables supporting Sydney Harbour Bridge

15.3 Response of solids to changes in length

We have already seen that for small compressive or tensile (elongating) forces, we expect the resulting linear strain to be proportional to the applied stress. Knowing how particular materials respond to forces that act to stretch or compress them is of great importance in construction; for example, the footings of a building need to be able to withstand the load caused by the weight of the building without significant compression or fracture, as do columns and structural beams supporting above-ground storeys and roofs. Similarly, the cables of a suspension bridge such as the Sydney Harbour Bridge (Figure 15.11) must support the weight of the bridge and the traffic without breaking or even measurably stretching. We will now look at how we quantify the response of different materials to compressive and tensile forces.

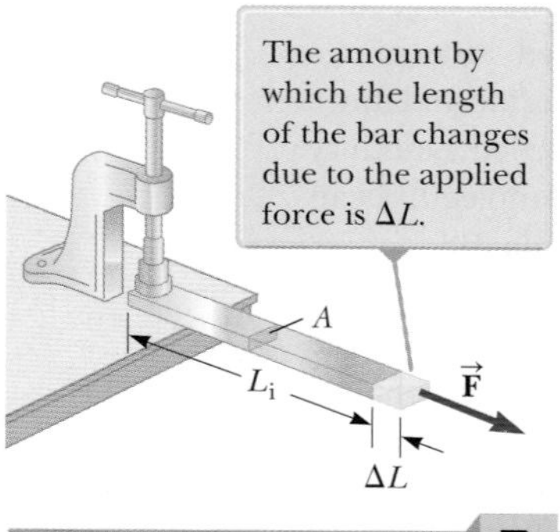
The amount by which the length of the bar changes due to the applied force is ΔL.
A
L_i
$\vec{F}$
ΔL

Active Figure 15.12
A force $\vec{F}$ is applied to the free end of a bar clamped at the other end.

Consider a long bar of cross-sectional area A and initial length L_i that is clamped at one end as in Active Figure 15.12. We define the **tensile stress** as the ratio of the magnitude of the external force F to the cross-sectional area A. Figure 15.13a shows the results of such an experiment where ΔL is measured as a function of applied force. Figure 15.13b shows a stress–strain curve for the same experiment. Stress–strain curves are the most common way of showing the relationship between stress and strain. Recalling that stress is force per unit area and tensile strain is fractional change in length, the two curves must be the same shape with the axes reversed.

For relatively small stresses, the bar returns to its initial length when the force is removed, and both curves are linear. Once the **proportional limit** is exceeded (point A on the graphs), the strain is no longer proportional to the stress and the curves are not linear. We can understand this by remembering that the intermolecular force is only approximately linear around the equilibrium position. In this region, the restoring forces in the material still balance the applied stress and so the material is still elastic. As the stress increases, the elastic limit (point B) is exceeded and the material is permanently distorted; the rapid extension for only a small additional load typical of ductile materials is visible in Figure 15.13a. Eventually, the **tensile breaking stress** is reached (point C).

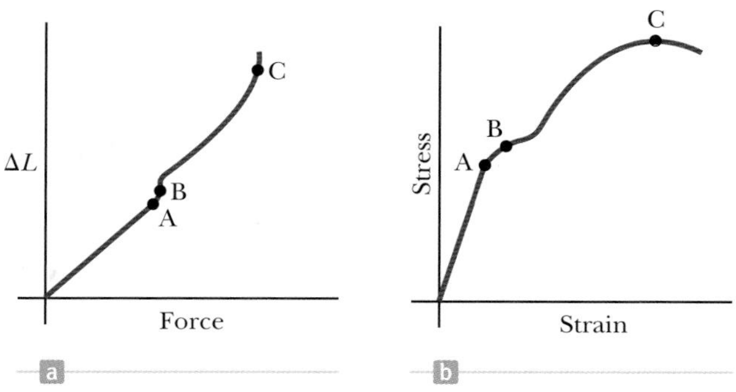

Figure 15.13
(a) Force versus extension and (b) stress versus strain for a ductile material

TRY THIS

Get a length of elastic and tie one end to a fixed object such as a door handle. Measure the length below the handle. Now tie a weight such as a cup to the other end and measure how much it has stretched. Add another weight and measure the length again. Does the elastic obey Hooke's law? Now try the same thing with a double strip of the elastic. How does the extension change with twice as much material being stretched? Can you exceed the elastic limit without breaking the elastic? Now try the same with another material, e.g. nylon fishing line.

Pitfall Prevention 15.1

Elastic materials do not necessarily show strain proportional to stress. A material is elastic so long as it returns to its original shape once the applied force is removed, which can continue after the proportional limit is exceeded.

As long as the stresses are small enough, we can define a constant of proportionality relating tensile or compressive stress to resulting tensile strain. We call it **Young's modulus**:

$$Y \equiv \frac{\text{tensile stress}}{\text{tensile strain}} = \frac{F/A}{\Delta L/L_0} \qquad (15.10)$$

◀ Young's modulus

Young's modulus is named after the British physicist Thomas Young. It is typically used to characterise a rod or wire stressed under either tension or compression. Because strain is a dimensionless quantity, Y has units of force per unit area, N/m^2, and dimensions $[ML^{-1}T^{-2}]$.

It can be shown that Young's modulus is related to the intermolecular force and the equilibrium separation by

$$Y \approx \left.\frac{dF}{dR}\right|_{r_0}\left(\frac{1}{r_0}\right)$$

If we assume the interaction is given by the Lennard-Jones potential, this gives

$$Y \approx \frac{72\epsilon}{r_0^{\,3}} \qquad (15.11)$$

Differentiation, including the differentiation of polynomials, is reviewed in Appendix B.7. Σ

Putting in typical values of the well depth ϵ (10^{-21}–10^{-20} J) and r_0 (0.1–1 nm) leads us to expect values of Y in the range 10^8–10^{12} N/m^2. Values for a range of different materials, including amorphous and crystalline substances, are given in Table 15.1. As you can see, Young's modulus is typically of the order of tens or hundreds of GN/m^2, indicating that our simple model provides a good estimate.

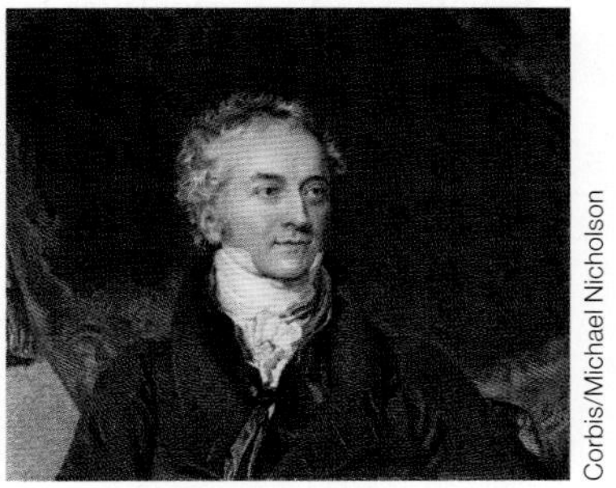

Thomas Young
British physicist (1773–1829)
Young made considerable contributions to various areas of physics including light and vision, the mechanics of solids and liquids (the Laplace equation in Chapter 13 (Equation 13.10) is also called the Young–Laplace equation). You may be familiar with his double-slit experiment, which helped demonstrate the wave nature of light. He was also a talented linguist and Egyptologist whose accomplishments include partial deciphering of the famous Rosetta stone hieroglyphs and the proposal of a new, universal alphabet.

Table 15.1
Some elastic moduli

Substance	Young's modulus (N/m²)	Bulk modulus (N/m²)	Shear modulus (N/m²)
Diamond	122×10^{10}	44.2×10^{10}	47.8×10^{10}
Tungsten	35×10^{10}	20×10^{10}	14×10^{10}
Steel	20×10^{10}	6×10^{10}	8.4×10^{10}
Copper	11×10^{10}	14×10^{10}	4.2×10^{10}
Brass	9.1×10^{10}	6.1×10^{10}	3.5×10^{10}
Gold	7.9×10^{10}	18.0×10^{10}	2.7×10^{10}
Aluminum	7.0×10^{10}	7.0×10^{10}	2.5×10^{10}
Glass	$6.5\text{–}7.8 \times 10^{10}$	$5.0\text{–}5.5 \times 10^{10}$	$2.6\text{–}3.2 \times 10^{10}$
Quartz	5.6×10^{10}	2.7×10^{10}	2.6×10^{10}
Rubber	$\sim 10^5$	$\sim 10^9$	$\sim 10^5$
Water	–	0.21×10^{10}	–
Mercury	–	2.8×10^{10}	

±?

Example 15.4

Stage design

In Example 7.2, we analysed a cable used to support an actor as he swung onto the stage. Now suppose the tension in the cable is 940 ± 30 N as the actor reaches the lowest point, where the uncertainty is related to the possible variations in the actor's weight over the time the play runs. What diameter should a 10.0 m steel cable have if we do not want it to stretch more than 0.50 cm under these conditions?

Solution

Conceptualise Look back at Example 7.2 to recall what is happening in this situation. We ignored any stretching of the cable there, but we wish to address this phenomenon in this example.

Model We model the steel cable as having a linear stress–strain curve for the applied tension. This allows us to apply **Equation 15.10**. We will also assume the cable has a circular cross section, and use the Young's modulus for steel from **Table 15.1**.

Analyse Solve **Equation 15.10** for the cross-sectional area of the cable:

$$A = \frac{FL_0}{Y\Delta L}$$

The cross-sectional area of the cable is

$$A = \frac{\pi d^2}{4} = \frac{FL_0}{Y\Delta L}$$

which gives

$$d = \sqrt{\frac{4FL_0}{\pi Y \Delta L}}$$

Check the dimensions to make sure they're consistent:

$$[\mathrm{L}] = \sqrt{\frac{[\mathrm{MLT^{-2}}][\mathrm{L}]}{[\mathrm{ML^{-1}T^{-2}}][\mathrm{L}]}} = [\mathrm{L}] \ ☺$$

Substitute the known values, using the maximum value of F:

$$d = \sqrt{\frac{4\,(970\ \mathrm{N})(10.0\ \mathrm{m})}{\pi(20.0 \times 10^{10}\ \mathrm{N/m^2})(0.005\ \mathrm{m})}} = 3.51 \times 10^{-3}\ \mathrm{m}$$

Finalise Our calculation suggests you could get away with a 3.51 mm diameter cable; however, we have assumed that the cable is completely free of flaws. To provide a larger margin of safety, you would probably use a cable made up of many smaller wires having a total cross-sectional area substantially greater than our calculated value.

Example 15.5

Measuring Young's modulus

One of the experiments in your first-year physics laboratory asks you to measure Young's modulus for a metal wire. The wire is suspended from a girder in the ceiling and you are given a series of weights to hang from the hook attached to the free end of the wire. You are given the length of the unextended wire as $L_i = (3.000 \pm 0.002)$ m and you measure its diameter to be $d = (0.40 \pm 0.05)$ mm. You take careful measurements of the extension of the wire as a function of the applied weight, using a Vernier scale, and record your results in a table in your logbook.

W (N)	0	10	20	30	40	50	60	70	80
$\Delta L \pm 0.01$ cm	0	0.11	0.19	0.34	0.45	0.56	0.68	0.83	0.97

What is your result for Young's modulus? Suggest a way of improving the precision of your measurement.

Solution

Conceptualise To find Young's modulus Y for the wire, we need to work out how our data relate to **Equation 15.10**, which tells us that tensile stress is equal to $Y \times$ tensile strain.

Model We model the metal as having a linear stress–strain curve for the strains applied during the experiment, so that we can apply **Equation 15.10**. We will also assume the wire has a circular cross section. Since we have a table of data showing how one quantity varies with another, we need to work out how to plot the data so that Y can be obtained from the graph.

Analyse First we consider **Equation 15.10**, which we can rearrange as

$$F = YA\left(\frac{\Delta L}{L_0}\right) = \left(\frac{Y\pi d^2}{4}\right)\left(\frac{\Delta L}{L_0}\right)$$

Although it is normal to plot a graph with the controlled variable on the x axis, we can see that in this case if we plot the applied force F as a function of the extension ΔL, we should obtain a straight line with gradient $\frac{YA}{L_0}$, as shown in **Figure 15.14**. The data closely follow a straight line, as expected if Hooke's law is obeyed. If the data are entered into spreadsheet software, a linear regression can be performed, which gives a gradient $g = 8.256 \times 10^3$ N/m. Combining this with our values of d and L_0, we obtain

$$Y = \frac{4gL_0}{\pi d^2} = \left(\frac{4}{\pi}\right)\left[\frac{(8.256 \times 10^3\ \text{N/m})(3.00\ \text{m})}{1.6 \times 10^{-7}\ \text{m}^2}\right] = 1.97 \times 10^{11}\ \text{N/m}^2$$

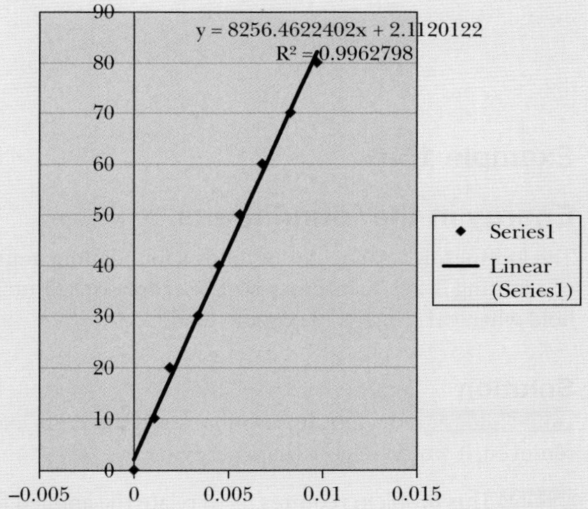

Figure 15.14
The force versus extension graph you plot in your logbook

Of course our result doesn't mean anything without an accompanying uncertainty. As usual, we examine the various contributions to the uncertainty to see if any dominate. The fractional uncertainty in the diameter is 12.5%, much bigger than the uncertainty in the unextended length or the likely uncertainty from the linear regression (this uncertainty is likely to be small, since the data lie very close to the calculated line of best fit). We therefore assume that the uncertainty in our value of Y is mainly due to the uncertainty in the diameter of the wire. Because the diameter is squared in our calculation, the fractional uncertainty in Y is related to the fractional uncertainty in d as

$$\frac{\Delta Y}{Y} = \frac{2\Delta d}{d} = 0.25$$

giving $\Delta Y = 0.5 \times 10^{11}$ N/m².

Finalise Our experiment has yielded a value of Young's modulus of $(2.0 \pm 0.5) \times 10^{11}$ N/m². The precision of our answer could be improved by making a more precise measurement of the diameter of the wire, since that dominated the uncertainty in the final value.

Courtesy of VintageVille.com

Figure 15.15
Magic jumping frog

Energy stored in a longitudinally stressed solid

Because the molecules in an unstressed material are at their lowest potential energy (they are at the minimum of the curves shown in **Figures 15.5** and **15.7**), we do work on them to change their separation. This means that a stressed material stores energy. You can see this when you compress a spring such as in a jumping frog toy: when the spring is released from tension, the toy frog hops high into the air, as the potential energy stored in the spring is converted first to kinetic, then gravitational potential, and then kinetic energy again.

Let's consider a wire with Young's modulus Y, cross-sectional area A and original length L_0. A load is attached to one end of the wire to produce an extension. As long as the proportional limit of the wire has not been exceeded **Equation 15.10** holds and we can write the applied force in terms of the extension x as

$$F = \frac{YAx}{L_0}$$

The work done to produce an extension ΔL is therefore

$$W = \int_0^{\Delta L} F\,dx = \frac{YA\Delta L^2}{2L_0}$$

Σ *Integration, including the integration of polynomials, is reviewed in Appendix B.7.*

As the energy stored in the wire is equal to the work done in stretching it, we can write

$$E_{\text{stored}} = \frac{YA\Delta L^2}{2L_0} \tag{15.12}$$

Example 15.6

Energy in the foundations

The footings in a particular building's foundations must each be able to bear a 100 tonne load. If the uncompressed height of a footing is 2.0 m, its cross-sectional area is 0.25 m^2 and its Young's modulus is 20 GPa, how much energy will a footing hold when carrying its maximum load?

Solution

Conceptualise The footing is compressed under its load and so its potential energy is increased – if the load were removed, it would release that energy.

Model This problem requires us to relate the applied force to the total stored energy; we will need to use **Equation 15.12**.

Analyse

Equation 15.12 tells us that

$$E_{\text{stored}} = \frac{YA\Delta L^2}{2L_0} \tag{1}$$

We do not know the compression ΔL, but we do know the applied force. Remembering that

$$Y = \frac{F/A}{\Delta L/L_0}$$

we can write

$$\Delta L = \frac{FL_0}{YA}$$

Substitute into **Equation (1)** to obtain

$$E_{\text{stored}} = \frac{F^2 L_0}{2YA}$$

Example 15.6 cont.

Check dimensions: $$[ML^2T^{-2}] = \frac{([MLT^{-2}]^2\,[L]}{([ML^{-1}\,T^{-2}][L]^2)} = [ML^2T^{-2}]$$ ☺

Substitute in the given values and assume $g = 9.8\ m/s^2$:

$$E = \frac{[(10^5\ kg \times 9.80\ m/s^2)^2(2.0\ m)]}{2(20\times 10^9\ N/m^2)(0.25\ m^2)} = 192\ J$$

Finalise This is about 1/5 of the food energy in a jam donut.

TRY THIS

Take a rubber band and stretch it to breaking point or tighten a guitar string until it breaks – be careful not to hurt yourself or others. Explain what you see. How does it relate to the idea of energy stored in a stressed wire?

We can also determine the energy stored in a wire from a graph of force against extension as shown in **Figure 15.13a**. As the energy stored in the wire is equal to the integral of the applied force, this is equivalent to saying that it is equal to the area under the force–extension curve.

Example 15.7

Catapult

Marcus is playing with a homemade catapult made of a 6 cm length of thick rubber band (cross-sectional area 10 mm^2) attached to a small forked branch. He loads a toy car (mass 200 g) into the catapult and stretches the rubber band to double its original length. With what velocity will the toy car leave the catapult when Marcus releases it?

Solution

Conceptualise Imagine firing a projectile using a rubber band. As the rubber band is stretched, the energy stored in it increases. When the rubber band is released, the stored energy is converted to kinetic energy of the projectile.

Model We will make some assumptions: (1) the rubber band does not pass its proportional limit, so **Equation 15.12** applies, and (2) all of the energy stored in the stretched band is converted to kinetic energy of the toy car. We can use **Equation 15.12** to calculate the energy stored in the rubber band and to relate that energy to the velocity of the car.

Analyse If the energy stored in the rubber band at its maximum elongation is entirely converted to kinetic energy and assuming that all of the energy is given to the toy car, the kinetic energy is

$$K = \frac{1}{2}mv^2 = \frac{YA\Delta L^2}{2L_0}$$

Rearrange for the velocity: $$v = \sqrt{\frac{YA\Delta L^2}{mL_0}} = \Delta L\sqrt{\frac{YA}{mL_0}}$$

Check dimensions: $$[LT^{-1}] = [L]\sqrt{\frac{[ML^{-1}T^{-2}][L^2]}{[M][L]}} = [LT^{-1}]$$ ☺

Substitute in the given values and the Young's modulus for rubber given in **Table 15.1**:

$$v = (0.06\ m)\sqrt{\frac{(10^5\ N/m^2)(10^{-5}\ m^2)}{(0.2\ kg)(0.06\ m)}} = 0.5\ m/s$$

Finalise Think about your own experience using catapults – does this seem like a reasonable estimate of the velocity to you? In fact, rubber does not respond linearly to applied stresses – that is, its Young's modulus is a function of stress so that the strain increases more slowly than stress. This means more energy is stored in the material than you obtain with **Equation 15.12** and objects released from a rubber band catapult can achieve considerably higher velocities.

Poisson's ratio

There is one last factor to consider in our description of the response of materials to tensile forces. If you did either of the *Try this* activities above, you may have noticed that as you stretched the elastic or rubber band, the diameter or cross-sectional area reduced. If the original diameter d decreases by Δd as a wire or bar is extended from an initial length L_i by an extension ΔL, we define **Poisson's ratio** as

$$\sigma = \left(\frac{\Delta L}{L_\text{i}}\right)\left(\frac{d}{\Delta d}\right) \tag{15.13}$$

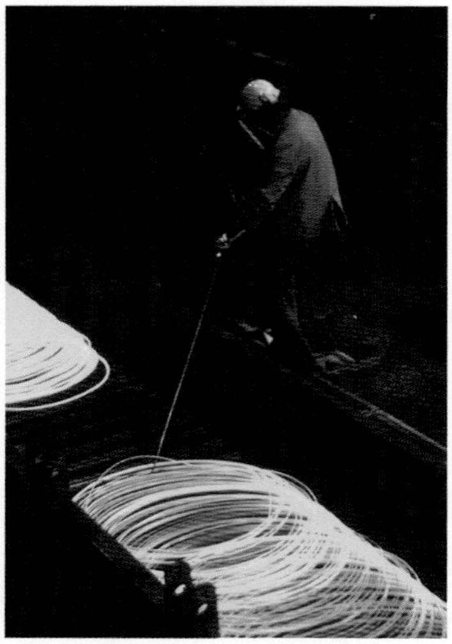

Alamy/Jim Pickerell

Figure 15.16
Drawing metal to make wire of the desired diameter

Poisson's ratio is the ratio of the fractional change in diameter to the fractional extension. This quantity is a constant for a given material and is another useful parameter for construction and engineering applications.

It can be shown (see Problem 47) that when Poisson's ratio is equal to ½, volume is conserved. Experiments show that $\sigma = 0.48$ for rubber, but is less than 0.3 for most metals. Thus rubber almost conserves volume when subjected to tensile stresses, but the volume of metals *increases* as they are stretched.

Of course, when a wire or elastic material is stretched, it is often the case that the change in diameter does not happen uniformly over the whole length. Instead, we often observe 'necking' – that is, the formation of narrow waists or necks at one or more points along the length. Necking is most often observed in ductile materials and some amorphous materials such as polymers. In a ductile metal, the narrower points occur where there are flaws in the material's microscopic structure, leading to easier movement of molecular planes with respect to each other. The elongation of a polymer is due to a somewhat different mechanism from that described above. Polymer molecules are long chains and tend to be kinked or coiled up when the material is not subject to tensile stress; a significant part of their elongation is due to the unfurling of molecules, rather than an increase in intermolecular separation. Necking in polymers is due to a combination of factors including softening of the material due to increased temperature as the work done on the molecules is dissipated in the form of heat. The formation of necks is exploited in the manufacture of polymer threads and metal wires as shown in **Figure 15.16**.

TRY THIS

Get some nylon thread or fishing line, a length of cotton and a long human hair (preferably one of your own, but ask someone else for one if you have short hair). Apply a tensile stress to each one in turn. Do they maintain a uniform cross section as they stretch? If you break any of them, look closely at the parts close to the broken ends.

15.4 Response of solids to changes in pressure

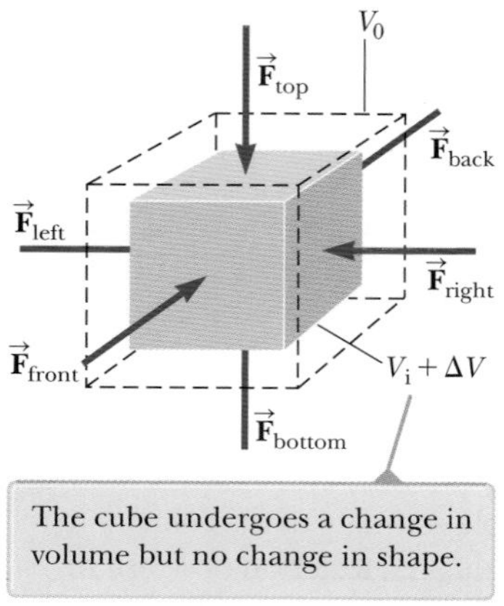

Active Figure 15.17
A cube under uniform pressure

If an object is moved from a lower-pressure environment to a higher-pressure environment, such as from land to deep sea, or vice versa, it experiences changes of uniform magnitude in the force applied perpendicularly over the entire surface. You can imagine that the response of different materials to such changes is particularly important in the design of submarines and aeroplanes and even more so in the design of satellites and spaceships, for which construction materials may experience significant changes of pressure.

Consider an object experiencing a uniform increase in pressure as shown in **Active Figure 15.17**. (We assume here the object is made of a single substance.) There are similarities here to the situation in which an object is subject to a linear compressive stress as discussed in the previous section. In this case, however, the compression or elongation occurs in all directions, so an object subject to this type of deformation undergoes a change in volume but no change in shape. We can characterise the degree to which a material resists deformation caused by changes in pressure by a new elastic modulus, called the **bulk modulus**.

To define the bulk modulus, we first define the **volume stress** as the ratio of the magnitude of the total force F exerted on a surface to the area A of the surface. It is therefore identical to the pressure. If the pressure on an object changes by an amount $\Delta P = \Delta F/A$, the object experiences a volume change ΔV. The **volume strain** is equal to the change in volume ΔV divided by the initial volume V_0. Therefore, from Equation 15.9, the bulk modulus is defined as

$$B \equiv \frac{\text{volume stress}}{\text{volume strain}} = -\frac{\Delta F/A}{\Delta V/V_0} = -\frac{\Delta P}{\Delta V/V_0} \quad (15.14)$$

◀ Bulk modulus

The negative sign indicates that an increase in pressure results in a decrease in volume and vice versa.

We might expect that the bulk modulus of a material is related to its Young's modulus, since in effect bulk volume change can be thought of as a three-dimensional extension of linear extension or compression, as long as stress is proportional to strain. If we consider the compression of a cube as in Active Figure 15.17, a reduction in volume by 1% only requires a reduction in side length of 0.3%. If the resistance to compression or expansion is primarily due to the decreasing or increasing separation of molecules interacting via potentials such as the Lennard-Jones potential, we might therefore expect the bulk modulus to be less than but similar to the Young's modulus (remembering that these moduli only make sense as constants of proportionality for very small deformations). In fact, it can be shown that the bulk modulus depends on the force between molecules (F), the number of nearest neighbours each molecule has (N) and the equilibrium separation (r_0) as

$$B = \frac{N}{18r_0}\left.\frac{dF}{dr}\right|_{r_0}$$

which for a Lennard-Jones potential is

$$B = \frac{4N\epsilon}{r_0^{\ 3}} \quad (15.15)$$

The number of nearest neighbours depends on how the molecules are packed within the solid and varies between about 6 for a relatively open-packed material up to 12 for a close-packed material. We might therefore expect the bulk modulus of a material to be between about 1/3 to 2/3 of the Young's modulus. Table 15.1 lists bulk moduli for some materials; you can see that for most (but not all) materials, the bulk modulus is indeed smaller than the Young's modulus. Note that if you look up such values in a different source, you may find the reciprocal of the bulk modulus listed. The reciprocal of the bulk modulus is called the **compressibility** of the material.

Notice from Table 15.1 that both solids and liquids have a bulk modulus. Neither shear modulus (see the next section) nor Young's modulus is given for liquids, however, because liquids flow in response to elongation and shear forces.

Example 15.8

Treasure at the bottom of the ocean

A pirate's chest filled with pieces of eight, gold coins each weighing 6.77 g, sinks with the ship during a storm. How much smaller are the coins when they are at the bottom of the ocean, where they experience a pressure of 2.00×10^7 Pa, than when they were on land, where air pressure was 1.0×10^5 Pa?

Solution

Conceptualise The increased pressure squeezes the gold and reduces its volume.

Model We will assume that the deformation is small, and apply Equation 15.14.

Analyse Solve Equation 15.14 for the volume change of a coin:

$$\frac{\Delta V}{V} = -\frac{\Delta P}{B}$$

Example 15.8 cont.

Substitute numerical values:

$$\frac{\Delta V}{V} = -\frac{(2.00 \times 10^7 - 1.0 \times 10^5 \text{ N/m}^2)}{(18 \times 10^{10} \text{ N/m}^2)} = 1.1 \times 10^{-4}$$

The negative sign indicates that the volume of the coin decreases. Thus the coin's new size is 0.011% less than its original size – not enough for anyone to notice, and small enough to justify our assumption of a small deformation.

15.5 Response of solids to shear forces

Another type of deformation occurs when an object is subjected to a force parallel to one of its faces while the opposite face is held fixed by another force (**Active Figure 15.18a**). This may seem at first sight to be similar to the linear compression described in Section 15.3, but in this case you can imagine the face against which the force acts as rotating through an angle around an axis defined by its contact with the bottom face, as shown in **Figure 15.19**. The stress in this case is called a shear stress. If you imagine the material as being composed of many planes of molecules, a shear force results in the relative movement of these planes. If the object is originally a rectangular block, a shear stress results in a shape whose cross section is a parallelogram. A book pushed sideways as shown in **Active Figure 15.18b** is an example of an object subjected to a shear stress. To a first approximation (for small distortions), no change in volume occurs with this deformation. The object does not collapse so long as the internal, intermolecular forces balance the applied shear force.

Shear stresses occur in many construction and engineering contexts. For example, the load on a roof beam due to the weight of the roofing materials produces a shear stress on the beam as shown in **Figure 15.20**. It is important to know whether the beam will fracture or break under likely loads and so it is useful to have a means of quantifying a material's resistance to shear stress.

We define the **shear stress** as F/A, the ratio of the tangential force to the area A of the face being sheared. The **shear strain** is defined as the ratio $\Delta x/h$, where Δx is the horizontal distance that the sheared face moves and h is the height of the object. In terms of these quantities, the **shear modulus** or **modulus of rigidity** is

Shear modulus ▶

$$S \equiv \frac{\text{shear stress}}{\text{shear strain}} = \frac{F/A}{\Delta x/h} \tag{15.16}$$

The shear strain $\Delta x/h$ is equal to the tangent of the angle the side makes with the normal, γ. If $\Delta x \ll h$, $\tan\gamma \approx \gamma$ (when γ is measured in radians) and we can write

$$S = \frac{F}{\gamma A} \tag{15.17}$$

Σ *Useful approximations, including the small angle approximation, are given in Appendix B.5.*

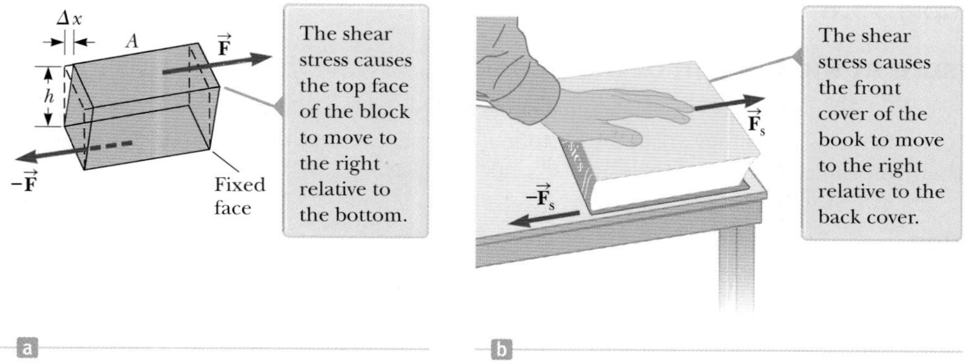
Δx
A
$\vec{F}$
h
$-\vec{F}$
Fixed face
The shear stress causes the top face of the block to move to the right relative to the bottom.
$\vec{F}_s$
$-\vec{F}_s$
The shear stress causes the front cover of the book to move to the right relative to the back cover.
a
b

Active Figure 15.18

(a) A shear deformation in which a rectangular block is distorted by two forces of equal magnitude but opposite directions applied to two parallel faces (b) A book is under shear stress when a hand placed on the cover applies a horizontal force away from the spine.

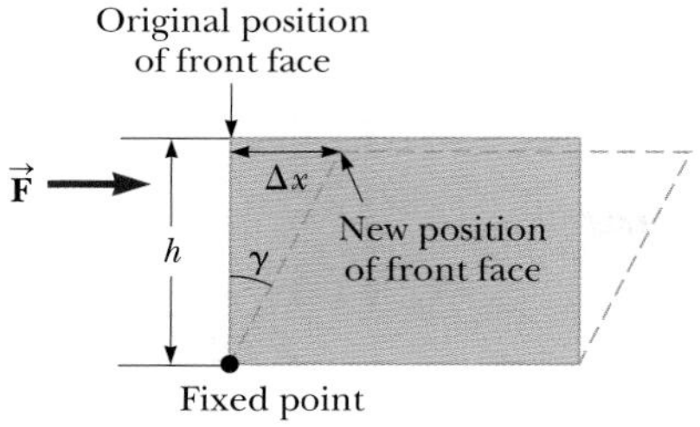

Figure 15.19
Side view of a rectangular block experiencing a shearing stress

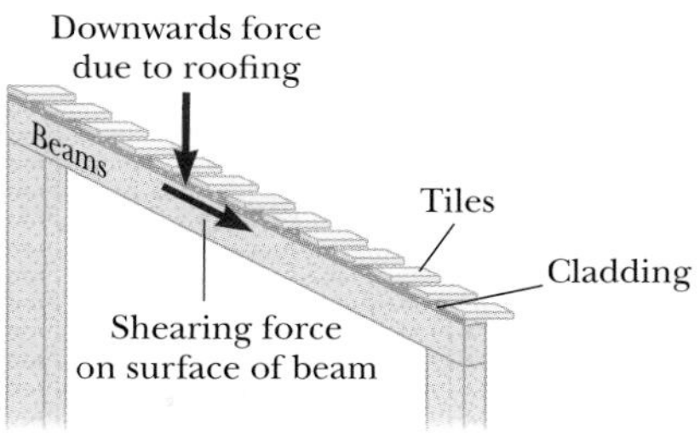

Figure 15.20
Shear stress due to the load carried by a roof beam

Values of the shear modulus for some materials are given in **Table 15.1**. Like Young's modulus, the unit of shear modulus is N/m^2. You can see that the shear modulus is usually a little lower than Young's modulus, but of the same order of magnitude.

Example 15.9

Shear gold?

A particular design of gold bracelet requires cuboid pieces of gold with two opposite side faces shaped like parallelograms. One way of creating the desired shape would be to start with small cubes of gold and subject to them to a shear force to create sloped faces on two of the sides. If the cube dimension is 4 mm, what force must be applied to create an angle to the normal of 15°?

Solution

Conceptualise Start by drawing a diagram to help visualise the problem.

Model The angle, 15°, is not a very small angle, so we will apply **Equation 15.16** rather than **Equation 15.17.**

Analyse Using **Equation 15.16**, and $\Delta x = h \tan \gamma$ gives:

$$S = \frac{F/A}{\Delta x/h} = \frac{F}{h^2 \tan\gamma}$$

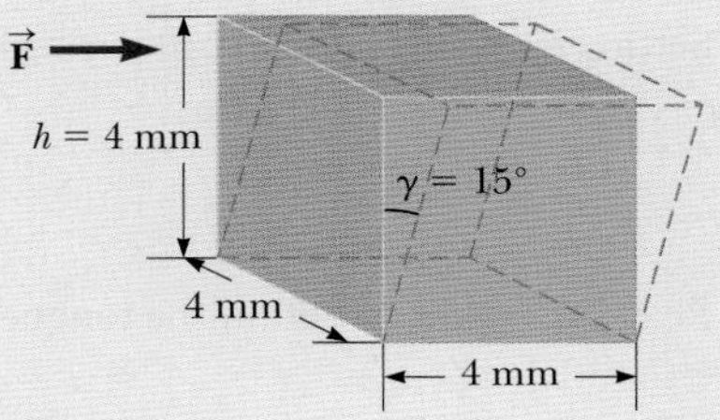

Figure 15.21
(Example 15.9)

Rearranging gives an expression for the required force:

$$F = Sh^2 \tan\gamma$$

Substitute the given values to obtain

$$F = (2.7 \times 10^{10}\ \mathrm{N/m^2})(0.004\ \mathrm{m})^2 \tan(15°) = 1.2 \times 10^5\ \mathrm{N}$$

Finalise The force required to shear the gold is huge, suggesting that this is not a sensible way to create the desired shapes. Instead, it would be better to cut the metal into the desired shape, or to cool molten gold in appropriately shaped moulds.

What If? What if the bracelet were to be made from titanium?

Answer Gold is a relatively soft metal and so we might expect that other metals would require even more force to produce the same shear angle. The shear modulus of titanium is 50% larger than that of gold so 50% more force would be needed.

Example 15.10

Tongue-and-groove joints

Tongue-and-groove timber joints need to be approximately as strong as the wood itself in order to carry useful loads. Consider two wooden beams of width w and height h, to be glued together with a tongue length d as shown in the diagram. The wood is known to break when subjected to tensile stresses at or above σ_{wood}. How long must the joint length d be for the joint to support the same maximum load as the wood itself?

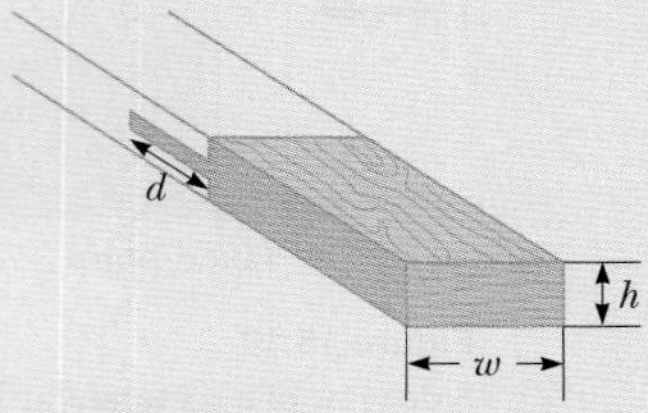

Figure 15.22 (Example 15.10) A tongue-and-groove joint

Solution

Conceptualise We use **Figure 15.22** to help us visualise the problem. An axial load (that is, one applied along a direction parallel to the axis of the two beams) will tend to shear the glue joint; we need to equate the maximum shear force the joint can withstand to the maximum load the wood can support without fracture.

Model This problem requires us to use the concepts of both shear and tensile stress.

Analyse The shear stress in the glue joint is the load divided by the area. Remembering that the shear stress will be applied over *both* sides of the joint, the total area is

$$A = 2wd$$

Let the load that causes the joint to break be F_{glue}. Then the shear stress experienced by the joint as it fails is

$$\sigma_{glue} = \frac{F_{glue}}{2wd}$$

Now the load needed to fracture the wood F_{wood} is equal to the maximum tensile strength multiplied by the cross-sectional area:

$$F_{wood} = \sigma_{wood} hw$$

Setting $F_{glue} = F_{wood}$ (that is, requiring the glue joint to break under the same load conditions as the wood itself) we can write

$$\sigma_{glue} = \frac{\sigma_{wood} hw}{2wd}$$

Rearrange for d to obtain the tongue length in terms of the other parameters:

$$d = \frac{\sigma_{wood} h}{2\sigma_{glue}}$$

σ_{wood} has the same dimensions as σ_{glue}, and h is a length, so our dimensions are okay. ☺

Finalise Note that the length of the glue joint is independent of the width of the beam. This makes sense, since the load is being applied perpendicular to it. Our result tells us that the thicker the wooden beams, the longer the joint has to be. If the characteristic tensile breaking stress of the wood is the same as the shear breaking stress of the glue joint, the tongue length needs to be half the height of the beam. As most wood glues are designed so that the joints are as strong as or stronger than the wood alone, this provides a good rule of thumb for calculating joint lengths.

Quick Quiz 15.5

For the three parts of this Quick Quiz, choose from the following choices the correct elastic modulus that describes the relationship between stress and strain for the system of interest, which is in italics: (a) Young's modulus (b) shear modulus (c) bulk modulus (d) none of those choices **(i)** A *block of iron* is sliding across a horizontal floor. The friction force between the block and the floor causes the block to deform. **(ii)** A trapeze artist swings through a circular arc. At the bottom of the swing, the *wires* supporting the trapeze are longer than when the trapeze artist simply hangs from the trapeze, due to the increased tension in them. **(iii)** A spherical steel satellite is launched into space from Earth. The lower pressure in space causes the radius of the satellite to increase.

15.6 Bending and twisting

Our descriptions of the three types of deformation above refer to relatively simple situations in which the objects have simple geometries and the stresses are of a single type. These examples allow us to develop models of the deformation and the material's response, but it is worth briefly considering some more complex types of deformation.

One very common way in which materials are stressed is **bending** or **flexure**. When a rod or beam is bent as shown in Figure 15.23, it is subject to a combination of tensile and compressive stresses, plus some shear stresses within the material. The shear stresses tend to be much smaller than the normal tensile and compressive stresses, and so can be neglected to first approximation.

Let's consider the stresses shown in Figure 15.23a. The inner arc of the rod is compressed while the outer arc is extended; both exert a restoring force to return the rod to its original position. The restoring force due to the material in the centre of the rod, which exhibits no resulting strain, is negligible. Thus most of the stress in a bent rod or bar is experienced in the outer layers of material. This is why hollow pipes are almost as difficult to bend as solid rods; the solid material in the middle of the rod exerts a smaller restoring force than the material at the edges and can be removed with little change to the object's resistance to bending. This provides a big advantage in construction work and other areas, since strong frameworks can be constructed out of relatively lightweight materials by the use of hollow pipes (for example as scaffolding) rather than solid bars.

When the end of a cantilevered beam is deflected from its equilibrium position due to the application of a load, it bounces back when the load is removed. For small deflections it can be shown that the restoring force is given by

$$F = \frac{-Ywt^3d}{4L^3} \quad (15.18)$$

where w, t and L are the width, thickness and depth of the beam respectively and d is the deflection as indicated in Figure 15.23b. You can see that this equation looks very much like Hooke's law, with the magnitude of the force proportional to but directed opposite to the displacement of the end of the beam from its equilibrium position. As we shall see in the next chapter, motion governed by forces that follow Hooke's law is usually *oscillatory*. Next time you go swimming, watch the motion of the diving board first as someone walks out to its end and then as they jump off.

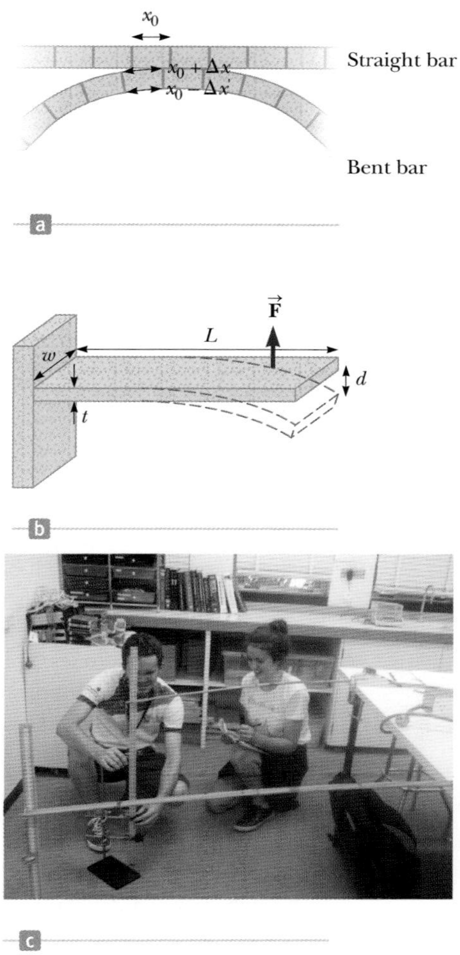

Figure 15.23
(a) A metal rod is bent. (b) Restoring force due to deflection of a beam (c) Students performing a bending beam experiment to measure Y

Twisting a solid material also results in strain. In solid mechanics, the twisting of an object due to an applied torque is called **torsion**. Let's consider a cylindrical wire or bar with one end fixed and the other subjected to a torque so that the whole wire twists, as illustrated in Figure 15.24.

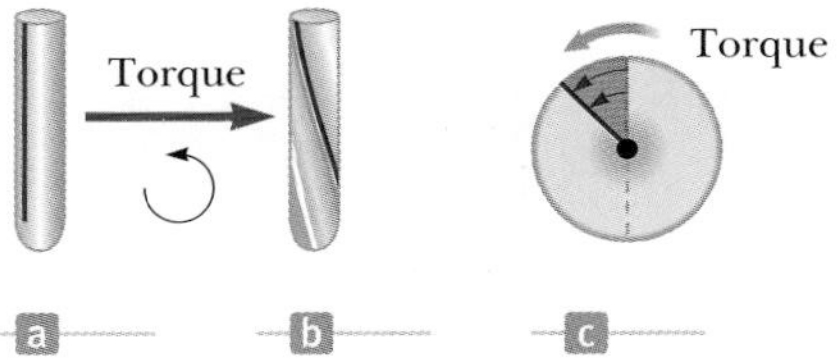

Figure 15.24
Twisting a cylindrical bar: the red line in (a) and (b) shows the displacement of points along the length of the wire caused by the applied torque. The cross section of the wire shown in (c) illustrates how points towards the outer surface of the wire are displaced further than those towards the centre.

If we imagine the wire to be constructed of a series of circular sections, like a stack of circular counters, the resultant shearing stress is perpendicular to the radius. For solid or hollow wires or bars of uniform circular cross section of radius R and length L of material with shear modulus S, an angle of rotation θ is produced by a torque τ:

$$\theta = \frac{\tau L}{JS} = \frac{\sigma L}{RS} \quad (15.19)$$

where σ is the applied shear stress and J is the torsion constant of the wire or bar, which for a solid cylinder is:

$$J = \frac{\pi R^4}{2}$$

For a hollow pipe of inner radius R_i and outer radius R_o,

$$J = \frac{\pi(R_o^{\,4} - R_i^{\,4})}{2}$$

Let's consider the implications of this. Equation 15.19 shows that for two cylindrical pipes, identical in all ways except that one is solid and the other is hollow, the ratio of the angular displacements produced by an applied torque will be

$$\frac{\theta_{\text{solid}}}{\theta_{\text{hollow}}} = \frac{J_{\text{hollow}}}{J_{\text{solid}}}$$

If the two pipes both have outer radius R, with the inner radius of the hollow pipe αR, we obtain

$$\frac{\theta_{\text{solid}}}{\theta_{\text{hollow}}} = \frac{(R^4 - \alpha^4 R^4)}{R^4}$$
$$= 1 - \alpha^4$$

This means that hollow pipes, although not as resistant to torsion as solid pipes, can still withstand significant twisting. For example, if the inner radius is half the outer radius (that is, $\alpha = 0.5$), the angle produced by an applied torque will be less than 7% greater than that produced by the same torque on a solid pipe of the same outer radius. This is consistent with our expectations based on Figure 15.24c that most of the stress is experienced in the outer part of the pipe and so hollow (and therefore relatively light) pipes can often be used instead of solid ones. This makes sense if we consider the stresses applied to the material being twisted; just as with bending, the molecules along the central axis of the bar experience the smallest displacement from their equilibrium positions, while those at the outside are twisted furthest.

Equation 15.19 also shows that the applied torque and the angular displacement are directly proportional to each other

$$\tau = \frac{k}{S}\theta$$

where k is a constant depending on the dimensions and geometry of the object being twisted. This should remind you once again of Hooke's law. We shall look at one application of torsion – the torsional pendulum – in Section 16.5.

End-of-chapter resources

Corbis/David Sailors

At the start of this chapter, we asked why concrete is often reinforced with steel rods, which apply a compressive stress to the concrete. We know that if the stress on a solid object exceeds a certain value, the object fractures. The maximum stress that can be applied before fracture occurs is called the *tensile strength*, *compressive strength* or *shear strength*, depending on whether the applied stress compresses, elongates or shears the material; these breaking stresses are characteristic of the material.

With no reinforcement or compressive stress, concrete has a tensile strength of about 2×10^6 N/m², a compressive strength of 20×10^6 N/m², and a shear strength of 2×10^6 N/m². Because it is much stronger under compression than under tension or shear, vertical columns of concrete can support very heavy loads, whereas horizontal beams of concrete tend to sag and crack as shown in Figure 15.25a. Unfortunately, concrete is most easily cast in thin sections or slabs, which are very brittle. They can be strengthened by the use of steel rods to reinforce the concrete as illustrated in Figure 15.25b. This results in an increase in the overall tensile strength of the material because the tensile strength of the steel (typically 200–300 $\times$ 10^6 N/m²) is much higher than that of concrete; however, there remains a danger that the concrete will crack or fracture even while the steel rods remain

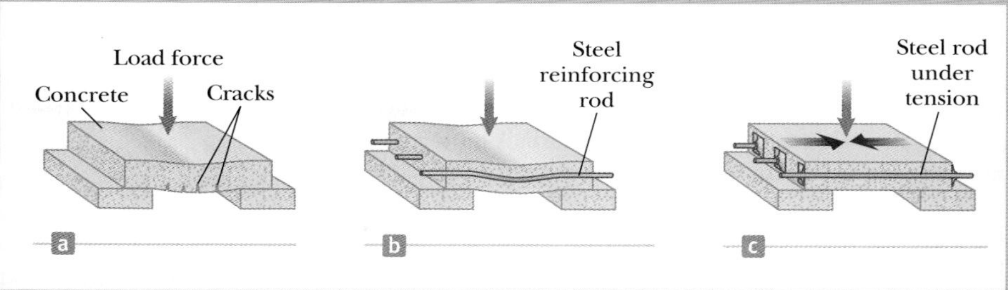

Figure 15.25
(a) A concrete slab with no reinforcement tends to crack under a heavy load. (b) The strength of the concrete is increased by using steel reinforcement rods. (c) The concrete is further strengthened by prestressing it with steel rods under tension.

undamaged. A significant increase in shear and tensile strength is achieved if the reinforced concrete is prestressed as shown in Figure 15.25c. As the concrete is being poured, the steel rods are held under tension by external forces. The external forces are released after the concrete cures; the result is a permanent tension in the steel and hence a compressive stress on the concrete. Any applied tensile stress now serves to counteract the existing compressive stress, resulting in little or no net stress on the concrete. In addition, the shear strength of prestressed concrete can be as much as 40 times larger than concrete alone. Prestressed concrete slabs can therefore support much heavier loads, allowing the use of thinner slabs. It is the main material used for upper-storey floors in skyscrapers and containment vessels in nuclear reactors.

The problems found in this chapter may be assigned online in Enhanced Web Assign.

ENHANCED WebAssign

Worked solutions to every fifth problem are available in the Student Solutions Manual. Register online at **www.cengagebrain.com** for access.

Summary

Definitions

We can describe the elastic properties resulting from intermolecular potentials using the concepts of stress and strain. **Stress** is a quantity proportional to the force producing a deformation; **strain** is a measure of the degree of deformation. Stress is proportional to strain, and the constant of proportionality is the **elastic modulus**:

$$\text{Elastic modulus} \equiv \frac{\text{stress}}{\text{strain}} \tag{15.9}$$

The response of a solid to tensile stress is characterised by Young's modulus:

$$Y \equiv \frac{\text{tensile stress}}{\text{tensile strain}} = \frac{F/A}{\Delta L/L_0} \tag{15.10}$$

The response of a solid to change in pressure is characterised by the bulk modulus or modulus of rigidity:

$$B \equiv \frac{\text{volume stress}}{\text{volume strain}} = -\frac{\Delta F/A}{\Delta V/V_0} = -\frac{\Delta P}{\Delta V/V_0} \tag{15.14}$$

The response of a solid to shear stress is characterised by the shear modulus:

$$S \equiv \frac{\text{shear stress}}{\text{shear strain}} = \frac{F/A}{\Delta x/h} \tag{15.16}$$

Concepts and principles

A crystalline solid is one in which constituent ions or molecules occupy regularly spaced, geometric patterns that can be described as sites on a lattice. Crystalline solids have sharply defined melting points.

An amorphous solid is one in which there is no short- or long-range order. Amorphous solids have no sharply defined melting point.

Neutral molecules interact via forces that are repulsive at short range and attractive at long range, giving rise to equilibrium separations where the net force on one molecule due to the other is zero.

An estimate of the average intermolecular spacing r_0 can be obtained from the molar mass and density of a substance:

$$r_0 \approx \left(\frac{M}{\rho N_A}\right)^{1/3} \tag{15.1}$$

The force on one molecule due to another can be written in the form

$$F = A\left[\left(\frac{r_0}{r}\right)^m - \left(\frac{r_0}{r}\right)^n\right] \tag{15.2}$$

The relative potential energy of two neutral molecules is reasonably well approximated by the Lennard-Jones potential

$$U_{LJ}(r) = \epsilon\left[\left(\frac{r_0}{r}\right)^{12} - 2\left(\frac{r_0}{r}\right)^6\right] \tag{15.4}$$

which has an associated force

$$F = \frac{12\epsilon}{r_0}\left[\left(\frac{r_0}{r}\right)^{13} - \left(\frac{r_0}{r}\right)^7\right] \tag{15.5}$$

The potential energy curve for ions takes a form such as

$$U_{\text{ionic}} = \frac{a}{r^9} - \frac{b}{r} \quad (15.6)$$

where b is positive if the ions have unlike charges.

The intermolecular forces give rise to the bulk properties of solid materials, including resistance to various types of deformation.

Materials that have been deformed but for which the elastic limit has not been exceeded store potential energy, which is released when the deforming stress is removed. Three common types of deformation are represented by (1) the resistance of a solid to elongation under a load, characterised by **Young's modulus** Y; (2) the resistance of a solid to the motion of internal planes sliding past each other, characterised by the **shear modulus** S; and (3) the resistance of a solid or fluid to a volume change, characterised by the **bulk modulus** B.

The energy stored in a linearly compressed or elongated solid of constant cross-sectional area A and unstressed length L_0 is

$$E_{\text{stored}} = \frac{YA\Delta L^2}{2L_0} \quad (15.12)$$

Chapter review quiz

To help you revise Chapter 15: Solids, complete the automatically graded Chapter review quiz at http://login.cengagebrain.com.

Conceptual questions

1. Very old glass window panes such as those in medieval cathedrals in Europe are often thicker at the bottom than at the top. Why?
2. What tests could you perform to find out whether a particular material is crystalline or amorphous?
3. Does the density of a material depend on the average intermolecular separation? Explain your answer.
4. Why do we introduce the concept of stress and not just describe the response of solid objects to applied force?
5. What do we mean when we say a material is elastic? Is elasticity an innate property of a material? Explain your answer.
6. Give examples of situations where it is important to be able to accurately predict the response of a given material to each of the three types of applied stress (tensile, bulk compressive and shear).
7. Explain the difference between the changes in microscopic structure under tensile and shear stresses.
8. Which is easier to break, a thin or a thick piece of the same kind of elastic? Why?
9. Can Poisson's ratio ever exceed ½? Explain your answer.
10. What kind of deformation does a cube of jelly exhibit when it jiggles?
11. When a cantilever beam is bent by the application of a load at the free end, do all parts of the beam produce a restoring force? Explain your answer.
12. Why is scaffolding made of hollow, rather than solid, pipes?

Problems

Section 15.1 Characteristics of solids and intermolecular potentials

1. Estimate the average distance between molecules or ions in (a) sodium chloride (table salt), (b) lead and (c) diamond. Look up any properties you need using a reputable source (which means that anything obtained from Wikipedia needs to be verified by another source).
2. Identify three amorphous solids, explaining why you think they are amorphous.
3. Identify three crystalline solids, explaining why you think they are crystalline.
4. Two molecules interact via a Lennard-Jones type of force, with well depth $\epsilon = 10^{-20}$ J and equilibrium separation $r_0 = 0.5$ nm. How much work must be done to increase their separation by 10%?
5. Two molecules interact via a Lennard-Jones type of force, with well depth $\epsilon = 0.5 \times 10^{-20}$ J and equilibrium separation $r_0 = 0.4$ nm. By how much does their relative potential energy change if their separation is reduced by 1%?
6. An alternative formulation of the intermolecular potential energy curve, as proposed by Richard A. Buckingham, is

$$U = \gamma\left[e^{-r/r_0} - \left(\frac{r_0}{r}\right)^6\right]$$

(a) Plot this as a function of the separation r. (b) What is the significance of γ? (c) Derive an expression for the force on one molecule due to another when their relative potential energy is described by this expression.

7. The bond between two ions in sodium chloride can be approximated by $U = ar^{-9} - br^{-1}$. (a) Sketch the restoring force as a function of r. (b) Find the equilibrium separation r_0 in terms of the two constants a and b.

Section 15.2 Elasticity, stress and strain

8. Explain how potentials such as the Lennard-Jones potential can be used to explain Hooke's law.
9. If the force at distances close to the equilibrium separation between two molecules can be written as $F = -k\Delta r$, what is the potential energy curve? Using sketches to illustrate your answer, explain how this relates to the Lennard-Jones potential energy curve.
10. In a certain physics lab, you are provided with samples of fishing line, horse hair, elastic and cotton, together with a retort stand, a ruler and a set of weights. You are asked to find out as much about the elastic properties of any one of these materials as you can. Describe what measurements you would take and how you would analyse them. How would you determine whether the elastic limit had been exceeded?

Section 15.3 Response of solids to changes in length

11. In the physics lab described in the previous question, you ±? choose to investigate the properties of fishing line. You obtain the following data. Can you determine Young's modulus for the

fishing line? If so, do. Is the proportional limit exceeded during the experiment?

Load (g) ± 10 g	0	200	400	600	800	1000	1500	2000	3000	5000
Length (m) ± 1 mm	1.000	1.001	1.002	1.003	1.004	1.006	1.011	1.021	1.058	broke

12. ±? A (200 ± 1) kg load is hung on a wire of length (4.000 ± 0.001) m, cross-sectional area $(0.20 \pm 0.01) \times 10^{-4}$ m², and Young's modulus $(8.00 \pm 0.05) \times 10^{10}$ N/m². What is its increase in length?
13. Assume Young's modulus for bone is 1.50×10^{10} N/m². The bone breaks if stress greater than 1.50×10^{8} N/m² is imposed on it. (a) What is the maximum force that can be exerted on the femur bone in the leg if it has a minimum effective diameter of 2.50 cm? (b) If this much force is applied compressively, by how much does the 25.0 cm bone shorten?
14. A steel wire of diameter 1 mm can support a tension of 0.2 kN. A steel cable to support a tension of 20 kN should have diameter of what order of magnitude?
15. A 2.00 m-long cylindrical steel wire with a cross-sectional diameter of 4.00 mm is placed over a light, frictionless pulley. An object of mass m_1 = 5.00 kg is hung from one end of the wire and an object of mass m_2 = 3.00 kg is hung from the other end as shown in Figure P15.15. The objects are released and allowed to move freely. Compared with its length before the objects were attached, by how much has the wire stretched while the objects are in motion?

Figure P15.15

16. A metal wire of diameter 1.0 mm has Young's modulus $Y = 2.0 \times 10^{11}$ N/m² and initial length L = 4.00 m. What load must be suspended from it to produce an extension of 1.0 mm? What is the energy stored in the wire under this tension?
17. Show that the energy stored per unit volume in a stretched wire is equal to half the applied stress multiplied by the resultant strain.
18. ±? A wire of initial length (5.00 ± 0.01) m and radius (1.00 ± 0.02) mm is extended by (1.5 ± 0.1) mm when subjected to a tension of 100 N. What is the stored energy per unit volume?

Section 15.4 Response of solids to changes in pressure

19. When water freezes, it expands by about 9.00%. What pressure increase occurs inside your domestic water pipes if the water inside freezes in winter? (The bulk modulus of ice is 2.00×10^{9} N/m².)
20. The deepest point in the ocean is in the Mariana Trench, about 11 km deep, in the Pacific. The pressure at this depth is huge, about 1.13×10^{8} N/m². (a) Calculate the change in volume of 1.00 m³ of seawater carried from the surface to this deepest point. (b) The density of seawater at the surface is 1.03×10^{3} kg/m³. Find its density at the bottom. (c) Explain whether or when it is a good approximation to think of water as incompressible.
21. The NASA spacecraft *Stardust* (monitored by the Deep Space Tracking Station at Tidbinbilla) used aerogel (sometimes known as *solid smoke*) to trap particles from the tail of comet Wild 2 and bring them back to Earth for analysis. Several different types of aerogel can be manufactured. These have different elastic properties, and NASA scientists had to carefully choose which type to use. Consider two types of aerogel, type A with bulk modulus 32.21 MPa and type B with bulk modulus 546 kPa. By what fraction will each expand when taken from the Earth's surface to outer space? Take the pressure in space to be approximately 10^{-15} Pa. Which do you think would be a better choice for the *Stardust* mission?

Section 15.5 Response of solids to shear forces

22. A rectangular besser block of length 30 cm, height 15 cm and width 15 cm is fixed to a concrete garden path with a layer of mortar 0.5 cm deep. If the mortar has a shear modulus of 2.00 GPa, what will the resulting shear angle be if the besser block is struck on its small end with a mallet producing an applied force of 500 N?
23. ±? A child slides across a floor in a pair of rubber-soled shoes. The friction force acting on each foot is (20.0 ± 0.5) N. The footprint area of each shoe sole is (14.0 ± 0.5) cm² and the thickness of each sole is (5.0 ± 0.1) mm. Find the horizontal distance by which the upper and lower surfaces of each sole are offset. The shear modulus of the rubber is 3.00 MN/m².
24. Assume that if the shear stress in steel exceeds about 4.00×10^{8} N/m² the steel ruptures. Determine the shearing force necessary to (a) shear a steel bolt 1.00 cm in diameter and (b) punch a 1.00 cm diameter hole in a steel plate 0.500 cm thick.
25. Two strips of metal are riveted together at their ends by four rivets, each of diameter 0.5 cm. What is the maximum tension that can be exerted by the riveted strip if the shearing stress on the rivets is not to exceed 600 MPa?

Section 15.6 Bending and twisting

26. A diver of mass 60 kg stands relaxed at the end of a diving board of length 2.0 m, width 30.0 cm and thickness 3.0 cm. The diving board is made of a material with Young's modulus 3.0×10^{9} Pa. What is the downwards deflection of the end of the board? When the diver launches herself from the board, she exerts an additional downwards force equal to her weight. What is the maximum deflection of the end of the board as she dives from it?
27. ±? The Young's modulus of a material can be measured by taking a thin sample of it, fixing one end firmly and applying a load to the other end, effectively creating a cantilevered beam. Sensitive measurements of the film's deflection can be obtained using interference techniques (see Chapter 18), which can measure deflections of the order of the wavelength of the light being used. The lab you work for has been provided with a sample of a mystery material of length 5.00 cm, width 2.00 cm and thickness 0.05 cm. You apply a load of (50.0 ± 0.5) g to the free end and measure a deflection of (0.98 ± 0.01) mm. What is the Young's modulus of the sample?
28. A torque of 100 N.m is applied to a copper cylinder of length 1.00 m and radius 3.0 cm. What is the resulting angle of twist?
29. Two cylindrical metal bars are identical except that one is hollow with an inner radius 2/3 of the outer radius. How much bigger is the angle through which the hollow bar is twisted compared to the solid bar, when the same torque is applied to each?

30. Two wires of cylindrical cross section, one with twice the diameter of the other, but both of the same length, are made from different materials. Twice the shear stress is applied to the thinner wire than the thicker, but the resulting twist angles are found to be the same. How are the shear moduli of the materials related?

Additional problems

31. A footbridge connecting two terminals at an airport is supported at numerous points along its edges by a vertical cable above each point and a vertical column underneath. The steel cable is 1.27 cm in diameter and is 5.75 m long before loading. The aluminium column is a hollow cylinder with an inside diameter of 16.14 cm, an outside diameter of 16.24 cm, and an unloaded length of 3.25 m. When the footbridge exerts a load force of 8500 N on one of the support points, how much does the point move down?

32. The lintel of pre-stressed reinforced concrete in Figure P15.32 is 1.50 m long. The concrete encloses one steel reinforcing rod with cross-sectional area 1.50 cm². The rod joins two strong end plates. The cross-sectional area of the concrete perpendicular to the rod is 50.0 cm². Young's modulus for the concrete is 30.0 × 10⁹ N/m². After the concrete cures and the original tension T_1 in the rod is released, the concrete is to be under compressive stress 8.00 × 10⁶ N/m². (a) By what distance will the rod compress the concrete when the original tension in the rod is released? (b) What is the new tension T_2 in the rod? (c) How much longer than the unstressed length will the rod then be? (d) By what extension distance should the rod have been stretched when the concrete was poured? (e) Find the required original tension T_1 in the rod.

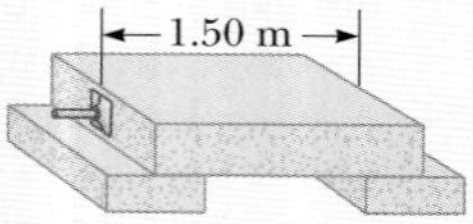

Figure P15.32

33. The Young's modulus of a material can be measured by taking a thin sample of it, fixing one end firmly and applying a load to the other end, effectively creating a cantilevered beam. Sensitive measurements of the film's deflection can be obtained using interference techniques (see Chapter 18), which can measure deflections of the order of the wavelength of the light being used. The lab you work for has been provided with a sample of mystery material of length 5.00 cm, width 1.00 cm and thickness 0.40 mm. You apply increasing loads to the end of the sample and measure the deflections, obtaining the data recorded in the table below. What is the Young's modulus of the material?

Load (g) ± 1 g	50	100	150	200	250	300	350
Deflection (mm) ± 0.005 mm	0.425	0.840	1.270	1.705	2.130	2.600	2.975

34. Review. A 30.0-kg hammer, moving with speed 20.0 m/s, strikes a steel spike 2.30 cm in diameter. The hammer rebounds with speed 10.0 m/s after 0.110 s. What is the average strain in the spike during the impact?

35. When a gymnast performing on the rings executes the *iron cross*, he maintains the position at rest shown in Figure P15.35a. In this manoeuvre, the gymnast's feet (not shown) are off the floor. The primary muscles involved in supporting this position are the latissimus dorsi and the pectoralis major. One of the rings exerts an upwards force $\vec{F}_h$ on a hand as shown in Figure P15.35b. The force $\vec{F}_s$ is exerted by the shoulder joint on the arm. The latissimus dorsi and pectoralis major muscles exert a total force $\vec{F}_m$ on the arm. (a) Using the information in the figure, find the magnitude of the force $\vec{F}_m$ for an athlete of weight 750 N. (b) Suppose an athlete in training cannot perform the iron cross but can hold a position similar to the figure in which the arms make a 45° angle with the horizontal rather than being horizontal. Why is this position easier for the athlete?

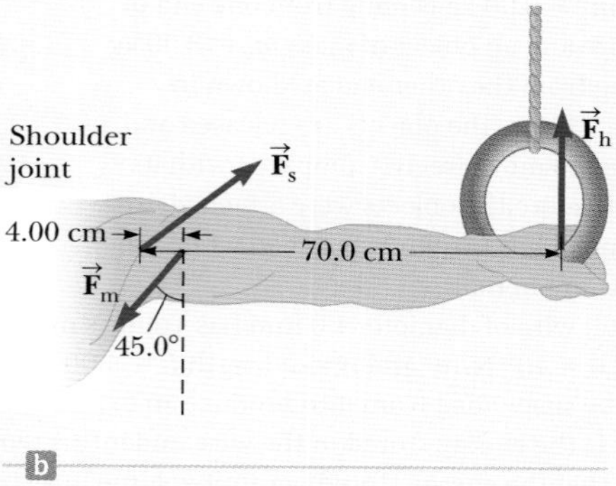

Figure P15.35

36. Figure P15.36 shows a light truss formed from three struts lying in a plane and joined by three smooth hinge pins at their ends. The truss supports a downwards force of $\vec{F}$ = 1000 N applied at the point *B*. The truss has negligible weight. The piers at *A* and *C* are smooth. (a) Given θ_1 = 30.0° and θ_2 = 45.0°, find n_A and n_C. (b) One can show that the force any strut exerts on a pin must be directed along the length of the strut as a force of tension or compression. Use that fact to identify the directions of the forces that the struts exert on the pins joining them. Find the force of tension or of compression in each of the three bars.

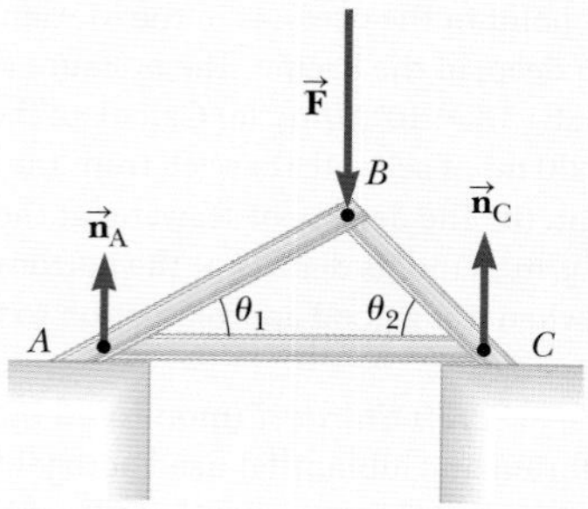

Figure P15.36

37. A wire of length L, Young's modulus Y, and cross-sectional area A is stretched elastically by an amount ΔL. By Hooke's law, the restoring force is $-k\,\Delta L$. (a) Show that $k = YA/L$. (b) Show that the work done in stretching the wire by an amount ΔL is

$$W = \frac{\frac{1}{2}YA\Delta L^2}{L}$$

38. A walkway suspended across a hotel lobby is supported at numerous points along its edges by a vertical cable above each point and a vertical column underneath. The steel cable is 1.27 cm in diameter and 5.75 m long before loading. The aluminium column is a hollow cylinder with an inside diameter of 16.14 cm, an outside diameter of 16.24 cm, and an unloaded length of 3.25 m. When the walkway exerts a load force of 8500 N on one of the support points, how much does the point move down?

39. ±? An aluminium wire is 0.850 m long and has a circular cross section of diameter 0.780 mm. Fixed at the top end, the wire supports a 1.20 kg object that swings in a horizontal circle. Determine the angular velocity of the object required to produce a strain of 1.00×10^{-3}.

Challenge problems

40. ±? Estimate the force with which a karate master strikes a board, assuming the hand's speed at the moment of impact is (10.00 ± 0.01) m/s and decreases to (1.00 ± 0.01) m/s during a (2.00 ± 0.001) ms time interval of contact between the hand and the board. The mass of his hand and arm is (1.0 ± 0.1) kg. (b) Estimate the shear stress, assuming this force is exerted on a (1.0 ± 0.1) cm-thick pine board that is (10.0 ± 0.1) cm wide. (c) If the maximum shear stress a pine board can support before breaking is 3.60×10^6 N/m^2, will the board break?

41. ±? In a factory, square plates of aluminium of side length (20.0 ± 0.1) cm and thickness (1.0 ± 0.1) cm are subjected to shear forces to produce diamond-shaped plates. If a shear angle of 12° is required, what force must be applied?

42. Show that Poisson's ratio is equal to $\frac{1}{2}$ when volume is conserved. (Hint: Consider infinitesimal changes to the length and width.)

43. A hungry bear weighing 700 N walks out on a beam in an attempt to retrieve a basket of goodies hanging at the end of the beam (Figure P15.43). The beam is uniform, weighs 200 N and is 6.00 m long, and is supported by a wire at an angle of $\theta = 60.0°$. The basket weighs 80.0 N. (a) Draw a force diagram for the beam. (b) When the bear is at $x = 1.00$ m, find the tension in the wire supporting the beam and the components of the force exerted by the wall on the left end of the beam. (c) **What If ?** If the wire can withstand a maximum tension of 900 N, what is the maximum distance the bear can walk before the wire breaks?

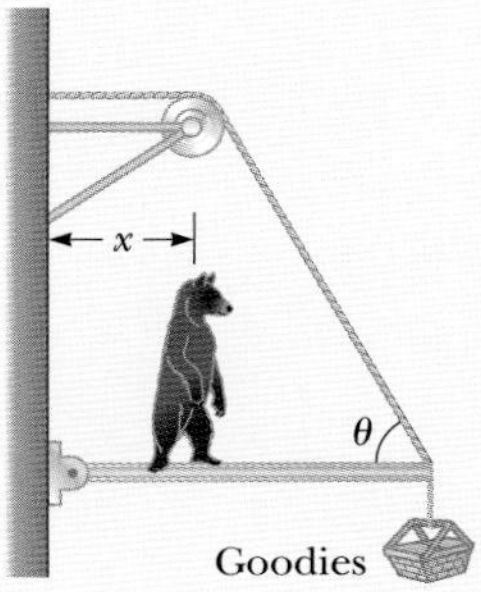

Figure P15.43

44. When you are walking rapidly downhill, the force on your knee cartilage can be up to eight times your body weight. Depending on the angle of the slope you are descending, this can result in a large and potentially damaging shear force on the cartilage. Typically, the cartilage has an area of about 11.0 cm^2 and a shear modulus of 12.0×10^6 N/m^2. If you weigh 80 kg and the force on the cartilage is directed at an angle of 15° to the normal, what will the resulting shear angle be?

Oscillations and mechanical waves

PART

3

Since the first public exhibition of the Foucault pendulum in Paris in 1851, more than 100 Foucault pendulums have been installed in universities, public libraries, schools and museums all over the world. The Foucault pendulum is a large pendulum that swings in a plane due to the Earth's rotation around its own axis. The plane of oscillation appears to rotate relative to its local surroundings. String length, force of gravity, degree of rotation and latitude are the variables that affect the period of the swing and the degree of rotation.

We now study a type of motion called *periodic* motion, the repeating motion of an object in which it continues to return to a given position after a fixed time interval, such as the pendulum shown in the photo. The repetitive movements of such an object are called *oscillations*. We will first focus our attention on a special case of periodic motion called *simple harmonic motion*. All periodic motions can be modelled as combinations of simple harmonic motions.

Simple harmonic motion forms the basis for our understanding of *mechanical waves*. Sound waves, seismic waves, waves on stretched strings and water waves are all produced by some source of oscillation. As a sound wave travels through the air, elements of the air oscillate back and forth; as a water wave travels across a pond, elements of the water oscillate up and down, and backwards and forwards.

To explain many other phenomena in nature, we must understand the concepts of oscillations and waves. For instance, although skyscrapers and bridges appear to be rigid, they actually oscillate, something the architects and engineers who design and build them must take into account. Much of what scientists have learned about atomic structure has come from information carried by waves. Finally, most of the information we obtain about the world around us, and most of our communication with other people, is carried by waves – light waves and sound waves.

chapter 16

Oscillatory motion

Taipei 101, with 101 floors above ground, was the world's tallest building until the completion of the Burj Kalifa in Dubai in 2010. This 660-tonne metal sphere, made of 41 steel plates, swings like a pendulum under computer control. This sphere is part of a damper near the top of the Taipei 101. What is the purpose of this system and how does it work?

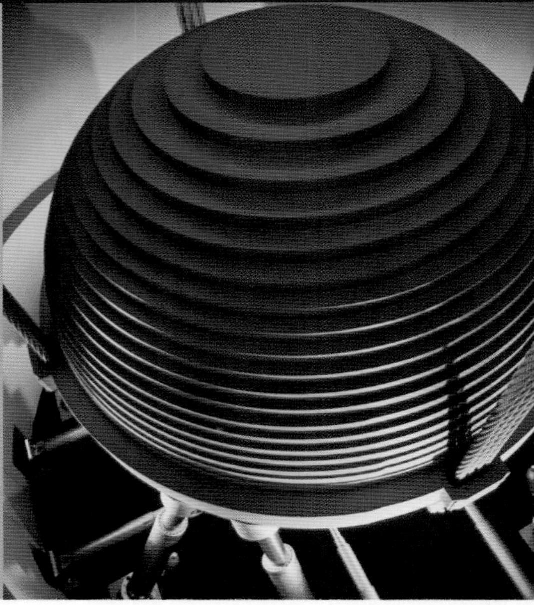

Alamy/Ranjit Doroszkeiwicz

On completing this chapter, students will understand:

- what simple harmonic motion is
- how simple harmonic motion is described mathematically
- that circular motion and simple harmonic motion are related
- how pendulums can be used for timekeeping
- that most oscillations are damped
- how driving oscillations increase the amplitude of oscillation
- that forced oscillations can result in resonance.

Students will be able to:

- describe simple harmonic motion
- identify systems undergoing simple harmonic motion
- apply the analysis model 'particle in simple harmonic motion'
- analyse oscillating systems using an energy approach
- calculate the period of simple and physical pendulums
- analyse the behaviour of oscillators subject to a damping and/or driving force.

Periodic motion is motion of an object that regularly returns to a given position after a fixed time interval. There are many examples of this: any object that swings like a pendulum, including the motion of your limbs as you walk, and any object that behaves as if it bounces on a spring, such as a car's suspension system. The molecules in a solid oscillate about their equilibrium positions; electromagnetic waves, such as light waves, are characterised by oscillating electric and magnetic field vectors. This motion can be modelled as *simple harmonic motion*. The simple harmonic motion model is very important and useful because of the huge number of situations it can be used to describe and analyse.

ENHANCED WebAssign

The problems found in this chapter may be assigned online in Enhanced Web Assign.

16.1 Motion of an object attached to a spring

Consider a block of mass m attached to the end of a spring, with the block free to move on a frictionless, horizontal surface (Active Figure 16.1). When the spring is neither stretched nor compressed, the block is at rest at the position called the **equilibrium position** of the system, which we identify as $x = 0$ (Active Figure 16.1b). The block oscillates back and forth if disturbed from its equilibrium position.

We can understand the oscillating motion of the block in Active Figure 16.1 qualitatively by first recalling that when the block is displaced to a position x, the spring exerts on the block a force that is proportional to the position and given by **Hooke's law** (see Section 6.3):

$$F_s = -kx \qquad (16.1)$$

◀ Hooke's law

We call F_s a **restoring force** because it is always directed toward the equilibrium position and hence *opposite* the displacement of the block from equilibrium. It acts to restore the block to the equilibrium position. That is, when the block is displaced to the right of $x = 0$ in Active Figure 16.1a, the position is positive and the restoring force is directed to the left. When the block is displaced to the left of $x = 0$ as in Figure 16.1c, the position is negative and the restoring force is directed to the right.

When the block is displaced from the equilibrium point and released, it is subject to a net force and consequently undergoes an acceleration. Applying Newton's second law to the motion of the block, with Equation 16.1 providing the net force in the x direction, we obtain

$$-kx = ma_x$$

$$a_x = -\frac{k}{m}x \qquad (16.2)$$

That is, the acceleration of the block is proportional to its position and the direction of the acceleration is opposite the direction of the displacement of the block from equilibrium. Systems that behave in this way exhibit **simple harmonic motion**. An object moves with simple harmonic motion whenever its acceleration is proportional to its position and is oppositely directed to the displacement from equilibrium.

◀ Simple harmonic motion

If the block in Active Figure 16.1 is displaced to a position $x = A$ and released from rest, its *initial* acceleration is $-kA/m$. When the block passes through the equilibrium position $x = 0$, its acceleration is zero. At this instant, its speed is a maximum because the acceleration changes sign. The block then continues to travel to the left of equilibrium with a positive acceleration and finally reaches $x = -A$, at which time its acceleration is $+kA/m$ and its speed is again zero as discussed in Sections 6.3 and 6.6. The block completes a full cycle of its motion by returning to the original position, again passing through $x = 0$ with maximum speed. Therefore, the block oscillates between the turning points $x = \pm A$. In the absence of friction, this idealised motion will continue forever because the force exerted by the spring is conservative. Real systems are generally subject to friction, so they do not oscillate forever. We shall explore the details of the situation with friction in Section 16.6.

Pitfall Prevention 16.1

All the results we discuss for the horizontal spring also apply to a vertical spring except that for the vertical spring we define the $x = 0$ position as the rest position of the block, as the weight of the block causes the spring to extend. Hence the equilibrium position is not the same as the unstretched length of the spring in this case.

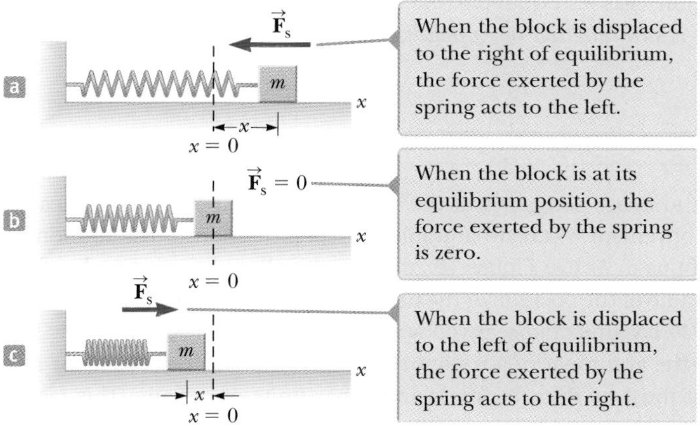
a
b
c
$\vec{F}_s$
$\vec{F}_s = 0$
m
x
$x = 0$
When the block is displaced to the right of equilibrium, the force exerted by the spring acts to the left.
When the block is at its equilibrium position, the force exerted by the spring is zero.
When the block is displaced to the left of equilibrium, the force exerted by the spring acts to the right.

Active Figure 16.1
A block attached to a spring moving on a frictionless surface

Quick Quiz 16.1

A block on the end of a spring is pulled to position $x = A$ and released from rest. In one full cycle of its motion, through what total distance does it travel? (a) $A/2$ (b) A (c) $2A$ (d) $4A$

TRY THIS

Get a rubber band and hang an object from the end. Stretch it downwards and release it, and then watch what happens. How does the rubber band's behaviour compare with your answer to the *Quick Quiz?*

Pitfall Prevention 16.2

The acceleration of a particle in simple harmonic motion is not constant. Equation 16.3 shows that its acceleration varies with position x. Therefore, we *cannot* apply the particle with constant acceleration analysis model in this situation.

16.2 Analysis model: particle in simple harmonic motion

The motion described in the preceding section occurs so often that we identify the **particle in simple harmonic motion** analysis model to analyse such situations. We will show that the model is a valid approximation when the displacement of the object from equilibrium is small, but even when this is not the case, or when the force is not exactly proportional to the displacement, the simple harmonic motion model can still often be used as a first approximation, as we shall see later in the quantum mechanics chapters.

We develop a mathematical representation for this model, starting by choosing x as the axis along which the oscillation occurs. Recall that, by definition, $a = dv/dt = d^2x/dt^2$, so we can express Equation 16.2 as

$$\frac{d^2x}{dt^2} = -\frac{k}{m}x \tag{16.3}$$

This is a second order differential equation. We seek a function whose second derivative is the same as the original function with a negative sign and multiplied by the ratio of the spring constant to the mass, k/m. The trigonometric functions sine and cosine exhibit this behaviour, so we can build a solution around one or both of them. The following cosine function is a solution to the differential equation:

Position versus time for a particle in simple harmonic motion ▶

$$x(t) = A\cos\left(\sqrt{\frac{k}{m}}t + \phi\right) \tag{16.4}$$

where A, k, m and ϕ are constants. Taking the derivatives:

$$\frac{dx}{dt} = A\frac{d}{dt}\cos\left(\sqrt{\frac{k}{m}}t + \phi\right) = -A\sqrt{\frac{k}{m}}\sin\left(\sqrt{\frac{k}{m}}t + \phi\right) \tag{16.5}$$

$$\frac{d^2x}{dt^2} = -A\sqrt{\frac{k}{m}}\frac{d}{dt}\sin\left(\sqrt{\frac{k}{m}}t + \phi\right) = -A\frac{k}{m}\cos\left(\sqrt{\frac{k}{m}}t + \phi\right) \tag{16.6}$$

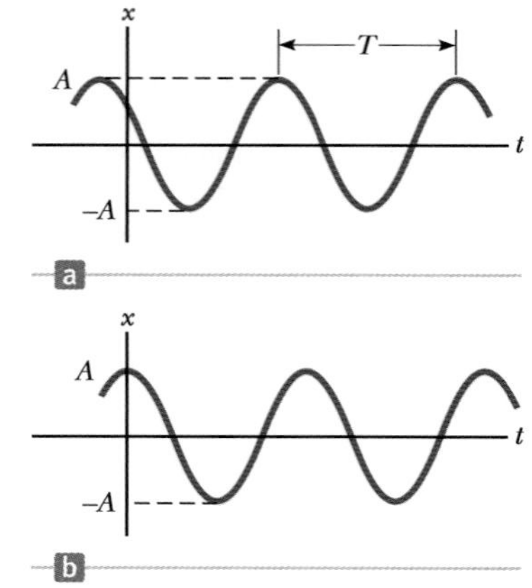

Active Figure 16.2
(a) An x–t graph for a particle undergoing simple harmonic motion. The amplitude of the motion is A and the period (defined below in Equation 16.9) is T. (b) The x–t graph for the special case in which $x = A$ at $t = 0$ and hence $\phi = 0$

We can see that this solution satisfies Equation 16.3.

Σ *Differential calculus is summarised in Appendix B.7.*

Quick Quiz 16.2

What are the units and dimensions of k/m?

The parameters A, k, m and ϕ are constants of the motion. To give physical significance to these constants, it is convenient to create a graphical representation of the motion by plotting x as a function of t as in Active Figure 16.2a. First, A, called the **amplitude** of the motion, is the maximum value of the position of the particle in either the positive or negative x direction. The factor $\sqrt{k/m}$ has units rad/s, as you will have seen from the *Quick Quiz*. This must be the case because the argument of a trigonometric function such as cosine or sine must be a pure number, it must be dimensionless and it has units of radians. This factor, appearing as it does in front of the variable t in our expression for x, controls the rate at which x varies in time.

If we denote the factor $\sqrt{k/m}$ with the symbol ω, then

$$\omega = \sqrt{\frac{k}{m}} \tag{16.7}$$

◀ Angular frequency

and **Equation 16.3** can be written in the form

$$\frac{d^2x}{dt^2} = -\omega^2 x \tag{16.8}$$

Pitfall Prevention 16.3

ω must be expressed in radians per second. The argument of a trigonometric function, such as sine or cosine, *must* be a pure number. The radian is a pure number because it is a ratio of lengths. Therefore, ω must be expressed in radians per second (and not, for example, in revolutions per second) if t is expressed in seconds.

The constant ω is the **angular frequency** and it has units of radians per second and dimensions T^{-1}. It is a measure of how rapidly the oscillations are occurring; the more oscillations per unit time, the higher the value of ω. We use the same symbol, ω, as we used for the angular speed in Chapter 9, because there is a close relationship between simple harmonic motion and circular motion, as we shall see in the next section.

The constant angle ϕ is called the **phase constant** or initial phase angle and, along with the amplitude A, is determined by the position and velocity of the particle at $t = 0$. If the particle is at its maximum position $x = A$ at $t = 0$, the phase constant is $\phi = 0$ and the graphical representation of the motion is as shown in **Active Figure 16.2b**. The quantity $(\omega t + \phi)$ is called the **phase** of the motion. Notice that the function $x(t)$ is periodic and its value is the same each time ωt increases by 2π radians.

Quick Quiz 16.3

Consider a graphical representation (**Figure 16.3**) of simple harmonic motion as described mathematically in **Equation 16.4**. When the particle is at point Ⓐ on the graph, what can you say about its position and velocity? **(a)** The position and velocity are both positive. **(b)** The position and velocity are both negative. **(c)** The position is positive and the velocity is zero. **(d)** The position is negative and the velocity is zero. **(e)** The position is positive and the velocity is negative. **(f)** The position is negative and the velocity is positive.

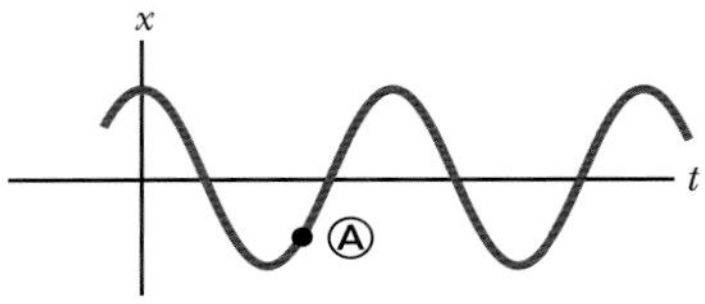

Figure 16.3
(Quick Quiz 16.3) An x–t graph for a particle undergoing simple harmonic motion. At a particular time, the particle's position is indicated by Ⓐ in the graph.

Quick Quiz 16.4

Figure 16.4 shows two curves representing particles undergoing simple harmonic motion. The correct description of these two motions is that the simple harmonic motion of particle B is **(a)** of larger angular frequency and larger amplitude than that of particle A, **(b)** of larger angular frequency and smaller amplitude than that of particle A, **(c)** of smaller angular frequency and larger amplitude than that of particle A, or **(d)** of smaller angular frequency and smaller amplitude than that of particle A.

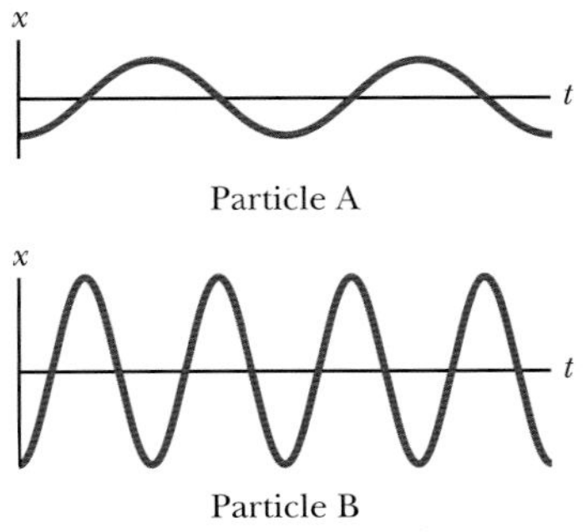

Figure 16.4
(Quick Quiz 16.4) Two x–t graphs for particles undergoing simple harmonic motion. The amplitudes and frequencies are different for the two particles.

The **period** T of the motion is the time interval required for the particle to go through one full cycle of its motion (**Active Figure 16.2a**). That is, the values of x and v for the particle at time t equal the values of x and v at time $t + T$. Because the phase increases by 2π radians in a time interval of T,

$$[\omega(t + T) + \phi] - (\omega t + \phi) = 2\pi$$

Simplifying this expression gives $\omega T = 2\pi$, or

$$T = \frac{2\pi}{\omega} \tag{16.9}$$

The inverse of the period is called the **frequency** f of the motion. The frequency is the number of oscillations the particle undergoes per unit time interval:

$$f = \frac{1}{T} = \frac{\omega}{2\pi} \tag{16.10}$$

The units of f are cycles per second, or **hertz** (Hz). Rearranging **Equation 16.10** gives

$$\omega = 2\pi f = \frac{2\pi}{T} \tag{16.11}$$

Recalling from **Equation 16.7** that $\omega = \sqrt{k/m}$, we can express the period and frequency in terms of the physical characteristics m and k of the system:

Period ▶

$$T = \frac{2\pi}{\omega} = 2\pi\sqrt{\frac{m}{k}} \tag{16.12}$$

Frequency ▶

$$f = \frac{1}{T} = \frac{1}{2\pi}\sqrt{\frac{k}{m}} \tag{16.13}$$

That is, the period and frequency depend *only* on the mass of the particle and the force constant of the spring and *not* on the parameters of the motion, such as A or ϕ.

Pitfall Prevention 16.4

We identify two kinds of frequency for a simple harmonic oscillator: f, called simply the *frequency*, is measured in hertz, and ω, the *angular frequency*, is measured in radians per second. Be sure you are clear about which frequency is being discussed or requested in a given problem.

TRY THIS

Get a few different rubber bands and hang an object such as a stapler from one of them. Pull it down slightly and note the frequency of its oscillations. Now hang the object from a different rubber band. How does this frequency compare to the first? What happens when you use two rubber bands together as a long chain? What if you use two rubber bands in parallel to suspend the object?

Quick Quiz 16.5

An object of mass m is hung from a spring and set into oscillation. The period of the oscillation is measured and recorded as T. The object of mass m is removed and replaced with an object of mass $2m$. When this object is set into oscillation, what is the period of the motion? (a) $2T$ (b) $\sqrt{2}\,T$ (c) T (d) $T/\sqrt{2}$ (e) $T/2$

We can obtain the velocity and acceleration of a particle undergoing simple harmonic motion from **Equations 16.5** and **16.6**:

Velocity of a particle in simple harmonic motion ▶

$$v = \frac{dx}{dt} = -\omega A\sin(\omega t + \phi) \tag{16.14}$$

Acceleration of a particle in simple harmonic motion ▶

$$a = \frac{d^2x}{dt^2} = -\omega^2 A\cos(\omega t + \phi) \tag{16.15}$$

From **Equation 16.14**, we see that because the sine and cosine functions oscillate between ± 1, the extreme values of the velocity v are $\pm\omega A$. Likewise, **Equation 16.15** shows that the extreme values of the acceleration a are $\pm\omega^2 A$. Therefore, the *maximum* values of the magnitudes of the velocity and acceleration are

Maximum magnitudes of velocity and acceleration in simple harmonic motion ▶

$$v_{\max} = \omega A = \sqrt{\frac{k}{m}}A \tag{16.16}$$

$$a_{\max} = \omega^2 A = \frac{k}{m}A \tag{16.17}$$

Figure 16.5a plots position versus time for an arbitrary value of the phase constant. The associated velocity–time and acceleration–time curves are illustrated in **Figures 16.5b** and **16.5c**, respectively. They show that the phase of the velocity differs from the phase of the position by $\pi/2$ rad, or 90°. That is, when x is a maximum or a minimum, the velocity is zero. Likewise, when x is zero, the speed is a maximum. The phase of the acceleration differs from the phase of the position by π radians, or 180°. For example, when x is a maximum, a has a maximum magnitude in the opposite direction.

Suppose a block is set into motion by pulling it from equilibrium by a distance A and releasing it from rest at $t = 0$ as in **Active Figure 16.6**. We must then require our solutions for $x(t)$ and $v(t)$ (**Equations 16.4** and **16.14**) to obey the initial conditions that $x(0) = A$ and $v(0) = 0$:

$$x(0) = A\cos\phi = A$$

$$v(0) = -\omega A\sin\phi = 0$$

These conditions are met if $\phi = 0$, giving $x = A\cos\omega t$ as our solution. To check this solution, notice that it satisfies the condition that $x(0) = A$ because $\cos 0 = 1$.

The position, velocity and acceleration of the block versus time are plotted in **Figure 16.7a** for this special case. The acceleration reaches extreme values of $\mp\omega^2 A$ when the position has extreme values of $\pm A$. The velocity has extreme values of $\pm\omega A$, which both occur at $x = 0$. Hence, the quantitative solution agrees with our qualitative description of this system.

Let's consider another possibility. Suppose the system is oscillating and we define $t = 0$ as the instant the block passes through the unstretched position of the spring while moving to the right (**Active Figure 16.8**). In this case, our solutions for $x(t)$ and $v(t)$ must obey the initial conditions that $x(0) = 0$ and $v(0) = v_i$:

$$x(0) = A\cos\phi = 0$$

$$v(0) = -\omega A\sin\phi = v_i$$

The first of these conditions tells us that $\phi = \pm\pi/2$. With these choices for ϕ, the second condition tells us that $A = \mp v_i/\omega$. Because the initial velocity is positive and the amplitude must be positive, we must have $\phi = -\pi/2$. Hence, the solution is

$$x = \frac{v_i}{\omega}\cos\left(\omega t - \frac{\pi}{2}\right)$$

The graphs of position, velocity and acceleration versus time for this choice of $t = 0$ are shown in **Figure 16.7b**. Notice that these curves are the same as those in **Figure 16.7a**, but shifted to the right by a quarter of a cycle. This shift is described mathematically by the phase constant $\phi = -\pi/2$, which is a quarter of a full cycle of 2π.

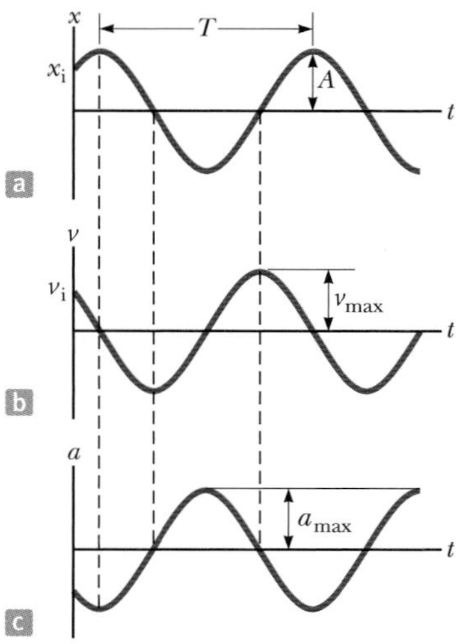

Figure 16.5
Graphical representation of simple harmonic motion (a) Position versus time (b) Velocity versus time (c) Acceleration versus time. Notice that at any specified time the velocity is 90° out of phase with the position and the acceleration is 180°out of phase with the position.

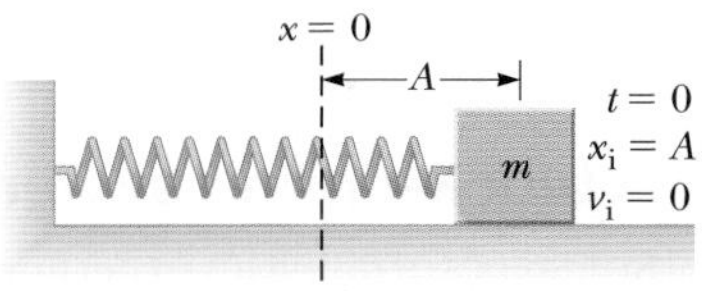

Active Figure 16.6
A block–spring system that begins its motion from rest with the block at $x = A$ at $t = 0$

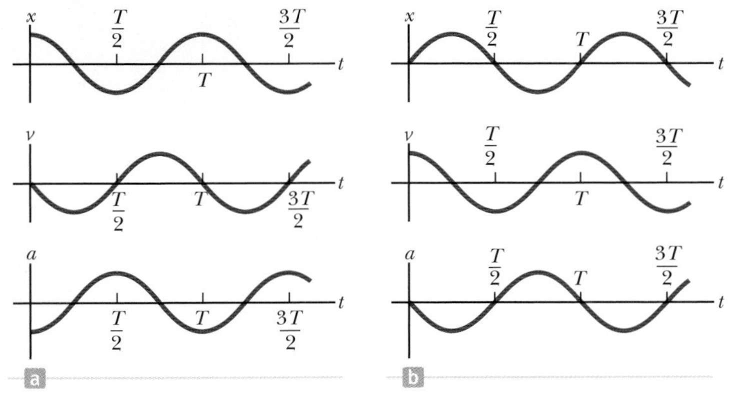

Figure 16.7
(a) Position, velocity, and acceleration versus time for the block in **Active Figure 16.6** under the initial conditions that at $t = 0$, $x(0) = A$, and $v(0) = 0$
(b) Position, velocity, and acceleration versus time for the block in **Active Figure 16.8** under the initial conditions that at $t = 0$, $x(0) = 0$, and $v(0) = v_i$

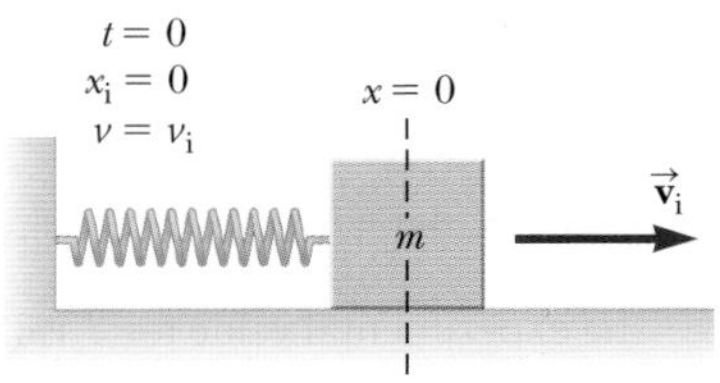

Active Figure 16.8
The block–spring system is undergoing oscillation, and $t = 0$ is defined at an instant when the block passes through the equilibrium position $x = 0$ and is moving to the right with speed v_i.

Analysis Model 16.1

Particle in simple harmonic motion

Imagine an object that is subject to a force that is proportional to the negative of the object's position, $F = -kx$. Such a force equation is known as Hooke's law, and it describes the force applied to an object attached to an ideal spring. The parameter k in Hooke's law is called the *spring constant* or the *force constant*. The position of an object acted on by a force described by Hooke's law is given by

$$x(t) = A\cos(\omega t + \phi) \quad (16.4)$$

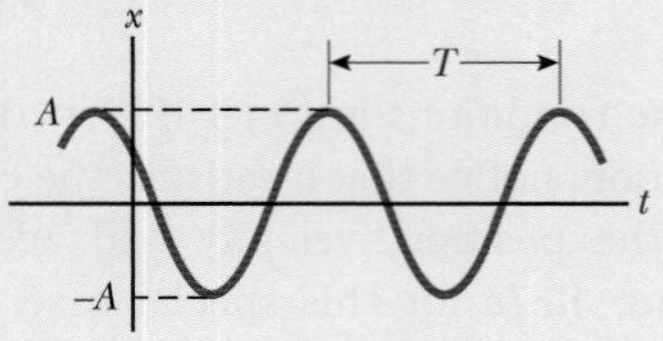

where A is the **amplitude** of the motion, ω is the **angular frequency,** and ϕ is the **phase constant.** The values of A and ϕ depend on the initial position and initial velocity of the particle.

The **period** of the oscillation of the particle is

$$T = \frac{2\pi}{\omega} = 2\pi\sqrt{\frac{m}{k}} \quad (16.12)$$

and the inverse of the period is the **frequency.**

Examples

- a bungee jumper hangs from a bungee cord and oscillates up and down
- a guitar string vibrates back and forth in a standing wave, with each element of the string moving in simple harmonic motion (Chapter 18)
- a piston in an engine oscillates up and down within the cylinder of the engine (Chapter 22)
- an atom in a diatomic molecule vibrates back and forth as if it is connected by a spring to the other atom in the molecule (Chapter 42)

±? Example 16.1

A block–spring system

Students are doing an experiment to measure the spring constant of a particular spring. A (200 ± 1) g block is connected to a light spring with an unknown spring constant, which is hanging vertically from a retort stand. The block is displaced 5.00 cm from the equilibrium position and released from rest. The students measure the period five times and find the following values: 1.26 s, 1.25 s, 1.25 s, 1.28 s, 1.27 s.

Find the spring constant of the spring.

Solution

Conceptualise This is much like **Active Figure 16.6** except that the block is moving vertically. You have already experimented with vertical systems like this if you did the *Try this* experiments, and you may also have done this same experiment at school or university.

Figure 16.9
A student measuring the spring constant of a spring

Model The block is modelled as a particle undergoing simple harmonic motion. We will also need to calculate an uncertainty in this problem.

Analyse Rearranging **Equation 16.12** for k: $T = 2\pi\sqrt{\frac{m}{k}} \Rightarrow k = \frac{4\pi^2 m}{T^2}$

Check dimensions: $[MT^{-2}] = [M]/[T^2] = [MT^{-2}]$ ☺

The experimental value for T is (1.262 ± 0.018) s, which is the average measured value with an uncertainty equal to the difference between this and the maximum value. We choose the maximum value in this case because this gives the larger estimate of the uncertainty.

So $$k = \frac{4\pi^2 m}{T^2} = \frac{4\pi^2(0.200\ \text{kg})}{(1.262\ \text{s})^2} = 4.958\ \text{kg/s}^2$$

with an uncertainty: $$\Delta k = k\left[\frac{\Delta m}{m} + 2\frac{\Delta T}{T}\right] = 4.958\ \text{kg/s}^2\left[\left(\frac{1\ \text{g}}{200\ \text{g}}\right) + 2\left(\frac{0.018\ \text{s}}{1.262\ \text{s}}\right)\right] = 0.17\ \text{kg/s}^2$$

Example 16.1 cont.

So we can write the experimental result for the spring constant as $k = (4.96 \pm 0.17)\ \text{kg/s}^2$.

Note that the unit kg/s^2 is the same as N/m.

Finalise Ideally, the students should collect more data and use a graphical method to find the spring constant. If they have a set of weights of various masses they can measure the period as a function of mass and use a graph to find the spring constant *k*. What should they plot such that the gradient of their graph is the spring constant? You can explore this further in problem 9 at the end of the chapter.

Example 16.2

Watch out for potholes!

A car with a mass of 1300 kg is constructed so that its frame is supported by four springs. Each spring has a force constant of 20 000 N/m. Two people riding in the car have a combined mass of 160 kg.

(A) Find the frequency of vibration of the car after it is driven over a pothole in the road.

Solution

Conceptualise Think about your experiences with cars. When you sit in a car, it moves downward a small distance because your weight is compressing the springs further. If you push down on the front bumper and release it, the front of the car oscillates a few times. Also recall the *Try this* example in which you connected springs in parallel and hung an object from them. Draw a diagram showing the physical situation and the model being used.

Model We model the car as a mass supported by a single spring, which allows us to apply the particle in simple harmonic motion analysis model.

Analyse For a given extension x of the springs, the combined force on the car is the sum of the forces from the individual springs.

Find an expression for the total force on the car:

$$F_{\text{total}} = \sum(-kx) = -\left(\sum k\right)x$$

In this expression, x has been factored from the sum because it is the same for all four springs. The effective spring constant for the combined springs is the sum of the individual spring constants.

Evaluate the effective spring constant:

$$k_{\text{eff}} = \sum k = 4 \times 20\,000\ \text{N/m} = 80\,000\ \text{N/m}$$

Use **Equation 16.13** to find the frequency of vibration:

$$f = \frac{1}{2\pi}\sqrt{\frac{k_{\text{eff}}}{m}} = \frac{1}{2\pi}\sqrt{\frac{80\,000\ \text{N/m}}{1460\ \text{kg}}} = 1.18\ \text{Hz}$$

Note that the mass we used here is that of the car plus the people because that is the total mass that is oscillating.

(B) What is the maximum vertical velocity and acceleration of the car and people if the amplitude of oscillation is 3 cm?

Solution

Analyse We use **Equations 16.16** and **16.17**:

$$v_{\text{max}} = \omega A = 2\pi f A = 2\pi(1.18\ \text{Hz})(0.03\ \text{m}) = 0.222\ \text{m/s}$$

$$a_{\text{max}} = \omega^2 A = 4\pi^2 f^2 A = 4\pi^2\,(1.18\ \text{Hz})^2(0.03\ \text{m}) = 1.65\ \text{m/s}^2$$

Finalise We have treated the car as a particle, and modelled the entire car as moving up and down together. A more realistic model is to treat the car as an extended body and model the oscillations as including rocking of the car, such that as the front goes up as the back goes down and vice versa.

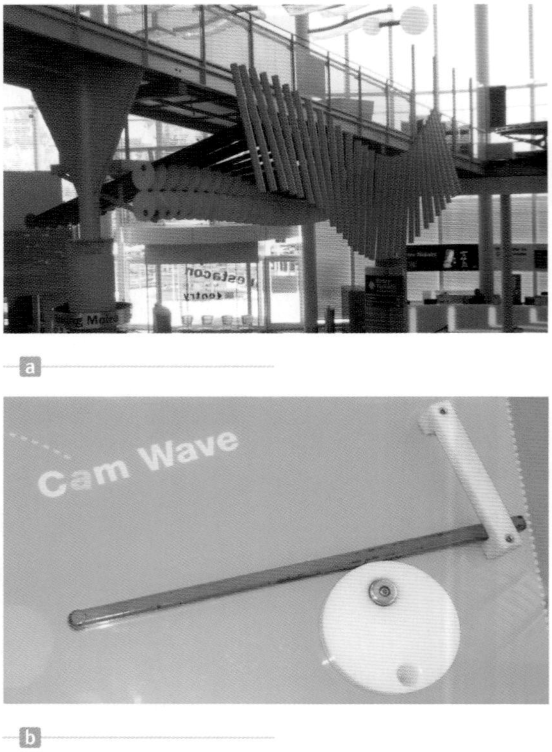
a
Cam Wave
b

Figure 16.10
(a) The 'cam wave' at Questacon in Canberra. Each camshaft has a series of offset cams attached to beams. As the camshafts rotate, the top edge of each cam goes up and down, making the beams oscillate up and down. (b) A single cam and beam, showing how each moves

16.3 Comparing simple harmonic motion to uniform circular motion

Some common devices in everyday life exhibit a relationship between oscillatory motion and circular motion, including the crankshaft in a car engine, which converts the up-and-down motion of the pistons to the rotational motion of the shaft, and the camshaft, which rotates and moves the valves up and down. Similar systems of cams are used in many machines, including printing presses and any device in which motors are used to produce translational motion.

Figure 16.10a shows a giant double camshaft at Questacon, with a beam attached to each cam. A cam is a disc or oval mounted so that the top edge of the cam moves up and down as the camshaft (a rod with cams mounted on it) rotates. Each cam is slightly offset from the previous one. **Figure 16.10b** shows a single cam and beam. As the camshaft rotates, each cam rotates with it, pushing on the beam and making it oscillate up and down. Hence circular motion is converted to oscillatory motion. In this section, we explore the relationship between these two types of motion.

Active Figure 16.11 is a view of an experimental arrangement that shows this relationship. A ball is attached to the rim of a turntable of radius A, which is illuminated from above by a lamp. The ball casts a shadow on a screen. As the turntable rotates with constant angular speed, the shadow of the ball moves back and forth in simple harmonic motion.

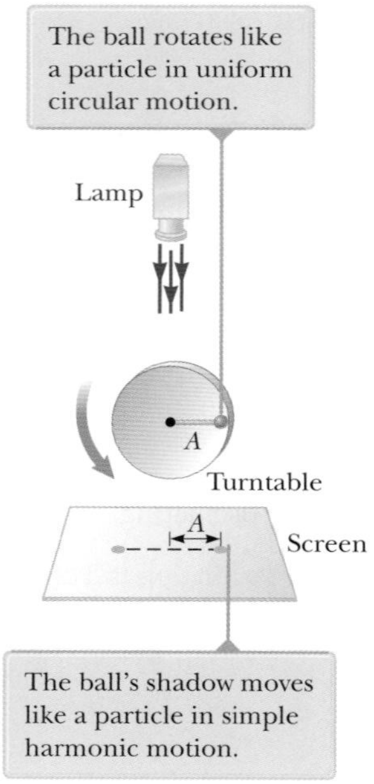
The ball rotates like a particle in uniform circular motion.
Lamp
A
Turntable
A
Screen
The ball's shadow moves like a particle in simple harmonic motion.

Active Figure 16.11
An experimental setup for demonstrating the connection between simple harmonic motion and uniform circular motion

Consider a particle located at point P on the circumference of a circle of radius A as in **Figure 16.12a**, with the line OP making an angle ϕ with the x axis at $t = 0$. We call this circle a *reference circle* for comparing simple harmonic motion with uniform circular motion, and we choose the position of P at $t = 0$ as our reference position. If the particle moves along the circle with constant angular speed ω until OP makes an angle θ with the x axis as in **Figure 16.12b**, at some time $t > 0$ the angle between OP and the x axis is $\theta = \omega t + \phi$. As the particle moves along the circle, the projection of P on the x axis, labelled point Q, moves back and forth along the x axis between the limits $x = \pm A$. Notice that points P and Q always have the same x coordinate. From the right triangle OPQ, we see that this x coordinate is

$$x(t) = A\cos(\omega t + \phi) \qquad \textbf{(16.18)}$$

This expression is the same as **Equation 16.4** and shows that the point Q moves with simple harmonic motion along the x axis. Therefore, simple harmonic motion along a straight line can be represented by the projection of uniform circular motion along a diameter of a reference circle.

The time interval for one complete revolution of the point P on the reference circle is equal to the period of motion T for simple harmonic motion between $x = \pm A$. Therefore, the angular speed ω of P is the same as the angular frequency ω of simple harmonic motion along the x axis, which is why we use the same symbol. The phase constant ϕ for simple harmonic motion corresponds to the initial angle OP makes with the x axis. The radius A of the reference circle equals the amplitude of the simple harmonic motion.

Because the relationship between linear and angular speed for circular motion is $v = r\omega$ (see **Equation 9.10**), the particle moving on the reference circle of radius A has a velocity of magnitude ωA. From the geometry in **Figure 16.12c**, we see that the x component of this

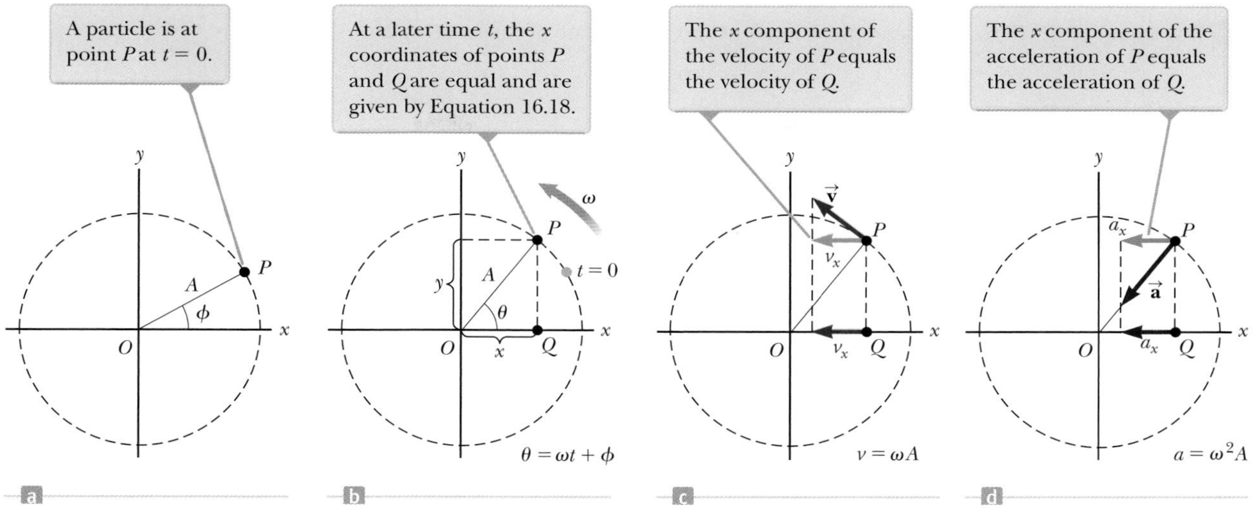
A particle is at point P at $t = 0$.
At a later time t, the x coordinates of points P and Q are equal and are given by Equation 16.18.
The x component of the velocity of P equals the velocity of Q.
The x component of the acceleration of P equals the acceleration of Q.
$\theta = \omega t + \phi$
$v = \omega A$
$a = \omega^2 A$
a
b
c
d

Figure 16.12
Relationship between the uniform circular motion of a point P and the simple harmonic motion of a point Q. A particle at P moves in a circle of radius A with constant angular speed ω.

velocity is $-\omega A \sin(\omega t + \phi)$. By definition, point Q has a velocity given by dx/dt. Differentiating **Equation 16.18** with respect to time, we find that the velocity of Q is the same as the x component of the velocity of P.

The acceleration of P on the reference circle is directed radially inwards toward O and has a magnitude $v^2/A = \omega^2 A$. From the geometry in **Figure 16.12d**, we see that the x component of this acceleration is $-\omega^2 A \cos(\omega t + \phi)$. This value is also the acceleration of the projected point Q along the x axis, as you can verify by taking the second derivative of **Equation 16.18**.

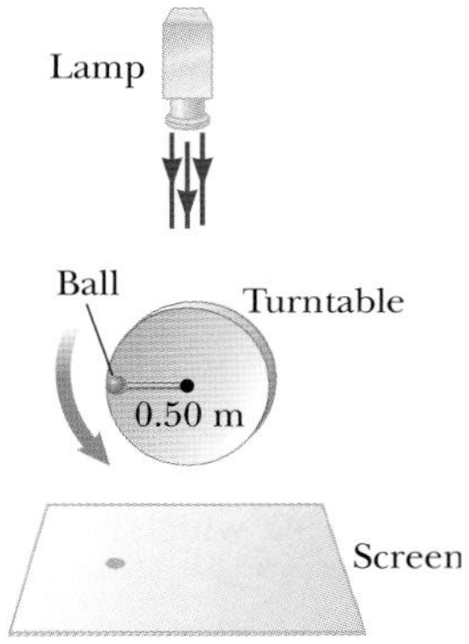
Lamp
Ball
Turntable
0.50 m
Screen

Figure 16.13
(Quick Quiz 16.6) An object moves in circular motion, casting a shadow on the screen below. Its position at an instant of time is shown.

Quick Quiz 16.6

Figure 16.13 shows the position of an object in uniform circular motion at $t = 0$. A light shines from above and projects a shadow of the object on a screen below the circular motion. What are the correct values for the *amplitude* and *phase constant* (relative to an x axis to the right) of the simple harmonic motion of the shadow? **(a)** 0.50 m and 0 **(b)** 1.00 m and 0 **(c)** 0.50 m and π **(d)** 1.00 m and π

Example 16.3

Circular motion with constant angular speed

The ball in **Active Figure 16.11** rotates anticlockwise in a circle of radius 3.00 m with a constant angular speed of 8.00 rad/s. At $t = 0$, its shadow has an x coordinate of 2.00 m and is moving to the right.

(A) Determine the x coordinate of the shadow as a function of time in SI units.

Solution

Conceptualise Study **Active Figure 16.11**. Notice that the shadow is *not* at its maximum position at $t = 0$. As we already have **Active Figure 16.11**, we do not need to draw another diagram, although it is good practice for you to do so.

Model The ball on the turntable can be modelled as a particle in uniform circular motion. The shadow is modelled as a particle in simple harmonic motion.

Example 16.3 cont.

Analyse Use **Equation 16.18** to write an expression for the x coordinate of the rotating ball:

$$x = A\cos(\omega t + \phi)$$

Solve for the phase constant:

$$\phi = \cos^{-1}\left(\frac{x}{A}\right) - \omega t$$

Substitute numerical values for the initial conditions:

$$\phi = \cos^{-1}\left(\frac{2.00\text{ m}}{3.00\text{ m}}\right) - 0 = \pm 48.2° = \pm 0.841\text{ rad}$$

If we were to take $\phi = +0.841$ rad as our answer, the shadow would be moving to the left at $t = 0$. Because the shadow is moving to the right at $t = 0$, we must choose $\phi = -0.841$ rad.

Write the x coordinate as a function of time:

$$x = 3.00\cos(8.00t - 0.841)$$

(B) Find the x components of the shadow's velocity and acceleration at any time t.

Solution

Analyse We differentiate the x coordinate with respect to time to find the velocity at any time in m/s:

$$v_x = \frac{dx}{dt} = -A\omega\sin(\omega t + \phi) = (-3.00\text{ m})(8.00\text{ rad/s})\sin(8.00t - 0.841)$$
$$= -24.0\sin(8.00t - 0.841)$$

Differentiate the velocity with respect to time to find the acceleration at any time in m/s:

$$a_x = \frac{dv_x}{dt} = -A\omega^2\cos(\omega t - \phi) = (-3.00\text{ m})(8.00\text{ rad/s})^2\cos(8.00t - 0.841)$$
$$= -192\cos(8.00t - 0.841)$$

Finalise These results are equally valid for the ball moving in uniform circular motion and the shadow moving in simple harmonic motion. The value of the phase constant puts the ball in the lower right quadrant of the xy coordinate system of **Figure 16.12**, which is consistent with the shadow having a positive value for x and moving towards the right.

16.4 Energy of the simple harmonic oscillator

What energy does a particle undergoing simple harmonic motion have? Consider the block–spring system in **Active Figure 16.1**. Because the surface is frictionless, the system is isolated and we expect the total mechanical energy of the system to be constant. We assume a massless spring, so the kinetic energy of the system corresponds only to that of the block. We can use **Equation 16.14** to express the kinetic energy of the block as

Kinetic energy of a simple harmonic oscillator ▶

$$K = \frac{1}{2}mv^2 = \frac{1}{2}m\omega^2A^2\sin^2(\omega t + \phi) \tag{16.19}$$

The elastic potential energy stored in the spring for any elongation x is given by $\frac{1}{2}kx^2$ (see **Equation 6.19**). Using **Equation 16.4** gives

Potential energy of a simple harmonic oscillator ▶

$$U = \frac{1}{2}kx^2 = \frac{1}{2}kA^2\cos^2(\omega t + \phi) \tag{16.20}$$

We see that K and U are *always* positive quantities or zero. Because $\omega^2 = k/m$, we can express the total mechanical energy of the simple harmonic oscillator as

$$E = K + U = \frac{1}{2}kA^2[\sin^2(\omega t + \phi) + \cos^2(\omega t + \phi)]$$

From the identity $\sin^2\theta + \cos^2\theta = 1$, we see that the quantity in square brackets is unity. Therefore, this equation reduces to

$$E = \frac{1}{2}kA^2 \tag{16.21}$$

◀ Total energy of a simple harmonic oscillator

Some useful trigonometric identities are given in Appendix B.4.

Σ

That is, the total mechanical energy of a simple harmonic oscillator is a constant of the motion and is proportional to the square of the amplitude. The total mechanical energy is equal to the maximum potential energy stored in the spring when $x = \pm A$ because $v = 0$ at these points and there is no kinetic energy. At the equilibrium position, where $U = 0$ because $x = 0$, the total energy, all in the form of kinetic energy, is again $\frac{1}{2}kA^2$.

Finally, we can obtain the velocity of the block at an arbitrary position by expressing the total energy of the system at some arbitrary position x as

$$E = K + U = \frac{1}{2}mv^2 + \frac{1}{2}kx^2 = \frac{1}{2}kA^2$$

$$v = \pm\sqrt{\frac{k}{m}(A^2 - x^2)} = \pm\omega\sqrt{A^2 - x^2} \tag{16.22}$$

◀ Velocity as a function of position for a simple harmonic oscillator

Equation 16.22 verifies that the speed is a maximum at $x = 0$ and is zero at the turning points $x = \pm A$.

Plots of the kinetic and potential energies versus time are shown in **Active Figure 16.14a**, where we have taken $\phi = 0$. At all times, the sum of the kinetic and potential energies is a constant equal to $\frac{1}{2}kA^2$, the total energy of the system.

The variations of K and U with the position x of the block are plotted in **Active Figure 16.14b**. Energy is continuously being transformed between potential energy stored in the spring and kinetic energy of the block.

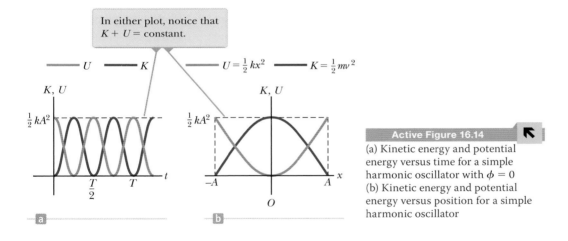

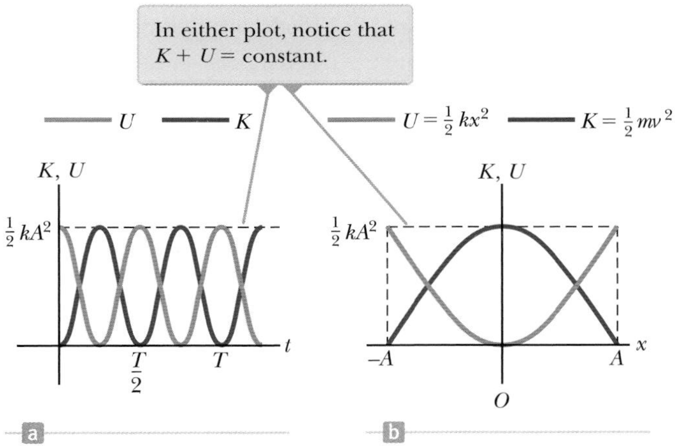

Active Figure 16.14
(a) Kinetic energy and potential energy versus time for a simple harmonic oscillator with $\phi = 0$
(b) Kinetic energy and potential energy versus position for a simple harmonic oscillator

Active Figure 16.15 illustrates the position, velocity, acceleration, kinetic energy, and potential energy of the block–spring system for one full period of the motion. Most of the ideas discussed so far are incorporated in this figure.

Quick Quiz 16.7

A weight is hanging from a vertical spring. If you pull the weight downwards, you increase the elastic potential energy of the system and decrease the gravitational potential energy. Which changes more and how do you know? Is the change in the total energy of the system positive or negative?

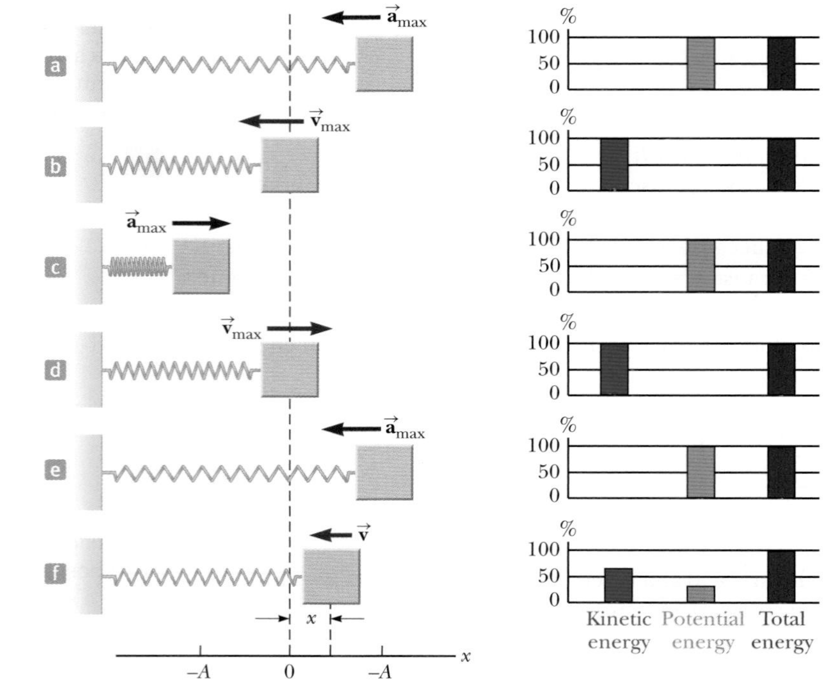

t	x	v	a	K	U
0	A	0	$-\omega^2 A$	0	$\frac{1}{2}kA^2$
$\frac{T}{4}$	0	$-\omega A$	0	$\frac{1}{2}kA^2$	0
$\frac{T}{2}$	$-A$	0	$\omega^2 A$	0	$\frac{1}{2}kA^2$
$\frac{3T}{4}$	0	ωA	0	$\frac{1}{2}kA^2$	0
T	A	0	$-\omega^2 A$	0	$\frac{1}{2}kA^2$
t	x	v	$-\omega^2 x$	$\frac{1}{2}mv^2$	$\frac{1}{2}kx^2$

Active Figure 16.15

(a) through (e) Several instants in the simple harmonic motion for a block–spring system. Energy bar graphs show the distribution of the energy of the system at each instant. The parameters in the table at the right refer to the block–spring system, assuming at $t = 0$, $x = A$; hence, $x = A \cos \omega t$. For these five special instants, one of the types of energy is zero. (f) An arbitrary point in the motion of the oscillator. The system possesses both kinetic energy and potential energy at this instant as shown in the bar graph.

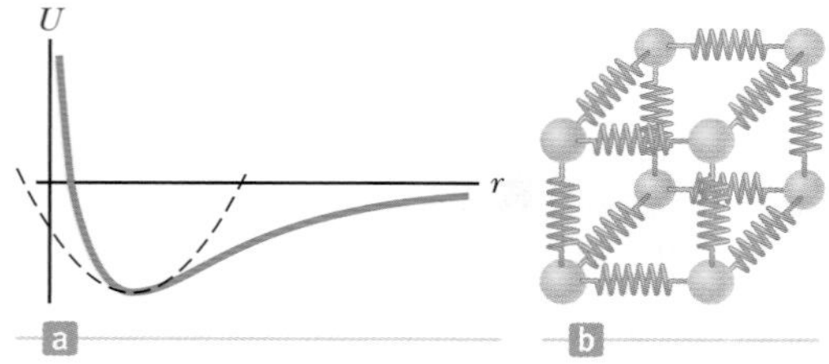

Figure 16.16

(a) If the atoms in a molecule do not move too far from their equilibrium positions, a graph of potential energy versus separation distance between atoms is similar to the graph of potential energy versus position for a simple harmonic oscillator (dashed black curve).
(b) The forces between atoms in a solid can be modelled by imagining springs between neighbouring atoms.

You may wonder why we are spending so much time studying simple harmonic oscillators. We do so because they are good models of a wide variety of physical phenomena. For example, recall the Lennard-Jones potential discussed in Example 6.8. This complicated function describes the forces holding atoms together. We will also use the simple harmonic oscillator model when we study quantum mechanics in Chapter 40.

Figure 16.16a shows that for small displacements from the equilibrium position, the potential energy curve for this function approximates a parabola, which represents the potential energy function for a simple harmonic oscillator. Therefore, we can model the complex atomic binding forces as being due to tiny springs as depicted in **Figure 16.16b**.

Example 16.4

Molecular vibrations

Consider an oxygen molecule modelled as two balls (the atoms) connected via a massless spring (the bond). The bond length is 1.21×10^{-10} m or 1.21 Å and the total vibrational energy is about 3×10^{-21} J.

(A) If the frequency of oscillation is 9×10^{12} Hz, find the spring constant of the bond.

Example 16.4 cont.

Solution

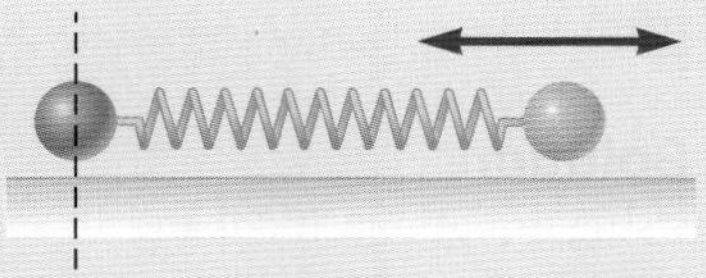

Conceptualise We considered the rotational energy of an oxygen molecule modelled as balls on a stick in Example 10.1. We use a similar idea here and begin by drawing a diagram. To make it simpler, we consider one of the two atoms as stationary and the other as vibrating towards and away from it as the spring (bond) is stretched and compressed.

Model The oxygen atom is modelled as a particle in simple harmonic motion.

Analyse Use **Equation 16.7** to find the spring constant: $\omega = \sqrt{\frac{k}{m}}$ so $k = m\omega^2$

The mass of an oxygen atom is 15.999u or $15.999 \times 1.66 \times 10^{-27}$ kg $= 2.66 \times 10^{-26}$ kg (see the periodic table in Appendix C and the back endpapers for unit conversions) and we are given $f = 9 \times 10^{12}$ Hz, so remembering to convert to rad/s we have $\omega = 2\pi f = 5.7 \times 10^{13}$ rad/s.

which gives us $k = m\omega^2 = (2.66 \times 10^{-26}\text{ kg})(5.7 \times 10^{13}\text{ rad/s})^2 = 85\text{ kg/s}^2$

Note that as we have only one significant figure in our value for frequency, we should express this to the same precision, as 90 kg/s^2.

(B) Find the amplitude of the oscillations.

Solution

Analyse Use **Equation 16.21** to express the total energy of the oscillator system: $E = \frac{1}{2}kA^2$

and rearrange for A: $A = \sqrt{\frac{2E}{k}} = \sqrt{\frac{2(3 \times 10^{-21}\text{ J})}{85\text{ kg/s}^2}} = 8 \times 10^{-12}\text{ m}$

This is about 7% of the bond length, which seems reasonable.

(C) What is the maximum velocity of the oxygen atom?

Solution

Using **Equation 16.21** again to relate the total energy to the maximum kinetic energy: $E = \frac{1}{2}kA^2 = \frac{1}{2}mv_{max}{}^2$

Solve for the maximum speed and substitute numerical values: $v_{max} = \sqrt{\frac{2E}{m}} = \sqrt{\frac{2(3 \times 10^{-21}\text{ J})}{2.66 \times 10^{-26}\text{ kg}}} = 500\text{ m/s}$

(D) What is the velocity of the oxygen atom when the position is $\frac{1}{2}A$?

Solution

Use **Equation 16.22**:

$$v = \pm\sqrt{\frac{k}{m}(A^2 - x^2)}$$

$$= \pm\sqrt{\frac{85\text{ N/m}}{2.66 \times 10^{-26}\text{ kg}}[(8.0 \times 10^{-12}\text{ m})^2 - (4.0 \times 10^{-12})^2]}$$

$$= \pm 400\text{ m/s}$$

The positive and negative signs indicate that the atom could be moving to either the right or the left at this instant.

Example 16.4 cont.

(E) What fraction of the total vibrational energy is present as kinetic and potential energy at this position?

Solution

Using **Equations 16.20** and **16.21**, we find that when $x = \frac{1}{2}A$:

$$U = \frac{1}{2}kx^2 = \frac{1}{2}k\left(\frac{1}{2}A\right)^2 = \frac{1}{2} \times \frac{1}{4}kA^2 = \frac{1}{4}E$$

Hence the potential energy is $\frac{1}{4}$ of the total energy and the kinetic energy must therefore be the remaining $\frac{3}{4}$ of the total energy.

Finalise This is a very simple model of a diatomic molecule and we shall explore a more physical model when we come to quantum mechanics in Chapter 39, where we shall see that the energy associated with vibrational motion (and rotational motion) is quantised. We will then introduce a new analysis model in Chapter 40, the quantum harmonic oscillator, which is an extension of the simple harmonic oscillator.

16.5 The pendulum

The **simple pendulum** is another mechanical system that exhibits periodic motion. It consists of a particle-like bob of mass m suspended by a light string of length L that is fixed at the upper end as shown in **Active Figure 16.17**. The motion occurs in the vertical plane and is driven by the gravitational force. Provided the angle θ is small (less than about 10° so that we can apply the small angle approximation), the motion is very close to that of a simple harmonic oscillator and we shall model it as such.

Σ *The small angle approximation, and other useful approximations, are given in Appendix B.5.*

The forces acting on the bob are the force $\vec{\mathbf{T}}$ exerted by the string and the gravitational force $m\vec{\mathbf{g}}$. The tangential component $mg\sin\theta$ of the gravitational force always acts towards $\theta = 0$, opposite the displacement of the bob from the lowest position. Therefore, the tangential component is a restoring force, and we can apply Newton's second law for motion in the tangential direction:

$$F_t = ma_t \rightarrow -mg\sin\theta = m\frac{d^2s}{dt^2}$$

where the negative sign indicates that the tangential force acts towards the equilibrium position and s is the bob's position measured along the arc. We have expressed the tangential acceleration as the second derivative of the position s. Because $s = L\theta$ (from the definition of radians) and L is constant, this equation reduces to

$$\frac{d^2\theta}{dt^2} = -\frac{g}{L}\sin\theta$$

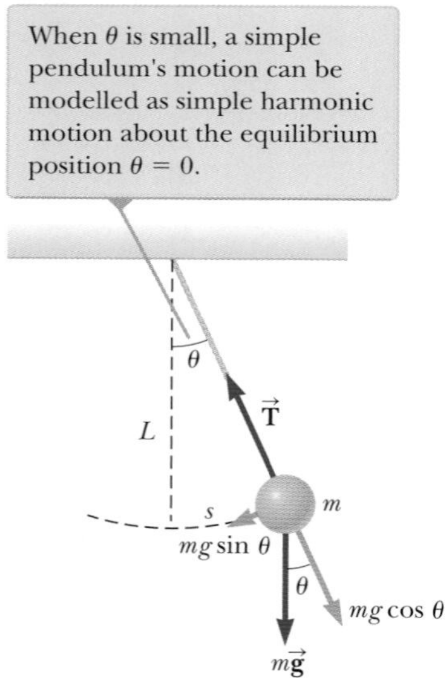
When θ is small, a simple pendulum's motion can be modelled as simple harmonic motion about the equilibrium position θ = 0.
θ
L
T
s
m
mg sin θ
θ
mg cos θ
mg

Active Figure 16.17
A simple pendulum

Considering θ as the position, this equation is similar to **Equation 16.3**, although the right side is proportional to $\sin\theta$ rather than to θ; hence, we would not expect simple harmonic motion because this expression is not of exactly the same mathematical form as **Equation 16.3**. If we assume θ is *small* (less than about 10° or 0.2 rad), however, we can use the **small angle approximation**, $\sin\theta \approx \theta$, where θ is measured in radians. As long as θ is less than approximately 10°, the angle in radians and its sine are the same to within an accuracy of better than 1.0%.

Therefore, for small angles, the equation of motion becomes

$$\frac{d^2\theta}{dt^2} = -\frac{g}{L}\theta \text{ (for small values of } \theta) \qquad (16.23)$$

Equation 16.23 has exactly the same mathematical form as **Equation 16.3**, so we conclude that the motion for small amplitudes of oscillation can be modelled as simple harmonic motion. Therefore, the solution of **Equation 16.23** is $\theta = \theta_{max}\cos(\omega t + \phi)$, where θ_{max} is the *maximum angular position* and the angular frequency ω is

Angular frequency for ▶ a simple pendulum

$$\omega = \sqrt{\frac{g}{L}} \qquad (16.24)$$

The period of the motion is

$$T = \frac{2\pi}{\omega} = 2\pi\sqrt{\frac{L}{g}} \qquad (16.25)$$

◀ Period of a simple pendulum

In other words, the period and frequency of a simple pendulum depend only on the length of the string and the acceleration due to gravity. Because the period is independent of the mass, we conclude that all simple pendulums that are of equal length and are at the same location (so that g is constant) oscillate with the same period.

So the simple pendulum can be used as a timekeeper because its period depends only on its length and the local value of g. It is also a convenient device for making precise measurements of the free-fall acceleration. Such measurements are important because variations in local values of g can provide information on the location of ore deposits and other valuable underground resources.

Pitfall Prevention 16.5

The pendulum *does not* exhibit true simple harmonic motion for *any* angle. If the angle is less than about 10°, the motion is close to and can be *modelled* as simple harmonic.

Quick Quiz 16.8

A grandfather clock depends on the period of a pendulum to keep correct time. **(i)** Suppose a grandfather clock is calibrated correctly and then a naughty child slides the bob of the pendulum downwards on the oscillating rod. Does the grandfather clock run (a) slow, (b) fast or (c) correctly? **(ii)** Suppose a grandfather clock is calibrated correctly at sea level and is then taken to the top of a very tall mountain. Does the grandfather clock now run (a) slow, (b) fast or (c) correctly?

TRY THIS

Make a simple pendulum using a piece of string and any handy object that you can tie to the string. Vary the length of the string and time the period (to be accurate, you should of course time 10 or more periods and then calculate the period). Find the length of string that gives you a one-second period.

Example 16.5

A connection between length and time

Christian Huygens (1629–1695), a great clockmaker and physicist, suggested that an international unit of length could be defined as the length of a simple pendulum having a period of exactly 1 s. How much shorter would our length unit be if his suggestion had been followed?

Solution

Conceptualise If you did the *Try this* example, you will already have a good idea of the length. If not, try it now so you do have a sense of what length is needed.

Model We will assume that any oscillations of the pendulum are small so that we can apply the small angle approximation and use the simple pendulum model described in this section.

Analyse Solve **Equation 16.25** for the length: $L = \dfrac{T^2 g}{4\pi^2}$

Check dimensions: $[L] = [T^2][LT^{-2}]/[] = [L]$ ☺

Substitute values: $L = \dfrac{(1\ \text{s})^2(9.80\ \text{m/s}^2)}{4\pi^2} = 0.248\ \text{m}$

Finalise The metre's length would be slightly less than a quarter of its current length. Note that the number of significant digits depends only on how precisely we know g because the time has been defined to be exactly 1 s.

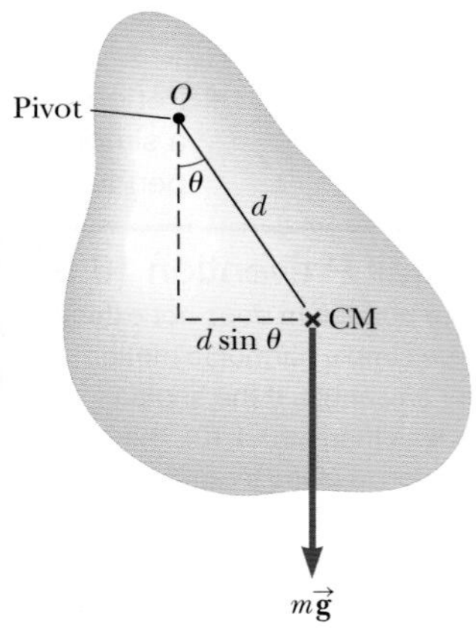

Figure 16.18
A physical pendulum pivoted at O

Physical pendulum

If a hanging object oscillates about a fixed axis that does not pass through its centre of mass and the object cannot be approximated as a point mass, we cannot treat the system as a simple pendulum. In this case, the system is called a **physical pendulum**.

Consider a rigid object pivoted at a point O that is a distance d from the centre of mass (**Figure 16.18**). The gravitational force provides a torque about an axis through O, and the magnitude of that torque is $mgd \sin\theta$, where θ is as shown in **Figure 16.18**. We model the object as a rigid object subject to a net torque and use the rotational form of Newton's second law, $\Sigma\tau_{ext} = I\alpha$, where I is the moment of inertia of the object about the axis through O. The result is

$$-mgd\sin\theta = I\frac{d^2\theta}{dt^2}$$

The negative sign indicates that the torque about O acts to decrease θ. That is, the gravitational force produces a restoring torque. If we again assume θ is small, we can use the small angle approximation, $\sin\theta \approx \theta$ and the equation of motion reduces to

$$\frac{d^2\theta}{dt^2} = -\left(\frac{mgd}{I}\right)\theta = -\omega^2\theta \tag{16.26}$$

Because this equation is of the same mathematical form as **Equation 16.3**, its solution is that of the simple harmonic oscillator. That is, the solution of **Equation 16.26** is given by $\theta = \theta_{max}\cos(\omega t + \phi)$, where θ_{max} is the maximum angular position and

$$\omega = \sqrt{\frac{mgd}{I}}$$

The period is

Period of a ▶ physical pendulum

$$T = \frac{2\pi}{\omega} = 2\pi\sqrt{\frac{I}{mgd}} \tag{16.27}$$

This result can be used to measure the moment of inertia of a flat, rigid object. If the location of the centre of mass – and hence the value of d – is known, the moment of inertia can be obtained by measuring the period.

Quick Quiz 16.9

What happens to **Equation 16.27** when $I = md^2$, that is, when all the mass is concentrated at the centre of mass?

TRY THIS

Balance a wire coat hanger so that the hook is supported by your extended index finger. When you give the hanger a small angular displacement with your other hand and then release it, it oscillates. Now stick a large lump of putty (Blu-tack or similar) to the coat hanger and observe its oscillations. What happens to the period if you move the putty? Where should the putty be put for the shortest period? What about the longest period?

Example 16.6

A swinging rod

A uniform rod of mass M and length L is pivoted about one end and oscillates in a vertical plane. Find the period of oscillation if the amplitude of the motion is small.

Solution

Conceptualise Imagine a rod swinging back and forth when pivoted at one end. Try it with a ruler. Draw a diagram showing the system, as in **Figure 16.19**.

Example 16.6 cont.

Model We cannot reasonably model a rod as a point particle, so we use the physical pendulum model. We will assume that any oscillations of the pendulum are small so that we can apply the small angle approximation.

Analyse In Chapter 9, we found that the moment of inertia of a uniform rod about an axis through one end is $\frac{1}{3}ML^2$. The distance d from the pivot to the centre of mass of the rod is $L/2$.

Substitute these quantities into **Equation 16.27**:

$$T = 2\pi\sqrt{\frac{\frac{1}{3}ML^2}{Mg(L/2)}} = 2\pi\sqrt{\frac{2L}{3g}}$$

Check dimensions: $[T] = ([L]/[LT^{-2}])^{1/2} = [T]$ ☺

Finalise The period of a rod is less than that of a simple pendulum by a factor of $\sqrt{2/3}$. In general, a physical pendulum will have a period less than that of a simple pendulum with the same total length and mass.

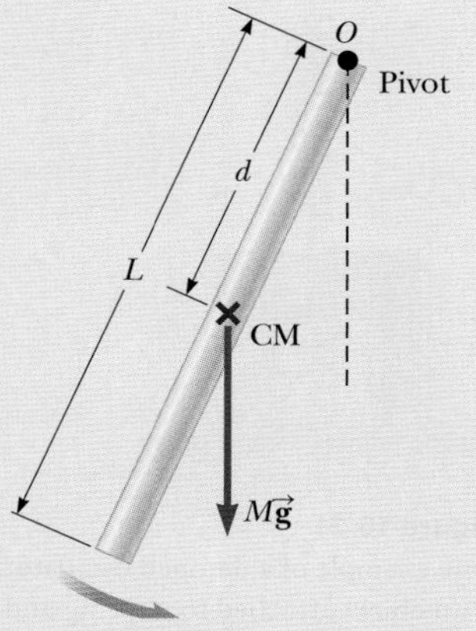
O
Pivot
d
L
CM
$M\vec{g}$

Figure 16.19
(Example 16.6) A rigid rod oscillating about a pivot through one end is a physical pendulum with $d = L/2$.

In one of the Moon landings, an astronaut walking on the Moon's surface had a belt hanging from his space suit and the belt oscillated as a physical pendulum. A scientist on the Earth observed this motion on television and used it to estimate the free-fall acceleration on the Moon using this same method and estimating the length of the belt.

TRY THIS

Estimate the period at which your arm will swing freely from your shoulder joint when released from rest at a small angle from hanging vertically downwards. Now do the experiment and measure it. What do you notice about the motion? You may need someone else to time it for you.

Torsional pendulum

Figure 16.20 shows a rigid object suspended by a wire attached at the top to a fixed support. When the object is twisted through some angle θ, the twisted wire exerts on the object a restoring torque that is proportional to the angular position. That is,

$$\tau = -\kappa\theta$$

where κ (Greek letter kappa) is called the *torsion constant* of the support wire and is a rotational analogue to the force constant κ for a spring. The value of κ can be obtained by applying a known torque to twist the wire through a measurable angle θ. Applying Newton's second law for rotational motion, we find that

$$\sum \tau = I\alpha \rightarrow -\kappa\theta = I\frac{d^2\theta}{dt^2}$$

$$\frac{d^2\theta}{dt^2} = -\frac{\kappa}{I}\theta \qquad (16.28)$$

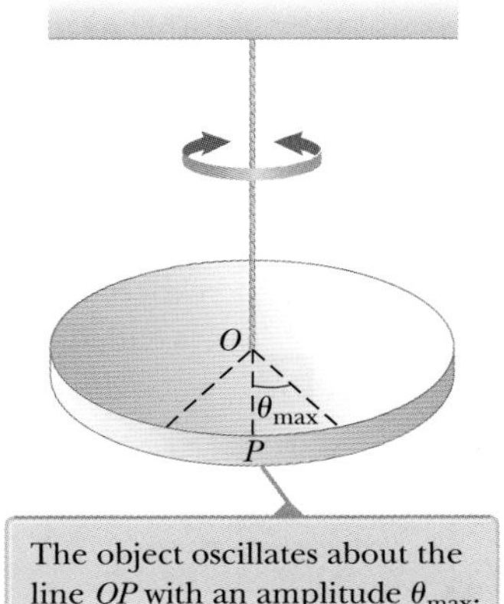
O
θ_{max}
P
The object oscillates about the line OP with an amplitude θ_{max}.

Figure 16.20
A torsional pendulum

Again, this result is the equation of motion for a simple harmonic oscillator, with $\omega = \sqrt{\kappa/I}$ and a period

$$T = 2\pi\sqrt{\frac{I}{\kappa}} \qquad (16.29)$$

◀ Period of a torsional pendulum

This system is called a *torsional pendulum*. There is no small-angle restriction in this situation as long as the elastic limit (as discussed in Chapter 15) of the wire is not exceeded.

Cavendish used a torsional pendulum to establish the value of the gravitational constant G in his famous experiment, as described in worked Example 11.1.

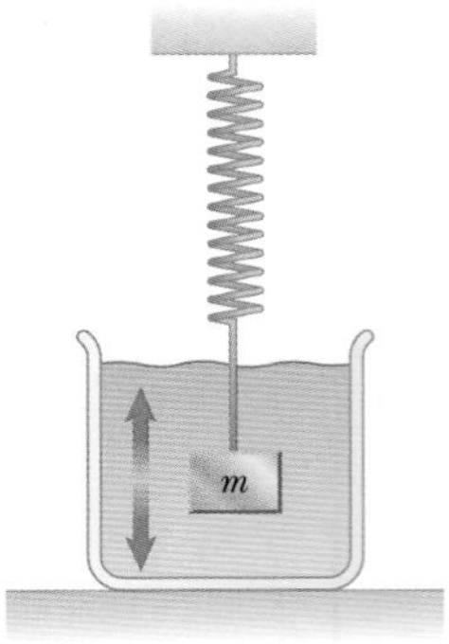
m

Figure 16.21
One example of a damped oscillator is an object attached to a spring and submerged in a viscous liquid.

16.6 Damped oscillations

The oscillatory motions we have considered so far have been for ideal systems, that is, systems that oscillate indefinitely with the action of only one force, a linear restoring force. In most real systems non-conservative forces such as friction and air resistance retard the motion. Consequently, the mechanical energy of the system diminishes in time, and the motion is said to be *damped*. The lost mechanical energy is transformed into internal energy in the object and the retarding medium. **Figure 16.21** depicts one such system: an object attached to a spring and submersed in a viscous liquid. The opening photograph for this chapter depicts a practical application of damped oscillations. The piston-like devices below the sphere are dampers that transform mechanical energy of the oscillating sphere into internal energy. A third example is your arm, as you would have noted in the *Try this* example – it will only swing back and forth a few times before coming to rest.

One common type of retarding force is that discussed in Section 5.2, where the force is proportional to the speed of the moving object and acts in the direction opposite the velocity of the object with respect to the medium. This retarding force is often observed when an object moves through air, for instance. Because the retarding force can be expressed as $\vec{\mathbf{R}} = -b\vec{\mathbf{v}}$ (where b is a constant called the *damping coefficient*) and the restoring force of the system is $-kx$, we can write Newton's second law as

$$\sum F_x = -kx - bv_x = ma_x$$

$$-kx - b\frac{dx}{dt} = m\frac{d^2x}{dt^2} \tag{16.30}$$

This is a second-order differential equation, with a first-order term. When the retarding force is small compared with the maximum restoring force – that is, when b is small – the solution to **Equation 16.30** is

$$x = Ae^{-(b/2m)t}\cos(\omega t + \phi) \tag{16.31}$$

where the angular frequency of the damped oscillation is

$$\omega = \sqrt{\frac{k}{m} - \left(\frac{b}{2m}\right)^2} \tag{16.32}$$

This result can be verified by substituting **Equation 16.31** into **Equation 16.30**. It is convenient to express the angular frequency of a damped oscillator in the form

$$\omega = \sqrt{\omega_0^{\,2} - \left(\frac{b}{2m}\right)^2}$$

where $\omega_0 = \sqrt{k/m}$ is the angular frequency in the absence of a retarding force (the undamped oscillator) and is called the **natural frequency** of the system.

Σ *Differential calculus is summarised in Appendix B.7.*

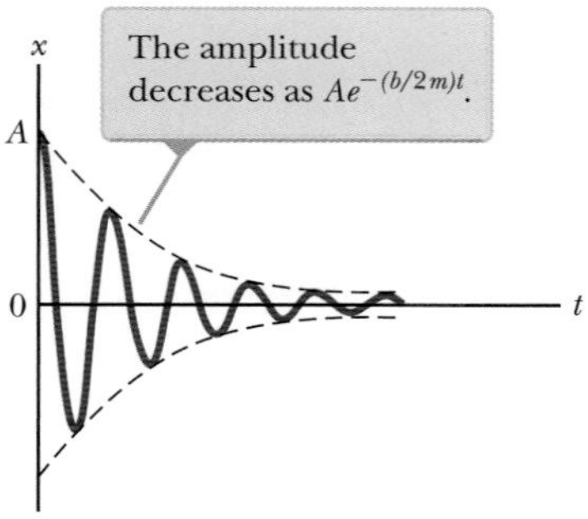
x
The amplitude decreases as $Ae^{-(b/2m)t}$.
A
0
t

Active Figure 16.22
Graph of position versus time for a damped oscillator

Quick Quiz 16.10

What are the units of the damping coefficient, b?

Active Figure 16.22 shows the position as a function of time for an object oscillating in the presence of a retarding force. When the retarding force is small, the oscillatory character of the motion is preserved but the amplitude decreases exponentially in time, with the result that the motion ultimately becomes undetectable. Any system that behaves in this way is known as a **damped oscillator**. The dashed black lines in **Active Figure 16.22**, which define the *envelope* of the oscillatory curve, represent the exponential factor in **Equation 16.31**. This envelope shows that the amplitude decays exponentially with time. The greater the retarding force, the more rapidly the oscillations decay.

When the magnitude of the retarding force is small such that $b/2m < \omega_0$, the system is said to be **underdamped**. The resulting motion is represented by the blue curve in **Figure 16.23**. A car's suspension system is typically underdamped. As the value of b increases, the amplitude of the oscillations decreases more and more rapidly. When b reaches a critical value b_c such that $b_c/2m = \omega_0$, the system does not oscillate and is said to be **critically damped**. In this case, the system, once released from rest at some non-equilibrium position, approaches but does not pass through the equilibrium position. The graph of position versus time for this case is the red curve in **Figure 16.23**.

If the retarding force is large compared with the restoring force, such that $b/2m > \omega_0$, because the medium is extremely viscous or some other means of damping such as large counterweights is used, the system is **overdamped**. Again, the displaced system, when free to move, does not oscillate but returns to its equilibrium position. As the damping increases, the time interval required for the system to approach equilibrium also increases, as indicated by the black curve in **Figure 16.23**. For critically damped and overdamped systems, there is no angular frequency ω and the solution in **Equation 16.31** is not valid. A swinging door that doesn't quite close may be critically damped or overdamped.

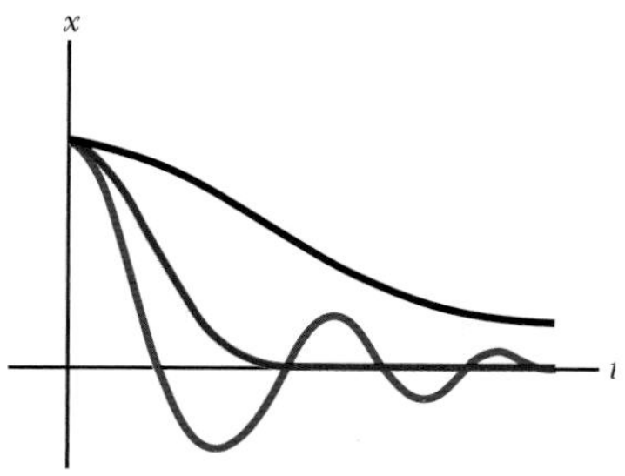

Figure 16.23
Graphs of position versus time for an underdamped oscillator (blue curve), a critically damped oscillator (red curve) and an overdamped oscillator (black curve)

TRY THIS

Hang a heavy waterproof object from a spring or rubber band. Pull it downwards and release it. Observe the oscillations and count how many it makes before coming to rest. Now submerge the object in water and repeat the experiment. How many oscillations does it make this time? Can you measure a difference in the period of oscillations in air and in water?

±? Example 16.7

A torsion pendulum experiment

A student is doing an experiment with a torsional pendulum, as shown. The pendulum bob has a mass of (2.0 ± 0.1) kg and is hung from a thin wire.

He measures the natural frequency of the pendulum in air and then lowers the pendulum so it is partly submerged in a container of oil. The period of the torsional pendulum is now measured to be (0.50 ± 0.01) s. He then records the maximum amplitude as a function of time and collects the following data:

$t \pm 0.1$ (s)	1.0	2.0	3.0	4.0	5.0	6.0	7.0	8.0	9.0	10.0
$\theta_{max} \pm 0.1$ (°)	6.4	4.0	2.5	1.7	1.0	0.7	0.4	0.3	0.3	0.1

Plot a graph to find the damping coefficient, b.

The torsion pendulum experiment at Monash University.

Solution

Conceptualise We are given a picture of the experiment, so we do not need to draw a diagram. If you were doing this experiment yourself, however, you would certainly draw a large clear diagram showing the apparatus. You may have done a similar experiment and already have analysed data like this to find a damping coefficient. We sketch roughly what the motion of the torsional pendulum looks like, marking the points at which our data was collected.

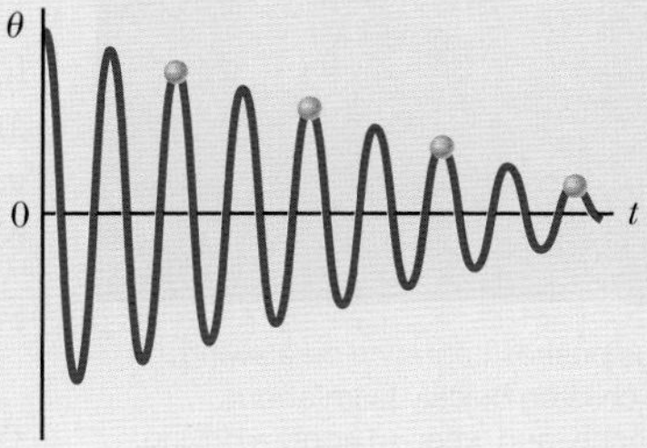

Model We model this system as a torsional pendulum undergoing damped simple harmonic motion. The amplitudes of oscillation are small, less than 10°. For a torsional pendulum we do not require that the angle is small because we do not need the small angle approximation; however, small oscillations are less likely to exceed the elastic limit of the wire as it twists. We will apply **Equation 16.31** and plot an appropriate graph.

Example 16.7 cont.

Analyse Equation 16.31 tells us that:

$$\theta = \theta_{max} e^{-(b/2m)t} \cos(\omega t + \phi)$$

where in this case the amplitude A is replaced by the angular amplitude θ_{max} and the linear displacement x with the angular displacement θ. The data we have are for when the amplitude is a maximum, every second cycle. At these times the term $\cos(\omega t + \phi)$ must be equal to one, so for the data given Equation 16.31 can be reduced to:

$$\theta = \theta_{max} e^{-(b/2m)t}$$

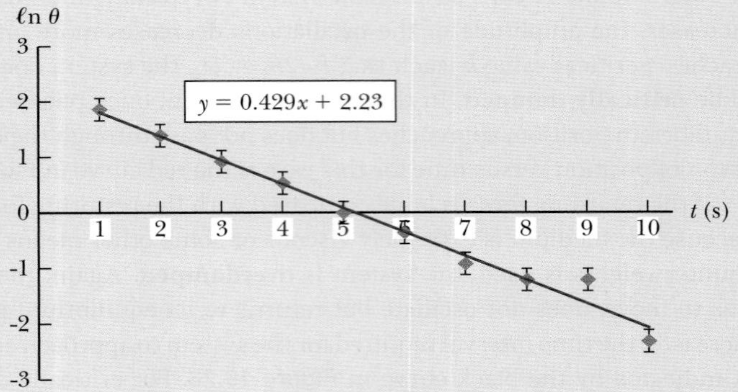

If we take the natural logarithm of both sides we find:

$$\ln\theta = \ln(\theta_{max} e^{-(b/2m)t}) = \ln\theta_{max} - (b/2m)t$$

Hence we plot a graph of $\ln\theta$ vs t, which will have a gradient of $b/2m$ and an intercept of $\ln\theta_{max}$.

From the graph we find that the gradient is $-(0.43 \pm 0.02)\text{s}^{-1}$ and the intercept is 2.2 ± 0.1.

So $b/2m = (0.43 \pm 0.02)\ \text{s}^{-1}$, and recalling that the mass is (2.0 ± 0.1) kg we have:

$$b = 2m\,(\text{gradient}) = 2(2.0\ \text{kg})(0.43\ \text{s}^{-1}) = 1.72\ \text{kg/s}$$

with an uncertainty of $\Delta b = b\left(\frac{\Delta m}{m} + \frac{\Delta\text{gradient}}{\text{gradient}}\right) = 1.72\ \text{kg/s}\left(\frac{0.1\ \text{kg}}{2.0\ \text{kg}} + \frac{0.02\ \text{s}^{-1}}{0.43\ \text{s}^{-1}}\right) = 0.166\ \text{kg.s}^{-1}$.

So our final experimental result for the damping coefficient is $b = (1.72 \pm 0.17)$ kg/s.

Finalise It is important to consider what to plot when analysing data. A graph is almost always useful if you can work out what to plot such that the gradient and intercept can give you useful data. Often the relationship between the things measured, in this case time and angle, is not linear, so some data manipulation is needed. A ln–ln graph is particularly useful when you do not already know what the relationship is between two variables. We can also find the initial angle in this case: In $\theta_{max} = 2.2$, so $\theta_{max} = 9°$. Note also that in this case all the fractional uncertainties were of the same order of magnitude, so we had to include all of them in our uncertainty analysis. Often this is not the case and one or more can be ignored, simplifying the analysis.

16.7 Forced oscillations

We have seen that the mechanical energy of a damped oscillator decreases in time as a result of the resistive force. It is possible to compensate for this energy decrease by applying a periodic external force that does positive work on the system. At any instant, energy can be transferred into the system by an applied force that acts in the direction of motion of the oscillator. For example, a child on a swing can be kept in motion by appropriately timed pushes. The amplitude of motion remains constant if the energy input per cycle of motion exactly equals the decrease in mechanical energy in each cycle that results from resistive forces.

Marcus pushes Laurence on a swing. Small children quickly learn about resonance, and when to apply a driving force, by playing on swings.

A common example of a forced oscillator is a damped oscillator driven by an external force that varies periodically, such as $F(t) = F_0 \sin\omega t$, where F_0 is a constant and ω is the angular frequency of the driving force. In general, the frequency ω of the driving force is variable, whereas the natural frequency ω_0 of the oscillator is fixed by the values of k and m. Newton's second law in this situation gives

$$\sum F_x = ma_x \rightarrow F_0 \sin\omega t - b\frac{dx}{dt} - kx = m\frac{d^2x}{dt^2} \tag{16.33}$$

After the driving force on an initially stationary object begins to act, the amplitude of the oscillation will increase. After a sufficiently long period of time, when the energy input per cycle from the driving force equals the amount of mechanical energy transformed to internal energy for each cycle, a **steady-state condition** is reached in which the oscillations continue with constant amplitude. In this situation, the solution of Equation 16.33 is

$$x = A\cos(\omega t + \phi) \tag{16.34}$$

where

$$A = \frac{F_0/m}{\sqrt{(\omega^2 - {\omega_0}^2)^2 + \left(\frac{b\omega}{m}\right)^2}} \tag{16.35}$$

◀ Amplitude of a driven oscillator

and where $\omega_0 = \sqrt{k/m}$ is the natural frequency of the undamped oscillator ($b = 0$).

Equations 16.34 and 16.35 show that the forced oscillator vibrates at the frequency of the driving force and that the amplitude of the oscillator is constant for a given driving force because it is being driven in steady-state by an external force. For small damping, the amplitude is large when the frequency of the driving force is near the natural frequency of oscillation, or when $\omega \approx \omega_0$. The dramatic increase in amplitude near the natural frequency is called **resonance** and the natural frequency ω_0 is also called the **resonance frequency** of the system.

The reason for large-amplitude oscillations at the resonance frequency is that energy is being transferred to the system at a large rate. We can better understand this by taking the first time derivative of x in Equation 16.34, which gives an expression for the velocity of the oscillator. We find that v is proportional to $\sin(\omega t + \phi)$, which is the same trigonometric function as that describing the driving force. Therefore, the applied force $\vec{\mathbf{F}}$ is in phase with the velocity. The rate at which work is done on the oscillator by $\vec{\mathbf{F}}$ equals the dot product $\vec{\mathbf{F}} \cdot \vec{\mathbf{v}}$; this rate is the power delivered to the oscillator. Because the product $\vec{\mathbf{F}} \cdot \vec{\mathbf{v}}$ is a maximum when $\vec{\mathbf{F}}$ and $\vec{\mathbf{v}}$ are in phase, we conclude that at resonance the applied force is in phase with the velocity and the power transferred to the oscillator is a maximum.

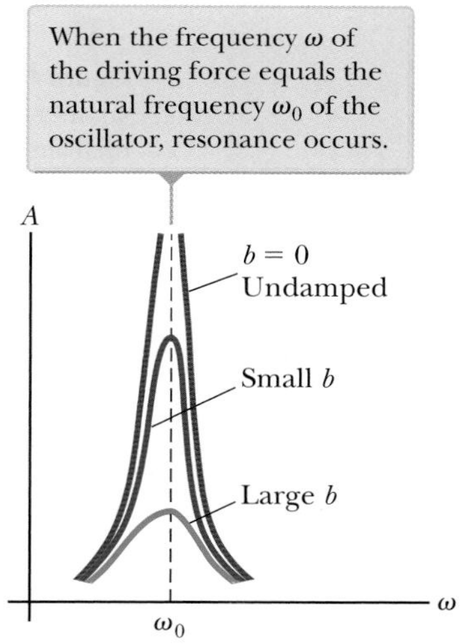

When the frequency ω of the driving force equals the natural frequency ω₀ of the oscillator, resonance occurs.
A
b = 0
Undamped
Small b
Large b
ω₀
ω

Figure 16.24
Graph of amplitude versus frequency for a damped oscillator when a periodic driving force is present. Notice that the shape of the resonance curve depends on the size of the damping coefficient b.

Figure 16.24 is a graph of amplitude as a function of driving frequency for a forced oscillator with and without damping. Notice that the amplitude increases with decreasing damping ($b \to 0$) and that the resonance curve broadens as the damping increases. In the absence of a damping force ($b = 0$), we see from Equation 16.35 that the steady-state amplitude approaches infinity as ω approaches ω_0. In other words, if there are no losses in the system and we continue to drive an initially motionless oscillator with a periodic force that is in phase with the velocity, the amplitude of motion builds without limit (see the red-brown curve in Figure 16.24). This limitless building does not occur in practice because some damping is always present in reality.

Resonance is important in many mechanical systems, including musical instruments that rely on standing waves (Chapter 17) and wire fences that 'sing' in the wind. Resonance can also be undesirable. A bridge has natural frequencies that can be set into resonance by an appropriate driving force. A dramatic example of such resonance occurred in 1940 when the Tacoma Narrows Bridge in the USA was destroyed by resonant vibrations. Although the winds were not particularly strong on that occasion, the motion of the wind across the roadway provided a periodic driving force whose frequency matched that of the bridge. The resulting oscillations of the bridge caused it to ultimately collapse (Figure 16.25) because the bridge design had inadequate built-in safety features.

TRY THIS

Take a friend to a playground with a swing. Get them to give you a push at the natural frequency of the you–swing system. Note how quickly your swinging gains amplitude when they push at the correct frequency. Does the period vary with amplitude? How quickly do you lose amplitude when they stop pushing? Now try to damp your motion by pumping your legs in the opposite way that you would do if you were making yourself swing – try this while your friend is pushing you and see how much it affects the energy transfer from them to you. Try it when they stop pushing. How quickly do you lose amplitude this time?

Figure 16.25
(a) In 1940, turbulent winds set up torsional vibrations in the Tacoma Narrows Bridge, causing it to oscillate at a frequency near one of the natural frequencies of the bridge structure. (b) Once established, this resonance condition led to the bridge's collapse. (Some mathematicians and physicists are currently challenging this interpretation.)

16.8 Vibrations in crystals

So far in this chapter we have been looking at oscillations on a macroscopic scale, but oscillations are also important on a microscopic scale. In the previous chapter we saw that there is a potential energy associated with the bonds between atoms, and how the shape and depth of the 'potential well' relates to the material's response to an applied stress. As well as the potential energy due to interactions with their neighbours, the molecules or ions in solids have thermal energy due to their interactions with their environment (see Chapter 19). At relatively low temperatures, this thermal energy is entirely in the form of kinetic energy – the molecules are constantly in motion, vibrating around their equilibrium positions.

We shall now show how to calculate the frequency of these vibrations. We have already seen that for small displacements from equilibrium, the forces on a molecule obey Hooke's law – that is, they are directly proportional but oppositely directed to the displacement:

$$F = -k(r - r_0) \tag{16.36}$$

We will simplify the notation by setting r_0 as the origin and so allowing ourselves to write

$$F = -kr$$

As we have seen earlier in this chapter, systems experiencing this kind of force exhibit *simple harmonic motion* with frequency

$$f = \frac{1}{2\pi}\sqrt{\frac{k}{m}}$$

where m is the mass in motion, in this case the mass of the molecule. We want to find out what this frequency is for a molecule in a Lennard-Jones potential (see Chapter 15) or similar. To do that, we need to evaluate the spring constant k. We know that

$$k = -\frac{dF}{dr}$$

and we showed in Example 15.3 that the force associated with the Lennard-Jones potential is

$$F(r) = \frac{12\epsilon}{r_0}\left[\left(\frac{r_0}{r}\right)^{13} - \left(\frac{r_0}{r}\right)^{7}\right] \tag{16.37}$$

where ϵ is the depth of the potential well and r_0 is the equilibrium spacing between molecules.

Differentiating this gives a function of r; but because we are only considering small-amplitude vibrations, we can evaluate it at r_0 to find k, i.e.

$$k = -\left.\frac{dF}{dr}\right|_{r_0} = \frac{72\epsilon}{r_0^{\,2}} \tag{16.38}$$

Differentiation of polynomials is described in Appendix B.7. Σ

Now imagine a line of three molecules in a solid as shown in **Figure 16.26**. We assume the two molecules at the sides are fixed and only the central one is allowed to vibrate around its equilibrium position. When the central molecule moves to the left, it moves further away from the one on the right but closer to the one on the left. As a result, it will experience forces due to both molecules – a repulsive force from the molecule on the left and an attractive one from the molecule on the right – which both act to draw it back towards its equilibrium position.

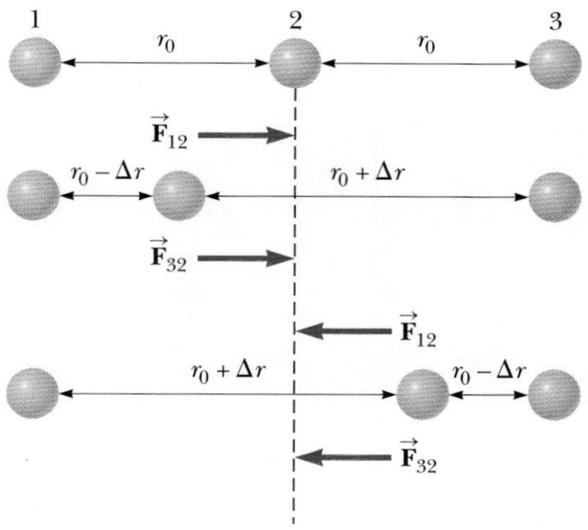

Figure 16.26
Three molecules in a crystal lattice

The total force on the molecule is thus twice the force given by **Equation 16.38**, and the constant k is

$$k = -\left.\frac{dF}{dr}\right|_{r_0} = \frac{144\epsilon}{r_0^{\,2}}$$

Of course, molecules in solids do not only have two neighbours to interact with. The number of neighbours that need to be counted depends on how closely packed the material is, so we shall represent it as N. You might think that k will then just be given by **Equation 16.38** with an additional factor of N. However, because the neighbours are not all the same distance away, a better approximation includes a factor of 3 that takes account of this, and we find that

$$k = -\left.\frac{dF}{dr}\right|_{r_0} = \frac{72N\epsilon}{3r_0^{\,2}} \tag{16.39}$$

This gives a frequency

$$f_E = \frac{1}{2\pi}\sqrt{\frac{k}{m}} = \frac{1}{2\pi r_0}\sqrt{\frac{24N\epsilon}{m}} \tag{16.40}$$

◀ Einstein frequency

which is called the Einstein frequency. We saw in Section 15.4 that we can write the bulk modulus in terms of the same constants (**Equation 15.15**), so we can relate the Einstein frequency of molecular vibration to this elastic property:

$$f_E = \frac{1}{2\pi}\sqrt{\frac{6Br_0}{m}} \tag{16.41}$$

In Chapter 17, we will see that the speed of sound in a solid is also proportional to the square root of the bulk modulus of the material. It can be shown that the Einstein frequency is the frequency of sound waves with a wavelength about half the length of the molecular separation.

As we shall see in Part 4 of this book, the Einstein frequency can also be related to the specific heat of insulators at high temperatures. This model assumes that molecules oscillate with a single frequency; improvements to the single-oscillator model allow for a range of fixed allowed frequencies.

End of chapter resources

Ranjit Doroszkeiwicz/Alamy

We began this chapter with an image of a pendulum that is part of the damping system for Taipei 101, and asked what the purpose of it is and how it works. The damping system of Taipei 101 is designed specifically to prevent oscillations of the building, a tower over 500 m tall, from becoming too large. Strong winds could cause the tower to oscillate, as in the case of the Tacoma Narrows Bridge. Taipei is also subject to earthquakes, the shaking from which is periodic in nature, as we shall see in the next chapter. It is possible that a building may be shaken at its resonant frequency by an earthquake, setting up extremely large vibrations in the building and doing a great deal of damage. Hence the need for the damping system. The gigantic pendulum oscillates in the opposite direction to the building, damping any vibrations due to wind or movement of the ground, and so preventing the building itself from oscillating substantially. This is similar to the damping you did with your legs on the swing if you did the *Try this* example.

The problems found in this chapter may be assigned online in Enhanced Web Assign.

ENHANCED WebAssign

Worked solutions to every fifth problem are available in the Student Solutions Manual. Register online at **www.cengagebrain.com** for access.

Summary

Concepts and principles

The kinetic energy and potential energy for an object of mass m oscillating at the end of a spring of force constant k vary with time and are given by

$$K = \frac{1}{2}mv^2 = \frac{1}{2}m\omega^2A^2\sin^2(\omega t + \phi) \quad (16.19)$$

$$U = \frac{1}{2}kx^2 = \frac{1}{2}kA^2\cos^2(\omega t + \phi) \quad (16.20)$$

The total energy of a simple harmonic oscillator is a constant of the motion and is given by

$$E = \frac{1}{2}kA^2 \quad (16.21)$$

A **simple pendulum** of length L moves in simple harmonic motion for small angular displacements from the vertical. Its period is

$$T = 2\pi\sqrt{\frac{L}{g}} \quad (16.25)$$

A **physical pendulum** is an extended object that, for small angular displacements, moves in simple harmonic motion about a pivot that does not go through the centre of mass. The period of this motion is

$$T = 2\pi\sqrt{\frac{I}{mgd}} \quad (16.27)$$

where I is the moment of inertia of the object about an axis through the pivot and d *is* the distance from the pivot to the centre of mass of the object.

If an oscillator experiences a damping force $\vec{\mathbf{R}} = -b\vec{\mathbf{v}}$, its position for small damping is described by

$$x = Ae^{-(b/2m)t}\cos(\omega t + \phi) \quad (16.31)$$

where

$$\omega = \sqrt{\frac{k}{m} - \left(\frac{b}{2m}\right)^2} \quad (16.32)$$

If an oscillator is subject to a sinusoidal driving force that is described by $F(t) = F_0 \sin\omega t$, it exhibits **resonance,** in which the amplitude is largest when the driving frequency ω matches the natural frequency $\omega_0 = \sqrt{k/m}$ of the oscillator.

The atoms in a solid vibrate about their equilibrium position due to thermal energy. The frequency of these vibrations is related to the bulk modulus of the material and is called the Einstein frequency:

$$f_E = \frac{1}{2\pi}\sqrt{\frac{6Br_0}{m}} \quad (16.41)$$

Analysis model for problem solving

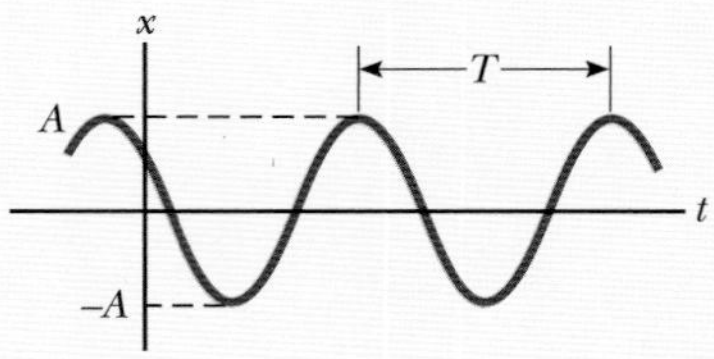

Particle in simple harmonic motion If a particle is subject to a force of the form of Hooke's law $F = -kx$, the particle exhibits **simple harmonic motion.** Its position is described by

$$x(t) = A\cos(\omega t + \phi) \quad (16.4)$$

where A is the **amplitude** of the motion, ω is the **angular frequency,** and ϕ is the **phase constant.** The value of ϕ depends on the initial position and initial velocity of the oscillator. The **period** of the oscillation is

$$T = \frac{2\pi}{\omega} = 2\pi\sqrt{\frac{m}{k}} \tag{16.12}$$

and the inverse of the period is the **frequency.**

Chapter review quiz

To help you revise Chapter 16: Oscillatory motion, complete the automatically graded Chapter review quiz at http://login.cengagebrain.com.

Conceptual questions

1. Consider a simple harmonic oscillator. (a) Can the quantities position and velocity have the same sign? (b) Can velocity and acceleration have the same sign? (c) Can position and acceleration have the same sign?
2. Equations 2.12 and 2.14 to 2.16 give position as a function of velocity and time, velocity as a function of time, and velocity as a function of position for an object moving in a straight line with constant acceleration. The quantity v_{xi} appears in every equation. (a) Do any of these equations apply to an object moving in a straight line with simple harmonic motion? (b) Make a table of equations to describe simple harmonic motion. Include equations giving acceleration as a function of time and acceleration as a function of position. State the equations in such a form that they apply equally to a block–spring system, to a pendulum and to other vibrating systems. (c) What quantity appears in every equation?
3. (a) If the coordinate of a particle varies as $x = -A \cos \omega t$, what is the phase constant in Equation 16.4? (b) At what position is the particle at $t = 0$?
4. Figure CQ16.4 shows graphs of the potential energy of four different systems versus the position of a particle in each system. A particle is set into motion with a push at an arbitrarily chosen location. Describe its subsequent motion in each case (a), (b), (c) and (d).

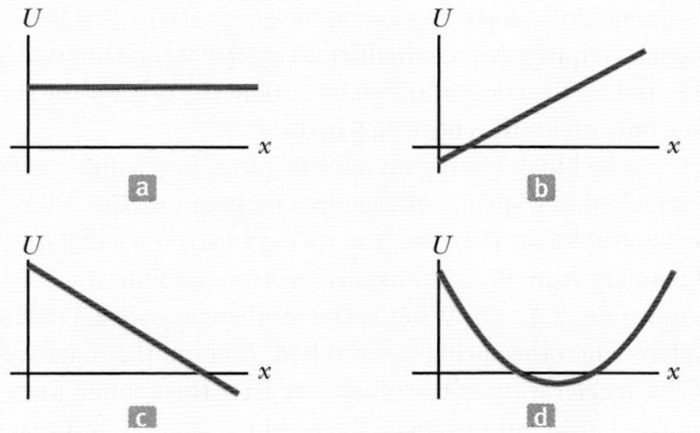

Figure CQ16.4

5. The mechanical energy of an undamped block–spring system is constant, as kinetic energy transforms to elastic potential energy and vice versa. For comparison, explain what happens to the energy of a damped oscillator in terms of the mechanical, potential and kinetic energies.
6. (a) Is it possible to have damped oscillations when a system is at resonance? (b) Will damped oscillations occur for all possible values of b and k? Explain your answers.
7. If a pendulum clock keeps perfect time at the base of a mountain, will it also keep perfect time when it is moved to the top of the mountain? Explain your answer.
8. A pendulum bob is a sphere filled with water. What would happen to the frequency of vibration of this pendulum if there were a hole in the sphere that allowed the water to leak out slowly? Sketch a graph of the period of the pendulum as a function of time.
9. Consider the simplified single-piston engine in Figure CQ16.9. Assuming the wheel rotates with constant angular speed, explain why the piston rod oscillates in simple harmonic motion.

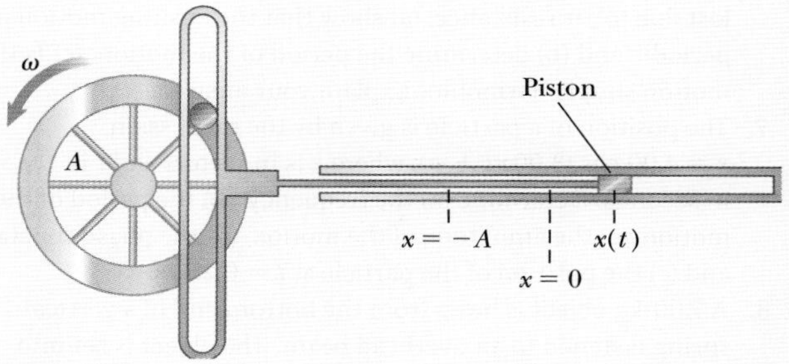

Figure CQ16.9

10. A simple pendulum is suspended from the ceiling of a stationary lift, and the period is determined. (a) What happens to the period when the lift is accelerating upwards? (b) What happens to the period when the lift is accelerating downwards? (c) What happens to the period when the lift is moving upwards at constant velocity? Explain your answer in each case.
11. What limits the amplitude of motion of a real vibrating system that is driven at one of its resonant frequencies?
12. Despite a reasonably steady hand, a person often spills his coffee when carrying it to his seat. Discuss resonance as a possible cause of this difficulty and devise a means for preventing the spills.

Problems

Section 16.1 Motion of an object attached to a spring

1. A 0.60 kg block attached to a spring with force constant 130 N/m is free to move on a frictionless, horizontal surface as in Active Figure 16.1. The block is released from rest when the spring is stretched 0.13 m. At the instant the block is released, find (a) the force on the block and (b) its acceleration.
2. A 65 kg person sits down on the edge of a bed with an inner-spring mattress, such that their weight is taken by a single spring. The spring compresses a distance of 4.5 cm. What is the force constant of the spring?

Section 16.2 Analysis model: simple harmonic motion

3. A vertical spring stretches 3.9 cm when a 10 g object is hung from it. The object is replaced with a block of mass 25 g that oscillates up and down in simple harmonic motion. Calculate the period of motion.

4. In an engine, a piston oscillates with simple harmonic motion so that its position varies according to the expression

$$x = 5.00\cos\left(2t + \frac{\pi}{6}\right)$$

where x is in centimetres and t is in seconds. At $t = 0$, find (a) the position of the particle, (b) its velocity and (c) its acceleration. Find (d) the period and (e) the amplitude of the motion.

5. A piston in an engine exhibits simple harmonic motion. The engine is running at the rate of 3600 rev/min. Taking the extremes of its position relative to its centre point as ± 5.00 cm, find the magnitudes of (a) the maximum velocity and (b) the maximum acceleration of the piston.

6. A ball dropped from a height of 4.00 m makes an elastic collision with the ground. Assuming no mechanical energy is lost due to air resistance, (a) show that the ensuing motion is periodic and (b) determine the period of the motion. (c) Is the motion simple harmonic? Explain your answer.

7. The position of a particle is given by the expression $x = 4.00\cos(3.00\pi t + \pi)$, where x is in meters and t is in seconds. Determine (a) the frequency, (b) the period of the motion, (c) the amplitude of the motion, (d) the phase constant and (e) the position of the particle at $t = 0.250$ s.

8. A 7.00-kg object is hung from the bottom end of a vertical spring fastened to an overhead beam. The object is set into vertical oscillations having a period of 2.60 s. Find the force constant of the spring.

9. ±? A group of students is doing an experiment using a spring. They hang weights from a spring using a holder and measure the position x of the end of the spring as they add additional weights, to give a total mass of m. They collect the following data:

x (cm) ± 0.1 cm	10.0	12.1	14.2	16.4	18.5	20.6	22.6	24.5	26.2	27.3
m (g) ± 1 g	50	100	150	200	250	300	350	400	450	500

(a) Plot a graph to find the spring constant for this spring. Explain your working, including any decisions you make on how to find the spring constant from the graph.

In a second part to the experiment the students measure the period T of oscillation as a function of mass hanging from the spring. They collect the following data:

T (s) ± 0.01 s	0.28	0.40	0.49	0.56	0.63	0.70	0.75
m (g) ± 1 g	50	100	150	200	250	300	350

(b) Plot a graph to find the spring constant for this spring in this second part of the experiment. Does it agree with the value found in the first part? (c) Why have the students collected fewer data points in the second part of the experiment? (It is not because they are running out of time.)

Section 16.3 Comparing simple harmonic motion with uniform circular motion

10. While driving behind a car travelling at 3.00 m/s, you notice that one of the car's tyres has an object stuck to it, as shown in Figure P16.10. (a) Explain why the object, from your viewpoint behind the car, executes simple harmonic motion. (b) If the radii of the car's tyres are 0.300 m, what is the object's period of oscillation?

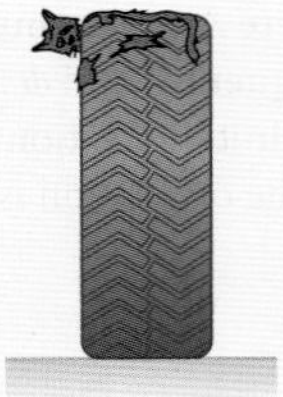

Figure P16.10

Section 16.4 Energy of the simple harmonic oscillator

11. ±? To test the resiliency of its bumper during low-speed collisions, a (1000 ± 50) kg car is driven into a brick wall. The car's bumper behaves like a spring with a force constant $(5.0 \pm 0.1) \times 10^6$ N/m and compresses (3.16 ± 0.05) cm as the car is brought to rest. What was the speed of the car before impact, assuming no mechanical energy is transformed or transferred away during impact with the wall?

12. A 50.0-g object connected to a spring with a force constant of 35.0 N/m oscillates with an amplitude of 4.00 cm on a frictionless, horizontal surface. Find (a) the total energy of the system and (b) the speed of the object when its position is 1.00 cm. Find (c) the kinetic energy and (d) the potential energy when its position is 3.00 cm.

13. A simple harmonic oscillator of amplitude A has a total energy E. Determine (a) the kinetic energy and (b) the potential energy when the position is one-third the amplitude. (c) For what values of the position does the kinetic energy equal one-half the potential energy? (d) Are there any values of the position where the kinetic energy is greater than the maximum potential energy? Explain your answer.

14. A 65.0 kg bungee jumper steps off a bridge with a light bungee cord tied to her body and to the bridge. The unstretched length of the cord is 11.0 m. The jumper reaches the bottom of her motion 36.0 m below the bridge before bouncing back. We wish to find the time interval between her leaving the bridge and her arriving at the bottom of her motion. Her overall motion can be separated into an 11.0 m free fall and a 25.0 m section of simple harmonic oscillation. (a) For the free-fall part, what is the appropriate analysis model to describe her motion? (b) For what time interval is she in free fall? (c) For the simple harmonic oscillation part of the plunge, is the system of the bungee jumper, the spring and the Earth isolated or non-isolated? (d) From your response in part (c) find the spring constant of the bungee cord. (e) What is the location of the equilibrium point where the spring force balances the gravitational force exerted on the jumper? (f) What is the angular frequency of the oscillation? (g) What time interval is required for the cord to stretch by 25.0 m? (h) What is the total time interval for the entire 36.0 m drop?

15. A 0.250 kg block resting on a frictionless, horizontal surface is attached to a spring whose force constant is 83.8 N/m. A horizontal force $\vec{\mathbf{F}}$ causes the spring to stretch a distance of 5.46 cm from its equilibrium position. (a) Find the magnitude of $\vec{\mathbf{F}}$. (b) What is the total energy stored in the system when the spring is stretched? (c) Find the magnitude of the acceleration of the block just after the applied force is removed. (d) Find the speed of the block when it first reaches the equilibrium position. (e) If the surface is not frictionless but the block still reaches the equilibrium position, would your answer to part (d) be larger or smaller? (f) What other information would you need to know to find the actual answer to part (d) in this case? (g) What is the largest value of the coefficient of friction that would allow the block to reach the equilibrium position?

Section 16.5 The pendulum

16. A 'seconds pendulum' is one that moves through its equilibrium position once each second. The length of a seconds pendulum is 0.9927 m at Tokyo, Japan and 0.9942 m at Cambridge, England. What is the ratio of the free-fall accelerations at these two locations?

17. A simple pendulum makes 120 complete oscillations in 3.00 min at a location where $g = 9.80\ \text{m/s}^2$. Find (a) the period of the pendulum and (b) its length.

18. A particle of mass m slides without friction inside a hemispherical bowl of radius R. Show that if the particle starts from rest with a small displacement from equilibrium, it moves in simple harmonic motion with an angular frequency equal to that of a simple pendulum of length R. That is, $\omega = \sqrt{g/R}$.

19. ±? A student is doing an experiment to find the moment of inertia of a physical pendulum in the form of a planar object. She measures the frequency of oscillation to be (0.450 ± 0.005) Hz, the mass to be (2.20 ± 0.01) kg, and the pivot to be located (0.350 ± 0.005) m from the centre of mass. Determine the moment of inertia of the pendulum about the pivot point.

20. **Review.** A simple pendulum is 5.00 m long. What is the period of small oscillations for this pendulum if it is located in an elevator (a) accelerating upward at $5.00\ \text{m/s}^2$? (b) accelerating downward at $5.00\ \text{m/s}^2$? (c) What is the period of this pendulum if it is placed in a truck that is accelerating horizontally at $5.00\ \text{m/s}^2$?

21. A very light rigid rod of length 0.500 m extends straight out from one end of a metre stick. The combination is suspended from a pivot at the upper end of the rod as shown in **Figure P16.21**. The combination is then pulled out by a small angle and released. (a) Determine the period of oscillation of the system. (b) By what percentage does the period differ from the period of a simple pendulum 1.00 m long?

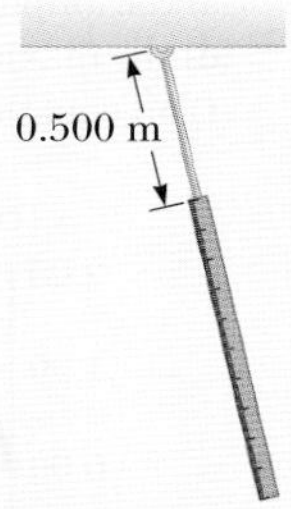

Figure P16.21

22. A watch balance wheel (**Figure P16.22**) has a period of oscillation of 0.250 s. The wheel is constructed so that its mass of 20.0 g is concentrated around a rim of radius 0.500 cm. Determine (a) the wheel's moment of inertia and (b) the torsion constant of the attached spring.

© Cengage Learning/George Semple

Figure P16.22

23. A simple pendulum is used in a harmonic motion experiment. Its period is measured for small angular displacements and three lengths. For lengths of 1.000 m, 0.750 m and 0.500 m, total time intervals for 50 oscillations of 99.8 s, 86.6 s and 71.1 s are measured with a stopwatch. (a) Determine the period of motion for each length. (b) Determine the mean value of g obtained from these three independent measurements and compare it with the accepted value. (c) Plot T^2 versus L and obtain a value for g from the gradient of your best-fit straight-line graph. (d) Compare the value found in part (c) with that obtained in part (b).

Section 16.6 Damped oscillations

24. A pendulum with a length of 1.00 m is released from an initial angle of 15.0°. After 1000 s, its amplitude has been reduced by friction to 5.50°. What is the value of $b/2m$?

25. Show that the time rate of change of mechanical energy for a damped, undriven oscillator is given by $dE/dt = -bv^2$ and hence is always negative. Hint: Differentiate the expression for the mechanical energy of an oscillator and use **Equation 16.30**.

26. A 10.6 kg object oscillates at the end of a vertical spring that has a spring constant of 2.05×10^4 N/m. The effect of air resistance is represented by the damping coefficient $b = 3.00$ N.s/m. (a) Calculate the frequency of the damped oscillation. (b) By what percentage does the amplitude of the oscillation decrease in each cycle? (c) Find the time interval that elapses while the energy of the system drops to 5.00% of its initial value.

27. Show that **Equation 16.31** is a solution of **Equation 16.30** provided that $b^2 < 4mk$.

28. For the damped torsional pendulum in Example 16.7, find the value of k for this pendulum and the undamped natural frequency ω_0.

Section 16.7 Forced oscillations

29. Marcus is pushing Laurence on a swing. The length of chain holding the swing seat and Laurence is 3.2 m. (a) Assuming the mass of the chains is small compared to the mass of the seat plus Laurence, with what frequency does Marcus need to push to have the greatest effect? (b) How would your answer change if you included the mass of the chains? (c) How would your answer change if Laurence was trying to help by pumping his legs but doing it at the wrong time?

30. Elton is bouncing up and down in his cot. His mass is 12.5 kg, and the cot mattress can be modelled as a light spring with force constant 700 N/m. (a) Elton soon learns to bounce with maximum amplitude and minimum effort by bending his knees. At what frequency does he do this? (b) When he uses the mattress as a trampoline – losing contact with it for part of each cycle – what minimum amplitude of oscillation does he require?

31. A 2.00 kg object attached to a spring moves without friction ($b = 0$) and is driven by an external force given by the expression $F = 3.00 \sin(2\pi t)$, where F is in newtons and t is in seconds. The force constant of the spring is 20.0 N/m. Find (a) the resonance angular frequency of the system, (b) the angular frequency of the driven system and (c) the amplitude of the motion.

32. ±? Damping is negligible for a (0.150 ± 0.005) kg object hanging from a light, (6.3 ± 0.1) N/m spring. A sinusoidal force with an amplitude of (1.70 ± 0.01) N drives the system. At what frequency will the force make the object vibrate with an amplitude of 0.440 m?

33. Considering an undamped, forced oscillator, show that **Equation 16.34** is a solution of **Equation 16.33**, with an amplitude given by **Equation 16.35**.

Section 16.8 Vibrations in crystals

34. Given that the speed of sound in a material is equal to $\sqrt{B/\rho}$, show that the Einstein frequency is the frequency of sound waves with wavelengths approximately equal to twice the average intermolecular spacing.

35. The packing of molecules in ice is fairly open, so that the number of nearest neighbours for any given molecule can be taken to be 6. Estimate the Einstein frequency of vibration for water molecules in an ice lattice. State your sources for the values of any additional quantities you use. Is this frequency within the range of human hearing?

Additional problems

36. A metronome used to keep time for a musician is a physical pendulum made from a thin rod with a large fixed weight (mass M) at the bottom, a distance L below the pivot, and a smaller movable weight (mass m) above the pivot point, as shown in Figure P16.36. The frequency of 'ticking' is adjusted by sliding the smaller weight up or down the rod, varying the distance ℓ between the small weight and the pivot point.

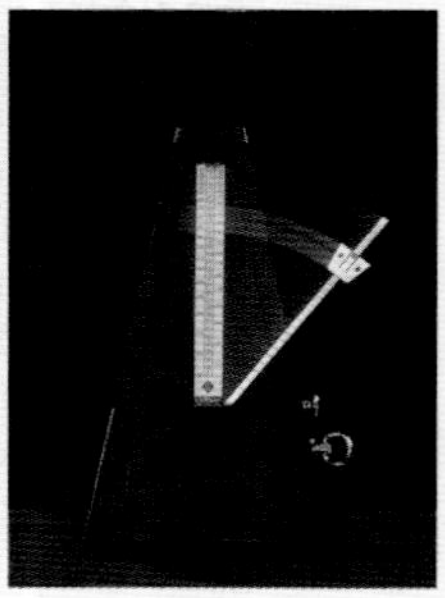

Science Photo Library/Martyn F. Chillmaid

Figure P16.36

(a) Write an expression for the period of a metronome in terms of M, m, L and ℓ and any constants required. (b) Sketch a graph showing how the period varies with ℓ.

37. The free-fall acceleration on Mars is 3.7 m/s². (a) What length of pendulum has a period of 1.0 s on Earth? (b) What length of pendulum would have a 1.0-s period on Mars? An object is suspended from a spring with force constant 10 N/m. Find the mass suspended from this spring that would result in a period of 1.0 s (c) on Earth and (d) on Mars.

38. Consider the physical pendulum of Figure 16.18. (a) Represent its moment of inertia about an axis passing through its centre of mass and parallel to the axis passing through its pivot point as I_{CM}. Show that its period is

$$T = 2\pi\sqrt{\frac{I_{CM} + md^2}{mgd}}$$

where d is the distance between the pivot point and the centre of mass. (b) Show that the period has a minimum value when d satisfies $md^2 = I_{CM}$.

39. The mass of the deuterium molecule (D_2) is twice that of the hydrogen molecule (H_2). If the vibrational frequency of H_2 is 1.30×10^{14} Hz, what is the vibrational frequency of D_2? Assume the 'spring constant' of attracting forces is the same for the two molecules.

40. This problem extends the reasoning of problem 42 in Chapter 8. Two gliders are set in motion on an air track. Glider 1 has mass $m_1 = 0.240$ kg and moves to the right with speed 0.740 m/s. It will have a rear-end collision with glider 2, of mass $m_2 = 0.360$ kg, which initially moves to the right with speed 0.120 m/s. A light spring of force constant 45.0 N/m is attached to the back end of glider 2 as shown in Figure P8.42. When glider 1 touches the spring, superglue instantly and permanently makes it stick to the end of the spring. (a) Find the common speed of the two gliders when the spring is at maximum compression. (b) Find the maximum spring compression distance. The motion after the gliders become attached consists of a combination of (1) the constant-velocity motion of the centre of mass of the two-glider system found in part (a) and (2) simple harmonic motion of the gliders relative to the centre of mass. (c) Find the energy of the centre-of-mass motion. (d) Find the energy of the oscillation.

41. A small ball of mass M is attached to the end of a uniform rod of equal mass M and length L that is pivoted at the top (Figure P16.41). Determine the tension in the rod (a) at the pivot and (b) at the point P when the system is stationary. (c) Calculate the period of oscillation for small displacements from equilibrium and (d) determine this period for $L = 2.00$ m.

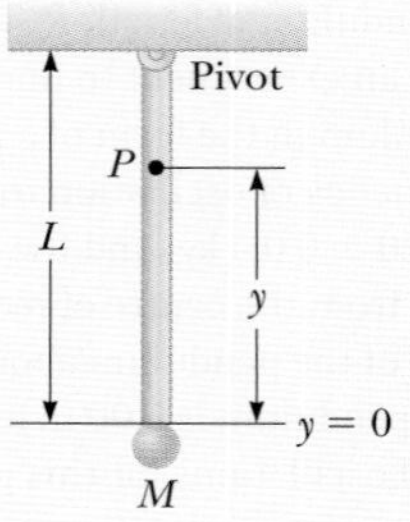

Figure P16.41

42. An object attached to a spring vibrates with simple harmonic motion as described by Figure P16.42. For this motion, find (a) the amplitude, (b) the period, (c) the angular frequency, (d) the maximum speed, (e) the maximum acceleration and (f) an equation for its position x as a function of time.

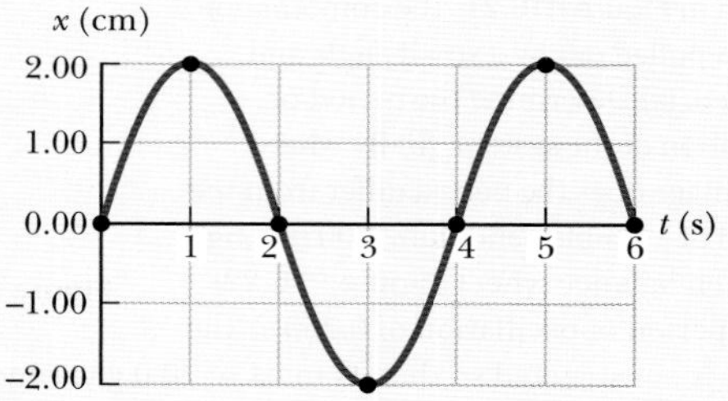

Figure P16.42

43. A large block P attached to a light spring executes horizontal, simple harmonic motion as it slides across a frictionless surface with a frequency $f = 1.50$ Hz. Block B rests on it as shown in Figure P16.43 and the coefficient of static friction between the two is $\mu_s = 0.600$. What maximum amplitude of oscillation can the system have if block B is not to slip?

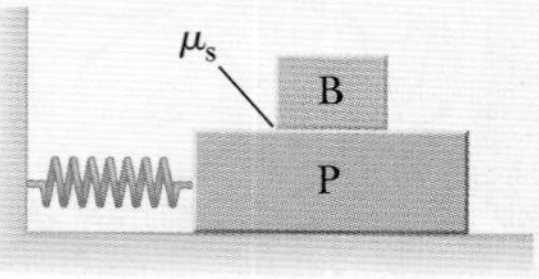

Figure P16.43

44. A one-person seesaw at a playground consists of a horizontal plank of mass M and length L pivoted at one end and supported by a spring of force constant k at the other end (Figure P16.44). (a) Find the angular frequency at which the plank oscillates when it is pushed down and released. What assumptions have you made in finding your answer and under what conditions are they valid? (b) How would the frequency change if a child of mass m was sitting at the end of the plank directly over the spring?

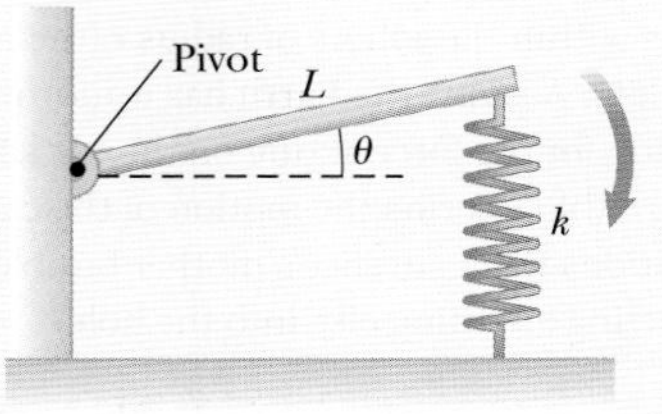

Figure P16.44

45. A particle of mass 4.00 kg is attached to a spring with a force constant of 100 N/m. It is oscillating on a frictionless, horizontal surface with an amplitude of 2.00 m. A 6.00 kg object is dropped vertically on top of the 4.00 kg object as it passes through its equilibrium point. The two objects stick together. (a) What is the new amplitude of the vibrating system after the collision? (b) By what factor has the period of the system changed? (c) By how much does the energy of the system change as a result of the collision? (d) Account for the change in energy.

46. A simple pendulum with a length of 2.23 m and a mass of 6.74 kg is given an initial speed of 2.06 m/s at its equilibrium position. Assume it undergoes simple harmonic motion. Determine (a) its period, (b) its total energy and (c) its maximum angular displacement.

47. One end of a light spring with force constant $k = 100$ N/m is attached to a vertical wall. A light string is tied to the other end of the horizontal spring. As shown in **Figure P16.47**, the string changes from horizontal to vertical as it passes over a pulley of mass M in the shape of a solid disc of radius $R = 2.00$ cm. The pulley is free to turn on a fixed, smooth axle. The vertical section of the string supports an object of mass $m = 200$ g. The string does not slip at its contact with the pulley. The object is pulled downwards a small distance and released. (a) What is the angular frequency ω of oscillation of the object in terms of the mass M? (b) What is the highest possible value of the angular frequency of oscillation of the object? (c) What is the highest possible value of the angular frequency of oscillation of the object if the pulley radius is doubled to $R = 4.00$ cm?

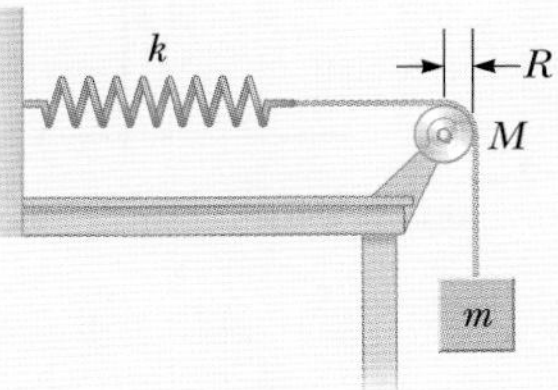

Figure P16.47

48. People who ride motorcycles and bicycles learn to look out for bumps in the road and especially for corrugations or washboarding, a condition in which many equally spaced ridges are worn into the road. What is so bad about corrugations? A motorcycle has several springs and shock absorbers in its suspension, but you can model it as a single spring supporting a block. You can estimate the force constant by thinking about how far the spring compresses when a heavy rider sits on the seat. A motorcyclist travelling at speed must be particularly careful of corrugations that are a certain distance apart. What is the order of magnitude of their separation?

49. A ball of mass m is connected to two rubber bands of length L, each under tension T, as shown in **Figure P16.49**. The ball is displaced by a small distance y perpendicular to the length of the rubber bands. Assuming the tension does not change, show that (a) the restoring force is $-(2T/L)y$ and (b) the system exhibits simple harmonic motion with an angular frequency $\omega = \sqrt{2T/mL}$.

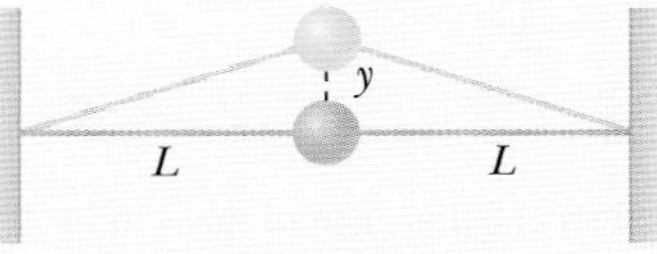

Figure P16.49

50. When a block of mass M, connected to the end of a spring of mass m and force constant k, is set into simple harmonic motion, the period of its motion is:

$$T = 2\pi\sqrt{\frac{M + (m/3)}{k}}$$

A two-part experiment is conducted with the use of blocks of various masses suspended vertically from a spring with mass $m = 7.40$ g. In the first part of the experiment students measure the static extension as a function of mass M and collect the following data:

M (g)	20.0	40.0	50.0	60.0	70.0	80.0
x (cm)	17.0	29.3	35.3	41.3	47.1	49.3

(a) Plot an appropriate graph and from the gradient of your graph determine a value for k for this spring. The system is now set into simple harmonic motion, and the total time interval required for 10 oscillations is measured with a stopwatch. The following data is collected:

M (g)	20.0	40.0	50.0	60.0	70.0	80.0
10T (s)	7.03	9.62	10.67	11.67	12.52	13.41

(b) Plot an appropriate graph and use the gradient of the graph to find a second value for k. (c) Compare this value of k with that obtained in part (a). (d) Obtain a value for m from your graph and compare it with the given value of 7.40 g.

51. Consider the damped oscillator illustrated in **Figure 16.21**. The mass of the object is 375 g, the spring constant is 100 N/m, and $b = 0.100$ N.s/m. (a) Over what time interval does the amplitude drop to half its initial value? (b) **What If?** Over what time interval does the mechanical energy drop to half its initial value? (c) Show that, in general, the fractional rate at which the amplitude decreases in a damped harmonic oscillator is one-half the fractional rate at which the mechanical energy decreases.

52. A block of mass m is connected to two springs of force constants k_1 and k_2 in two ways as shown in **Figure P16.52**. In both cases, the block moves on a frictionless table after it is displaced from equilibrium and released. Show that in the two cases the block exhibits simple harmonic motion with periods

$$\text{(a) } T = 2\pi\sqrt{\frac{m(k_1+k_2)}{k_1k_2}} \quad \text{and} \quad \text{(b) } T = 2\pi\sqrt{\frac{m}{k_1+k_2}}$$

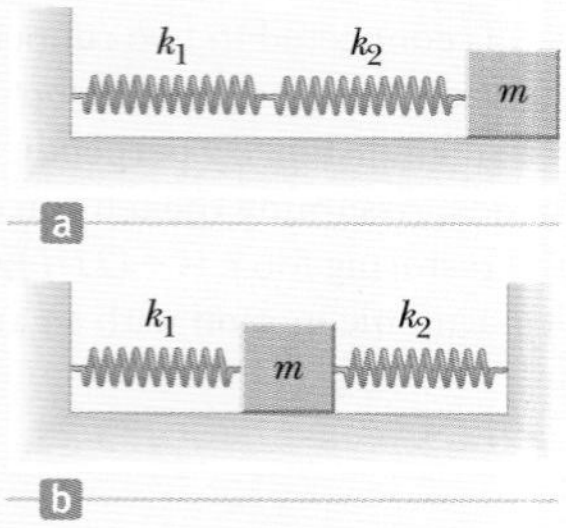

Figure P16.52

53. Two identical steel balls, each of mass (67.4 ± 0.1) g, are ±? moving in opposite directions at (5.0 ± 0.2) m/s. They collide head-on and bounce apart elastically. By squeezing one of the balls in a vice while precise measurements are made of the resulting amount of compression, you find that Hooke's law is a good model of the ball's elastic behaviour. A force of (16.0 ± 0.1) kN exerted by each jaw of the vice reduces the diameter by (0.20 ± 0.01) mm. Model the motion of each ball, while the balls are in contact, as one-half of a cycle of simple harmonic motion. Calculate the time interval for which the balls are in contact. (If you solved problem 36 in Chapter 6, compare your results from this problem with your results from that one.)

Challenge problems

54. A small disc of radius r and mass m is attached rigidly to the face of a second larger disc of radius R and mass M as shown in Figure P16.54. The centre of the small disc is located at the edge of the large disc. The large disc is mounted at its centre on a frictionless axle. The assembly is rotated through a small angle θ from its equilibrium position and released. (a) Show that the speed of the centre of the small disc as it passes through the equilibrium position is

$$v = 2\left[\frac{Rg(1-\cos\theta)}{(M/m)+(r/R)^2+2}\right]^{1/2}$$

(b) Show that the period of the motion is

$$T = 2\pi\left[\frac{(M+2m)R^2+mr^2}{2mgR}\right]^{1/2}$$

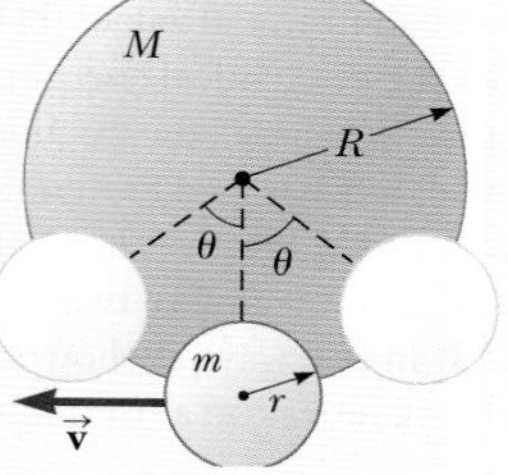

Figure P16.54

55. Imagine that a tunnel has been dug from one side of the Earth to the other and Laurence, having tired of dropping things into the toilet, drops his mother's car keys into the hole (Figure P16.55). An object, such as a set of car keys, at a distance r from the centre of the Earth is pulled toward the centre of the Earth *only* by the mass within the sphere of radius r (the reddish region in Figure P16.55). Assume the Earth has uniform density. (a) Find an expression for the force acting on the car keys as a function of position, r. (b) Describe the motion of the car keys. (c) Under what conditions will Laurence's mother be able to get them back without putting anything else into the hole?

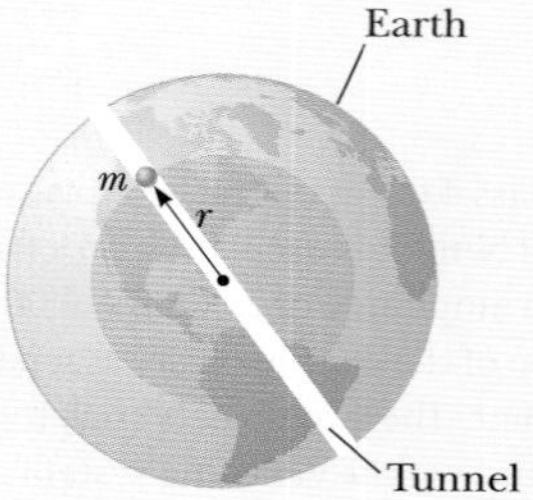

Figure P16.55

56. A block of mass M is connected to a spring of mass m and oscillates in simple harmonic motion on a frictionless, horizontal track. The force constant of the spring is k and the equilibrium length is ℓ. Assume all portions of the spring oscillate in phase and the velocity of a segment of the spring of length dx and mass $dm = (m/\ell)dx$ is proportional to the distance x from the fixed end; that is, $v_x = (x/\ell)v$. Find (a) the kinetic energy of the system when the block has a speed v and (b) the period of oscillation.

57. A system consists of a spring with force constant $k = 1250$ N/m, length $L = 1.50$ m, and an object of mass $m = 5.00$ kg attached to the end (Figure P16.57). The object is placed at the level of the point of attachment with the spring unstretched, at position $y_i = L$, and then it is released so that it swings like a pendulum. (a) Find the y position of the object at the lowest point. (b) Will the pendulum's period be greater or less than the period of a simple pendulum with the same mass m and length L? Explain your answer.

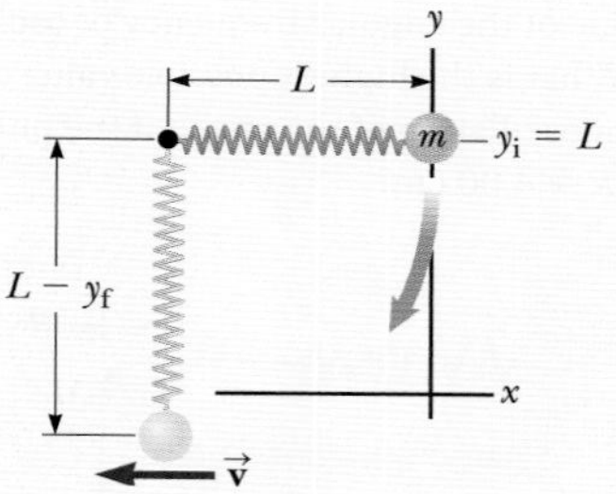

Figure P16.57

58. A light, cubic container of volume a^3 is initially filled with a liquid of mass density ρ. The cube is supported by a light string to form a simple pendulum of length L_i, measured from the centre of mass of the filled container, where $L_i \gg a$. The liquid is allowed to flow from the bottom of the container at a constant rate (dM/dt). At any time t, the level of the liquid in the container is h and the length of the pendulum is L (measured relative to the instantaneous centre of mass). (a) Find the period of the pendulum as a function of time. (b) What is the period of the pendulum after the liquid completely runs out of the container?

chapter 17

Alamy/DIZ Muenchen GmbH, Sueddeutsche Zeitung Photo

Wave motion

On the 30 June 1908, a meteor burned up and exploded in the atmosphere above the Tunguska River valley in Siberia. It knocked down trees over thousands of square kilometres. A witness reported that he saw a moving light in the sky and felt his face become warm. He felt the ground shake and was thrown about a metre away, then he heard a very loud rumbling. How can we explain these observations and the order in which they happened?

On completing this chapter, students will understand:

- how mechanical waves propagate
- that waves transfer energy but not matter
- that sound waves are longitudinal waves
- that waves may be described by a differential equation called the wave equation
- how the speed of a wave is determined by the properties of the medium
- why the observed frequency of a wave depends on the movement of the source and the observer of the wave
- that the intensity of spherical waves decreases with the square of the distance from the source.

Students will be able to:

- describe a wave using a wave function
- distinguish between transverse and longitudinal waves
- apply the travelling wave analysis model to analyse wave motion
- identify solutions to the wave equation
- calculate wave speeds for transverse and longitudinal waves in different media
- calculate the observed frequency and wavelength of a wave when the source and/or observer are in motion
- express the intensity of a sound using the decibel scale.

The world is full of waves, the two main types being *mechanical* waves and *electromagnetic* waves. Most of the information that we get about the world around us comes from waves – our ears detect sound waves and our eyes detect electromagnetic waves. Sound waves are a type of mechanical wave, which is the subject of this chapter. Later we shall look at light and other electromagnetic waves.

ENHANCED WebAssign

The problems found in this chapter may be assigned online in Enhanced Web Assign.

The ripples in the water make the leaf move vertically but there is no net displacement.

17.1 Propagation of waves

Imagine throwing a stone into a pond. At the point where the stone hits the water's surface, circular waves are created. These waves move outwards from the creation point in expanding circles until they reach the shore. If you were to examine carefully the motion of a leaf floating on the disturbed water, you would see that the leaf moves up and down about its original position but does not undergo any net displacement away from or towards the point at which the stone hit the water. The water *wave* moves from the point of origin to the shore, but the *water* is not carried with it. There is no net movement of the medium when a wave travels through it. However, the wave has caused the leaf to move at one point in the water by a stone dropping at another location. The leaf has gained kinetic energy, so energy must have transferred from the point at which the stone is dropped to the position of the leaf. This feature is central to wave motion: ***energy* is transferred over a distance but *matter* is not**.

All mechanical waves require (1) some source of disturbance, (2) a medium containing elements that can be disturbed and (3) some physical mechanism through which elements of the medium can influence each other. One way to demonstrate wave motion is to flick one end of a long string that is under tension and has its opposite end fixed as shown in Figure 17.1. In this manner, a single *pulse* is formed and travels along the string with a definite speed, while the string itself has no overall displacement. Figure 17.1 represents four consecutive 'snapshots' of the creation and propagation of the travelling pulse. The hand is the source of the disturbance. The string is the medium through which the pulse travels – individual elements of the string are disturbed from their equilibrium position. Furthermore, the elements are connected together so they influence each other. The pulse has a definite height and a definite speed of propagation along the medium. The shape of the pulse changes very little as it travels along the string.

Transverse and longitudinal waves

As the pulse in Figure 17.1 travels, each disturbed element of the string moves in a direction *perpendicular* to the direction of propagation. Figure 17.2 illustrates this point for one particular element, labelled *P*. Notice that no part of the string moves in the direction of the propagation. *A travelling wave or pulse that causes the elements of the disturbed medium to move perpendicular to the direction of propagation is called a* ***transverse wave.***

Compare this wave with another type of pulse, one moving down a long, stretched spring as shown in Figure 17.3. The left end of the spring is pushed briefly to the right and then pulled briefly to the left. This movement creates a sudden compression of a region of the coils. The compressed region travels along the spring (to the right in Figure 17.3). Notice that the direction of the displacement of the coils is *parallel* to the direction of propagation of the compressed region. *A travelling wave or pulse that causes the elements of the medium to move parallel to the direction of propagation is called a* **longitudinal wave**.

Sound waves are another example of longitudinal waves. The disturbance in a sound wave is a series of high-pressure and low-pressure regions that travel through air.

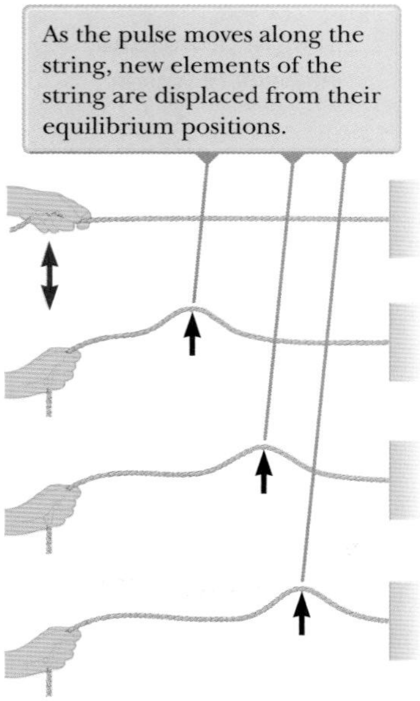

As the pulse moves along the string, new elements of the string are displaced from their equilibrium positions.

Figure 17.1
A hand moves the end of a stretched string up and down once (red arrow), causing a pulse to travel along the string.

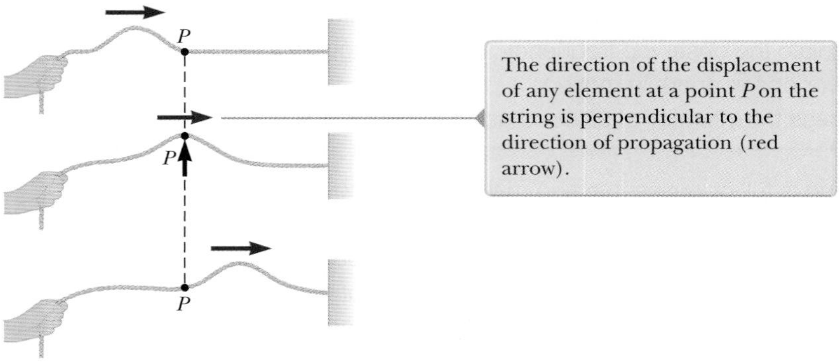

P
P
P
The direction of the displacement of any element at a point P on the string is perpendicular to the direction of propagation (red arrow).

Figure 17.2
The displacement of a particular string element for a transverse pulse travelling on a stretched string

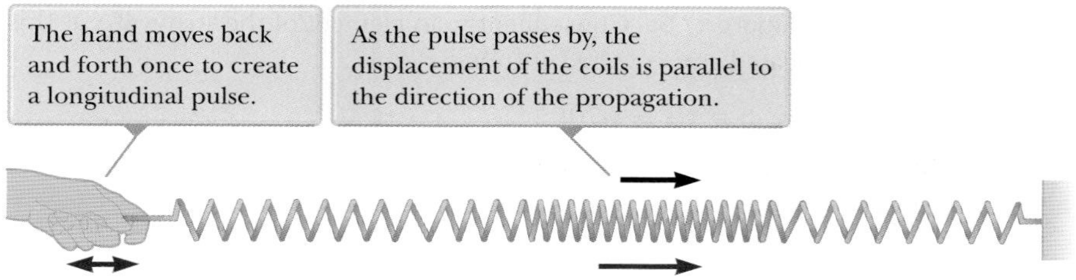
The hand moves back and forth once to create a longitudinal pulse.
As the pulse passes by, the displacement of the coils is parallel to the direction of the propagation.

Figure 17.3
A longitudinal pulse along a stretched spring

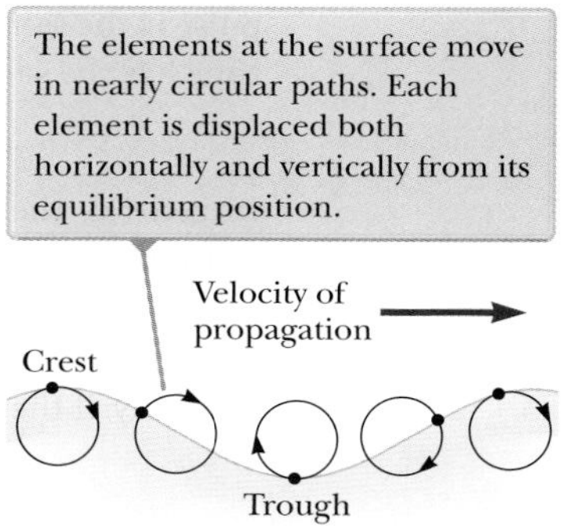
The elements at the surface move in nearly circular paths. Each element is displaced both horizontally and vertically from its equilibrium position.
Velocity of propagation
Crest
Trough

Active Figure 17.4
The motion of water elements on the surface of deep water in which a wave is propagating is a combination of transverse and longitudinal displacements.

TRY THIS

You can make both transverse and longitudinal waves using a large spring (such as a large slinky toy). Get a friend to hold one end still then send pulses along the spring towards them by either moving your end of the spring up and down or moving it quickly towards yourself then back towards your friend. Another way of sending longitudinal pulses is to hold the end of the spring with one hand and use the other hand to compress the section of the spring closest to you and then release it. Try to observe the speed of the two different types of pulses – are they the same or different? What can you do to make them faster or slower?

Some waves in nature exhibit a combination of transverse and longitudinal displacements. Surface water waves are a good example. When a water wave travels on the surface of deep water, elements of water at the surface move in nearly circular paths as shown in **Active Figure 17.4**. The disturbance has both transverse and longitudinal components. The transverse displacements seen in **Active Figure 17.4** represent the variations in vertical position of the water elements. The longitudinal displacements represent elements of water moving back and forth in a horizontal direction.

The three-dimensional waves that travel out from a point under the Earth's surface at which an earthquake occurs are of both types, transverse and longitudinal. The longitudinal waves are the faster of the two, travelling at speeds of 7 to 8 km/s near the surface. They are called **P waves**, with 'P' standing for *primary*, because they travel faster than the transverse waves and arrive first at a seismograph (a device used to detect waves due to earthquakes). The slower transverse waves, called **S waves**, with 'S' standing for *secondary*, travel through the Earth at 4 to 5 km/s near the surface. By recording the time interval between the arrivals of these two types of waves at a seismograph, the distance from the seismograph to the point of origin of the waves can be determined.

Quick Quiz 17.1

In a long line of people waiting to buy tickets, the first person leaves and a pulse of motion occurs as people step forwards to fill the gap. As each person steps forwards, the gap moves through the line. Is the propagation of this gap (a) transverse or (b) longitudinal?

Consider a pulse travelling to the right on a long string as shown in **Figure 17.5**. **Figure 17.5a** represents the shape and position of the pulse at time $t = 0$. At this time, the shape of the pulse, whatever it may be, can be represented by some mathematical function that we will write as $y(x, 0) = f(x)$. This function describes the transverse position y of the element of the string located at each value of x at time $t = 0$. Because the speed of the pulse is v, the pulse has travelled to the right a distance vt at the time t (**Figure 17.5b**). We assume the shape of the pulse does not change with time. Therefore, at time t, the shape of the

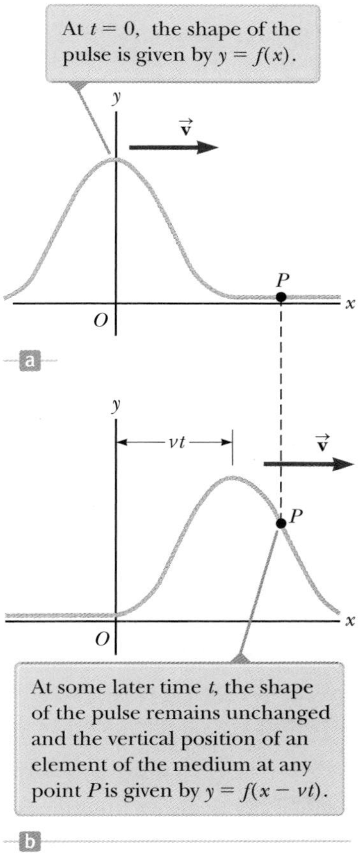
At $t = 0$, the shape of the pulse is given by $y = f(x)$.
y
$\vec{v}$
P
x
O
a
y
vt
$\vec{v}$
P
x
O
At some later time t, the shape of the pulse remains unchanged and the vertical position of an element of the medium at any point P is given by $y = f(x - vt)$.
b

Figure 17.5
A one-dimensional pulse travelling to the right on a string with a speed v

pulse is the same as it was at time $t = 0$ as in **Figure 17.5a**. Consequently, an element of the string at x at this time has the same y position as an element located at $x - vt$ had at time $t = 0$:

$$y(x, t) = y(x - vt, 0)$$

In general, then, we can represent the transverse position y for all positions and times, measured in a stationary frame with the origin at O, as

$$y(x, t) = f(x - vt) \tag{17.1}$$

Similarly, if the pulse travels to the left, the transverse positions of elements of the string are described by

$$y(x, t) = f(x + vt) \tag{17.2}$$

The function y, sometimes called the **wave function**, depends on both of the two variables, x and t.

Consider an element of the string at point P in **Figure 17.5**, identified by a particular value of its x coordinate. As the pulse passes through P, the y coordinate of this element increases, reaches a maximum and then decreases to zero. The wave function $y(x, t)$ represents the y coordinate – the transverse position – of any element located at position x at any time t. Furthermore, if t is fixed (as, for example, in the case of taking a snapshot of the pulse), the wave function $y(x)$, sometimes called the **waveform**, defines a curve representing the geometric shape of the pulse at that time.

TRY THIS

In your next physics lecture (or some other large lecture), start a 'Mexican wave'. You will probably have to ask some friends in advance to participate to make sure enough people do it or you could ask the lecturer to request people to participate. Try to estimate how fast the disturbance moves around the lecture room. Is this a transverse or longitudinal wave?

Example 17.1

A Mexican wave

A lecturer asks her class of physics students to make a Mexican wave, starting at the left-hand side of the lecture theatre. Her students do so and a pulse moving across the lecture theatre is represented by the wave function:

$$y(x, t) = \frac{2}{(x - 3.0t)^2 + 1}$$

where x and y are measured in metres and t is measured in seconds. Find expressions for the wave function at $t = 0$, $t = 1.0$ s and $t = 2.0$ s and draw the pulse at each of these times.

Getty Images

A Mexican wave

Solution

Conceptualise If you have ever been part of a Mexican wave or watched one, for example at the cricket, you will be familiar with the pulse of movement as people raise their arms and then lower them again. This is a transverse pulse, as the pulse moves horizontally but the medium (the people) move vertically.

Model We model the Mexican wave as a transverse wave. We need to interpret the mathematical representation, the wave function, given in the question.

Analyse The wave function is of the form $y = f(x - vt)$. Inspection of the expression for $y(x, t)$ and comparison with **Equation 17.1** reveal that the wave speed is $v = 3.0$ m/s to the right. Furthermore, by letting $x - 3.0t = 0$, we find that the maximum value of y is given by $A = 2.0$ m. This implies that the students are either exceptionally tall, or are standing up as part of the wave.

Example 17.1 cont.

Write the wave function expression at $t = 0$: $$y(x, 0) = \frac{2}{x^2 + 1}$$

Write the wave function expression at $t = 1.0$ s: $$y(x, 1.0) = \frac{2}{(x - 3.0)^2 + 1}$$

Write the wave function expression at $t = 2.0$ s: $$y(x, 2.0) = \frac{2}{(x - 6.0)^2 + 1}$$

For each of these expressions, we can substitute various values of x and plot the wave function. This procedure yields the wave functions shown in the three parts of **Figure 17.6**.

Finalise These snapshots show that the Mexican wave moves to the right without changing its shape and that it has a constant speed of 3.0 m/s.

What If? What if the wave function were $$y(x, t) = \frac{2}{(x + 3.0t)^2 + 1}$$

How would that change the situation?

Answer The new feature in this expression is the plus sign in the denominator rather than the minus sign. This represents a pulse with the same shape as that in **Figure 17.6**, but moving to the left as time progresses. This could be the Mexican wave travelling back in the opposite direction.

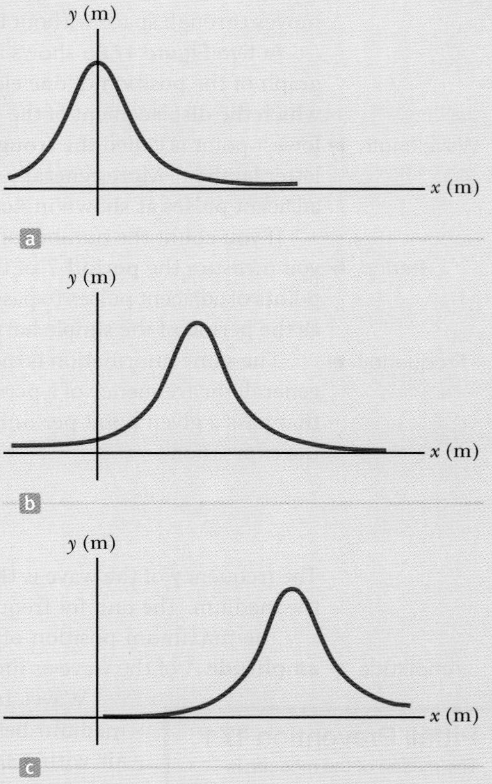

Figure 17.6
(Example 17.1) Graphs of the function $y(x, t) = 2/[(x - 3.0t)^2 + 1]$ at (a) $t = 0$, (b) $t = 1.0$ s and (c) $t = 2.0$ s

17.2 Analysis model: travelling wave

In this section we introduce an important wave function whose shape is shown in **Active Figure 17.7**. The wave represented by this curve is called a **sinusoidal wave** because the curve is the same as that of the function $\sin\theta$ plotted against θ. A sinusoidal wave could be established on the rope in **Figure 17.1** by shaking the end of the rope up and down in simple harmonic motion.

The sinusoidal wave is the simplest example of a periodic continuous wave and can be used to build more complex waves (see Section 18.7). The brown curve in **Active Figure 17.7** represents a snapshot of a travelling sinusoidal wave at $t = 0$ and the blue curve represents a snapshot of the wave at some later time t. There are two types of motion that occur. First, the entire waveform in **Active Figure 17.7** moves to the right so that the brown curve moves towards the right and eventually reaches the position of the blue curve. This movement is the motion of the *wave*. If we focus on one element of the medium, such as the element at $x = 0$, we see that each element moves up and down along the y axis in simple harmonic motion. This movement is the motion of the *elements of the medium*. It is important to differentiate between the motion of the wave and the motion of the elements of the medium. Remember the example of the leaf on the pond – the leaf goes up and down but does not travel in the direction of the wave as it spreads out from the point where the stone hit the water.

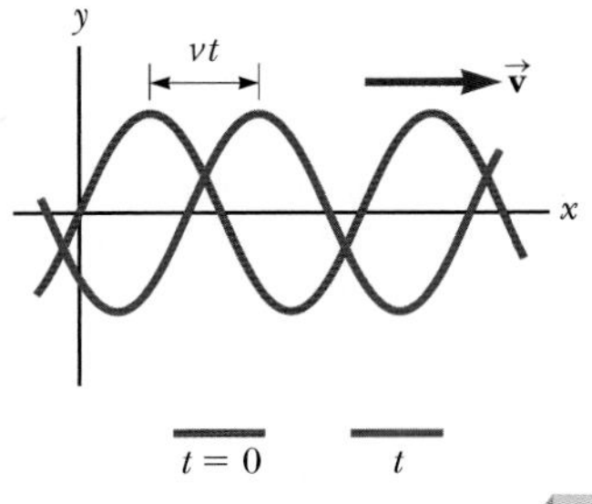

Active Figure 17.7
A one-dimensional sinusoidal wave travelling to the right with a speed v. The brown curve represents a snapshot of the wave at $t = 0$, and the blue curve represents a snapshot at some later time t.

In the early chapters of this book, we developed several analysis models based on three approximations: the particle, the system and the rigid object. We now develop the principal features and mathematical representations of the analysis model of a **travelling wave**. This model is used in situations in which a wave moves through space without interacting with other waves.

Active Figure 17.8a shows a snapshot of a wave moving through a medium. Active Figure 17.8b shows a graph of the position of one element of the medium as a function of time. A point in Active Figure 17.8a at which the displacement of the element from its normal position is highest is called the **crest** of the wave. The lowest point is called the **trough**. The distance from one crest to the next is called the **wavelength**, λ (Greek letter lambda). More generally, the wavelength is the minimum distance between any two identical points on adjacent pulses as shown in Active Figure 17.8a.

Wavelength ▶

If you count the number of seconds between the arrivals of two adjacent crests at a given point in space, you measure the **period** T of the wave. In general, the period is the time interval required for two identical points of adjacent pulses to pass by a point as shown in Active Figure 17.8b. The period of the wave is the same as the period of the simple harmonic oscillation of one element of the medium.

Period ▶

The same information is more often given by the inverse of the period, which is called the **frequency** f. In general, the frequency of a periodic wave is the number of crests (or troughs, or any other point on the wave) that pass a given point per unit time interval. The frequency of a sinusoidal wave is related to the period by the expression

Frequency ▶

$$f = \frac{1}{T} \tag{17.3}$$

The frequency of the wave is the same as the frequency of the simple harmonic oscillation of any element of the medium. The unit for frequency is s^{-1}, or **hertz** (Hz); the unit for T is seconds.

The maximum position of an element of the medium relative to its equilibrium position is called the **amplitude** A of the wave as indicated in Active Figure 17.8. The SI unit of amplitude is metres, m.

Amplitude ▶

Pitfall Prevention 17.1

The motion of the medium is not the same as the motion of the wave. When a transverse wave passes through a medium, the elements of the medium move in simple harmonic motion perpendicular to the motion of the wave. The elements of the medium do not travel along with the wave. Do not confuse the two motions.

Waves travel with a specific speed and this speed depends on the properties of the medium being disturbed. For instance, sound waves travel through room-temperature air with a speed of about 343 m/s whereas they travel through most solids with a much greater speed.

Consider the sinusoidal wave in Active Figure 17.8a, which shows the position of the wave at $t = 0$. Because the wave is sinusoidal, we expect the wave function at this instant to be expressed as $y(x, 0) = A \sin ax$, where A is the amplitude and a is a constant to be determined. At $x = 0$, we see that $y(0, 0) = A \sin a(0) = 0$, consistent with Active Figure 17.8a. The next value of x for which y is zero is $x = \lambda/2$. Therefore,

$$y\left(\frac{\lambda}{2}, 0\right) = A \sin\left(a\frac{\lambda}{2}\right) = 0$$

For this equation to be true, we must have $a\lambda/2 = \pi$, or $a = 2\pi/\lambda$. Therefore, the function describing the positions of the elements of the medium through which the sinusoidal wave is travelling can be written

$$y(x, 0) = A \sin\left(\frac{2\pi}{\lambda}x\right) \tag{17.4}$$

Pitfall Prevention 17.2

Notice the visual similarity between Active Figures 17.8a and 17.8b. The shapes are the same, but (a) is a graph of vertical position versus horizontal position, whereas (b) is vertical position versus time. Active Figure 17.8a is a pictorial representation of the wave *for a series of elements of the medium*; it is what you would see at an instant of time. Active Figure 17.8b is a graphical representation of the position of *one element of the medium* as a function of time. That both figures have the identical shape represents Equation 17.1: a wave is the *same* function of both x and t.

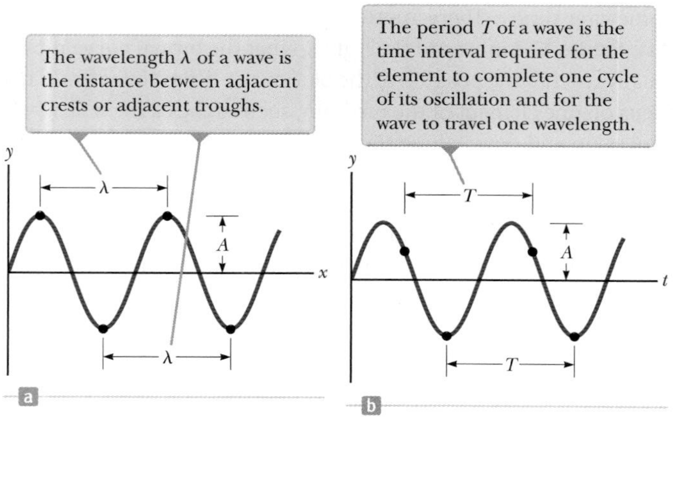
The wavelength λ of a wave is the distance between adjacent crests or adjacent troughs.
The period T of a wave is the time interval required for the element to complete one cycle of its oscillation and for the wave to travel one wavelength.
y
λ
A
x
λ
a
y
T
A
t
T
b

Active Figure 17.8
(a) A snapshot of a sinusoidal wave
(b) The position of one element of the medium as a function of time

where the constant A represents the wave amplitude and the constant λ is the wavelength. Notice that the vertical position of an element of the medium is the same whenever x is increased by an integral multiple of λ. Based on our discussion of **Equation 17.1**, if the wave moves to the right with a speed v, the wave function at some later time t is

$$y(x, t) = A \sin\left[\frac{2\pi}{\lambda}(x - vt)\right] \tag{17.5}$$

If the wave were travelling to the left, the quantity $(x - vt)$ would be replaced by $(x + vt)$ as we found when we developed **Equations 17.1** and **17.2**.

By definition, the wave travels through a displacement Δx equal to one wavelength λ in a time interval Δt of one period T. Therefore, the wave speed, wavelength and period are related by the expression

$$v = \frac{\Delta x}{\Delta t} = \frac{\lambda}{T} \tag{17.6}$$

Substituting this expression for v into **Equation 17.5** gives

$$y = A \sin\left[2\pi\left(\frac{x}{\lambda} - \frac{t}{T}\right)\right] \tag{17.7}$$

This form of the wave function shows the *periodic* nature of y. Note that we will often use y rather than $y(x, t)$ as a shorthand notation. At any given time t, y has the *same* value at the positions x, $x + \lambda$, $x + 2\lambda$, and so on. Furthermore, at any given position x, the value of y is the same at times t, $t + T$, $t + 2T$, and so on.

We can express the wave function in a convenient form by defining two other quantities, the **angular wave number** k (usually called simply the **wave number**) and the **angular frequency** ω:

$$k \equiv \frac{2\pi}{\lambda} \tag{17.8}$$

◀ Angular wave number

$$\omega \equiv \frac{2\pi}{T} = 2\pi f \tag{17.9}$$

◀ Angular frequency

Using these definitions, **Equation 17.7** can be written in the more compact form

$$y = A \sin(kx - \omega t) \tag{17.10}$$

◀ Wave function for a sinusoidal wave

Using **Equations 17.3, 17.8**, and **17.9**, the wave speed v originally given in **Equation 17.6** can be expressed in the following alternative forms:

$$v = \frac{\omega}{k} \tag{17.11}$$

◀ Speed of a sinusoidal wave

$$v = \lambda f \tag{17.12}$$

Equations 17.11 and **17.12** are known as *dispersion relations* because they relate the frequency to the wavelength. Dispersion will be described in detail in Chapter 37.

The wave function given by **Equation 17.10** assumes the vertical position y of an element of the medium is zero at $x = 0$ and $t = 0$. That need not be the case. If it is not, we generally express the wave function in the form

$$y = A \sin(kx - \omega t + \phi) \tag{17.13}$$

◀ General expression for a sinusoidal wave

where ϕ is the **phase constant**, as in Chapter 16. This constant can be determined from the initial conditions. The primary equations in the mathematical representation of the travelling wave analysis model are **Equations 17.3, 17.10**, and **17.12**.

TRY THIS, AGAIN

Using a long stretched spring such as a slinky toy, create transverse waves by shaking one end up and down in simple harmonic motion. Does the frequency with which you shake the end of the spring change the wave speed? What about the wavelength? Now pull it back and forth to create longitudinal waves. Does changing the frequency affect the wave speed now? What effect does changing the amplitude have?

Pitfall Prevention 17.3

Wave speed is not determined by frequency or wavelength. The speed of a wave depends only on the properties of the medium. The frequency depends on the source of the wave. The speed and frequency are related by the wavelength, as shown in Equation 17.12.

Pitfall Prevention 17.4

Do not confuse v, the speed of the wave, with v_y, the transverse velocity of a point on the string. The speed v is constant for a uniform medium, and depends on the properties of the medium. v_y varies sinusoidally and depends on the frequency and amplitude of the wave.

Quick Quiz 17.2

A sinusoidal wave of frequency f is travelling along a stretched string. The string is brought to rest and a second travelling wave of frequency $2f$ is established on the string. **(i)** What is the wave speed of the second wave? (a) twice that of the first wave (b) half that of the first wave (c) the same as that of the first wave (d) impossible to determine **(ii)** From the same choices, describe the wavelength of the second wave. **(iii)** From the same choices, describe the amplitude of the second wave.

Sinusoidal waves on strings

Active Figure 17.9 represents snapshots of a wave created using an oscillating blade attached to a taut string at intervals of $T/4$. Because the end of the blade oscillates in simple harmonic motion, each element of the string, such as that at P, also oscillates vertically with simple harmonic motion. We make the approximation that there is no horizontal movement as this would cause the tension in the string to vary and make the analysis very complex.

Therefore, every element of the string can be treated as a simple harmonic oscillator vibrating with a frequency equal to the frequency of oscillation of the blade. Notice that although each element oscillates in the y direction, the wave travels in the x direction with a speed v. (This is the definition of a transverse wave.)

If we define $t = 0$ as the time for which the configuration of the string is as shown in Active Figure 17.9a, the wave function can be written as

$$y = A \sin(kx - \omega t)$$

We can use this expression to describe the motion of any element of the string. An element at point P (or any other element of the string) moves only vertically and so its x coordinate remains constant. Therefore, the **transverse speed** v_y (not to be confused with the wave speed v) and the **transverse acceleration** a_y of elements of the string are

$$v_y = \left.\frac{dy}{dt}\right|_{x=\text{constant}} = \frac{\partial y}{\partial t} = -\omega A \cos(kx - \omega t) \tag{17.14}$$

$$a_y = \left.\frac{dv_y}{dt}\right|_{x=\text{constant}} = \frac{\partial v_y}{\partial t} = -\omega^2 A \sin(kx - \omega t) \tag{17.15}$$

These expressions use **partial derivatives** because y depends on both x and t. When we take a partial derivative we hold all variables constant except the one we are taking the derivative with respect to. In the operation $\partial y/\partial t$ in Equation 17.14 we take a derivative with respect to t while holding x constant.

Σ *Differential calculus is summarised in Appendix B.7.*

The maximum magnitudes of the transverse speed and transverse acceleration are:

$$v_{y,\max} = \omega A \tag{17.16}$$

$$a_{y,\max} = \omega^2 A \tag{17.17}$$

The transverse speed and transverse acceleration of elements of the string do not reach their maximum values simultaneously. The transverse speed reaches its maximum value (ωA) when $y = 0$, whereas the magnitude of the transverse acceleration reaches its maximum value ($\omega^2 A$) when $y = \pm A$. Finally, Equations 17.16 and 17.17 are identical in mathematical form to the corresponding equations for simple harmonic motion, Equations 16.16 and 16.17.

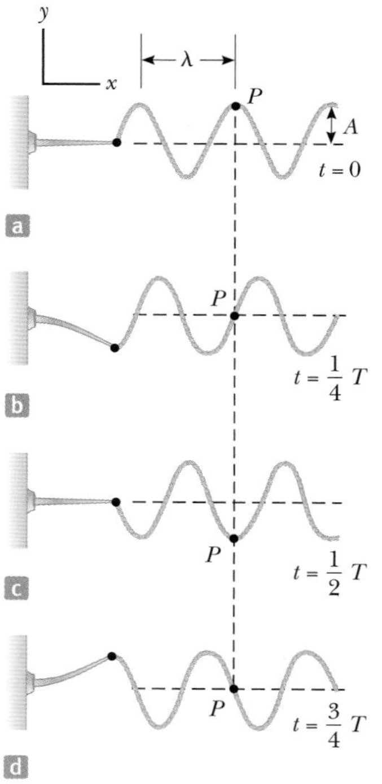

Active Figure 17.9

The left end of the string is connected to a blade that is set into oscillation. Every element of the string, such as that at point P, oscillates with simple harmonic motion in the vertical direction.

Quick Quiz 17.3

The amplitude of a wave is doubled, with no other changes made to the wave. As a result of this doubling, which of the following statements is correct? **(a)** The speed of the wave changes. **(b)** The frequency of the wave changes. **(c)** The maximum transverse speed of an element of the medium changes. **(d)** Statements (a) through (c) are all true. **(e)** None of statements (a) through (c) is true.

Analysis Model 17.1

Travelling wave

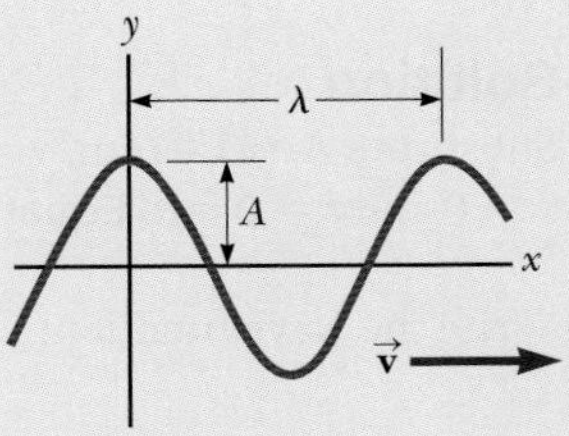

Imagine a source vibrating such that it influences the medium that is in contact with the source. Such a source creates a disturbance that propagates through the medium. If the source vibrates in simple harmonic motion with period T, sinusoidal waves propagate through the medium at a speed given by

$$v = \frac{\lambda}{T} = \lambda f \qquad (17.6, 17.12)$$

where λ is the **wavelength** of the wave and f is its **frequency.** A sinusoidal wave can be expressed as

$$y = A \sin (kx - \omega t) \qquad (17.10)$$

where A is the **amplitude** of the wave, k is its **wave number,** and ω is its **angular frequency.**

Examples

- a vibrating blade sends a sinusoidal wave down a string attached to the blade
- a loudspeaker vibrates back and forth, emitting sound waves into the air
- a guitar body vibrates, emitting sound waves into the air (Chapter 18)
- a vibrating electric charge creates an electromagnetic wave that propagates into space at the speed of light (Chapter 34)

Example 17.2

A wave on a fence wire

A cicada sits on a strand of wire fence and buzzes with a frequency of 6.0 kHz, producing a sinusoidal wave travelling in the positive x direction with an amplitude of 1.0 mm, at a speed of 550 m/s. Assume that at $x = 0$ and $t = 0$ the displacement y is equal to the amplitude.

(A) Find the wavelength λ, wave number k, period T and angular frequency ω of the wave.

Solution

Conceptualise We begin as usual by drawing a diagram.

Model We model the wave as a transverse travelling wave and use the travelling wave analysis model.

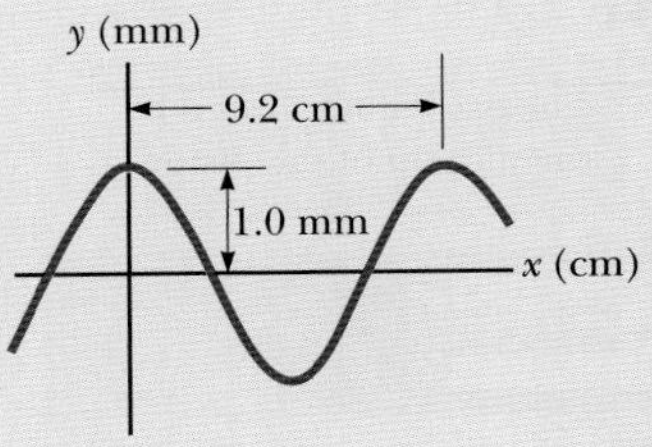

Figure 17.10 (Example 17.2) A sinusoidal wave of wavelength $\lambda = 9.2$ cm and amplitude $A = 1.0$ mm

Analyse Evaluate the wavelength from Equation 17.12:

$$\lambda = \frac{v}{f} = \frac{550 \text{ m/s}}{6000 \text{ s}^{-1}} = 0.0917 \text{ m} = 9.2 \text{ cm}$$

Evaluate the wave number from Equation 17.8:

$$k = \frac{2\pi}{\lambda} = \frac{2\pi \text{ rad}}{0.0917 \text{ m}} = 69 \text{ m}^{-1}$$

Evaluate the period of the wave from Equation 17.3:

$$T = \frac{1}{f} = \frac{1}{6000 \text{ s}^{-1}} = 1.67 \times 10^{-4} \text{ s} = 0.17 \text{ ms}$$

Evaluate the angular frequency of the wave from Equation 17.9:

$$\omega = 2\pi f = 2\pi(6000 \text{ s}^{-1}) = 38\,000 \text{ rad/s}$$

Example 17.2 cont.

(B) Determine the phase constant ϕ and write a general expression for the wave function.

Solution

Substitute $A = 1.0$ mm, $y = 1.0$ mm, $x = 0$ and $t = 0$ into **Equation 17.13**:

$$1.0 = (1.0)\sin\phi \rightarrow \sin\phi = 1 \rightarrow \phi = \frac{\pi}{2}\text{ rad}$$

Write the wave function:

$$y = A\sin\left(kx - \omega t + \frac{\pi}{2}\right) = A\cos(kx - \omega t)$$

Substitute the values for A, k and ω in SI units into this expression:

$$y = 0.001\cos(69.0x - 38\,000t)$$

17.3 Sound waves

Sound is a longitudinal wave, moving through a medium such that particles in the medium move back and forth in simple harmonic motion in the same direction as the motion of the wave. We most commonly experience sound waves in air, but in fact any longitudinal wave moving through a medium, whether solid, liquid or gas, is a sound wave.

We begin our description of sound waves by considering the motion of a one-dimensional longitudinal pulse moving through a long tube containing a compressible gas as shown in **Figure 17.11**. A piston at the left end can be quickly moved to the right to compress the gas and create the pulse. Before the piston is moved, the gas is undisturbed and of uniform density as represented by the uniformly shaded region in **Figure 17.11a**. When the piston is pushed to the right (**Figure 17.11b**), the gas just in front of it is compressed (as represented by the more heavily shaded region); the pressure and density in this region are now higher than they were before the piston moved. When the piston comes to rest (**Figure 17.11c**), the compressed region of the gas continues to move to the right, corresponding to a longitudinal pulse travelling through the tube with speed v.

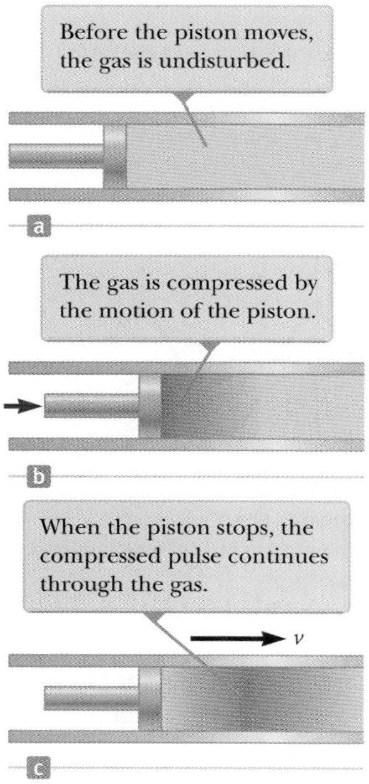

Before the piston moves, the gas is undisturbed.
a
The gas is compressed by the motion of the piston.
b
When the piston stops, the compressed pulse continues through the gas.
v
c

Figure 17.11
Motion of a longitudinal pulse through a compressible gas. The compression (darker region) is produced by the moving piston.

One can produce a one-dimensional *periodic* sound wave in the tube of gas in **Figure 17.11** by moving the piston in simple harmonic motion. The results are shown in **Active Figure 17.12**. The darker parts of the coloured areas in this figure represent regions in which the gas is compressed, and the density and pressure are above their equilibrium values. A compressed region is formed whenever the piston is pushed into the tube. This compressed region, called a **compression**, moves through the tube, continuously compressing the region just in front of itself. When the piston is pulled back, the gas in front of it expands, and the pressure and density in this region fall below their equilibrium values (represented by the lighter parts of the coloured areas in **Active Figure 17.12**). These low-pressure regions, called **rarefactions**, also propagate along the tube, following the compressions.

As the piston oscillates sinusoidally, regions of compression and rarefaction are continuously set up. The distance between two successive compressions (or two successive rarefactions) equals the wavelength, λ, of the sound wave. Because the sound wave is longitudinal, as the compressions and rarefactions travel through the tube, any small element of the gas moves with simple harmonic motion parallel to the direction of the wave. If $s(x, t)$ is the position of a small element relative to its equilibrium position, we can express this harmonic position function as

$$s(x, t) = s_{max}\cos(kx - \omega t) \tag{17.18}$$

where s_{max} is the maximum position of the element relative to equilibrium. This parameter is often called the **displacement amplitude** of the wave. The parameter k is the wave number and ω is the angular frequency of the wave. Notice that the displacement of the element is along x, in the direction of propagation of the sound wave, hence we use $s(x, t)$ here instead of $y(x, t)$ because the displacement of elements of the medium is not perpendicular to the x direction.

The variation in the gas pressure ΔP measured from the equilibrium value, called the **acoustic pressure** of the sound wave, is also periodic with the same wave number and angular frequency as for the displacement in **Equation 17.18**, but the variations in displacement and pressure are out of phase, as we shall show. Therefore, we can write

$$\Delta P = \Delta P_{max} \sin(kx - \omega t) \quad (17.19)$$

where the **pressure amplitude** ΔP_{max} is the maximum change in pressure from the equilibrium value.

We have expressed the displacement with a cosine function and the pressure with a sine function. We can justify this by considering the piston–tube arrangement of **Figure 17.11**. In **Figure 17.13a**, we focus our attention on a small cylindrical element of undisturbed gas of length Δx and area A. The volume of this element is $V_i = A\Delta x$.

Figure 17.13b shows this element of gas after a sound wave has moved it to a new position. The disc's two flat faces move through different distances s_1 and s_2. The change in volume ΔV of the element in the new position is equal to $A\Delta s$, where $\Delta s = s_1 - s_2$.

From the definition of bulk modulus (see **Equation 15.14**), we express the pressure variation in the element of gas as a function of its change in volume:

$$\Delta P = -B\frac{\Delta V}{V_i}$$

We substitute for the initial volume and the change in volume of the element:

$$\Delta P = -B\frac{A\Delta s}{A\Delta x}$$

Let the thickness Δx of the disc approach zero so that the ratio $\Delta s/\Delta x$ becomes a partial derivative:

$$\Delta P = -B\frac{\partial s}{\partial x} \quad (17.20)$$

Substitute the position function given by **Equation 17.18**:

$$\Delta P = -B\frac{\partial}{\partial x}[s_{max}\cos(kx - \omega t)] = Bs_{max}k\sin(kx - \omega t)$$

From this result, we see that a displacement described by a cosine function leads to a pressure described by a sine function. We also see that the displacement and pressure amplitudes are related by

$$\Delta P_{max} = Bs_{max}k \quad (17.21)$$

This procedure applies equally as well to solids and liquids, and **Equation 17.21** tells us that the pressure amplitude of a wave through any medium is related to the bulk modulus of that material. As we shall see in the next section, wave speed also depends on the bulk modulus of a material. This discussion also shows that a sound wave may be described equally well in terms of either pressure or displacement.

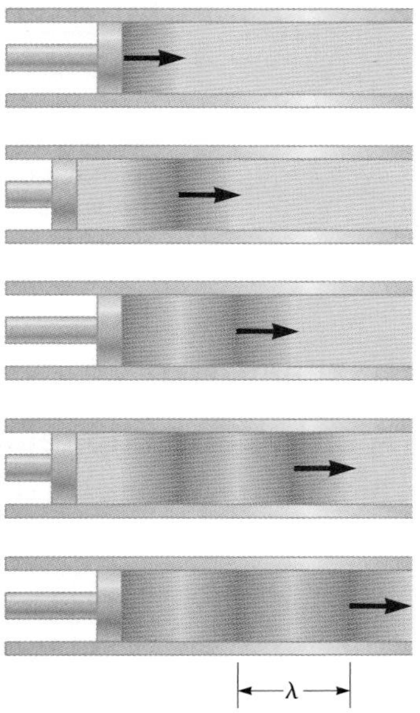

Active Figure 17.12
A longitudinal wave propagating through a gas-filled tube. The source of the wave is an oscillating piston at the left.

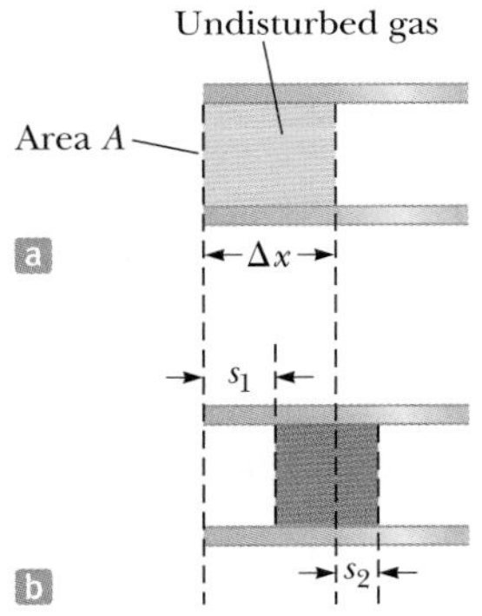

Figure 17.13
(a) An undisturbed element of gas of length Δx in a tube of cross-sectional area A. (b) When a sound wave propagates through the gas, the element is moved to a new position and has a different length. The parameters s_1 and s_2 describe the displacements of the ends of the element from their equilibrium positions.

Quick Quiz 17.4

If you blow across the top of an empty soft-drink bottle, a pulse of sound travels down through the air in the bottle. At the moment the pulse reaches the bottom of the bottle, what is the correct description of the displacement of elements of air from their equilibrium positions and the pressure of the air at this point? **(a)** The displacement and pressure are both at a maximum. **(b)** The displacement and pressure are both at a minimum. **(c)** The displacement is zero, and the pressure is a maximum. **(d)** The displacement is zero, and the pressure is a minimum.

TRY THIS

Get a pair of polystyrene or similar disposable cups, and a piece of string or twine several metres long. Put a small hole in the bottom of each cup so that you can just push a string through it. Push one end of the string through the hole in the bottom of one cup and tie a knot in the end of the string so that it cannot be pulled back though the hole. The knot should be on the inside of the cup. Now do the same with the other end of the string and the second cup. Get a friend to hold one cup and walk away as far as the string will reach, so that it is held taut. Now you can talk to each other using your cup and string 'telephone', with no charges!

Example **17.3**

A string and cup telephone

Two small boys have made themselves a telephone so that they can talk to each other between different rooms without annoying their mother. They make a small hole in the bottom of each of two plastic cups and connect them by feeding one end of a string through each hole and tying a knot to secure it. Marcus sits in the cubby house with his telephone and screams into it, producing a clear note with a frequency of 12 kHz. This sound wave takes 0.13 s to reach Laurence, who is 10 m away.

Find values for the wave number k and the angular frequency ω, and use these to write expressions for the pressure and displacement waves in the string.

Solution

Conceptualise You have probably never played with a string and cup telephone, unless you did the *Try this* example above. Try making one as described in the *Try this* example and experiment with it. You will find that the sound carried by the string if the string is held taut is louder and clearer than you might expect.

Model We model the wave as a longitudinal travelling wave (sound wave) and use the travelling wave analysis model from the previous section, as well as the equations for sound waves developed in this section.

Analyse Evaluate the angular frequency of the wave from **Equation 17.9**:

$$\omega = 2\pi f = 2\pi\,(12\,000\text{ s}^{-1}) = 75\,400\text{ rad/s} = 75\,000\text{ rad/s}$$

Evaluate the wave number from **Equations 17.8** and **17.12**:

$$k = \frac{2\pi}{\lambda},\ \lambda = \frac{v}{f}\quad\text{and}\quad v = \frac{\Delta x}{\Delta t}$$

so

$$k = \frac{2\pi f \Delta t}{\Delta x} = \frac{\omega \Delta t}{\Delta x} = \frac{(75\ 400\text{ rad/s})(0.13\text{ s})}{10\text{ m}}$$

$$= 980\text{ m}^{-1}$$

We can now write the expressions for Δs and ΔP using **Equations 17.18** and **17.19**:

$$s(x, t) = s_{max}\cos(kx - \omega t) = s_{max}\cos(980x - 75\ 000t)$$

$$\Delta P = \Delta P_{max}\sin(kx - \omega t) = \Delta P_{max}\sin(980x - 75\ 000t)$$

(B) Draw a graph of pressure and displacement as a function of position for a small section of string.

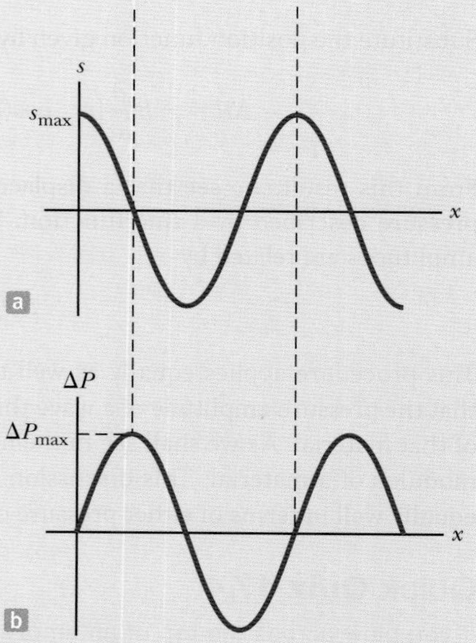

Figure 17.14
(Example 17.3) (a) Displacement amplitude and (b) pressure amplitude versus position for a sinusoidal longitudinal wave

Solution

We can choose any small section we like, so we do not need to worry about the phase angle. The two graphs will be 90° out of phase as shown in **Figure 17.14**.

Finalise The total string length is 10 m and the wavelength is only about 6 mm, which is why we have been asked to draw the graph for only a short section of string – if we drew it for the entire string we would have to draw more than 1000 peaks and troughs!

17.4 The linear wave equation

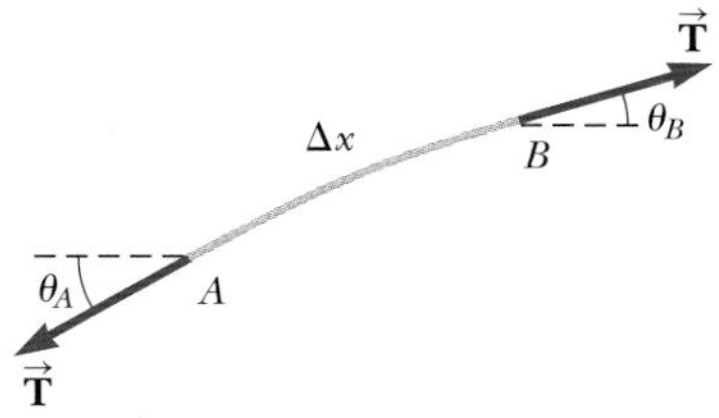

Figure 17.15
An element of a string under tension T.

In Section 17.1, we introduced the concept of the wave function to represent waves travelling on a string. All wave functions $y(x, t)$ represent solutions of an equation called the *linear wave equation*. This equation gives a complete description of the wave motion and from it one can derive an expression for the wave speed, which we shall look at in more detail in the next section. We shall derive the linear wave equation as applied to waves on strings.

Suppose a travelling wave is propagating along a string that is under a tension T. Consider one small string element of length Δx (**Figure 17.15**). The ends of the element make small angles θ_A and θ_B with the x axis. The net force acting on the element in the vertical direction is

$$\sum F_y = T\sin\theta_B - T\sin\theta_A = T(\sin\theta_B - \sin\theta_A)$$

Because the angles are small, we can use the approximation $\sin\theta \approx \tan\theta$ to express the net force as

$$\sum F_y \approx T(\tan\theta_B - \tan\theta_A) \qquad (17.22)$$

Σ

The small angle approximation is described in Appendix B.5.

Imagine undergoing an infinitesimal displacement outwards from the right end of the rope element in **Figure 17.15** along the blue line representing the force $\vec{\mathbf{T}}$. This displacement has infinitesimal x and y components and can be represented by the vector $dx\hat{\mathbf{i}} + dy\hat{\mathbf{j}}$. The tangent of the angle with respect to the x axis for this displacement is dy/dx. As we evaluate this tangent at a particular instant of time, we must express it in partial derivative form as $\partial y/\partial x$. Substituting for the tangents in **Equation 17.22** gives

$$\sum F_y \approx T\left[\left(\frac{\partial y}{\partial x}\right)_B - \left(\frac{\partial y}{\partial x}\right)_A\right] \qquad (17.23)$$

Now let's apply Newton's second law to the element, with the mass of the element given by $m = \mu\Delta x$, where μ (Greek letter mu) is the mass per unit length:

$$\sum F_y = ma_y = \mu\Delta x\left(\frac{\partial^2 y}{\partial t^2}\right) \qquad (17.24)$$

Combining **Equation 17.23** with **Equation 17.24** gives

$$\mu\Delta x\left(\frac{\partial^2 y}{\partial t^2}\right) = T\left[\left(\frac{\partial y}{\partial x}\right)_B - \left(\frac{\partial y}{\partial x}\right)_A\right]$$

$$\frac{\mu}{T}\frac{\partial^2 y}{\partial t^2} = \frac{(\partial y/\partial x)_B - (\partial y/dx)_A}{\Delta x} \qquad (17.25)$$

The right side of **Equation 17.25** can be expressed in a different form if we note that the partial derivative of any function is defined as

$$\frac{\partial f}{\partial x} \equiv \lim_{\Delta x \to 0}\frac{f(x+\Delta x) - f(x)}{\Delta x}$$

> **Pitfall Prevention 17.5**
> Equation 17.27 is the linear wave equation. Sometimes Equation 17.12 is mistakenly called the wave equation because it is one of the first equations students learn when studying waves. Equation 17.12 is *not* a wave equation, it is a dispersion relation.

Associating $f(x + \Delta x)$ with $(\partial y/\partial x)_B$ and $f(x)$ with $(\partial y/\partial x)_A$, we see that, in the limit $\Delta x \to 0$, **Equation 17.25** becomes

$$\frac{\mu}{T}\frac{\partial^2 y}{\partial t^2} = \frac{\partial^2 y}{\partial x^2} \qquad (17.26)$$

◀ Linear wave equation for a string

This expression is the linear wave equation as it applies to waves on a string. As we shall see in the next section, the speed of a wave on a string is given by $v = \sqrt{\frac{T}{\mu}}$, so **Equation 17.26** can be written as

$$\frac{\partial^2 y}{\partial x^2} = \frac{1}{v^2}\frac{\partial^2 y}{\partial t^2} \qquad (17.27)$$

◀ Linear wave equation in general

which is the general form of the linear wave equation.

Equation 17.27 applies in general to various types of travelling waves. For waves on strings, y represents the vertical position of elements of the string. For sound waves propagating through a gas, y corresponds to the longitudinal position of elements of the gas from equilibrium or variations in either the pressure or the density of the gas. In the case of electromagnetic waves, y corresponds to electric or magnetic field components.

Example 17.4

A solution to the wave equation

Show that $y = A\sin(kx - \omega t)$ is a solution of the wave equation.
Differential calculus is summarised in Appendix B.7.

Solution

Conceptualise This is a simple mathematical problem, involving taking partial derivatives.

Analyse We will need to take partial derivatives of the function given, and substitute these into the wave Equation 17.27.

Considering the left-hand side first, we see that:

$$\frac{\partial y}{\partial x} = Ak\cos(kx - \omega t)$$

$$\frac{\partial^2 y}{\partial x^2} = -Ak^2\sin(kx - \omega t)$$

Now the right-hand side:

$$\frac{\partial y}{\partial t} = -A\omega\cos(kx - \omega t)$$

$$\frac{\partial^2 y}{\partial x^2} = -A\omega^2\sin(kx - \omega t)$$

$$\left(\frac{1}{v^2}\right)\frac{\partial^2 y}{\partial x^2} = -\left(\frac{1}{v^2}\right)A\omega^2\sin(kx - \omega t)$$

Setting the left-hand side equal to the right-hand side:

$$-Ak^2\sin(kx - \omega t) = -\left(\frac{1}{v^2}\right)A\omega^2\sin(kx - \omega t)$$

So if $v = \omega/k = 2\pi f/(2\pi/\lambda) = f\lambda$ (as stated in Equation 17.12 and using the definitions of angular frequency and wave number), the function $A\sin(kx - \omega t)$ is a solution of the wave function.

Finalise What other functions can you find that are solutions of the wave equation?

Although we do not prove it here, the linear wave equation is satisfied by *any* wave function of the form $y = f(x \pm vt)$. Furthermore, we have seen that the linear wave equation is a direct consequence of Newton's second law applied to any element of a string carrying a travelling wave.

17.5 The speed of waves

Imagine plucking a taut string, such as a fence wire or guitar string, so that you send a transverse pulse along it. The pulse moves to the right with a uniform speed v measured relative to a stationary frame of reference as shown in Figure 17.16a. Instead of staying in this reference frame, it is more convenient to choose a different inertial reference frame that moves along with the pulse with the same speed as the pulse so that the pulse is at rest within the frame. This change of reference frame is permitted because Newton's laws are valid in any stationary frame or one that moves with constant velocity.

In our new reference frame, shown in the magnified view of **Figure 17.16b**, all elements of the string move to the left: a given element of the string initially to the right of the pulse moves to the left, rises up and follows the shape of the pulse, and then continues to move to the left. Both parts of **Figure 17.16** show such an element at the instant it is located at the top of the pulse.

The small element of the string of length Δs forms an approximate arc of a circle of radius R. In the moving frame of reference (which moves to the right at a speed v along with the pulse), the shaded element moves to the left with a speed v. This element has a centripetal acceleration equal to v^2/R, which is supplied by components of the force $\vec{\mathbf{T}}$, whose magnitude is the tension in the string. The force $\vec{\mathbf{T}}$ acts on both sides of the element and is tangent to the arc as shown in **Figure 17.16b**. The horizontal components of $\vec{\mathbf{T}}$ cancel and each vertical component $T \sin \theta$ acts downwards. Hence, the total force on the element is $2T \sin \theta$ towards the arc's centre. Because the element is small, θ is small and we can use the small-angle approximation $\sin \theta \approx \theta$. So, the total radial force is

$$F_r = 2T \sin \theta \approx 2T\theta$$

The small angle approximation, and other useful approximations, are given in Appendix B.5. Σ

The element has a mass $m = \mu \Delta s$, where μ is the mass per unit length of the string. Because the element forms part of a circle and subtends an angle 2θ at the centre, $\Delta s = R(2\theta)$, and hence

$$m = \mu \Delta s = 2\mu R\theta$$

Applying Newton's second law to this element in the radial direction gives

$$F_r = ma = \frac{mv^2}{R}$$

$$2T\theta = \frac{2\mu R\theta v^2}{R}$$

therefore:

$$v = \sqrt{\frac{T}{\mu}} \tag{17.28}$$

◀ Speed of a wave on a stretched string

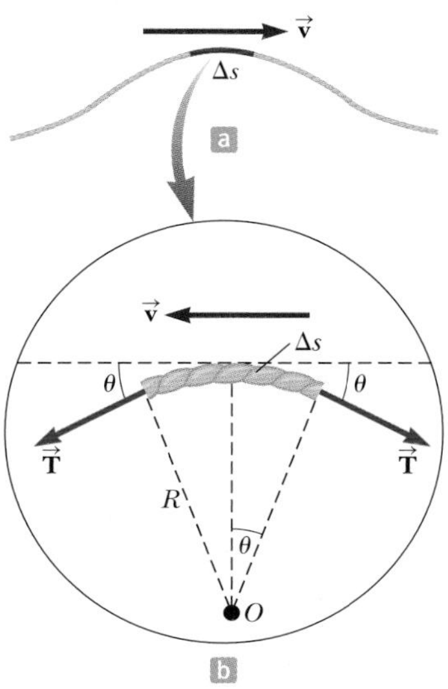

Figure 17.16
(a) In the reference frame of the Earth, a pulse moves to the right on a string with speed v. (b) In a frame of reference moving to the right with the pulse, the small element of length Δs moves to the left with speed v.

Notice that this derivation is based on the assumption that the pulse height is small relative to the length of the string. Using this assumption, we were able to use the small angle approximation $\sin\theta \approx \theta$. Furthermore, the model assumes the tension T is not affected by the presence of the pulse; therefore, T is the same at all points on the string. Finally, this derivation does *not* assume any particular shape for the pulse. Therefore, a pulse of *any shape* travels along the string with speed $v = \sqrt{T/\mu}$ without any change in pulse shape.

Quick Quiz 17.5

A lecture demonstration consists of two very long, identical rubber tubes. One end of each tube is tied to the wall of the lecture theatre and the lecturer stretches them out so they are taut. He then holds the free end of each tube in his hand and sends a pulse down them. One pulse reaches the far wall much sooner than the other. The 'trick' is that one tube is filled with water. Which one?

TRY THIS, AGAIN

Using a long stretched spring such as a slinky toy, with one end attached to a wall or held firmly by a friend, create transverse pulses by moving the free end up and then down. What can you do to increase the wave speed? Try changing the speed at which you move your hand up and down and the amplitude. Try stretching the spring more or less to change the tension. Now pull it back and forth to create longitudinal waves and repeat the experiment. Do the same factors affect speed in a longitudinal wave?

Pitfall Prevention 17.6

Do not confuse the T in Equation 17.28 for the tension with the symbol T used in this chapter for the period of a wave. The context of the equation should help you identify which quantity is meant and if you are still unsure use the units (or dimensions) of the other quantities in the equation to tell you what the dimensions of T are. In Equation 17.28, T must have units of kg.m/s^2, or N.

Example **17.5**

Rescuing the hiker

An 80.0 kg hiker is trapped on a mountain ledge after a storm. A helicopter rescues the hiker by hovering above him and lowering a cable to him. The mass of the cable is 8.00 kg and its length is 15.0 m. A sling of mass 70.0 kg is attached to the end of the cable. The hiker attaches himself to the sling and the helicopter then accelerates upwards. Realising he has left his mobile phone on the ledge, the hiker tries to signal the pilot by jerking on the cable, sending transverse pulses up the cable. A pulse takes 0.250 s to travel the length of the cable. What is the acceleration of the helicopter? Assume the tension in the cable is uniform.

Solution

Conceptualise Imagine the effect of the acceleration of the helicopter on the cable. The greater the upwards acceleration, the larger the tension in the cable. In turn, the larger the tension, the higher the speed of pulses on the cable. Draw a diagram to help you visualise the situation and add the data given in the problem. Show the forces acting on the hiker/sling combination.

Model We model the hiker and sling as a particle subject to a net force, with constant acceleration. We model the wave as a transverse travelling wave and make the approximation that the tension is constant in the rope, even though the rope is not massless.

Analyse Use the time interval for the pulse to travel from the hiker to the helicopter to find the speed of the pulses on the cable:

$$v = \frac{\Delta x}{\Delta t} = \frac{15.0\text{ m}}{0.250\text{ s}} = 60.0\text{ m/s}$$

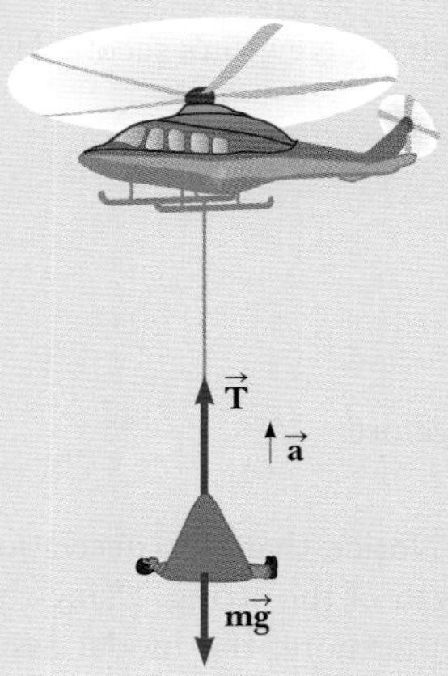

Solve **Equation 17.28** for the tension in the cable: $v = \sqrt{\frac{T}{\mu}} \rightarrow T = \mu v^2$

Model the hiker and sling as a particle subject to a net force, noting that the acceleration of this particle of mass m is the same as the acceleration of the helicopter: $\sum F = ma \rightarrow T - mg = ma$

Solve for the acceleration: $a = \frac{T}{m} - g = \frac{\mu v^2}{m} - g = \frac{m_{\text{cable}} v^2}{l_{\text{cable}} m} - g$

Check dimensions (noting that g is an acceleration and hence has the same units as a): $[LT^{-2}] = [M][LT^{-1}]^2/[L][M] = [LT^{-2}]$ ☺

Substitute numerical values: $a = \frac{(8.00\text{ kg})(60.0\text{ m/s})^2}{(15.0\text{ m})(150.0\text{ kg})} - 9.80\text{ m/s} = 3.00\text{ m/s}^2$

Finalise A real cable has stiffness in addition to tension, as discussed in Chapter 6. The stiffness increases the restoring force and hence increases the wave speed. Consequently, for a real cable, the speed of 60.0 m/s that we determined is most likely associated with a smaller acceleration of the helicopter. We have also assumed that the tension is constant in the rope, even though the rope is not massless. The tension will increase slightly down the length of the rope as it must support its own weight as well as the hiker and sling, giving a varying wave speed along its length.

Speed of sound waves

We have found that the speed of a transverse wave on a string depends upon the linear mass density of the string and the restoring force within the string that acts to return elements of the string to their equilibrium positions, which is the tension in the string. Now consider a sound wave travelling through air. What factors determine the speed of a sound wave? We will now derive an expression for the speed of sound in a gas.

Consider the cylindrical element of gas between the piston and the dashed line in **Figure 17.17a**. This element of gas is in equilibrium under the influence of forces of equal magnitude, from the piston on the left and from the rest of the gas on the right. The magnitude of these forces is PA, where P is the pressure in the gas and A is the cross-sectional area of the tube.

Figure 17.17b shows the situation after a time interval Δt during which the piston moves to the right at a constant speed v_x due to a force from the left on the piston that has increased in magnitude to $(P + \Delta P)A$. The movement of the piston creates a pressure pulse that moves through the gas, in other words, a sound wave. By the end of the time interval Δt, every bit of gas in the element is moving with speed v_x. That will not be true in general for a macroscopic element of gas, but it will become true if we shrink the length of the element to an infinitesimal value.

The length of the undisturbed element of gas is chosen to be $v\Delta t$, where v is the speed of sound in the gas and Δt is the time interval between the configurations in **Figures 17.17a** and **17.17b**. Therefore, at the end of the time interval Δt, the sound wave will just reach the right end of the cylindrical element of gas. The gas to the right of the element is undisturbed because the sound wave has not reached it yet.

We can model the element of gas as a non-isolated system in terms of momentum. The force from the piston has provided an impulse to the element that, in turn, exhibits a change in momentum. Therefore, we evaluate both sides of the impulse–momentum theorem:

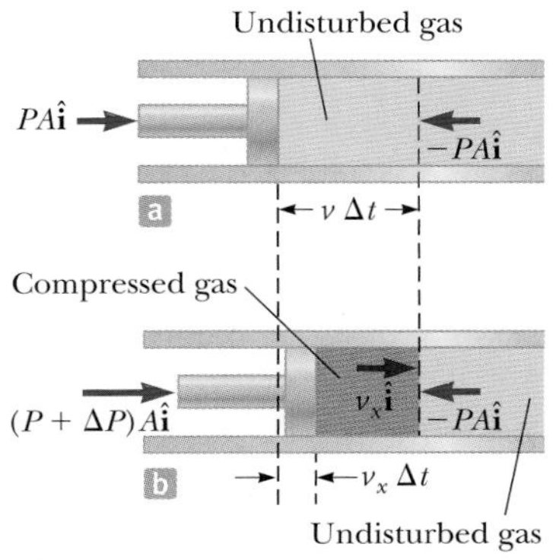

Figure 17.17
(a) An undisturbed element of gas of length $v\,\Delta t$ in a tube of cross-sectional area A. The element is in equilibrium between forces on either end. (b) When the piston moves inward at constant velocity v_x due to an increased force on the left, the element also moves with the same velocity.

$$\vec{\mathbf{I}} = \Delta\vec{\mathbf{p}} \tag{17.29}$$

On the left, the impulse is provided by the constant force due to the increased pressure on the piston:

$$\vec{\mathbf{I}} = \sum \vec{\mathbf{F}}\Delta t = (A\Delta P\Delta t)\hat{\mathbf{i}}$$

The pressure change ΔP on the right-hand side can be related to the volume change and then to the speeds v and v_x through the bulk modulus:

$$\Delta P = -B\frac{\Delta V}{V_i} = -B\frac{(-v_x A\Delta t)}{vA\Delta t} = B\frac{v_x}{v}$$

Therefore, the impulse becomes

$$\vec{\mathbf{I}} = \left(AB\frac{v_x}{v}\Delta t\right)\hat{\mathbf{i}} \tag{17.30}$$

On the right-hand side of the impulse–momentum theorem, **Equation 17.29**, the change in momentum of the element of gas of mass m is as follows:

$$\Delta\vec{\mathbf{p}} = m\Delta\vec{\mathbf{v}} = (\rho V_i)(v_x\hat{\mathbf{i}} - 0) = (\rho v v_x A\Delta t)\hat{\mathbf{i}} \tag{17.31}$$

Substituting **Equations 17.30** and **17.31** into **Equation 17.29**, we find

$$AB\frac{v_x}{v}\Delta t = \rho v v_x A\Delta t$$

which reduces to an expression for the speed of sound in a gas:

$$v = \sqrt{\frac{B}{\rho}} \tag{17.32}$$

◀ Speed of sound in a gas

It is interesting to compare this expression with **Equation 17.28** for the speed of transverse waves on a string, $v = \sqrt{T/\mu}$. In both cases, the wave speed depends on an elastic property of the medium (bulk modulus B or string tension T) and on an inertial property of the medium (volume density ρ or linear density μ). In fact, the speed of all mechanical waves follows an expression of the general form

General expression for speed of mechanical waves ►

$$v = \sqrt{\frac{\text{elastic property}}{\text{inertial property}}} \qquad (17.33)$$

For longitudinal sound waves in a solid rod of material, the speed of sound depends on Young's modulus, Y, and the density, ρ. **Table 17.1** lists the speed of sound (longitudinal waves) in several different materials.

Table 17.1
Speed of sound in various media

Gases	***v* (m/s)**
Hydrogen (0°C)	1286
Helium (0°C)	972
Air (20°C)	343
Air (0°C)	331
Oxygen (0°C)	317
Liquids at 25°C	
Glycerine	1904
Seawater	1533
Water	1493
Mercury	1450
Carbon tetrachloride	926
Solids[a]	
Pyrex glass	5640
Iron	5950
Copper	5010
Lead	1960
Rubber	1600

[a]Values given are for propagation of longitudinal waves in bulk media.

We see from **Equation 17.32** that the speed of sound depends on the density of the medium. Therefore anything that changes the density, such as temperature, will change the speed of sound in that medium. For sound travelling through air, the relationship between wave speed and air temperature is

$$v = 331\sqrt{1 + \frac{T_C}{273}} \qquad (17.34)$$

where v is in metres/second, 331 m/s is the speed of sound in dry air at 0°C, and T_C is the air temperature in degrees Celsius. Using this equation, one finds that at 20°C, the speed of sound in air is approximately 343 m/s. Humidity will also affect the speed of sound because humid air is less dense than dry air. Hence the speed of sound in Darwin is higher than that in Auckland.

As we have an expression (**Equation 17.32**) for the speed of sound, we can now express the relationship between pressure amplitude and displacement amplitude for a sound wave (**Equation 17.21**) as

$$\Delta P_{max} = Bs_{max}k = (\rho v^2)s_{max}\left(\frac{\omega}{v}\right) = \rho v \omega s_{max} \qquad (17.35)$$

This expression is sometimes more useful than **Equation 17.21** because the density of a gas is generally more readily available, or easily measured, than the bulk modulus.

In Chapter 15, we saw that the way a material responds to a stress depends on the type of stress – compression, tension or shear. For a transverse wave in a solid medium, the wave speed depends upon the shear modulus and the density. Therefore measuring the wave speed through a material and the density of a material is a simple way of measuring the shear modulus. As you might expect from the discussion of strengths of materials in Chapter 15, the speed of transverse waves through solids is generally much smaller than the speed of longitudinal waves.

Quick Quiz 17.6

We have talked about longitudinal waves in solids and gases; they also propagate through liquids. Why have we only discussed transverse waves in solids and liquids?

Example 17.6

As discussed in Section 17.1, an earthquake produces two waves: a longitudinal wave called the primary or P wave, and a transverse wave called the secondary or S wave. The speed of an S wave in a particular area where the ground is primarily dry sandstone is about 2.4 km/s and the speed of a P wave about 3.8 km/s. The bulk modulus of the sandstone is 35 GPa.

What is the shear modulus of the sandstone?

Solution

Conceptualise The two waves are travelling through the same medium, but with very different speeds. From **Equation 17.33**, wave speed is given by $v = \sqrt{\text{elastic property}/\text{inertial property}}$. The inertial property (density) relates to the medium, and so it is the same for both types of wave; it must be that the elastic properties (shear modulus for the S wave and bulk modulus for the P wave) are different.

Example 17.6 cont.

Model We assume that the medium (the ground) is uniform and we use the travelling wave model for the two waves. We model the transverse wave as causing only shearing of the material, and the longitudinal wave as causing only compression and tension.

Analyse

$$v = \sqrt{\frac{\text{elastic property}}{\text{inertial property}}}$$

For the transverse (*S*) wave:

$$v_S = \sqrt{\frac{S}{\rho}} \quad (1)$$

For the longitudinal (*P*) wave:

$$v_P = \sqrt{\frac{B}{\rho}} \quad (2)$$

From **Equation (2)** we see that $\rho = B/v_P^2$. Substituting this into **Equation (1)** and rearranging for *S*:

$$S = B\left(\frac{v_S^2}{v_P^2}\right)$$

We can see that this is dimensionally correct because shear modulus, S, and bulk modulus, B, have the same dimensions and the quantity in brackets is a dimensionless ratio. ☺

Substitute the numbers:

$$S = 35\text{ GPa}\left[\frac{(2.4\text{ km/s})^2}{(3.8\text{ km/s})^2}\right] = 14\text{ GPa}$$

Note that we have not needed to convert our units as our answer reduces to a ratio. Had we substituted numbers earlier, rather than arriving at this algebraic expression, we would have needed to convert the units to ensure we did not end up with an incorrect numerical answer.

Finalise As we expect from our knowledge of the mechanical properties of materials, the shear modulus is much smaller than the bulk modulus.

What If? What if the sandstone was saturated with water? How would this change the two wave speeds?

Answer In fact the transverse wave speed would not change very much. The water filling up the pores in the stone would increase the density of the stone, decreasing the wave speed slightly, but would have very little effect on the shear modulus. In contrast, the *P* wave would travel much faster because water has a very large bulk modulus, making it almost incompressible. *P* waves from earthquakes are observed to speed up when they enter the watertable; the *S* waves do not.

17.6 The Doppler effect

In this section we consider the effects of relative motion on waves. Perhaps you have noticed how the sound of a vehicle's horn or the siren from an emergency vehicle changes as the vehicle moves past you. The frequency of the sound you hear as the vehicle approaches you is higher than the frequency you hear as it moves away from you. This experience is one example of the **Doppler effect**, named after Austrian physicist Christian Johann Doppler (1803–1853). In 1842, he predicted the effect for both sound waves and light waves.

Imagine you are riding your bicycle on the road and fail to notice a parked truck. The driver sounds the horn at you to wake you up. If you stop, the waves travel past you at speed *v*, and you hear a frequency given by $f = v/\lambda$, which is just the frequency of sound emitted by the horn. But what happens if you continue to ride towards the parked truck?

In this case you, the observer *O*, are moving and the sound source, *S*, is stationary. For simplicity, we assume the air is also stationary and the observer moves directly toward the source, as in **Active Figure 17.18**. The observer moves with a speed v_O toward a stationary point source ($v_S = 0$), where *stationary* means at rest with respect to the medium, air.

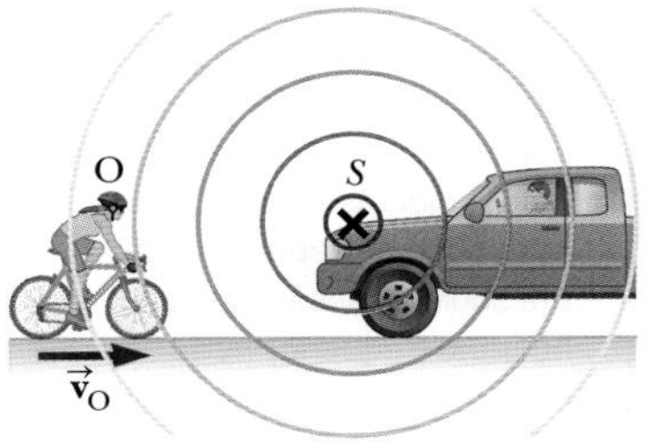

Active Figure 17.18
An observer *O* (the cyclist) moves with a speed v_O toward a stationary point source *S*, the horn of a parked truck. The observer hears a frequency *f′* that is greater than the source frequency.

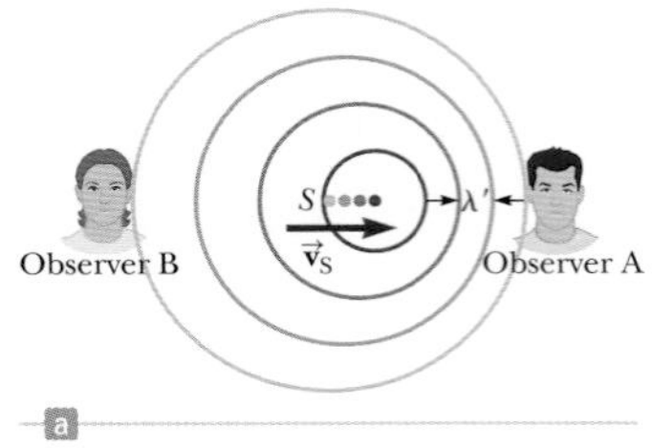

Active Figure 17.19

A source S moving with a speed v_S towards stationary observer A and away from stationary observer B. Observer A hears an increased frequency, and observer B hears a decreased frequency.

When the observer moves towards the source, the speed of the waves relative to the observer is $v' = v + v_O$, but the wavelength λ is unchanged. Hence, using **Equation 17.12** $v = \lambda f$, we can say that the frequency f' heard by the observer is *increased* and is given by

$$f' = \frac{v'}{\lambda} = \frac{v + v_O}{\lambda}$$

Because $\lambda = v/f$, we can express f' as

$$f' = \left(\frac{v + v_O}{v}\right) f \quad \text{(observer moving towards source)} \qquad \textbf{(17.36)}$$

If the observer is moving away from the source, the speed of the wave relative to the observer is $v' = v - v_O$. The frequency heard by the observer in this case is *decreased* and is given by

$$f' = \left(\frac{v - v_O}{v}\right) f \quad \text{(observer moving away from source)} \qquad \textbf{(17.37)}$$

These last two equations can be reduced to a single equation by adopting a sign convention. Whenever an observer moves with a speed v_O relative to a stationary source, the frequency heard by the observer is given by **Equation 17.36**, with v_O interpreted as follows: a positive value is substituted for v_O when the observer moves towards the source and a negative value is substituted when the observer moves away from the source.

Now suppose the source is in motion and the observer is at rest. If the source moves directly towards observer A in **Active Figure 17.19**, each new wave is emitted from a position to the right of the origin of the previous wave. As a result, the wave fronts heard by observer A are closer together than they would be if the source were not moving. (**Active Figure 17.20** shows this effect for waves moving on the surface of water.) As a result, the wavelength λ' measured by observer A is shorter than the wavelength λ of the source. During each vibration, which lasts for a time interval T (the period), the source moves a distance $v_S T = v_S/f$ and the wavelength is *shortened* by this amount. Therefore, the observed wavelength λ' is

$$\lambda' = \lambda - \Delta\lambda = \lambda - \frac{v_S}{f}$$

Because $\lambda = v/f$, the frequency f' heard by observer A is

$$f' = \frac{v}{\lambda'} = \frac{v}{\lambda - (v_S/f)} = \frac{v}{(v/f) - (v_S/f)}$$

$$f' = \left(\frac{v}{v - v_S}\right) f \quad \text{(source moving towards observer)} \qquad \textbf{(17.38)}$$

That is, the observed frequency is *increased* whenever the source is moving towards the observer.

When the source moves away from a stationary observer, as is the case for observer B in **Active Figure 17.19**, the observer measures a wavelength λ' that is *greater* than λ and hears a *decreased* frequency:

$$f' = \left(\frac{v}{v + v_S}\right) f \quad \text{(source moving away from observer)} \qquad \textbf{(17.39)}$$

We can express the general relationship for the observed frequency when a source is moving and an observer is at rest as **Equation 17.38**, with the same sign convention applied to v_S as was applied to v_O: a positive value is substituted for v_S when the source moves towards the observer and a negative value is substituted when the source moves away from the observer.

Pitfall Prevention 17.7

Doppler effect does not depend on distance. Although the *intensity* of a sound varies as the distance changes, the apparent *frequency* depends only on the relative speed of the source and the observer. As you listen to an approaching source, you will detect increasing intensity but constant frequency. As the source passes, you will hear the frequency suddenly drop to a new constant value and the intensity begin to decrease.

Finally, combining **Equations 17.36** and **17.38** gives the following general relationship for the observed frequency that includes all four conditions described by **Equations 17.36** through **17.39**:

$$f' = \left(\frac{v + v_O}{v - v_S}\right)f \qquad (17.40)$$

◀ General Doppler-shift expression

In this expression, the signs for the values substituted for v_O and v_S depend on the direction of the velocity. A positive value is used for motion of the observer or the source *towards* the other (associated with an *increase* in observed frequency), and a negative value is used for motion of one *away from* the other (associated with a *decrease* in observed frequency).

Although the Doppler effect is most typically experienced with sound waves, it is a phenomenon common to all waves. For example, the relative motion of source and observer produces a frequency shift in light waves. The Doppler effect is used in police radar systems to measure the speeds of motor vehicles and in SONAR (SOund NAvigation and Ranging) systems, for example in submarine navigation and detection.

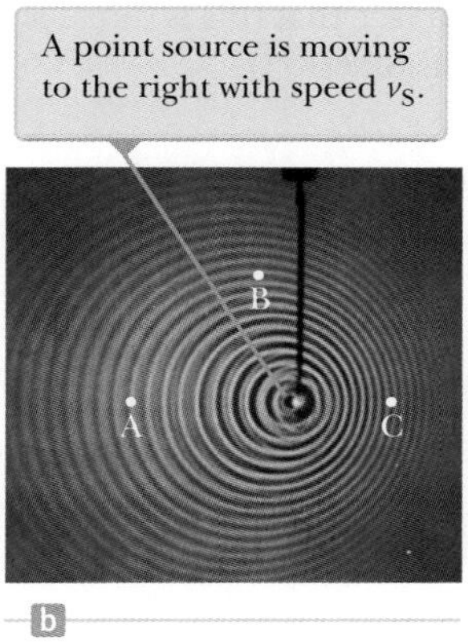
A point source is moving to the right with speed v_S.
B
A
C
b

Courtesy of the Educational Development Center, Newton, MA

Active Figure 17.20
The Doppler effect in water, observed in a ripple tank. Letters shown in the photo refer to Quick Quiz 17.7.

Quick Quiz 17.7

Consider detectors of water waves at three locations A, B and C in **Active Figure 17.20**. Which of the following statements is true? **(a)** The wave speed is highest at location A. **(b)** The wave speed is highest at location C. **(c)** The detected wavelength is largest at location B. **(d)** The detected wavelength is largest at location C. **(e)** The detected frequency is highest at location C. **(f)** The detected frequency is highest at location A.

TRY THIS

Get a small object that makes a constant sound, such as a small buzzer. (Small battery-powered children's toys often make annoying noises and are very suitable for this experiment.) Tie it securely to a length of string and swing it at constant speed in a horizontal circle above your head. What do you notice about the sound? Ask a friend to listen as you swing it around – what do they hear? Now try swinging it in a vertical circle or swap with your friend. What do you hear now? *Caution: Make sure the object is tied securely and do not hit anyone with the object!*

Example 17.7

The broken alarm clock

Your alarm clock awakens you with a steady and irritating sound of frequency 600 Hz. One morning, it malfunctions and cannot be turned off. In frustration, you drop the clock out of your fourth-floor window, 15.0 m from the ground. Assume the speed of sound is 343 m/s. As you listen to the falling alarm clock, what frequency do you hear just before you hear it strike the ground?

Solution

Conceptualise The speed of the alarm clock increases as it falls. Therefore, it is a source of sound moving away from you with an increasing speed so the frequency you hear should be less than 600 Hz.

Model We combine the particle with constant acceleration model for the falling alarm clock with our understanding of the frequency shift of sound due to the Doppler effect.

Analyse Use **Equation 2.12** to express the speed of the source of sound:

$$v_S = v_i + a_y t = 0 - gt = -gt \qquad (1)$$

From **Equation 2.15**, find the time at which the clock strikes the ground:

$$y_f = y_i + v_{yi}t - \frac{1}{2}gt^2 = 0 + 0 - \frac{1}{2}gt^2 \rightarrow t = \sqrt{-\frac{2y_f}{g}}$$

Substitute into **Equation (1)**:

$$v_S = (-g)\sqrt{-\frac{2y_f}{g}} = -\sqrt{-2gy_f}$$

Example 17.7 cont.

Check dimensions: $[LT^{-1}] = ([LT^{-2}][L])^{1/2} = [LT^{-1}]$ ☺

Use **Equation 17.40** to determine the Doppler-shifted frequency heard from the falling clock:

$$f' = \left[\frac{v+0}{v-(-\sqrt{-2gy_f})}\right]f = \left(\frac{v}{v+\sqrt{-2gy_f}}\right)f$$

We can see that this is dimensionally correct because the expression in the brackets is a ratio of velocities and hence dimensionless. ☺

Substitute numerical values:

$$f' = \left[\frac{343\text{ m/s}}{343\text{ m/s} + \sqrt{-2(9.80\text{ m/s}^2)(-15.0\text{ m})}}\right](600\text{ Hz})$$

$$= 571\text{ Hz}$$

Finalise The frequency is lower than the actual frequency of 600 Hz because the clock is moving away from you. If it were to fall from a higher floor so that it passes below $y = -15.0$ m, the clock would continue to accelerate and the frequency would continue to drop. What would you hear if you threw it upwards and listened as it rose and fell again? Try it with a small noise-making toy – just be careful it doesn't hit you!

Shock waves

Now consider what happens when the speed v_S of a source *exceeds* the wave speed v. This situation is depicted graphically in **Figure 17.21a**. The circles represent spherical wave fronts emitted by the source at various times during its motion. At $t = 0$, the source is at S_0 and moving toward the right. At later times, the source is at S_1 and then S_2, and so on. At the time t, the wave front centred at S_0 reaches a radius of vt. In this same time interval, the source travels a distance $v_S t$. Notice in **Figure 17.21a** that a straight line can be drawn tangent to all the wave fronts generated at various times. The envelope of these wave fronts is a cone whose apex half-angle θ (the 'Mach angle') is given by

$$\sin\theta = \frac{vt}{v_S t} = \frac{v}{v_S}$$

The ratio v_S/v is referred to as the *Mach number* and the conical wave front produced when $v_S > v$ (supersonic speeds) is known as a *shock wave.*

Jet aeroplanes travelling at supersonic speeds produce shock waves, which are responsible for the loud 'sonic boom' one hears. The shock wave carries a great deal of energy concentrated on the surface of the cone, with correspondingly great pressure variations. Such shock waves are unpleasant to hear and can cause damage to buildings when aircraft fly supersonically at low altitudes.

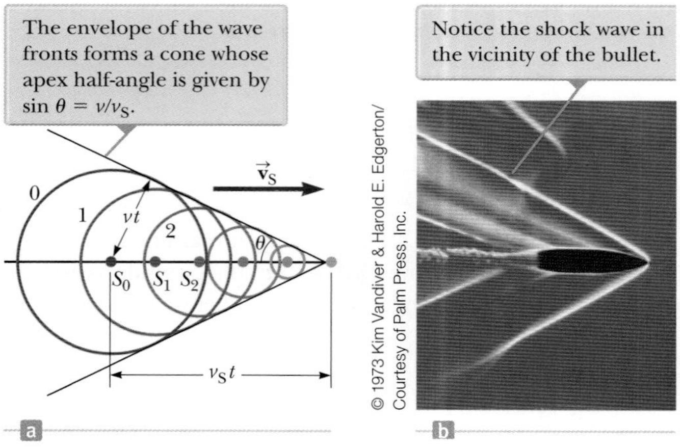
The envelope of the wave fronts forms a cone whose apex half-angle is given by sin θ = v/vS.
vS
0
1
vt
2
θ
S0
S1
S2
vSt
a
Notice the shock wave in the vicinity of the bullet.
b

Figure 17.21
(a) A representation of a shock wave produced when a source moves from S_0 to the right with a speed v_S that is greater than the wave speed v in the medium (b) A stroboscopic photograph of a bullet moving at supersonic speed through the hot air above a candle

Quick Quiz 17.8

An aeroplane flying with a constant velocity moves from a cold air mass into a warm air mass. Does the Mach number **(a)** increase, **(b)** decrease or **(c)** stay the same?

17.7 Energy, power and intensity of waves

We began this chapter by considering a leaf floating on a pond. As a ripple passes the leaf it is momentarily lifted up, so the Earth–leaf system gains potential energy. Waves transport energy through a medium, even though there is no net movement of material.

Consider a sinusoidal wave travelling on a string (**Figure 17.22**). The source of the energy is some external agent at the left end of the string. We can consider the string to be a non-isolated system. As the external agent performs work on the end of the string, moving it up and down, energy enters the system of the string and propagates along its length. Now consider an infinitesimal element of the string of length dx and mass dm. Each such element moves vertically in simple harmonic motion. Therefore, we can model each element of the string as a simple harmonic oscillator, with the oscillation in the y direction. We assume that all elements have the same angular frequency ω and the same amplitude A.

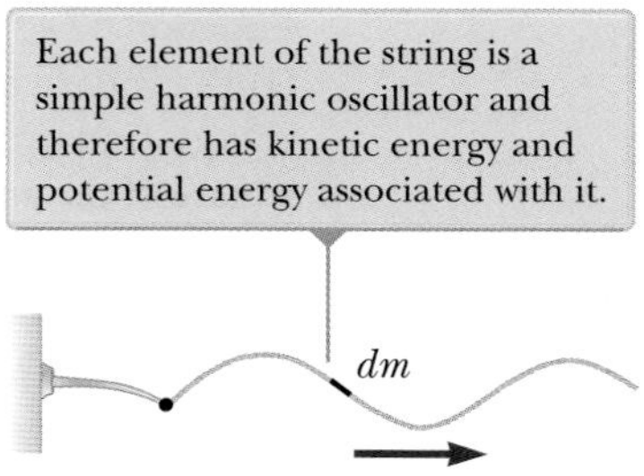

Each element of the string is a simple harmonic oscillator and therefore has kinetic energy and potential energy associated with it.
dm

Figure 17.22
A sinusoidal wave travelling along the x axis on a stretched string

The kinetic energy K associated with a moving particle is $K = \frac{1}{2}mv^2$. The kinetic energy dK associated with the up and down motion of this element is

$$dK = \frac{1}{2}(dm)v_y^{\,2}$$

where v_y is the transverse speed of the element. If μ is the mass per unit length of the string, the mass dm of the element of length dx is equal to $\mu\, dx$. Hence, the kinetic energy of an element of the string is

$$dK = \frac{1}{2}(\mu\, dx)v_y^{\,2} \qquad (17.41)$$

Substituting for the general transverse speed of an element of the medium using **Equation 17.14** gives

$$dK = \frac{1}{2}\mu[-\omega A\cos(kx - \omega t)]^2\,dx = \frac{1}{2}\mu\omega^2A^2\cos^2(kx - \omega t)dx$$

If we take a snapshot of the wave at time $t = 0$, the kinetic energy of a given element is

$$dK = \frac{1}{2}\mu\omega^2A^2\cos^2 kx\,dx$$

Integrating this expression over all the string elements in a wavelength of the wave gives the total kinetic energy K_λ in one wavelength:

$$\begin{aligned}K_\lambda &= \int dK = \int_0^\lambda \frac{1}{2}\mu\omega^2A^2\cos^2 kx\,dx = \frac{1}{2}\mu\omega^2A^2\int_0^\lambda \cos^2 kx\,dx\\ &= \frac{1}{2}\mu\omega^2A^2\left[\frac{1}{2}x + \frac{1}{4k}\sin 2kx\right]_0^\lambda = \frac{1}{2}\mu\omega^2A^2\left[\frac{1}{2}\lambda\right]\\ &= \frac{1}{4}\mu\omega^2A^2\lambda\end{aligned}$$

Appendix B.8 summarises integral calculus and includes a table of useful integrals.

In addition to kinetic energy, there is potential energy associated with each element of the string due to its displacement from the equilibrium position and the restoring forces from neighbouring elements. A similar analysis to that above for the total potential energy U_λ in one wavelength gives exactly the same result:

$$U_\lambda = \frac{1}{4}\mu\omega^2A^2\lambda$$

The total energy in one wavelength of the wave is the sum of the potential and kinetic energies:

$$E_\lambda = U_\lambda + K_\lambda = \frac{1}{2}\mu\omega^2A^2\lambda \qquad (17.42)$$

As the wave moves along the string, this amount of energy passes by a given point on the string during a time interval of one period of the oscillation. Therefore, the power P, or rate of energy transfer T_{MW} associated with the mechanical wave, is

$$P = \frac{T_{MW}}{\Delta t} = \frac{E_\lambda}{T} = \frac{\frac{1}{2}\mu\omega^2A^2\lambda}{T} = \frac{1}{2}\mu\omega^2A^2\left(\frac{\lambda}{T}\right)$$

$$P = \frac{1}{2}\mu\omega^2A^2v \qquad (17.43)$$

◀ Power of a wave

Equation 17.43 shows that the rate of energy transfer by a sinusoidal wave on a string is proportional to (1) the square of the frequency, (2) the square of the amplitude and (3) the wave speed. In fact, the rate of energy transfer in *any* sinusoidal wave is proportional to the square of the angular frequency and to the square of the amplitude.

Quick **Quiz 17.9**

Which of the following, taken by itself, would be most effective in increasing the rate at which energy is transferred by a wave travelling along a string? **(a)** reducing the linear mass density of the string by one half **(b)** doubling the wavelength of the wave **(c)** doubling the tension in the string **(d)** doubling the amplitude of the wave

±? Example **17.8**

Power supplied to a vibrating string

A dhunaki is an ancient Indian device for carding cotton which uses a vibrating taut string to loosen the fibres, before it is spun into threads. A particular dhunaki string has $\mu = (5.00 \pm 0.05) \times 10^{-2}$ kg/m and is under a tension of 80 ± 1 N. How much power must be supplied to the string to generate sinusoidal waves at a frequency of 60 Hz and a minimum amplitude of 0.60 cm?

Getty Images/SSPL Science Museum

Solution

Conceptualise Consider **Active Figure 17.9** again and notice that the vibrating blade supplies energy to the string at a certain rate. This energy then propagates to the right along the string.

Model We model the waves on the dhunaki as travelling waves and assume constant tension in the string and constant wave speed and amplitude on the string.

Analyse Use **Equation 17.43** to evaluate the power: $P = \frac{1}{2}\mu\omega^2 A^2 v$

Use **Equations 17.9** and **17.28** to substitute for ω and v: $P = \frac{1}{2}\mu(2\pi f)^2 A^2\left(\sqrt{\frac{T}{\mu}}\right) = 2\pi^2 f^2 A^2\sqrt{\mu T}$

Check dimensions: $[\mathrm{ML^2T^{-3}}] = [\mathrm{T^{-1}}]^2[\mathrm{L}]^2([\mathrm{ML^{-1}}][\mathrm{MLT^{-2}}])^{1/2} = [\mathrm{ML^2T^{-3}}]$ ☺

Substitute numerical values to find the minimum power required: $P_{min} = 2\pi^2(60\text{ Hz})^2(0.006\text{ m})^2\sqrt{(0.0495\text{ kg/m})(79\text{ N})} = 5.05\text{ W}$

Substitute numerical values to find the maximum power required: $P_{max} = 2\pi^2(60\text{ Hz})^2(0.006\text{ m})^2\sqrt{(0.0505\text{ kg/m})(81\text{ N})} = 5.17\text{ W}$

In order to ensure that the amplitude of oscillation reached at least 6 mm, you would choose a power of 5.17 W in this case.

Intensity of sound waves

Sound waves also transfer energy. When you sit in a physics lecture there is no transfer of air from the lecturer's mouth to your ears, but there is a transfer of energy (and information).

Consider again the element of gas acted on by the piston in **Figure 17.17**. Imagine that the piston is moving back and forth in simple harmonic motion at angular frequency ω. Imagine also that the length of the element becomes very small so that the entire element moves with the same velocity as the piston. Then we can model the element as a particle on which the piston is doing work. The rate at which the piston is doing work on the element at any instant of time is given by Equation 6.25:

$$\text{Power} = \vec{\mathbf{F}} \cdot \vec{\mathbf{v}}_x$$

where we have used *Power* rather than P so that we don't confuse power P with pressure P. The force $\vec{\mathbf{F}}$ on the element of gas is related to the pressure, and the velocity $\vec{\mathbf{v}}_x$ of the element is the derivative of the displacement function, so we find

$$\begin{aligned}\text{Power} &= [\Delta P(x,t)A]\hat{\mathbf{i}} \circ \frac{\partial}{\partial t}[s(x,t)\hat{\mathbf{i}}] \\ &= [\rho v\omega A s_{max} \sin(kx - \omega t)]\left\{\frac{\partial}{\partial t}[s_{max} \cos(kx - \omega t)]\right\} \\ &= \rho v\omega A s_{max} \sin(kx - \omega t)][\omega s_{max} \sin(kx - \omega t)] \\ &= \rho v\omega^2 A s_{max}^2 \sin^2(kx - \omega t)\end{aligned}$$

Appendix B.7 summarises differential calculus, and the vector dot product is described in Appendix B.6. Σ

We now find the time average power over one period of the oscillation. For any given value of x, which we can choose to be $x = 0$, the average value of $\sin^2 (kx - \omega t)$ over one period T is

$$\frac{1}{T}\int_0^T \sin^2(0 - \omega t)\,dt = \frac{1}{T}\int_0^T \sin^2 \omega t\, dt = \frac{1}{T}\left(\frac{t}{2} + \frac{\sin 2\omega t}{2\omega}\right)\Bigg|_0^T = \frac{1}{2}$$

Therefore,

$$\text{Power}_{avg} = \frac{1}{2}\rho v\omega^2 A s^2_{max}$$

We define the **intensity** I of a wave, or the power per unit area, as the rate at which the energy transported by the wave transfers through a unit area A perpendicular to the direction of travel of the wave:

$$I \equiv \frac{\text{Power}_{avg}}{A} \tag{17.44}$$

◀ Intensity of a sound wave

In this case, the intensity is therefore

$$I = \frac{1}{2}\rho v(\omega s_{max})^2$$

Hence, the intensity of a periodic sound wave is proportional to the square of the displacement amplitude and to the square of the angular frequency. This expression can also be written in terms of the pressure amplitude ΔP_{max}; in this case, we use Equation 17.35 to obtain

$$I = \frac{\Delta P^2_{max}}{2\rho v} \tag{17.45}$$

The waves on strings that we studied are constrained to move along the one-dimensional string and the sound waves we have studied so far have been constrained to move in one dimension along the length of the tube. However, we do not usually experience sound in a piston but rather in open air.

Consider the case of a point source emitting sound waves equally in all directions. If the air around the source is perfectly uniform, the power radiated in all directions is the same and the speed of sound in all directions is the same. The result in this situation is called a **spherical wave**. Figure 17.23 shows these spherical waves as a series of circular arcs concentric with the source. Each arc represents a surface over which the phase of the wave is constant. We call such a surface of constant phase a **wave front**. The radial distance between adjacent wave fronts that have the same phase is the wavelength λ of the wave. The radial lines pointing outwards from the source, representing the direction of propagation of the waves, are called **rays**.

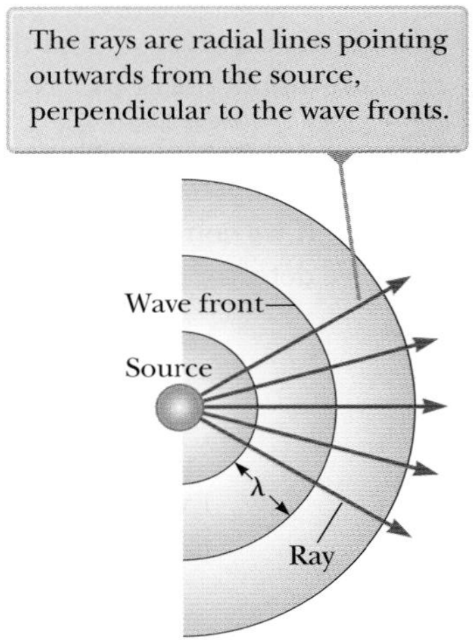
The rays are radial lines pointing outwards from the source, perpendicular to the wave fronts.
Wave front
Source
λ
Ray

Figure 17.23
Spherical waves emitted by a point source. The circular arcs represent the spherical wave fronts that are concentric with the source.

The average power emitted by the source must be distributed uniformly over each spherical wave front of area $4\pi r^2$. Hence, the wave intensity at a distance r from the source is

$$I = \frac{\text{Power}_{\text{avg}}}{A} = \frac{\text{Power}_{\text{avg}}}{4\pi r^2} \tag{17.46}$$

The intensity decreases as the square of the distance from the source. This inverse-square law is characteristic of anything that originates from a point source or can be modelled as such – for example, the behaviour of gravity as we saw in Chapter 11.

Example 17.9

Hearing limits

The faintest sounds the human ear (of a typical young person) can detect at a frequency of 1000 Hz correspond to an intensity of about 1.00×10^{-12} W/m², which is called the *threshold of hearing*. The loudest sounds the ear can tolerate at this frequency correspond to an intensity of about 1.00 W/m², the *threshold of pain*. Determine the pressure amplitude and displacement amplitude associated with these two limits.

Solution

Conceptualise Think about the quietest environment you have ever experienced. It is likely that the intensity of sound in even this quietest environment is higher than the threshold of hearing.

Model We use the expressions for pressure and displacement amplitude developed in this section. We will assume that the medium is air at normal temperature and pressure so that the sound wave is travelling at 343 m/s and the air has density 1.20 kg/m³.

Analyse Use Equation 17.45:

$$\begin{aligned}\Delta P_{\text{max}} &= \sqrt{2\rho v I} \\ &= \sqrt{2(1.20\ \text{kg/m}^3)(343\ \text{m/s})(1.00 \times 10^{-12}\ \text{W/m}^2)} \\ &= 2.87 \times 10^{-5}\ \text{N/m}^2\end{aligned}$$

Calculate the corresponding displacement amplitude using Equation 17.35, recalling that $\omega = 2\pi f$ (Equation 17.9):

$$\begin{aligned}s_{\text{max}} &= \frac{\Delta P_{\text{max}}}{\rho v \omega} = \frac{2.87 \times 10^{-5}\ \text{N/m}^2}{(1.20\ \text{kg/m}^3)(343\ \text{m/s})(2\pi \times 1000\ \text{Hz})} \\ &= 1.11 \times 10^{-11}\ \text{m}\end{aligned}$$

In a similar manner, we find that the loudest sounds the human ear can tolerate (the threshold of pain) correspond to a pressure amplitude of 28.7 N/m² and a displacement amplitude equal to 1.11×10^{-5} m.

Finalise Because atmospheric pressure is about 10^5 N/m², the result for the pressure amplitude tells us that the ear is sensitive to pressure fluctuations as small as 3 parts in 10^{10}! The displacement amplitude is also a remarkably small number. If we compare this result for s_{max} to the size of an atom (about 10^{-10} m), we see that the ear is an extremely sensitive detector of sound waves.

Sound level in decibels

Example 17.9 illustrates the wide range of intensities the human ear can detect. Because this range is so wide, it is convenient to use a logarithmic scale, where the **sound level** β (Greek letter beta) is defined by the equation

Sound level ▶ in decibels

$$\beta \equiv 10 \log\left(\frac{I}{I_0}\right) \tag{17.47}$$

The constant I_0 is the *reference intensity*, and is taken to be at the threshold of hearing ($I_0 = 1.00 \times 10^{-12}$ W/m^2), and I is the intensity in watts per square metre to which the sound level β corresponds, where β is measured in **decibels** (dB). The unit *bel* is named after the inventor of the telephone, Alexander Graham Bell (1847–1922). The prefix *deci-* is the SI prefix that stands for 10^{-1}.

On the dB scale the threshold of pain ($I = 1.00$ W/m^2) corresponds to a sound level of $\beta = 120$ dB and the threshold of hearing corresponds to $\beta = 0$ dB.

Prolonged exposure to high sound levels may seriously damage the human ear. Ear plugs are recommended whenever sound levels exceed 90 dB. Recent evidence suggests that 'noise pollution' may be a contributing factor to high blood pressure, anxiety and stress. Table 17.2 gives some typical sound levels.

Table 17.2
Sound levels

Source of sound	β (dB)
Nearby jet aeroplane	150
Jackhammer	130
Rock concert	120
Lawn mower	100
Busy traffic	80
Vacuum cleaner	70
Normal conversation	60
Mosquito buzzing	40
Whisper	30
Rustling leaves	10
Threshold of hearing	0

Quick Quiz 17.10

Increasing the intensity of a sound by a factor of 100 causes the sound level to increase by what amount? (a) 100 dB (b) 20 dB (c) 10 dB (d) 2 dB

Example 17.10

The loud boys

A physicist (one of the authors) who is the mother of twin boys, Marcus and Laurence, suspects that she is going deaf from her babies' screaming. She borrows a sound level meter and discovers that the sound level when Marcus is screaming is 126 dB at a distance of 20 cm from his mouth, about the distance between his mouth and his mother's ear when she holds him while he is screaming.

(A) What is the intensity of sound at this distance?

Solution

Conceptualise Imagine holding a screaming baby, producing a sound level well above that allowed in the workplace. A second baby may start screaming at any moment.

Model We are given a sound level, so we will use Equation 17.47, assuming that Marcus produces a scream of approximately 1000 Hz.

Analyse Use Equation 17.47 to calculate the sound intensity at the ear:

$$\beta_M = 10\log\left(\frac{I}{I_0}\right)$$

$$I = I_0(10^{0.1\beta}) = (1.00 \times 10^{-12}\ \text{W/m}^2)(10^{0.1(126\ \text{dB})}) = 3.98\ \text{W/m}^2$$

(B) If we model Marcus as a point source, what is the average power he puts into his screaming?

Solution

We now use Equation 17.46:

$$I = \frac{\text{Power}_{\text{avg}}}{A} = \frac{\text{Power}_{\text{avg}}}{4\pi r^2}$$

Rearranging for power and substituting gives:

$$\text{Power}_{\text{avg}} = 4\pi r^2 I = 4\pi(0.20\ \text{m})^2(3.98\ \text{W/m}^2) = 2.0\ \text{W}$$

This very modest power explains why it is possible for Marcus to scream continuously for many hours.

(C) Find the sound level heard by the physicist if Laurence also starts screaming and she holds both screaming babies at approximately the same distance.

Solution

Use Equation 17.47 to calculate the sound level assuming the intensity has doubled from 3.98 W/m^2 to 7.96 W/m^2:

$$\beta_{L+M} = 10\log\left(\frac{I}{I_0}\right) = 10\log\left(\frac{7.96\ \text{W/m}^2}{1.00 \times 10^{-12}\ \text{W/m}^2}\right) = 129\ \text{dB}$$

Finalise These results show that when the intensity is doubled, the sound level increases by only 3 dB. This 3-dB increase is independent of the original sound level.

Loudness is a psychological response to a sound. It depends on both the intensity and the frequency of the sound. As a rule of thumb, a doubling in loudness is approximately associated with an increase in sound level of 10 dB. (This rule of thumb is relatively inaccurate at very low or very high frequencies.) A doubling in loudness corresponds to an increase in intensity by a factor of 10. In other words, two screaming babies is subjectively only slightly louder than one, it takes 10 screaming babies to be (subjectively) twice as loud as one! This is why we use a logarithmic scale to describe sounds, rather than a linear one.

Loudness and frequency

The discussion of sound level in decibels relates to a *physical* measurement of the strength of a sound. Let us now extend our discussion of the *psychological* 'measurement' of the strength of a sound.

We have stated that the threshold intensity is 10^{-12} W/m^2, corresponding to an intensity level of 0 dB. In reality, this value is the threshold only for a sound of frequency 1000 Hz, which is a standard reference frequency in acoustics. If we perform an experiment to measure the threshold intensity at other frequencies, we find a distinct variation of this threshold as a function of frequency. For example, at 100 Hz, a barely audible sound must have an intensity level of about 30 dB! Unfortunately, there is no simple relationship between physical measurements and psychological 'measurements'.

By using test subjects, the human response to sound has been studied, and the results are shown in the white area of **Figure 17.24** along with the approximate frequency and sound-level ranges of other sound sources. The lower curve of the white area corresponds to the threshold of hearing. Notice that humans are sensitive to frequencies ranging from about 20 Hz to about 20 000 Hz. The upper bound of the white area is the threshold of pain. Here the boundary of the white area appears straight because the psychological response is relatively independent of frequency at this high sound level.

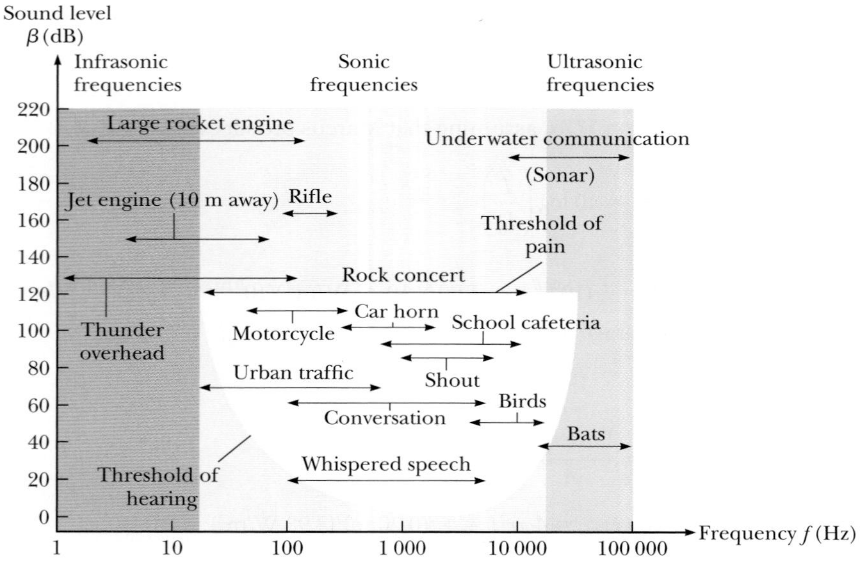

Figure 17.24
Approximate ranges of frequency and sound level of various sources and that of normal human hearing, shown by the white area. (From Reese, R. L., 2000, *University Physics*, Pacific Grove, Brooks/Cole.)

End-of-chapter resources

So what can we now say about the experiences of the witness to the Tunguska event? He recalled a moving light in the sky, brighter than the Sun, and felt his face become warm. What he saw was the light from the meteor, which was burning up due to friction (air resistance). The light and heat that he felt were both electromagnetic waves, travelling at the speed of light. He felt the ground shake and then an invisible agent picked him up and dropped him about a metre from where he had been seated. The initial shaking was the longitudinal wave or *P* wave travelling through the ground, after the collision of the remains of the meteor with the ground. The *P* wave moves much faster than the transverse or *S* wave because the bulk modulus of rock is greater than the shear modulus. The transverse or *S* wave, which arrived next, caused the ground to move up and down, throwing him upwards. He heard a very loud, protracted rumbling, as the sound wave in air, the slowest of the waves he experienced, finally arrived.

Alamy/DIZ Muenchen GmbH, Sueddeutsche Zeitung Photo

The problems found in this chapter may be assigned online in Enhanced Web Assign.

Worked solutions to every fifth problem are available in the Student Solutions Manual. Register online at **www.cengagebrain.com** for access.

Summary

Definitions

A one-dimensional **sinusoidal wave** is one for which the positions of the elements of the medium vary sinusoidally. A sinusoidal wave travelling to the right can be expressed with a **wave function**

$$y(x, t) = A\sin\left[\frac{2\pi}{\lambda}(x - vt)\right] \qquad (17.5)$$

where A is the **amplitude**, λ is the **wavelength**, and v is the **wave speed**.

The **angular wave number** k and **angular frequency** ω of a wave are defined as:

$$k \equiv \frac{2\pi}{\lambda} \qquad (17.8)$$

$$\omega \equiv \frac{2\pi}{T} = 2\pi f \qquad (17.9)$$

where T is the **period** of the wave and f is its **frequency**.

A **transverse wave** is one in which the elements of the medium move in a direction *perpendicular* to the direction of propagation.

A **longitudinal wave** is one in which the elements of the medium move in a direction *parallel* to the direction of propagation. Sound is a longitudinal wave.

The **sound level** of a sound wave in decibels is

$$\beta \equiv 10\log\left(\frac{I}{I_0}\right) \qquad (17.47)$$

where I_0 is a reference intensity, the threshold of hearing (1.00×10^{-12} W/m^2), and I is the intensity of the sound wave in watts per square metre.

Concepts and principles

Any one-dimensional wave travelling with a speed v in the x direction can be represented by a wave function of the form

$$y(x, t) = f(x \pm vt) \qquad (17.1, 17.2)$$

where the positive sign applies to a wave travelling in the negative x direction and the negative sign applies to a wave travelling in the positive x direction.

For sound waves the variation in pressure from the equilibrium value is

$$\Delta P = \Delta P_{max} \sin(kx - \omega t) \qquad (17.19)$$

where ΔP_{max} is the **pressure amplitude** or **acoustic pressure**. The pressure wave is 90° out of phase with the displacement wave. The relationship between s_{max} and ΔP_{max} is

$$\Delta P_{max} = \rho v \omega s_{max} \qquad (17.35)$$

Wave functions are solutions to a differential equation called the **linear wave equation**:

$$\frac{\partial^2 y}{\partial x^2} = \frac{1}{v^2}\frac{\partial^2 y}{\partial t^2} \qquad (17.27)$$

The speed of a wave depends upon the properties of the medium:

$$v = \sqrt{\frac{\text{elastic property}}{\text{inertial property}}} \qquad (17.33)$$

For a transverse wave on a string:

$$v = \sqrt{\frac{T}{\mu}} \qquad (17.28)$$

For a sound wave:

$$v = \sqrt{\frac{B}{\rho}} \qquad (17.32)$$

The **power** transmitted by a sinusoidal wave on a stretched string is

$$P = \frac{1}{2}\mu\omega^2 A^2 v \qquad (17.43)$$

The **intensity** of a periodic sound wave, which is the power per unit area, is

$$I \equiv \frac{\text{Power}_{\text{avg}}}{A} = \frac{\Delta P_{\text{max}}^2}{2\rho v} \qquad (17.44, 17.45)$$

The change in frequency heard by an observer whenever there is relative motion between a source of sound waves and the observer is called the **Doppler effect**. The observed frequency is

$$f' = \left(\frac{v + v_O}{v - v_S}\right) f \qquad (17.40)$$

The signs for the values substituted for v_O and v_S depend on the direction of the velocity.

Analysis model for problem solving

Travelling wave The wave speed of a sinusoidal wave is

$$v = \frac{\lambda}{T} = \lambda f \qquad (17.6, 17.12)$$

A sinusoidal wave can be expressed as

$$y = A \sin(kx - \omega t) \qquad (17.10)$$

Chapter review quiz

To help you revise Chapter 17: Wave motion, complete the automatically graded Chapter review quiz at http://login.cengagebrain.com.

Conceptual questions

1. (a) How would you create a longitudinal wave in a stretched spring? (b) Would it be possible to create a transverse wave in a spring?
2. Does the vertical speed of an element of a horizontal, taut string, through which a wave is travelling, depend on the wave speed? Explain your answer.
3. (a) If a long rope is hung from a ceiling and waves are sent up the rope from its lower end, why does the speed of the waves change as they ascend? (b) Does the speed of the ascending waves increase or decrease? Explain your answer.
4. Why is a solid substance able to transport both longitudinal waves and transverse waves but a homogeneous fluid is able to transport only longitudinal waves?
5. In Example 17.5 a hiker sent transverse pulses up the cable being used by a helicopter to lift him to safety. What happens if the hiker begins to swing on the end of the cable like a pendulum? Sketch the speed of the pulses on the cable as a function of time and explain the variations.
6. Explain how the distance to a lightning bolt can be determined by counting the seconds between the flash and the sound of thunder. Approximately how many seconds is equivalent to one kilometre in distance?
7. You are driving toward a cliff and honk your horn. Is there a Doppler shift of the sound when you hear the echo? If so, is it like a moving source or a moving observer? What if the reflection occurs not from a cliff but from the front edge of a huge alien spacecraft moving towards you as you drive along a deserted outback road?
8. How can an object move with respect to an observer so that the sound from it is not shifted in frequency?
9. A student taking a quiz finds these two equations on a reference sheet:

$$f = \frac{1}{T} \quad \text{and} \quad v = \sqrt{\frac{T}{\mu}}$$

She has forgotten what T represents in each equation. (a) Use dimensional analysis to determine the units required for T in each equation. (b) Explain how you can identify the physical quantity each T represents from the units.

10. In Chapter 12 we studied relativity, which states it is only the *relative* velocities of observers that is important. Is this true for the Doppler effect? In other words, does the observed frequency depend on whether it is the source that is moving or the observer, or only the relative motion? Explain your answer and any apparent conflict with the theory of relativity.

Problems

Section 17.1 Propagation of waves

1. At $t = 0$, a transverse pulse in a wire is described by the function

$$y = \frac{6.00}{x^2 + 3.00}$$

where x and y are in metres. If the pulse is travelling in the positive x direction with a speed of 4.50 m/s, write the function $y(x, t)$ that describes this pulse.

2. Ocean waves with a crest-to-crest distance of 10.0 m can be described by the wave function

$$y(x, t) = 0.800 \sin[0.628(x - vt)]$$

where x and y are in metres, t is in seconds and $v = 1.20$ m/s. (a) Sketch $y(x, t)$ at $t = 0$. (b) Sketch $y(x, t)$ at $t = 2.00$ s. (c) Compare the graph in part (b) with that in part (a) and explain similarities and differences. (d) How has the wave moved between graph (a) and graph (b)?

3. ±? A seismographic station receives S and P waves from an earthquake, separated in time by (17.3 ± 0.1) s. Assume the waves have travelled over the same path at speeds of 4.50 km/s and 7.80 km/s. Find the distance from the seismograph to the focus of the quake.

Section **17.2** Analysis model: travelling wave

4. The wave function for a travelling wave on a taut string is (in SI units)

$$y(x, t) = 0.350 \sin\left(10\pi t - 3\pi x + \frac{\pi}{4}\right)$$

(a) What are the speed and direction of travel of the wave? (b) What is the vertical position of an element of the string at $t = 0$, $x = 0.100$ m? What are (c) the wavelength and (d) the frequency of the wave? (e) What is the maximum transverse speed of an element of the string?

5. A certain uniform string is held under constant tension. (a) Draw a side-view snapshot of a sinusoidal wave on a string as shown in diagrams in the text. (b) Immediately below your diagram for part (a), draw the same wave at a moment later by one-quarter of the period of the wave. (c) Then, draw a wave with an amplitude 1.5 times larger than the wave in diagram (a). (d) Next, draw a wave differing from the one in your diagram for part (a) just by having a wavelength 1.5 times larger. (e) Finally, draw a wave differing from that in the diagram for part (a) just by having a frequency 1.5 times larger.

6. A sinusoidal wave is travelling along a rope. The oscillator that generates the wave completes (80 ± 1) vibrations per minute. A given crest of the wave is measured to travel (425 ± 5) cm along the rope in (10.0 ± 0.1) s. What is the wavelength of the wave?

7. A wave is described by $y = 0.0200 \sin(kx - \omega t)$, where $k = 2.11$ rad/m, $\omega = 3.62$ rad/s, x and y are in metres, and t is in seconds. Determine (a) the amplitude, (b) the wavelength, (c) the frequency and (d) the speed of the wave.

8. The string shown in **Figure P17.8** is driven at a frequency of 5.00 Hz. The amplitude of the motion is $A = 12.0$ cm and the wave speed is $v = 20.0$ m/s. Furthermore, the wave is such that $y = 0$ at $x = 0$ and $t = 0$. Determine (a) the angular frequency and (b) the wave number for this wave. (c) Write an expression for the wave function. Calculate (d) the maximum transverse speed and (e) the maximum transverse acceleration of an element of the string.

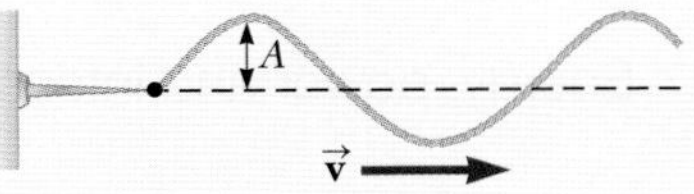

Figure P17.8

9. When a particular wire is vibrating with a frequency of 4.00 Hz, a transverse wave of wavelength 60.0 cm is produced. Determine the speed of waves along the wire.

Section **17.3** Sound waves

10. Write an expression that describes the pressure variation as a function of position and time for a sinusoidal sound wave in air. Assume the speed of sound is 343 m/s, $\lambda = 0.100$ m and $\Delta P_{max} = 0.200$ Pa.

11. A sound wave in air has a pressure amplitude equal to 4.00×10^{-3} Pa. Calculate the displacement amplitude of the wave at a frequency of 10.0 kHz.

12. A sinusoidal sound wave moves through a medium and is described by the displacement wave function

$$s(x, t) = 2.00 \cos(15.7x - 858t)$$

where s is in micrometres, x is in metres and t is in seconds. Find (a) the amplitude, (b) the wavelength and (c) the speed of this wave. (d) Determine the instantaneous displacement from equilibrium of the elements of the medium at the position $x = 0.0500$ m at $t = 3.00$ ms. (e) Determine the maximum speed of the element's oscillatory motion.

Section **17.4** The linear wave equation

13. Show that the wave function $y = e^{b(x-vt)}$ is a solution of the linear wave equation (**Equation 17.27**), where b is a constant.

14. Show that the wave function $y = \ln[b(x - vt)]$ is a solution to **Equation 17.27**, where b is a constant.

15. (a) Show that the function $y(x, t) = x^2 + v^2t^2$ is a solution to the wave equation. (b) Show that the function in part (a) can be written as $f(x + vt) + g(x - vt)$ and determine the functional forms for f and g. (c) Repeat parts (a) and (b) for the function $y(x, t) = \sin(x)\cos(vt)$.

Section **17.5** The speed of waves

16. A piano string having a mass per unit length equal to 5.00×10^{-3} kg/m is under a tension of 1350 N. Find the speed with which a wave travels on this string.

17. Transverse waves travel with a speed of 20.0 m/s on a string under a tension of 6.00 N. What tension is required for a wave speed of 30.0 m/s on the same string?

18. The elastic limit of a steel wire is 2.70×10^8 Pa. What is the maximum speed at which transverse wave pulses can propagate along this wire without exceeding this stress? (The density of steel is 7.86×10^3 kg/m^3.)

19. A group of students set up an experiment in which tension is maintained in a string by hanging a weight from one end; the other end is fixed. The measured wave speed is v = (24.0 ± 0.5) m/s when the suspended mass is m = (3.00 ± 0.01) kg. (a) What is the mass per unit length of the string? (b) What is the wave speed when the suspended mass is $m = 2.00$ kg?

20. A steel wire of length 30.0 m and a copper wire of length 20.0 m, both with 1.00 mm diameters, are connected end to end and stretched to a tension of 150 N. During what time interval will a transverse wave travel the entire length of the two wires?

21. An experimenter wishes to generate in air a sound wave that has a displacement amplitude of 5.50×10^{-6} m. The pressure amplitude is to be limited to 0.840 Pa. What is the minimum wavelength the sound wave can have?

22. Earthquakes at fault lines in the Earth's crust create seismic waves, which are longitudinal (P waves) or transverse (S waves). The P waves have a speed of about 7 km/s. Estimate the average bulk modulus of the Earth's crust given that the density of rock is about 2500 kg/m^3.

23. An Ethernet cable is 4.00 m long. The cable has a mass of 0.200 kg. A transverse pulse is produced by plucking one end of the taut cable. The pulse makes four trips down and back along the cable in 0.800 s. What is the tension in the cable?

24. A sound wave propagates in air at 27°C with frequency 4.00 kHz. It passes through a region where the temperature gradually changes and then moves through air at 0°C. Give numerical answers to the following questions to the extent possible and state your reasoning about what happens to the wave physically. (a) What happens to the speed of the wave? (b) What happens to its frequency? (c) What happens to its wavelength?

25. A sound wave in air has a pressure amplitude equal to 4.00×10^{-3} Pa. Calculate the displacement amplitude of the wave at a frequency of 10.0 kHz.

26. The speed of sound in air (in metres per second) depends on temperature according to the approximate expression

$$v = 331.5 + 0.607T_C$$

where T_C is the Celsius temperature. In dry air, the temperature decreases about 1°C for every 150 m rise in altitude. (a) Assume this change is constant up to an altitude of 9000 m. What time interval is required for the sound from an aeroplane flying at 9000 m to reach the ground on a day when the ground temperature is 30°C? (b) **What If?** Compare your answer with the time interval required if the air were uniformly at 30°C. Which time interval is longer?

27. A hammer strikes one end of a thick iron rail of length 8.50 m. A microphone located at the opposite end of the rail detects two pulses of sound, one that travels through the air and a longitudinal wave that travels through the rail. (a) Which pulse reaches the microphone first? (b) Find the separation in time between the arrivals of the two pulses.

Section 17.6 The Doppler effect

28. A driver travels northbound on a highway at a speed of 25.0 m/s. A police car, travelling southbound at a speed of 40.0 m/s, approaches with its siren producing sound at a frequency of 2500 Hz. (a) What frequency does the driver hear as the police car approaches? (b) What frequency does the driver detect after the police car passes him? (c) Repeat parts (a) and (b) for the case when the police car is behind the driver and travels northbound.

29. When high-energy charged particles move through a transparent medium with a speed greater than the speed of light in that medium, a shock wave, or bow wave, of light is produced. This phenomenon is called the *Cerenkov effect*. When a nuclear reactor is shielded by a large pool of water, Cerenkov radiation can be seen as a blue glow in the vicinity of the reactor core due to high-speed electrons moving through the water (**Figure P17.29**). In a particular case, the Cerenkov radiation produces a wave front with an apex half-angle of 53.0°. Calculate the speed of the electrons in the water. The speed of light in water is 2.25×10^8 m/s.

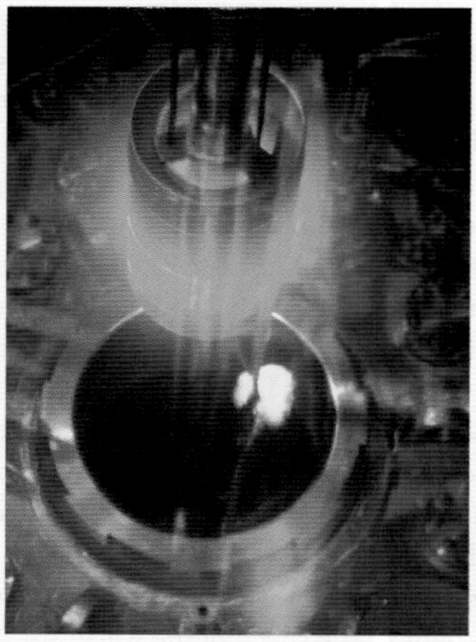

B.Benoit/Photo researchers, Inc.

Figure P17.29

30. Standing at a crossing, you hear a frequency of 560 Hz from the siren of an approaching ambulance. After the ambulance passes, the observed frequency of the siren is 480 Hz. Determine the ambulance's speed from these observations.

31. A tuning fork vibrating at 512 Hz falls from rest and accelerates at 9.80 m/s². How far below the point of release is the tuning fork when waves of frequency 485 Hz reach the release point?

32. Tristan is swinging a buzzer attached to a cord around his head, as directed by a *Try this* example in his physics textbook. The buzzer produces a sound with frequency 5.00 kHz and Tristan swings it so that it travels in a horizontal circle with velocity 10.0 m/s. (a) What are the maximum and minimum frequencies heard by Tristan? (b) What are the maximum and minimum frequencies heard by Jason, who is standing a short distance away? (c) If the Tristan suddenly lets go of the string, what frequency will he hear? State any approximations that you make.

33. A supersonic jet travelling at Mach 3.00 at an altitude of h = 20 000 m is directly over a person at time t = 0 as shown in **Figure P17.33**. Assume the average speed of sound in air is 335 m/s over the path of the sound. (a) At what time will the person encounter the shock wave due to the sound emitted at t = 0? (b) Where will the plane be when this shock wave is heard?

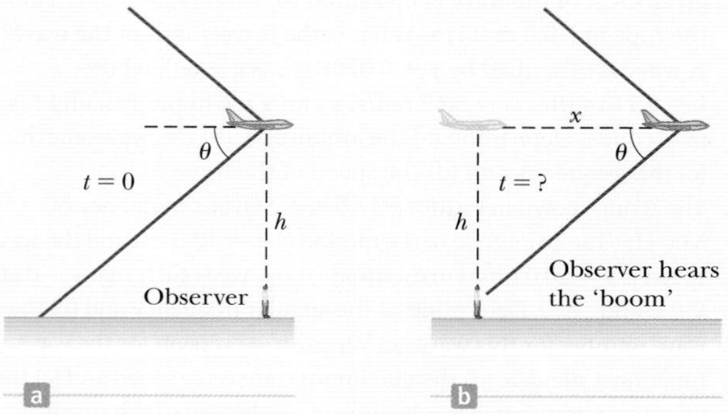

Figure P17.33

Section 17.7 Energy, power and intensity of waves

34. Sinusoidal waves 5.00 cm in amplitude are to be transmitted along a string that has a linear mass density of 4.00×10^{-2} kg/m. The source can deliver a maximum power of 300 W and the string is under a tension of 100 N. What is the highest frequency f at which the source can operate?

35. A sinusoidal wave on a string is described by the wave function

$$y = 0.15 \sin (0.80x - 50t)$$

where x and y are in metres and t is in seconds. The mass per unit length of this string is 12.0 g/m. Determine (a) the speed of the wave, (b) the wavelength, (c) the frequency and (d) the power transmitted by the wave.

36. In a region far from the epicentre of an earthquake, a seismic wave can be modelled as transporting energy in a single direction without absorption, just as a string wave does. Suppose the seismic wave moves from granite into mud with similar density but with a much smaller bulk modulus. Assume the speed of the wave gradually drops by a factor of 25.0, with negligible reflection of the wave. (a) Explain whether the amplitude of the ground shaking will increase or decrease. (b) Does it change by a predictable factor? This phenomenon was an important factor in the February 2011 magnitude 6.3 earthquake that devastated Christchurch. Christchurch is on soft sedimentary rock and soil, surrounded by hard volcanic basalt. The city was subjected to much greater shaking than would normally be expected from even such a large quake.

37. A two-dimensional water wave spreads in circular ripples. Show that the amplitude A at a distance r from the initial disturbance is proportional to $1/\sqrt{r}$. *Hint:* Consider the energy carried by one outward-moving ripple.

38. The area of a typical eardrum is about 5.00×10^{-5} m². (a) Calculate the average power of sound incident on an eardrum at the threshold of pain, which corresponds to an intensity of 1.00 W/m². (b) How much energy is transferred to an eardrum exposed to this sound for 1.00 min?

39. Calculate the sound level (in decibels) of a sound wave that has ±? an intensity of 4.00 μW/m². If the uncertainty in the intensity is 0.01 μW/m², what is the uncertainty in the sound level? What is the fractional uncertainty in each?

40. A train sounds its horn as it approaches an intersection. The horn can just be heard at a level of 50 dB by an observer 10 km away. (a) What is the average power generated by the horn? (b) What intensity level of the horn's sound is observed by someone waiting at an intersection 50 m from the train? Treat the horn as a point source and neglect any absorption of sound by the air.

41. A person wears a hearing aid that uniformly increases the sound level of all audible frequencies of sound by 30.0 dB. The hearing aid picks up sound having a frequency of 250 Hz at an intensity of 3.0×10^{-11} W/m². What is the intensity delivered to the eardrum?

42. The power output of a certain public-address speaker is ±? (6.00 ± 0.05) W. Suppose it broadcasts equally in all directions. (a) Within what distance from the speaker would the sound be painful to the ear? (b) At what distance from the speaker would the sound be barely audible?

43. The most soaring vocal melody is in Johann Sebastian Bach's *Mass in B Minor*. In one section, the basses, tenors, altos and sopranos carry the melody from a low D to a high A. In concert pitch, these notes are now assigned frequencies of 146.8 Hz and 880.0 Hz. Find the wavelengths of (a) the initial note and (b) the final note. Assume the chorus sings the melody with a uniform sound level of 75.0 dB. Find the pressure amplitudes of (c) the initial note and (d) the final note. Find the displacement amplitudes of (e) the initial note and (f) the final note.

44. Show that the difference between decibel levels β_1 and β_2 of a sound is related to the ratio of the distances r_1 and r_2 from the sound source by $\beta_2 - \beta_1 = 20 \log\left(\frac{r_1}{r_2}\right)$.

Additional problems

45. A sinusoidal wave is described by the wave function $y = 0.25 \sin (0.30x - 40t)$ where x and y are in metres and t is in seconds. Determine for this wave (a) the amplitude, (b) the angular frequency, (c) the angular wave number, (d) the wavelength, (e) the wave speed, and (f) the direction of motion.

46. A sinusoidal wave in a rope is described by the wave function

$$y = 0.20 \sin (0.75\pi x + 18\pi t)$$

where x and y are in metres and t is in seconds. The rope has a linear mass density of 0.250 kg/m. The tension in the rope is provided by an arrangement like the one illustrated in **Figure P17.46**. What is the mass of the suspended object?

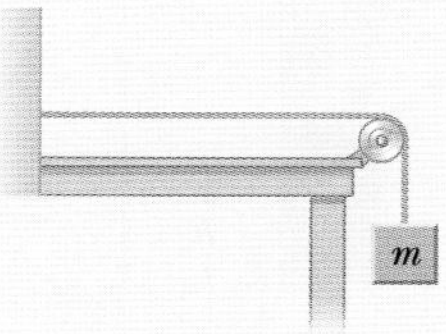

Figure P17.46

47. A taut rope has a mass of 0.180 kg and a length of 3.60 m. What power must be supplied to the rope so as to generate sinusoidal waves having an amplitude of 0.100 m and a wavelength of 0.500 m and travelling with a speed of 30.0 m/s?

48. A block of mass M hangs from a rubber cord. The block is supported so that the cord is not stretched. The unstretched length of the cord is L_0, and its mass is m, much less than M. The spring constant for the cord is k. The block is released and stops momentarily at the lowest point. (a) Determine the tension in the cord when the block is at this lowest point. (b) What is the length of the cord in this 'stretched' position? (c) If the block is held in this lowest position, find the speed of a transverse wave in the cord.

49. The highest note written for a singer in a published score was F-sharp above high C, 1.480 kHz, for Zerbinetta in the original version of Richard Strauss's opera *Ariadne auf Naxos*. (a) Find the wavelength of this sound in air. (b) Suppose people in the fourth row of seats hear this note with level 81.0 dB. Find the displacement amplitude of the sound. (c) **What If?** In response to complaints, Strauss later transposed the note down to F above high C, 1.397 kHz. By what increment did the wavelength change?

50. Assume a 150 W loudspeaker broadcasts sound equally in all directions and produces sound with a level of 103 dB at a distance of 1.60 m from its centre. (a) Find its sound power output. If a salesman claims the speaker is rated at 150 W, he is referring to the maximum electrical power input to the speaker. (b) Find the efficiency of the speaker, that is, the fraction of input power that is converted into useful output power.

51. The tensile stress in a thick copper bar is 99.5% of its elastic breaking point of 13.0×10^{10} N/m². If a 500 Hz sound wave is transmitted through the material, (a) what displacement amplitude will cause the bar to break? (b) What is the maximum speed of the elements of copper at this moment? (c) What is the sound intensity in the bar?

52. A flowerpot is knocked off a window ledge from a height $d = 20.0$ m above the sidewalk as shown in **Figure P17.52**. It falls towards an unsuspecting man of height $h = 1.75$ m who is standing below. Assume the man requires a time interval of $\Delta t = 0.300$ s to respond to the warning. How close to the sidewalk can the flowerpot fall before it is too late for a warning shouted from the balcony to reach the man in time?

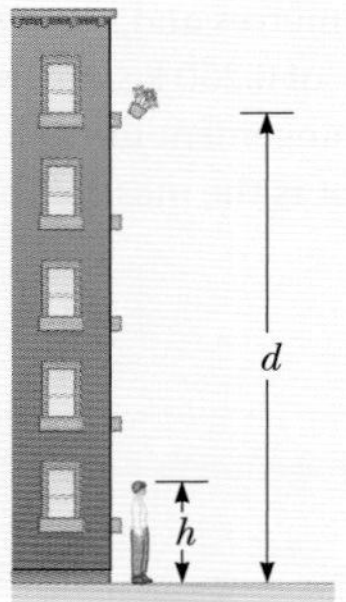

Figure P17.52

53. For a certain type of steel, stress is always proportional to strain with Young's modulus 20×10^{10} N/m^2. The steel has density 7.86×10^3 kg/m^3. It will fail by bending permanently if subjected to compressive stress greater than its yield strength $\sigma_y = 400$ MPa. A rod 80.0 cm long, made of this steel, is fired at 12.0 m/s straight at a very hard wall. (a) The speed of a one-dimensional compressional wave moving along the rod is given by

$$v = \sqrt{Y/\rho}$$

where Y is Young's modulus for the rod and ρ is the density. Calculate this speed. (b) After the front end of the rod hits the wall and stops, the back end of the rod keeps moving as described by Newton's first law until it is stopped by excess pressure in a sound wave moving back through the rod. What time interval elapses before the back end of the rod receives the message that it should stop? (c) How far has the back end of the rod moved in this time interval? Find (d) the strain and (e) the stress in the rod. (f) If it is not to fail, what is the maximum impact speed a rod can have in terms of σ_y, Y and ρ?

54. An undersea earthquake or a landslide can produce an ocean wave of short duration carrying great energy, called a tsunami. When its wavelength is large compared to the ocean depth d, the speed of a water wave is given approximately by $v = \sqrt{gd}$. Assume an earthquake occurs all along a tectonic plate boundary running north to south and produces a straight tsunami wave crest moving everywhere to the west. (a) What physical quantity can you consider to be constant in the motion of any one-wave crest? (b) Explain why the amplitude of the wave increases as the wave approaches shore. (c) If the wave has amplitude 1.80 m when its speed is 200 m/s, what will be its amplitude where the water is 9.00 m deep? (d) Explain why the amplitude at the shore should be expected to be still greater, but cannot be meaningfully predicted by your model.

55. With particular experimental methods, it is possible to produce and observe in a long, thin rod both a transverse wave whose speed depends primarily on tension in the rod and a longitudinal wave whose speed is determined by Young's modulus and the density of the material according to the expression $v = \sqrt{Y/\rho}$. The transverse wave can be modelled as a wave in a stretched string. A particular metal rod is 150 cm long and has a radius of 0.200 cm and a mass of 50.9 g. Young's modulus for the material is 6.80×10^{10} N/m^2. What must the tension in the rod be if the ratio of the speed of longitudinal waves to the speed of transverse waves is 8.00?

56. A rope of total mass m and length L is suspended vertically. Waves above a short distance from the free end of the rope can be represented to a good approximation by the linear wave equation. Find an expression for the wave speed at any point a distance x from the lower end by considering the rope's tension as resulting from the weight of the segment below that point and sketch the wave speed as a function of distance, x.

57. An aluminium wire is held between two clamps under zero tension at room temperature. Reducing the temperature, which results in a decrease in the wire's equilibrium length, increases the tension in the wire. Taking the cross-sectional area of the wire to be 5.00×10^{-6} m^2, the density to be 2.70×10^3 kg/m^3, and Young's modulus to be 7.00×10^{10} N/m^2, what strain ($\Delta L/L$) results in a transverse wave speed of 100 m/s?

58. A pulse travelling along a string of linear mass density μ is described by the wave function

$$y = [A_0 e^{-bx}] \sin (kx - \omega t)$$

where the factor in brackets is said to be the amplitude. (a) What is the power $P(x)$ carried by this wave at a point x? (b) What is the power $P(0)$ carried by this wave at the origin? (c) Calculate the ratio $P(x)/P(0)$.

59. A bat moving at 5.00 m/s is chasing a flying insect. If the bat emits a 40.0 kHz chirp and receives back an echo at 40.4 kHz, (a) what is the speed of the insect? (b) Will the bat be able to catch the insect? Explain your answer.

60. A large meteoroid enters the Earth's atmosphere at a speed of 20.0 km/s and is not significantly slowed before entering the ocean. (a) What is the Mach angle of the shock wave from the meteoroid in the lower atmosphere? (b) If we assume the meteoroid survives the impact with the ocean surface, what is the (initial) Mach angle of the shock wave the meteoroid produces in the water?

Challenge problems

61. The Doppler equation presented in the text is valid when the motion between the observer and the source occurs on a straight line so that the source and observer are moving either directly towards or directly away from each other. (a). Find an expression for the observed frequency when the direction of travel of the observer and source are moving at angles θ_O and θ_S respectively to the direction of propagation of the wave. Use your expression to solve the following problem: A train moves at a constant speed of $v = 25.0$ m/s towards an intersection with a level crossing. A car is stopped near the crossing, 30.0 m from the tracks. The road runs perpendicular to the train tracks. The train's horn emits a frequency of 500 Hz when the train is 40.0 m from the intersection. (b) What is the frequency heard by the passengers in the car? (c) If the train emits this sound continuously and the car is stationary at this position long before the train arrives until long after it leaves, what range of frequencies do passengers in the car hear? (d) Suppose the car is foolishly trying to beat the train to the intersection and is travelling at 40.0 m/s towards the tracks. When the car is 30.0 m from the tracks and the train is 40.0 m from the intersection, what is the frequency heard by the passengers in the car now?

62. In Section 17.5, we derived the speed of sound in a gas using the impulse–momentum theorem applied to the cylinder of gas in **Figure 17.17**. Let us find the speed of sound in a gas using a different approach based on the element of gas in **Figure 17.13**. Proceed as follows. (a) Draw a force diagram for this element showing the forces exerted on the left and right surfaces due to the pressure of the gas on either side of the element. (b) By applying Newton's second law to the element, show that

$$-\frac{(\partial \Delta P)}{\partial x} A \Delta x = \rho A \,\Delta x \frac{\partial^2 s}{\partial t^2}$$

(c) By substituting $\Delta P = -(B\,\partial s/\partial x)$ (**Equation 17.20**), derive the following wave equation for sound:

$$\frac{B}{\rho}\frac{\partial^2 s}{\partial x^2} = \frac{\partial^2 s}{\partial t^2}$$

(d) Substitute into the wave equation the trial solution $s(x, t) = s_{max} \cos (kx - \omega t)$. Show that this function satisfies the wave equation, provided $\omega/k = v = \sqrt{B/\rho}$.

63. Assume an object of mass M is suspended from the bottom of the rope of mass m and length L in problem 56. (a) Show that the time interval for a transverse pulse to travel the length of the rope is

$$\Delta t = 2\sqrt{\frac{L}{mg}}\left(\sqrt{M+m} - \sqrt{M}\right)$$

(b) Find an expression for the wave speed at any point x from the lower end. (c) Show that for $m \ll M$, the expression in part (a) reduces to

$$\Delta t = \sqrt{\frac{mL}{Mg}}$$

64. If a loop of chain is spun at high speed, it can roll along the ground like a circular hoop without collapsing. Consider a chain of uniform linear mass density μ whose centre of mass travels to the right at a high speed v_0 as shown in **Figure P17.64**. (a) Determine the tension in the chain in terms of μ and v_0. Assume the weight of an individual link is negligible compared to the tension. (b) If the loop rolls over a small bump, the resulting deformation of the chain causes two transverse pulses to propagate along the chain, one moving clockwise and one moving anticlockwise. What is the speed of the pulses travelling along the chain? (c) Through what angle does each pulse travel during the time interval over which the loop makes one revolution?

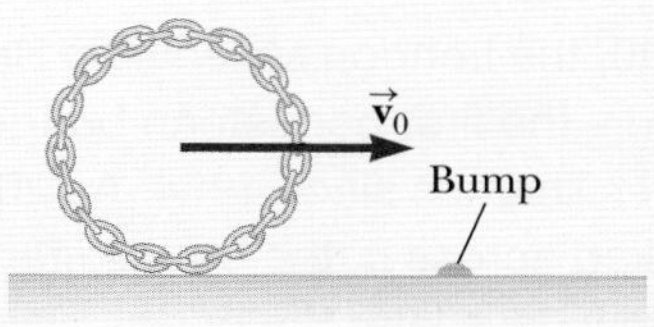

Figure P17.64

65. A string on a musical instrument is held under tension T and extends from the point $x = 0$ to the point $x = L$. The string is overwound with wire in such a way that its mass per unit length $\mu(x)$ increases uniformly from μ_0 at $x = 0$ to μ_L at $x = L$. (a) Find an expression for $\mu(x)$ as a function of x over the range $0 \le x \le L$. (b) Find an expression for the time interval required for a transverse pulse to travel the length of the string.

chapter 18

Superposition and interference

In June 2000, the London Millennium Bridge was opened – the first pedestrian bridge built over the river Thames since 1984. As 2000 people walked across it, the bridge started to wobble. A few days later the bridge was closed. Was it the synchronous movement from the pedestrians' steps, the load or the wind that caused the bridge to resonate?

Shutterstock.com/piotreknik

On completing this chapter, students will understand:

- how waves behave at boundaries
- when a reflected wave is inverted and when it is not
- that waves superimpose when they occupy the same space
- how interference results from superposition of waves
- how constraining a wave under boundary conditions leads to quantisation
- that standing waves are the result of interference
- that waves can interfere spatially and temporally
- why beats occur.

Students will be able to:

- analyse situations in which multiple waves are present
- apply the interference of waves analysis model to analyse interference
- calculate the frequencies and wavelengths of standing waves
- distinguish between spatial and temporal interference
- analyse temporal interference involving beats
- interpret a Fourier spectrum graph
- describe how different instruments produce different sounds
- explain how a synthesiser produces the sounds of different instruments.

ENHANCED WebAssign

The problems found in this chapter may be assigned online in Enhanced Web Assign.

The wave model for mechanical waves was introduced in the previous chapter. In this chapter we shall look at what happens when a wave is incident on an interface between two media. We will also look at what happens when more than one wave is present in a medium – the phenomenon of interference. We shall introduce the idea of quantisation, which is important in understanding the way in which many musical instruments work and which we will meet again when we study quantum physics in Chapter 39.

18.1 Reflection and transmission

The travelling wave model in Chapter 17 describes waves travelling through a uniform medium without interacting with anything along the way. We now consider how a travelling wave is affected when it encounters a change in the medium. For example, consider a pulse travelling on a string that is rigidly attached to a support at one end, as in **Active Figure 18.1**. When the pulse reaches the support, a severe change in the medium occurs: the string ends. As a result, the pulse undergoes **reflection**, that is, the pulse moves back along the string in the opposite direction.

Notice that the reflected pulse is *inverted*. This inversion can be explained as follows. When the pulse reaches the fixed end of the string, the string produces an upwards force on the support. By Newton's third law, the support must exert an equal-magnitude and oppositely directed (downwards) reaction force on the string. This downwards force causes the pulse to invert upon reflection.

Now consider another case. This time, the pulse arrives at the end of a string that is free to move vertically as in **Active Figure 18.2**. The tension at the free end is maintained because the string is tied to a ring of negligible mass that is free to slide vertically on a smooth post without friction. Again, the pulse is reflected, but this time it is not inverted. When it reaches the post, the pulse exerts a force on the free end of the string, causing the ring to accelerate upwards. The ring rises as high as the incoming pulse, and then the downwards component of the tension force pulls the ring back down. This movement of the ring produces a reflected pulse that is not inverted and that has the same amplitude as the incoming pulse.

Finally, consider a situation in which the boundary is intermediate between these two extremes. In this case, part of the energy in the incident pulse is reflected and part undergoes **transmission**; that is, some of the energy passes through the boundary. For instance, suppose a light string is attached to a heavier string as in **Active Figure 18.3**. When a pulse travelling on the light string reaches the boundary between the two strings, part of the pulse is reflected and inverted and part is transmitted to the heavier string. The reflected pulse is inverted for the same reasons described earlier in the case of the string rigidly attached to a support.

The reflected pulse has a smaller amplitude than the incident pulse. In Section 17.7, we showed that the energy carried by a wave is related to its amplitude. According to the principle of conservation of energy, when the pulse breaks up into a reflected pulse and a transmitted pulse at the boundary, the sum of the energies of these two pulses must equal the energy of the incident pulse. Because the reflected pulse contains only part of the energy of the incident pulse, its amplitude must be smaller.

When a pulse travelling on a heavy string strikes the boundary between the heavy string and a lighter one as in **Active Figure 18.4**, again part is reflected and part is transmitted. In this case, the reflected pulse is not inverted.

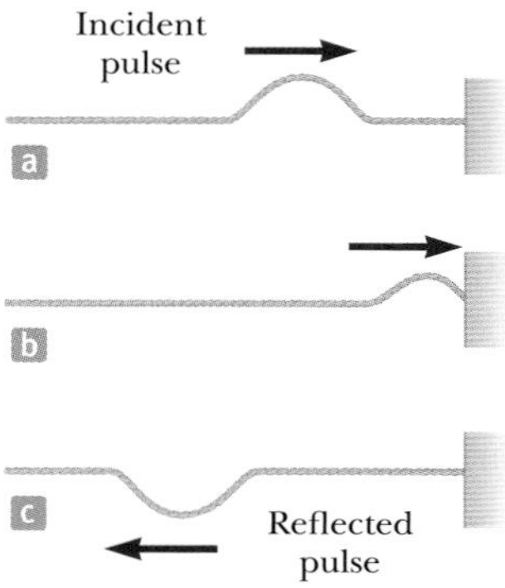

Active Figure 18.1

The reflection of a travelling pulse at the fixed end of a stretched string. The reflected pulse is inverted, but its shape is otherwise unchanged.

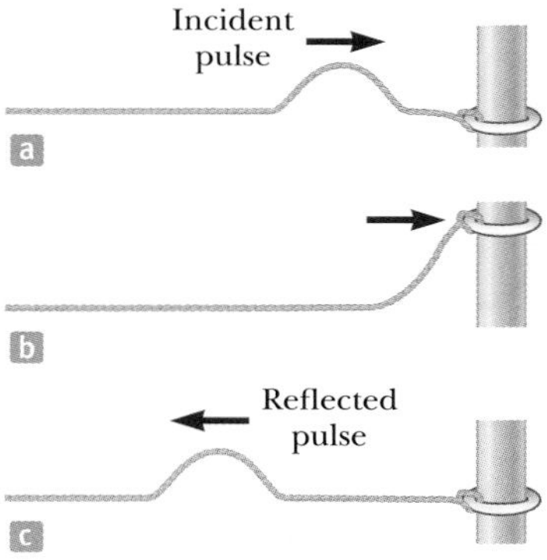

Active Figure 18.2

The reflection of a travelling pulse at the free end of a stretched string. The reflected pulse is not inverted.

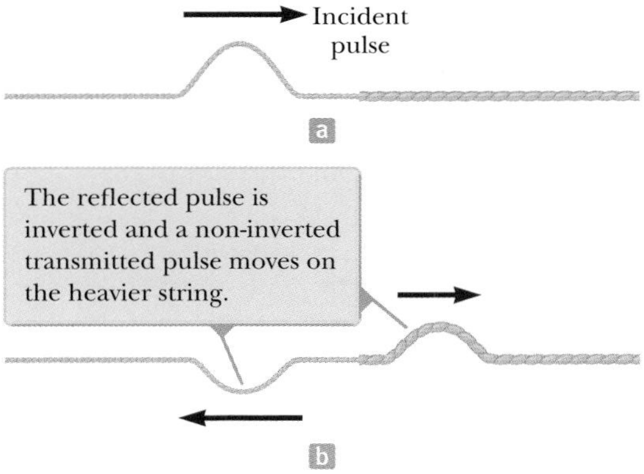

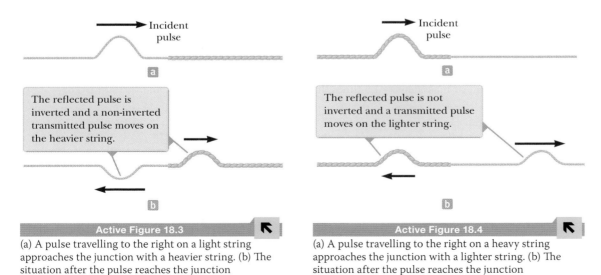

Active Figure 18.3

(a) A pulse travelling to the right on a light string approaches the junction with a heavier string. (b) The situation after the pulse reaches the junction

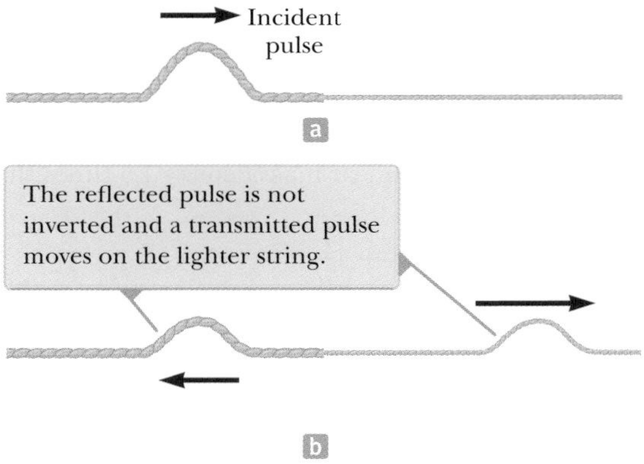

Active Figure 18.4

(a) A pulse travelling to the right on a heavy string approaches the junction with a lighter string. (b) The situation after the pulse reaches the junction

In both cases, the relative heights of the reflected and transmitted pulses depend on the relative densities of the two strings. If the strings are identical, there is no discontinuity at the boundary and no reflection takes place.

According to **Equation 17.28**, a wave travels more slowly on a heavy string than on a light string if both are under the same tension. The following general rules apply to reflected waves:

Determining when a reflected wave is inverted ▶

When a wave travels from medium A to medium B and $v_A > v_B$ (that is, when B is denser than A), it is inverted upon reflection. When a wave or pulse travels from medium A to medium B and $v_A < v_B$ (that is, when A is denser than B), it is not inverted upon reflection. The transmitted wave is never inverted.

TRY THIS

Hold one end of a large spring such as a slinky and get a friend to hold the other end steady and not allow it to move. Send a transverse pulse towards them and watch what happens when it reflects back towards you. Try attaching one end of the spring to a second, different spring or a length of rope and repeat the experiment. What effects can you observe now?

Pitfall Prevention 18.1

Don't forget that wave speed is determined by the properties of the medium, *not* by the frequency and wavelength of the wave. Be careful how you apply Equation 17.12, $v = f\lambda$.

Quick Quiz 18.1

Two lengths of rope are tied together. They are made of the same material, but one has twice the diameter of the other. The tension in both is the same. A wave is sent along from the thinner to the thicker section. **(i)** By what factor does the wave speed decrease? **(a)** $\sqrt{2}$ **(b)** 2 **(c)** 4 **(d)** it does not change. **(ii)** Which wave is inverted? **(a)** the transmitted wave **(b)** the reflected wave **(c)** both **(d)** neither

Example 18.1

A damaged guitar string

A steel guitar string has a steel inner core that is wound around with a chromed steel wire. A particular string has become worn and the winding has come off it so that part of the string has a density one and a half times that of the other part. Consider a wave travelling along such a string, and coming to the interface between the damaged and undamaged sections. If the wavelength, speed and frequency in the undamaged section are λ_o, v_o and f_o respectively, what are the wavelength, speed and frequency of the transmitted part of the wave in the damaged section?

Solution

Conceptualise Begin by drawing a diagram (**Figure 18.5**). The wave moves from an area of higher linear mass density to lower linear mass density, so we expect the wave speed to increase for the transmitted part of the wave. We also expect part of the wave to be reflected.

Model We assume the tension is the same in both parts of the string and we model the pulse as a travelling wave so that we can apply the travelling wave analysis model from Chapter 17. We will also need to use ideas from this chapter.

Analyse The wave will be partly transmitted and partly reflected. The transmitted part will have a different speed to the incident wave as the medium has a linear mass density 1.5 times that in the damaged section. From **Equation 17.28**:

$$v = \sqrt{\frac{T}{\mu}}$$

We see that if $\mu \rightarrow 2/3\mu$, $v \rightarrow \sqrt{3T/2\mu}$ so the new speed is

$$v_{\text{transmitted}} = \sqrt{\frac{3}{2}}\, v_{\text{incident}}$$

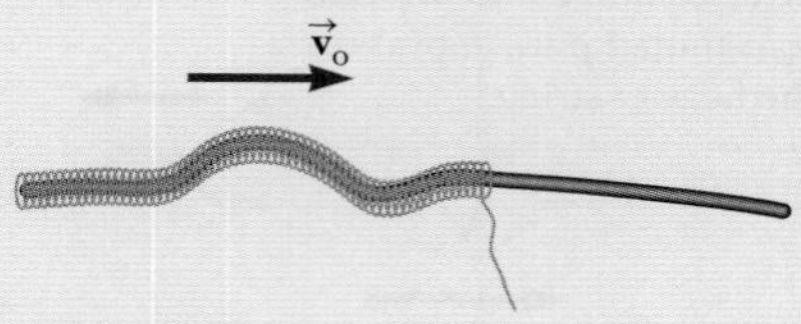

Figure 18.5
(Example 18.1)

Example 18.1 cont.

The frequency of the transmitted wave must be equal to that of the incident wave, as it is the oscillations of the undamaged part of the string that act as the source of the wave in the damaged part. Hence $f_{\text{transmitted}} = f_{\text{incident}}$.

The wavelength of the transmitted wave will be determined by the frequency and velocity of the transmitted wave, according to **Equation 17.12**, $v = \lambda f$:

$$\lambda_{\text{transmitted}} = (v_{\text{transmitted}})(f_{\text{transmitted}}) = \left(\sqrt{\frac{3}{2}}\, v_{\text{incident}}\right)(f_{\text{incident}}) = \sqrt{\frac{3}{2}}\,(v_{\text{incident}})(f_{\text{incident}})$$

Hence $$\lambda_{\text{transmitted}} = \sqrt{\frac{3}{2}}\,\lambda_{\text{incident}}$$

Finalise The transmitted wave has the same frequency as the incident wave, because this is determined by the frequency of oscillation of the source of the wave. The incident wave is the source of oscillation at the boundary of the new medium, so both waves must have the same frequency. However, the change in wave speed in the new medium causes a change in wavelength proportional to the change in speed.

18.2 Analysis model: interference of waves

Many interesting wave phenomena in nature cannot be described by a single travelling wave. Instead, one must analyse these phenomena in terms of a combination of travelling waves. As noted in the introduction, waves have a remarkable difference from particles in that waves can be combined at the *same* location in space. To analyse such wave combinations, we make use of the **superposition principle**:

> If two or more travelling waves are moving through a medium, the resultant value of the wave function at any point is the algebraic sum of the values of the wave functions of the individual waves.

◄ Superposition principle

Waves that obey this principle are called *linear waves*. In the case of mechanical waves, linear waves are generally characterised by having amplitudes much smaller than their wavelengths. Waves that violate the superposition principle are called *non-linear waves* and are often characterised by large amplitudes. We will deal only with linear waves.

One consequence of the superposition principle is that two travelling waves can pass through each other without being destroyed or even altered. For instance, when two pebbles are thrown into a pond and hit the surface at different locations, the expanding circular surface waves from the two locations simply pass through each other with no permanent effect. The resulting complex pattern can be viewed as two independent sets of expanding circles.

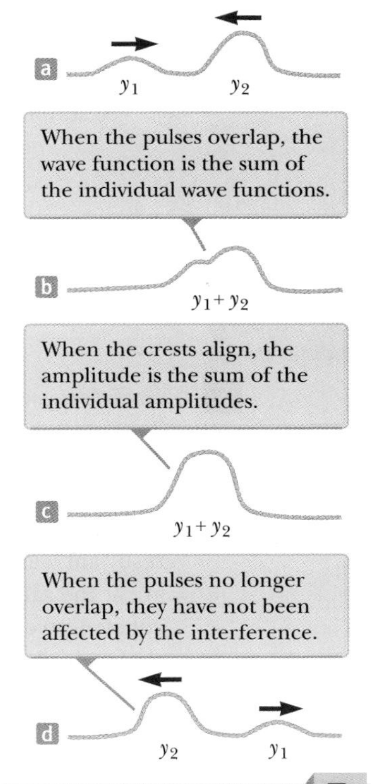

a
y_1
y_2
When the pulses overlap, the wave function is the sum of the individual wave functions.
b
$y_1 + y_2$
When the crests align, the amplitude is the sum of the individual amplitudes.
c
$y_1 + y_2$
When the pulses no longer overlap, they have not been affected by the interference.
d
y_2
y_1

Active Figure 18.6
Constructive interference. Two positive pulses travel on a stretched string in opposite directions and overlap.

Constructive and destructive interference

Active Figure 18.6 is a pictorial representation of the superposition of two pulses. The wave function for the pulse moving to the right is y_1, and the wave function for the pulse moving to the left is y_2. The pulses have the same speed but different shapes and the displacement of the elements of the medium is in the positive y direction for both pulses. When the waves overlap (**Active Figure 18.6b**), the wave function for the resulting complex wave is given by $y_1 + y_2$. When the crests of the pulses coincide (**Active Figure 18.6c**), the resulting wave given by $y_1 + y_2$ has a larger amplitude than that of the individual pulses. The two pulses finally separate and continue moving in their original directions (**Active Figure 18.6d**). Notice that the pulse shapes remain unchanged after the interaction, as if the two pulses had never met!

The combination of separate waves in the same region of space to produce a resultant wave is called **interference**. For the two pulses shown in **Active Figure 18.6**, the displacement of the elements of the medium is in the positive y direction for both pulses and the resultant pulse (created when the individual pulses overlap) exhibits an amplitude greater than that of either individual pulse. Because the displacements caused by the two pulses are in the same direction, we refer to their superposition as **constructive interference**.

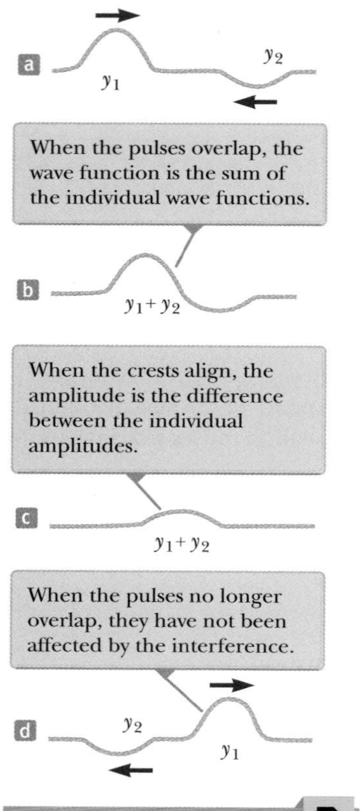
a
y_2
y_1
When the pulses overlap, the wave function is the sum of the individual wave functions.
b
$y_1 + y_2$
When the crests align, the amplitude is the difference between the individual amplitudes.
c
$y_1 + y_2$
When the pulses no longer overlap, they have not been affected by the interference.
d
y_2
y_1

Active Figure 18.7
Destructive interference. Two pulses, one positive and one negative, travel on a stretched string in opposite directions and overlap.

Now consider two pulses travelling in opposite directions on a taut string where one pulse is inverted relative to the other as shown in **Active Figure 18.7**. When these pulses begin to overlap, the resultant pulse is given by $y_1 + y_2$, but the values of the function y_2 are negative. Again, the two pulses pass through each other; because the displacements caused by the two pulses are in opposite directions; however, we refer to their superposition as **destructive interference**. Note that neither wave is destroyed in destructive interference – both continue on unchanged once they have passed each other.

The superposition principle is central to the analysis model called **interference of waves**. In many situations, both in acoustics and optics, waves combine according to this principle and exhibit interesting phenomena with many practical applications.

Quick Quiz 18.2

Two pulses move in opposite directions on a string and are identical in shape except that one has positive displacements of the elements of the string and the other has negative displacements. At the moment the two pulses completely overlap on the string, what happens? **(a)** The energy associated with the pulses has disappeared. **(b)** The string forms a straight line. **(c)** The pulses have vanished and will not reappear.

Superposition of sinusoidal waves

Let us now apply the principle of superposition to two sinusoidal waves travelling in the same direction in a linear medium. If the two waves are travelling to the right and have the same frequency, wavelength and amplitude but differ in phase, we can express their individual wave functions as

$$y_1 = A\sin(kx - \omega t) \quad \text{and} \quad y_2 = A\sin(kx - \omega t + \phi)$$

where, as usual, $k = 2\pi/\lambda$, $\omega = 2\pi f$, and ϕ is the phase constant as discussed in Section 17.2. Hence, the resultant wave function y is

$$y = y_1 + y_2 = A\left[\sin(kx - \omega t) + \sin(kx - \omega t + \phi)\right]$$

To simplify this expression, we use the trigonometric identity

$$\sin a + \sin b = 2\cos\left(\frac{a-b}{2}\right)\sin\left(\frac{a+b}{2}\right)$$

Letting $a = kx - \omega t$ and $b = kx - \omega t + \phi$, we find that the resultant wave function y reduces to

Resultant ▶ of two travelling sinusoidal waves

$$y = 2A\cos\left(\frac{\phi}{2}\right)\sin\left(kx - \omega t + \frac{\phi}{2}\right)$$

Σ *Useful trigonometric identities are given in Appendix B.4.*

This result has several important features. The resultant wave function y is also sinusoidal and has the same frequency and wavelength as the individual waves because the sine function incorporates the same values of k and ω that appear in the original wave functions. The amplitude of the resultant wave is $2A\cos(\phi/2)$ and its phase is $\phi/2$. If the phase constant ϕ equals 0, then $\cos(\phi/2) = \cos 0 = 1$ and the amplitude of the resultant wave is $2A$, twice the amplitude of either individual wave. In this case, the crests of the two waves are at the same locations in space and the waves are said to be everywhere *in phase* and therefore interfere constructively. The individual waves y_1 and y_2 combine to form the red-brown curve y of amplitude $2A$ shown in **Active Figure 18.8a**. Because the individual waves are in phase, they are indistinguishable in **Active Figure 18.8a**, where they appear as a single blue curve. In general, constructive interference occurs when $\cos(\phi/2) = \pm 1$. That is true, for example, when $\phi = 0, 2\pi, 4\pi, \ldots$ rad, that is, when ϕ is an *even* multiple of π.

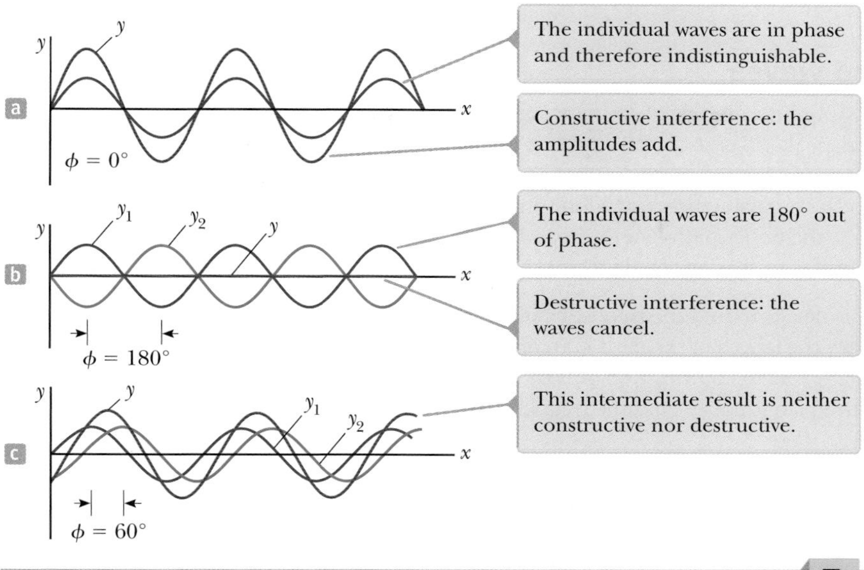
a
y
y
x
$\phi = 0°$
The individual waves are in phase and therefore indistinguishable.
Constructive interference: the amplitudes add.
b
y
y_1
y_2
y
x
$\phi = 180°$
The individual waves are 180° out of phase.
Destructive interference: the waves cancel.
c
y
y
y_1
y_2
x
$\phi = 60°$
This intermediate result is neither constructive nor destructive.

Active Figure 18.8

The superposition of two identical waves y_1 and y_2 (blue and green, respectively) to yield a resultant wave (red-brown)

When ϕ is equal to π rad or to any *odd* multiple of π, then $\cos(\phi/2) = \cos(\pi/2) = 0$ and the crests of one wave occur at the same positions as the troughs of the second wave (**Active Figure 18.8b**). Therefore, as a consequence of destructive interference, the resultant wave has *zero* amplitude everywhere as shown by the straight red-brown line in **Active Figure 18.8b**. Finally, when the phase constant has an arbitrary value other than 0 or an integer multiple of π rad (**Active Figure 18.8c**), the resultant wave has an amplitude whose value is somewhere between 0 and $2A$.

In the more general case in which the waves have the same wavelength but different amplitudes, the results are similar with the following exceptions. In the in-phase case, the amplitude of the resultant wave is not twice that of a single wave, but rather is the sum of the amplitudes of the two waves. When the waves are π radians out of phase, they do not completely cancel as in **Active Figure 18.8b**. The result is a wave whose amplitude is the difference in the amplitudes of the individual waves.

Analysis Model 18.1

Waves in interference

Imagine two waves travelling in the same location through a medium. The displacement of elements of the medium is affected by both waves. According to the **principle of superposition**, the displacement is the sum of the individual displacements that would be caused by each wave. When the waves are in phase, **constructive interference** occurs and the resultant displacement is larger than the individual displacements. **Destructive interference** occurs when the waves are out of phase.

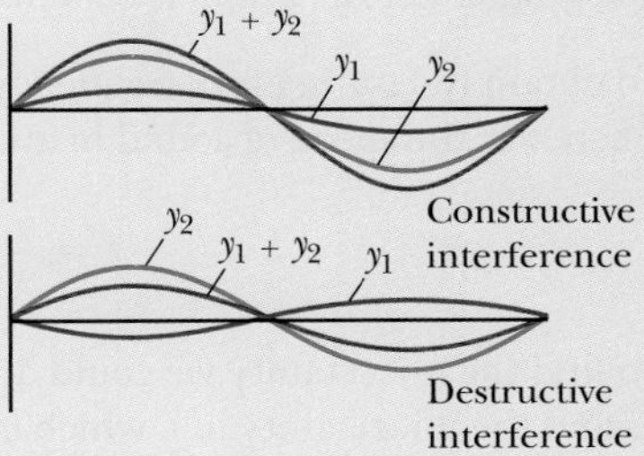
$y_1 + y_2$
y_1
y_2
Constructive interference
y_2
$y_1 + y_2$
y_1
Destructive interference

Examples

- a piano tuner listens to a piano string and a tuning fork vibrating together and notices beats (Section 18.6)
- light waves from two coherent sources combine to form an interference pattern on a screen (Chapter 37)
- a thin film of oil on top of water shows swirls of color (Chapter 37)
- x-rays passing through a crystalline solid combine to form a Laue pattern (Chapter 38)

Example 18.2

Interference of sound waves

A simple device for demonstrating interference of sound waves is illustrated in **Figure 18.9**. Sound from a loudspeaker *S* is sent into a tube at point *P*, where there is a T-shaped junction. Half the sound energy travels in one direction, along a path of length r_1, and half travels in the opposite direction along a path of length r_2. The lower path length r_1 is fixed, but the upper path length r_2 can be varied by sliding the U-shaped tube, which is similar to that on a slide trombone.

A group of students is using this device to find the frequency of a tuning fork of unknown pitch. They slowly vary the length of the path r_2 and find they get maxima in the sound heard at *R* when $r_2 = 1.32$ m, 2.07 m, 2.86 m, 3.64 m and 4.43 m. They forget to write down the measured length of r_1. Find the frequency of the tuning fork.

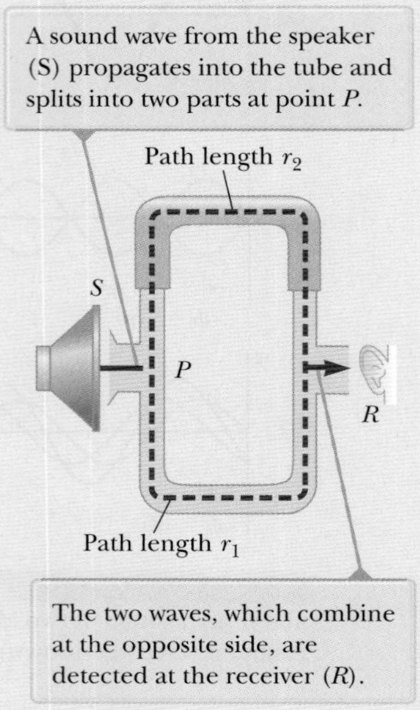
A sound wave from the speaker (S) propagates into the tube and splits into two parts at point P.
Path length r_2
S
P
R
Path length r_1
The two waves, which combine at the opposite side, are detected at the receiver (R).

Figure 18.9
An acoustical system for demonstrating interference of sound waves. The upper path length r_2 can be varied by sliding the upper section.

Solution

Conceptualise In **Figure 18.9**, a sound wave enters a tube and is split into two different paths before recombining at the other end. If the waves are in phase, the sound waves will interfere constructively and the sound will be loud. Examine **Active Figure 18.8** again to help you understand the situation.

Model Because the sound waves from two separate sources combine, we apply the waves in interference analysis model.

Analyse **Figure 18.9** shows the physical arrangement of the system. The sound will be loud when the waves travelling along the two different paths recombine in phase – that is, when the path lengths are different by a whole number of wavelengths so that the peaks can 'line up' when the sound recombines. In other words, for constructive interference we require the path difference $r_2 - r_1 = \Delta r = 0, \lambda, 2\lambda, \ldots n\lambda$. A graph of Δr vs n would have a gradient of λ; however, we do not have a value for r_1, nor can we be sure that the data begins with a path difference of 0. In fact this doesn't matter if we use a graphical method, because a plot of r_2 vs n will still have a gradient of λ, although it will be displaced by (have an intercept of) $(r_1 + m\lambda)$ where m is the number of maxima between the zero path difference and the first maximum recorded by the students. A plot of r_2 vs n is shown below, taking the first measurement as $n = 1$ and labelling the data assuming no maxima were missed by the students.

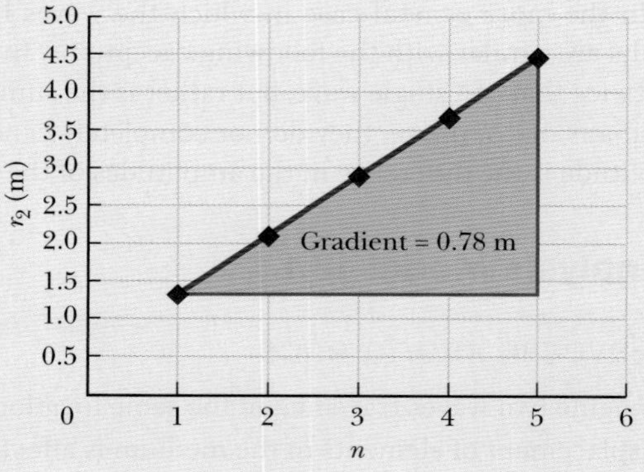
r_2 (m)
5.0
4.5
4.0
3.5
3.0
2.5
2.0
1.5
1.0
0.5
0
1
2
3
4
5
6
n
Gradient = 0.78 m

The gradient of the graph is 0.78 m.

To obtain the tuning fork frequency, use **Equation 17.12**, $v = \lambda f$, where v is the speed of sound in air, 343 m/s:

$$f = \frac{v}{\lambda} = \frac{343 \text{ m/s}}{0.78 \text{ m}} = 440 \text{ Hz, which is the note A above middle C.}$$

To find the uncertainty we could draw lines of maximum and minimum gradient to get a range of gradients, then use this to find the uncertainty in f, which is what the students will have done to ensure they get good marks for the experiment.

Finalise The distance along any path from speaker to receiver is called the path length, r. When the difference in the path lengths $\Delta r = |r_2 - r_1|$ is either zero or some integer multiple of the wavelength λ (that is, $\Delta r = n\lambda$, where $n = 0, 1, 2, 3, \ldots$) the two waves reaching the receiver at any instant are in phase and interfere constructively as shown in **Active Figure 18.8a**. For this case, a maximum in the sound intensity is detected at the receiver. If the path length r_2 is adjusted such that the path difference $\Delta r = \lambda/2, 3\lambda/2, \ldots, n\lambda/2$ (for n odd), the two waves are exactly π rad, or 180°, out of phase at the receiver and hence cancel each other. In this case of destructive interference, no sound is detected at the receiver. This simple experiment demonstrates that a phase difference may arise between two waves generated by the same source when they travel along paths of unequal lengths.

Interference occurs whenever the wave from a single source is split and recombined or when multiple sources are driven in or out of phase. When the sources all produce the same frequency and are in phase then a **path difference** of 0, λ, 2λ, . . . , $n\lambda$ gives constructive interference. A path difference of $\frac{1}{2}\lambda$, $(1 + \frac{1}{2})\,\lambda$, $(2 + \frac{1}{2})\,\lambda$, . . ., $(n + \frac{1}{2})\,\lambda$ gives destructive interference. If the sources do not have the same frequency the situation is more complex, and the waves move in and out of phase giving rise to effects such as beats, as we shall see later in this chapter.

18.3 Standing waves

Consider connecting two speakers to a single sound source. Suppose we turn the speakers so that they face each other and then have them emit sound of the same frequency and amplitude. In this situation, two identical waves travel in opposite directions in the same medium as in Figure 18.10. These waves combine in accordance with the interference of waves model.

We can analyse such a situation by considering wave functions for two transverse sinusoidal waves having the same amplitude, frequency and wavelength but travelling in opposite directions in the same medium:

$$y_1 = A\sin(kx - \omega t) \quad \text{and} \quad y_2 = A\sin(kx + \omega t)$$

$\vec{v}$
$\vec{v}$

Figure 18.10
Two identical loudspeakers emit sound waves towards each other. When they overlap, identical waves travelling in opposite directions will combine to form standing waves.

where y_1 represents a wave travelling in the positive x direction and y_2 represents one travelling in the negative x direction. Adding these two functions gives the resultant wave function y:

$$y = y_1 + y_2 = A\sin(kx - \omega t) + A\sin(kx + \omega t)$$

Using the trigonometric identity $\sin(a \pm b) = \sin a \cos b \pm \cos a \sin b$, this reduces to

$$y = (2A\sin kx)\cos(\omega t) \qquad (18.1)$$

Σ

Useful trigonometric identities are given in Appendix B.4.

Equation 18.1 represents the wave function of a **standing wave**. A standing wave such as the one on a string shown in Figure 18.11 is an oscillation pattern *with a stationary outline* that results from the superposition of two identical waves travelling in opposite directions.

Notice that Equation 18.1 does not contain a function of $(kx - \omega t)$. Therefore, it is not an expression for a single travelling wave. When you observe a standing wave, there is no sense of motion in the direction of propagation of either original wave. Comparing Equation 18.1 with Equation 16.4, we see that it describes a special kind of simple harmonic motion. Every element of the medium oscillates in simple harmonic motion with the same angular frequency ω (according to the $\cos\omega t$ factor in the equation). The amplitude of the simple harmonic motion of a given element (given by the factor $2A\sin kx$, the coefficient of the cosine function) depends on the location x of the element in the medium.

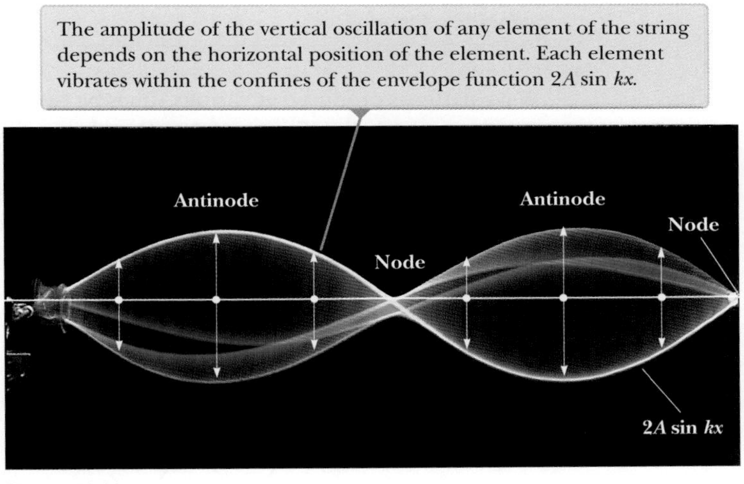

The amplitude of the vertical oscillation of any element of the string depends on the horizontal position of the element. Each element vibrates within the confines of the envelope function 2*A* sin *kx*.
Antinode
Antinode
Node
Node
2*A* sin *kx*

Figure 18.11
Multiflash photograph of a standing wave on a string. The time behaviour of the vertical displacement from equilibrium of an individual element of the string is given by $\cos\omega t$. That is, each element vibrates at an angular frequency ω.

TRY THIS

Get a large spring such as a slinky again and get a friend to hold one end of it. Shake your end of the slinky up and down and adjust your shaking frequency until every coil of the spring is moving up at the same time and then down. Now try gradually increasing the frequency at which you shake the coil. You should be able to find other standing wave patterns. You can also do this by plucking the centre of a taut wire such as a fence wire, although it is difficult to get more than one standing wave pattern in this case.

In the *Try this* example a standing wave is formed from the combination of waves moving away from your hand and reflected from the fixed end back towards your hand. Notice that there is no sense of travelling along the spring like there was for the pulse. You only see up-and-down motion of the elements of the spring.

Equation 18.1 shows that the amplitude of the simple harmonic motion of an element of the medium has a minimum value of zero when x satisfies the condition $\sin kx = 0$, that is, when

$$kx = 0,\ \pi,\ 2\pi,\ 3\pi,\ \ldots$$

Because $k = 2\pi/\lambda$, these values for kx give

Positions of ▶ nodes

$$x = 0, \frac{\lambda}{2}, \lambda, \frac{3\lambda}{2}, \ldots \frac{n\lambda}{2} \qquad n = 0, 1, 2, 3, \ldots \tag{18.2}$$

These points of zero amplitude are called **nodes**.

The elements of the medium with the *greatest* possible displacement from equilibrium have an amplitude of $2A$, which we define as the amplitude of the standing wave. The positions in the medium at which this maximum displacement occurs are called **antinodes**. The antinodes are located at positions for which the coordinate x satisfies the condition $\sin kx = \pm 1$, that is, when

$$kx = \frac{\pi}{2}, \frac{3\pi}{2}, \frac{5\pi}{2}, \ldots$$

Therefore, the positions of the antinodes are given by

Positions of ▶ antinodes

$$x = \frac{\lambda}{4}, \frac{3\lambda}{4}, \frac{5\lambda}{4}, \ldots \frac{n\lambda}{4} \qquad n = 1, 3, 5, \ldots \tag{18.3}$$

Two nodes and two antinodes are labelled in the standing wave in **Figure 18.11**. The light blue curve labelled $2A \sin kx$ in **Figure 18.11** represents one wavelength of the travelling waves that combine to form the standing wave.

Quick Quiz 18.3

In terms of λ, what is the distance between adjacent antinodes? What is the distance between a node and an adjacent antinode?

Wave patterns produced at various times by two transverse travelling waves moving in opposite directions are shown in **Active Figure 18.12**. The blue and green curves are the wave patterns for the individual travelling waves, and the red-brown curves are the wave patterns for the resultant standing wave. At $t = 0$ (**Active Figure 18.12a**), the two travelling waves are in phase, giving a wave pattern in which each element of the medium is experiencing its maximum displacement from equilibrium. One-quarter of a period later, at $t = T/4$ (**Active Figure 18.12b**), the travelling waves have moved one-quarter of a wavelength (one to the

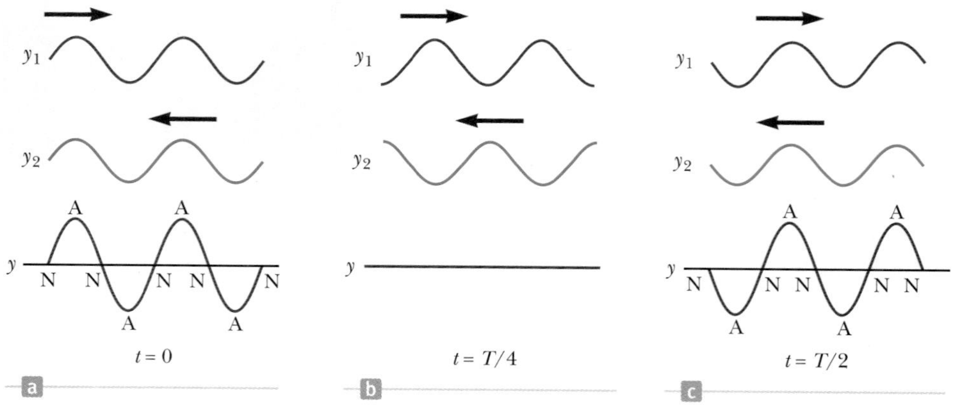

Active Figure 18.12

Standing-wave patterns produced at various times by two waves of equal amplitude travelling in opposite directions. For the resultant wave y, the nodes (N) are points of zero displacement and the antinodes (A) are points of maximum displacement.

right and the other to the left). At this time, the travelling waves are out of phase, and each element of the medium is passing through the equilibrium position in its simple harmonic motion. The result is zero displacement for elements at all values of x; that is, the wave pattern is a straight line. At $t = T/2$ (**Active Figure 18.12c**), the travelling waves are again in phase, producing a wave pattern that is inverted relative to the $t = 0$ pattern. In the standing wave, the elements of the medium alternate in time between the extremes shown in **Active Figures 18.12a** and **18.12c**.

Pitfall Prevention 18.2

We need to distinguish carefully here between the amplitude of the individual waves, which is A; the amplitude of the standing wave, which is $2A$; and the amplitude of the simple harmonic motion of the elements of the medium, which is $2A \sin kx$. A given element in a standing wave vibrates within the constraints of the *envelope* function $2A \sin kx$, where x is that element's position in the medium. Such vibration is in contrast to travelling sinusoidal waves, in which all elements oscillate with the amplitude A of the wave.

Quick Quiz 18.4

Consider the waves in **Active Figure 18.12** to be waves on a stretched string. Define the velocity of elements of the string as positive if they are moving upwards in the figure. (i) At the moment the string has the shape shown by the red-brown curve in **Active Figure 18.12a**, what is the instantaneous velocity of elements along the string? **(a)** zero for all elements **(b)** positive for all elements **(c)** negative for all elements **(d)** varies with the position of the element **(ii)** From the same choices, at the moment the string has the shape shown by the red-brown curve in **Active Figure 18.12b**, what is the instantaneous velocity of elements along the string?

Example 18.3

Formation of a standing wave

Two waves travel in opposite directions along a stretched slinky to produce a standing wave. The individual wave functions are

$$y_1 = 4.0 \sin(3.0x - 2.0t)$$

$$y_2 = 4.0 \sin(3.0x + 2.0t)$$

where x and y are measured in centimetres and t is in seconds.

(A) Find the amplitude of the simple harmonic motion of the element of the medium located at $x = 2.3$ cm.

Solution

Conceptualise You may have already experimented with producing such waves if you did the *Try this* example. We can represent the waves graphically by the blue and green curves in **Active Figure 18.12**.

Model The two waves are identical except for their direction of travel, so they will combine to form a standing wave. We can model this standing wave using the equations developed in this section.

Analyse From the equations for the waves, we see that $A = 4.0$ cm, $k = 3.0$ rad/cm, and $\omega = 2.0$ rad/s.

Use **Equation 18.1** to write an expression for the standing wave:

$$y = (2A \sin kx) \cos(\omega t) = 8.0 \sin(3.0x) \cos(2.0t)$$

Find the amplitude of the simple harmonic motion of the element at the position $x = 2.3$ cm by evaluating the coefficient of the cosine function at this position:

$$y_{\max} = (8.0 \text{ cm}) \sin 3.0x\big|_{x=2.3}$$
$$= (8.0 \text{ cm}) \sin(6.9 \text{ rad}) = 4.6 \text{ cm}$$

(B) Find the positions of the nodes and antinodes if one end of the string is at $x = 0$.

Solution

Find the wavelength of the travelling waves:

$$k = \frac{2\pi}{\lambda} = 3.0 \text{ rad/cm} \rightarrow \lambda = \frac{2\pi}{3.0} \text{ cm}$$

Use **Equation 18.2** to find the locations of the nodes:

$$x = n\frac{\lambda}{2} = n\left(\frac{\pi}{3.0}\right) \text{cm} \quad n = 0, 1, 2, 3, \ldots$$

Use **Equation 18.3** to find the locations of the antinodes:

$$x = n\frac{\lambda}{4} = n\left(\frac{\pi}{6.0}\right) \text{cm} \quad n = 1, 3, 5, 7, \ldots$$

18.4 Analysis model: waves under boundary conditions

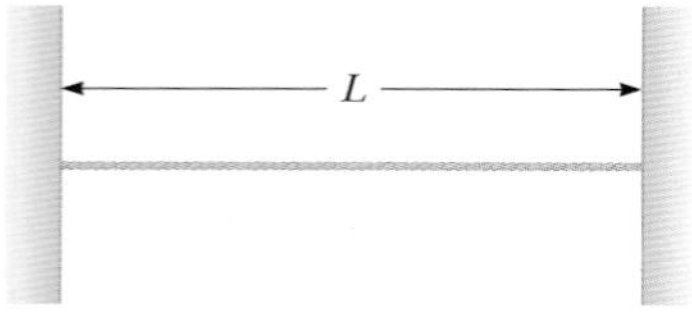

Figure 18.13
A string of length L fixed at both ends

Consider a string of length L fixed at both ends as shown in **Figure 18.13**. Waves can travel in both directions on the string. Therefore, standing waves can be set up in the string by a continuous superposition of waves incident on and reflected from the ends. Notice that there is a *boundary condition* for the waves on the string. Because the ends of the string are fixed, they must necessarily have zero displacement and are therefore nodes by definition. This boundary condition results in the string having a number of discrete natural patterns of oscillation, called **normal modes**, each of which has a characteristic frequency that is easily calculated. This situation in which only certain frequencies of oscillation are allowed is called **quantisation**. Quantisation is a common occurrence when waves are subject to boundary conditions and is a central feature of quantum physics as we shall see in Chapter 39. Notice in **Active Figure 18.12** that there are no boundary conditions, so standing waves of *any* frequency can be established; there is no quantisation without boundary conditions. Because boundary conditions occur so often for waves, we identify an analysis model called **waves under boundary conditions** for the discussion that follows.

The normal modes of oscillation for the string in **Figure 18.13** can be found by imposing the boundary conditions that the ends be nodes. We also know that the distance between a node and adjacent antinode is $\frac{1}{4}\lambda$. The first normal mode that is consistent with these requirements, shown in **Active Figure 18.14a**, has nodes at its ends and one antinode in the middle. This normal mode is the longest-wavelength mode that is consistent with our boundary conditions. The first normal mode occurs when the wavelength λ_1 is equal to twice the length of the string, or $\lambda_1 = 2L$. The section of a standing wave from one node to the next node is called a *loop*. In the first normal mode, the string is vibrating in one loop. In the second normal mode (see **Active Figure 18.14b**), the string vibrates in two loops and the wavelength λ_2 is equal to the length of the string, $\lambda_2 = L$. The third normal mode (see **Active Figure 18.14c**) has three loops and $\lambda_3 = 2L/3$. In general, the wavelengths of the various normal modes for a string of length L fixed at both ends are

Wavelengths of ▶ normal modes

$$\lambda_n = \frac{2L}{n} \qquad n = 1, 2, 3, \ldots \tag{18.4}$$

where the index n refers to the nth normal mode of oscillation. These modes are the *possible* modes of oscillation for the string. The *actual* modes that are excited on a string are discussed shortly.

The natural frequencies associated with the modes of oscillation are obtained from the relationship $f = v/\lambda$, where the wave speed v is the same for all frequencies. Using **Equation 18.4**, we find that the natural frequencies f_n of the normal modes are

Frequencies of ▶ normal modes

$$f_n = \frac{v}{\lambda_n} = n\frac{v}{2L} \qquad n = 1, 2, 3, \ldots \tag{18.5}$$

These natural frequencies are also called the *quantised frequencies* associated with the vibrating string fixed at both ends.

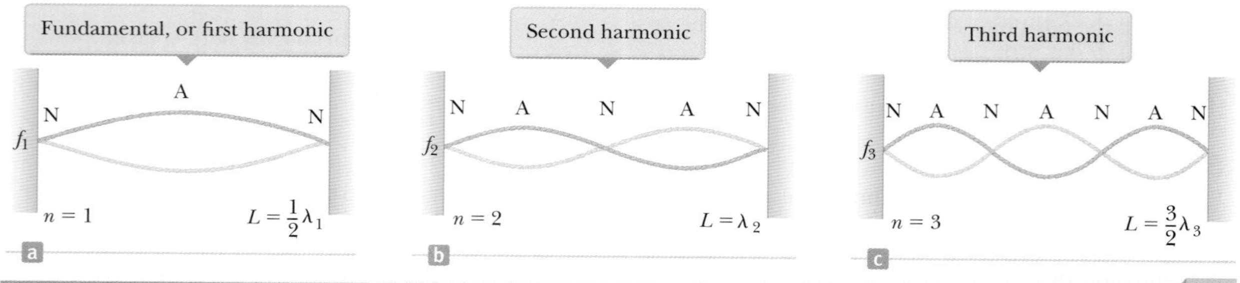

Active Figure 18.14
The normal modes of vibration of the string in **Figure 18.9** form a harmonic series. The string vibrates between the extremes shown.

Because $v = \sqrt{T/\mu}$ (see **Equation 17.28**) for waves on a string, where T is the tension in the string and μ is its linear mass density, we can also express the natural frequencies of a taut string as

$$f_n = \frac{n}{2L}\sqrt{\frac{T}{\mu}} \qquad n = 1, 2, 3, \ldots \tag{18.6}$$

The lowest frequency f_1, which corresponds to $n = 1$, is called either the **fundamental** or the **fundamental frequency** and is given by

$$f_1 = \frac{1}{2L}\sqrt{\frac{T}{\mu}} \tag{18.7}$$

◄ Fundamental frequency of a taut string

The frequencies of the other normal modes are integer multiples of the fundamental frequency. Frequencies of normal modes that exhibit such an integer multiple relationship form a **harmonic series** and the normal modes are called **harmonics**. The fundamental frequency f_1 is the frequency of the first harmonic, the frequency $f_2 = 2f_1$ is the frequency of the second harmonic, and the frequency $f_n = nf_1$ is the frequency of the nth harmonic. Other oscillating systems, such as oscillating sheets of material, for example drumheads, exhibit normal modes, but the frequencies are not related as integer multiples of a fundamental. Therefore, we do not use the term *harmonic* in association with those types of systems.

Let us examine further how the various harmonics are created in a string. To excite only a single harmonic, the string must be distorted into a shape that corresponds to that of the desired harmonic. After being released, the string vibrates at the frequency of that harmonic. This manoeuvre is difficult to perform, however, and is not how a string of a musical instrument is excited. If the string is distorted such that its shape is not that of just one harmonic, the resulting vibration includes a combination of various harmonics. Such a distortion occurs in musical instruments when the string is plucked (as in a guitar), bowed (as in a cello) or struck (as in a piano). When the string is distorted into a non-sinusoidal shape, only waves that satisfy the boundary conditions can persist on the string. These waves are the harmonics.

The frequency of a string that defines the musical note that it plays is that of the fundamental. The string's frequency can be varied by changing either the string's tension or its length. For example, the tension in guitar and violin strings is varied by a screw adjustment mechanism or by tuning pegs located on the neck of the instrument. As the tension is increased, the frequency of the normal modes increases in accordance with **Equation 18.6**. Once the instrument is 'tuned', players vary the frequency by moving their fingers along the neck, thereby changing the length of the oscillating portion of the string. As the length is shortened, the frequency increases because, as **Equation 18.6** specifies, the normal-mode frequencies are inversely proportional to string length.

Quick Quiz 18.5

When a standing wave is set up on a string fixed at both ends, which of the following statements is true? **(a)** The number of nodes is equal to the number of antinodes. **(b)** The wavelength is equal to the length of the string divided by an integer. **(c)** The frequency is equal to the number of nodes times the fundamental frequency. **(d)** The shape of the string at any instant shows a symmetry about the midpoint of the string.

TRY THIS, AGAIN

Using a large spring such as a slinky with a friend holding one end of it still, shake your end of the slinky up and down and adjust your shaking frequency to find standing wave patterns. How many harmonics can you find?

TRY THIS TOO

Get some rubber bands of different lengths and thicknesses, and stretch them over an empty tissue box to make a homemade toy guitar. What do you notice about the sound made when you pluck the different 'strings' above the hole in the box? Which ones make a high-pitched sound and which ones make a low-pitched sound?

Analysis Model 18.2

Waves under boundary conditions

Imagine a wave that is not free to travel throughout all space as in the travelling wave model. If the wave is subject to boundary conditions, such that certain requirements must be met at specific locations in space, the wave is limited to a set of **normal modes** with quantised wavelengths and quantised natural frequencies.

For waves on a string fixed at both ends, the natural frequencies are

$$f_n = \frac{n}{2L}\sqrt{\frac{T}{\mu}} \quad n = 1, 2, 3, \ldots \tag{18.6}$$

where T is the tension in the string and μ is its linear mass density.

Examples

- waves travelling back and forth on a guitar string combine to form a standing wave
- sound waves travelling back and forth in a clarinet combine to form standing waves (Section 18.5)
- a microscopic particle confined to a small region of space is modelled as a wave and exhibits quantised energies (Chapter 40)
- the Fermi energy of a metal is determined by modelling electrons as wave-like particles in a box (Chapter 42)

Example 18.4

Give me a C!

The middle C string on a piano has a fundamental frequency of 262 Hz, and the string for the first A above middle C has a fundamental frequency of 440 Hz.

(A) Calculate the frequencies of the next two harmonics of the C string.

Solution

Conceptualise Remember that the harmonics of a vibrating string have frequencies that are related by integer multiples of the fundamental. Draw a diagram to help you, or refer to **Figure 18.13**.

Model We can apply the waves under boundary conditions analysis model.

Analyse As $f_n = nf_1$, and the fundamental frequency is $f_1 = 262$ Hz:

$$f_2 = 2f_1 = 524 \text{ Hz}$$
$$f_3 = 3f_1 = 786 \text{ Hz}$$

(B) If the A and C strings have the same linear mass density μ and length L, determine the ratio of tensions in the two strings.

Solution

Analyse Use **Equation 18.7** to write expressions for the fundamental frequencies of the two strings:

$$f_{1\text{A}} = \frac{1}{2L}\sqrt{\frac{T_\text{A}}{\mu}} \quad \text{and} \quad f_{1\text{C}} = \frac{1}{2L}\sqrt{\frac{T_\text{C}}{\mu}}$$

Divide the first equation by the second and solve for the ratio of tensions:

$$\frac{f_{1\text{A}}}{f_{1\text{C}}} = \sqrt{\frac{T_\text{A}}{T_\text{C}}} \rightarrow \frac{T_\text{A}}{T_\text{C}} = \left(\frac{f_{1\text{A}}}{f_{1\text{C}}}\right)^2 = \left(\frac{440}{262}\right)^2 = 2.82$$

Finalise If the frequencies of piano strings were determined solely by tension, this result suggests that the ratio of tensions from the lowest string to the highest string on the piano would be enormous. Such large tensions would make it difficult to design a frame to support the strings. In reality, the frequencies of piano strings vary due to additional parameters, including the mass per unit length and the length of the string, hence this is not a good assumption.

Resonance

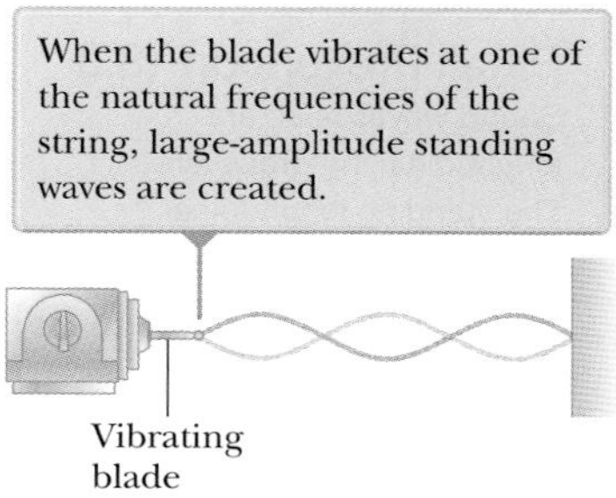
When the blade vibrates at one of the natural frequencies of the string, large-amplitude standing waves are created.
Vibrating blade

Figure 18.15
Standing waves are set up in a string when one end is connected to a vibrating blade.

Suppose we drive a string with a vibrating blade as in **Figure 18.15**. The fixed end is a node, and the end connected to the blade is very nearly a node because the amplitude of the blade's motion is small compared with that of the elements of the string. As the blade oscillates, transverse waves sent down the string are reflected from the fixed end. When the frequency of the blade equals one of the natural frequencies of the string, standing waves are produced and the string oscillates with a large amplitude. In this resonance case, the wave generated by the oscillating blade is in phase with the reflected wave and the string rapidly absorbs energy from the blade. This phenomenon, known as *resonance*, was discussed in Section 16.7 where we looked at driven oscillators. Although a block–spring system or a simple pendulum has only one natural frequency, standing-wave systems have a whole set of natural frequencies, such as that given by **Equation 18.6** for a string. Because an oscillating system exhibits a large amplitude when driven at any of its natural frequencies, these natural frequencies are often referred to as **resonance frequencies**. If the string is driven at a frequency that is not one of its natural frequencies, the oscillations are of low amplitude and exhibit no stable pattern because the transfer of energy to the string happens at a much lower rate.

18.5 Standing waves in pipes, rods and membranes

Resonance can occur in any system that can oscillate, or support transverse or longitudinal waves. For example, resonance is very important in the excitation of musical instruments based on air columns. The waves under boundary conditions model can be applied to sound waves in a column of air such as that inside an organ pipe or a clarinet. Standing waves in this case are the result of interference between longitudinal sound waves travelling in opposite directions.

In a pipe closed at one end, the closed end is a **displacement node** because the rigid barrier at this end does not allow longitudinal motion of the air. Because the pressure wave is 90° out of phase with the displacement wave, the closed end of an air column corresponds to a **pressure antinode**. The open end of an air column is approximately a **pressure node** and a **displacement antinode**. We can understand why no pressure variation occurs at an open end by noting that the end of the air column is open to the atmosphere; therefore, the pressure at this end must remain constant at atmospheric pressure.

You may wonder how a sound wave can reflect from an open end because there may not appear to be a change in the medium at this point: the medium through which the sound wave moves is air both inside and outside the pipe. Sound can be represented as a pressure wave, however, and a compression region of the sound wave is constrained by the sides of the pipe as long as the region is inside the pipe. As the compression region exits at the open end of the pipe, the constraint of the pipe is removed and the compressed air is free to expand into the atmosphere. Therefore, there is a change in the *character* of the medium between the inside of the pipe and the outside even though there is no change in the *material* of the medium. This change in character is sufficient to allow some reflection.

Strictly speaking, the open end of an air column is not exactly a displacement antinode. A compression reaching an open end does not reflect until it passes beyond the end. For a tube of circular cross section, an end correction equal to approximately $0.6R$, where R is the tube's radius, must be added to the length of the air column. Hence, the effective length of the air column is longer than the true length L. We shall assume that the length of the pipe is large compared to the radius and ignore this end correction in this discussion.

With the boundary conditions of nodes or antinodes at the ends of the air column, we have a set of normal modes of oscillation as is the case for the string fixed at both ends. Therefore, the air column has quantised frequencies. The first three normal modes of oscillation of a pipe open at both ends are shown in **Figure 18.16a**. Notice that both ends are displacement antinodes (approximately). In the first normal mode, the standing wave extends between two adjacent antinodes, which is a distance of half a wavelength. Therefore, the wavelength is twice the length of the pipe and the fundamental frequency is $f_1 = v/2L$. As **Figure 18.16a** shows, the frequencies of the higher harmonics are $2f_1$, $3f_1$, . . .

Pitfall Prevention 18.3
Sound waves in air are longitudinal, not transverse. The standing longitudinal waves are drawn as transverse waves in Figure 18.16. As they are in the same direction as the propagation, it is difficult to draw longitudinal displacements. Therefore, it is best to interpret the red-brown curves in Figure 18.16 as a graphical representation of the waves (our diagrams of string waves are pictorial representations), with the vertical axis representing the horizontal displacement $s(x, t)$ of the elements of the medium.

In a pipe open at both ends, the ends are displacement antinodes and the harmonic series contains all integer multiples of the fundamental.

In a pipe closed at one end, the open end is a displacement antinode and the closed end is a node. The harmonic series contains only odd integer multiples of the fundamental.

First harmonic
$$\lambda_1 = 2L$$
$$f_1 = \frac{v}{\lambda_1} = \frac{v}{2L}$$

Second harmonic
$$\lambda_2 = L$$
$$f_2 = \frac{v}{L} = 2f_1$$

Third harmonic
$$\lambda_3 = \frac{2}{3} L$$
$$f_3 = \frac{3v}{2L} = 3f_1$$

First harmonic
$$\lambda_1 = 4L$$
$$f_1 = \frac{v}{\lambda_1} = \frac{v}{4L}$$

Third harmonic
$$\lambda_3 = \frac{4}{3} L$$
$$f_3 = \frac{3v}{4L} = 3f_1$$

Fifth harmonic
$$\lambda_5 = \frac{4}{5} L$$
$$f_5 = \frac{5v}{4L} = 5f_1$$

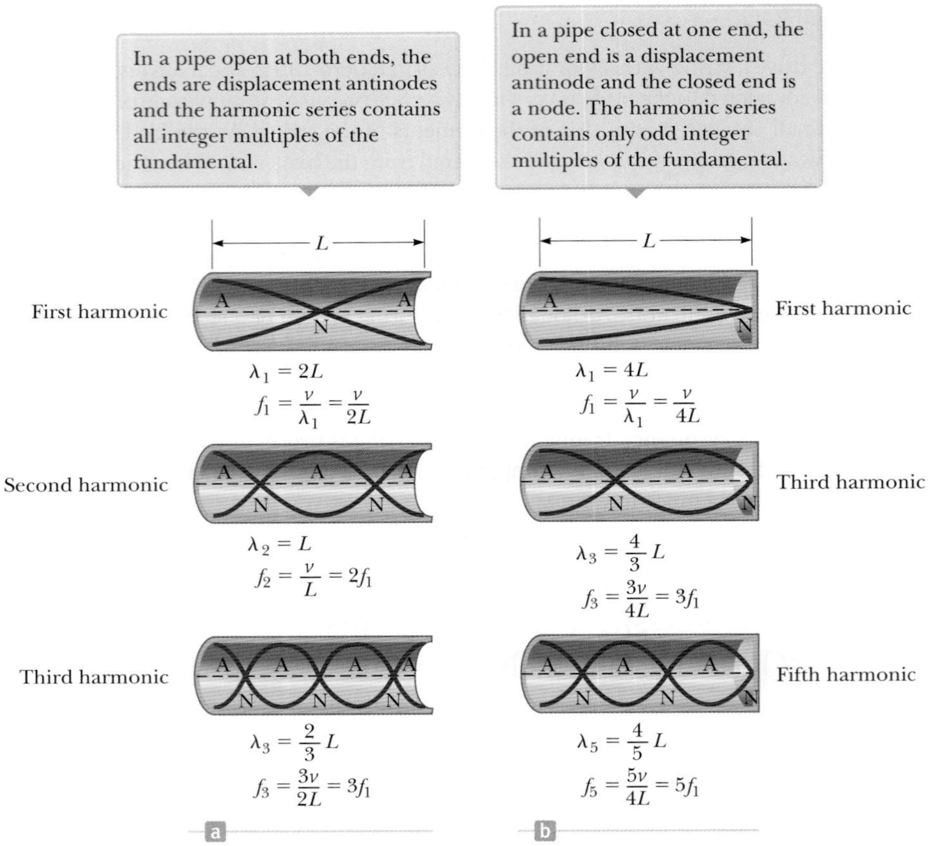

Figure 18.16
Graphical representations of the motion of elements of air in standing longitudinal waves in (a) a column open at both ends and (b) a column closed at one end

Because all harmonics are present and because the fundamental frequency is given by the same expression as that for a string (see **Equation 18.5**), we can express the natural frequencies of oscillation as

▶ Natural frequencies of a pipe open at both ends

$$f_n = n\frac{v}{2L} \qquad n = 1, 2, 3, \ldots \qquad (18.8)$$

where v is the speed of sound in air.

If a pipe is closed at one end and open at the other, the closed end is a displacement node (see **Figure 18.16b**). In this case, the standing wave for the fundamental mode extends from an antinode to the adjacent node, which is one-quarter of a wavelength. Hence, the wavelength for the first normal mode is $4L$ and the fundamental frequency is $f_1 = v/4L$. As **Figure 18.16b** shows, the higher-frequency waves that satisfy the boundary conditions are those that have a node at the closed end and an antinode at the open end; hence, the higher harmonics have frequencies $3f_1, 5f_1, \ldots$ In this case only odd integral multiples of the fundamental frequency are present in the harmonic series. We express this result mathematically as

▶ Natural frequencies of a pipe closed at one end

$$f_n = n\frac{v}{4L} \qquad n = 1, 3, 5, \ldots \qquad (18.9)$$

Musical instruments based on air columns are generally excited by resonance. The air column is presented with a sound wave that is rich in many frequencies. The air column then responds with a large-amplitude oscillation to the frequencies that match the quantised frequencies in its set of harmonics. In many woodwind instruments, the initial rich sound is provided by a vibrating reed. In brass instruments, this excitation is provided by the sound coming from the vibration of the player's lips.

TRY THIS

Get an empty bottle with a narrow neck and blow across the top of it to produce a sound. The sound of the air rushing across the bottle opening has many frequencies, including one that sets the air cavity in the bottle into resonance, producing a much louder sound than the blowing. How can you change the sound produced?

TRY THIS TOO

Make your own didjeridu (didgeridoo) using a length of pipe such as the pipe from a vacuum cleaner or a cardboard mailing tube. Purse your lips as you blow into one end of it. What do you need to do with your mouth to get a sound? What can you do to vary the sound? If you can find more than one tube, note the differences in the sounds they produce.

Quick Quiz 18.6

A pipe open at both ends resonates at a fundamental frequency f_{open}. When one end is covered and the pipe is again made to resonate, the fundamental frequency is f_{closed}. Which of the following expressions describes how these two resonant frequencies compare? **(a)** $f_{\text{closed}} = f_{\text{open}}$ **(b)** $f_{\text{closed}} = \frac{1}{2}f_{\text{open}}$ **(c)** $f_{\text{closed}} = 2f_{\text{open}}$ **(d)** $f_{\text{closed}} = \frac{3}{2}f_{\text{open}}$

Quick Quiz 18.7

Balboa Park in San Diego has an outdoor organ. When the air temperature increases, the fundamental frequency of one of the organ pipes **(a)** stays the same, **(b)** goes down, **(c)** goes up or **(d)** is impossible to determine.

Example 18.5

Wind in a culvert

A section of drainage pipe in a culvert is 1.23 m in length and makes a howling noise when the wind blows across its open ends.

(A) Determine the frequencies of the first three harmonics of the pipe. Take $v = 343$ m/s as the speed of sound in air.

Solution

Conceptualise The sound of the wind blowing across the end of the pipe contains many frequencies and the pipe responds to the sound by vibrating at the natural frequencies of the air column.

Model We model the air column as open at both ends and apply the waves under boundary conditions analysis model for a pipe open at both ends.

Analyse Use **Equation 18.8** to find the frequency of the first harmonic:

$$f_1 = \frac{v}{2L} = \frac{343 \text{ m/s}}{2(1.23 \text{ m})} = 139 \text{ Hz}$$

Find the next two harmonics by multiplying by integers:

$$f_2 = 2f_1 = 279 \text{ Hz}$$

$$f_3 = 3f_1 = 418 \text{ Hz}$$

(B) What are the three lowest natural frequencies of the culvert if it is blocked at one end?

Solution

Model We now model the air column as closed at one end and apply the waves under boundary conditions analysis model for a pipe closed at one end.

Analyse Use **Equation 18.9** to find the first harmonic:

$$f_1 = \frac{v}{4L} = \frac{343 \text{ m/s}}{4(1.23 \text{ m})} = 69.7 \text{ Hz}$$

Find the next two harmonics by multiplying by odd integers:

$$f_3 = 3f_1 = 209 \text{ Hz}$$

$$f_5 = 5f_1 = 349 \text{ Hz}$$

Example 18.6

Measuring the frequency of a tuning fork

A simple apparatus for demonstrating resonance in an air column is shown in **Figure 18.17**.

A vertical pipe open at both ends is partially submerged in water and a tuning fork vibrating at an unknown frequency is placed near the top of the pipe. The length L of the air column can be adjusted by moving the pipe vertically. For a certain pipe and tuning fork, the smallest value of L for which resonance occurs and a loud sound can be heard is 9.0 cm.

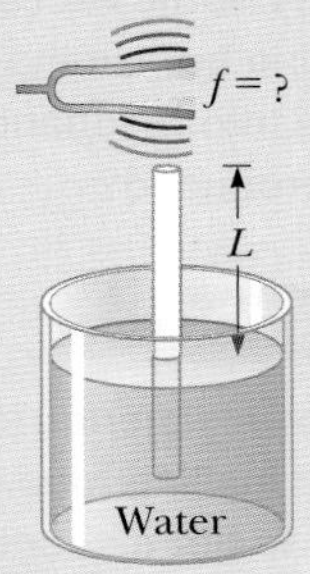

Figure 18.17
(Example 18.6) Apparatus for demonstrating the resonance of sound waves in a pipe closed at one end. The length L of the air column is varied by moving the pipe vertically while it is partially submerged in water.

(A) What is the frequency of the tuning fork?

Solution

Conceptualise In the culvert in the preceding example, the pipe length was fixed and the air column was presented with a mixture of frequencies. The pipe in this example is presented with one single frequency from the tuning fork and the length of the pipe is varied until resonance is achieved. We have a diagram of the apparatus, but it will be helpful to draw a diagram showing the wave patterns produced.

Model Although the pipe is open at its lower end to allow the water to enter, the water's surface acts as a barrier. Therefore, this setup can be modelled as an air column closed at one end.

Analyse Use **Equation 18.9** to find the fundamental frequency for $L = 0.0900$ m:

$$f_1 = \frac{v}{4L} = \frac{343 \text{ m/s}}{4(0.0900 \text{ m})} = 953 \text{ Hz}$$

Because the tuning fork causes the air column to resonate at this frequency, this frequency must also be that of the tuning fork.

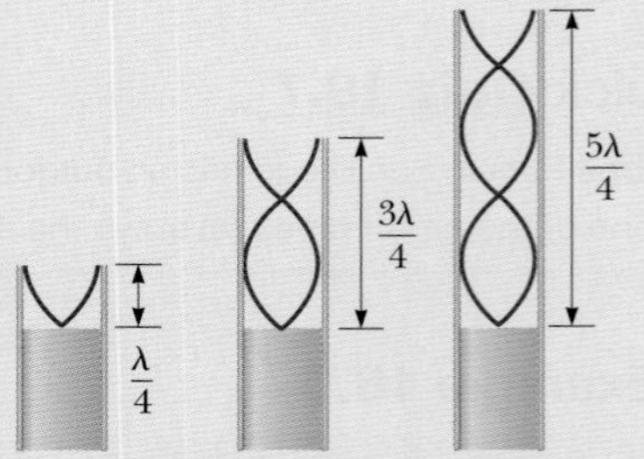

Figure 18.18
(Example 18.6) The first three normal modes of the system (displacement wave)

(B) What are the values of L for the next two resonance conditions?

Solution

Model We now model the wave produced by the tuning fork as a travelling wave and apply the travelling wave analysis model.

Analyse Use **Equation 17.12** to find the wavelength of the sound wave from the tuning fork:

$$\lambda = \frac{v}{f} = \frac{343 \text{ m/s}}{953 \text{ Hz}} = 0.360 \text{ m}$$

Notice from **Figure 18.18** that the length of the air column for the second resonance is $3\lambda/4$:

$$L = \frac{3\lambda}{4} = 0.270 \text{ m}$$

Notice from **Figure 18.18** that the length of the air column for the third resonance is $5\lambda/4$:

$$L = \frac{5\lambda}{4} = 0.450 \text{ m}$$

Rods and membranes

Standing waves can also be set up in rods and membranes. Musical instruments that depend on transverse standing waves in rods include triangles, marimbas, xylophones, glockenspiels, chimes and vibraphones. **Figure 18.19** shows two small boys playing with the 'lithophone' at Questacon in Canberra, which is a large xylophone made from stone blocks.

The positions at which the bars are clamped to the frame determine the standing wave patterns that are possible. For example, if a bar is clamped at the centre then the fundamental mode is shown in **Figure 18.20a**. If it is clamped at a position one-quarter along its length, the mode is that shown in **Figure 18.20b**.

A rod clamped in the middle, as in **Figure 18.20a**, and stroked parallel to the rod at one end also oscillates as depicted in **Figure 18.20a**, but in this case the oscillations of the elements of the rod are longitudinal and so the red-brown curves in **Figure 18.20** represent *longitudinal* displacements of various parts of the rod. The oscillations in this setup are analogous to those in a pipe open at both ends. The red-brown lines in **Figure 18.20a** represent the first normal mode, for which the wavelength is $2L$ and the frequency is $f = v/2L$, where v is the speed of longitudinal waves in the rod.

Figure 18.19
The lithophone is a musical instrument made from large stone blocks. The length of each block determines the note that it plays.

TRY THIS

Hold a ruler firmly by one end against a table and 'twang' the other end. What happens to the sound produced if you change the length of ruler that is free to vibrate by holding more of the ruler firmly against the table? If you have a clamp and a metal rod, try setting up the situation shown in **Figure 18.20** and see if you can get the rod to 'sing' by stroking it. You can also make a wine glass sing by gently rubbing around the rim.

Two-dimensional oscillations can be set up in a flexible membrane stretched over a circular hoop such as that in a drumhead. As the membrane is struck at some point, waves that arrive at the fixed boundary are reflected many times. The resulting sound is not harmonic because the standing waves have frequencies that are *not* related by integer multiples.

Whereas nodes are *points* in one-dimensional standing waves on strings and in air columns, a two-dimensional oscillator has *curves* along which there is no displacement of the elements of the medium. **Figure 18.21** shows examples of nodes for different harmonics on Chladni plates. These thin metal plates vibrate when the edge is 'bowed' using a violin bow.

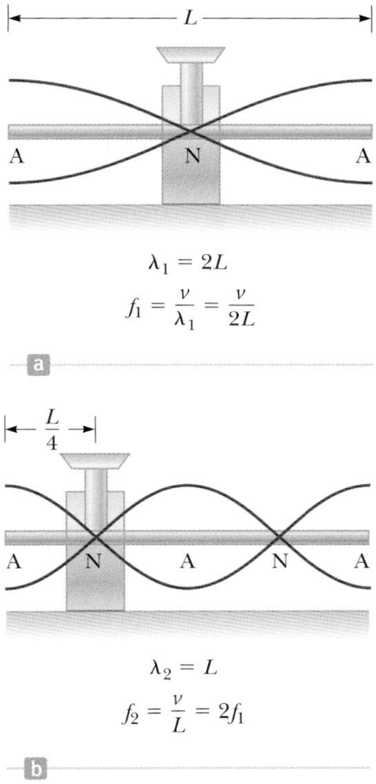

Figure 18.20
Normal-mode longitudinal vibrations of a rod of length L (a) clamped at the middle to produce the first normal mode and (b) clamped at a distance $L/4$ from one end to produce the second normal mode

TRY THIS

Stretch some elastic fabric, such as some lycra used for a swimsuit, across a container such as a large bowl or lunch box to make a drum. Tap it and see what sounds it makes. You can vary the sound by stretching the fabric more or less tightly. Now sprinkle a little salt or caster sugar over the surface and tap it gently. How many patterns can you make by tapping very gently and rapidly in different places on the surface? This also works very well with a clamped metal plate excited by bowing with a violin bow at the edge (but this is not very good for the bow).

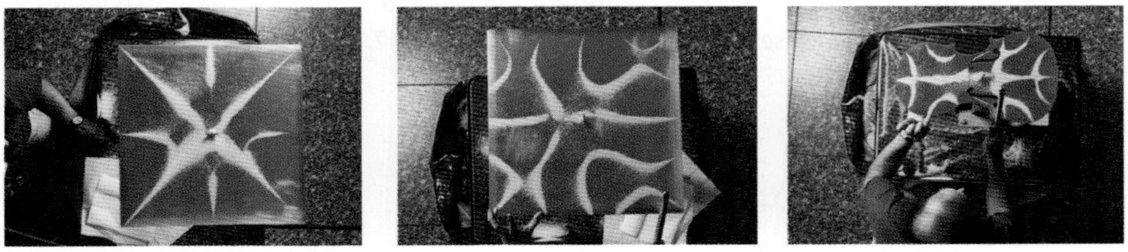

Figure 18.21
Patterns produced in the sand on a square Chladni plate and a viola-shaped plate when the plate is 'bowed' and vibrated at different frequencies (Figures courtesy of Questacon)

Conceptual Example 18.7

In and out of tune

It is interesting to investigate what happens to the frequencies of instruments during a concert as the temperature of the instrument rises as it is played. What sort of instruments become flat (play a lower frequency note than desired) and what sort become sharp (play a higher frequency note than that desired)?

Conceptual Example 18.7 cont.

Solution

First consider the sound emitted by instruments that use a column of air. As the air inside gets warmer, the density of the air will decrease and the speed of sound will increase. The higher speed leads to a higher frequency, from $v = f\lambda$, because the wavelength is fixed by the length of the air column. (The instrument itself does not heat up enough to change length significantly.) We can see this from **Equations 18.8** and **18.9**. Hence any woodwind or brass instruments, such as flutes and clarinets, will become very slightly sharp during a long concert.

Now consider the stringed instruments, such as violins and cellos. As the strings thermally expand, the expansion causes their tension to decrease. Linear mass density does not change significantly, so the speed must decrease. The length and hence wavelength for the standing waves does not change, resulting in a decrease in frequency. Hence the sound produced by a violin, piano or other stringed instrument becomes flat. What do you think would happen to a xylophone as it warms up?

18.6 Beats: interference in time

The interference phenomena we have studied so far involve the superposition of waves having the same frequency. The amplitude of oscillation of elements of the medium varies with the position in space, so we refer to the phenomenon as *spatial interference*. Standing waves in strings and pipes are common examples of spatial interference.

Another type of interference results from the superposition of two waves having slightly *different* frequencies. In this case, when the two waves are observed at a point in space, they are periodically in and out of phase. That is, there is a *temporal* (time) alternation between constructive and destructive interference. As a consequence, we refer to this phenomenon as *interference in time* or *temporal interference*. For example, if two tuning forks of slightly different frequencies are struck, one hears a sound of periodically varying amplitude. This phenomenon is called **beating**.

Definition of beating ▶

Beating is the periodic variation in amplitude at a given point due to the superposition of two waves having slightly different frequencies.

The number of amplitude maxima one hears per second, or the *beat frequency*, equals the difference in frequency between the two sources as we shall show below. The maximum beat frequency that the human ear can detect is about 20 beats/s. When the beat frequency exceeds this value, the beats blend indistinguishably with the sounds producing them.

Consider two sound waves of equal amplitude and slightly different frequencies f_1 and f_2 travelling through a medium. We use equations similar to **Equation 17.13** to represent the wave functions for these two waves at a point that we identify as $x = 0$. We also choose the phase angle in **Equation 17.13** as $\phi = \pi/2$, which enables us to use the cosine rule below:

$$y_1 = A\sin\left(\frac{\pi}{2} - \omega_1 t\right) = A\cos(2\pi f_1 t)$$

$$y_2 = A\sin\left(\frac{\pi}{2} - \omega_2 t\right) = A\cos(2\pi f_2 t)$$

Using the superposition principle, we find that the resultant wave function at this point is

$$y = y_1 + y_2 = A(\cos 2\pi f_1 t + \cos 2\pi f_2 t)$$

The trigonometric identity $\cos a + \cos b = 2\cos\left(\frac{a-b}{2}\right)\cos\left(\frac{a+b}{2}\right)$ allows us to write the expression for y as

Resultant of two waves of different frequencies but equal amplitude ▶

$$y = \left[2A\cos 2\pi\left(\frac{f_1 - f_2}{2}\right)t\right]\cos 2\pi\left(\frac{f_1 + f_2}{2}\right)t \quad (18.10)$$

Σ *Appendix B.4 contains a table of useful trigonometric identities.*

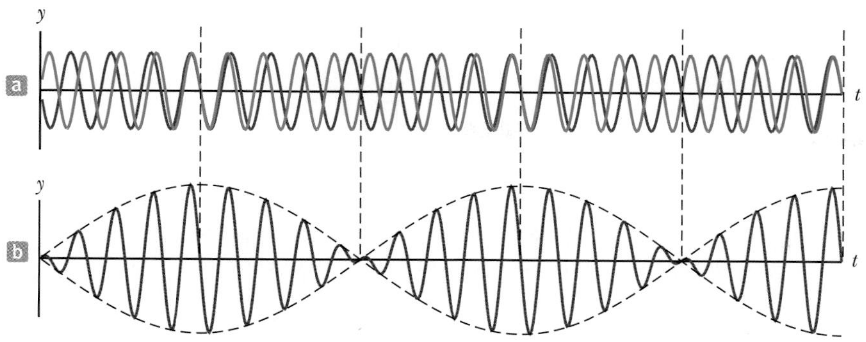

Active Figure 18.22
Beats are formed by the combination of two waves of slightly different frequencies. (a) The individual waves (b) The combined wave. The envelope wave (dashed line) represents the beating of the combined sounds.

Graphs of the individual waves and the resultant wave are shown in **Active Figure 18.22**. From **Equation 18.10**, we see that the resultant wave has an effective frequency equal to the average frequency $(f_1 + f_2)/2$. This wave is multiplied by an envelope wave given by the expression in the brackets:

$$y_{\text{envelope}} = 2A \cos 2\pi \left(\frac{f_1 - f_2}{2} \right) t \qquad (18.11)$$

That is, the amplitude and therefore the intensity of the resultant sound vary in time. The dashed black line in **Active Figure 18.22b** is a graphical representation of the envelope wave in **Equation 18.11** and is a sinusoidal wave varying with frequency $(f_1 - f_2)/2$.

A maximum in the amplitude of the resultant sound wave is detected whenever

$$\cos 2\pi \left(\frac{f_1 - f_2}{2} \right) t = \pm 1$$

Hence, there are *two* maxima in each period of the envelope wave. Because the amplitude varies with frequency as $(f_1 - f_2)/2$, the number of beats per second, or the beat frequency f_{beat}, is twice this value. That is,

$$f_{\text{beat}} = |f_1 - f_2| \qquad (18.12)$$

◀ Beat frequency

Quick Quiz 18.8

One tuning fork vibrates at 438 Hz and a second one vibrates at 442 Hz. What does a listener hear? **(i)** Does the resultant sound have a frequency of **(a)** 220 Hz, **(b)** 438 Hz, **(c)** 440 Hz, **(d)** 442 Hz or **(e)** 880 Hz? **(ii)** Is the beat frequency **(a)** 2 Hz, **(b)** 4 Hz, **(c)** 8 Hz or **(d)** 440 Hz?

Example 18.8

The mistuned piano strings

Two identical piano strings of length 0.750 m are each tuned exactly to 440 Hz. The tension in one of the strings is then increased by 1.0%. If they are now struck, what is the beat frequency between the fundamentals of the two strings?

Solution

Conceptualise As the tension in one of the strings is changed, its fundamental frequency changes. Therefore, when both strings are played, they will have different frequencies and beats will be heard.

Model We must combine our understanding of the waves under boundary conditions model for strings with our new knowledge of beats.

Analyse Set up a ratio of the fundamental frequencies of the two strings using **Equation 18.5**:

$$\frac{f_2}{f_1} = \frac{(v_2/2L)}{(v_1/2L)} = \frac{v_2}{v_1}$$

Use **Equation 17.28** to substitute for the wave speeds on the strings:

$$\frac{f_2}{f_1} = \frac{\sqrt{T_2/\mu}}{\sqrt{T_1/\mu}} = \sqrt{\frac{T_2}{T_1}}$$

Example 18.8 cont.

Incorporate that the tension in one string is 1.0% larger than the other, that is, $T_2 = 1.010T_1$:

$$\frac{f_2}{f_1} = \sqrt{\frac{1.010T_1}{T_1}} = 1.005$$

Solve for the frequency of the tightened string:

$$f_2 = 1.005f_1 = 1.005(440 \text{ Hz}) = 442 \text{ Hz}$$

Find the beat frequency using **Equation 18.12**:

$$f_{\text{beat}} = 442 \text{ Hz} - 440 \text{ Hz} = 2 \text{ Hz}$$

Finalise Notice that a 1.0% mistuning in tension leads to an easily audible beat frequency of 2 Hz. A piano tuner can use beats to tune a stringed instrument by 'beating' a note against a reference tone of known frequency. The tuner can then adjust the string tension until the frequency of the sound it emits equals the frequency of the reference tone. The tuner does so by tightening or loosening the string until the beats produced by it and the reference source become too infrequent to notice.

18.7 Fourier analysis and synthesis

It is relatively easy to distinguish the sounds coming from a violin and a trumpet even when they are both playing the same note; however, the same note played on a clarinet and an oboe sound quite similar. We can use the pattern of the sound waves from various sources to explain these effects.

The wave patterns produced by a musical instrument are the result of the superposition of frequencies that are integer multiples of a fundamental. This superposition results in the corresponding richness of musical tones. The human perceptive response associated with various mixtures of harmonics is the *quality* or *timbre* of the sound. For instance, the sound of the trumpet is perceived to have a 'brassy' quality (that is, we have learned to associate the adjective *brassy* with that sound); this quality enables us to distinguish the sound of the trumpet from that of the violin, whose quality is often perceived as 'bright' or 'rich'. The clarinet and oboe, however, both contain air columns excited by reeds; because of this similarity, they have similar mixtures of frequencies and it is more difficult for the human ear to distinguish them on the basis of their sound quality.

The sound wave patterns produced by the majority of musical instruments are non-sinusoidal. Characteristic patterns produced by a tuning fork, a flute and a clarinet, each playing the same note, are shown in **Figure 18.23**. Each instrument has its own characteristic pattern. Notice that despite the differences in the patterns, each pattern is periodic. This point is important for our analysis of these waves. If the wave pattern is periodic, it can be represented as closely as desired by the combination of a sufficiently large number of sinusoidal waves that form a harmonic series. In fact, we can represent any periodic function as a series of sine and cosine terms by using a mathematical technique based on **Fourier's theorem**, developed by

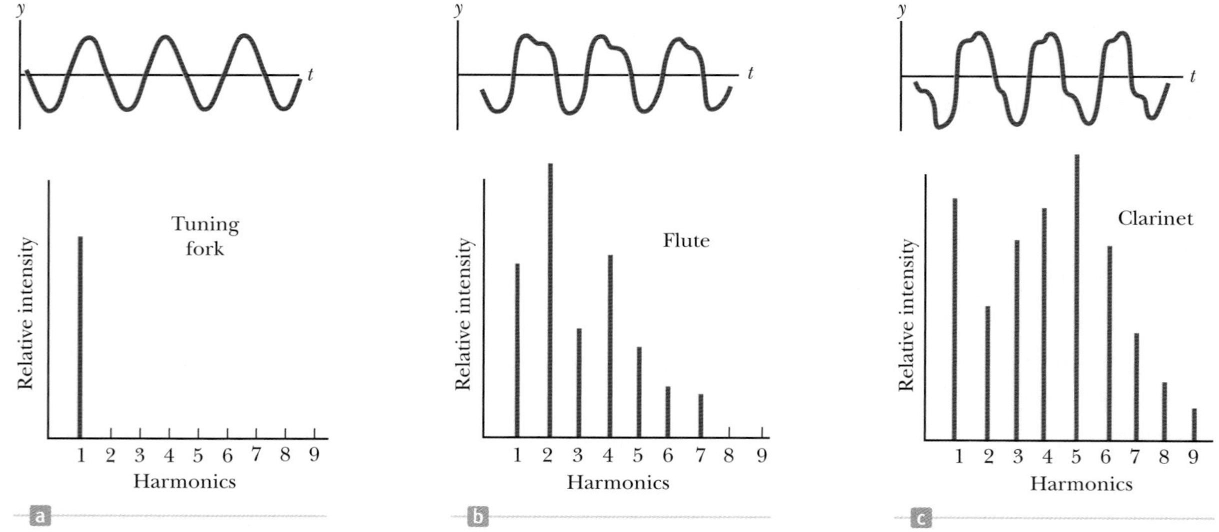

Figure 18.23
(top) The sound wave patterns produced by (a) a tuning fork, (b) a flute and (c) a clarinet, and (below) the harmonics of each wave pattern. The wave patterns can be constructed using Fourier synthesis from the patterns of harmonics.

Jean Baptiste Joseph Fourier (1786–1830). The corresponding sum of terms that represents the periodic wave pattern is called a **Fourier series**. Let $y(t)$ be any function that is periodic in time with period T such that $y(t + T) = y(t)$. Fourier's theorem states that this function can be written as

$$y(t) = \sum (A_n \sin 2\pi f_n t + B_n \cos 2\pi f_n t) \quad (18.13)$$

◀ Fourier's theorem

where the lowest frequency is $f_1 = 1/T$. The higher frequencies are integer multiples of the fundamental, $f_n = nf_1$, and the coefficients A_n and B_n represent the amplitudes of the various waves.

Figure 18.23 also represents a harmonic analysis of the wave patterns shown, called a **Fourier spectrum**. Each bar in the graph represents one of the terms in the series in Equation 18.13 up to $n = 9$. Notice that a struck tuning fork produces only one harmonic (the first), whereas the flute and clarinet produce the first harmonic and many higher ones. A pure sound, such as that represented in Figure 18.23a does not sound very musical and many pure notes sound quite unpleasant. It is very difficult to sing a pure note, as the students in Figure 18.24 are trying to do. However, a clear whistle is very close to a single pure frequency. The variation in relative intensity of the various harmonics for the flute and the clarinet is the reason for their differing timbres. In general, any musical sound consists of a fundamental frequency f plus other frequencies that are integer multiples of f, all having different intensities.

Figure 18.24
Students at the Australian National University observe the waves created when they sing and whistle into a microphone connected to an oscilloscope.

The analysis of a sound using Fourier's theorem involves determining the coefficients of the harmonics in Equation 18.13 from a knowledge of the wave pattern. The reverse process, called *Fourier synthesis*, can also be performed. In this process, the various harmonics are added together to form a resultant wave pattern. As an example of Fourier synthesis, consider the building of a square wave as shown in Active Figure 18.25. The symmetry of the square wave results in only odd multiples of the fundamental frequency combining in its synthesis. In Active Figure 18.25a, the blue curve shows the combination of f and $3f$. In Active Figure 18.25b, we have added $5f$ to the combination and obtained the green curve. Notice how the general shape of the square wave is approximated, even though the upper and lower portions are not flat as they should be. Active Figure 18.25c shows the result of adding odd frequencies up to $9f$. This approximation (red-brown curve) to the square wave is better than the approximations in Active Figures 18.25a and 18.25b. To approximate the square wave more closely, we must add a large number of odd multiples of the fundamental frequency.

Musical sounds can be **synthesised** electronically by mixing different amplitudes of any number of harmonics. This is how a **synthesiser** or electronic keyboard works, combining sinusoidal waves to produce sounds approximating those of different real instruments.

Fourier analysis and Fourier synthesis have many applications beyond music. Fourier analysis is useful in the analysis of any repeating signal and has many applications in engineering, in particular signal processing. It allows a signal to be analysed to find periodicity, which allows encoding to save computational and transmission time, and memory usage.

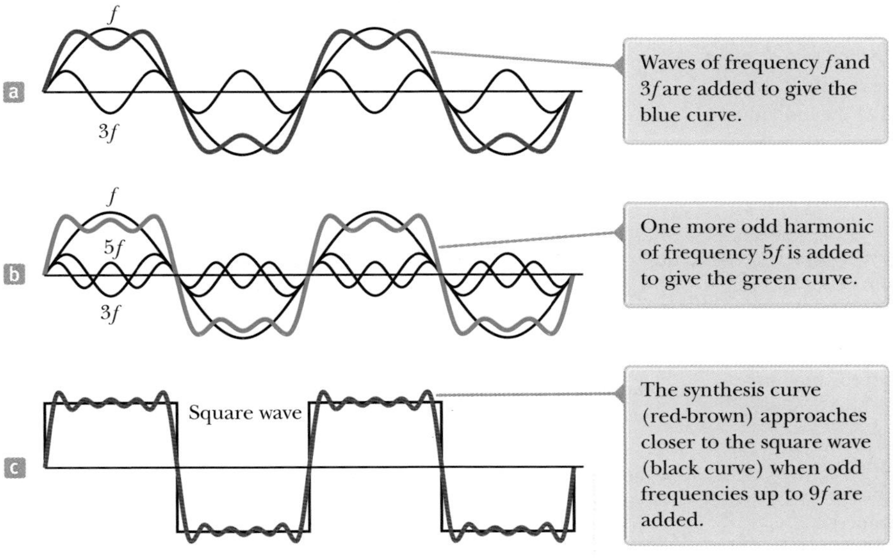
a
f
3f
Waves of frequency f and 3f are added to give the blue curve.
b
f
5f
3f
One more odd harmonic of frequency 5f is added to give the green curve.
c
Square wave
The synthesis curve (red-brown) approaches closer to the square wave (black curve) when odd frequencies up to 9f are added.

Active Figure 18.25
Fourier synthesis of a square wave, represented by the sum of odd multiples of the first harmonic, which has frequency f

End-of-chapter resources

Shutterstock.com/piotreknik

On its opening day, the London Millennium Bridge exhibited vertical movements within the limits of normal practice as predicted in design, but greater lateral vibrations that had not been anticipated during design. These strong lateral vibrations were caused by resonance, which made walking difficult for a significant number of pedestrians (Section 18.4). For more than a year, engineering designers and researchers from UK universities conducted an extensive analysis on the resonant vibrational modes. One possible explanation for this phenomenon is that the bridge's natural vibrating frequency was amplified by an identical frequency caused by the synchronous lateral excitation from pedestrians' steps. For example, the lateral motion at the first lateral mode of the bridge was in the range of 0.5 to 1.0 Hz, which coincides with the frequencies of a normal walking pace. New experimental studies have been developed and tested. Using the Fourier's theorem (Section 18.7), a new theoretical model confirms that people will naturally synchronise their footsteps with each other and with the movement of the bridge, so that the bridge vibration increases to a maximum.

The problems found in this chapter may be assigned online in Enhanced Web Assign.

Worked solutions to every fifth problem are available in the Student Solutions Manual. Register online at **www.cengagebrain.com** for access.

Summary

Concepts and principles

A wave is totally or partially reflected when it reaches the end of the medium in which it propagates or when it reaches a boundary where its speed changes discontinuously. If a wave travelling on a string meets a fixed end, the wave is reflected and inverted. If the wave reaches a free end, it is reflected but not inverted.

The **superposition principle** specifies that when two or more waves move through a medium, the value of the resultant wave function equals the algebraic sum of the values of the individual wave functions.

Standing waves are formed from the combination of two sinusoidal waves having the same frequency, amplitude and wavelength but travelling in opposite directions. The resultant standing wave is described by the wave function:

$$y = (2A \sin kx) \cos (\omega t) \qquad (18.1)$$

Hence, the amplitude of the standing wave is $2A$, and the amplitude of the simple harmonic motion of any element of the medium varies according to its position as $2A \sin kx$. The points of zero amplitude (called **nodes**) occur at $x = n\lambda/2$ ($n = 0, 1, 2, 3, \ldots$). The maximum amplitude points (called **antinodes**) occur at $x = n\lambda/4$ ($n = 1, 3, 5, \ldots$). Adjacent antinodes are separated by a distance $\lambda/2$. Adjacent nodes also are separated by a distance $\lambda/2$.

The phenomenon of **beating** is the periodic variation in intensity at a given point due to the superposition of two waves having slightly different frequencies.

Fourier analysis can be used to analyse and mathematically represent any periodic signal as a sum of harmonic waves with different amplitudes. **Fourier synthesis** is used to produce a periodic signal by adding a series of harmonic waves with different amplitudes.

Analysis models for problem solving

Waves in interference When two travelling waves having equal frequencies superimpose, the resultant wave has an amplitude that depends on the phase angle ϕ between the two waves. **Constructive interference** occurs when the two waves are in phase, corresponding to $\phi = 0, 2\pi, 4\pi, \ldots$ rad. **Destructive interference** occurs when the two waves are 180° out of phase, corresponding to $\phi = \pi, 3\pi, 5\pi, \ldots$ rad.

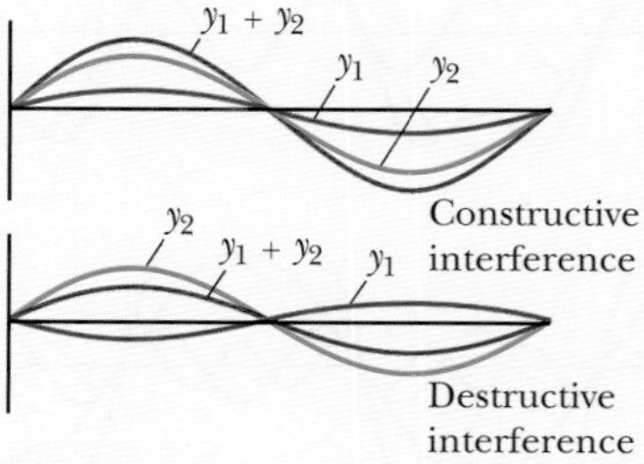

Waves under boundary conditions When a wave is subject to boundary conditions, only certain natural frequencies are allowed; we say that the frequencies are **quantised**.

For waves on a string fixed at both ends, the natural frequencies are

$$f_n = \frac{n}{2L}\sqrt{\frac{T}{\mu}} \qquad n = 1, 2, 3, \ldots \qquad (18.6)$$

where T is the tension in the string and μ is its linear mass density.

For sound waves with speed v in an air column of length L open at both ends, the natural frequencies are

$$f_n = n\frac{v}{2L} \qquad n = 1, 2, 3, \ldots \tag{18.8}$$

If an air column is open at one end and closed at the other, only odd harmonics are present and the natural frequencies are

$$f_n = n\frac{v}{4L} \qquad n = 1, 3, 5, \ldots \tag{18.9}$$

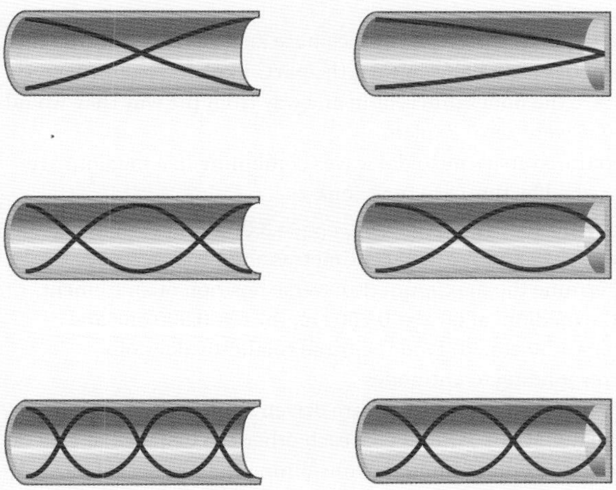

Chapter review quiz

To help you revise Chapter 18: Superposition and interference, complete the automatically graded chapter review quiz at http://login.cengagebrain.com.

Conceptual questions

1. When a pulse travels on a taut string, does it always invert upon reflection? Explain your answer.
2. Does the phenomenon of wave interference apply only to sinusoidal waves?
3. When two waves interfere constructively or destructively, is there any gain or loss in energy in the system of the waves? Explain your answer.
4. Explain how a musical instrument such as a piano may be tuned by using the phenomenon of beats.
5. An aeroplane mechanic notices that the sound from a twin-engine aircraft rapidly varies in loudness when both engines are running. What could be causing this variation from loud to soft?
6. A tuning fork by itself produces a faint sound. Explain how each of the following methods can be used to obtain a louder sound from it. Explain also any effect on the time interval for which the fork vibrates audibly. (a) holding the edge of a sheet of paper against one vibrating tine (b) pressing the handle of the tuning fork against a chalkboard or a tabletop (c) holding the tuning fork above a column of air of properly chosen length as in Example 18.6 (d) holding the tuning fork close to an open slot cut in a sheet of foam plastic or cardboard (with the slot similar in size and shape to one tine of the fork and the motion of the tines perpendicular to the sheet)
7. A soft-drink bottle resonates as air is blown across its top. What happens to the resonance frequency as the level of fluid in the bottle decreases? Try it yourself!
8. A crude model of the human throat is that of a pipe open at both ends with a vibrating source to introduce the sound into the pipe at one end. Assuming the vibrating source produces a range of frequencies, discuss the effect of changing the pipe's length.
9. The vocal cords in your throat are a pair of vibrating membranes, which produce a sound wave. The sound wave is then modified in your throat, mouth and nasal cavities to produce sounds including speech. Try screaming loudly at a high pitch (in the middle of a maths or chemistry lecture is a good time to try this). Now try singing a deep bass note. Explain what you do differently with you vocal cords in each case and how this affects the sound produced. Why do you think people are more likely to make a high-pitched scream when suddenly frightened rather than a low-pitched yell?
10. Marcus and Laurence receive a small, cheap keyboard as a Christmas present. It has buttons to make it sound somewhat like different instruments. How does the device do this? As soon as they start fighting over it their mother gets rid of it because the sound it makes is unbearably annoying and 'tinny'. What could be the problem with the way it synthesises the different instruments?

Problems

Note: Unless otherwise specified, assume the speed of sound in air is 343 m/s, its value at an air temperature of 20.0°C. At any other temperature T_C, the speed of sound in air is described by

$$v = 331\sqrt{1 + \frac{T_C}{273}}$$

where v is in m/s and T is in °C.

Section 18.1 Reflection and transmission

1. One of the reasons Christchurch was so badly affected by the February 2011 earthquake is that it is situated on sedimentary rock and soil, with much of the surrounding area being volcanic basalt. Consider a transverse or S seismic wave travelling in sedimentary rock where the wave speed is around 1.5 km/s, incident on basalt, in which the S wave speed is approximately 3.5 km/s. The incident wave has frequency f_i and wavelength λ_i. (a) Describe the transmitted wave in terms of frequency and wavelength, and whether it is inverted relative to the incident wave. (b) Describe the reflected wave in terms of frequency and wavelength, and whether it is inverted relative to the incident wave.

Section 18.2 Analysis model: interference of waves

2. Two waves on one string are described by the wave functions

$$y_1 = 3.0\cos(4.0x - 1.6t) \text{ and } y_2 = 4.0\sin(5.0x - 2.0t)$$

where x and y are in centimetres and t is in seconds. Find the superposition of the waves $y_1 + y_2$ at the points (a) $x = 1.00$, $t = 1.00$; (b) $x = 1.00$, $t = 0.500$; and (c) $x = 0.500$, $t = 0$. *Note:* Remember that the arguments of the trigonometric functions are in radians.

3. Two sinusoidal waves are travelling in the same direction along a stretched string. The waves are 90.0° out of phase. Each wave has an amplitude of 4.00 cm/s. Find the amplitude of the resultant wave.

4. Two pulses of different amplitudes approach each other, each having a speed of $v = 1.00$ cm/s. **Figure P18.4** shows the positions of the pulses at time $t = 0$. (a) Sketch the resultant wave at $t = 2.00$ s, 4.00 s, 5.00 s and 6.00 s. (b) **What If?** If the pulse on the right is inverted so that it is upright, how would your sketches of the resultant wave change?

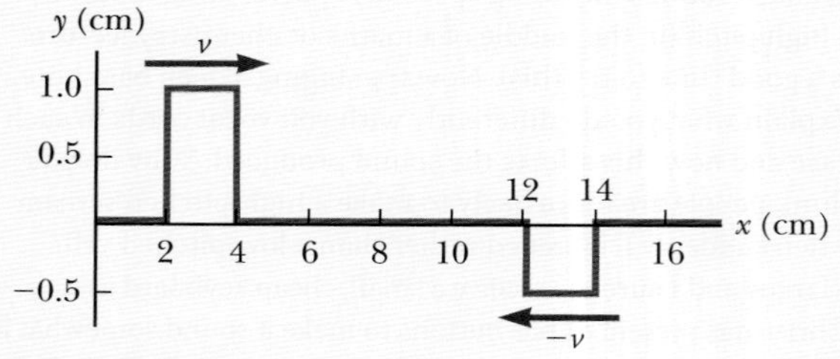

Figure P18.4

5. Two pulses travelling on the same string are described by

$$y_1 = \frac{5}{(3x - 4t)^2 + 2} \quad \text{and} \quad y_2 = \frac{-5}{(3x + 4t - 6)^2 + 2}$$

(a) In which direction does each pulse travel? (b) At what instant do the two cancel everywhere? (c) At what point do the two pulses always cancel?

6. Two travelling sinusoidal waves are described by the wave functions

$$y_1 = 5.00\sin[\pi(4.00x - 1200t)]$$

$$y_2 = 5.00\sin[\pi(4.00x - 1200t - 0.250)]$$

where x, y_1 and y_2 are in metres and t is in seconds. (a) What is the amplitude of the resultant wave function $y_1 + y_2$? (b) What is the frequency of the resultant wave function?

7. Two identical loudspeakers are placed on a wall 2.00 m apart. A listener stands 3.00 m from the wall directly in front of one of the speakers. A single oscillator is driving the speakers at a frequency of 300 Hz. (a) What is the phase difference in radians between the waves from the speakers when they reach the observer? (b) **What If?** What is the frequency closest to 300 Hz to which the oscillator may be adjusted such that the observer hears minimal sound?

8. Two sinusoidal waves in a string are defined by the wave functions

$$y_1 = 2.00\sin(20.0x - 32.0t) \text{ and } y_2 = 2.00\sin(25.0x - 40.0t)$$

where x, y_1 and y_2 are in centimetres and t is in seconds. (a) What is the phase difference between these two waves at the point $x = 5.00$ cm at $t = 2.00$ s? (b) What is the positive x value closest to the origin for which the two phases differ by $\pm\pi$ at $t = 2.00$ s? (At that location, the two waves add to zero.)

Section 18.3 Standing waves

9. Two sinusoidal waves travelling in opposite directions interfere to produce a standing wave with the wave function

$$y = 1.50\sin(0.400x)\cos(200t)$$

where x and y are in metres and t is in seconds. Determine (a) the wavelength, (b) the frequency and (c) the speed of the interfering waves.

10. Two waves simultaneously present on a long string have a phase difference ϕ between them so that a standing wave formed from their combination is described by

$$y(x, t) = 2A\sin\left(kx + \frac{\phi}{2}\right)\cos\left(\omega t - \frac{\phi}{2}\right)$$

(a) Despite the presence of the phase angle ϕ, is it still true that the nodes are a half wavelength apart? Explain your answer. (b) Are the nodes different in any way from the way they would be if ϕ were zero? Explain.

11. Two transverse sinusoidal waves combining in a medium are described by the wave functions

$$y_1 = 3.00\sin\pi(x + 0.600t) \qquad y_2 = 3.00\sin\pi(x - 0.600t)$$

where x, y_1, and y_2 are in centimetres and t is in seconds. Determine the maximum transverse position of an element of the medium at (a) $x = 0.250$ cm, (b) $x = 0.500$ cm, and (c) $x = 1.50$ cm. (d) Find the three smallest values of x corresponding to antinodes.

12. Two identical loudspeakers are driven in phase by a common oscillator at 800 Hz and face each other at a distance of 1.25 m. Locate the points along the line joining the two speakers where relative minima of sound pressure amplitude would be expected.

Section 18.4 Analysis model: waves under boundary conditions

13. A certain vibrating string on a piano has a length of 74.0 cm and forms a standing wave having two antinodes. (a) Which harmonic does this wave represent? (b) Determine the wavelength of this wave. (c) How many nodes are there in the wave pattern? (d) Draw the wave pattern.

14. A string that is 30.0 cm long and has a mass per unit length of 9.00×10^{-3} kg/m is stretched to a tension of 20.0 N. Find (a) the fundamental frequency and (b) the next three frequencies that could cause standing-wave patterns on the string.

15. The A string on a cello vibrates in its first normal mode with a frequency of 220 Hz, and may not vary by more than 2 Hz. The vibrating segment is (70.0 ± 0.2) cm long and has a mass of 1.20 g. (a) Find the allowable range of tension of the string. (b) Determine the frequency of vibration when the string vibrates in three segments.

16. In the arrangement shown in **Figure P18.16**, an object can be hung from a string (with linear mass density $\mu = 0.002\ 00$ kg/m) that passes over a light pulley. The string is connected to a vibrator (of constant frequency f), and the length of the string between point P and the pulley is $L = 2.00$ m. When the mass m of the object is either 16.0 kg or 25.0 kg, standing waves are observed; no standing waves are observed with any mass between these values, however. (a) What is the frequency of the vibrator? *Note:* The greater the tension in the string, the smaller the number of nodes in the standing wave. (b) What is the largest object mass for which standing waves could be observed?

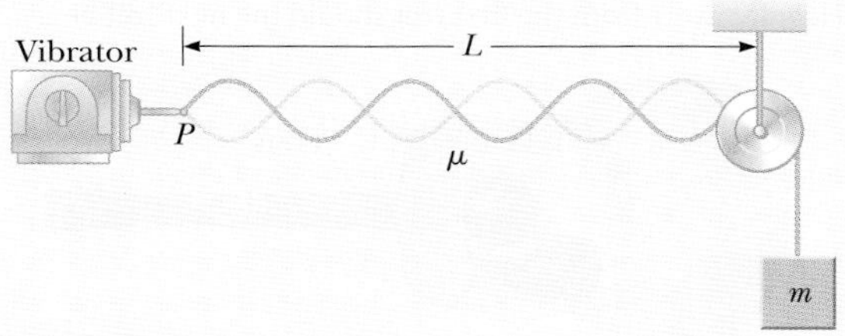

Figure P18.16

17. ±? A violin string has a length of 0.350 m and is tuned to concert G, with $f_G = 392$ Hz. (a) How far from the end of the string must the violinist place her finger to play concert A, with $f_A = 440$ Hz? (b) If this position is to remain correct to half the width of a finger (that is, to within 0.600 cm), what is the maximum allowable percentage change in the string tension?

18. The Bay of Fundy, Nova Scotia, Canada, has the highest tides in the world. Assume in mid-ocean and at the mouth of the bay the water surface oscillates with an amplitude of a few centimetres and a period of 12 h 24 min. At the head of the bay, the amplitude is several metres. Assume the bay has a length of 210 km and a uniform depth of 36.1 m. The speed of long-wavelength water waves is given by $v = \sqrt{gd}$, where d is the water's depth. Argue for or against the proposition that the tide is magnified by standing-wave resonance.

19. High-frequency sound can be used to produce standing-wave vibrations in a wine glass. A standing-wave vibration in a wine glass is observed to have four nodes and four antinodes equally spaced around the 20.0 cm circumference of the rim of the glass. If transverse waves move around the glass at 900 m/s, an opera singer would have to produce a high harmonic with what frequency to shatter the glass with a resonant vibration as shown in **Figure P18.19**?

Steve Bronstein/Stone/Getty Images

Figure P18.19

Section **18.5** Standing waves in pipes, rods and membranes

20. Two adjacent natural frequencies of an organ pipe are determined to be 550 Hz and 650 Hz. Calculate (a) the fundamental frequency and (b) the length of this pipe.

21. The overall length of a piccolo is 32.0 cm. The resonating air column is open at both ends. (a) Find the frequency of the lowest note a piccolo can sound. (b) Opening holes in the side of a piccolo effectively shortens the length of the resonant column. Assume the highest note a piccolo can sound is 4000 Hz. Find the distance between adjacent antinodes for this mode of vibration.

22. ±? A tube that is open at one end and closed at the other by a movable piston is used in an experiment to measure the speed of sound in air, as shown in **Figure P18.22**, on a warm day in a lab with no air conditioning. A 384 Hz tuning fork is held at the open end. Resonance is heard when the piston is at a distance $d_1 = (22.8 \pm 0.2)$ cm from the open end and again when it is at a distance $d_2 = (68.3 \pm 0.4)$ cm from the open end. (a) What speed of sound is implied by these data? (b) How far from the open end will the piston be when the next resonance is heard?

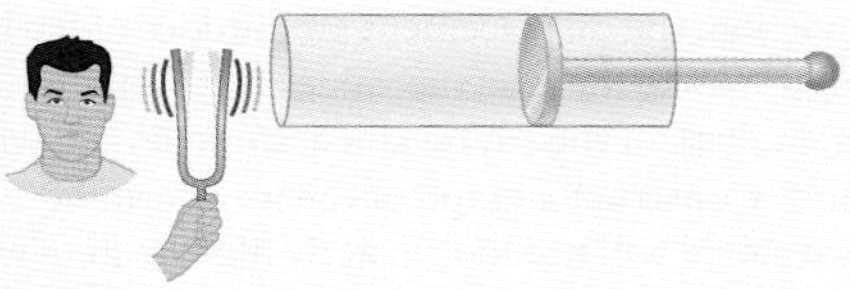

Figure P18.22

23. A pipe open at both ends has a fundamental frequency of 300 Hz when the temperature is 0°C. (a) What is the length of the pipe? (b) What is the fundamental frequency at a temperature of 30.0°C?

24. A shower cubicle has dimensions 86.0 cm × 86.0 cm × 210 cm. Assume the cubicle acts as a pipe closed at both ends. Assume singing voices range from 130 Hz to 2000 Hz and the speed of sound in the hot humid air is 355 m/s. For someone singing in this shower, which frequencies would sound the richest?

25. As shown in **Figure P18.25**, water is pumped into a tall, vertical cylinder at a volume flow rate R. The radius of the cylinder is r and at the open top of the cylinder a tuning fork is vibrating with a frequency f. As the water rises, what time interval elapses between successive resonances?

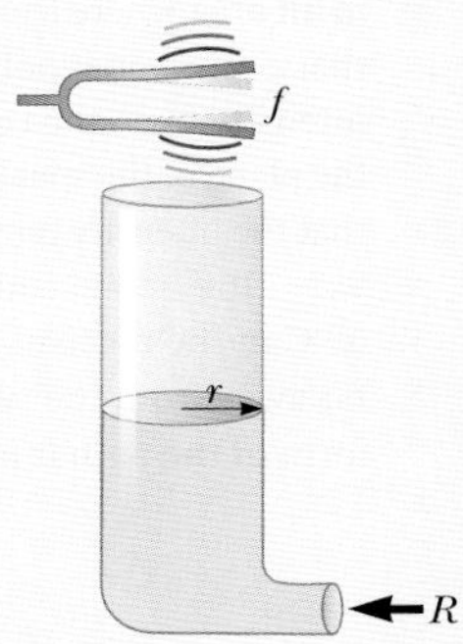

Figure P18.25

26. An aluminium rod 1.60 m long is held at its centre. It is stroked with a rosin-coated cloth to set up a longitudinal vibration. The speed of sound in a thin rod of aluminium is 5100 m/s. (a) What is the fundamental frequency of the waves established in the rod? (b) What harmonics are set up in the rod held in this manner? (c) **What If?** What would be the fundamental frequency if the rod were copper, in which the speed of sound is 3560 m/s?

Section **18.6** Beats: interference in time

27. In certain ranges of a piano keyboard, more than one string is tuned to the same note to provide extra loudness. For example, the note at 110 Hz has two strings at this frequency. If one string slips from its normal tension of 600 N to 540 N, what beat frequency is heard when the hammer strikes the two strings simultaneously?

28. While attempting to tune the note C at 523 Hz, a piano tuner hears 2.00 beats/s between a reference oscillator and the string. (a) What are the possible frequencies of the string? (b) When she tightens the string slightly, she hears 3.00 beats/s. What is the frequency of the string now? (c) By what percentage should the piano tuner now change the tension in the string to bring it into tune?

29. A student holds a tuning fork oscillating at 256 Hz. He walks towards a wall at a constant speed of 1.33 m/s. (a) What beat frequency does he observe between the tuning fork and its echo? (b) How fast must he walk away from the wall to observe a beat frequency of 5.00 Hz?

Section 18.7 Fourier analysis and synthesis

30. Suppose a flautist plays a 523 Hz C note with first harmonic displacement amplitude $A_1 = 100$ nm. From Figure 18.23 read, by proportion, the displacement amplitudes of harmonics 2 through 7. Take these as the values A_2 through A_7 in the Fourier analysis of the sound and assume $B_n = 0$ where $n = 1$ to 7. Construct a graph of the waveform of the sound. Your waveform will not look exactly like the flute waveform in Figure 18.23, because you simplify by ignoring cosine terms; nevertheless, it produces the same sensation to human hearing.

Additional problems

31. Two identical loudspeakers 10.0 m apart are driven by the same oscillator with a frequency of $f = 21.5$ Hz (Figure P18.31) in an area where the speed of sound is 344 m/s. (a) Show that a receiver at point A records a minimum in sound intensity from the two speakers. (b) If the receiver is moved in the plane of the speakers, show that the path it should take so that the intensity remains at a minimum is along the hyperbola $9x^2 - 16y^2 = 144$ (shown in red-brown in Figure P18.31). (c) Can the receiver remain at a minimum and move very far away from the two sources? If so, determine the limiting form of the path it must take. If not, explain how far it can go.

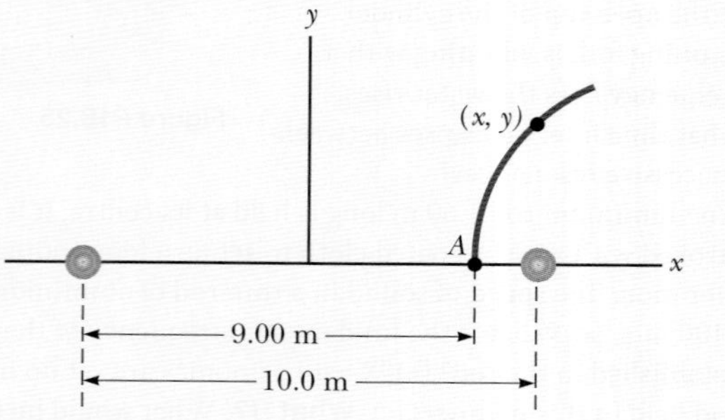

Figure P18.31

32. Verify by direct substitution that the wave function for a standing wave given in Equation 18.1,

$$y = (2A \sin kx) \cos(\omega t)$$

is a solution of the general linear wave equation, Equation 17.27:

$$\frac{\partial^2 y}{\partial x^2} = \frac{1}{v^2}\frac{\partial^2 y}{\partial t^2}$$

33. In the arrangement shown in Figure P18.16, an object of mass $m = 5.00$ kg hangs from a cord around a light pulley. The length of the cord between point P and the pulley is $L = 2.00$ m. (a) When the vibrator is set to a frequency of 150 Hz, a standing wave with six loops is formed. What must be the linear mass density of the cord? (b) How many loops (if any) will result if m is changed to 45.0 kg? (c) How many loops (if any) will result if m is changed to 10.0 kg?

34. The fret closest to the bridge on a guitar is 21.4 cm from the bridge as shown in Figure P18.34. When the thinnest string is pressed down at this first fret, the string produces the highest frequency that can be played on that guitar, 2349 Hz. The next lower note that is produced on the string has frequency 2217 Hz. How far away from the first fret should the next fret be?

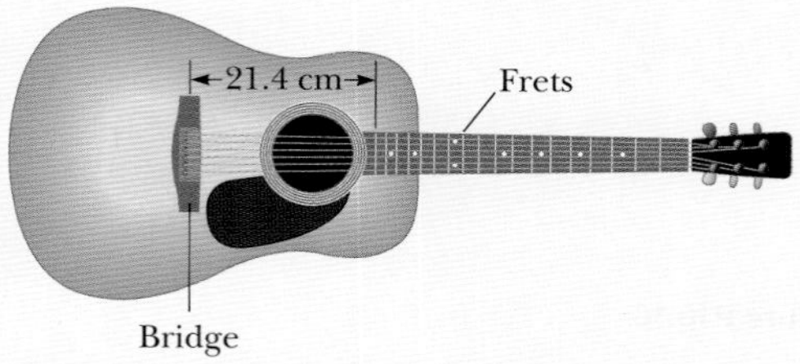

Figure P18.34

35. A string fixed at both ends and having a mass of 4.80 g, a length of 2.00 m and a tension of 48.0 N vibrates in its second ($n = 2$) normal mode. (a) Is the wavelength in air of the sound emitted by this vibrating string larger or smaller than the wavelength of the wave on the string? (b) What is the ratio of the wavelength in air of the sound emitted by this vibrating string and the wavelength of the wave on the string?

36. A quartz watch contains a crystal oscillator in the form of a block of quartz that vibrates by contracting and expanding. An electric circuit feeds in energy to maintain the oscillation and also counts the voltage pulses to keep time. Two opposite faces of the block, 7.05 mm apart, are antinodes, moving alternately towards each other and away from each other. The plane halfway between these two faces is a node of the vibration. The speed of sound in quartz is equal to 3.70×10^3 m/s. Find the frequency of the vibration.

37. ±? As part of an experiment, one end of a horizontal string is attached to a vibrating blade and the other end passes over a pulley as in Figure P18.37a. A sphere of mass (2.00 ± 0.05) kg hangs on the end of the string. The string is vibrating in its second harmonic. A container of water is raised under the sphere so that the sphere is completely submerged. In this configuration, the string vibrates in its fifth harmonic as shown in Figure P18.37b. What is the radius of the sphere?

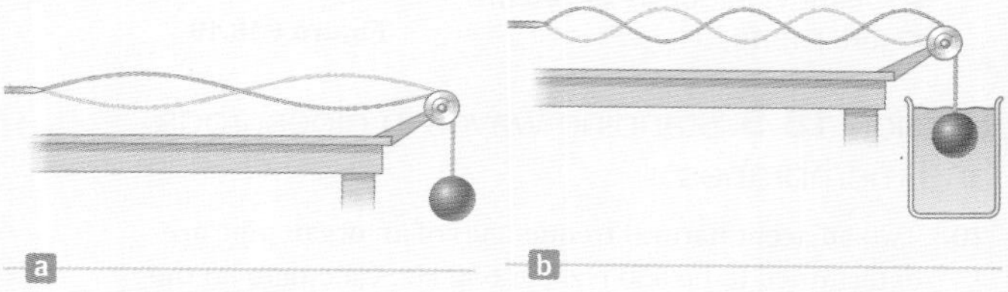

Figure P18.37
Problems 37 and 38

38. Consider the apparatus shown in **Figure P18.37** and described in problem 37. Suppose the number of antinodes in **Figure P18.37b** is an arbitrary value n. (a) Find an expression for the radius of the sphere in the water as a function of only n. (b) What is the minimum allowed value of n for a sphere of non-zero size? (c) What is the radius of the largest sphere that will produce a standing wave on the string? (d) What happens if a larger sphere is used?

39. The top end of a yo-yo string is held stationary. The yo-yo itself is much more massive than the string. It starts from rest and moves down with constant acceleration 0.800 m/s^2 as it unwinds from the string. The rubbing of the string against the edge of the yo-yo excites transverse standing-wave vibrations in the string. Both ends of the string are nodes even as the length of the string increases. Consider the instant 1.20 s after the motion begins from rest. (a) Show that the rate of change with time of the wavelength of the fundamental mode of oscillation is 1.92 m/s. (b) Is the rate of change of the wavelength of the second harmonic also 1.92 m/s at this moment? Explain your answer.

40. On a marimba (**Figure P18.40**), the wooden bar that sounds a tone when struck vibrates in a transverse standing wave having three antinodes and two nodes. The lowest-frequency note is 87.0 Hz, produced by a bar 40.0 cm long. (a) Find the speed of transverse waves on the bar. (b) A resonant pipe suspended vertically below the centre of the bar enhances the loudness of the emitted sound. If the pipe is open at the top end only, what length of the pipe is required to resonate with the bar in part (a)?

Figure P18.40

41. Two train whistles have identical frequencies of 180 Hz. When one train is at rest in the station and the other is moving nearby, a commuter standing on the station platform hears beats with a frequency of 2.00 beats/s when the whistles operate together. What are the two possible speeds and directions the moving train can have?

42. Two wires are welded together end to end. The wires are made of the same material, but the diameter of one is twice that of the other. They are subjected to a tension of 4.60 N. The thin wire has a length of 40.0 cm and a linear mass density of 2.00 g/m. The combination is fixed at both ends and vibrated in such a way that two antinodes are present, with the node between them being right at the weld. (a) What is the frequency of vibration? (b) What is the length of the thick wire?

43. A standing wave is set up in a string of variable length and tension by a vibrator of variable frequency. Both ends of the string are fixed. When the vibrator has a frequency f, in a string of length L and under tension T, n antinodes are set up in the string. (a) If the length of the string is doubled, by what factor should the frequency be changed so that the same number of antinodes is produced? (b) If the frequency and length are held constant, what tension will produce $n + 1$ antinodes? (c) If the frequency is tripled and the length of the string is halved, by what factor should the tension be changed so that twice as many antinodes are produced?

44. Two waves are described by the wave functions

$$y_1(x, t) = 5.00 \sin(2.00x - 10.0t)$$

$$y_2(x, t) = 10.0 \cos(2.00x - 10.0t)$$

where x, y_1 and y_2 are in metres and t is in seconds. (a) Show that the wave resulting from their superposition can be expressed as a single sine function. (b) Determine the amplitude and phase angle for this sinusoidal wave.

45. Marcus and Laurence have been given flutes for Christmas and are diligently practising, one in his room and the other outside (to prevent them fighting). Laurence, who is inside, plays a middle C (frequency of 261.6 Hz) by covering all the holes. (a) Consider the flute as a pipe that is open at both ends. Find the length of the flute, assuming middle C is the fundamental, and it is 20°C in the room. (b) Marcus, standing just outside the window to annoy his brother, also plays middle C on his identical flute. A beat frequency of 3.00 Hz is produced. What is the temperature outside?

Challenge problems

46. A string of linear density 1.60 g/m is stretched between clamps 48.0 cm apart. The string does not stretch appreciably as the tension in it is steadily raised from 15.0 N at $t = 0$ to 25.0 N at $t = 3.50$ s. The string is vibrating in its fundamental mode throughout this process. Find the number of oscillations it completes during the 3.50 s interval.

47. Consider the apparatus shown in **Figure P18.47a**, in which the hanging object has mass M and the string is vibrating in its second harmonic. The vibrating blade at the left maintains a constant frequency. The wind begins to blow to the right, applying a constant horizontal force $\vec{F}$ on the hanging object. What is the magnitude of the force the wind must apply to the hanging object so that the string vibrates in its first harmonic as shown in **Figure P18.47b**?

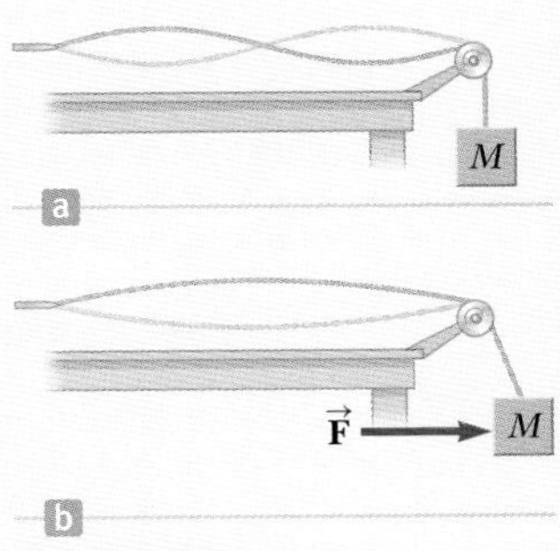

Figure P18.47

48. In Active Figures 18.25a and 18.25b, notice that the amplitude of the component wave for frequency f is large, that for $3f$ is smaller and that for $5f$ smaller still. How do we know exactly how much amplitude to assign to each frequency component to build a square wave? This problem helps us find the answer to that question. Let the square wave in Active Figure 18.25c have an amplitude A and let $t = 0$ be at the extreme left of the figure. So, one period T of the square wave is described by

$$y(t) = \begin{cases} A & 0 < t < \dfrac{T}{2} \\ -A & \dfrac{T}{2} < t < T \end{cases}$$

Express Equation 18.13 with angular frequencies:

$$y(t) = \sum_n (A_n \sin n\omega t + B_n \cos n\omega t)$$

Now proceed as follows. (a) Multiply both sides of Equation 18.13 by $\sin m\omega t$ and integrate both sides over one period T. Show that the left-hand side of the resulting equation is equal to 0 if m is even and is equal to $4A/m\omega$ if m is odd. (b) Using trigonometric identities, show that all terms on the right-hand side involving B_n are equal to zero. (c) Using trigonometric identities, show that all terms on the right-hand side involving A_n are equal to zero *except* for the one case of $m = n$. (d) Show that the entire right-hand side of the equation reduces to $\frac{1}{2} A_m T$. (e) Show that the Fourier series expansion for a square wave is

$$y(t) = \sum_n \frac{4A}{n\pi} \sin n\omega t$$

case study 4

The alphorn

Nicoleta Gaciu

The alphorn is a traditional long wooden instrument made of young spruce or pine tree bark that has been used by shepherds all over Europe since primeval times. While in the past the alphorn was used as an instrument for communication, today it is mostly used as a musical instrument. The instrument has a simple conical shape, which can be modified by adding a curved 'bell' at the open end. To produce a pure note while playing an alphorn is a great challenge. Its geometry and the skill of the musician determine the number of distinctive notes that can be played. In a straight alphorn, the sound wave is reflected at more or less the opening of the tube. The sound waves are created by blowing air into the mouthpiece and the vibrations are reflected back and forth between the player's lips, the mouthpiece and the open end of the alphorn. An audible tone is produced when the original wave constructively interferes with the reflected one. If the alphorn is longer, then the wavelength is longer and the frequency lower. The alphorn can play a harmonic series or sequence of harmonics (see Section 18.4). The frequencies of these sounds are integer multiples of the fundamental frequency.

The 'bell' end and the mouthpiece play a significant role with respect to the acoustics of the alphorn and affect the values of the frequency, increasing the low-frequency power transmission even more. This frequency shift is due to the change of the reflection point: where the wave is reflected back in the bell. If the bell end is the same size as the wavelength, it becomes difficult to support the standing wave because almost no wave is reflected. Many questions regarding the resonance shift and changes in the radiation of the alphorn with different bell-end shapes and lengths remain unanswered.

Shutterstock.com/Luca Lorenzelli

In classical music, many composers have incorporated an alphorn theme or a solo alphorn in their composition. For example, an alphorn theme was included by Johannes Brahms in the fourth movement of his Symphony No.1 in C minor and by Ludwig van Beethoven in the last movement of his Symphony No. 6 in F major (Pastoral Symphony). In 1755 Leopold Mozart composed *Sinfonia Pastorella* for alphorn and strings. In the same period, Johann Sebastian Bach wrote a piece of music for the 'lituus', a forgotten long wooden musical instrument, with a sound quality somewhere between a trumpet and an alphorn. In 2009, scientists at the Musical Acoustics Laboratory in Edinburgh developed and used an integrated software package for designing, testing and optimising brass instruments.[1] The result was a 2.7-metre-long horn with a flared bell at the end. Since then, the Schola Cantorum Basiliensis, a Swiss-based music conservatorium specialising in early music, has used the 'Edinburgh design' to build two identical examples of the long-lost instrument.

For musicians and musical instrument manufacturers, the optimisation of air-column instruments has been very important as new methods of designing these instruments have been introduced. In 2009 a new design of 'brass' air-column instruments was introduced by a group of researchers from the University of Nottingham, which included structural and acoustical interactions in the optimisation process.[2] Since 2012, the team of engineers and numerical-methods specialists from the Acoustics and Audio Group at the University of Edinburgh have been exploring numerical techniques for the simulation of a variety of instrument families, in order to generate synthetic sound.[3]

The human audible frequency range is between 20 Hz and 20000 Hz. Assume you want to design an alphorn for the fundamental frequency, and the first characteristic of the instrument you have to take into consideration is its length. To calculate the length of the alphorn, we can use the equation

$$L = n\frac{v}{2f_n}, \quad n = 1$$

where f is the frequency and v is the speed of sound in air for temperatures close to room temperature. In our case, the length of the simple canonical alphorn, whose first harmonic frequency is 27.5 Hz, is

$$L = \frac{332 \text{ m/s}}{2 \times 27.5 \text{ Hz}} = 6.04 \text{ m}$$

The traditional melodies played with a typical alphorn, which is about 6 metres or more in length, use at least two to eight of the harmonics.

The power of the alphorn lies in how far we can communicate with it. Assume that we measure the intensity of the sound at the end of the alphorn described earlier and it is $I_1 = 0.1 \text{ mW/m}^2$. The standard threshold of hearing intensity detectable by the ear is $I_0 = 10^{-12} \text{ W/m}^2$. The sound intensity is inversely proportional to the distance of the point of measurement from the source and is described by the following equation:

$$I \sim \frac{1}{r^2}$$

The maximum distance at which we can hear the alphorn sound will be:

$$\frac{I_0}{I_1} = \frac{r_1^2}{r_0^2} \rightarrow r_1 = r_0\sqrt{\frac{I_0}{I_1}} = 6.04 \times \sqrt{\frac{0.1 \times 10^{-2}}{10^{-12}}} = 60.4 \times 10^3 \text{ m} = 60.4 \text{ km}$$

References

1 http://www.ph.ed.ac.uk/news/bachs-long-lost-horn-reconstructed-24-09-09

2 Brackett, D. J., Ashcroft, I. A. and Hague, R. J. M., 2009, Multi-physics optimisation of 'brass' instruments—a new method to include structural and acoustical interactions, *Structural and Multidisciplinary Optimization*, 40 (1–6): 611–24.

3 http://www.ness-music.eu/publications

Nicoleta Gaciu has a PhD in Physics from the University of Surrey, a Master in Physics from Stuttgart University and a BSc in Physics from Babes-Bolyai University. She has taught physics, mathematics and computing to undergraduate and postgraduate students, and is a member of the Institute of Physics and a Fellow of the Higher Education Academy. Nicoleta has published research on spatio-temporal dynamics of bio and optoelectronic systems, twin-stripe and broad area semiconductor lasers, managerial statistics, physics and leadership, and band effects in magnetostrictive properties of different compounds. She is now a Senior Lecturer at Oxford Brookes University and her main areas of research are physics education, metacognition and advanced research methods.

Thermodynamics

PART 4

A bubble in one of the many mud pots in Rotorua, New Zealand, is caught just at the moment of popping. A mud pot is a pool of bubbling hot mud that demonstrates the existence of thermodynamic processes below the Earth's surface.

Have you ever wondered why a burn to your skin from steam seems to hurt more than a burn from boiling water? What happens to the kinetic energy of a moving object when the object comes to rest? How a refrigerator is able to cool its contents? How usable energy is generated in a power plant or in the engine of your car? In the following chapters we study thermodynamics, the field of physics that provides us with a way to answer these and other questions. Thermodynamics provides a model that describes situations in which the temperature or state (solid, liquid, gas) of a system changes due to the transfer of energy to or from the system. It also explains how changes in the bulk properties of matter correlate with the mechanics of atoms and molecules.

The historical development of thermodynamics parallelled the development of the atomic theory of matter. By the 1820s, chemical experiments had provided solid evidence for the existence of atoms. At that time, scientists recognised that a connection between thermodynamics and the mechanics of atoms and molecules must exist. In 1827, botanist Robert Brown reported that grains of pollen suspended in a liquid move erratically from one place to another as if under constant agitation. In 1905, Albert Einstein used kinetic theory to explain the cause of this erratic motion, known today as *Brownian motion*. Einstein explained this phenomenon by assuming the grains are under constant bombardment by 'invisible' molecules in the liquid, which themselves move erratically and whose average speed depends on the liquid's temperature. A connection was thus forged between the everyday world and the tiny, invisible building blocks that make up this world.

chapter 19

Heat and temperature

Shutterstock.com/Bjorn Stefanson

In this photograph we see evidence of water in all three phases. The water in the lake is liquid. Solid water appears on the mountains in the form of snow. The clouds in the sky consist of liquid water droplets that have condensed from the gaseous water vapour in the air. What causes the water to take on these different forms?

On completing this chapter, students will understand:

- that *heat* can refer to either the process of energy transfer between systems caused by a difference in temperatures or to the energy transferred in this process
- that temperature is a measure of the kinetic energy of atoms and molecules in a material
- why solids and liquids expand when they are heated
- why both the volume and pressure of gases vary with temperature
- that specific heat characterises a system's thermal insensitivity to heat transfer
- that latent heat characterises the change in internal energy required to produce a phase change in a given material.

Students will be able to:

- distinguish between heat, internal energy, thermal energy and temperature
- identify when two objects are in thermal equilibrium
- calculate volume expansions for materials undergoing temperature changes
- use the ideal gas law to relate pressure, volume and temperature
- describe how thermometers work and how they are calibrated
- calculate how much energy is required to change the temperature of a sample of a given material and to produce a change in the material's phase.

Our quantitative study of mechanics required us to define such concepts as *mass, force* and *kinetic energy.* A quantitative description of thermal phenomena requires us to define additional concepts such as *heat*, *internal energy* and *temperature*. We start this chapter by defining these terms. We then focus on the physical observables that tell us whether heat has been transferred to or from a system, such as changes in size, changes in temperature and changes in phase.

19.1 Heat, internal energy and temperature

It is worth starting our discussion of thermodynamics by defining what we mean by heat when we use the term scientifically. When we walk barefoot on sand on a hot, sunny day we feel the sensation of hotness on the soles of our feet, whereas if we wade into the sea in winter we are likely to experience a sensation of coldness. However, the terms hotness and coldness are rather vague, and as we shall see in the next chapter sensations of hotness and coldness depend not only on the temperature of the object being touched but also on its thermal properties (in particular, its thermal conductivity) and the temperature of the skin. Hotness and coldness are thus not suitable terms to use in a scientific analysis of what is happening in these circumstances. We also know that the sand on the beach is heated by the sun and cools down at night. We therefore see that the concept of heat is connected with the idea of a transfer of something (which in this case produces a change in temperature) to or from an object. Although early theories imagined that the something being transferred was a fluid, we now know that the something is a form of energy.

TRY THIS

Fill one bowl or bucket with cold (but not freezing) water and another with warm (but not hot) water. Put one hand in each bowl and wait for a minute. Now swap your hands between the buckets. Do the temperatures of the water feel the same as they did before you swapped?

Scottish physicist James Clerk Maxwell (1831–1879) gave one of the first modern definitions of heat in 1871. According to his definition, heat is a form of energy that may be transferred from one object to another. Modern definitions allow two uses of the word heat:

1. Heat is the term used to describe the *energy* transferred from a system at higher temperature to a system at lower temperature; and
2. Heat is also the term used to describe the *flow* of energy between systems caused by differences in temperature.

◀ Definition of heat

Some physicists only accept the second definition, with heat used as a verb, as a valid definition. However, the use of heat as a noun, to describe the amount of energy transferred, is so common in engineering that we will also accept and use that definition in this text.

It is important to note that heat is not stored within a system. You may recall that work (measured in units of energy) describes only the energy added to or removed from a system to change its kinetic or potential energy. Similarly, heat (also measured in units of energy) describes only the energy that is transferred from one system to another, or between a system and its surroundings due to thermal processes. When energy in the form of heat is added to a system, it is stored as **internal energy**. Internal energy is all the energy of a system that is associated with its microscopic components – atoms and molecules – when viewed from a reference frame at rest with respect to the centre of mass of the system. The condition on the reference frame of the observer ensures that kinetic energy due to the overall motion of the system is not included in internal energy. Internal energy includes all the kinetic energy of the random translational, rotational and vibrational motion of molecules; vibrational potential energy associated with forces between atoms in molecules; and electric potential energy associated with forces between molecules. It is often useful to relate internal energy to the temperature of an object, but as we shall see in Section 19.4, this relationship is limited, since internal energy changes can also occur without associated changes in temperature.

◀ Internal energy

Pitfall Prevention 19.1
Heat is not stored in a system. It is a common misconception that heat can be stored in a system, for example in a solar hot water tank or in a heat pack used to reduce muscle pain. In such cases, energy is stored in the system in the form of internal energy; heat is the energy transferred to or from the water or heat pack as its temperature changes.

As heat, like work, is a form of energy, it is measured in units of energy, which in the SI system means joules, J ($kg.m^2/s^2$). However, you may also come across the calorie (cal), which is defined as the amount of energy transfer necessary to raise the temperature of 1 g of water from 14.5°C to 15.5°C and is related to the joule as 1 cal = 4.186 J. Although this unit was defined before the equivalence between heat and other forms of energy became clear, it is still sometimes used, especially in giving the energy content of foods (where the kilocalorie, written as a capital 'C' may be given instead of kilojoules).

We have already used the concept of temperature in our description of heat and internal energy. However, just as with the term heat, we need to give a scientific definition of temperature. In thermodynamics, the temperature of an object increases when the average speed of its constituent atoms or molecules increases. In fact, the temperature depends on the average speed squared, and hence on the average kinetic energy:

$$T \propto K = \frac{1}{2}mv_{avg}^2$$

Thermal energy

Temperature can be seen as a measure of the total kinetic energy of the constituent atoms or molecules, sometimes referred to as the *thermal energy*. When energy is transferred to a system in the form of heat, the internal energy of the system increases; however, this can result in an increase in the thermal energy or an increase in the intermolecular potential energy (or both). Because temperature is a measure of the thermal energy only, it is *not* a measure of heat.

Pitfall Prevention 19.2

It is important to distinguish between internal energy and thermal energy. Thermal energy is that part of the internal energy associated with random motion of molecules and is therefore directly related to temperature. However, internal energy can also be stored in a material by increasing the average intermolecular distances and hence increasing the intermolecular potential energy or bond energy. Therefore, internal energy = thermal energy + bond energy.

TRY THIS

Get two glasses and some food colouring. Fill one glass with hot water and the other with cold. Put a drop of food colouring in each glass and observe what happens. Can you explain what you see in terms of the relationship between temperature and molecular speeds discussed above?

We observe experimentally that two objects at different initial temperatures eventually reach some intermediate temperature when placed in contact with each other. For example, if you half-fill the sink with hot water and then add water from the cold tap, energy is transferred from the hotter water to the colder water and the final temperature of the mixture is somewhere between the initial temperatures of the water emerging from the taps. On the other hand, if you half-fill the sink with water from the cold tap and then add more water from the same tap, no energy is exchanged and no temperature change is observed, except for that due to the interaction of the water with the environment (the metal of the sink and the air in the room).

Now, imagine that two objects A and B are placed in an insulated container such that they interact with each other but not with the environment; the two objects are said to be in **thermal contact**. If the objects are at different temperatures, energy will be transferred between them, as is the case when hot and cold water mix. If no heat energy is exchanged between them, as is the case with mixing cold water with more cold water, they are said to be in **thermal equilibrium**. Figure 19.1 shows a practical example of this situation. If a third object (C) is added into the container and no heat flow occurs between it and objects A and B, then we say that it is in thermal equilibrium with both of them. We summarise this in a statement known as the **zeroth law of thermodynamics** (the law of equilibrium):

Zeroth law of thermodynamics

If objects A and B are separately in thermal equilibrium with a third object C, then A and B are in thermal equilibrium with each other.

Temperature

This statement enables us to define temperature without needing to measure average speeds or kinetic energies of constituent molecules. We can think of temperature as the property that determines whether an object is in thermal equilibrium with other objects. **Two objects in thermal equilibrium with each other are at the same temperature**. Conversely, if two objects have different temperatures, they are not in thermal equilibrium with each other.

iStockphoto/Steve Snyder

Figure 19.1
Two objects in thermal equilibrium, isolated from the environment

Quick Quiz 19.1

Two objects with different sizes, masses and temperatures are placed in thermal contact. In which direction does the energy travel? **(a)** Energy travels from the larger object to the smaller object. **(b)** Energy travels from the object with more mass to the one with less mass. **(c)** Energy travels from the object at higher temperature to the object at lower temperature.

This definition of temperature allows us to compare the relative temperatures of different objects but does not define a scale. In practice, we define temperature scales through variations in the observable physical properties of materials. On the **Celsius temperature scale**, a mixture of ice and water in thermal equilibrium is defined to have a temperature of zero degrees Celsius, which is written as 0°C; this temperature is called the *ice point* of water. The temperature of a mixture of water and steam in thermal equilibrium at atmospheric pressure is defined as 100°C; this is the *steam point* of water. As described in the next section, variations in physical properties such as the volume of mercury or a monatomic gas between these two points can then be used to establish degrees of temperature inside and outside the range of temperature at which water is liquid. An alternative temperature scale which is commonly used in thermodynamics is the **absolute** or **Kelvin scale**. On this scale, a change in temperature of one kelvin (1 K) is equal to a change in temperature of 1°C. The ice and steam points of water are 273.15 K and 373.15 K respectively. We shall discuss the origin of the absolute temperature scale in Section 19.4. Temperature is a basic quantity, and we cannot express the dimensions of temperature in terms of length L, mass M, and time T. Temperature has its own dimension which we write as **Θ**, which is the capital Greek letter theta.

Pitfall Prevention 19.3
Notations for temperatures in the Kelvin scale do not use the degree sign. The unit for a Kelvin temperature is simply 'kelvins' and not 'degrees Kelvin'.

19.2 Thermal expansion of liquids and solids

So far we have discussed heat and temperature without considering what experimental observables allow us to know that an object's temperature has changed (or by how much). Although we know that we can sometimes sense changes in an object's temperature due to changes in our physical sensation of hotness, this sensation does not provide a reliable, reproducible and quantifiable way for us to measure temperature. In the following, we consider one of the effects of temperature changes that provides us with a means to quantify such changes. This is the phenomenon whereby a substance's volume increases as its temperature increases. Such **thermal expansion** plays an important role in numerous engineering applications. For example, thermal-expansion joints such as those shown in Figure 19.2 must be included in concrete highways, railway tracks, brick walls and bridges to compensate for expansion of the materials as the temperature changes.

Thermal expansion is a consequence of the change in the *average* separation between the atoms in an object. To understand this concept, let's model the atoms as being connected by stiff springs as discussed in Section 16.4 and shown in Figure 16.16b. At ordinary temperatures, the atoms in a solid oscillate about their equilibrium positions with an amplitude of approximately 10^{-11} m and a frequency of approximately 10^{13} Hz. The average spacing between the atoms is about 10^{-10} m. As the temperature of the solid increases, the atoms oscillate with greater amplitudes; as a result, the average separation between them increases. Consequently, the object expands. More precisely, thermal expansion arises from the *asymmetrical* nature of the potential energy curve for the atoms in a solid as shown in Figure 16.16a. If the oscillators were truly simple harmonic, the average atomic separations would not change regardless of the amplitude of vibration.

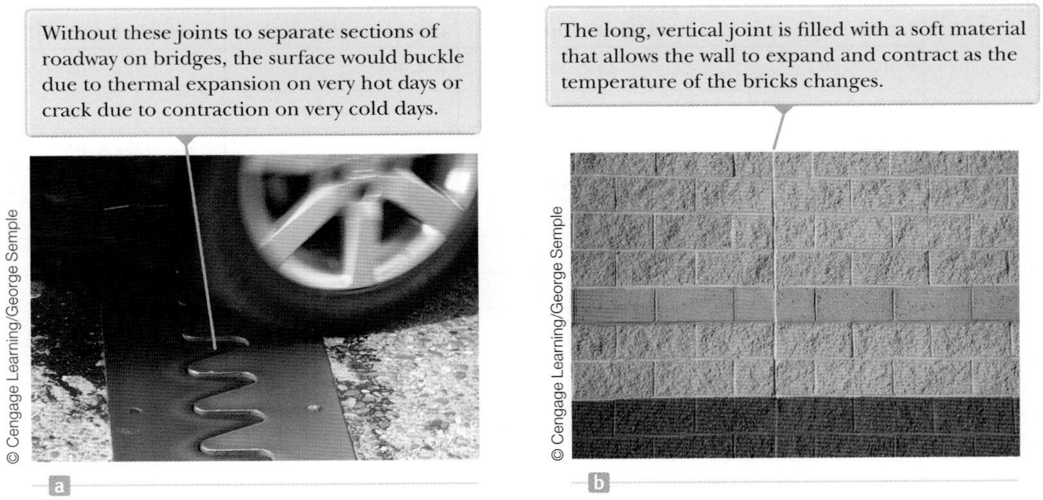

Without these joints to separate sections of roadway on bridges, the surface would buckle due to thermal expansion on very hot days or crack due to contraction on very cold days.
© Cengage Learning/George Semple
a
The long, vertical joint is filled with a soft material that allows the wall to expand and contract as the temperature of the bricks changes.
© Cengage Learning/George Semple
b

Figure 19.2
Thermal-expansion joints in (a) a bridge and (b) a wall

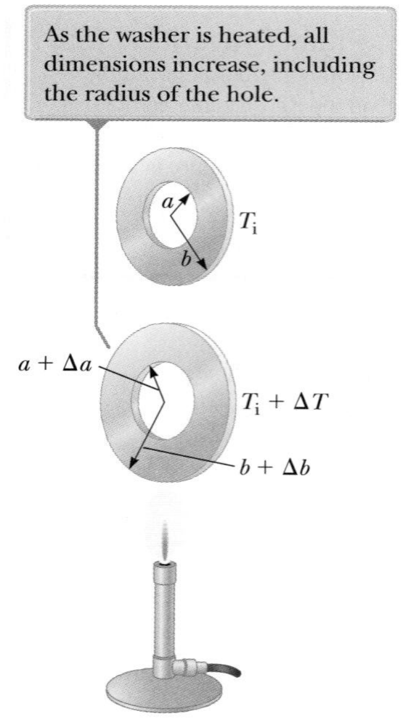

Active Figure 19.3
Thermal expansion of a homogeneous metal washer (The expansion is exaggerated in this figure.)

If thermal expansion is sufficiently small relative to an object's initial dimensions, the change in any dimension is, to a good approximation, proportional to the first power of the temperature change. Suppose an object has an initial length L_i along some direction at some temperature and the length increases by an amount ΔL for a change in temperature ΔT. Because it is convenient to consider the fractional change in length per degree of temperature change, we define the **average coefficient of linear expansion** as

$$\alpha \equiv \frac{\Delta L / L_i}{\Delta T}$$

Experiments show that α is constant for small changes in temperature. For purposes of calculation, this equation is usually rewritten as

$$\Delta L = \alpha L_i \Delta T \quad (19.1)$$

or as

$$L_f - L_i = \alpha L_i (T_f - T_i) \quad (19.2)$$

where L_f is the final length, T_i and T_f are the initial and final temperatures respectively, and the proportionality constant α is the average coefficient of linear expansion for a given material and has units of $(°C)^{-1}$ and dimensions Θ^{-1}. **Equation 19.1** can be used for both thermal expansion, when the temperature of the material increases, and thermal contraction, when its temperature decreases.

It may be helpful to think of thermal expansion as an effective magnification of an object. For example, as a metal washer is heated (**Active Figure 19.3**), all dimensions, including the radius of the hole, increase according to **Equation 19.1**. A cavity in a piece of material expands in the same way as if the cavity were filled with the material. You may have seen an illustration of this in a lecture demonstration in which a ball can pass through a metal ring after the ring is heated but not before.

Table 19.1 lists the average coefficients of linear expansion for various materials. For these materials, α is positive, indicating an increase in length with increasing temperature. That is not always the case, however. Some substances, for example calcite ($CaCO_3$), expand along one dimension (positive α) and contract along another (negative α) as their temperatures are increased.

A simple mechanism called a *bimetallic strip*, found in practical devices such as thermostats, uses the difference in coefficients of expansion for different materials. It consists of two thin strips of dissimilar metals bonded together. As the temperature of the strip increases, the two metals expand by different amounts and the strip bends as shown in **Figure 19.4**.

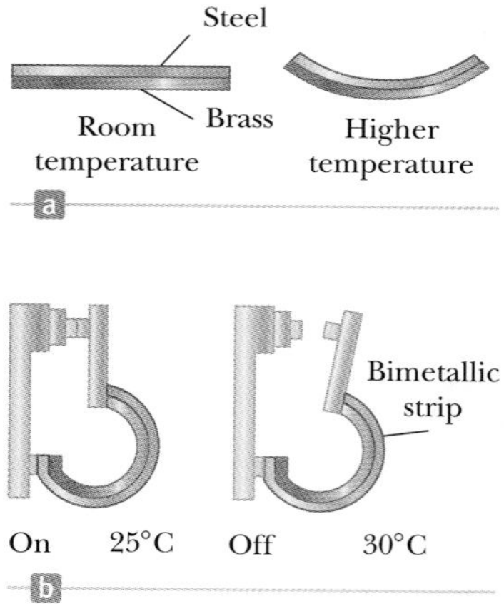

Figure 19.4
(a) A bimetallic strip bends as the temperature changes because the two metals have different expansion coefficients. (b) A bimetallic strip used in a thermostat to make or break electrical contact

Table 19.1
Average expansion coefficients for some materials near room temperature

Material (solids)	**Average linear expansion coefficient (α) $(°C)^{-1}$**	**Material (liquids and gases)**	**Average volume expansion coefficient (β) $(°C)^{-1}$**
Aluminium	24×10^{-6}	Acetone	1.5×10^{-4}
Brass and bronze	19×10^{-6}	Alcohol, ethyl	1.12×10^{-4}
Concrete	12×10^{-6}	Benzene	1.24×10^{-4}
Copper	17×10^{-6}	Petrol	9.6×10^{-4}
Glass (ordinary)	9×10^{-6}	Glycerine	4.85×10^{-4}
Glass (Pyrex)	3.2×10^{-6}	Mercury	1.82×10^{-4}
Invar (Ni–Fe alloy)	0.9×10^{-6}	Turpentine	9.0×10^{-4}
Lead	29×10^{-6}	Air[a] at 0°C	3.67×10^{-3}
Steel	11×10^{-6}	Helium[a]	3.665×10^{-3}

[a]Gases do not have a specific value for the volume expansion coefficient because the amount of expansion depends on the type of process through which the gas is taken. The values given here assume the gas undergoes an expansion at constant pressure.

TRY THIS

Screw the metal lid tightly onto a glass jam jar at room temperature and then put it in the fridge overnight. In the morning, try to unscrew the lid. Repeat your experiment, but this time run the lid under hot water for a minute before you try to open the jar. Can you explain why it is easier to remove the lid the second time round?

Pitfall Prevention 19.4

Do holes become larger or smaller? When an object's temperature is raised, every linear dimension increases in size. That includes any holes in the material, which expand in the same way as if the hole were filled with the material, as shown in **Active Figure 19.3**.

Because the linear dimensions of an object change with temperature, it follows that surface area and volume change as well. The change in volume is proportional to the initial volume V_i and to the change in temperature according to the relationship

$$\Delta V = \beta V_i \Delta T \tag{19.3}$$

where β is the **average coefficient of volume expansion**. To find the relationship between β and α, assume the average coefficient of linear expansion of the solid is the same in all directions; that is, assume the material is *isotropic*. Consider a solid box of dimensions l, w and h. Its volume at some temperature T_i is $V_i = lwh$. If the temperature changes to $T_i + \Delta T$, its volume changes to $V_i + \Delta V$, where each dimension changes according to **Equation 19.1**. Therefore,

$$\begin{aligned} V_i + \Delta V &= (l + \Delta l)(w + \Delta w)(h + \Delta h) \\ &= (l + \alpha l \Delta T)(w + \alpha w \Delta T)(h + \alpha h \Delta T) \\ &= lwh(1 + \alpha \Delta T)^3 \\ &= V_i\left[1 + 3\alpha\Delta T + 3(\alpha\Delta T)^2 + (\alpha\Delta T)^3\right] \end{aligned}$$

Dividing both sides by V_i and isolating the term $\Delta V/V_i$, we obtain the fractional change in volume:

$$\frac{\Delta V}{V_i} = 3\alpha\Delta T + 3(\alpha\Delta T)^2 + (\alpha\Delta T)^3$$

Because $\alpha\Delta T \ll 1$ for typical values of ΔT ($< \sim 100°C$), we can neglect the terms $3(\alpha\Delta T)^2$ and $(\alpha\Delta T)^3$. When we make this approximation, we see that

$$\frac{\Delta V}{V_i} = 3\alpha\Delta T \rightarrow \Delta V = 3\alpha V_i \Delta T$$

Comparing this expression to **Equation 19.3** shows that

$$\beta = 3\alpha$$

In a similar way, you can show that the change in area of a rectangular plate is given by $\Delta A = 2\alpha A_i \Delta T$ (see problem 45).

The mathematical model for thermal expansion presented above is based on three approximations: that the material is isotropic (expansion is the same in all directions), that α is constant for the temperature range, and that the terms $(\alpha\Delta T)^2$ and $(\alpha\Delta T)^3$ are small. While these approximations are often reasonable, this is not always the case, particularly for non-isotropic materials such as calcite. As with all models, we need to remember that it is an approximation, and that the real situation may be more complicated.

Quick Quiz 19.2

Two spheres are made of the same metal and have the same radius, but one is hollow and the other is solid. The spheres are taken through the same temperature increase. Which sphere expands more? **(a)** The solid sphere expands more. **(b)** The hollow sphere expands more. **(c)** They expand by the same amount. **(d)** There is not enough information to say.

Example 19.1

Expansion of a railway track

A segment of steel railway track has a length of 30.000 m when the temperature is 0.0°C.

(A) What is its length when the temperature is 40.0°C?

Solution

Conceptualise As the rail is relatively long, we expect to obtain a measurable increase in length for a 40°C temperature increase.

Model We will assume that the material of the track is isotropic, and that α is constant over the range of temperatures. This allows us to apply the model developed in this section.

Analyse Use **Equation 19.1** and the value of the coefficient of linear expansion from **Table 19.1**:

$$\Delta L = \alpha L_i \Delta T = (11 \times 10^{-6}\ (°\text{C})^{-1})(30.000\ \text{m})(40.0°\text{C}) = 0.013\ \text{m}$$

Find the new length of the track:

$$L_f = 30.000\ \text{m} + 0.013\ \text{m} = 30.013\ \text{m}$$

(B) Suppose the ends of the rail are rigidly clamped at 0.0°C so that expansion is prevented. What is the thermal stress set up in the rail if its temperature is raised to 40.0°C?

Solution

Conceptualise The thermal stress is the same as the tensile stress in the situation in which the rail expands freely and is then compressed by a mechanical force F back to its original length.

Model We now need to use the model for tensile stress and strain developed in Chapter 15, and make the assumption that the material is in the linear region of the stress-strain curve where the stress is small.

Analyse Find the tensile stress from **Equation 15.10** using Young's modulus for steel from **Table 15.1**:

$$\text{Tensile stress} = \frac{F}{A} = Y\frac{\Delta L}{L_i}$$

$$\frac{F}{A} = (20 \times 10^{10}\ \text{N/m}^2)\left(\frac{0.013\ \text{m}}{30.000\ \text{m}}\right) = 8.7 \times 10^7\ \text{N/m}^2$$

Finalise The expansion in part (A) is 1.3 cm. This expansion is indeed measurable as predicted in the *Conceptualise* step. The thermal stress in part (B) can be avoided by leaving small expansion gaps between the rails.

What If? What if the temperature drops to −40.0°C? What is the length of the unclamped segment?

Answer The expression for the change in length in **Equation 19.1** is the same whether the temperature increases or decreases. Therefore, if there is an increase in length of 0.013 m when the temperature increases by 40°C, there is a decrease in length of 0.013 m when the temperature decreases by 40°C. (We assume α is constant over the entire range of temperatures.) The new length at the colder temperature is 30.000 m − 0.013 m = 29.987 m.

The unusual behaviour of water

Liquids generally increase in volume with increasing temperature and have average coefficients of volume expansion about 10 times greater than those of solids. Cold water is an exception to this rule, as you can see from its density–temperature curve shown in **Figure 19.5**. As the temperature increases from 0°C to 4°C, water contracts and its density therefore increases. Above 4°C, water expands with increasing temperature and so its density decreases. Therefore, the density of water reaches a maximum value of 1.000 g/cm^3 at 4°C. Cold water is an example of a material for which the approximations made in our model are not appropriate, and hence the model does not predict the behaviour of the material. From **Figure 19.5** we can see that the model is still appropriate between approximately 10°C and 80°C, but outside this range α cannot be treated as constant.

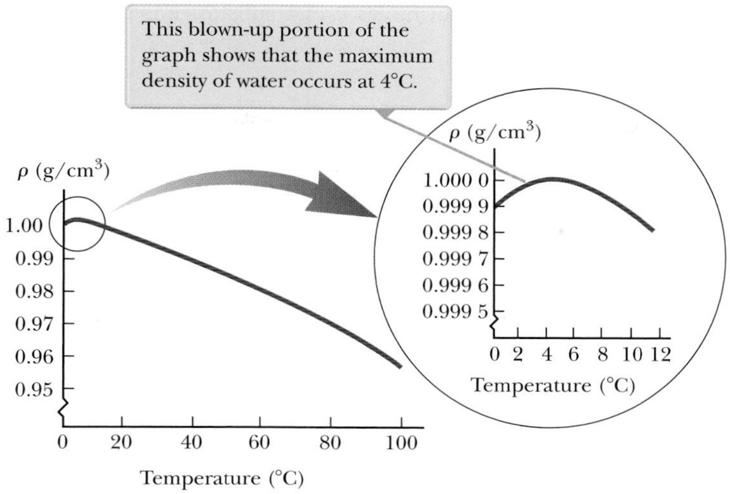

Figure 19.5
The variation in the density of water at atmospheric pressure with temperature

We can use this unusual thermal-expansion behaviour of water to explain why a pond begins freezing at the surface rather than at the bottom. When the air temperature drops from, for example, 7°C to 6°C, the surface water also cools and consequently decreases in volume. The surface water is denser than the water below it, which has not cooled and decreased in volume. As a result, the surface water sinks, and warmer water from below moves to the surface. When the air temperature is between 4°C and 0°C, however, the surface water expands as it cools, becoming less dense than the water below it. The mixing process stops and eventually the surface water freezes. As the water freezes, the ice remains on the surface because ice is less dense than water. The ice continues to build up at the surface, while water near the bottom remains at 4°C. If that were not the case, fish and other forms of marine life would not survive.

19.3 The ideal gas law

The volume expansion equation $\Delta V = \beta V_i \Delta T$ is based on the assumption that the material has an initial volume V_i before the temperature change occurs. Such is the case for solids and liquids because they have a fixed volume at a given temperature.

The case for gases is completely different. The interatomic forces within gases are very weak and, in many cases, we can imagine these forces to be non-existent and still make very good approximations. Therefore, *there is no equilibrium separation* for the atoms and no 'standard' volume at a given temperature; the volume depends on the size of the container. As a result, we cannot express changes in volume ΔV in a process on a gas with Equation 19.3 because we have no defined volume V_i at the beginning of the process. Equations involving gases contain the volume V, rather than a *change* in the volume from an initial value, as a variable.

For a gas, it is useful to know how the quantities volume V, pressure P and temperature T are related for a sample of gas of mass m. In general, the equation that relates these quantities, called the *equation of state*, is very complicated. If the gas is maintained at a very low pressure (or low density), however, the equation of state is quite simple and can be determined from experimental results. Such a low-density gas is commonly referred to as an **ideal gas**.

◀ Ideal gas model

The assumptions in the **ideal gas model** are that (1) the temperature of the gas is not too low (the gas must not condense) and (2) the pressure is low. This implies that the gas molecules do not interact except via collisions and that the molecular volume is negligible compared with the volume of the container.

In reality, an ideal gas does not exist. The concept of an ideal gas is nonetheless very useful because real gases at low pressures are well modelled as ideal gases.

It is convenient to express the amount of gas in a given volume in terms of the number of moles n. One **mole** of any substance is the amount of the substance that contains **Avogadro's number** $N_A = 6.022 \times 10^{23}$ of constituent particles (atoms or molecules); n is a pure number, it has units *mol* but no

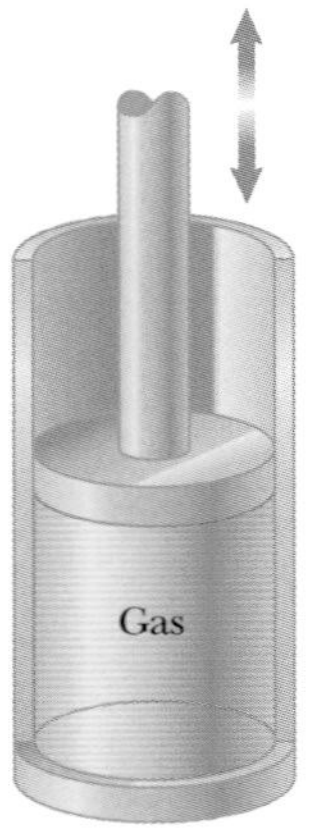

Active Figure 19.6

An ideal gas confined to a cylinder whose volume can be varied by means of a movable piston

dimensions. The number of moles n of a substance is related to its mass m through the expression

$$n = \frac{m}{M} \tag{19.4}$$

where M is the molar mass of the substance. The molar mass of each chemical element is the atomic mass (from the periodic table; see Appendix C) expressed in grams per mole. For example, the mass of one He atom is 4.00 u (atomic mass units), so the molar mass of He is 4.00 g/mol.

Now suppose an ideal gas is confined to a cylindrical container whose volume can be varied by means of a movable piston as in **Active Figure 19.6**. If we assume the cylinder does not leak, the mass (or the number of moles) of the gas remains constant. For such a system, experiments provide the following information.

- When the gas is kept at a constant temperature, its pressure is inversely proportional to the volume. (This behaviour is described historically as Boyle's law.)
- When the pressure of the gas is kept constant, the volume is directly proportional to the temperature. (This behaviour is described historically as Charles' law.)
- When the volume of the gas is kept constant, the pressure is directly proportional to the temperature. (This behaviour is described historically as Gay-Lussac's law.)

These observations are summarised by the **equation of state for an ideal gas**:

Equation of state for an ideal gas ▶

$$PV = nRT \tag{19.5}$$

In this expression, also known as the **ideal gas law**, n is the number of moles of gas in the sample and R is a constant. Experiments on numerous gases show that as the pressure approaches zero, the quantity PV/nT approaches the same value R for all gases. For this reason, R is called the **universal gas constant**. In SI units, in which pressure is expressed in pascals (1 Pa = 1 N/m^2) and volume in cubic metres, the product PV has units of newton.metres, or joules, and R has the value

$$R = 8.314 \text{ J/mol.K} \tag{19.6}$$

If the pressure is expressed in atmospheres and the volume in litres (1 L = 10^3 cm^3 = 10^{-3} m^3), then R has the value

$$R = 0.082\,06 \text{ L.atm/mol.K}$$

Using this value of R, **Equation 19.5** suggests that the volume occupied by 1 mol of *any* gas at atmospheric pressure and at 0°C (273 K) is 22.4 L. Although the ideal gas law assumes low pressure, it provides a good approximation of the behaviour of gases at atmospheric pressures.

Pitfall Prevention 19.5

There are a variety of physical quantities for which the letter k is used. Two we have seen previously are the force constant for a spring (Chapter 6) and the wave number for a mechanical wave (Chapter 16). Boltzmann's constant is another k, and we will see k used for thermal conductivity in Chapter 20 and for an electrical constant in Chapter 23. To make some sense of this confusing state of affairs, we use a subscript 'B' for Boltzmann's constant to help us recognise it. In this book, you will see Boltzmann's constant as k_B, but you may see Boltzmann's constant in other resources as simply k.

Quick Quiz 19.3

What are the dimensions of R?

The ideal gas law is often expressed in terms of the total number of molecules N. Because the number of moles n equals the ratio of the total number of molecules and Avogadro's number N_A, we can write **Equation 19.5** as

$$PV = nRT = \frac{N}{N_A}RT$$

$$PV = Nk_BT \tag{19.7}$$

where k_B is **Boltzmann's constant**, which has the value

Boltzmann's constant ▶

$$k_B = \frac{R}{N_A} = 1.38 \times 10^{-23} \text{ J/K} \tag{19.8}$$

It is common to call quantities such as P, V and T the **thermodynamic variables** of an ideal gas. If the equation of state is known, one of the variables can always be expressed as some function of the other two.

TRY THIS

Inflate a balloon and measure its circumference at its widest point by noting the length of thread or string you can tie around it. Estimate the temperature of the air in the balloon and the temperature in your freezer. Put the balloon in the freezer for a couple of hours. Calculate the change in the balloon's volume you expect if the pressure due to the balloon's surface is constant. When you remove it, carefully measure the circumference again. Do your observations match your predictions?

Quick Quiz 19.4

On a winter day, you turn on the heating and the temperature of the air inside your home increases. Assume your home has the normal amount of leakage between inside air and outside air. Is the number of moles of air in your room at the higher temperature **(a)** larger than before, **(b)** smaller than before or **(c)** the same as before?

Example 19.2

Heating a spray can

A spray can containing a propellant gas at twice atmospheric pressure (202 kPa) and having a volume of 125.00 cm^3 is in thermal equilibrium with its surroundings in the back of a rubbish truck at 22°C. The rubbish is transferred from the truck into an incinerator. When the temperature of the gas in the can reaches 195°C, what is the pressure inside the can?

Solution

Conceptualise Intuitively, you should expect that the pressure of the gas in the container increases because of the increasing temperature.

Model The ideal gas law is a good model for gases at low temperatures. In this case, the gas is not at low temperature, and it is likely that there will be interactions between molecules other than elastic collisions. However, we can use the ideal gas law, **Equation 19.5**, to get a first approximation to the behaviour of the gas in the can. We will also make the approximation that the can itself does not expand significantly.

Analyse Rearrange **Equation 19.5**:

$$\frac{PV}{T} = nR \qquad (1)$$

No air escapes during the compression, so n, and therefore nR, remains constant. Hence, set the initial value of the left side of **Equation (1)** equal to the final value:

$$\frac{P_i V_i}{T_i} = \frac{P_f V_f}{T_f} \qquad (2)$$

Because the initial and final volumes of the gas are assumed to be equal, cancel the volumes:

$$\frac{P_i}{T_i} = \frac{P_f}{T_f} \qquad (3)$$

Solve for P_f:

$$P_f = \left(\frac{T_f}{T_i}\right)P_i = \left(\frac{468\text{ K}}{295\text{ K}}\right)(202\text{ kPa}) = 320\text{ kPa}$$

Finalise The higher the temperature, the higher the pressure exerted by the trapped gas as expected. If the pressure increases sufficiently, the can may explode. **This is why you should never dispose of spray cans in a fire**.

What If? Suppose we include a volume change due to thermal expansion of the steel can as the temperature increases. Does that alter our answer for the final pressure significantly?

Answer Because the thermal expansion coefficient of steel is very small, we do not expect much of an effect on our final answer.

Find the change in the volume of the can using **Equation 19.3** and the value for α for steel from **Table 19.1**:

$$\Delta V = \beta V_i \Delta T = 3\alpha V_i \Delta T$$
$$= 3[11 \times 10^{-6}(°\text{C})^{-1}](125.00\text{ cm}^3)(173°\text{C}) = 0.71\text{ cm}^3$$

Example 19.2 cont.

Start from **Equation (2)** again and find an equation for the final pressure:

$$P_f = \left(\frac{T_f}{T_i}\right)\left(\frac{V_i}{V_f}\right)P_i$$

We can see that both sides have dimensions of pressure, $[ML^{-1}T^{-2}]$, as the two fractions are dimensionless. ☺

This result differs from **Equation (3)** only in the factor V_i/V_f. Evaluate this factor:

$$\frac{V_i}{V_f} = \frac{125.00\ \text{cm}^3}{(125.00\ \text{cm}^3 + 0.71\ \text{cm}^3)} = 0.994 = 99.4\%$$

Therefore, the final pressure will differ by only 0.6% from the value calculated without considering the thermal expansion of the can. Taking 99.4% of the previous final pressure, the final pressure including thermal expansion is 318 kPa.

±? Example **19.3**

Inflating crisp packets

You buy a packet of potato crisps on a hot summer's day, when the temperature is (35 ± 1)°C and you are at sea level, where atmospheric pressure is $(1.013 \pm 0.001) \times 10^5$ Pa. When you first buy them, you estimate that the air in the packet occupies a volume of 0.50 ± 0.02 litres. You have bought the crisps to eat on a plane journey. During the journey, the cabin pressure reaches a minimum of $(0.753 \pm 0.001) \times 10^5$ Pa and the temperature a minimum of (18 ± 1)°C. What is the volume occupied by the air in the packet at this point?

Solution

Conceptualise We expect the volume to increase due to the change in pressure but to decrease due to the change in temperature.

Model We model the air in the packet as ideal and use the ideal gas law to calculate the new volume.

Analyse Rearranging **Equation 19.5** gives an expression for volume in terms of pressure and temperature

$$V = \frac{nRT}{P}$$

Assuming no air leaks from the packet, the quantity nR is constant throughout this problem. We know the initial conditions are

$$V_i = \frac{nRT_i}{P_i}$$

and the final conditions are

$$V_f = \frac{nRT_f}{P_f}$$

Assuming that the pressure and temperature inside the crisp packet are the same as the pressure and temperature of its environment, take the ratio of these and rearrange to obtain

$$V_f = \left(\frac{T_f P_i}{T_i P_f}\right)V_i$$

We can see that both sides have dimensions of volume, $[L^3]$, as the fraction is dimensionless. ☺
Substitute the values given in the question (remembering to use temperatures on the Kelvin scale):

$$V_f = \frac{(291\ \text{K}) \times (1.013 \times 10^5\ \text{Pa})}{(307\ \text{K}) \times (0.753 \times 10^5\ \text{Pa})} \times 0.50\ \text{L} = 0.64\ \text{L}$$

Finalise We cannot give a final answer without calculating the uncertainty. Looking at the equation for the final volume we derived in the solution, we see that the final volume depends on the initial and final temperatures, the initial and final pressures and the initial volume in a way that means that all of the associated fractional uncertainties must be added

Example 19.3 cont.

in quadrature. As usual, we check to see whether any of these fractional uncertainties will dominate. It is important to remember that the uncertainty of ±1°C in the temperatures is equal to an uncertainty of ±1 K when we convert to kelvins, which is about 0.3%. The fractional uncertainties in the pressures are even smaller. Thus the fractional uncertainty in the initial volume (2/50 or 4%) is much larger than any of the other fractional uncertainties and will dominate the uncertainty in the final answer. We therefore estimate the uncertainty simply as 4% of the final volume, giving $V_f = (0.64 \pm 0.03)$ L.

19.4 Thermometers

Thermometers are devices used to measure the temperature of a system. All thermometers are based on the principle that some physical property of a system changes as the system's temperature changes. As we have already seen, the volumes of liquids and solids increase as temperature increases; the volume of a gas increases with temperature at constant pressure, and the pressure of a gas increases with temperature at constant volume. Other physical changes that can form the basis on which thermometers function include the electric resistance of a conductor, which decreases with increasing temperature (see Chapter 27), and the colour of an object.

A common thermometer in everyday use consists of a mass of liquid – usually mercury or alcohol – that expands into a glass capillary tube when heated. Such thermometers depend on the linear expansion of the liquid with increasing temperature, so that any temperature change in the range of the thermometer can be defined as being proportional to the change in length of the liquid column. Such thermometers can be calibrated using the ice and steam points of water.

Unfortunately, thermometers calibrated in this way present problems when extremely accurate readings are needed. For instance, the readings given by an alcohol thermometer calibrated at the ice and steam points of water might agree with those given by a mercury thermometer only at the calibration points. For example, since mercury and alcohol have different thermal expansion properties, when one thermometer reads a temperature of 50°C, the other may indicate a slightly different value. The discrepancies between thermometers are especially large when the temperatures to be measured are far from the calibration points.

Quick Quiz 19.5

If you are asked to make a very sensitive glass thermometer, which of the following working liquids would you choose? (a) mercury (b) alcohol (c) petrol (d) glycerine

The constant-volume gas thermometer and the Kelvin temperature scale

An additional practical problem of any thermometer is the limited range of temperatures over which it can be used. A mercury thermometer, for example, cannot be used below the freezing point of mercury, which is −39°C, and an alcohol thermometer is not useful for measuring temperatures above 85°C, the boiling point of alcohol. To surmount this problem, we need a universal thermometer whose readings are independent of the substance used in it. The gas thermometer approaches this requirement. One version of a gas thermometer is the constant-volume apparatus shown in **Figure 19.7**. The physical change exploited in this device is the variation of pressure of a fixed volume of gas with temperature as described by the ideal gas law. The flask is immersed in an ice-water bath and mercury reservoir *B* is raised or lowered until the top of the mercury in column *A* is at the zero point on the scale. The height *h*, the difference between the mercury levels in reservoir *B* and column *A*, indicates the pressure in the flask at 0°C by means of **Equation 13.4**, $P = P_0 + \rho g h$.

The flask is then immersed in water at the steam point. Reservoir *B* is re-adjusted until the top of the mercury in column *A* is again at zero on the scale, which ensures that the volume of the gas is the same as it was when the flask was in the ice bath (hence the designation 'constant volume'). This adjustment of reservoir *B* gives a value for the

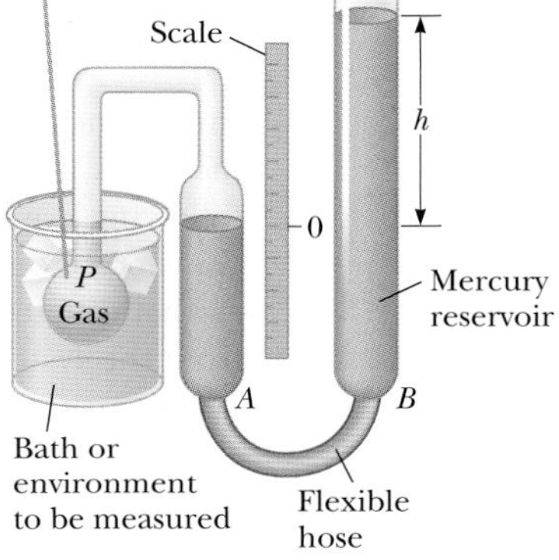
The volume of gas in the flask is kept constant by raising or lowering reservoir B to keep the mercury level in column A constant.
Scale
h
0
P
Gas
Mercury reservoir
A
B
Bath or environment to be measured
Flexible hose

Figure 19.7
A constant-volume gas thermometer measures the pressure of the gas contained in the flask immersed in the bath.

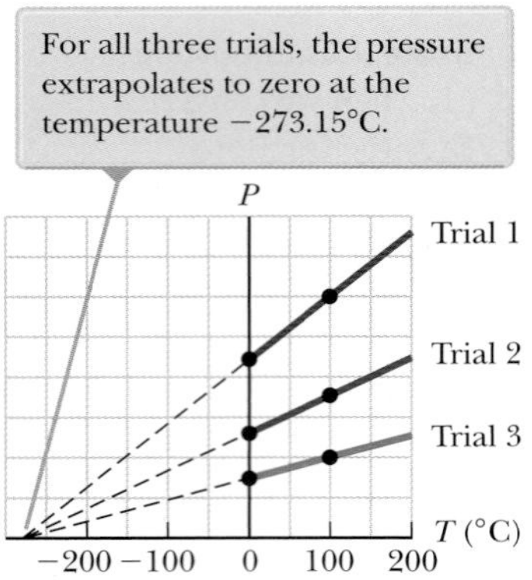

Figure 19.8
Pressure versus temperature for experimental trials in which gases have different pressures in a constant-volume gas thermometer

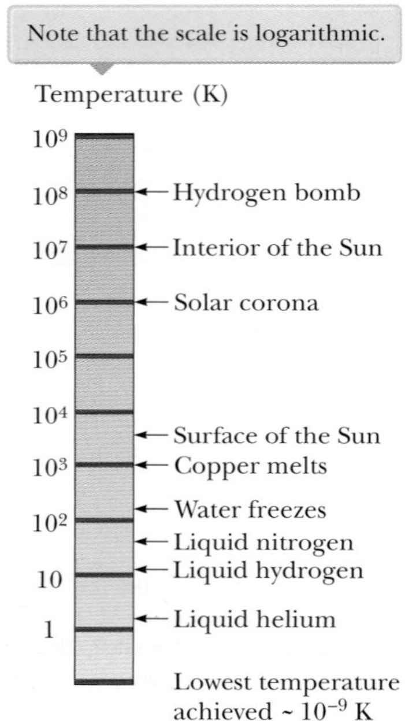

Figure 19.9
Absolute temperatures at which various physical processes occur

gas pressure at 100°C. These two pressure measurements provide the calibration points, and the ideal gas law indicates that the relationship between pressure and temperature between and beyond these points should be linear. To measure the temperature of a substance, the gas flask of **Figure 19.7** is placed in thermal contact with the substance and allowed to reach equilibrium and the height of reservoir *B* is adjusted until the top of the mercury column in *A* is at zero on the scale. The height of the mercury column in *B* indicates the pressure of the gas; knowing the pressure, the temperature of the substance can then be found.

Now suppose temperatures of different gases at different initial pressures are measured with gas thermometers. Experiments show that the thermometer readings are nearly independent of the type of gas used as long as the gas pressure is low and the temperature is well above the point at which the gas liquefies (**Figure 19.8**). The agreement among thermometers using various gases improves as the pressure is reduced.

If we extend the straight lines in **Figure 19.8** toward negative temperatures, we find a remarkable result: **in every case, the pressure is zero when the temperature is −273.15°C!** This finding suggests some special role that this particular temperature must play. It is used as the basis for the absolute temperature scale introduced in Section 19.1, which has −273.15°C as its zero point. This temperature is often referred to as **absolute zero**. It is indicated as a zero because at a lower temperature, the pressure of the gas would become negative, which is meaningless.

Because the ice and steam points are experimentally difficult to duplicate and depend on atmospheric pressure, an absolute temperature scale based on two new fixed points was adopted in 1954 by the International Committee on Weights and Measures. The first point is absolute zero. The second reference temperature for this new scale was chosen as the **triple point of water**, which is the single combination of temperature and pressure at which liquid water, gaseous water and ice (solid water) coexist in equilibrium. This triple point occurs at a temperature of 0.01°C and a pressure of 4.58 mm of mercury. On the new scale, which uses the unit *kelvin*, the temperature of water at the triple point was set at 273.16 K. This choice was made so that the old absolute temperature scale based on the ice and steam points would agree closely with the new scale based on the triple point.

Figure 19.9 gives the absolute temperature for various physical processes and structures. The temperature of absolute zero (0 K) cannot be achieved, although laboratory experiments have come very close, reaching temperatures of less than one nanokelvin.

19.5 Specific and latent heats

In the previous sections, we have seen how the thermal expansion of an object depends on the material it is composed of, so that different materials expand by different amounts when heated through the same temperature change. In Section 19.1, we discussed how the internal energy of an object increases when energy is transferred to it and we made the distinction between increases in thermal energy, which are measured as increases in temperature, and increases in intermolecular potential energy, which do not show up as increases in temperature. In the following, we shall see how the fraction of the energy transferred to an object in the form of heat (energy transferred due to a temperature difference) that is converted to thermal energy depends on the material, so that different materials undergo different changes in temperature for the same transfer of energy to or from the system.

Specific heat

When energy is transferred to a system in the form of heat and the materials in the system stay in the same phase (solid, liquid or gas), the temperature of the system usually rises. Experimentally, we find that the quantity of energy required to raise the temperature of a given mass by some amount varies from one substance to another. For example, the quantity of energy required to raise the temperature of 1 kg of water by 1°C is 4186 J, but the quantity of energy required to raise the temperature of 1 kg of copper by 1°C is only 387 J.

Experiments show that for a given substance, the amount of energy needed to raise a sample of the substance of mass m through a temperature change ΔT is directly proportional to both the mass and temperature change:

$$Q \propto m\Delta T \tag{19.9}$$

We call the constant of proportionality the **specific heat** or **specific heat capacity** c

$$c \equiv \frac{Q}{m\Delta T} \tag{19.10}$$

◄ Specific heat

Specific heat capacity is a measure of how thermally insensitive a substance is to the addition of energy. The greater a material's specific heat, the more energy must be added to a given mass of the material to cause a particular temperature change. Table 19.2 lists some specific heats. Specific heat (or specific heat capacity) has units of J/kg.°C which is the same as J/kg.K as it is only the temperature difference that matters and this is the same in both Celsius and kelvin. The dimensions of specific heat are $L^2\,T^{-2}\Theta^{-1}$.

Table 19.2
Specific heats of some substances at 25°C and atmospheric pressure

Substance	Specific heat (J/kg.°C)
Solids	
Aluminium	900
Copper	387
Gold	129
Iron	448
Lead	128
Silicon	703
Silver	234
Other solids	
Brass	380
Glass	837
Ice (−5°C)	2110
Wood	1700
Liquids	
Alcohol, ethyl	2400
Mercury	140
Water (15°C)	4186
Gas	
Steam (100°C)	2010

TRY THIS

Fill a small cup or metal container with water and take the water's temperature. Light a candle under the cup and let it burn for a few minutes, keeping note of the time. Measure the temperature of the water again. How long would you need to burn the candle to raise the water's temperature until it was hot enough to have a bath in? How many candles would you need to heat enough water to have a bath?

Combining Equations 19.9 and 19.10, we can see that the energy Q transferred between a sample of mass m of a material and its surroundings resulting in a temperature change ΔT is

$$Q = mc\Delta T \tag{19.11}$$

For example, the energy required to raise the temperature of 0.500 kg of water by 3.00°C is $Q = (0.500\text{ kg})(4186\text{ J/kg.°C})(3.00\text{°C}) = 6.28 \times 10^3$ J. Notice that when the temperature increases, Q and ΔT are taken to be positive and energy transfers into the system. When the temperature decreases, Q and ΔT are negative and energy transfers out of the system.

We can identify $mc\Delta T$ as the change in internal energy of the system if we ignore any thermal expansion or contraction of the system. (Thermal expansion or contraction would result in a very small amount of work being done on (or by) the system by (or on) the surrounding air.) Then, Equation 19.11 is a reduced form of the conservation of energy equation, Equation 7.2: $\Delta E_{int} = Q$. The internal energy of the system can be changed by transferring energy into the system by any mechanism. For example, if the system is a baked potato in an oven, Equation 7.2 reduces to the following analogue to Equation 19.11: $\Delta E_{int} = T_{ER} = mc\Delta T$, where T_{ER} is the energy transferred to the potato from the oven. If the system is the air in a bicycle pump, which becomes hot when the pump is operated, Equation 7.2 reduces to the following analogue to Equation 19.11: $\Delta E_{int} = W = mc\Delta T$, where W is the work done on the pump by the operator. By identifying $mc\Delta T$ as ΔE_{int}, we have taken a step towards a better understanding of temperature: temperature is related to the energy of the molecules of a system. We will learn more details of this relationship in Chapter 21.

Pitfall Prevention 19.6

The symbol Q represents the amount of energy transferred, but keep in mind that the energy transfer in Equation 19.11 could be by *any* of the methods introduced in Chapter 7; it does not have to be heat. For example, repeatedly bending a rubber eraser raises the temperature at the bending point by *mechanical work*.

Specific heat varies with temperature. If, however, temperature intervals are not too great, the temperature variation can be ignored and c can be treated as a constant.[1] For example, the specific heat of water varies by only about 1% from 0°C to 100°C at atmospheric pressure. More generally, if c varies over the temperature range of interest, $Q = m\int_{T_i}^{T_f} c\,dT$. Unless stated otherwise, we shall neglect such variations.

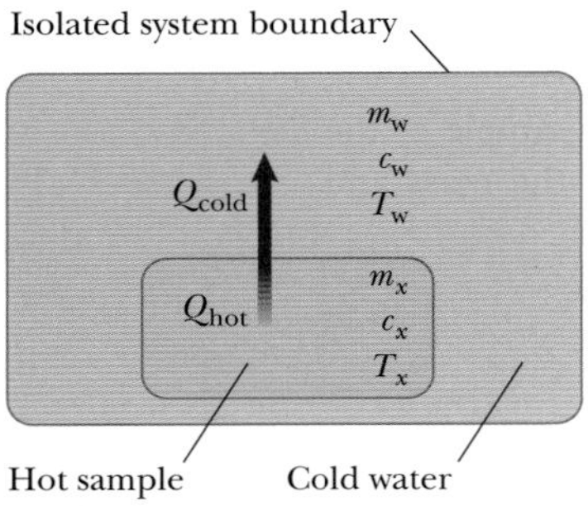

Figure 19.10
In a calorimetry experiment, a hot sample whose specific heat is unknown is placed in cold water in a container that isolates the system from the environment.

Quick Quiz 19.6

Imagine you have 1 kg each of iron, glass and water, and all three samples are at 10°C. **(a)** Rank the samples from highest to lowest temperature after 100 J of energy is added to each sample. **(b)** Rank the samples from greatest to least amount of energy transferred by heat if each sample increases in temperature by 20°C.

> **Pitfall Prevention 19.7**
> The negative sign in the equation is necessary for consistency with our sign convention for energy transfer. The energy transfer Q_{hot} has a negative value because energy is leaving the hot substance. The negative sign in the equation ensures that the right side is a positive number, consistent with the left side, which is positive because energy is entering the cold water.

Notice from **Table 19.2** that water has the highest specific heat of common materials. This high specific heat is in part responsible for the moderate climates found near large bodies of water. As the temperature of a body of water decreases during the winter, energy is transferred from the cooling water to the air by heat, increasing the internal energy of the air. Because of the high specific heat of water, a relatively large amount of energy is transferred to the air for even modest temperature changes of the water.

A simple technique for measuring specific heat involves heating a sample to some known temperature T_x, placing it in a vessel containing water of known mass and temperature $T_w < T_x$, and measuring the temperature of the water after equilibrium has been reached. This technique is called **calorimetry**, and devices in which this energy transfer occurs are called **calorimeters**. Calorimeters are generally made such that the material in contact with the water has a low specific heat and is a good insulator. This is so that energy is not transferred by heat to the surroundings, and only a very small amount of energy is used in any changes of temperature of the vessel. **Figure 19.10** shows the hot sample in the cold water and the resulting energy transfer by heat from the high-temperature part of the system to the low-temperature part. If the system of the sample and the water is isolated, the principle of conservation of energy requires that the amount of energy Q_{hot} that leaves the sample (of unknown specific heat) equals the amount of energy Q_{cold} that enters the water. Conservation of energy allows us to write the mathematical representation of conservation of energy in this context as

$$Q_{cold} = -Q_{hot} \quad \textbf{(19.12)}$$

Suppose m_x is the mass of a sample of some substance whose specific heat we wish to determine. Let's call its specific heat c_x and its initial temperature T_x as shown in **Figure 19.10**. Likewise, let m_w, c_w and T_w represent corresponding values for the water. If T_f is the final temperature after the system comes to equilibrium, **Equation 19.11** shows that the energy transfer to the water is $m_w c_w(T_f - T_w)$, which is positive because $T_f > T_w$, and that the energy transfer from the sample of unknown specific heat is $m_x c_x(T_f - T_x)$, which is negative. Substituting these expressions into **Equation 9.12** gives

$$m_w c_w(T_f - T_w) = -m_x c_x(T_f - T_x)$$

This equation can be solved for the unknown specific heat c_x.

±? Example 19.4

Cooling a hot ingot

In a chemistry experiment, a 0.0500-kg ingot of an unknown metal is heated to (200.0 ± 0.1)°C and then dropped into a calorimeter containing 0.400 kg of water initially at (20.0 ± 0.1)°C. The final equilibrium temperature of the mixed system is (22.4 ± 0.1)°C. Find the specific heat of the metal.

Solution

Conceptualise Imagine the process occurring in the isolated system of **Figure 19.10**. Energy leaves the hot ingot and goes into the cold water, so the ingot cools off and the water warms up. Once both are at the same temperature, the energy transfer stops.

Model We will make the approximation that all of the energy transferred from the hot ingot by heat results in an increase of the thermal energy and hence temperature of the water. We will also assume that there is no change of phase (boiling) of the water.

Example 19.4 cont.

Analyse Use **Equation 19.11** to evaluate each side of **Equation 19.12**:

$$m_w c_w (T_f - T_w) = -m_x c_x (T_f - T_x)$$

Solve for c:
$$c_x = \frac{m_w c_w (T_f - T_w)}{m_x (T_x - T_f)}$$

We can see by inspection that the dimensions of both sides of the equation are that for *c*, as the fraction on the right hand side is otherwise dimensionless. ☺

Substitute numerical values:
$$c_x = \frac{(0.400\ \text{kg})(4186\ \text{J/kg.°C})(22.4°\text{C} - 20.0°\text{C})}{(0.0500\ \text{kg})(200.0°\text{C} - 22.4°\text{C})}$$
$$= 453\ \text{J/kg.°C}$$

As we are not given uncertainties in the masses of water or ingot, we will assume that the uncertainty in the experiment is dominated by the uncertainty in the temperature measurements.

To calculate the uncertainty, we first need to find the uncertainty in the temperature changes. As this involves a subtraction, we simply add the absolute uncertainties in each temperature to get the uncertainty in the change:

$$\Delta(T_f - T_w) = \Delta T_f + \Delta T_w = 0.1°\text{C} + 0.1°\text{C} = 0.2°\text{C}$$

The uncertainty in $(T_x - T_f)$ is also 0.2°C.

So
$$\Delta c_x = c_x \left(\frac{\Delta(T_f - T_w)}{(T_f - T_w)} + \frac{\Delta(T_f - T_x)}{(T_x - T_f)} \right)$$
$$= 453\ \text{J/kg.°C} \left(\frac{0.2°\text{C}}{(22.4°\text{C} - 20.0°\text{C})} \right) + \left(\frac{0.2°\text{C}}{(200°\text{C} - 22.4°\text{C})} \right) = 38\ \text{J/kg.°C}$$

Our final answer is thus
$$c_x = (450 \pm 40)\ \text{J/kg.°C}$$

Finalise The ingot is most likely iron as you can see by comparing this result with the data given in **Table 19.2**. The temperature of the ingot is initially above the steam point. Therefore, some of the water may vaporise when the ingot is dropped into the water. We assume the system is sealed and this steam cannot escape. Because the final equilibrium temperature is lower than the steam point, any steam that does result recondenses back into water.

What If? Suppose you are performing an experiment in the laboratory that uses this technique to determine the specific heat of a sample and you wish to decrease the overall uncertainty in your final result for c_x. Of the data given in this example, changing which value would be most effective in decreasing the uncertainty?

Answer The largest experimental uncertainty is associated with the small difference in temperature of 2.4°C for the water. For this temperature difference to be larger experimentally, the most effective change is to decrease the amount of water.

Latent heat

In some situations, the transfer of energy in the form of heat to an object does not result in a change in temperature, but the material instead undergoes a **phase change**, for example when the substance boils, melts or undergoes a change in its crystalline structure. All such phase changes involve a change in the system's internal energy but no change in its temperature. The increase in internal energy in boiling, for example, results in the breaking of bonds between molecules in the liquid state; this bond breaking allows the molecules to move farther apart in the gaseous state, with a corresponding increase in intermolecular potential energy. Different substances respond differently to the addition or removal of energy as they change phase because their internal molecular arrangements vary.

Consider a system containing a substance in two phases in equilibrium, such as water and ice. When discussing two phases of a material, we will use the term *higher-phase material* to mean the material existing at the higher temperature. So, in this example, water is the higher-phase material, whereas steam would be the higher-phase material if we were considering a system of steam and water. Now imagine that energy Q enters the system. As a result, the final amount of water is increased due to the melting of some of the ice.

Experimentally we observe that the amount of energy needed to increase the amount of higher-phase material by an amount Δm is directly proportional to Δm:

Energy transferred to a substance during a phase change ▶

$$Q = L\Delta m \tag{19.13}$$

That is

$$L \equiv \frac{Q}{\Delta m} \tag{19.14}$$

Latent heat ▶

We call the constant of proportionality L the **latent heat** for this phase change. The value of L for a substance depends on the nature of the phase change as well as on the properties of the substance. **Latent heat of fusion** L_f is the term used when the phase change is from solid to liquid (*to fuse* means 'to combine by melting') and **latent heat of vaporisation** L_v is the term used when the phase change is from liquid to gas (the liquid 'vaporises'). Latent heat has units of has units of J/kg, and hence dimensions $L^2\,T^{-2}$.

When energy enters a system, causing melting or vaporisation, the amount of the higher-phase material increases, so Δm is positive and Q is positive, consistent with our sign convention. When energy is extracted from a system, causing freezing or condensation, the amount of the higher-phase material decreases, so Δm is negative and Q is negative, again consistent with our sign convention. Keep in mind that Δm in Equations 19.13 and 19.14 always refers to the higher-phase material. If the entire amount of the lower-phase material undergoes a phase change, the change in mass Δm of the higher-phase material is equal to the initial mass of the lower-phase material. The amount of energy released when a gas condenses or a liquid freezes is exactly the same as that required for the reverse phase change and so the latent heat of fusion is also the latent heat of solidification and the latent heat of vaporisation is also the latent heat of condensation.

The latent heats of various substances vary considerably as data in Table 19.3 show.

TRY THIS

Take an ice cube out of the freezer and use a candle flame to melt it. By timing how long it takes to melt the ice and estimating the volume of water in the ice cube, estimate the power (energy per second) obtained from the candle. Now melt a square of chocolate with the same candle flame. Can you estimate the latent heat of chocolate using this method?

To understand the role of latent heat in phase changes, consider the energy required to convert a 1.00-g cube of ice at −30.0°C to steam at 120.0°C at constant pressure. Figure 19.11 indicates the experimental results obtained when energy is gradually added to the ice. Initially, as energy is added to the ice, the temperature of the ice increases. Once the temperature reaches 0.0°C, the addition of more energy produces a phase change, as ice melts to water, with no change in temperature. Once all the ice has been converted to water, the addition

Pitfall Prevention 19.8

Signs are critical. Sign errors occur very often when students apply calorimetry equations. For phase changes, remember that Δm in Equation 19.13 is always the change in mass of the higher-phase material. In Equation 19.11, be sure your ΔT is *always* the final temperature minus the initial temperature. In addition, you must *always* include the negative sign on the right side of Equation 19.12.

Table 19.3
Latent heats of fusion and vaporisation

Substance	Melting point (°C)	Latent heat of fusion	Boiling point (°C)	Latent heat of vaporisation (J/kg)
Helium	−269.65	5.23×10^3	−268.93	2.09×10^4
Oxygen	−218.79	1.38×10^4	−182.97	2.13×10^5
Nitrogen	−0.97	2.55×10^4	−195.81	2.01×10^5
Alcohol, ethyl	114	1.04×10^5	78	8.54×10^5
Water	0.00	3.33×10^5	100.00	2.26×10^6
Sulfur	119	3.81×10^4	444.60	3.26×10^5
Lead	327.3	2.45×10^4	1750	8.70×10^5
Aluminium	660	3.97×10^5	2450	1.14×10^7
Silver	960.80	8.82×10^4	2193	2.33×10^6
Gold	1063.00	6.44×10^4	2660	1.58×10^6
Copper	1083	1.34×10^5	1187	5.06×10^6

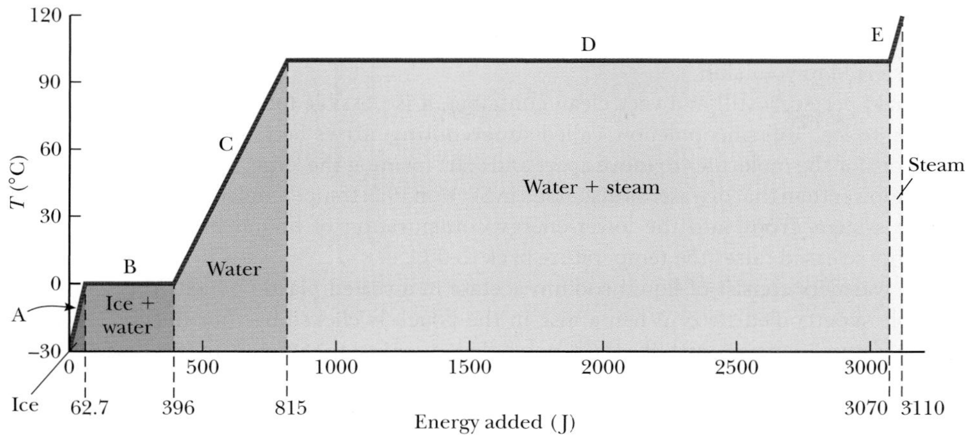

Figure 19.11
A plot of temperature versus energy added when 1.00 g of ice initially at −30.0°C is converted to steam at 120.0°C

of more energy starts to increase the temperature of the water up to 100°C. At this point, the addition of energy again produces a phase change, as water at 100°C is converted to steam at the same temperature. Finally, once the sample has all been vaporised, the addition of energy again produces an increase in temperature.

The results are presented as a graph of temperature of the system of the ice cube versus energy added to the system (**Figure 19.11**). Let's examine each portion of the red-brown curve, which is divided into parts A through E.

Part A. On this portion of the curve, the temperature of the ice changes from −30.0°C to 0.0°C. **Equation 19.11** indicates that the temperature varies linearly with the energy added, so the experimental result is a straight line on the graph. Because the specific heat of ice is 2090 J/kg.°C, we can calculate the amount of energy added by using **Equation 19.11**.

$$Q = m_i c_i \Delta T = (1.00 \times 10^{-3}\ \text{kg})(2090\ \text{J/kg.°C})(30.0°\text{C}) = 62.7\ \text{J}$$

Part B. When the temperature of the ice reaches 0.0°C, the ice–water mixture remains at this temperature – even though energy is being added – until all the ice melts. From **Equation 19.13**, the energy required to melt 1.00 g of ice at 0.0°C is

$$Q = L_f \Delta m_w = L_f m_i = (3.33 \times 10^5\ \text{J/kg})(1.00 \times 10^{-3}\ \text{kg}) = 333\ \text{J}$$

At this point, we have moved to the 396 J (= 62.7 J + 333 J) mark on the energy axis in **Figure 19.11**.

Part C. Between 0.0°C and 100.0°C, no phase change occurs, so all energy added to the water is used to increase its temperature. The amount of energy necessary to increase the temperature from 0.0°C to 100.0°C is

$$Q = m_w c_w \Delta T = (1.00 \times 10^{-3}\ \text{kg})(4.19 \times 10^3\ \text{J/kg.°C})(100.0°\text{C}) = 419\ \text{J}$$

Part D. At 100.0°C, another phase change occurs as the water changes from water at 100.0°C to steam at 100.0°C. Similar to the ice–water mixture in part B, the water–steam mixture remains at 100.0°C – even though energy is being added – until all the liquid has been converted to steam. The energy required to convert 1.00 g of water to steam at 100.0°C is

$$Q = L_v \Delta m_s = L_v m_w = (2.26 \times 10^6\ \text{J/kg})(1.00 \times 10^{-3}\ \text{kg}) = 2.26 \times 10^3\ \text{J}$$

Part E. On this portion of the curve, as in parts A and C, no phase change occurs; therefore, all energy added is used to increase the temperature of the steam. The energy that must be added to raise the temperature of the steam from 100.0°C to 120.0°C is

$$Q = m_s c_s \Delta T = (1.00 \times 10^{-3}\ \text{kg})(2.01 \times 10^3\ \text{J/kg.°C})(20.0°\text{C}) = 40.2\ \text{J}$$

The total amount of energy that must be added to change 1 g of ice at −30.0°C to steam at 120.0°C is the sum of the results from all five parts, which is 3.11×10^3 J. Conversely, to cool 1 g of steam at 120.0°C to ice at −30.0°C, we must remove 3.11×10^3 J of energy.

Notice in **Figure 19.11** the relatively large amount of energy that is transferred into the water to vaporise it to steam. Imagine reversing this process, with a large amount of energy transferred out of steam to condense it into water. That is why a burn to your skin from steam at 100°C is more damaging than a splash on your skin of water at 100°C. A very large amount of energy enters your skin from the steam and the steam remains at 100°C for a long time while it condenses, whereas the temperature of water at 100°C immediately starts to drop as energy is transferred to your skin.

If liquid water is held perfectly still in a very clean container, it is possible for the water to drop below 0°C without freezing into ice. This phenomenon, called **supercooling**, arises because the water requires a disturbance of some sort for the molecules to move apart and start forming the large, open ice structure that makes the density of ice lower than that of water as discussed in Section 19.2. If supercooled water is disturbed, it suddenly freezes. The system drops into the lower-energy configuration of bound molecules of the ice structure and the energy released raises the temperature back to 0°C.

Commercial hand warmers consist of liquid sodium acetate in a sealed plastic pouch. The solution in the pouch is in a stable supercooled state. When a disc in the pouch is clicked by your fingers, the liquid solidifies and the temperature increases, just like the supercooled water just mentioned. In this case, however, the freezing point of the liquid is higher than body temperature, so the pouch feels warm to the touch. To reuse the hand warmer, the pouch must be boiled until the solid liquefies. Then, as it cools, it passes below its freezing point into the supercooled state.

It is also possible to create **superheating**. For example, clean water in a very clean cup placed in a microwave oven can sometimes rise in temperature beyond 100°C without boiling because the formation of a bubble of steam in the water requires scratches in the cup or some type of impurity in the water to serve as a nucleation site. When the cup is removed from the microwave oven, the superheated water can become explosive as bubbles form immediately and the hot water is forced upwards out of the cup.

Example 19.5

Cooling the steam

A barista froths and heats milk by passing steam through it. What mass of steam initially at 130°C is needed to warm 200 g of milk in a 100-g glass container from 20.0°C to 50.0°C?

Solution

Conceptualise Imagine placing cool water and steam together in a closed insulated container. The system eventually reaches a uniform state of water with a final temperature of 50.0°C.

Model We will make the approximation that no energy is transferred to the environment by heat during the process, so we model the glass, milk and steam as an isolated system. We will also make the approximation that milk has the same heat capacity as water, as milk is mostly water.

Analyse Write **Equation 19.12** to describe the process:

$$Q_{\text{cold}} = -Q_{\text{hot}} \quad (1)$$

The steam undergoes three processes: first a decrease in temperature to 100°C, then condensation into liquid water, and finally a decrease in temperature of the water to 50.0°C. Find the energy transfer in the first process using the unknown mass m_s of the steam:

$$Q_1 = m_s c_s \Delta T_s$$

Find the energy transfer in the second process:

$$Q_2 = L_v \Delta m_s = L_v (0 - m_s) = -m_s L_v$$

Find the energy transfer in the third process:

$$Q_3 = m_s c_w \Delta T_{\text{hot water}}$$

Add the energy transfers in these three stages:

$$Q_{\text{hot}} = Q_1 + Q_2 + Q_3 = m_s(c_s \Delta T_s - L_v + c_w \Delta T_{\text{hot water}}) \quad (2)$$

The 20.0°C water and the glass undergo only one process, an increase in temperature to 50.0°C. Find the energy transfer in this process:

$$Q_{\text{cold}} = m_w c_w \Delta T_{\text{cold water}} + m_g c_g \Delta T_{\text{glass}} \quad (3)$$

Example 19.5 cont.

Substitute **Equations (2)** and **(3)** into **Equation (1)**:

$$m_w c_w \Delta T_{\text{cold water}} + m_g c_g \Delta T_{\text{glass}} = -m_s(c_s \Delta T_s - L_v + c_w \Delta T_{\text{hot water}})$$

Solve for m_s:

$$m_s = \frac{m_w c_w \Delta T_{\text{cold water}} + m_g c_g \Delta T_{\text{glass}}}{c_s \Delta T_s - L_v + c_w \Delta T_{\text{hot water}}}$$

Check dimensions:

$$[\text{M}] = \frac{[\text{M}][\text{L}^2\text{T}^{-2}\Theta^{-1}][\Theta]}{[\text{L}^2\text{T}^{-2}]} = [\text{M}]\ ☺$$

Substitute numerical values:

$$m_s = \frac{(0.200\text{ kg})(4186\text{ J/kg.°C})(50.0°\text{C} - 20.0°\text{C}) + (0.100\text{ kg})(837\text{ J/kg.°C})(50.0°\text{C} - 20.0°\text{C})}{(2010\text{ J/kg.°C})(100°\text{C} - 130°\text{C}) - (2.26 \times 10^6\text{ J/kg}) + (4186\text{ J/kg.°C})(50.0°\text{C} - 100°\text{C})}$$

$$= 1.09 \times 10^{-2}\text{ kg} = 10.9\text{ g}$$

What If? What if the final state of the system is water at 100°C? Would we need more steam or less steam? How would the analysis above change?

Answer More steam would be needed to raise the temperature of the water and glass to 100°C instead of 50.0°C. There would be two major changes in the analysis. First, we would not have a term Q_3 for the steam because the water that condenses from the steam does not cool below 100°C. Second, in Q_{cold}, the temperature change would be 80.0°C instead of 30.0°C. For practice, show that the result is a required mass of steam of 31.8 g.

End-of-chapter resources

At the beginning of this chapter, we asked what causes the water to take on the different forms shown in the photograph. We now know that the phase of the water depends on its temperature, which in turn depends on how much energy has been transferred into or away from the system. The water in the lake and the snow on the ground are both in thermal contact with the ground below and the air above. We know that liquid water and ice can coexist when the temperature is 0°C. On a still winter's day, it is likely that they are close to being in thermal equilibrium with the ground, as the temperature of the ground varies quite slowly, especially when covered by layers of water or snow, as we shall see in the following chapter. The temperature of the air is also likely to be about 0°C, but it may be slightly lower or higher depending on the time of day and the thickness of the cloud cover. If the air temperature is also 0°C, then the whole system may be in thermal equilibrium. If the air temperature is below 0°C, it may be in thermal equilibrium with the snow but not with the water, so that the water in the lake may cool and a layer of ice form on its surface. If the air temperature is above 0°C, some of the snow may start to melt (without changing temperature).

Water in all three phases

Shutterstock.com/Bjorn Stefanson

So far in this chapter, we have only considered the possibility of materials existing in their gaseous form if they are heated through their boiling points. Yet we know the air can contain water vapour at much lower temperatures than 100°C, as is evident from condensation of water out of the air onto a cold glass on a hot day or the formation of clouds at high altitudes where the temperature is very low. We shall see in the next chapter how this happens.

The problems found in this chapter may be assigned online in Enhanced Web Assign.

Worked solutions to every fifth problem are available in the Student Solutions Manual. Register online at **www.cengagebrain.com** for access.

Summary

Definitions

The term *heat* is used in two related ways:

1. Heat is the term used to describe the *energy* transferred from a system at higher temperature to a system at lower temperature; and
2. Heat is also the term used to describe the *flow* of energy between systems caused by differences in temperature.

The symbol Q represents the amount of energy transferred as heat.

Internal energy is all a system's energy that is associated with the system's microscopic components. Internal energy includes kinetic energy of random translation, rotation and vibration of molecules, vibrational potential energy within molecules, and potential energy between molecules.

Two objects are in **thermal equilibrium** with each other if there is no net transfer of energy between them when they are in thermal contact.

Temperature is the property that determines whether an object is in thermal equilibrium with other objects. Two objects in thermal equilibrium with each other are at the same temperature. The SI unit of absolute temperature is the **kelvin**, which is defined to be 1/273.16 of the difference between absolute zero and the temperature of the triple point of water.

The **specific heat** c of a substance is the amount of heat required to raise the substance's temperature through 1 K per kg:

$$c \equiv \frac{Q}{m\Delta T} \qquad (19.10)$$

The **latent heat** of a substance is defined as the ratio of the energy input to a substance to the change in mass of the higher-phase material:

$$L \equiv \frac{Q}{\Delta m} \qquad (19.14)$$

Concepts and principles

The **zeroth law of thermodynamics** states that if objects A and B are separately in thermal equilibrium with a third object C, then objects A and B are in thermal equilibrium with each other.

When the temperature of an object is changed by an amount ΔT, its length changes by an amount ΔL that is proportional to ΔT and to its initial length L_i:

$$\Delta L = \alpha L_i \Delta T \qquad (19.1)$$

where the constant α is the average coefficient of linear expansion. The average coefficient of volume expansion β for a solid is approximately equal to 3α.

An **ideal gas** is one for which PV/nT is constant. An ideal gas is described by the **equation of state**

$$PV = nRT \qquad (19.5)$$

where n equals the number of moles of the gas, P is its pressure, V is its volume, R is the universal gas constant (8.314 J/mol.K) and T is the absolute temperature of the gas. A real gas behaves approximately as an ideal gas if it has a low density.

The energy Q required to change the temperature of a mass m of a substance by an amount ΔT is

$$Q = mc\Delta T \qquad (19.11)$$

where c is the specific heat of the substance.

The energy required to change the phase of a pure substance is

$$Q = L\Delta m \qquad (19.13)$$

where L is the latent heat of the substance, which depends on the nature of the phase change and the substance, and Δm is the change in mass of the higher-phase material.

Chapter review quiz

To help you revise Chapter 19: Heat and temperature, complete the automatically graded Chapter review quiz at http://login.cengagebrain.com.

Conceptual questions

1. What is wrong with the following statement? 'Given any two bodies, the one with the higher temperature contains more heat.'
2. Is it possible for two objects to be in thermal equilibrium if they are not in contact with each other? Explain your answer.
3. A piece of copper is dropped into a beaker of water. (a) If the water's temperature rises, what happens to the temperature of the copper? (b) Under what conditions are the water and copper in thermal equilibrium?
4. Use a periodic table of the elements (see Appendix C) to determine the number of grams in one mole of (a) hydrogen, which has diatomic molecules, (b) helium and (c) carbon monoxide.
5. What does the ideal gas law predict about the volume of a sample of gas at absolute zero? (b) Why is this prediction incorrect?
6. A car radiator is filled to the brim with water when the engine is cool. (a) What happens to the water when the engine is running and the water has been raised to a high temperature? (b) What do modern cars have in their cooling systems to prevent the loss of coolants?
7. When the metal ring and metal sphere in Figure CQ19.7 are both at room temperature, the sphere can barely be passed through the ring. (a) After the sphere is warmed in a flame, it cannot be passed through the ring. Explain why this happens. (b) **What If?** What if the ring is warmed and the sphere is left at room temperature? Does the sphere pass through the ring?

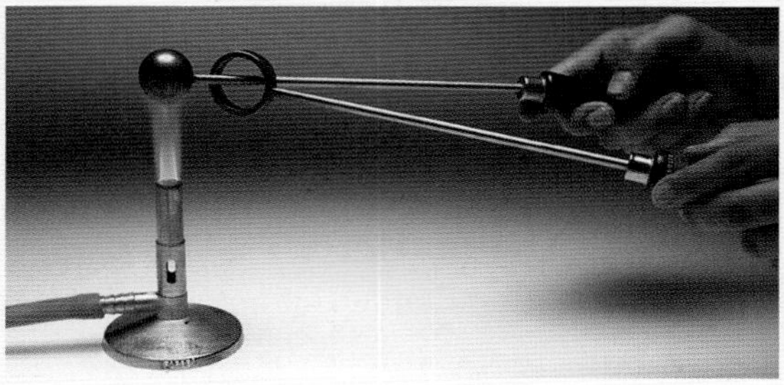

Figure CQ19.7

8. Some thermometers are made of a mercury column in a glass tube. Based on the operation of these thermometers, which has the larger coefficient of linear expansion, glass or mercury? (Don't answer the question by looking at a table.)
9. Two identical cans of soft drink are cooled from room temperature, one in a closed, insulated container of water at zero degrees and one in a similar container of ice at zero degrees. How will the final temperatures of the cans compare? Explain your answer.
10. When a solid is heated through a temperature change but doesn't undergo a phase change, which form of internal energy has increased the most: the average kinetic energy of the molecules or the average intermolecular potential energy?
11. When a solid is melted but its temperature does not change, which form of internal energy has increased the most, the average kinetic energy of the molecules or the average intermolecular potential energy?
12. Which process requires the biggest transfer of energy to or from the system: condensing 1 litre of water vapour or melting 1 kg of ice?

Problems

Section 19.1 Heat, internal energy and temperature

1. A student eats a 920 kJ doughnut. (a) How many steps must the student climb on a very tall stairway to change the gravitational potential energy of the student–Earth system by a value equivalent to the food energy in the doughnut? Assume the height of a single stair is 15.0 cm. (c) If the human body is only 25.0% efficient in converting chemical potential energy to mechanical energy, how many steps must the student climb to expend the same amount of energy as he obtained from the doughnut?
2. Two objects made of different materials are in thermal equilibrium. (a) What can you say about their internal energies? (b) What can you say about the average speeds of the molecules in the two materials? (c) What can you say about the average kinetic energies of the molecules?

Section 19.2 Thermal expansion of liquids and solids

3. ±? A copper telephone wire has essentially no sag between poles (35.0 ± 0.2) m apart on a winter day when the temperature is 2.0°C. How much longer is the wire on a summer day when the temperature is 35.0°C?
4. The concrete sections of a highway are designed to have a length of 25.0 m. The sections are poured and cured at 10.0°C. What minimum spacing should the engineer leave between the sections to avoid buckling if the concrete is to reach a temperature of 50.0°C?
5. The active element of a laser is made of a glass rod 30.0 cm long and 1.50 cm in diameter. Assume the average coefficient of linear expansion of the glass is equal to $9.00 \times 10^{-6}\ (°C)^{-1}$. If the temperature of the rod increases by 65.0°C, what is the increase in (a) its length, (b) its diameter and (c) its volume?
6. Inside the wall of a house, an L-shaped section of hot-water pipe consists of three parts: a straight, horizontal piece $h = 28.0$ cm long; an elbow; and a straight, vertical piece $\ell = 134$ cm long (Figure P19.6). A stud and a second-storey floorboard hold the ends of this section of copper pipe stationary. Find the magnitude and direction of the displacement of the pipe elbow when the water flow is turned on, raising the temperature of the pipe from 18.0°C to 46.5°C.

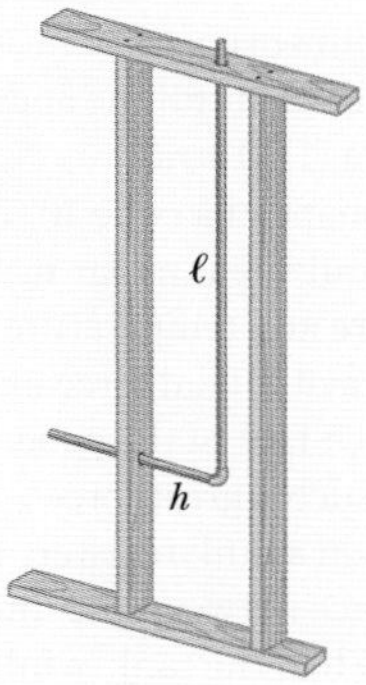

Figure P19.6

7. At 20.0°C, an aluminium ring has an inner diameter of 5.000 cm and a brass rod has a diameter of 5.050 cm. (a) If only the ring is warmed, what temperature must it reach so that it will just slip over the rod? (b) **What If?** If both the ring and the rod are warmed together, what temperature must they both reach so that the ring barely slips over the rod? (c) Would this latter process work? Explain your answer.
8. A volumetric flask made of Pyrex is calibrated at 20.0°C. It is filled to the 100 mL mark with 35.0°C acetone. After the flask is filled, the acetone cools and the flask warms so that the combination of acetone and flask reaches a uniform temperature of 32.0°C. The combination is then cooled back to 20.0°C. (a) What is the volume of the acetone when it cools to 20.0°C? (b) At the temperature of 32.0°C, does the level of acetone lie above or below the 100 mL mark on the flask? Explain.
9. ±? On a day on which the temperature is (20 ± 2)°C, a concrete pavement is poured in such a way that the ends of the pavement are unable to move. Take Young's modulus for concrete to be 7.00×10^9 N/m² and the compressive strength to be 2.00×10^9 N/m². (a) What is the stress in the cement on a hot day when the temperature of the pavement reaches 40°C? (b) Does the concrete fracture?
10. The Golden Gate Bridge in San Francisco has a main span of length 1.28 km, one of the longest in the world. Imagine that a steel wire with this length and a cross-sectional area of 4.00×10^{-6} m² is laid in a straight line on the bridge deck with its ends attached to the towers of the bridge. On a summer day the temperature of the wire is 35.0°C. (a) In winter the towers stay the same distance apart and the bridge deck keeps the same shape as its expansion joints open. When the temperature drops to −10.0°C, what is the tension in the wire? Take Young's modulus for steel to be 20.0×10^{10} N/m². (b) Permanent deformation occurs if the stress in the steel exceeds its elastic limit of 3.00×10^8 N/m². At what temperature would the wire reach its elastic limit? (c) **What If?** Explain how your answers to parts (a) and (b) would change if the Golden Gate Bridge were twice as long.

Section 19.3 Ideal gas law

11. ±? Gas is confined in a tank at a pressure of (11.0 ± 0.5) atm and a temperature of (25.0 ± 0.1)°C. If two-thirds of the gas is withdrawn and the temperature is raised to (75.0 ± 0.1)°C, what is the pressure of the gas remaining in the tank?

12. In state-of-the-art vacuum systems, pressures as low as 1.00×10^{-9} Pa are being attained. Calculate the number of molecules in a 1.00-m^3 vessel at this pressure and a temperature of 27.0°C.
13. An auditorium has dimensions $10.0 \text{ m} \times 20.0 \text{ m} \times 30.0$ m. How many molecules of air fill the auditorium at 20.0°C and a pressure of 101 kPa (1.00 atm)?
14. A container in the shape of a cube 10.0 cm on each edge contains air (with equivalent molar mass 28.9 g/mol) at atmospheric pressure and temperature 300 K. Find (a) the mass of the gas, (b) the gravitational force exerted on it and (c) the force it exerts on each face of the cube. (d) Why does such a small sample exert such a great force?
15. The pressure gauge on a tank registers the gauge pressure, which is the difference between the interior pressure and exterior pressure. When the tank is full of oxygen (O_2), it contains 12.0 kg of the gas at a gauge pressure of 40.0 atm. Determine the mass of oxygen that has been withdrawn from the tank when the pressure reading is 25.0 atm. Assume the temperature of the tank remains constant.
16. To measure how far below the ocean surface a bird dives to catch a fish, a scientist uses a method originated by Lord Kelvin. He dusts the interiors of plastic tubes with powdered sugar and then seals one end of each tube. He captures the bird at night time in its nest and attaches a tube to its back. He then catches the same bird the next night and removes the tube. In one trial, using a tube 6.50 cm long, water washes away the sugar over a distance of 2.70 cm from the open end of the tube. Find the greatest depth to which the bird dived, assuming the air in the tube stayed at constant temperature.
17. A car tyre is inflated with air originally at 10.0°C and normal atmospheric pressure. During the process, the air is compressed to 28.0% of its original volume and the temperature is increased to 40.0°C. (a) What is the tyre pressure? (b) After the car is driven at high speed, the tyre's air temperature rises to 85.0°C and the tyre's interior volume increases by 2.00%. What is the new tyre pressure (absolute)?
18. The mass of a hot-air balloon and its cargo (not including the air inside) is 200 kg. The air outside is at 10.0°C and 101 kPa. The volume of the balloon is 400 m³. To what temperature must the air in the balloon be warmed before the balloon will lift off? (Air density at 10.0°C is 1.244 kg/m³.)
19. At (25.0 ± 0.1) m below the surface of the sea, where the
±? temperature is (5.00 ± 0.05)°C, a diver exhales an air bubble having a volume of (1.000 ± 0.001) cm³. If the surface temperature of the sea is (20.0 ± 0.1)°C, what is the volume of the bubble just before it breaks the surface?
20. The pressure gauge on a cylinder of gas registers the gauge pressure, which is the difference between the interior pressure and the exterior pressure P_0. Let's call the gauge pressure P_g. When the cylinder is full, the mass of the gas in it is m_i at a gauge pressure of P_{gi}. Assuming the temperature of the cylinder remains constant, show that the mass of the gas *remaining in* the cylinder when the pressure reading is P_{gf} is given by

$$m_f = m_i\left(\frac{P_{gf} + P_0}{P_{gi} + P_0}\right)$$

Section 19.4 Thermometers

21. The boiling point of liquid hydrogen is 20.3 K at atmospheric pressure. What is this temperature on the Celsius scale?
22. Liquid nitrogen has a boiling point of −195.81°C at atmospheric pressure. Express this temperature in kelvins.
23. In an experiment in the undergraduate physics labs, a constant-volume gas thermometer is calibrated in dry ice (−78.5°C) and in boiling ethyl alcohol (78.0°C). The separate pressures are 0.900 atm and 1.635 atm. (a) What value of absolute zero in degrees Celsius does the calibration yield? What pressures would be found at (b) the freezing and (c) the boiling points of water? Hint: Use the linear relationship $P = A + BT$, where A and B are constants.

Section 19.5 Specific and latent heats

24. The highest waterfall in the world is the Salto Angel Falls in Venezuela. Its longest single falls has a height of 807 m. If water at the top of the falls is at 15.0°C, what is the maximum temperature of the water at the bottom of the falls? Assume all the kinetic energy of the water as it reaches the bottom goes into raising its temperature.
25. What mass of water at (25.0 ± 0.1)°C must be allowed to
±? come to thermal equilibrium with a 1.85 kg cube of aluminium initially at (150.0 ± 0.1)°C to lower the temperature of the aluminium to (65.0 ± 0.1)°C?
26. The temperature of a silver bar rises by 10.0°C when it absorbs 1.23 kJ of energy by heat. The mass of the bar is 525 g. Determine the specific heat of silver from these data.
27. In cold climates in the southern hemisphere, a house can be built with very large windows facing north to take advantage of solar heating. Sunlight shining in during the daytime is absorbed by the floor, interior walls and objects in the room, raising their temperature to 28.0°C. If the house is well insulated, you may model it as losing energy by heat steadily at the rate 6000 W on a day when the average exterior temperature is 4°C and when the conventional heating system is not used at all. During the period between 5.00 p.m. and 7.00 a.m., the temperature of the house drops and a sufficiently large 'thermal mass' is required to keep it from dropping too far. The thermal mass can be a large quantity of stone (with specific heat 850 J/kg.°C) in the floor and the interior walls exposed to sunlight. What mass of stone is required if the temperature is not to drop below 18.0°C overnight?
28. A 1.50 kg iron horseshoe initially at 600°C is dropped into a bucket containing 20.0 kg of water at 25.0°C. What is the final temperature of the water–horseshoe system? Ignore the heat capacity of the container and assume a negligible amount of water boils away.
29. A (3.00 ± 0.01) g copper coin at (25.0 ± 0.1)°C drops
±? (50.0 ± 0.1) m to the ground. (a) Assuming 60.0% of the change in gravitational potential energy of the coin–Earth system goes into increasing the internal energy of the coin, determine the coin's final temperature. (b) **What If?** Does the result depend on the mass of the coin? Explain your answer.
30. Two thermally insulated vessels are connected by a narrow tube fitted with a valve that is initially closed as shown in Figure P19.30. One vessel of volume 16.8 L contains oxygen at a temperature of 300 K and a pressure of 1.75 atm. The other vessel of volume 22.4 L contains oxygen at a temperature of

450 K and a pressure of 2.25 atm. When the valve is opened, the gases in the two vessels mix and the temperature and pressure become uniform throughout. (a) What is the final temperature? (b) What is the final pressure?

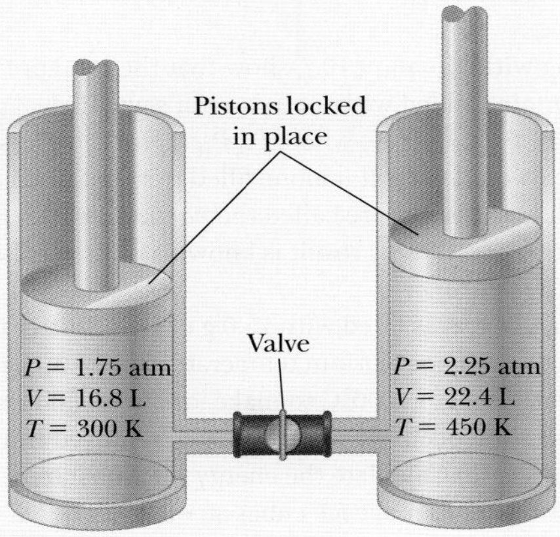

Figure P19.30

31. How much energy is required to change a 40.0 g ice cube from ice at −10.0°C to steam at 110°C?

32. A 3.00 g lead bullet at 30.0°C is fired at a speed of 240 m/s into a large block of ice at 0°C, in which it becomes embedded. What mass of ice melts?

33. A 1.00-kg block of copper at 20.0°C is dropped into a large vessel of liquid nitrogen at 77.3 K. How many kilograms of nitrogen boil away by the time the copper reaches 77.3 K? (The specific heat of copper is 0.092 0 cal/g. °C, and the latent heat of vaporization of nitrogen is 48.0 cal/g.)

34. A car has a mass of 1500 kg and its aluminium brakes have an overall mass of 6.00 kg. (a) Assume all the mechanical energy that transforms into internal energy when the car stops is deposited in the brakes and no energy is transferred out of the brakes in the form of heat. The brakes are originally at 20.0°C. How many times can the car be stopped from 25.0 m/s before the brakes start to melt? (b) Identify some effects ignored in part (a) that are important in a more realistic assessment of the warming of the brakes.

Additional problems

35. ±? A steel beam being used in the construction of a skyscraper has a length of 35.000 m when delivered on a cold day at a temperature of (5.0 ± 0.1)°C. What is the length of the beam when it is being installed later on a warm day when the temperature reaches (35.0 ± 0.1)°C?

36. ±? A bicycle tyre is inflated to a gauge pressure of (2.50 ± 0.01) atm when the temperature is 15.0°C. While a man rides the bicycle, the temperature of the tyre rises to 45.0°C. Assuming the volume of the tyre does not change, find the gauge pressure in the tyre at the higher temperature.

37. An aluminium calorimeter with a mass of 100 g contains 250 g of water. The calorimeter and water are in thermal equilibrium at 10.0°C. Two metallic blocks are placed into the water. One is a 50.0-g piece of copper at 80.0°C. The other has a mass of 70.0 g and is originally at a temperature of 100°C. The entire system stabilises at a final temperature of 20.0°C. (a) Determine the specific heat of the unknown sample. (b) Using the data in **Table 19.1**, can you make a positive identification of the unknown material? Can you identify a possible material? (c) Explain your answers for part (b).

38. The density of petrol is 730 kg/m³ at 0°C. Its average coefficient of volume expansion is $9.60 \times 10^{-4}\ (°C)^{-1}$. How many extra kilograms of petrol would you receive if you bought 50.0 L of petrol at 0°C rather than at 25.0°C from a pump that is not temperature compensated?

39. A mercury thermometer is constructed as shown in **Figure P19.39**. The glass capillary tube has a diameter of 0.004 00 cm, and the bulb has a diameter of 0.250 cm. Find the change in height of the mercury column that occurs with a temperature change of 30.0°C.

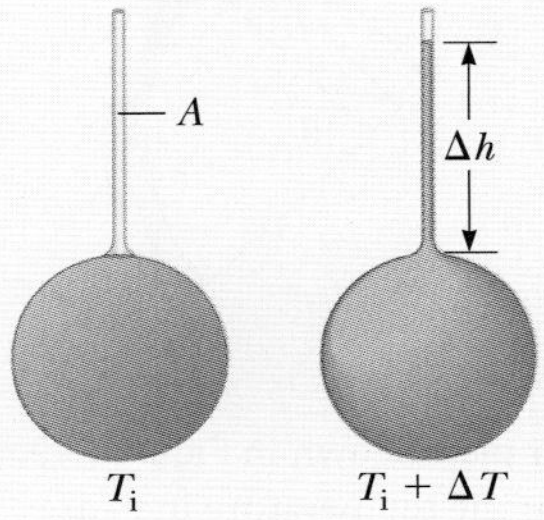

Figure P19.39
Problems 39 and 40

40. A liquid with a coefficient of volume expansion b just fills a spherical shell of volume V (**Figure P19.39**). The shell and the open capillary of area A projecting from the top of the sphere are made of a material with an average coefficient of linear expansion α. The liquid is free to expand into the capillary. Assuming the temperature increases by ΔT, find the distance Δh the liquid rises in the capillary.

41. A liquid has a density ρ. (a) Show that the fractional change in density for a change in temperature ΔT is $\Delta\rho/\rho = -\beta\Delta T$. (b) What does the negative sign signify? (c) Fresh water has a maximum density of 1.0000 g/cm³ at 4.0°C. At 10.0°C, its density is 0.9997 g/cm³. What is β for water over this temperature interval? (d) At 0°C, the density of water is 0.9999 g/cm³. What is the value for β over the temperature range 0°C to 4.00°C?

42. Take the definition of the coefficient of volume expansion to be

$$\beta = \frac{1}{V}\frac{dV}{dT}\bigg|_{P=\text{constant}} = \frac{1}{V}\frac{\partial V}{\partial T}$$

Use the equation of state for an ideal gas to show that the coefficient of volume expansion for an ideal gas at constant pressure is given by $\beta = 1/T$ where T is the absolute temperature. (b) What value does this expression predict for β at 0°C? State how this result compares with the experimental values for (c) helium and (d) air in **Table 19.1**. Note: These values are much larger than the coefficients of volume expansion for most liquids and solids.

43. A clock with a brass pendulum has a period of 1.000 s at 20.0°C. If the temperature increases to 30.0°C, (a) by how much does the period change and (b) how much time does the clock gain or lose in one week?

44. A bimetallic strip of length L is made of two ribbons of different metals bonded together. (a) First assume the strip is originally straight. As the strip is warmed, the metal with the greater average coefficient of expansion expands more than the other, forcing the strip into an arc with the outer radius having a greater circumference (**Figure P19.44**). Derive an expression for the angle of bending θ as a function of the initial length of the strips, their average coefficients of linear expansion, the change in temperature, and the separation of the centres of the strips ($\Delta r = r_2 - r_1$). (b) Show that the angle of bending decreases to zero when ΔT decreases to zero and also when the two average coefficients of expansion become equal. (c) **What If?** What happens if the strip is cooled?

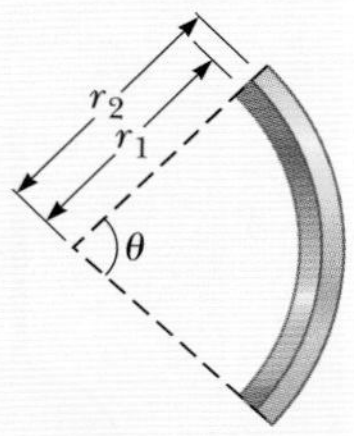

Figure P19.44

45. The rectangular plate shown in **Figure P19.45** has an area A_i equal to ℓw. If the temperature increases by ΔT, each dimension increases according to **Equation 19.1**, where α is the average coefficient of linear expansion. (a) Show that the increase in area is $\Delta A = 2\alpha A_i \Delta T$. (b) What approximation does this expression assume?

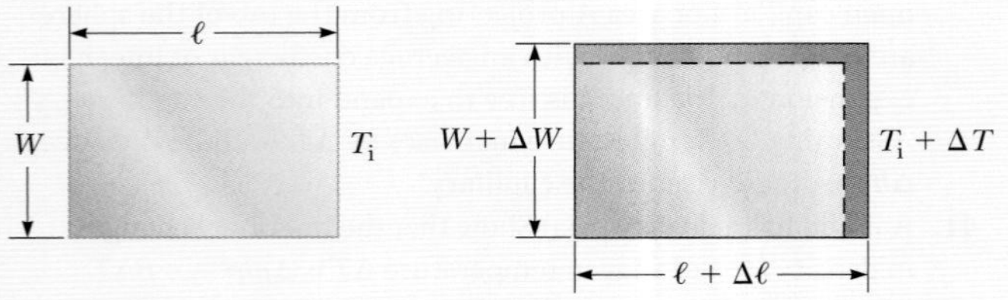

Figure P19.45

46. Your younger brother is confused. His toy water cannon has a tank with a capacity of 5.00 L. He pours 4.00 L of water into the tank and seals it, so it also contains air at atmospheric pressure. Next, he uses a hand-operated pump to inject more air until the absolute pressure in the tank reaches 2.40 atm. Now he uses the cannon to spray out water – not air – until the stream becomes feeble, which it does when the pressure in the tank reaches 1.20 atm. He finds that to spray out all the water, he must pump up the tank three times. Here is the puzzle: most of the water sprays out after the second pumping. The first and the third pumping-up processes seem just as difficult as the second but result in a much smaller amount of water coming out. Use your understanding of fluids and the ideal gas law to explain why this happens.

47. Water in an electric teakettle is boiling. The power absorbed by the water is 1.00 kW. Assuming the pressure of vapour in the kettle equals atmospheric pressure, determine the speed of effusion of vapour from the kettle's spout if the spout has a cross-sectional area of 2.00 cm^2. Model the steam as an ideal gas.

48. Starting with **Equation 19.7**, show that the total pressure P in a container filled with a mixture of several ideal gases is $P = P_1 + P_2 + P_3 + \cdots$, where $P_1, P_2, \ldots$ are the pressures that each gas would exert if it alone filled the container. (These individual pressures are called the *partial pressures* of the respective gases). This result is known as *Dalton's law of partial pressures*.

49. An ice-cube tray is filled with 75.0 g of water. After the filled tray reaches an equilibrium temperature of 20.0°C, it is placed in a freezer set at −8.00°C to make ice cubes. (a) Describe the processes that occur as energy is being removed from the water to make ice. (b) Calculate the energy that must be removed from the water to make ice cubes at −8.00°C.

50. A *flow calorimeter* is an apparatus used to measure the specific heat of a liquid. The technique of flow calorimetry involves measuring the temperature difference between the input and output points of a flowing stream of the liquid while energy is added by heat at a known rate. A liquid of density ρ flows through the calorimeter with volume flow rate R. At steady state, a temperature difference ΔT is established between the input and output points when energy is supplied at the rate P. What is the specific heat of the liquid?

51. A 75.0-kg cross-country skier glides over snow as in **Figure P19.51**. The coefficient of friction between skis and snow is 0.200. Assume all the snow beneath her skis is at 0°C and that all the internal energy generated by friction is added to snow, which sticks to her skis until it melts. How far would she have to ski to melt 1.00 kg of snow?

iStockphoto.com/technotr

Figure P19.51

52. A student measures the following data in a calorimetry ±? experiment designed to determine the specific heat of aluminium:

Initial temperature of water and calorimeter	(70.0 ± 0.5)°C
Mass of water	0.400 kg
Mass of calorimeter	0.040 kg
Specific heat of calorimeter	0.630 kJ/kg.°C
Initial temperature of aluminium	(27.0 ± 0.5)°C
Mass of aluminium	0.200 kg
Final temperature of mixture	(66.3 ± 0.5)°C

Use these data to determine the specific heat of aluminium. Is your result consistent with the value listed in **Table 19.2**?

Challenge problems

53. A cylinder is closed by a piston connected to a spring of constant 2.00×10^3 N/m (see **Figure P19.53**). With the spring relaxed, the cylinder is filled with 5.00 L of gas at a pressure of 1.00 atm and a temperature of 20.0°C (a) If the piston has a cross-sectional area of 0.0100 m² and negligible mass, how high will it rise when the temperature is raised to 250°C? (b) What is the pressure of the gas at 250°C?

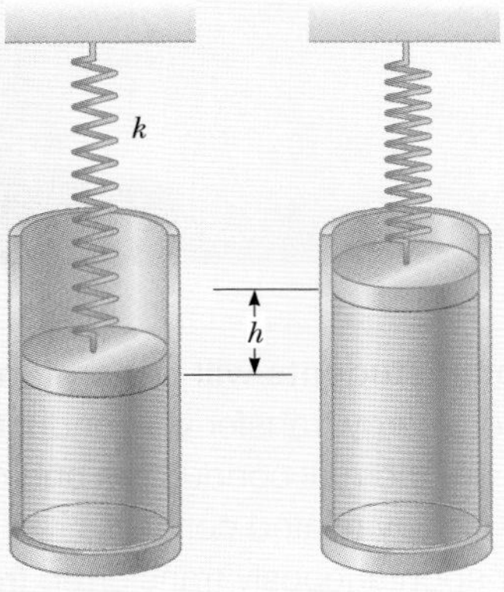

Figure P19.53

54. A steel guitar string with a diameter of 1.00 mm is stretched between supports 80.0 cm apart. The temperature is 0.0°C. (a) Find the mass per unit length of this string. (Use the value 7.86×10^3 kg/m³ for the density.) (b) The fundamental frequency of transverse oscillations of the string is 200 Hz. What is the tension in the string? Next, the temperature is raised to 30.0°C. Find the resulting values of (c) the tension and (d) the fundamental frequency. Assume both the Young's modulus of 20.0×10^{10} N/m² and the average coefficient of expansion $\alpha = 11.0 \times 10^{-6}$ (°C)$^{-1}$ have constant values between 0.0°C and 30.0°C.

55. A 1.00 km steel railroad rail is fastened securely at both ends when the temperature is 20.0°C. As the temperature increases, the rail buckles, taking the shape of an arc of a vertical circle. Find the height h of the centre of the rail when the temperature is 25.0°C. (Hint: You will need to solve a transcendental equation.)

chapter 20

Energy transfer processes and thermodynamics

Hot Spot® imaging can be used in cricket to show whether the ball hit the bat or not. How does it work?

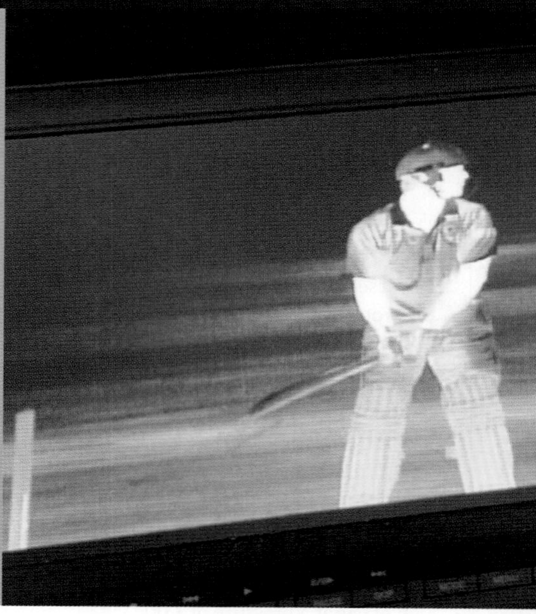

Corbis/Craig Shepheard

On completing this chapter, students will understand:

- the microscopic basis of energy transfer via conduction, convection and radiation
- why some substances are thermal conductors and others thermal insulators
- that work and heat are closely related concepts
- that energy is only ever spontaneously transferred from a hotter to a colder body
- that the first law of thermodynamics is a statement of the conservation of energy.

Students will be able to:

- model thermal conduction in simple systems
- calculate the net rate of energy loss through radiation from an object to its environment
- qualitatively describe convection and its role in some atmospheric phenomona
- distinguish between heat and work
- apply the first law of thermodynamics to analyse energy transfers.

Until about 1850, thermodynamics and mechanics were considered to be two distinct branches of science; conservation of energy seemed to apply to only certain kinds of mechanical systems. However, once experiments began to show a strong connection between the transfer of energy in thermal and mechanical processes the concept of energy was generalised from mechanics to include internal energy, and the principle of conservation of energy emerged as a universal law of nature. In this chapter, we examine heat and its connection to work. We introduce the first law of thermodynamics, which describes systems in which only internal energy changes, and transfers of energy are by heat and work. Finally, we look at some important applications of this law.

ENHANCED WebAssign

The problems found in this chapter may be assigned online in Enhanced Web Assign.

20.1 Energy transfer mechanisms in thermal processes

We often associate the concept of temperature with how hot or cold an object feels when we touch it: we know the sand on a sunny beach is at a higher temperature during the day than during the night. However, we cannot rely on our senses to determine whether two objects made of different materials are at the same temperature. For example, if you stand barefoot with one foot on the bathmat and the other on the tiled floor, the tile feels colder than the bathmat *even though both are at the same temperature.* The two objects feel different because the tile transfers energy by heat at a higher rate than the mat does. Your skin 'measures' the rate of energy transfer by heat rather than the actual temperature.

In Chapter 7, we introduced a global approach to the energy analysis of physical processes through **Equation 7.1**, $\Delta E_{\text{system}} = \Sigma T$, where T represents energy transfer, which can occur by several mechanisms. In Chapter 19, we looked at the effects of transferring energy in the form of heat to or from a system, but we did not consider the nature of the transfer processes themselves. In this section, we explore how energy can be transferred in the form of heat through three different processes – conduction, convection and radiation.

Thermal conduction

The process of thermal conduction is typically the dominant mechanism for transferring heat energy through solids. In this process, the transfer can be represented on an atomic scale as an exchange of kinetic energy between microscopic particles – molecules, atoms, and free electrons – in which particles with lower energy gain energy in collisions with particles with higher energy. For example, if you hold one end of a long metal bar and insert the other end into a flame, you will find that the temperature of the metal in your hand soon increases. The energy reaches your hand by means of conduction. Before the rod is inserted into the flame, the microscopic particles in the metal are vibrating about their equilibrium positions. As the flame raises the temperature of the rod, the particles in the rod near the flame begin to vibrate with greater and greater amplitudes. These particles, in turn, collide with their neighbors and transfer some of their energy in the collisions. The free electrons in the material close to the flame also gain energy, and move faster within the material, undergoing collisions as well as spreading throughout the material. Slowly, the amplitudes of vibration of metal atoms and electrons further and further from the flame increase until eventually those in the metal near your hand are affected. This increased vibration is detected by an increase in the temperature of the metal and of your potentially burned hand.

The rate of thermal conduction depends on the properties of the substance being heated. For example, it is possible to hold a piece of asbestos in a flame indefinitely, which implies that very little energy is conducted through the asbestos. In general, metals are good thermal conductors and materials such as asbestos, cork, paper and fibreglass are poor conductors. Gases also are poor conductors because the separation distance between the particles is so great. Metals are good thermal conductors because they contain large numbers of electrons that are relatively free to move through the metal and so can transport energy over large distances. Therefore, in a good conductor such as copper, conduction takes place by means of both the vibration of atoms and the motion of free electrons. As we shall see in Chapter 27, materials that are good thermal conductors are generally also good electrical conductors, as both forms of conduction depend on the presence of relatively free electrons.

TRY THIS

Get a wooden spoon and a metal spoon. Pour some water into a container from a freshly boiled kettle. Hold one end of each spoon in your hands and immerse the other end in the hot water. Explain your observations in terms of thermal conductivity.

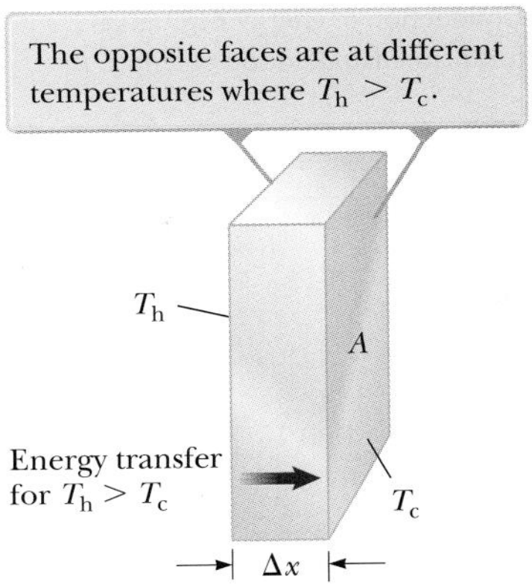
The opposite faces are at different temperatures where $T_h > T_c$.
T_h
A
Energy transfer for $T_h > T_c$
T_c
Δx

Figure 20.1
Energy transfer through a conducting slab with cross-sectional area A and thickness Δx

Conduction occurs only if there is a difference in temperature between two parts of the conducting medium. Consider a slab of material of thickness Δx and cross-sectional area A. One face of the slab is at a temperature T_c, and the other face is at a temperature $T_h > T_c$ (**Figure 20.1**). Experimentally, it is found that energy Q transfers in a time interval Δt from the hotter face to the colder one. The rate $P = Q/\Delta t$ at which this energy transfer

Table 20.1
Thermal conductivities

Substance	Thermal conductivity (W/m.°C)
Metals (at 25°C)	
Aluminium	238
Copper	397
Gold	314
Iron	79.5
Lead	34.7
Silver	427
Non-metals (approximate values)	
Asbestos	0.08
Brick	0.7
Concrete	0.8
Diamond	2300
Glass	0.8
Ice	2
Plasterboard	0.174
Rubber	0.2
Wall sheathing	2.0
Water	0.6
Wood	0.08
Gases (at 20°C)	
Air	0.0234
Helium	0.138
Hydrogen	0.172
Nitrogen	0.0234
Oxygen	0.0238

occurs is found to be proportional to the cross-sectional area and the temperature difference $\Delta T = T_h - T_c$, and inversely proportional to the thickness:

$$P = \frac{Q}{\Delta t} \propto A\frac{\Delta T}{\Delta x}$$

Notice that P has units of watts when Q is in joules and Δt is in seconds. For a slab of infinitesimal thickness dx and temperature difference dT, we can write the **law of thermal conduction** as

$$P = kA\left|\frac{dT}{dx}\right| \qquad (20.1)$$

where the proportionality constant k is the **thermal conductivity** of the material and $|dT/dx|$ is the **temperature gradient** (the rate at which temperature varies with position). The thermal conductivity k must have units of W/m.°C, and hence dimensions $MLT^{-3}\Theta^{-1}$ for Equation 20.1 to be dimensionally correct.

Substances that are good thermal conductors have large thermal conductivity values, whereas good thermal insulators have low thermal conductivity values. Table 20.1 lists thermal conductivities for various substances. Notice that metals are generally better thermal conductors than non-metals.

TRY THIS

Go into the kitchen and take a wooden spoon and a metal spoon out of the drawer. Does one feel colder than the other? Why?

Suppose a long, uniform rod of length L is thermally insulated so that energy cannot escape by heat from its surface except at the ends as shown in Figure 20.2. One end is in thermal contact with an energy reservoir at temperature T_c, and the other end is in thermal contact with a reservoir at temperature $T_h > T_c$. (If you tried the first *Try this* above, the hot water was one reservoir and you were the other.)

When a steady state has been reached, the temperature at each point along the rod is constant in time. In this case, if we assume k is not a function of temperature, the temperature gradient is the same everywhere along the rod and is

$$\left|\frac{dT}{dx}\right| = \frac{T_h - T_c}{L}$$

Therefore, the rate of energy transfer by conduction through the rod is

$$P = kA\left(\frac{T_h - T_c}{L}\right) \qquad (20.2)$$

For a compound slab containing several materials of thicknesses $L_1, L_2, \ldots$ and thermal conductivities $k_1, k_2, \ldots$, the rate of energy transfer through the slab at steady state is

$$P = \frac{A(T_h - T_c)}{\sum_i (L_i/k_i)} \qquad (20.3)$$

where T_h and T_c are the temperatures of the outer surfaces (which are held constant) and the summation is over all slabs. Example 20.1 shows how Equation 20.3 results from a consideration of two thicknesses of materials.

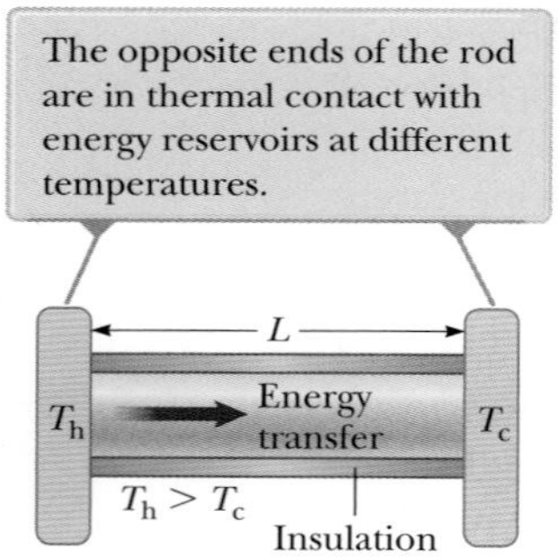

Figure 20.2
Conduction of energy through a uniform, insulated rod of length L

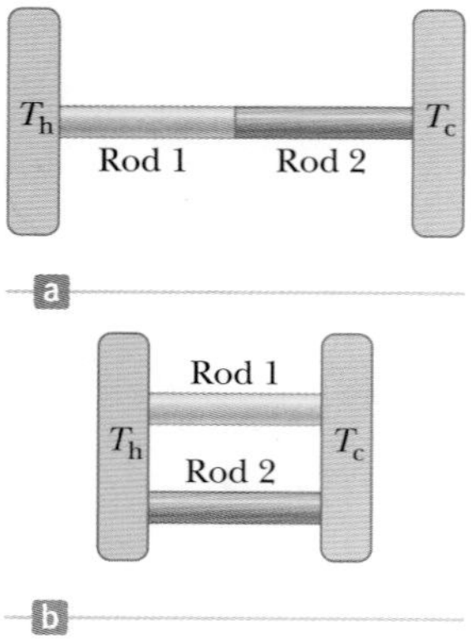

Figure 20.3
(Quick Quiz 20.1) In which case is the rate of energy transfer larger?

Quick Quiz 20.1

You have two rods of the same length and diameter, but they are formed from different materials. The rods are used to connect two regions at different temperatures so that energy transfers through the rods by heat. They can be connected in series as in Figure 20.3a or in parallel as in Figure 20.3b. In which case is the rate of energy transfer by heat larger? (a) The rate is larger when the rods are in series. (b) The rate is larger when the rods are in parallel. (c) The rate is the same in both cases.

Example 20.1

Energy transfer through two slabs

Two slabs of different roofing materials, of thickness L_1 and L_2 and thermal conductivities k_1 and k_2, are in thermal contact with each other as shown in Figure 20.4. The temperatures of their outer surfaces are T_c and T_h respectively, and $T_h > T_c$. Determine the temperature at the interface and the rate of energy transfer by conduction through an area A of the slabs in the steady-state condition.

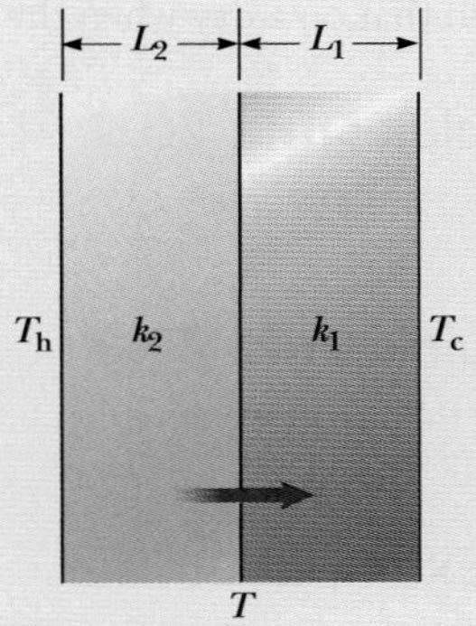

Figure 20.4 (Example 20.1) Energy transfer by conduction through two slabs in thermal contact with each other. At steady state, the rate of energy transfer through slab 1 equals the rate of energy transfer through slab 2.

Solution

Conceptualise Imagine an insulated roof on a hot day. The outer surface is at T_H and the inner surface is at T_C. There will be a transfer of energy by heat from the outside to the inside. Hence, the temperature varies with position in the two slabs, most likely at different rates in each part of the compound slab.

Model We are told that the system is in the steady-state condition. This tells us that the rate of energy transfer is constant, and the temperature at any given position within the slabs is also constant. Hence the power transferred through each slab must be the same, or the temperature at the interface would vary. We will use the model for energy transfer by conduction developed in this section.

Analyse Use Equation 20.2 to express the rate at which energy is transferred through an area A of slab 1:

$$P_1 = k_1 A\left(\frac{T - T_c}{L_1}\right) \quad (1)$$

Express the rate at which energy is transferred through the same area of slab 2:

$$P_2 = k_2 A\left(\frac{T_h - T}{L_2}\right) \quad (2)$$

Set these two rates as equal to represent the steady-state situation:

$$k_1 A\left(\frac{T - T_c}{L_1}\right) = k_2 A\left(\frac{T_h - T}{L_2}\right)$$

Solve for T:

$$T = \frac{k_1 L_2 T_c + k_2 L_1 T_h}{k_1 L_2 + k_2 L_1} \quad (3)$$

Substitute Equation (3) into either Equation (1) or Equation (2):

$$P = \frac{A(T_h - T_c)}{(L_1/k_1) + (L_2/k_2)} \quad (4)$$

Check dimensions:

$$[\mathrm{ML^2T^{-3}}] = \frac{[\mathrm{L^2}][\Theta]}{[\mathrm{L}]/[\mathrm{MLT^{-3}\Theta^{-1}}]} = [\mathrm{ML^2T^{-3}}]$$ ☺

Finalise Extension of this procedure to several slabs of materials leads to Equation 20.3.

What If? Suppose you are building an insulated container with two layers of insulation and the rate of energy transfer determined by Equation (4) turns out to be too high. You have enough room to increase the thickness of one of the two layers by 20%. How would you decide which layer to choose?

Answer To decrease the power as much as possible, you must increase the denominator in Equation (4) as much as possible. Whichever thickness you choose to increase, L_1 or L_2, you increase the corresponding term L/k in the denominator by 20%. For this percentage change to represent the largest absolute change, you want to take 20% of the larger term. Therefore, you should increase the thickness of the layer that has the larger value of L/k, and hence the smaller value of k.

In engineering practice, the term L/k for a particular substance is referred to as the ***R*-value** of the material. Therefore, Equation 20.3 reduces to

$$P = \frac{A(T_h - T_c)}{\sum_i R_i} \quad (20.4)$$

where $R_i = L_i/k_i$.

At any vertical surface open to the air, a very thin stagnant layer of air adheres to the surface. One must consider this layer when determining the *R*-value for a wall. The thickness of this stagnant layer on an outside wall depends on the speed of the wind. Energy transfer through the walls of a house on a windy day is greater than that on a day when the air is calm.

Quick Quiz 20.2

What are the units of the *R*-value of a material? **(a)** K/m **(b)** W/m².K or **(c)** m².K/W?

Quick Quiz 20.3

For the most effective insulation, should an *R*-value be **(a)** low or **(b)** high?

Example 20.2

The *R*-value of a typical wall

Calculate the total *R*-value for a wall constructed as shown in **Figure 20.5a**. Starting outside the house (towards the front in the figure) and moving inwards, the wall consists of 10.0 cm of brick, 1.2 cm of sheathing, an air space 2.0 cm thick and 1.2 cm of plasterboard. Assume that the layers of stagnant air inside and outside the home are each 1 cm thick.

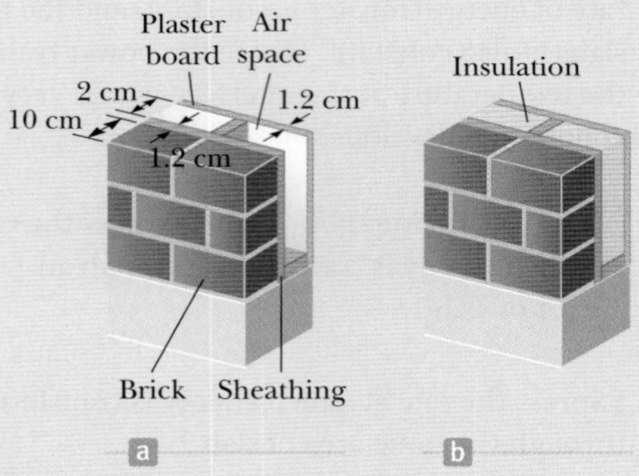

Figure 20.5
(Example 20.2) An exterior house wall containing (a) an air space and (b) insulation

Solution

Conceptualise Use **Figure 20.5** to help conceptualise the structure of the wall. Imagine energy being transferred from the outside to the inside on a hot day. Each layer decreases the rate at which energy is transferred, and hence acts to increase the total *R*-value.

Model We will use the model for thermal conductivity developed in this section. We need to find the *R*-value for each layer of material, which we can do using the data in **Table 20.1**.

Analyse Using the definition of R as $R = L/k$ and the data from **Table 20.1**:

$$R_1 \text{ (outside stagnant air layer)} = \frac{L_1}{k_{\text{air}}} = \frac{0.01 \text{ m}}{0.0234 \text{ W/m.K}} = 0.43 \text{ m}^2\text{.K/W}$$

$$R_2 \text{ (brick)} = \frac{L_2}{k_{\text{brick}}} = \frac{0.10 \text{ m}}{0.70 \text{ W/m.K}} = 0.143 \text{ m}^2\text{.K/W}$$

$$R_3 \text{ (sheathing)} = \frac{L_3}{k_{\text{sheathing}}} = \frac{0.012 \text{ m}}{2.00 \text{ W/m.K}} = 0.0060 \text{ m}^2\text{.K/W}$$

$$R_4 \text{ (air space)} = \frac{L_4}{k_{\text{air}}} = \frac{0.02 \text{ m}}{0.0234 \text{ W/m.K}} = 0.85 \text{ m}^2\text{.K/W}$$

$$R_5 \text{ (plasterboard)} = \frac{L_5}{k_{\text{plasterboard}}} = \frac{0.012 \text{ m}}{0.174 \text{ W/m.K}} = 0.071 \text{ m}^2\text{.K/W}$$

$$R_6 \text{ (inside stagnant air layer)} = \frac{L_6}{k_{\text{air}}} = \frac{L_1}{k_{\text{air}}} = R_1 = 0.43 \text{ m}^2\text{.K/W}$$

Add the *R*-values to obtain the total *R*-value for the wall:

$$R_{\text{total}} = R_1 + R_2 + R_3 + R_4 + R_5 + R_6 = 1.9 \text{ m}^2\text{.K/W}$$

Example 20.2 cont.

Finalise This R-value of just under 2 is fairly typical for the walls of many homes in Australia; however, it is lower than that recommended for many places, particularly those with a cool climate. In contrast, a strawbale wall has an R-value of around 10. Note also that the single largest contribution to the total R-value comes from the air space. As we shall see in the following section, the presence of air allows for energy to be transported through convection, as well as conduction. This reduces the *effective* R-value of the air layer inside the wall, so that in practice it only contributes a smaller amount to the wall's insulating properties.

Convection

Look back at the photograph of a burning candle at the start of Chapter 14. Immediately above the candle, the smoke rises straight upwards. In this situation, the air directly above the hot end of the candle is heated and expands. As a result, the density of this air decreases and the air rises. This hot air carries the smoke particles with it, allowing you to 'see' the air's motion. Energy transferred by the movement of a warm substance is said to have been transferred by **convection**, which is a form of matter transfer, T_{MT} in Equation 7.2. When resulting from differences in density, as with air around a fire, the process is referred to as *natural convection*. Airflow at a beach is an example of natural convection, as is the mixing that occurs as surface water in a lake cools and sinks. When the heated substance is forced to move by a fan or pump, as in some hot-air and hot-water heating systems, the process is called *forced convection*.

It is important to note that when convection occurs there is a transfer of both energy *and* matter. Recall from Chapter 7 that energy can be transferred in several different ways, including the transfer of matter, as is the case when you fill your car up with petrol. Convection is different to the two other mechanisms of heat transfer we describe here, conduction and radiation, in that matter as well as energy is transferred. In heat transfer by conduction and radiation there is no matter transferred, only energy. However, in all three cases, there is a transfer of energy due to a temperature difference.

TRY THIS, AGAIN

Fill two glasses with water, one with hot and one with cold. Add a drop of food colouring to each glass. Explain why the food colouring spreads faster through the hot water than the cold.

TRY THIS

Fill a glass with cold water. Pack ice against one side of the glass and use a candle to warm the bottom of the glass on the opposite side to the ice. Very carefully, drop a droplet of food colouring (or dark cordial) into the glass on the cold side. What do you observe?

If it were not for convection currents, it would be very difficult to boil water. As water is heated in a kettle, the lower layers are warmed first. This water expands and rises to the top because its density is lowered. At the same time, the denser, cool water at the surface sinks to the bottom of the kettle and is heated. If the warmer layers did not rise, the relatively lower thermal conductivity of water would mean that the bottom layer would reach boiling point while the surface remained relatively cool.

The same process occurs when a room is heated by a radiator. The hot radiator warms the air in the lower regions of the room. The warm air expands and rises to the ceiling because of its lower density. The denser, cooler air from above sinks, and the continuous air current pattern shown in Figure 20.6 is established.

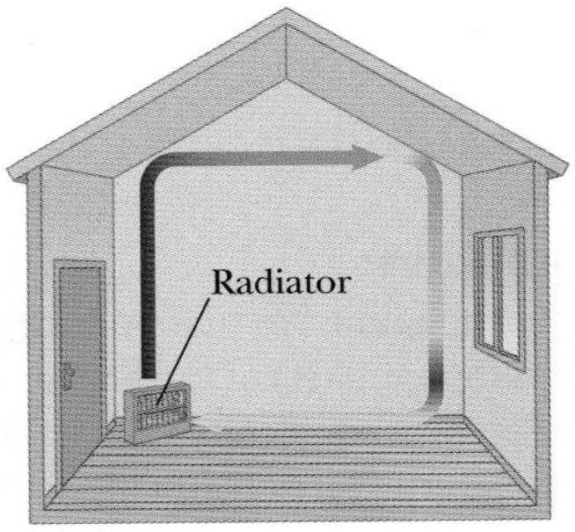

Figure 20.6
Convection currents are set up in a room warmed by a radiator.

Convective currents form an extremely important part of the way energy is transferred in the Earth's environment. In the following, we consider two qualitative examples; although a quantitative description of convection is beyond the scope of this book, they offer powerful illustrations of why convection is an important phenomenon.

Conceptual Example 20.3

Sea breezes

If you live close to the coast, you may have noticed that breezes frequently develop during the day when the wind blows in from the sea towards the land. Similarly, you may have noticed breezes blowing in the opposite direction at night. Why do these phenomena occur?

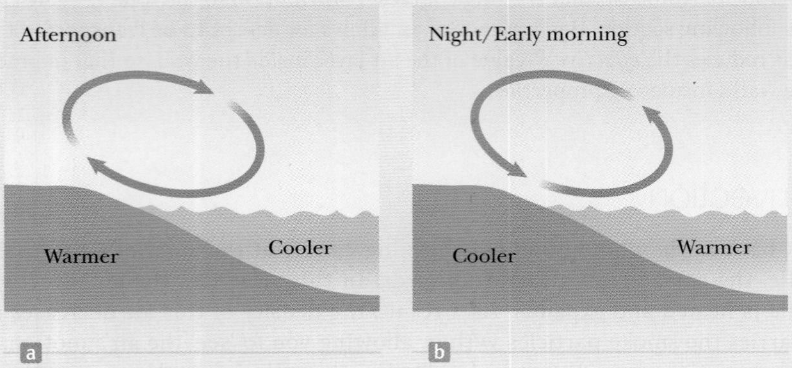

Figure 20.7
(Conceptual Example 20.3) Convection creating (a) a sea breeze and (b) a land breeze

Solution

The sea and land breezes in the day and night are the result of convective currents in the air, driven by temperature differences inland and at sea. These temperature differences are the result of the different specific heat capacities (see Section 19.5) of water and land. The sea has a greater specific heat than land, so the surface of the sea warms up more slowly than the surface of the land, despite the fact that they both receive approximately the same energy per unit area per unit time from the sun. As the temperature of the surface of the land rises, the land heats the air above it. The warm air is less dense and so it rises, reducing the sea-level air pressure over the land. The cooler air above the sea is now at higher pressure than that over the land, so it flows towards the land, creating a sea breeze. This process is illustrated in Figure 20.7a.

At night, the lower specific heat capacity of the land means it cools faster than the sea. If the temperature of the surface of the land drops below that of the sea surface temperature, convective currents are set up which circulate in the opposite direction to those which drove the sea breeze during the day.

Conceptual Example 20.4

Thermohaline circulation in the oceans

The Gulf Stream is a powerful ocean current that brings a stream of relatively warm water from the tropics far north into the Atlantic. It is responsible for the relatively warm climate of Ireland and the west coast of the United Kingdom, which would otherwise experience temperatures closer to those of countries at similar latitudes, such as Denmark, southern Sweden and Lithuania. An image of the Gulf Stream taken by NASA is shown in Figure 20.8. The Gulf Stream is part of a more complex network of ocean currents driven by thermohaline circulation. What causes these currents to flow?

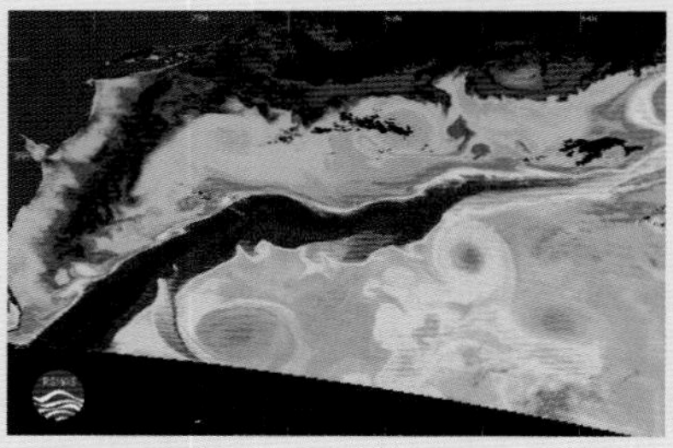
NASA

Figure 20.8
(Conceptual Example 20.4) Thermal imaging shows the Gulf Stream as a stream of relatively warmer (yellow) water flowing through the ocean.

Solution

Thermohaline circulation is driven by differences in the density of seawater in different parts of the ocean. Some of the density difference is due to differences in salinity (the 'haline' part of thermohaline), but differences in temperature are an equally significant cause. The Sun warms the oceans more at the equator than at the poles. The warmer water is less dense than the colder water so floats above it, making it more likely to be driven by currents in the air (wind). Once a current of warm water is set up moving away from the equator, the relatively low thermal conductivity of water and the relatively high specific heat capacity mean that the water can flow significant distances

Conceptual Example 20.4 cont.

while retaining some of the thermal energy acquired at the equator. Once the water nears the poles in the North Atlantic, however, surface winds result in rapid cooling, so that the surface water becomes colder than the deeper water and sinks downwards, displacing the deeper water. Together, these effects set up convective currents sometimes known as the Atlantic Conveyor Belt, of which the Gulf Stream is a part.

It has been suggested that melting Arctic sea ice due to changes in the Earth's climate may ultimately disturb thermohaline circulation to the point where the Atlantic Conveyor Belt slows or even stops, potentially lowering average temperatures along the west coasts of Great Britain and Norway by tens of degrees.

Warming of the Earth's climate can also act to increase convection in the atmosphere and seas. As the energy balance changes, and more energy is absorbed by the Earth's atmosphere and surface, more energy will be transferred within the atmosphere and seas by convection, leading to more severe weather events such as storms and cyclones, and longer-term weather patterns including droughts.

Radiation

The third means of energy transfer we shall discuss is **thermal radiation**, T_{ER} in Equation 7.2. All objects radiate energy continuously in the form of electromagnetic waves (see Chapter 34) produced by thermal vibrations of the molecules. You are likely to be familiar with thermal electromagnetic radiation in the form of the orange glow from an electric hotplate or the coils of a toaster.

The rate at which an object radiates energy is proportional to the fourth power of its absolute temperature. Known as **Stefan's law**, this behaviour is expressed mathematically as

$$P = \sigma A e T^4 \qquad (20.5)$$

◀ Stefan's law

where P is the power in watts of electromagnetic waves radiated from the surface of the object; σ is a constant, called the Stefan-Boltzmann constant, equal to 5.6696×10^{-8} W/m^2.K^4; A is the surface area of the object in square metres; e is a dimensionless constant called the **emissivity**; and T is the surface temperature in kelvins. The value of e can vary between zero and one depending on the properties of the surface of the object. The emissivity is equal to the **absorptivity**, which is the fraction of the incoming radiation that the surface absorbs. To understand why emissivity and absoptivity are equal, consider an object in thermal equilibrium with its environment that receives and emits energy only through radiation, such as a satellite in space. For the satellite to be in thermal equilibrium with its environment, it must emit and absorb energy at the same rate. Thus its absorptivity and emissivity factors must be equal.

A mirror has very low absorptivity because it reflects almost all incident light. Therefore, a mirror surface also has a very low emissivity. At the other extreme, a black surface has high absorptivity and high emissivity. An **ideal absorber** is defined as an object that absorbs all the energy incident on it and, for such an object, $e = 1$. An object for which $e = 1$ is often referred to as a **black body**. We shall investigate experimental and theoretical approaches to radiation from a black body in Chapter 39.

Every second, approximately 1370 J of electromagnetic radiation from the Sun passes perpendicularly through each 1 m^2 at the top of the Earth's atmosphere. This radiation is primarily visible and infra-red light accompanied by a significant amount of ultraviolet radiation. We shall study these types of radiation in detail in Chapter 34. Enough energy arrives at the surface of the Earth each day to supply all our energy needs on this planet hundreds of times over, if only it could be captured and used efficiently. The growth in the use of solar energy throughout the world reflects the increasing efforts being made to use this abundant energy.

What happens to the atmospheric temperature at night is another example of the effects of energy transfer by radiation. If there is cloud cover above the Earth, the water vapour in the clouds absorbs part of the infra-red radiation emitted by the Earth and re-emits it back to the surface. Consequently, temperature levels at the surface remain moderate. In the absence of this cloud cover, there is less in the way to prevent this radiation from escaping into space; therefore, the temperature decreases more on a clear night than on a cloudy one.

As an object radiates energy at a rate given by Equation 20.5, it also absorbs electromagnetic radiation from the surroundings, which consist of other objects that radiate energy. If the latter process did not occur, an object would eventually radiate all its energy and its temperature would reach absolute zero. If an object is at a temperature T and its surroundings are at an average temperature T_0, the net rate of energy gained or lost by the object as a result of radiation is

Stefan–Boltzmann equation ▶

$$P_{net} = \sigma Ae(T^4 - T_0^4) \tag{20.6}$$

When an object is in equilibrium with its surroundings, it radiates and absorbs energy at the same rate and its temperature remains constant. When an object is hotter than its surroundings, it radiates more energy than it absorbs and its temperature decreases.

TRY THIS

Take a book with a light cover and a book with a dark cover outside on a sunny day and leave them in the sun for the same length of time. Go back and touch their surfaces – do they feel the same temperature? Explain your observations. Now bring the two books inside out of the sun. Which reaches room temperature faster?

If you did the *Try this* above, you would have noted that the dark-coloured book absorbed more energy, and became hotter than the light-coloured book. The Earth's temperature is determined by the amount of radiation that it absorbs from the sun and emits into space. The amount of radiation incident on the Earth is approximately constant, and to maintain constant temperature, or energy balance, the amount reflected and radiated needs to match the amount absorbed. The amount absorbed depends on the emissivity of the surface, with surfaces such as polar ice having a low emissivity (or high albedo) and hence reflecting most of the incident energy. As polar ice melts, more energy is absorbed, leading to more ice melting, and so on. The Earth's surface also radiates energy, although very little of that radiated energy escapes directly into space; most is absorbed and re-radiated by the atmosphere, particularly by water vapour and greenhouse gases such as carbon dioxide. This makes the Earth a less efficient radiator, resulting in increasing temperatures, as has been observed. This is why reducing carbon dioxide and other greenhouse gas emissions is the main focus in addressing climate change.

Example 20.5

Cooling coffee

On a cold winter day when the air temperature is 10°C, you fill an insulated coffee mug (inner diameter 7.5 cm and depth 10.0 cm) with black coffee at 85°C but forget to put the lid on.

(A) At what rate does the coffee initially lose energy to the environment by radiation?

Solution

Conceptualise We start by drawing a diagram to help visualise the problem.

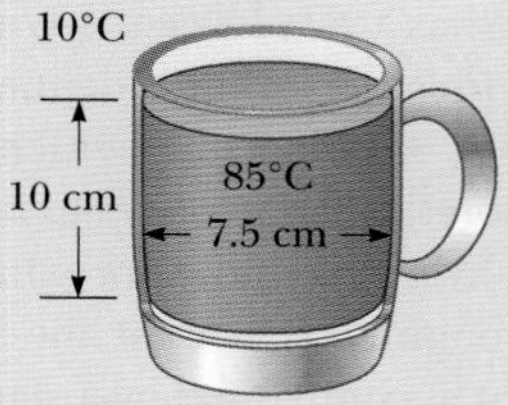

Figure 20.9
(Example 20.5) Cooling coffee

Model We shall assume that the sides of the coffee mug are perfect insulators, so that heat is only lost from the top surface. We will also make the approximation that as the coffee is black we can treat it as a black body with $e = 1$. We can then use the Stefan-Boltzmann equation to find the rate at which energy is lost by radiation.

Analyse The circular surface of the coffee has an area

$$A = \pi d^2/4$$

Substituting the given values into **Equation 20.6** and remembering to use kelvins to give the temperatures, we obtain

$$P_{net} = \sigma Ae(T^4 - T_0^4) = (5.6696 \times 10^{-8}\ \text{W/m}^2.\text{K}^4)\left(\frac{\pi 0.075^2}{4}\right)(1)(358.15^4 - 283.15^4\ \text{K})$$

$$= 2.5\ \text{W}$$

Example 20.5 cont.

(B) By how much does the coffee's temperature decrease in the first second?

Solution

Model We will make the approximation that the coffee has the same density and specific heat as water. We will also assume that the temperature change is small, so the rate at which energy is transferred is approximately constant over the one second.

Analyse The answer to part A tells us that the energy lost in one second is 2.5 J. The specific heat capacity of water (see **Table 19.2**) is 4186 J/kg.K. The volume of coffee in the mug is

$$V = Ah = \frac{\pi d^2 h}{4}$$

so the mass of coffee in the mug is

$$m = V\rho = \frac{\pi d^2 h\rho}{4}$$

We know that the change in temperature is related to the energy removed from the system through

$$Q = mc\Delta T$$

Rearranging for ΔT gives

$$\Delta T = \frac{Q}{mc} = \frac{4Q}{\pi d^2 h\rho c}$$

Check dimensions:

$$[\Theta] = \frac{[\mathrm{ML^2T^{-2}}]}{[\mathrm{L^2}][\mathrm{L}][\mathrm{ML^{-3}}][\mathrm{L^2T^{-2}\Theta^{-1}}]} = [\Theta]$$ ☺

Substituting in the numbers given in the question and the standard value for the density of water we obtain

$$\Delta T = \frac{4 \times 2.5\ \mathrm{J}}{\pi \times 0.075^2 \times 0.1\ \mathrm{m}^3 \times 1000\ \mathrm{kg/m^3} \times 4186\ \mathrm{J/kg.K}} = 0.0014\ \mathrm{K}$$

As this is not a substantial change in temperature, the factor $(T^4 - T_0^4)$ does not change significantly over this period. It is therefore reasonable to assume that the energy loss is constant over the first second. We have only considered cooling by radiation. However, in reality, convective cooling due to air at the surface of the coffee is likely to be an important cooling factor.

The Dewar flask

The *Dewar flask* (invented by Sir James Dewar) is a container designed to minimise energy transfers by conduction, convection and radiation. Such a container is used to store cold or hot liquids for long periods of time. (An insulated bottle or coffee mug is a common household equivalent of a Dewar flask.) The standard construction (**Figure 20.10**) consists of a double-walled Pyrex glass vessel with silvered walls. The space between the walls is evacuated to minimise energy transfer by conduction and convection. The silvered surfaces minimise energy transfer by radiation because silver is a very good reflector and has very low emissivity.

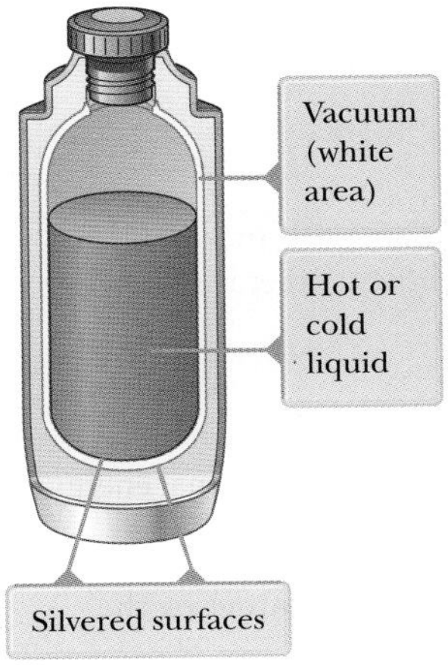

Figure 20.10
(a) A cross-sectional view of a Dewar flask, which is used to store hot or cold substances and (b) an insulated container holding liquid nitrogen at temperatures below 77 K

James Prescott Joule
British physicist (1818–1889)
Joule received some formal education in mathematics, philosophy and chemistry from John Dalton but was in large part self-educated. Joule's study of the quantitative relationships between electrical, mechanical and chemical effects of heat led him to recognise the equivalence of mechanical work and heat, and led to the establishment of the principle of conservation of energy.

A further reduction in energy loss is obtained by reducing the size of the neck. Dewar flasks are commonly used to store liquid nitrogen (boiling point 77 K) and liquid oxygen (boiling point 90 K).

20.2 Work and internal energy

In Chapters 6 and 7, we found that whenever friction is present in a mechanical system, the mechanical energy in the system decreases; in other words, mechanical energy is not conserved in the presence of non-conservative forces. The principle of conservation of energy tells us that this mechanical energy does not simply disappear; experimentally we find it is transformed into internal energy.

TRY THIS

Find a nail and touch its head to get a sense of its temperature. Now hammer it into a scrap piece of wood. What has happened to all the kinetic energy of the hammer once you have driven the nail into the wood? Touch the head of the nail again. Is it the same temperature as it was when you started? Explain your observations.

James Prescott Joule first established the equivalence of the decrease in mechanical energy and the increase in internal energy. A schematic diagram of Joule's experiment is shown in Figure 20.11.

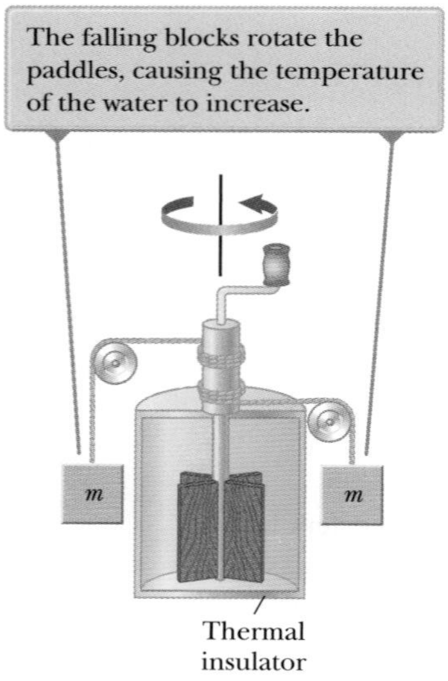
The falling blocks rotate the paddles, causing the temperature of the water to increase.
m
m
Thermal insulator

Figure 20.11
Joule's experiment for determining the mechanical equivalent of heat

The system of interest is the Earth, the two blocks and the water in a thermally insulated container. Work is done within the system on the water by a rotating paddle wheel, which is driven by heavy blocks falling at a constant speed. If the energy transformed in the bearings and the energy passing through the walls by heat are neglected, the decrease in potential energy of the system as the blocks fall equals the work done by the paddle wheel on the water and, in turn, the increase in internal energy of the water. If the two blocks fall through a distance h, the decrease in potential energy is $2mgh$, where m is the mass of one block; this energy causes the temperature of the water to increase. By varying the conditions of the experiment, Joule found that the decrease in mechanical energy is proportional to the product of the mass of the water and the increase in water temperature. The proportionality constant was found to be approximately 4.18 J/g.°C. Hence, 4.18 J of mechanical energy raises the temperature of 1 g of water by 1°C. More precise measurements taken later demonstrated the proportionality to be 4.186 J/g.°C when the temperature of the water was raised from 14.5°C to 15.5°C. Thus Joule's experiment showed that mechanical work, like heat transfer, can result in change in an object's temperature.

±? Example **20.6**

Temperature increase through deceleration

Marcus throws a lump of putty of mass 100.0 g at his brother. It hits a window with a speed of (20.0 ± 0.1) m/s and sticks to it. Assuming all the internal energy generated by the impact remains within the putty, what is the temperature change of the putty? Take the specific heat of the putty to be 377 J/kg.°C.

Solution

Conceptualise The lump of putty has kinetic energy when it is moving: All of that energy is transformed into internal energy.

No work is done on the putty because the force from the window moves through no displacement. This example is similar to the *Try this* with a skateboarder pushing off a wall in Section 8.7. There, no work is done on the skateboarder by the

Example 20.6 cont.

wall and potential energy stored in the body from previous meals is transformed to kinetic energy. Here, no work is done by the window on the putty and kinetic energy is transformed to internal energy.

Model We model the putty as an isolated system. We assume that all of the kinetic energy of the putty is transformed into thermal energy, raising the temperature of the putty, and ignore any energy transfer to the window.

Analyse Reduce the conservation of energy equation, Equation 7.2, to the appropriate expression for the system of the putty:

$$\Delta K + \Delta E_{int} = 0 \quad (1)$$

The change in the putty's internal energy is related to its change in temperature:

$$\Delta E_{int} = mc\Delta T \quad (2)$$

Substitute Equation (2) into Equation (1):

$$\left(0 - \frac{1}{2}mv^2\right) + mc\Delta T = 0$$

Solve for ΔT:

$$\Delta T = \frac{\frac{1}{2}mv^2}{mc} = \frac{v^2}{2c} \quad (3)$$

Check dimensions:

$$[\Theta] = \frac{[LT^{-1}]^2}{[L^2T^{-2}\Theta^{-1}]} = [\Theta] \text{ ☺}$$

Substitute values:

$$\Delta T = \frac{(20\text{m/s})^2}{2(377\text{J/kg.°C})} = 0.531\text{°C}$$

To find the uncertainty, note that ΔT is proportional to v^2, and assume that as no uncertainty in c is given, it is negligible compared to that in v. So:

$$\Delta(\Delta T) = 2\left(\frac{\Delta v}{v}\right)\Delta T = 2\left(\frac{0.1\text{ m/s}}{20.0\text{ m/s}}\right)(0.531\text{°C}) = 0.005\text{°C}$$

So our final answer is $\Delta T = 0.531\text{°C} \pm 0.005\text{°C}$.

Finalise Notice that the result does not depend on the mass of the putty.

20.3 Work and heat in thermodynamic processes

In thermodynamics, we describe the *state* of a system using such variables as pressure, volume, temperature and internal energy. As a result, these quantities belong to a category called **state variables**. For any given configuration of the system, we can identify values of the state variables. (For mechanical systems, the state variables include kinetic energy K and potential energy U.) A state of a system can be specified only if the system is in thermal equilibrium internally. In the case of a gas in a container, internal thermal equilibrium requires that every part of the gas be at the same pressure and temperature.

A second category of variables in situations involving energy is **transfer variables**. These variables are those that appear on the right side of the conservation of energy equation, Equation 7.2. Such a variable has a non-zero value if a process occurs in which energy is transferred across the system's boundary. The transfer variable is positive or negative, depending on whether energy is entering or leaving the system. Because a transfer of energy across the boundary represents a change in the system, transfer variables are not associated with a given state of the system, but rather with a *change* in the state of the system.

In the previous sections, we discussed heat as a transfer variable. In this section, we study another important transfer variable for thermodynamic systems: work. Work performed on particles was studied extensively in Chapter 6 and here we investigate the work done on a deformable system, a gas. Consider a gas contained in a cylinder fitted with a movable piston (Figure 20.12). At equilibrium, the gas occupies a volume V and exerts a uniform pressure P on the cylinder's walls and on the piston. If the piston has a cross-sectional area A, the force exerted by the gas on the piston is $F = PA$. Now let's assume we push the piston inwards and compress the gas **quasi-statically**, that is, slowly

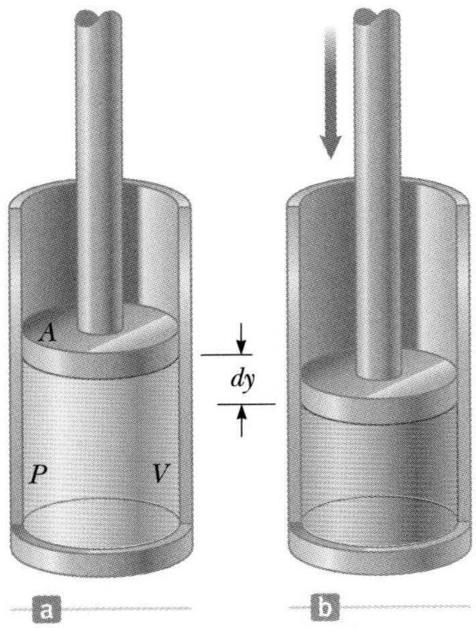

Figure 20.12
Work is done on a gas contained in a cylinder at a pressure P as the piston is pushed downwards so that the gas is compressed.

enough to allow the system to remain essentially in internal thermal equilibrium at all times. As the piston is pushed downwards by an external force $\vec{\mathbf{F}} = -F\hat{\mathbf{j}}$ through a displacement of $d\vec{\mathbf{r}} = dy\hat{\mathbf{j}}$ (**Figure 20.12b**), the work done on the gas is, according to our definition of work in Chapter 6,

$$dW = \vec{\mathbf{F}} \cdot d\vec{\mathbf{r}} = -F\hat{\mathbf{j}} \cdot dy\hat{\mathbf{j}} = -Fdy = -PAdy$$

where the magnitude F of the external force is equal to PA because the piston is always in equilibrium between the external force and the force from the gas. The mass of the piston is assumed to be negligible in this discussion. Because $A\,dy$ is the change in volume of the gas dV, we can express the work done on the gas as

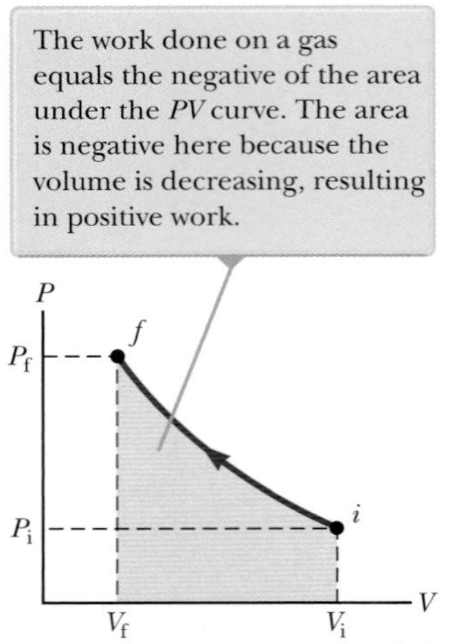
The work done on a gas equals the negative of the area under the *PV* curve. The area is negative here because the volume is decreasing, resulting in positive work.
P
f
P_f
P_i
i
V
V_f
V_i

Active Figure 20.13

A gas is compressed quasi-statically (slowly) from initial state i to final state f. An outside agent must do positive work on the gas to compress it.

$$dW = -P\,dV \quad (20.7)$$

If the gas is compressed, dV is negative and the work done on the gas is positive. If the gas expands, dV is positive and the work done on the gas is negative. If the volume remains constant, the work done on the gas is zero. The total work done on the gas as its volume changes from V_i to V_f is given by the integral of **Equation 20.7**:

Work done on a gas ▶

$$W = -\int_{V_i}^{V_f} P\,dV \quad (20.8)$$

Σ *Integration is summarised in Appendix B.7.*

To evaluate this integral, you must know how the pressure varies with volume during the process.

In general, the pressure is not constant during a process followed by a gas, but depends on the volume and temperature. If the pressure and volume are known at each step of the process, the state of the gas at each step can be plotted on a graph called a ***PV* diagram** as in **Active Figure 20.13**. This type of diagram allows us to visualise a process through which a gas is progressing. The curve on a *PV* diagram is called the *path* taken between the initial and final states.

Notice that the integral in **Equation 20.8** is equal to the area under a curve on a *PV* diagram. Therefore, we can identify an important use for *PV* diagrams:

The work done on a gas in a quasi-static process that takes the gas from an initial state to a final state is the negative of the area under the curve on a *PV* diagram, evaluated between the initial and final states.

For the process of compressing a gas in a cylinder, the work done depends on the particular path taken between the initial and final states. To illustrate this important point, consider several different paths connecting i and f (**Active Figure 20.14**). In the process shown in **Active Figure 20.14a**, the volume of the gas is first reduced from V_i to V_f at constant pressure P_i and the pressure of the gas then increases from P_i to P_f by heating at constant volume V_f. The work done on the gas along this path is $-P_i(V_f - V_i)$. In **Active Figure 20.14b**, the pressure of the gas is increased from P_i to P_f at constant volume V_i and then the volume of the gas is reduced from V_i to V_f at constant pressure P_f. The work done on the gas is $-P_f(V_f - V_i)$. This value

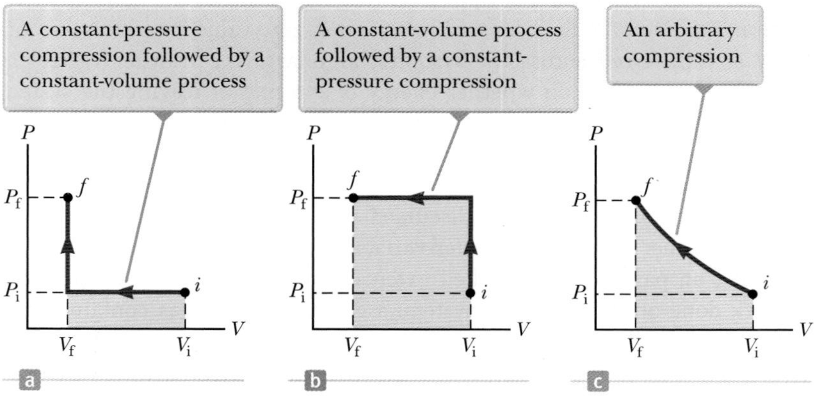
A constant-pressure compression followed by a constant-volume process
A constant-volume process followed by a constant-pressure compression
An arbitrary compression
P
P_f
P_i
f
i
V
V_f
V_i
a
b
c

Active Figure 20.14

The work done on a gas as it is taken from an initial state to a final state depends on the path between these states.

is greater than that for the process described in **Active Figure 20.14a** because the piston is moved through the same displacement by a larger force. Finally, for the process described in **Active Figure 20.14c**, where both P and V change continuously, the work done on the gas has some value between the values obtained in the first two processes. To evaluate the work in this case, the function $P(V)$ must be known so that we can evaluate the integral in **Equation 20.8**.

TRY THIS

Slowly pump a bicycle pump 20 times. Touch the barrel of the pump. Can you sense any change in temperature? Now pump the bicycle pump as rapidly as you can, also 20 times. Can you sense any change in temperature of the barrel now? Explain your observations.

The energy transfer Q into or out of a system by heat also depends on the process. Consider the situations depicted in **Figure 20.15**. In each case, the gas has the same initial volume, temperature and pressure, and is assumed to be ideal. In **Figure 20.15a**, the gas is thermally insulated from its surroundings except at the bottom of the gas-filled region, where it is in thermal contact with an energy reservoir. An *energy reservoir* is a source of energy that is considered to be so great that a finite transfer of energy to or from the reservoir does not change its temperature. The piston is held at its initial position by an external agent such as a hand. When the force holding the piston is reduced slightly, the piston rises very slowly to its final position shown in **Figure 20.15b**. Because the piston is moving upwards, the gas is doing work on the piston. During this expansion to the final volume V_f, just enough energy is transferred by heat from the reservoir to the gas to maintain a constant temperature T_i.

Now consider the completely thermally insulated system shown in **Figure 20.15c**. When the membrane is broken, the gas expands rapidly into the vacuum until it occupies a volume V_f and is at a pressure P_f. The final state of the gas is shown in **Figure 20.15d**. In this case, the gas does no work because it does not apply a force; no force is required to expand into a vacuum. Furthermore, no energy is transferred by heat through the insulating wall.

As we discuss in Section 20.4, experiments show that the temperature of the ideal gas does not change in the process indicated in **Figures 20.15c** and **20.15d**. Therefore, the initial and final states of the ideal gas in **Figures 20.15a** and **20.15b** are identical to the initial and final states in **Figures 20.15c** and **20.15d**, but the paths are different. In the first case, the gas does work on the piston and energy is transferred slowly to the gas by heat. In the second case, no energy is transferred by heat and the value of the work done is zero.

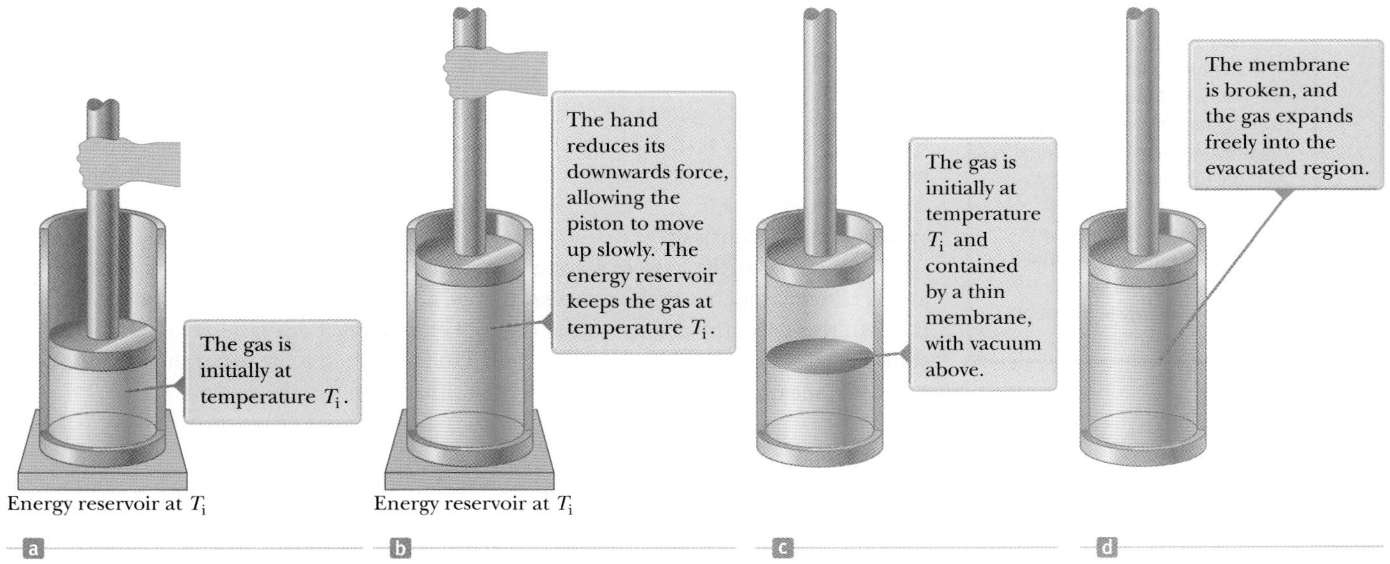

Figure 20.15
Gas in a cylinder (a) The gas is in contact with an energy reservoir. The walls of the cylinder are perfectly insulating, but the base in contact with the reservoir is conducting. (b) The gas expands slowly to a larger volume. (c) The gas is contained by a membrane in half of a volume, with vacuum in the other half. The entire cylinder is perfectly insulating. (d) The gas expands freely into the larger volume.

Therefore, energy transfer by heat, like work done, depends on the initial, final and intermediate states of the system. In other words, because heat and work depend on the path, neither quantity is determined solely by the endpoints of a thermodynamic process.

Quick Quiz 20.4

Which of the compression processes shown in **Figure 20.16** takes more work to do?

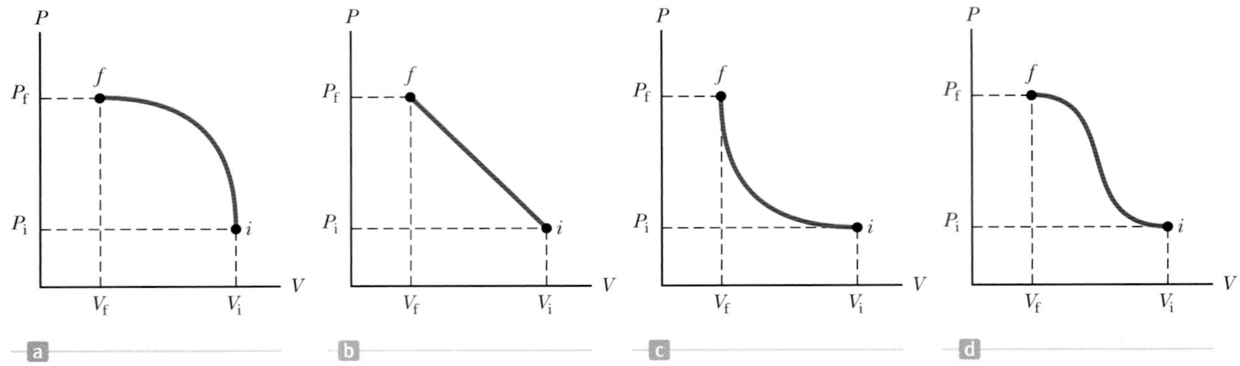

Figure 20.16
(Quick Quiz 20.4)

20.4 The first law of thermodynamics

When we introduced the law of conservation of energy in Chapter 7, we stated that the change in the energy of a system is equal to the sum of all transfers of energy across the system's boundary. The **first law of thermodynamics** is a special case of the law of conservation of energy that describes processes in which only the internal energy changes and the only energy transfers are by heat and work:

First law of thermodynamics ▶

$$\Delta E_{int} = Q + W \tag{20.9}$$

> **Pitfall Prevention 20.1**
> It is an unfortunate accident of history that the traditional symbol for internal energy is *U*, which is also the traditional symbol for potential energy as introduced in Chapter 6. To avoid confusion between potential energy and internal energy, we use the symbol E_{int} for internal energy in this book.

An important consequence of the first law of thermodynamics is that there exists a quantity known as internal energy whose value is determined by the state of the system. The internal energy is therefore a state variable like pressure, volume and temperature.

Let us investigate some special cases in which the first law can be applied. First, consider an *isolated system*, that is, one that does not interact with its surroundings. In this case, no energy transfer by heat takes place and the work done on the system is zero; hence, the internal energy remains constant. That is, because $Q = W = 0$, it follows that $\Delta E_{int} = 0$; therefore, $E_{int,i} = E_{int,f}$. We conclude that the internal energy E_{int} of an isolated system remains constant.

Next, consider the case of a system that can exchange energy with its surroundings and is taken through a **cyclic process**, that is, a process that starts and ends at the same state. In this case, the change in the internal energy must again be zero because E_{int} is a state variable; therefore, the energy Q added to the system must equal the negative of the work W done on the system during the cycle. That is, in a cyclic process

$$\Delta E_{int} = 0 \quad \text{and} \quad Q = -W \quad \text{(cyclic process)}$$

On a *PV* diagram, a cyclic process appears as a closed curve. (The processes described in **Active Figure 20.14** are represented by open curves because the initial and final states differ.) In a cyclic process, the net work done on the system per cycle equals the area enclosed by the path representing the process on a *PV* diagram.

We shall now use the first law of thermodynamics to analyse various processes through which a gas is taken. As a model, let's consider the sample of gas contained in the piston–cylinder apparatus in **Active Figure 20.17**. This figure shows work being done on the gas and energy transferring in by heat, so the internal energy of the gas is rising. In the following discussion of various processes, refer back to this figure and mentally alter the directions of the transfer of energy to reflect what is happening in the process.

Before we apply the first law of thermodynamics to specific systems, it is useful to first define some idealised thermodynamic processes. An **adiabatic process** is one during which no energy enters or leaves the system by heat, that is, $Q = 0$. An adiabatic process can be achieved either by thermally insulating the walls of the system or by performing the process rapidly so that there is negligible time for energy to transfer by heat. Applying the first law of thermodynamics to an adiabatic process gives

$$\Delta E_{\text{int}} = W \qquad (20.10)$$

◀ Adiabatic process

This result shows that if a gas is compressed adiabatically such that W is positive (work is done on the gas), then ΔE_{int} is positive and the temperature of the gas increases. Conversely, the temperature of a gas decreases when the gas expands (does work on its environment) adiabatically.

Adiabatic processes are very important in engineering practice. Some common examples are the expansion of hot gases in an internal combustion engine, the liquefaction of gases in a cooling system, and the compression stroke in a diesel engine.

The process described in Figures 20.15c and 20.15d, called an **adiabatic free expansion**, is unique. The process is adiabatic because it takes place in an insulated container. Because the gas expands into a vacuum, it does not apply a force on a piston as does the gas in Figures 20.15a and 20.15b, so no work is done on or by the gas. Therefore, in this adiabatic process, both $Q = 0$ and $W = 0$. As a result, $\Delta E_{\text{int}} = 0$ for this process as can be seen from the first law. That is, the initial and final internal energies of a gas are equal in an adiabatic free expansion. As we shall see in Chapter 21, the internal energy of an ideal gas depends only on its temperature. Therefore, we expect no change in temperature during an adiabatic free expansion. This prediction is in accord with the results of experiments performed at low pressures. Experiments performed at high pressures show a slight change in temperature after the expansion due to intermolecular interactions. This change in temperature is not predicted by the ideal gas model because it assumes that there are no interactions other than perfectly elastic collisions between gas molecules.

Active Figure 20.17

The first law of thermodynamics equates the change in internal energy E_{int} in a system to the net energy transfer to the system by heat Q and work W. In the situation shown here, the internal energy of the gas increases.

A process that occurs at constant pressure is called an **isobaric process**. In Active Figure 20.17, an isobaric process could be established by allowing the piston to move freely so that it is always in equilibrium between the net force from the gas pushing upwards and the weight of the piston plus the force due to atmospheric pressure pushing downwards. The first process in Active Figure 20.14a and the second process in Active Figure 20.14b are both isobaric.

In such a process, the values of the heat and the work are both usually non-zero. The work done on the gas in an isobaric process is simply

$$W = -P(V_{\text{f}} - V_{\text{i}}) \qquad (20.11)$$

◀ Isobaric process

where P is the constant pressure of the gas during the process.

A process that takes place at constant volume is called an **isovolumetric process**. In Active Figure 20.17, clamping the piston at a fixed position would ensure an isovolumetric process. The constant-volume processes in Active Figure 20.14a and 20.14b are both isovolumetric.

Because the volume of the gas does not change in such a process, the work given by Equation 20.8 is zero. Hence, from the first law we see that in an isovolumetric process, because $W = 0$,

$$\Delta E_{\text{int}} = Q \qquad (20.12)$$

◀ Isovolumetric process

This expression specifies that if energy is added by heat to a system kept at constant volume, all the transferred energy remains in the system as an increase in its internal energy. For example, when a can of spray paint is thrown into a fire, energy enters the system (the gas in the can) by heat through the metal walls of the can. Consequently, the temperature, and therefore the pressure, in the can increases until the can possibly explodes.

Pitfall Prevention 20.2

Some physics and engineering books present the first law as $\Delta E_{\text{int}} = Q - W$, with a minus sign between the heat and work. The reason is that work is defined in these treatments as the work done *by* the gas rather than *on* the gas, as in our treatment. The equivalent equation to Equation 20.8 in these treatments defines work as $W = \int_{V_i}^{V_f} P\,dV$. Therefore, if positive work is done by the gas, energy is leaving the system, leading to the negative sign in the first law.

In your studies in chemistry or engineering courses, or in your reading of other physics books, be sure to note which sign convention is being used for the first law.

Isothermal process

A process that occurs at constant temperature is called an **isothermal process**. This process can be established by immersing the cylinder in Active Figure 20.17 in an ice-water bath or by putting the cylinder in contact with some other constant-temperature reservoir. A plot of P versus V at constant temperature for an ideal gas yields a hyperbolic curve called an *isotherm*. The internal energy of an ideal gas is a function of temperature only. Hence, in an isothermal process involving an ideal gas, $\Delta E_{int} = 0$. For an isothermal process, we conclude from the first law that the energy transfer Q must be equal to the negative of the work done on the gas; that is, $Q = -W$. Any energy that enters the system by heat is transferred out of the system by work or vice versa; as a result, no change in the internal energy of the system occurs in an isothermal process.

Pitfall Prevention 20.3

$Q \neq 0$ in an isothermal process. Do not fall into the common trap of thinking there must be no transfer of energy by heat if the temperature does not change as is the case in an isothermal process. Because the cause of temperature change can be either heat *or* work, the temperature can remain constant even if energy enters the gas by heat, which can only happen if the energy entering the gas by heat leaves by work.

Quick Quiz 20.5

Complete the last three columns of the following table by adding the correct signs (−, +, or 0) for Q, W, and ΔE_{int}. For each situation, the system to be considered is identified.

Situation	System	Q	W	ΔE_{int}
(a) Rapidly pumping up a bicycle tyre	Air in the pump			
(b) Saucepan of room-temperature water sitting on a hot stove	Water in the pan			
(c) Air quickly leaking out of a balloon	Air originally in the balloon			

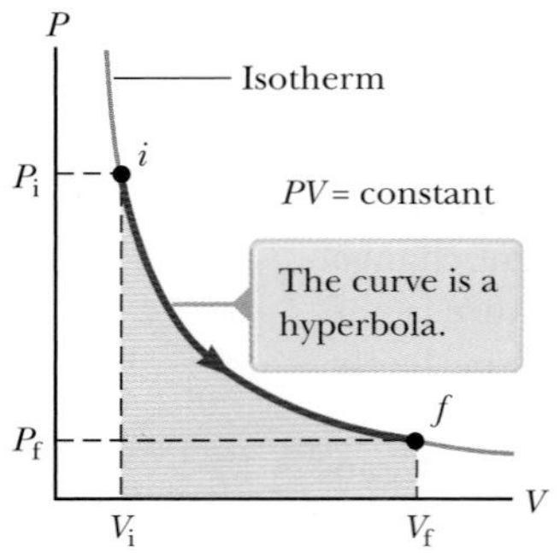

Figure 20.18
The PV diagram for an isothermal expansion of an ideal gas from an initial state to a final state

Isothermal expansion of an ideal gas

Suppose an ideal gas is allowed to expand quasi-statically at constant temperature. This process is described by the PV diagram shown in Figure 20.18. The curve is a hyperbola, and the ideal gas law (Equation 19.5) with T constant indicates that the equation of this curve is $PV = nRT =$ constant.

Let's calculate the work done on the gas in the expansion from state i to state f. The work done on the gas is given by Equation 20.8. Because the gas is ideal and the process is quasi-static, the ideal gas law is valid for each point on the path. Therefore

$$W = -\int_{V_i}^{V_f} P\,dV = -\int_{V_i}^{V_f} \frac{nRT}{V}\,dV$$

Because T is constant in this case, it can be removed from the integral along with n and R:

$$W = -nRT\int_{V_i}^{V_f} \frac{dV}{V} = -nRT \ln V\Big|_{V_i}^{V_f}$$

Evaluating the result at the initial and final volumes gives

$$W = nRT \ln\left(\frac{V_i}{V_f}\right) \qquad (20.13)$$

Curves including hyperbolae are described in Appendix B.3 and a table of integrals is given in Appendix B.7.

Numerically, this work W equals the negative of the shaded area under the PV curve shown in Figure 20.18. Because the gas expands, $V_f > V_i$ and the value for the work done on the gas is negative as we expect. If the gas is compressed, then $V_f < V_i$ and the work done on the gas is positive.

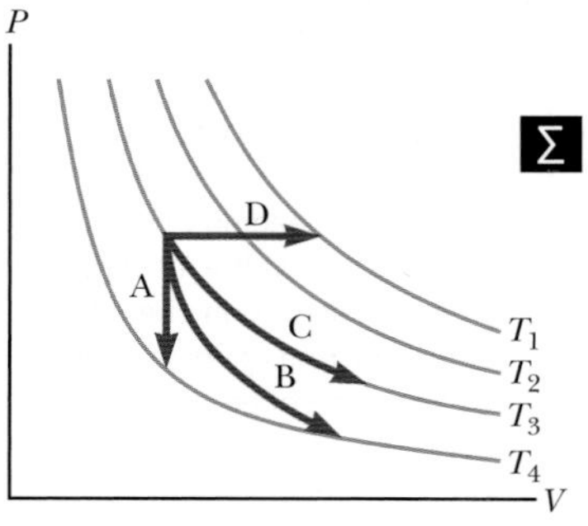

Figure 20.19
(Quick quiz 20.6) Identify the nature of paths A, B, C, and D.

Quick Quiz 20.6

Characterise the paths A, B, C and D in Figure 20.19 as isobaric, isovolumetric, isothermal or adiabatic. For path B, $Q = 0$. The blue curves are isotherms.

Example **20.7**

An isothermal expansion

A 1.0 mol sample of an ideal gas is kept at 0.0°C during an expansion from 3.0 L to 10.0 L.

(A) How much work is done on the gas during the expansion?

Solution

Conceptualise Run the process in your mind: the cylinder in **Active Figure 20.17** is immersed in an ice-water bath and the piston moves outwards so that the volume of the gas increases. You can also use the graphical representation in **Figure 20.18** to conceptualise the process.

Model We will assume the gas behaves as an ideal gas. Because the temperature of the gas is fixed, the process is isothermal.

Analyse Use **Equation 20.13**:

$$W = nRT\ln\left(\frac{V_\mathrm{i}}{V_\mathrm{f}}\right)$$

$$= (1.0\ \mathrm{mol})(8.31\ \mathrm{J/mol.K})(273\ \mathrm{K})\ln\left(\frac{3.0\ \mathrm{L}}{10.0\ \mathrm{L}}\right)$$

$$= -2.7\times10^{3}\ \mathrm{J}$$

(B) How much energy transfer by heat occurs between the gas and its surroundings in this process?

Solution

Find the heat from the first law:

$$\Delta E_\mathrm{int} = Q + W$$

$$0 = Q + W$$

$$Q = -W = 2.7\times10^{3}\ \mathrm{J}$$

(C) If the gas is returned to the original volume by means of an isobaric process, how much work is done on the gas?

Solution

Use **Equation 20.11**: $W = -P(V_\mathrm{f} - V_\mathrm{i})$

We can use the ideal gas law to write the pressure, noting that as the pressure is constant, so:

$$P = \frac{nRT}{V_i}$$

and:

$$W = \frac{nRT}{V_i}(V_f - V_i)$$

Check dimensions:

$$[\mathrm{ML^2T^{-2}}] = \frac{([\,][\mathrm{ML^2T^{-2}\Theta^{-1}}\,][\Theta]}{\mathrm{L^3}}[\mathrm{L^3}] = [\mathrm{ML^2T^{-2}}]$$ ☺

$$W = -\frac{(1.0\ \mathrm{mol})(8.31\ \mathrm{J/mol.K})(273\ \mathrm{K})}{10.0\times10^{-3}\ \mathrm{m^3}}(3.0\times10^{-3}\ \mathrm{m^3} - 10.0\times10^{-3}\ \mathrm{m^3})$$

$$W = 1.6\times10^{3}\ \mathrm{J}$$

Finalise We used the initial temperature and volume to calculate the work done because the final temperature was unknown. The work done on the gas is positive because the gas is being compressed.

Example **20.8**

Boiling water

Suppose 1.00 g of water vaporises isobarically at atmospheric pressure (1.013×10^5 Pa). Its volume in the liquid state is $V_\mathrm{i} = V_\mathrm{liquid} = 1.00\ \mathrm{cm^3}$ and its volume in the vapour state is $V_\mathrm{f} = V_\mathrm{vapour} = 1671\ \mathrm{cm^3}$. Find the work done in the expansion and the change in internal energy of the system.

Example 20.8 cont.

Solution

Conceptualise Imagine a saucepan with boiling water in it. There is a phase change occurring as the water evaporates to steam.

Model Recall that during a phase change there is no temperature change, and we will assume that the steam maintains a constant temperature during the process. We will model the expansion as the steam 'pushing the air out of the way' and ignore any mixing of steam with air. As the expansion takes place at constant pressure, we model the process as isobaric.

Analyse Use **Equation 20.11** to find the work done on the system as the air is pushed out of the way:

$$W = -P(V_f - V_i) = -(1.013 \times 10^5 \text{ Pa})(1671 \times 10^{-6} \text{ m}^3 - 1.00 \times 10^{-6} \text{ m}^3) = -169 \text{ J}$$

Use **Equation 19.13** and the latent heat of vaporisation for water to find the energy transferred into the system by heat, recalling that water has a density of 1.00 g/cm³.

$$Q = L_v \Delta m_s = m_s L_v = (1.00 \times 10^{-3} \text{ kg})(2.26 \times 10^6 \text{ J/kg}) = 2260 \text{ J}$$

Use the first law to find the change in internal energy of the system:

$$\Delta E_{int} = Q + W = 2260 \text{ J} + (-169 \text{ J}) = 2.09 \text{ kJ}$$

The positive value for ΔE_{int} indicates that the internal energy of the system increases. The largest fraction of the energy (2090 J/2260 J = 93%) transferred to the liquid goes into increasing the internal energy of the system. The remaining 7% of the energy transferred leaves the system as work done by the steam on the surrounding atmosphere.

±? Example **20.9**

Heating a solid

A (1.0 ± 0.1) kg bar of copper is heated at atmospheric pressure so that its temperature increases from 20°C to 50°C.

(A) What is the work done on the copper bar by the surrounding atmosphere?

Solution

Conceptualise This example involves a solid, whereas the preceding two examples involved liquids and gases. For a solid, the change in volume due to thermal expansion is very small.

Categorise As the expansion takes place at constant atmospheric pressure, we categorise the process as isobaric.

Analyse Find the work done on the copper bar using **Equation 20.11**:

$$W = -P\Delta V$$

Express the change in volume using **Equation 19.3** and assume that the volume expansion coefficient β can be approximated as three times the linear expansion coefficient, 3α:

$$W = -P(\beta V_i \Delta T) = -P(3\alpha V_i \Delta T) = -3\alpha P V_i \Delta T$$

Substitute for the volume in terms of the mass and density of the copper:

$$W = -3\alpha P\left(\frac{m}{\rho}\right)\Delta T$$

Check the dimensions to make sure they are consistent.

$$[\text{ML}^2\text{T}^{-2}] = [\Theta^{-1}][\text{ML}^{-1}\text{T}^{-2}]\frac{[\text{M}]}{[\text{ML}^{-3}]}[\Theta] = [\text{ML}^2\text{T}^{-2}] \ ☺$$

Substitute numerical values:

$$W = -3\left[1.7 \times 10^{-5} \text{ (°C)}^{-1}\right](1.013 \times 10^5 \text{ N/m}^2)\left(\frac{1.0 \text{ kg}}{8.92 \times 10^3 \text{ kg/m}^3}\right)(50\text{°C} - 20\text{°C})$$

$$= -1.7 \times 10^{-2} \text{ J}$$

Example 20.9 cont.

Note that the work is directly proportional to the mass, so the fractional uncertainty in the work will be the same as that in the mass, i.e.

$$\Delta W = \left(\frac{\Delta m}{m}\right)W = \left(\frac{0.1\text{ kg}}{1.0\text{ kg}}\right)1.7\times10^{-2}\text{J} = 0.17\times10^{-2}\text{ J}$$

and so the final answer is

$$W = -(1.7 \pm 0.2)\times10^{-2}\text{ J}$$

As this work is negative, work is done *by* the copper bar on the atmosphere.

(B) How much energy is transferred to the copper bar by heat?

Solution

Use **Equation 19.11** and the specific heat of copper from **Table 19.2**:

$$Q = mc\Delta T = (1.0\text{ kg})(387\text{ J/kg.°C})(50°\text{C} - 20°\text{C})$$
$$= 1.2\times10^4\text{ J}$$

Again, the energy is directly proportional to the mass and so the uncertainty in the energy can be treated in the same way as the uncertainty in the work, giving a final answer of

$$Q = (1.2 \pm 0.1)\times10^4\text{ J}$$

(C) What is the increase in internal energy of the copper bar?

Solution

Use the first law of thermodynamics:

$$\Delta E_{\text{int}} = Q + W = 1.2\times10^4\text{ J} + (-1.7\times10^{-2}\text{ J})$$
$$= 1.2\times10^4\text{ J}$$

The uncertainty in ΔE_{int} is the sum of the uncertainties in Q and W. This is dominated by the uncertainty in Q, which is six orders of magnitude greater than that in W. So our final answer is:

$$\Delta E_{\text{int}} = (1.2 \pm 0.1)\times10^4\text{ J}$$

Finalise Most of the energy transferred into the system by heat goes into increasing the internal energy of the copper bar. The amount of energy used to do work on the surrounding atmosphere is only about 10^{-6}. Hence, when the thermal expansion of a solid or a liquid is analysed, the small amount of work done on or by the system is usually ignored.

End-of-chapter resources

Corbis/Craig Shepheard

At the start of this chapter, we asked how the Hot Spot® imaging used in some cricket matches works. Hot Spot imaging uses cameras that are sensitive to infra-red radiation. As we shall see in Chapter 34, infra-red radiation is electromagnetic radiation with wavelengths somewhat longer than those of visible light; it is this type of radiation that dominates in the transfer of energy as heat through electromagnetic radiation. When the cricket ball strikes the bat or the batsman's body or pads, work is done in changing the direction of the ball and some of the energy of the collision is dissipated as heat. The main process through which energy is dissipated is friction, although some is dissipated through the small amount of compression and subsequent expansion of the wood, the pad or the batsman's leg. This energy is emitted in the form of infra-red radiation, which is detected by cameras positioned at either end of the pitch.

The problems found in this chapter may be assigned online in Enhanced Web Assign.

Worked solutions to every fifth problem are available in the Student Solutions Manual. Register online at **www.cengagebrain.com** for access.

Summary

Concepts and principles

Conduction can be viewed as an exchange of kinetic energy between colliding molecules or electrons. The rate of energy transfer by conduction through a slab of area A is

$$P = kA\left|\frac{dT}{dx}\right| \quad (20.1)$$

where k is the **thermal conductivity** of the material from which the slab is made and $|dT/dx|$ is the **temperature gradient**.

In **convection**, a warm substance transfers energy from one location to another due to a temperature difference. Both matter and energy are transferred by the process of convection.

All objects emit **thermal radiation** in the form of electromagnetic waves at the rate

$$P = \sigma A e T^4 \quad (20.5)$$

The **work** done on a gas as its volume changes from some initial value V_i to some final value V_f is

$$W = -\int_{V_i}^{V_f} P\,dV \quad (20.8)$$

where P is the pressure of the gas, which may vary during the process. To evaluate W, the process must be fully specified; that is, P and V must be known during each step. The work done depends on the path taken between the initial and final states.

The **first law of thermodynamics** states that when a system undergoes a change from one state to another, the change in its internal energy is

$$\Delta E_{int} = Q + W \quad (20.9)$$

where Q is the energy transferred into the system by heat and W is the work done on the system. Although Q and W both depend on the path taken from the initial state to the final state, the quantity ΔE_{int} does not depend on the path.

In a **cyclic process** (one that originates and terminates at the same state), $\Delta E_{int} = 0$ and therefore $Q = -W$. That is, the energy transferred into the system by heat equals the negative of the work done on the system during the process.

In an **adiabatic process**, no energy is transferred by heat between the system and its surroundings ($Q = 0$). In this case, the first law gives $\Delta E_{int} = W$. In the **adiabatic free expansion** of a gas, $Q = 0$ and $W = 0$, so $\Delta E_{int} = 0$. That is, the internal energy of the gas does not change in such a process.

An **isobaric process** is one that occurs at constant pressure. The work done on a gas in such a process is $W = -P(V_f - V_i)$.

An **isovolumetric process** is one that occurs at constant volume. No work is done in such a process, so $\Delta E_{int} = Q$.

An **isothermal process** is one that occurs at constant temperature. The work done on an ideal gas during an isothermal process is

$$W = nRT\ln\left(\frac{V_i}{V_f}\right) \quad (20.13)$$

Chapter review quiz

To help you revise Chapter 20: Energy transfer processes and thermodynamics, complete the automatically graded Chapter review quiz at http://login.cengagebrain.com.

Conceptual questions

1. In usually warm climates that experience a hard frost, fruit growers will spray the fruit trees with water, hoping that a layer of ice will form on the fruit. Why would such a layer be advantageous?
2. In describing his upcoming trip to the Moon, astronaut Jim Lovell said, 'I'll be walking in a place where there's a 400-degree difference between sunlight and shadow.' Suppose an astronaut standing on the Moon holds a thermometer in his gloved hand. (a) Is the thermometer reading the temperature of the vacuum at the Moon's surface? (b) Does it read any temperature? If so, what object or substance has that temperature?
3. Why is a person able to remove a piece of dry aluminium foil from a hot oven with bare fingers, whereas a burn results if there is moisture on the foil?
4. Using the first law of thermodynamics, explain why the *total* energy of an isolated system is always constant.
5. Is it possible to convert internal energy to mechanical energy? Use examples to explain your answer.
6. It is the morning of a day that will become hot. You just purchased drinks for a picnic and are loading them, with ice, into a chest in the back of your car. (a) You wrap a wool blanket around the chest. Does doing so help to keep the drinks cool or should you expect the wool blanket to warm them up? Explain your answer. (b) Your younger sister suggests you wrap her up in another wool blanket to keep her cool on the hot day like the ice chest. Explain your response to her.
7. You need to pick up a very hot cooking pot in your kitchen. You have a pair of cotton oven mitts. To pick up the pot most comfortably, should you soak them in cold water or keep them dry?
8. Suppose you pour hot coffee for your guests and one of them wants it with milk. He wants the coffee to be as warm as possible several minutes later when he drinks it. To have the warmest coffee, should the person add the milk just after the coffee is poured or just before drinking? Explain your answer.
9. When camping in a canyon on a still night, a camper notices that as soon as sunlight strikes the surrounding peaks, a breeze begins to stir. What causes the breeze?
10. In 1801, Humphry Davy rubbed together pieces of ice inside an icehouse. He made sure that nothing in the environment was at a higher temperature than the rubbed pieces. He observed the production of drops of liquid water. Make a table listing this and other experiments or processes to illustrate each of the following situations. (a) A system can absorb energy by heat, increase in internal energy and increase in temperature. (b) A system can absorb energy by heat and increase in internal energy without an increase in temperature. (c) A system can absorb energy by heat without increasing in temperature or in internal energy. (d) A system can increase in internal energy and in temperature without absorbing energy by heat. (e) A system can increase in internal energy without absorbing energy by heat or increasing in temperature.

Problems

Section 20.1 Energy transfer mechanisms in thermal processes

1. ±? A glass windowpane in a home is (0.62 ± 0.01) cm thick and has dimensions (1.000 ± 0.002) m × (2.000 ± 0.002) m. On a certain day, the temperature of the interior surface of the glass is 25.0°C and the exterior surface temperature is 0°C. (a) What is the rate at which energy is transferred by heat through the glass? (b) How much energy is transferred through the window in one day, assuming the temperatures on the surfaces remain constant?

2. ±? A student is trying to decide what to wear. His bedroom is at (20 ± 1)°C. His skin temperature is (35 ± 1)°C. The area of his exposed skin is 1.50 m^2. People all over the world have skin that is dark in the infra-red, with emissivity about 0.900. Find the net energy transfer from his body by radiation in 10.0 min.

3. A copper rod and an aluminium rod of equal diameter are joined end to end in good thermal contact. The temperature of the free end of the copper rod is held constant at 100°C and that of the far end of the aluminium rod is held at 0°C. If the copper rod is 0.150 m long, what must be the length of the aluminum rod so that the temperature at the junction is 50.0°C?

4. The tungsten filament of a 100 W lightbulb radiates 2.00 W of light. (The other 98 W is carried away by convection and conduction.) The filament has a surface area of 0.250 mm^2 and an emissivity of 0.950. Find the filament's temperature. (The melting point of tungsten is 3683 K.)

5. At noon on a sunny day, the Sun delivers 1000 W to each square metre of a road. If the hot asphalt transfers energy only by radiation, what is its steady-state temperature?

6. At our distance from the Sun, the intensity of solar radiation is 1370 W/m^2. The temperature of the Earth is affected by the *greenhouse effect* of the atmosphere. This phenomenon describes the effect of absorption of infra-red light emitted by the surface so as to make the surface temperature of the Earth higher than if it were airless. For comparison, consider a spherical object of radius r with no atmosphere at the same distance from the Sun as the Earth. Assume its emissivity is the same for all kinds of electromagnetic waves and its temperature is uniform over its surface. (a) Explain why the projected area over which it absorbs sunlight is πr^2 and the surface area over which it radiates is $4\pi r^2$. (b) Calculate its steady-state temperature. Is it chilly?

7. For bacteriological testing of water supplies and in medical clinics, samples must routinely be incubated for 24 h at 37°C. Peace Corps volunteer and MIT engineer Amy Smith invented a low-cost, low-maintenance incubator. The incubator consists of a foam-insulated box containing a waxy material that melts at 37.0°C interspersed among tubes, dishes or bottles containing the test samples and growth medium (bacteria food). Outside the box, the waxy material is first melted on a stove or solar energy collector. Then the waxy material is put into the box to keep the test samples warm as the material solidifies. The heat of fusion of the phase-change material is 205 kJ/kg. Model the insulation as a panel with surface area 0.490 m^2, thickness 4.50 cm, and conductivity 0.0120 W/m.°C. Assume the exterior temperature is 23.0°C for 12.0 h and 16.0°C for 12.0 h. (a) What mass of the waxy material is required to conduct the bacteriological test? (b) Explain why your calculation can be done without knowing the mass of the test samples or of the insulation.

8. A bar of gold (Au) is in thermal contact with a bar of silver (Ag) of the same length and area (**Figure P20.8**). One end of the compound bar is maintained at 80.0°C and the opposite end is at 30.0°C. When the energy transfer reaches steady state, what is the temperature at the junction?

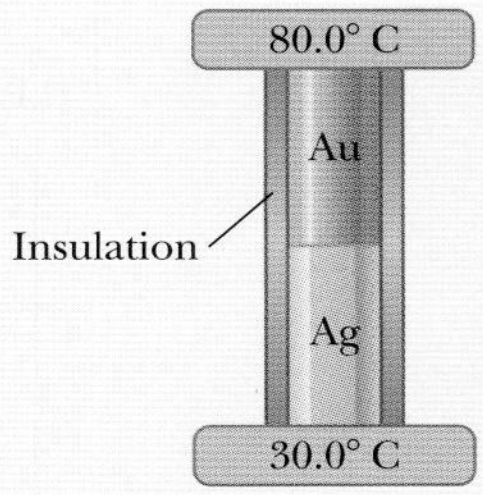

Figure P20.8

Section 20.2 Work and internal energy, and Section 20.3 Work and heat in thermodynamic processes

9. ±? An ideal gas is enclosed in a cylinder that has a movable piston on top. The piston has a mass $m \pm \Delta m$ and an area $A \pm \Delta A$ and is free to slide up and down, keeping the pressure of the gas constant. How much work is done on the gas as the temperature of n mol of the gas is raised from T_1 to T_2?

10. An ideal gas is taken through a quasi-static process described by $P = \alpha V^2$, with $\alpha = 5.00$ atm/m^6, as shown in **Figure P20.10**. The gas is expanded to twice its original volume of 1.00 m^3. How much work is done on the expanding gas in this process?

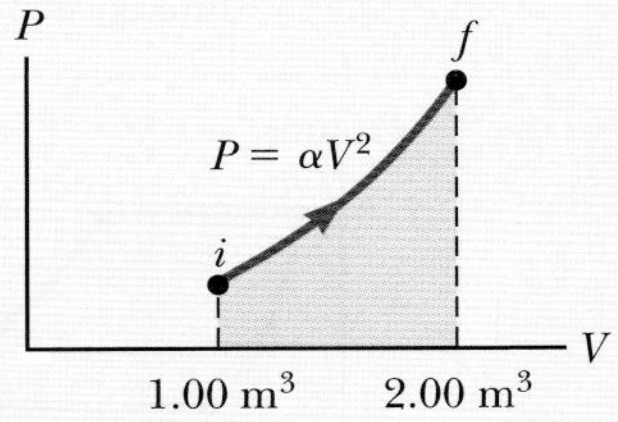

Figure P20.10

11. Determine the work done on a gas that expands from i to f as indicated in **Figure P20.11**. (b) **What If?** How much work is done on the gas if it is compressed from f to i along the same path?

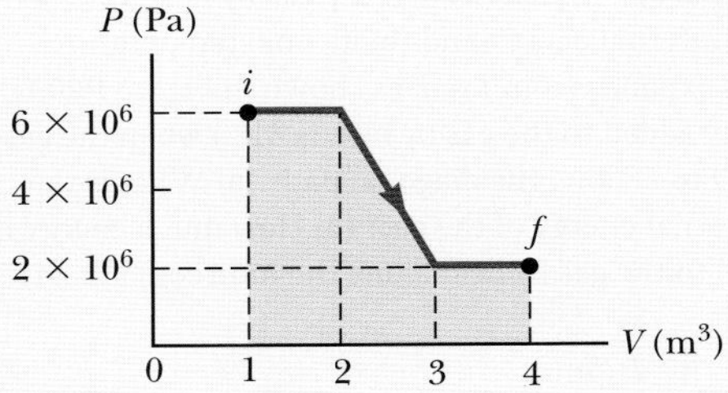

Figure P20.11

12. One mole of an ideal gas is warmed slowly so that it goes from the PV state (P_i, V_i) to $(3P_i, 3V_i)$ in such a way that the pressure of the gas is directly proportional to the volume. (a) How much work is done on the gas in the process? (b) How is the temperature of the gas related to its volume during this process?

Section 20.4 The first law of thermodynamics

13. A gas is taken through the cyclic process described in **Figure P20.13**. (a) Find the net energy transferred to the system by heat during one complete cycle. (b) **What If?** If the cycle is reversed – that is, the process follows the path *ACBA* – what is the net energy input per cycle by heat?

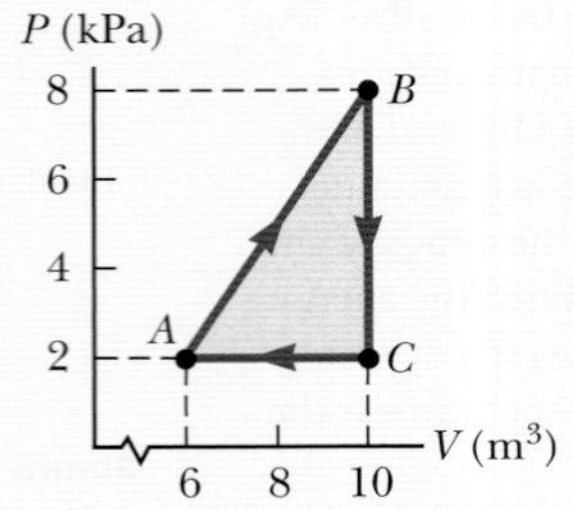

Figure P20.13

14. ±? A thermodynamic system undergoes a process in which its internal energy decreases by (500 ± 30) J. Over the same time interval, (220 ± 10) J of work is done on the system. Find the energy transferred from it by heat.

15. A sample of an ideal gas goes through the process shown in **Figure P20.15**. From *A* to *B*, the process is adiabatic; from *B* to *C*, it is isobaric with 100 kJ of energy entering the system as heat; from *C* to *D*, the process is isothermal; and from *D* to *A*, it is isobaric with 150 kJ of energy leaving the system as heat. Determine the difference in internal energy $E_{\text{int, B}} - E_{\text{int, A}}$.

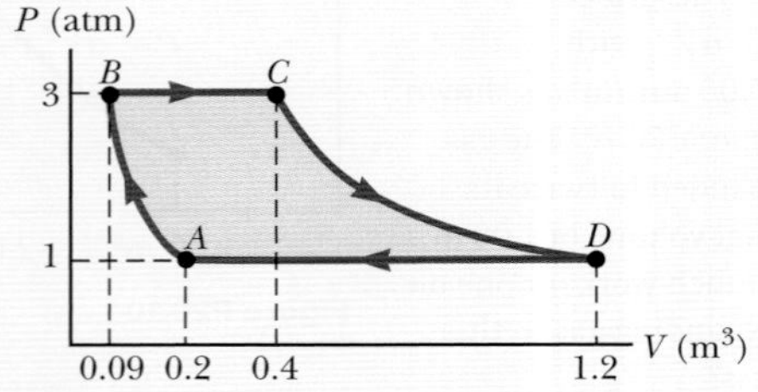

Figure P20.15

16. An ideal gas initially at 300 K undergoes an isobaric expansion at 2.50 kPa. If the volume increases from 1.00 m³ to 3.00 m³ and 12.5 kJ is transferred to the gas by heat, find (a) the change in its internal energy and (b) its final temperature.

17. A gas expands from *I* to *F* as shown in **Figure P20.17**. The energy added to the gas by heat is 418 J when the gas goes from *I* to *F* along the diagonal path. (a) What is the change in internal energy of the gas? (b) How much energy must be added to the gas by heat along the indirect path *IAF*?

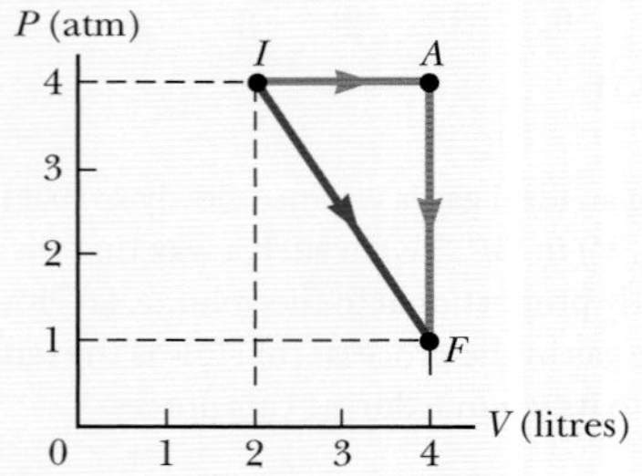

Figure P20.17

18. ±? A 2.00 mol sample of helium gas initially at (300 ± 10) K and (0.400 ± 0.001) atm is compressed isothermally to (1.20 ± 0.01) atm as it is pumped into a storage cylinder. Noting that the helium behaves as an ideal gas, find (a) the final volume of the gas, (b) the work done on the gas and (c) the energy transferred by heat.

19. One mole of an ideal gas does 3000 J of work on its surroundings as it expands isothermally to a final pressure of 1.00 atm and final volume of 25.0 L. Determine (a) the initial volume and (b) the temperature of the gas.

20. An ideal gas initially at P_i, V_i and T_i is taken through a cycle as shown in **Figure P20.20**. (a) Find the net work done on the gas per cycle. (b) What is the net energy added by heat to the system per cycle?

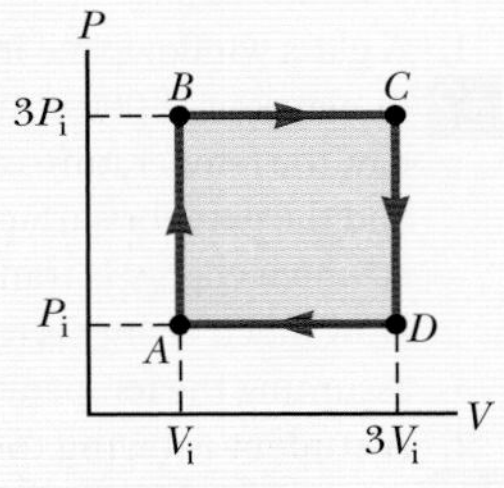

Figure P20.20

21. ±? On a hot day, a 1.00 kg aluminium window frame is warmed at atmospheric pressure so that its temperature increases from (22.0 ± 0.1)°C to (40.0 ± 0.1)°C. Find (a) the work done on the aluminium, (b) the energy added to it by heat and (c) the change in its internal energy.

Additional problems

22. An aluminium rod 0.500 m in length and with a cross-sectional area of 2.50 cm² is inserted into a thermally insulated vessel containing liquid helium at 4.20 K. The rod is initially at 300 K. (a) If half of the rod is inserted into the helium, how many litres of helium boil off by the time the inserted half cools to 4.20 K? Assume the upper half does not yet cool. (b) If the circular surface of the upper end of the rod is maintained at 300 K, what is the approximate boil-off rate of liquid helium in litres per second after the lower half has reached 4.20 K? (Aluminum has thermal conductivity of 3100 W/m.K at 4.20 K; ignore its temperature variation. The density of liquid helium is 125 kg/m³.)

23. One mole of an ideal gas is contained in a cylinder with a movable piston. The initial pressure, volume, and temperature are P_i, V_i and T_i respectively. Find the work done on the gas in the following processes. In operational terms, describe how to carry out each process and show each process on a *PV* diagram: (a) an isobaric compression in which the final volume is half the initial volume (b) an isothermal compression in which the final pressure is four times the initial pressure (c) an isovolumetric process in which the final pressure is three times the initial pressure.

24. Water boils in a kettle. The power absorbed by the water is 1.00 kW. Assuming the pressure of vapour in the kettle equals atmospheric pressure, determine the speed of effusion of vapour from the kettle's spout if the spout has a cross-sectional area of 2.00 cm². Model the steam as an ideal gas.

25. ±? In air at 0°C, a 1.60 kg copper block at 0°C is set sliding at (2.50 ± 0.01) m/s over a sheet of ice at 0°C. Friction brings the block to rest. (a) Find the mass of the ice that melts. (b) As the block slows down, identify its energy input *Q*, its change in internal energy ΔE_{int} and the change in mechanical energy for the block–ice system. (c) For the ice as a system, identify its energy input *Q* and its change in internal energy ΔE_{int}. (d) A 1.60 kg block of ice at 0°C is set sliding at 2.50 m/s over a sheet of copper at 0°C. Friction brings the block to rest. Find the mass of the ice that melts. (e) Evaluate *Q* and ΔE_{int} for the block of ice as a system and ΔE_{mech} for the block–ice system. (f) Evaluate *Q* and ΔE_{int} for the metal sheet as a system. (g) A thin, 1.60 kg slab of copper at 20°C is set sliding at 2.50 m/s over an identical stationary slab at the same temperature. Friction quickly stops the motion. Assuming no energy is transferred to the environment by heat, find the change in temperature of both objects. (h) Evaluate Q and ΔE_{int} for the sliding slab and ΔE_{mech} for the two-slab system. (i) Evaluate Q and ΔE_{int} for the stationary slab.

26. During periods of high activity, the Sun has more sunspots than usual. Sunspots are cooler than the rest of the luminous layer of the Sun's atmosphere (the photosphere). Paradoxically, the total power output of the active Sun is not lower than average but is the same or slightly higher than average. Work out the details of the following crude model of this phenomenon. Consider a patch of the photosphere with an area of 5.10×10^{14} m^2. Its emissivity is 0.965. (a) Find the power it radiates if its temperature is uniformly 5800 K, corresponding to the quiet Sun. (b) To represent a sunspot, assume 10.0% of the patch area is at 4800 K and the other 90.0% is at 5890 K. Find the power output of the patch. (c) State how the answer to part (b) compares with the answer to part (a). (d) Find the average temperature of the patch. Note that this cooler temperature results in a higher power output. (The sunspot cycle has a period of about 11 years, with the most recent maximum having occurred in 2013 and the next maximum predicted for 2024.)

27. Gas in a container is at a pressure of 1.50 atm and a volume of 4.00 m^3. What is the work done on the gas (a) if it expands at constant pressure to twice its initial volume, and (b) if it is compressed at constant pressure to one-quarter its initial volume?

Challenge problems

28. (a) The inside of a hollow cylinder is maintained at a temperature T_a, and the outside is at a lower temperature T_b (**Figure P20.28**). The wall of the cylinder has a thermal conductivity k. Ignoring end effects, show that the rate of energy conduction from the inner surface to the outer surface in the radial direction is

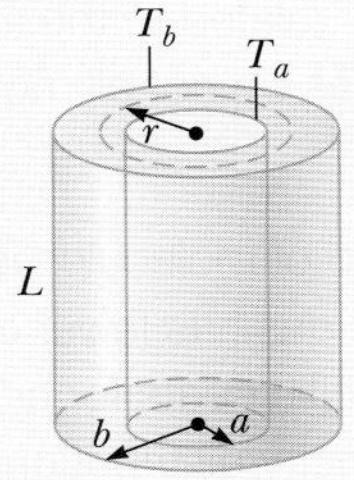

Figure P20.28

$$\frac{dQ}{dt} = 2\pi Lk\left[\frac{T_a - T_b}{\ln(b/a)}\right]$$

Suggestions: The temperature gradient is dT/dr. A radial energy current passes through a concentric cylinder of area $2\pi rL$. (b) The passenger section of a jet airliner is in the shape of a cylindrical tube with a length of 35.0 m and an inner radius of 2.50 m. Its walls are lined with an insulating material 6.00 cm in thickness and having a thermal conductivity of 4.00×10^{-5} cal/s.cm.°C. A heater must maintain the interior temperature at 25.0°C while the outside temperature is −35.0°C. What power must be supplied to the heater?

29. Consider the piston–cylinder apparatus shown in **Figure P20.29**. The bottom of the cylinder contains 2.00 kg of water at just under 100.0°C. The cylinder has a radius of $r = 7.50$ cm. The piston of mass $m = 3.00$ kg sits on the surface of the water. An electric heater in the cylinder base transfers energy into the water at a rate of 100 W. Assume the cylinder is much taller than shown in the figure, so we don't need to be concerned about the piston reaching the top of the cylinder. (a) Once the water begins boiling, how fast is the piston rising? Model the steam as an ideal gas. (b) After the water has completely turned to steam and the heater continues to transfer energy to the steam at the same rate, how fast is the piston rising?

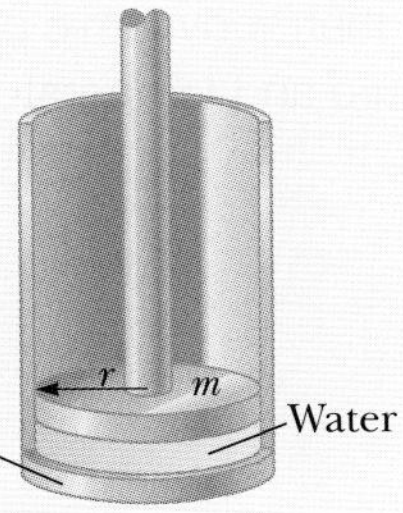

Figure P20.29

30. A spherical shell has inner radius 3.00 cm and outer radius 7.00 cm. It is made of material with thermal conductivity $k = 0.800$ W/m.°C. The interior is maintained at temperature 5°C and the exterior at 40°C. After an interval of time, the shell reaches a steady state with the temperature at each point within it remaining constant in time. (a) Explain why the rate of energy transfer P must be the same through each spherical surface, of radius r, within the shell and must satisfy

$$\frac{dT}{dr} = \frac{P}{4\pi kr^2}$$

(b) Prove that

$$\int_5^{40} dT = \frac{P}{4\pi k}\int_{0.03}^{0.07} r^{-2}dr$$

where T is in degrees Celsius and r is in metres. (c) Find the rate of energy transfer through the shell. (d) Prove that

$$\int_5^T dT = 1.84\int_{0.03}^{r} r^{-2}\,dr$$

where T is in degrees Celsius and r is in metres. (e) Find the temperature within the shell as a function of radius. (f) Find the temperature at $r = 5.00$ cm, halfway through the shell.

31. A cylinder is closed by a piston connected to a spring of constant 2.00×10^3 N/m (see **Figure P20.31**). With the spring relaxed, the cylinder is filled with 5.00 L of gas at a pressure of 1.00 atm and a temperature of 20.0°C. (a) If the piston has a cross-sectional area of 0.0100 m^2 and negligible mass, how high will it rise when the temperature is raised to 250°C? (b) What is the pressure of the gas at 250°C?

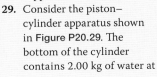

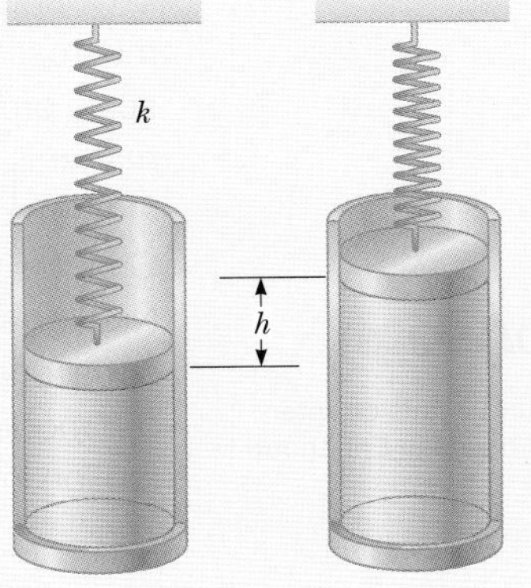

$T = 20.0$°C $\quad T = 250$°C

Figure P20.31

chapter 21

The kinetic theory of gases

As the boy pumps air into his bicycle tyre with the hand-operated pump, the gas is compressed; after vigorously pumping for a minute or two, the boy will find that the pump feels warm. How can we model the compression and heating of the gas inside the pump?

On completing this chapter, students will understand:

- why temperature is related to the average kinetic energy of molecules in a gas
- what forms of motion contribute to that kinetic energy
- how specific heat can be related to the microscopic properties of a gas
- that the molecules in a gas do not all move with the same speed.

Students will be able to:

- calculate the mean speed of molecules in a gas
- explain why a monatomic gas has fewer degrees of freedom than a polyatomic gas
- calculate the ratio of specific heats for an ideal gas
- calculate energy and velocity distributions for molecules in a gas
- distinguish between rms, average and most probable speeds.

In Chapter 19, we discussed the properties of an ideal gas by using such macroscopic variables as pressure, volume and temperature. Such large-scale properties can be related to a description on a microscopic scale, where matter is treated as a collection of molecules. In Chapter 20 we began to look at thermodynamic processes in gases, using the state variables described in Chapter 19. In this chapter, we develop a microscopic model of gases. Applying Newton's laws of motion in a statistical manner to a collection of particles provides a reasonable description of thermodynamic processes. We shall begin to develop our model by relating pressure and temperature directly to the details of molecular motion in a sample of gas.

ENHANCED WebAssign

The problems found in this chapter may be assigned online in Enhanced Web Assign.

21.1 Molecular model of an ideal gas

The idea of the ideal gas model was introduced in Chapter 19, where we used the mathematical representation $PV = nRT$ (**Equation 19.5**) to describe it. So far we have considered the macroscopic state of a gas only, and not been concerned about the microscopic states of the individual molecules. In the following, we develop a microscopic model of the thermal properties of an ideal gas, called **kinetic theory**. This model will allow us to make predictions about the distribution of speeds of molecules in different types of gases at different temperatures. It will also allow us to make estimates of the specific heat capacity of some gases. Such calculations form the basis of many important fields of science, for example models of planetary atmospheres including that of the Earth.

In developing our microscopic model, we make the following assumptions.

◀ Assumptions of the microscopic model of an ideal gas

1. **The number of molecules in the gas is large and the average separation between them is large compared with their dimensions.** In other words, the molecules occupy a negligible volume in the container. That is consistent with the ideal gas model, in which we model the molecules as particles.
2. **The molecules obey Newton's laws of motion but as a whole they move randomly.** By 'randomly,' we mean that any molecule can move in any direction with any speed.
3. **The molecules interact only by short-range forces during elastic collisions.** This is consistent with the ideal gas model, in which the molecules exert no long-range forces on each other.
4. **The molecules make elastic collisions with the walls.** These collisions lead to the macroscopic pressure on the walls of the container.
5. The gas under consideration is a pure substance, that is, **all molecules are identical.** Although we often picture an ideal gas as consisting of single atoms, the behaviour of molecular gases approximates that of ideal gases well at low pressures. Usually, molecular rotations or vibrations have no effect on the motions considered here.

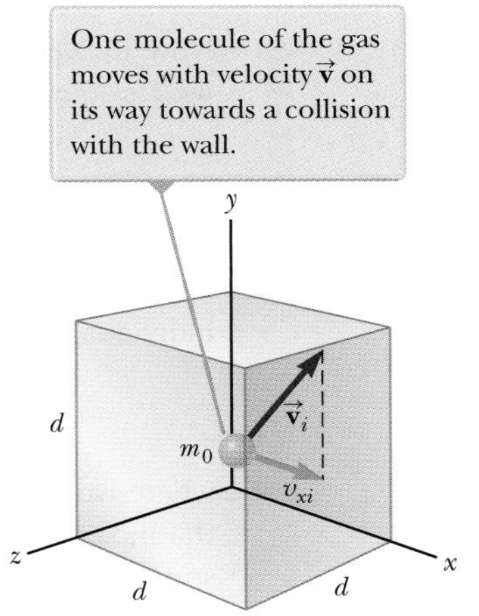
One molecule of the gas moves with velocity $\vec{\mathbf{v}}$ on its way towards a collision with the wall.
y
d
m_0
$\vec{\mathbf{v}}_i$
v_{xi}
z
d
d
x

Figure 21.1
A box with sides of length d containing an ideal gas

For our first application of kinetic theory, let us relate the macroscopic variable of pressure P to microscopic quantities. This process will allow us to make a key link between the molecular and large-scale worlds.

Consider a collection of N molecules of an ideal gas in a container of volume V. The container is a cube with edges of length d (**Figure 21.1**). We shall first focus our attention on one of these molecules of mass m_0 and assume it is moving so that its component of velocity in the x direction is v_{xi} as in **Active Figure 21.2**. The subscript i here refers to the ith molecule in the collection, not to an initial value. We will combine the effects of all the molecules shortly. As the molecule collides elastically with any wall (assumption 4), its velocity component perpendicular to the wall is reversed because the mass of the wall is far greater than the mass of the molecule. We model the molecule as a non-isolated system for which the impulse from the wall causes a change in the molecule's momentum. Because the momentum component p_{xi} of the molecule is $m_0 v_{xi}$ before the collision and $-m_0 v_{xi}$ after the collision, the change in the x component of the momentum of the molecule is

$$\Delta p_{xi} = -m_0 v_{xi} - (m_0 v_{xi}) = -2m_0 v_{xi}$$

Because the molecules obey Newton's laws (assumption 2), we can apply the impulse–momentum theorem (**Equation 8.10**) to the molecule to give

$$\overline{F}_{i,\text{ on molecule}} \Delta t_{\text{collision}} = \Delta p_{xi} = -2m_0 v_{xi}$$

where $\overline{F}_{i,\text{ on molecule}}$ is the x component of the average force the wall exerts on the molecule during the collision and $\Delta t_{\text{collision}}$ is the duration of the collision. For this discussion, we use a bar over a variable to represent the average value of the variable, such as $\overline{F}$ for the average force, rather than the subscript 'avg' that we have used before. This notation is to save confusion because we already have a number of subscripts on variables.

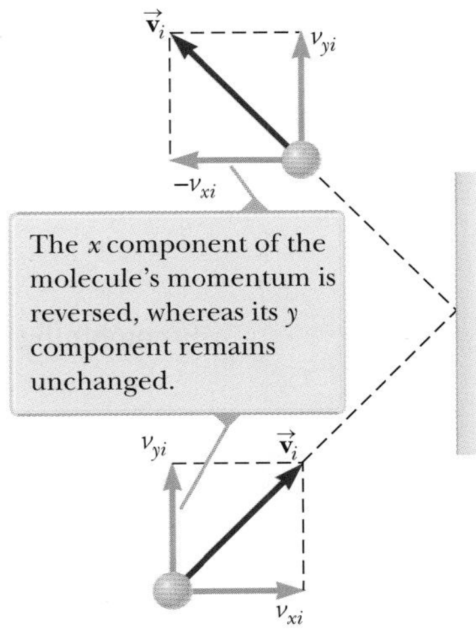
$\vec{\mathbf{v}}_i$
v_{yi}
$-v_{xi}$
The x component of the molecule's momentum is reversed, whereas its y component remains unchanged.
v_{yi}
$\vec{\mathbf{v}}_i$
v_{xi}

Active Figure 21.2
A molecule makes an elastic collision with the wall of the container. In this construction, we assume the molecule moves in the xy plane.

For the molecule to make another collision with the same wall after this first collision, it must travel a distance of $2d$ in the x direction (across the container and back). Therefore, the time interval between two collisions with the same wall is

$$\Delta t = \frac{2d}{v_{xi}}$$

The force that causes the change in momentum of the molecule in the collision with the wall occurs only during the collision. We can, however, average the force over the time interval for the molecule to move across the cube and back. Sometime during this time interval the collision occurs, so the change in momentum for this time interval is the same as that for the short duration of the collision. Therefore, we can rewrite the impulse–momentum theorem as

$$\overline{F}_i \Delta t = -2m_0 v_{xi}$$

where $\overline{F}_i$ is the average force component over the time interval for the molecule to move across the cube and back. Because exactly one collision occurs for each such time interval, this result is also the long-term average force on the molecule over long time intervals containing any number of multiples of Δt.

This equation and the preceding one enable us to express the x component of the long-term average force exerted by the wall on the molecule as

$$\overline{F}_i = -\frac{2m_0 v_{xi}}{\Delta t} = -\frac{2m_0 v_{xi}^2}{2d} = -\frac{m_0 v_{xi}^2}{d}$$

Now, by Newton's third law, the x component of the long-term average force exerted by the *molecule* on the *wall* is equal in magnitude and opposite in direction:

$$\overline{F}_{i,\text{on wall}} = -\overline{F}_i = -\left(-\frac{m_0 v_{xi}^2}{d}\right) = \frac{m_0 v_{xi}^2}{d}$$

The total average force $\overline{F}$ exerted by the gas on the wall is found by adding the average forces exerted by the individual molecules. Adding terms such as that above for all molecules gives

$$\overline{F} = \sum_{i=1}^{N} \frac{m_0 v_{xi}^2}{d} = \frac{m_0}{d} \sum_{i=1}^{N} v_{xi}^2$$

where we have factored out the length of the box and the mass m_0 because assumption 5 tells us that all the molecules are the same. We now impose assumption 1, that the number of molecules is large. For a small number of molecules, the actual force on the wall would vary with time. It would be non-zero during the short interval of a collision of a molecule with the wall and zero when no molecule happens to be hitting the wall. For a very large number of molecules such as Avogadro's number, however, these variations in force are smoothed out so that the average force given above is the same over *any* time interval. Therefore, the *constant* force F on the wall due to the molecular collisions is

$$F = \frac{m_0}{d} \sum_{i=1}^{N} v_{xi}^2$$

To proceed further, let's consider how to express the average value of the square of the x component of the velocity for N molecules. The average of a set of values is the sum of the values over the number of values:

$$\overline{v_x^2} = \frac{\sum_{i=1}^{N} v_{xi}^2}{N}$$

By combining the two expressions the total force on the wall can be written as:

$$F = \frac{m_0}{d} N\overline{v_{xi}^2} \tag{21.1}$$

Σ *The Pythagorean theorem is given in Appendix B.4.*

Now let's focus again on one molecule with velocity components v_{xi}, v_{yi} and v_{zi}. The Pythagorean theorem (see Appendix B.4) relates the square of the speed of the molecule to the squares of the velocity components:

$$v_i^2 = v_{xi}^2 + v_{yi}^2 + v_{zi}^2$$

Hence, the average value of v^2 for all the molecules in the container is related to the average values of v_x^2, v_y^2 and v_z^2 according to the expression

$$\overline{v^2} = \overline{v_x^2} + \overline{v_y^2} + \overline{v_z^2}$$

Because the motion is completely random (assumption 2), the average values $\overline{v_x^2}$, $\overline{v_y^2}$ and $\overline{v_z^2}$ are equal to one another. Using this fact and the preceding equation, we find that

$$\overline{v^2} = 3\overline{v_x^2}$$

Therefore, from **Equation 21.1**, the total force exerted on the wall is

$$F = \frac{1}{3}N\frac{m_0\overline{v^2}}{d}$$

Using this expression, we can find the total pressure exerted on the wall:

$$P = \frac{F}{A} = \frac{F}{d^2} = \frac{1}{3}N\frac{m_0\overline{v^2}}{d^3} = \frac{1}{3}\left(\frac{N}{V}\right)m_0\overline{v^2}$$

$$P = \frac{2}{3}\left(\frac{N}{V}\right)\left(\frac{1}{2}m_0\overline{v^2}\right) \quad (21.2)$$

◀ Relationship between pressure and molecular kinetic energy

This result indicates that the pressure of a gas is proportional to (1) the number of molecules per unit volume and (2) the average translational kinetic energy of the molecules, $\frac{1}{2}m_0\overline{v^2}$. In analysing this simplified model of an ideal gas, we obtain an important result that relates the macroscopic quantity of pressure to a microscopic quantity, the average value of the square of the molecular speed.

Notice that **Equation 21.2** verifies some features of pressure with which you are already familiar from Chapter 19. One way to increase the pressure inside a container is to increase the number of molecules per unit volume, N/V, in the container. That is what you do when you add air to a tyre. The pressure in the tyre can also be raised by increasing the average translational kinetic energy of the air molecules in the tyre. That can be accomplished by increasing the temperature of that air, which is why the pressure inside a tyre increases as the tyre warms up during long road trips. The continuous flexing of the tyre as it moves along the road surface results in work done on the rubber as parts of the tyre distort, causing an increase in internal energy of the rubber. The increased temperature of the rubber results in the transfer of energy by heat into the air inside the tyre. This transfer increases the air's temperature and this increase in temperature in turn produces an increase in pressure.

Molecular interpretation of temperature

Let's now consider another macroscopic variable, the temperature T of the gas, and see how it relates to microscopic variables at the molecular level. We can gain some insight into the meaning of temperature by first writing **Equation 21.2** in the form

$$PV = \frac{2}{3}N\left(\frac{1}{2}m_0\overline{v^2}\right)$$

Let's now compare this expression with the equation of state for an ideal gas (**Equation 19.7**):

$$PV = Nk_BT$$

Recall that the equation of state is based on experimental observations concerning the macroscopic behaviour of gases. Equating the right sides of these expressions gives

$$T = \frac{2}{3k_B}\left(\frac{1}{2}m_0\overline{v^2}\right) \quad (21.3)$$

◀ Relationship between temperature and molecular kinetic energy

This result tells us that temperature is a direct measure of average molecular kinetic energy. By rearranging **Equation 21.3**, we can relate the translational molecular kinetic energy to the temperature:

$$\frac{1}{2}m_0\overline{v^2} = \frac{3}{2}k_BT \quad (21.4)$$

◀ Average kinetic energy per molecule

That is, the average translational kinetic energy per molecule is $\frac{3}{2}k_BT$. Because $\overline{v_x^2} = \frac{1}{3}\overline{v^2}$, it follows that

$$\frac{1}{2}m_0\overline{v_x^2} = \frac{1}{2}k_BT \quad (21.5)$$

In a similar manner, for the y and z directions,

Theorem of equipartition of energy ▶

$$\frac{1}{2}m_0\overline{v_y^2} = \frac{1}{2}k_BT \quad \text{and} \quad \frac{1}{2}m_0\overline{v_z^2} = \frac{1}{2}k_BT$$

Therefore, each translational degree of freedom contributes an equal amount of energy, $\frac{1}{2}k_BT$, to the gas. (In general, a 'degree of freedom' refers to an independent means by which a molecule can possess energy.) A generalisation of this result, known as the **theorem of equipartition of energy**, is as follows:

> Each degree of freedom contributes $\frac{1}{2}k_BT$ to the energy of a system, where possible degrees of freedom are those associated with translation, rotation and vibration of molecules.

The total translational kinetic energy of N molecules of gas is simply N times the average energy per molecule, which is given by **Equation 21.4**:

Total translational kinetic energy of N molecules ▶

$$K_{\text{tot trans}} = N\frac{1}{2}m_0\overline{v^2} = \frac{3}{2}Nk_BT = \frac{3}{2}nRT \tag{21.6}$$

where we have used $k_B = R/N_A$ and $n = N/N_A$. If the gas molecules possess only translational kinetic energy, **Equation 21.6** represents the internal energy of the gas. This result implies that the internal energy of an ideal gas depends *only* on the temperature. We will follow up on this point in Section 21.2.

The square root of $\overline{v^2}$ is called the **root-mean-square (rms) speed** of the molecules. From **Equation 21.4**, we find that the rms speed is

Root-mean-square speed ▶

$$v_{\text{rms}} = \sqrt{\overline{v^2}} = \sqrt{\frac{3k_BT}{m_0}} = \sqrt{\frac{3RT}{M}} \tag{21.7}$$

where M is the molar mass in kilograms per mole and is equal to m_0N_A. (We shall use rms values again when we look at AC circuits in Chapter 33.) This expression shows that, at a given temperature, lighter molecules move faster on the average than do heavier molecules. For example, at a given temperature, hydrogen molecules, whose molar mass is 2.02×10^{-3} kg/mol, have an average speed approximately four times that of oxygen molecules, whose molar mass is 32.0×10^{-3} kg/mol. Typical rms speeds for gases at room temperature range from about 200 m/s to about 2000 m/s; the rms speeds of the two lightest gases, hydrogen and helium, actually exceed the escape velocity required to escape Earth's gravitational field.

Pitfall Prevention 21.1

Taking the square root of $\overline{v^2}$ does not 'undo' the square because we have taken an average between squaring and taking the square root. Although the square root of $\overline{v}^2$ is $\overline{v} = v_{\text{avg}}$ because the squaring is done after the averaging, the square root of $\overline{v^2}$ is *not* v_{avg}, but rather v_{rms}.

Quick Quiz 21.1

Two containers hold an ideal gas at the same temperature and pressure. Both containers hold the same type of gas, but container B has twice the volume of container A. (i) What is the average translational kinetic energy per molecule in container B? (a) twice that of container A (b) the same as that of container A (c) half that of container A (d) impossible to determine (ii) From the same choices, describe the internal energy of the gas in container B.

Here we have characterised the behaviour of the particles in terms of their speeds. As we shall see in Chapter 27 when we study current, we can also describe the particles in terms of their mean free path and mean free time, which are the average distances and times between collisions. Mean free path and time also depend on temperature, and are important in determining the conductivity of a material.

Example 21.1

A tank of helium

A tank used for filling helium balloons has a volume of 0.300 m³ and contains 2.00 mol of helium gas at 20.0°C.

(A) What is the total translational kinetic energy of the gas molecules?

Example 21.1 cont.

Solution

Conceptualise Start by drawing a diagram. Imagine a microscopic model of a gas in which you can watch the molecules move about the container more rapidly as the temperature increases.

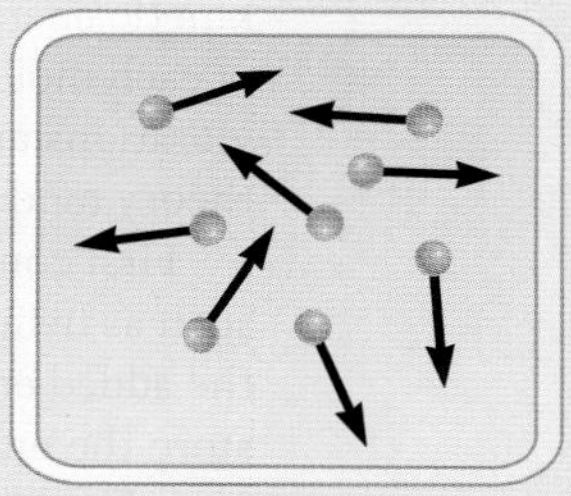

Figure 21.3 (Example 21.1) Helium atoms in a tank

Model We model the helium as an ideal gas, and use the equations developed in this section.

Analyse Use Equation 21.6 with $n = 2.00$ mol and $T = 293$ K:

$$K_{\text{tot trans}} = \frac{3}{2}nRT = \frac{3}{2}(2.00 \text{ mol})(8.31 \text{ J/mol.K})(293 \text{ K})$$
$$= 7.30 \times 10^3 \text{ J}$$

(B) What is the average kinetic energy per molecule?

Solution

Use Equation 21.4:

$$\frac{1}{2}m_0\overline{v}^2 = \frac{3}{2}k_BT = \frac{3}{2}(1.38 \times 10^{-23} \text{ J/K})(293 \text{ K})$$
$$= 6.07 \times 10^{-21} \text{ J}$$

What If? What if the temperature is raised from 20.0°C to 40.0°C? Because 40.0 is twice as large as 20.0, is the total translational energy of the molecules of the gas twice as large at the higher temperature?

Answer The expression for the total translational energy depends on the temperature, and the value for the temperature must be expressed in kelvins, not in degrees Celsius. Therefore, the ratio of 40.0 to 20.0 is *not* the appropriate ratio. Converting the Celsius temperatures to kelvins, 20.0°C is 293 K and 40.0°C is 313 K. Therefore, the total translational energy increases by a factor of only 313 K/293 K = 1.07.

21.2 Molar specific heat of an ideal gas

Consider an ideal gas undergoing several processes such that the change in temperature is $\Delta T = T_f - T_i$ for all processes. The temperature change can be achieved by taking a variety of paths from one isotherm to another as shown in Figure 21.4. Because ΔT is the same for each path, the change in internal energy ΔE_{int} is the same for all paths. The work W done on the gas (the negative of the area under the curves) is different for each path. Therefore the heat associated with a given change in temperature does *not* have a unique value as $Q = \Delta E_{\text{int}} - W$, and hence the specific heat of a gas does not have a unique value.

We can address this difficulty by defining specific heats for two special processes: isovolumetric and isobaric. Because the number of moles n is a convenient measure of the amount of gas, we define the **molar specific heats** associated with these processes as follows:

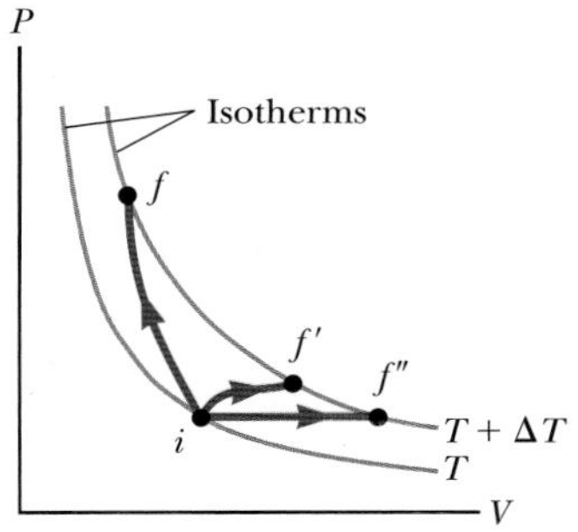

Figure 21.4 An ideal gas is taken from one isotherm at temperature T to another at temperature $T + \Delta T$ along three different paths.

$$Q = nC_V \Delta T \text{ (constant volume)} \qquad (21.8)$$

$$Q = nC_P \Delta T \text{ (constant pressure)} \qquad (21.9)$$

where C_V is the **molar specific heat at constant volume** and C_P is the **molar specific heat at constant pressure**. When energy is added by heat to a gas at constant volume, no work is done on or by the gas, so $\Delta E_{\text{int}} = Q$. When energy is added to a gas by heat at constant pressure, not only does the internal energy of the gas increase, but (negative) work is done on the gas because of the change in volume required to keep the pressure constant. Therefore, the heat Q in Equation 21.9 must account for both the increase in internal energy and the transfer of energy out of the system by work.

For this reason, Q is greater in Equation 21.9 than in Equation 21.8 for given values of n and ΔT. Therefore, C_P is greater than C_V.

In the previous section, we found that the temperature of a gas is a measure of the average translational kinetic energy of the gas molecules. This kinetic energy is associated with the motion of the centre of mass of each molecule. It does not include the energy associated with the internal motion of the molecule, namely, vibrations and rotations about the centre of mass. That should not be surprising because the simple kinetic theory model assumes a structureless molecule.

First consider the simplest case of an ideal monatomic gas, that is, a gas containing one atom per molecule such as helium, neon or argon. When energy is added to a monatomic gas in a container of fixed volume, all the added energy goes into increasing the translational kinetic energy of the atoms. There is no other way to store the energy in a monatomic gas. Therefore, from **Equation 21.6**, we see that the internal energy E_{int} of N molecules (or n mol) of an ideal monatomic gas is

Internal energy ▶ of an ideal monatomic gas

$$E_{int} = K_{tot\ trans} = \frac{3}{2}Nk_BT = \frac{3}{2}nRT \tag{21.10}$$

For a monatomic ideal gas, E_{int} is a function of T only and the functional relationship is given by **Equation 21.10**. In general, the internal energy of any ideal gas is a function of T only and the exact relationship depends on the type of gas.

If energy is transferred as heat to a system at constant volume, no work is done on the system. That is, $W = -\int P\,dV = 0$ for a constant-volume process. Hence, from the first law of thermodynamics,

$$Q = \Delta E_{int} \tag{21.11}$$

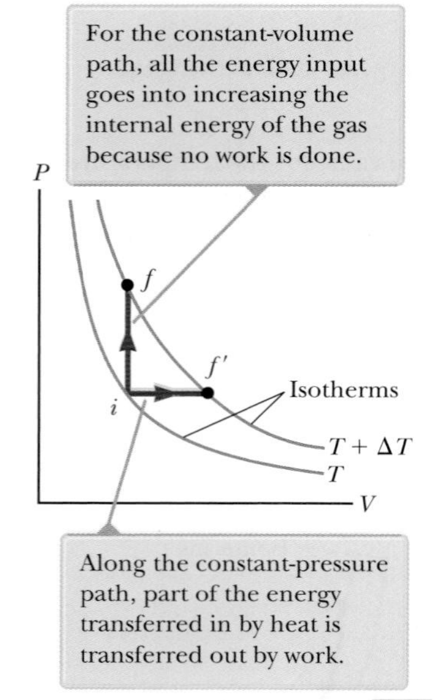
For the constant-volume path, all the energy input goes into increasing the internal energy of the gas because no work is done.
P
f
f′
i
Isotherms
T + ΔT
T
V
Along the constant-pressure path, part of the energy transferred in by heat is transferred out by work.

Active Figure 21.5
Energy is transferred by heat to an ideal gas in two ways.

In other words, all the energy transferred as heat goes into increasing the internal energy of the system. A constant-volume process from i to f for an ideal gas is described in **Active Figure 21.5**, where ΔT is the temperature difference between the two isotherms. Substituting the expression for Q given by **Equation 21.8** into **Equation 21.11**, we obtain

$$\Delta E_{int} = nC_V\,\Delta T \tag{21.12}$$

This equation applies to all ideal gases, those gases having more than one atom per molecule as well as monatomic ideal gases. In the limit of infinitesimal changes, we can use **Equation 21.12** to express the molar specific heat at constant volume as

$$C_V = \frac{1}{n}\frac{dE_{int}}{dT} \tag{21.13}$$

Let's now apply the results of this discussion to a monatomic gas. Substituting the internal energy from **Equation 21.10** into **Equation 21.13** gives

$$C_V = \frac{3}{2}R \tag{21.14}$$

This expression predicts a value of $C_V = \frac{3}{2}R = 12.5$ J/mol.K for *all* monatomic gases. This prediction is in excellent agreement with measured values of molar specific heats for such gases as helium, neon, argon and xenon over a wide range of temperatures (**Table 21.1**). Small variations in **Table 21.1** from the predicted values are because real gases are not ideal gases. In real gases, weak intermolecular interactions occur, which are not addressed in our ideal gas model.

Now suppose the gas is taken along the constant-pressure path $i \rightarrow f'$ shown in **Active Figure 21.5**. Along this path, the temperature again increases by ΔT. The energy that must be transferred by heat to the gas in this process is $Q = nC_P\,\Delta T$. Because the volume changes in this process, the work done on the gas is $W = -P\Delta V$, where P is the constant pressure at which the process occurs. Applying the first law of thermodynamics to this process, we have

$$\Delta E_{int} = Q + W = nC_P\,\Delta T + (-P\Delta V) \tag{21.15}$$

In this case, part of the energy added to the gas leaves the system by work (for example, the gas moves a piston through a displacement) and the remainder appears as an increase in the internal energy of the gas. The change in internal energy for the process $i \rightarrow f'$, however, is equal to that for the process $i \rightarrow f$ because E_{int} depends only on temperature for an ideal gas and ΔT is the same for both processes. In addition, because

Table 21.1

Molar specific heats of various gases

	Molar specific heat (J/mol.K)[a]			
Gas	C_P	C_V	$C_P - C_V$	$\gamma = C_P/C_V$
Monatomic gases				
He	20.8	12.5	8.33	1.67
Ar	20.8	12.5	8.33	1.67
Ne	20.8	12.7	8.12	1.64
Diatomic gases				
H_2	28.8	20.4	8.33	1.41
O_2	29.4	21.1	8.33	1.40
Cl_2	34.7	25.7	8.96	1.35
Polyatomic gases				
CO_2	37.0	28.5	8.50	1.30
H_2O	35.4	27.0	8.37	1.30
CH_4	35.5	27.1	8.41	1.31

[a] All values except that for water were obtained at 300 K.

$PV = nRT$, note that for a constant-pressure process, $P\Delta V = nR\Delta T$. Substituting this value for $P\Delta V$ into **Equation 21.15** with $\Delta E_{int} = nC_V\Delta T$ (**Equation 21.12**) gives

$$nC_V\Delta T = nC_P\Delta T - nR\Delta T$$

$$C_P - C_V = R \qquad \textbf{(21.16)}$$

This expression applies to *any* ideal gas. It predicts that the molar specific heat of an ideal gas at constant pressure is greater than the molar specific heat at constant volume by an amount R, the universal gas constant (which has the value 8.31 J/mol.K). This expression is a good approximation for real monatomic gases as the data in **Table 21.1** show.

Because $C_V = \frac{3}{2}R$ for a monatomic ideal gas, **Equation 21.16** predicts a value $C_P = \frac{5}{2}R = 20.8$ J/mol.K for the molar specific heat of a monatomic gas at constant pressure. The ratio of these molar specific heats is a dimensionless quantity γ (Greek letter gamma):

$$\gamma = \frac{C_P}{C_V} = \frac{5R/2}{3R/2} = \frac{5}{3} = 16.7 \qquad \textbf{(21.17)}$$

◀ Ratio of molar specific heats for a monatomic ideal gas

Theoretical values of C_V, C_P and γ are in excellent agreement with experimental values obtained for monatomic gases, but they are in serious disagreement with the values for the more complex gases (see **Table 21.1**). That is not surprising; the value $C_V = \frac{3}{2}R$ was derived for a monatomic ideal gas, and we expect some additional contribution to the molar specific heat from the internal structure of the more complex molecules. In Section 21.4, we describe the effect of molecular structure on the molar specific heat of a gas. The internal energy – and hence the molar specific heat – of a complex gas must include contributions from the rotational and the vibrational motions of the molecule.

In the case of solids and liquids heated at constant pressure, very little work is done because the thermal expansion is small. Consequently, C_P and C_V are approximately equal for solids and liquids.

Quick Quiz 21.2

(i) How does the internal energy of an ideal gas change as it follows path $i \rightarrow f$ in Active **Figure 21.5**? (a) E_{int} increases. (b) E_{int} decreases. (c) E_{int} stays the same. (d) There is not enough information to determine how E_{int} changes. (ii) From the same choices, how does the internal energy of an ideal gas change as it follows path $f \rightarrow f'$ along the isotherm labeled $T + \Delta T$ in Active **Figure 21.5**?

Σ Example 21.2

Heating a cylinder of helium

A cylinder of helium gas for filling balloons contains (3.00 ± 0.01) mol of gas at a temperature of (300 ± 1) K.

(A) If the gas is heated at constant volume, how much energy Q_1 must be transferred by heat to the gas for its temperature to increase to (500 ± 1) K?

Solution

Conceptualise Run the process in your mind with the help of the piston–cylinder arrangement in **Active Figure 19.6**. Imagine that the piston is clamped in position to maintain the constant volume of the gas so no work is done.

Model We model the helium as an ideal monatomic gas, and the process as isovolumetric. We can use the value for C_V from **Table 21.1**.

Analyse Use **Equation 21.8** to find the energy transfer: $Q_1 = nC_V\,\Delta T$

Substitute the given values: $Q_1 = (3.00\text{ mol})(12.5\text{ J/mol.K})(500\text{ K} - 300\text{ K}) = 7.50 \times 10^3\text{ J}$

The uncertainty is:

$$\Delta Q_1 = Q_1\left(\frac{\Delta(\Delta T)}{\Delta T} + \frac{\Delta n}{n}\right) = 7.5 \times 10^3\text{ J}\left(\frac{2\text{ K}}{200\text{ K}} + \frac{0.01\text{ mol}}{3.00\text{ mol}}\right) = 0.1 \times 10^3\text{ J}$$

where the uncertainty in ΔT is 2K because we must add the absolute uncertainties in the final and initial temperatures. Hence, our final answer is:

$$Q_1 = (7.5 \pm 0.1) \times 10^3\text{ J}$$

(B) If the cylinder is opened, the gas may be heated at constant pressure. How much energy must be transferred to the gas at constant pressure to raise the temperature to 500 K?

Solution

Model We now model the process as isobaric.

Analyse Use **Equation 21.9** to find the energy transfer: $Q_2 = nC_P\,\Delta T$

Substitute the given values: $Q_2 = (3.00\text{ mol})(20.8\text{ J/mol.K})(500\text{ K} - 300\text{ K}) = 12.5 \times 10^3\text{ J}$

The uncertainty is:

$$\Delta Q_2 = Q_2\left(\frac{\Delta(\Delta T)}{\Delta T} + \frac{\Delta n}{n}\right) = 12.5 \times 10^3\text{ J}\left(\frac{2\text{ K}}{200\text{ K}} + \frac{0.01\text{ mol}}{3.00\text{ mol}}\right) = 0.16 \times 10^3\text{ J}$$

$$Q_2 = (12.5 \pm 0.2) \times 10^3\text{ J}$$

This value is larger than Q_1 because of the transfer of energy out of the gas by work as the gas expands against its environment.

21.3 Adiabatic processes for an ideal gas

An **adiabatic process** is one in which no energy is transferred by heat between a system and its surroundings: $Q = 0$. For example, if a gas is compressed (or expanded) rapidly, very little energy is transferred out of (or into) the system by heat, so the process is nearly adiabatic. Such processes occur in the cycle of a petrol engine, which is discussed in detail in Chapter 22. Another example of an adiabatic process is the slow expansion of a gas that is thermally insulated from its surroundings. All three variables in the ideal gas law – P, V and T – change during an adiabatic process.

Let's imagine an adiabatic gas process involving an infinitesimal change in volume dV and an accompanying infinitesimal change in temperature dT. The work done on the gas is $-PdV$. Because the internal energy of an ideal gas depends only on temperature, the change in the internal energy in an adiabatic process is the same as that for an isovolumetric process between the same temperatures, $dE_{int} = nC_V dT$ (**Equation 21.12**). Hence, the first law of thermodynamics, $\Delta E_{int} = Q + W$, with $Q = 0$, becomes the infinitesimal form

$$dE_{int} = nC_V dT = -P dV$$

Taking the total differential of the equation of state of an ideal gas, $PV = nRT$, gives

$$P dV + V dP = nR dT$$

Eliminating dT from these two equations, we find that

$$P dV + V dP = -\frac{R}{C_V} P dV$$

Substituting $R = C_P - C_V$ and dividing by PV gives

$$\frac{dV}{V} + \frac{dP}{P} = -\left(\frac{C_P - C_V}{C_V}\right)\frac{dV}{V} = (1-\gamma)\frac{dV}{V}$$

$$\frac{dP}{P} + \gamma\frac{dV}{V} = 0$$

Integrating this expression, we have

$$\ln P + \gamma \ln V = \text{constant}$$

which is equivalent to

$$PV^{\gamma} = \text{constant} \qquad (21.18)$$

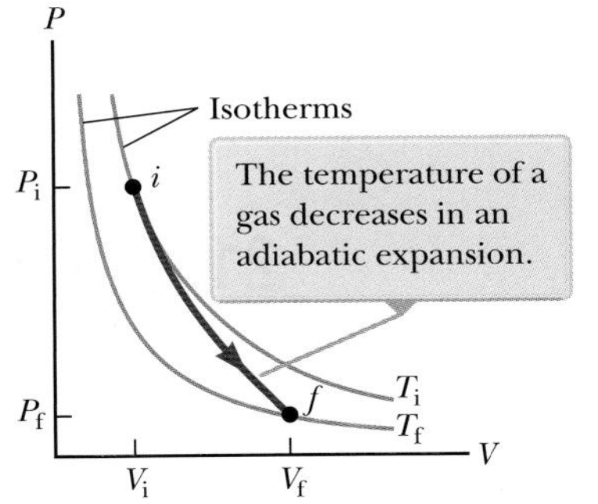

Figure 21.6
The *PV* diagram for an adiabatic expansion of an ideal gas

Σ *Differential calculus is summarised in Appendix B.7. A table of useful integrals is given in Appendix B.8.*

The *PV* diagram for an adiabatic expansion is shown in **Figure 21.6**. Because $\gamma > 1$, the *PV* curve is steeper than it would be for an isothermal expansion. By the definition of an adiabatic process, no energy is transferred by heat into or out of the system. Hence, from the first law, we see that ΔE_{int} is negative (work is done *by* the gas, so its internal energy decreases) and so ΔT also is negative. Therefore, the temperature of the gas decreases ($T_f < T_i$) during an adiabatic expansion. Conversely, the temperature increases if the gas is compressed adiabatically. Applying **Equation 21.18** to the initial and final states, we see that

$$P_i V_i^{\gamma} = P_f V_f^{\gamma} \qquad (21.19)$$

Using the ideal gas law, we can express **Equation 21.18** as

$$TV^{\gamma-1} = \text{constant} \qquad (21.20)$$

Example 21.3

A diesel engine cylinder

Air at 20.0°C in the cylinder of a diesel engine is compressed from an initial pressure of 1.00 atm and volume of 800.0 cm^3 to a volume of 60.0 cm^3. Find the final pressure and temperature of the air.

Solution

Conceptualise Imagine what happens if a gas is compressed into a smaller volume. Our discussion above and **Figure 21.6** tell us that the pressure and temperature both increase.

Model We model the air as an ideal gas with $\gamma = 1.4$, and the compression as adiabatic because it happens very rapidly.

Analyse Use **Equation 21.19** to find the final pressure:

$$P_f = P_i\left(\frac{V_i}{V_f}\right)^{\gamma} = 1.00\text{ atm}\left(\frac{800.0\text{ cm}^3}{60.0\text{ cm}^3}\right)^{1.40}$$

$$= 37.6\text{ atm}$$

Example 21.3 cont.

Use the ideal gas law to find the final temperature:

$$\frac{P_i V_i}{T_i} = \frac{P_f V_f}{T_f}$$

$$T_f = \frac{P_f V_f}{P_i V_i} T_i = \frac{(37.6\text{ atm})(60.0\text{ cm}^3)}{(1.00\text{ atm})(800.0\text{ cm}^3)}(293\text{ K})$$

$$= 826\text{ K} = 553°\text{ C}$$

Finalise The temperature of the gas increases by a factor of 826 K/293 K = 2.82. The high compression in a diesel engine raises the temperature of the gas enough to cause the combustion of fuel without the use of spark plugs.

21.4 The equipartition of energy

Predictions based on our model for molar specific heat agree quite well with the behaviour of monatomic gases, but not with the behaviour of more complex gases (see **Table 21.1**). The value predicted by the model for the quantity $C_P - C_V = R$, however, is the same for all gases. This similarity is not surprising because this difference is the result of the work done on the gas, which is independent of its molecular structure.

To understand the variations in C_V and C_P in gases more complex than monatomic gases we need to consider the origin of molar specific heat. So far, we have assumed the sole contribution to the internal energy of a gas is the translational kinetic energy of the molecules. The internal energy of a gas, however, includes contributions from the translational, vibrational and rotational motion of the molecules. The rotational and vibrational motions of molecules can be changed by collisions and therefore are 'coupled' to the translational motion of the molecules. The branch of physics known as *statistical mechanics* has shown that, for a large number of particles obeying the laws of Newtonian mechanics, the available energy is, on average, shared equally by each independent degree of freedom. Recall from Section 21.1 that the equipartition theorem states that, at equilibrium, each degree of freedom contributes $\frac{1}{2}k_B T$ of energy per molecule.

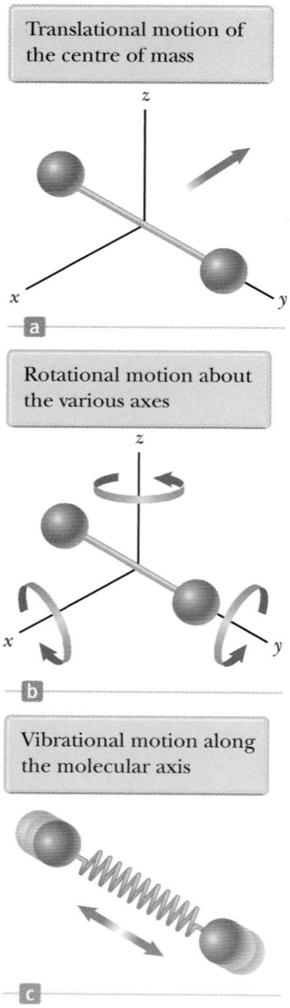

Figure 21.7
Possible motions of a diatomic molecule

TRY THIS

Put a blob of putty or Blu-Tack on each end of a pen or pencil. Spin it around the three axes shown in **Figure 21.7b**. How hard is it to spin it around the long axis compared to the short axes? Can you explain your observations?

Let's consider a diatomic gas whose molecules have the shape of a dumbbell (**Figure 21.7**). In this model, the centre of mass of the molecule can translate in the *x*, *y*, and *z* directions (**Figure 21.7a**). In addition, the molecule can rotate about three mutually perpendicular axes (**Figure 21.7b**). The rotation about the *y* axis can be neglected because the molecule's moment of inertia I_y and its rotational energy $\frac{1}{2}I_y\omega^2$ about this axis are negligible compared with those associated with the *x* and *z* axes. (If the two atoms are modelled as particles, then I_y is zero.) Therefore, there are five degrees of freedom for translation and rotation: three associated with the translational motion and two associated with the rotational motion. Because each degree of freedom contributes, on average, $\frac{1}{2}k_B T$ of energy per molecule, the internal energy for a system of *N* molecules, ignoring vibration for now, is

$$E_{\text{int}} = 3N\left(\frac{1}{2}k_B T\right) + 2N\left(\frac{1}{2}k_B T\right) = \frac{5}{2}Nk_B T = \frac{5}{2}nRT$$

We can use this result and **Equation 21.13** to find the molar specific heat at constant volume:

$$C_V = \frac{1}{n}\frac{dE_{\text{int}}}{dT} = \frac{1}{n}\frac{d}{dT}\left(\frac{5}{2}nRT\right) = \frac{5}{2}R \qquad (21.21)$$

From **Equations 21.16** and **21.17**, we find that

$$C_P = C_V + R = \frac{7}{2}R$$

$$\gamma = \frac{C_P}{C_V} = \frac{\frac{7}{2}R}{\frac{5}{2}R} = \frac{7}{5} = 1.40$$

These results agree quite well with most of the data for diatomic molecules given in **Table 21.1**. That is rather surprising because we have not yet accounted for the possible vibrations of the molecule.

In our model for vibration, the two atoms are joined by an imaginary spring (see **Figure 21.7c**). The vibrational motion adds two more degrees of freedom, which correspond to the kinetic energy and the potential energy associated with vibrations along the length of the molecule. Hence, a model that includes all three types of motion predicts a total internal energy of

$$E_{\text{int}} = 3N\left(\frac{1}{2}k_{\text{B}}T\right) + 2N\left(\frac{1}{2}k_{\text{B}}T\right) + 2N\left(\frac{1}{2}k_{\text{B}}T\right) = \frac{7}{2}Nk_{\text{B}}T = \frac{7}{2}nRT$$

and a molar specific heat at constant volume of

$$C_V = \frac{1}{n}\frac{dE_{\text{int}}}{dT} = \frac{1}{n}\frac{d}{dT}\left(\frac{7}{2}nRT\right) = \frac{7}{2}R \tag{21.22}$$

This value is inconsistent with experimental data for molecules such as H_2 and N_2 (see **Table 21.1**) and suggests a breakdown of our model, which is based on classical physics.

It might seem that our model is a failure for predicting molar specific heats for diatomic gases. We can claim some success for our model, however, if measurements of molar specific heat are made over a wide temperature range rather than at the single temperature that gives us the values in **Table 21.1. Figure 21.8** shows the molar specific heat of hydrogen as a function of temperature. The remarkable feature about the three plateaus in the graph's curve is that they are at the values of the molar specific heat predicted by **Equations 21.14, 21.21**, and **21.22**! For low temperatures, the diatomic hydrogen gas behaves like a monatomic gas. As the temperature rises to room temperature, its molar specific heat rises to a value for a diatomic gas, consistent with the inclusion of rotation but not vibration. For high temperatures, the molar specific heat is consistent with a model including all types of motion.

Before addressing the reason for this behaviour, we make some brief remarks about polyatomic gases. For molecules with more than two atoms, the vibrations are more complex than for diatomic molecules and the number of degrees of freedom is even larger. The result is an even higher predicted molar specific heat, which is in qualitative agreement with experiment. The molar specific heats for the polyatomic gases in **Table 21.1** are higher than those for diatomic gases. The more degrees of freedom available to a molecule, the more 'ways' there are to store energy, resulting in a higher molar specific heat.

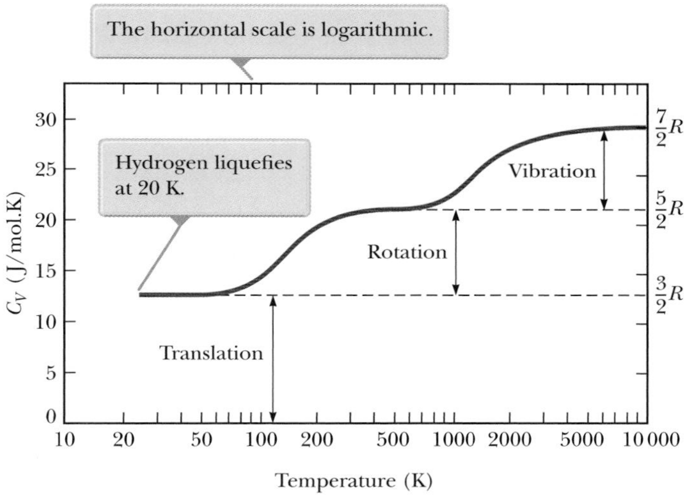

Figure 21.8
The molar specific heat of hydrogen as a function of temperature

Energy quantisation

Our model for molar specific heats has been based so far on purely classical notions. It predicts a value of the specific heat for a diatomic gas that, according to **Figure 21.8**, only agrees with experimental measurements made at high temperatures. To explain why this value is only true at high temperatures and why the plateaus in **Figure 21.8** exist, we must go beyond classical physics and introduce some quantum physics into the model. In Chapter 18, we discussed quantisation of frequency for vibrating strings and air columns in which only certain frequencies of standing waves can exist. That is a natural result whenever waves are subject to boundary conditions.

Quantum physics shows that atoms and molecules can be described by the physics of waves under boundary conditions. Consequently, these waves have quantised frequencies. Furthermore, in quantum

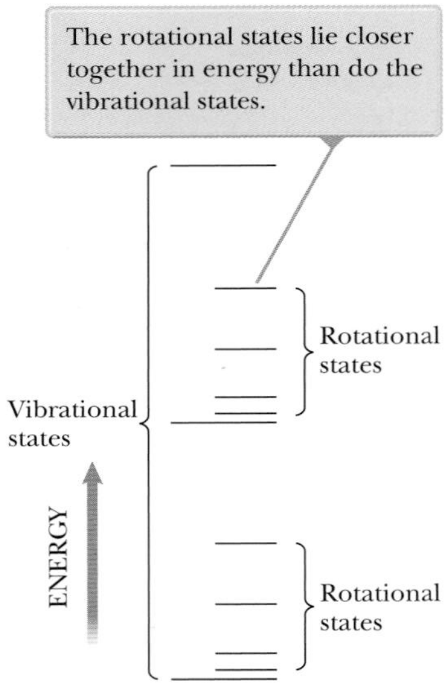
The rotational states lie closer together in energy than do the vibrational states.
Rotational states
Vibrational states
ENERGY
Rotational states

Figure 21.9
An energy-level diagram for vibrational and rotational states of a diatomic molecule

physics, the energy of a system is proportional to the frequency of the wave representing the system. Hence, **the energies of atoms and molecules are quantised**.

For a molecule, quantum physics tells us that the rotational and vibrational energies are quantised. **Figure 21.9** shows an **energy-level diagram** for the rotational and vibrational quantum states of a diatomic molecule. The lowest allowed state is called the **ground state**. Notice that vibrational states are separated by larger energy gaps than are rotational states.

At low temperatures, the energy a molecule gains in collisions with its neighbours is generally not large enough to raise it to the first excited state of either rotation or vibration. Therefore, even though rotation and vibration are allowed according to classical physics, they do not occur in reality at low temperatures. All molecules are in the ground state for rotation and vibration. The only contribution to the molecules' average energy is from translation, and the specific heat is that predicted by **Equation 21.14**.

As the temperature is raised, the average energy of the molecules increases. In some collisions, a molecule may have enough energy transferred to it from another molecule to excite the first rotational state. As the temperature is raised further, more molecules can be excited to this state. The result is that rotation begins to contribute to the internal energy and the molar specific heat rises. At about room temperature in **Figure 21.8**, the second plateau has been reached and rotation contributes fully to the molar specific heat. The molar specific heat is now equal to the value predicted by **Equation 21.21**.

There is no contribution at room temperature from vibration because the molecules are still in the ground vibrational state. The temperature must be raised even further to excite the first vibrational state, which happens in **Figure 21.8** between 1000 K and 10 000 K. At 10 000 K on the right side of the figure, vibration is contributing fully to the internal energy and the molar specific heat has the value predicted by **Equation 21.22**.

The predictions of this model are supportive of the theorem of equipartition of energy. In addition, the inclusion in the model of energy quantisation from quantum physics allows a full understanding of **Figure 21.8**. Quantisation is discussed more fully in Chapters 39 to 44.

University of Vienna, courtesy AIP Emilio Segre Visual Archives

Ludwig Boltzmann
Austrian physicist (1844–1906)
Boltzmann made many important contributions to the development of the kinetic theory of gases, electromagnetism and thermodynamics. His pioneering work in the field of kinetic theory led to the branch of physics known as statistical mechanics.

Quick Quiz 21.3

The molar specific heat of a diatomic gas is measured at constant volume and found to be 29.1 J/mol.K. What are the types of energy that are contributing to the molar specific heat? (a) translation only (b) translation and rotation only (c) translation and vibration only (d) translation, rotation and vibration

Quick Quiz 21.4

The molar specific heat of a gas is measured at constant volume and found to be $11R/2$. Is the gas most likely to be (a) monatomic, (b) diatomic or (c) polyatomic?

Pitfall Prevention 21.2
The distribution function $n_V(E)$ is defined in terms of the number of molecules with energy in the range E to $E + dE$ rather than in terms of the number of molecules with energy E. Because the number of molecules is finite and the number of possible values of the energy is infinite, the number of molecules with an *exact* energy E may be zero.

21.5 Distribution of molecular speeds

Thus far, we have considered only average values of the energies of molecules in a gas and have not addressed the distribution of energies among molecules. In reality, the motion of the molecules is extremely chaotic. Any individual molecule collides with others at an enormous rate, typically a billion times per second. Each collision results in a change in the speed and direction of motion of each of the participant molecules. **Equation 21.7** shows that rms molecular speeds increase with increasing temperature.

What is the relative number of molecules that possess some characteristic such as energy within a certain range?

We shall address this question by considering the **number density** $n_V(E)$. This quantity, called a *distribution function*, is defined such that $n_V(E)\,dE$ is the number of molecules per unit volume with energy between E and $E + dE$. The ratio of the number of molecules that have the desired characteristic to the total number of molecules is the probability that a particular molecule has that characteristic. In general, the number density is found from statistical mechanics to be

$$n_V(E) = n_0 e^{-E/k_BT} \quad (21.23)$$

◀ Boltzmann distribution law

where n_0 is defined such that $n_0 dE$ is the number of molecules per unit volume having energy between $E = 0$ and $E = dE$. This equation, known as the **Boltzmann distribution law**, is important in describing the statistical mechanics of a large number of molecules. It states that the probability of finding the molecules in a particular energy state varies exponentially as the negative of the energy divided by k_BT. All the molecules would fall into the lowest energy level if the thermal agitation at a temperature T did not excite the molecules to higher energy levels.

Example 21.4

Thermal excitation of atomic energy levels

Quantum mechanics tells us that atoms can occupy only certain discrete energy levels. Consider a gas at a temperature of 2500 K whose atoms can occupy only two energy levels separated by 1.50 eV, where 1 eV (electron volt) is an energy unit equal to 1.60×10^{-19} J (**Figure 21.10**). Determine the ratio of the number of atoms in the higher energy level to the number in the lower energy level.

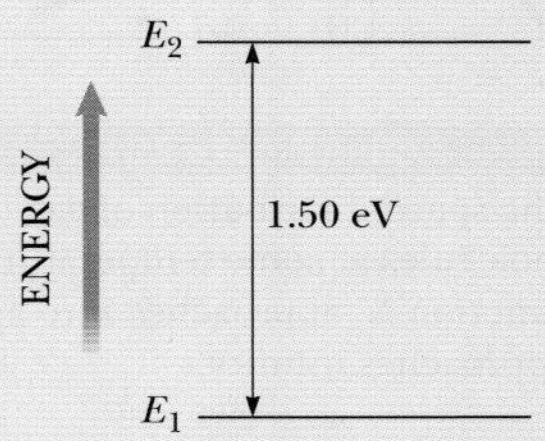

Figure 21.10
(Example 21.4) Energy-level diagram for a gas whose atoms can occupy two energy states

Solution

Conceptualise **Figure 21.10** helps you visualise the two states on an energy-level diagram. In this case, the atom has two possible energies, E_1 and E_2, where $E_1 < E_2$.

Model We shall assume that the Boltzmann distribution law describes this quantised system.

Analyse Set up the ratio of the number of atoms in the higher energy level to the number in the lower energy level and use **Equation 21.23** to express each number:

$$\frac{n_V(E_2)}{n_V(E_1)} = \frac{n_0 e^{-E_2/k_BT}}{n_0 e^{-E_1/k_BT}} = e^{-(E_2 - E_1)/k_BT} \quad (1)$$

Evaluate k_BT in the exponent:

$$k_BT = (1.38 \times 10^{-23}\ \text{J/K})(2500\ \text{K})\left(\frac{1\ \text{eV}}{1.60 \times 10^{-19}\ \text{J}}\right) = 0.216\ \text{eV}$$

Substitute this value into **Equation (1)**:

$$\frac{n_V(E_2)}{n_V(E_1)} = e^{-1.50\ \text{eV}/0.216\ \text{eV}} = e^{-6.96} = 9.52 \times 10^{-4}$$

Finalise This result indicates that at $T = 2500$ K, only a small fraction of the atoms are in the higher energy level. In fact, for every atom in the higher energy level, there are about 1000 atoms in the lower level. The number of atoms in the higher level increases at even higher temperatures, but the distribution law specifies that at equilibrium there are always more atoms in the lower level than in the higher level.

What If? What if the energy levels in **Figure 21.10** were closer together in energy? Would that increase or decrease the fraction of the atoms in the upper energy level?

Answer If the excited level is lower in energy than that in **Figure 21.10**, it would be easier for thermal agitation to excite atoms to this level and the fraction of atoms in this energy level would be larger. We can see from **Equation (1)** that as E_2 approaches E_1 the (negative) exponent on the right-hand side gets smaller, and so the ratio on the left-hand side gets larger.

We now consider the distribution of molecular speeds. In 1860, James Clerk Maxwell derived an expression that describes the distribution of molecular speeds. His work and subsequent developments by other scientists were highly controversial because direct detection of molecules could not be achieved experimentally at that time. About 60 years later, however, experiments were devised that confirmed Maxwell's predictions.

Let's consider a container of gas whose molecules have some distribution of speeds. Suppose we want to determine how many gas molecules have a speed in the range from, for example, 400 to 401 m/s. Intuitively, we expect the speed distribution to depend on temperature. Furthermore, we expect the distribution to peak in the

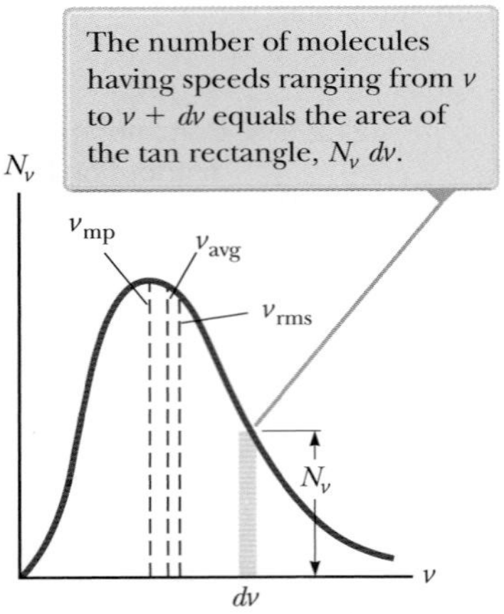
The number of molecules having speeds ranging from v to $v + dv$ equals the area of the tan rectangle, $N_v\,dv$.
N_v
v_{mp}
v_{avg}
v_{rms}
N_v
dv
v

Active Figure 21.11

The speed distribution of gas molecules at some temperature. The function N_v approaches zero as v approaches infinity.

vicinity of v_{rms}. That is, few molecules are expected to have speeds much less than or much greater than v_{rms} because these extreme speeds result only from an unlikely chain of collisions.

The observed speed distribution of gas molecules in thermal equilibrium is shown in **Active Figure 21.11**. The quantity N_v, called the **Maxwell–Boltzmann speed distribution function**, is defined as follows. If N is the total number of molecules, the number of molecules with speeds between v and $v + dv$ is $dN = N_v\,dv$. This number is equal to the area of the shaded rectangle in **Active Figure 21.11**. The fraction of molecules with speeds between v and $v + dv$ is $(N_v\,dv)/N$. This fraction is also equal to the probability that a molecule has a speed in the range v to $v + dv$.

The fundamental expression that describes the distribution of speeds of N gas molecules is

$$N_v = 4\pi N\left(\frac{m_0}{2\pi k_B T}\right)^{3/2} v^2 e^{-m_0 v^2/2k_B T} \tag{21.24}$$

where m_0 is the mass of a gas molecule, k_B is Boltzmann's constant and T is the absolute temperature.

As indicated in **Active Figure 21.11**, the average speed is somewhat lower than the rms speed. The *most probable speed* v_{mp} is the speed at which the distribution curve reaches a peak. Using **Equation 21.24**, we find that

$$v_{rms} = \sqrt{\overline{v^2}} = \sqrt{\frac{3k_B T}{m_0}} = 1.73\sqrt{\frac{k_B T}{m_0}} \tag{21.25}$$

$$v_{avg} = \sqrt{\frac{8k_B T}{\pi m_0}} = 1.60\sqrt{\frac{k_B T}{m_0}} \tag{21.26}$$

$$v_{mp} = \sqrt{\frac{2k_B T}{m_0}} = 1.41\sqrt{\frac{k_B T}{m_0}} \tag{21.27}$$

Equation 21.25 has previously appeared as **Equation 21.7**. From these equations, we see that

$$v_{rms} > v_{avg} > v_{mp}$$

Active Figure 21.12 represents speed distribution curves for nitrogen, N_2. The curves were obtained by using **Equation 21.24** to evaluate the distribution function at various speeds and at two temperatures. Notice that the peak in the curve shifts to the right as T increases, indicating that the average speed increases with increasing temperature, as expected. Because the lowest speed possible is zero and the upper classical limit of the speed is infinity, the curves are asymmetrical. (In Chapter 12, we showed that the actual upper limit is the speed of light.)

Equation 21.24 shows that the distribution of molecular speeds in a gas depends both on mass and on temperature. At a given temperature, the fraction of molecules with speeds exceeding a fixed value increases as the mass decreases. Hence, lighter molecules such as H_2 and He escape into space more readily from the Earth's atmosphere than do heavier molecules such as N_2 and O_2. (See the discussion of escape speed in Chapter 11.)

The speed distribution curves for molecules in a liquid are similar to those shown in **Active Figure 21.12**. We can understand the phenomenon of evaporation of a liquid from this distribution in speeds, given that some molecules in the liquid are more energetic than others. Some of the faster-moving molecules in the liquid penetrate the surface and even leave the liquid at temperatures well below the boiling point. The molecules that escape the liquid by evaporation are those that have sufficient energy to overcome the attractive forces of the molecules in the liquid phase. Consequently, the molecules left behind in the liquid phase have a lower average kinetic energy; as a result, the temperature of the liquid decreases. Hence, evaporation is a cooling process. For example, an alcohol-soaked cloth can be placed on a feverish head to cool and comfort a patient.

TRY THIS

Soak a facecloth or tea towel in water at room temperature and wring it out so that it is thoroughly damp but not dripping wet. Use a thermometer to take the temperature of the wet cloth, remembering to wait long enough for the thermometer and cloth to be in thermal equilibrium with each other. Now shake the cloth vigorously for a few minutes. Take the temperature of the cloth again. Can you explain your observations using the ideas discussed above?

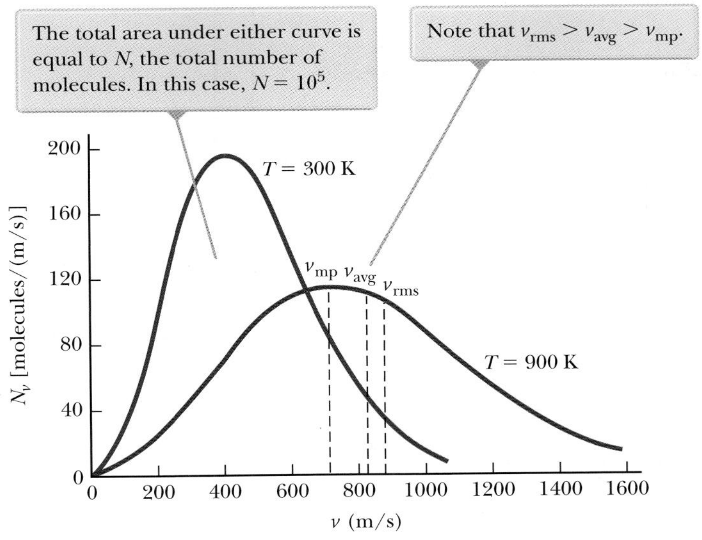

Active Figure 21.12

The speed distribution function for 10^5 nitrogen molecules at 300 K and 900 K

Example 21.5

A system of nine particles

Nine particles have speeds of 5.00, 8.00, 12.0, 12.0, 12.0, 14.0, 14.0, 17.0 and 20.0 m/s.

(A) Find the average speed of the particles.

Solution

Conceptualise Imagine a small number of particles moving in random directions with the speeds listed.

Model As we are dealing with a small number of particles, we can calculate the average speed directly.

Analyse Find the average speed of the particles by dividing the sum of the speeds by the total number of particles:

$$v_{avg} = \frac{(5.00 + 8.00 + 12.0 + 12.0 + 12.0 + 14.0 + 14.0 + 17.0 + 20.0)\text{ m/s}}{9}$$

$$= 12.7\text{ m/s}$$

(B) What is the rms speed of the particles?

Solution

Find the average speed squared of the particles by dividing the sum of the speeds squared by the total number of particles:

$$\overline{v^2} = \frac{5.00^2 + 8.00^2 + 12.0^2 + 12.0^2 + 12.0^2 + 14.0^2 + 14.0^2 + 17.0^2 + 20.0^2\text{ m/s}}{9}$$

$$= 178\text{ m}^2/\text{s}^2$$

Find the rms speed of the particles by taking the square root:

$$v_{rms} = \sqrt{\overline{v^2}} = \sqrt{178\text{ m}^2/\text{s}^2} = 13.3\text{ m/s}$$

(C) What is the most probable speed of the particles?

Solution

Three of the particles have a speed of 12.0 m/s, two have a speed of 14.0 m/s and the remaining four have different speeds. Hence, the most probable speed v_{mp} is 12.0 m/s.

Finalise Compare this example, in which the number of particles is small and we know the individual particle speeds, with the next example.

Example 21.6

Molecular speeds in a hydrogen gas

A 0.500 mol sample of hydrogen gas is at 300 K.

(A) Find the average speed, the rms speed and the most probable speed of the hydrogen molecules.

Solution

Conceptualise Imagine a huge number of particles in a real gas, all moving in random directions with different speeds.

Model We cannot calculate the averages as was done in Example 21.5 because the individual speeds of the particles are not known. We are dealing with a very large number of particles, however, so we can use the Maxwell–Boltzmann speed distribution function.

Analyse Use **Equation 21.26** to find the average speed:

$$v_{avg} = 1.60\sqrt{\frac{k_B T}{m_0}} = 1.60\sqrt{\frac{(1.38 \times 10^{-23}\ \text{J/K})(300\ \text{K})}{2(1.67 \times 10^{-27}\ \text{kg})}}$$
$$= 1.78 \times 10^3\ \text{m/s}$$

Use **Equation 21.25** to find the rms speed:

$$v_{rms} = 1.73\sqrt{\frac{k_B T}{m_0}} = 1.73\sqrt{\frac{(1.38 \times 10^{-23}\ \text{J/K})(300\ \text{K})}{2(1.67 \times 10^{-27}\ \text{kg})}}$$
$$= 1.93 \times 10^3\ \text{m/s}$$

Use **Equation 21.27** to find the most probable speed:

$$v_{mp} = 1.41\sqrt{\frac{k_B T}{m_0}} = 1.41\sqrt{\frac{(1.38 \times 10^{-23}\ \text{J/K})(300\ \text{K})}{2(1.67 \times 10^{-27}\ \text{kg})}}$$
$$= 1.57 \times 10^3\ \text{m/s}$$

(B) Find the number of molecules with speeds between 400 m/s and 401 m/s.

Solution

Use **Equation 21.24** to evaluate the number of molecules in a narrow speed range between v and $v + dv$:

$$N_v dv = 4\pi N\left(\frac{m_0}{2\pi k_B T}\right)^{3/2} v^2 e^{-m_0 v^2/2k_B T}\,dv \qquad (1)$$

Evaluate the constant in front of v^2:

$$4\pi N\left(\frac{m_0}{2\pi k_B T}\right)^{3/2} = 4\pi n N_A\left(\frac{m_0}{2\pi k_B T}\right)^{3/2}$$
$$= 4\pi(0.500\ \text{mol})(6.02 \times 10^{23}\ \text{mol}^{-1})\left[\frac{2(1.67 \times 10^{-27}\ \text{kg})}{2\pi(1.38 \times 10^{-23}\ \text{J/K})(300\ \text{K})}\right]^{3/2}$$
$$= 1.74 \times 10^{14}\ \text{s}^3/\text{m}^3$$

Evaluate the exponent of e that appears in **Equation (1)**:

$$-\frac{m_0 v^2}{2k_B T} = -\frac{2(1.67 \times 10^{-27}\ \text{kg})(400\ \text{m/s})^2}{2(1.38 \times 10^{-23}\ \text{J/K})(300\ \text{K})} = -0.0645$$

Evaluate $N_v\,dv$ using **Equation (1)**:

$$N_v\,dv = (1.74 \times 10^{14}\ \text{s}^3/\text{m}^3)(400\ \text{m/s})^2 e^{-0.0645}(1\ \text{m/s})$$
$$= 2.61 \times 10^{19}\ \text{molecules}$$

Finalise In this evaluation, we could calculate the result without integration because $dv = 1$ m/s is much smaller than $v = 400$ m/s. Had we sought the number of particles between, say, 400 m/s and 500 m/s, we would need to integrate **Equation (1)** between these speed limits.

End-of-chapter resources

At the begining of this chapter, we asked what happens when the boy pumps air into his bicycle tyre. We can start to answer this question by modelling air, which is mostly nitrogen, as an ideal diatomic gas. The downstroke (pushing the piston into the pump) compresses the air adiabatically, so that it reaches a significantly higher pressure than atmospheric pressure before entering the tyre. Because the compression is adiabatic, work is done on the gas by the piston and the internal energy of the gas increases as a result. Because we are dealing with a gas rather than a liquid, the increase in internal energy is entirely in the form of increased kinetic energy, which may come from rotational, vibrational or translational motion. At the temperatures occurring a typical bike pump, the translational and rotational modes will be important in determining the temperature of the air in the pump.

The problems found in this chapter may be assigned online in Enhanced Web Assign.

ENHANCED WebAssign

Worked solutions to every fifth problem are available in the Student Solutions Manual. Register online at **www.cengagebrain.com** for access.

Summary

Concepts and principles

The pressure of N molecules of an ideal gas contained in a volume V is

$$P = \frac{2}{3}\left(\frac{N}{V}\right)\left(\frac{1}{2}m_0\overline{v^2}\right) \quad (21.2)$$

The average translational kinetic energy per molecule of a gas, $\frac{1}{2}m_0\overline{v^2}$, is related to the temperature T of the gas through the expression

$$\frac{1}{2}m_0\overline{v^2} = \frac{3}{2}k_BT \quad (21.4)$$

where k_B is Boltzmann's constant. Each translational degree of freedom (x, y or z) has $\frac{1}{2}k_BT$ of energy associated with it.

The internal energy of N molecules (or n mol) of an ideal monatomic gas is

$$E_{int} = \frac{3}{2}Nk_BT = \frac{3}{2}nRT \quad (21.10)$$

The change in internal energy for n mol of any ideal gas that undergoes a change in temperature ΔT is

$$\Delta E_{int} = nC_V\,\Delta T \quad (21.12)$$

where C_V is the **molar specific heat at constant volume**.

The molar specific heat of an ideal monatomic gas at constant volume is $C_V = \frac{3}{2}R$; the molar specific heat at constant pressure is $C_P = \frac{5}{2}R$. The ratio of specific heats is given by $\gamma = C_P/C_V = \frac{5}{3}$.

The theorem of equipartition of energy states that at equilibrium each degree of freedom of a molecule contributes $\frac{1}{2}k_BT$ of energy.

Monatomic gases have only translational degrees of freedom, diatomic and polyatomic gases have rotational and vibrational degrees of freedom also. Hence diatomic and polyatomic gases have higher internal energy for a given temperature, and higher specific heat.

If an ideal gas undergoes an adiabatic expansion or compression, the first law of thermodynamics, together with the equation of state, shows that

$$PV^{\gamma} = \text{constant} \quad (21.18)$$

The **Boltzmann distribution law** describes the distribution of particles among available energy states. The relative number of particles having energy between E and $E + dE$ is $n_V(E)\,dE$, where

$$n_V(E) = n_0e^{-E/k_BT} \quad (21.23)$$

The **Maxwell–Boltzmann speed distribution function** describes the distribution of speeds of molecules in a gas:

$$N_v = 4\pi N\left(\frac{m_0}{2\pi k_BT}\right)^{3/2} v^2e^{-m_0v^2/2k_BT} \quad (21.24)$$

Equation 21.24 enables us to calculate the **root-mean-square speed**, the **average speed**, and the **most probable speed** of molecules in a gas:

$$v_{rms} = \sqrt{\overline{v^2}} = \sqrt{\frac{3k_BT}{m_0}} = 1.73\sqrt{\frac{k_BT}{m_0}} \quad (21.25)$$

$$v_{avg} = \sqrt{\frac{8k_BT}{\pi m_0}} = 1.60\sqrt{\frac{k_BT}{m_0}} \quad (21.26)$$

$$v_{mp} = \sqrt{\frac{2k_BT}{m_0}} = 1.41\sqrt{\frac{k_BT}{m_0}} \quad (21.27)$$

Chapter review quiz

To help you revise Chapter 21: The kinetic theory of gases, complete the automatically graded Chapter review quiz at http://login.cengagebrain.com.

Conceptual questions

1. Dalton's law of partial pressures states that the total pressure of a mixture of gases is equal to the sum of the pressures that each gas in the mixture would exert if it were alone in the container. Give a convincing argument for this law based on the kinetic theory of gases.
2. One container is filled with helium gas and another with argon gas. Both containers are at the same temperature. Which molecules have the higher rms speed? Explain your answer.
3. When alcohol is rubbed on your body, it lowers your skin temperature. Explain this effect.
4. What happens to a helium-filled latex balloon released into the air? Does it expand or contract? Does it stop rising at some height?
5. Which is denser: dry air or air saturated with water vapour? Explain your answer. This has impications for aircraft – why?
6. Why does a diatomic gas have a greater energy content per mole than a monatomic gas at the same temperature?
7. Hot air rises, so why does it generally become cooler as you climb a mountain? Note: Air has low thermal conductivity.
8. When the gas in a helium-filled balloon escapes, it leaves the Earth's atmosphere. Why?

Problems

Section 21.1 Molecular model of an ideal gas

1. A cylinder contains a mixture of helium and argon gas in equilibrium at 150°C. (a) What is the average kinetic energy for each type of gas molecule? (b) What is the rms speed of each type of molecule?
2. ±? In a 30.0 s interval, (500 ± 10) hailstones strike a glass window of area 0.600 m^2 at an angle of 45.0° to the window surface. Each hailstone has a mass of 5.00 g and a speed of 8.00 m/s. Assuming the collisions are elastic, find (a) the average force and (b) the average pressure on the window during this interval.
3. A spherical balloon of volume 4.00×10^3 cm^3 contains helium at a pressure of 1.20×10^5 Pa. How many moles of helium are in the balloon if the average kinetic energy of the helium atoms is 3.60×10^{-22} J?
4. A 2.00 mol sample of oxygen gas is confined to a 5.00 L vessel at a pressure of 8.00 atm. Find the average translational kinetic energy of the oxygen molecules under these conditions.
5. ±? How many atoms of helium gas fill a spherical balloon of diameter (30.0 ± 0.1) cm at (20.0 ± 0.1)°C and (1.00 ± 0.01) atm? (b) What is the average kinetic energy of the helium atoms? (c) What is the rms speed of the helium atoms?
6. In a period of 1.00 s, 5.00×10^{23} nitrogen molecules strike a wall with an area of 8.00 cm^2. Assume the molecules move with a speed of 300 m/s and strike the wall head-on in elastic collisions. What is the pressure exerted on the wall? Note: The mass of one N_2 molecule is 4.65×10^{-26} kg.

Section 21.2 Molar specific heat of an ideal gas

Note: Here we define a 'monatomic ideal gas' to have molar specific heats $C_V = \frac{3}{2}R$ and $C_P = \frac{5}{2}R$ and a 'diatomic ideal gas' to have $C_V = \frac{5}{2}R$ and $C_P = \frac{7}{2}R$.

7. A sample of a diatomic ideal gas has pressure P and volume V. When the gas is warmed, its pressure triples and its volume doubles. This warming process includes two steps, the first at constant pressure and the second at constant volume. Determine the amount of energy transferred to the gas by heat.
8. ±? In a constant-volume process, 209 J of energy is transferred by heat to 1.00 mol of an ideal monatomic gas initially at (300 ± 5) K. Find (a) the work done on the gas, (b) the increase in internal energy of the gas and (c) its final temperature.
9. A 1.00 mol sample of hydrogen gas is heated at constant pressure from 300 K to 420 K. Calculate (a) the energy transferred to the gas by heat, (b) the increase in its internal energy and (c) the work done on the gas.
10. A vertical cylinder with a heavy piston contains air at 300 K. The initial pressure is 2.00×10^5 Pa and the initial volume is 0.350 m^3. Take the molar mass of air as 28.9 g/mol and assume $C_V = \frac{5}{2}R$. (a) Find the specific heat of air at constant volume in units of J/kg.°C. (b) Calculate the mass of the air in the cylinder. (c) Suppose the piston is held fixed. Find the energy input required to raise the temperature of the air to 700 K. (d) **What If?** Assume again the conditions of the initial state and assume the heavy piston is free to move. Find the energy input required to raise the temperature to 700 K.

Section 21.3 Adiabatic processes for an ideal gas

11. ±? During the compression stroke of a particular petrol engine, the pressure increases from (1.00 ± 0.05) atm to (20.0 ± 0.2) atm. If the process is adiabatic and the air–fuel mixture behaves as a diatomic ideal gas, (a) by what factor does the volume change and (b) by what factor does the temperature change? Assuming the compression starts with 0.0160 mol of gas at 27.0°C, find the values of (c) Q, (d) ΔE_{int} and (e) W that characterise the process.
12. A 2.00 mol sample of a diatomic ideal gas expands slowly and adiabatically from a pressure of 5.00 atm and a volume of 12.0 L to a final volume of 30.0 L. Draw a PV diagram for this process. (a) What is the final pressure of the gas? (b) What are the initial and final temperatures? Find (c) Q, (d) ΔE_{int} and (e) W for the gas during this process.
13. Air in a thundercloud expands as it rises. If its initial temperature is 300 K and no energy is lost by thermal conduction on expansion, what is its temperature when the initial volume has doubled?
14. How much work is required to compress 5.00 mol of air at 20.0°C and 1.00 atm to one-tenth of the original volume (a) by an isothermal process? (b) **What If?** How much work is required to produce the same compression in an adiabatic process? (c) What is the final pressure in part (a)? (d) What is the final pressure in part (b)?
15. An ideal gas with specific heat ratio γ confined to a cylinder is put through a closed cycle. Initially, the gas is at P_i, V_i and T_i. First, its pressure is tripled under constant volume. It then expands adiabatically to its original pressure and finally is compressed isobarically to its original volume. (a) Draw a PV diagram of this cycle. (b) Determine the volume at the end of the adiabatic expansion. Find (c) the temperature of the gas at the start of the adiabatic expansion and (d) the temperature at the end of the cycle. (e) What was the net work done on the gas for this cycle?

16. Air (a diatomic ideal gas) at 27.0°C and atmospheric pressure is drawn into a bicycle pump that has a cylinder with an inner diameter of 2.50 cm and length 50.0 cm. The down-stroke adiabatically compresses the air, which reaches a gauge pressure of 8.00×10^5 Pa before entering the tyre. We wish to investigate the temperature increase of the pump. (a) What is the initial volume of the air in the pump? (b) What is the number of moles of air in the pump? What is the absolute pressure of the compressed air? What is the volume of the compressed air? (e) What is the temperature of the compressed air? (f) What is the increase in internal energy of the gas during the compression? (g) **What If?** The pump is made of steel that is 2.00 mm thick. Assume 4.00 cm of the cylinder's length is allowed to come to thermal equilibrium with the air and the pump is compressed once. After the adiabatic expansion, conduction results in the energy increase in part (f) being shared between the gas and the 4.00 cm length of steel. What will be the increase in temperature of the steel after one compression?

Section 21.4 The equipartition of energy

17. A certain molecule has f degrees of freedom. Show that an ideal gas consisting of such molecules has the following properties: (a) its total internal energy is $fnRT/2$, (b) its molar specific heat at constant volume is $fR/2$, (c) its molar specific heat at constant pressure is $(f + 2)R/2$, and (d) its specific heat ratio is $\gamma = C_p/C_V = (f + 2)/f$.

18. In a crude model of a rotating diatomic chlorine molecule (Cl_2), the two Cl atoms are 2.00×10^{-10} m apart and rotate about their centre of mass with angular speed $\omega = 2.00 \times 10^{12}$ rad/s. What is the rotational kinetic energy of one molecule of Cl_2, which has a molar mass of 70.0 g/mol?

19. Consider a sample containing 2.00 mol of an ideal diatomic gas. Assuming the molecules rotate but do not vibrate, find (a) the total heat capacity of the sample at constant volume and (b) the total heat capacity at constant pressure. (c) **What If?** Repeat parts (a) and (b), assuming the molecules both rotate and vibrate.

Section 21.5 Distribution of molecular speeds

20. Fifteen identical particles have various speeds: one has a speed of 2.00 m/s, two have speeds of 3.00 m/s, three have speeds of 5.00 m/s, four have speeds of 7.00 m/s, three have speeds of 9.00 m/s and two have speeds of 12.0 m/s. Find (a) the average speed, (b) the rms speed and (c) the most probable speed of these particles.

21. One cubic metre of atomic hydrogen at 0°C at atmospheric pressure contains approximately 2.70×10^{25} atoms. The first excited state of the hydrogen atom has an energy of 10.2 eV above that of the lowest state, called the ground state. Use the Boltzmann factor to find the number of atoms in the first excited state (a) at 0°C and (b) at (1.00×10^4)°C.

22. At what temperature would the average speed of helium atoms equal (a) the escape speed from the Earth, 1.12×10^4 m/s, and (b) the escape speed from the Moon, 2.37×10^3 m/s? Note: The mass of a helium atom is 6.64×10^{-27} kg.

23. Consider a container of nitrogen gas molecules at (900 ± 10) K. ±? Calculate (a) the most probable speed, (b) the average speed and (c) the rms speed for the molecules. (d) State how your results compare with the values displayed in **Active Figure 21.12**.

24. Assume the Earth's atmosphere has a uniform temperature of 20.0°C and uniform composition, with an effective molar mass of 28.9 g/mol. (a) Show that the number density of molecules depends on height y above sea level according to

$$n_V(y) = n_0 e^{-m_0 g y / k_B T}$$

where n_0 is the number density at sea level (where $y = 0$). This result is called the *law of atmospheres.*
(b) Commercial jetliners typically cruise at an altitude of 11.0 km. Find the ratio of the atmospheric density at this level to the density at sea level.

25. From the Maxwell–Boltzmann speed distribution, show that the most probable speed of a gas molecule is given by **Equation 21.27**.

26. The law of atmospheres states that the number density of molecules in the atmosphere depends on height y above sea level according to

$$n_V(y) = n_0 e^{-m_0 g y / k_B T}$$

where n_0 is the number density at sea level (where $y = 0$). The average height of a molecule in the Earth's atmosphere is given by

$$y_{avg} = \frac{\int_0^\infty y n_V(y)\,dy}{\int_0^\infty n_V(y)\,dy} = \frac{\int_0^\infty y e^{-m_0 g y / k_B T}\,dy}{\int_0^\infty e^{-m_0 g y / k_B T}\,dy}$$

(a) Prove that this average height is equal to $k_B T / m_0 g$.
(b) Evaluate the average height, assuming the temperature is 10.0°C and the molecular mass is 28.9 u, and both are uniform throughout the atmosphere.

Additional problems

27. A small oxygen tank at a gauge pressure of 125 atm has a ±? volume of 6.88 L at 21.0°C. (a) If an athlete breathes oxygen from this tank at the rate of (8.5 ± 0.1) L/min when measured at atmospheric pressure and the temperature remains at 21.0°C, how long will the tank last before it is empty? (b) At a particular moment during this process, what is the ratio of the rms speed of the molecules remaining in the tank to the rms speed of those being released at atmospheric pressure?

28. A 1.00-L insulated bottle is full of tea at 90.0°C. You pour out one cup of tea and immediately screw the stopper back on the bottle. Make an order-of-magnitude estimate of the change in temperature of the tea remaining in the bottle that results from the admission of air at room temperature. State the quantities you take as data and the values you measure or estimate for them.

29. The *mean free path* ℓ of a molecule is the average distance that a molecule travels before colliding with another molecule. It is given by

$$\ell = \frac{1}{\sqrt{2}\pi d^2 N_V}$$

where d is the diameter of the molecule and N_V *is* the number of molecules per unit volume. The number of collisions that a molecule makes with other molecules per unit time, or *collision frequency f*, is given by

$$f = \frac{v_{avg}}{\ell}$$

(a) If the diameter of an oxygen molecule is 2.00×10^{-10} m, find the mean free path of the molecules in a scuba tank that has a volume of 12.0 L and is filled with oxygen at a gauge pressure of 100 atm at a temperature of 25.0°C. (b) What is the average time interval between molecular collisions for a molecule of this gas?

30. In a sample of a solid metal, each atom is free to vibrate about some equilibrium position. The atom's energy consists of kinetic energy for motion in the x, y and z directions plus elastic potential energy associated with the Hooke's law forces exerted by neighbouring atoms on it in the x, y and z directions. According to the theorem of equipartition of energy, assume the average energy of each atom is $\frac{1}{2}k_BT$ for each degree of freedom. (a) Prove that the molar specific heat of the solid is $3R$. The *Dulong–Petit law* states that this result generally describes pure solids at sufficiently high temperatures. (You may ignore the difference between the specific heat at constant pressure and the specific heat at constant volume.) (b) Evaluate the specific heat c of iron. Compare this to the accepted value at 25°C and atmospheric pressure, 448 J/kg.°C. (c) Evaluate the specific heat capacity of gold. Compare this to the accepted value at 25°C and atmospheric pressure, 129 J/kg.°C, and the value you obtained for iron.

31. In a cylinder, n moles of an ideal gas undergo an adiabatic process. (a) Starting with the expression $W = -\int P\,dV$ and using the condition $PV^\gamma =$ constant, show that the work done on the gas is

$$W = \left(\frac{1}{\gamma - 1}\right)(P_fV_f - P_iV_i)$$

(b) Starting with the first law of thermodynamics, show that the work done on the gas is equal to $nC_V(T_f - T_i)$. (c) Are these two results consistent with each other? Explain your answer.

32. The compressibility κ of a substance is defined as the fractional change in volume of that substance for a given change in pressure:

$$\kappa = -\frac{1}{V}\frac{dV}{dP}$$

(a) Explain why the negative sign in this expression ensures κ is always positive. (b) Show that if an ideal gas is compressed isothermally, its compressibility is given by $\kappa_1 = 1/P$. (c) **What If?** Show that if an ideal gas is compressed adiabatically, its compressibility is given by $\kappa_2 = 1/(\gamma P)$. Determine values for (d) κ_1 and (e) κ_2 for a monatomic ideal gas at a pressure of 2.00 atm.

33. Model air as a diatomic ideal gas with $M = 28.9$ g/mol. A cylinder with a piston contains 1.20 kg of air at 25.0°C and 2.00×10^5 Pa. Energy is transferred by heat into the system as it is permitted to expand, with the pressure rising to 4.00×10^5 Pa. Throughout the expansion, the relationship between pressure and volume is given by

$$P = CV^{1/2}$$

where C is a constant. Find (a) the initial volume, (b) the final volume, (c) the final temperature, (d) the work done on the air and (e) the energy transferred by heat.

34. As a sound wave passes through a gas, the compressions are either so rapid or so far apart that thermal conduction is prevented by a negligible time interval or by effective thickness of insulation. The compressions and rarefactions are adiabatic. (a) Show that the speed of sound in an ideal gas is

$$v = \sqrt{\frac{RT}{M}}$$

where M is the molar mass. The speed of sound in a gas is given by **Equation 17.32**; use that equation and the definition of the bulk modulus from Section 15.4. (b) Calculate the theoretical speed of sound in air at 20.0°C and state how it compares with the value in **Table 17.1**. Take $M = 28.9$ g/mol. (c) Show that the speed of sound in an ideal gas is

$$v = \sqrt{\frac{k_BT}{m_0}}$$

where m_0 is the mass of one molecule. (d) State how the result in part (c) compares with the most probable, average and rms molecular speeds.

35. The latent heat of vaporisation for water at room temperature is 2430 J/g. Consider one particular molecule at the surface of a glass of liquid water, moving upwards with sufficiently high speed that it will be the next molecule to join the vapour. (a) Find its translational kinetic energy. (b) Find its speed. Now consider a thin gas made only of molecules like that one. (c) What is its temperature? (d) Why are you not burned by water evaporating from a vessel at room temperature?

36. For a Maxwellian gas, use a computer or programmable calculator to find the numerical value of the ratio $N_v(v)/N_v(v_{mp})$ for the following values of v: (a) $v = (v_{mp}/50.0)$, (b) $(v_{mp}/10.0)$, (c) $(v_{mp}/2.00)$, (d) v_{mp}, (e) $2.00v_{mp}$, (f) $10.0v_{mp}$ and (g) $50.0v_{mp}$. Give your results to three significant figures. Now plot a graph showing how the ratio varies with v.

37. A triatomic molecule can have a linear configuration, as does CO_2 (**Figure P21.37a**) or it can be non-linear, like H_2O (**Figure P21.37b**). Suppose the temperature of a gas of triatomic molecules is sufficiently low that vibrational motion is negligible. What is the molar specific heat at constant volume, expressed as a multiple of the universal gas constant, (a) if the molecules are linear and (b) if the molecules are non-linear? At high temperatures, a triatomic molecule has two modes of vibration, and each contributes $\frac{1}{2}R$ to the molar specific heat for its kinetic energy and another $\frac{1}{2}R$ for its potential energy. Identify the high-temperature molar specific heat at constant volume for a triatomic ideal gas of (c) linear molecules and (d) non-linear molecules. (e) Explain how specific heat data can be used to determine whether a triatomic molecule is linear or non-linear. Are the data in **Table 21.1** sufficient to make this determination?

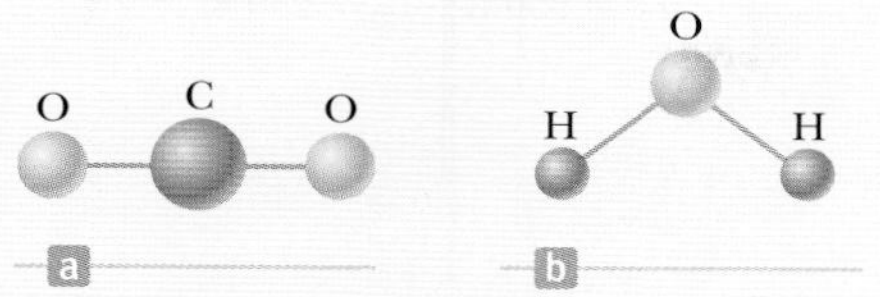

Figure P21.37

38. When a small particle is suspended in a fluid, bombardment by molecules makes the particle jitter about at random. Robert Brown discovered this motion in 1827 while studying plant fertilisation and the motion has become known as *Brownian motion*. The particle's average kinetic energy can be taken as $\frac{3}{2}k_BT$, the same as that of a molecule in an ideal gas. Consider a spherical particle of density 1.00×10^3 kg/m³ in water at (20.0 ± 0.5)°C. (a) For a particle of diameter d, evaluate the rms speed. (b) The particle's actual motion is a random walk, but imagine that it moves with constant velocity equal in magnitude to its rms speed. In what time interval would it move by a distance equal to its own diameter? (c) Evaluate the rms speed and the time interval for a particle of diameter 3.00 μm. (d) Evaluate the rms speed and the time interval for a sphere of mass 70.0 kg, modelling your own body.

39. A sample of a monatomic ideal gas occupies 5.00 L at atmospheric pressure and 300 K (point A in **Figure P21.39**). It is warmed at constant volume to 3.00 atm (point B). Then it is allowed to expand isothermally to 1.00 atm (point C) and at last compressed isobarically to its original state. (a) Find the number of moles in the sample. Find (b) the temperature at point B, (c) the temperature at point C and (d) the volume at point C. (e) Now consider the processes $A \to B$, $B \to C$, and $C \to A$. Describe how to carry out each process experimentally. (f) Find Q, W and ΔE_{int} for each of the processes. (g) For the whole cycle $A \to B \to C \to A$, find Q, W and ΔE_{int}.

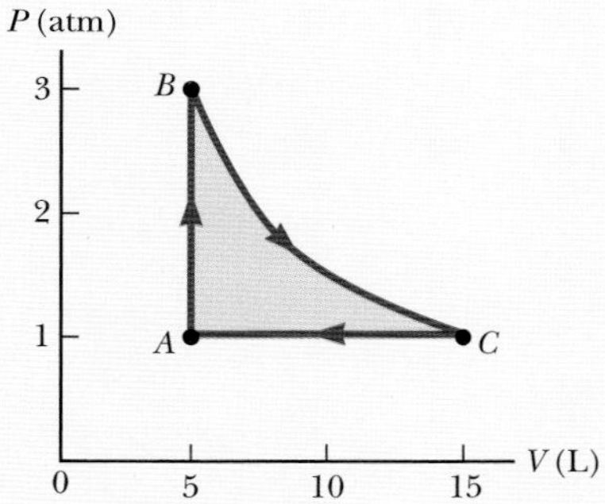

Figure P21.39

40. Consider the particles in a gas centrifuge, a device used to separate particles of different mass by whirling them in a circular path of radius r at angular speed ω. The force acting on a gas molecule towards the centre of the centrifuge is $m_0\omega^2 r$. (a) Discuss how a gas centrifuge can be used to separate particles of different mass. (b) Suppose the centrifuge contains a gas of particles of identical mass. Show that the density of the particles as a function of r is

$$n(r) = n_0 e^{m_0 r^2 \omega^2 / 2k_B T}$$

41. Using the Maxwell–Boltzmann speed distribution function, verify **Equations 21.25** and **21.26** for (a) the rms speed and (b) the average speed of the molecules of a gas at a temperature T. The average value of v^n is

$$\overline{v^n} = \frac{1}{N}\int_0^\infty v^n N_v \, dv$$

Use the table of integrals B.6 in Appendix B.

42. On the PV diagram for an ideal gas, one isothermal curve and one adiabatic curve pass through each point as shown in **Figure P21.42**. Prove that the gradient of the adiabatic curve is steeper than the gradient of the isotherm at that point by the factor γ.

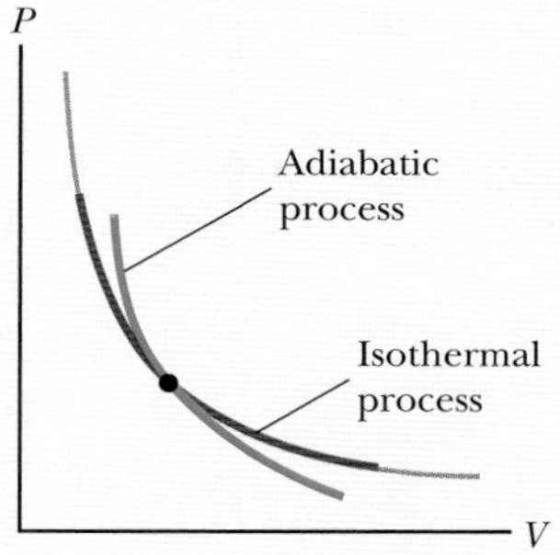

Figure P21.42

43. If it has enough kinetic energy, a molecule at the surface of the Earth can 'escape the Earth's gravitation' in the sense that it can continue to move away from the Earth forever as discussed in Section 11.4. Using the principle of conservation of energy, show that the minimum kinetic energy needed for 'escape' is $m_0 g R_E$, where m_0 is the mass of the molecule, g *is* the free-fall acceleration at the surface, and R_E is the radius of the Earth. (b) Calculate the temperature for which the minimum escape kinetic energy is 10 times the average kinetic energy of an oxygen molecule.

44. Using multiple laser beams, physicists have been able to cool and trap sodium atoms in a small region. In one experiment, the temperature of the atoms was reduced to 0.240 mK. (a) Determine the rms speed of the sodium atoms at this temperature. The atoms can be trapped for about 1.00 s. The trap has a linear dimension of roughly 1.00 cm. (b) Over what approximate time interval would an atom wander out of the trap region if there were no trapping action?

Challenge problems

45. **Equations 21.25** and **21.26** show that $v_{rms} > v_{avg}$ for a collection of gas particles, which turns out to be true whenever the particles have a distribution of speeds. Let us explore this inequality for a two-particle gas. Let the speed of one particle be $v_1 = av_{avg}$ and the other particle have speed $v_2 = (2 - a)v_{avg}$. (a) Show that the average of these two speeds is v_{avg}. (b) Show that

$$v_{rms}^2 = v_{avg}^2(2 - 2a + a^2)$$

(c) Argue that the equation in part (b) proves that, in general, $v_{rms} > v_{avg}$. (d) Under what special condition will $v_{rms} = v_{avg}$ for the two-particle gas?

46. A cylinder is closed at both ends and has insulating walls. It is divided into two compartments by an insulating piston that is perpendicular to the axis of the cylinder as shown in **Figure P21.46a**. Each compartment contains 1.00 mol of oxygen that behaves as an ideal gas with $\gamma = 1.40$. Initially, the two compartments have equal volumes and their temperatures are (550 ± 10) K and (250 ± 10) K. The piston is then allowed to move slowly parallel to the axis of the cylinder until it comes to rest at an equilibrium position (**Figure P21.46b**). Find the final temperatures in the two compartments.

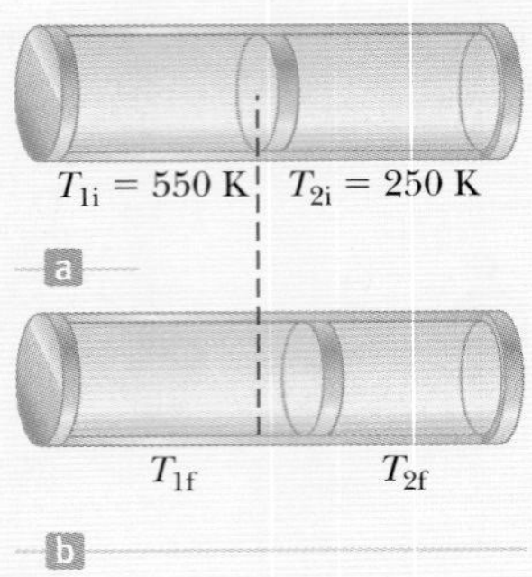

Figure P21.46

chapter 22

The second law of thermodynamics, heat engines and entropy

The device pictured uses the heat energy radiating from a cup of tea to produce mechanical energy, thereby spinning the flywheel. How does it work?

On completing this chapter, students will understand:

- that heat transfer from a colder to a hotter body cannot occur without the input of work
- how heat pumps and refrigerators work
- that real thermodynamic processes are irreversible
- why the Carnot efficiency is the theoretical limit on the efficiency of a heat engine
- how petrol and diesel engines work
- what entropy is.

Students will be able to:

- identify the reservoirs and engine in a variety of heat engines
- calculate the efficiency and coefficient of performance of a heat engine
- describe the main differences between Stirling, steam and petrol engines
- distinguish between reversible and irreversible processes
- distinguish between macro- and microstates
- calculate changes in entropy.

In this chapter, we study the second law of thermodynamics. The first law of thermodynamics is a statement of conservation of energy, stating that a change in internal energy in a system can occur as a result of energy transfer by heat, by work or by both. Although the first law of thermodynamics is very important, it makes no distinction between processes that occur spontaneously and those that do not. Only certain types of energy conversion and energy transfer processes actually take place in nature, however, and we find that many processes are only observed to proceed spontaneously in one direction – that is, they are *irreversible*. The second law of thermodynamics and the concept of entropy help us to understand this observation.

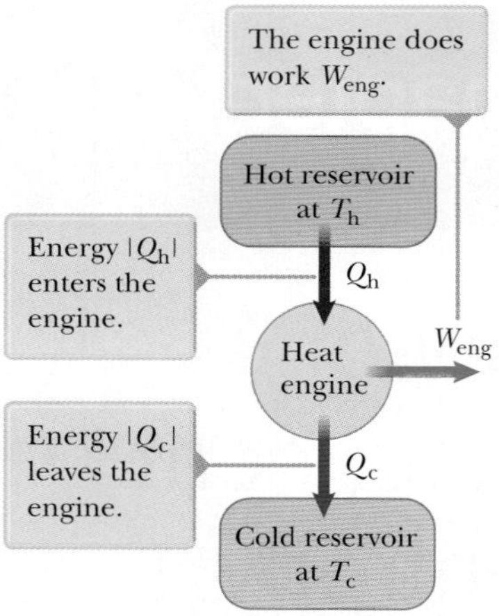

The engine does work W_{eng}.
Hot reservoir at T_h
Energy $|Q_h|$ enters the engine.
Q_h
Heat engine
W_{eng}
Energy $|Q_c|$ leaves the engine.
Q_c
Cold reservoir at T_c

Active Figure 22.1

Schematic representation of a heat engine

> **Pitfall Prevention 22.1**
> Be careful to use the total energy input to the engine Q_h when calculating an engine's efficiency, not the net energy transferred to it Q_{net}.

22.1 The second law of thermodynamics and heat engines

In the previous chapter we looked at processes that occur when gases expand, contract and change temperature due to changes in internal energy as energy is transferred by heat and work. In this chapter we will look at useful applications of these processes in engines. We begin by introducing an important new model, the heat engine. A **heat engine** is a device that takes in energy by heat and, operating in a cyclic process, expels a fraction of that energy by means of work. For instance, in a typical process by which a power plant produces electricity, a fuel such as coal is burned and the high-temperature gases produced are used to convert liquid water to steam. This steam is directed at the blades of a turbine, setting it into rotation. The mechanical energy associated with this rotation is used to drive an electric generator. Another device that can be modelled as a heat engine is the internal combustion engine in a car. This device uses energy from a burning fuel to perform work on pistons that results in the motion of the car.

A heat engine carries some working substance through a cyclic process during which (1) the working substance absorbs energy by heat from a high-temperature energy reservoir, (2) work is done by the engine and (3) energy is expelled by heat to a lower-temperature reservoir.

It is useful to represent a heat engine schematically as in **Active Figure 22.1**. The engine absorbs a quantity of energy $|Q_h|$ from the hot reservoir. For the mathematical discussion of heat engines, we use absolute values to make all energy transfers by heat positive, and the direction of transfer is indicated with an explicit positive or negative sign. The engine does work W_{eng} (so that *negative work* $W = -W_{eng}$ is done *on* the engine) and then gives up a quantity of energy $|Q_c|$ to the cold reservoir. Because the working substance goes through a cycle, its initial and final internal energies are equal: $\Delta E_{int} = 0$. Hence, from the first law of thermodynamics, $\Delta E_{int} = Q + W = Q - W_{eng} = 0$, and the net work W_{eng} done by a heat engine is equal to the net energy Q_{net} transferred to it. As you can see from **Active Figure 22.1**, $Q_{net} = |Q_h| - |Q_c|$; therefore,

$$W_{eng} = |Q_h| - |Q_c| \tag{22.1}$$

Thermal ▶ efficiency of a heat engine

The **thermal efficiency** ϵ of a heat engine is defined as the ratio of the net work done by the engine during one cycle to the energy input at the higher temperature during the cycle:

$$\epsilon \equiv \frac{W_{eng}}{|Q_h|} = \frac{|Q_h| - |Q_c|}{|Q_h|} = 1 - \frac{|Q_c|}{|Q_h|} \tag{22.2}$$

You can think of the efficiency as the ratio of what you gain (work) to what you give (energy transfer at the higher temperature). Efficiency is dimensionless. In practice, all heat engines expel only a fraction of the input energy Q_h by mechanical work; consequently, their efficiency is always less than 100%. For example, a good car engine has an efficiency of about 20% and diesel engines have efficiencies ranging from 35% to 40%.

Equation 22.2 shows that a heat engine has 100% efficiency ($\epsilon = 1$) only if $|Q_c| = 0$, that is, if no energy is expelled to the cold reservoir. In other words, a heat engine with perfect efficiency would have to expel all the input energy by work. Because efficiencies of real engines are well below 100%, the **Kelvin–Planck form of the second law of thermodynamics** states the following:

> It is impossible to construct a heat engine that, operating in a cycle, produces no effect other than the input of energy by heat from a reservoir and the performance of an equal amount of work.

This statement of the second law means that during the operation of a heat engine, W_{eng} can never be equal to $|Q_h|$ or, alternatively, that some energy $|Q_c|$ must be rejected to the environment. Every heat engine *must* have some energy exhaust.

J-L Charmet/Science Photo Library/ Photo Researchers, Inc.

Lord Kelvin
British physicist and mathematician (1824–1907)
Born William Thomson in Belfast, Kelvin was the first to propose the use of an absolute scale of temperature. Kelvin collaborated closely with Joule on the interpretation of Joule's experiments relating mechanical work to heat. His work in thermodynamics led to the idea that energy cannot pass spontaneously from a colder object to a hotter object, encapsulated in the second law of thermodynamics and the concept of entropy.

Quick Quiz 22.1

The energy input to an engine is 3.00 times greater than the work it performs. **(i)** What is its thermal efficiency? (a) 3.00 (b) 1.00 (c) 0.333 (d) impossible to determine **(ii)** What fraction of the energy input is expelled to the cold reservoir? (a) 1/3 (b) 2/3 (c) 1 (d) impossible to determine

Example 22.1

The efficiency of an engine

An engine transfers 2.00×10^3 J of energy from a hot reservoir during a cycle and transfers 1.50×10^3 J as exhaust heat to a cold reservoir.

(A) Find the efficiency of the engine.

Solution

Conceptualise Review **Active Figure 22.1**; think about energy going into the engine from the hot reservoir and part coming out by work and part by heat into the cold reservoir.

Model We model the engine as a heat engine as shown in **Active Figure 22.1**.

Analyse Find the efficiency of the engine from **Equation 22.2**:

$$\epsilon = 1 - \frac{|Q_c|}{|Q_h|} = 1 - \frac{1.50 \times 10^3 \text{ J}}{2.00 \times 10^3 \text{ J}} = 0.250, \text{ or } 25.0\%$$

(B) How much work does this engine do in one cycle?

Solution

Find the work done by the engine by taking the difference between the input and output energies:

$$W_{eng} = |Q_h| - |Q_c| = 2.00 \times 10^3 \text{ J} - 1.50 \times 10^3 \text{ J}$$
$$= 5.0 \times 10^2 \text{ J}$$

What If? Suppose you were asked for the power output of this engine. Do you have sufficient information to answer this question?

Answer No, you do not have enough information. The power of an engine is the *rate* at which work is done by the engine. You know how much work is done per cycle, but you have no information about the time interval associated with one cycle. If you were told that the engine operates at 2000 rpm (revolutions per minute), however, you could relate this rate to the period of rotation T of the mechanism of the engine. Assuming there is one thermodynamic cycle per revolution, the power is

$$P = \frac{W_{eng}}{T} = \frac{5.0 \times 10^2 \text{ J}}{\left(\frac{1}{2000} \text{ min}\right)}\left(\frac{1 \text{ min}}{60 \text{ s}}\right) = 1.7 \times 10^4 \text{ W}$$

22.2 Heat pumps and refrigerators

In a heat engine, the direction of energy transfer is from the hot reservoir to the cold reservoir, which is the natural direction. The role of the heat engine is to process the energy from the hot reservoir so as to do useful work. What if we wanted to transfer energy from the cold reservoir to the hot reservoir? Because that is not the natural direction of energy transfer, we must put some energy into a device to be successful. Devices that perform this task are called **heat pumps** and **refrigerators**. For example, homes in summer can be cooled using heat pumps called *air conditioners*. The air conditioner transfers energy from the cool room in the home to the warm air outside.

In a refrigerator or a heat pump, the engine takes in energy $|Q_c|$ from a cold reservoir and expels energy $|Q_h|$ to a hot reservoir (**Active Figure 22.2**), which can be accomplished only if work is done *on* the engine. From the first law, we know that the energy given up to the hot reservoir must equal the sum of the work done and the energy taken in from the

Pitfall Prevention 22.2

Notice the distinction between the first and second laws of thermodynamics. If a gas undergoes a *one-time isothermal process*, then $\Delta E_{int} = Q + W = 0$ and $W = -Q$. Therefore, the first law allows *all* energy input by heat to be expelled by work. In a heat engine, however, in which a substance undergoes a *cyclic* process, only a *portion* of the energy input by heat can be expelled by work according to the second law.

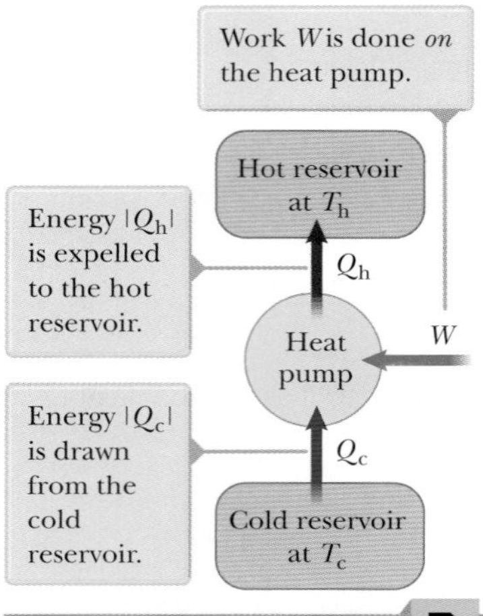

Work W is done *on* the heat pump.
Hot reservoir at T_h
Energy $|Q_h|$ is expelled to the hot reservoir.
Q_h
Heat pump
W
Energy $|Q_c|$ is drawn from the cold reservoir.
Q_c
Cold reservoir at T_c

Active Figure 22.2
Schematic representation of a heat pump

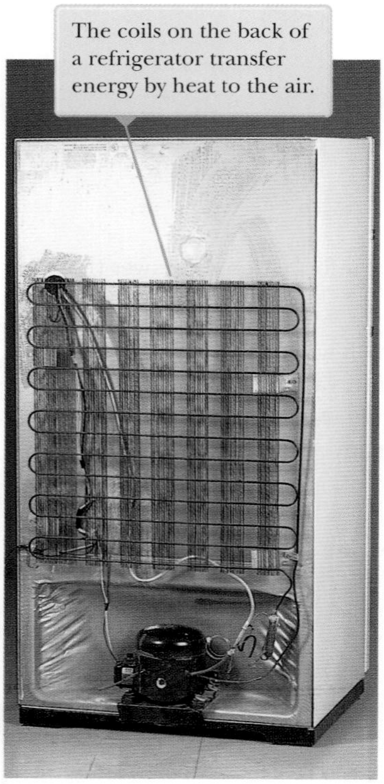

The coils on the back of a refrigerator transfer energy by heat to the air.

Figure 22.3
The back of a household refrigerator. The air surrounding the coils is the hot reservoir.

cold reservoir. Therefore, the refrigerator or heat pump transfers energy from a colder body (for example, the contents of a kitchen refrigerator or the winter air outside a building) to a hotter body (the air in the kitchen or a room in the building). In practice, it is desirable to carry out this process with a minimum of work. If the process could be accomplished without doing any work, the refrigerator or heat pump would be 'perfect'. The **Clausius statement** of the second law of thermodynamics, first formulated by Rudolf Clausius (1822–1888), is:

> It is impossible to construct a cyclical machine whose sole effect is to transfer energy continuously by heat from one object to another object at a higher temperature without the input of energy by work.

In simpler terms, energy does not transfer spontaneously by heat from a cold object to a hot object. Work input is required to run a refrigerator.

The Clausius and Kelvin–Planck statements of the second law of thermodynamics appear at first sight to be unrelated, but in fact they are equivalent in all respects. Although we do not prove so here, if either statement is false, so is the other.

In practice, a heat pump includes a circulating fluid that passes through two sets of metal coils that can exchange energy with the surroundings. The fluid is cold and at low pressure when it is in the coils located in a cool environment, where it absorbs energy by heat. The resulting warm fluid is then compressed and enters the other coils as a hot, high-pressure fluid. There it releases its stored energy to the warm surroundings. In an air conditioner, energy is absorbed into the fluid in coils located in a building's interior; after the fluid is compressed, energy leaves the fluid through coils located outdoors. In a refrigerator, the external coils are behind or underneath the unit (**Figure 22.3**). The internal coils are in the walls of the refrigerator and absorb energy from the contents of the refrigerator.

TRY THIS

Look at your own refrigerator or freezer. Can you identify all the elements of the heat pump shown in **Active Figure 22.2**?

The effectiveness of a heat pump is described in terms of the **coefficient of performance** (COP). The COP is similar to the thermal efficiency for a heat engine in that it is a ratio of what you gain (energy transferred to or from a reservoir) to what you give (work input). For a heat pump operating in the cooling mode, 'what you gain' is energy removed from the cold reservoir. The most effective refrigerator or air conditioner is one that removes the greatest amount of energy from the cold reservoir in exchange for the least amount of work. For these devices operating in the cooling mode, we define the COP in terms of $|Q_c|$:

$$\text{COP (cooling mode)} = \frac{\text{energy transferred at low temperature}}{\text{work done on heat pump}} = \frac{|Q_c|}{W} \tag{22.3}$$

Like efficiency, COP is dimensionless because it a ratio of energies. A good refrigerator should have a high COP, typically 5 or 6.

In addition to cooling applications, heat pumps are becoming increasingly popular for heating purposes. The energy-absorbing coils for a heat pump are located outside a building, in contact with the air or buried in the ground. The other set of coils are in the building's interior. The circulating fluid flowing through the coils absorbs energy from the outside and releases it to the interior of the building from the interior coils.

In the heating mode, the COP of a heat pump is defined as the ratio of the energy transferred to the hot reservoir to the work required to transfer that energy:

$$\text{COP (heating mode)} = \frac{\text{energy transferred at high temperature}}{\text{work done on heat pump}} = \frac{|Q_h|}{W} \tag{22.4}$$

If the outside temperature is −4°C or higher, a typical value of the COP for a heat pump is about 4. That is, the amount of energy transferred to the building is about four times greater than the work done by the motor in the heat pump. As the outside temperature decreases, however, it becomes more difficult for the heat pump to extract sufficient energy from the air and so the COP decreases. Therefore, the use of heat pumps that extract energy from the air, although satisfactory in moderate climates, is not appropriate in areas where winter temperatures are very low. It is possible to use heat pumps in colder areas by burying the external coils deep in the ground. In that case, the energy is extracted from the ground, which tends to be warmer than the air in the winter.

Quick Quiz 22.2

The energy entering an electric heater by electrical transmission can be converted to internal energy with an efficiency of 100%. By what factor does the cost of heating your home change when you replace your electric heating system with an electric heat pump that has a COP of 4.00? Assume the motor running the heat pump is 100% efficient. **(a)** 4.00 **(b)** 2.00 **(c)** 0.500 **(d)** 0.250

±?

Example 22.2

Freezing water

A refrigerator has a COP of 5.00. When the refrigerator is running, its power input is 500 W. A sample of water of mass 500 g and temperature (20 ± 1)°C is placed in the freezer compartment. How long does it take to freeze the water to ice at 0°C? Assume all other parts of the refrigerator stay at the same temperature and there is no leakage of energy from the exterior, so the operation of the refrigerator results only in energy being extracted from the water.

Solution

Conceptualise Energy leaves the water, reducing its temperature and then freezing it to ice. The time interval required for this entire process is related to the rate at which energy is withdrawn from the water, which, in turn, is related to the power input of the refrigerator.

Model We are told to assume that that all energy extracted from the inside of the fridge comes from the water. Therefore we model the system as a heat pump with the water as the cold reservoir. We will also need to use our understanding of temperature changes and phase changes from Chapter 19.

Analyse Use the power rating of the refrigerator to find the time interval Δt required for the freezing process to occur:

$$P = \frac{W}{\Delta t} \rightarrow \Delta t = \frac{W}{P}$$

Use **Equation 22.3** to relate the work W done on the heat pump to the energy $|Q_c|$ extracted from the water:

$$\Delta t = \frac{|Q_c|}{P(\text{COP})}$$

Use **Equations 19.11 and 19.13** to substitute the amount of energy $|Q_c|$ that must be extracted from the water of mass m:

$$\Delta t = \frac{|mc\Delta T + L_f \Delta m|}{P(\text{COP})}$$

Recognise that the amount of water that freezes is $\Delta m = -m$ because all the water freezes:

$$\Delta t = \frac{|m(c\Delta T - L_f)|}{P(\text{COP})}$$

Check dimensions:

$$[\text{T}] = \frac{[\text{M}]([\text{L}^2\text{T}^{-2}\Theta^{-1}][\Theta] - [\text{L}^2\ \text{T}^{-2}])}{[\text{ML}^2\ \text{T}^{-3}]} = \frac{[\text{M}][\text{L}^2\ \text{T}^{-2}]}{[\text{ML}^2\ \text{T}^{-3}]} = \text{T} ☺$$

Subsitute numerical values:

$$\Delta t = \frac{|(0.500\ \text{kg})[4186\ \text{J/kg.°C})(-20.0°\text{C}) - 3.33 \times 10^5\ \text{J/kg}]|}{(500\ \text{W})(5.00)}$$

$$= 83.3\ \text{s}$$

Example 22.2 cont.

To calculate the uncertainty in the time, we note that the time is directly proportional to the change in temperature, which is the only variable for which we have been given an uncertainty. The fractional uncertainty in the time will therefore be equal to that in the change in temperature (1/20), giving a final answer of

$$\Delta t = (83 \pm 4)\ \text{s}$$

Finalise In reality, the time interval for the water to freeze in a refrigerator is much longer than (83 ± 4) s, which suggests that the assumptions of our model are not valid. Only a small part of the energy extracted from the refrigerator interior in a given time interval comes from the water. Energy must also be extracted from anything else the fridge contains, including air and the container in which the water is placed, and energy that continuously enters the fridge as heat from the warmer environment must be continuously extracted.

22.3 Reversible and irreversible processes

In the next section, we shall discuss a theoretical heat engine that is the most efficient possible. To understand its nature, we must first examine the meaning of reversible and irreversible processes. In a **reversible** process, the system undergoing the process can be returned to its initial conditions along the same path on a *PV* diagram and every point along this path is an equilibrium state. A process that does not satisfy these requirements is **irreversible**. Consider dropping a ball. It will bounce a few times and then come to a stop on the ground as the mechanical energy in the system (the ball and the Earth) is dissipated due to friction and ends up as internal energy of the ball and the ground. The reverse of this process, a stationary ball on the ground 'gathering in' internal energy from the ground and spontaneously bouncing up on its own, does not violate the first law of thermodynamics as energy can be conserved in this process. However, this process does *not* occur – a dropped ball bouncing and coming to a stop is an irreversible process.

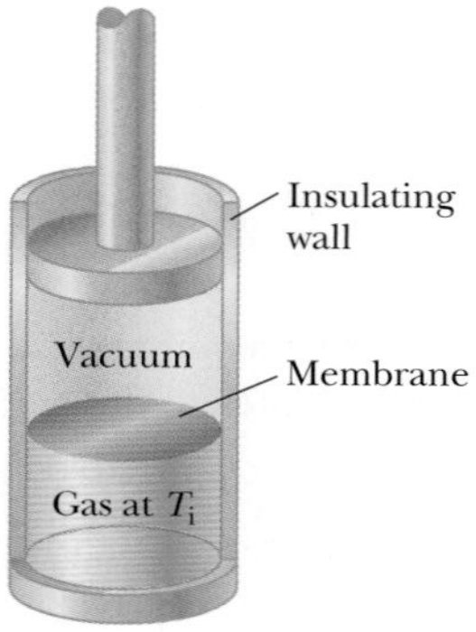

Figure 22.4
Adiabatic free expansion of a gas

All natural processes are known to be irreversible. Let's examine the adiabatic free expansion of a gas, which was discussed in Section 21.3, and show that it cannot be reversible. Consider a gas in a thermally insulated container as shown in **Figure 22.4**. A membrane separates the gas from a vacuum. When the membrane is punctured, the gas expands freely into the vacuum. As a result of the puncture, the system has changed because it occupies a greater volume after the expansion. Because the gas does not exert a force through a displacement, it does no work on the surroundings as it expands. In addition, no energy is transferred to or from the gas by heat because the container is insulated from its surroundings. Therefore, in this adiabatic process, the system has changed but the surroundings have not.

TRY THIS

Inflate a balloon and pinch it closed without tying it. Attach a second balloon to the first so that its neck is slipped over the neck of the inflated balloon and tape them firmly together so that no air can escape. Now release the first balloon's pinched neck. What happens? Could the process spontaneously reverse? Why not?

For this process to be reversible, we must return the gas to its original volume and temperature without changing the surroundings. Imagine trying to reverse the process by compressing the gas to its original volume. To do so, we fit the container with a piston and use an engine to force the piston inwards. During this process, the surroundings change because work is being done by an outside agent on the system. In addition, the system changes because the compression increases the temperature of the gas. The temperature of the gas can be lowered by allowing it to come into contact with an external energy reservoir. Although this step returns the gas to its original conditions, the surroundings are again affected because energy is being added to the surroundings from the gas. If this energy could be used to drive the engine that compressed the gas, the net energy transfer to the surroundings would be zero. In this way, the system and its surroundings could be returned to their initial conditions and we could identify the process as reversible. The Kelvin–Planck statement of the second law, however, specifies that the energy removed from the gas to return the temperature to its original value cannot be completely converted

Pitfall Prevention 22.3
All processes are irreversible. The reversible process at the macroscopic scale is an idealisation; all real processes are irreversible.

to mechanical energy in the form of the work done by the engine in compressing the gas. Therefore, we must conclude that the process is irreversible.

We could also argue that the adiabatic free expansion is irreversible by relying on the portion of the definition of a reversible process that refers to equilibrium states. For example, during the sudden expansion, significant variations in pressure occur throughout the gas. Therefore, there is no well-defined value of the pressure for the entire system at any time between the initial and final states. In fact, the process cannot even be represented as a path on a *PV* diagram. The *PV* diagram for an adiabatic free expansion would show the initial and final conditions as points, but these points would not be connected by a path. Therefore, because the intermediate conditions between the initial and final states are not equilibrium states, the process is irreversible.

Although all real processes are irreversible, some are almost reversible. If a real process occurs very slowly such that the system is always very nearly in an equilibrium state, the process can be approximated as being reversible. Suppose a gas is compressed isothermally in a piston–cylinder arrangement in which the gas is in thermal contact with an energy reservoir and we continuously transfer just enough energy from the gas to the reservoir to keep the temperature constant. For example, imagine that the gas is compressed very slowly by dropping grains of sand onto a frictionless piston as shown in Figure 22.5. As each grain lands on the piston and compresses the gas a small amount, the system deviates from an equilibrium state, but it is so close to one that it achieves a new equilibrium state in a relatively short time interval. Each grain added represents a change to a new equilibrium state, but the differences between states are so small that the entire process can be approximated as occurring through continuous equilibrium states. The process can be reversed by slowly removing grains from the piston.

The gas is compressed slowly as individual grains of sand drop onto the piston.
Energy reservoir

Figure 22.5
A method for compressing a gas in a reversible isothermal process

A general characteristic of a reversible process is that no dissipative effects (such as turbulence or friction) that convert mechanical energy to internal energy can be present. Such effects are impossible to eliminate completely. Hence, it is not surprising that real processes in nature are irreversible.

22.4 The Carnot engine

In 1824, Sadi Carnot described a theoretical engine, now called a **Carnot engine**, that is of great importance from both practical and theoretical viewpoints. He showed that a heat engine operating in an ideal, reversible cycle – called a **Carnot cycle** – between two energy reservoirs is the most efficient engine possible. Such an ideal engine establishes an upper limit on the efficiencies of all other engines. That is, the net work done by a working substance taken through the Carnot cycle is the greatest amount of work possible for a given amount of energy supplied to the substance at the higher temperature.

J-L Charmet/Science Photo Library/ Photo Researchers, Inc.

Sadi Carnot
French engineer (1796–1832)
Carnot was the first to show the quantitative relationship between work and heat. In 1824, he published his only work, *Reflections on the Motive Power of Heat*, which reviewed the industrial, political and economic importance of the steam engine. In it, he defined work as 'weight lifted through a height'.

To understand why this is the case, we examine the consequences if a more efficient heat engine were possible. Imagine two heat engines A and B operating between the *same* energy reservoirs. Engine A is a Carnot engine with efficiency ϵ_C. Engine B is a hypothetical engine with efficiency $\epsilon > \epsilon_C$. Because the cycle in the Carnot engine is reversible, engine A can operate in reverse as a refrigerator. The hypothetical engine B is used to drive the Carnot engine as a Carnot refrigerator. The output by work of engine B is matched to the input by work of the Carnot refrigerator (engine A). For the *combination* of the engine and refrigerator, no exchange by work with the surroundings occurs. Because we have assumed engine B is more efficient than the refrigerator, the net result of the combination is a transfer of energy from the cold to the hot reservoir without work being done on the combination. According to the Clausius statement of the second law, this process is impossible. Hence, the assumption that $\epsilon > \epsilon_C$ must be false.

This result is summarised in **Carnot's theorem**, which can be stated as follows:

No real heat engine operating between two energy reservoirs can be more efficient than a Carnot engine operating between the same two reservoirs.

Pitfall Prevention 22.4
The Carnot engine is a theoretical model, it is an idealisation like a massless inextensible rope, it is **not** a type of real engine. We study it because it provides a useful ideal model that highlights some of the important features of real engines.

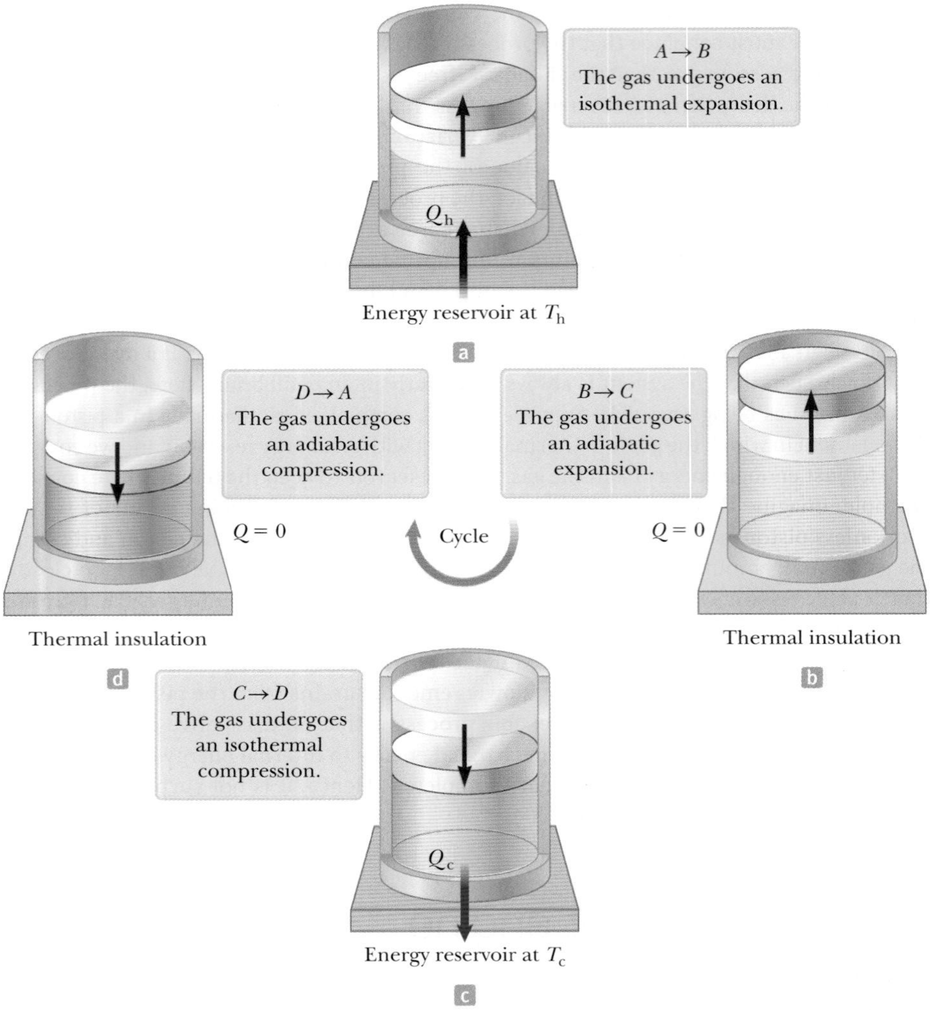

Active Figure 22.6

The Carnot cycle. The letters *A*, *B*, *C* and *D* refer to the states of the gas shown in **Active Figure 22.7**. The arrows indicate the direction of the piston during each process.

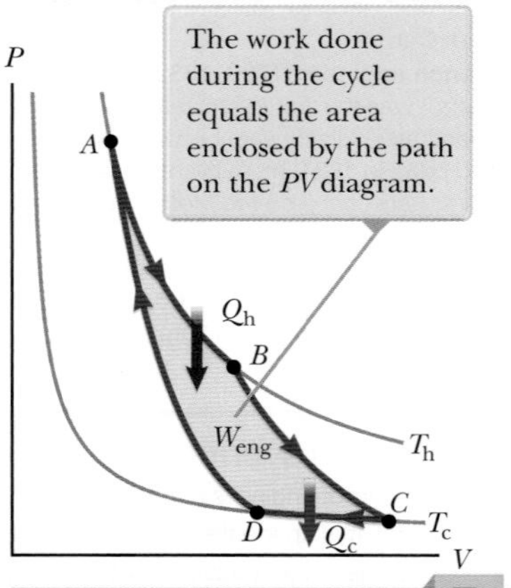

Active Figure 22.7

PV diagram for the Carnot cycle. The net work done W_{eng} equals the net energy transferred into the Carnot engine in one cycle, $|Q_h| - |Q_c|$.

All real engines are less efficient than the Carnot engine because they do not operate through a reversible cycle. The efficiency of a real engine is further reduced by such practical difficulties as friction and energy losses by conduction.

To describe the Carnot cycle taking place between temperatures T_c and T_h, let's assume the working substance is an ideal gas contained in a cylinder fitted with a movable piston at one end. The cylinder's walls and the piston are thermally non-conducting. Four stages of the Carnot cycle are shown in **Active Figure 22.6** and the *PV* diagram for the cycle is shown in **Active Figure 22.7**. The Carnot cycle consists of two adiabatic processes and two isothermal processes, all reversible:

1. Process $A \rightarrow B$ (**Active Figure 22.6a**) is an isothermal expansion at temperature T_h. The gas is placed in thermal contact with an energy reservoir at temperature T_h. During the expansion, the gas absorbs energy $|Q_h|$ from the reservoir through the base of the cylinder and does work W_{AB} in raising the piston.
2. In process $B \rightarrow C$ (**Active Figure 22.6b**), the base of the cylinder is replaced by a thermally non-conducting wall and the gas expands adiabatically; that is, no energy enters or leaves the system by heat. During the expansion, the temperature of the gas decreases from T_h to T_c and the gas does work W_{BC} in raising the piston.

3. In process $C \rightarrow D$ (**Active Figure 22.6c**), the gas is placed in thermal contact with an energy reservoir at temperature T_c and is compressed isothermally at temperature T_c. During this time, the gas expels energy $|Q_c|$ to the reservoir and the work done by the piston on the gas is W_{CD}.
4. In the final process $D \rightarrow A$ (**Active Figure 22.6d**), the base of the cylinder is replaced by a non-conducting wall and the gas is compressed adiabatically. The temperature of the gas increases to T_h and the work done by the piston on the gas is W_{DA}.

The thermal efficiency of the engine is given by **Equation 22.2**:

$$\epsilon = 1 - \frac{|Q_c|}{|Q_h|}$$

In Example 22.3, we show that for a Carnot cycle,

$$\frac{|Q_c|}{|Q_h|} = \frac{T_c}{T_h} \tag{22.5}$$

Hence, the thermal efficiency of a Carnot engine is

$$\epsilon_C = 1 - \frac{T_c}{T_h} \tag{22.6}$$

◀ Efficiency of a Carnot engine.

This result indicates that all Carnot engines operating between the same two temperatures have the same efficiency. For the processes in the Carnot cycle to be reversible, they must be carried out infinitesimally slowly. Therefore, although the Carnot engine is the most efficient engine possible, it has zero power output because it takes an infinite time interval to complete one cycle! For a real engine, the short time interval for each cycle results in the working substance reaching a high temperature lower than that of the hot reservoir and a low temperature higher than that of the cold reservoir. The Curzon–Ahlborn efficiency $\epsilon_{C\text{–}A} = 1 - (T_c/T_h)^{1/2}$ provides a closer approximation to the efficiencies of real engines than does the Carnot efficiency.

Equation 22.6 can be applied to any working substance operating in a Carnot cycle between two energy reservoirs. According to this equation, the efficiency is zero if $T_c = T_h$, as one would expect. The efficiency increases as T_c is lowered and T_h is raised. The efficiency can be unity (100%), however, only if $T_c = 0$ K. Such reservoirs are not available; therefore, the maximum efficiency is always less than 100%. In most practical cases, T_c is near room temperature, which is about 300 K. Therefore, one usually strives to increase the efficiency by raising T_h.

Theoretically, a Carnot-cycle heat engine run in reverse constitutes the most effective heat pump possible and it determines the maximum COP for a given combination of hot and cold reservoir temperatures. Using **Equations 22.1** and **22.4**, we see that the maximum COP for a heat pump in its heating mode is

$$\text{COP}_C \text{ (heating mode)} = \frac{|Q_h|}{W}$$

$$= \frac{|Q_h|}{|Q_h| - |Q_c|} = \frac{1}{1 - \frac{|Q_c|}{|Q_h|}} = \frac{1}{1 - \frac{T_c}{T_h}} = \frac{T_h}{T_h - T_c}$$

The Carnot COP for a heat pump in the cooling mode is

$$\text{COP}_C \text{ (cooling mode)} = \frac{T_c}{T_h - T_c}$$

As the difference between the temperatures of the two reservoirs approaches zero in this expression, the theoretical COP approaches infinity. In practice, the low temperature of the cooling coils and the high temperature at the compressor limit the COP to values below 10.

Quick Quiz 22.3

Three engines operate between reservoirs separated in temperature by 300 K. The reservoir temperatures are as follows: Engine A: $T_h = 1000$ K, $T_c = 700$ K; Engine B: $T_h = 800$ K, $T_c = 500$ K; Engine C: $T_h = 600$ K, $T_c = 300$ K. Rank the engines in order of theoretically possible efficiency from highest to lowest.

Example 22.3

Efficiency of the Carnot engine

Show that the ratio of energy transfers by heat in a Carnot engine is equal to the ratio of reservoir temperatures, as given by **Equation 22.5**.

Solution

Conceptualise Make use of **Active Figures 22.6 and 22.7** to help you visualise the processes in the Carnot cycle.

Model We model the processes in the cycle as isothermal and adiabatic, and the entire process as reversible.

Analyse For the isothermal expansion (process $A \rightarrow B$ in **Active Figure 22.6**), find the energy transfer by heat from the hot reservoir using **Equation 20.13** and the first law of thermodynamics:

$$|Q_h| = |\Delta E_{int} - W_{AB}| = |0 - W_{AB}| = nRT_h \ln\frac{V_B}{V_A}$$

In a similar manner, find the energy transfer to the cold reservoir during the isothermal compression $C \rightarrow D$:

$$|Q_c| = |\Delta E_{int} - W_{CD}| = |0 - W_{CD}| = nRT_c \ln\frac{V_C}{V_D}$$

Divide the second expression by the first:

$$\frac{|Q_c|}{|Q_h|} = \frac{T_c \ln(V_C/V_D)}{T_h \ln(V_B/V_A)} \quad (1)$$

Apply **Equation 21.20** to the adiabatic processes $B \rightarrow C$ and $D \rightarrow A$:

$$T_h V_B^{\gamma-1} = T_c V_C^{\gamma-1}$$

$$T_h V_A^{\gamma-1} = T_c V_D^{\gamma-1}$$

Divide the first equation by the second:

$$\left(\frac{V_B}{V_A}\right)^{\gamma-1} = \left(\frac{V_C}{V_D}\right)^{\gamma-1}$$

$$\frac{V_B}{V_A} = \frac{V_C}{V_D} \quad (2)$$

Substitute **Equation (2)** into **Equation (1)**:

$$\frac{|Q_c|}{|Q_h|} = \frac{T_c \ln(V_C/V_D)}{T_h \ln(V_B/V_A)} = \frac{T_c \ln(V_C/V_D)}{T_h \ln(V_C/V_D)} = \frac{T_c}{T_h}$$

We can see that our final expression is dimensionally correct because both sides are dimensionless ratios. ☺

Finalise This last equation is **Equation 22.5**, the one we set out to prove.

SSPL/The Image Works

Figure 22.8
A Stirling engine from the early 19th century

22.5 Real engines

The Carnot cycle describes the operation of an ideal engine. In practice, real engines deviate from the Carnot cycle in ways that make them less efficient. In the following we consider three types of real engine: Stirling engines, which operate on a closed cycle with external hot and cold reservoirs driving the expansion and contraction of a working fluid; steam engines, which are driven by external combustion; and petrol or diesel engines, which are driven by internal combustion.

Stirling engines

A Stirling engine such as that shown in **Figure 22.8** operates on a cycle that can be closely related to the ideal Carnot cycle. The working fluid (air) is sealed within the engine and undergoes successive heating–expansion and cooling–contraction cycles, driving pistons with the help of a flywheel. In most cases, the cold reservoir is simply the engine's environment

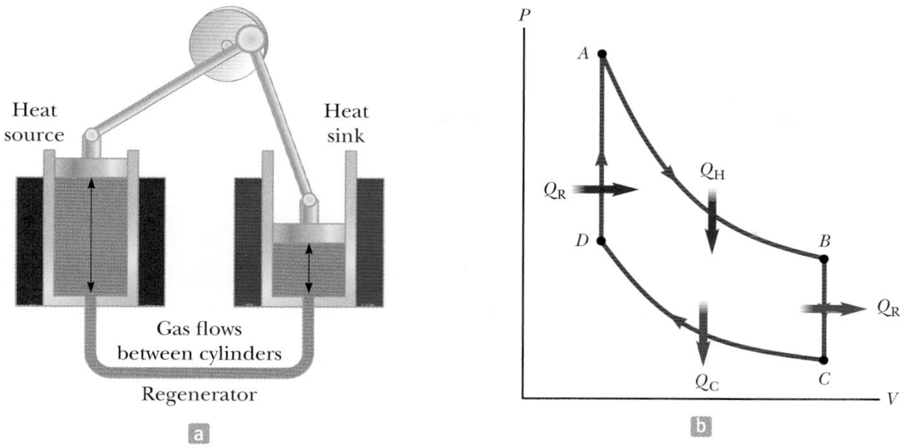

Figure 22.9
(a) The pistons in a Stirling engine
(b) PV diagram for an idealised Stirling engine. Q_R is the energy transferred by heat to and from the regenerator. A real Stirling engine has a PV curve shaped more like an ellipse, with the sharp corners rounded off.

(usually air or water). The heat source may be in the form of a flame, a resistor, which heats when current is passed through it; radiation from the Sun; or even your body heat, as very small Stirling engines have been constructed which will run from the heat from the palm of your hand alone.

The schematic illustration shown in **Figure 22.9** helps us to understand the Stirling engine's cycle. As the cylinder on the left is heated, the air inside expands and pushes against a piston, causing it to move. This is the part of the Stirling cycle that generates usable power. At this point in the cycle, most of the gas is in the hot cylinder; as its pressure increases, it flows through the *regenerator* into the cold cylinder, pushing against the piston in this cylinder. The gas is now at its maximum volume. Because the two pistons are coupled, this in turn causes the piston in the hot cylinder to push against the gas, moving most of it into the cold cylinder, where it cools and compresses. The cold piston, which is powered by the momentum of the flywheel (shown in **Figure 22.8**), further compresses the gas and pushes some back into the hot cylinder. The cycle then begins again. The idealised Stirling cycle shown in **Figure 22.9b** consists of four processes:

1. Process $A \rightarrow B$: an isothermal expansion in which energy Q_H is transferred from the hot reservoir to the gas and work is done by the gas. The work done is the area under the $A \rightarrow B$ curve.
2. Process $B \rightarrow C$: Isovolumetric (isochoric) heat removal. The gas is passed through the regenerator, cooling the gas and decreasing its pressure, and transferring energy Q_R to the regenerator for use in the next cycle. No work is done in this part of the cycle.
3. Process $C \rightarrow D$: Isothermal compression. Work is done on the gas, given by the area under the $C \rightarrow D$ curve, and energy Q_C is lost to the cold reservoir.
4. Process $D \rightarrow A$: Isovolumetric (isochoric) heat addition. The gas flows back through the regenerator where heat Q_R is transferred back in to the gas on the way back to the hot cylinder. No work is done in this part of the cycle.

If the heat transferred to and from the regenerator is the same in steps 2 and 4, then the net heat transfer in each cycle is $Q_H - Q_C$, and the total work done is the area enclosed in the PV curve. Note that the curve shown in **Figure 22.9b** is idealised. In reality, the PV diagram for a Stirling engine has a curved shape more like an ellipse, roughly like that shown in **Figure 22.9b** but with the edges rounded off.

Stirling engines are generally very inefficient and so have not played a significant role in modern industry or transport. However, there is increasing interest in the use of Stirling engines driven by direct solar heating, since in regions where solar energy is abundant they provide a low-cost alternative to fossil fuels.

Steam engines

The steam engine is a type of external combustion engine in which the movement of pistons, and hence the generation of power, is driven by the expansion and contraction of a working fluid (steam), heated by an external energy source. The operation of a steam engine (**Figure 22.10**) can also be compared to the ideal Carnot cycle. In contrast to a Stirling engine, a steam engine exploits the

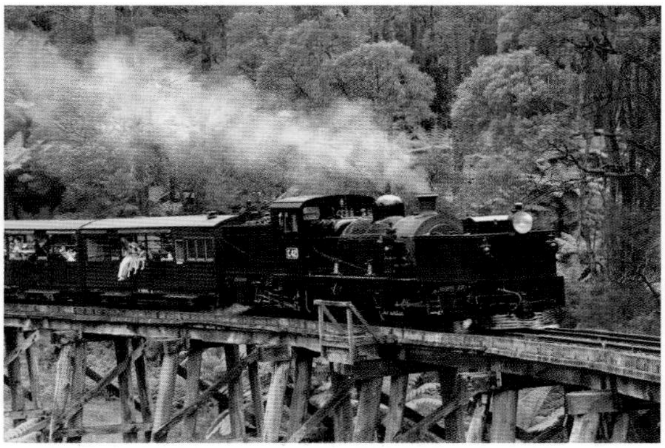

Figure 22.10
Puffing Billy in Victoria is a steam-driven locomotive. It obtains its energy by burning wood or coal. This energy vaporises water into steam, which powers the locomotive.

expansion of the working fluid as it changes phase from liquid to gas in its cycle. Heat is provided to water in a boiler from burning fuel. The water in the boiler evaporates to steam, which then does work by expanding against a piston. After the steam cools and condenses, the liquid water produced returns to the boiler and the cycle repeats.

Example 22.4

The steam engine

A steam engine has a boiler that operates at 500 K. The energy from the burning fuel changes water to steam and this steam then drives a piston. The cold reservoir's temperature is that of the outside air, approximately 300 K. What is the maximum thermal efficiency of this steam engine?

Solution

Conceptualise A steam engine does not operate with the efficiency of a Carnot engine, but we can use the Carnot cycle to find the maximum theoretical efficiency of any engine, including a steam engine, operating between two temperatures. Imagine that the gas (the working fluid) in **Figure 22.6** is steam.

Model We model the process as a Carnot cycle to find the maximum theoretical efficiency.

Analyse Substitute the reservoir temperatures into **Equation 22.6**:

$$\epsilon_C = 1 - \frac{T_c}{T_h} = 1 - \frac{300\text{ K}}{500\text{ K}} = 0.400 \text{ or } 40.0\%$$

This result is the highest *theoretical* efficiency of the engine. In practice, the efficiency is considerably lower.

What If? Suppose we wished to increase the theoretical efficiency of this engine. This increase can be achieved by raising T_h by ΔT or by decreasing T_c by the same ΔT. Which would be more effective?

Answer A given ΔT would have a larger fractional effect on a smaller temperature, so you would expect a larger change in efficiency if you alter T_c by ΔT. Let's test that numerically. Raising T_h by 50 K, corresponding to $T_h = 550$ K, would give a maximum efficiency of

$$\epsilon_C = 1 - \frac{T_c}{T_h} = 1 - \frac{300\text{K}}{550\text{K}} = 0.455$$

Decreasing T_c by 50 K, corresponding to $T_c = 250$ K, would give a maximum efficiency of

$$\epsilon_C = 1 - \frac{T_c}{T_h} = 1 - \frac{250\text{K}}{500\text{K}} = 0.500$$

Although changing T_c is *mathematically* more effective, changing T_h is generally more *practically* feasible.

Internal combustion engines

We shall now look at engines which are modified so that the fuel and the working fluid are combined. In such *internal combustion engines*, the fuel acting as the working fluid ignites, giving an even more powerful expansion stroke, but meaning that the working fluid is expended rather than regenerated, as in the closed cycle Stirling and steam engines. The petrol and diesel engines that provide power to vehicles such as cars, trucks, motorbikes and ships are examples of this type of engine.

First, we consider the processes that occur in each cycle of a petrol engine; these are shown in **Active Figure 22.11**.

We consider the interior of the cylinder above the piston to be the system that is taken through repeated cycles in the engine's operation. For a given cycle, the piston moves up and down twice, which represents a four-stroke cycle consisting of two upstrokes and two downstrokes. The processes in the cycle can be approximated by the **Otto cycle** shown in the *PV* diagram in **Active Figure 22.12**. In the following discussion, refer

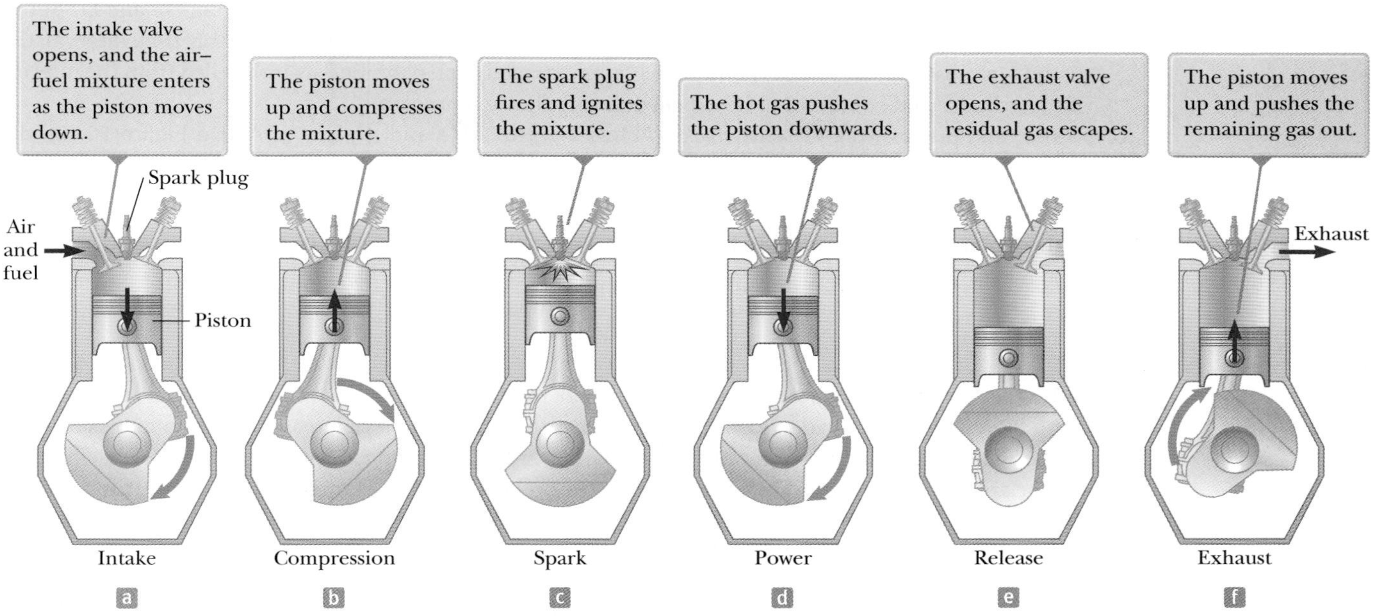
The intake valve opens, and the air–fuel mixture enters as the piston moves down.
The piston moves up and compresses the mixture.
The spark plug fires and ignites the mixture.
The hot gas pushes the piston downwards.
The exhaust valve opens, and the residual gas escapes.
The piston moves up and pushes the remaining gas out.
Spark plug
Air and fuel
Piston
Exhaust
Intake
Compression
Spark
Power
Release
Exhaust
a
b
c
d
e
f

Active Figure 22.11

The four-stroke cycle of a conventional petrol engine. The arrows on the piston indicate the direction of its motion during each process.

to **Active Figure 22.11** for the pictorial representation of the strokes and **Active Figure 22.12** for the significance on the *PV* diagram of the letter designations below:

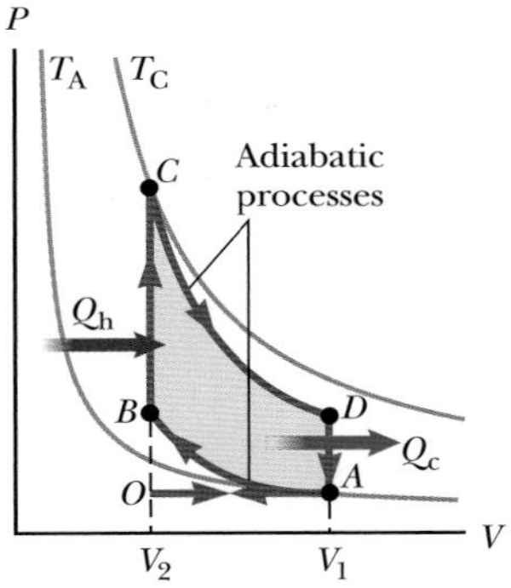
P
T_A
T_C
C
Adiabatic processes
Q_h
B
D
Q_c
O
A
V
V_2
V_1

Active Figure 22.12

PV diagram for the Otto cycle, which approximately represents the processes occurring in an internal combustion engine.

1. During the *intake stroke* (**Active Figure 22.11a** and $O \to A$ in **Active Figure 22.12**), the piston moves downwards and a mixture of air and fuel is drawn into the cylinder at approximately atmospheric pressure. That is the energy input part of the cycle: energy enters the system (the interior of the cylinder) by matter transfer as potential energy stored in the fuel. In this process, the volume increases from V_2 to V_1. This apparent backwards numbering is based on the compression stroke (process 2 below), in which the air–fuel mixture is compressed from V_1 to V_2.
2. During the *compression stroke* (**Active Figure 22.11b** and $A \to B$ in **Active Figure 22.12**), the piston moves upwards, the air–fuel mixture is compressed adiabatically from volume V_1 to volume V_2, and the temperature increases from T_A to T_B. The work done on the gas is positive and its value is equal to the negative of the area under the curve *AB* in **Active Figure 22.12**.
3. Combustion occurs when the spark plug fires (**Active Figure 22.11c** and $B \to C$ in **Active Figure 22.12**). That is not one of the strokes of the cycle because it occurs in a very short time interval while the piston is at its highest position. The combustion represents a rapid energy transformation from potential energy stored in chemical bonds in the fuel to internal energy associated with molecular motion, which is related to temperature. During this time interval, the mixture's pressure and temperature increase rapidly, with the temperature rising from T_B to T_C. The volume, however, remains approximately constant because of the short time interval. As a result, approximately no work is done on or by the gas. We can model this process in the *PV* diagram (**Active Figure 22.12**) as that process in which the energy $|Q_h|$ enters the system. (In reality, however, this process is a *conversion* of energy already in the cylinder from process $O \to A$.)
4. In the *power stroke* (**Active Figure 22.11d** and $C \to D$ in **Active Figure 22.12**), the gas expands adiabatically from V_2 to V_1. This expansion causes the temperature to drop from T_C to T_D. Work is done by the gas in pushing the piston downwards, and the value of this work is equal to the area under the curve *CD*.

5. Release of the residual gases occurs when an exhaust valve is opened (**Active Figure 22.11e** and $D \to A$ in **Active Figure 22.12**). The pressure suddenly drops for a short time interval. During this time interval, the piston is almost stationary and the volume is approximately constant. Energy is expelled from the interior of the cylinder and continues to be expelled during the next process.
6. In the final process, the *exhaust stroke* (**Active Figure 22.11e** and $A \to O$ in **Active Figure 22.12**), the piston moves upwards while the exhaust valve remains open. Residual gases are exhausted at atmospheric pressure and the volume decreases from V_1 to V_2. The cycle then repeats.

If the air–fuel mixture is assumed to be an ideal gas, the efficiency of the Otto cycle is

$$\epsilon = 1 - \frac{1}{(V_1/V_2)^{\gamma-1}} \text{ (Otto cycle)} \tag{22.7}$$

where V_1/V_2 is the **compression ratio** and γ is the ratio of the molar specific heats C_p/C_V for the air–fuel mixture. **Equation 22.7**, which is derived in Example 22.5, shows that the efficiency increases as the compression ratio increases. For a typical compression ratio of 8 and with $\gamma = 1.4$, **Equation 22.7** predicts a theoretical efficiency of 56% for an engine operating in the idealised Otto cycle. This value is much greater than that achieved in real engines (15% to 20%) because of such effects as friction, energy transfer by conduction through the cylinder walls, and incomplete combustion of the air–fuel mixture.

Diesel engines operate on a cycle similar to the Otto cycle, but they do not employ a spark plug. The compression ratio for a diesel engine is much greater than that for a petrol engine. Air in the cylinder is compressed to a very small volume and, as a consequence, the cylinder temperature at the end of the compression stroke is very high. At this point, fuel is injected into the cylinder. The temperature is high enough for the air–fuel mixture to ignite without the assistance of a spark plug. Diesel engines are more efficient than petrol engines because of their greater compression ratios and resulting higher combustion temperatures.

Example 22.5

Efficiency of the Otto cycle

Show that the thermal efficiency of an engine operating in an idealised Otto cycle (see **Active Figures 22.11** and **22.12**) is given by **Equation 22.7**. Treat the working substance as an ideal gas.

Solution

Conceptualise Study **Active Figures 22.11** and **22.12** to make sure you understand the working of the Otto cycle.

Model As seen in **Active Figure 22.12**, we model the processes in the Otto cycle as isovolumetric and adiabatic.

We model the energy input and output as occurring by heat in processes $B \to C$ and $D \to A$. (In reality, most of the energy enters and leaves by matter transfer as the air–fuel mixture enters and leaves the cylinder.)

Analyse Use **Equation 21.8** to find the energy transfers by heat for these processes, which take place at constant volume:

$$B \to C\text{: } |Q_h| = nC_V(T_C - T_B)$$

$$D \to A\text{: } |Q_c| = nC_V(T_D - T_A)$$

Substitute these expressions into **Equation 22.2**:

$$\epsilon = 1 - \frac{|Q_c|}{|Q_h|} = 1 - \frac{T_D - T_A}{T_C - T_B} \tag{1}$$

Apply **Equation 21.20** to the adiabatic processes $A \to B$ and $C \to D$:

$$A \to B\text{: } T_A V_A^{\gamma-1} = T_B V_B^{\gamma-1}$$

$$C \to D\text{: } T_C V_C^{\gamma-1} = T_D V_D^{\gamma-1}$$

Example 44.5 cont.

Solve these equations for the temperatures T_A and T_D, noting that $V_A = V_D = V_1$ and $V_B = V_C = V_2$:

$$T_A = T_B\left(\frac{V_B}{V_A}\right)^{\gamma-1} = T_B\left(\frac{V_2}{V_1}\right)^{\gamma-1} \quad (2)$$

$$T_D = T_C\left(\frac{V_C}{V_D}\right)^{\gamma-1} = T_C\left(\frac{V_2}{V_1}\right)^{\gamma-1} \quad (3)$$

Subtract Equation (2) from Equation (3) and rearrange:

$$\frac{T_D - T_A}{T_C - T_B} = \left(\frac{V_2}{V_1}\right)^{\gamma-1} \quad (4)$$

Substitute Equation (4) into Equation (1):

$$\epsilon = 1 - \frac{1}{(V_1/V_2)^{\gamma-1}}$$

Both sides are dimensionless as required ☺

Finalise This final expression is Equation 22.7.

22.6 Entropy

The zeroth law of thermodynamics involves the concept of temperature and the first law involves the concept of internal energy. Temperature and internal energy are both state variables; that is, the value of each depends only on the thermodynamic state of a system, not on the process that brought it to that state. Another state variable – this one related to the second law of thermodynamics – is *entropy S*. In this section, we define entropy on a macroscopic scale as it was first expressed by Clausius in 1865.

Entropy was originally formulated as a useful concept in thermodynamics. Its importance grew, however, as the field of statistical mechanics developed, because the analytical techniques of statistical mechanics provide an alternative means of interpreting entropy and give a more global significance to the concept. In statistical mechanics, the behaviour of a substance is described in terms of the statistical behaviour of its atoms and molecules.

To understand entropy we first need to distinguish between *microstates* and *macrostates* of a system. A **microstate** is a particular configuration of the individual constituents of the system. A **macrostate** is a description of the system's conditions from a macroscopic point of view. For a thermodynamic system, macrostates are described by macroscopic variables such as pressure, density and temperature. It is often the case that there are many microstates that correspond to the same macrostate. For example, the molecules in the gas in the container may have many different velocity configurations for the same pressure and temperature; the precise configuration of velocities specifices the microstate, while the pressure and temperature alone specify the macrostate.

For any given macrostate of the system, a number of microstates are possible. For example, the macrostate of a 4 on a pair of dice can be formed from the possible microstates 1–3, 2–2, and 3–1. The macrostate of 2 has only one microstate, 1–1. We assume that all microstates are equally probable.

We can compare these two macrostates in three ways:

(1) *Uncertainty*: If we know that a macrostate of 4 exists, there is some uncertainty as to the microstate that exists, because there are multiple microstates that will result in a 4. In comparison, there is lower uncertainty (in fact, *zero* uncertainty) for a macrostate of 2 because there is only one microstate.

(2) *Choice*: There are more choices of microstates for a 4 than for a 2.

(3) *Probability*: The macrostate of 4 has a higher probability than a macrostate of 2 because there are more ways (microstates) of achieving a 4.

The notions of uncertainty, choice, and probability are central to the concept of entropy, as we discuss below.

Pitfall Prevention 22.5

Although we begin our discussion of entropy with macroscopic non-thermodynamic systems such as cards and dice, entropy is a thermodynamic variable and should only be applied to thermodynamic systems.

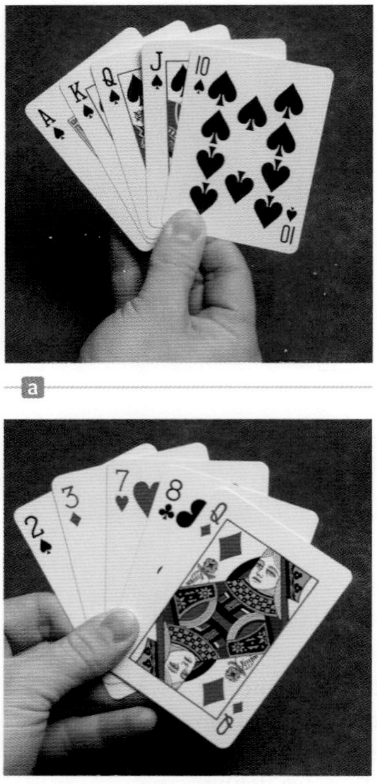

Figure 22.13
(a) A Royal Flush has a low probability of occurring. (b) A worthless poker hand, one of many

Let's look at another example related to a poker hand. There is only one microstate associated with the macrostate of a royal flush of five spades, laid out in order from ten to ace (**Figure 22.13a**). **Figure 22.13b** shows another poker hand. The macrostate here is 'worthless hand'. The *particular* hand (the microstate) in **Figure 22.13b** and the hand in **Figure 22.13a** are equally probable. There are, however, *many* other hands similar to that in **Figure 22.13b**; that is, there are many microstates that also qualify as worthless hands. If you, as a poker player, are told your opponent holds a macrostate of a royal flush in spades, there is *zero uncertainty* as to what five cards are in the hand, only *one choice* of what those cards are, and a *low probability* that the hand actually occurred. In contrast, if you are told that your opponent has the macrostate of 'worthless hand', there is *high uncertainty* as to what the five cards are, *many choices* of what they could be, and a *high probability* that a worthless hand occurred. Another variable in poker, of course, is the value of the hand, related to the probability: the higher the probability, the lower the value. The important point to take away from this discussion is that uncertainty, choice, and probability are related in these situations: if one is high, the others are high, and vice versa.

Another way of describing macrostates is by means of 'missing information.' For high-probability macrostates with many microstates, there is a large amount of missing information, meaning we have very little information about what microstate actually exists. For a macrostate of a 2 on a pair of dice, we have no missing information; we *know* the microstate is 1–1. For a macrostate of a worthless poker hand, however, we have lots of missing information, related to the large number of choices we could make as to the actual hand that is held.

For thermodynamic systems, the variable **entropy** S is used to represent the level of uncertainty, choice, probability, or missing information in the system.

Entropy is also often described in terms of order and disorder, with higher entropy being associated with lower levels of order. Consider again the example of the two poker hands. The royal flush is a very ordered hand in that it has a particular sequence of cards, while the worthless hand is a very disordered set. In general, more highly ordered hands, such as pairs and runs, are less likely to occur than less ordered hands, and hence are of lower value in poker.

Uncertainty, choice, probability and missing information are related to disorder: if one is high, the others are high, and vice versa.

Quick Quiz 22.4

(a) Suppose you select four cards at random from a standard deck of playing cards and end up with a macrostate of four twos. How many microstates are associated with this macrostate? (b) Suppose you pick up two cards and end up with a macrostate of two aces. How many microstates are associated with this macrostate?

TRY THIS

Take a coin and flip it. How many macrostates are there for this system? How many microstates? Now take two coins of the same denomination and flip them. How many macrostates are there now? How many microstates?

We can also imagine ordered macrostates and disordered macrostates in physical processes, not just in games of dice and poker. The result of a dice throw or a poker hand stays fixed once the dice are thrown or the cards are dealt. Most physical systems, however, are in a constant state of flux, changing from moment to moment from one microstate to another. Based on the relationship between the probability of a macrostate and the number of associated microstates, we therefore see that the probability of a system moving in time from an ordered macrostate to a disordered macrostate is far greater than the probability of the reverse because there are more microstates in a disordered macrostate.

Pitfall Prevention 22.6
The number of microstates for a system is always equal to or greater than the number of macrostates.

The original formulation of entropy in thermodynamics involves the transfer of energy by heat during a reversible process. Consider any infinitesimal process in which a system changes from one equilibrium state to another. If dQ_r is the amount of energy transferred by heat when the system follows a reversible path between the states, the change in entropy dS is equal to this amount of energy for the reversible process divided by the absolute temperature of the system:

$$dS = \frac{dQ_r}{T} \qquad (22.8)$$

◀ Change in entropy for an infinitesimal process

We can see from **Equation 22.8** that entropy must have units of J/K, and dimensions of $ML^2T^{-2}\Theta^{-1}$. We have assumed the temperature is constant because the process is infinitesimal. Because entropy is a state variable, the change in entropy during a process depends only on the endpoints and therefore is independent of the actual path followed. Consequently, the entropy change for an irreversible process can be determined by calculating the entropy change for a *reversible* process that connects the same initial and final states.

The subscript r on the quantity dQ_r is a reminder that the transferred energy is to be measured along a reversible path even though the system may actually have followed some irreversible path. When energy is absorbed by the system, dQ_r is positive and the entropy of the system increases. When energy is expelled by the system, dQ_r is negative and the entropy of the system decreases. Notice that **Equation 22.8** does not define entropy but rather the *change* in entropy. Hence, the meaningful quantity in describing a process is the *change* in entropy.

To calculate the change in entropy for a *finite* process, first recognise that T is generally not constant during the process. Therefore, we must integrate **Equation 22.8**:

$$\Delta S = \int_i^f dS = \int_i^f \frac{dQ_r}{T} \qquad (22.9)$$

◀ Change in entropy for a finite process

Integral calculus is summarised in Appendix B.8. Σ

As with an infinitesimal process, the change in entropy ΔS of a system going from one state to another has the same value for *all* paths connecting the two states. That is, the finite change in entropy ΔS of a system depends only on the properties of the initial and final equilibrium states. Therefore, we are free to choose any particular reversible path over which to evaluate the entropy in place of the actual path as long as the initial and final states are the same for both paths. Hence we can choose a path that makes the integral easier to calculate. This is explored further in Section 22.7.

Quick Quiz 22.5

An ideal gas is taken from an initial temperature T_i to a higher final temperature T_f along two different reversible paths. Path A is at constant pressure and path B is at constant volume. What is the relation between the entropy changes of the gas for these paths? **(a)** $\Delta S_A > \Delta S_B$ **(b)** $\Delta S_A = \Delta S_B$ **(c)** $\Delta S_A < \Delta S_B$

Example 22.6

Change in entropy: melting

A solid that has a latent heat of fusion L_f melts at a temperature T_m. Calculate the change in entropy of this substance when a mass m of the substance melts.

Solution

Conceptualise Imagine placing the substance in a warm environment so that energy enters the substance by heat. The process can be reversed by placing the substance in a cool environment so that energy leaves the substance by heat. The mass m of the substance that melts is equal to Δm, the change in mass of the higher-phase (liquid) substance.

Model Because the melting takes place at a fixed temperature, we model the process as isothermal.

Analyse Use **Equation 19.13** in **Equation 22.9**, noting that the temperature remains fixed:

$$\Delta S = \int \frac{dQ_r}{T} = \frac{1}{T_m}\int dQ_r = \frac{Q_r}{T_m} = \frac{L_f \Delta m}{T_m} = \frac{L_f m}{T_m}$$

Example 26.6 cont.

Finalise Notice that Δm is positive so that ΔS is positive, representing that energy is added to the ice cube.

What If? Suppose you did not have Equation 22.9 available to calculate an entropy change. How could you argue from the statistical description of entropy that the changes in entropy should be positive?

Answer When a solid melts, its entropy increases because the molecules are much more disordered in the liquid state than they are in the solid state. The positive value for ΔS also means that the substance in its liquid state does not spontaneously transfer energy from itself to the warm surroundings and freeze because to do so would involve a spontaneous increase in order and a decrease in entropy.

Let's consider the changes in entropy that occur in a Carnot heat engine that operates between the temperatures T_c and T_h. In one cycle, the engine takes in energy $|Q_h|$ from the hot reservoir and expels energy $|Q_c|$ to the cold reservoir. These energy transfers occur only during the isothermal portions of the Carnot cycle; therefore, the constant temperature can be brought out in front of the integral sign in Equation 22.9. The integral then simply has the value of the total amount of energy transferred by heat. Therefore, the total change in entropy for one cycle is

$$\Delta S = \frac{|Q_h|}{T_h} - \frac{|Q_c|}{T_c}$$

where the minus sign indicates that energy is leaving the engine. In Example 22.3, we showed that for a Carnot engine,

$$\frac{|Q_c|}{|Q_h|} = \frac{T_c}{T_h}$$

Using this result in the previous expression for ΔS, we find that the total change in entropy for a Carnot engine operating in a cycle is *zero*:

$$\Delta S = 0$$

Now consider a system taken through an arbitrary (non-Carnot) reversible cycle. Because entropy is a state variable – and hence depends only on the properties of a given equilibrium state – we conclude that $\Delta S = 0$ for *any* reversible cycle. In general, we can write this condition as

$$\Delta S = \oint \frac{dQ_r}{T} = 0 \quad \text{(reversible cycle)} \qquad (22.10)$$

where the symbol $\oint$ indicates that the integration is over a closed path.

For any real engine, the change in entropy will be greater than zero; the processes used to generate power (for example the combustion of fuel) are not reversible.

22.7 Entropy and the second law

A calculation of the change in entropy for a system requires information about a reversible path connecting the initial and final equilibrium states. To calculate changes in entropy for real (irreversible) processes, remember that entropy (like internal energy) depends only on the *state* of the system. That is, entropy is a state variable, and so the change in entropy depends only on the initial and final states.

You can calculate the entropy change in some irreversible process between two equilibrium states by devising a reversible process (or series of reversible processes) between the same two states and calculating $\Delta S = \int dQ_r/T$ for the reversible process. In irreversible processes, it is important to distinguish between Q, the actual energy transfer in the process, and Q_r, the energy that would have been transferred by heat along a reversible path. Only Q_r is the correct value to be used in calculating the entropy change.

If we consider a system and its surroundings to include the entire Universe, the Universe is always moving towards a higher-probability macrostate, corresponding to the continuous spreading of energy and increasing disorder. An alternative way of stating this behaviour is as follows:

The entropy of the Universe increases in all real processes.

◀ Entropy statement of the second law of thermodynamics

This statement is yet another wording of the second law of thermodynamics that can be shown to be equivalent to the Kelvin–Planck and Clausius statements.

When dealing with a system that is not isolated from its surroundings, remember that the increase in entropy described in the second law is that of the system *and* its surroundings. When a system and its surroundings interact in an irreversible process, the increase in entropy of one is greater than the decrease in entropy of the other. Hence, the change in entropy of the Universe must be greater than zero for an irreversible process and equal to zero for a reversible process. Ultimately, because real processes are irreversible, the entropy of the Universe should increase steadily and eventually reach a maximum value. At this value, the Universe will be in a state of uniform temperature and density. All physical, chemical and biological processes will have ceased at this time because a state of maximum entropy implies that no energy is available for doing work. This gloomy state of affairs is sometimes referred to as the *heat death* of the Universe.

We began our discussion of entropy by looking at probability and uncertainty, and then defined entropy changes in terms of heat. So we know that the concept of entropy is linked to that of energy, although they are not the same thing.

One way of conceptualizing a change in entropy is to relate it to *energy spreading*. A natural tendency is for energy to undergo spatial spreading in time, representing an increase in entropy. Remember our example of dropping a ball as an irreversible process. If a ball is dropped onto a floor, it bounces several times and eventually comes to rest. The initial gravitational potential energy in the ball–Earth system has been transformed to internal energy in the ball and the floor. That energy spreads outwards by heat into the air and into regions of the floor farther from the drop point. In addition, some of the energy spreads throughout the room by sound. It would be unnatural for energy in the room and floor to reverse this motion and concentrate into the stationary ball so that it spontaneously begins to bounce again. Such a process would result in a decrease in the entropy of the universe, and hence is not observed.

Quick **Quiz 22.6**

True or False: The entropy change in an adiabatic process must be zero because $Q = 0$.

Entropy change in thermal conduction

Let's now consider a system consisting of a hot reservoir and a cold reservoir that are in thermal contact with each other and isolated from the rest of the Universe. A process occurs during which energy Q is transferred from the hot reservoir at temperature T_h to the cold reservoir at temperature T_c. The process as described is irreversible (energy would not spontaneously flow from cold to hot), so we must find an equivalent reversible process to allow us to calcualte the change in entropy ΔS. Because the temperature of a reservoir does not change during the process, we can replace the real process for each reservoir with a reversible, isothermal process in which the same amount of energy is transferred. Consequently, for a reservoir, the entropy change does not depend on whether the process is reversible or irreversible.

Because the cold reservoir absorbs energy Q, its entropy increases by Q/T_c. At the same time, the hot reservoir loses energy Q, so its entropy change is $-Q/T_h$. Since $T_h > T_c$, the increase in entropy of the cold reservoir is greater than the decrease in entropy of the hot reservoir. Therefore, the change in entropy of the system (and of the Universe) is greater than zero:

$$\Delta S_U = \frac{Q}{T_c} + \frac{-Q}{T_h} > 0$$

This increase is consistent with our interpretation of entropy changes as representing the spreading of energy. In the initial configuration, the hot reservoir has excess internal energy relative to the cold reservoir. The process that occurs spreads the energy into a more equitable distribution between the two reservoirs.

Recall from Chapter 20 that the heat capacity of a diatomic or polyatomic gas increases with increasing temperature as more quantum states corresponding to different modes of vibration and rotation become available as the energy of the system increases. As entropy is related to the number of possible microstates we can conclude that the minimum possible entropy of a system corresponds to the 0K configuration. As we shall see in Chapter 40, it is not possible to actually reach 0K, although it is possible to get arbitrarily close.

Entropy change in a free expansion

Let's again consider the adiabatic free expansion of a gas occupying an initial volume V_i (**Figure 22.14**). In this situation, a membrane separating the gas from an evacuated region is broken and the gas expands to a volume V_f. This process is irreversible; the gas would not spontaneously crowd into half the volume after filling the entire volume. What are the changes in entropy of the gas and of the Universe during this process? The process is neither reversible nor quasi-static. As shown in Section 20.4, the initial and final temperatures of the gas are the same.

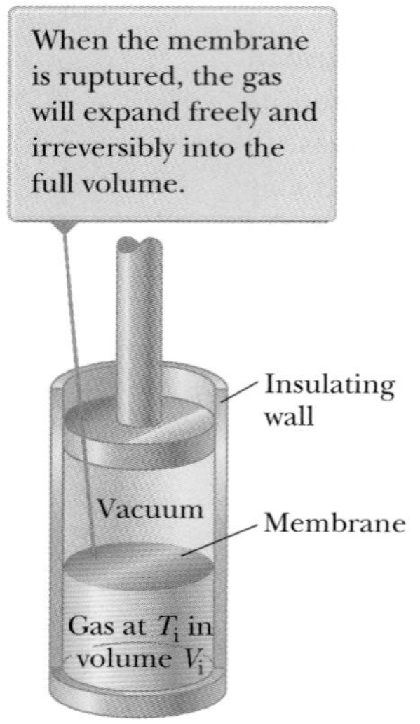
When the membrane is ruptured, the gas will expand freely and irreversibly into the full volume.
Insulating wall
Vacuum
Membrane
Gas at T_i in volume V_i

Figure 22.14
Adiabatic free expansion of a gas. The container is thermally insulated from its surroundings; therefore, $Q = 0$.

To apply **Equation 22.9**, we cannot take $Q = 0$, the value for the irreversible process, but must instead find Q_r; that is, we must find an equivalent reversible path that shares the same initial and final states. A simple choice is an isothermal, reversible expansion in which the gas pushes slowly against a piston while energy enters the gas by heat from a reservoir to hold the temperature constant. Because T is constant in this process, **Equation 22.9** gives

$$\Delta S = \int_i^f \frac{dQ_r}{T} = \frac{1}{T}\int_i^f dQ_r$$

For an isothermal process, the first law of thermodynamics specifies that $\int_i^f dQ_r$ is equal to the negative of the work done on the gas during the expansion from V_i to V_f, which is given by **Equation 20.13**. Using this result, we find that the entropy change for the gas is

$$\Delta S = nR\ln\left(\frac{V_f}{V_i}\right) \qquad (22.11)$$

Pitfall Prevention 22.6
Don't confuse the model with reality. Remember that the isothermal, reversible expansion is only a *replacement* process used to calculate the entropy change for the gas; it is not the *actual* process.

Because $V_f > V_i$, we conclude that ΔS is positive. This positive result indicates that the entropy of the gas *increases* as a result of the irreversible, adiabatic expansion.

It is easy to see that the energy has spread due to the expansion. Instead of being concentrated in a relatively small space, the molecules and the energy associated with them are scattered over a larger region. We can also see that the system is now more disordered.

Because the free expansion takes place in an insulated container, no energy is transferred by heat from the surroundings. (Remember that the isothermal, reversible expansion is only a *replacement* process used to calculate the entropy change for the gas; it is not the *actual* process.) Therefore, the free expansion has no effect on the surroundings and the entropy change of the surroundings is zero.

22.8 Entropy on a microscopic scale

As we have seen, entropy can be approached by relying on macroscopic concepts. Entropy can also be treated from a microscopic viewpoint through statistical analysis of molecular motions. Let's use a microscopic model to investigate once again the free expansion of an ideal gas, which was discussed from a macroscopic point of view in Section 22.7.

In the kinetic theory of gases, gas molecules are represented as particles moving randomly. Suppose the gas is initially confined to the volume V_i shown in **Figure 22.14**. When the membrane is removed, the molecules eventually are distributed throughout the greater volume V_f of the entire container. For a given uniform distribution of gas in the volume, there are a large number of equivalent microstates and the entropy of the gas can be related to the number of microstates corresponding to a given macrostate. This is similar to the idea of there being many different possible ways of obtaining a worthless hand in poker, compared to only a few ways of getting a valuable hand.

Let's count the number of microstates by considering the variety of molecular locations available to the molecules. Let's assume each molecule occupies some microscopic volume V_m. The total number of possible locations of a single molecule in a macroscopic initial volume V_i is the ratio $w_i = V_i/V_m$, which is a huge number. We use w_i here to represent either the number of *ways* the molecule can be placed in the initial volume or the number of microstates, which is equivalent to the number of available locations. We assume the probabilities of a molecule occupying any of these locations are equal.

As more molecules are added to the system, the number of possible ways the molecules can be positioned in the volume multiplies. For example, if you consider two molecules, for every possible placement of the first, all possible placements of the second are available. Therefore, there are w_i ways of locating the first molecule and, for each way, there are w_i ways of locating the second molecule. The total number of ways of locating the two molecules is $w_i w_i = w_i^2$.

Neglecting the very small probability of having two molecules occupying the same location, each molecule may go into any of the V_i/V_m locations, and so the number of ways of locating N molecules in the volume becomes $W_i = w_i^N = (V_i/V_m)^N$. W_i is a pure number, representing the number of microstates possible for a given macrostate, and not to be confused with work. Similarly, when the volume is increased to V_f, the number of ways of locating N molecules increases to $W_f = w_f^N = (V_f/V_m)^N$. The ratio of the number of ways of placing the molecules in the volume for the initial and final configurations is

$$\frac{W_f}{W_i} = \frac{(V_f/V_m)^N}{(V_i/V_m)^N} = \left(\frac{V_f}{V_i}\right)^N$$

Taking the natural logarithm of this equation and multiplying by Boltzmann's constant gives

$$k_B \ln\left(\frac{W_f}{W_i}\right) = k_B \ln\left(\frac{V_f}{V_i}\right)^N = nN_A k_B \ln\left(\frac{V_f}{V_i}\right)$$

where we have used the equality $N = nN_A$. We know from **Equation 19.8** that $N_A k_B$ is the universal gas constant R; therefore, we can write this equation as

$$k_B \ln W_f - k_B \ln W_i = nR\ln\left(\frac{V_f}{V_i}\right) \qquad \textbf{(22.12)}$$

From **Equation 22.11**, we know that when a gas undergoes a free expansion from V_i to V_f, the change in entropy is

$$S_f - S_i = nR\ln\left(\frac{V_f}{V_i}\right) \qquad \textbf{(22.13)}$$

Notice that the right sides of **Equations 22.12** and **22.13** are identical. Therefore, from the left sides, we make the following important connection between entropy and the number of microstates for a given macrostate:

$$S \equiv k_B \ln W \qquad \textbf{(22.14)}$$

◀ Entropy (microscopic definition)

The more microstates there are that correspond to a given macrostate, the greater the entropy of that macrostate. Although our discussion used the specific example of the free expansion of an ideal gas, a more rigorous development of the statistical interpretation of entropy would lead us to the same conclusion.

We have assumed that individual microstates are equally probable. As there are far more microstates associated with a disordered macrostate than with an ordered macrostate, a disordered macrostate is much more probable than an ordered one. Therefore **Equation 22.14** agrees with our earlier statement that entropy is a measure of the disorder in a system.

Let's explore this concept by considering 100 molecules in a container. At any given moment, the probability of one molecule being in the left part of the container shown in **Active Figure 22.15a** as a result of random motion is $\frac{1}{2}$. If there are two molecules as shown in **Active Figure 22.15b**, the probability of both being in the left part is $\left(\frac{1}{2}\right)^2$, or 1 in 4. If there are three molecules (**Active Figure 22.15c**), the probability of them all being in the left portion at the same moment is $\left(\frac{1}{2}\right)^3$, or 1 in 8. For 100 independently moving molecules, the probability that the 50 fastest ones will be found in the left part at any moment is $\left(\frac{1}{2}\right)^{50}$. Likewise, the probability that the remaining 50 slower molecules will be found in the right part at any moment is $\left(\frac{1}{2}\right)^{50}$. Therefore, the probability of finding this fast–slow separation as a result of random motion is the product $\left(\frac{1}{2}\right)^{50}\left(\frac{1}{2}\right)^{50} = \left(\frac{1}{2}\right)^{100}$, which corresponds to about 1 in 10^{30}. When this calculation is extrapolated from 100 molecules to the number in 1 mol of gas (6.02×10^{23}), the ordered arrangement is found to be *extremely* improbable.

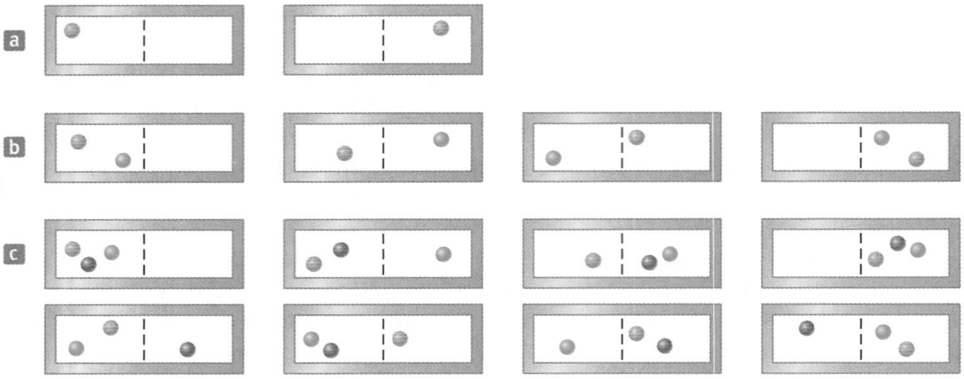

Active Figure 22.15

(a) One molecule in a container has a 1-in-2 chance of being on the left side. (b) Two molecules have a 1-in-4 chance of being on the left side at the same time. (c) Three molecules have a 1-in-8 chance of being on the left side at the same time.

Conceptual Example 22.7

Marbles

You have a bag containing an equal number of red and green marbles. You are allowed to draw four marbles from the bag according to the following rules. Draw one marble, record its color and return it to the bag. Shake the bag and then draw another marble. Continue this process until you have drawn and returned four marbles. What are the possible macrostates for this set of events? What is the most likely macrostate? What is the least likely macrostate?

Solution

Because each marble is returned to the bag before the next one is drawn and the bag is then shaken, the probability of drawing a red marble is always the same as the probability of drawing a green one. All the possible microstates and macrostates are shown in Table 22.1. As this table indicates, there is only one way to draw a macrostate of four red marbles, so there is only one microstate for that macrostate. There are, however, four possible microstates that correspond to the macrostate of one green marble and three red marbles, six microstates that correspond to two green marbles and two red marbles, four microstates that correspond to three green marbles and one red marble, and one microstate that corresponds to four green marbles. The most likely macrostate – two red marbles and two green marbles – corresponds to the largest number of microstates. The least likely macrostates – four red marbles or four green marbles – correspond to the smallest number of microstates.

Table 22.1

Possible results of drawing four marbles from a bag

Macrostate	Possible microstates	Total number of microstates
All R	RRRR	1
1G, 3R	RRRG, RRGR, RGRR, GRRR	4
2G, 2R	RRGG, RGRG, GRRG, RGGR, GRGR, GGRR	6
3G, 1R	GGGR, GGRG, GRGG, RGGG	4
All G	GGGG	1

Example 22.8

Adiabatic free expansion: one last time

Let's verify that the macroscopic and microscopic approaches to the calculation of entropy lead to the same conclusion for the adiabatic free expansion of an ideal gas. Suppose an ideal gas expands to four times its initial volume. As we have seen for this process, the initial and final temperatures are the same.

(A) Using a macroscopic approach, calculate the entropy change for the gas.

Solution

Conceptualise Look back at Figure 22.14, which is a diagram of the system before the adiabatic free expansion. Imagine breaking the membrane so that the gas moves into the evacuated area. The expansion is irreversible.

Model We can replace the irreversible process with a reversible isothermal process between the same initial and final states. This approach is macroscopic, so we use a thermodynamic variable, in particular, the volume V.

Analyse Use **Equation 22.11** to evaluate the entropy change:

$$\Delta S = nR\ln\left(\frac{V_f}{V_i}\right) = nR\ln\left(\frac{4V_i}{V_i}\right) = nR\ln 4$$

(B) Using statistical considerations, calculate the change in entropy for the gas and show that it agrees with the answer you obtained in part (A).

Solution

Model This approach is microscopic, so we use variables related to the individual molecules. We make the assumption that all microstates are equally probable.

Analyse The number of microstates available to a single molecule in the initial volume V_i is $w_i = V_i/V_m$. Use this number to find the number of available microstates for N molecules:

$$W_i = w_i^N = \left(\frac{V_i}{V_m}\right)^N$$

Find the number of available microstates for N molecules in the final volume $V_f = 4V_i$:

$$W_f = \left(\frac{V_f}{V_m}\right)^N = \left(\frac{4V_i}{V_m}\right)^N$$

Use **Equation 22.14** to find the entropy change:

$$\Delta S = k_B \ln W_f - k_B \ln W_i = k_B \ln\left(\frac{W_f}{W_i}\right)$$

$$= k_B \ln\left(\frac{4V_i}{V_i}\right)^N = k_B \ln(4^N) = Nk_B \ln 4 = nR\ln 4$$

Finalise The answer is the same as that for part (A), which dealt with macroscopic parameters.

What If? In part (A), we used **Equation 22.11**, which was based on a reversible isothermal process connecting the initial and final states. Would you arrive at the same result if you chose a different reversible process?

Answer You *must* arrive at the same result because entropy is a state variable. For example, consider the two-step process in **Figure 22.16**: a reversible adiabatic expansion from V_i to $4V_i$ ($A \rightarrow B$) during which the temperature drops from T_1 to T_2 and a reversible isovolumetric process ($B \rightarrow C$) that takes the gas back to the initial temperature T_1. During the reversible adiabatic process, $\Delta S = 0$ because $Q_r = 0$.

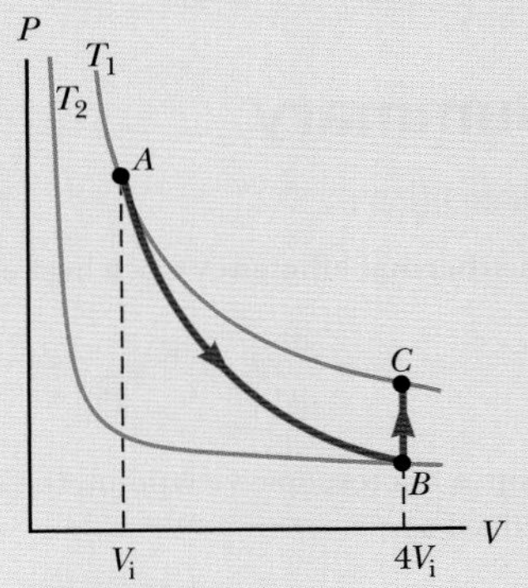

Figure 22.16 (Example 22.8) A gas expands to four times its initial volume and back to the initial temperature by means of a two-step process.

For the reversible isovolumetric process ($B \rightarrow C$), use **Equation 22.9**:

$$\Delta S = \int_i^f \frac{dQ_r}{T} = \int_{T_2}^{T_1} \frac{nC_V dT}{T} = nC_V \ln\left(\frac{T_1}{T_2}\right)$$

Find the ratio of temperature T_1 to T_2 from **Equation 21.20** for the adiabatic process:

$$\frac{T_1}{T_2} = \left(\frac{4V_i}{V_i}\right)^{\gamma-1} = (4)^{\gamma-1}$$

Substitute to find ΔS:

$$\Delta S = nC_V \ln(4)^{\gamma-1} = nC_V(\gamma - 1)\ln 4$$

$$= nC_V\left(\frac{C_P}{C_V} - 1\right)\ln 4 = n(C_P - C_V)\ln 4 = nR\ln 4$$

You do indeed obtain the exact same result for the entropy change.

End-of-chapter resources

At the start of this chapter we saw a device we now know to be a Stirling engine, which uses the heat coming from a cup of tea to drive a flywheel. In this case the heat sink for this Stirling engine is the air above the teacup.

To the right of the shaft of the flywheel we can see a grey 'power piston' connected by a rod to the flywheel. Beneath this we see a blue 'displacer piston' near to the heat source (tea) that is connected to the flywheel by the rod behind the flywheel.

When the power piston is down and the displacer piston is up, the heat from the tea increases the pressure in the air-tight vessel of the engine that acts to raise the power piston. The connection to the flywheel then lowers the displacer piston.

With the displacer piston lowered, the air inside is cut off from its heat source on the bottom plate and starts losing heat through the top plate to the air. The cooled gas is then compressed by the motion of the flywheel (which continues moving due to conservation of angular momentum) raising the displacer piston and lowering the power piston back to their initial position from which the cycle repeats.

The problems found in this chapter may be assigned online in Enhanced Web Assign.

Worked solutions to every fifth problem are available in the Student Solutions Manual. Register online at **www.cengagebrain.com** for access.

Summary

Definitions

The **thermal efficiency** ϵ of a heat engine is

$$\epsilon \equiv \frac{W_{eng}}{|Q_h|} = \frac{|Q_h| - |Q_c|}{|Q_h|} = 1 - \frac{|Q_c|}{|Q_h|} \quad (22.2)$$

From a microscopic viewpoint, the **entropy** of a given macrostate is defined as

$$S \equiv k_B \ln W \quad (22.14)$$

where k_B is Boltzmann's constant and W is the number of microstates of the system corresponding to the macrostate.

In a **reversible** process, the system can be returned to its initial conditions along the same path on a *PV* diagram, and every point along this path is an equilibrium state. A process that does not satisfy these requirements is **irreversible**.

Concepts and principles

A **heat engine** is a device that takes in energy by heat and, operating in a cyclic process, expels a fraction of that energy by means of work. The net work done by a heat engine in carrying a working substance through a cyclic process ($\Delta E_{int} = 0$) is

$$W_{eng} = |Q_h| - |Q_c| \quad (22.1)$$

where $|Q_h|$ is the energy taken in from a hot reservoir and $|Q_c|$ is the energy expelled to a cold reservoir.

Two ways the **second law of thermodynamics** can be stated are as follows:

- It is impossible to construct a heat engine that, operating in a cycle, produces no effect other than the input of energy by heat from a reservoir and the performance of an equal amount of work (the Kelvin–Planck statement).
- It is impossible to construct a cyclical machine whose sole effect is to transfer energy continuously by heat from one object to another object at a higher temperature without the input of energy by work (the Clausius statement).

Carnot's theorem states that no real heat engine operating irreversibly between the temperatures T_c and T_h can be more efficient than a hypothetical engine operating reversibly in a Carnot cycle between the same two temperatures.

The thermal efficiency of a heat engine operating in the Carnot cycle is

$$\epsilon_C = 1 - \frac{T_c}{T_h} \quad (22.6)$$

The second law of thermodynamics states that when real (irreversible) processes occur, the entropy of the system plus the surroundings increases. Therefore, yet another way the second law can be stated is as follows:

- The entropy of the Universe increases in all real processes.

Entropy is a measure of the spread of energy in a system, which is related to the number of possible microstates that can result in the observed macrostate of the system. Entropy is related to probability and uncertainty, and to the disorder of a system. The **change in entropy** dS of a system during a process between two infinitesimally separated equilibrium states is

$$dS = \frac{dQ_r}{T} \quad (22.8)$$

where dQ_r is the energy transfer by heat to the system for a reversible process that connects the initial and final states.

The change in entropy of a system during an arbitrary process between an initial state and a final state is

$$\Delta S = \int_i^f \frac{dQ_r}{T} \quad (22.9)$$

The value of ΔS for the system is the same for all paths connecting the initial and final states. The change in entropy for a system undergoing any reversible, cyclic process is zero and when such a process occurs the entropy of the Universe remains constant.

Chapter review quiz

To help you revise Chapter 22: The second law of thermodynamics, heat engines and entropy, complete the automatically graded Chapter review quiz at http://login.cengagebrain.com.

Conceptual questions

1. A steam-driven turbine is one major component of many electricity-generating plants. Why is it advantageous to have the temperature of the steam as high as possible?
2. Does the second law of thermodynamics contradict the first law? Explain your answer.
3. 'The first law of thermodynamics says you can't really win and the second law says you can't even break even.' Explain how this statement applies to a particular device or process; alternatively, argue against the statement.
4. Is it possible to construct a heat engine that creates no thermal pollution? Explain your answer.
5. (a) Give an example of an irreversible process that occurs in nature. (b) Give an example of a process in nature that is nearly reversible.
6. The device shown in **Figure CQ22.6**, called a thermoelectric converter, uses a series of semiconductor cells to transform

Photos courtesy of PASCO Scientific Company

Figure CQ22.6

internal energy to electric potential energy, which we shall study in Chapter 25. In the photograph on the left, both legs of the device are at the same temperature and no electric potential energy is produced. When one leg is at a higher temperature than the other as shown in the photograph on the right, however, electric potential energy is produced as the device extracts energy from the hot reservoir and drives a small electric motor. (a) Why is the difference in temperature necessary to produce electric potential energy in this demonstration? (b) In what sense does this intriguing experiment demonstrate the second law of thermodynamics?
7. Discuss three different common examples of natural processes that involve an increase in entropy. Be sure to account for all parts of each system under consideration.
8. Discuss the change in entropy of a gas that expands (a) at constant temperature and (b) adiabatically.
9. Suppose your housemate cleans and tidies up your messy room after a big party. Because she is creating more order, does this process represent a violation of the second law of thermodynamics?
10. If you shake a jar full of jelly beans of different sizes, the larger beans tend to appear near the top and the smaller ones tend to fall to the bottom. Why? (b) Does this process violate the second law of thermodynamics?
11. The energy exhaust from a coal-fired electricity-generating station is carried by 'cooling water' into a lake. The water is warm from the viewpoint of plants and animals in the lake. Some of them congregate around the outlet port and can impede the water flow. (a) Use the theory of heat engines to explain why this action can reduce the electric power output of the station. (b) An engineer says that the electric output is reduced because of 'higher back pressure on the turbine blades'. Comment on the accuracy of this statement.
12. A heat pump can add more energy to a house than it consumes in electrical energy. How is this possible?
13. Describe some of the factors that affect the effeciency of petrol engines.

Problems

Section 22.1 The second law of thermodynamics and heat engines

1. An engine absorbs 1.70 kJ from a hot reservoir at 277°C and expels 1.20 kJ to a cold reservoir at 27°C in each cycle. (a) What is the engine's efficiency? (b) How much work is done by the engine in each cycle? (c) What is the power output of the engine if each cycle lasts (0.30 ± 0.01) s?
2. A heat engine takes in 360 J of energy from a hot reservoir and performs 25.0 J of work in each cycle. Find (a) the efficiency of the engine and (b) the energy expelled to the cold reservoir in each cycle.
3. A particular heat engine has a mechanical power output of 5.00 kW and an efficiency of 25.0%. The engine expels 8.00×10^3 J of exhaust energy in each cycle. Find (a) the energy taken in during each cycle and (b) the time interval for each cycle.
4. A multicylinder petrol engine in an aeroplane, operating at 2.50×10^3 rev/min, takes in 7.89×10^3 J of energy and exhausts 4.58×10^3 J for each revolution of the crankshaft. (a) How many litres of fuel does it consume in 1.00 h of operation if the heat of combustion of the fuel is equal to 4.03×10^7 J/L? (b) What is the mechanical power output of the engine? Ignore friction and express the answer in horsepower. (c) What is the torque exerted by the crankshaft on the load? (d) What power must the exhaust and cooling system transfer out of the engine?

Section 22.2 Heat pumps and refrigerators

5. A refrigerator has a coefficient of performance equal to 5.00. The refrigerator takes in 120 J of energy from a cold reservoir in each cycle. Find (a) the work required in each cycle and (b) the energy expelled to the hot reservoir.
6. During each cycle, a refrigerator ejects (625 ± 5) kJ of energy to a high-temperature reservoir and takes in (550 ± 5) kJ of energy from a low-temperature reservoir. Determine (a) the work done on the refrigerant in each cycle and (b) the coefficient of performance of the refrigerator.

7. A freezer has a coefficient of performance of 6.30. It is advertised as using electricity at a rate of 457 kWh/yr. (a) On average, how much energy does it use in a single day? (b) On average, how much energy does it remove from the refrigerator in a single day? (c) What maximum mass of water at 20.0°C could the freezer freeze in a single day? Note: One kilowatt-hour (kWh) is an amount of energy equal to running a 1 kW appliance for one hour.
8. A domestic heat pump has a coefficient of performance equal to 4.20 and requires a power of 1.75 kW to operate. (a) How much energy does the heat pump add to a home in one hour? (b) If the heat pump is reversed so that it acts as an air conditioner in the summer, what would be its coefficient of performance?

Section 22.4 The Carnot engine

9. One of the most efficient heat engines ever built is a coal-fired steam turbine in the Ohio River valley, operating between 1870°C and 430°C. (a) What is its maximum theoretical efficiency? (b) The actual efficiency of the engine is 42.0%. How much mechanical power does the engine deliver if it absorbs 1.40×10^5 J of energy each second from its hot reservoir?
10. A Carnot engine has a power output of 150 kW. The engine operates between two reservoirs at 20.0°C and 500°C. (a) How much energy enters the engine by heat per hour? (b) How much energy is exhausted by heat per hour?
11. A heat engine is being designed to have a Carnot efficiency of ±? 65.0% when operating between two energy reservoirs. (a) If the temperature of the cold reservoir is (20 ± 1)°C, what must be the temperature of the hot reservoir? (b) Can the actual efficiency of the engine be equal to 65.0%? Explain.
12. An ideal refrigerator or ideal heat pump is equivalent to a Carnot engine running in reverse. That is, energy $|Q_c|$ is taken in from a cold reservoir and energy $|Q_h|$ is rejected to a hot reservoir. (a) Show that the work that must be supplied to run the refrigerator or heat pump is

$$W = \frac{T_h - T_c}{T_c}|Q_c|$$

(b) Show that the coefficient of performance (COP) of the ideal refrigerator is

$$\text{COP} = \frac{T_c}{T_h - T_c}$$

13. What is the maximum possible coefficient of performance of a heat pump that brings energy from outdoors at −3.00°C into a 22.0°C house? Note: The work done to run the heat pump is also available to warm the house.
14. How much work does an ideal Carnot refrigerator require to remove 1.00 J of energy from liquid helium at 4.00 K and expel this energy to a room-temperature (293 K) environment?
15. A heat engine operates in a Carnot cycle between 80.0°C and 350°C. It absorbs 21 000 J of energy per cycle from the hot reservoir. The duration of each cycle is 1.00 s. (a) What is the mechanical power output of this engine? (b) How much energy does it expel in each cycle by heat?
16. An ideal gas is taken through a Carnot cycle. The isothermal expansion occurs at 250°C, and the isothermal compression takes place at 50.0°C. The gas takes in 1.20×10^3 J of energy from the hot reservoir during the isothermal expansion. Find (a) the energy expelled to the cold reservoir in each cycle and (b) the net work done by the gas in each cycle.
17. You are part of a team of engineers designing an electric power plant that makes use of the temperature gradient in the ocean. The system is to operate between 20.0°C (surface-water temperature) and 5.00°C (water temperature at a depth of about 1 km). (a) What is the maximum efficiency of such a system? (b) If the electric power output of the plant is 75.0 MW, how much energy is taken in from the warm reservoir per hour? (c) In view of your answer to part (a), explain whether you think such a system is worthwhile. Note that the 'fuel' is free.
18. Suppose you build a two-engine device with the exhaust energy output from one heat engine supplying the input energy for a second heat engine. We say that the two engines are running *in series*. Let ϵ_1 and ϵ_2 represent the efficiencies of the two engines. (a) The overall efficiency of the two-engine device is defined as the total work output divided by the energy put into the first engine by heat. Show that the overall efficiency ϵ is given by

$$\epsilon = \epsilon_1 + \epsilon_2 - \epsilon_1\epsilon_2$$

What If? For parts (b) through (e) that follow, assume the two engines are Carnot engines. Engine 1 operates between temperatures T_h and T_i. The gas in engine 2 varies in temperature between T_i and T_c. In terms of the temperatures, (b) what is the efficiency of the combination engine? (c) Does an improvement in net efficiency result from the use of two engines instead of one? (d) What value of the intermediate temperature T_i results in equal work being done by each of the two engines in series? (e) What value of T_i results in each of the two engines in series having the same efficiency?
19. An electricity-generating station is designed to have an electric output power of 1.40 MW using a turbine with two-thirds the efficiency of a Carnot engine. The exhaust energy is transferred by heat into a cooling tower at 110°C. (a) Find the rate at which the station exhausts energy by heat as a function of the fuel combustion temperature T_h. (b) If the firebox is modified to run hotter by using more advanced combustion technology, how does the amount of energy exhaust change? (c) Find the exhaust power for T_h = 800°C. (d) Find the value of T_h for which the exhaust power would be only half as large as in part (c). (e) Find the value of T_h for which the exhaust power would be one-fourth as large as in part (c).
20. A heat pump used for heating, shown in **Figure P22.20**, is ±? essentially an air conditioner installed backwards. It extracts energy from colder air outside and deposits it in a warmer room. Suppose the ratio of the actual energy entering the room to the work done by the device's motor is 10.0% of the theoretical maximum ratio. Determine the energy entering the room per joule of work done by the motor given that the inside temperature is (20.0 ± 0.5)°C and the outside temperature is −(5 ± 1)°C.

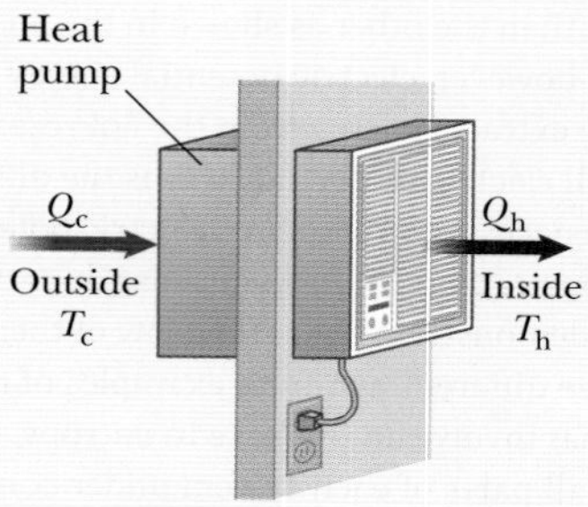

Figure P22.20

Section 22.5 Real engines

Note: For problems in this section, assume the gas in the engine is diatomic with $\gamma = 1.40$.

21. In a cylinder of a car engine, immediately after combustion the gas is confined to a volume of 50.0 cm^3 and has an initial pressure of 3.00×10^6 Pa. The piston moves outwards to a final volume of 300 cm^3 and the gas expands without energy transfer by heat. (a) What is the final pressure of the gas? (b) How much work is done by the gas in expanding?

22. A petrol engine has a compression ratio of 6.00. (a) What is the efficiency of the engine if it operates in an idealised Otto cycle? (b) **What If?** If the actual efficiency is 15.0%, what fraction of the fuel is wasted as a result of friction and energy transfers by heat that could be avoided in a reversible engine? Assume complete combustion of the air–fuel mixture.

23. An idealised diesel engine operates in a cycle known as the *air-standard diesel cycle* shown in Figure P22.23. Fuel is sprayed into the cylinder at the point of maximum compression, *B*. Combustion occurs during the expansion $B \to C$, which is modelled as an isobaric process. Show that the efficiency of an engine operating in this idealised diesel cycle is

$$\epsilon = 1 - \frac{1}{\gamma}\left(\frac{T_D - T_A}{T_C - T_B}\right)$$

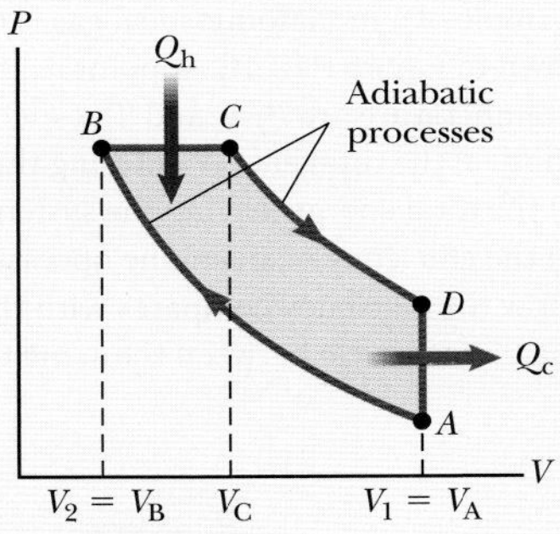

Figure P22.23

Section 22.7 Entropy and the second law

24. An ice tray contains 500 g of liquid water at 0°C. Calculate the change in entropy of the water as it freezes slowly and completely at 0°C.

25. ±? A polystyrene cup holding (125 ± 5) g of hot water at 100°C cools to room temperature, 20.0°C. What is the change in entropy of the room? Neglect the specific heat of the cup and any change in temperature of the room.

26. Two 1.00×10^3 kg cars both travelling at 20.0 m/s undergo a head-on collision and stick together. Find the change in entropy of the surrounding air resulting from the collision if the air temperature is 23.0°C. Ignore the energy carried away from the collision by sound.

27. A 1.00 mol sample of H_2 gas is contained in the left side of the container shown in Figure P22.27, which has equal volumes on the left and right. The right side is evacuated. When the valve is opened, the gas streams into the right side. (a) What is the entropy change of the gas? (b) Does the temperature of the gas change? Assume the container is so large that the hydrogen behaves as an ideal gas.

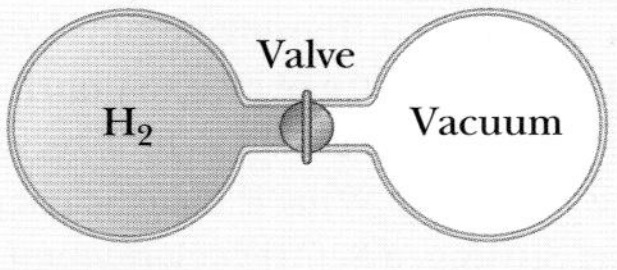

Figure P22.27

28. A 70.0-kg log falls from a height of 25.0 m into a lake. If the log, the lake, and the air are all at 300 K, find the change in entropy of the air during this process.

29. The temperature at the surface of the Sun is approximately 5800 K and the temperature at the surface of the Earth is approximately 290 K. What entropy change of the Universe occurs when 1.00×10^3 J of energy is transferred by radiation from the Sun to the Earth?

Section 22.8 Entropy on a microscopic scale

30. In a game in which you roll two dice, macrostates can be defined by the score calculated as the sum of the two sides of the dice facing upwards. (a) What are the macrostates with the smallest number of microstates? (b) How much more probable is it that you will score 7 than 12?

31. (a) Prepare a table like Table 22.1 for the following occurrence. You toss four coins into the air simultaneously and then record the results of your tosses in terms of the numbers of heads (H) and tails (T) that result. For example, HHTH and HTHH are two possible ways in which three heads and one tail can be achieved. (b) On the basis of your table, what is the most probable result recorded for a toss? In terms of entropy, (c) what is the most ordered macrostate and (d) what is the most disordered?

Additional problems

32. The energy absorbed by an engine is three times greater than the work it performs. (a) What is its thermal efficiency? (b) What fraction of the energy absorbed is expelled to the cold reservoir?

33. A steam engine is operated in a cold climate where the exhaust temperature is 0°C. (a) Calculate the theoretical maximum efficiency of the engine using an intake steam temperature of 100°C. (b) If, instead, superheated steam at 200°C is used, find the maximum possible efficiency.

34. Energy transfers by heat through the exterior walls and roof of a house at a rate of 5.00×10^3 J/s = 5.00 kW when the interior temperature is 22.0°C and the outside temperature is −5.00°C. (a) Calculate the electric power required to maintain the interior temperature at 22.0°C if the power is used in electric resistance heaters that convert all the energy transferred in by electrical transmission into internal energy. (b) **What If?** Calculate the electric power required to maintain the interior temperature at 22.0°C if the power is used to drive an electric motor that operates the compressor of a heat pump that has a coefficient of performance equal to 60.0% of the Carnot-cycle value.

35. ±? An airtight freezer holds *n* moles of air at (25.0 ± 0.5)°C and 1.00 atm. The air is then cooled to −(18.0 ± 0.5)°C. (a) What is the change in entropy of the air if the volume is held constant? (b) What would the entropy change be if the pressure were maintained at 1.00 atm during the cooling?

36. Suppose a heat engine is connected to two energy reservoirs, one a pool of molten aluminium (660°C) and the other a block of solid mercury (−38.9°C). The engine runs by freezing 1.00 g of aluminium and melting 15.0 g of mercury during each cycle. The heat of fusion of aluminium is 3.97×10^5 J/kg; the heat of fusion of mercury is 1.18×10^4 J/kg. What is the efficiency of this engine?

37. In 1816, Robert Stirling, a Scottish clergyman, patented the *Stirling engine* (see **Figure 22.8**). Fuel is burned externally to warm one of the engine's two cylinders. A fixed quantity of inert gas moves cyclically between the cylinders, expanding in the hot one and contracting in the cold one. **Figure P22.37** represents a model for its thermodynamic cycle. Consider n moles of an ideal monatomic gas being taken once through the cycle, consisting of two isothermal processes at temperatures $3T_i$ and T_i and two constant-volume processes. Let us find the efficiency of this engine. (a) Find the energy transferred by heat into the gas during the isovolumetric process AB. (b) Find the energy transferred by heat into the gas during the isothermal process BC. (c) Find the energy transferred by heat into the gas during the isovolumetric process CD. (d) Find the energy transferred by heat into the gas during the isothermal process DA. (e) Identify which of the results from parts (a) through (d) are positive and evaluate the energy input to the engine by heat. (f) From the first law of thermodynamics, find the work done by the engine. (g) From the results of parts (e) and (f), evaluate the efficiency of the engine. A Stirling engine is easier to manufacture than an internal combustion engine or a turbine. It can run on burning garbage. It can run on the energy transferred by sunlight and produce no material exhaust. Stirling engines are not currently used in cars due to long startup times and poor acceleration response.

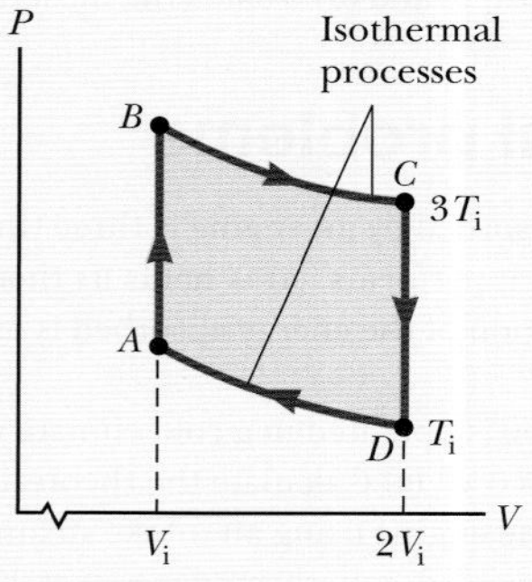

Figure P22.37

38. This problem complements problem 41 in Chapter 10. In the operation of a single-cylinder internal combustion piston engine, one charge of fuel explodes to drive the piston outwards in the *power stroke*. Part of its energy output is stored in a turning flywheel. This energy is then used to push the piston inwards to compress the next charge of fuel and air. In this compression process, assume an original volume of 0.120 L of a diatomic ideal gas at atmospheric pressure is compressed adiabatically to one-eighth of its original volume. (a) Find the work input required to compress the gas. (b) Assume the flywheel is a solid disc of mass 5.10 kg and radius 8.50 cm, turning freely without friction between the power stroke and the compression stroke. How fast must the flywheel turn immediately after the power stroke? This situation represents the minimum angular speed at which the engine can operate without stalling. (c) When the engine's operation is well above the point of stalling, assume the flywheel puts 5.00% of its maximum energy into compressing the next charge of fuel and air. Find its maximum angular speed in this case.

39. ±? A biology laboratory is maintained at a constant temperature of (7.0 ± 0.1)°C by an air conditioner, which is vented to the air outside. On a typical hot summer day, the outside temperature reaches 27°C and the air-conditioning unit emits energy to the outside at a rate of 10.0 kW. Model the unit as having a coefficient of performance (COP) equal to 40.0% of the COP of an ideal Carnot device. (a) At what rate does the air conditioner remove energy from the laboratory? (b) Calculate the power required for the work input. (c) Find the change in entropy of the Universe produced by the air conditioner in 1.00 h. (d) **What If?** The outside temperature increases to 32.0°C. Find the fractional change in the COP of the air conditioner.

40. Argon enters a turbine at a rate of 80.0 kg/min, a temperature of 800°C, and a pressure of 1.50 MPa. It expands adiabatically as it pushes on the turbine blades and exits at pressure of 300 kPa. (a) Calculate its temperature at exit. (b) Calculate the (maximum) power output of the turning turbine. (c) The turbine is one component of a model closed-cycle gas turbine engine. Calculate the maximum efficiency of the engine.

41. A 1.00 mol sample of a monatomic ideal gas is taken through the cycle shown in **Figure P22.41**. At point A, the pressure, volume and temperature are P_i, V_i and T_i respectively. In terms of R and T_i, find (a) the total energy entering the system by heat per cycle, (b) the total energy leaving the system by heat per cycle, and (c) the efficiency of an engine operating in this cycle. (d) Explain how the efficiency compares with that of an engine operating in a Carnot cycle between the same temperature extremes.

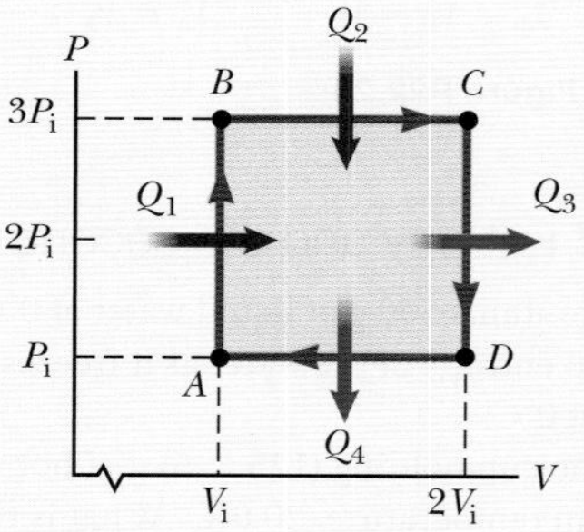

Figure P22.41

42. A sample consisting of n moles of an ideal gas undergoes a reversible isobaric expansion from volume V_i to volume $3V_i$. Find the change in entropy of the gas by calculating $\int_i^f dQ/T$, where $dQ = nC_P\,dT$.

43. An athlete whose mass is 70.0 kg drinks 0.5 L of refrigerated water at 3°C. (a) Ignoring the temperature change of the body that results from the water intake (so that the body is regarded as a reservoir always at 37°C), find the entropy increase of the entire system. (b) **What If?** Assume the entire body is cooled by the drink and the average specific heat of a person is equal to the specific heat of liquid water. Ignoring any other energy transfers by heat and any metabolic energy release, find the athlete's temperature after he drinks the cold water given an initial body temperature of 37°C. (c) Under these assumptions, what is the entropy increase of the entire system? (d) State how this result compares with the one you obtained in part (a).

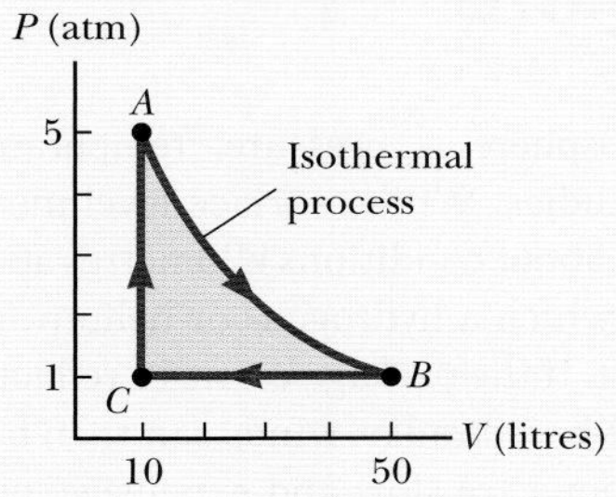

Figure P22.44

44. A 1.00 mol sample of an ideal monatomic gas is taken through the cycle shown in **Figure P22.44**. The process $A \rightarrow B$ is a reversible isothermal expansion. Calculate (a) the net work done by the gas, (b) the energy added to the gas by heat, (c) the energy exhausted from the gas by heat and (d) the efficiency of the cycle. (e) Explain how the efficiency compares with that of a Carnot engine operating between the same temperature extremes.

45. A 1500-kg car is moving at 20.0 m/s. The driver brakes to a stop. The brakes cool off to the temperature of the surrounding air, which is nearly constant at 20.0°C. What is the total entropy change?

46. A sample of an ideal gas expands isothermally, doubling in volume. (a) Show that the work done on the gas in expanding is $W = -nRT \ln 2$. (b) Because the internal energy E_{int} of an ideal gas depends solely on its temperature, the change in internal energy is zero during the expansion. It follows from the first law that the energy input to the gas by heat during the expansion is equal to the energy output by work. Does this process have 100% efficiency in converting energy input by heat into work output? (c) Does this conversion violate the second law? Explain.

Challenge problems

47. A 1.00 mol sample of an ideal gas ($\gamma = 1.40$) is carried through the Carnot cycle described in **Active Figure 22.7**. At point A, the pressure is 25.0 atm and the temperature is 600 K. At point C, the pressure is 1.00 atm and the temperature is 400 K. (a) Determine the pressures and volumes at points A, B, C and D. (b) Calculate the net work done per cycle.

48. The compression ratio of an Otto cycle as shown in **Active Figure 22.12** is $V_A/V_B = 8.00$. At the beginning A of the compression process, 500 cm^3 of gas is at 100 kPa and 20.0°C. At the beginning of the adiabatic expansion, the temperature is $T_C = 750$°C. Model the working fluid as an ideal gas with $\gamma = 1.40$. (a) Fill in this table to follow the states of the gas:

	T (K)	**P (kPa)**	**V (cm³)**
A	293	100	500
B			
C	1023		
D			

(b) Fill in this table to follow the processes:

	Q	**W**	ΔE_{int}
$A \rightarrow B$			
$B \rightarrow C$			
$C \rightarrow D$			
$D \rightarrow A$			
ABCDA			

(c) Identify the energy input $|Q_h|$, (d) the energy exhaust $|Q_c|$ and (e) the net output work W_{eng}. (f) Calculate the thermal efficiency. (g) Find the number of crankshaft revolutions per minute required for a one-cylinder engine to have an output power of 1.00 kW = 1.34 hp. Note: The thermodynamic cycle involves four piston strokes.

case study 5

David Low

Remote sensing the atmosphere

Information about the thermodynamic state of the atmosphere – pressure, temperature, wind and humidity – is required for weather prediction and climate studies. While surface-level measurements are relatively easy to make, it is significantly harder to acquire data about conditions kilometres above the ground. The standard technique is the balloon-borne radiosonde, in which a hydrogen- or helium-filled balloon of about 1 m diameter carries a small instrumented package aloft. If the package carries a reflector, it is called a rawinsonde (*ra*dio *win*d *sonde*), and can be tracked by ground-based radar: winds can then be inferred from the rate of change of position of the package. However, balloons are slow and expensive, which limits the number that can be launched every day.

Remote-sensing technology allows for continuous observation of upper-level winds. The idea is to send a pulse of sound or a burst of electromagnetic energy into the sky, usually in a few different directions. A tiny part of this energy is reflected or scattered by turbulent inhomogeneities in the atmosphere; technically speaking, small fluctuations in temperature or humidity cause the *refractive index* (a measure of wave speed in a medium, discussed in Chapter 35) to vary, and the wave energy can 'echo' from these variations. As the air is moving – wind – the echo experiences a change in frequency due to the Doppler effect (discussed in Chapter 17). If the energy sent up is acoustic (i.e. sound waves), the instrument is called a *sodar* (*so*und *d*etection *a*nd *r*anging); if the energy is electromagnetic (e.g. radio waves), the instrument is called a *radar* (*ra*dio *d*etection *a*nd *r*anging). An equivalent device that uses a laser to probe the atmosphere is called a *lidar* (*li*ght *d*etection *a*nd *r*anging).

What about the other meteorological variables? A device known as a radio acoustic sounding system (RASS) combines two physical processes to determine profiles of temperature. First, an acoustic source emits a pulse: high-frequency pulses near 2000 Hz might sound like a 'ping' or 'beep', while low-frequency pulses of about 100 Hz can sound like a mooing cow! This sound wave is a pressure wave, and as it travels upwards it creates regions of compression and rarefaction in the atmosphere. The perturbation to the atmosphere caused by the acoustic pulse is much, much larger than the natural turbulence-induced background atmospheric perturbations. Thus, if a radar located next to the acoustic source emits an electromagnetic pulse, it will echo more from the artificial acoustic perturbation than from the background. These stronger echoes are Doppler shifted due to the acoustic perturbation, which is travelling at the speed of sound, and the speed of sound depends on temperature. In Chapter 17, we saw how the speed of a wave in a medium depends on the bulk modulus of that medium (**Equation 17.32**). Combining the definition of bulk modulus (**Equation 15.14**) with the relationship between pressure and volume for an adiabatic process in an ideal gas (**Equation 21.18**) and the ideal gas law (**Equation 19.5**), we can show that the speed of sound is given by:

$$c_s = \sqrt{\frac{B}{\rho}} = \sqrt{\frac{\gamma P}{\rho}} = \sqrt{\frac{\gamma RT}{M_d}} = K_d\sqrt{T}$$

where K_d is about 20 m/s/$K^{0.5}$. Thus if you know how the speed of sound varies as you go up through the atmosphere, you can determine how temperature varies. And hence (in kelvins):

$$T \approx \frac{c_s^{\ 2}}{400\ \text{m}^2/\text{s}^2}$$

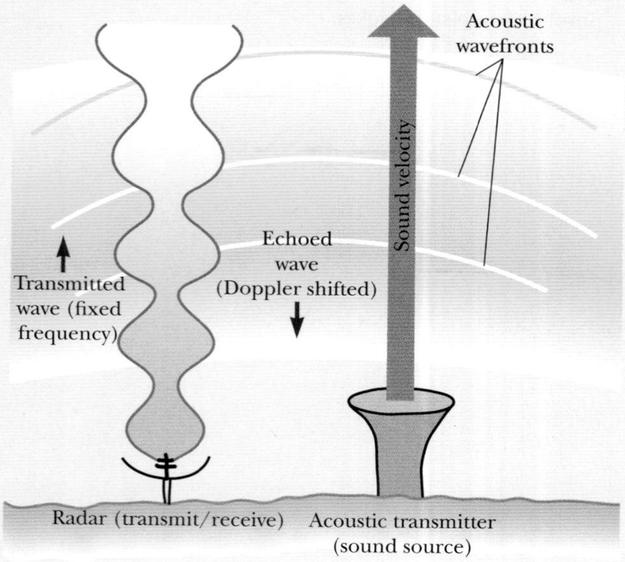

Figure CS5.1
A schematic of RASS technique

Thus, by measuring the Doppler shift of the radar echo, we can determine the speed of sound at different levels in the atmosphere, and hence we can determine how temperature varies vertically through the atmosphere!

The largest (and 'noisiest') RASS in the world is in Japan. It has measured temperature profiles from the surface up to 20-km altitude.

Atmospheric-profiling systems have been used to study weather features such as cold fronts moving across southern Australia, and sea breezes colliding across the Cape York Peninsula. These features change rapidly in time, so the ability to reliably measure profiles of atmospheric variables on a timescale of minutes is invaluable. Radars, sodars and RASS provide data that help us understand how meteorological features form, move and evolve, which contributes toward improved forecasts. And who doesn't want a better weather forecast?

Figure CS5.2
Acoustic profiler (sodar) at Weipa (Cape York Peninsula, Queensland) is part of a study on tropical convection.

(Michael Reeder, Monash University, Australia; Michael.Reeder@monash.edu, +61-3-9905-4464).

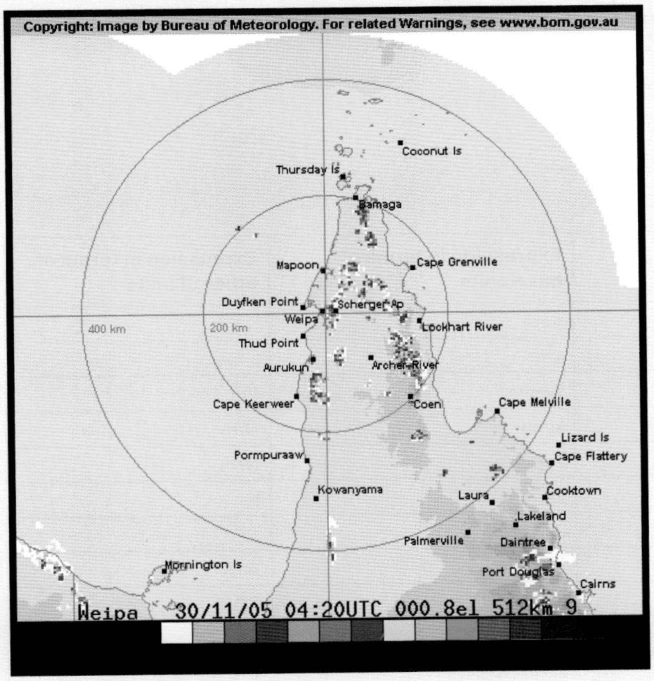

Figure CS5.3
Sea breezes from the eastern and western coasts of Cape York Peninsula can collide, causing uplift and precipitation (seen as specked blue/yellow regions).

(Original radar image copyright by Australian Bureau of Meteorology.)

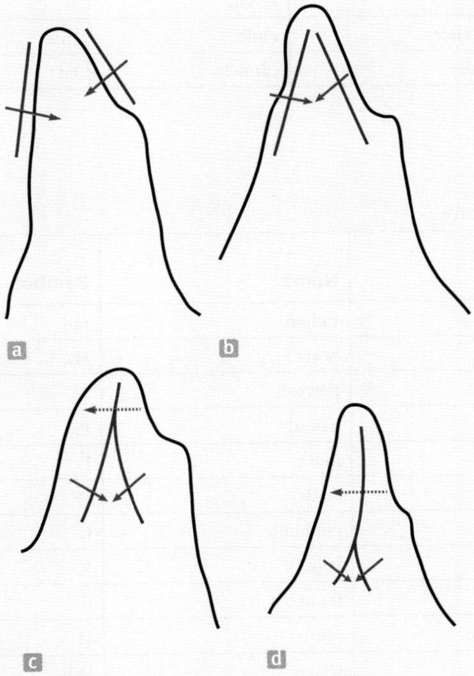

Figure CS5.4
(a, b) Sea breeze fronts ——— form on both the eastern and western coasts of the Cape York Peninsula, and propagate inland ⟶. (c, d) When these fronts collide, they reinforce. The joint front usually moves westward ⋯⋯▸ – why?

David Low is an academic at the University of New South Wales at the Australian Defence Force Academy. He has been studying atmospheric processes since he was an undergraduate, starting at about 100-km altitude and working his way down. He now probes the lowest 1000 m or so of the atmosphere, and sometimes wonders if he may eventually become a geographer... or a geologist!

appendix A

SI units

Table A.1
SI units

Base quantity	SI base unit	
	Name	**Symbol**
Length	metre	m
Mass	kilogram	kg
Time	second	s
Electric current	ampere	A
Temperature	kelvin	K
Amount of substance	mole	mol
Luminous intensity	candela	cd

Table A.2
Some derived SI units

Quantity	Name	Symbol	Expression in terms of base units	Expression in terms of other SI units
Plane angle	radian	rad	m/m	
Frequency	hertz	Hz	s^{-1}	
Force	newton	N	$kg\cdot m/s^2$	J/m
Pressure	pascal	Pa	$kg/m\cdot s^2$	N/m^2
Energy	joule	J	$kg\cdot m^2/s^2$	$N\cdot m$
Power	watt	W	$kg\cdot m^2/s^3$	J/s
Electric charge	coulomb	C	$A\cdot s$	
Electric potential	volt	V	$kg\cdot m^2/A\cdot s^3$	W/A
Capacitance	farad	F	$A^2\cdot s^4/kg\cdot m^2$	C/V
Electric resistance	ohm	Ω	$kg\cdot m^2/A^2\cdot s^3$	V/A
Magnetic flux	weber	Wb	$kg\cdot m^2/A\cdot s^2$	$V\cdot s$
Magnetic field	tesla	T	$kg/A\cdot s^2$	
Inductance	henry	H	$kg\cdot m^2/A^2\cdot s^2$	$T\cdot m^2/A$

Mathematics review

B.1 Scientific notation

Many quantities used by scientists have very large or very small values. The speed of light, for example, is about 300 000 000 m/s, and the ink required to make the dot over an *i* in this textbook has a mass of about 0.000 000 001 kg. It is very cumbersome to read, write and keep track of such numbers. We avoid this problem by using a method incorporating powers of the number 10:

$$10^0 = 1$$

$$10^1 = 10$$

$$10^2 = 10 \times 10 = 100$$

$$10^3 = 10 \times 10 \times 10 = 1000$$

and so on. The number of zeros corresponds to the power to which 10 is raised, called the **exponent** of 10. For example, the speed of light, 300 000 000 m/s, can be expressed as 3.00×10^8 m/s.

In this method, some representative numbers smaller than unity are the following:

$$10^{-1} = \frac{1}{10} = 0.1$$

$$10^{-2} = \frac{1}{10 \times 10} = 0.01$$

$$10^{-3} = \frac{1}{10 \times 10 \times 10} = 0.001$$

In these cases, the number of places the decimal point is to the left of the digit 1 (including the digit 1 itself) equals the value of the (negative) exponent. Numbers expressed as some power of 10 multiplied by another number between one and 10 are said to be in **scientific notation**. For example, the scientific notation for 5 943 000 000 is 5.943×10^9 and that for 0.000 083 2 is 8.32×10^{-5}.

When numbers expressed in scientific notation are being multiplied, the following general rule is very useful:

$$10^n \times 10^m = 10^{n+m} \tag{B.1}$$

where n and m can be *any* numbers (not necessarily integers). For example, $10^2 \times 10^5 = 10^7$. The rule also applies if one of the exponents is negative: $10^3 \times 10^{-8} = 10^{-5}$. When dividing numbers expressed in scientific notation, note that

$$\frac{10^n}{10^m} = 10^n \times 10^{-m} = 10^{n-m} \tag{B.2}$$

B.2 Algebra

Powers

When powers of a given quantity x are multiplied, the following rule applies:

$$x^n \times x^m = x^{n+m} \tag{B.3}$$

For example, $x^2x^4 = x^{2+4} = x^6$.

When dividing the powers of a given quantity, the rule is

$$\frac{x^n}{x^m} = x^{n-m} \tag{B.4}$$

Table B.1
Rules of exponents

$x^0 = 1$
$x^1 = x$
$x^n x^m = x^{n+m}$
$\frac{x^n}{x^m} = x^{n-m}$
$x^{1/n} = \sqrt[n]{x}$
$(x^n)^m = x^{nm}$

For example, $x^8/x^2 = x^{8-2} = x^6$.

A power that is a fraction, such as $\frac{1}{3}$, corresponds to a root as follows:

$$x^{1/n} = \sqrt[n]{x} \tag{B.5}$$

For example, $4^{1/3} = \sqrt[3]{4} = 1.5874$.

Finally, any quantity x^n raised to the mth power is

$$(x^n)^m = x^{nm} \tag{B.6}$$

Table B.1 summarises the rules of exponents.

Factoring

Some useful formulas for factoring an equation are:

$ax + ay + az = a(x + y + z)$	common factor
$a^2 + 2ab + b^2 = (a + b)^2$	perfect square
$a^2 - b^2 = (a + b)(a - b)$	differences of squares

Quadratic equations

The general form of a quadratic equation is:

$$ax^2 + bx + c = 0 \tag{B.7}$$

where x is the unknown quantity and a, b and c are numerical factors referred to as **coefficients** of the equation. This equation has two roots, given by

$$x = \frac{-b \pm \sqrt{b^2 - 4ac}}{2a} \tag{B.8}$$

If $b^2 \geq 4ac$, the roots are real.

Example B.1

The equation $x^2 + 5x + 4 = 0$ has the following roots corresponding to the two signs of the square-root term:

$$x_- = \frac{-5 \pm \sqrt{5^2 - (4)(1)(4)}}{2(1)} = \frac{-5 \pm \sqrt{9}}{2} = \frac{-5 \pm 3}{2}$$

$$x_+ = \frac{-5 + 3}{2} = -1 \quad x_- = \frac{-5 - 3}{2} = -4$$

where x_+ refers to the root corresponding to the positive sign and x_- refers to the root corresponding to the negative sign.

Linear equations

A linear equation has the general form

$$y = mx + b \tag{B.9}$$

where m and b are constants. This equation is referred to as linear because the graph of y versus x is a straight line as shown in **Figure B.1**. The constant b, called the ***y*-intercept**, represents the value of y at which the straight line intersects the y axis. The constant m is equal to the **gradient** or slope of the straight line. If any two points on the straight line are specified by the coordinates (x_1, y_1) and (x_2, y_2), as in **Figure B.1**, the slope of the straight line can be expressed as:

$$\text{Gradient} = \frac{y_2 - y_1}{x_2 - x_1} = \frac{\Delta y}{\Delta x} \tag{B.10}$$

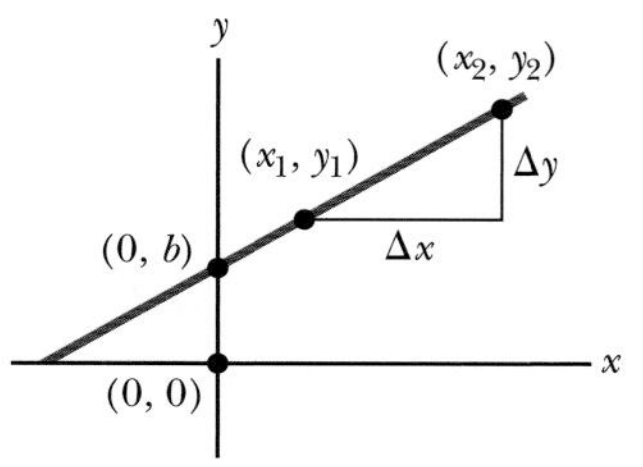

Figure B.1
A straight line graphed on an xy coordinate system. The gradient of the line is the ratio of Δy to Δx.

Note that m and b can have either positive or negative values. If $m > 0$, the straight line has a *positive* gradient, as in **Figure B.1**. If $m < 0$, the straight line has a *negative* gradient. In **Figure B.1**, both m and b are positive. Three other possible situations are shown in **Figure B.2**.

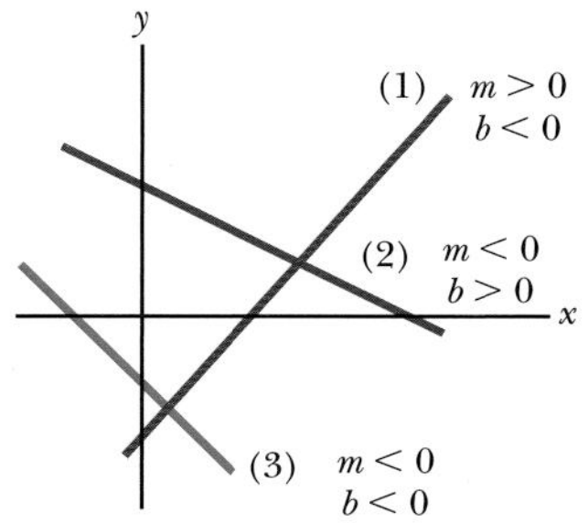

Figure B.2
The brown line has a positive gradient and a negative y-intercept. The blue line has a negative gradient and a positive y-intercept. The green line has a negative gradient and a negative y-intercept.

Solving simultaneous linear equations

Consider the equation $3x + 5y = 15$, which has two unknowns, x and y. Such an equation does not have a unique solution. For example, $(x = 0, y = 3)$, $(x = 5, y = 0)$, and $(x = 2, y = \frac{9}{5})$ are all solutions to this equation.

If a problem has two unknowns, a unique solution is possible only if we have *two* pieces of information, such as two equations. In general, if a problem has n unknowns, its solution requires n equations. To solve two simultaneous equations involving two unknowns, x and y, we solve one of the equations for x in terms of y and substitute this expression into the other equation.

In some cases, the two pieces of information may be (1) one equation and (2) a condition on the solutions. For example, suppose we have the equation $m = 3n$ and the condition that m and n must be the smallest positive nonzero integers possible. Then, the single equation does not allow a unique solution, but the addition of the condition gives us that $n = 1$ and $m = 3$.

Example B.2

Solve the two simultaneous equations

(1) $5x + y = -8$

(2) $2x - 2y = 4$

Solution

From **Equation (2)**, $x = y + 2$. Substitution of this equation into **Equation (1)** gives:

$$5(y + 2) + y = -8$$
$$6y = -18$$
$$y = -3$$
$$x = y + 2 = -1$$

Alternative solution Multiply each term in **Equation (1)** by the factor 2 and add the result to **Equation (2)**:

$$\begin{array}{r} 10x + 2y = -16 \\ 2x - 2y = 4 \\ \hline 12x = -12 \\ x = -1 \\ y = x - 2 = -3 \end{array}$$

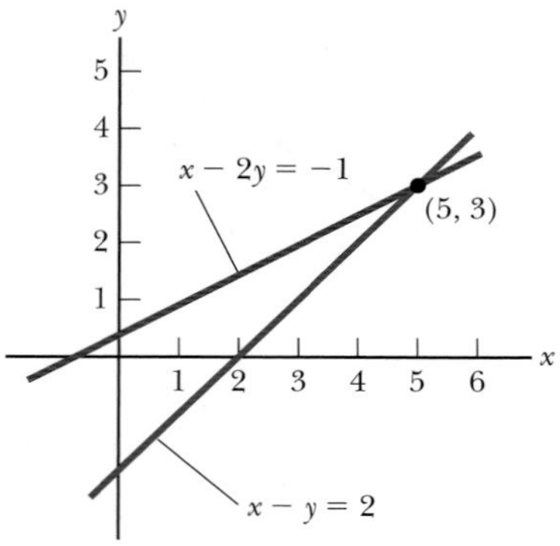

Figure B.3
A graphical solution for two linear equations

Two linear equations containing two unknowns can also be solved by a graphical method. If the straight lines corresponding to the two equations are plotted in a conventional coordinate system, the intersection of the two lines represents the solution. For example, consider the two equations

$$x - y = 2$$
$$x - 2y = -1$$

These equations are plotted in **Figure B.3**. The intersection of the two lines has the coordinates $x = 5$ and $y = 3$, which represents the solution to the equations. You should check this solution by the analytical technique discussed earlier.

Logarithms

Suppose a quantity x is expressed as a power of some quantity a:

$$x = a^y \quad \text{(B.11)}$$

The number a is called the **base** number. The **logarithm** of x with respect to the base a is equal to the exponent to which the base must be raised to satisfy the expression $x = a^y$:

$$y = \log_a x \quad \text{(B.12)}$$

Conversely, the **antilogarithm** of y is the number x:

$$x = \text{antilog}_a y \quad \text{(B.13)}$$

In practice, the two bases most often used are base 10 and base e = 2.718 282 (Euler's constant), which is called the *natural* logarithm base. When base 10 is used:

$$y = \log_{10} x \quad \text{or} \quad x = 10^y \quad \text{(B.14)}$$

When natural logarithms are used:

$$y = \ln x \quad \text{or} \quad x = e^y \quad \text{(B.15)}$$

For example, $\log_{10} 52 = 1.716$, so $\text{antilog}_{10} 1.716 = 10^{1.716} = 52$. Likewise, $\ln 52 = 3.951$, so antiln $3.951 = e^{3.951} = 52$.

In general, note you can convert between base 10 and base e using the equality

$$\ln x = 2.302\,585 \log_{10} x \quad \text{(B.16)}$$

Finally, here are some useful properties of logarithms:

$$\left.\begin{aligned} \log ab &= \log a + \log b \\ \log a/b &= \log a - \log b \\ \log a^n &= n \log a \end{aligned}\right\} \text{any base}$$

$$\ln e = 1$$

$$\ln e^a = a$$

$$\ln\left(\frac{1}{a}\right) = -\ln a$$

B.3 Geometry

The **distance** d between two points having coordinates (x_1, y_1) and (x_2, y_2) is

$$d = \sqrt{(x_2 - x_1)^2 + (y_2 - y_1)^2} \quad \text{(B.17)}$$

Two angles are equal if their sides are perpendicular, right side to right side and left side to left side. For example, the two angles marked θ in **Figure B.4** are the same because the sides of the angles are perpendicular.

To distinguish the left and right sides of an angle, imagine standing at the apex of the angle and facing into the angle.

Similar triangles: Two triangles are *similar* if their corresponding angles are the same. Their side lengths will not necessarily match, but will be in the same proportion. Each of the following properties applies to similar triangles, and can be used to show that two triangles are similar:

- both triangles have two matching angles
- a pair of sides has the same length ratio for each triangle, **and** the included angle is the same
- the ratios of side lengths are the same for both triangles.

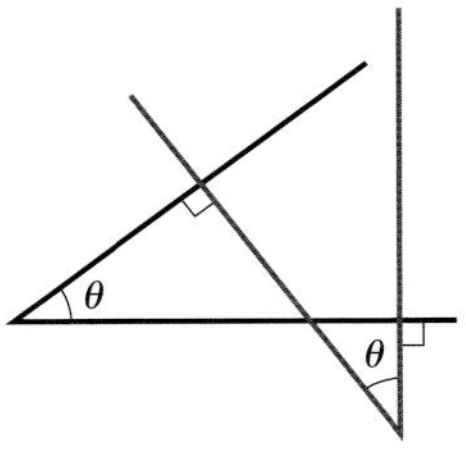

Figure B.4
The angles are equal because their sides are perpendicular.

Radian measure: The arc length s of a circular arc (**Figure B.5**) is proportional to the radius r for a fixed value of θ (in radians):

$$s = r\theta$$
$$\theta = \frac{s}{r} \quad \text{(B.18)}$$

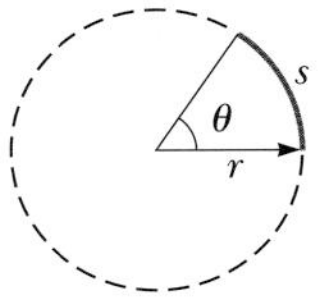

Figure B.5
The angle θ in radians is the ratio of the arc length s to the radius r of the circle.

Table B.2 gives the **areas** and **volumes** for several geometric shapes used throughout this text.

Table B.2
Useful information for geometry

Shape	Area or Volume	Shape	Area or Volume
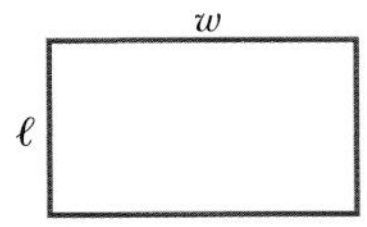 Rectangle	Area $= \ell w$	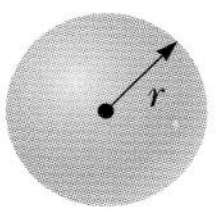 Sphere	Surface area $= 4\pi r^2$ Volume $= \frac{4\pi r^3}{3}$
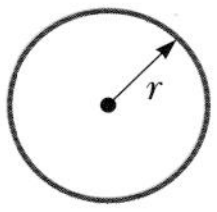 Circle	Area $= \pi r^2$ Circumference $= 2\pi r$	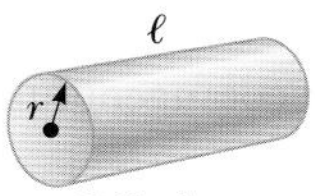 Cylinder	Lateral surface area $= 2\pi r\ell$ Volume $= \pi r^2\ell$
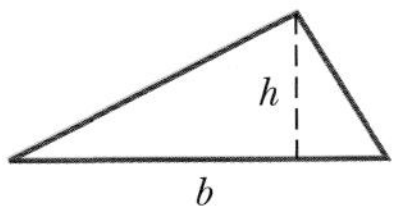 Triangle	Area $= \frac{1}{2}bh$	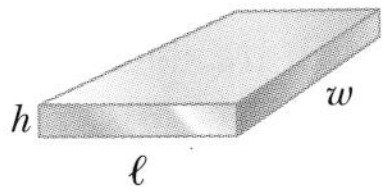 Rectangular box	Surface area $= 2(\ell h + \ell w + hw)$ Volume $= \ell wh$

The equation of a **straight line** (**Figure B.6**) is

$$y = mx + b \quad \text{(B.19)}$$

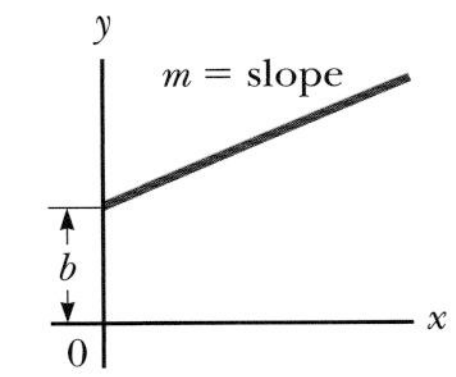

Figure B.6
A straight line with a gradient of m and a y-intercept of b

where b is the y-intercept and m is the gradient of the line.

The equation of a **circle** of radius R centred at the origin is

$$x^2 + y^2 = R^2 \quad \text{(B.20)}$$

The equation of an **ellipse** having the origin at its centre (**Figure B.7**) is

$$\frac{x^2}{a^2} + \frac{y^2}{b^2} = 1 \quad \text{(B.21)}$$

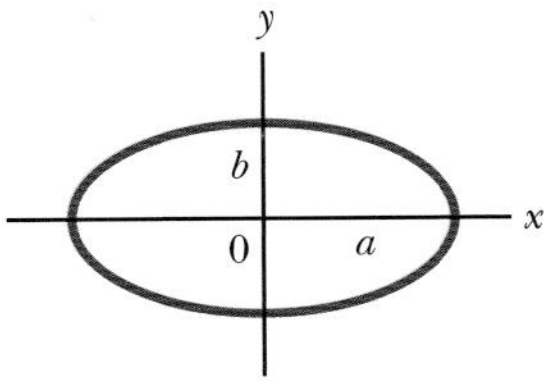

Figure B.7
An ellipse with semimajor axis a and semiminor axis b

where a is the length of the semimajor axis (the longer one) and b is the length of the semiminor axis (the shorter one).

The equation of a **parabola** the vertex of which is at $y = b$ (**Figure B.8**) is

$$y = ax^2 + b \quad \text{(B.22)}$$

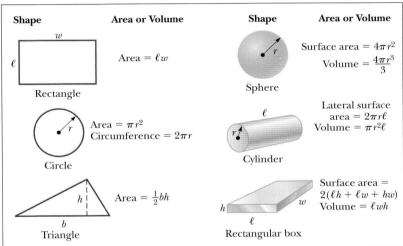

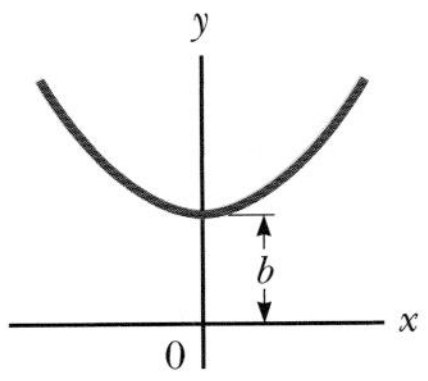

Figure B.8
A parabola with its vertex at $y = b$

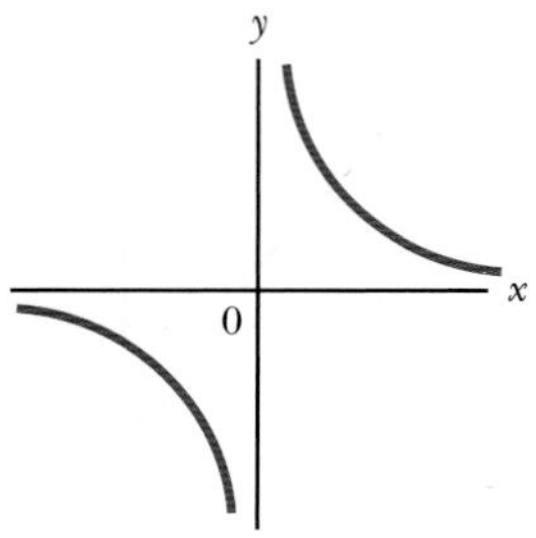

Figure B.9
A rectangular hyperbola

The equation of a **rectangular hyperbola** (Figure B.9) is

$$xy = \text{constant} \tag{B.23}$$

B.4 Trigonometry

That portion of mathematics based on the special properties of the right triangle is called trigonometry. By definition, a right triangle is a triangle containing a 90° angle. Consider the right triangle shown in Figure B.10, where side a is opposite the angle θ, side b is adjacent to the angle θ, and side c is the hypotenuse of the triangle. The three basic trigonometric functions defined by such a triangle are the sine (sin), cosine (cos), and tangent (tan). In terms of the angle θ, these functions are defined as follows:

$$\sin\theta = \frac{\text{side opposite }\theta}{\text{hypotenuse}} = \frac{a}{c} \tag{B.24}$$

$$\cos\theta = \frac{\text{side adjacent to }\theta}{\text{hypotenuse}} = \frac{b}{c} \tag{B.25}$$

$$\tan\theta = \frac{\text{side opposite }\theta}{\text{side adjacent to }\theta} = \frac{a}{b} \tag{B.26}$$

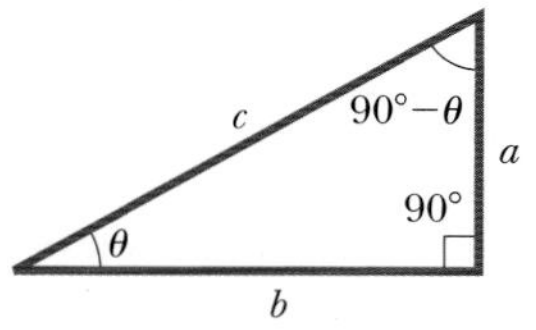

Figure B.10
A right triangle, used to define the basic functions of trigonometry

The Pythagorean theorem provides the following relationship among the sides of a right triangle:

$$c^2 = a^2 + b^2 \tag{B.27}$$

From the preceding definitions and the Pythagorean theorem, it follows that $\sin^2\theta + \cos^2\theta = 1$.

$$\tan\theta = \frac{\sin\theta}{\cos\theta}$$

The cosecant, secant, and cotangent functions are defined by

$$\csc\theta = \frac{1}{\sin\theta} \qquad \sec\theta = \frac{1}{\cos\theta} \qquad \cot\theta = \frac{1}{\tan\theta}$$

The following relationships are derived directly from the right triangle shown in Figure B.10:

$$\sin\theta = \cos(90° - \theta)$$
$$\cos\theta = \sin(90° - \theta)$$
$$\cot\theta = \tan(90° - \theta)$$

Some properties of trigonometric functions are the following:

$$\sin(-\theta) = -\sin\theta$$
$$\cos(-\theta) = \cos\theta$$
$$\tan(-\theta) = -\tan\theta$$

The following relationships apply to *any* triangle as shown in Figure B.11:

$$\alpha + \beta + \gamma = 180°$$

$$\text{Law of cosines} \quad \begin{cases} a^2 = b^2 + c^2 - 2bc\cos\alpha \\ b^2 = a^2 + c^2 - 2ac\cos\beta \\ c^2 = a^2 + b^2 - 2ab\cos\gamma \end{cases}$$

$$\text{Law of sines} \quad \frac{a}{\sin\alpha} = \frac{b}{\sin\beta} = \frac{c}{\sin\gamma}$$

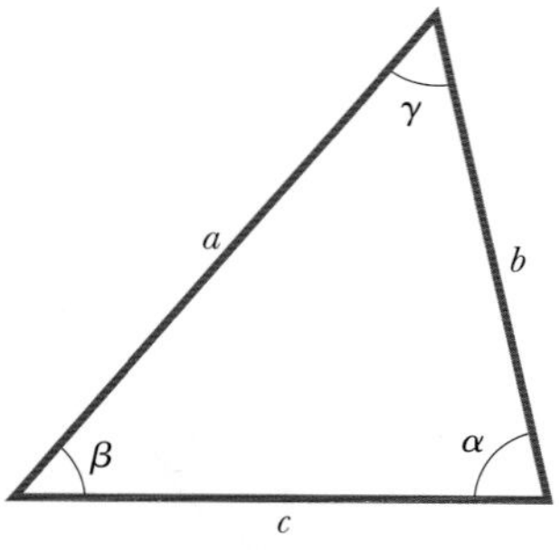

Figure B.11
An arbitrary, non-right triangle

Table B.3 lists some additional useful trigonometric identities.

Table B.3
Some trigonometric identities

$\sin^2\theta + \cos^2\theta = 1$	$\csc^2\theta = 1 + \cot^2\theta$
$\sec^2\theta = 1 + \tan^2\theta$	$\sin^2\frac{\theta}{2} = \frac{1}{2}(1 - \cos\theta)$
$\sin 2\theta = 2\sin\theta\cos\theta$	$\cos^2\frac{\theta}{2} = \frac{1}{2}(1 + \cos\theta)$
$\cos 2\theta = \cos^2\theta - \sin^2\theta$	$1 - \cos\theta = 2\sin^2\frac{\theta}{2}$
$\tan 2\theta = \frac{2\tan\theta}{1 - \tan^2\theta}$	$\tan\frac{\theta}{2} = \sqrt{\frac{1 - \cos\theta}{1 + \cos\theta}}$
$\sin(A \pm B) = \sin A\cos B \pm \cos A\sin B$	
$\cos(A \pm B) = \cos A\cos B \mp \sin A\sin B$	
$\sin A \pm \sin B = 2\sin[\frac{1}{2}A \pm B]\cos[\frac{1}{2}A \mp B]$	
$\cos A + \cos B = 2\cos[\frac{1}{2}A + B]\cos[\frac{1}{2}A - B]$	
$\cos A - \cos B = 2\sin[\frac{1}{2}A + B]\sin[\frac{1}{2}B - A]$	

Example B.3

Consider the right triangle in Figure B.12 in which $a = 2.00$, $b = 5.00$, and c is unknown. From the Pythagorean theorem, we have

$$c^2 = a^2 + b^2 = 2.00^2 + 5.00^2 = 4.00 + 25.0 = 29.0$$

$$c = \sqrt{29.0} = 5.39$$

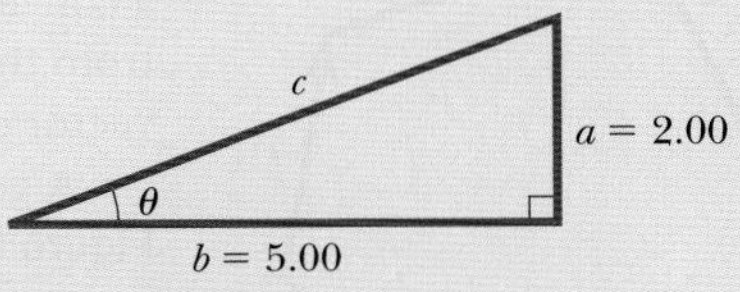

Figure B.12
(Example B.3)

To find the angle θ, note that:

$$\tan\theta = \frac{a}{b} = \frac{2.00}{5.00} = 0.400$$

Hence

$$\theta = \tan^{-1} 0.400 = 21.8°$$

where $\tan^{-1}(0.400)$ is the notation for 'angle whose tangent is 0.400', sometimes written as arctan (0.400).

B.5 Series expansions

$$(a + b)^n = a^n + \frac{n}{1!}a^{n-1}b + \frac{n(n-1)}{2!}a^{n-2}b^2 + \cdots$$

$$(1 + x)^n = 1 + nx + \frac{n(n-1)}{2!}x^2 + \cdots$$

$$e^x = 1 + x + \frac{x^2}{2!} + \frac{x^3}{3!} + \cdots$$

$$\ln(1 \pm x) = \pm x - \tfrac{1}{2}x^2 \pm \tfrac{1}{3}x^3 - \cdots$$

$$\left.\begin{aligned} \sin x &= x - \frac{x^3}{3!} + \frac{x^5}{5!} - \cdots \\ \cos x &= 1 - \frac{x^2}{2!} + \frac{x^4}{4!} - \cdots \\ \tan x &= x + \frac{x^3}{3} + \frac{2x^5}{15} + \cdots |x| < \frac{\pi}{2} \end{aligned}\right\} x \text{ in radians}$$

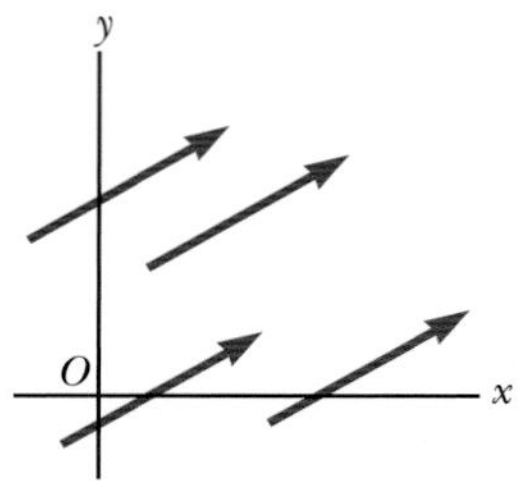

Figure B.13
These four vectors are equal because they have equal lengths and point in the same direction.

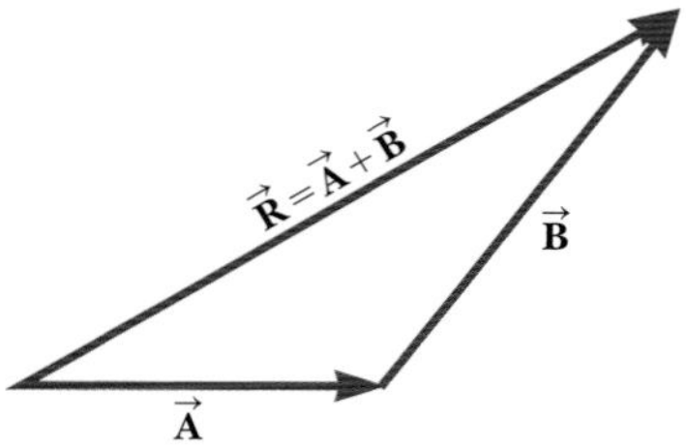

Active Figure B.14

When vector $\vec{\mathbf{B}}$ is added to vector $\vec{\mathbf{A}}$, the resultant $\vec{\mathbf{R}}$ is the vector that runs from the tail of $\vec{\mathbf{A}}$ to the tip of $\vec{\mathbf{B}}$.

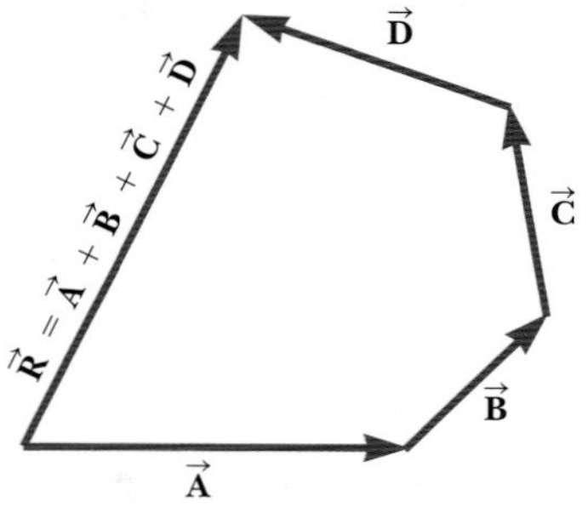

Figure B.15
Geometric construction for summing four vectors. The resultant vector $\vec{\mathbf{R}}$ is, by definition, the one that completes the polygon.

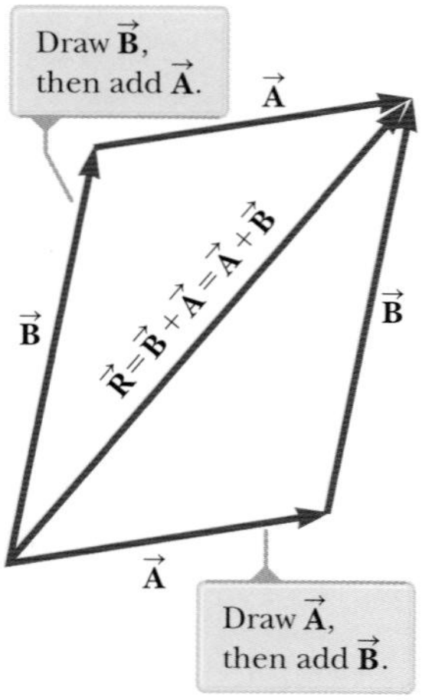

Figure B.16
This construction shows that $\vec{\mathbf{A}} + \vec{\mathbf{B}} = \vec{\mathbf{B}} + \vec{\mathbf{A}}$ or, in other words, that vector addition is commutative.

For $x \ll 1$, the *small angle approximations* can be used:

$$
\begin{array}{ll}
(1 + x)^n \approx 1 + nx & \sin x \approx x \\
e^x \approx 1 + x & \cos x \approx 1 \\
\ln(1 \pm x) \approx \pm x & \tan x \approx x
\end{array}
$$

B.6 Vectors

In our study of physics, we often need to work with physical quantities that have both numerical and directional properties. Quantities of this nature are **vector quantities**. **Scalar quantities** are those that have only a numerical value and no associated direction.

Equality of two vectors

For many purposes, two vectors $\vec{\mathbf{A}}$ and $\vec{\mathbf{B}}$ may be defined to be equal if they have the same magnitude and if they point in the same direction. That is, $\vec{\mathbf{A}} = \vec{\mathbf{B}}$ only if $A = B$ and if $\vec{\mathbf{A}}$ and $\vec{\mathbf{B}}$ point in the same direction along parallel lines. For example, all the vectors in Figure B.13 are equal even though they have different starting points. This property allows us to move a vector to a position parallel to itself in a diagram without affecting the vector.

Adding vectors

The rules for adding vectors are conveniently described by a graphical method. To add vector $\vec{\mathbf{B}}$ to vector $\vec{\mathbf{A}}$, first draw vector $\vec{\mathbf{A}}$ on graph paper, with its magnitude represented by a convenient length scale, and then draw vector $\vec{\mathbf{B}}$, to the same scale, with its tail starting from the tip of $\vec{\mathbf{A}}$, as shown in Active Figure B.14. The **resultant vector** $\vec{\mathbf{R}} = \vec{\mathbf{A}} + \vec{\mathbf{B}}$ is the vector drawn from the tail of $\vec{\mathbf{A}}$ to the tip of $\vec{\mathbf{B}}$.

A geometric construction can also be used to add more than two vectors as shown in Figure B.15 for the case of four vectors. The resultant vector $\vec{\mathbf{R}} = \vec{\mathbf{A}} + \vec{\mathbf{B}} + \vec{\mathbf{C}} + \vec{\mathbf{D}}$ is the vector that completes the polygon. In other words, $\vec{\mathbf{R}}$ is the vector drawn from the tail of the first vector to the tip of the last vector. This technique for adding vectors is often called the 'head to tail method'.

When two vectors are added, the sum is independent of the order of the addition. This property, which can be seen from the geometric construction in Figure B.16, is known as the **commutative law of addition**:

$$\vec{\mathbf{A}} + \vec{\mathbf{B}} = \vec{\mathbf{B}} + \vec{\mathbf{A}} \tag{B.28}$$

When three or more vectors are added, their sum is independent of the way in which the individual vectors are grouped together. A geometric proof of this rule for three vectors is given in Figure B.17. This property is called the **associative law of addition**:

$$\vec{\mathbf{A}} + (\vec{\mathbf{B}} + \vec{\mathbf{C}}) = (\vec{\mathbf{A}} + \vec{\mathbf{B}}) + \vec{\mathbf{C}} \tag{B.29}$$

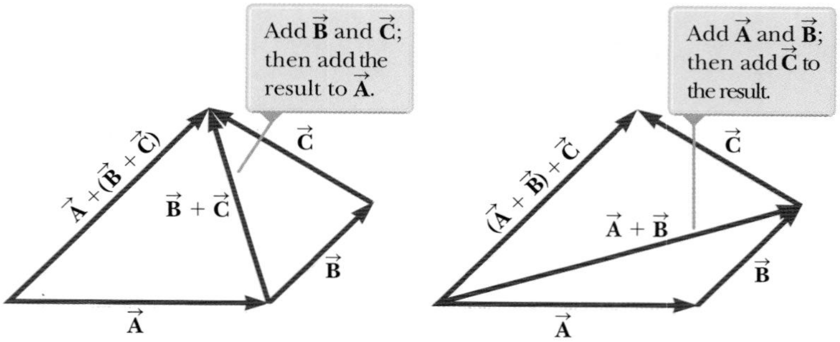

Figure B.17
Geometric constructions for verifying the associative law of addition

Negative of a vector

The negative of the vector $\vec{\mathbf{A}}$ is defined as the vector that when added to $\vec{\mathbf{A}}$ gives zero for the vector sum. That is, $\vec{\mathbf{A}} + (-\vec{\mathbf{A}}) = 0$. The vectors $\vec{\mathbf{A}}$ and $-\vec{\mathbf{A}}$ have the same magnitude but point in opposite directions.

Subtracting vectors

The operation of vector subtraction makes use of the definition of the negative of a vector. We define the operation $\vec{\mathbf{A}} - \vec{\mathbf{B}}$ as vector $-\vec{\mathbf{B}}$ added to vector $\vec{\mathbf{A}}$:

$$\vec{\mathbf{A}} - \vec{\mathbf{B}} = \vec{\mathbf{A}} + (-\vec{\mathbf{B}}) \quad \text{(B.30)}$$

The geometric construction for subtracting two vectors in this way is illustrated in **Figure B.18a**.

Another way of looking at vector subtraction is to notice that the difference $\vec{\mathbf{A}} - \vec{\mathbf{B}}$ between two vectors $\vec{\mathbf{A}}$ and $\vec{\mathbf{B}}$ is what you have to add to the second vector to obtain the first. In this case, as **Figure B.18b** shows, the vector $\vec{\mathbf{A}} - \vec{\mathbf{B}}$ points from the tip of the second vector to the tip of the first.

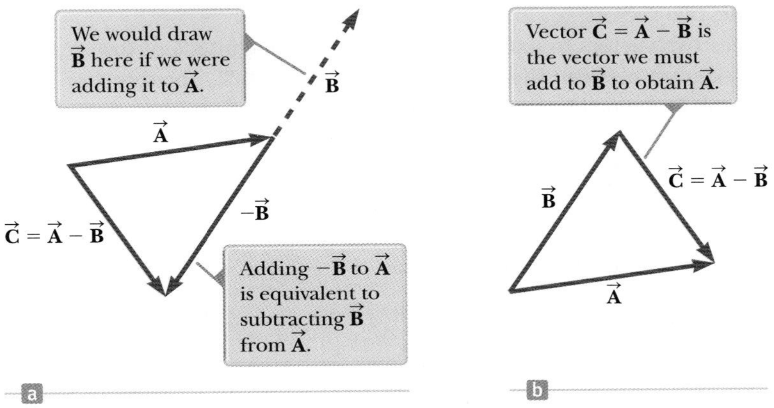

Figure B.18
(a) Subtracting vector $\vec{\mathbf{B}}$ from vector $\vec{\mathbf{A}}$. The vector $-\vec{\mathbf{B}}$ is equal in magnitude to vector $\vec{\mathbf{B}}$ and points in the opposite direction. (b) A second way of looking at vector subtraction

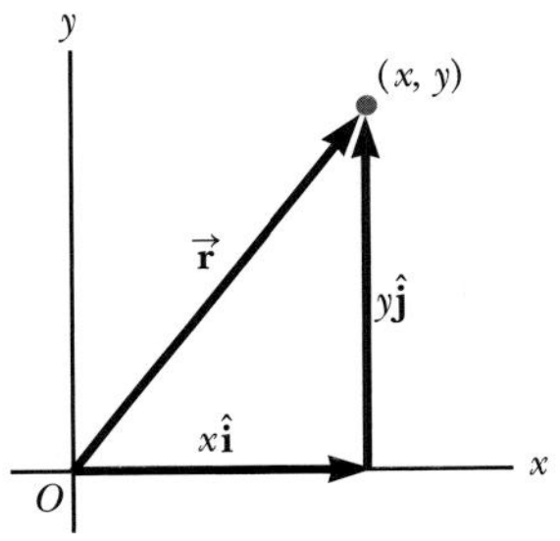

Figure B.19
The point whose Cartesian coordinates are (x, y) can be represented by the position vector $\vec{\mathbf{r}} = x\hat{\mathbf{i}} + y\hat{\mathbf{j}}$.

Adding and subtracting vectors using components

Often, the graphical method is not sufficiently accurate so we use an alternative approach. Suppose we wish to add vector $\vec{\mathbf{B}}$ to vector $\vec{\mathbf{A}}$. If $\vec{\mathbf{A}}$ and $\vec{\mathbf{B}}$ both have x, y, and z components, they can be expressed in the form

$$\vec{\mathbf{A}} = A_x\hat{\mathbf{i}} + A_y\hat{\mathbf{j}} + A_z\hat{\mathbf{k}} \quad \text{(B.31)}$$

$$\vec{\mathbf{B}} = B_x\hat{\mathbf{i}} + B_y\hat{\mathbf{j}} + B_z\hat{\mathbf{k}} \quad \text{(B.32)}$$

where $\hat{\mathbf{i}}$, $\hat{\mathbf{j}}$, and $\hat{\mathbf{k}}$ are vectors of unit length ('unit vectors') in the x, y and z directions respectively. **Figure B.19** shows an example in two dimensions.

The sum of $\vec{\mathbf{A}}$ and $\vec{\mathbf{B}}$ is

$$\vec{\mathbf{R}} = (A_x + B_x)\hat{\mathbf{i}} + (A_y + B_y)\hat{\mathbf{j}} + (A_z + B_z)\hat{\mathbf{k}} \quad \text{(B.33)}$$

If a vector $\vec{\mathbf{R}}$ has x, y and z components, the magnitude of the vector is

$$R = \sqrt{R_x^2 + R_y^2 + R_z^2} \quad \text{(B.34)}$$

The angle θ_x that $\vec{\mathbf{R}}$ makes with the x axis is found from the expression

$$\cos\theta_x = \frac{R_x}{R} \quad \text{(B.35)}$$

with similar expressions for the angles with respect to the y and z axes.

The extension of our method to adding more than two vectors is also straightforward. For example, $\vec{\mathbf{A}} + \vec{\mathbf{B}} + \vec{\mathbf{C}} = (A_x + B_x + C_x)\hat{\mathbf{i}} + (A_y + B_y + C_y)\hat{\mathbf{j}} + (A_z + B_z + C_z)\hat{\mathbf{k}}$.

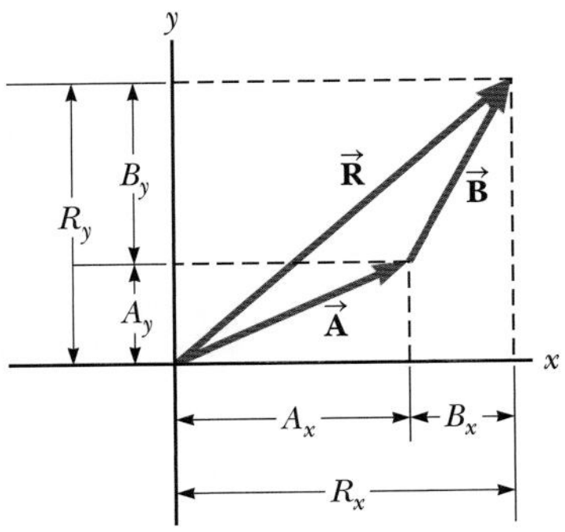

Figure B.20
This geometric construction for the sum of two vectors shows the relationship between the components of the resultant $\vec{\mathbf{R}}$ and the components of the individual vectors.

Multiplying a vector by a scalar

If vector $\vec{\mathbf{A}}$ is multiplied by a positive scalar quantity m, the product $m\vec{\mathbf{A}}$ is a vector that has the same direction as $\vec{\mathbf{A}}$ and magnitude mA.

If vector $\vec{\mathbf{A}}$ is multiplied by a negative scalar quantity $-m$, the product $-m\vec{\mathbf{A}}$ is directed opposite $\vec{\mathbf{A}}$.

Dot and cross products

The **dot product** combines two vectors to produce a scalar quantity (i.e. a number).

The dot product is defined algebraically as

$$\vec{\mathbf{A}} \bullet \vec{\mathbf{B}} = A_x B_x + A_y B_y + A_z B_z \tag{B.36}$$

Geometrically, the dot product is the projection of one vector onto the other, as shown in Figure B.21, and can also be written as

$$\vec{\mathbf{A}} \bullet \vec{\mathbf{B}} = AB\cos\theta \tag{B.37}$$

where θ is the angle between the two vectors.

The **cross product** combines two vectors to make a new vector. The magnitude of the cross product is equal to the area of the parallelogram between them, as shown in Figure B.22. The direction of the cross product is perpendicular to the plane defined by the original vectors.

The cross product is defined as

$$\vec{\mathbf{A}} \times \vec{\mathbf{B}} = AB\sin\theta\,\hat{\mathbf{n}} \tag{B.38}$$

where $\hat{\mathbf{n}}$ is the unit vector perpendicular to the plane defined by $\vec{\mathbf{A}}$ and $\vec{\mathbf{B}}$.

In terms of components, the cross product can be written as

$$\vec{\mathbf{A}} \times \vec{\mathbf{B}} = (A_y B_z - A_z B_y)\hat{\mathbf{i}} + (A_z B_x - A_x B_z)\hat{\mathbf{j}} + (A_x B_y - A_y B_x)\hat{\mathbf{k}} \tag{B.39}$$

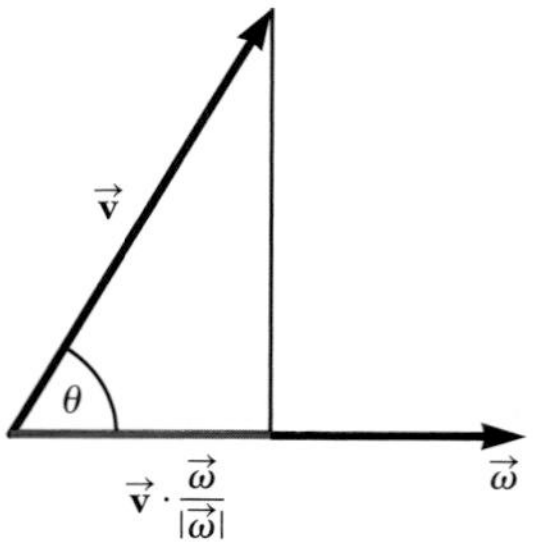

Figure B.21
The dot product describes the projection of one vector onto another.

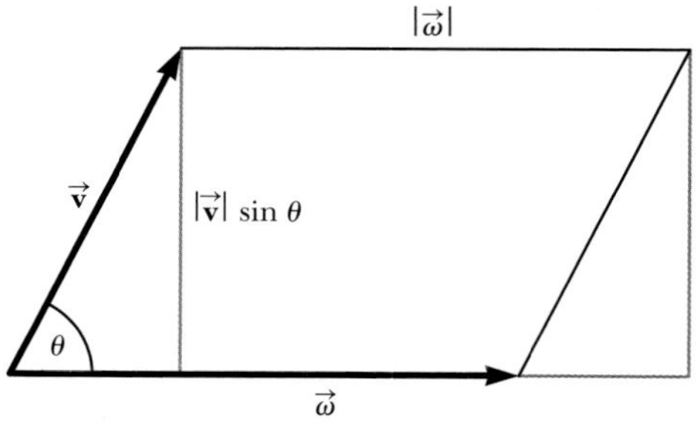

Figure B.22
The magnitude of the cross product is the area of the parallelogram between two vectors.

B.7 Differential calculus

In various branches of science, it is sometimes necessary to use the basic tools of calculus, invented by Newton, to describe physical phenomena. The use of calculus is fundamental in physics.

First, a **function** must be specified that relates one variable to another (e.g., a coordinate as a function of time). Suppose one of the variables is called y (the dependent variable), and the other x (the independent variable). We might have a function relationship such as

$$y(x) = ax^3 + bx^2 + cx + d$$

If a, b, c and d are specified constants, y can be calculated for any value of x. We usually deal with continuous functions, that is, those for which y varies 'smoothly' with x.

The **derivative** of y with respect to x is defined as the limit as Δx approaches zero of the slopes of chords drawn between two points on the y versus x curve. Mathematically, we write this definition as

$$\frac{dy}{dx} = \lim_{\Delta x \to 0} \frac{\Delta y}{\Delta x} = \lim_{\Delta x \to 0} \frac{y(x + \Delta x) - y(x)}{\Delta x} \tag{B.40}$$

where Δy and Δx are defined as $\Delta x = x_2 - x_1$ and $\Delta y = y_2 - y_1$ (**Figure B.23**). Note that dy/dx *does not* mean dy divided by dx, but rather is simply a notation of the limiting process of the derivative as defined by **Equation B.40**.

A useful expression to remember when $y(x) = ax^n$, where a is a *constant* and n is *any* positive or negative number (integer or fraction), is

$$\frac{dy}{dx} = nax^{n-1} \quad \text{(B.41)}$$

If $y(x)$ is a polynomial or algebraic function of x, we apply **Equation B.41** to *each* term in the polynomial and take $d[\text{constant}]/dx = 0$. In Examples B.4 to B.6, we evaluate the derivatives of several functions.

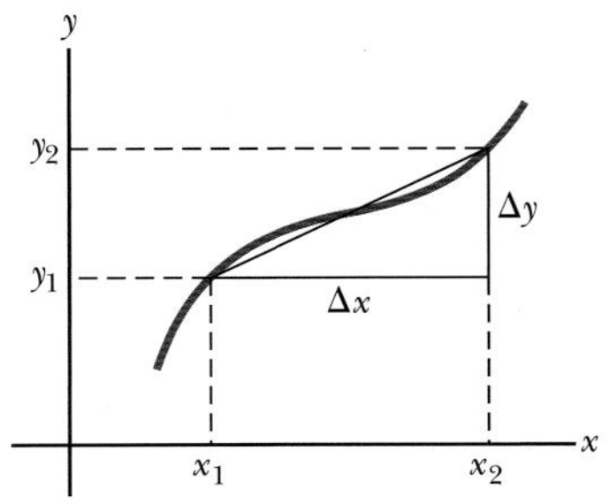

Figure B.23
The lengths Δx and Δy are used to define the derivative of this function at a point.

Example B.4

Suppose $y(x)$ is given by

$$y(x) = ax^3 + bx + c$$

where a and b are constants. It follows that

$$y(x + \Delta x) = a(x + \Delta x)^3 + b(x + \Delta x) + c$$
$$= a(x^3 + 3x^2\Delta x + 3x\Delta x^2 + \Delta x^3) + b(x + \Delta x) + c$$

so

$$\Delta y = y(x + \Delta x) - y(x) = a(3x^2\Delta x + 3x\Delta x^2 + \Delta x^3) + b\Delta x$$

Substituting this into **Equation B.51** gives

$$\frac{dy}{dx} = \lim_{\Delta x \to 0} \frac{\Delta y}{\Delta x} = \lim_{\Delta x \to 0} [3ax^2 + 3ax\Delta x + a\Delta x^2] + b$$

$$\frac{dy}{dx} = 3ax^2 + b$$

Example B.5

Find the derivative of

$$y(x) = 8x^5 + 4x^3 + 2x + 7$$

Solution

Applying **Equation B.41** to each term independently and remembering that d/dx (constant) $= 0$, we have:

$$\frac{dy}{dx} = 8(5)x^4 + 4(3)x^2 + 2(1)x^0 + 0$$

$$\frac{dy}{dx} = 40x^4 + 12x^2 + 2$$

Example B.6

Find the derivative of $y(x) = x^3/(x + 1)^2$ with respect to x.

Solution

We can rewrite this function as $y(x) = x^3(x + 1)^{-2}$ and apply **Equation B.42**:

$$\frac{dy}{dx} = (x + 1)^{-2}\frac{d}{dx}(x^3) + x^3\frac{d}{dx}(x + 1)^{-2}$$
$$= (x + 1)^{-2}3x^2 + x^3(-2)(x + 1)^{-3}$$

$$\frac{dy}{dx} = \frac{3x^2}{(x+1)^2} - \frac{2x^3}{(x+1)^3} = \frac{x^2(x+3)}{(x+1)^3}$$

Special properties of the derivative

A. Derivative of the product of two functions If a function $f(x)$ is given by the product of two functions – say, $g(x)$ and $h(x)$ – the derivative of $f(x)$ is defined as

$$\frac{d}{dx}f(x) = \frac{d}{dx}[g(x)h(x)] = g\frac{dh}{dx} + h\frac{dg}{dx} \tag{B.42}$$

B. Derivative of the sum of two functions If a function $f(x)$ is equal to the sum of two functions, the derivative of the sum is equal to the sum of the derivatives:

$$\frac{d}{dx}f(x) = \frac{d}{dx}[g(x) + h(x)] = \frac{dg}{dx} + \frac{dh}{dx} \tag{B.43}$$

C. Chain rule of differential calculus If $y = f(x)$ and $x = g(z)$, then dy/dz can be written as the product of two derivatives:

$$\frac{dy}{dx} = \frac{dy}{dx}\frac{dx}{dz} \tag{B.44}$$

D. The second derivative The second derivative of y with respect to x is defined as the derivative of the function dy/dx (the derivative of the derivative). It is usually written as

$$\frac{d^2y}{dx^2} = \frac{d}{dx}\left(\frac{dy}{dx}\right) \tag{B.45}$$

E. Partial derivative The partial derivative of a function of more than one variable is taken with respect to one of the variables, with the other variables considered to be constant. For example, the partial derivatives of a function $f(x, y)$ are written

$$\frac{\partial f}{\partial x} \text{ and } \frac{\partial f}{\partial y} \tag{B.46}$$

Some of the more commonly used derivatives of functions are listed in Table B.4.

Table B.4
Derivatives for several functions

$\frac{d}{dx}a = 0$
$\frac{d}{dx}ax^n = nax^{n-1}$
$\frac{d}{dx}e^{ax} = ae^{ax}$
$\frac{d}{dx}\cos ax = -a\sin ax$
$\frac{d}{dx}\tan ax = a\sec^2 ax$
$\frac{d}{dx}\cot ax = -a\csc^2 ax$
$\frac{d}{dx}\sec x = \tan x\sec x$
$\frac{d}{dx}\csc x = -\cot x\csc x$
$\frac{d}{dx}\ln ax = \frac{1}{x}$
$\frac{d}{dx}\sin^{-1} ax = \frac{a}{\sqrt{1 - a^2x^2}}$
$\frac{d}{dx}\cos^{-1} ax = \frac{-a}{\sqrt{1 - a^2x^2}}$
$\frac{d}{dx}\tan^{-1} ax = \frac{a}{1 + a^2x^2}$
Note: The symbols a and n represent constants.

B.8 Integral calculus

We think of integration as the inverse of differentiation. As an example, consider the expression

$$f(x) = \frac{dy}{dx} = 3ax^2 + b \quad \text{(B.47)}$$

which was the result of differentiating the function

$$y(x) = ax^3 + bx + c$$

in Example B.4. We can write **Equation B.47** as $dy = f(x)\,dx = (3ax^2 + b)dx$ and obtain $y(x)$ by 'summing' over all values of x. Mathematically, we write this inverse operation as

$$y(x) = \int f(x)dx$$

For the function $f(x)$ given by **Equation B.34**, we have

$$y(x) = \int (3ax^2 + b)dx = ax^3 + bx + c$$

where c is a constant of the integration. This type of integral is called an *indefinite integral* because its value depends on the choice of c. A general **indefinite integral** $I(x)$ is defined as

$$I(x) = \int f(x)dx \quad \text{(B.48)}$$

where $f(x)$ is called the *integrand* and $f(x) = dI(x)/dx$.

For a *general continuous* function $f(x)$, the integral can be described as the area under the curve bounded by $f(x)$ and the x axis, between two specified values of x, say, x_1 and x_2, as in **Figure B.24**.

The area of the blue element in **Figure B.15** is approximately $f(x_i)\ \Delta x_i$. If we sum all these area elements between x_1 and x_2 and take the limit of this sum as $\Delta x_i \to 0$, we obtain the *true* area under the curve bounded by $f(x)$ and the x axis, between the limits x_1 and x_2:

$$\text{Area} = \lim_{x_{\Delta i} \to 0} \sum_i f(x_i)\Delta x_i = \int_{x_1}^{x_2} f(x)dx \quad \text{(B.49)}$$

Integrals of the type defined by **Equation B.49** are called **definite integrals**.

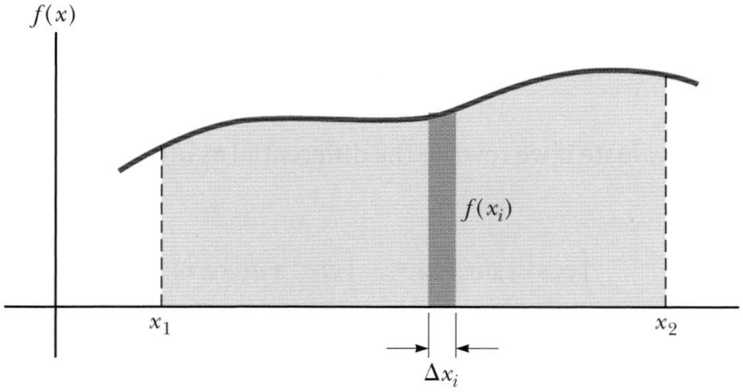

Figure B.24
The definite integral of a function is the area under the curve of the function between the limits x_1 and x_2.

One common integral that arises in practical situations has the form

$$\int x^n\,dx = \frac{x^{n+1}}{n+1} + c \quad (n \neq -1) \tag{B.50}$$

If the limits of the integration are known, this integral becomes a *definite integral* and is written

$$\int_{x_1}^{x_2} x^n\,dx = \left.\frac{x^{n+1}}{n+1}\right|_{x_1}^{x_2} = \frac{x_2^{n+1} - x_1^{n+1}}{n+1} \quad (n \neq -1) \tag{B.51}$$

Integration by parts

Sometimes it is useful to apply the method of *integrating by parts*. This method uses the property

$$\int u\,dv = uv - \int v\,du \tag{B.52}$$

where u and v are *carefully* chosen so as to reduce a complex integral to a simpler one. In many cases, several reductions have to be made. Consider the function

$$I(x) = \int x^2 e^x\,dx$$

which can be evaluated by integrating by parts twice. First, if we choose $u = x^2$, $v = e^x$, we obtain

$$\int x^2 e^x\,dx = \int x^2\,d(e^x) = x^2 e^x - 2\int e^x x\,dx + c_1$$

Now, in the second term, choose $u = x$, $v = e^x$, which gives

$$\int x^2 e^x\,dx = x^2 e^x - e^x + 2\int e^x\,dx + c_1$$

or

$$\int x^2 e^x\,dx = x^2 e^x - 2xe^x + 2e^x + c_2$$

Integration by substitution

In other cases, it is useful to look for a change of variable such that the differential of the function is the differential of the independent variable appearing in the integrand. For example, consider the integral

$$I(x) = \int \cos^2 x \sin x\,dx$$

This integral becomes easy to evaluate if we rewrite the differential as $d(\cos x) = -\sin x\,dx$. The integral then becomes

$$\int \cos^2 x \sin x\,dx = -\int \cos^2 x\,d(\cos x)$$

If we now change variables, letting $y = \cos x$, we obtain

$$\int \cos^2 x \sin x\,dx = -\int y^2\,dy = -\frac{y^3}{3} + c = -\frac{\cos^3 x}{3} + c$$

Table B.5 lists some useful indefinite integrals. **Table B.6** gives Gauss's probability integral and other definite integrals. More complete lists can be found in mathematics textbooks.

Table B.5

Some indefinite integrals (An arbitrary constant should be added to each of these integrals.)

$$\int x^n\,dx = \frac{x^{n+1}}{n+1} \quad \text{(provided } n \neq 1)$$

$$\int \frac{dx}{x} = \int x^{-1}\,dx = \ln x$$

$$\int \frac{dx}{a+bx} = \frac{1}{b}\ln(a+bx)$$

$$\int \frac{x\,dx}{a+bx} = \frac{x}{b} - \frac{a}{b^2\ln(a+bx)}$$

$$\int \frac{dx}{x(x+a)} = -\frac{1}{a}\ln\frac{x+a}{x}$$

$$\int \frac{dx}{(a+bx)^2} = -\frac{1}{b(a+bx)}$$

$$\int \frac{dx}{a^2+x^2} = \frac{1}{a}\tan^{-1}\frac{x}{a}$$

$$\int \frac{dx}{a^2-x^2} = \frac{1}{2a}\ln\frac{a+x}{a-x}(a^2-x^2>0)$$

$$\int \frac{dx}{x^2-a^2} = \frac{1}{2a}\ln\frac{x-a}{x+a}(x^2-a^2>0)$$

$$\int \frac{x\,dx}{a^2\pm x^2} = \pm\tfrac{1}{2}\ln(a^2\pm x^2)$$

$$\int \frac{dx}{\sqrt{a^2-x^2}} = \sin^{-1}\frac{x}{a} = -\cos^{-1}\frac{x}{2}(a^2-x^2>0)$$

$$\int \frac{dx}{\sqrt{x^2\pm a^2}} = \ln(x+\sqrt{x^2\pm a^2})$$

$$\int \frac{x\,dx}{\sqrt{a^2-x^2}} = -\sqrt{a^2-x^2}$$

$$\int \frac{x\,dx}{\sqrt{x^2\pm a^2}} = \sqrt{x^2\pm a^2}$$

$$\int \sqrt{a^2-x^2}\,dx = \tfrac{1}{2}\left(x\sqrt{a^2-x^2} + a^2\sin^{-1}\frac{x}{|a|}\right)$$

$$\int x\sqrt{a^2-x^2}\,dx = -\tfrac{1}{3}(a^2-x^2)^{3/2}$$

$$\int \sqrt{x^2\pm a^2}\,dx = \tfrac{1}{2}\left[x\sqrt{x^2\pm a^2} \pm a^2\ln(x=\sqrt{x^2\pm a^2})\right]$$

$$\int x\left(\sqrt{x^2\pm a^2}\right)dx = \tfrac{1}{3}(x^2\pm a^2)^{3/2}$$

$$\int e^{ax}\,dx = \frac{1}{a}e^{ax}$$

$$\int \ln ax\,dx = (x\ln ax) - x$$

$$\int xe^{zx}\,dx = \frac{d^{ax}}{a^2}(ax-1)$$

$$\int \frac{dx}{a+be^{cx}} = \frac{x}{a} - \frac{1}{ac}\ln(a+be^{cx})$$

$$\int \sin ax\,dx = -\frac{1}{a}\cos ax$$

$$\int \cos ax\,dx = \frac{1}{a}\sin ax$$

$$\int \tan ax\,dx = -\frac{1}{a}\ln(\cos ax) = \frac{1}{a}\ln(\sec ax)$$

$$\int \cot ax\,dx = \frac{1}{a}\ln(\sin ax)$$

$$\int \sec ax\,dx = \frac{1}{a}\ln(\sec ax) + (\tan ax) = \frac{1}{a}\ln\left[\tan\left(\frac{ax}{2}+\frac{\pi}{4}\right)\right]$$

$$\int \csc ax\,dx = \frac{1}{a}\ln(\csc ax - \cot ax) = \frac{1}{a}\ln\left(\tan\frac{ax}{2}\right)$$

$$\int \sin^2 ax\,dx = \frac{x}{2} - \frac{\sin 2ax}{4a}$$

$$\int \cos^2 ax\,dx = \frac{x}{2} + \frac{\sin 2ax}{4a}$$

$$\int \frac{dx}{\sin^2 ax} = -\frac{1}{a}\cot ax$$

$$\int \frac{dx}{\cos^2 ax} = \frac{1}{a}\tan ax$$

$$\int \tan^2 ax\,dx = \frac{1}{a}(\tan ax) - x$$

$$\int \cot^2 ax\,dx = -\frac{1}{a}(\cot ax) - x$$

$$\int \sin^{-1} ax\,dx = x(\sin^{-1} ax) + \frac{\sqrt{1-a^2x^2}}{a}$$

$$\int \cos^{-1} ax\,dx = x(\cos^{-1} ax) - \frac{\sqrt{1-a^2x^2}}{a}$$

$$\int \frac{dx}{(x^2+a^2)^{3/2}} = \frac{x}{a^2\sqrt{x^2+a^2}}$$

$$\int \frac{x\,dx}{(x^2+a^2)^{3/2}} = -\frac{1}{\sqrt{x^2+a^2}}$$

Table B.6
Gauss's probability integral and other definite integrals

$$\int_0^\infty x^n e^{-ax} dx = \frac{n!}{a^{n+1}}$$

$$I_0 = \int_0^\infty e^{-ax^2} dx = \frac{1}{2}\sqrt{\frac{\pi}{a}} \qquad \text{(Gauss's probability integral)}$$

$$I_1 = \int_0^\infty x e^{-ax^2} dx = \frac{1}{2a}$$

$$I_2 = \int_0^\infty x^2 e^{-ax^2} dx = -\frac{dI_0}{da} = \frac{1}{4}\sqrt{\frac{\pi}{a^3}}$$

$$I_3 = \int_0^\infty x^3 e^{-ax^2} dx = -\frac{dI_1}{da} = \frac{1}{2a^2}$$

$$I_4 = \int_0^\infty x^4 e^{-ax^2} dx = \frac{d^2 I_0}{da^2} = \frac{3}{8}\sqrt{\frac{\pi}{a^5}}$$

$$I_5 = \int_0^\infty x^5 e^{-ax^2} dx = \frac{d^2 I_1}{da^2} = \frac{1}{a^3}$$

$\vdots$

$$I_{2n} = (-1)^n \frac{d^n}{da^n} I_0$$

$$I_{2n+1} = (-1)^n \frac{d^n}{da^n} I_1$$

B.9 Complex numbers

A **complex** number z is written in general as

$$z = a + bi$$

where a and b are **real** numbers, and i is the **imaginary** unit such that $i^2 = -1$.
The quantity a is called the real part of z, and b is the imaginary part.
The **complex conjugate** of a complex number, written as z^*, is obtained by replacing i with $-i$;

$$z = a + bi \quad z^* = a - bi$$

The **absolute square** of a complex number is the product of the number and its complex conjugate

$$|z|^2 = zz^* = (a + bi)(a - bi) = a^2 + b^2$$

The absolute square is a positive real number.
The above definitions for complex conjugate and absolute square also apply to **complex functions**.
A useful connection between complex numbers and trigonometry is given by **Euler's formula**:

$$e^{i\theta} = \cos\theta + i\sin\theta$$

appendix C

Tables of data

Table C.1
Conversion factors

Length						
	m	**cm**	**km**	**in.**	**ft**	**mi**
1 metre	1	10^2	10^{-3}	39.37	3.281	6.214×10^{-4}
1 centimetre	10^{-2}	1	10^{-5}	0.3937	3.281×10^{-2}	6.214×10^{-6}
1 kilometre	10^3	10^5	1	3.937×10^4	3.281×10^3	0.6214
1 inch	2.540×10^{-2}	2.540	2.540×10^{-5}	1	8.333×10^{-2}	1.578×10^{-5}
1 foot	0.3048	30.48	3.048×10^{-4}	12	1	1.894×10^{-4}
1 mile	1609	1.609×10^5	1.609	6.336×10^4	5280	1

Mass				
	kg	**g**	**slug**	**u**
1 kilogram	1	10^3	6.852×10^{-2}	6.024×10^{26}
1 gram	10^{-3}	1	6.852×10^{-5}	6.024×10^{23}
1 slug	14.59	1.459×10^4	1	8.789×10^{27}
1 atomic mass unit	1.660×10^{-27}	1.660×10^{-24}	1.137×10^{-28}	1
Note: 1 metric tonne = 1000 kg.				

Time					
	s	**min**	**h**	**day**	**yr**
1 second	1	1.667×10^{-2}	2.778×10^{-4}	1.157×10^{-5}	3.169×10^{-8}
1 minute	60	1	1.667×10^{-2}	6.994×10^{-4}	1.901×10^{-6}
1 hour	3600	60	1	4.167×10^{-2}	1.141×10^{-4}
1 day	8.640×10^4	1440	24	1	2.738×10^{-5}
1 year	3.156×10^7	5.259×10^5	8.766×10^3	365.2	1

Speed				
	m/s	**cm/s**	**ft/s**	**mi/h**
1 metre per second	1	10^2	3.281	2.237
1 centimetre per second	10^{-2}	1	3.281×10^{-2}	2.237×10^{-2}
1 foot per second	0.304 8	30.48	1	0.681 8
1 mile per hour	0.447 0	44.70	1.467	1
Note: 1 mi/min = 60 mi/h = 88 ft/s.				

Force		
	N	**lb**
1 newton	1	0.2248
1 pound	4.448	1

(Continued)

Table C.1
Conversion factors (*continued*)

Energy, energy transfer			
	J	**ft.lb**	**eV**
1 joule	1	0.737 6	6.242×10^{18}
1 foot-pound	1.356	1	8.464×10^{18}
1 electron volt	1.602×10^{-19}	1.182×10^{-19}	1
1 calorie	4.186	3.087	2.613×10^{19}
1 British thermal unit	1.055×10^{3}	7.779×10^{2}	6.585×10^{21}
1 kilowatt-hour	3.600×10^{6}	2.655×10^{6}	2.247×10^{25}
	cal	**Btu**	**kWh**
1 joule	0.2389	9.481×10^{-4}	2.778×10^{-7}
1 foot-pound	0.3239	1.285×10^{-3}	3.766×10^{-7}
1 electron volt	3.827×10^{-20}	1.519×10^{-22}	4.450×10^{-26}
1 calorie	1	3.968×10^{-3}	1.163×10^{-6}
1 British thermal unit	2.520×10^{2}	1	2.930×10^{-4}
1 kilowatt-hour	8.601×10^{5}	3.413×10^{2}	1

Pressure

	Pa	**atm**	
1 pascal	1	9.869×10^{-6}	
1 atmosphere	1.013×10^{5}	1	
1 centimetre mercury[a]	1.333×10^{3}	1.316×10^{-2}	
1 pound per square inch	6.895×10^{3}	6.805×10^{-2}	
1 pound per square foot	47.88	4.725×10^{-4}	
	cm Hg	**lb/in.^{2}**	**lb/ft^2**
1 pascal	7.501×10^{-4}	1.450×10^{-4}	2.089×10^{-2}
1 atmosphere	76	14.70	2.116×10^{3}
1 centimetre mercury[a]	1	0.194 3	27.85
1 pound per square inch	5.171	1	144
1 pound per square foot	3.591×10^{-2}	6.944×10^{-3}	1

[a] At 0°C and at a location where the free-fall acceleration has its 'standard' value, 9.806 65 m/s^2

Table C.2
Symbols, dimensions, and units of physical quantities

Quantity	Common symbol	Unit[a]	Dimensions[b]	Unit in terms of base SI units
Acceleration	$\vec{a}$	m/s^2	L/T^2	m/s^2
Amount of substance	n	mole		mol
Angle	θ, ϕ	radian (rad)	1	
Angular acceleration	$\vec{\alpha}$	rad/s^2	T^{-2}	s^{-2}
Angular frequency	ω	rad/s	T^{-1}	s^{-1}
Angular momentum	$\vec{L}$	kg•m^2/s	ML^2/T	kg•m^2/s
Angular velocity	$\vec{\omega}$	rad/s	T^{-1}	s^{-1}
Area	A	m^2	L^2	m^2
Atomic number	Z			
Capacitance	C	farad (F)	Q^{2}T^2/ML^2	A^2•s^4/kg•m^2
Charge	q, Q, e	coulomb (C)	Q	A•s

(*Continued*)

Table C.2
Symbols, dimensions and units of physical quantities (*continued*)

Quantity	Common symbol	Unit[a]	Dimensions[b]	Unit in terms of base SI units
Charge density				
Line	λ	C/m	Q/L	A•s/m
Surface	σ	C/m^2	Q/L^2	A•s/m^2
Volume	ρ	C/m^3	Q/L^3	A•s/m^3
Conductivity	σ	1/Ω•m	Q^{2}T/ML^3	A^2•s^3/kg•m^3
Current	I	**ampere**	Q/T	A
Current density	J	A/m^2	Q/TL^2	A/m^2
Density	ρ	kg/m^3	M/L^3	kg/m^3
Dielectric constant	κ			
Electric dipole moment	$\vec{\mathbf{p}}$	C•m	QL	A•s•m
Electric field	$\vec{\mathbf{E}}$	V/m	ML/QT^2	kg•m/A•s^3
Electric flux	Φ_E	V•m	ML^3/QT^2	kg•m^3/A•s^3
Electromotive force	ϵ	volt (V)	ML^2/QT^2	kg•m^2/A•s^3
Energy	E, U, K	joule (J)	ML^2/T^2	kg•m^2/s^2
Entropy	S	J/K	ML^2/T^{2}K	kg•m^2/s^2•K
Force	$\vec{\mathbf{F}}$	newton (N)	ML/T^2	kg•m/s^2
Frequency	f	hertz (Hz)	T^{-1}	s^{-1}
Heat	Q	joule (J)	ML^2/T^2	kg•m^2/s^2
Inductance	L	henry (H)	ML^2/Q^2	kg•m^2/A^2•s^2
Length	l, L	**metre**	L	m
Displacement	$\Delta x, \Delta \vec{\mathbf{r}}$			
Distance	d, h			
Position	$x, y, z, \vec{\mathbf{r}}$			
Magnetic dipole moment	$\vec{\boldsymbol{\mu}}$	N•m/T	QL^2/T	A•m^2
Magnetic field	$\vec{\mathbf{B}}$	tesla (T) (= Wb/m^2)	M/QT	kg/A•s^2
Magnetic flux	Φ_B	weber (Wb)	ML^2/QT	kg•m^2/A•s^2
Mass	m, M	**kilogram**	M	kg
Molar specific heat	C	J/mol•K		kg•m^2/s^2•mol•K
Moment of inertia	I	kg•m^2	ML^2	kg•m^2
Momentum	$\vec{\mathbf{p}}$	kg•m/s	ML/T	kg•m/s
Period	T	s	T	s
Permeability of free space	μ_0	N/A^2 (= H/m)	ML/Q^2	kg•m/A^2•s^2
Permittivity of free space	ϵ_0	C^2/N•m^2(= F/m)	Q^{2}T^2/ML^3	A^2•s^4/kg•m^3
Potential	V	volt (V)(= J/C)	ML^2/QT^2	kg•m^2/A•s^3
Power	P	watt (W)(= J/s)	ML^2/T^3	kg•m^2/s^3
Pressure	P	pascal (Pa)(= N/m^2)	M/LT^2	kg/m•s^2
Resistance	R	ohm (Ω)(= V/A)	ML^2/Q^{2}T	kg•m^2/A^2•s^3
Specific heat	c	J/kg•K	L^2/T^{2}K	m^2/s^2•K
Speed	v	m/s	L/T	m/s
Temperature	T	**kelvin**	K	K
Time	t	**second**	T	s
Torque	$\vec{\boldsymbol{\tau}}$	N•m	ML^2/T^2	kg•m^2/s^2
Velocity	$\vec{\mathbf{v}}$	m/s	L/T	m/s
Volume	V	m^3	L^3	m^3
Wavelength	λ	m	L	m
Work	W	joule (J)(= N•m)	ML^2/T^2	kg•m^2/s^2

[a]The base SI units are given in **bold** letters.

[b]The symbols M, L, T, K and Q denote mass, length, time, temperature and charge, respectively.

Periodic table of the elements

Symbol — **Ca** 20 — Atomic number
Atomic mass† — 40.078
$4s^2$ — Electron configuration

Group I	Group II	Transition elements						
H 1 1.007 9 $1s$								
Li 3 6.941 $2s^1$	**Be** 4 9.0122 $2s^2$							
Na 11 22.990 $3s^1$	**Mg** 12 24.305 $3s^2$							
K 19 39.098 $4s^1$	**Ca** 20 40.078 $4s^2$	**Sc** 21 44.956 $3d^14s^2$	**Ti** 22 47.867 $3d^24s^2$	**V** 23 50.942 $3d^34s^2$	**Cr** 24 51.996 $3d^54s^1$	**Mn** 25 54.938 $3d^54s^2$	**Fe** 26 55.845 $3d^64s^2$	**Co** 27 58.933 $3d^74s^2$
Rb 37 85.468 $5s^1$	**Sr** 38 87.62 $5s^2$	**Y** 39 88.906 $4d^15s^2$	**Zr** 40 91.224 $4d^25s^2$	**Nb** 41 92.906 $4d^45s^1$	**Mo** 42 95.94 $4d^55s^1$	**Tc** 43 (98) $4d^55s^2$	**Ru** 44 101.07 $4d^75s^1$	**Rh** 45 102.91 $4d^85s^1$
Cs 55 132.91 $6s^1$	**Ba** 56 137.33 $6s^2$	57–71*	**Hf** 72 178.49 $5d^26s^2$	**Ta** 73 180.95 $5d^36s^2$	**W** 74 183.84 $5d^46s^2$	**Re** 75 186.21 $5d^56s^2$	**Os** 76 190.23 $5d^66s^2$	**Ir** 77 192.2 $5d^76s^2$
Fr 87 (223) $7s^1$	**Ra** 88 (226) $7s^2$	89–103**	**Rf** 104 (261) $6d^27s^2$	**Db** 105 (262) $6d^37s^2$	**Sg** 106 (266)	**Bh** 107 (264)	**Hs** 108 (277)	**Mt** 109 (268)

*Lanthanide series	**La** 57 138.91 $5d^16s^2$	**Ce** 58 140.12 $5d^14f^16s^2$	**Pr** 59 140.91 $4f^36s^2$	**Nd** 60 144.24 $4f^46s^2$	**Pm** 61 (145) $4f^56s^2$	**Sm** 62 150.36 $4f^66s^2$
**Actinide series	**Ac** 89 (227) $6d^17s^2$	**Th** 90 232.04 $6d^27s^2$	**Pa** 91 231.04 $5f^26d^17s^2$	**U** 92 238.03 $5f^36d^17s^2$	**Np** 93 (237) $5f^46d^17s^2$	**Pu** 94 (244) $5f^67s^2$

Note: Atomic mass values given are averaged over isotopes in the percentages in which they exist in nature.
† For an unstable element, mass number of the most stable known isotope is given in parentheses.

			Group III	Group IV	Group V	Group VI	Group VII	Group 0
							H 1 1.007 9 $1s^1$	**He** 2 4.002 6 $1s^2$
			B 5 10.811 $2p^1$	**C** 6 12.011 $2p^2$	**N** 7 14.007 $2p^3$	**O** 8 15.999 $2p^4$	**F** 9 18.998 $2p^5$	**Ne** 10 20.180 $2p^6$
			Al 13 26.982 $3p^1$	**Si** 14 28.086 $3p^2$	**P** 15 30.974 $3p^3$	**S** 16 32.066 $3p^4$	**Cl** 17 35.453 $3p^5$	**Ar** 18 39.948 $3p^6$
Ni 28 58.693 $3d^84s^2$	**Cu** 29 63.546 $3d^{10}4s^1$	**Zn** 30 65.41 $3d^{10}4s^2$	**Ga** 31 69.723 $4p^1$	**Ge** 32 72.64 $4p^2$	**As** 33 74.922 $4p^3$	**Se** 34 78.96 $4p^4$	**Br** 35 79.904 $4p^5$	**Kr** 36 83.80 $4p^6$
Pd 46 106.42 $4d^{10}$	**Ag** 47 107.87 $4d^{10}5s^1$	**Cd** 48 112.41 $4d^{10}5s^2$	**In** 49 114.82 $5p^1$	**Sn** 50 118.71 $5p^2$	**Sb** 51 121.76 $5p^3$	**Te** 52 127.60 $5p^4$	**I** 53 126.90 $5p^5$	**Xe** 54 131.29 $5p^6$
Pt 78 195.08 $5d^96s^1$	**Au** 79 196.97 $5d^{10}6s^1$	**Hg** 80 200.59 $5d^{10}6s^2$	**Tl** 81 204.38 $6p^1$	**Pb** 82 207.2 $6p^2$	**Bi** 83 208.98 $6p^3$	**Po** 84 (209) $6p^4$	**At** 85 (210) $6p^5$	**Rn** 86 (222) $6p^6$
Ds 110 (271)	**Rg** 111 (272)	**Cn** 112 (285)	113†† (284)	**Fl** 114 (289)	115†† (288)	**Lv** 116 (292)	117†† (294)	118†† (294)

Eu 63 151.96 $4f^76s^2$	**Gd** 64 157.25 $4f^75d^16s^2$	**Tb** 65 158.93 $4f^85d^16s^2$	**Dy** 66 162.50 $4f^{10}6s^2$	**Ho** 67 164.93 $4f^{11}6s^2$	**Er** 68 167.26 $4f^{12}6s^2$	**Tm** 69 168.93 $4f^{13}6s^2$	**Yb** 70 173.04 $4f^{14}6s^2$	**Lu** 71 174.97 $4f^{14}5d^16s^2$
Am 95 (243) $5f^77s^2$	**Cm** 96 (247) $5f^76d^17s^2$	**Bk** 97 (247) $5f^86d^17s^2$	**Cf** 98 (251) $5f^{10}7s^2$	**Es** 99 (252) $5f^{11}7s^2$	**Fm** 100 (257) $5f^{12}7s^2$	**Md** 101 (258) $5f^{13}7s^2$	**No** 102 (259) $5f^{14}7s^2$	**Lr** 103 (262) $5f^{14}6d^17s^2$

††Elements 113, 115, 117, and 118 have not yet been officially named. Only small numbers of atoms of these elements have been observed. *Note*: For a description of the atomic data, visit *http://physics.nist.gov/PhysRefData/Elements/per_text.html.*

Answers to Quick Quizzes

Chapter 1

1.1. (a)
1.2. False
1.3. 10^{-3} m, 100 m, 10^6 m
1.4. 2, 3, 6

Chapter 2

2.1. (c)
2.2. (b)
2.3. Velocity is negative, acceleration is positive
2.4. (c)
2.5. (a)–(e), (b)–(d), (c)–(f)
2.6. **(i)** acceleration (e), **(ii)** speed (d)

Chapter 3

3.1. vectors: (b), (c); scalars: (a), (d), (e)
3.2. (b)
3.3. (a)
3.4. **(i)** (b) **(ii)** (a)
3.5. 15°, 30°, 45°, 60°, 75°
3.6. **(i)** (d) **(ii)** (b)
3.7. **(i)** (b) **(ii)** (d)

Chapter 4

4.1. (d)
4.2. (a) The force required to lift the unit against gravity will be the same in both cases, as the weight depends upon the local gravity. Accelerating the unit horizontally will require a force proportional to its mass, not its weight, which will be different on Earth and the Moon.
4.3. (d)
4.4. **(a)**
4.5. **(c)** The force of the cushion on you. Action–reaction pairs are $F_{\text{a on b}} = F_{\text{b on a}}$.
4.6. **(i)** (c) **(ii)** (a)
4.7. Above the horizontal (b), below the horizontal (a)
4.8. (c)

Chapter 5

5.1. (b)
5.2. (b) pulling at an angle above the horizontal. This decreases the frictional force acting on the bag by decreasing the normal force exerted by the floor.
5.3. (a)
5.4. **(i)** (a) **(ii)** (b)
5.5. (b)
5.6. **(a)** Because the speed is constant, the only direction the force can have is that of the centripetal acceleration. The force is larger at Ⓒ than at Ⓐ because the radius at Ⓒ is smaller. There is no force at Ⓑ because the wire is straight. **(b)** In addition to the forces in the centripetal direction in part (a), there are now tangential forces to provide the tangential acceleration. The tangential force is the same at all three points because the tangential acceleration is constant.

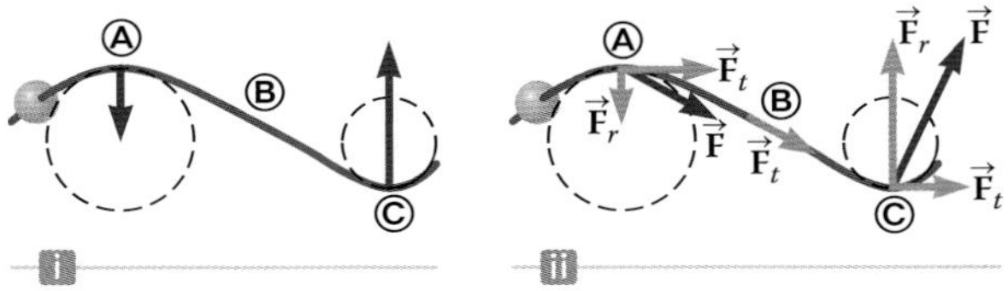

Chapter 6

6.1. (a)
6.2. (c), (a), (d), (b)
6.3. kg.s^{-2}, L/T^2
6.4. (a)
6.5. (b)
6.6. (c)
6.7. **(i)** (c) **(ii)** (a)

Chapter 7

7.1. (a) For the television set, energy enters by electrical transmission (through the power cord). Energy leaves by heat (from hot surfaces into the air), mechanical waves (sound from the speaker), and electromagnetic radiation (from the screen). (b) For the petrol-powered lawn mower, energy enters by matter transfer (petrol). Energy leaves by work (on the blades of grass), mechanical waves (sound), and heat (from hot surfaces into the air). (c) For the hand-cranked pencil sharpener, energy enters by work (from your hand turning the crank). Energy leaves by work (done on the pencil), mechanical waves (sound), and heat due to the temperature increase from friction.
7.2. **(i)** (b) **(ii)** (b) **(iii)** (a)
7.3. (a)
7.4. $v_1 = v_2 = v_3$
7.5. (c)
7.6. (c)

Chapter 8

8.1. (d)
8.2. (b), (c), (a)
8.3. When he pushes for the same distance, he does the same amount of work on each, as $W = Fd$, and this will cause the change in the kinetic energy of each boy. Hence (e) is true. Twin 2 is heavier, so will have a smaller velocity, $p = 2K/v$, so (a) is also true.
If he pushes for the same time and with the same force, at the end of the time each twin will have the same momentum, as $\Delta p = F\Delta t$, hence in this case (b) is true. Twin 1 will be moving faster as he has the smaller mass, so in this case he will have the greater kinetic energy, so (f) is also true.

8.4 (a) all are equal (b) F(dashboard) > F (seat belt) > F(air bag)
8.5 (a)
8.6 (b)
8.7 **(i)** (a) **(ii)** (b)

Chapter 9

9.1. **(i)** (c) **(ii)** (b)
9.2. (b)
9.3. **(i)** (b) **(ii)** (a)
9.4. $-\hat{\mathbf{k}}, -\hat{\mathbf{i}}, -\hat{\mathbf{j}}$
9.5. **(i)** (b) **(ii)** (a)
9.6. (b)
9.7. (a)
9.8. (b)

Chapter 10

10.1. (b)
10.2. **(i)** (a) **(ii)** (c)
10.3. (b)
10.4. (a)

Chapter 11

11.1. $m^3/(kg.s^2)$, $L^3M^{-1}T^{-2}$
11.2. At the surface of the Earth, $F = (GmM_E)/(R_E{}^2) = mg$, so $g = (GM_E)/(R_E{}^2)$
11.3. (e)
11.4. (c)
11.5. The potential energy is proportional to the integral of the force, so we expect the potential energy to vary with $1/r$.
11.6. (a) closest (b) furthest (c) closest (d) all points
11.7.

Body	T^2/r^3 (s^2/m^3)
Mercury	2.98×10^{-19}
Venus	2.99×10^{-19}
Earth	2.97×10^{-19}
Mars	2.98×10^{-19}
Jupiter	2.97×10^{-19}
Saturn	2.95×10^{-19}
Uranus	2.97×10^{-19}
Neptune	2.94×10^{-19}

The value for the solar system is from this data is $(2.97 \pm 0.02) \times 10^{-19}$ s^2/m^3.

Chapter 12

12.1. (d)
12.2. (d)
12.3. (a)
12.4. (d)
12.5. **(i)** (c) **(ii)** (a)
12.6. **(a)** $v_1 = v_3 < v_2$
(b) $K_1 < K_2 = K_3$
(c) $m_1 = m_2 < m_3$

Chapter 13

13.1. (a)
13.2. (a)
13.3. **(i)**(b) **(ii)**(a)
13.4. (c)
13.5. (b)
13.6. The density of saltwater is greater than the density of freshwater.
13.7. The larger bubble will expand and the smaller bubble contract until it disappears.

Chapter 14

14.1. (a)
14.2. Yes

Chapter 15

15.1. (c)
15.2. (a)
15.3. (c)
15.4. (a)
15.5. **(i)** (b) **(ii)** (a) **(iii)** (c)

Chapter 16

16.1. (d)
16.2. Units: $kg.rad^2/s^2$, dimensions: MT^{-2}.
16.3. (f)
16.4. (a)
16.5. (b)
16.6. (c)
16.7. The change in the elastic potential energy is proportional to the square of the displacement, and the change is positive, while the gravitational potential energy decreases linearly with the change in position (a negative change). Further, when you release the weight after pulling it down, the weight and spring oscillate, so the net change in energy must be positive. When the weight passes through the equilibrium position it has the same elastic potential energy as before you pulled down on it, but now it also has additional kinetic energy – a net increase in the total energy of the system. Hence, the increase in elastic potential energy must have been greater than the decrease in gravitational potential energy.
16.8. **(i)** (a) **(ii)** (a)
16.9. If $I = md^2$, then Equation 16.27 reduces to Equation 16.25, with the length L equal to d.
16.10. Units: kg/s

Chapter 17

17.1. (b)
17.2. **(i)** (c) **(ii)** (b) **(iii)** (d)
17.3. (c)
17.4. (c)
17.5. The pulse will move more slowly along the tube containing the water because it has a much greater mass per unit length.
17.6. A transverse wave cannot propagate through a gas because there is no force holding the elements of the gas together at some equilibrium distance. When a shear force is applied, the molecules simply move but are not 'pulled back' towards their previous position because there is no restoring force between them.
17.7. (e)
17.8. (b)
17.9. (d)
17.10. (b)

Chapter 18

18.1. **(i)** If the diameter increases by a factor of two then the area, and hence the linear mass density, increases by a factor of 4. The wave speed is inversely proportional to the square root of the mass density, so the velocity decreases by a factor of 2 (answer (b)).
(ii) When the speed decreases, the reflected wave is inverted (answer (b)).
18.2. (b)
18.3. **(i)** λ **(ii)** $\lambda/2$
18.4. **(i)** (a) **(ii)** (d)
18.5. (d)
18.6. (b)
18.7. (c)
18.8. **(i)** The listener hears (c), the average frequency of the two, 440 Hz.
(ii) The beat frequency is equal to the difference in the frequencies, so (b) 4 Hz.

Chapter 19

19.1. (c)
19.2. (c)
19.3. $[ML^2T^{-2}\Theta^{-1}]$
19.4. (b)
19.5. (c)
19.6. **(a)** iron, glass, water **(b)** water, glass, iron

Chapter 20

20.1. (b)
20.2. (b)
20.3. (b)
20.4. (a)
20.5.

Situation	System	Q	W	ΔE_{int}
(a) Rapidly pumping up a bicycle tyre	Air in the pump	0	–	–
(b) Pan of room-temperature water sitting on a hot stove	Water in the pan	0	–	–
(c) Air quickly leaking out of a balloon	Air originally in the balloon	0	–	–

20.6. Path A is isovolumetric, path B is adiabatic, path C is isothermal, and path D is isobaric.

Chapter 21

21.1. **(i)** (b) **(ii)** (a)
21.2. **(i)** (a) **(ii)** (c)
21.3. (d)
21.4. (c)

Chapter 22

22.1. **(i)** (c) **(ii)** (b)
22.2. (d)
22.3. C, B, A
22.4. **(a)** one **(b)** six
22.5. (a)
22.6. False

Answers to odd-numbered problems

Chapter 1

1. (a) Making d three times larger with d^2 in the bottom of the fraction makes Δt nine times smaller
   (b) Δt is inversely proportional to the square of d.
   (c) Plot Δt on the vertical axis and $1/d^2$ on the horizontal axis.
   (d) From the last version of the equation, the slope is $4QL / k\pi(T_h - T_c)$. Note that this quantity is constant as both Δt and d vary.
3. (a) $5.52 \times 10^3 \text{kg/m}^3$ (b) Between the tabulated densities of aluminium and iron
5. 7.69cm
7. (a) Incorrect (b) Correct
9. 871 m^2
11. 151 μm
13. 2.86cm
15. 1.67×10^6
17. (a) $5.12 \times 10^{29} \approx$ bacteria (b) 10^{14} kg
19. $P = (62.2 \pm 0.6)$ cm
21. The boy's height at the first measurement is (1.45 ± 0.01) m or between 1.44m and 1.46m. At the second measurement the height is (1.46 ± 0.01) or between 1.45m and 1.47m. As these ranges overlap, it is possible that boy has not grown in the time between measurements, as the results are not significantly different.
23. (a) $R = (15.47 \pm 0.05)\ \Omega$ (b) $\rho = (2.5 \pm 0.2) \times 10^{-4}\ \Omega$.m
25. (a) 796 (b) 1.1 (c) 17.66
27. 316m
29. 10^{11} stars
31. 3.41 m

Chapter 2

1. (a) 5m/s (b) 1.2m/s (c) −2.5m/s (d) −3.3m/s (e) 0m/s
3. (a) 3.75m/s (b) 0
5. (a) 27.0m (b) $27.0\text{ m} + (18.0\text{ m/s})\Delta t + (3.00\text{ m/s}^2)(\Delta t)^2$
   (c) 18.0m/s
7. (a) 2.80 h (b) 218 km
9. (a)

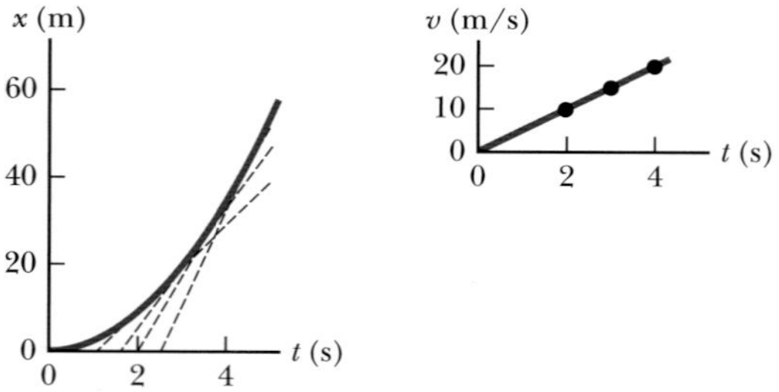

   (b) At $t = 5.0$ s: 23 m/s; At $t = 4.0$ s: 18 m/s; At $t = 3.0$ s: 14 m/s; At $t = 2.0$ s: 9.0 m/s (c) 4.6 m/s² (d) zero
11. (a) 20m/s; 5m/s (b) 263m
15. (a) 6.61m/s (b) −0.448 m/s²
17. As in the algebraic solution to Example 2.8, we let t represent the time the police officer has been moving. We graph $x_{car} = 45 + 45t$ and $x_{police\ officer} = 1.5t^2$.

   They intersect at $t = 31$ s

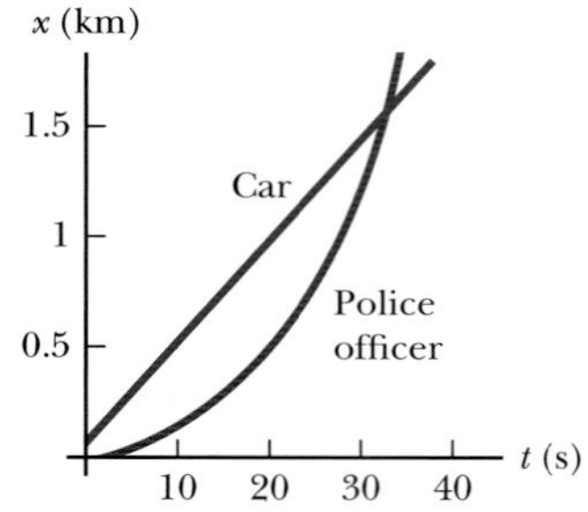

19. 3.10 m/s
21. (a) 1.88km (b) 1.46km (c) 3.3 m/s²; $a_2 = 0$; −5.0 m/s²
   (d) $x_1 = (1.67\text{ m/s}^2)t^2$; $x_2 = (50\text{ m/s})t - 375\text{ m}$;
   $x_3 = (250\text{ m/s})t - (2.5\text{ m/s}^2)t^2 - 4\,375\text{ m}$ (e) 37.5m/s
23. (a & b) The rock does reach the top of the wall with $v_f = 3.69$m/s (c) 2.39m/s (d) Does not agree (e) The upward-moving rock spends more time in flight because its average speed is smaller than the downward-moving rock, so the rock has more time to change its speed.
25. (a) 7.82m (b) 0.782s
27. (a) $a_x(t) = a_{xi} + Jt$; $v_x(t) = v_{xi} + a_{xi}t + \frac{1}{2}Jt^2$;
   $x(t) = x_i + v_{xi}t + \frac{1}{2}a_{xi}t^2 + \frac{1}{6}Jt^3$ (b) $a_x^2 = a_{xi}^2 + 2J(v_x - v_{xi})$
29. (a) 4.00m/s (b) 1.00ms (c) 0.816m
31. (a) 3.00s (b) 15.3m/s (c) 31.4m/s; 34.8m/s
33. (a) In order for the trailing athlete to be able to catch the leader, his speed (v_1) must be greater than that of the leading athlete (v_2), and the distance between the leading athlete and the finish line must be great enough to give the trailing athlete sufficient time to make up the deficient distance, d
   (b) $\dfrac{d}{(v_1 - v_2)}$ (c) $d_2 = \dfrac{v_2 d}{v_1 - v_2}$
35. (a) $v \cot\theta$ (b) As A moves toward the origin, θ goes from $\sim 0° \rightarrow 90°$, $\cot\theta$ decreases from a very large value to zero, so v_B increases from to zero. At $\theta = 45°$, $v_B = v$.
37. (a) Laura: 5.32 m/s²; Helen: 3.75 m/s² (b) Laura: 10.6 m/s; Helen: 11.2 m/s (c) Laura is ahead by (53.19 m − 50.56 m) = 2.63 m (d) In the time interval from the 2.00s mark to the 3.00s mark, the distance between them will be the greatest; Laura is ahead of Helen by 4.47m.
39 (a) Intuitively, path D is longer but has greater acceleration, so which path is faster will depend on whether the length or the acceleration decreases more.
   (b) D is vertical, so $D = \frac{1}{2}gt^2$; Rearranging for t: $t = \sqrt{\dfrac{2D}{g}}$

(c and d) We need to define some angles:

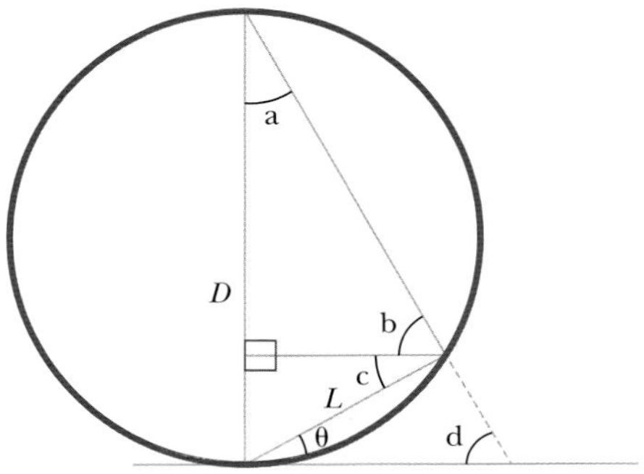

Now we can see that θ = c, and b = d. Using the triangles: 180° – (b + c) = (θ + d) = (b + c). So 180° = 2(b + c), therefore b + c = 90°. Hence the angle within the triangle in the circle is a right angle. For path L: $L = D\sin(a)$, and the component of gravity in the direction of L is: $g_L = g\sin(a)$.

So $D\sin(a) = \frac{1}{2} g \sin(a) t^2$ And we again have $t = \sqrt{\frac{2D}{g}}$.

Hence the time is the same for both paths. (e) So was your guess correct?

Chapter 3

1. -2.75 m; -4.76 m
3. (a) 47.2 units; 122°
5. (a) 4.87 km at 28.6°S of W (b) 23.3 m/s (c) 13.5 m/s along R
7. (a) $\vec{\mathbf{r}} = 18.0t\hat{\mathbf{i}} + (4.00t - 4.90t^2)\hat{\mathbf{j}}$, where $\vec{\mathbf{r}}$ is in metres and t is in seconds
(b) $\vec{\mathbf{v}} = 18.0\hat{\mathbf{i}} + [4.00 - (9.80)t]\hat{\mathbf{j}}$, where $\vec{\mathbf{v}}$ is in metres per second and t is in seconds
(c) $\vec{\mathbf{a}} = -9.80\hat{\mathbf{j}}$ m/s^2
(d) $(54.0\,\text{m})\hat{\mathbf{i}} - (32.1\,\text{m})\hat{\mathbf{j}}$; $(18.0\,\text{m/s})\hat{\mathbf{i}} - (25.4\,\text{m/s})\hat{\mathbf{j}}$; $(-9.80\,\text{m/s}^2)\hat{\mathbf{j}}$
9. (a) 0.800 m/s^2; -0.300 m/s^2 (b) 339° from $+x$ axis (c) 360 m; -72.7 m; $-15.2°$
11. (a) 2.81 m/s (b) The mug's velocity is 60.2° below the horizontal when it strikes the ground
13. (a) 3.96 m/s (b) Porpoising reduces the time interval by 9.6% (c) The same on every planet
15. $x = 7.23 \times 10^3$ m; $y = 1.68 \times 10^3$ m
17. 1.21 s
19. $v = 7.58 \times 10^3$ m/s; $T = 96.7$ min
21. $a = (340\,280 \pm 50)$ m/s^2
23. (a) 13.0 m/s^2 (b) 5.70 m/s (c) 7.50 m/s^2
25. (a) 15 km/h, east (b) 15 km/h, west (c) 60 s
27. (a) 57.7 km/h at 60.0° west of vertical (b) 28.9 km/h downward
29. (a) 101 m/s (b) 3.27×10^4 ft (c) 20.6 s
31. (a) 1.4 m (b) 2.3 m
33. $x = 18.8$ m; $y = -17.3$ m
35. 1.69 km/s (b) 1.80 h
37. (a) $\Delta x = (6.8 \pm 0.4)$ km (b) 3000 m directly above the bomb
39. (a) 20.0 m/s (b) 5.00 s (c) 31.5 m/s at 59.4° below the horizontal (d) 6.53 s (e) 24.5 m
41. (a) $\Delta t_1 = \frac{L}{c+v} + \frac{L}{c-v} = \frac{2L/c}{1 - v^2/c^2}$
(b) $\Delta t_2 = \frac{2L}{\sqrt{c^2 - v^2}} = \frac{2L/c}{\sqrt{1 - v^2/c^2}}$
(c) Since the term $\left(1 - \frac{v^2}{c^2}\right) < 1$, $\Delta t_1 > \Delta t_2$, so Sarah, who swims cross-stream, returns first
43 $\theta = \tan^{-1}\left(\frac{v_{yf}}{v_{xf}}\right) = \tan^{-1}\left(\frac{\sqrt{2gh}}{v}\right)$

Chapter 4

1. (a) -4.47×10^{-15} m/s^2 (b) $+2.09 \times 10^{-10}$ N
3. 8.71 N
5. (a) Force exerted by spring on hands, to the left and right (b) Force exerted by wagon on handle, downward to the left. Force exerted by wagon on planet, upward. Force exerted by wagon on ground, downward (c) Force exerted by football on player, downward to the right. Force exerted by football on planet, upward (d) Force exerted by small-mass object on large-mass object, to the left (e) Force exerted by negative charge on positive charge, to the left. (f) Force exerted by iron on magnet, to the left
7. (a) $t = \sqrt{\frac{2h}{g}}$ (b) $\frac{F}{m}$ (c) $\frac{Fh}{mg}$ (d) $\sqrt{\left(\frac{F}{m}\right)^2 + g^2}$
9. (a) $a = 0.49$ m/s^2, $\mathbf{a} = (0.40 \pm 0.02)\mathbf{i} + (0.28 \pm 0.02)\mathbf{j}$
(b) $a = 0.59$ m/s^2, $\mathbf{a} = (0.54 \pm 0.02)\mathbf{i} + (0.24 \pm 0.01)\mathbf{j}$
(c) (0.68 ± 0.03)m/s^2 or (0.12 ± 0.02)m/s^2
11. (a) â is at 181° (b) 11.2 kg (c) 37.5 m/s (d) $(-37.5\,\hat{\mathbf{i}} - 0.893\,\hat{\mathbf{j}})$ m/s
13. (a) 7.0 m/s^2 horizontally to the right (b) 21 N (c) 14 N horizontally to the right
15. (b) -2.54 m/s^2 (c) 3.18 m/s
17. (b) 3000N (c) In this case, the weight of the wire will be comparable to that of the bird, so approximating the wire as an ideal (massless) rope is not a good approximation. Mains wire typically has a mass/length of about 50kg/km.
19. (a) 8.00 kg or 78.4 N (b) 105 N
21. (b) 6.30 m/s^2 (c) 31.5N
23. (a) $m(g + a)$, $2m(g + a)$ (b) Upper string breaks first (c) 0; 0
25. (a) 706 N (b) 814 N (c) 706 N (d) 648 N
27. (a) 3.60 m/s^2 (b) $T = 0$ (c) Someone in the car (non-inertial observer) claims that the forces on the mass along x are T and a fictitious force ($-Ma$) (d) Someone at rest outside the car (inertial observer) claims that T is the only force on M in the x-direction
29. (a) 491 N (b) 50.1 kg (c) 2.00 m/s^2
31. (a) $T = \left(M + \frac{m}{L}\ell\right)g$
33. (b) 0.408 m/s^2 (c) 83.3 N
35. (b) $T_1 = T_2 = T_3 = \frac{Mg}{2}$; $T_4 = \frac{3Mg}{2}$; $T_5 = Mg$ (c) $F = \frac{Mg}{2}$
37. $F = (M + m_1 + m_2)\frac{m_1 g}{m_2}$
39. (a) 30.7° (b) 0.843 N
41. (a) $T_1 = \frac{2mg}{\sin\theta_1}$; $\frac{2mg}{\tan\theta_1} = T_3$; $T_2 = \frac{mg}{\sin[\tan^{-1}(\frac{1}{2}\tan\theta_1)]}$
(b) $\theta_2 = \tan^{-1}\left(\frac{\tan\theta_1}{2}\right)$
43. $R = mg\cos\theta\sin\theta$ to the right + $(M + m\cos^2\theta)g$ upward

Chapter 5

1. (a)

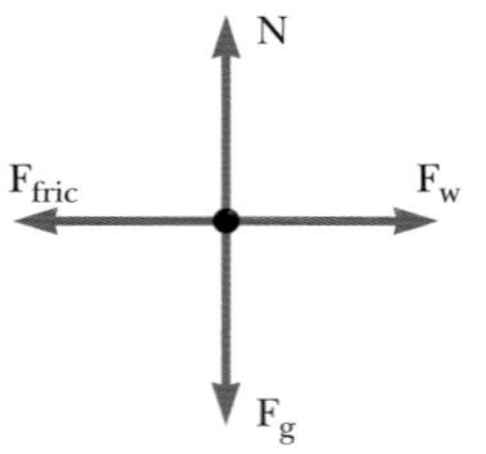

$N = -F_g = mg$

$F_{fric} = \mu_R m$

(b) $F_{friction}$ of the marble surface on the student and the frictional force of the student on the marble; F_w, the force of the air on the student, and the force of the student on the air; N_1, the normal force of the marble on the student and the normal (downwards) force of the student on the marble; the gravitational attraction of the Earth on the student and the gravitational attraction of the student on the Earth (c) $v(x) = \sqrt{\frac{2x}{m}(F_w - \mu_k m)}$

3. (a) $\mu_s = 0.31 \pm 0.01$ (b) $\mu_k = 0.24 \pm 0.01$

5. (a) 4.18 (b) Time would increase as the wheels would skid and only kinetic friction would act; or perhaps the car would flip over.

7. (a) The free-body diagrams for each object appear at right (b) 2.31 m/s^2, down for m_1, left for m_2, and up for m_3 (c) $T_{12} = 30.0$ N; $T_{23} = 24.2$ N (d) T_{12} decreases; T_{23} increases (friction disappears ($\mu_k = 0$), and so the acceleration would become larger)

9. (a) 48.6 N; 31.7 N (b) If $P > 48.6$ N, the block slides up the wall If $P < 31.7$ N, the block slides down the wall. (c) 62.7 N

11. 0.0600 m

13.

$$-k(t - 0) = \left.\frac{v^{-1}}{-1}\right|_{v_i}^{v} = -\frac{1}{v} + \frac{1}{v_i}$$

$$\frac{1}{v} = \frac{1}{v_i} + kt = \frac{1 + v_i kt}{v_i}$$

$$v = \frac{v_i}{1 + v_i kt}$$

15. (a) $b = (1.47 \pm 0.16)$N.s/m (b) $t = (2.0 \pm 0.2) \times 10^{-3}$s (c) $R = (2.94 \pm 0.02) \times 10^{-2}$N

17. 37 m/s or 133 km/h

19. 10^1 N

21. (a) 8.33×10^{-8} N inward (b) 9.15×10^{22} m/s^2 inward

23. (a) -0.233 m/s$^2\,\hat{\mathbf{i}} + 0.163$ m/s$^2\,\hat{\mathbf{j}}$ (b) 6.53m/s (c) -0.181 m/s$^2\hat{\mathbf{i}} + 0.181$ m/s$^2\hat{\mathbf{j}}$

25. 14.3m/s

27. (a) The gravitational force and the contact force exerted on the water by the bucket. (b) Contact force exerted by the bucket. (c) 3.13m/s (d) The water would follow the parabolic path of a projectile.

29. (a) 2.49×10^4 N (b) 12.1m/s

31. (a) 6.17m (b) Mg, downward (c) 6.05 m/s^2 (d) The normal force would have to point away from the centre of the curve. Unless the riders have belts, they will fall from the cars. In a teardrop-shaped loop, the radius of curvature r decreases, causing the centripetal acceleration to increase. The speed would decrease as the car rises (because of gravity), but the overall effect is that the required centripetal force increases, meaning the normal force increases. There is less danger if not wearing a seatbelt.

33. (b) $(\cos\theta = \mu_s \sin\theta) > 0 \rightarrow \mu_s \tan\theta < 1 \rightarrow \tan\theta < \frac{1}{\mu_s}$ If this condition is not met, no value of P can move the crate

35. $T_2 = 781$ N

37. $\frac{n\sin\theta}{n\cos\theta} = \frac{mv^2/r}{gr} \rightarrow \tan\theta = \frac{v^2}{gr} \rightarrow v^2 = gr\tan\theta$

$\rightarrow v^2 = g(L\cos\theta)\tan\theta \rightarrow v = (gL\sin\theta)^{1/2}$

39. (a) 217N (b) 283N (c) From above, $T_2 > T$ always, so string 2 will break first.

41. (a) 0.0162kg/m (b) $\frac{1}{2}D\rho A$ (c) 0.778 (d) For stacked coffee filters falling at terminal speed, a graph of air resistance force as a function of squared speed demonstrates that the force is proportional to the speed squared. This proportionality agrees with that described by the theoretical equation $R = \frac{1}{2}D\rho A v^2$ The value of the constant slope of the graph implies that the drag coefficient for coffee filters is $D = 0.78$.

43. (a) 2.63 m/s² (b) 201 m (c) 17.7 m/s

45. (a) 5.19 m/s (c) 444 N

47. (a)

t(s)	*d(m)*
1.00	4.88
2.00	18.9
3.00	42.1
4.00	43.8
5.00	112
6.00	154
7.00	199
8.00	246
9.00	296
10.0	347
11.0	399
12.0	452
13.0	505
14.0	558
15.0	611
16.0	664
17.0	717
18.0	770
19.0	823
20.0	876

(c) 53.0 m/s

49. 12.8 N

51. 0.0928°

53. (a) $x = \frac{1}{k}\ln(1 + v_i kt)$ (b) $v_i e^{-kx} = v$

55. (a) $v = 0.0132$ m/s (b) 1.03 m/s (c) 6.87 m/s

Chapter 6

1. (a) 31.9 J (b & c) 0 (d) 31.9 J

3. (a) 472 J (b) 2.76 kN

5. -470×10^3 J; or -4.70 kJ

7. (a) 7.50 J (b) 15.0 J (c) 7.50 J (d) 30.0 J

9. (a) $y_2 = (0.94 \pm 0.04)$cm (b) work $= (1.25 \pm 0.05)$J

11. (a)

F (N)	L (mm)	F (N)	L (mm)
0.00	0.00	12.0	98.0
2.00	15.0	14.0	112
4.00	32.0	16.0	126
6.00	49.0	18.0	149
8.00	64.0	20.0	175
10.0	79.0	22.0	190

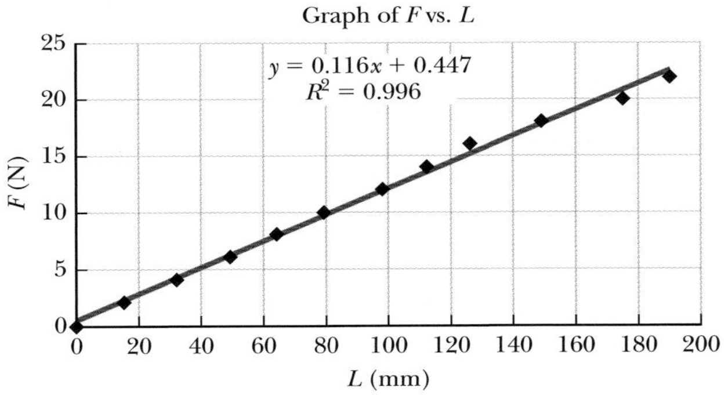

(b) 116 N/m (c) To draw the straight line we use all the points listed and also the origin. If the coils of the spring touched each other, a bend or nonlinearity could show up at the bottom end of the graph. If the spring were stretched 'too far', a nonlinearity could show up at the top end. But there is no visible evidence for a bend in the graph near either end (d) $k = 116$ N/m (e) 12.2 N

13. 50.0 J
15. 0.299 m/s
17. (a) 1.20 J (b) 5.00 m/s (c) 6.30 J
19. 8.78×10^5 N; the force on the pile driver is upward
21. (a) 55.5 J (b) 64.5 J
23. (a) $U_g = (800 \pm 80)$ J (b) $U_g = (107 \pm 11)$ J (c) $U_g = 0$
27. $(7 - 9x^2y)\hat{\mathbf{i}} - 3x^3\hat{\mathbf{j}}$
31. $P = 1.23$ kW
33. (a) $P = 1.02 \times 10^4$ W (b) $P = 1.06 \times 10^4$ W (c) 5.82×10^6 J
35. (a) $1.80 = b$; $4.01 \times 10^4\,\text{N/m}^{1.8} = a$ (b) 295 J
37. (a) $(3x^2 - 4x - 3)\hat{\mathbf{i}}$ (b) 1.87 and -0.535 (c) The stable point is at $x = -0.535$, point of minimum $U(x)$; the unstable point is at $x = 1.87$, maximum in $U(x)$
39. $$x = \frac{g\sin\theta + \sqrt{(g\sin\theta)^2 + \left(k/m\right)[v^2 + 2(g\sin\theta)d]}}{k/m}$$
41. 5×10^5 N
43. (a) $-2kx\left(1 - \dfrac{L}{\sqrt{x^2+L^2}}\right)\hat{\mathbf{i}}$ (b) $kx^2 + 2kL\left(L - \sqrt{x^2+L^2}\right)$ (c) $x = 0$ (d) $v = 0.823$ m/s

Chapter 7

1. $h = 10.2$ m
3. (a) $v = \sqrt{3.00gR}$ (b) 0.098 0 N downward
5. (a) -196 J (b) -196 J (c) -196 J (d) The results should all be the same since the gravitational force is conservative
7. (a) Yes, the child-Earth system is isolated because the only force that can do work on the child is her weight. The normal force from the slide can do no work because it is always perpendicular to her displacement. The slide is frictionless, and we ignore air resistance (b) No, because there is no friction. (c) $E = mgh$ (d) $E = \frac{1}{2}mv_i^2 + \frac{mgh}{5}$ (e) $\frac{1}{2}mv_{xi}^2 + mgy_{max}$ (f) $v_i = \sqrt{\frac{8gh}{5}}$ (g) $h_{max} = \left(1 - \frac{4}{5}\cos^2\theta\right)$ (h) No. If friction is present, the mechanical energy of the system would *not* be conserved, so her kinetic energy at all points after leaving the top of the waterslide would be reduced when compared with the frictionless case. Consequently, her launch speed, maximum height reached, and final speed would be reduced as well
9. (a) $v = 1.40$ m/s (b) 4.60 cm from the start (c) $v = 1.79$ m/s
11. (a) 0.791 m/s (b) 0.531 m/s
13. (a) $v_B = 1.65\,\text{m/s}^2$ (b) The red bead slides a greater distance along the curved path, so friction transforms more of its potential energy into internal energy. There is less of the red bead's original potential energy in the form of kinetic energy when it arrives at point B. The result is that the green bead arrives at point B at higher speed
15. $v = (3.7 \pm 0.1)$ m/s
17. (a) 24.5 m/s (b) Yes. This is too fast for safety (c) 206 m (d) The air drag is proportional to the square of the skydiver's speed, so it will change quite a bit. It will be larger than her 784 N weight only after the chute is opened. It will be nearly equal to 784 N before she opens the chute and again before she touches down whenever she moves near terminal speed
19. (a) 168 J
21. (a) $d = 0.378$ m (b) 2.30 m/s (c) $D = 1.08$ m
23. (a) 0.403 m or $-$ 0.357 m (b) From a perch at a height of 2.80 m above the top of a pile of mattresses, a 46.0 kg child jumps upward at 2.40 m/s. The mattresses behave as a linear spring with force constant 19.4 kN/m. Find the maximum amount by which they are compressed when the child lands on them. (c) 0.0232 m (d) This result is the distance by which the mattresses compress if the child just stands on them. It is the location of the equilibrium position of the oscillator
25. (a) 2.17×10^3 W (b) 58.6 kW
27. 236 s (3.93 minutes)
29. $v = 1.24$ m/s
31. (a) Use conservation of energy
33. (a) (627 N)y (b) $\frac{1}{2}(81\,\text{N/m})(39.2\,\text{m} - y)^2$ (c) (627 N)y; (40.5 N/m)y^2 $-$ (2550 N)y + 62 200 J (e) 10.0 m (f) stable equilibrium; 31.5 m (g) 24.1 m/s
35. (b) 7.42 m/s

Chapter 8

1. (a) $K = \frac{1}{2}mv^2 = \frac{1}{2}m\left(\frac{p}{m}\right)^2$ (b) $\sqrt{2mK}$ (c) 10%
3. $\mu_k = 0.0409 \pm 0.0007$
5. 40.5 g
7. 10^{-23} m/s
9. (a) 9.60×10^{-2} s (b) 3.65×10^5 N (c) 26.5 g
11. (a) $K_i + U_{xi} = K_f + U_{sf}$: $0 + \frac{1}{2}kx^2 = \frac{1}{2}mv^2 + 0$ $v = x\sqrt{\frac{k}{m}}$ (b) $I = |\vec{\mathbf{p}}_f - \vec{\mathbf{p}}_i| = mv_f - 0 = mx\sqrt{\frac{k}{m}} = x\sqrt{km}$ (c) No difference
13. (a) 981 N.s, up (b) $v = (3.43 \pm 0.03)$ m/s (c) 3.83 m/s, up (d) $y_{top} = (0.75 \pm 0.4)$ m
15. (a) 20.9m/s east (b) -8.68×10^3 J (c) Most of the energy was transformed to internal energy with some being carried away by sound

17. (a) 0.284 (b) 1.15×10^{-3} J; 4.54×10^{-14}J

19. $v_C = 91.2$ m/s

21. 0.556 m

23. $(3.00\hat{\mathbf{i}} - 1.20\hat{\mathbf{j}})$ m/s

25. 2.50 m/s at $-60.0°$

27. $\vec{\mathbf{r}}_{CM} = (0\hat{\mathbf{i}} + 1.00\hat{\mathbf{j}})$ m

29. (a) 15.9 g (b) 0.153 m

31. 7.00 m

33. (a) Yes. $18.0\hat{\mathbf{i}}$ kg.m/s (b) No. The friction force exerted by the floor on each stationary bit of caterpillar tread acts over no distance, so it does zero work (c) Yes, we could say that the final momentum of the card came from the floor or from the Earth through the floor (d) No. The kinetic energy came from the original gravitational potential energy of the Earth-elevated load system in the amount $KE = \left(\frac{1}{2}\right)(6\text{ kg})(3\text{ m/s})^2 = 27.0$ J (e) Yes. The acceleration is caused by the static friction force exerted by the floor that prevents the wheels from slipping backward

35. (a) Yes (b) No. (c) 103 kg · m/s up (d) Yes. You could say that it came from the planet, that gained momentum 103 kg m/s down, but it came through the force exerted by the floor over a time interval on the person, so it came through the floor or from the floor through direct contact. (e) 88.2 J (f) No. The energy came from chemical energy in the person's leg muscles. The floor did not work on the person

37. (a) 3.90×10^7 N (b) 3.20 m/s^2

39. (a) 2.67 m/s; 10.7 m/s (b) 5.33 m/s; 2.67 m/s (c) part (a): 7.11×10^{-3} J; part (b): 2.84×10^{-2} J

41. (a) $\dfrac{K_B}{K_A} = \dfrac{m_1}{m_1 + m_2}$ (b) 1.00

43. 3.6 m

45. $\left(\dfrac{M+m}{m}\sqrt{\dfrac{gd^2}{2h}}\right)$

47. (a) 0; inelastic (b) $(-0.250\hat{\mathbf{i}} + 0.750\hat{\mathbf{j}} - 2.00\hat{\mathbf{k}})$m/s; perfectly inelastic (c) $(-2.67 - 0.333a)\hat{\mathbf{k}}$ m/s; $-3.58\hat{\mathbf{k}}$ m/s, $a = 2.74$; $-0.419\hat{\mathbf{k}}$ m/s, $a = -6.74$

49. 0.179 m/s

51. 0.0635 L

53. (a) 3.75 N (b) 3.75 N (c) 3.75 N (d) 2.81 J (e) 1.41 J/s (f) One-half of the work input becomes kinetic energy of the moving sand and the other half becomes additional internal energy. The internal energy appears when the sand does not elastically bounce under the hopper, but has friction eliminate its horizontal motion relative to the belt. By contrast, all of the impulse input becomes momentum of the moving sand

Chapter 9

1. 7.27×10^5 rad/s

3. (a) 5.00 rad; 10.0 rad/s; 4.00 rad/s^2 (b) 53.0 rad; 22.0 rad/s; 4.00 rad/s^2

5. (a) 5.24 s (b) 27.4 rad

7. $\theta = (30 \pm 6)$ revolutions

9. $\sim 10^7$ rev/yr

11. (a) 8.00 rad/s (b) 8.00 m/s (c) 64.1 m/s^2 (d) 9.00 rad

13. (a) 3.47 rad/s (b) 1.74 m/s (c) 2.78 s (d) 1.02 rotations

15. $\dfrac{a}{g}\sqrt{1+\pi^2}$

17. 168 N.m (clockwise)

19. (b) This is a class 2 lever (c) The screwdriver has a mechanical advantage of 4, as this is the ratio of the radius at which the effort is applied to the radius at which the force is applied to the load (d) $\tau_{\text{friction}} = \tau_{\text{app}}$

$F_{\text{friction}} = 4F_{\text{app}} = 4(50\text{ N}) = 200\text{ N}.$

21. 1.28 kg.m^2

23. (a) 1.03 s (b) 10.3 rev

25. (a) 21.6 kg.m^2 (b) 3.60 N.m (c) 52.5 revolutions

27. (a) 1.95 s (b) If the pulley were massless, the acceleration would be larger by a factor 35/32.5 and the time shorter by the square root of the factor 32.5/35. That is, the time would be reduced by 3.64%

29. (a) $m_1 = 9.00$ g (b) $m_2 = 52.5$ g (c) $m_3 = 49.0$ g

31. (a) The wall is frictionless, but it does exert a horizontal normal force, n_w. $\sum F_x = f - n_w = 0$ $\sum F_y = n_g - 800\text{ N} - 500\text{ N} = 0$ Taking torques about an axis at the foot of the ladder, $(800\text{ N})(4.00\text{ m})\sin 30.0° + (500\text{ N})(7.50\text{ m})\sin 30.0° - n_w (15.0\text{ cm})\cos 30.0° = 0$

33. (a) 859 N (b) 36.9° to the left and upward

35. $\dfrac{11}{12}mL^2$

37. (a) $\sqrt{\dfrac{3g}{L}}$ (b) $\dfrac{3g}{2L}$ (c) $\vec{\mathbf{a}} = -\dfrac{3}{2}g\hat{\mathbf{i}} - \dfrac{3}{4}g\hat{\mathbf{j}}$

(d) $\vec{\mathbf{F}} = M\vec{\mathbf{a}} = -\dfrac{3}{2}Mg\hat{\mathbf{i}} + \dfrac{1}{4}Mg\hat{\mathbf{j}}$

39. (a) $0 \le 35.3°$ (b) 0.184 m from the moving end

41. $x = (0.90 \pm 0.03)$ m

43. 177 kg

45. (a) $T = 2.71$ kN(b) 2.65 kN (c) You should lift 'with your knees' rather than "with your back." In this situation, with a load weighing only 200 N, you can make the compressional force in your spine about ten times smaller by bending your knees and lifting with your back as straight as possible. (d) In this situation, you can make the compressional force in your spine about ten times smaller by bending your knees and lifting with your back as straight as possible.

47. (a) 2.88 s (b) 12.8 s

49. (b) 60.0°

Chapter 10

1. (a) 143 kg.m^2 (b) 2.57×10^3 J

3. 1.04×10^{-3} J

5. $I = mr^2\left(\dfrac{2gh}{v^2} - 1\right)$

7. (a) 74.3 W (b) 401 W

9. (a) $\dfrac{2}{3}g\sin\theta$ (b) $a_{\text{hoop}} = \dfrac{mgR^2\sin\theta}{2mR^2} = \dfrac{1}{2}g\sin\theta$. The acceleration of the hoop is smaller than that of the disk (c) $\dfrac{1}{3}\tan\theta$

11. (a) 2.38 m/s (b) The centripetal acceleration at the top is $\dfrac{v_2^2}{r} = \dfrac{(2.38\text{ m/s})^2}{0.450\text{ m}} = 12.6\text{ m/s}^2 > g$ Thus, the ball must be in contact with the track, with the track pushing downward on it (c) 4.31 m/s (d) $\sqrt{-140\text{ m}^2/\text{s}^2}$ (e) Never makes it to the top of the loop

13. $(-22.0\text{ kg.m}^2/\text{s})\hat{\mathbf{k}}$

15. $(60.0\text{ kg.m}^2/\text{s})\hat{\mathbf{k}}$

17. (a) 0.433 kg.m^2/s (b) 1.73 kg.m^2/s

19. (a) zero (b) $\dfrac{-mv_i^3 \sin^2\theta\cos\theta}{2g}\hat{\mathbf{k}}$ (c) $\dfrac{-2mv_i^3 \sin\theta\sin\theta}{g}\hat{\mathbf{k}}$ (d) The downward force of gravity exerts a torque in the $-z$ direction

21. (a) $1.57 \times 10^8\,\text{kg.m}^2/\text{s}$ (b) $6.26 \times 10^3\,\text{s} = 1.74$ h

23. (a) 1.91 rad/s (b) 2.53 J/6.44 J

25. (a) $\sum \tau_{ext} = 0$, $L_i = mv\ell$, so $L_{fi} = L_i = mv\ell$ (b) $\dfrac{M}{M+m}$

27. (a) $\omega = \dfrac{2mv_i d}{(M+2m)R^2}$ (b) No; some mechanical energy of the system (the kinetic energy of the clay) changes into internal energy (c) The linear momentum of the system is not constant. The axle exerts a backward force on the cylinder when the clay strikes

29. 5.45×10^{22} N.m

31. (a) 4.00 J (b) 1.60 s (c) 0.800 m

33. $\omega = \sqrt{\dfrac{10}{7}\cdot\dfrac{(R-r)(1-\cos\theta)g}{r^2}}$

35. (a) 0 (b) Since the total angular momentum of the system is initially zero, the total angular momentum remains zero, so the monkey and bananas move upward with the same speed at any instant. (c) The monkey will not reach the bananas. The motions of the monkey and bananas are identical, so the bananas remain out of the monkey's reach—until they get tangled in the pulley.

37. (a) 11.1 m/s (b) $5.32 \times 10^3\ \text{kg.m}^2/\text{s}$ (c) The wheels on his skateboard prevent any tangential force from acting on him. Then no torque about the axis of the channel acts on him and his angular momentum is constant. His legs convert chemical into mechanical energy. They do work to increase his kinetic energy. The normal force acts in the upward direction, perpendicular to the direction of motion of the skateboarder (d) 12.0 m/s (e) 1.08 kJ

39. (a) $\vec{\omega} = 2.11\hat{\mathbf{j}}$ rad/s (b) We take the x axis east, the y axis up, and the z axis south. Marcus has moment of inertia 0.730 kg.m² about the axis of the stool and is originally turning counterclockwise at 2.40 rad/s. At a point 0.350 m to the east of the axis, he catches a 0.120 kg apple moving toward the south at 4.30 m/s. He continues to hold the apple in his outstretched arm. Find his final angular velocity (c) Yes, with the left-hand side representing the final situation and the right-hand side representing the original situation, the equation describes the process of Marcus throwing the apple to Laurence.

41. The flywheel can be shaped like a cup or open barrel, 9.00 cm in outer radius and 7.68 cm in inner radius, with its wall 6 cm high, and with its bottom forming a disk 2.00 cm thick and 9.00 cm in radius. It is mounted to the crankshaft at the centre of this disk and turns about its axis of symmetry. Its mass is 7.27 kg. If the disk were made somewhat thinner and the barrel wall thicker, the mass could be smaller

43. $P = 3/8F_g$

45. $\dfrac{M}{m}\sqrt{3ga\left(\sqrt{2}-1\right)}$

Chapter 11

1. (a) 2.50×10^{-7} N
(b) 2.45 m from the 500-kg object toward the smaller object.

3. $F = (7.41 \pm 0.16) \times 10^{-10}$ N

5. $m_1 = 3.00\,\text{kg}$, so $m_2 = 2.00\,\text{kg}$

7. $\Delta g = 2.62 \times 10^{12}$ N/kg

9. $\dfrac{\rho M}{\rho E} = \dfrac{2}{3}$

11. 2.82×10^9 J

13. -2.08×10^{13} J

15. 1.66×10^4 m/s

17. $\sqrt{2}v$

19. 15.6 km/s

21. (a) 0.71 yr (b) This trip cannot be taken at just any time. The departure must be timed so that the spacecraft arrives at the aphelion when the target planet is located there

23. 35.2 AU

25. $\omega = 1.63 \times 10^4$ rad/s

27. $x = 3.06 \times 10^{-8}$ m

29. (a) $v_i = \left(\dfrac{GM_E}{r}\right)^{1/2}$ (b) $\dfrac{5}{4}\left(\dfrac{GM_E}{r}\right)^{1/2}$ (c) $r_f = \dfrac{25r}{7}$

31. $\dfrac{\Delta g_M}{g} = 2.25 \times 10^{-7}$

33. (a) $\dfrac{dg}{dr} = -\dfrac{2GM_E}{R_E^3}$ (b) $|\Delta g| = \dfrac{2GM_E h}{R_E^3}$ (c) $1.85 \times 10^{-5}\,\text{m/s}^2$

35. (a) 2×10^8 yrs (b) About 10^{41} kg (c) On the order of 10^{11}

37. (a) 7.79×10^3 m/s (b) 7.85×10^3 m/s (c) -3.04×10^9 J (d) -3.08×10^9 J (e) 4.69×10^7 J (f) One component of the gravitational force pulls forward on the satellite

39. $v_i = \sqrt{\dfrac{GM_E}{4R_E}}$

41. 1.82×10^{-2} s

43. To three-digit precision, the solution is 1.48×10^9 m

Chapter 12

1. (a) 0.436 m (b) less than 0.436 m

3. 0.917 c

5. (a) 20.0 m (b) 19.0 m (c) 0.312 c

7. 0.800 c

9. 1.55 ns

11. (a) $v = 0.943c$ (b) 2.55×10^3 m The later pulse is to the left of the origin

13. Event B occurred earlier. The time elapsed between the events was 444 ns

15. $0.960c$

17. (a) 2.83×10^8 m/s (b) The result would be the same

19. (a) $0.141c$ (b) $0.436c$

21. 3.74×10^5 MeV

23. (a) 2.92×10^{-14} J or 183 keV (b) 7.24×10^{-13} J or 4.47 MeV

25. 1.62×10^3 MeV/c

27. (a) 4.08 MeV (b) 29.6 MeV

29. (a) Isolated (b) Isolated system: conservation of energy, and isolated system: conservation of momentum (c) 6.22, 2.01 (d) $3.09m_1 + m_2 = 1.66 \times 10^{-27}$ kg (e) $m_2 = 3.52m_1$ (f) $m_1 = 2.51 \times 10^{-28}$ kg, $m_2 = 8.84 \times 10^{-28}$ kg

31. (a) smaller (b) 3.18×10^{-12} kg (c) It is too small a fraction of 9.00 g to be measured

33. 4.28×10^9 kg/s

35. (a) $(1 - 1.12 \times 10^{-10})c$ (b) 6.00×10^{27} J (c) \$$1.83 \times 10^{20}$

37. (a) $\sim 10^2$ s or 10^3 s (b) $\sim 10^8$ km

39. (a) 0.905 MeV (b) 0.394 MeV (c) 3.99×10^{-22} kg·m/s (d) 65.4°

41. (a) $\dfrac{2d}{c+v}$ (b) $\dfrac{2a}{c}\sqrt{\dfrac{c-v}{c+v}}$

43. (a) 76.0 minutes (b) 52.1 minutes

45. (a) Use conservation of energy and momentum, remembering to use the quadratic expression for total energy (b) Remember the total energy is the sum of the kinetic and rest energies

Chapter 13

1. (a) -4×10^{17} kg/m^3 (b) Atoms are mostly empty space, and therefore all matter (solids and liquids as well as gases) is mostly empty space

3. They push as much air out as possible by squeezing the suction cups. This allows the outside air pressure, which is higher, to stick the cups to the surface. This is because more air molecules are pushing down compared to those pushing out

5. (a) $(5.88 \pm 0.06) \times 10^6$ N down (b) 196 ± 2 kN outward (c) 196 ± 2 kN outward

7. 225 N

9. (a) 65.1 N (b) $F = 275$ N

11. (a) 29.4kN (to the right) (b) $\tau = 16.3$ kN.m counterclockwise

13. 0.986×10^5 Pa

15. (a) $P = P_0 + \rho gh$. (b) Mg/A

17. (a) $m_{\text{payload}} = 444$ kg (b) $m_{\text{payload}} = 480$ kg

19. (a) $F_{\text{top}} = P_{\text{top}} A = 1.0179 \times 10^3$ N and $F_{\text{bot}} = 1.0297 \times 10^3$ N (b) 86.2 N (c) 11.8 N

21. $m = 2.67 \times 10^3$ kg

23. (a) 3.7 kN (b) 1.9 kN (c) Atmospheric pressure at this high altitude is much lower than at Earth's surface, so the balloons expanded and eventually burst

25. (a) $L = 11.6$ cm (b) $\rho_0 = 0.963$ g/cm^3 (c) No; the density r is not linear in h

27. Easier

29. (a) As R gets bigger, ΔP decreases. (b) $P_{2.5 \times 10^{-3}\,\text{m}} = 58.24$ Pa; $P_{2.5 \times 10^{-2}\,\text{m}} = \frac{1}{10} P_{5 \times 10^{-3}\,\text{m}} = 5.824$ Pa

31. (a) $P = 1.02 \times 10^7$ Pa (b) 6.61×10^5 N

33. (a) 6.70 cm (b) 5.74 cm

35. $F_g = F'_g + \left(V - \frac{F'_g}{\rho g}\right)\rho_{\text{air}}\, g$

37. 1.28×10^4 m^2. The acceleration of gravity does not affect the answer.

39. (a) $\tau = \int_0^H y\left[\rho g(H - y)w\,dy\right] = \frac{1}{6}\rho gwH^3$

(b) $\frac{1}{6}\rho\, gwH^3 = y_{\text{eff}}\left[\frac{1}{2}\rho\, gwH^2\right]$ and $y_{\text{eff}}\ \frac{1}{3}H$

41 $\ln\left(\frac{P}{P_0}\right) = -\frac{\rho_0 gy}{P_0}$ where $\alpha = \frac{\rho_0 g}{P_0}$, $P = P_0 e^{-\alpha y}$

Chapter 14

1. $\frac{A_{\text{tot}}}{A_1} = 4.3 \times 10^{14}$ capillaries

3. (a) 0.471 m/s (b) 4.24 m/s

5. (a) 0.638 m/s (b) 2.55 m/s (c) 1.80×10^3 m^3/s

7. $d = 0.247$ cm

9. 2.25 m (above the level where the water emerges)

13. (a) 1.91 ± 0.06 m/s (b) $(8.6 \pm 0.3) \times 10^{-4}$ m^3/s

15. =740 N

17. 0.3 Pa s.

19. a = 2, b = 1, c = 2

21. $\Delta P = 662$ Pa

23. 4480 Pa

27. (a) 4.43 m/s (b) 10.1 m

29. At high speed, the flow becomes turbulent, resulting in flow in directions other than opposite the desired direction of motion

31. (a) 6.80×10^4 Pa (b) Higher

33. 347 m/s

35. (a) 8.04 m/s (b) Gravitational force and buoyant force (c) The net upward force on the ball brings its downward motion to a stop; 4.18 m (d) 8.04 m/s (e) The time intervals are equal (f) With friction present, Δt_{down} is less than Δt_{up}. The magnitude of the ball's acceleration on the way down is greater than its acceleration on the way up. The two motions cover equal distances and both have zero speed at one end point, so the downward trip with larger-magnitude acceleration must take less time

37. (a) 0.533 s (b) 14.5 m/s (c) 0.145 m/s (d) $P_2 = 1.013 \times 10^5$ Pa (e) $P_1 = 2.06 \times 10^5$ Pa (f) 33.0 N

39.

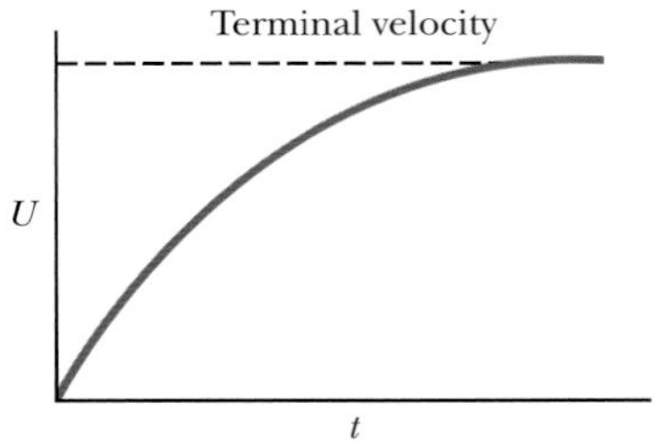

41. $t = \frac{Ah}{v_2 A'} = \frac{Ah}{A'\sqrt{2gd}}$

43. (a) 1.25 cm (b) 14.3 m/s

Chapter 15

1. (a) 0.35 nm (b) 0.31 nm (c) 0.17 nm

3. e.g. Carbon 60, Ice, NaCl

5. 1.9×10^{-21} J

7. (a) $F(r) = -\frac{d}{dr}U(r) = -9ar^{-10} - 6r^{-2}$ (b) $\left(\frac{9a}{b}\right)^{\frac{1}{8}}$

9. $U = k\Delta r^2$

11. Plot a graph of $\Delta\ell$ versus mg. In principle, we can determine Y from the slope, however, we need to know the cross-section of the fishing line. The proportional limit is exceeded around 1000 g, above which the proportionality changes

13. (a) 118 kN (b) 2.5 mm

15. $x_{\text{CG}} = \frac{\sum m_i x_i}{\sum m_i} = \frac{(72.0\,\text{cm}^2)(2.00\,\text{cm}) + (32.0\,\text{cm}^2)(8.00\,\text{cm})}{72.0\,\text{cm}^2 + 32.0\,\text{cm}^2} =$ 3.85 cm;

$y_{\text{CG}} = \frac{\sum m_i y_i}{\sum m_i} = \frac{(72.0\,\text{cm}^2)(9.00\,\text{cm}) + (32.0\,\text{cm}^2)(2.00\,\text{cm})}{104\,\text{cm}^2} =$ 6.85 cm

17. $E = YA\Delta L^2 / 2L_{il}$; use small extension approximation

19. 1.65×10^8 N/m^2

21. 18.56%

23. 0.024 ± 0.001 mm

25. 47000 N

27. $(1.00 \pm 0.01) \times 10^{11}$ Pa

29. 24.6% larger

31. 8.60×10^{-4} m

33. $(8.91 \pm 0.05) \times 10^9$ Pa

35. (a) 9.28 kN (b) 6.56 kN

37. (a) $\frac{YA}{L_i}$ (b) $YA\frac{(\Delta L)^2}{2L_i}$

39. 5.73 rad/s

41. $(1.06 \pm 0.11) \times 10^{11}$ N

43. (b) 343 N, 171 N, 683 N (c) 5.14 m

Chapter 16

1. (a) 17N to the left (b) $a = 28\,\text{m/s}^2$ to the left
3. 0.63s
5. (a) 18.8 m/s (b) 7.11 km/s^2
7. (a) $f = 1.50$ Hz (b) 0.667 s
   (c) 4.00 m (d) π rad (e) 2.83 m
9. (a) 24.7 kg.s^{-2} The uncertainty is small and will be around 1% from the uncertainties in the raw data
   (b) $k = \dfrac{4\pi^2}{\text{gradient}} = 24.4\,\text{kg/s}^2$ with an uncertainty of around 4%. Hence these two values do agree when uncertainties are taken into account (c) The spring does not only obey Hooke's law (it departs from linear behaviour) for masses above 400 g, hence the students have chosen not to collect data in the non-linear region in the second part of the experiment
13. (a) $\frac{8}{9}E$ (b) $\frac{1}{9}E$ (c) $\pm\sqrt{\frac{2}{3}}A$ (d) No
15. (a) 4.58 N (b) 0.125 J (c) 18.3 m/s^2 (d) 1.00 m/s (e) Smaller (f) The coefficient of kinetic friction between the block and surface (g) 0.934
17. (a) 1.50 s (b) 0.559 m
19. $I = (0.94 \pm 0.04)\,\text{kg.m}^2$
21. (a) 2.09 s (b) 4.08%
23. (a)

Length, L (m)	1.000	0.750	0.500
Period, T (s)	2.00	1.73	1.42

   (b) Thus, $g_{\text{ave}} = 9.85\,\text{m/s}^2$ This agrees with the accepted value of $g = 9.80\,\text{m/s}^2$. Within 0.5% (c) 9.94 m/s^2
25. $\dfrac{dE}{dt} = -bv^2 < 0$
27. Take the first and second derivatives, substitute them into the differential equation and compare the coefficients. This will show that the expression given is a solution to the equation
29. (a) $f = \dfrac{1}{T} = \sqrt{\dfrac{g}{L}} = \sqrt{\dfrac{9.8\,\text{m/s}^2}{3.2\,\text{m}}} = 1.75\,\text{s}^{-1}$ (b) If we included the mass of the chains, this would lift the centre of mass of the system, giving a smaller effective length and increasing the frequency (c) If Laurence tried to help, but pumped his legs ineffectively, this would act as damping on the system, decreasing the frequency
31. (a) 3.16 s^{-1} (b) 6.28 s^{-1} (c) 5.09 cm
33. $A = \dfrac{F_0/m}{\omega^2 - \omega_0^2}$, where $\omega_0 = \sqrt{\dfrac{k}{m}}$
35. 1.7×10^{12} Hz
37. (a) 25 cm (b) 9.4 cm (c and d) 0.25 kg
39. 0.919×10^{14} Hz
41. (a) $2Mg$ (b) $Mg\left(1 + \dfrac{y}{L}\right)$ (c) $\dfrac{4\pi}{3}\sqrt{\dfrac{2L}{g}}$ (d) 2.68 s
43. 6.62 cm
45. (a) 1.26 m (b) 1.58 (c) Energy decreased by 120 J (d) Mechanical energy is transformed into internal energy in the perfectly inelastic collision
47. (a) $\omega = \sqrt{\dfrac{200}{0.400 + M}}$ where ω is in s^{-1} and M is in kilograms.
   (b) 22.4 s^{-1} (c) $f = 22.4\,\text{s}^{-1}$
49. (a) $\sin\theta \approx \tan\theta$ and $\sum \mathbf{F} = \dfrac{-2Ty}{L}\hat{\mathbf{j}}$ (b) $\omega = \sqrt{\dfrac{k}{m}} = \sqrt{\dfrac{2T}{mL}}$
51. (a) 5.20 s (b) 2.60 s (c) $\dfrac{(dA/dt)}{A} = \dfrac{1}{2}\dfrac{dE/dt}{E}$
53. Time is $(9.1 \pm 0.3) \times 10^{-5}$ s
55. (a) $F_g = -G\dfrac{M_{\text{closer than } r}\,m}{r^2} = -G\dfrac{m}{r^2}\cdot\dfrac{\left(\frac{4}{3}\pi r^3\right)}{\left(\frac{4}{3}\pi R_E^3\right)}M_E = -G\dfrac{M_E m}{R_E^3}r$
   (b) The 'spring constant' for this motion is $k = G\dfrac{M_E m}{R_E^3}$
   (c) Ignoring air resistance, the keys will return to the position they were dropped from. However, if there is any air in the hole, the drag will prevent the keys coming back all the way
57. (a) $y_f = -0.110$ m (b) Its period will be longer

Chapter 17

1. $y = \dfrac{6}{[(x - 4.50t)^2 + 3]}$
3. 184 km
7. (a) 2.00 cm (b) 2.98 m (c) 0.576 Hz (d) 1.72 m/s
9. 2.40 m/s
11. 1.55×10^{-10} m
13. Take the second derivative of $e^{b(x - vt)}$ with respect to time and with respect to position, then substitute into $\dfrac{\partial^2 y}{\partial t^2} = v^2\dfrac{\partial^2 y}{\partial x^2}$, to show that it is a solution.
15. (a) By substitution, we must test $2 = \dfrac{1}{v^2}2v^2$ and this is true, so the wave function does satisfy the wave equation.
   (b) $f(x + vt) = \dfrac{1}{2}(x + vt)^2$ and $g(x - vt) = \dfrac{1}{2}(x - vt)^2$
   (c) $f(x + vt) = \dfrac{1}{2}\sin(x + vt)$ and $g(x - vt) = \dfrac{1}{2}\sin(x - vt)$
17. 13.5 N
19. (a) 0.0510 kg/m (b) 19.6 m/s
21. 5.81 m
23. 80.0 N
25. 1.55×10^{-10} m
27. (a) The two pulses travel the same distance, and so the one that travels at the highest velocity will arrive first. Because the speed of sound in air is 343 m/s and the speed of sound in the iron rod is 5950 m/s, the pulse travelling through the iron rail will arrive first. (b) 23.4 ms
29. 2.82×10^8 m/s
31. 19.7 m
33. (a) 56.3 s (b) 56.6 km
35. (a) 62.5 m/s (b) 7.85 m (c) 7.96 Hz (d) 21.1 W
37. The power-per-width across the wave front $\dfrac{P}{2\pi r}$ is proportional to amplitude squared so amplitude is proportional to $\sqrt{\dfrac{P}{2\pi r}}$
39. 66.0 dB
41. 3.0×10^{-8} W/m^2
43. (a) 2.34 m (b) 0.390 m (c and d) 0.61 Pa (e) 4.25×10^{-7} m (f) 7.09×10^{-8} m
45. (a) 0.250 m (b) 40.0 rad/s (c) 0.300 rad/m (d) 20.9 m (e) 133 m/s (f) The wave moves to the right, in the + × direction
47. 1.07 kW
49. (a) 0.232 m (b) 8.41×10^{-8} m (c) 13.8 mm
51. (a) 4.63 mm (b) 14.5 m/s (c) 4.73×10^9 W/m^2
53. (a) 5.04×10^3 m/s (b) 1.59×10^{-4} s (c) 1.90 mm
   (d) 2.38×10^{-3} (e) $\sigma = Y\left(\dfrac{\Delta L}{L}\right) = (20.0 \times 10^{10}\,\text{N/m}^2)$
   $(2.38 \times 10^{-3}) = 4.76 \times 10^8\,\text{N/m}^2$ (f) $\dfrac{\sigma_y}{\sqrt{\rho Y}}$
55. 1.34×10^4 N

57. 3.86×10^{-4}
59. The bat is gaining on its prey at 1.71 m/s
61. (a) $f' = 531$ Hz (b) 539 Hz to 466 Hz (c) 568 Hz
63. (a) $t = 2\sqrt{\frac{L}{mg}}\left(\sqrt{M+m} - \sqrt{M}\right)$ (b) $2\sqrt{\frac{L}{g}}$ (c) $\sqrt{\frac{mL}{Mg}}$
65. (a) $\mu(x) = \frac{(\mu_L - \mu_0)x}{L} + \mu$ (b) $\Delta t = \frac{2L}{3\sqrt{T}(\mu_L - \mu_0)}\left(\mu_L^{3/2} - \mu_0^{3/2}\right)$

Chapter 18

3. 5.66 cm
5. (a) wave 1 $+x$ direction wave 2 $-x$ direction (b) $t = 0.750$ s (c) $x = 1.00$ m
7. (a) 3.33 rad (b) 283
9. (a) 15.7 m (b) 31.8 Hz (c) 500 m/s
11. (a) 4.24 cm (b) 6.00 cm (c) 6.00 cm (d) 0.500 cm, 1.50 cm, 2.50 cm
13. (a) second harmonic (b) 74.0 cm (c) 3
15. (a) 160 N to 166 N (b) 660 Hz
17. (a) 31.2 cm from bridge (b) 3.85 %
19. 9.00 kHz
21. (a) 536 Hz (b) 42.9 mm
23. (a) 0.552 m (b) 316 Hz
25. $\frac{\pi r^2 v}{2Rf}$
27. $\Delta f = 110/\text{s} - 104.4/\text{s} = 5.64$ beats/s
29. (a) 1.99 Hz (b) 3.38 m/s
31. (a) Point A is one-half wavelength farther from one speaker than from the other. The waves from the two sources interfere destructively, so the receiver records a minimum in sound intensity. (b) 144 (c) Yes; the limiting form of the path is two straight lines through the origin with slope ± 0.75
33. (a) 4.90×10^{-3} kg/m (b) 2 (c) no standing wave will form
35. (a) larger (b) 2.43
37. (7.37 ± 0.06)cm
39. (a) 1.92 m/s (b) 0.960 m/s half as much as for the first harmonic
41. 3.85 m/s away from the station or 3.77 m/s toward the station
43. (a) Frequency should be halved (b) $T' = \left[\frac{n}{n+1}\right]^2 T$ (c) $\frac{T'}{T} = \frac{9}{16}$
45. (a) 0.656 m (b) 13.5°C
47. $F = \sqrt{15}Mg$

Chapter 19

1. (a) 1.14×10^4 stairs (b) 2.85×10^3 stairs
3. $+3.27 \pm 0.02$ cm
5. (a) 0.176 mm (b) 8.78μm (c) 0.0930 cm^3
7. (a) $T = 437$°C (b) $T = 2.1 \times 10^3$°C (c) No. Aluminium melts at 660°C (Table 19.2). Also, although it is not in Table 19.2, Internet research shows that brass (an alloy of copper and zinc) melts at about 900°C
9. (a) $(2.5 \pm 0.3) \times 10^6$ N/m^2 (b) The concrete will not fracture
11. 4.28 atm
13. 1.50×10^{29} molecules
15. 4.39 kg
17. (a) $P_f = 3.95$ atm $= 4.00 \times 10^5$Pa (b) 4.49×10^5 Pa
19. 3.68 ± 0.15 cm^3
21. -253°C
23. (a) -270°C (b) $1.2\dot{7}$ atm (c) 1.74 atm
25. 0.845 ± 0.012 kg
27. 1.78×10^4 kg
29. (a) $T = 25.8$°C (b) As shown above, the symbolic result from part (a) shows no dependence on mass. Both the change in gravitational potential energy and the change in internal energy of the system depend on the mass, so the mass cancels
31. $Q_{\text{needed}} = 1.22 \times 10^5$ J
33. 0.415 kg
35. 35.016 m. The uncertainty is smaller than the least significant figure.
37. (a) 1.82×10^3 J/kg.°C (b) We cannot make definite identification. It might be beryllium. (c) The material might be an unknown alloy or a material not listed in the table.
39. 3.37 cm
41. (a) $\rho = \frac{m}{V}$ and $d\rho = -\frac{m}{V^2}dV$. Assume very small changes. (b) As the temperature increases, the density decreases. (c) 5×10^{-5}°C^{-1} (d) -25×10^{-5}°C^{-1}
43. 57.5 s lost
45. (a) $\Delta A = 2\alpha A \Delta T$ (b) $\alpha \Delta T \ll 1$
47. 3.76 m/s
49. (a) First, energy must be removed from the liquid water to cool it to 0°C. Next, energy must be removed from the water at 0°C to freeze it, which corresponds to a liquid-to-solid phase transition. Finally, once all the water has frozen, additional energy must be removed from the ice to cool it from 0°C to -8.00°C. (b) 32.5 kJ
51. 2.27 km
53. (a) 0.169 m (b) 1.35×10^5 Pa
55. Yields $h = 4.54$ m, a remarkably large value compared to $\Delta L = 5.50$ cm.

Chapter 20

1. (a) $(6.45 \pm 0.10) \times 10^3$ W (b) $(5.57 \pm 0.09) \times 10^8$ J
3. 9.00 cm
5. 364 K
7. (a) 0.964 kg or more (b) The test samples and the inner surface of the insulation can be pre-warmed to 37.0°C as the box is assembled. Then nothing changes in temperature during the test period and the masses of the test samples and insulation make no difference.
9. $-nR(T_2 - T_1)$
11. (a) $W_{i\to f} = -12.0$ MJ (b) $W_{f\to i} = +12.0$ MJ
13. (a) 12.0 kJ (b) $Q = -W = -12.0$ kJ
15. 4.29×10^4 J
17. (a) -88.5 J (b) 722 J
19. (a) 0.007 65 m^3 (b) 305 K
21. (a) -0.0486 ± 0.0005 J (b) 16.2 ± 0.2 kJ (c) 16.2 ± 0.2 kJ
23. (a) $+\frac{P_iV_i}{2}$ (b) $+1.39P_iV_i$ (c) $W = 0$
25. (a) 1.50×10^{-5} kg – the uncertainty is smaller than the least significant figure (b) -5.00 ± 0.03 J (c) 5.00 ± 0.03 J (d) 15.0 mg (e) 5.00 ± 0.03 J, -5.00 ± 0.03 J (f) $\Delta E_{\text{int}} = 0$ (g) $4.04 \pm 0.02 \times 10^{-3}$ °C (h) $Q = 0$ $\Delta E_{\text{int}} = 2.50$ J (i) $Q = 0$ $\Delta E_{\text{int}} = 2.50$ J
27. (a) -6.08×10^5 J (b) 4.56×10^5 J
29. (a) 4.19 mm/s (b) 12.6 mm/s
31. (a) 0.169 m (b) $P' = 1.35 \times 10^5$ Pa

Chapter 21

1. (a) 8.76×10^{-21}J (b) helium $v_{\text{rms}} = 1.62$km/s, argon $v_{\text{rms}} = 514$m/s
3. 3.32mol

5. (a) $N = (3.64 \pm 0.04) \times 10^{23}$ atoms (b) 6.07×10^{-27} J (c) 1.35 km/s
7. 1.35 PV
9. (a) 3.46 kJ (b) 2.45 kJ (c) –1.01 kJ
11. (a) 0.118 ± 0.005 (b) 235 ± 0.16 (c) $Q = 0$ (d) 135 ± 16 J (e) $+135 \pm 16$ J
13. 227 K
15. (b) $(3^{1/\gamma})V_i$ (c) $3T_i$ (d) T_i (e) $-P_iV_i\left[\left(\frac{1}{\gamma-1}\right)(1-3^{1/\gamma})+(1-3^{1/\gamma})\right]$
17. (a) $E_{\text{int}} = Nf\left(\frac{k_BT}{2}\right) = f\left(\frac{nRT}{2}\right)$ (b) $C_V = \frac{1}{n}\left(\frac{dE_{\text{int}}}{dT}\right) = \frac{1}{2}fR$
(c) $C_P = C_V + R = \frac{1}{2}(f+2)R$ (d) $C_P = C_V + R = \frac{1}{2}(f+2)R$
19. (a) $nC_V = 41.6$ J/K (b) $nC_P = 58.2$ J/K (c) 58.2 J/K, 74.8 J/K
21. (a) almost all of the time no atom is excited (b) 2.70×10^{20}
23. (a) 731 m/s (b) 825 m/s (c) 895 m/s (d) the graph appears to be drawn correctly within about 10 m/s
27. (a) 1.7 ± 0.2 h (b) 1.00
29. (a) 2.26×10^{-9} m (b) 5.09×10^{-12} seconds
31. (c) The expressions are equal because $PV = nRT$ and $\gamma = (C_V + R)/CV = 1 + R/C_V$ give $R = (\gamma - 1)C_V$, so $PV = n(\gamma - 1)C_VT$ and $PV/(\gamma - 1) = nC_VT$.
33. (a) 0.514 m^3 (b) 2.06 m^3 (c) 2.38×10^3 K (d) −480 KJ (e) 2.28 MJ
35. (a) 7.27×10^{-20} J/molecule (b) 2.20 km/s (c) 3.51×10^3 K (d) The evaporating particles emerge with much less kinetic energy, as negative work is performed on them by restraining forces as they leave the liquid. Much of the initial kinetic energy is used up in overcoming the latent heat of vaporisation. There are also very few of these escaping at any moment in time.
37. (a) $\frac{5}{2}R$ (b) $3R$ (c) $\frac{9}{2}R$ (d) $5R$ (e) Measure the constant-volume specific heat of the gas as a function of temperature and look for plateaus on the graph. If the first jump goes from $\frac{3}{2}R$ to $\frac{5}{2}R$, the molecules can be diagnosed as linear. If the first jump goes from $\frac{3}{2}R$ to $3R$, the molecules must be nonlinear. The tabulated data at one temperature are insufficient for the determination. At room temperature some of the heavier molecules appear to be vibrating.
39. (a) 0.203 mol (b) 900 K (c) 900 K (d) 15.0 L (e) A→B: lock the piston in place and put the cylinder into an oven at 900 K, gradually heating the gas. B→C: keep the sample in the oven while gradually letting the gas expand to lift a load on the piston as far as it can. C→A: carry the cylinder back into the room at 300 K and let the gas gradually cool and contract without touching the piston.
(f) for $A \to B$: $W = 0$, $\Delta E_{\text{int}} = 1.52$ kJ, $Q = 1.52$ kJ; for $B \to C$: $\Delta E_{\text{int}} = 0$, $W = -1.67$ kJ, $Q = 167$ kJ; for $C \to A$: $\Delta E_{\text{int}} = -1.52$ kJ, $W = 1.01$ kJ, $Q = -2.53$ kJ
(g) $Q_{ABCA} = 0.656$ kJ, $W_{ABCA} = -0.656$ kJ, $\Delta E_{\text{int}} = 0$
41. (a) $\sqrt{\frac{3k_BT}{m_0}}$ (b) $\sqrt{\frac{8K_BT}{\pi m_0}}$
43. (a) m_0gR_E (b) 1.60×10^4 K
45. (d) The only possibility for $v_{\text{rms}} = v_{\text{avg}}$ is $a = 1$

Chapter 22

1. (a) 0.294 (or 29.4%) (b) 5.00×10^2 J (c) 1.67 ± 0.06 kW
3. (a) 10.7 kJ (b) 0.533 s
5. (a) 24.0 J (b) 144 J
7. (a) 4.51×10^5 J (b) 284×10^7 J
9. (a) 67.2% (b) 58.8 kW
11. (a) 837 ± 3 K (b) No. A real engine will always have an efficiency *less* than the Carnot efficiency because it operates in an irreversible manner
13. 11.8
15. (a) 9.10 kW (b) 11.9 kJ
17. (a) 5.12% (b) 5.27 Tj/h (c) As fossil fuel prices rise, this way to use solar energy will become a good buy
19. (a) $\frac{Q_c}{\Delta t} = 1.40\left(\frac{0.5T_h + 383}{T_h - 383}\right)$ where $Q_c/\Delta t$ is in megawatts and T is in kelvins (b) The exhaust power decreases as the firebox temperature increases (c) 1.87 MW (d) 3.84×10^3 K (e) No answer exists as the energy exhaust cannot be that small
21. (a) 244 kPa (b) 192 J
25. 143 ± 6 J/K
27. (a) 5.76 J/K (b) no change in temperature
29. 3.28 J/K
31. (a)

Result	Possible Combinations	Total
All heads	HHHH	1
3H, 1T	THHH, HTHH, HHTH, HHHT	4
2H, 2T	TTHH, THTH, THHT, HTTH, HTHT, HHTT	6
1H, 3T	HTTT, THTT, TTHT, TTTH	4
All tails	TTTT	1

(b) 2 heads and 2 tails (c) either all heads or all tails (d) 2 heads and 2 tails
33. (a) 0.268 (b) 0.423
35. (a) $-0.390nR$ (b) $-0.545nR$
37. (a) $3nRT_i$ (b) $3nRT_i \ln 2$ (c) $-3nRT_i$ (d) $-nRT_i \ln 2$ (e) $3nRT_i(1 + \ln 2)$ (f) $2nRT_i \ln 2$ (g) 0.273
39. (a) 8.5 ± 0.5 kW (b) 1.5 ± 0.5 kW (c) $(1.1 \pm 0.6) \times 10^4$ J/K (d) 20.0 %
41. (a) $10.5nRT_i$ (b) $8.50nRT_i$ (c) 0.190 (d) 0.833, the Carnot efficiency is much higher
43. (a) 13.4 J/K (b) 310 K (c) 13.3 J/K (d) smaller by less than 1%
45. 1.02 kJ/K
47. (a) $P_A = 25.0$ atm, $P_B = 4.14$ atm, $P_C - 1.00$ atm, $P_D = 6.03$ atm, $V_A = 1.97 \times 10^{-3}$ m^3, $V_B = 11.9 \times 10^{-3}$ m^3a, $V_C = 32.8 \times 10^{-3}$ m^3 $V_D = 5.44 \times 10^{-3}$ m^3 (b) 2.99 kJ

Index

Table of symbols

Symbol	Description	SI units	Section	Page
☺	the expression is dimensionally correct		1.3	8
a, $\mathbf{a}$	acceleration	$m \cdot s^{-2}$	2.4	30
a	semi-major axis length of an ellipse	m	11.4	333
a	slit width	m	38.1	1061
$\mathbf{a}_{AB}$	acceleration of A relative to B	$m \cdot s^{-2}$	3.7	74
a_c	centripetal acceleration	$m \cdot s^{-2}$	3.5	70
$\mathbf{a}_{CM}$	acceleration of the centre of mass	$m \cdot s^{-2}$	8.7	235
a_o	Bohr radius	m	41.1	1140
a_r, $\mathbf{a}_r$	radial acceleration	$m \cdot s^{-2}$	3.5	72
a_t, $\mathbf{a}_t$	tangential acceleration	$m \cdot s^{-2}$	3.5	72
$\mathbf{A}$, A	arbitrary vector in Cartesian coordinates, a component of the vector		3.1	57
A	area, cross sectional area	m^2	5.2	132
A	amplitude of oscillation or wave	m	16.2	466
A	constant associated with attractive force between atoms	$J \cdot m^{-n}$	42.1	1169
A	mass number	-	43.1	1201
b	resistive constant for movement in a fluid	$kg \cdot s^{-1}$	5.2	128
b	damping coefficient	$kg \cdot s^{-1}$	16.6	482
b	semi-minor axis length of an ellipse	m	11.5	337
$\mathbf{B}$, B	arbitrary vector in Cartesian coordinates, a component of the vector		6.2	157
B, $\mathbf{B}$	buoyant force	N ($kg \cdot m \cdot s^{-2}$)	13.5	388
B	bulk modulus	Pa ($kg \cdot m^{-1} \cdot s^{-2}$)	15.4	450
B, $\mathbf{B}$	magnetic field	T ($kg \cdot C^{-1} \cdot s^{-1}$)	29.2	816
B	constant associated with repulsive force between atoms	$J \cdot m^{-m}$	42.1	1169
B	Baryon number	-	44.5	1249
B_{ext}, $\mathbf{B}_{ext}$	external magnetic field	T ($kg \cdot C^{-1} \cdot s^{-1}$)	30.6	861
c	speed of light in vacuum	$m \cdot s^{-1}$	11.4	336
c	specific heat capacity	$J \cdot kg^{-1} \cdot K^{-1}$ ($m^2 \cdot s^{-2} \cdot K^{-1}$)	19.5	575
$\mathbf{C}$, C	arbitrary vector in Cartesian coordinates, a component of the vector		9.4	257
C	capacitance	F ($C^2 \cdot s^2 \cdot kg^{-1} \cdot m^{-2}$)	26.3	745
C	centre of curvature		36.2	1001
C	charm quantum number	-	44.7	1254
C_{eq}	equivalent or total capacitance	F ($C^2 \cdot s^2 \cdot kg^{-1} \cdot m^{-2}$)	28.5	798
C_p	molar specific heat at constant pressure	$J \cdot mol^{-1} \cdot K^{-1}$ ($kg \cdot m^2 \cdot s^{-2} \cdot K^{-1}$)	21.2	617
C_V	molar specific heat at constant volume	$J \cdot mol^{-1} \cdot K^{-1}$ ($kg \cdot m^2 \cdot s^{-2} \cdot K^{-1}$)	21.2	617
COP	coefficient of performance	-	22.2	638
d	differential change in a quantity, differential element		2.2	25

Symbol	Description	SI units	Section	Page
d	distance	m	12.3	354
d	distance between slits, grooves, planes of atoms	m	37.2	1039
d_{effort}	distance from pivot to applied force	m	9.4	259
d_{load}	distance from pivot to load	m	9.4	259
D	drag coefficient	-	5.2	131
D	distance from centre of mass to axis of interest	m	9.5	268
D	diameter of a lens or aperture	m	36.7	1024
e	emissivity	-	20.1	595
e	electron charge	C	23.1	670
E	energy	J ($kg \cdot m^2 \cdot s^{-2}$)	6.7	174
E, $\mathbf{E}$	electric field	$N \cdot C^{-1}$ ($kg \cdot m \cdot s^{-2} \cdot C^{-1}$)	23.3	676
E_b	binding energy	J ($kg \cdot m^2 \cdot s^{-2}$)	43.1	1204
E_F	Fermi energy	J or eV ($kg \cdot m^2 \cdot s^{-2}$)	42.4	1181
E_g	energy (band) gap	J or eV ($kg \cdot m^2 \cdot s^{-2}$)	42.6	1187
E_H	Hall field	$N \cdot C^{-1}$ ($kg \cdot m \cdot s^{-2} \cdot C^{-1}$)	30.5	859
E_{ind}	induced electric field	$N \cdot C^{-1}$ ($kg \cdot m \cdot s^{-2} \cdot C^{-1}$)	26.2	743
E_{mech}	mechanical energy (kinetic and potential) of a system	J ($kg \cdot m^2 \cdot s^{-2}$)	7.2	184
E_o	external electric field	$N \cdot C^{-1}$ ($kg \cdot m \cdot s^{-2} \cdot C^{-1}$)	26.2	743
E_R	rest energy	J ($kg \cdot m^2 \cdot s^{-2}$)	12.5	368
E_{system}	energy of a system	J ($kg \cdot m^2 \cdot s^{-2}$)	7.1	184
emf	electromotive force	V ($kg \cdot m^2 \cdot s^{-2} \cdot C^{-1}$)	27.1	761
f	frequency of periodic motion or wave	Hz (s^{-1})	16.2	467
f	focal length	m	36.2	1001
f'	observed frequency	Hz (s^{-1})	17.6	513
f_1	fundamental (first harmonic) frequency	Hz (s^{-1})	18.4	541
f_{beat}	beat frequency	Hz (s^{-1})	18.6	549
f_c	cut-off frequency	Hz (s^{-1})	39.3	1098
f_E	Einstein frequency	Hz (s^{-1})	16.8	487
$\mathbf{f_k}$, f_k	force of kinetic friction	N ($kg \cdot m \cdot s^{-2}$)	5.1	122
f_n	n^{th} harmonic frequency	Hz (s^{-1})	18.4	541
$\mathbf{f_s}$, f_s	force of static friction	N ($kg \cdot m \cdot s^{-2}$)	5.1	121
$f(E)$	Fermi-Dirac distribution function	-	42.4	1181
$\mathbf{F}$, F	force, magnitude of force	N ($kg \cdot m \cdot s^{-2}$)	4.5	90
F	Focal point		36.2	1001
$\mathbf{F}_{12}$, F_{12}	force of object 1 on object 2	N ($kg \cdot m \cdot s^{-2}$)	4.6	94
$\mathbf{F_{app}}$, F_{app}	applied (external) force, magnitude of applied force	N ($kg \cdot m \cdot s^{-2}$)	6.5	169
F_B, $\mathbf{F}_B$	magnetic force	N ($kg \cdot m \cdot s^{-2}$)	30.1	844
F_e, $\mathbf{F}_e$	electrostatic force	N ($kg \cdot m \cdot s^{-2}$)	23.3	676

Symbol	Description	SI units	Section	Page
F_{effort}	applied (external) force on a lever	N (kg • m • s^{-2})	9.4	259
$\mathbf{F}_{fric}$, F_{fric}	frictional force	N (kg • m • s^{-2})	5.1	123
$\mathbf{F}_g$, F_g	gravitational force (weight)	N (kg • m • s^{-2})	4.4	87
F_{load}	load (force) on a lever	N (kg • m • s^{-2})	9.4	259
F_P, $\mathbf{F}_\mathbf{P}$	force due to pressure gradient	N (kg • m • s^{-2})	14.4	421
F_r, $\mathbf{F}_\mathbf{r}$	radial force	N (kg • m • s^{-2})	5.4	142
F_s, $\mathbf{F}_\mathbf{s}$	force exerted by a spring	N (kg • m • s^{-2})	6.3	161
F_t, $\mathbf{F}_\mathbf{t}$	tangential force	N (kg • m • s^{-2})	5.4	142
F_v, $\mathbf{F}_\mathbf{v}$	viscous force	N (kg • m • s^{-2})	14.4	421
g, $\mathbf{g}$	acceleration due to gravity	m • s^{-2}	3.4	63
$g(E)$	density of states function	-	42.4	1182
$\mathbf{g}$	gravitational field	N • kg^{-1} (m • s^{-2})	11.2	327
G	universal gravitational constant	N • m^2 • kg^{-2} (m^3 • s^{-2} • kg^{-1})	11.1	324
h	height, object height	m	3.4	64
h	Planck's constant	J • s (kg • m^2 • s^{-1})	29.6	831
$\hbar$	h-bar, Planck's constant divide by 2π	J • s (kg • m^2 • s^{-1})	40.2	1115
h'	image height	m	36.1	999
H	Hubble's constant	m • s^{-1} • ly^{-1} (s^{-1})	44.9	1260
i, j, k	unit vectors in Cartesian coordinates		3.3	60
I, $\mathbf{I}$	impulse, change in momentum	kg • m • s^{-1}	8.3	219
I	moment of inertia	kg • m^2	9.5	263
I	intensity	W • m^{-2} (kg • s^{-3})	17.7	519
I	threshold of hearing, reference intensity	W • m^{-2} (kg • s^{-3})	17.7	520
I	current	A (C • s^{-1})	27.1	761
I	image		36.2	1001
I	nuclear angular momentum quantum number	-	43.2	1208
I_{CM}	moment of inertia about the centre of mass	kg • m^2	9.5	268
I_P	moment of inertia axis through point P	kg • m^2	10.2	296
J	polar moment of inertia (second moment of area)	m^4	15.6	456
J	current density	A • m^{-2} (C • s^{-1} • m^{-2})	27.2	765
J	rotational quantum number	-	42.1	1172
k	spring constant	N • m^{-1} (kg • s^{-2})	6.3	161
k	angular wave number (wave number)	radians • m^{-1} (m^{-1})	17.2	501
k	thermal conductivity	W • m^{-1} • $°C^{-1}$ (kg • m • s^{-3} • K^{-1})	20.1	590
K	kinetic energy	J (kg • m^2 • s^{-2})	6.4	165
k_B	Boltzmann's constant	J • K^{-1} (kg • m^2 • s^{-2} • K^{-1})	19.3	570
k_e	Coulomb constant	N • m^2 • C^{-2} (kg • m^3 • s^{-2} • C^{-2})	23.2	673
K_R	rotational kinetic energy	J (kg • m^2 • s^{-2})	10.1	290

Symbol	Description	SI units	Section	Page
l	length	m	14.4	421
l	orbital angular momentum quantum number	-	41.2	1143
L, [L]	dimensions of length		1.3	8
L, $\mathbf{L}$	angular momentum, orbital angular momentum	$kg \cdot m^2 \cdot s^{-1}$	10.3	300
L	length	m or ly	1.3	8
L	distance between slits and screen	m	37.2	1039
L	latent heat for a phase change	$J \cdot kg^{-1}$ (m^2s^{-2})	19.5	578
L	inductance	H ($kg \cdot m^2 \cdot C^{-2}$)	32.1	901
L_e, L_μ, L_τ	electron, muon and tau lepton numbers	-	44.5	1250
L_f	latent heat of fusion	$J \cdot kg^{-1}$ (m^2s^{-2})	19.5	578
l_o, L_o	equilibrium length	m	15.2	443
L_p	proper length	m or ly	12.3	358
$\mathbf{L}_{tot}$	total angular momentum of system	$kg \cdot m^2 \cdot s^{-1}$	10.4	304
L_v	latent heat of vaporisation	$J \cdot kg^{-1}$ (m^2s^{-2})	19.5	578
L_z	projection of orbital angular momentum in z direction	$kg \cdot m^2 \cdot s^{-1}$	41.4	1149
m	mass	kg	4.3	86
m	angular magnification	-	36.7	1025
m	order number	-	37.2	1039
m_l	orbital magnetic quantum number	-	41.2	1144
m_p	proper mass	kg	12.5	369
m_s	spin quantum number	-	41.4	1150
M, [M]	dimensions of mass		1.3	8
M	total mass in a system	kg	8.6	232
M	magnetisation	$A \cdot m^{-1}$ ($C \cdot m^{-1} \cdot s^{-1}$)	29.6	833
M	mutual inductance	H ($kg \cdot m^2 \cdot C^{-2}$)	32.4	908
M	magnification	-	36.1	999
M_E	mass of Earth	kg	11.1	325
M_P	mass of a planet	kg	11.5	338
M_S	mass of the Sun	kg	11.5	338
$\mathbf{n}$, n	normal force	N ($kg \cdot m \cdot s^{-2}$)	4.6	95
n	number of moles	mol	19.3	569
n	number of turns per unit length in a solenoid or toroid	m^{-1}	29.4	826
n	number of charged particles per unit volume	m^{-3}	30.3	853
n	refractive index	-	35.4	982
n	principle quantum number	-	39.1	1091
N	number of items, e.g., particles, loops	-	19.3	570
N	number per unit volume	m^{-3}	42.4	1182
N	neutron number	-	43.1	1201

Symbol	Description	SI units	Section	Page
N_A	Avogadro's number	-	15.1	436
n_o	number density in volume V in lowest energy state	m^{-3}	21.5	624
n_V	number density in volume V	m^{-3}	21.5	624
O, O'	observers in inertial reference frames S, s′, respectively		12.3	353
O	object		36.2	1001
$\mathbf{p}, p$	momentum (linear)	$kg \cdot m \cdot s^{-1}$	8.1	214
$\mathbf{p}, p$	electric dipole moment	$C \cdot m$	26.2	741
$\mathbf{p}_{tot}$	total momentum of a system	$kg \cdot m \cdot s^{-1}$	8.7	235
p	object distance from lens or mirror	m	36.2	1001
P	power, rate of energy transfer	W ($kg \cdot m^2 \cdot s^{-3}$)	6.7	174
P	pressure	Pa ($kg \cdot m^{-1} \cdot s^{-2}$)	13.1	381
P	power of a lens	dioptre (m^{-1})	36.6	1022
P_o	atmospheric pressure	Pa ($kg \cdot m^{-1} \cdot s^{-2}$)	13.2	384
$P(x)$	probability of finding particle at x	-	40.1	1113
Q	energy transferred by heat	J ($kg \cdot m^2 \cdot s^{-2}$)	7.1	184
Q	volume rate of flow	$m^3 \cdot s^{-1}$	14.1	409
q, Q	charge	C	23.1	669
Q	quality factor	-	33.6	940
Q	disintegration energy	eV or J ($kg \cdot m^2 \cdot s^{-2}$)	43.4	1211
q	image distance from lens or mirror	m	36.2	1001
r	radius of circle or curved path	m	3.1	55
$\mathbf{r}$	position vector	m	3.2	58
$\mathbf{r}_{AB}$	position of A relative to B	m	3.7	74
$\mathbf{r}_{CM}$	position of centre of mass	m	10.6	313
r_n	radius of nth atomic orbit	m	41.1	1140
r_o	equilibrium separation between particles	m	15.1	438
R	horizontal range of a projectile	m	3.4	64
$\mathbf{R}, R$	resistive force in a fluid	N ($kg \cdot m \cdot s^{-2}$)	5.2	128
R	radius of circle, disc or sphere	m	10.2	295
R	ideal gas constant	$J \cdot mol^{-1} \cdot K^{-1}$ ($kg \cdot m^2 \cdot s^{-2} \cdot mol^{-1} \cdot K^{-1}$)	19.3	569
R	R-value for thermal materials	$°C \cdot m^2\ W^{-1} \cdot (s^3 \cdot K \cdot kg^{-1})$	20.1	591
R	resistance	Ω ($kg \cdot m^2 \cdot s^{-1} \cdot C^{-2}$)	27.2	765
R	reflection coefficient	-	40.7	1129
R	decay rate or activity	Bq (s^{-1})	43.4	1217
R_{eq}	equivalent or total resistance	Ω ($kg \cdot m^2 \cdot s^{-1} \cdot C^{-2}$)	28.3	792
R_E	radius of Earth	m	11.1	325
R_H	Hall coefficient	$m^3 \cdot C^{-1}$	30.5	860

Symbol	Description	SI units	Section	Page
R_H	Rydberg constant	m^{-1}	39.2	1094
R_s	Schwarzschild radius of a black hole	m	11.4	336
Re	Reynolds number	-	14.5	423
RBE	relative biological effectiveness	-	43.7	1232
rms	root mean square		33.1	928
s	distance travelled	m	2.1	23
$\mathbf{s}$	displacement vector	m	25.2	717
s	arc length	m	3.1	55
s	spin quantum number	-	41.4	1151
S, S'	inertial reference frame (labels)		12.2	351
S	shear modulus or modulus of rigidity	Pa ($kg \cdot m^{-1} \cdot s^{-2}$)	15.5	452
S	entropy	$J \cdot K^{-1}$ ($kg \cdot m^2 \cdot s^{-2} \cdot K^{-1}$)	22.6	650
S, $\mathbf{S}$	spin angular momentum	$J \cdot s$ ($kg \cdot m^2 \cdot s^{-1}$)	29.6	831
$\mathbf{S}$	Poynting vector	$W \cdot m^{-2}$ ($kg \cdot s^{-3}$)	34.3	959
S	strangeness quantum number	-	44.6	1252
S_z	projection of spin angular momentum vector in z direction	$J \cdot s$ ($kg \cdot m^2 \cdot s^{-1}$)	41.4	1151
t	time	s	1.3	8
t, t'	time in S and S' inertial frames (respectively)	s	12.2	351
t	thickness	m	37.4	1045
T, [T]	dimensions of time		1.3	8
$\mathbf{T}$, T	tension in, or exerted by, a cable	N ($kg \cdot m \cdot s^{-2}$)	4.6	99
T	energy transferred into a system	J ($kg \cdot m^2 \cdot s^{-2}$)	7.1	184
T	period of orbit, oscillation, other periodic motion	s	11.5	339
T	temperature	K	19.1	563
T	Transmission coefficient	-	40.7	1129
$T_{1/2}$	half life	s	43.4	1212
T_c	air temperature, cold reservoir temperature	°C	17.5	512
T_c	critical temperature	K	42.8	1193
T_{ER}	energy transferred into a system by electromagnetic radiation	J ($kg \cdot m^2 \cdot s^{-2}$)	7.2	185
T_{ET}	energy transferred into a system by electrical transmission	J ($kg \cdot m^2 \cdot s^{-2}$)	7.2	185
T_{MT}	energy transferred into a system by matter transfer	J ($kg \cdot m^2 \cdot s^{-2}$)	7.2	185
T_{MW}	energy transferred into a system by mechanical waves	J ($kg \cdot m^2 \cdot s^{-2}$)	7.2	185
$\mathbf{u}_{AB}$	velocity of A measured by (relative to) to B	$m \cdot s^{-1}$	3.7	74
u, u'	speed measured in S and S' inertial reference frames, respectively	$m \cdot s^{-1}$	12.2	351
u_E	energy density in an electric field	$J \cdot m^{-3}$ ($kg \cdot m^{-1} \cdot s^{-2}$)	26.1	740
u_B	energy density in a magnetic field	$J \cdot m^{-3}$ ($kg \cdot m^{-1} \cdot s^{-2}$)	32.3	907
U	potential energy	J ($kg \cdot m^2 \cdot s^{-2}$)	7.2	185
U_C	potential energy stored in a capacitor	J ($kg \cdot m^2 \cdot s^{-2}$)	32.5	912

Symbol	Description	SI units	Section	Page
U_g	gravitational potential energy	J ($kg \cdot m^2 \cdot s^{-2}$)	6.5	169
U_L	potential energy stored in an inductor	J ($kg \cdot m^2 \cdot s^{-2}$)	32.5	912
$\mathbf{U_{LJ}}$	Lennard-Jones potential	J ($kg \cdot m^2 \cdot s^{-2}$)	15.1	439
$\mathbf{v}$, v	velocity or speed	$m \cdot s^{-1}$	2.1	22
$\mathbf{v_{AB}}$	velocity of A relative to B	$m \cdot s^{-1}$	3.7	74
v_{avg}	average speed	$m \cdot s^{-1}$	21.5	626
$\mathbf{v_{CM}}$	velocity of the centre of mass	$m \cdot s^{-1}$	8.7	235
v_d	drift velocity	$m \cdot s^{-1}$	27.1	762
v_{esc}	escape speed	$m \cdot s^{-1}$	11.4	335
v_{mp}	most probable speed	$m \cdot s^{-1}$	21.5	626
v_{rms}	root –mean-square speed	$m \cdot s^{-1}$	21.1	616
V	volume	m^3	13.1	381
V	electric potential	V ($kg \cdot m^2 \cdot s^{-2} \cdot C^{-1}$)	25.1	714
V_g	gravitational potential	$J \cdot kg^{-1}$ ($m^2 \cdot s^{-2}$)	11.3	331
V_o	equilibrium volume	m^3	15.4	451
w	velocity gradient	s^{-1}	14.3	417
W	work done by a force	J ($kg \cdot m^2 \cdot s^{-2}$)	6.2	156
W	number of microstates	-	22.8	654
W_{app}	work done on a system by an applied (external) force	J ($kg \cdot m^2 \cdot s^{-2}$)	6.3	163
W_{eng}	work done on a system by an engine	J ($kg \cdot m^2 \cdot s^{-2}$)	22.1	636
W_{ext}	work done on a system by an external force	J ($kg \cdot m^2 \cdot s^{-2}$)	6.3	160
W_{int}	work done in a system by an internal force	J ($kg \cdot m^2 \cdot s^{-2}$)	7.2	185
W_g	work done by the gravitational force	J ($kg \cdot m^2 \cdot s^{-2}$)	6.3	165
W_s	work done by a spring	J ($kg \cdot m^2 \cdot s^{-2}$)	6.3	162
x	position in the x direction in Cartesian coordinates	m	3.1	55
x, y, z	positions in the x, y, z directions (Cartesian coordinates) in inertial reference frame S	m	12.2	351
x', y', z'	positions in the x, y, z directions (Cartesian coordinates) in inertial reference frame S'	m	12.2	351
X_C, X_L	capacitive reactance, inductive reactance	Ω ($kg \cdot m^2 \cdot s^{-1} \cdot C^{-2}$)	33.2	931
x_{CM}	centre of mass position in the x direction in Cartesian coordinates	m	8.6	232
y	position in the y direction in Cartesian coordinates	m	3.1	55
Y	Young's modulus	Pa ($kg \cdot m^{-1} \cdot s^{-2}$)	15.3	445
y_{CM}	centre of mass position in the y direction in Cartesian coordinates	m	8.6	232
Z	impedance of a circuit	Ω ($kg \cdot m^2 \cdot s^{-1} \cdot C^{-2}$)	33.4	935
Z	atomic number	-	41.5	1154
z_{CM}	centre of mass position in the z direction in Cartesian coordinates	m	8.6	232
α	angular acceleration	radians $\cdot s^{-2}$	9.1	250
α	coefficient of linear thermal expansion	K^{-1}	19.2	566

Symbol	Description	SI units	Section	Page
α	temperature coefficient of resistivity	K^{-1}	27.3	771
α	Madelung constant	-	42.3	1179
β	compressibility	Pa^{-1} ($m \cdot s^2 \cdot kg^{-1}$)	14.1	408
β	sound level in dB	-	17.7	520
β	coefficient of volume thermal expansion	K^{-1}	19.2	566
χ	magnetic susceptibility	-	30.6	862
δ	path difference	m	37.2	1039
Δ	uncertainty in a quantity		1.6	12
Δ	change in a quantity		2.1	22
Δt_p	proper time interval	s	12.3	354
Δv	time varying potential difference	V ($kg \cdot m^2 \cdot s^{-2} \cdot C^{-1}$)	32.4	910
ΔV	potential difference	V ($kg \cdot m^2 \cdot s^{-2} \cdot C^{-1}$)	25.1	714
ΔV_H	Hall voltage	V ($kg \cdot m^2 \cdot s^{-2} \cdot C^{-1}$)	30.5	859
ε	efficiency	-	7.3	199
ε	electromotive force (emf)	V ($kg \cdot m^2 \cdot s^{-2} \cdot C^{-1}$)	28.2	786
ε_L	self-induced electromotive force (emf)	V ($kg \cdot m^2 \cdot s^{-2} \cdot C^{-1}$)	32.1	901
ϵ	equilibrium inter-molecular potential energy	J ($kg \cdot m^2 \cdot s^{-2}$)	15.1	439
ϵ	efficiency of a heat engine	-	22.1	636
ϵ	permittivity of a material	$C^2 \cdot N^{-1} \cdot m^{-2}$ ($C^2 \cdot s^2 \cdot kg^{-1} m^{-3}$)	26.2	743
ϵ_C	efficiency of a Carnot engine	-	22.4	641
ϵ_0	permittivity of free space	$C^2 \cdot N^{-1} \cdot m^{-2}$ ($C^2 \cdot s^2 \cdot kg^{-1} m^{-3}$)	23.2	673
ϕ	phase angle, phase shift or phase difference	radians	16.2	466
ϕ	work function of a metal	J or eV ($kg \cdot m^2 \cdot s^{-2}$)	39.3	1097
Φ_B	magnetic flux	Wb ($kg \cdot m^2 \cdot s^{-1} \cdot C^{-1}$)	29.5	827
Φ_E	electric flux	$N \cdot m^2 \cdot C^{-1}$ ($kg \cdot m^3 \cdot s^{-2} \cdot C^{-1}$)	24.1	695
γ	Lorentz factor	-	12.3	355
γ	surface tension	$J \cdot m^{-2}$ ($kg \cdot s^{-2}$)	13.6	392
γ	angle of shear	radians	15.5	453
γ	ratio of C_p to C_V	-	21.2	619
γ_{LS}	surface tension for liquid – solid interface	$J \cdot m^{-2}$ ($kg \cdot s^{-2}$)	13.6	396
γ_{LG}	surface tension for liquid – gas interface	$J \cdot m^{-2}$ ($kg \cdot s^{-2}$)	13.6	396
η	viscosity	$Pa \cdot s$ ($kg \cdot m^{-1} \cdot s^{-1}$)	14.3	417
κ	torsion constant	$N \cdot m \cdot rad^{-1}$ ($kg \cdot m^2 \cdot s^{-2}$)	16.5	481
κ	dielectric constant (relative permittivity)	-	26.2	743
λ	wavelength	m	17.2	500
λ'	observed wavelength	m	17.6	514
λ	linear charge density	$C \cdot m^{-1}$	23.4	680
λ	decay constant	s^{-1}	43.4	1217

Symbol	Description	SI units	Section	Page
λ_c	cut-off wavelength	m	39.3	1098
μ	linear mass density	$kg \cdot m^{-1}$	17.4	507
μ, $\boldsymbol{\mu}$	magnetic dipole moment	$A \cdot m^2$ ($C \cdot m^2 \cdot s^{-1}$)	29.2	822
μ	permeability of a material	$T \cdot m \cdot A^{-1}$ ($kg \cdot m \cdot C^{-2}$)	30.6	862
μ	reduced mass	kg	42.2	1172
μ_0	permeability of free space	$T \cdot m \cdot A^{-1}$ ($kg \cdot m \cdot C^{-2}$)	29.2	817
μ_B	Bohr magneton	$A \cdot m^2$ ($C \cdot m^2 \cdot s^{-1}$)	29.6	831
μ_k	coefficient of kinetic friction	-	5.1	122
μ_n	nuclear magneton	$A \cdot m^2$ ($C \cdot m^2 \cdot s^{-1}$)	43.3	1209
μ_s	coefficient of static friction	-	5.1	121
ν	kinematic viscosity	$m^2 \cdot s^{-1}$	14.5	424
θ	angle, angle subtended by an arc	radians	3.1	55
θ	angular displacement	radians	9.1	249
θ	contact angle	radians	13.6	396
θ	Mach angle or apex half angle	radians	17.6	516
θ_c	critical angle	radians	5.1	124
θ_p	Brewster's angle, polarising angle	radians	38.5	1075
ρ	density of material	$kg \cdot m^{-3}$	5.2	132
ρ	volume charge density	$C \cdot m^{-3}$	23.4	680
ρ	resistivity	$\Omega \cdot m$ ($kg \cdot m^3 \cdot s^{-1} \cdot C^{-2}$)	27.2	765
σ	Poisson's ratio	-	15.3	450
σ	Stefan-Boltzmann constant	$W \cdot m^2 \cdot K^{-4}$ ($kg \cdot s^{-3} \cdot K^{-4}$)	20.1	595
σ	surface charge density	$C \cdot m^{-2}$	23.4	680
σ	conductivity	$\Omega^{-1} \cdot m^{-1}$ ($s \cdot C^2 \cdot kg^{-1} \cdot m^{-3}$)	27.3	768
σ_{ind}	induced surface charge density	$C \cdot m^{-2}$	26.5	752
σ_x	standard deviation in x	m	40.1	1114
τ	time constant, time taken for a process to be 63% complete	s	5.1	129
$\boldsymbol{\tau}$, τ	torque	$N \cdot m$ ($kg \cdot m^2 \cdot s^{-2}$)	9.4	256
τ	mean free time	s	27.3	769
$\boldsymbol{\tau}_{ext}$, τ_{ext}	external torque	$N \cdot m$ ($kg \cdot m^2 \cdot s^{-2}$)	9.7	273
ω	angular speed or angular frequency	radians $\cdot s^{-1}$	9.1	250
ω_d	angular frequency of a damped oscillator	radians $\cdot s^{-1}$	16.6	482
ω_o	natural frequency or resonance frequency	radians $\cdot s^{-1}$	16.6	482
ω_p	angular speed of precession, precessional frequency	radians $\cdot s^{-1}$	9.1	250
ψ	spatial component of wave function	$m^{-1/2}$	40.1	1112
Ψ	wave function	$m^{-1/2}$ or $m^{-3/2}$	40.1	1112
∇, ∇^2	del operator, del squared or Laplacian operator		41.2	1142

Standard abbreviations and symbols for units

Symbol	Unit	Symbol	Unit
A	ampere	K	kelvin
u	atomic mass unit	kg	kilogram
atm	atmosphere	kmol	kilomole
Btu	British thermal unit	L	litre
C	coulomb	lb	pound
°C	degree Celsius	ly	light-year
cal	calorie	m	metre
d	day	min	minute
eV	electron volt	mol	mole
°F	degree Fahrenheit	N	newton
F	farad	Pa	pascal
ft	foot	rad	radian
G	gauss	rev	revolution
g	gram	s	second
H	henry	T	tesla
h	hour	V	volt
hp	horsepower	W	watt
Hz	hertz	Wb	weber
in.	inch	yr	year
J	joule	Ω	ohm

Mathematical symbols used in the text and their meaning

Symbol	Meaning
$=$	is equal to
$\equiv$	is defined as
$\neq$	is not equal to
$\propto$	is proportional to
$\sim$	is of the order of
$>$	is greater than
$<$	is less than
$\gg$ ($\ll$)	is much greater (less) than
$\approx$	is approximately equal to
Δx	the change in x
$\sum_{i=1}^{N} x_i$	the sum of all quantities x_i from $i = 1$ to $i = N$
$\lvert x \rvert$	the absolute value of x (always a non-negative quantity)
$\Delta x \to 0$	Δx approaches zero
$\frac{dx}{dt}$	the derivative of x with respect to t
$\frac{\partial x}{\partial t}$	the partial derivative of x with respect to t
$\int$	integral